2019 22nd European Microelectronics and Packaging Conference & Exhibition (EMPC 2019)

Pisa, Italy
16-19 September 2019

IEEE Catalog Number:	CFP1954H-POD
ISBN:	978-1-7281-6291-1

Copyright © 2019, IMAPS-Europe
All Rights Reserved

*** *This is a print representation of what appears in the IEEE Digital Library. Some format issues inherent in the e-media version may also appear in this print version.*

IEEE Catalog Number: CFP1954H-POD
ISBN (Print-On-Demand):   978-1-7281-6291-1
ISBN (Online):   978-0-9568086-6-0

Additional Copies of This Publication Are Available From:

Curran Associates, Inc
57 Morehouse Lane
Red Hook, NY 12571 USA
Phone: (845) 758-0400
Fax: (845) 758-2633
E-mail: curran@proceedings.com
Web: www.proceedings.com

2019 22nd European Microelectronics and Packaging Conference & Exhibition (EMPC)
Contents of Proceedings
Tuesday 16th September to Thursday 19th September
Released for IEEE-Xplore: ISBN 978-0-9568086-6-0

The Papers in these Proceedings are organised into Topics, corresponding with the Topics of the EMPC-2019 Conference.

Conference Session Topics
Advanced Packaging for Electronics and Photonics - AP
Applications - Apl
Embedded Electronics & Smart Textiles - EST
Green Electronics - GrE
Interconnection Technologies - InT
Manufacturing Technologies - MT
MEMS
Modelling - Mod
Nano Technologies - NT
Poster-Orals - Post
Power & Thermal Management - P&T
Reliability & Quality - R&Q
Substrate Technologies - Sub

Conference Session Papers

AP = Advanced Packaging

AP-01, Mike Tsai, **Advanced Antenna Integration of 3D System in Package Solutions for IoT and 5G Application 1**

AP-02, Jaber Derakhshandeh, **Novel embedded microbump approach for die-to-die and wafer-to-wafer interconnects with variable microbump diameters and down to 5 μm interconnect pitch scaling 7**

AP-03, Jean-Pol Delrue, **Glass Wafer Level Packaging Enabled by Laser Induced Deep Etching of Closed Cavities 13**

AP-04, Sheng-Chi Hsieh, **mmWave Antenna Design in Advanced Fan-Out Technology for 5G Application 17**

AP-05, Heidi Lundén, **Hermetic and Radiation Tolerant Glass Package for VCSELs Using Novel Micro Bonding Method 21**

AP-06, Mark Shaw, **High Capacity Silicon Photonics Packaging 26**

AP-07, Friedrich-Leonhard Schein, **High density fan-out panel level packaging of multiple dies embedded in IC substrates 34**

Apl = Applications

Apl-01, Outi Rusanen, **Injection Molded Structural Electronics Brings Surfaces to Life 40**

Apl-02, Daniel Ciprian Vasile, **Innovative Authentication Method for IoT Devices 47**

Apl-03, Susan Trulli, **Multi-material Printed Microwave Element for Phased Array Applications 53**

Apl-04, Alexander Schulz, **Laser structured passive Components and RF Filters in LTCC Technology with Operating Frequencies up to 40 GHz focusing on 5G Mobile Applications 59**

EST = Embedded Electronics & Smart Textiles

EST-01, Heike Bartsch, **LTCC as substrate - enabling semiconductor and packaging integration 64**

EST-02, Carmen Meuser, **Printed Functional Applications: Batteries, Communication Elements, Antennas and Conductive Paths on Technical Textiles 68**

EST-03, Dionysios Manessis, **Embedding Technologies for the Manufacturing of Advanced Miniaturised Modules toward the Realisation of Compact and Environmentally Friendly Electronics Devices 75**

EST-04, Mamta Pradhan, **Ultra-thin Capacitors in Silicon for 3D-Integration and Flexible Electronics 83**

GrE = Green Electronics

GrE-01, Benjamin Schellscheidt, **Life-Cycle Assessment for Power Electronics Module Manufacturing 89**

GrE-02, Karsten Schischke, **Embedding as a key Board-Level Technology for Modularization and Circular Design of Smart Mobile Products: Environmental Assessment 93**

GrE-03, Marek Koscielski, **Recovery of valuable BGA components from used electronic mobile devices and their application in new electronics products 101**

GrE-04, Attila Géczy, **Challenges of SMT assembling on biodegradable PCB substrates 105**

InT = Interconnection Technologies

InT-01, James Su, **Flip Chip Assembly on Coreless Substrate Challenge with Die Bond Solution 110**

InT-02, Francesco Bertocci, **Corrosion of Tin-Indium Solder during the Manufacturing Process of Biomedical Ultrasound Transducers 115**

InT-03, Sandy Klengel, **Influence of copper wire material to corrosion resistant packages and systems for high temperature applications 123**

InT-04, Tamás Hurtony, **Investigation of the Microstructure of Mn-doped Tin-Silver-Copper Solder Alloys Solidified with Different Cooling Rates 130**

InT-05, Anna Steenmann, **TLPB Improves Solder Connections by On Chip Creation of Intermetallic Phase Precursors 135**

InT-06, Masahiro Inoue, **Design of Interfacial Chemistry for Inducing Low Temperature Sintering of Silver Micro-fillers within Epoxy-based Binders 140**

InT-07, Jakob Schober, **Copper Pillar Bumps on Acoustic Wave Components 145**

InT-08, Andreas Schneider, **Stencil Printing and Flip-Chip Bonding for Assembly of Pixelated X-ray Detectors using PCB-type Interposer and Flexible Printed Circuit Boards 153**

InT-09, Knut E. Aasmundtveit, **High-energy X-ray Tomography for 3D Void Characterization in Au–Sn Solid-Liquid Interdiffusion (SLID) Bonds 161**

InT-10, Sri Krishna Bhogaraju, **Hybrid Cu particle paste with surface-modified particles for high temperature electronics packaging 168**

InT-11, Martin Ihle. **Functional Printing of MMIC-Interconnects on LTCC Packages for sub-THz applications 176**

InT-12, David Stenzel, **Characterization of Alternative Sinter Materials for Power Electronics** 180

InT-13, Daniel Ernst, **Heterogeneous Integration of Surface Mounted Devices for Bendable Electronics** 188

InT-14, Olivier Mailliart, **Assembly of very fine pitches Infrared focal plane array with indium micro balls** 193

InT-15, Luca Del Carro, **Sintering of oxide-free copper pastes for the attachment of SiC power devices** 199

MEMS = MEMS

MEMS-01, Lara Natta, **Aluminium Nitride based bio-MEMS for vascular graft monitoring** 205

MEMS-02, Michael Fischer, **Active cooling using fluid channels in a silicon-ceramic composite substrate** 211

MEMS-03, Adrian Goldberg, **Monolithic Ceramic IR-Emitter in Multilayer Technology** 216

Mod = Modelling

Mod-01, Xiangyang Shi, **Electrical Modeling and Analysis of Through-Silicon-Via Crosstalk Based on Scalable Physical Lumped Circuit Model for 3D Packaging** 222

Mod-02, Duyen Nu Bich Do, **New Encapsulation Concepts for Medical Ultrasound Probes – A Heat Transfer Simulation Study** 226

Mod-03, Haksan Jeong, **Thermomechanical Properties of Fan-Out Wafer Level Package Fabricated with Various Epoxy Mold Compounds** 233

Mod-04, Stoyan Stoyanov, **Packaging Challenges and Reliability Performance of Compound Semiconductor Focal Plane Arrays** 238

Mod-05, Zhaorong Li, **Evaluation Method of Electronic Performance Margins for SiP Based on Field-Circuit and Multiple Physical Cooperative Simulation** 246

Mod-06, Ghanshyam Gadhiya, **Assessment of FOWLP process dependent wafer warpage using parametric FE study** 251

MT = Manufacturing Technologies

MT-01, Daniel Wright, **Development of mechanically compliant flip chip interconnect using single metal coated polymer spheres** 258

MT-02, Yusuke Fumita, **A novel DAF mount method for SDBG process** 265

MT-03, Jan Bickel, **Increasing the productivity of the novel atmospheric pressure sputtering technology for 3D chip interconnection** 270

MT-04, Rainer Dohle, **New Assembly Technology for VCSEL arrays comprising ultra-thin diodes** 276

MT-05, Ruben Kahle, **Evaluation of adaptive processes for the embedding of bare dies in IC substrates** 282

MT-06, Silvia Altenbockum, **High throughput two-stage bonding technique for advanced wafer level packaging** 286

MT-07, Mona Bakr, **Effect of overmolding process on the integrity of electronics circuits** 292

MT-08, Mathilde Billaud, **Process Flow and Cost Modelling for Fan-Out Panel Level Packaging** 300

MT-09, Ulrike Passlack, **Hybrid Systems-in-Foil (HySiF) – Low Stress CFP Process for Biomedical Application** 306

MT-10, Eric Cadalen, **Dispensing Improvements with Drop on Demand Technology** 312

MT-11, Matthias Hunstig, **Thermosonic wedge-wedge bonding using dosed tool heating** 320

MT-12, Hiroshi Nishikawa, **Effect of bonding temperature on shear strength of joints using micro-sized Ag particles for high temperature packaging technology** 327

MT-13, Aurélien Griffart, **Opto-electronics flip-chip bonding Automation and *in-situ* quality monitoring** 331

NT = Nano-Technologies

NT-01, Igor Kadija, **Flexible Electronics Printing by Electroplating** 338

NT-02, Abdelhafid Zehri, **Graphene-coated copper nanoparticles for thermal conductivity enhancement in water-based nano-fluid** 345

Post = Poster Orals

Post-01, Yong-Sung Eom, **Advanced Interconnection technology with Laser Assisted Bonding Process for PET Substrate** 381

Post-02, Jiri Hlina, **Advanced Application Capabilities of Thick Printed Copper Technology** 386

Post-03, Jae Hak Lee, **Study on the Embedded Flexible Hybrid Stack Package using Polymer Elastic Bump Interconnection Method** 391

Post-04, Cristina Somma, **Digital Core Supply in Automotive FC-BGA Package: a Smart Modularity Solution for Design and Validation** 396

Post-05, Marco Rovitto, **Transfer Molding Simulation to Predict Filling Flaws and Optimize Package Design** 402

Post-06, Adil Shehzad, **New approach to apply 1,2,3-benzotriazole as a capping layer on UBMs for 3D TCB stacking and investigation of oxidation protection and solder wetting** 407

Post-07, Georg Lorenz, **Investigation of mechanical and microstructural properties of a new, corrosion resistant gold-palladium coated copper bond wire** 415

Post-08, Vladimir Cherman, **High spatial resolution measurements of thermo-mechanical stress effects in flip-chip packages** 421

Post-10, Hoang-Vu Nguyen, **Reworkable Anisotropic Conductive Adhesive for Assembly of Medical Devices** 375

Post-11, Fabien Piallat, **ALD Coatings to Mitigate Against Tin Whiskers and Upgrade the Environmental Durability of Electronic Circuit Boards** 427

Post-12, Martin Hirman, **Comparison of QFN Chips Glued by ACA and NCA Adhesives on the Flexible Substrate** 432

14.52-14.57: Post-13, Jessica Richter, **Tilting Behavior and Phase Formation of Sn-Cu Composite Solder for Large Area Baseplate Solder Joints** 439

Post-15, Adam Yuile, **Simulation of temperature profiles in reflow ovens for soldering area array components** 445

Post-16, Vlad Serea, **A Finite Element Modelling Approach of Test Setups for Multilayer Ceramic Capacitors 450**

Post-17, Kalyan Mitra, **Manufacturing of bio-compatible & degradable devices using inkjet technology for transient electronics 454**

Post-18, Graham Wilson, **Substrate to Baseplate Attach: A Novel Solder Solution with an Embedded Metal Matrix for Enhanced Reliability 462**

Post-19, Sarthak Acharya, **Realization of Embedded Passives using an additive Covalent bonded metallization approach 466**

Post-20, Bozena Matuskova, **Ultra-thin polymer spray coating for advanced adhesive bonding applications 472**

Post-21, Steffen Wiese, **Temperature evaluation of solder joints for adjusting reflow profiles 477**

Post-22, Dionysios Manessis, **High Frequency Substrate Technologies for the Realisation of Software Programmable Metasurfaces on PCB Hardware Platforms with Integrated Controller Nodes 484**

Post-23, Nan Wang, **Highly Thermal Conductive and Light-weight Graphene-based Heatsink 491**

Post-24, Amedeo Maierna, **Electronic Packaging for MEMS Infrared Sensor With Filtered Optical Window 495**

Post-25, Minoru Ueshima, **Ag and Si particles sintering technology for SiC power device 503**

Post-26, Zeynep Gökdeniz, **Behaviour of Silver-Sintered Joints by Cyclic Mechanical Loading and Influence of Temperature 508**

Post-27, Arne Neiser, **The Challenge of a Self-Soldering PCB 514**

Post 28, Rebecca Petrich, **Investigation of ScAlN for piezoelectric and ferroelectric applications 518**

Post-29, Christian Schwarzer, **Investigation of Pressureless Sintered Interconnections on Plasma Based Additive Copper Metallization for 3-Dimentional Ceramic Substrates in High Temperature Applications 523**

Post-30, Dirk Seehase, **Alternative Heating Methods for Printed Circuit Boards and a Practical Comparison of Direct Resistance and Inductive Heating Processes 531**

Post-31 Julien Vieilledent, **Mechanical behaviour of SAC305 lead-free alloy 539**

Post-32 Ali Roshanghias, **Low-temperature fine-pitch flip-chip bonding by using snap cure adhesives and Au stud bumps 546**

P&T = Power & Thermal Management

P&T-01, Suiying Ye, **Anisotropic Composite Core Material for Inductor-based Fully Integrated Voltage Regulator 352**

P&T-02, Madalin Vasile Moise, **Implementation of a prototype embedded system for in-car multipoint temperature measuring 360**

P&T-03, Kathrin Reinhardt, **PrintPOWER – Paste systems for multifunctional copper power modules 364**

P&T-04, Krzysztof Stojek, **Metalization impact on heat transfer through sintered nanosilver based thermal joints 371**

R&Q= Reliability & Quality Assurance

R&Q-01, Greg Caswell, **Can Electrolytic Capacitors Meet the Demands of High Reliability Applications? 550**

R&Q-02, Akira Saito, **Tin Whisker Growth Mechanism on Tin Plating of MLCCs Mounted with Sn-3.5Ag-8In-0.5Bi Solder in 30°C60%RH**   555

R&Q-03, Takahashi Furuhashi, **Characterization of Impact of Vertical Stress on FinFETs**   559

R&Q-04, Balázs Illés, **Characterization of Tin Pest Phenomenon in a Low Ag Content SAC Solder Alloy**   563

R&Q-05, Dan Wargulski, **Paving the way for the replacement of solder interconnections in power electronics by silver-sinter using pulsed infrared thermography**   568

R&Q-06, Vladimír Sítko, **Optical Method for Validation of Changes In The Cleaning Process and Cleaning Process Optimization**   576

R&Q-07, Elisabeth Kolbinger, **An approach for failure prediction in H³TRB-tests**   584

R&Q-08, Kazuaki Ano, **Risk Assessment Study of Package Warpage by using Bayesian Networks**   589

R&Q-09, Erik Wiss, **Flex Cracking of Multilayer Ceramic Capacitors: Experiments on Fracture Propagation**   597

R&Q-10, Chengcheng Chen, **Investigating the Overdischarge Failure on Copper Dendritic Phenomenon of Lithium ion Batteries in Portable Electronics**   601

R&Q-11, Nadine Pflügler, **Time-dependent Behaviour of Bonded Silicon Dies under Subcritical, Constant Shear Loading**   606

R&Q-12, Enrico Galbiati, **Acceleration factors for reliability analysis of electronics assemblies**   613

R&Q-13, Yeong K. Kim, **High frequency effects on the PBGA stress developments at random vibration**   620

R&Q-14, Bjoern Boehme, **Chip Package Interaction with Cu Pillar Interconnects – Impact of Die Warpage**   624

R&Q-15, Felix Wuest, **Integrated Condition Monitoring by Measuring the Delay of Gate Turn-off**   630

R&Q-16, René Metasch, **Numerical study on the influence of material models for tin-based solder alloys on reliability statements**   635

R&Q-17, Dong-Kil Shin, **Effect of Test Fixture on the Drop Reliability of Solid State Drive**   642

Sub = Substrate Technologies

Sub-01, Erick Merle. Spory, **High-Temperature Inter-Cavity Silicon Interposer Substrate for Multi-Chip Module Assembly**   646

Sub-02, Messaoud Bedjaoui, **Laser Frit Sealing Approach for Ultrathin Glass**   652

Sub-03, Nam Gutzeit, **Picosecond Laser Structuring Technology for LTCC – the Improvement of Fine Line Structuring**   658

2019 22nd European Microelectronics and Packaging Conference & Exhibition (EMPC)

Technical Papers

Advanced Antenna Integration of 3D System in Package Solutions for IoT and 5G Application

Mike Tsai*, Ryan Chiu, Eric He, J. Y. Chen, Frank Chu, Jensen Tsai,
Yu-Po Wang, Shunyu Jian and Simon Chen

Siliconware Precision Industries Co. Ltd.
No. 153, Sec. 3, Chung-Shan Rd. Tantzu, Taichung 427, Taiwan, R.O.C.
*Email: miketsai@spil.com.tw ; Tel: 886-4-25341525 ext. 6718

Abstract

Recently, the coming fast and large growing opportunity market will be the Internet of Things (IoT) and fifth generation (5G) connectivity application. Particularly for 5G, need to design by beamforming antennas to send Radio Frequency (RF) signals to different network & client devices. By using MIMO (Multi-Input Multi-Out) technology which uses more transmitters and receivers to ensure each antenna of data transmission performance. To meet those new application requirements & challenges, high-gain antennas have been proposed for wireless connectivity products such as high data transmission, very low latency and low power consumption requirements.

A 3D System in Package (3D SiP) including different approach, such as the double side assembly technology and antenna in package (AiP) which is a combination solutions for these requirements. In this paper, SPIL provided an alternative 3D antenna SiP technologu which used surface mount technology (SMT) and 3D antenna integration were designed to shrink the package size, the calculation of package size can be shrunk around 55% area and package size reduced from 10 x 8mm to 6 x6mm. This new solution could be integrated an antenna inside SiP module as AiP technology to get small form factor and additional major benefites to address cost, performance, and time-to-market.

The characterization analysis will utilize typical reliability testing (Temperature Cycle Test, High Temperature Storage Test, unbias HAST) results and verify the package structure. Finally, this paper will find out the suitable 3D SiP with advanced antenna integration solution for future IoT and 5G Connectivity devices application.

Key words: IoT, 5G, mmWave, 3D SiP, Heterogeneous Integration, Antenna in Package (AiP)

Introduction

From mobile network generation, the 5G applications such as high communication quality and power conservation requirements become insufficient for higher and higher data rates comparing with 4G mobile networks [1]. The System in Package (SiP) is a combination with different function IC and passive components that integration into a small package. The 3D stacking structure on SiP module could offer a small form factor and a low cost system solution for many wireless connectivity applications. For example from power consumption and compact form factor, such as Bluetooth Low Energy (BLE), and wireless connectivity module such as Wireless Fidelity (WiFi) will become the major heterogeneous integration requirement in near marketing [2]. The package solution for IoT product which including light-weighted, small form factor, low-profile, low cost and good electrical performances. The major process solution such as package surface coating, to provide

electromagnetic interference (EMI) to prevent the noise during different working function area. This structure could be provided shorter development time to achieve time-to-market requirement [3]. The mm-wave electrical performance of 3D glass structure was provided good electrical performance and reduction in the overall package foot-print [4]. The antenna pattern such as 4 x 4 array with patch antennas for Small Cells application. This application will require large antenna array technology such as Massive MIMO and Beam-forming [5]. The design of antenna array had more challenges regarding integration and module (IC, package and antenna) co-design and the SiP module concepts to provide high integration with antenna with industrial assembly constraints to enable volume production [6]. Difference antenna technology such as 77GHz automotive radar systems have been apply into ADAS application with mmWave transceiver on flip antenna in package solution [7]. The 3D antenna was designed with a metal lid structure to connect with substrate to connectivity IC and processing the connectivity data for redio frequency (RF) transport and receive for stacking die on passive SiP module [8].

In this paper, the 3D antenna SiP module with antenna integration was studied and demonstrated on the 6 x 6mm module size. By using a metal lid as 3D structure to process the connectivity data for redio frequency (RF) transport and receive. For example, frequency was separated to three groups as Low-band, Mid-band and High-band for RF connectivity application [9], refer to Figure 1 for different frequency band application.

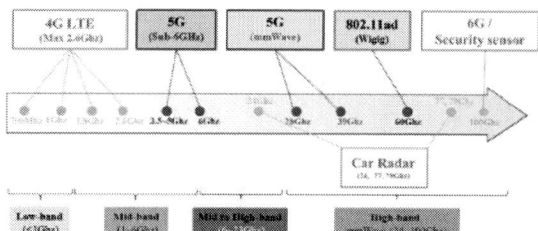

Figure 1: Different frequency band application

Test Vehicle Decription

From the high integration package trend and smaller form factor requirement, this test vehicle (TV) was designed with from single side PCB antenna to 3D AiP antenna structure technology and shrunk from 10 x 8mm to 6 x 6mm package size by using two major new assembly technology as below items.

✓   High integration SMT integration technology
✓   3D antenna in package integration technology

In order to shink the package size, all IC chips and components were designed into SiP module with fine component to component spacing and also integrated a 3D antenna technology, the basic comparison information was shown in Table 1.

Table 1: 3D antenna SiP package structure concept comparison

Item	Traditional SiP	New TV SiP
Dimension (mm)	10 x 8	6 x 6
Package Structure	2D Antenna	3D Antenna
Antenna Integration	2D Antenna	3D Antenna

In order to reduce the package size, this 3D antenna integration technology which uses metal lid structure comparing with traditional PCB antenna as 2D structure (snake-shape trace layout). The 3D antenna provided the high ultization for product design layout requirement. The calculation result of overall package size can be reduced about 55% and get the small form factor and benifites. From package

size reduction benefits which can be saved both of assembly and substrate cost. The basic information comparison between traditional AiP structure & new 3D antenna SiP module was shown in Table 2.

Table 2: Basic information comparison between traditional & new SiP module structure

Item	Traditional AiP	3D AiP
Dimension (mm)	10 x 8	6 x 6
3D Structure Photo	2D Antenna	3D Antenna
Placement Area (mm²)	80	36
Size Ratio	100%	45%

Process Flow

First, the major assembly procedure and process flow sequences were defined in Figure 2. The 3D antenna SiP was manufactured by high speed SMT process and also used components by tape & reel format for SMT machine preparation. The detail SMT process includes the solder printing, solder paste inspection, component attachment, automated optical inspection, thermal reflow for solder melting and de-flux cleaning process; the process flow shows in step (1) ~ (2) of Figure 2.

Process Flow   **Sequences**

	Process Flow	
(1)	Substrate in-coming	
(2)	Solder printing/ SMT attached	Passive
(3)	3D Antenna attached	3D Antenna
(4)	Die attached & Reflow	
(5)	Molding	
(6)	Marking & Ball attached	EMI coating
(7)	Singulation & Sputter coating	
(8)	Testing & finished	

Figure 2: 3D SiP of process flow

From the step (3) of process flow, the 3D metal lid antenna is manufacturing by mechanical punching process. This standard punching process which uses a force through into raw metal frame material to specific shape as 3D antenna structure. By using punching process to offer the cheapest method and high manufacturing speed to low down the overall package assembly cost. Before the die attached process, general preparation the silicon wafer will be proceed wafer grinding process to achieve thinner wafer thickness and sawing to separate the each of dies. During the die attached process, die will be picked up and dipping with flux material. The die bonding machine has high bonding accuracy to control the alignment to substrate side, and go through reflow for solder melting for components, 3D antenna and die together; refer to step (4) of Figure 2. The bending the 3D antenna by metal lid parts with 90 degrees for the footprint requirement to connect on substrate surface. The 3D antenna finished parts attached on substrate side result on Figure 3.

From the molding process in step (5) ~ (6) of Figure 2, we will use epoxy molding compound (EMC) as non-conductive material and protection purpose. This paper also uses partial coating on IC location (non-antenna area) to get good EMI shielding performance. During the sputter coating process, the 3D antenna SiP module will put on specific cover tooling to prevent the sputter coating material accrete on molding surface, refer to step (7) of Figure 2.

Finally, this 3D antenna SiP module will finish other standard process such as laser marking, ball attachment, package singulation process and testing for all assembly process; full process flow shows in step (8) of Figure 2.

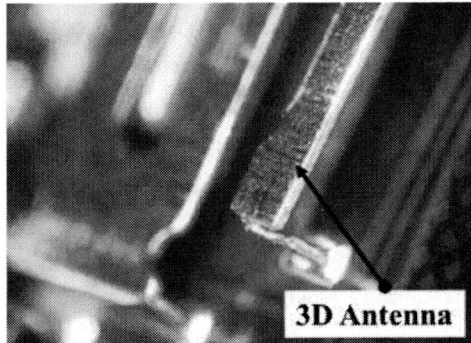

Figure 3: 3D antenna attached on substrate result

And also 3D metal antenna x-section image as shown in Figure 4, the metal antenna was adhered to substrate of bonding pad by solder paste material. Comparing with PCB antenna, the 3D metal antenna could be provided high speed attached by SMT process, cost effective by mechanical punching process and mature assembly process requirement. This metal frame structure could be used for different approach such as signal antenna purpose, EMI shielding between two function area purpose and high thermal requirement as heat-sink purpose.

Figure 4: X-section image of 3D metal antenna

Experiment Result

(1) Electrical Simulation for 3D Antenna SiP

Typically, the traditional PCB antenna design occupies a lot of design space for SiP layout due to the 2D side by side antenna structure. By using the 3D antenna integration technology with metal frame structure to get the small form factor for product design. This paper of TV working frequency was design at 2.4GHz for IoT and 5G product applications. To compare simulation of return loss (S11) with different signal data from connectivity quality requirementm. As shown in Figure 5, the simulation result is shown acceptable for 2.4GHz frequency range. Based on standard antenna return loss requirement which was set on -10dB, the 3D metal antenna design could provided wider bandwidth to cover from 2.25GHz ~ 2.5GHz of working frequency (around 0.25GHz bandwidth). Comparing with 2D PCB antenna struacture which provided frequency bandwidth to cover from 2.32GHz ~ 2.48GHz of working frequency (around 0.16GHz bandwidth). The simulation of return loss result shown frequency bendwidth had around 1.50X enhancement.

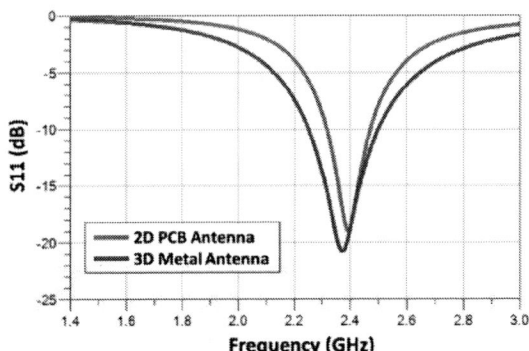

Figure 5: S11 simulation comparison between 2D PCB antenna and 3D metal antenna

(2) Molding Flow Verification for 3D Antenna SiP

The molding flow simulation is proceed through Finite Element Method (FEM) model to create module structure by using Moldex3D software for mold flowability simulation and the mold flow simulation with quarter symmetric structure of 3D sub-model was shown in Figure 6. This test vehicle uses transfer molding process and the mold thickness is 1.0mm for flowability verification study.

Figure 6: Quarter 4 block strip simulation model

The simulation analysis result of the mold flow was shown in Table 3. To compare different transfer rate for both 75% and 100%, base on 75% transfer rate, there is a balance mold pressure to push the EMC as wave from mold chase of air vent site through mold chase. When 100% transfer rate, the simulation shown void free is found during simulation result and no found mold incomplete. We can use this 3D metal antenna design and straucture for actual molding process verification.

The 3D antenna SiP of transfer molding appearance result was shown in Figure 7. By using molded underfill (MUF) technology, there are no metal antenna exposed and also no mold incomplete fill phenomenon from mold edge. The mold surface result also shows void free. The 3D antenna module can provide the clearance space between die area and component area for good molding flowability.

Table 3: Molding flowability simulation by different transfer rate status

Transfer rate(time)	75%	100%
Molding flowability simulation		Void Free

Figure 7: Demonstration of molding process result with antenna structure

(3) Warpage Performance for 3D Antenna SiP

The one of key challenge from this study was warpage performance for 3D antenna SiP structure. We need to consider warpage performance affects by EMC material and substrate including antenna structure inside when strip level warpage performance at 25℃ and 260℃. In this study, the difference material property is measured by the TMA (CTE) & DMA (modulus). Those two different compound material properties was measured by TMA & DMA curve in house as shown in Figure 8.

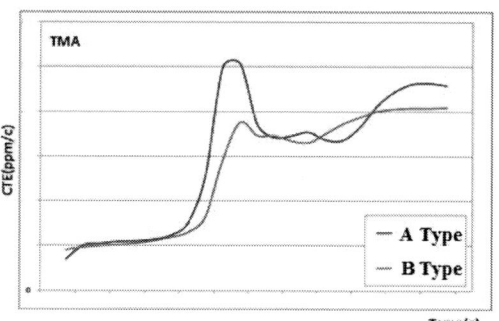

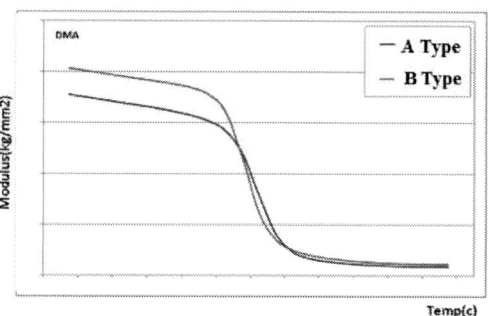

Figure 8: Difference EMC of TMA and DMA curve chart comparison

The purpose of strip warpage simulation which was used to check assembly workability after molding process. The output result was shown strip warpage ratio comparing with two EMC type in Table 4. The type B EMC got better strip warpage result than type A EMC material. The 3D metal antenna structure can provided good supporting inside molding. Due to low CTE EMC can provided warpage mis-match and also 3D metal antenna can provided structure balance during assembly process. The 3D metal antenna is not only provided RF connectivity purpose, but also provided warpage enhancement comparing with 2D antenna structure.

Table 4: Warpage simulation comparison result of EMC and antenna factor

Simulation Leg#	1	2	3	4
EMC type	Type A		Type B	
CTE	High		Low	
Antenna	w/	w/o	w/	w/o
Strip WPG Ratio (Smiling)	1.00x	1.19x	0.73x	0.90x

To proceed the actual unit level warpage measurement by Shadow Moiré as shown in Figure 9. We used sample from Leg1 and Leg3 to comform the unit warpage which the type B EMC had low CTE to reduce EMC expending to smiling face (-) shape comparing with type A EMC result. Both of type A and Type B of room temperature warpage shown as -28um & -18um and high temperature shown as 43um and 35um, respectively. Both EMC legs are within JEDEC warpage requirement (Max. 80um).

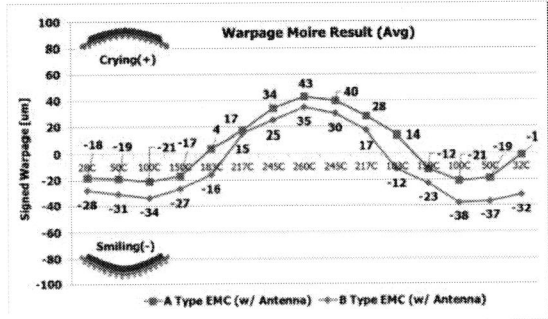

Figure 9: Shadow Moiré measurement result

Reliability Result

First, this paper had demonstrated the 3D antenna SiP into 6 x 6 mm module size, the demonstrated photo as shown in Figure 10.

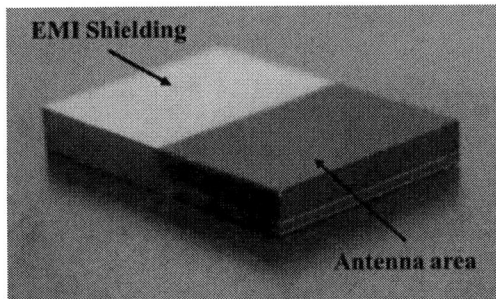

Figure 10: Demonstration result of 3D antenna SiP

As shown in Figure 11 for x-section of 3D metal antenna was used SMT process to mount on substrate. The L-shape metal to provide robust structure during assembly handling process and prevent deformation.

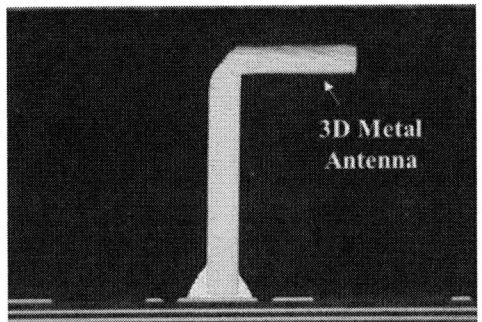

Figure 11: 3D metal antenna of x-section result

Refer to Figure 12, the 3D metal antenna was molded by EMC material for protection and also prevent oxidation. The result also shows good soldering on 3D metal antenna of side wall and Cu substate layer without any non-wetting problem.

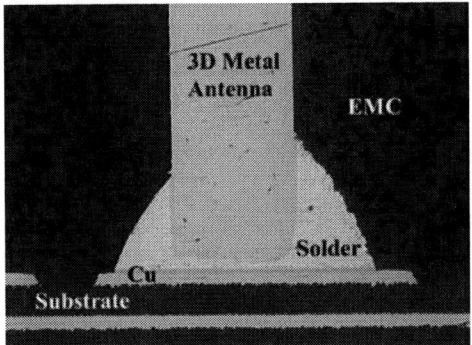

Figure 12: 3D metal antenna of solder joint x-section SEM result

The reliability test items follows standard JEDEC condition includes moisture soaking level 3, temperature cycling test, un-bias HAST and high temperature storage in Table 5 as detail temperature requirement. Each reliability test uses O/S testing and SAT inspecation to confirm the quality performance and got the passed result for all reliability test items including T0, MSL 3, TCT1000, u-HAST96 and HTS1000.

Table 5: Reliability test result for 3D antenna SiP

Reliability Test Items	Read Point	Sample size	O/S Test Result
Time Zero	T0	0 / 231pcs	All Pass
MSL3	Precon	0 / 154pcs	All Pass
TCT (-55 ℃ ~+125 ℃)	1000 Cycles	0 / 77pcs	All Pass
u-HAST (130℃/85%RH)	96 Hours	0 / 77pcs	All Pass
HTS (150℃)	1000 Hours	0 / 77pcs	All Pass

Conclusions

In this paper, we demonstrated the 3D antenna SiP technology included structure and feasibility data for IoT and 5G application. 3D antenna integration can shrink the overall package size to get low cost and high integration requirement. The 3D metal antenna design could provided wider bandwidth comparing with traditional 2D PCB antenna design structure, the simulation of return loss result shown frequency bendwidth had around 1.50X enhancement. This structure provided package structure balance during thermal assembly process for strip and unit warpage enhancement. Completed package level reliability test for following test conditions, passed MSL 3, TCT1000, u-HAST96 and HTS1000. This paper provided the 3D SiP with advanced antenna integration solution for future product requirement.

Acknowledgements

Authors would like to thank you for members of SPIL SiP technology development team (PIE/TD/PE) and reliability team of SPIL for their help in sample preparation, also thanks for CRD team for material simulation and experimental setup, and substrate suppler partners for test substrate manufacturing, and also thanks metal shielding vendor for antenna manufacturing.

References

[1] Ying-Wei Lu, Bo-Siang Fang, Hsuan-Hao Mi and Kuan-Ta Chen, "Mm-Wave Antenna in Package (AiP) Design Applied to 5th Generation (5G) Cellular User Equipment Using Unbalanced Substrate", in Proc. 68th Electronic Components and Technol. Conf. (ECTC), pp.208-213, 2018.

[2] Mike Tsai, Albert Lan, Chi Liang Shih, Terence Huang, Ryan Chiu, S. L. Chung, J. Y. Chen, Frank Chu, Cheng Kai Chang, Sheng Ming Yang, Daniel Chen, Nicholas Kao, "Alternative 3D Small form Factor Methodology of System in Package for IoT and Wearable Devices Application", in Proc. 67th Electronic Components and Technol. Conf. (ECTC), pp.1541-1546, 2017.

[3] Chih-Chun Hsu, Sheng-Mou Lin, Ying-Chih Chen, Wen-Zhou Wu, Che-Ya Chou, Charles Nan-Cheng Chen, Gavin Gao and Chen-Chao Wang, "3-D Antenna-in-Package Design with Metallic Coating Technique", in Proc. 68th Electronic Components and Technol. Conf. (ECTC), pp.1472-1478, 2018.

[4] Atom O. Watanabe, Tong-Hong Lin, P. Markondeya Raj,Venky Sundaram, Manos M. Tentzeris, Rao R. Tummala and Tomonori Ogawa, "Leading-Edge and Ultra-Thin 3D Glass-Polymer 5G Modules with Seamless Antenna-to-Transceiver Signal Transmissions", in Proc. 68th Electronic Components and Technol. Conf. (ECTC), pp.2020-2025, 2018.

[5] Chung-Yi Hsu, Chia-Ling Chiang, Lih-Tyng Hwang, and Fa-Shian Chang, "Design of a Compact Broadband Butler Matrix and its Application in Organic Beam-former at the 5 GHz Band", in Proc. 68th Electronic Components and Technol. Conf. (ECTC), pp.2121-2127, 2018.

[6] Xiaoxiong Gu, Duixian Liu, Christian Baks, Jean-Olivier Plouchart, Wooram Lee and Alberto Valdes-Garcia, "An Enhanced 64-Element Dual-Polarization Antenna Array Package for W-band Communication and Imaging Applications", in Proc. 68th Electronic Components and Technol. Conf. (ECTC), pp.197-201, 2018.

[7] Cheng-Yu Ho, Sheng-Chi Hsieh, Ming-Fong Jhong, Chen-Chao Wang and Chun-Yen Ting, "A 77GHz Antenna-in-Package with Low-Cost Solution for Automotive Radar Applications", in Proc. 68th Electronic Components and Technol. Conf. (ECTC), pp.191-196, 2018.

[8] Mike Tsai, Ryan Chiu, Eric He, J. Y. Chen, Royal Chen, Jensen Tsai and Yu-Po Wang, "Innovative Packaging Solutions of 3D System in Package with Antenna Integration for IoT and 5G Application", in Proc. 20th Electronics Packaging Technology Conf. (EPTC), 2018.

[9] 3GPP "NR; User Equipment (UE) radio transmission and reception; Part 2: Range 2 Standalone (Release 15)" 3GPP TS 38.101-2 V15.2.0.

[10] JEDEC Solid State Technology Association. JESD22-A104C: Temperature Cycling; 2005

Novell embedded microbump approach for die-to-die and wafer-to-wafer interconnects with variable microbump diameters and down to 5 μm interconnect pitch scaling

J. Derakhshandeh, E. Beyne, G. Capuz, F. Inoue, V. Cherman, I. De Preter, F. Duval, J. Slabbekoorn, C. Gerets, C. Heyvaert, F. Beirnaert, T. Cochet, P. Bex, L. Hou, M. Lofrano, G. Jamieson, N. Heylen, S. Suhard, M. Honore, T. Webers, J. Bertheau, A. Phommahaxay, G. Van der Plas, K. J. Rebibis, A. Miller and G. Beyer

Imec, Leuven, Belgium, Email: jaber.derakhshandeh@imec.be

Abstract— In this paper a novel solder-based die-to-die or wafer-to-wafer interconnect approach is introduced. This technique allows for microbump interconnects with different diameters on a single die and allows for pitch scaling down to 5μm. A metal damascene process is used to create the metal pad layers for soldering. Solder μbumps are created by semi-additive electroplating. After embedding the solder μbumps in a partially cured polymer, the wafer surface is planarized and the solder surface is exposed. This results in flat die surfaces of the die before bonding. Using a thermo-compression bonding process these flat surfaces can be aligned, the solder reacts with the metal pad to form a solder joint and the polymer can bond and cure to ensure a void less underfill layer for mechanical strength and increased reliability. The selection of suitable metals and polymers as well as the different process steps together with reliability data will be discussed in detail. Electrical yield and quality of bonding is demonstrated using imec 5um pitch PTCU/W test vehicle for die to wafer bonding.

Key words: 3D STACKING, EMBEDDED BUMPS, FINE PITCH MICROBUMPS, TCB, IMC GROWTH RATE

I. INTRODUCTION

Advanced 2.5D, 3D and heterogeneous die stacking technologies require increasingly denser interconnects with reduced die separation, improved thermal performance and high current carrying capabilities [1]. The novel technology proposed in this paper addresses some of these challenges by allowing scaling μbump pitches down to 5μm while, at the same time, allowing for larger solder bump connections, up to 40μm pitch. It also reduces the die-to-die separation distance to approximately 4μm, which greatly reduces the electrical μbump parasitic inductance and resistance as well as the die-to-die thermal resistance.

Conventional μbump technology uses semi-additive electroplating for realizing small μbump metal pads and μbump solder bumps or pillars. Our approach for scaling, which involves conventional μbumps from 40 to 20μm pitches is illustrated in figure 1a top. Using a wafer-level applied underfill (WLUF) and thermocompression bonding, high electrical yield and good reliability can be obtained [2]. Scaling this approach to 10μm pitch and below however, becomes increasingly difficult as placement tolerances become smaller and processing of the semi-additive metal and solder μbumps becomes challenging, as we reported earlier in [3].

Furthermore, the risk of solder μbumps sliding in between metal μbumps during assembly increase the risks of shorting connections [3].

II. A NOVEL PROCESS FOR 5μM PITCH MICROBUMPS

In order to increase the tolerance to misalignment and avoid the observed problems with μbump sliding and interlocking during thermo-compression bonding (TCB) we introduce the use of metal-damascene solder pads. Instead of electroplating a μbump on the die surface, a metal pad is created in the chip passivation layer, resulting in a flat wafer. This allows for the fabrication of metal landing pads with very narrow pad-to-pad spacings, which increases the tolerance to TCB misalignment as illustrated in figure 1b. To realize the solder bump, a pure solder pillar or metal/solder pillar is electroplated on the similarly planarized chip passivation.

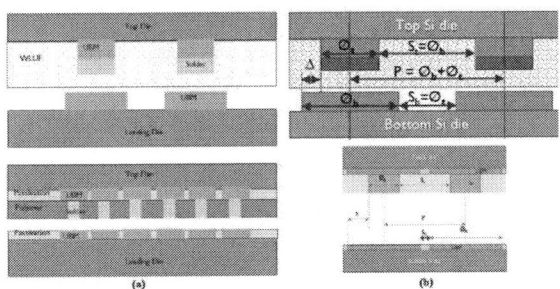

Fig. 1. Comparison of conventional (top) vs embedded bumps (bottom), (a): bonding (b) TCB overlay

When scaling μbump pitches, the diameter of these solder bumps scales down significantly, down to 2μm ∅ for 5μm pitch connections. As shown in Fig. 2 in conventional bumping the required accuracy of TCB tool for 10um pitch is 1um which current TCB tools can not make it. Using damascene pad process, since bottom pads can be larger with narrower space, the required overlay accuracy for TCB is 3.5um for 10um pitch, which can be handled by current TCB tools. For 5um pitch a 1um accuracy is needed when using embedded bumps.

After processing these μbumps, they are embedded in a polymer layer that will act as a wafer-level applied underfill and provide mechanical integrity during the assembly. Next the surface of the solder μbumps is exposed by surface planarization or chemical-mechanical polishing, resulting in a flat wafer surface. The stacking process consists of aligning and bonding flat surfaces. This approach can be applied for face-to-face (F2F) and back-to-face (B2F) die-to-die or die-to-wafer (D2W) stacking as well as wafer-to-wafer (W2W) bonding. The embedded μbump bonding approach is less sensitive to surface defects and small particles as compared to direct hybrid bonding (Cu/oxide to Cu/oxide) due to the use of a polymer bonding layer and solder bumping.

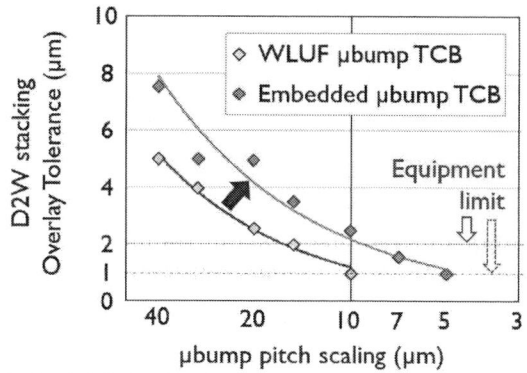

Fig. 2. (a): Calculated overlay tolerance of TCB for conventional and embedded microbumps.

III. SELECTION OF UBM METALS

For below 10um pitch since UBM and solder volume are limited, there is a possibility to convert all solder or UBM to IMC layer thus reliability issues can happen. The ultimate solution for stacking below 10um pitch is to use hybrid bonding where Cu to Cu and oxide to oxide bonding take place without any solder. However, in die to wafer bonding, having clean dies and controlling or removing particles is a challenge which needs development of clean and vacuum operating die to wafer TCB tools. Any trapped particle will prevent electrical connection of two flat dies. Using the introduced approach of polymer embedded bumps, particle issue can be solved because the polymer is compliant material. Since this approach is using a damascene process for creating metal pads and under-bump metal (UBM) pads, the thickness of the passivation layer should match the anticipated UBM metal consumption during solder intermetallic compound (IMC) formation. When using Cu as UBM metal, a passivation layer thickness of 2-3 μm must be used to ensure reliability life of product. In order to avoid the need for such thick passivation layers, different metals with lower UBM consumption should be used. We propose the use of Ni or Co damascene metallization. Fig.3 displays the measured IMC growth rate versus temperature for Cu/Sn, Co/Sn and Ni/Sn systems in both reaction and diffusion modes. Below empirical power law equation describes IMC growth rate as a function of time and temperature:

$$x(t,T) = x_o + At^n e^{\frac{-Q}{RT}} \qquad (1)$$

Where x is the IMC thickness, x_o is the initial IMC thickness, A is the interdiffusion coefficient, n is the power factor, t is the aging time, Q is the activation energy, R is the universal gas constant ($8.314 \frac{J}{mol.K}$) and T is the ageing temperature.

In reaction mode power factor is close to 1 which means IMC thickness has linear relation with time, while in diffusion mode which happens during device working time, power factor is 0.5 meaning that IMC thickness is proportional to square root of time. At temperatures below 160°C, Cu/Sn IMC (mainly Cu_6Sn_5 phase) grows very fast while Co/Sn ($CoSn_3$ phase) and Ni/Sn (Ni_3Sn_4 phase) IMC growth rates are very slow. With further ageing, dissolution of Ni and Co is rather limited and a thin UBM is sufficient.

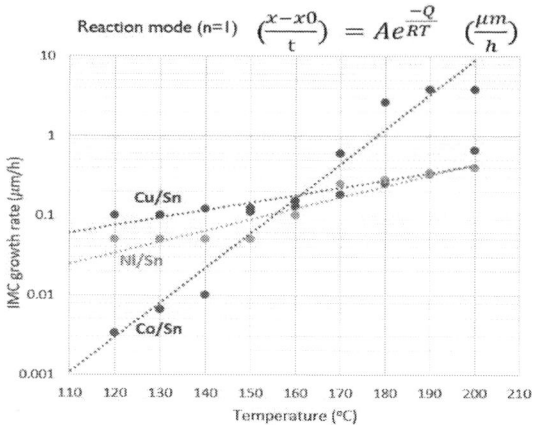

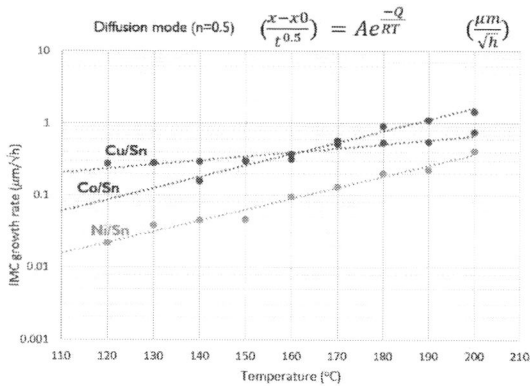

Fig. 3. IMC growth rate vs temperature in reaction (top) and diffusion (bottom) modes for Cu/Sn, Ni/Sn and Co/Sn systems

Table 1 summaries the extracted IMC growth parameters in both reaction and diffusion modes. Lower activation energy can be seen for Cu/Sn which means IMC grows easier in Cu/Sn system while for Co/Sn specially at the beginning of bonding which is reaction mode, activation energy is 5 times higher than Co. Therefore, there will be no reaction between Co and Sn unless we increase the temperature to overcome the activation energy. Since interdiffusion coefficient of Co/Sn system is very high when passing the barrier of activation energy, IMC will grow very fast as shown in the plots of Fig.3. Therefore, a barrier is needed in between Co/Sn and Ni/Sn to prevent further IMC formation and solder consumption. In general, for embedded bumps and high-density interconnects Cu/Sn must be replaced by Co/Sn and Ni/Sn systems.

8

Table 1: Extracted growth rate parameters

Diffusion couple		Interdiffusion coefficient (A)	Activation energy Q (kJ/mol)
Cu/Sn	Reaction mode	1645.9 µm/h	32.73
	Diffusion mode	99.25 µm/√h	19.84
Ni/Sn	Reaction mode	107279 µm/h	49.08
	Diffusion mode	302117 µm/√h	53.91
Co/Sn	Reaction mode	10^{18} µm/h	155.22
	Diffusion mode	6×10^6 µm/√h	59.72

Cu has one advantage that creates fast IMC layer which is good for bonding thus to have minimum UBM consumption and fast IMC formation during TCB one should use the proposed metallurgy system in Fig. 4 for optimum performance. In top wafer, thin layer of Cu creates an IMC layer which acts as barrier for IMC formation between Ni (or Co) and Sn. A thin layer of Cu finish can be used at bottom wafer to obtain fast IMC during TCB.

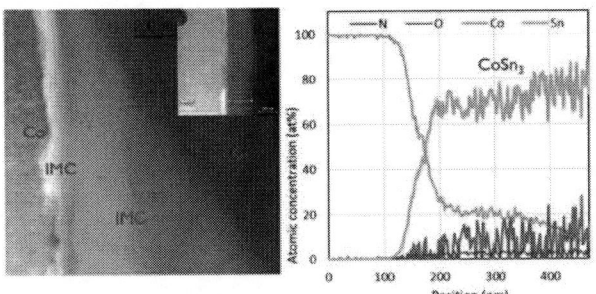

Fig. 4. Metallurgy options for UBM and solder in top and bottom dies

Table 2 shows other properties of IMC layers such as thermal conductivity, CTE and hardness. Ni_3Sn_4 and $CoSn_3$ IMCs have single phase IMC and almost similar electrical resistivity, thermal conductivity and CTE values.

Table 2: Other parameters of IMC layers

	Electrical resistivity (µΩcm)	Thermal conductivity (W/K.m)	CTE (10^{-6}/°C)	Young's modulus (GPa)	Hardness (GPa)
Cu	1.7	401	17	130	1.2
Ni	7	90.9	13	210	7.8
Co	6.2	100	13	209	3.3
Sn	12.3	78	22	53	0.15
Cu_3Sn	8.9	70.4	18.2	134	6.9
Cu_6Sn_5	17.5	34.1	18.3	120	7
Ni_3Sn_4	30	19.6	14.6	211	8.5
$CoSn_3$	39	19.6	15	98.5	4

$CoSn_3$ IMC is twice softer than Ni_3Sn_4 layer. It can also be seen that Cu/Sn has two IMC phases of Cu_6Sn_5 and Cu_3Sn. Cu_6Sn_5 IMC grows immediately after plating and transforms to Cu_3Sn IMC where with ageing the well-known Kirkendall

voids can be formed. This is another limitation of Cu UBM which cannot be used in fine pitch microbumps. In our proposed system at Fig. 4, Cu thickness if very little thus only Cu_6Sn_5 IMC phase can be made in top die to prevent solder consumption and at bottom die for fast bonding [2].

Fig. 5 shows TEM and EDX mapping element analysis of Co/Sn sample verifying single phase $CoSn_3$ IMC layer.

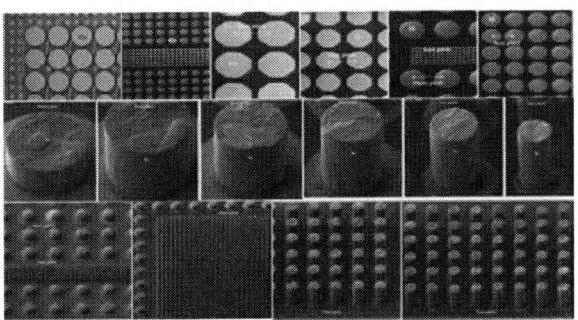

Fig. 5. (a): TEM and EDX analysis shows single phase $CoSn_3$ IMC

Fig. 6 top shows optical and SEM images of processed Cu, Ni and Co UBM metal pads by damascene process on 300mm Si wafers. For Co and Ni damascene process, special slurry and seed/barrier layers are needed to prevent the corrosion of pads.

Fig. 6. Top: UBM (Cu, Ni and Co) damascene process development, Middle/bottom: Plated Sn bumps with different bump sizes in same chip.

IV. PROCESSING SOLDER µBUMPS

On top wafers after UBM CMP, Sn bumps were formed by electroplating inside patterned resist followed by seed and barrier wet etching. The Sn µbump diameter scales with reducing µbump pitch. For 5µm pitch, 2µm diameter Sn µbumps are plated. The plating thickness varies between 5.5 and 6.5 µm for bumps ranging from 2 to 16 µm, respectively. This is acceptable as the µbumps are planarized in a later step. The seed layer was etched without creating any undercut in µbump as shown in Fig.6 middle and bottom SEM images.

V. POLYMER SELECTION AND PROCESSING

In this technology the polymer layer is used as a supporting material for the high aspect ratio Sn µbumps as well as an underfill (UF) material during thermocompression bonding to ensure mechanical stability and good reliability of the µbump assemblies. After spin coating, the polymer is soft baked to remove any solvents and only partially cured to obtain a stable film for planarization and still reactive as "glue" layer during stacking. The process to planarize the polymer and "reveal" the Sn μbumps is shown schematically in Fig. 7a. Mainly two methods are considered: chemical-mechanical polishing (CMP) or mechanical diamond bit cutting (surface planer). Surface planer is a mechanical method for planarization which offers good planarization over the wafer for different pitches and applicable to most polymers. CMP offers a cleaner surface, some protrusion of Sn bumps but suffers from height differences (erosion) in the dense structures. After planarization, all μbumps have the same height, independent on their size and pitch. As shown in Fig. 7b surface planer gives smoother polymer and Sn surface while CMP has made some scratches on polymer and Sn bumps. Depending on polymer type and soft bake condition a suitable slurry is needed while surface planer is almost independent on polymer type.

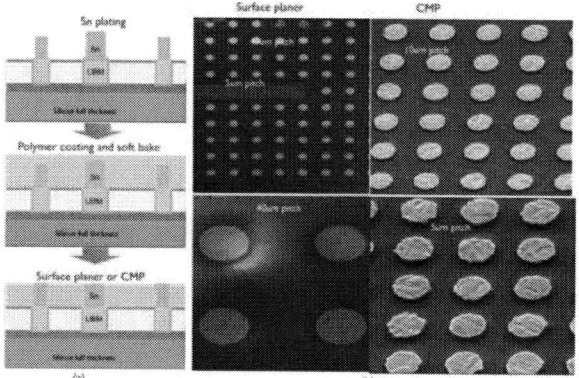

Fig. 7. Revealing Sn bumps by surface planer or CMP.

Several polymers were tested for this application and for every polymer the soft bake condition had to be optimized to obtain good planarization (no surface scratches and no damage on Sn bumps), good die adhesion, less voiding and good IMC formation during TCB. Fig. 8 compares the surface of 4 different types of polymers after planarization by surface planer. In all tested polymers there is no bump deformation after planarization. The surface of wafers is smooth for first 3 polymers and it is rough in polymer 4. An optimization of the polymer bake condition before surface planarization was necessary for this polymer in order to obtain a smooth surface.

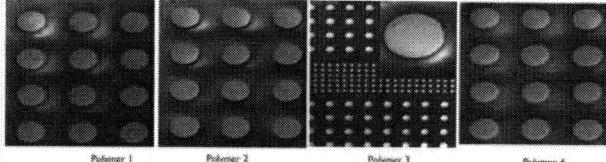

Fig. 8. SEM images after surface planer for 4 different types of polymers.

The surface topography, measured by AFM, in the revealed μbumps by surface planer using polymer 1 are shown in Fig.9 for different designed pitches. 40um pitch bumps have around 250nm protrusion while 5um pitch bumps have 150nm protrusion. Such a small protrusion is desired for the assembly process to have good electrical yield.

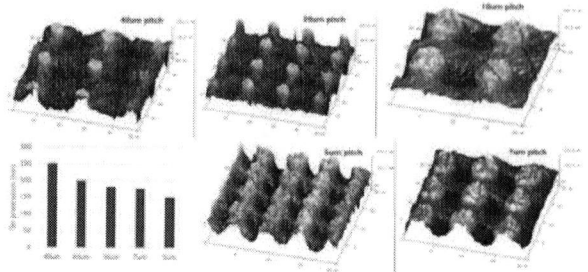

Fig. 9. AFM images of Sn bumps after revealing by surface planer. Larger bumps have 100nm more protrusion. (polymer 1)

In polymer 4 which is polyimide-based polymer, surface of polymer is rough due to high softbake condition and amount of bump protrusion is shown in Fig. 10. As it can be observed the protrusion is less than polymer 1 therefore higher TCB force is needed for polymer 4 to obtain a proper electrical connection.

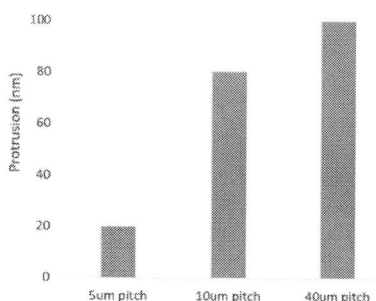

Fig. 10. Measured bump protrusion by AFM for polymer 4

For multi-die stacking, planarization by CMP or surface planer may fail due to thin Si wafer thickness variation over the wafer and the fact that TBM (temporary bond material) is not a uniform layer between device wafer and carrier wafer. In this condition the best way to reveal the bumps is to use photo-patternable polymers. Fig. 11 shows SEM images of a photo patterned wafer in different microbumps pitches all in same wafer. Polymer resolution should be good enough with 2um opening and 3um space CDs for 5um pitch structures.

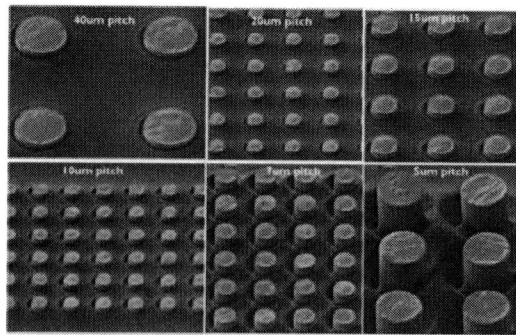

Fig. 11. Revealing Sn bumps using photo-patternable polymer

VI. RESULTS AND DISCUSSIONS

The thermocompression bonding was performed using a "TC advanced" bonding tool from Besi. This tool is a fully automated high precision, dual bond head thus high throughput TCB tool. The die placement accuracy is about 2μm. This should enable 10 μm pitch connections. With a tool placement accuracy of 1μm; a 5μm pitch connection is possible. A 10 second TCB profile cycle and a bonding force of 1 to 4kgf was used. Fig. 12 top left shows electrical yield of single bump kelvin structures and in top right the yield of daisy chains for different pitches with a stacking force of 4kgf measured on more than 100 structures of 40, 20, 15, 10, 7 and 5um pitches using polymer 3, are plotted. Y-axis is cdf (cumulative distribution function) percentage and x-axis is the resistance values in Ohm. Daisy chains have 219 and 73 bumps in each structure. As plotted in Fig. 12 bottom, with increasing stacking force from 1kgf to 4kgf, the electrical yield is improved to 75% for 5um pitch, 90% for 10um pitch and 100% for larger pitches.

conventional bumps where pressure on the bumps can be as high as 50MPa range. This is another advantage of this method which means there will be no stress on low K dielectric and no change on transistors performance after 3D stacking.

Cross section SEM images of daisy chains are shown in Fig. 13 for two stacking forces of 1kgf and 4kgf. It can be observed that in all pitches there is very good soldering or IMC formation between top and bottom dies. The height of Sn after planarization is 4um which can be seen in 1kgf samples. In 4kgf samples Sn is deformed where Sn and UBM thickness are reduced to 2.5um and 0.5um, respectively. It shows that there is a good IMC formation at higher forces. No voiding is seen in the polymer layer. TCB alignment is very good with 1kgf while there is slight misalignment when increasing force to 4kgf. For this specific die size (7.2mm by 7.2mm) 2kgf or 3kgf will be the optimum TCB forces.

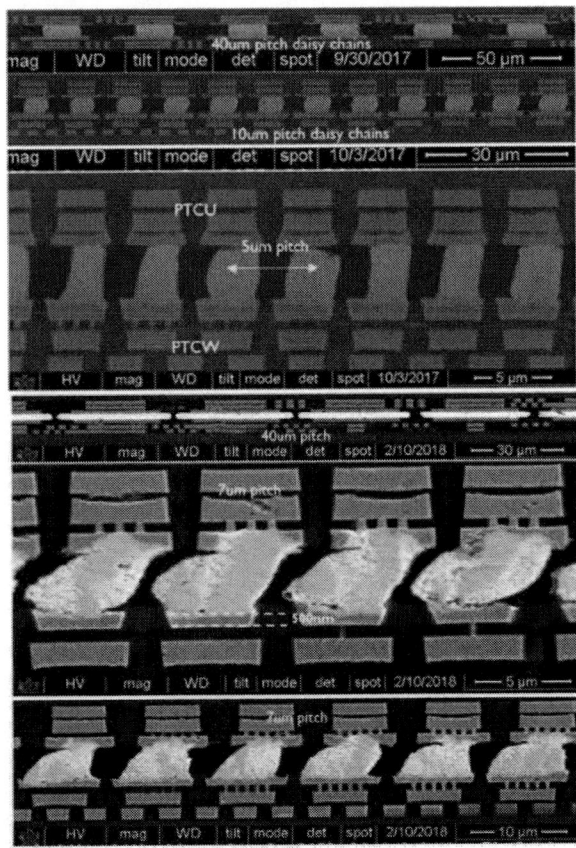

Fig. 13. SEM images of joint and IMC formation for different pitches with 1kgf (top) and 4kgf (bottom) TCB D2W using polymer 3

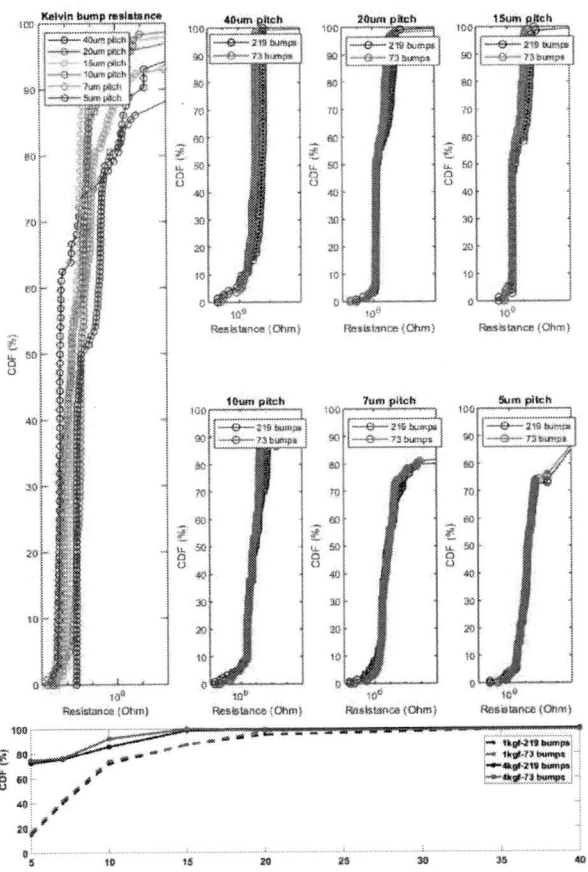

Fig. 12. Top left: Electrical yield of single bump kelvin structures, Top right: Electrical yield of daisy chains for different pitches at 4kgf stacking force, Bottom: Electrical yield vs pitch of daisy chains for 1kgf and 4kgf TCB forces. Polymer 3 was used in this example.

Applied TCB 4kgf force can be converted to 0.75MPa pressure on bumps which is quite low compared with

Fig. 14 shows the measured die shear force of 16 stacked samples at 1kgf using different polymers. Stacked samples using first 3 polymers have die shear force higher than 120kgf while it is 3 times less in polymer 4. The reason is that polymer 4 was baked at higher temperature after coating to remove the solvents therefore polymer tackiness is lower. In the tested samples with higher die shear force top die was broken while in polymer 4 top die was sheared off showing weak adhesion. One can use extra backing to cure polymer completely after die population on full wafer. These results show that snap cured polymers are preferred for this application.

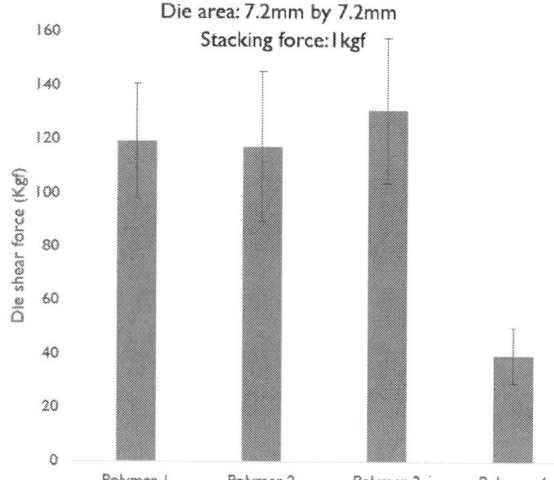

Fig. 14. Measured die shear force of 16 stacked samples per polymer

Furthermore, stacking using dry film (wafer level underfill, WLUF or non-conductive film, NCF) laminated wafer was also performed. Based on viscosity of NCF layer, the lamination condition must be optimized to make sure that different pitch bumps will be coved by the film. Fig. 15 shows cross section SEM images of bonded samples before and after MSL test (moisture sensitivity level). It can be seen that the electrical connection is good, but some voiding can be seen. Since this voiding does not grow after MSL test, this type of boding can still pass reliability tests successfully.

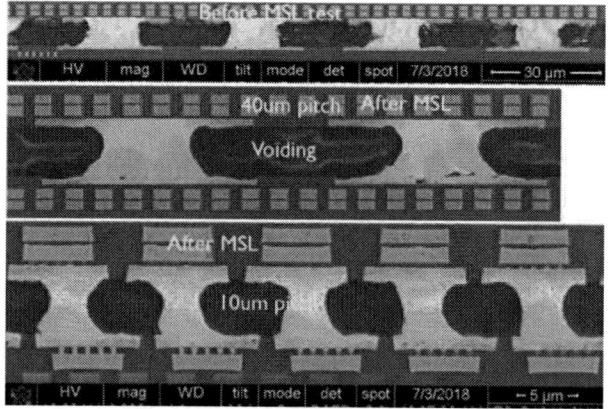

Fig. 15. SEM images of boned samples using NCF dry film before and after MSL test

Fig. 16 shows SAM images of stacked 16 dies before (left) and after (right) MSL test for dry film. No voiding seen in the stacks before and after MSL test.

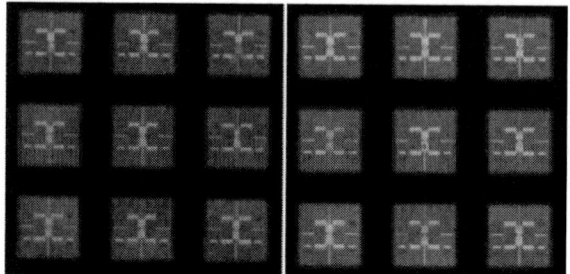

Fig. 16. SAM images of the stacked 16 dies before (left) and after (right) MSL test using dry NCF film

Fig. 17 shows cross section SEM images of bonded samples using polymer 4 before and after MSL test. No voiding can be seen between the bumps. Electrical connection is good for large pitches but for small pitches it is marginal. Electrical yield of this polymer was low for 5um and 7um pitches due to harder polymer. Increasing stacking force or lowering the baking condition can improve the connection of small pitch bumps.

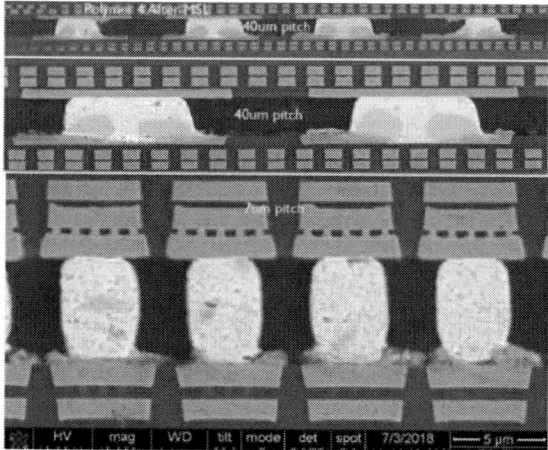

Fig. 17. SEM images of bonded samples using polymer 4 before and after MSL test

VII. CONCLUSIONS

In this paper for first time a novel approach was introduced which enables microbumps pitch scaling to 5um and integrating different microbumps diameter in one design. Both top and bottom dies or wafers are flat. UBM is formed using damascene process and solder is embedded in a polymer and planarized. Electrical yield of 75% and 90% were obtained for 5um and 10um pitches for D2W stacking. SEM images show very good IMC formation between top and bottom dies. It was shown that Ni and Co should be used as UBM layer instead of Cu in this technology.

REFERENCES

[1] E. Beyne, "The 3-D Interconnect Landscape", IEEE Design & Test, 2016, Vol.33, no. 3, pp.8-20.
[2] I. De Preter et al, "3D stacking of Co and Ni based microbumps", *ESTC 2016*
[3] J. Derakhshandeh et al, "3D stacking using bump-less process for sub 10um pitch interconnects", *ECTC 2016*

Glass Wafer Level Packaging Enabled by Laser Induced Deep Etching of Closed Cavities

Jean-Pol Delrue, Roman Ostholt, and Norbert Ambrosius

LPKF Laser & Electronics AG, Osteriede 7, 30827 Garbsen, Germany

+49-513170950, and info@vitrion.com

Abstract

Glass capping wafers are regularly used in wafer level packaging -(WLP) applications. There are several reasons to use glass in these applications such as high surface quality, chemical inertia, hermeticity and among others silicon matched CTE, high thermal stability, transparency and low cost. The objective of this paper is to demonstrate the capability to manufacture closed pockets for glass capping wafers based on Laser Induced Deep Etching (LIDE). Today's glass capping wafers are usually manufactured by masked isotropic wet etching. Isotropic wet etching is associated with under etching which results in pockets showing pronounced fillet at the bottom of the pocket. Obviously, to be packaged, a device requires a pocket with a depth bigger than the height of the device. As the radius of the fillet at the bottom of the pocket equals the depth of the pocket, the device pitch in wafer level packaging using today's glass capping wafers is ultimately limited by the manufacturing process of the capping wafers. Authors will present closed cavities with rectangular shaped profile in glass made by LIDE technology. The demonstrated cavities have a depth of >300 μm and a surface roughness of Ra ≈ 0.5μm.

Key words: wafer Level Packaging, closed pockets for glass capping wafers, Laser induced deep etching.

Laser Induced Deep Etching

LPKF developed a technology called Laser Induced Deep Etching (LIDE). It consists of a two-step process using first a fast laser beam modifying the glass, then followed by a chemical etching of the glass. The LIDE technology is capable to manufacture deep microfeatures in glass with unprecedent quality and precision. It is a fully digital-mask free process on panel sizes up to 20" X 20". As shown by the authors before [1], LIDE technology is capable to generate high aspect and high quality microfeatures in glass. As such, the technology is regularly used in applications like through glass vias (TGV), glass spacer wafers and microfluidics [2].

WLP with glass capping wafer

Wafer-level packaging is an established low-cost and high-volume production technology [3]. Glass capping wafers are usually made with an etch mask (typically chromium). Glass wafers are wet etched. This is associated with under etching of the etch mask, resulting in capping wafers with closed cavities with a pronounced fillet at the bottom (Figure 1).

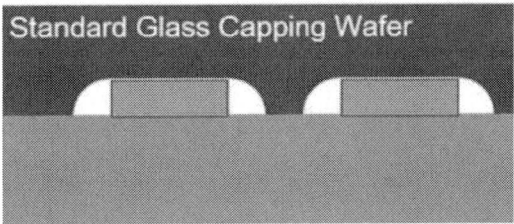

Figure 1: Wafer Level Packaging with standard glass capping wafer

Glass Capping Wafers made by LIDE

LIDE enables the manufacturer of cavities with steep sidewalls. For this purpose, the process is controlled in such a way that the modifications are intentionally not introduced over the entire thickness of the substrate. Usually, the modifications are arranged in a hexagonal pattern to produce a surface that is as flat as possible.

A SEM image of a closed cavity made by LIDE is shown in figure 2. The sidewalls have a tilt angle of approximately 1°.

Figure 2: Closed glass cavity made by LIDE

The advantages of using a capping wafer manufactured by LIDE for wafer level packaging essentially result from the favorable cross-section profile.

This profile fits better to the also rectangular silicon dies and enables higher population densities as it can be seen in the schematic figure 3 below compared to the schematic of figure 1.

Figure 3: Wafer Level Packaging with LIDE processed glass capping wafer

Experimental design

The goal of this paper is to characterize closed cavities manufactured by Laser Induced Deep Etching technology in terms of:

- Position tolerance
- Size tolerance of cavities
- Depth control
- Roughness

All the results shown below are generated with the help of a specialized laser machine equipment from LPKF Laser & Electronics AG. The array of modifications intentionally does not penetrate the full substrate.

Tests are carried out using φ 150 mm Schott Borofloat 33 glass wafers with an initial thickness of 500 μm. The test layout represented in figure 4 consists of 2 mm x 1 mm cavities with a pitch of 7 mm which leads to 332 cavities over one panel in total.

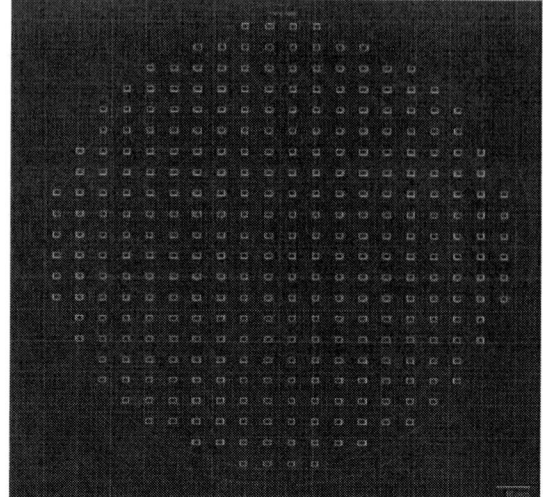

Figure 4: Test Layout

To eliminate the glass thickness tolerances during the laser modification, the position of the wafer surface is measured prior to the modification. The deviation is corrected accordingly during the laser modification.

After the laser modifications, the wafers are etched in a hydrofluoric containing etchant to generate cavities with a target depth of 180 μm.

In order to check the capability of the glass capping wafers made by LIDE, the measurements of the height and width were done with an Automatic Optical Inspection device (AOI). In this specific case, we used a Werth Videocheck HA600 equipment. This measurement tool has an accuracy of 0.5 μm + L/900 with a reproducibility of <1 μm. A type 1 study of the Videocheck HA600 shows a standard deviation of the diameter measurement of 150 nm. For the material alignment crosses were machined in each corner of the glass wafers.

Variation of laser pulse pitch

The roughness of the cavity mainly depends on the distance between the individual laser modifications of the glass.

To evaluate the cavity roughness, the pitch of the laser modifications was progressively changed. A schematic visualization of the Laser modifications and the resulting etched profile are shown in figure 5. Figure 6 shows different SEM pictures of the closed cavities with different roughness's. Roughness measurements of the bottom of the cavities will be addressed later in the measurement results paragraph where we will show that a longer etching time leads to a better bottom cavity roughness.

Figure 5: below shows the schematic of the cavity.

Figure 6: SEM pictures of closed cavities

To measure the cavity depth, the glass panel was cut with a diamond scribing wheel. The cutting facet was then inspected with a Keyence VK X210 microscope and the glass thickness and cavity depth were measured and shown (Figure 7).

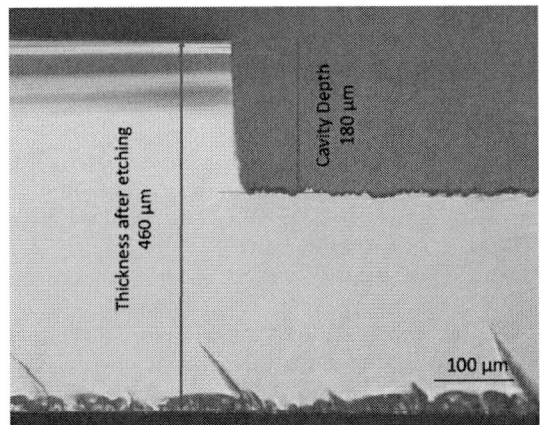

Figure 7: Light microscopy image of a cross section of a cavity made by LIDE.

Measurement results

The main results regarding the cavities, e.g. their size tolerances, their depth control and their position tolerance are presented respectively in figures 8, 9 and 10 below.

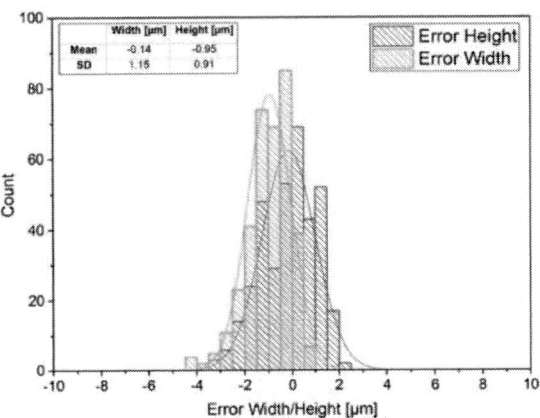

Figure 8: Measurement results on the cavity size

The mean size error of the cavity height and width is -0,14 µm in the x-direction and -0,95 µm in the y-direction. The standard deviation was measured with a value of 1,15 µm in the x-direction and of 0,91 µm in the y-direction.

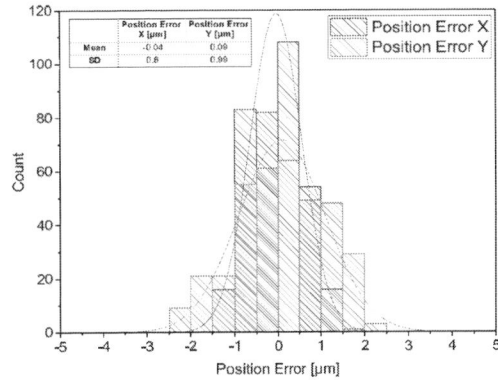

Figure 9: Measurement results on the cavity depth

The mean position error of the cavities is of -0,04 µm in the x-direction and of 0,09 µm in the y-direction with respect to alignment features, the standard deviation is of 0,6 µm in the x-direction and of 0,99 µm in the y-direction.

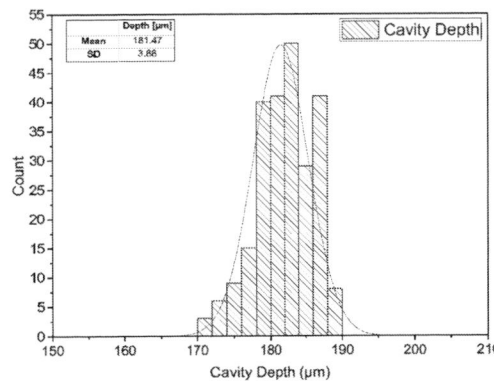

Figure 10: Measurement results on position error

The measured mean cavity depth is 181,47 µm whereas the targeted value is 180 µm. The standard deviation is 3,88 µm.

Roughness Measurement

As it is shown in figure 11, if etched longer the spikes at the bottom of the cavities are reduced and so the roughness as well. So, the etching time parameter can be used to achieve the requested roughness down to a Ra of ≈ 0.5µm.

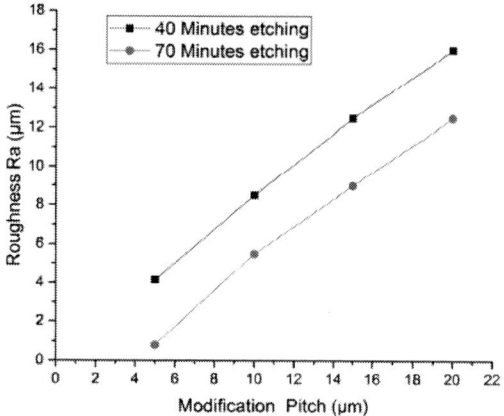

Figure 11: Measurement results of the roughness of the bottom cavities versus the Laser pitch applied for two different etching times

Discussion

The positional accuracy of closed cavities made by LIDE is within the expectation as the laser tool specification is within +/-5 µm (cpk>1.33).

Width and height of the cavities are defined by the same direct writing process. As such, the size tolerance equals the positional tolerance values.

Smallest roughness obtained by LIDE is approximately Ra ≈ 0.5 µm. Although the bottom surface of the pockets created by LIDE does not meet optical standard requirements, most capping applications do not require such a low roughness.

Cavity depth shows a standard deviation of 3.9 µm and is thus significantly larger than the other tolerances. The reason for this is suspected to be insufficient control of the mechanically guided processing head of the laser machine relative to the substrate. Further investigations will show whether and to what extent better cavity depth tolerance can be achieved.

Cost contributions

To evaluate a new manufacturing process, it is essential to consider its cost and, in particular, to understand its influencing factors. Even though LIDE is a very fast manufacturing process, it remains a direct writing technology. The pulse pitch is the main contributor to the overall cost. Fortunately, there is a tradeoff between an acceptable bottom surface roughness and a manufacturing cost to choose an optimal pulse pitch. In addition, wet etching used in LIDE is typically a batch process and does not contribute significantly to the overall process costs.

Conclusions

Glass capping wafers with steep side walls have been achieved with the help of the Laser Induced Deep Etching technology. The size tolerances of the cavities, their depth value (of >300 µm) as well as their position tolerance have been measured. They are within specifications to accommodate WLP packaging. The bottom pocket surface roughness is of Ra≈ 0.5µm which is also more than enough to accommodate the dies.

In conclusion, we believe that LIDE technology is a good candidate to manufacture closed cavities for Wafer Level Packaging, taking into account the fact that the steep side walls achieved allow higher population densities of the die packing.

References

[1] Ostholt, R. (2017), "Novel Method for High-Volume Via Formation in Solid –Core Glass for IC substrates" Chip Scale Review, Sep./Oct 2017, Vol. 21/5

[2] Ostholt, R. (2018). "Laser Induced Deep Etching of Glass (LIDE) and its Contribution to Heterogeneous Integration". Semi 3D Summit 2019, Dresden, Germany

[3] Z.Zhang, C.P. Wong, "Design, Process and Reliability of Wafer Level Packaging" from a book called "Micro- and Opto- Electronics Materials and Structures: Physics, Mechanics, Design, Reliability, Packaging. (pp. B135 – B150), January 2007.

mmWave Antenna Design in Advanced Fan-Out Technology for 5G Application

Sheng-Chi Hsieh, Fu-Chen. Chu, Cheng-Yu Ho and Chen-Chao Wang

ASE Group. No. 26, Chin 3rd Road Nantze Export Processing Kaohsiung 811, Taiwan

Email: Ricky_Hsieh@aseglobal.com

Abstract

In this paper, the fabrication process of AiP is based on Fan-Out substrate. This solution brings short interconnection between die to die and die to antenna for excellent electrical performance. They tend to offer the lowest transition loss due to circuitry directly connected to the antenna. In this study, the package height is slightly less than 0.75 mm, with three RDL layers, two mold layers for antenna design, mold thickness of 250um, and ball height of 200um. Its benefits include a smaller package footprint compared to conventional organic substrate or laminate packages, higher maximum connection density, as well as desirable electrical and thermal performance. In this work, the stacked patch antenna is a wideband deisgn to improve the bandwitdh. Based on design theory of antenna, proper design requires careful optimization of the thickness, feed point and width of patch atenn. By choosing the antenna parameters properly, the proposed stacked patch antenna on fan-out package has better than 10dB return loss in 26.8-32.5 GHz range, with ~5.7GHz bandwidth and provides a high-gain (above ~10.3 dBi) radiation pattern with 2x2 patch antenna array. Beamforming array is a very important technique in 5G application. The proposed thin profile AiP solution can achieve both broadband and high directivity characteristics, which meet the requirement of 5G systems in the allocated frequency bands at 28 GHz applications.

Key words: 28 GHz 5G systems; flip-chip ball grid array (FCBGA) package; flip chip chip scale (FCCSP) package; Fan out wafer level chip scale package (Fan-out WLP); Advanced single sided substrates (fan-out package); Antenna in package (AiP)

Introduction

The recent trends have significantly increased in demands for millimeter-wave wireless communication systems. The potential applications include 28/39-GHz 5th mobile generation 60-GHz high-speed wireless data link, 77-GHz automotive radar and 94-GHz radar imaging, etc [1]-[2]. In the microwave frequency range, the physical wavelength is typically too large to allow antennas be placed on PCB substrate or phone case of reasonable sizes. Hybrid circuits with the antennas on PCB also often are used at low frequencies due to low conductive loss. Connectors and RF cables have generally been used to interface the antenna and RF module. However, as the operating frequency of the mobile systems increases to millimeter-wave, it is often severely degraded. Thus, the complexities of the packages and interconnections are raised since it is required to integrate antenna and RF chips to improve performance by reducing interconnected distance. The best solution is the millimeter wave antenna in package (AiP) to integrate antenna with RF ICs into chip-scale package devices. As shown in Fig. 1, the physical wavelength in millimeter-wave range is smaller than current commercial systems including GSM, LTE, and Wi-Fi. The more compact antenna size makes it suitable for further system in

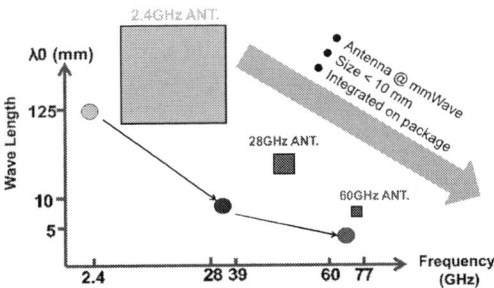

Figure 1: Comparison of wavelength in air at 28 GHz, 60GHz and 77 GHz mmWave applications

package integration. AiP offers a solution for interconnect and antenna implementation problem by integrating the adaptive antennas to the RF chips in the package. A possible high gain antenna can be achieved by antenna array designed with directivity radiators. An increased number of antenna elements exacerbate the problem of finding compact, low cost solutions for the implementation of the antenna and high frequency interconnects. One of the effective solutions is implemented on organic substrate [3]-[5].

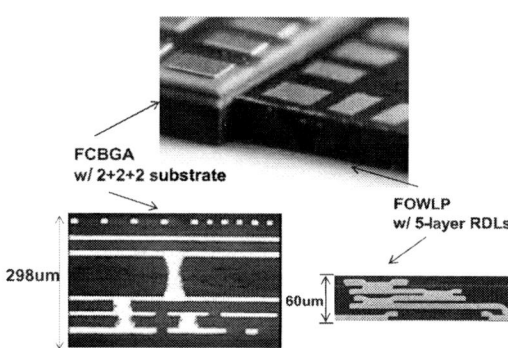

Figure 2: Cross-section view of the mmWave transceiver on flip chip ball grid array (FCBGA) AiP

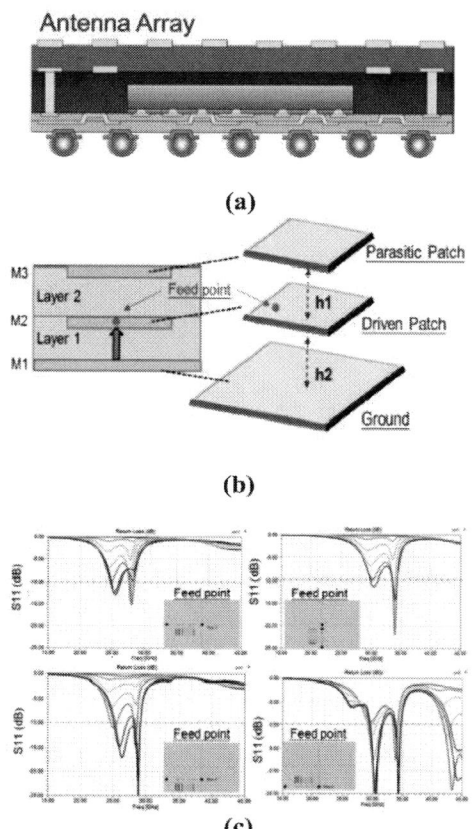

Figure 3: Photograph of FCBGA and Fan-out WLP

It is a low-cost and low loss multilayer organic substrate for flip chip ball grid array (FCBGA) package as shown in Fig. 2. However, the thick substrate thickness is not easy to mount in thin mobile phone case. In addition, 5G mmWave systems create significant challenges for packaging engineers since the power consumption caused by high date rate at mmWave is obviously from active device. The thermal issue of interface on PCB is a very serious challenge to the 5G millimeter-wave systems create significant challenges for packaging engineers since the power consumption caused by high date rate at mmWave is is obviously from active device. The thermal issue of interface on PCB is a very serious challenge to the 5G millimeter-wave systems. A novel solution for advanced packaging is the Fan out wafer level chip scale package (Fan-out WLP). Fan-out WLP is a new generation platform that will support die to die integration, particularly for heterogeneous high-density IO system in a package [6]-[7]. Its benefits include a smaller package footprint compared to conventional lead-frame or laminate packages, maximum connection density, as well as desirable electrical and thermal performance. Fig. 3 shows the comparison of package thickness between FCBGA and FOWLP. After that, fan-out WLP offers small form factor, excellent electrical and thermal performance for mmWave AiP into mobile devices. In addition, foundry-like process is another approach

Figure 4: Fanout antenna in package design (a) Geometry of the fan-out Antenna in Package (b) The Schematic of propose stacking patch antenna (c) Varrying return loss by adjusting the position of the feed point

to reduce process variation and increase yield. A stacking patch antenna with broadband frequency has been widely used in wireless application. A slot tuning has been proposed in the antenna performance improvement. This work proposes a stacking patch antenna with slot structure to increase the bandwidth and directivity characteristics, which meets the requirement of 5G millimeter-wave application.

Fanout Antenna in Package(AiP) Design

The proposed fan-out structure is shown in Fig 4(a). The packaging technology has a number of important features. Package height is slightly less than 0.75 mm, with three RDL layers. It has two mold layers for antenna design, with a mold thickness of 250um, and ball height of 200um. This work proposed a patch antenna design in fan-out package for 28-GHz band. Patch antenna consists of a patch of metal layer with a large ground. There are several advantages to employ patch antenna for mobile applications including low profile, high gain and high compatibility for packaging with IC. However, there is a problem in patch antenna for

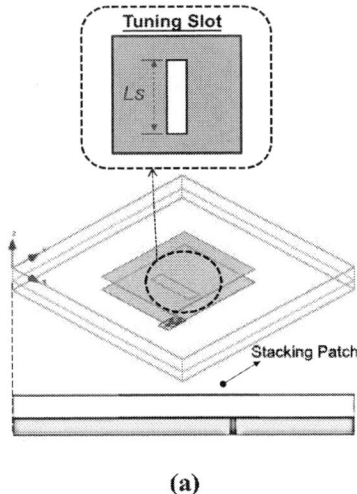

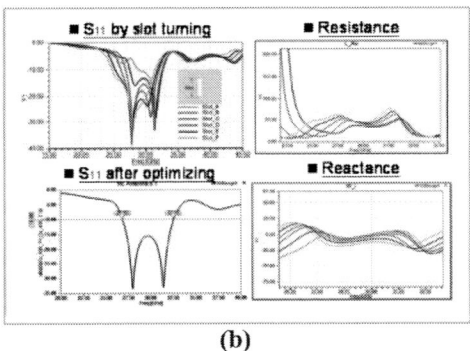

(a)

(b)

Figure 5: Simulation of notching slot structure (a) Geometry of the tuning slot (b) S11 result of various return loss by adjusting the dimension

wireless application. It is well-known that patch antenna have very narrow band. In this paper, we propose stacking patch to obtain broadband antenna design as shown in Fig. 4(b). In the past years, several ideas have been published to increase the bandwidth. One effective idea to improve this issue is by adding a parasitic element above the main patch. In general theory of stacking patches, it is realized by electromagnetic coupling to add bandwidth. In this design, the parasitic element is located on top mold

layer above main patch with a thickness of mold layer 250 μm. The large ground is located at bottom of mold layer. Based on design theory of a patch antenna, proper design requires careful optimization of the feeding point, and width of antenna. The design of the feeding structure is directly connected to main patch. However, the feeding point has to match the antenna impedance. The input impedance varies obviously with resonance frequency depending on different feeding point of patch antenna. A rectangular patch of dimensions is around 2.5mm by 2.5mm. The dielectric constant of mold layers is

εr=3.6. A well feeding point can obtain required bandwidth to cover operating frequency ranges. The return loss of our antenna is a function of frequency which we simulated from HFSS simulation. Fig. 4(c) shows varying return loss by adjusting the position of the feed point. The return loss depends on feeding location which can achieve good impedance matching between the plate and the feeding point. The current loop will introduce the inductance to match the antenna impedance for capacitive loading. Optimization of feeding point for patch antenna is an important initial step to determine antenna structure and to improve impedance bandwidth. In this case, the antenna bandwidth can achieve up to 3.6-GHz after optimization by adjusting the position of the feed point. Futhermore, the system required the antenna bandwidth still needs to be enhanced for 28-GHz application. In this paper, two techniques are employed to improve the bandwidth for patch antenna. One is adding a stacking patch to obtain the broadband antenna design. The other way of enhancing the bandwidth of patch antenna is to add a tuning slot. The proposed tuning slot for patch antenna is as shown in Fig 5(a). The tuning slot is designed on main patch near the feeding strip that can be used to achieve good matching with the inductance. By having the slot design being closed to resonance allows an impedance response similar to adding parasitic to improve the bandwidth. In this case, we choose the simple way of enhancing the bandwidth without increasing the complexity of fan-out structure by adding a tuning slot for antenna impedance matching. Fig. 5(b) shows varrying return loss by adjusting the dimension of notching slot structure. As shown in Fig. 5(b), the return loss of < -10 dB bandwidth is from 26.8 to 32.5 GHz. By combining both techniques, the patch antenna can achieve 5.7 GHz bandwidth which can fit the requirements of 28-GHz for wireless communication applications.

Patch Antenna Array Design

They provide a solution to increase the antenna gain and improve the performance over a single antenna. The patch antenna array allows better control of direction and can be easily controlled for beam direction with phase array technology. In order to obtain best directivity and linearity, the linear arrays are designed to half-wave spacing to give identical results. In this work, we performed a full EM simulation of the 1x2 and 2x2 wideband antenna arrays with identical feeding network. The 3D schematic model of a 1x2 and 2x2 patch antenna is shown in Fig. 6. The 2x2 antenna array has dimensions of around 10mm by 10mm with 3mm spacing for each antenna. The four antenna elements are made of stacked patch antenna structure with same feeding location.

Conclusion

Fan-out WLP minimizes 28-GHz interconnection losses of transition from chip to fan-out package to PCB. Small form factor AiP solutions are important prerequisites for 5G systems market. The proposed AiP solution on fan-out package has a broadband and high-gain (above 5 dBi w/ single antenna) performance. The return loss of < - 10 dB bandwidth is from 26 to 33 GHz, which covers the requirements of 28GHz systems. Finally, this work also demonstrates the simulation for antenna array and radiation pattern of AiP on fan-out package. The maximum gain of the proposed 1x2 and 2x2 antenna arrays are 8.1 dBi and 10.3 dBi.. This work provides a thin AiP approach and is suitable for mmWave applications.

References

[1] J.P. John, "A Great New Invention", U.S. Patent 2,825,028, September 21, 2017.

[2] A. Fischer, Z. Tong, A. Hamidipour, L. Maurer and A. Atelzer, "77-GHz Multi-Channel Radar Transceiver With Antenna in Package," IEEE Trans Antennas Propag., vol. 62, no. 3, pp. 1386 – 1394, March 2014.

[3] J. Hasch, E. Topak, R. Schnabel, T. Zwick, R. Weigel and C. Waldschmidt, "Millimeter-Wave Technology for Automotive Radar Sensors in the 77 GHz Frequency Band," IEEE Trans. Microw. Theory Techn., vol. 62, no. 3, pp. 1386 – 1394, March 2014.

[4] A. Fischer, Z. Tong, A. Hamidipour, L. Maurer, and A. Stelzer, "A 28-GHz antenna in package," in 41st Europ. Microwave Conference, vol. 57, no. 11, pp. 1316–1319, Oct. 2011.

[5] C.-Y. Ho, S.-C. Hsieh, M.-F. Jhong, C.-C. Wang and C.-Y. Ting, "A 77GHz Antenna-in-Package with Low-Cost Solution for Automotive Radar Applications," in Proc. IEEE Electronic Components and Technology Conference (ECTC), pp. 191-196, 2018.

[6] C.-Y. Ho, M.-F. Jhong, P.-C Pan, C.-Y. Huang, C.-C. Wang and C.-Y. Ting, "Integrated Antenna-in-Package on Low-Cost Organic Substrate for Millimeter-Wave Wireless Communication Applications," in Proc. IEEE Electronic Components and Technology Conference (ECTC), pp. 242-247, 2017.

[7] C.-T. Wang, and Douglas C.-H. Yu, "Signal and Power Integrity Analysis on Integrated Fan-out PoP(InFO_PoP) Technology for Next Generation Mobile Applications," in Proc. IEEE Electronic Components and Technology Conference (ECTC), pp. 380-385, 2016.

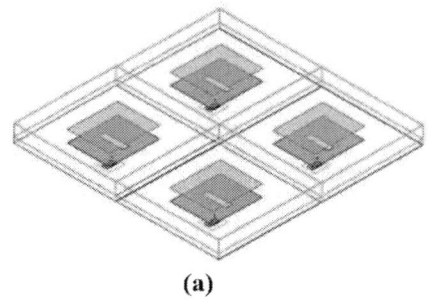

(a)

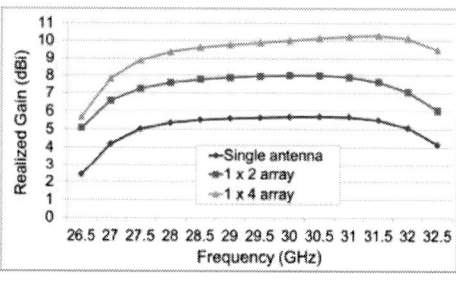

(b)

Figure 6: 2x2 Antenna array Design (a) Geometry of the 1x2 and 2x2 patch antenna array (b) Simulation result of antenna gain

Hermetic and Radiation Tolerant Glass Package for VCSELs Using Novel Micro Bonding Method

H. Lundén*, A. Määttänen*, and I. Oxtoby**

*) SCHOTT Primoceler Oy & Vesiroineenkatu 3, 33720 Tampere, Finland

**) Optocap Ltd & 5 Bain Square EH54 7DQ Livingston, Scotland, UK

+358400807616, heidi.lunden@schott.com, www.schott.com/primoceler

Abstract

In this study, we use novel glass micro bonding method to manufacture hermetic and radiation tolerant all-in-glass packages for vertical cavity surface emitting lasers (VCSEL) suitable for space applications. Thermal analysis demonstrates that the properties of the designed Leadless Glass Package (LGP) are well-suited for the intended conditions, even at higher temperatures than required. Evaluation tests show excellent device functionality and robust package structure. The same design could easily be adapted to other glass types, such as Borofloat 33, and end applications such as MEMS and active implants. The assembly design and technologies are well-suited for commercial volume scale manufacturing.

Key words: Glass, hermetic, VCSEL, all-in-glass packaging

Introduction

Early applications for Vertical Cavity Surface Emitting Lasers (VCSELs) were focused on data communication. However, changes in industry needs have created new markets for VCSELs [1]. Functionalities such as gesture sensing and 3D scanning are being implemented in new mobile devices. The same features are also found in other industry areas. In the automotive and space sector, Light Detection and Ranging (LIDAR) is used in autonomous vehicles as a leading example. These 3D scanning and LIDAR applications often utilize high power measurement lasers, including and especially VCSELs. Augmented Reality (AR) and Virtual Reality (VR) are also notable VCSEL applications [2].

The use of VCSELs in these applications is driven by several key benefits. Production costs are low, enabling efficient manufacturing in high quantities. This makes VCSELs especially suitable for consumer applications. VCSELs are highly reliable and their compact size makes them ideal for products in which miniaturization is required [1, 2].

Currently, the most common packages for VCSEL devices are Transistor Outline (TO) cans. Additionally, various leaded or leadless ceramics and plastic packages are used (LCC and PLCC). Figure 1 presents the different package solutions used for VCSELs [3].

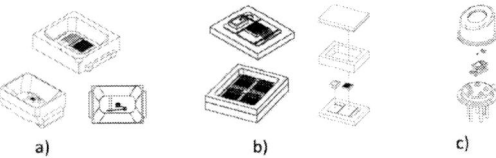

Figure 1: Several types of packages for VCSELs. a) PLCC b) ceramic substrate c) TO-46 can [4].

Quad-flat packages (QFP) and quad-flat no-leads (QFN) type packages are also widely used for VCSELs, especially for arrays and devices having 10 or more pins [3].

To enable the functionality of an optoelectronic part that is emitting or receiving light, a transparent lid is required. In addition, the optoelectronic part must be protected from the surrounding environment and various adverse conditions such as moisture, particulates, and mechanical stresses. Hermeticity is a crucial factor that enables sustained stable conditions inside the package [5]. Poor hermeticity exposes the package to moisture. In combination with temperature, moisture is one of the main ageing acceleration elements that causes damage in microelectronic devices. Moisture is a significant reliability risk since it can cause or facilitate a variety of different failure mechanisms, such as electrical leakage, metal migration, and corrosion of wires, metallization, and bondpads [6-8].

Conventionally, optoelectronic packages have only a glass lid and the other package parts are metal or ceramic. However, glass can also be used to construct the entire package. For example, in medical implants and Micro Electro Mechanical Systems (MEMS), all-in-glass packages have been introduced. Advantages of glass include transparency that allows for transmission of both visible light and radio frequencies. This enables enhanced functionalities, such as wireless data and energy transfer [9, 10]. When compared to ceramics, glass also offers more affordable prototype manufacturing [11].

In this paper, we design, model, manufacture, and test a leadless glass package (LGP) for VCSEL arrays. Novel glass micro bonding is used to construct the package. The package must be suitable for use in space applications.

Experimental

Several design rules were set for the package. The requirements are listed in Table 1.

Table 1: Design rules

#	Requirement	Details
1	Hermetic	-Hermetic Through Glass Via (TGV) technology -Hermetic sealing method
2	All glass	-Sealing technique enabling all-in-glass structure
3	Radiation tolerant	-Selection of suitable glass type
4	Volume manufacturing capability	-Commercially available materials, components -Manufacturing techniques suitable for wafer scale production

The package was designed to provide a hermetic protective seal for a VCSEL as well as I/O interconnection to connect the sensor package using a standard Ball Grid Array (BGA) connection. The all-in-glass package was designed with a three-layer construction using a BK7G18 glass wafer with hermetically sealed-in FeNi42 through-glass vias (TGV). To enable wire bonding and die attachment, top platings were placed on the TGV wafer. Three-layer construction is presented in Figure 2.

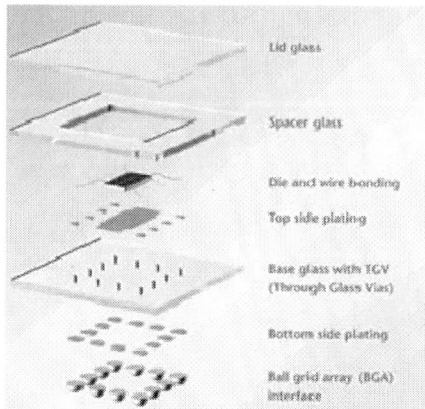

Figure 2: Three-layer all-in-glass package structure. (SCHOTT Primoceler Oy)

A standard package size and interface was used for the design. The final dimensions of the package were 5×5×1.5 mm with a cavity volume of 4.7 mm^3. The sample was designed in a way that enabled both single and wafer level manufacturing.

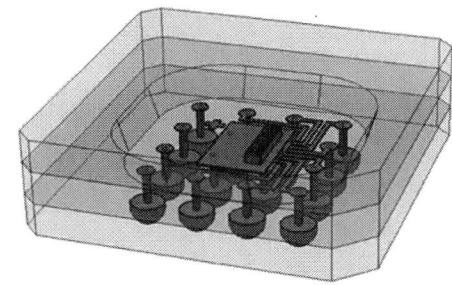

Figure 3: All-in-glass package design. (SCHOTT Primoceler Oy/Optocap Ltd)

VCSELs are known as a power device, which necessitated the use of thermal modelling to verify thermal performance of the package design. Thermal modelling was performed by Optocap. Models were generated in SolidWorks 3D CAD and the thermal flow using Solidworks Simulation software. An 850 nm VCSEL array with eight connections was used for the study. It was assumed that all four channels are running simultaneously, and no forced cooling was used. Based on VCSEL limitations, the operating temperature range was set from 5 °C to 45 °C. However, modelling was done for a wider temperature range for consideration of other conditions, such as storage environments. Modelling was performed for the glass package with the VCSEL and wirings encapsulated, but not containing metallisation platings or BGA balling.

Known and proven technologies were used for the assembly phases executed by Optocap. The die was attached using epoxy and Au wire was used for wire bonding. SCHOTT Primoceler Oy's glass micro bonding method was used to construct the package. The principle of the bonding process is illustrated in Figure 4.

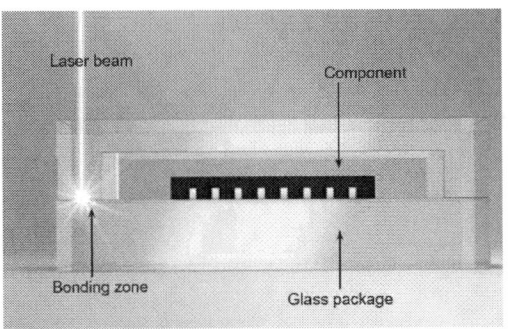

Figure 4: Principle of the bonding process. (SCHOTT Primoceler Oy)

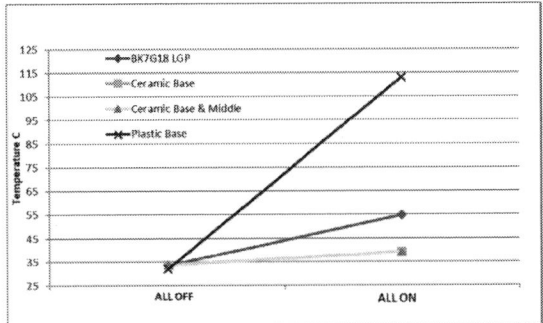

Figure 5: Comparison of thermal performance. (Optocap Ltd)

The bonding process is laser-based. No additive materials or layers are required, which results in excellent gap control and retaining optical properties. Contaminants at material interfaces are also avoided. The process workflow is simplified because no wait time for curing is required. The heat load is extremely low, which enables package miniaturization and the need for only minimal safety zones.

A large evaluation test campaign was executed to test package properties. Samples were exposed to pre- and post-stress tests. Electro-optical behaviour and hermeticity were investigated before and after thermal and mechanical stress tests. Additionally, baseline and post-stress Residual Gas Analysis (RGA) was performed to research the stability of cavity content. BGA balling using lead-free SAC was done to study all-in-glass package mountability onto a PCB. A lead-free solution was also selected to demonstrate the suitability for non-space applications which may not accept lead. Visual inspection was performed throughout the entire campaign.

Results and discussion

As a result of thermal modelling, it was noted that the material properties of BK7G18 are sufficient for the intended conditions of an operating temperature range from 5 to 45 °C. Therefore, the package design fulfills the set requirements. Thermal modelling shows that the package performs at even higher temperatures than required.

Comparison of thermal performance of the glass package design to ceramic and plastic packages is shown in Figure 5.

When compared to plastics, such as PLCC packages, glass has much better thermal performance. Ceramic delivers the highest thermal performance of all compared materials. However, already few simple design amendments improve thermal performance of the glass package. For example, adding metallized pads on both sides of the base glass will significantly improve the properties. Additionally, integrating the glass package onto a PCB would improve the results: PCB works as a heat sink.

Manufacturing was completed using 50×50 mm glass sheets. Figure 6 shows an assembled and bonded glass sheet. Single devices were diced from this glass sheet. Figure 7 shows manufactured single devices.

Figure 6: Assembled and bonded 50×50 mm glass sheet. (SCHOTT Primoceler Oy)

Figure 7: Manufactured all-in-glass packages for VCSELs. (SCHOTT Primoceler Oy)

After manufacturing, an evaluation test campaign that included pre- and post-stress E/O and hermeticity tests was performed. Detailed test descriptions are shown in Table 2. "Pass" indicates that no visual defects or changes in the device functionality were observed.

Table 2: Test methods

Test	Reference	Result
Thermal cycling	MIL-STD 750, method 1051 Condition B, 100 cycles -55 °C/+125 °C	Pass
Moisture resistance	JEDEC STD Method A101B 240 h or TBD at 85 °C/85%	Pass
Resistance to glass cracking	MIL-STD-750, method 1057 Condition A, 10 cycles 0 °C/+100 °C	Pass
Mechanical shock	MIL-STD 750, method 2016 5 shocks, 1500 g - 0,5 ms - 1/2 sine, 3 axes (×2 directions)	Pass
Mechanical vibration	MIL-STD 750, method 2056 4 cycles, 20 g, 80 to 2000 Hz, 0.06 inch 20 to 80, 3 axes	Pass
Hermeticity	Krypton 85 Standard limit 5×10^{-8} atmcm^3/s air, fine and gross	Leak rate limit of 6.0×10^{-12} atmcm^3/s Kr-85 ($1,1 \times 10^{-11}$ atmcm^3/s air)

No abnormal deviations were observed in the E/O tests performed by Optocap. No failures were detected during visual inspection at the test house or at SCHOTT Primoceler Oy. Pre- and post-stress optical pictures of bond seams are shown in Figure 8.

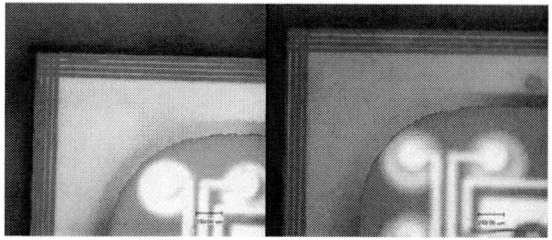

Figure 8: Left: Pre-stress; Right: post-stress. Good quality bonding seams. No changes caused by environmental stresses. (SCHOTT Primoceler Oy)

To inspect the package in more detail, cross-sectional samples of some stressed devices were examined. A cross-sectional image of the sample is shown in Figure 9. Three-layer structure with through glass vias, VCSEL, and wiring can be seen.

Figure 9: Cross-sectional image of the package. Simple visual inspection can be performed. (SCHOTT Primoceler Oy)

Krypton 85 method was used for hermeticity testing. Leak rate was set higher than required in the MIL standard: 6.0×10^{-12} atmcm^3/s Kr-85 (1.1×10^{-11} atmcm^3/s air) instead of standard limit 5×10^{-8} atmcm^3/s. Both fine and gross leak were tested. All the tested samples passed both fine and gross leak tests. Pre- and post-stress RGA results were in line with hermeticity results; no indication of leakage was detected.

BGA balling followed by PCB mounting using lead-free solder was performed successfully. PCB mounted all-in-glass package is shown in Figure 10.

Figure 10: PCB mounted glass package. (SCHOTT Primoceler Oy)

X-ray inspection of PCB mounted devices exposed to thermal cycling showed no signs of noticeable solder joint degradation.

Conclusions

To summarize, hermetic and radiation tolerant all-in-glass packages for VCSELs have been designed, modelled, and manufactured using novel glass micro bonding method. Thermal modelling proved that the design and materials are suitable for the intended conditions. Environmental tests were performed successfully. Excellent hermeticity, exceeding the standard requirements, was demonstrated 6.0×10^{-12} atmcm^3/s Kr-85 (1.1×10^{-11} atmcm^3/s air). Also, BGA balling and mounting the devices onto a PCB was successfully performed. Glass micro bonding proved to be a suitable method for the designed package.

The design and package solution could easily be adapted to other glass types such as Borofloat 33 and other end applications, such as active implants, MEMS, and automotive uses. The design is suitable for volume manufacturing.

The novel glass micro bonding technique offers several benefits. Risk of surface and internal contamination, such as outgassing, is reduced as no additive materials are used for bonding. Presented low temperature bonding technique can be beneficial when encapsulating sensitive components and reducing safety zones, enabling miniaturization possibilities.

Acknowledgements

We acknowledge Liam Murphy from the European Space Agency (ESA) for his expertise and assistance with this project. ESA is to provide for and promote, for exclusively peaceful purposes, cooperation among European states in space research and technology and their space applications.

References

[1] Osram Opto Semiconductors. Vertical take-off for VCSEL technology in IR applications. Available: https://www.osram.com/os/news-and-events/spotlights/vertical-take-off-for-vcsel-technology-in-ir-applications.jsp 20.5.2019

[2] H. Moench, G. Carpaiji, S. Gronenborn, R. Gudde, J. Hellmig, J. Kolb, A. Lee. "VCSEL based sensors for distance and velocity. Proceedings of SPIE Vol. 9766, 97660A. A1-A10.

[3] Vixar. VCSEL Packages datasheet: VCSEL Standard Product Packaging Options. Available: https://www.laser2000.es/out/groupdownloads/Vixar_VCSEL_PackagesDS_30SEP2014_datasheet_Laser_2000.pdf 20.5.2019

[4] Vixar. VCSEL Modules. Available: http://vixarinc.com/products/vcsel-modules 19.5.2019

[5] S. Costello, M. P. Y. Desmulliez, "Hermeticity testing of MEMS and microelectronic packages", Artech House, Norwood, 2013, 195 p.

[6] T. J. Rossiter, "Moisture Analysis: History, Sampling and Case Studies." JEDEC Meeting, 13 Jan 2004. Available: http://www.orslabs.com/pdf/Moisture%20Analysis%20History%20Sampling%20and%20Case%20Studies.pdf 17.5.2019

[7] H. Greenhouse, R. Lowry, B. Romenesko. Hermeticity of Electronic Packages. Second Edition. William Andrew, 2011. 360 p.

[8] Failure mechanisms and models for semiconductor devices, JEDEC Solid State Technology Association, Oct 2011

[9] Y.-H. Joung, "Development of Implantable Medical Devices: From an Engineering Perspective" International Neurourolocigal Journal. 2013 Sep; 17(3): 98–106. Published online 2013 Sep 30.

[10] J. Bibin, "Smart system for invasive measurement of biomedical parameters", Logos Verlag Berlin GmbH, 2017, pp. 5-13.

[11] H. Lundén, A. Määttänen, I. Thanasopoulos. Room temperature bonding for packaging CMOS image sensors; direct sapphire ceramic sealing. Proceedings of the Nordic Conference on Microelectronics Packaging (Nordpac), Gothenburg, Sweden, June 18-20. 2017.

High Capacity Silicon Photonics Packaging

Marco Binda[1], Antonio Canciamilla, Alessio Daverio[1], Antonio Fincato[2], Piero Gambini[1]
Luca Maggi[1], Piero Orlandi, Luca Ramini, Matteo Repossi[2], Angelica Simbula[3], Mark Shaw[1]

[1]STMicroelectronics Agrate, Via Camillo Olivetti 2, Agrate Brianza, 20864, Italy
[2]STMicroelectronics Castelletto, Via Tolomeo 1, Cornaredo, 20010, Italy
[3]Dipartimento di Ingegneria Industriale e dell'Informazione, Università di Pavia, Pavia, 27100, Italy

Mark.Shaw@st.com

Abstract

A large amount of progress has been made in the industrialization of Silicon Photonics fabricated in CMOS Fabs, enabling the adoption of 100G QSFP modules. Further progress has now increased the transmission capacity of Silicon Photonics devices. In this paper, we outline the package design and evaluation of high capacity Silicon Photonics devices from simulation of the package performance to the prototype packaging results.

Key words: Silicon Photonics, Packaging

1. Introduction

Silicon Photonics-based pluggable modules based on the QSFP and COBO formats (Fig 1) with 4 parallel Tx and 4 parallel Rx fibres at 25Gbs are now widely deployed in data centres. These devices produced on 300mm wafers exploit traditional CMOS fabrication methods and materials [1] and flip chip bonding of the electrical driver chip onto the optical device with Cu Pillar technology. The use of gratings for vertical optical coupling between silicon and Single Mode Fibre (SMF) facilitates wafer-level testing [2] and interconnection via Low loss low profile couplers [3].

Fig 1a 3D 4-channel transceiver chip mounted on QSFP demonstration board with optical fibre

Fig 1b On Board Optics demonstrator

To increase overall capacity of data centre optical links we can increase the bit rate, increase the number of fibres or increase the bandwidth of each single fibre connection. In this paper, we will discuss the implications on packaging and demonstrate results of each of these options.

2. Increased Bitrate

High speed interconnection between Integrated Circuit (IC) and printed circuit board (PCB) represents a critical aspect when considering the assembly of Silicon Photonic (SiPho) IC, typically a 3D chip-on-chip structure composed by Photonic IC (PIC) and Electric IC (EIC). The most reliable and simple solution is standard wire bonding technology, which however poses limitations on high speed signals beyond 28 Gbaud [4].

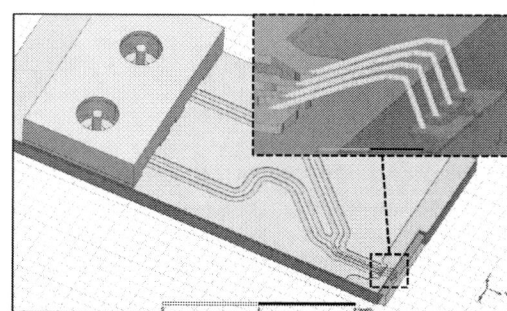

Fig 2 HFSS simulation of an assembly of PCB and Photonic IC with wirebondings for 56 Gbaud data rates

As a reference testcase, we consider the assembly of a 3D SiPho IC on a multilayer PCB (see Fig 2), including the signal path on the PCB and a portion of the PIC including the pad with a short segment of matched transmission line. The whole structure has been simulated with Ansys HFSS and the results are shown in Fig 3.

A rapid degradation of the frequency response in air is evident at frequencies higher than 15GHz, especially in terms of Return Loss (RL) and Insertion Loss (IL). The Time Domain reflectometry step response in air Fig.4b clearly shows that a large impedance discontinuity is the root cause of the phenomenon, essentially due to wirebondings, acting as large value inductors. Generally speaking, the discontinuity can be approximated by a segment of transmission line whose characteristic impedance can be written as

$$Z_{wb} = \sqrt{\frac{L_{wb}}{C_{wb} + C_{pads}}} \quad (1)$$

where L_{wb} represents the wires inductance, C_{wb} is the capacitance between the wires and C_{pads} is the aggregate capacitance of the pads on PCB and PIC substrates.

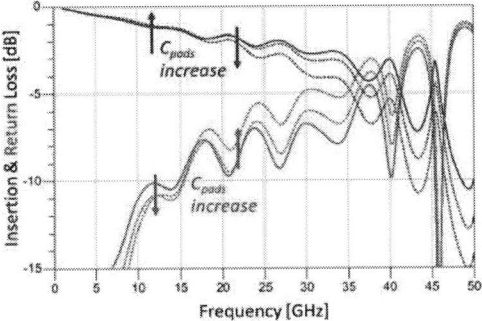

Fig 3 Insertion and Return Losses for the signal path shown in Fig2 (solid line)

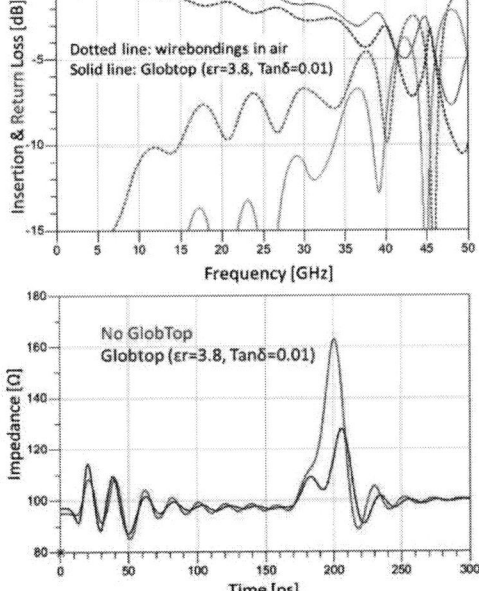

Fig 4 Improvment of the frequency response with GlobTop (top a) of the transmission path by compensating the impedance step (bottom b) caused by the wirebonds

L_{wb} is proportional to the length of the wires and their pitch, but a reduction of metal-to-edge distance on PCB and PIC is difficult due to physical tolerances.

According to Eq.1, Z_{wb} can also be reduced by increasing ($C_{pads} + C_{wb}$). The effect of an increase of C_{pads} has been simulated: Fig 3 shows that the

improvement is limited to frequencies below 15-17 GHz, i.e. for datarates below 25-28 Gbaud. Alternatively, we can increase C_{wb} by adopting a Glob-Top resin to cover the assembly, replacing the air around the wirebondings with a medium featuring an apropriate dielectric constant: the results are shown in Fig 4. This approach does not impact on the PCB and PIC fabrication and assembly except for the additional step required to apply the resin. Devices assembled with this glob top have also fully passed package qualification tests demonstrating its suitability for use for production devices.

The result of this approach has been implemented in the design of a 400G (4x100G) transceiver (Fig 5a-5b).

Fig 5a 400G (4 channel x 100 G) transceiver chip mounted on demonstrator board with glob top on wirebond connections

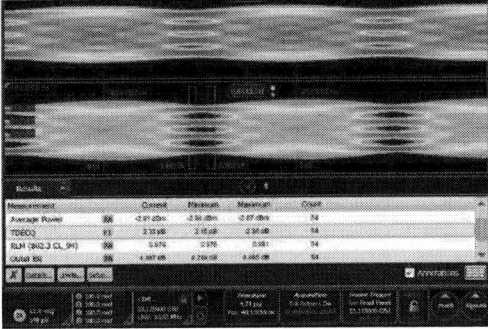

Fig 5b 100Gbps (56Gbaud PAM4) eye diagram

3.1 Broad band coupling device characterisation

The coupling efficiency of grating coupler technology is limited in bandwidth and sensitive to temperature variations. This hinders the successful development of integrated Silicon Photonics chips for transmission standards such as Coarse WDM and Coherent, but also for applications requiring a temperature range larger than 70 °C.

In this work we demonstrate broadband coupling by using a two stage adiabatic coupler. The first stage couples the Si waveguides to a SiN waveguide [5,6], and the second stage couples the SiN waveguide to an interposer, which is then connected to the optical fibre. The presence of an intermediate SiN layer is functional not only for the adiabatic coupling with the interposer, but also for the realization of additional optical elements such as an Optical Mux and Demux [7]. In order to verify and optimise the various elements of the Si-SiN–interface, tests were performed using edge coupling to the SiN waveguides, enabling measurements to be independent of the efficiency of the SiN-Interposer transition. To do this, a good quality edge facet is required. Standard dicing techniques such as scribe and cleave have difficulty in achieving repeatable results on 12'' wafers with the crack not propagating uniformly, and even after optimization standard blade dicing cannot produce repeatable low loss results.

Fig 6 Stealth dicing showing modified layers and 'cleaved' optical quality area

A stealth dicing process was therefore developed ensuring that the laser modified layers within the dicing area start just below the optical waveguides (Fig 6): this ensures that during the expansion stage the cleave is uniformly propagated. This technique is shown to provide repeatable low loss facets within devices and across the wafer.

The optical devices were characterized injecting laser light from one edge of the chip by means of a polarization maintaining lensed fibre and collecting the transmitted light at the other edge. Thanks to a polarization maintaining optical switch it was possible to tune the input state of polarization between TE and TM; at the output side, the presence of a polarization splitter made possible the analysis and the behaviour of the DUT under different states. The spot size of the lensed fibre was 2μm and the working distance 12μm. The coupling loss with the input and output waveguides was about 2dB, mainly due to mode size mismatch. In order to be able to characterize devices with low loss a cut-back technique was used. Several structures were designed where the same DUT was cascaded multiple times. The loss of the single device was obtained linearly interpolating the results.

As an example, we report the data obtained over the whole O-band of a Polarization Splitter and Rotator (PSR), a device that splits the random polarization at its input into TE-state at the first output and TM-state at the second output. Three different structures were designed with 1, 5 and 9 cascaded PSR respectively and measured (Fig 7).

TE and TM losses were then calculated by interpolating at each wavelength the set of three points. In this way, it was possible to determine which design among a Design of Experiments performed better, independently of the coupling losses or the propagation losses into the chip: in fact these contributions are common to all the optical structures and didn't affect the interpolated values. The interpolated values for the PSR are shown in Fig 9.

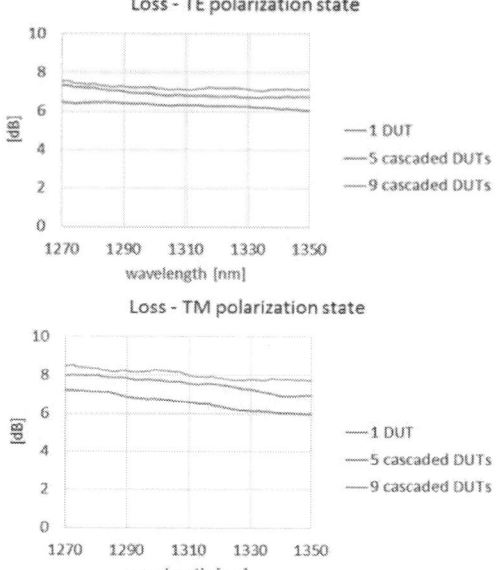

Fig 7 Insertion loss of the optical structures with multiple cascaded devices under test when laser polarization state is selected to TE (Top) and TM (Bottom)

3.2. Coupling from SiN to Si and Polarization management

To enable signal transmission from the silicon layer to the SiN layer (and vice-versa), a vertical adiabatic transition has been designed by keeping the SiN waveguide width constant around 0.7 μm along the propagation direction, while tapering the underlying Si strip waveguide width from 250 to 60 nm. Insertion loss (IL) and Polarisation dependent Loss (PDL) of identical devices with the mentioned design parameters, measured on nine different chips, are reported in Fig 8 and 9. Maximum losses are flat and below 0.25 dB (IL+PDL) over the O-band.

Moreover, to enable also polarization diversity schemes (beside polarization insensitive ones), a polarization splitter and rotator (PSR) is required. The high degree of flexibility of the developed platform can be exploited to design an integrated PSR in different ways: two optical layers (SiN and Si) and different waveguide structures (strip or rib-based) can be suitably combined to optimize the PSR according to target specifications. In this summary, we report the results obtained for a PSR designed in Si rib technology by exploiting mode-evolution principles [8] using the interpolation method to characterize the devices. IL measured for TE input polarization path is reported in the upper chart of Fig 9. Values measured over eight different chips are always lower than 0.55 dB. TM polarization, which is rotated to TE polarization at the second output of the device, shows IL values always below 0.7 dB (see Fig. 9 lower chart) over the O-band.

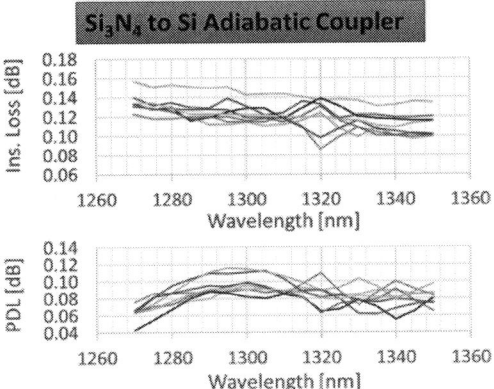

Fig 8 Measured IL (upper chart) and PDL (lower chart) of identical SiN to Si adiabatic couplers over nine different chips

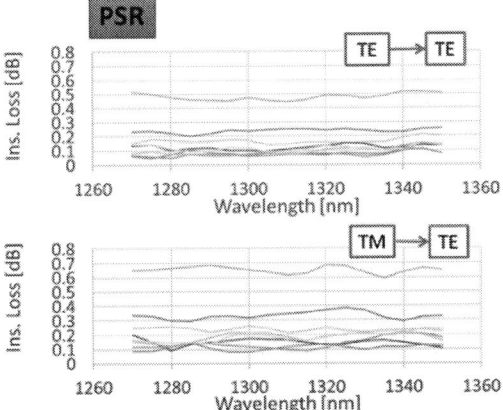

Fig 9 Measured IL for TE (upper chart) and TM (lower chart) input polarization through the designed PSR for eight different chips

3.3. Broadband coupling technology with a glass interposer

Fig. 10 shows a side-view sketch of the fiber-to-silicon broadband coupling scheme. A glass-based interposer fabricated by means of field-assisted Ion-Exchange process [9], can be assembled to the Silicon Photonics chip in correspondence with a cavity etched at wafer level in the back-end of line (BEOL). Using the Ion exchange process we can obtain a waveguide with a 3D variable shape: exposed and narrower at the chip end area to optimize

the adiabatic coupling to a SiN waveguide, buried and wider at the other end to optimize the coupling to a SMF [10]. To optimize the coupling with SiN waveguide, the refractive index profile of the Ion exchange waveguide is asymmetric, with the maximum value close to the bottom surface, as shown in the theoretical index profile in Fig 11.

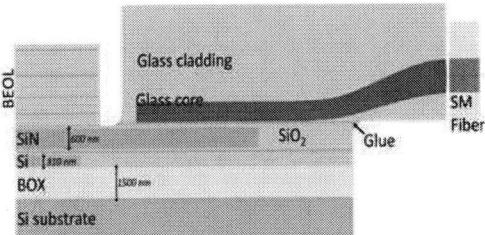

Fig 10 Side-view sketch of the broadband coupling scheme

The FWHM of the gaussian refractive index distribution is determined by the width of the mask aperture used for the diffusion process during the fabrication of the Ion Exchange waveguide.

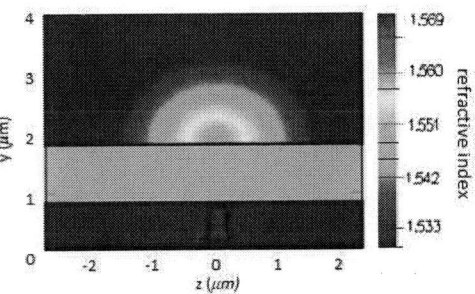

Fig. 11 Refractive index profile of the transversal cross-section of an Ion-Exhange interposer with mask width 1.7μm, on top of the glue layer (light green) and the SiN waveguide (red)

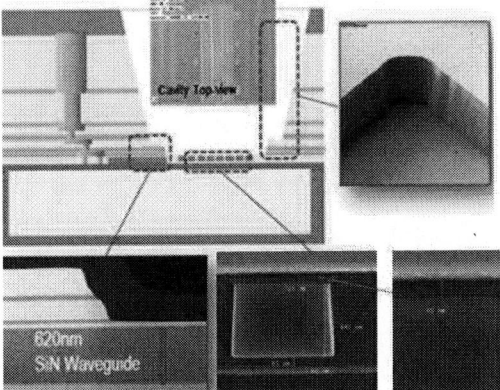

Fig 12 Schematic etched cavity with insets showing SEM image of sidewalls and details of the SiN waveguide at the start of the transition.

The etching process of the cavity is optimised in order to obtain a large area of 8.75 mm^2, with only few tens of nanometres of SiO$_2$ (see Fig 12) above the SiN waveguide.

To obtain adiabatic coupling between the SiN layer and the interposer, a linearly tapered SiN strip waveguide is used. Design parameters as well as assembly specifications were determined by performing extensive simulations with a commercial-grade simulator eigenmode solver and propagator [11]. SiN waveguide maximum and minimum widths were set to respectively 380 and 180 nm to guarantee efficient coupling of both TE and TM polarization over the O-band. Preliminary studies of the glass interposer highlighted that the tolerance to glue thickness depends both on the choice of glue refractive index, η and the glass waveguide width.

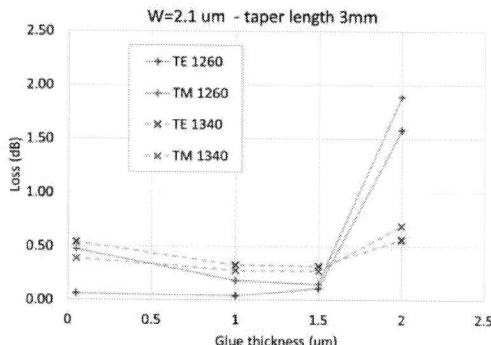

Fig 13 Loss vs glue thickness simulation for a glass waveguide width of 2.1μm, η=1.548

The glass waveguide width is determined by the opening in the mask aperture used for the ion diffusion fabrication process. We identified glass waveguides with a mask aperture width between 1.5 and 2.5 μm to be the most suitable. Then, setting a maximum acceptable coupling penalty due to the assembly process to 1 dB, the following optimal parameters were determined: a length of 3 mm, a maximum glue thickness of 1 μm (Fig 13) and a maximum lateral misalignment of 2 μm (fig 14).

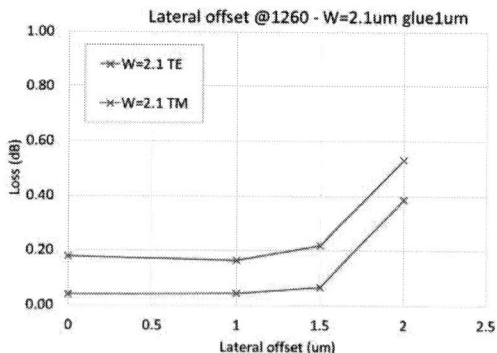

Fig 14 Loss Vs Lateral Offset, η=1.548

To test the functionality of these devices, we designed three optical loops with different lengths, each loop being composed of two SiN to glass adiabatic couplers and a 180° SiN bend (see Cavity Top view Fig 12).

The fabricated, etched and diced Silicon Photonics chips were assembled with the optical interposers (see Fig 15) using a high precision SET150 flip chip bonder and fixed with a UV epoxy. It must be noted that the choice of the epoxy depends not only on an appropriate refractive index, but also the viscosity for dispensing and curing properties must be considered.

Fig 15 Glass interposer assembled Silicon Photonics Device with Cavity Etched BEOL

At this stage, the interposer was pre-assembled with fibres in a polished V-groove attached to the interposer. It is anticipated that, in a production environment, the interposer will be first passively assembled with standard flip chip assembly equipment followed by an active fibre attach as is the case with current optical modules. Further improvements to the assembly process can be easily implemented, such as better alignment markers and tailored pick up tools, guaranteeing high process throughput and large production volumes without impacting the cost of the assembly line.

The measurements for four different chips with 3mm long SiN to fibre adiabatic couplers for IL and PDL are reported in Fig 16. For each wavelength, IL corresponds to maximum transmission power and PDL to the difference between maximum and minimum transmission power over all possible polarization states. It can be seen that maximum IL is lower than 1.9 dB over 100 nm bandwidth, while PDL is kept always below 0.4 dB. Maximum loss including fibre coupling loss is always below 2 dB.

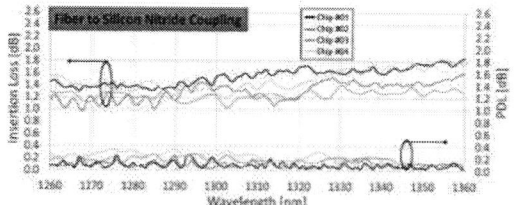

Fig 16 Measured fibre to SiN waveguide adiabatic coupler IL and PDL of four assembled chips..

An initial device showed high PDL with very high loss in the TM polarization state, with additional variation depending on the wavelength (Table 1). This can be explained by a tilt in the interposer relative to the waveguide and the different overlap

30

lengths required for the different polarisation states and wavelengths (see Fig 18).

Table 1 Coupling loss of device with tilt for 3mm taper

Wavelength and polarisation	Loss in dB
1260 TE	3
1340 TE	2
1260 TM	10
1340 TM	6
PDL 1260	7
PDL 1340	4

Sectioning this device along the coupling region (Fig 18), we observe a variation in gap or glue thickness along the waveguide. Simulations demonstrate that an inclination between 0.065 and 0.07 degrees (fig 19) corresponding to a glue thickness at the end of the 3mm taper of about 3.5 microns, is compatible with the observed PDL.

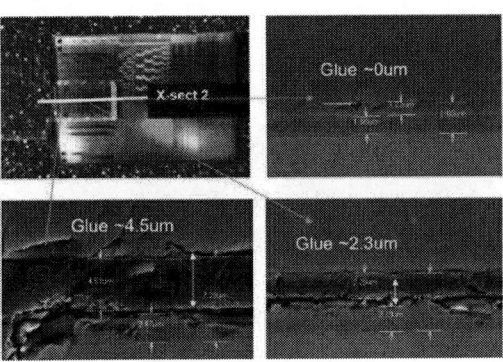

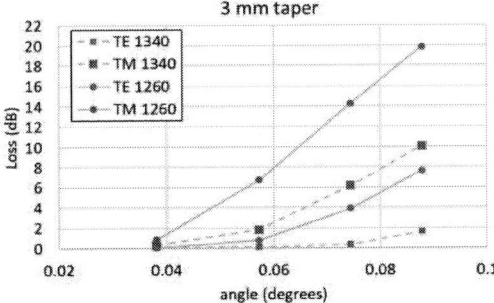

Fig 18 Cross section of device with tilt

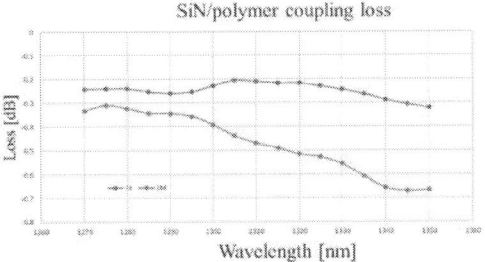

Fig 19 simulated dependence of loss vs inclination for different wavelengths and polarisation

After monitoring of the 1260nm TM polarization was introduced during assembly, devices were screened and only devices with low loss fixed in place with the UV epoxy. The high loss device with tilt is attributed to the fact the assembly was performed in a non clean room environment. In fact, after cleaning devices with high loss, they were realigned and assembled with low loss. It is therefore anticipated that resolving the contamination problem would eliminate the need

for the in-process loss check used during the development.

3.4. Direct write Polymer

An alternative to attaching an Interposer to the silicon Photonic chip is the direct deposition of the polymer waveguides on the waveguide area [12].

Fig 20 Polymer waveguides fabricated on top of SiN waveguides

Direct writing can then be used to produce the waveguides, again fiducials placed on the silicon photonic chip are used to align the laser relative to the SiN waveguides (see Fig 20).

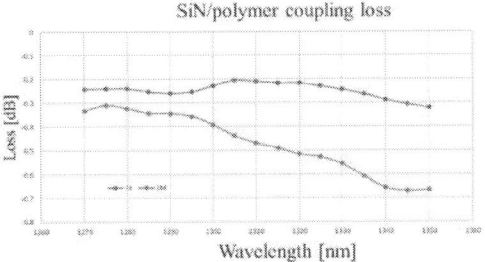

Fig 21 Measured TE and TM coupling SiN to polymer waveguides

The results demonstrate that low loss for the SiN/polymer transition can be achieved (IL < 0.8 dB), as shown in Fig 21,. This is in line with the value resulting from simulation which needs to be added to the fiber-to-polymer loss (0.5dB) to obtain the overall coupling loss.

3.5. Comparison with the state-of-the-art and grating couplers

Table 2. Fibre to silicon coupling comparison

Parameter	Minimum IL	1-dB BW	100nm BW penalty
Glass interposer	1.25-1.55 dB	> 100 nm	0.6-0.7 dB
Direct write polymer	1.45-1.6 dB	> 100 nm	0.3 dB
Metamaterial [13]	1.3 dB	> 100 nm	0.8 dB
Polymer [13]	1.6 dB	> 100 nm	0.6 dB
Grating SPGC [1]	2.2 dB	18-25 nm	>15dB

A performance comparison with other Silicon Photonics I/Os is presented in Table 2. For this work, Silicon coupling loss has been estimated by adding maximum measured SiN to Si IL (i.e. 0.25 dB) to the measured SiN to fibre IL. The proposed solution has performance comparable with other state-of-the-art broadband coupling approaches, the glass interposer has the advantages of a well-established supply chain and robust component assembly, the direct writing process has lower loss but requires the writing process within the cavity and a standard edge coupling attach. If compared with current Single Polarization Grating Coupler (SPGC) this work demonstrates comparable minimum IL, but provides a much larger operational bandwidth over both polarisations. The cost is a more complex assembly flow, including an additional interposer chip.

To estimate also the penalty of coupling in a polarization diversity receiver, PSR IL can be added to the values reported in the first column of the table. These values allows for a 1 dB better performance with respect to peak performance of current Polarization Splitting Grating Couplers, (PSGC), (2-2.3 dB vs 3.2 dB) [1]. Moreover, efficient coupling is guaranteed, as before, over a much larger operational bandwidth and temperature range.

3.6. Wafer level testing

Wafer level testing of the devices can be provided by continuing the SiN waveguides into the scribe line and using grating couplers in unused areas of the device or a sacrificial test chip. Because of the restricted bandwidth of the grating couplers, if testing is to be performed over the full waveglength range, additional multiplexer structures are also included.

3.7. Cavity Chip dicing

The standard dicing process for the 3D Silicon photonics chip with the EIC assembled requires laser grooving of the back end of line (BEOL) layers in the scribe line of the Photonics chip before blade dicing in order to prevent chipping and delamination at the blade dicing or subsequent reliability testing. If this is performed on cavity devices, the ablated material from the laser grooving will cause a tilt in an assembled interposer, similar to what has been seen with contamination during assembly (Fig 22)

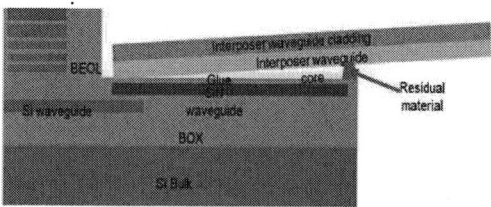

Fig 22 Devices with tilt from laser ablation

We have developed a stop/start laser grooving process where the laser beam is interrupted at the start of the cavity, and restarted at the end of the cavity,

ensuring that no debris remains in the area of the interposer bonding, Fig 23.

Trials have also been performed assembling 12'' cavity photonic wafers with the electrical IC using Cu pillar technology. The parts were visually inspected pre and post assembly verifying that no residues/contamination occurred with the cavity due to the various bumping and assembly steps.

Fig 23 Laser grooving with stop start in cavity

Singulation of polymer waveguides formed within a cavity can be performed using the same interrupted laser grooving process and a two blade dicing process, which has been shown to produce optical quality facets using the techniques outlined in section 3.1.

4. Higher fibre count.

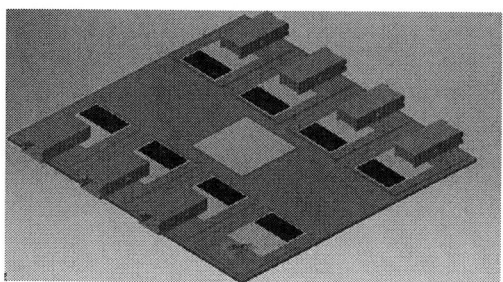

Fig 24 schematic of ASIC with optical submodules with one of the submodules shown without a housing to demonstrate the glass optical interposer

To reduce signal integrity issues in the electrical connections between ASIC and optical Interfaces, a Multichip module combining the two functions is proposed for ultra high capacity switches. Targeting 25Tbs with single wavelength would require 100G channels and 8 modules per ASIC requires 32Tx, 32RX, and 8 Polarisation maintaing (PM) fibre connections for the laser sources per module. The PM fibres are required as the laser sources are packaged separatly due to the operating temperature (Fig 24). If we increase the number of fibres interconnects, we are limited by the minimum possible pitch with the standard fibre coating diameter of 125μm. Clearly, this is not practical with a standard pitch. In addition, standard fibre coatings do not withstand soldering temperatures, so if the individual chip/modules are to be assembled with high yield, some form of connectorised device would be preferable. The proposed solution is to use a direct write glass interposer (Fig 25); such an interposer can support singlemode or PM interconnects, and can interface either through an appropriately mirrored angled facet to grating couplers or via adiabatic coupling. In order to interface with a large number of fibres with only one connector the ouput uses multicore fibre.

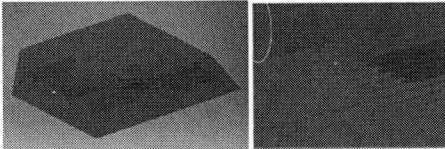

Fig 25 Glass optical interposer with laser written waveguides, and detail of multiple waveguide

The direct write waveguides form a fan out from a reduced pitch in one dimension on the silicon photonics device [14] to match the multicore and PM fibre positions at the connector interface (Fig 25). A standard MPO interface cannot be used due to the large number of connections, as the insertion force required to maintain a low loss connection between the multicore fibre MPO and the glass interposer would be transmitted directly to the joint between the optical fibre and silicon. To resolve this an alternative is to use an airgap connector [15] so that the interposer and the connector ferrule are not in physical contact. In this way a high capacity solderable sub module can be produced and scalable multiple connections with no additional insertion loss can be achieved.

5. Conclusion

In this paper we have outlined several different packaging technologies that can be applied in high capacity Datacom links and beyond, in Automotive, Sensing and High Performance Computing.

6. Acknowledgements: The authors would like to thank: Adamo Bazzotti and Donato Gallo for the QSFP, On Board Optics development, STMicroelectronics Crolles France for the Wafer process development and fabrication, Disco Germany and Japan for the stealth and ablation dicing development, Florian Gardillou of Teem Photonics for the glass interposers. Gianni Preve, Davide Rotta and Marco Chiesa from Inphotec Pisa, Aina Serrano Rodrigo and Jose Angel Ayucar from Universidad Politécnica de Valencia, Valencia Nanophotonics Technology Center for the assembly of the Interposers to the Silicon Photonics devices, Variooptics for the polymer waveguide fabbrication, Optoscribe and Sumitomo Electric Industries for the multicore optical interposer and air gap connector information. Piero Orlandi, Luca Ramini and Antonio Cianciamilla are no longer with STMicroelectronics.

7. References

[1] F. Boeuf et al., "Silicon Photonics R&D and Manufacturing on 300-mm Wafer Platform Developments in 300mm silicon photonics using traditional CMOS fabrication methods and materials", Journal of Lightwave Technology 34, 286-295 (2016).

[2] Bazzotti et al., "Silicon Photonics Assembly Industrialisation" Proc EMPC 2015

[3] Kumagai et al., "Low-Loss and Highly Reliable Low-Profile Coupler for Silicon Photonics," 2019 Optical Fiber Communications Conference (OFC), San Diego, CA, USA, 2019, pp. 1-3.

[4] Dong Gun Kam et al.," Packaging a 40-Gbps serial link using a wire-bonded plastic ball grid array", IEEE Design & Test of Computers, 2006 Volume 23, Issue 3, pp. 212-219

[5] C. Baudot et al., "Developments in 300mm silicon photonics using traditional CMOS fabrication methods and materials," in 2017 IEEE International Electron Devices Meeting (IEDM, San Francisco, CA, 2017), pp. 34.3.1-34.3.4.

[6] C.Baudot et al "Integrated SiN on SOI dual photonic platform for advanced datacom solutions" Proc. SPIE 10686, Silicon Photonics: From Fundamental Research to Manufacturing.

[7] S. Guerber et al., "Design and integration of an O-band silicon nitride AWG for CWDM applications," 2017 IEEE 14th International Conference on Group IV Photonics (GFP), Berlin, 2017, pp. 133-134.

[8] Yuan et al., "Mode-evolution-based polarization rotator splitter design via simple fabrication process", Optics Express 20, 10163-10169 (2012).

[9] Teem Photonics. https://www.teemphotonics.com/integrated-optics/waft-interface-products/

[10] Grelin et al, " Study of deeply buried waveguides: A way towards 3D integration", Materials Science and Engineering: B, Volume 149, Issue 2, 25 March 2008, Pages 185-189

[11] Lumerical Inc. http://www.lumerical.com/tcad-products/mode/

[12] E. Zgraggen et al., "Laser Direct Writing of Single-Mode Polysiloxane Optical Waveguides and Devices," in Journal of Lightwave Technology, vol. 32, no. 17, pp. 3036-3042, 1 Sept.1, 2014.

[13] T. Barwicz et al., "A Novel Approach to Photonic Packaging Leveraging Existing High-Throughput Microelectronics Facilities", IEEE Journal of Selected Topics in Quantum Electronics 22, 455-466 (2016).

[14] Psaila "3D laser direct writing for advanced photonic integration", Proc. SPIE 10924, Optical Interconnects XIX, 109240U (4 March 2019)

[15] Arao et al "Small Footprint Air-gap Multi Fiber Connector with Low Loss and Low Mating Force" Proc OFC 2018

High density fan-out panel level packaging of multiple dies embedded in IC substrates

Friedrich-Leonhard Schein[1], Ruben Kahle[2], Marc Kunz[3], and Andreas Ostmann[2]

1: Technical University of Berlin, 13355 Berlin, Germany 2: Fraunhofer Institute for Reliability and Microintegration (IZM), 13355 Berlin, Germany 3: Schmoll Maschinen GmbH, 63322 Roedermark, Germany

Phone: +49 30 46403-7937, e-mail: f.schein@tu-berlin.de

Abstract

The ongoing roadmaps of miniaturization and functional heterogenity in electronics packaging are pushing the demand for advanced substrate technologies. In this paper we show the embedding in core cavity (EiCC) process running with 5 µm L/S and chips with 50 µm bump pitch. Two 6x6 mm² dies are symmetrically embedded into an organic laminate matrix. A PCB core (100 µm thickness) with very low coefficient of thermal expansion (CTE) containing laser-cut cavities acts as a frame layer. Besides mechanical and handling stability the usage of such a frame offers the advantage of pre-integrating additional features like local fiducials, through vias or power lines by conventional PCB processes. Within that frame the dies are embedded by lamination of an organic build-up film. The chip contacts are then revealed in process based on plasma etching. After measuring chip positions the first redistribution layer (RDL) is formed in a semi-additive process (SAP) utilizing sputtering technique and adaptive laser direct imaging (LDI). Therefore, a newly developed LDI machine is used to write structures in a 7 µm photoresist. Subsequently a second RDL formation can be done. In this step high aspect ratio blind microvias with 20 µm diameter and up to 80 µm depth are drilled by UV-laser and filled in the following plating process. Altogether, with the combination of high density 5 µm L/S interconnects, high aspect ratio (2.5:1) blind microvias and 50 µm fine bump pitch on large panel formats we will give an outlook to upcoming challenges and possibilities in FO PLP.

Key words: panel level packaging, adaptive imaging, direct imaging, 3D system in package, IC substrate

Introduction

The scaling from wafer-level to panel-level is expected to be the next evolution of fan-out (FO) packaging [1]. However, there is still a technical gap between wafer back-end technologies and the PCB infrastructure. One key element to close this gap is to significantly lower the structure size of lines and spaces (L/S). While L/S become smaller the panel size will increase up to 600 mm x 600 mm. To address these upcoming challenges of panel-level packaging (PLP) a consortium of German partners from industry and research develops advanced technology building blocks for PLP. The project aims for an integrated process flow with multi-die embedding in core cavities (EiCC).

Feasibility of bare die embedding in FOPLP technique has been demonstrated in recent publications [2, 3]. However, with downscaling to finer contact pitches on chips (below 60 µm) the (mis)alignment challenge gets even more difficult to overcome. In order to stick with reasonable production yield and time to market we are pursuing an adaptive process flow approach [4]. Instead of costly accuracy improvements throughout the process chain the actual die positions are measured and the RDL wiring is adapted appropriately.

Demonstrator

The demonstrator package of 21 mm x 18 mm size contains two equal Si dies embedded into a PCB core cavity. Core and dies have 100 µm thickness. Die size is 6 mm x 6 mm each. Cu bumps of 25 µm diameter and 15 µm height act as chip contacts. Three rows with a pad pitch of 50 µm and pairwise on-chip interconnects for daisy-chain structures are placed at the outer chip area. The I/O count is 1200. In the inner chip area single bumps are arranged in cross-shape to support recognition in assembly machines. The fan-out is designed with 5 µm lines and spaces and pad openings of 15 µm diameter for contacting Cu pillars. This requires a via to pad tolerance of less than +/- 5µm.

Figure 1 depicts package and chip design. The three different colours in the zoomed package region encode different purposes. Blue and orange structures are both fixed, whereby the blue structures can be shifted according to the true chip position, while the orange fan-out image is stationary. Therefore, a distortion in length and angle (like a rubber band) is necessary for the connections drawn here in red. The inter-line distance for this adaptive part is designed to be 12 µm to fulfill the constraint of at least 5 µm space between Cu traces after distortion. The adaptive length is 100 µm.

Figure 1: Package and chip design

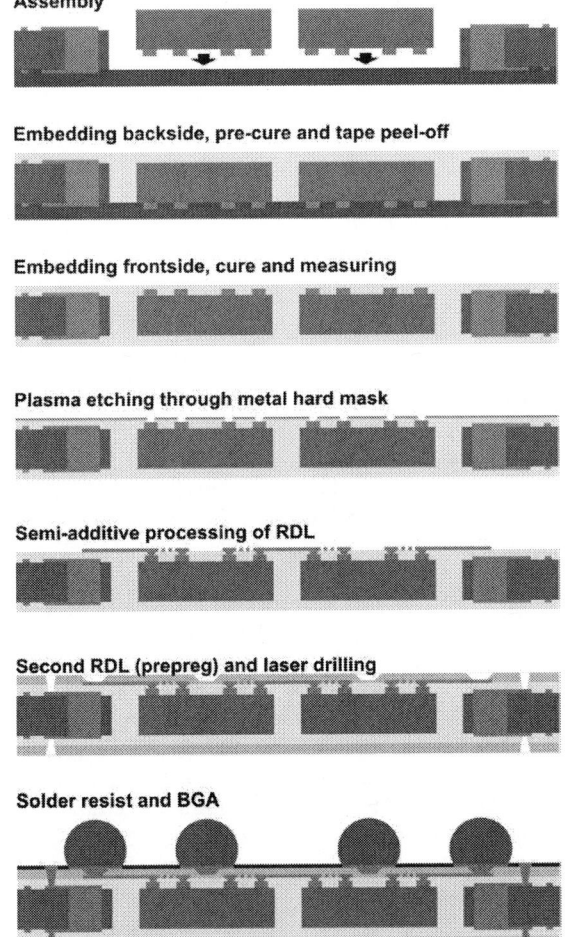

Figure 2: Process flow

The landing pads for vertical interconnects between first and second RDL have a diameter of 100 µm. For final BGA 150µm pads are provided with 300 µm pitch.

For process evaluation a panel size of 303 mm x 227 mm was used. Later extension to 510 mm x 515 mm is envisaged.

Package Fabrication and Results

A sketch of the general process actions is shown in Fig. 2. Parts of this process are described in the following paragraphs.

1. Core

At first, an organic glass fiber reinforced PCB core is structured with standard print and etch technology. To mimimize CTE mismatch between materials we used a core having a CTE xy of 4-6 ppm/K and a CTE z of 8-13 ppm/K (MCL-E-770G from Hitachi Chemicals). Its glass transition temperature Tg is above 260 °C.

Besides Fiducials also electrical test elements are defined. The change in lateral expansion of the core construct due to the modified Cu area has been determined beforehand and considered for the respective lithography layout. Then cavities are cut by laser for later die positions. A surrounding gap of 40 µm larger than chip x-y dimensions is left.

2. Assembly

Assembly preparation starts with the attachment of a single-side adhesive tape to the core substrate. Accordingly cavities are now equipped with a sticky bottom. Here we used a UV-releasable tape which has a 50 µm adhesive layer to allow full penetration of Cu bumps in the face-down assembly process.

Chip placement was performed with a Datacon EVO 2200 for 303 mm x 227 mm panel size and a SiPlace CA3 for larger panels. Since we follow an adaptive approach the assembly accuracy is not a crucial criterium here. The test chips have been fabricated with inhouse equipment.

3. Embedding

First step in embedding is the lamination of an Ajinomoto build-up film (ABF, here GX-T61, 25 µm thickness) onto the backside of the chip/core-construct in a Dynachem 7224-HP7 vacuum applicator for 90s at 90°C stage temperature with 0.6 Nm/mm² pressure. This fills the gap between chip and core and results in an even backside. Next, a prebake step of 120 °C for 30 min until epoxy gelation was applied. The pre-heating inhibits remelting of ABF in subsequent steps and reduces the amount of embedded die repositioning as previous experiments have shown [2, 4].

Since dies are now fixed in cavities the adhesive tape can be released by UV exposure. After that, samples were treated with an oxygen plasma in

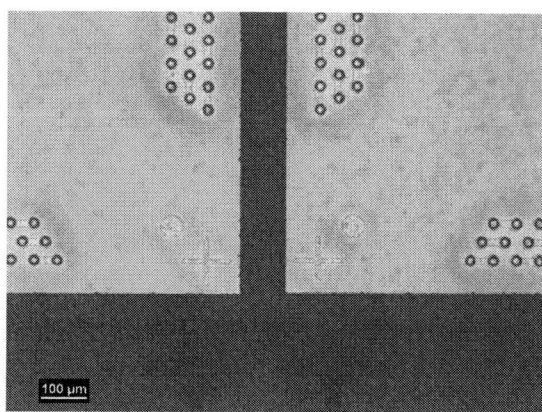

Figure 3: Chips embedded into PCB core cavity. The irregularly appearing surface is caused by a PET protection film.

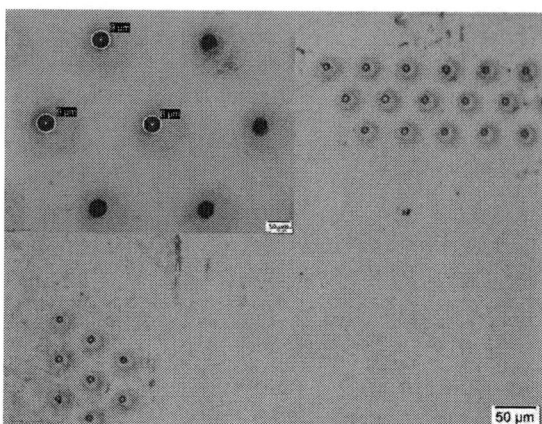

Figure 4: Openings in photo resist for structuring Ti/Cu hard mask for subsequent plasma etching.

a Nordson MARCH PCB-800 plasma system (2.7 kW, 185 mTorr, 2 slm O2, 75 °C). Frontside lamination of a 15 μm thin ABF film (approximately equal to bump height) was then carried out analogue to previous one but at 100°C. For final curing of the nano-filler dielectric a PCB lamination press Lauffer (RMV7) was used. Settings were 1 Nm/mm², vacuum of 1 mbar, and a temperature of 130 °C for 30 min followed by 170 °C for 30 min. Chips are now fixed and embedded in core cavities (Fig. 3).

4. Adaptive processing

After all embedding steps the die positions have to be measured. Therefore we used a Rudolph Firefly S1200 coordinate measuring machine (CMM). The optical transparent ABF layer allows recognition of core fiducials and chip bumps. Hence, a file with real displacement vectors (x, y and rotation angle) could be created.

Since we have a setup with two dies which moved independently it is not sufficient to shift, rotate and scale the RDL design only, because the angle and the direction of the die (re)positioning exceed the pad tolerances of 5 μm. This is why we used a chip surrounding region in which connection traces are treated as flexible like a rubber band (red lines in Fig. 1). Now by combining measured CMM and ideal CAM data the following lithography images were recalculated to fit to the respective panel [4].

5. Formation of Via to Chip

Next step is to reveal the Cu bump surface for subsequent electrical contact. Nowadays the conventional technique for via openings to chip contacts is laser drilling. However, the miniaturization leads to tight via-to-pad tolerances (here +/- 5 μm). This requires adequate drill hole diameters and position precision. The task to realize laser drill holes with diameters smaller than 15 μm is feasible but still in development [5, 6]. Also material ablation selectivity is limited.

Therefore we have taken a different approach based on plasma etching. In first experiments we observed that the homogeneity of plasma etching of few microns in z-direction over panel area of several 100 millimeters is challenging to control. Also the adhesion of metal layers on etched ABF might be not reproducible.

Thus we make use of a lithographic defined metal hard mask to open contact positions only. The plasma etch process can then be designed to selectively remove the dielectric. First, a Ti/Cu (100 nm / 300 nm) layer is sputtered onto the panel frontside in a CreaVac Creamet 600 Cl2 S3 deposition system. Then a 7 μm thin dry film photo resist (Hitachi RY5107UT) is applied by vacuum lamination. For lithography an Orbotech Paragon Ultra 200 Laser Direct Imaging (LDI) machine was used. Since the bump positions are known from a CMM an adapted image is applied for opening the resist only at the appropriate locations. Figure 4 shows a corresponding image after resist development.

Afterwards, by chemical etching of Cu with a $Na_2S_2O_8/H_2SO_4$ solution (micro etching) and subsequent resist stripping a Cu mask for the Ti layer is defined. Now a H_2O_2 based solution is used for etching Ti and thus revealing the ABF surface only over Cu bump positions. The opening diameter of nominal 10 μm is smaller than the bump diameter of 25 μm. The ABF is then etched utilizing a Nordson March PCB-800 plasma system. For all plasma steps a power of 4.5 kW and a total gas flow of 2.5 slm was applied. After warm-up with O_2/N_2 gases at 300 mTorr a sequenced two-step process at 180 mTorr with O_2/Ar first and $Ar/CF_4/O_2$ second was used for ABF etching. Residues are removed in a wet-chemical desmear process (sweller 60 °C for 5 min and etching at 80 °C for 15 min).

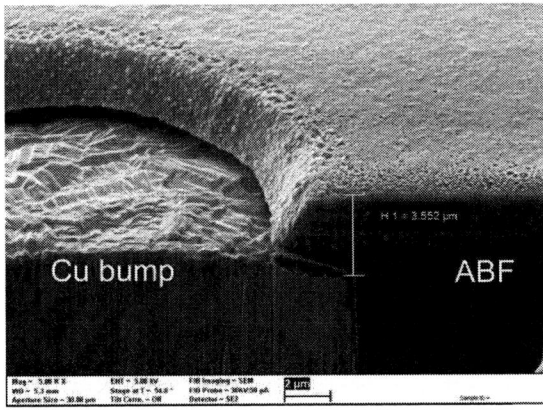

Figure 5: Plasma etched opening (after removal of Ti/Cu hard mask) over Cu bump embedded in ABF dielectric. Cracks/undercutting around bump is problematic.

In Fig. 5 a corresponding scanning electron microscope (SEM) image of a focused ion beam (FIB) cut after this process is shown. About 5 µm of the dielectric have been selectively etched. One can see filler particles of the ABF on the inside of the opening. The ABF surface surrounding the opening appears unaffected since it was covered by the metallic mask. No residues on the Cu bump are observed. The conical shape supports side wall metal coverage and electrolyte flow in further processes. However, there is an undercutting effect at the edge of the Cu bump. Concerning reliability this is not an acceptable result. We assume that the electric field of the plasma in the vicinity of the Cu is modified in a way that enhances ABF etch rate. Desmear chemistry may play a role here, too. Further investigations will be carried out to clarify the causes. In future experiments also a designated reactive-ion etching machine (parallel plate reactor) will be used instead of the March system made for PCB processes.

6. RDL formation

The redistribution layer is formed in a semi-additive process. Analogue to the hard mask a Ti/Cu (100 nm barrier and 300 nm seed) layer stack is deposited by sputtering. Again, the RY5107UT resist is vacuum laminated.

For exposure we now use a Schmoll MDI-ST-Ultra direct imaging system. It is equipped with photoheads carrying a digital micromirror device (DMD) with 2560 x 800 mirrors each. The light of a 405 nm photo diode is reflected by switchable mirrors and guided through an optical system to project an image onto the photo resist. The optical system with a sub-pixel size of 360 nm is designed for a lateral resolution of structures down to 4 µm. Due to the functional principle the photoheads move in strips over substrates with up to 620 mm x 620 mm size. The total exposure time for a panel depends on the number of photoheads (up to 6) and the required exposure dose of the resist.

Figure 6: Microscopic image of Cu RDL after SAP process.

Here we used one photohead, a dose of 550 mJ/cm² and a substrate size of 303 mm x 227 mm resulting in about 10 min of exposure time. Only four global fiducials per panel are required for registration due to the adaptive process flow. After exposure and 1 h hold time the resist development was performed in a horizontal Schmid CombiLine system in a carbonate solution.

To minimize resist foot an oxygen descum plasma was run afterwards. For subsequent Cu plating an electrolyte from Schlötter was used. With a current density of 1.0 A/dm² Cu traces of about 6.5 µm thickness have been plated with a growth rate of about 0.25 µm/min. Then the resist was stripped. Finally by differential etching the initial Cu seed and Ti barrier layer were removed using the chemicals described above. Figure 5 and 6 show the demonstrator fan-out design with 5 µm line/space after SAP process. The roughened Cu surface is due to differential etching chemistry.

After finishing the first RDL a prepreg can be laminated symmetrically on both panel sides to build up subsequent layers. High aspect ratio vias as an element for vertical interconnects are possible.

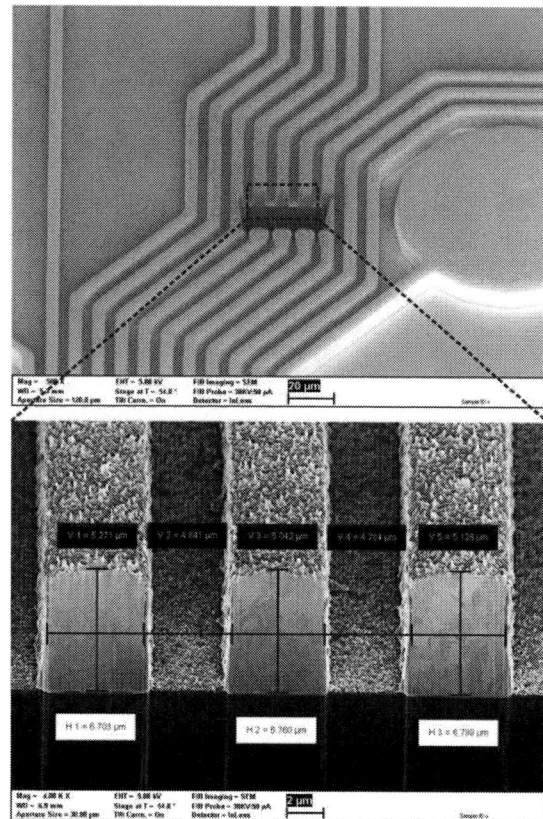

Figure 7: FIB section and SEM view of Cu RDL after SAP process.

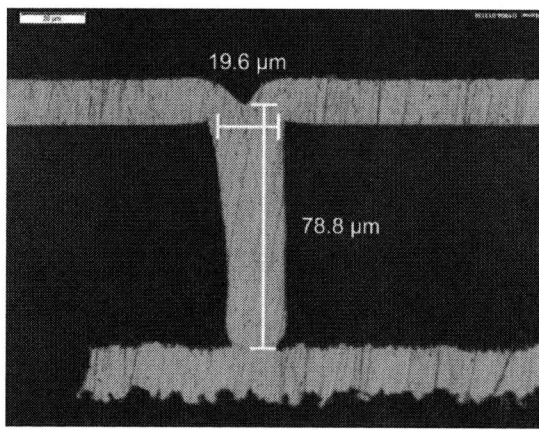

Figure 8: Cross-section of Cu filled blind microvia in ABF with aspect ratio 4:1.

Availibility of appropriate laser drilling technique and plating chemistry for filling such vias has been demonstrated [5]. Figure 8 shows a Cu filled microvia with a top diameter of 20 µm in a 79 µm thick ABF dielectric corresponding to an aspect ratio of 4:1.

Summary

Dies with a pitch of 50 µm (3 rows) were symmetrically embedded into cavities of a PCB core. First investigations on the concept of plasma etching to open buried bumps have been demonstrated. Intermediate results show its feasibility in principle but still non-negligible undercutting.

The fan-out redistribution layer with 5 µm L/S Cu traces were grown in a semi-additive process with sputtered Ti/Cu seed layer on organic substrates. As lithography tool a digital micromirror based direct imaging machine from Schmoll was used. With mask-less technology the build-up images can be recalculated and adjusted to individual, actual die position (adaptive imaging).

Next steps will not only include advanced building blocks like high aspect ratio vias (up to 4:1) but also the optimization of process capability (fine line SAP, plasma etching) for panel sizes up to 600 mm x 600 mm.

Acknowledgements

This work was financially supported by the German Ministry of Education and Research within the project "PEKOS" (No. 16ES0718K).

The authors thank Dimitri Mun from Schmoll Maschinen GmbH for supporting with direct imaging processing.

We deeply acknowledge Michael Gerberich, Lutz Gerhold, Malte Spanier, Christian Voigt, Lars Böttcher, Evelyn Wegner, and Eugen Formin from Fraunhofer IZM for experimental support and organizational contributions.

References

[1] Report "Status of Advanced Substrates 2019", Yole Développement

[2] K. Kikuchi, "Warpage Analysis with Newly Molding Material of Fan-Out Panel Level Packaging and the Board Level Reliability Test Results", 2018 IEEE 68th Electronic Components and Technology Conference (ECTC), May 21 – June 1, 2018, DOI: 10.1109/ECTC.2018.0014

[3] Y. Han, "Process Feasibility and Reliability Performance of Fine Pitch Si Bare Chip Embedded in Through Cavity of Substrate Core", IEEE Trans. Compon. Packag. Manuf. Technol, Vol. 5, No. 4, 551-561 April 2015, DOI: 10.1109/TCPMT.2015.2406880

[4] R. Kahle, "Evaluation of adaptive processes for the embedding of bare dies in IC substrates", 22nd Microelectronics and Packaging Conference (EMPC), Pisa, Italy, Sept. 16-19, 2019

[5] A. Ostmann, "High Density Interconnect Processes for Panel Level Packaging", 7th

Electronics System-Integration Technology Conference (ESTC) 2018, Dresden, Germany, Sept. 18-21, DOI:10.1109/ESTC.2018.854643

[6] W. Zhao, "Microdrilling of Through-Holes in Flexible Printed Circuits Using Picosecond Ultrashort Pulse Laser", Polymers, Vol. 10 (12), 1390, 2018, DOI: 10.3390/polym10121390

Injection Molded Structural Electronics Brings Surfaces to Life

Outi Rusanen, Tomi Simula, Paavo Niskala, Ville Lindholm and Mikko Heikkinen

TactoTek, Automaatiotie 1, 90460 Oulunsalo, Finland

E-mail addresses are in the form of firstname.lastname@tactotek.com

Abstract

This paper introduces IMSE technology enabling smart molded structures. (IMSE stands for Injection Molded Structural Electronics.) Smart molded structures are made by integrating and encapsulating printed electronics and standard electronic components within durable, 3D injection-molded plastics. IMSE technology and manufacturing differ from conventional electronics. Thus, TactoTek uses its internal IMSE Technology Validation process for certification of components, surface mounting adhesives and other materials. IMSE Technology Validation process includes also extensive reliability testing to ensure that IMSE solutions are reliable.

Key words: Structural electronics, printed electronics, validation process, reliability

Introduction to IMSE Technology

IMSE technology enables design innovation by integrating electronic functions into 3-dimensional injection molded plastic structures. Features, such as controls, sensors, illumination and communications, are embedded in thin 3D structures with plastic, wood and other surfaces.

The structures are light, thin and durable. In conventional use cases, such as an in-vehicle control panel, a single part replaces a multi-part conventional electronics structure and eliminates labor-intensive electro-mechanical assembly. The part also weighs less and is significantly thinner. TactoTek has demonstrated structural electronic designs with 70% weight and 90% thickness reduction when compared with conventional multi-part assemblies (Figure 1).

Introduction to IMSE Manufacturing

Typical IMSE solution consists of electronic film on the bottom carrying conductive inks, dielectrics, and surface mounted electronic components. On the top, there can be either surface film with decorations or a natural surface such as wood. Everything is injection molded to one-piece assembly. Figure 2 shows an illustration of 2-film design.

Depending on the use case and functionality, IMSE part can be also a 1-film design. One-film design has an electric film on the back or front side of the part. When using only back-side-film, we create visual surface by post process on top of injection molded resin. When using only front-side-film, we print both graphics and electronics onto the same film.

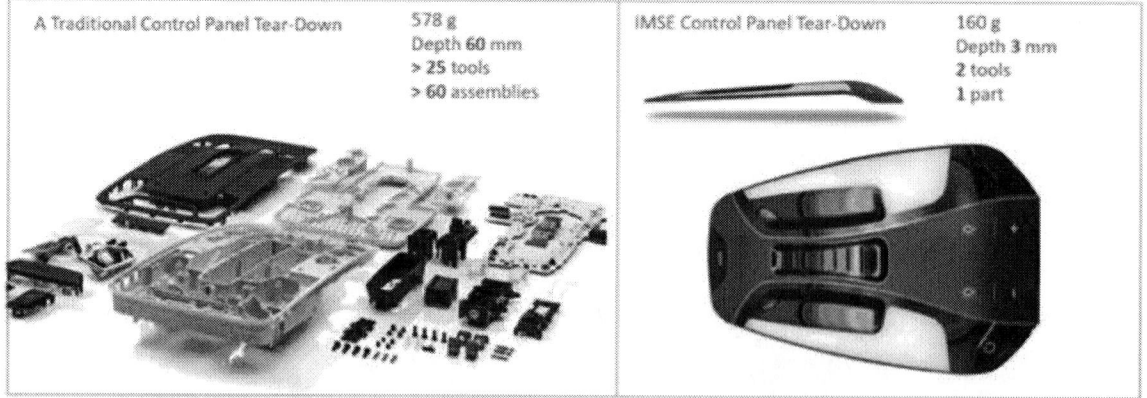

Figure 1: TactoTek has demonstrated IMSE designs with 70 % weight and 90 % thickness reduction when compared with conventional multi-part assemblies.

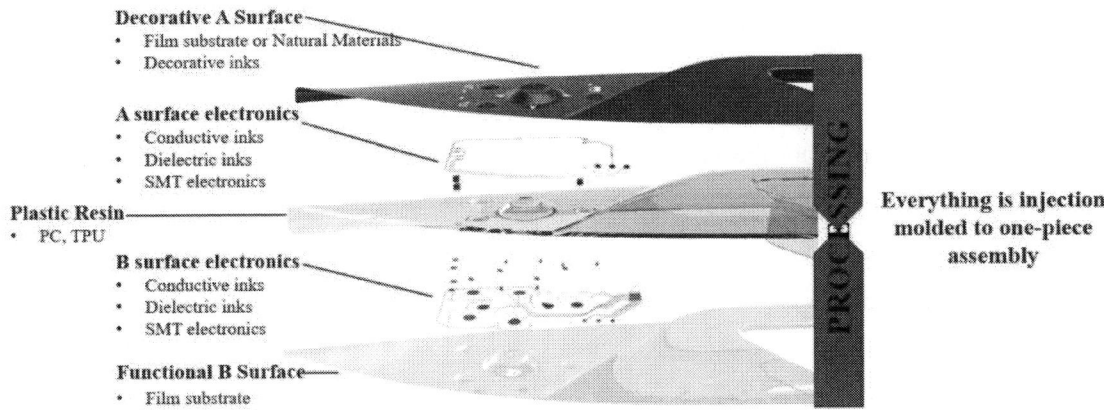

Figure 2: Illustration of IMSE part structure in 2-film design

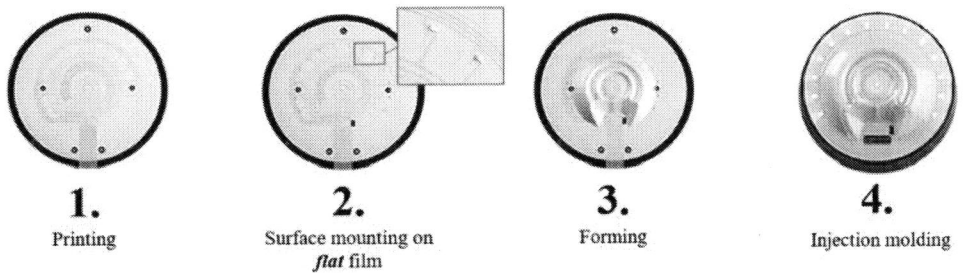

1.
Printing

2.
Surface mounting on *flat* film

3.
Forming

4.
Injection molding

Figure 3: Core manufacturing processes for IMSE

Core manufacturing processes for IMSE are printing, surface mounting, forming and injection molding (Figure 3). Taken individually, these processes are mature and TactoTek uses standard equipment suitable for mass production. However, the standard processes are combined in a unique way during manufacturing of IMSE.

Printing is the first core manufacturing process and TactoTek uses screen printing technology. Screen printing is suitable for mass-production and enables appropriate layer thickness (range of 10 μm). Electronics (functional inks) and decorations (graphic inks) are screen printed onto plastic film or another suitable substrate material. Electronics are typically printed using silver (Ag) conductive inks and dielectric inks to insulate between layers of circuitry.

Surface mounting is the second core process. Components are placed and bonded, mechanically and electrically, onto electronic films. IMSE structures use isotropically conductive and structural adhesives for component bonding. The output is two-dimensional film substrate with components.

Forming is the third core process and TactoTek uses a high-pressure thermoforming process. Two-dimensional electric and graphic films are thermoformed into three-dimensional shape and trimmed as needed. Outputs are 3D electric films with components and 3D graphic films. Forming is not used in conventional electronics manufacturing. During forming, component packages are subjected to elevated temperatures and pressures. The maximum temperature depends on the polymer film and is typically below 150 C. Maximum pressure is typically below 8 MPa (80 bar).

Injection molding is the fourth core manufacturing process. Three-dimensional electric films and 3D graphic films are used as inserts in an injection molding tool. Resin, such as polycarbonate is injected between the films resulting in a single molded part. The output is a strong and durable structure in which electronics are encapsulated within the molded plastic. Injection molding is not used in conventional electronics manufacturing, either. Some molding temperatures are higher than peak temperature during reflow soldering of SAC-solders. In addition, heat transfer from hot resin to electric films is through conduction. Thus, heat transfer to components and other materials is more efficient than during reflow soldering. Maximum pressures during injection molding are around 100 MPa (1000 Bar).

IMSE manufacturing often includes also pre-assembly and final assembly of control electronics. These processes are similar to conventional electronics components.

Requirements for an Ideal Component

The 2-film design can be illustrated also as a material stack, Figure 4. By material stack we mean the combination of films, inks, electronic components, surface mounting adhesives and injection molding resin.

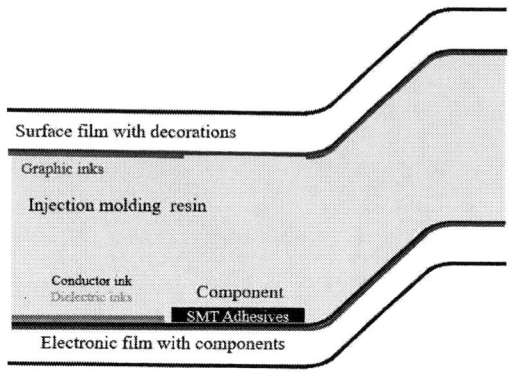

Figure 4: Illustration of a material stack in a 2-film design

Currently most electronic components are optimized for conventional electronics manufacturing that does not include the temperature and pressure exposure of thermoforming and injection molding processes. TactoTek has therefore determined requirements for an ideal component package [1]. By ideal component package we mean that it does not lower IMSE manufacturing yield, product reliability or impose significant design constraints. The requirements for the ideal

component package are listed below and partly illustrated in Figure 5.

1. Minimum contact spacing is 500 μm.
2. Minimum contact area is 500 μm x 300 μm.
3. Contact surfaces are clean. For example, there are no mold release agents or oils.
4. Component package substrate is structurally strong, such as glass fiber laminate or copper/bronze baseplate.
5. No package materials are sensitive to moisture. The moisture sensitivity level (MSL) is 0 or 1.
6. Component package does not have any cavities or hollow parts. The use of porous materials, such as low-fired ceramics, is also minimized.
7. Maximum package overall height is 1.0 mm.
8. Bottom of component package is flat and has room for bonding. Structural adhesives, such as epoxies, have good adhesion to bottom of component package.
9. Package shapes are simple, such as cuboids, cylinders or domes. All corners are rounded.
10. No wire bonding is exposed.
11. Maximum package size is 16 mm^2.
12. Package material has good adhesion to injected polymer resin, such as polycarbonate (PC), PET or TPU.
13. Maximum amount of contact pads is 16. (This requirement is a combination from requirements 1, 2 and 11.)

IMSE Technology Validation Process

IMSE Technology Validation process includes certification processes for conductive inks, electronic components and surface mounting adhesives. In addition, the total material stack is validated in Material Stack Platform (MSP). MSP test structures include all IMSE materials and go undergo all IMSE core processes. In this paper, we present the certification processes for components and surface mounting adhesives.

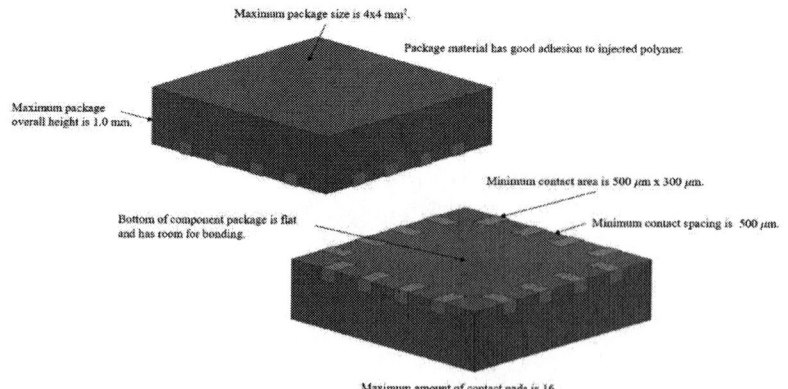

Figure 5: Requirements for ideal component package

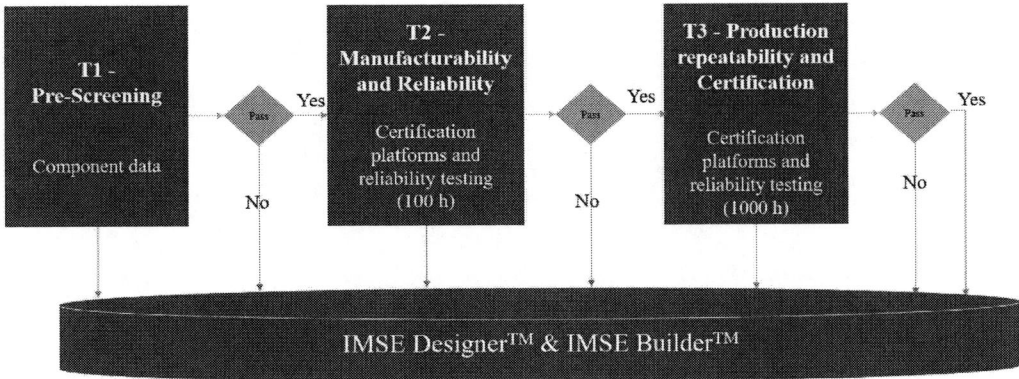

Figure 6: Component certification process

Component Certification Process

TactoTek certifies all electronic components that are embedded inside injection molded polymers. Certification has three steps, Figure 6. During Pre-Screening component data is compared with the ideal package. The component package does not need to fulfill all requirements to pass this step. However, there are some items that cause failure at Pre-Screening. For example, a package with moisture sensitivity level (MSL) of 4 or higher fails.

If a component passes the pre-screening (T1), TactoTek manufactures certification platforms with that component using internal company standards for layout and material stack. The T1 manufacturing lot includes a total of 200 components, or more. Components undergo surface mounting, forming and injection molding and they are tested after each process step. The certification platforms are subjected to reliability testing, as well. Typical environmental loads are change of temperature as well as elevated temperature and elevated temperature-humidity. Testing time is 100 hours or more.

If component passes T2 step, TactoTek manufactures more certification platforms. The T3 manufacturing lot includes at least 10 000 components so that TactoTek can assess production repeatability. Some of the certification platforms are subjected to reliability testing. Tests are the same as during T2 and testing time is increased to 1000 hours or more. Based on manufacturing yield, reliability testing and physical failure analysis, component can pass T3 step and receive IMSE certification.

Surface Mounting Adhesives Certification Process

By surface mounting adhesives we mean isotropically conductive and structural adhesives used for electrical and mechanical bonding of components, Figure 7. Typically, they are optimized for conventional electronics manufacturing that does not include the temperature and pressure exposure of thermoforming and injection molding processes. TactoTek has therefore identified critical-to-quality requirements for surface mounting adhesives. They are listed below:

1. Adhesive is compatible with typical TactoTek stack materials, see Table 1.
2. Adhesive can be applied by stencil printing, pin transfer or by (jet) dispensing. It is an advantage if adhesive is suitable for many application methods.
3. Adhesive cures during TactoTek reflow, typical peak temperature is 130 °C and profile length is about 3 minutes.
4. Adhesive working life is 72 hours or more.
5. Adhesive storage life is 3 months or more.
6. Isotropically conductive adhesive has resistivity less than 5 mΩ·cm.
7. It is an advantage if adhesive is solvent free.

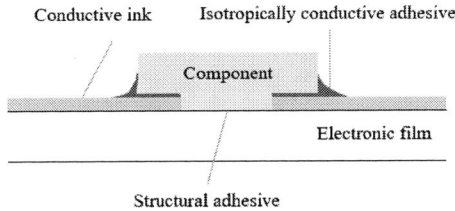

Figure 7: TactoTek uses isotropically conductive and structural adhesives for surface mounting

Table 1: Typical TactoTek stack materials

Film substrate	Polycarbonate films are most common materials
Graphic inks	Film-Insert-Molding ink families from Pröll, Nazdar and Marabu
Functional inks	In-Mold-Electronics families from Dupont, Henkel, Sun Chemical and Nagase

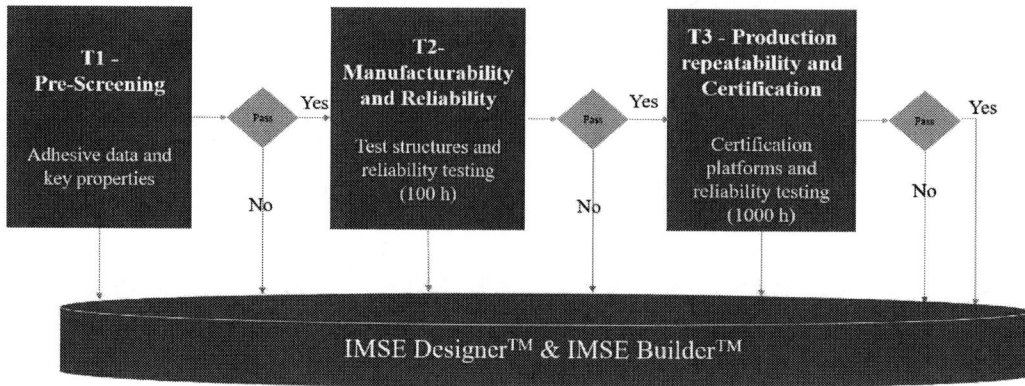

Figure 8: Surface mounting adhesive certification process

Steady-State Temperature-Humidity 85C/85%RH (JEDEC 22-A101) for 1000 hours	High Temperature Ageing at 110 C for 1000 hours	Rapid Change of Temperature between -40 C and +85 C * (IEC 60068-2-14) for 1000 cycles
26 MSPs were tested, • 18 were all powered on • 8 were powered off	19 MSPs were tested, they were all powered off	20 MSPs were tested, they were all powered off

Figure 9: Summary of reliability testing in MSP validation

TactoTek certifies surface mounting adhesives, as well. The certification is similar to component certification process and has three steps, Figure 8. During pre-screening, adhesive data is compared with critical-to-quality requirements. We also make quick trials to ensure that adhesive is compatible with typical stack materials and that it cures in TactoTek curing process (130 C for about 3 min).

If adhesives pass T1, TactoTek manufactures test platforms using 0 Ω - resistors. They undergo surface mounting, forming and injection molding. The resistance values are measured after each process step. The test structures are subjected to reliability testing, as well. Typically, tests are same as in component certification. Testing time is 100 hours or more.

If adhesives pass T2 step, TactoTek manufactures certification platforms using components, such as LEDs. They undergo surface mounting, forming and injection molding. Some of the certification platforms are subjected to reliability testing. The testing time is increased to at least 1000 hours. Based on manufacturing yield, reliability testing and physical failure analysis, adhesives can pass T3 step and receive IMSE certification.

Reliability Testing Results - Material Stack Platforms

The results presented here are from validation of Material Stack Platforms (MSPs). In this case, the components were two different types of LEDs.

Type1-LED is top-shooting and Type2-LED is side-shooting. MSPs included also structures for measuring resin transparency.

A total of 65 MSPs were tested in three different reliability tests, Figure 9. We data-logged LED forward voltage (V_f) for one LED in each MSP that was powered on, Figure 10. We also measured LED brightness (i.e. maximum and mean intensity) as well as LED color coordinates before and after reliability testing.

Figure 10: Photo of the MSPs that were powered on during Steady-State Temperature-Humidity (85/85) testing

All of the tested LEDs remained functional during testing and LED forward voltage (V_f) remained constant, as well. In addition, the color coordinates before and after testing had nearly the same values. We did, however, see a luminance decrease in both types of LEDs. For Type1-LEDs, it was most noticeable in High Temperature Aging.

Their luminance values after testing were about seven percent lower than before testing. For Type2-LEDs, the luminance values decreased significantly after Steady-State Temperature-Humidity test. The transparency of the injection molding resin had not changed during testing. Thus, we assume that the luminance decrease is caused by delamination between LED encapsulant and injection molding resin. We are currently making root-cause analysis to confirm this assumption.

Reliability Testing results - Extended Testing

Testing until failure is useful for better understanding reliability. That is why we tested one of the company demonstrator products for 3000 cycles in the change of temperature test. The demonstrator product, shown in Figure 11, contains 20 pieces of Type2-LEDs. Figure 12 shows summary of the tests and sample sizes.

Figure 11: Photo of a demonstrator product that was subjected to extended reliability testing

Change of Temperature
between -40 C and +85 C *
(IEC 60068-2-14) for 3000 cycles

20 pcs of Type2-LEDs were tested,
they were all powered on (I=15 mA).

*Test cycles had 1 h exposure (i.e. dwell) times and 1 hour ramp times. One test cycle lasted for 4 hours.

Figure 12: Summary of extended reliability test

Two Type2-LEDs failed during 3000 cycles. Failures occurred at 702 and at 2988 cycles. When we fit this data into a Weibull distribution, we predict 50 percent failure point at over 10000 cycles, Figure 13. Such a high value demonstrates the reliability of the technology.

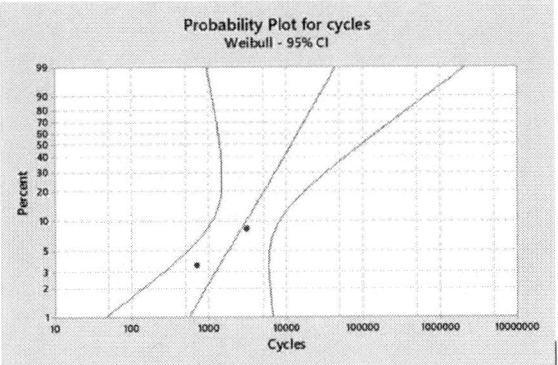

Figure 13: Extended reliability testing failures in Weibull-distribution

We performed cross-section and scanning-electron-microscope (SEM) analysis on a failed Type2-LED. It showed that the failure mode is a fracture in the conductive adhesive used for surface mounting, Figure 14. The failure mechanism is conductive adhesive creep caused by thermo-mechanical stresses during thermal cycling.

However, the number of tested components was small in the extended reliability testing. Thus, we have tested another demonstrator product with 20 Type2-LEDs. After 1362 cycles, all LEDs have remained functional and the test continues.

We have shared also reliability testing results from component certification platforms [2]. Tests and test durations were similar to MSP reliability testing. Components endured thermo-mechanical stresses and elevated temperature-humidity without failures as long as designs were according to TactoTek design guidelines.

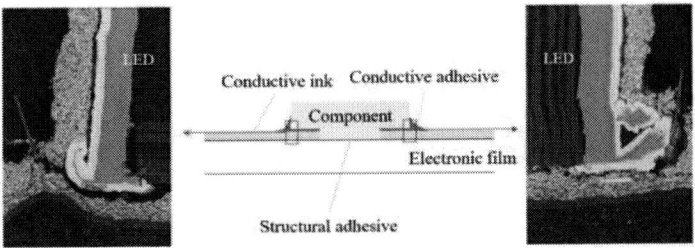

Figure 14: SEM-images of the cross-section show fracture in the conductive adhesive

Conclusion

TactoTek has demonstrated IMSE designs with 70 % weight and 90 % thickness reduction when compared with conventional multi-part assemblies. Because IMSE manufacturing is different from conventional electronics, TactoTek uses its internal IMSE Technology Validation process. It includes certification of conductive inks, electric components and surface mounting adhesives. TactoTek uses also Material Stack Platforms (MSPs) to certify injection molding resins.

The results from MSP reliability testing show that LED components pass tests such as 1000 hours in Steady-State Temperature Humidity, High Temperature Testing and Rapid Change of Temperature. Injection molding resin strengthens the structures and also protects electronics from environmental conditions, such as moisture, and thermo-mechanical stresses.

Acknowledgements

The authors thank the following TactoTek colleagues and teams and for their much-valued contribution to this work: Tuomas Kallio, Pasi Korhonen, Johannes Soutukorva, Advanced Engineering Team and Production Team.

References

[1] T. Simula, et al., "Component Packages for IMSE (Injection Molded Structural Electronics)," in Proceedings of NordPac, Oulu, Finland, 2018.

[2] O. Rusanen, et al. "Smart Molded Structures Bring Surfaces to Life", in Proceedings of IPC APEX EXPO Technical Conference, San Diego, 2019.

Innovative Authentication Method for IoT Devices

Daniel C. Vasile, Paul M. Svasta

UPB-CETTI, 313 Splaiul Independentei, Sector 6,
060042, Bucharest, ROMANIA

0040723283387, ciprian.vasile@cetti.ro

Abstract

This work provides an innovative method for a very important cryptographic function used in securing the communications in an IoT network: authenticating a device based on the identity obtained from a unique physical characteristic. An authentication procedure requires a secret key that is usually protected by an anti-tamper mechanism. IoT devices can use tamper detection circuits to protect secret data involved in cryptographic functions as: encryption, decryption, authentication, key exchange, etc. These functions are implemented in electronic security circuits. The tamper detection circuit is independently powered by a battery in order to permanently protect the secret data, even if the equipment is not supplied with power. This paper presents a solution to obtain a unique key from the probing process of the conductive mesh that protects the electronic security circuit.

Key words: mesh, anti-tamper, authentication, security, criptography

Introduction

IoT devices are spread in many areas of activities and are capable of sensing or actuation functions. Usually they are communicating through the Internet and they are in a large number. Even if they are used in home appliances, industrial automation or medical devices, their data storage and communications must be protected against cryptanalytic attacks. Generally, the IoT devices are small electronic circuits with limited computation power. In some applications they are installed in remote places where they must operate with very little energy and witout interruption and maintenance. Due to these reasons, IoT devices don't enjoy the facilities offered by an operating system. They run bare-metal programs that are optimized for memory and processing resources. Usually they are implemented in microcontrollers or security processors. The challenges for assuring the security of IoT devices are:

- They are deployed in critical environments.
- One type of device is replicated in thousands of identical devices, meaning that, if one of them can be hacked, the attack can be fulfilled on all others.
- They are hard to upgrade because this capability requires storage memory and software procedures.
- They have a long operating period, usually a few years, in which they must respond to security challenges.
- They are installed in places that are not protected by specialized equipment that can be found in enterprise facilities.

So, IoT devices must protect themselves in risks environments with minimum energy.

The protection of IoT devices is achieved through cryptographic functions and physical security.

Cryptographic functions protect data communications between devices. Therefore, no unauthorized entity has access to the transmitted information. Cryptographic functions cover all aspects of data security, as: confidentiality, integrity, non-repudiation and authentication.

Physical security is responsible for data protection inside the device. Some cryptographic functions rely on the secret data (keys) that are stored inside the device's memory. The secret data are important to the same extent as the cryptographic functions. The physical security of the IoT devices consists in detecting any intrusion attempt and reacting promptly (erasing the secret data).

This paper focuses on a new approach for the authentication functions used in cryptographic protocols. Authentication relies on a secret key known by the both sides involved in communication. The method for obtaining the secret key, provided by this work, consists in acquiring an electronic pattern (in form of a collection of numbers) of the conductive mesh that covers the entire security electronic circuit of the IoT device. The conductive mesh is used to detect physical intrusions and is connected to an active tamper detection circuit (ATDC). Its response, at predefined frequencies, provides an electrical pattern of the cryptographic key. The conductive mesh has a special construction consisting in three layers with conductive traces. The first one (nearest to the electronic circuit) is a conductive layer connected to the circuit's ground signal. The second layer contains very thin traces that cover the whole surface. It is provided with an input and an output port. The third layer (the outermost layer) contains traces, organized as areas that are similar to those on the second layer. Each area is short-circuited. If the traces on the third layer are short circuited (between adjacent traces) or if the circuits are opened, the response on the second layer is modified and the authentication function based on this response will fail.

This paper continues the work presented in the papers [1] and [2], by adding new features.

The interaction of the conductive layers can be modeled as two inductively coupled circuits, as presented in [3], and is depicted in Figure 1:

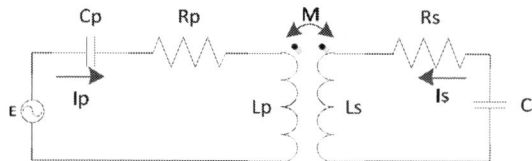

Figure 1: Equivalent circuit of the coupled layers 2 and 3

The input impedance of the trace on layer 2, function of the impedance of the trace on layer 3, can be expressed, according to [3], as:

$$Z_{in} = Z_p + \frac{(\omega M)^2}{Z_s} \qquad (1)$$

where $Z_s = R_s + j\omega L_s - j\frac{1}{\omega C_s}$ and $Z_p = R_p + j\omega L_p - j\frac{1}{\omega C_p}$. Z_{in} is the impedance seen by the signal source E, Z_p is the open-circuit impedance of the primary circuit, Z_s is the impedance of the secondary circuit and M is the mutual reactance of the coupled circuits. A change in the impedance of the secondary circuit (layer 3) will produce a change in the impedance of the primary circuit (layer 2), and so, the signal used to probe the conductive trace on layer 2 will be modified (phase and amplitude). Thus, physical intrusions attempts can be detected before even penetrating all the layers of the anti-tamper conductive mesh.

Design of the anti-tamper conductive mesh

The anti-tamper conductive mesh (ATCM) is made of PCB (Printed Circuit Board) or other printing technologies that are suitable for creating thin conductive traces on stackable layers. It is made of three layers of conductive traces with isolation layers between them. The structure of the ATCM is depicted in figure 2.

The layers have the following description:
- The first layer contains a conductive layer spread all over the surface (grid surface in figure 2). It is connected to the ground signal of the detection circuit. Thus, it acts as a reference for the signal layers 2 and 3 and as an electromagnetic shield between the electronic circuit and these two layers.

This layer can be a copper foil or can have a hashed design in order to ease the bending of the ATCM.
- The second layer contains conductive traces that follow a pattern like meanders (red trace in figure 2). It has an input port, used to connect a signal source (V_1 in figure 2), and an output port connected to a load (R_1 in figure 2). The signal is acquired at the output port in order to analyze the modifications in the signal parameters. This layer is the probing layer.
- The third layer, exposed to the outside, has conductive traces with the same pattern as those in the layer 2. Small areas covered by these traces are short-circuited (green trace in figure 2). Traces on layer 3 and 2 are capacitively and inductively coupled, hence they influence each other. The third layer is a detection layer because any intervention on this layer (trace interruption or shorts between traces) will cause a modification to the signal parameters in the second layer.

The ATCM covers both the ATDC and the protected security electronic circuit. The only slot provided by this structure is for a small connector for power supply and digital signals.

The design presented in figure 2 coresponds to a small area of the ATCM, delimited by the trace on the layer 3 that is short-circuited. The ATCM is made up of many such areas, side by side, with no gaps between them. The smaller the traces are the more efficient the ATCM is. The width of the conductive traces and the spaces between them depend on the technology used to make the ATCM.

In order to evaluate the efficiency of the ATCM, the mesh was modeled and simulated. The simulation results were compared to the correspondent signals from a fabricated test vehicle.

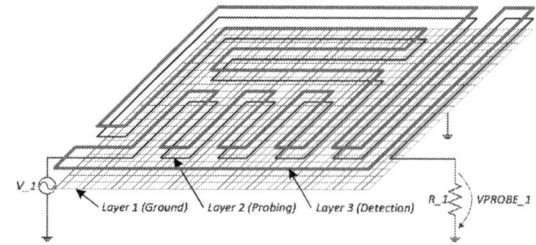

Figure 2: Structure of the anti-tamper conductive mesh

Simulation of the ATCM

For modeling and simulations it was used the ANSYS ® SIwave 2016.2.0 program and a structure that is similar to the one presented in figure 2. It is composed of two square areas, side by side, with traces connected in series, on layer 2, and with traces that are short-circuited on layer 3.

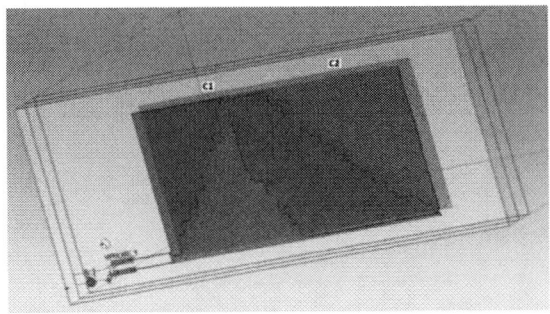

Figure 3: Simulation of the ATCM structure

Each area is a square of 30x30mm and they are noted C1 and C2. The signal source V_1 has an internal impedance of 50Ω and generates a sinusoidal signal with amplitude of 1V. The loading resistor has a resistance of 50Ω. The simulated structure is presented in figure 3.

The simulations were performed according to two structures made of different materials:

- The first one is a PCB that has 0.3mm FR4 dielectric layers with copper traces. This is a rigid structure.
- The second type is made of flexible PES (polyester) with thickness of 0.1mm. The traces were printed using the SW180 conductive paste, made by Tatsuta Electric Wire & Cable Co. Ltd.

There are some uses of the conductive adhesive pastes in anti-tamper designs, as presented in [4]. This paper presents an innovative design based on a mesh printed with the SW180 paste.

The assessments consisted in measuring the level of the signal on the R_1 resistor while sweeping the frequency of the V_1 source signal in the 0MHz ÷ 300MHz domain. The scope of this paper is to find measurable differences between the untampered and tampered states of the mesh. The acquired levels will not allow an electronic device to authenticate in case of tampering.

For each type of material, PCB and PES, the assessments tried to reproduce the tampering action and there resulted the following cases:

- not tampered (case noted as NT);
- C1 is tampered by opening the circuit (case noted as TC1_O);
- C1 is tampered by shorting the circuit between traces (case noted as TC1_S);
- C2 is tampered by opening the circuit (case noted as TC2_O);
- C2 is tampered by shorting the circuit between traces (case noted as TC2_S).

The opening and shorting of the circuits are related to layer 3 (detection layer), and they are depicted in figure 4, **a)** and **b)**.

The results are presented in figures 5 and 6 for the two constructive types. The green traces represent the results when the ATCM structures are not tampered. The other traces represent each case of tampering: TC1_O, TC1_S, TC2_O and TC2_S (as

explained above). As can be observed in figure 5, representing the simulation results for the PCB ATCM with 0.3mm isolating layers, starting from the frequency of 20MHz, the tampering action can be measured with high accuracy using a standard 10 bits analog to digital converter, usually found in most microcontrollers. A practical implementation of such an ATCM can use a signal generator with discrete frequencies, spread inside the active domain (20÷300MHz).

In case of PES foil, the results presented in figure 6 show signals that are much more attenuated compared to those signals presented in figure 5. The resistivity of conductive paste is much greater than the resistivity of copper traces, resulting in a higher attenuation. The usable frequency domain for the PES foil starts from 30MHz.

Each type of ATCM has extended frequency domains that can be used to detect physical intrusions. The tampering of the detection layer 3 (by shorting or opening the circuit) will produce a modification in the propagation of the signals in the signal layer (2). Therefore, the amplitude (and phase) of the probing signals will have values that will differ from the expected ones, at certain discrete frequencies.

At low frequencies, the short-circuit traces (TC1_S, TC2_S) show small differences from the untampered trace. This is due to the fact that a short-circuit on the third layer will split the trace of a specific area (C1 or C2) into two traces (short-circuited), one of them that is very small. The tampering, in this case, is detectable at higher frequencies.

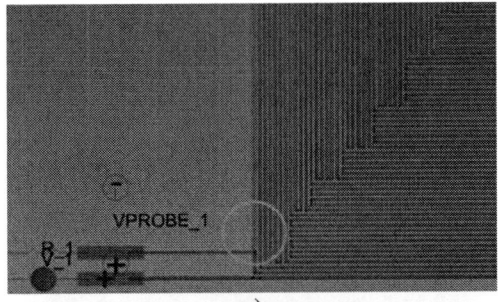

a)

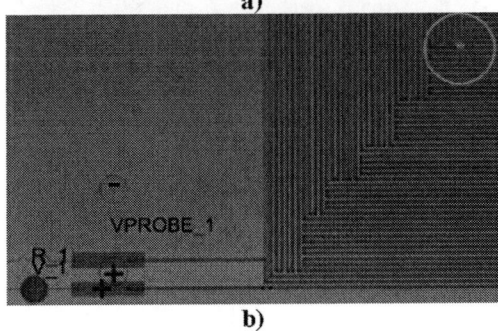

b)

Figure 4: Tampering the ATCM: a) open circuit, b) short circuit (marked with an orange circle)

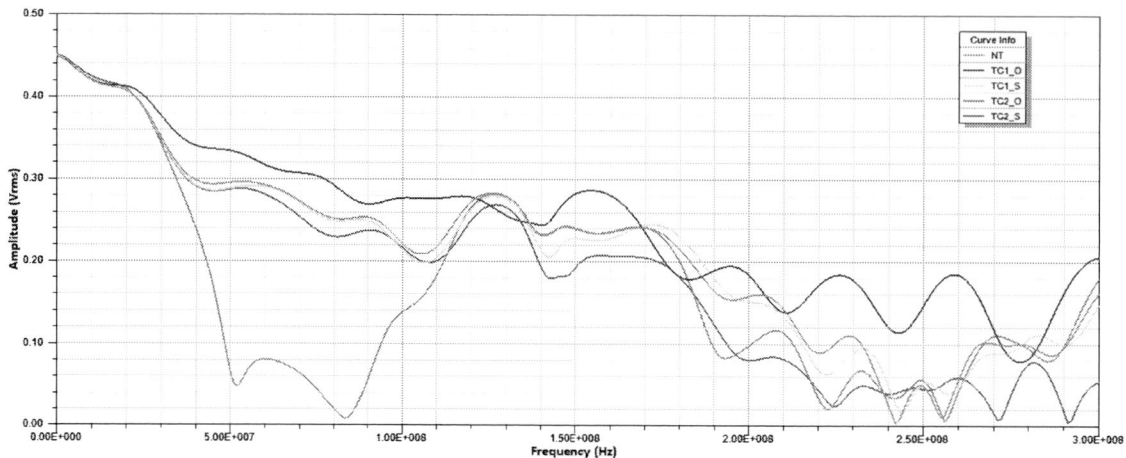

Figure 5: Simulation results for the ATCM structure made of PCB with 0.3mm isolating layers

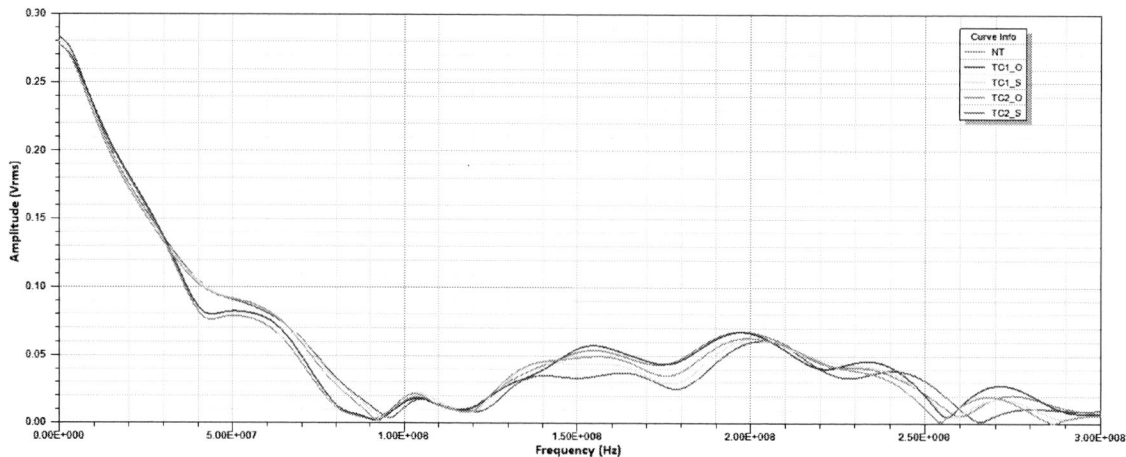

Figure 6: Simulation results for the ATCM structure made of PES with 0.1mm isolating layers

Practical test meshes

In order to validate the effect of tampering the ATCM, both structures used in simulation were made practically. The 0.1mm dielectric layers PES structure was printed using the SW180 conductive paste. The capacitance and the resistance of the traces on the second layer (signal layer) are presented in table 1. As expected, the ATCM with dielectric of 0.1mm and traces made of conductive paste has greater capacitance and resistance. This implies a high attenuation of the signal.

The ATCM structures, with the same design used in simulations, are presented in figure 7.

Measurements were performed using the Agilent N9310A signal generator and the Rohde&Schwarz HMO3004 oscilloscope. The signal generator has the output impedance of 50Ω. It was connected to the input port of the ATCM (similar to V_1 in figure 2). A 50Ω resistor was connected at the output port of the ATCM (similar to R_1 in figure 2). The observations proved that there are no abrupt modifications in magnitude for a frequency step smaller than 5MHz, therefore the frequency step was chosen to be of 5MHz. Measurements are presented in figures 8 and 9.

As expected, at low frequencies both ATCM structures cannot detect tamper attempts because of the small differences in signal amplitudes. Starting from 35MHz, for the PCB structure, and from 80MHz, for the PES structure, the differences in signals amplitudes can be measured and so the intrusions can be detected.

Both ATCM structures show a resonance domain at about 165MHz for PCB and 155MHz for PES, where the amplitude of the signal at the output port exceeds the amplitude of the signal provided by the source. At these frequencies, the ATCM structures behave like a parallel oscillating circuit connected in parallel to input and output ports.

Type of the mesh structure	Capacitance (between signal layer and ground)	Resistance of the signal trace
0.3mm PCB	162pF	21.2Ω
0.1mm PES	400pF	404Ω

Table 1. Parameters of the test mesh structures

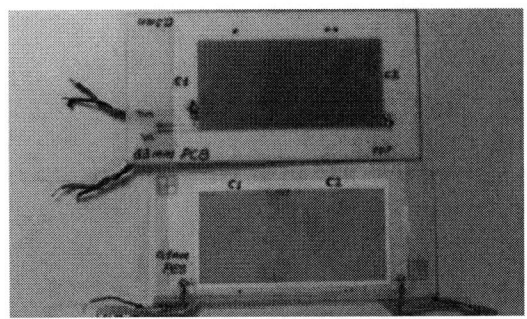

Figure 7: PCB and PES ATCM structures

Even if the practical results, presented in figures 8 and 9, do not match to those provided by the simulations, presented in figures 5 and 6, the tampering actions are detectable in the same extent. Simulations don't take in consideration all aspects of a practical assessment like the influences of measuring probes, electromagnetic couplings and the differences in the values of the parameters that characterize the materials.

Despite the fact that PES structure has a high resistance, compared to the PCB structure, in figure 9 can be observed that it shows a good conductivity at high frequencies, making this structure a very good candidate for anti-tamper mesh structures.

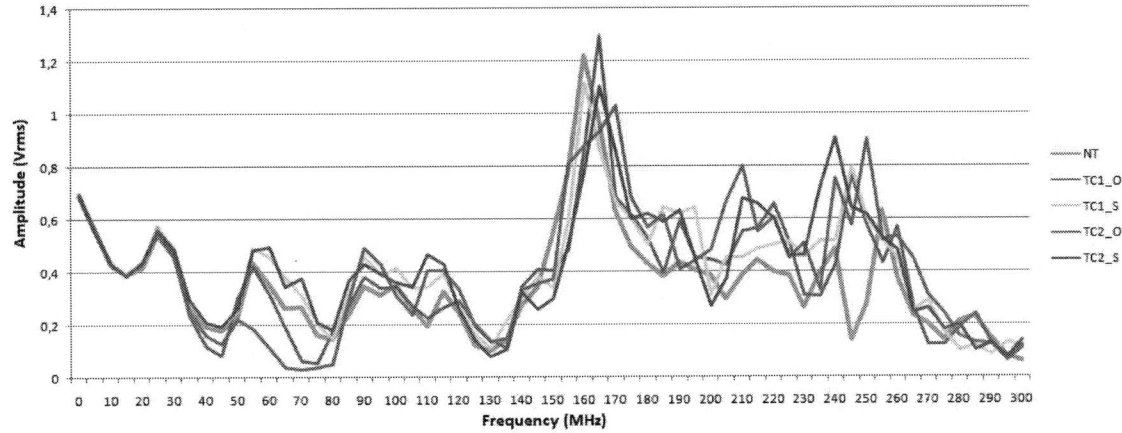

Figure 8: Test results for the ATCM structure made of PCB with 0.3mm isolating layers

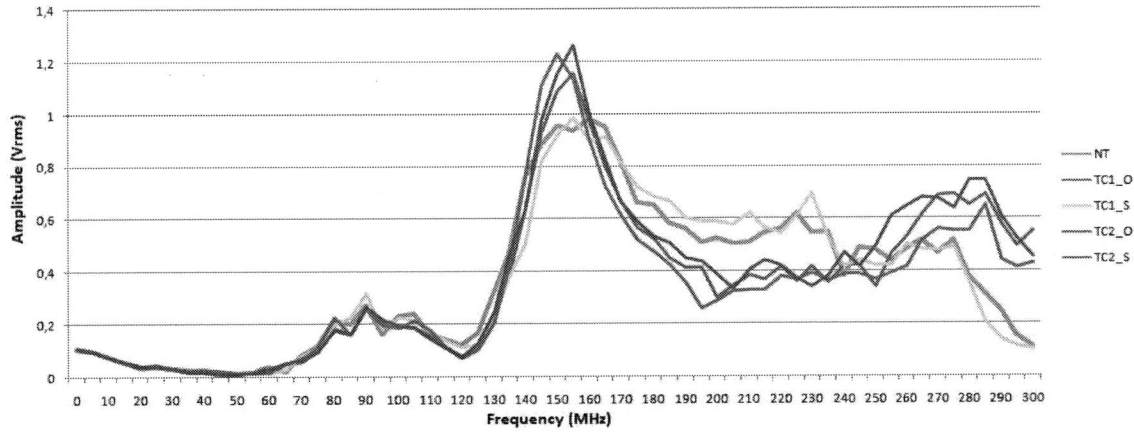

Figure 9: Test results for the ATCM structure made of PES with 0.1mm isolating layers

Authentication based on the ATCM characteristics

Every device in an IoT network must authenticate itself in order to establish a secure communication with other devices [5]. Each part of the communication must demonstrate its identity. There are many modern methods that are developed to achieve a secure authentication. Many of them are based on an identification data, a passphrase or a string. Identification data can be obtained from the

frequency response of the ATCM. Probing the mesh at different frequencies will provide the necessary data for authentication. Each response must be truncated to the bits sequences that are stable, eliminating the least significant bits (LSB) that are flipping when sampling the signal.

Unique identification of the IoT devices implies that the identification data, obtained from amplitude-frequency characteristic, to be unique.

In order to obtain a unique identification data for each device there can be used the following methods:

- Designing a unique mesh for each device.
- Applying a hash function (an injective function used to map an arbitrary size data vector onto a fixed size vector) to the numeric sequence obtained after probing the ATCM. Even if the same design of the ATCM is used to different devices, the probing results will differ in at least one bit. In this case, a hash function will transform this difference in a multiple different bits [5].
- Probing the ATCM at different frequencies for each device.

There may be used a combination of these methods in order to obtain the unique identification data.

Before using an IoT device that implements this identification procedure, it is registered to the authority responsible for the security of communications by sending the hash of the identification data. Based on this information, every device in the network can authenticate before starting a secure communication.

Conclusions

This paper proposes a solution that solves two problems of the security electronic circuits found in IoT devices: tamper detection and authentication.

The method proposed in this paper analyzes the response of a specially designed mesh that covers the electronic security circuit. Any attempt to make intrusions through this mesh will produce a detectable response when the mesh is probed with proper signals. The mesh is composed of three conductive layers, each having an important role. The first layer, near the electronic circuit, is connected to the ground signal and acts as reference plane and electromagnetic screen. The second layer, the middle one, contains thin traces that are probed with signals. The third layer, the outermost one, also contains thin traces with detection purpose. The signal layer (2) is probed with signals at discrete frequencies in order to detect variations of the signal's level at the output port. This structure was simulated and practically tested using two types of materials: PCB with 0.3mm dielectric layers and PES foil with 0.1mm dielectric layers.

Both simulations and practical assessments proved the efficiency of these structures in detecting physical intrusions. Starting from a specific frequency, each type of anti-tamper conductive mesh proves measurable levels in case of tampering.

The PES mesh shows very good conductive properties at high frequencies, even if it has a relatively high resistance. This makes the PES structure a good choice for an anti-tamper conductive mesh.

The ATCM together with an active tamper detection circuit compose a protection system that keeps the secrets of a security electronic circuit safe.

The amplitude-frequency characteristic can derive a unique identification data that is used in authentication functions and protocols. This is the second important use of this anti-tamper structure. As long as the response of the mesh is correct, the IoT device can authenticate and communicate in a network. When tampering occurs, the response of the mesh is altered and the IoT device can no longer communicate with other devices.

The designed anti-tamper mesh that is presented in this paper can protect against physical intrusions (tamper attempts) and can be also used as unique identification data in authentication protocols.

Acknowledgements

The practical part of this paper was technically supported by the following companies: **Tatsuta Electric Wire & Cable Co. Ltd.** and **ASM Assembly Systems GmbH.** We thank them for their availability, the technical assistance and the materials used in experiments.

References

[1] D. C. Vasile, P. M. Svasta, "Innovative Conductive Mesh Structure for the Protection of Security Electronic Circuits", Electronics System-Integration Technology Conference (ESTC), Dresden, Germany, September 18-21, 2018.

[2] D. C. Vasile, P. M. Svasta, "Antitamper Conductive Mesh Used for Securing Cryptographic Modules", 2018 IEEE 24th International Symposium for Design and Technology in Electronic Packaging (SIITME), Iaşi, Romania, October 25-28, 2018.

[3] K. Weidner, A. Wimmer, "Hardware Protection System For Deep-Drawn Printed Circuit Boards, As Half-Shells", U.S. Patent US20090109024A1, June 30, 2005.

[4] M. Nahvi and J. A. Edminister, "Schaum's Outlines Electric Circuits", 6th ed., McGraw-Hill Education, pp. 369-370, 2014.

[5] A. J. Menezes, P. C. Oorschot, S. A. Vanstone, "Handbook of Applied Cryptography", CRC Press, 1997.

Multi-material Printed Microwave Element for Phased Array Applications

S. Trulli[1], C. Laighton[1], E. Harper[1], D. Volkov[2], E. Engel[1], A.Akyurtlu[2], C. Armiento[1,2]

(1) Raytheon Integrated Defense Systems, Andover, MA 01810
(2) Raytheon UML Research Institute, University of Massachusetts, Lowell, MA 01854, USA

978-684-8667
susan_c_trulli@raytheon.com

Abstract

To realize additive manufacturing for microwave applications, full component solutions integrated with MMIC amplifiers are needed. A key obstacle is the utilization of multiple materials within integrated 3D printed electronics. The aim of this work is to demonstrate feasibility of printed multi-material structures that perform well at microwave frequencies. To show the efficacy of additive processing and materials for microwave applications, we designed and fabricated a C-Band phased array single element as a demonstration vehicle. We present design, simulated and measured results of the fully printed passive networks as well as characterization of materials and manufacturing processes required to realize a printed C- Band phased array single element. Passive networks are integrated with MMIC amplifiers within the channel. The technical challenges that were overcome include elimination of wirebonds through printed interconnects, fabricating 3-dimensional RF component structures that reduce the x-y footprint for the intended functionality, and creating tunable structures that allow for adaptation during the additive manufacturing process.

Key words: *printed; additive; multi-material; ink; interconnects; RF/microwave*

Introduction

Additively printed electronics (PE) has the potential to allow rapid prototyping, manufacturing of high mix, customizable and tunable designs, and the ability to create electronics in nontraditional form factors. These functionalized structures may include flexible, lightweight, conformable or wearable electronics with vertically or laterally graded mechanical or electrical properties. Developing printed electronics for the RF/microwave domain is particularly challenging since the materials, fabrication, device characterization and design methodologies require higher performance and greater precision than low frequency applications. Printing microwave electronics has the potential to benefit defense systems particularly for low cost expendable and attritable applications and for commercial products such as IoT applications which require customization.

Current additive manufacturing (AM) technology has limitations in achieving this potential. Materials are limited and material providers do not typically characterize printable inks for high frequency performance. High aspect ratio connections on a variety of materials are needed. An understanding of the impact of print parameters, substrate properties and ink properties on key print characteristics including line width control, surface roughness and print height of multi-material structures is needed. The different processing methods for creating the structures have the potential to affect the microwave performance. For printed electronics to gain adoption and realize their potential, these effects must be understood and a set of guidelines for printing microwave structures established. To this end, a C-Band Phased Array single element was designed as a demonstration vehicle. This paper will present data on the successful realization of this demonstrator and the design, characterization and fabrication approaches used to achieve this result.

Design and Fabrication

Our objective was to advance multi-material printing of integrated 3D electronics and non-planar structures for microwave applications. Commercially available materials, printers and processes were characterized for RF and microwave performance needs and integrated into the design solution. A C-Band Phased Array single element was designed as a demonstration vehicle and used to measure the effectiveness of the approach in terms of performance and manufacturability. C Band was chosen for its relevance to defense applications and due to requisite geometries needed for C band being realizable with current commercial materials and printers. The demonstrator design consisted of a phase match tuner, quadrature couplers, band pass filter, antenna, and two monolithic microwave integrated circuit (MMIC) gallium arsenide low noise amplifiers (GaAs LNA). See Figure 1.

This document does not contain technology or technical data controlled under either the U.S. International Traffic in Arms Regulations or the U.S. Export Administration Regulations.

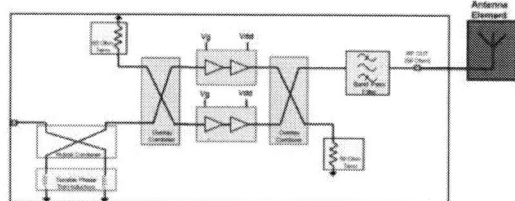

Figure 1: Schematic of C Band single element channel with patch antenna designed as the demonstration vehicle.

To meet project objectives, preliminary designs of the individual passive networks for the demonstrator were completed. Due to the three dimensional structure of key components such as the quadrature or overlay coupler, Ansys HFSS 3D was used to analyze, design and tune. From this, key needs for material sets were quantified. For example, for the quadrature coupler a preliminary design highlighted the conductor line widths required in the critical coupling region as well as the dielectric thickness and tolerance required for that region. This example is illustrated in Figure 2.

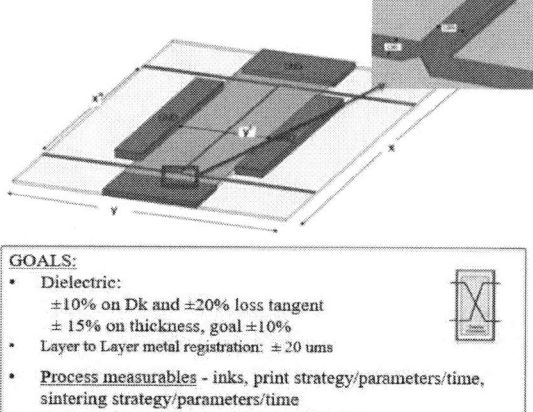

Figure 2: Example of preliminary component design (quadrature coupler) developed to identify material & print requirements shown. Once material is characterized for electrical and print properties, the design is iterated with measured data.

Based on this type of information driven by the demonstrator needs, literature and vendor searches were conducted to identify best material options. Once candidate materials were reviewed and agreed upon by the team, common methods were needed to characterize materials across laboratories with common test techniques so results could be verified. To this end, conductor and dielectric test beds were developed to enable efficient microwave characterization of materials and to examine the effect of process methods such as cure options on relevant microwave characteristics. For conductors, the focus was on insertion loss, while for dielectrics it was on evaluating dielectric constant and loss tangent at the relevant frequencies. A test bed consisting of a transmission line test coupon was developed as the standard to evaluate conductive inks for microwave applications. Copper 50 ohm transmission lines in microstrip, grounded coplanar waveguide (CPW) and ungrounded CPW lines of approximately one inch and three inch lengths were provided along with Through-Reflect-Line (TRL) calibration coupons for de-embedding S-parameter data. On the same substrate, Cu contact pads separated by the same distance as the 50 ohm lines are provided. The conductive ink to be tested was printed as a 50 ohm trace from pad to pad. This was then measured on a network analyzer and compared to the copper through lines and/or fully de-embedded using the TRL calibration. The test coupon is shown in Figure 3.

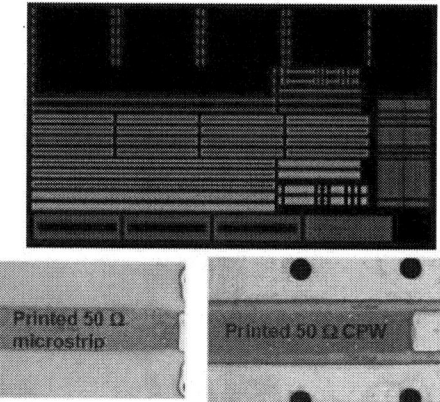

Figure 3: Insertion loss test bed used to characterize microwave performance of printed conductive inks and examples of printed 50Ω lines are shown.

Dielectric characterization was also critical and more problematic. Material vendors do not provide data at the required frequencies. Well-known techniques such as the waveguide method can be difficult as the sample preparation required for accurate measurements is very difficult to do with the thin printed materials. Samples typically fracture or small air gaps between the sample and the waveguide yield highly optimistic dielectric constant and loss tangent results. Ring resonator methods can be used but are only narrow band. To efficiently screen dielectric inks for suitability for microwave structures over a wide frequency range, we employed a novel method developed by UML [1] [2]. In this method, a dielectric ink is dispensed into a cylindrical capacitor on printed circuit board (PCB) and characterized for dielectric permittivity and loss tangent. This method is particularly well suited for thin, low viscosity inks.

Non-Export Controlled – See Sheet 1

For higher viscosity materials where the level of dielectric fill may be non-uniform, alternate capacitive structures were used. Figure 4 shows the test bad for the cylindrical capacitor on PCB used for dielectric ink characterization of dielectric constant and loss tangent over frequency and the alternate structures for higher viscosity inks.

a

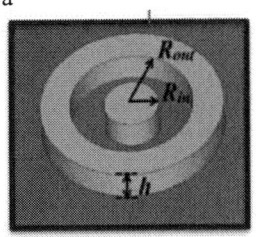 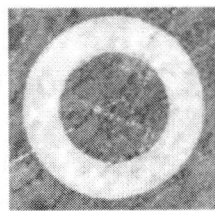

a.

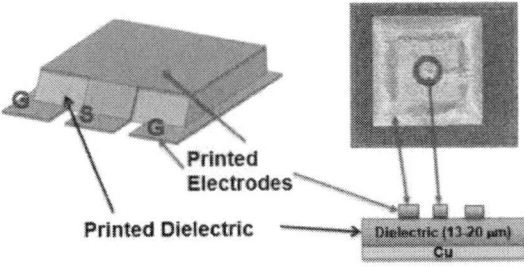

b.

Figure 4: Dielectric characterization for printed inks: a: A dielectric ink is dispensed into a cylindrical capacitor on PCB and characterized for dielectric permittivity and loss tangent. The left shows the empty test fixture and the right is filled with dielectric.

b: Alternate capacitive test structures are shown that are more suitable for dielectric characterization of viscous inks.

After materials were characterized from 500 MHz to at 20 GHz, designs were simulated with measured material performance. Printing and process design of experiments could then begin. In the course of development, materials were evaluated on a variety of printed platforms including the Optomec Aerosol Jet 5 Axis and 2D system, the multi-head nScrypt system, the Nordson dispensing system, and the LPKF system for via filling. Cure options such as the effect of thermal vs photonic vs laser cure were evaluated at the material characterization level with focus on the electrical performance difference due to sinter quality.

As print performance was developed, the first round of circuits were fabricated, tested and measured data was compared to simulations. First iteration microwave circuits were originally circuit elements added to a partially etched microstrip circuit as printing and process experience was developing. This was referred to as the 'hybrid approach" with printing on a circuit partially fabricated by traditional means.

For example, a five mil layer of Rogers RO3035 dielectric with copper on both sides was the initial starting point. The backside copper served as ground plane and the topside copper was etched for some portion of the first conductor metal layer. Required conductors and dielectrics were printed on top of this partially processed circuit. For each circuit, performance, fabrication and yield issues were considered and each circuit was optimized for a second iteration. These were then fabricated, tested and again compared to simulation. After the initial passives were fabricated and tested, they were used in the assembly of the phased array single element. After the success of that demonstrator, designs were optimized for more fully printed circuits. For the second iteration, single clad RO3035 was used as the dielectric base substrate and ground plane but all other layers including ground vias were additively manufactured.

Results

The transmission line test coupons structures proved to be very useful and an efficient way to screen conductors for microwave use. Independent testing by Raytheon of select university prepared samples validated the measurements. Conductive inks were effectively ranked for insertion loss across laboratories and test stands and an ink was identified that had insertion loss comparable to electroless nickel-immersion gold (ENIG) plated copper through X Band. Transmission line testing with these coupons was also effective in evaluating the effect of processing methods on electrical performance. The same inks were tested with thermal cure, photonic cure and laser cure. As can be seen in Figure 5, Ink A shows better results than Ink B (0.2 -0.3 dB), photonic curing gives comparable performance to oven curing for Ink A and photonic curing for Ink B improves loss. Since photonic curing is minutes vs hours, this information enabled shorter fabrication times in realizing final circuits.

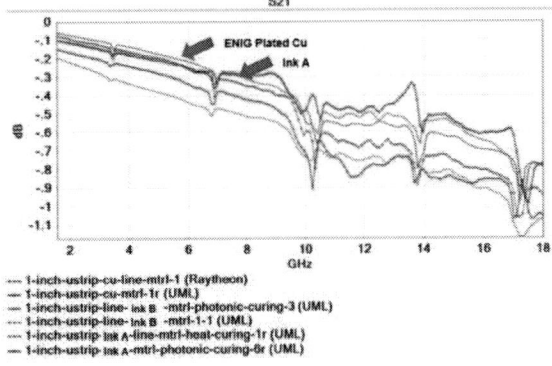

Figure 5: Conductive Ink characterization: Measured insertion loss of 1 inch long 50 Ω microstrip lines is shown comparing two printed inks to ENIG plated Cu and comparing cure methods for each ink.

Non-Export Controlled – See Sheet 1

Four different critical microwave components for insertion in the C Band phased array single element were successfully fabricated. These are a quadrature coupler, phase match tuner, band pass filter and antenna. To highlight the differences between the first and second circuit iteration, the phase match tuner is further described [4]. The phase match tuner was designed to use AM to provide fine adjustments in phase over a wide frequency range. The two port device is shown in Figure 6 and the phase of the device increases with the increase in the printed line length. Printing conductive ink to connect stub to ground effectively changes the electrical length and phase angle of the hybrid combiner.

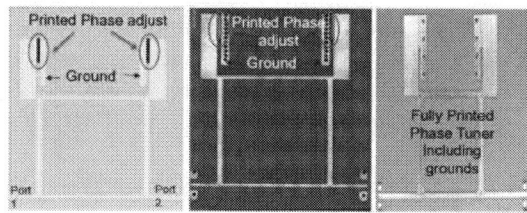

a.

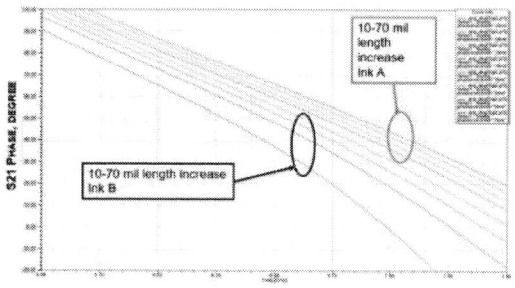

b.

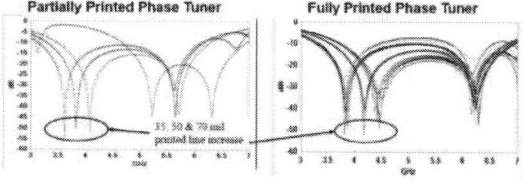

c.

Figure 6. Phase Match Tuner: a.) The circuit layout (left), the first iteration fabricated circuit (center) using the hybrid approach adding printed elements to a PCB and the second iteration (right) circuit with all topside circuit elements and ground via additively manufactured are shown.

b.) Measured performance of two inks used for phase adjust are compared for similar print lengths.

c.) The reference S paramaters of the hybrid approach (left) vs the fully printed approach (right) are presented.

A summary of key differences between the first and second iteration design and fabrication are highlighted in Table 1.

Table 1. Summary of key differences of the 1st iteration phase match tuner to the second iteration

1ST ITERATION	2ND ITERATION
Partially printed (tuning element only)	Fully printed circuit
No am of vias (pcb fabrication)	AM of ground vias, all locations
Cure time: themal cure of 3 hrs	Cure time: reduced by photonic or laser cure to < 15 mins
Center below target frequency	Retuned to center at 6.5 GHz; performance met with # of vias reduced by 50%
	High yielding: Print to Design target of -5,+10% achieved

A similar process of design, characterize materials, simulate, fabricate and optimize for performance and yield was completed for each circuit element. Once circuits were complete, integration to form the C Band phased array single element was initiated. All RF interconnects between circuits and from circuit to MMICs were printed to enable a coplanar transition using printed dielectrics and conductors. For this interconnect, an HFSS model was developed and measured results compared to the model as well as to a ribbon bonded interconnect for a similar geometry. Results are shown in Figure 7. [5.]

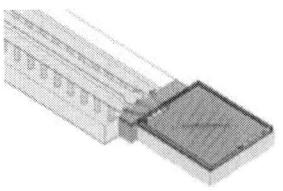

a)

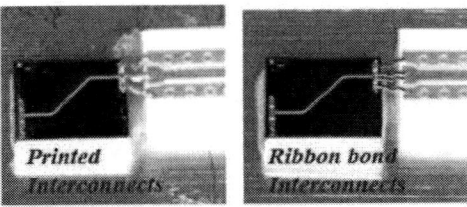

b.

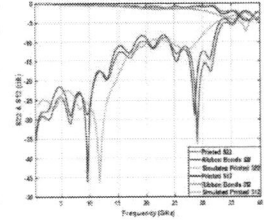

c.

Figure 7: a.) Schematic of die and alumina with printed CPW interconnects (light & dark green: printed dielectric) shown for HFSS model.

b.) Printed Interconnect vs ribbon bonded interconnect of GaAs through to alumina are fabricated for test.

c.) S parameters of measured vs simulated printed interconnects are plotted and compared to ribbon bonded interconnects.

All printed circuits were integrated onto a common metal ground using silver epoxy. RF interconnects were then printed. In this iteration, DC bias was bought from buss lines to the device with wirebonds as jumper wires were needed. Figure 8 shows the final printed C Band single element with patch antenna and printed RF interconnects.

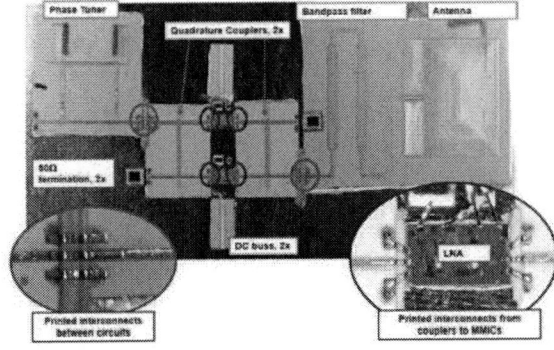

Figure 8. Printed C Band Phased Array single element with patch antenna and printed RF interconnects assembled for test.

In order to measure the gain on the amplifier section, an assembly was fabricated without the patch antenna and a two port measurement made from the phase match tuner input to the output of the band pass filter. This measured data was compared to the simulation of the full channel less antenna. It was also compared to a composite model of the individually measured components. The measured data for these components were combined in the ADS model and the LNA performance attached to the model. All three data sets are shown in Figure 9.

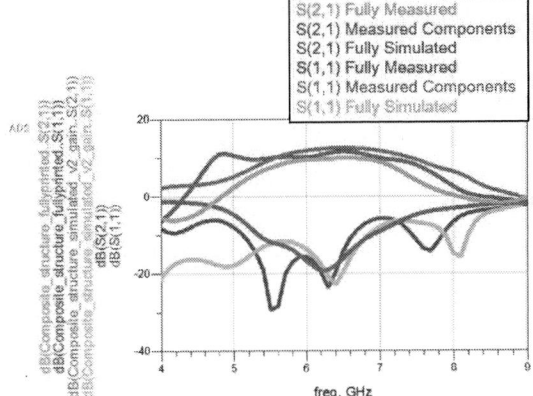

Figure 9. Results of C-Band phased array single element less antenna to measure gain are plotted. Three data sets are shown: fully simulated, fully measured and measured components joined with MMIC data in ADS composite simulation.

Conclusion

A printed C Band phased array single channel element with patch antenna was successfully demonstrated with very good correlation to simulations. To achieve this result, fully printed multi-layer structures on couplers, fully printed circuits on all other passive elements, drill and printed ground vias for all circuits, active MMIC integration, AM printed interconnects from MMIC to circuit and between circuit elements were demonstrated. Test beds were created to evaluated inks and processes and characterize for RF performance and used to inform design simulations.

Acknowledgements

The authors would like to thank America Makes, the National Center for Defense Manufacturing and Machining (NCDMM) and Raytheon Integrated Defense Systems for their support of this research. We also wish to acknowledge the members of Raytheon U Mass Lowell Research Institute (RURI) for their support as well as other collaborators in the America Makes Multi-material 3D Printing of Electronics and Structures Program, specifically, Roger Corporation for providing substrate material, nScrypt for printing support, and USF for independent material characterization.

Non-Export Controlled – See Sheet 1

References

[1] M. Haghzadeh, C. Armiento, A. Akyurtlu, "Microwave dielectric characterization of flexible plastic films using printed electronics", 2016 87th ARFTG Microwave Measurement Conference (ARFTG), pp. 1-4, 2016.

[2] E. Harper, M. Haghzadeh, E. Hajisaeid, C. Armiento and A. Akyurtlu, "Broadband microwave dielectric characterization method for printable dielectric inks," 2017 89th ARFTG Microwave Measurement Conference (ARFTG), Honolulu, HI, 2017, pp. 1-4.

[3] C. Laighton, S. C. Trulli, E. Harper, "Quadrature Coupler", U.S. Patent pending, 20180358676, June 13, 2017.

[4] C. Laighton, S. C. Trulli, E. Harper, " Phase shifter including a branchline coupler having phase adjusting sections formed by connectable conductive pads", U.S. Patent 10243246, March 26, 2019.

[5] S. C. Trulli, C. Laighton, E. Harper, "Interconnect Structure for Electrical Connecting a Pair of Microwave Transmission Lines Formed on a Pair of Spaced Structure Members", U.S. Patent 9978698, May 22, 2018.

[6] A. M. Nicolson, G. F. Ross, "Measurement of the Intrinsic Properties of Materials by Time-Domain Techniques", IEEE Transactions on Instrumentation and Measurement, vol. 19, no. 4, pp. 377-382, Nov. 1970.

[7] P.G. Bartley, S. B. Begley, "A new technique for the determination of the complex permittivity and permeability of materials", IEEE Instrumentation & Measurement Technology Conference Proceedings, pp. 54-57, 2010.

[8] N. Arnal, T. Ketterl, Y. Vega, J. Stratton, C. Perkowski, P. Deffenbaugh, et al., "3D multi-layer additive manufacturing of a 2.45 GHz RF front end," 2015 IEEE MTT-S International Microwave Symposium, Phoenix, AZ, 2015, pp. 1-4.

[9] J. O'Brien et al., "Miniaturization of microwave components and antennas using 3D manufacturing," 2015 9th European Conference on Antennas and Propagation (EuCAP), Lisbon, 2015, pp. 1-4.

Laser structured passive Components and RF Filter in LTCC Technology with Operating Frequencies up to 40 GHz focusing on 5G Mobile Applications

Alexander Schulz, Nam Gutzeit and Jens Müller

Ilmenau University of Technology
Faculty of Electrical Engineering and Information Technology
Department of Electronic Technologies
Institute of Micro- and Nanotechnologies MacroNano®

Gustav-Kirchhoff-Str. 1
98693 Ilmenau, Germany

Phone: +49 3677 693376

Fax: +49 3677 693379

alexander.schulz@tu-ilmenau.de

Abstract

This paper presents passive components and embedded low-pass filter (LPF) using Low Temperature Co-fired Ceramic (LTCC) materials and advanced picosecond laser structuring. These components and the LPF are well suited for multilayer System-in-Package (SiP) and Multi-Chip-Module (MCM) applications, e.g. wireless mobile applications focusing on 5G frequency bands. The passive components were measured, working up to 40 GHz and offering good RF performances. The LPF structure, consists of these passives, achieved a measured insertion loss lower than 0.6 dB up to 20 GHz, and a return loss of less than -13 dB in the pass-band of the LPF. The 3 dB frequency was 27.4 GHz. The whole LPF occupies a substrate area of 1.3 x 0.7 x 0.09 mm³ without probe tip ports and transitions. The LPF structures in a LTCC package suit well for 5G-band applications due to the compact dimensions and the good RF performance.

Key Words: LTCC, low-pass filter, 5G band, embedded filter, picosecond laser structuring

1. Introduction

Regulations on 5G network frequencies worldwide up to 60 GHz and offered bandwidth was followed directly by broader investigations in hardware development for high data rate mobile communication systems. Rapidly rising data rates in digital communication networks are essential in order to handle the growing amount of information which results in higher bandwidth requirements mainly. Real time and high-speed data transmissions are also key aspects for future mobile communication. Therefore, two wide-band frequency spectra from 24.25 GHz to 27.5 GHz and 31.8 GHz to 33.4 GHz are focused in the European Union for future 5G mobile applications. Commercial micro- and millimeter wave systems have to ensure compactness, high performance and low manufacturing costs. LTCC technology offers high-grade electrical performance up to millimeter wave frequencies and 3D capability. Moreover, passive components, Monolithic Microwave Integrated Circuits (MMICs) and antennas can be easily integrated which enables compact high performance RF systems in one LTCC package, also well suited for future 5G mobile devices.

Therefore, the implementation of passive components and structures on the surface as well as embedded in the multilayer which are electrically small related to the wavelength becomes more important. They should offer operating frequencies up to 40 GHz or higher regarding to the 5G specifications. Such passive structures can be combined to establish small RF circuits, e. g. filter, matching circuits and transformation networks. These circuits can be realized using distributed structures (transmission lines) as well as lumped elements [1, 2]. Concentrated or lumped components, e.g. coils and capacitors, which are manufactured by classical screen printing and conventional LTCC via technologies, do usually not fulfil these requirements since their large dimensions typically limit the maximum operating frequency. This contribution presents simulation results and the technological realization of small lumped components (2.5D coils, parallel plate and stacked interdigital capacitors) usable for RF circuits targeting on 5G applications. Conventional LTCC technology in combination with picosecond laser structuring enabled the tight requirements in structural quality and tolerances. These structures were characterized up to 40 GHz.

Moreover, a small RF low-pass filter (LPF) circuit used as functional technology demonstrator based on these structures was simulated, fabricated and fully electrically characterized by a network analyzer.

2. Low-pass Filter Design

A Chebyshew prototype low-pass filter is chosen as technology demonstrator. Aspects of the theory of filter design are described in literature [3, 4]. A five-element low-pass filter prototype (0.1 dB ripple) was calculated as well as transformed to the 1st 5G (lower) frequency band given above using normalized element values [4]. The schematic of the five-element LPF is depicted in Fig. 1.

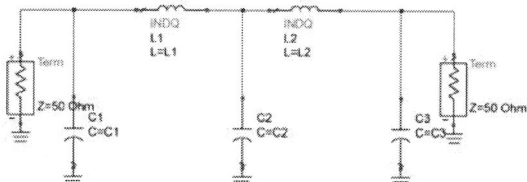

Figure 1: Schematic of the five-element LPF

The calculated values are verified as well as further optimized by circuit simulations using ADS (Advanced Design System) software from Keysight (1400 Fountaingrove Parkway Santa Rosa, CA 95403-1738). Table 1 contains the optimized inductance and capacitance values of an ideal five-element low-pass circuit shown in Fig. 1.

Table 1: Optimized lumped element values

	L1=L2 [nH]	C1=C3 [fF]	C2 [fF]
Values	0.47	135	240

3D electromagnetic field simulations were done to calculate and optimize the lumped component dimensions according to meet the inductance and capacitance values given in Table 1. The procedure is described in detail in the following paragraph.

These passive components were combined to a 3D five-element low-pass filter model which was further optimized by 3D field simulations to achieve the required filter behaviour (cf. chapter 2B).

A. Simulation of Passive Components

Modelling of the passive components were fulfilled under strictly consideration of the design rules of the applied laser structuring process which allows lines and spaces down to 25 µm and laser drilled as well as fillable via diameter ≥ 30 µm. The layer-to-layer distances are fixed to 45 µm according to the selected LTCC material system Du Pont 951 C2.

Geometrical dimensions of the designed passive components (2.5D coil and parallel plate as well as stacked interdigital capacitors, cf. Fig. 2a-c) selected for the introduced RF low-pass filter circuit were determined in HFSS (ANSYS, Inc., Southpointe 2600 ANSYS Drive, Canonsburg, PA 15317) by applying parameter sweeps of component dimensions until the component values given in Table 1 were achieved. These geometries were used to initiate the 3D simulations of the LPF structure described in the next paragraph. The interdigital capacitor configuration for C1 and C3 was chosen to demonstrate the capability of the laser-based fine structuring process. Nevertheless, a parallel plate capacitor could also be optionally used for C1 and C3 to achieve a similar electrical performance.

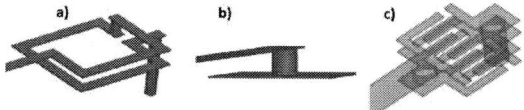

Figure 2: HFSS model of a) 2.5D coil, b) parallel plate and c) stacked interdigital capacitor

The main advantage of this 3D electromagnetic field simulation approach is the realistic prediction of the component behaviour which includes also parasitic effects. In a real coil, as depicted in the HFSS model in Fig. 2a, the parallel capacitance between coil windings has to be considered in the circuit design. It typically reduces the 1st resonant frequency of the coil. Similarly, a small series inductance must be added to the capacitance of the interdigital capacitor shown in Fig. 2c. The one-port simulated scattering parameters were converted into complex impedances which were used to compute the inductance and capacitance over frequency up to 40 GHz. Prior to this, line de-embedding in HFSS was applied to all structures to reduce the influence of the feeding lines. The simulated frequency dependent inductance value of the 2.5D coil (1.5 turns) and the capacitance values of the parallel plate and stacked interdigital capacitor are depicted in Fig. 5 and Fig. 6 after final optimizations of the LPF (see chapter 2B) including measurement results.

B. Simulation of the Low-pass Filter

The five-element low-pass filter 3D model was designed in HFSS consisting of the three separately optimized passive components. Additional parameter sweeps were necessary to include and compensate the parasitics related to the individual components and among them. The final HFSS 3D model is shown in Fig. 3. Optimizations of the LPF performance were done by varying each component dimensions based on parameter sweeps.

Figure 3: 5-element low-pass filter HFSS model

The calculated dimensions of the 2.5D coil (Fig. 2a and 3) are 0.15 x 0.12 mm^2 having a line width and via diameter of 30 µm. The one-layer parallel plate capacitor (Fig. 2b and 3) require lateral dimensions of 0.272 x 0.272 mm^2. The stacked interdigital capacitor consists of 6 fingers and 2-layers (Fig. 2c and 3) and has lines and spaces of 25 µm.

The simulated scattering parameters of the optimized 5-element LPF are depicted in Fig. 7 including measurement results. The simulated insertion loss is lower than 0.55 dB (up to 20 GHz) and the return loss is less than -17 dB in the passband. The simulated 3 dB cut-off frequency is at 27.5 GHz. The LPF structure is connected with a 50 Ω microstrip line (MSL). In order to measure the filter and the components using coplanar wave probes, a grounded coplanar waveguide (CPWg) to MSL transition was added to the test structure. It is compatible to coplanar probes with ground-signal-ground (GSG) pitches of 200 µm and 250 µm (see X-ray image in Fig. 4).

3. Fabrication and Measurement

The three-layer structured five-element LPF requiring dimensions of 1.3 x 0.7 x 0.09 mm^3 was manufactured using the LTCC material Du Pont 951 C2 and the cofire gold paste Du Pont 5740A. Including probe tip port and CPWg to MSL transition, a substrate area of 2.8 x 1.0 mm^2 is occupied. Three Du Pont 951 PT dielectric layers were added to the layer stack to enhance the mechanical stability of the substrate. A punched via diameter of 75 µm was used for vias of the interdigital capacitors. The signal micro vias of the 2.5D coils were laser drilled. The fired mean diameter of these micro vias is 35 µm. The shielding via fence of the GSG probe tip port was made with a 90 µm punch tool. The vias were filled with the gold via paste Du Pont 5738 using an extrusion bladder via filler PTC VF-1000 (Pacific Trinetics Corporation, 10542 Calle Lee, Suite 114, Los Alamitos, CA 90720).

All designed passive components require high structural resolution as well as low tolerances. The poor accuracy and the repeatability of 25-30 µm lines made by screen printing are insufficient for the series production of such small passive components. The combination of standard screen printing and high precision laser structuring of screen printed thick film pastes on LTCC using a picosecond UV laser is a promising solution to meet these requirements. It allows the manufacturing of structures with lines and spaces as small as 25 µm with high reproducibility.

The applied picosecond laser offers very short pulses (<10 ps), a small laser spot diameter (7 µm) and a high laser pulse energy (up to 70 µJ), which leads to a very small heat affected zone around the ablation areas and allows for a selective ablation process. Therefore, laser structuring of screen printed and dried gold thick films on unfired LTCC tapes with high precision and repeatability is possible [5-7]. This high-resolution patterning process can therefore be applied on any layer of the LTCC substrate.

By applying a special sequential structuring and lamination process a layer-to-layer misalignment of 5 µm or below was obtained [6, 7].

Standard screen printing of thick film paste was applied on each layer fully covering the component areas. All RF features on the inner layers (LPF, including transitions, passive components) were sequentially laser structured in the unfired state. Lamination and sintering were done according to the material supplier's recommendations. The laser structuring of the top layer was performed after firing. The X-ray image in Fig. 4 shows a manufactured LPF structure including CPWg to MSL transitions.

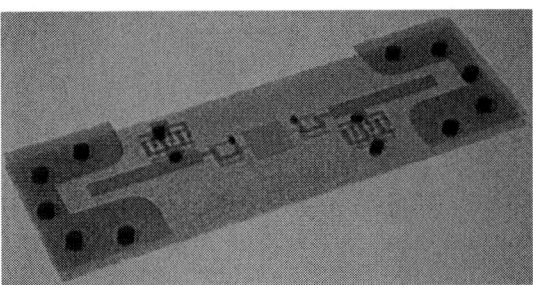

Figure 4: X-ray image of LPF including CPWg to MSL transitions

Fabricated RF structures (LPF, passives, de-embedding lines) were measured with a network analyzer (Agilent HP 8510) and 250 µm pitch coplanar probes up to 40 GHz on a wafer probe station. Calibration was made on substrate level using a SOLT calibration substrate from Cascade.

An influence analysis of manufacturing tolerances on RF measurement results of passive components was investigated. Nine samples of the introduced 2.5D Coil L_s and stacked interdigital capacitor C_F as well as a parallel plate capacitor C_P, having dimensions of 0.245 x 0.245 mm^2, featuring similar capacitances, were measured separately by network analyzer. The measured scattering parameters were converted into complex impedances which allow computing the mean values of inductance and capacitances including relative standard deviations (RSD) given in Table 2 for selected frequencies.

Table 2: Mean values of inductance and capacitances including relative standard deviations

Frequency	Ls [nH]	CP [fF]	CF [fF]
1 GHz	0.47 ± 2.7 %	141 ± 2.7 %	151 ± 4.9 %
10 GHz	0.49 ± 1.5 %	138 ± 2.1 %	148 ± 4.4 %
20 GHz	0.63 ± 4.1 %	135 ± 3.0 %	144 ± 3.5 %

The relative standard deviations of the computed inductance and capacitances based on s-parameter measurements are below 5 % up to 20 GHz for all three passives which verify high accuracy and reproducibility of the applied manufacturing processes. Higher RSD were determined during RF measurements for the more complex coil and finger capacitor design compared to the parallel plate capacitor.

Fig. 5 and Fig. 6 display the simulated and measured RF performance up to 40 GHz of each single passive component used for the LPF structure. Differences of about 15 % between simulation and measurement are probably related to manufacturing tolerances as well as to the line de-embedding procedure in HFSS mainly.

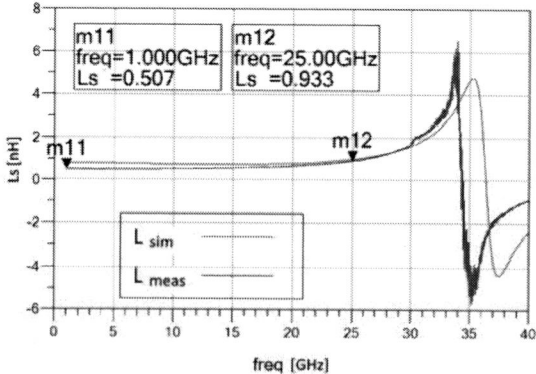

Figure 5: Simulated and measured inductance of the 2.5D coil

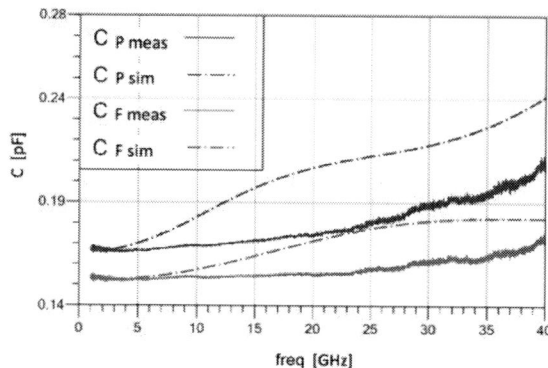

Figure 6: Simulated and measured capacitances of parallel plate (C_P) and stacked interdigital capacitor (C_F)

The simulated (dotted lines) and measured (dashed lines) scattering parameters, including line de-embedding, of the manufactured LPF are depicted in Fig. 7. The measured insertion loss is lower than 0.6 dB (up to 20 GHz) and the return loss is less than -13 dB in the filter passband. The 3 dB cut-off frequency is at 27.4 GHz which is slightly lower than simulated. The measured S11 and S22 parameter values are also higher than simulated. The insertion loss behaviour in the stop-band is even better than predicted. Nevertheless, the measured RF performance is in a good agreement according to the simulation of the LPF. Manufacturing tolerances are mainly responsible for the small differences compared to simulation results.

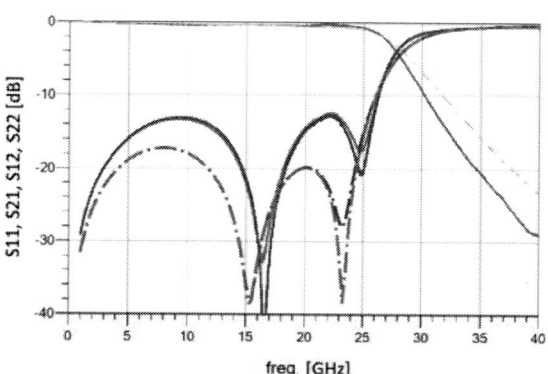

Figure 7: Simulated (dotted lines) and measured (dashed lines) scattering parameters of the manufactured LPF

4. Conclusion

Small passive components were designed, simulated, fabricated and measured up to 40 GHz. A five-element low-pass Filter (LPF) structure consisting of these passives was optimized, manufactured and characterized by scattering parameter measurements. Low insertion losses, a return loss of less than -13 dB in the filter passband and a 3 dB cut-off frequency at 27.4 GHz were achieved. Slight differences between measurement and simulation of the LPF are attributed to manufacturing tolerances mainly. Nevertheless, the combination of standard screen printing and advanced laser structuring is very promising for the technological realization of physically small passives as well as RF circuits targeting on 5G applications. The small LPF dimensions and the 3D capability can reduce the size of a RF system.

5. Acknowledgements

The authors are grateful to Ina Koch (Department of Electronics Technology and Institute of Micro- and Nanotechnologies (IMN) of Ilmenau University of Technology (TU Ilmenau) for the technical support in the manufacturing of the LTCC substrates. Additional thanks to the RF and Microwave Research Lab of TU Ilmenau, mainly to Michael Huhn, for measuring the RF structures. This work has been funded by the Carl Zeiss Foundation (funding code: 0563-2.8/581/2).

References

[1] Zhengwei Wang, Yin Tian, Shirong Bu, Zhengxiang Luo, "Study of L-band miniature lumped element LTCC band-pass filter", IEEE International Conference on Microwave and Millimeter Wave Technology, Chengdu, China, pp. 864-866, ISBN: 978-1-4244-5708-3 and 978-1-4244-5705-2, DOI: 10.1109/ICMMT. 2010.55 25175, May 8-11, 2010;

[2] Ruben Perrone and Jens Müller, "Compact RF-filter-modules with lumped elements in LTCC for applications up to 10GHz", IEEE 2nd Electronics System-Integration Technology Conference, Greenwich, UK, pp. 491-495, Print ISBN: 978-1-4244-2813-7, DOI: 10.1109/ESTC.2008.4684397, Sept. 1-4, 2008;

[3] Pozar M. David, "Microwave Engineering", New York: John Wiley & Sons, third edition, 2005;

[4] Matthaei George, Young Leo, Jones E. M. T, "Microwave Filters, Impedance-Matching Networks, and Coupling Structures", Norwood, MA: Artech House, first edition, 1980;

[5] Nam Gutzeit, Tilo Welker, Karl-Heinz Drüe and Jens Müller, "High resolution LTCC laser processing in the green and fired state for future technologies"; 12th International Conference and Exhibition on Ceramic Interconnect and Ceramic Microsystems Technologies (CICMT), Denver, USA, April 19-21, 2016;

[6] Nam Gutzeit, Alexander Schulz, Tilo Welker, Christoph Wagner, Eric Schäfer and Jens Müller, "High-precision picosecond laser structuring on LTCC for silicon chip assembly with high electrical contact density"; European Microelectronics and Packaging Conference (EMPC), Warschau, Polen, Sept. 10-13, 2017;

[7] Nam Gutzeit, Alexander Schulz, Jens Müller, Torsten Thelemann, "Miniaturized laser structured components in LTCC for 5G applications", 14th International Conference and Exhibition on Ceramic Interconnect and Ceramic Microsystems Technologies (CICMT), Aveiro, Portugal, April 18-20, 2018;

LTCC as substrate - enabling semiconductor and packaging integration

Heike Bartsch[1], Jörg Pezoldt[1], Francico M. Morales Sanchez[2], Juan J. Jimenez Rios[2], Jose M. Mánuel Delgado[2], Jonas Breiling[1], Jens Müller[1]

[1]Technische Universität Ilmenau, Institute of Micro-and Nanotechnologies MacroNano, Ilmenau, Germany
[2]Universidad de Cádiz, Instituto Universitario de Investigación en Microscopía Electrónica y Materiales (IMEYMAT), Cádiz, Spain

+49 3677 69 3452, heike.bartsch@tu-ilmenau.de

Abstract

Gallium nitride (GaN) is commonly used in high-power and high-frequency devices but one limiting factor for its wide application is the availability of cost-effective substrates. The approach of "engineered substrates" is consequently followed up in this work. Low temperature cofired ceramics are used as carrier substrates in this work. The flexible material system allows the creation of a thermo-mechanically compatible base for GaN components. Sputtered aluminum nitride (AlN) layer form the buffer layer for the GaN deposition. The surface quality of these buffer layers reaches an arithmetic mean height R_a of 10 nm after chemical mechanical polishing and forms thus a surface quality which is adequate for thin film processing. GaN layers grown on these substrates can have both, tensile and compressive stress levels, while such grown on (100) silicon for comparison purpose show exclusively tensile stress. The growth of the GaN layer and resulting morphology is assumed to be mainly influenced by the process conditions during molecular beam epitaxy. The polycrystalline GaN layers show pronounced c-axis orientation at deposition temperatures of 700°C and 800°C. These first results encourage a further development of this new substrate architecture for integrated GaN systems.

Key words: Low Temperature Cofired Ceramics, LTCC, wide bandgap semiconductors, engineered substrates, layer properties; Substrates Design and Technologies; Co-firing

Introduction

The wide bandgap of gallium nitride (GaN) of 3.4 eV qualifies it for the use in optoelectronic, high-power and high-frequency devices. Because of its very high breakdown voltages, high electron mobility and saturation velocity it is an ideal candidate for high-power and high-temperature microwave applications, e.g. RF power amplifiers at microwave frequencies or high-voltage switching devices for power grids. In this fields, GaN components compete with lateral double-diffused metal oxide field effect transistors (LDMOS) and gallium arsenide (GaAs) transistors.

One limiting factor for the breaktrough of GaN transistors is the availability of cost-effective substrates with low defect density. An interesting approach is the use of so-called "engineered substrates", which means that a thin GaN layer is applied on a carrier substrate. [1]. The quality of the GaN layer depends on the original material and the carrier determines the thermomechanical performance. Beside of lattice mismatch, the mismatch of the coefficient of thermal expansion (CTE) plays a crucial role for the forming of cracks and defects [2].

Low temperature Cofired Ceramics (LTCC) are composite materials with adaptable dielectric and thermo-mechanical properties [3]. They are widely used in RF applications and allow a high packaging density [4]. However, the surface quality of the pristine ceramic after firing influences the crystallinity of deposited layers on the materials. Sputtered AlN layers showed an inclined orientation of the c-axis and their electrical properties were influenced [5]. Nevertheless, AlN could provide an excellent buffer layer on ceramic substrates for GaN processes: It has good thermal conductivity [6] and high insulation resistance make them interesting candidates for heat spreaders on LTCC [5], which both are valuable properties for GaN system integration. Such layers can be sputtered with low thermal stress and high thickness on LTCC substrates [7].

These prerequisites encourage a new architecture for engineered substrates, based on LTCC as with adapted thermo-mechanical properties as carrier substrate and sputtered AlN as buffer layer for further deposition of GaN layers by the use of molecular beam epitaxy (MBE).

Substrate preparation

The LTCC tape 9k7V (DuPont, Nemours) is used as material for the carrier substrate. 10 layers without vias and printing were stacked and laminated at 70°C for 10 minutes at a pressure of 210 bar in an isostatic press. The laminates are then fired at 850°C for 30 minutes in a multipurpose fast ramping furnace (PEO 603, ATV GmbH, Vaterstetten). The obtained substrates were diced in pieces of 35 x 35 mm² using a wafer saw (ADT 7100 ProVectus 2, Advanced Dicing Technologies Ltd., Yokneam, Israel) and lapped with boron carbide slurry (B₄C). All lapping and polishing steps use the Logitech PM5 Lapper (Logitech Ltd. Glasgow, UK). After this treatment, the surface was polished with 1 μm diamond slurry for 90 minutes. On this surface, AlN was sputtered with pulsed reactive magnetron sputtering (CS400ES, VON ARDENNE GmbH, Dresden, Germany). Layers of 8 μm thickness were deposited at 300°C with a deposition rate of 0.2 nm/s. The N : Ar ratio was 40 : 10 Sccm. The sputtered surface was polished again with 1 μm diamond slurry for 90 minutes and finally, chemical mechanical polishing was carried out with a colloidal silica slurry (SF1 0CON-140, Logitech Ltd. Glasgow, UK). The preparation sequence is illustrated in Figure 1.

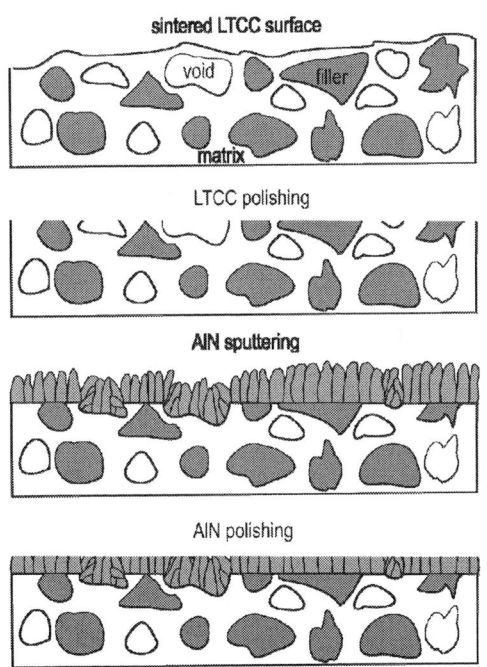

Figure 1: substrate preparation sequence 1) LTCC carrier substrate, 2) after polishing: pores are opened, 3) AlN sputtering: pores are filled, 4) AlN polishing: leveling of the sputtered layer.

Substrate Properties

The substrate geometry was measured along both diagonal lines with a profilometer (MicroProf® 200 with CWL 600 sensor in TTV mode, FRT GmbH, Bergisch Gladbach, Germany). The total thickness variance of the substrates after polishing is less than 3 μm. The arithmetic mean height (Sa) after polishing of the LTCC substrate was determined by laser scanning microscopy (OLS 4100, Olympus) considering a region of interest of 128 μm x 128 μm. It amounts to 345 nm after LTCC

polishing. The value decreases to 42 nm after sputtering and polishing of the AlN layer with 1 μm diamond slurry. A further decrease is obtained after CMP polishing, the value amounts to 10 nm after the process. Atomic force measurements were carried out at a Veeco-Bruker Dimension microscope working on tapping mode with a 8-nm, Si tip (Bruker Corporation, Massachusetts, US) of the surface encompassing an area of 5 x 5 μm revealed a root mean square height of 5.6 nm.

Figure 2 depicts the measured XRD θ-2θ plot, using a Siemens XRD 5000 difractometer (Siemens AG, Berlin, Germany) working on Bragg-Brentano symmetric scan conditions, with an angle step of 0.02°.

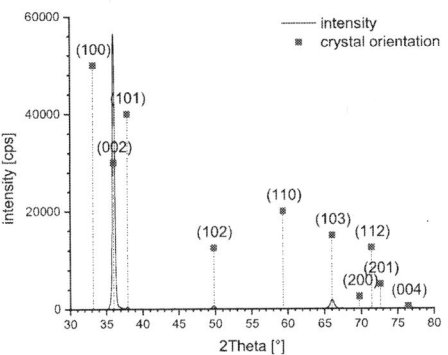

Figure 2: XRD of sputtered AlN on polished LTCC.

The sputtered AlN layers exposed a pronounced c-axis orientation. The grain structure is depicted in the SEM micrograph in Figure 3, obtained at a Field Effect GeminiSEM (Carl Zeiss GmbH, Jena, Germany), operating at 2-5 kV.

Figure 3: Sputtered AlN on polished LTCC, small pores are closed. Columnar structure is visible.

The columnar structure perpendicular to the LTCC surface is clearly visible. Small pores, which are layed open by the polishing step are completely filled up by the sputtered AlN.

MBE growth of GaN on LTCC substrates

These LTCC substrates were used for GaN growth, carried out by means of MBE. The substrates were heated up to 900°C in the chamber and the temperature was held for 4h, approximately. The base pressure was measured before starting the deposition. It amounts to 1.6×10^{-10} mbar. The substrate temperature was set between 600°C and 900°C and a gallium flux (I_{Ga}) from 2.0×10^{-5} mbar to 4.0×10^{-5} mbar was used. The deposited layers were investigated with a JEOL 2100 TEM and a JEOL 2010-Field-Effect-Gun Scanning-TEM microscopes (JEOL Ltd., Tokio, Japan), both working at 200 kV, and the same XRD diffractometer utilized for the LTCC characterization. Figure 4 depicts an image, recorded with a JEOL 2100 microscope, of a cross-section TEM (XTEM) preparation of a GaN layer, deposited on the AlN buffer layer, from the sample grown at 600 °C and under a pressure of 2.0×10^{-5} mbar. The GaN layer has a columnar, polycrystalline structure. The image reveals the grain orientation of both, AlN layer and GaN layer on top. The grain orientation of the GaN layer, in this region, is more randomly distributed that in the case of the AlN.

Figure 4: XTEM micrograph of a GaN layer grown at 600 °C and p = 2x10⁻⁵ mbar on sputtered and polished AlN buffer layer.

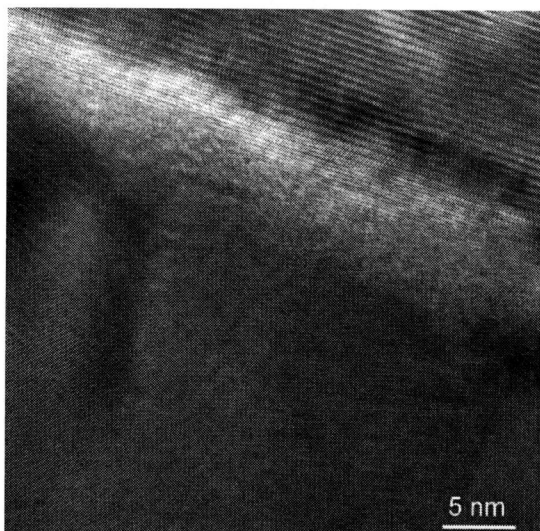

Figure 5: HighResolution XTEM image of the interface region for sample obtained using I_{Ga} = 4x10⁻¹¹A, T = 800°C and p = 2x10⁻⁵ mbar.

Figure 5 shows a high-resolution X-TEM image of the sample, obtained in a JEOL 2010-FEG microscope. It exposes details of the transition zone between AlN (bottom) and GaN (top). There is a thin amorphous layer with a thickness of 1-2 nm,

which forms a boundary between AlN and GaN grains. It is evident that the GaN morphology distinguishes from that of the AlN buffer layer. Nevertheless, although both structures are polycrystalline, they also present a hexagonal symmetry. It can be assumed that the GaN growth dependents foremostly on MBE process conditions rather than an imprinted crystal orientation of the AlN buffer layer. A more detailed (S)TEM study of the properties is presented in [8].

Best c-axis orientation reached layers which were grown at 700°C and 800°C. The relative texture coefficient was 80 %, approximately. The lowest roughness achieved layers grown at 800°C and a gallium flux of 6×10^{-11}A. It amounts to 8 nm, measured on an area of $1 \times 1\mu m$ (AFM).

Layer stress

Figure 6 depicts the lattice constants obtained from the XRD measurements. Additional data of layers grown on (100) silicon substrates with sputterd AlN buffer layers deposited at 700°C were taken as comparison values. These data were available from previous experiments. The reference value [9] marked in the graph bases on a thick layer of epitaxially laterally overgrown GaN, separated from the substrate without bending. This value forms the partition between tensile and compressive stress.

Different deposition condition for layers grown on ceramic substrates can yield compressive and tensile stress as well. The variation of the temperature at I_{Ga} of 4×10^{-11}A shows an evident trend of increasing tensile stress with increasing substrate temperature. The lowest stress value amounts to 0.2 GPa. It was observed at a substrate temperature of 800°C and I_{Ga} of 6×10^{-11}A. The GaN stress at a constant temperature of 800°C decreases firstly with the GaN flux, but, at 8×10^{-11}A the trend inverts.

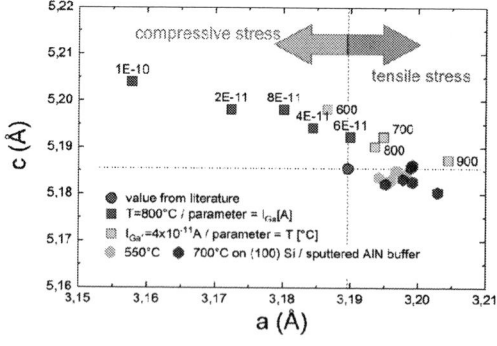

Figure 6: Lattice constant obtained from XRD data for different GaN flux and temperatures. Values obtained from GaN layers grown on (100) silicon with sputtered AlN buffer for comparison purpose.

The comparison samples grown on silicon with AlN buffer layer show tensile stress without exception, even though the substrate temperature is significantly lower. Stress values varied between 0.44 GPa and 2 GPa.

Conclusions

The present work introduces a substrate achitecture consisting of LTCC as carrier substrate and a surface finish of sputtered AlN. The sputtered layer smoothens the surface and CMP polishing creates a surface with a roughness R_a of 10 nm. This surface quality is adequate for thin film processing. MBE deposition of GaN layers on these substrates produces polycrystalline layers with columnar structure. A thin amorphous layer at the interface between AlN and GaN separates the grain structure of both materials and the GaN growth is assumed to be controlled by the MBE process.

In contrast to layers deposited on (100) silicon with AlN buffer layers, the layer stress of these GaN layers can take both, compressive and tensile values. These facts encourage the strategy to grow thick GaN layers on such substrates in the future, enabling high-quality layers with controlled stress level, which have adequate quality for electronic applications.

Acknowledgements

The authors would like to thank the support from the Alexander von Humboldt Foundation, through the "Humboldt Research Fellowship for Postdoctoral Researchers Programme" (Ref. 3.3-1 1S8421-ES-HFST-P), and from the University of Cádiz, through the "Programa de Fomento e Impulso de la Investigación de la Transferencia" (Proyectos de Investigación-UCA Jóvenes Investigadores PR2016-003 and Puente PR2016-094 and PR2016-042), the "Programa de ayudas a la realización de Tesis Doctoral del Plan Propio de Investigación y Transferencia" (Contratos predoctorales de Formación de Profesorado Universitario (fpUCA) 2016-060/PU/EPIF-FPU-CT/CP) and the Servicios Centrales de Investigación Científica y Tecnológica (SC-ICYT).

References

[1] Yole Developpement, "Free-Standing & Bulk GaN Substrates for Laser Diode, LED and Power Electronics: 2013 Report" [Online] Available: https://www.slideshare.net/Yole_Developpement/yole-bulk-ganoctober2013reportsample. Accessed on: Apr. 17 2019.

[2] K. Motoki, "Development of Gallium Nitride Substrates," SEI Technical review, Vol. 70, pp. 28–35, 2010.

[3] M. T. Sebastian and H. Jantunen, "Low loss dielectric materials for LTCC applications: A review," International Materials Reviews, Vol. 53, No. 2, pp. 57–90, 2013.

[4] H. Bartsch et al., "Applications of Microwave Dielectrics" John Wiley & Sons, first edition, Chichester, UK, Chapter 15, pp. 653–683, 2017.

[5] A. Bittner, A. Ababneh, H. Seidel, and U. Schmid, "Influence of the crystal orientation on the electrical properties of AlN thin films on LTCC substrates," Applied Surface Science, Vol. 257, No. 3, pp. 1088–1091, 2010.

[6] J. W. Lee, J. J. Cuomo, Y. S. Cho, and R. L. Keusseyan, "Aluminum Nitride Thin Films on an LTCC Substrate," J American Ceramic Society, Vol. 88, No. 7, pp. 1977–1980, 2005.

[7] H. Bartsch, R. Grieseler, J. Mánuel, J. Pezoldt, and J. Müller, "Magnetron Sputtered AlN Layers on LTCC Multilayer and Silicon Substrates," Coatings, Vol. 8, No. 8, p. 289, 2018.

[8] J. J. Jiménez et al., "Comprehensive (S)TEM characterization of polycrystalline GaN/AlN layers grown on LTCC substrates," Ceramics International, Vol. 45, No. 7, pp. 9114–9125, 2019.

[9] M. Yamaguchi et al., "Brillouin scattering study of bulk GaN," Journal of Applied Physics, Vol. 85, No. 12, pp. 8502–8504, 1999.

Printed Functional Applications: Batteries, Communication Elements, Antennas and Conductive Paths on Technical Textiles

Carmen Meuser, Andreas Willert and Ralf Zichner

Fraunhofer Institute for Electronic Nano Systems, Department of Printed Functionalities, Technologie-Campus 3, 09126 Chemnitz, Germany

Phone: +49 371 / 45001-443, carmen.meuser@enas.fraunhofer.de

Abstract

Today various applications are designed to make life much easier for the consumer. For example you want to be accessible always and everywhere, and detect many parameters of your own body functions, not only in the medical field, but also in the leisure area. In times when fitness and healthiness is getting higher and higher, with dozens of apps which recognize if I slept well, drank enough, or my sport unit is effective, we have to think about an easy and undisturbing way to collect those data and send them for example to our smartphone or computer. With printed functional applications we are able to collect those data without disturbing the consumer, because they are light weighted and flexible. In the medical field as well, non-perceptible sensors and functional applications are as desirable as possible to affect patients. The more common the patient feels, despite monitored by sensors, the faster he will recover and, for example, be discharged from the hospital. This reduction in the period of illness allows an immense reduction in the incidence of the illness and examination costs.

Key words: technical textile, functional printing, battery, antennas, conductive path

Introduction

Well-known printing technologies such as screen, inkjet or gravure are very common in the production of textiles for the process of graphic arts printing. Nowadays, an interesting pattern is no longer the only interest of a consumer. Various applications, especially in the field of functional devices, become increasingly more valuable as measuring the heart beat frequency, breath frequency, or body temperature are among simpler functions. The consumer wants to use this functional devices without disturbing his environment, so they have to be lightweight and flexible. State of the art rigid PCB (Printed Circuit Board) technology is not suitable for flexible applications. In contrast, printing technologies are able to generate flexible microelectronics, wireless sensors for example, on top of technical textiles. This means conductive paths, antennas and e.g. capacitive areas could be realized through printing.

All elements which are implemented in functional applications have their own challenges, in particular if they are applied onto flexible substrates. Technical textiles which are used as substrates increase these challenges enormously.

Research groups, all over the world, are concerned with the application of functionalities in textiles. However, it is seldom the case that the textiles are used as a substrate directly. The realisation of those applications normally takes place on flexible, but non-stretchable substrates, which are integrated in the textiles later on.

Especially in the medical field several research topics are possible, for example wound monitoring. This is intended to reduce the costs of wound care while reducing the incidence of complications caused by late recognized changes in the wound area. Kassal et al. [1] for example, describe a smart bandage for wound indication. In this case a uric acid biosensor is used to define the wound status. The sensor is screen printed onto the bandage encapsulated with an insulator. Afterwards a rigid, leight flexible PCB is implemented in the bandage, which is able to send the data versus RFID-technique (radio frequency identification), or by NFC (near-field communication). The restriction is the rigid PCB, that only allows a large bending radius, because the flexibility is not as high as that of the bandage. The same RFID technique is used in an optical sensor for pH values. [2] There are more examples for sensing biologic states but the RFID connection is also managed with a rigid PCB and a rigid coin cell is implemented as power source. [3] This affects the flexibility and reduce the wearing comfort for the user.

Moya et al. [4] describes the construction of inkjet-printed sensors. The most common materials which are used for electrochemical sensors and their needs in production, e.g. curig and sintering conditions, are mentioned.

Faraooqui and Shamim [5] even developed a monitoring system for wounds. The sensor is inkjet printed onto kapton and then implemented into a bandage. The kapton film has a dimension of 23.8 mm x 46.5 mm. In this area the flexibilty is restricted and there is no possibility of breathable acticity inside the material. Further a lithium battery (2 mm x 12 mm x 12.5 mm) is implemented into the system. This is also a rigid component, and the lithium battery is a security risk, because of cutting it with scissors or destroing it by removing the bandage.

Another combination of a medical sensor, a combination of antenna, sensor, conductive path and battery is described by Sun et al. [6]. In this case the system is not directly applied onto the textile. A rigid PCB and other mechanical elements are placed in a box which is fixed at the textile with snap buttons. This makes the use of that sensing system uncomfartable for the consumer.

Batteries

The range of applications is tremendous, while almost every applcation has its own requirements in battery shape, size, capacity and voltage. A wide range of rigid cells are used, e.g. in watches, notebooks, cameras, hearing aids, flashlights or even mobile phones, but all of them are rigid. For the functional devices on textiles, a light and flexible battery which is fully integrated into the system is required. A wide range of applications are enabled by printed, flexible batteries. Due to low costs, environmental friendliness, non-toxic and disposeable materials make them an attractive power supply. The development of primary printed batteries provides valuable information in the summary of Willert et al. [7] Printed batteries have recently emerged as a new battery system to address the issues on the flexibility and design diversity. [8] [9] Also first results and the challenges of the technical materials, like rouhgness, woven structue of the back sides and their influence of the batttery performance were presented from Willert et. al. [10]

Choi et al. [11] provides new insights into printed batteries. They mentioned that for the successful development of printed batteries, all components of the whole process have to be coordinated to each other. The development of the ink with well-tuned rheological properties and dispersion state combined with the suitable printing technology is only a small part. Due to the different substrates also the wetting behavior, the chemical interaction between surface and ink and the following sintering and curing conditions have to be adapted. In recent years, there have been many approaches to produce micro power sources. This includes different techniques like thin-film, stretchable and flexible batteries produced by a variety of methods. [12], [13], [14], [15], [16], [17]

High throughput printing methods offer the prospect of low-cost mass production capability and the potential for unprecedented levels of technical applications. This depends on the qualities if organic, inorganic and hybrid conductive materials which can be deposited on flexible substrates using additive high throughput printing methods. [18], [19], [20], [21]

In this work the low cost, environmental friendly, non-toxic Zn-carbon system is used to empower the different elements.

Printing of the whole system, including antennas, sensors, conductive paths and batteries offers numerous benefits, but there are still some challenges. All printing techniques have the same requirements to achieve good result, especially on technical textile, for example the surface roughness which influences the layer thickness and conductivity of printed functionalities, as well as the chemical composition of the textile surface are cruical. A second point is the thermal stability. Nowadays, the most metalized pastes and inks need for an effective functionalisation a temperature above 130 °C. For technical textiles the temperature is limited, because of the colour (some pigments change their colour at higher temperature) and the chemical ingredients. In the last coverage there are reactions, which destroy the technical textiles by using temperatures above 130 °C.

This work demonstrates antennas, communication elements, conductive paths and batteries which are screen printed on technical textiles. In terms of technical textiles, different surfaces, chemical compositions and woven backsides are inspected.

Results and discussion

The first task was the characterization of the used technical textiles. At the beginning five different textiles, with two different surface chemistries, PVC (polyvinyl chloride) and PU (polyurethane) were used. The surfaces of the technical textiles were characterized by scanning electron microscopy (SEM-Hitachi, TM-1000) and profilometry (Dektak 150, Veeco Instruments Inc.)

As shown in Figure 1, the PVC based technical textiles have a closed surface with small agglomerates and aggregates of melted particles. The surfaces look nearly similar, but there are big differences in the measured roughness.

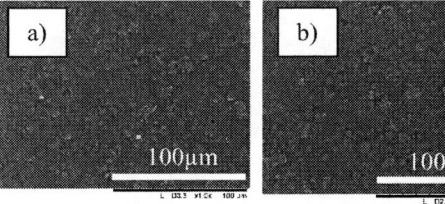

Figure 1: SEM images of two different PVC based technical textiles a) artificial leather b) mattress protective cover (hygienic underlay).

The roughness (R_a), calculated as an average out of three individual measurements for every textile were found to be $R_a = 2.5$ μm, while the height distribution could be up to 50 μm, see Figure 2 and Figure 3.

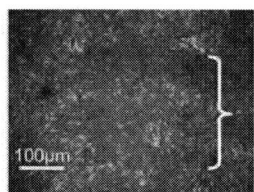

Height distribution up to 50 µm are visible within short sections.

Figure 2: Microscopic image of PVC based artificial leather.

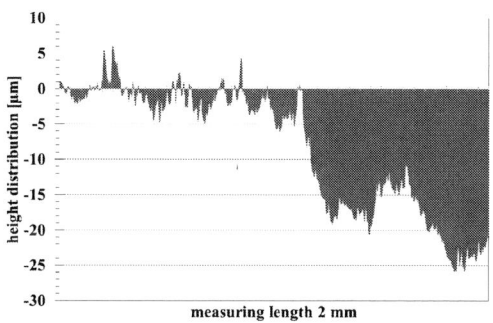

Figure 3: Height distribution of PVC based artificial leather.

The PU based hygienic materials show in contrast a structured and smooth surface without agglomerates and aggregates, only small round wells could be seen, Figure 4.

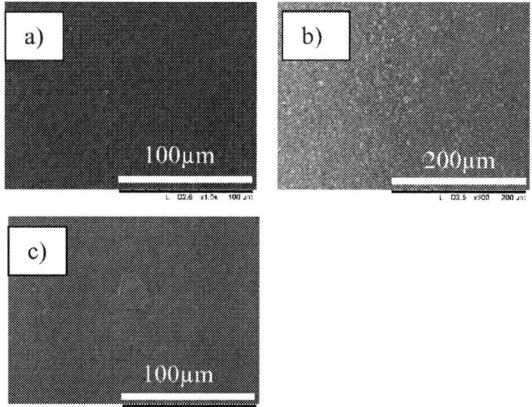

Figure 4: SEM images of three different PU based technical textiles. (hygienic underlays)

Due to the thin printed layer thickness, the large differences in the height are challenging. Also with regard to the subsequent bending and stretching of the technical textiles during the use, the height distribution of the technical textiles affects the conductive properties significantly.

Table 1: Roughness of the different textiles

material	R_a [µm]	R_t [µm]
PVC (a) artificial leather	1.5	9.7
PVC (b) hygienic underlay	0.7	3.4
PU (a) hygienic underlay	0.8	7.2
PU (b) hygienic underlay	2.4	13.2
PU (c) hygienic underlay	1.5	12.3

Battery fabrication

The battery is deposited layer by layer onto the technical textiles. As reference a battery on PET (polyethylene terephthalate) film (Melinex CW 401, thickness 100 µm) is printed. The carbon current collector is a screen printed pattern of Electrodag PF-407C (Henkel, Loctite). Cathodes and anodes were deposited with customized inks containing zinc (85 wt%) and manganese dioxide (64 wt%), respectively.

After printing, each layer was dried in a convection oven @ 110 °C for 10 min. A fiber based separator layer and a gelled aqueous based $ZnCl_2$ electrolyte complete the electrochemical system. The batteries were laminated using a chemical resistant double-sided glue tape.

Due to the surface roughness an inhomogeneous printed current collector was expected. The average thickness of the carbon current collector according to the different textiles is in the range from 15 µm up to 18 µm, see Table 2.

Table 2: Current collector thickness on different technical textiles

material	thickness of dry current collector [µm]
PVC (a) artificial leather	16
PVC (b) hygienic underlay	15
PU (a) hygienic underlay	15
PU (b) hygienic underlay	17
PU (c) hygienic underlay	18
PET reference	15

Due to the different woven backsides and the different thicknesses of the technical textiles a second impact in the printed layers could be recognized. In Figure 5 you could imagine the influence of the woven backside, especially for thin textiles. In this case the PU based material had a total thickness below 350 µm with only a 15 µm coating, so the structure of the woven backside could be seen in the functional structure. This did not adversely affect the functionality of the current collector nor the electrodes.

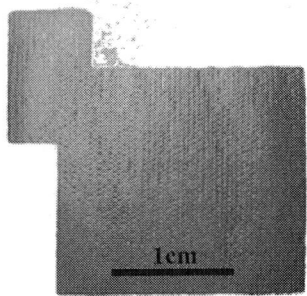

Figure 5: Screen printed current collector on PU based hygienic underlay. The structure of the woven backside could be recognized clearly.

The preparation of the anode and the cathode took part in two steps. First zinc-based anode was screen printed, followed from manganese dioxide-based cathode. Both printed electrodes have a rough surface by themselves, because the electrode structure shows a high porosity to ensure a maximum moistening with the electrolyte. For this reason no influence due to the roughness of the textiles is expected. The thickness of the whole electrode is about 80 µm for the zinc anodes and about 100 µm for the manganese dioxide cathodes. Figure 6 shows the printed current collector on a PU based hygienic underlay which is covered by the cathode (right) and anode (left).

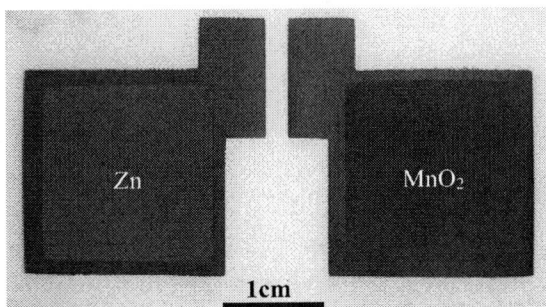

Figure 6: Printed electrodes on PU based hygienic underlay.

First measurements of the resistance, were performed using a dual channel source meter (Keithley 2636) and Suss PM5 Probe station (SussMicroTec) with a van-der-Pauw setup for resistance (5 mm x 5 mm) as well as a line setup. A higher line resistance for the electrodes on the PVC based technical textiles than the PU or PET based could be recognized, see Figure 7. The resistance values for the current collector without any electrode were significantly increased. On the PVC based textile the resistance is 36 times higher than on the PET reference, while on the PU it is raised only 1.5 times. This effect could be recognized for all technical textiles which were tested. Also the resistance for the zinc electrodes, which covered the current collector later on, decrease to a value which is 20 times higher on PVC and three times higher on PU.

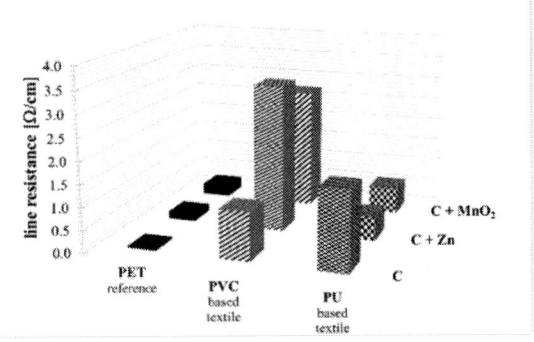

Figure 7: Line resistance of different electrode materials based on PET reference and PVC and PU based hygienic underlay.

Afterwards the battery was built in two steps, via application of a double-sided glue tape, implementation of the electrolyte solution and the separator. In the second step the separator was moistened with the other half of the electrolyte solution and then the opposite electrodes were applied. The batteries were tested intermediately with impedance and galvanostatic measurements by using a potentiostat (BioLogic Science Instruments, VMP3).

Five $3 V_{nom}$ batteries were built for each textile, which consist of the series connection of two single cells of $1.5 V_{nom}$ each, and the discharge characteristics were measured. The open circuit voltage (OCV) was about 3 V, see Figure 8, and discharge curves at voltage level about 2.5 V during a discharge with constant current of 200 µA. The working electrode area was 2.5 cm x 2.5 cm for each electrode. After 70 h and 72 h of discharge, the first cell of the battery discharged completely. The remaining cell discharged completely after 80 h and 96 h, respectively. By discharging the battery with 200 µA CC (constant current) down to the COV (cut of voltage) of 0.9 V, the battery on PET achieves an average area capacity of 3.1 mAh/cm², PVC based 3.0 mAh/cm² and PU based 2.4 mAh/cm².

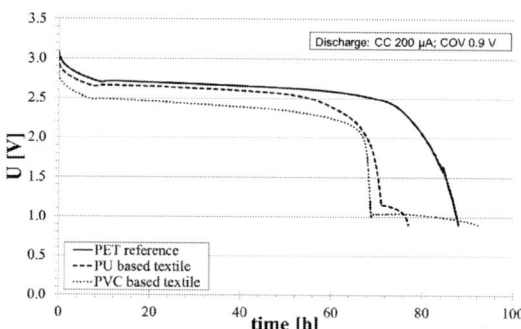

Figure 8: Discharge curves of 3 V_nom batteries based on PET, PU and PVC based technical textiles with constant current 200 μA.

Conductive patterns

Conductive patterns with different area sizes, e. g. 5 mm x 5mm and 10 mm x 10 mm were screen printed on the different technical textiles. Also lines with different lengths (10 mm up to 50 mm) and widths (1 mm up to 10 mm) were screen printed by two different silver inks, a standard and a flexible one, especially made for technical applications. Due to the temperature sensitivity of the technical textiles both inks were sintered at 110 °C in a convection oven. Giving a wide scope for application also the woven backsides of the technical textiles were imprinted, not only the front surface of the materials. In Figure 9 the difficulties of printing onto the woven backside are shown. No closed pattern is printable on the woven backsides, the silver ink covers only the yarn directly.

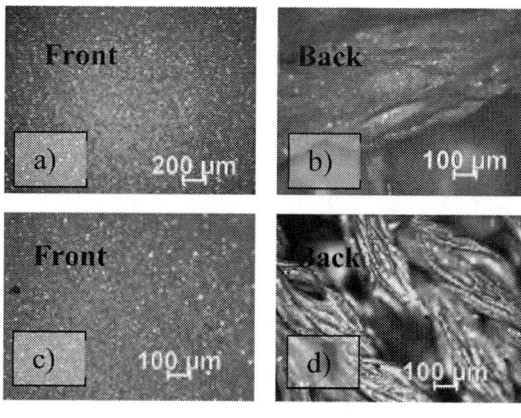

Figure 9: Screen printed patterns with standard silver ink on different technical textiles. a) front (PVC based), b) back (PVC based), c) front (PU based) and d) back (PU based).

Due to the effect, that there is no closed layer on the backside of the technical textiles, the further printing was done on the surface of the front side. The resistance of the silver patterns is dependent on the surface quality, the ink and the textile. The screen printed silver patterns and lines have a thickness of about 2 μm to 3 μm. The lowest specific conductivity

could be observed on the PVC based textiles, see Figure 10. A chemical reaction between the solvent of the ink and the finish coverage of the textiles takes place. Especially the chlorine effects the electrochemistry, also the conductivity, in a high range. On the PU based textiles the specific conductivity is higher and nearly 50 % of the PET reference.

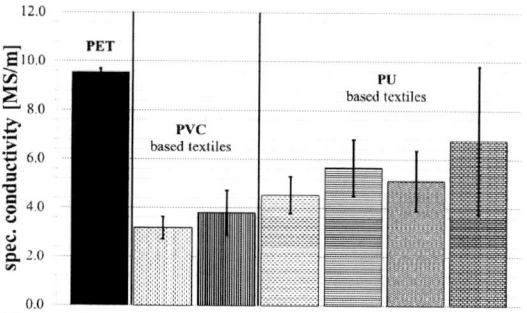

Figure 10: Specific conductivity of standard silver ink on PET, PVC and PU based technical textiles.

Communication elements

The goal was to print a 2.45 GHz Bluetooth antenna on different technical textiles, to develop a full electrical hybrid system.

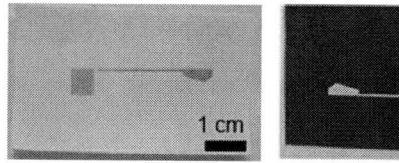

Figure 11: Screen printed silver Bluetooth antenna on PVC (white) and PU (blue) based textiles.

Due to the confirmed capacities, screen printed silver antennas could be applied to the PVC and PU based technical textiles, see Figure 11. In the first approach these antennas were printed individually and the conductivity, as well as their function was checked. Here, the antennas were connected with a coaxial measuring cable via two copper contact pads to a corresponding network analyzer (Vector Network Analyzer Rohde & Schwarz ZVL6) and then glued to the forearm of a test person to generate a corresponding everyday situation, see Figure 12.

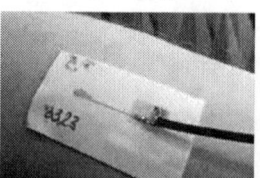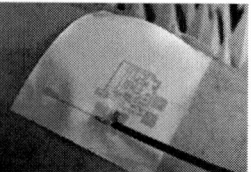

Figure 12: Bluetooth antennas on PU based hygienic underlay. a) flatbed screen printed and b) R2R screen printed Bluetooth antenna connected with hybrid system.

First the characteristic impedance of the antennas were measured in order to compensate the impedance influence of the different technical textiles. From this, impedance mismatch networks were generated, which were used in the functional test of the antennas on the different technical textiles. Figure 13 shows the resonance behavior of a part of the generated antennas. The functionality could be proven successfully, regardless of which technical textile (PVC or PU based) the antennas are applied.

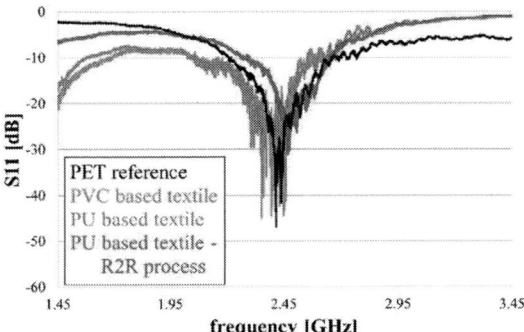

Figure 13: Resonance measurement of the antenna functionality at 2.45 GHz (Bluetooth-Low-Energy frequency).

Hybrid System

Subsequently, a hybrid system was developed, which contains not only the antenna and the connections for a printed body but also the most varied application sites for MCUs (Micro Controller Unit), sensors and other components such as resistors and coils.

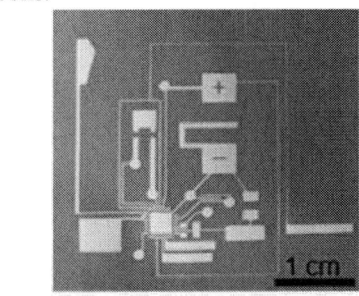

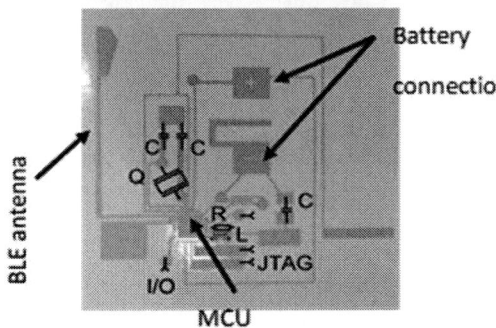

Figure 14: Screen printed hybrid systems with integrated Bluetooth antenna and different possibilities of interconnection, e.g. battery, MCU and sensors. Printed on a) PU based textile and b) transfer paper.

In Figure 14 the first flatbed screen printed hybrid system is shown on PU based hygienic underlay (red) and on transfer paper. The usage of the transfer paper should minimize the influence of the surface properties of the technical textile; in addition, the hybrid system printed on transfer paper could be transferred to any woven carrier with medium lamination. Another advantage is, that the hybrid system could be embedded in the multilayer assembly of the technical textile during the lamination and be protected from external influences.

The second approach was the printing of the hybrid system in a high throughput industrial R2R process. Due to the stretchabilty and high flexibility of the technical textiles, the printing conditions have to be adapted to the textiles respectively. In order to ensure a better assessment of the quality of the printed hybrid system, control marks according to IEC62899-301-2 and IEC62899-403-1 were implemented in the printing process, see Figure 15.

Figure 15: Control marks, implemented in the high throughput printing process, as quality control.

The R2R screen printed hybrid system of the second generation was further compressed and equipped with additional connection options, see Figure 16.

Figure 16: R2R screen printing of control marks, capacitive areas, conductive paths, antennas and hybrid systems on PU based hygienic underlay. In the enlargement the hybrid system 2nd generation.

Acknowledgements

We acknowledge the financial support by the Federal Ministry of Education and Research in the funding program »Unternehmen Region; Zwanzig20 – Partnerschaft für Innovation; consortium futureTEX; Projekt leiTEX. «.

References

[1] P. Kassal, J. Kim, R. Kumar, W. R. de Araujo, I. M. S. M. D. Steinberg and J. Wang, "Smart bandage with wireless connectivity for uric

acid biosensing as an indicator of wound status," *Electrochemistry Communications,* no. 56, pp. 6-10, 2015.

[2] P. Kassal, M. Zubak, G. Scheipl, G. J. Mohr, M. D. Steinberg and I. M. Steinberg, "Smart bandage with wireless connectivity for optical monitoring of pH," *Sensors and Actuators B: Chemical,* no. 246, pp. 455-460, 2017.

[3] P. Kassal, I. M. Steinberg and M. D. Steinberg, "Wireless smart tag with potentiometric input for ultra low-power chemical sensing," *Sensors and Actuators B: Chemical,* no. 184, pp. 254-259, 2013.

[4] A. Moya, G. Gabriel, R. Villa and F. J. del Campo, "Inkjet-printed electrochemical sensors," *Current Opinion in Electrochemistry,* vol. 1, no. 3, pp. 29-39, 2017.

[5] M. F. Farooqui and S. Atif, "Low Cost Inkjet Printed Smart Bandage for Wireless Monitoring of Chronic Wounds," *Scientific Reports,* no. 6, p. 28949, 2016.

[6] Y. Li, F. Sun, C. Yi and W. Li, "A wearable H-shirt for exerces ECG monitoring and individual lactate threshold computing," *Computers in industry,* vol. 92, no. 93, pp. 1-11, 2017.

[7] A. Willert and R. R. Baumann, "33rd International Conference on Digital Printing Technologies (NIP) - Printing for Fabrication 2017," Denver, 2017.

[8] R. E. Sousa, C. M. Costa and S. Lanceros-Mendez, "Advances and Future Challenges in Printed Batteries," *ChemSusChem,* no. 8, pp. 3539-3555, 2015.

[9] A. M. Gaikwad, A. C. Arias and D. A. Steingart, "Recent Progress on Flexible Batteries: Mechanical Challenges, Printing Technologies, and Future Prospects," *Energy Technology,* no. 3, pp. 305-328, 2015.

[10] A. Willert, C. Meuser and R. R. Baumann, "Printed batteries and conductive patterns in technical textiles," *Japanese Journal of Applied Physics,* vol. 57, no. 5S, 2018.

[11] K.-H. Choi, D. B. Ahn and S.-Y. Lee, "Current Status and Challenges in Printed Batteries: Toward Form Factor-Free, Monolithic Integrated Power Sources," *ACS Energy Leters,* no. 3, pp. 220-236, 2018.

[12] A. M. Gaikwad, D. A. Steingart, T. N. Ng, D. E. Schwartz and G. L. Whiting, "A flexible high potential printed battery for powering printed electronics," *Applied Physics Letters,* no. 102, p. 233302, 2013.

[13] N. Singh, C. Galande, A. Miranda, A. Mathkar, W. Gao, A. V. A. Reddy and P.

Aiayan, "Paintable Battery," *Scientific Report,* no. 2, p. 481, 2012.

[14] C. Meng, C. Liu, L. Chen, C. Hu and S. Fan, "Highly Flexible and All-Solid-State Paperlike Polymer Supercapacitors," *Nano Letters,* vol. 10, no. 10, pp. 4025-4031, 2010.

[15] V. L. M. Pushparaj, S. M., A. Kumar, S. Murugesan, R. Vajtai, R. J. Linhardt, O. Nalamasu and P. M. Ajayan, "Flexible energy storage devices based on nanocomposite paper," *Proceedings of the National Academy of Science of the United States of America,* vol. 34, no. 104, pp. 13574-13577, 2007.

[16] M. Peckerar, D. Zeynep, M. Dornajafti, N. Goldsman, Y. Ngut, R. B. Proctor, B. J. Krupsaw and D. A. Lowy, "A novel high energy density flexible galvanic cell," *Energy Environmental Science,* no. 4, pp. 1807-1812, 2011.

[17] P. Kritzer and J. A. Cook, "Nonwovens as Separators for Alkaline Batteries," *Journal of the Electrochemical Socitey,* vol. 154, no. 5, pp. A481-A494, 2007.

[18] S. Magdassi and A. Kamyshny, "Conductive nanomaterials for printed electonics," *Small,* no. 10, p. 3515, 14 01 2014.

[19] S. Park, M. Vosguerichian and Z. Bao, "A review of fabrication and applications of carbon nanotube film-based flexible electronics," *Nanoscale,* no. 5, pp. 1727-1255, 2013.

[20] J. Chang, T. Ge and E. Sanchez-Sinencio, "Challenges of printed electronics on flexible substrates," in *IEEE 55th International Midwest Symposium in Circuits and Systems (MWSCAS),* Boise, ID, USA, 2012.

[21] T. Ahmadraji, L. Gnzales-Macia, T. Ritvonen, A. Willert, S. Ylimaula, D. Donaghy, S. Tuurala, M. Suhonen, D. Smart, A. Morrin, V. Efremov, R. R. Baumann, M. Raja, A. Kemppainen and A. J. Killard, "Biomedical Diagnostics Enabled by Integrated Organic and Printed Electronics," *Analytical Chemistry,* vol. 89, no. 14, pp. 7447-7454, 2017.

Embedding Technologies for the Manufacturing of Advanced Miniaturised Modules toward the Realisation of Compact and Environmentally Friendly Electronic Devices

D. Manessis[1], K. Schischke[1], J. Pawlikowski[1], T. Krivec[2], G. Schulz[2], G. Podhradsky[3], R. Aschenbrenner[1], M. Schneider-Ramelow[1], A. Ostmann[1], and K-D. Lang[4]

[1]Fraunhofer Institute for Reliability and Microintegration (IZM), Gustav-Meyer-Alle 25, 13355 Berlin, Germany
[2]AT&S AG, Fabriksgasse 13, 8700 Leoben, Austria
[3] SPEECH, Speech Processing Solutions GmbH, Gutheil-Schoder-Gasse 8-12, 1100 Vienna, Austria
[4]Technical University of Berlin, Gustav-Meyer-Alle 25, 13355 Berlin, Germany

Corresponding Author: +49-30-46403788, Dionysios.Manessis@izm.fraunhofer.de

Abstract

The proposed work is performed in the frame of the EU project "sustainablySMART", which has undertaken research activities on "Eco-innovative approaches for advanced printed circuit boards" with the aim to demonstrate that embedding technologies are environmentally and economically beneficial since they save much surface space on main boards by embedding components in PCB layers. The main outcome is the manufacturing of robust and compact modules as sub-systems with specific functionalities. Based on this approach, the main board architecture of a voice recorder has been modified in order to be split in power, USB and DSP modules. This paper will describe the PCB embedding processes for the production of the digital signal processing (DSP) module, the power and the USB modules. In specific, for the DSP module, a 6-core layer with through vias and microvias is manufactured and then on its bottom side all the components are assembled which are going to be embedded. These components are the DSP BGA chip, voltage detector, bus buffer, etc. The components after embedding are routed to surface pads of the module. The rest of the components are assembled as SMT components on the surface of the DSP embedded module and these are the Flash memory as BGA package and 2-pad clock crystals. The DSP module (1.5cmx1.5cmx2.8mm) together with the other two embedded modules will be assembled on the main board of the voice recorder. This paper will elaborate on the new design architecture of the device backbone and the assembly of all embedded modules on the backbone.

Key words: embedding technologies, PCBs, miniaturisation, heterogeneous system integration, modularisation.

1. Introduction

Component embedding opens the door to new opportunities in the miniaturization of devices. The foot print of a device and the thickness can be reduced in one step. Considerable efforts have been devoted for the development of embedding technologies, mainly in two technology approaches [1-7]. The first one usually addresses thin semiconductor chips with various functionalities which should have copper metallisations. They are accurately assembled in PCBs with their active "face-up", subsequently are embedded in epoxy resins and are finally contacted with electroplated laser microvias [1-5]. The second approach addresses mainly pre-packaged chips as e.g. BGA packages and passives with various functionalities and thicknesses. A copper metallization is not required. The electronic component assembly uses existing assembly machines for conventional "face-down" placement, soldering of the components and

re-routing of the signals through copper electroplated laser microvias and drilled through-vias. It is an embedding technology that combines superior electronic assembly and state-of-the-art PCB technology solutions. It advances very fast with respect to via opening and PCB structuring capabilities with very fine line/spacings (L/S) even down to 8μm/12μm.

This approach offers the advantage that the supply chain is not required to provide thin dies with copper metallization, and in this way, alternatively pre-packaged components can be embedded. This is also the technique employed in this paper for advanced modularization. It offers a platform for heterogeneous integration of components with different functionalities and thicknesses which normally are soldered on top of mainboards as standalone SMT components. The "face-down" approach has been also used successfully for embedding of sensitive SoCs and MEMS components in very tiny medical microsystem modules used e.g in cardiac and cochlear implants and other hearing aid devices [6,7].

Main limitations for realization of modular concepts up to now are interfaces, upwards and downwards compatibility with other sub-system blocks, significantly larger form factors that require additional housing and connectors for the modules.

In this context, the objectives of this paper and also of the project is to develop and assess new and eco-innovative approaches of the PCB technology in a modular design, rethink the current functions of the Printed Circuit Board and invent new non-permanent interconnection techniques for module/module and module/backbone interlocks. These new innovative modularized PCB concepts intend to reduce interface tangling and make assembly and de-assembly processes very easy. Modules can be very easily reused. Concepts will be developed and assessed to reduce the needed amount of Printed Circuit Boards in electronic devices by merging different functions like HDI (high-density interconnect) and high-speed compatibility or enhanced thermal solution into one Printed Circuit Board. These effects are expected to have also a positive impact on the usage of resources.

The embedded modularization concept deployed in this paper will offer great miniaturization potential as it will be demonstrated with the digital voice recorder (DVR) of the Speech, where only the embedded digital signal processing (DSP) module saves 50% of space usage on board. In general, accurate electronic assembly of heterogeneous components in the inner layer of PCBs, their subsequent embedding in PCB layers and their signal re-routing through advanced PCB processes is the technology route of choice for advanced system or sub-system modularization with an imminent positive environmental impact, as will be shown in this paper.

2. Module Design & Embedding Concepts

The "Embedded module concept" will be mainly explored in the Sustainably SMART project as the main route for electronics miniaturization and will be described in this paper how it can be developed and utilized for an innovative redesign of the digital voice recorder (DVR) of the end user Speech. In this embedding approach, thin components with different functionalities will be soldered and embedded in one layer and others that can be thicker can be assembled on top of the modules [6,7]. In this way, a new embedding platform is created for heterogeneous integration of components that are conventionally assembled on top as SMT, toward the formation of an embedded module with specific functionality. In this paper, three module designs will be presented in order to demonstrate the project plans for modularization. In specific, IZM has developed the concept for the DSP

module, based on the concept of the heterogeneous integration of components. This concept combines very accurate assembly of components in inner layers and then innovative PCB technologies with microvias and through-vias for the redistribution of signals on the outer layers of the modules. The same embedding concept will be followed for the embedded USB and power modules that will be processed by AT&S.

The DSP module will be comprised of 8-copper layers with a 6-inner PCB and will have in total 68 components as embedded and SMT mounted on top of the module. Fig. 1 shows an Altium perspective of the design and how the layers are interconnected with though-vias. The DSP package and quartz crystals will be embedded and the flash memory BGA and passives (smallest 0201) will be assembled on top. The embedded components will be assembled with +/-10μm accuracy on the bottom of the 6-layer core.

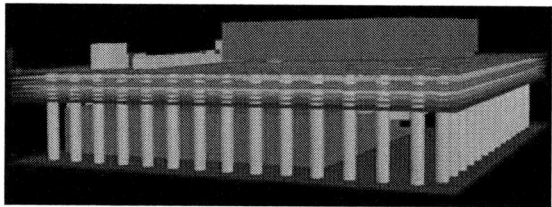

Figure 1: 3D-delineation of the 8-layer DSP board with components inside and Flash and SMT components assembled on top of the module.

The manufacturing of the embedded DSP module will save about 50% of space on the Speech backbone, as shown in Figure 2.

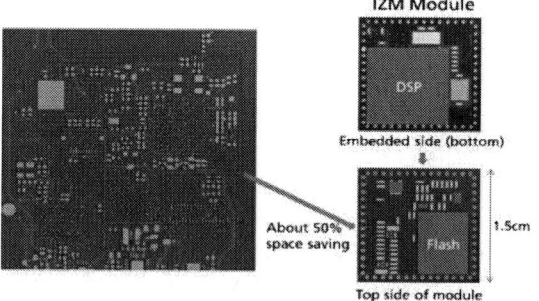

Figure 2: Design optimization through embedded module concept. By DSP module (size 1.5cmx1.5cm) a space saving of 50% will be achieved. Top right view of the embedded side and bottom right view of the SMT components mounted on the top of the DSP module.

Based on the same "embedded module concept", AT&S will also develop "base" modules which together with the functional DSP modules will substitute and innovate the electronic system of the digital voice recorder (DVR) of Speech. Based on "face-down" embedding technology, 2 modules have been designed, namely the USB module and the power module. The designs of the USB and power modules as well as the reconfiguration of the whole dvice PCB after assembly of the new 3 Modules are shown in Figure 3.

As shown in the bottom picture of Fig. 3, the components in the USB module building block will be completely embedded including the USB interface and the embedded part will be connected with the conventional tooth connector through a flex PCB segment in order to minimize any plugging and un-plugging stresses. The new embedded USB module then will be soldered to the Speech backbone PCB as a typical SMT component.

The power module will provide operational voltages to the DVR building blocks of 1.3V, 1.36V, 2.8V, 2.89V, 3.6V. It will be also embedded in order to minimize footprint and it will be connected to the backbone by contact spring/screw connection for very simple exchange.

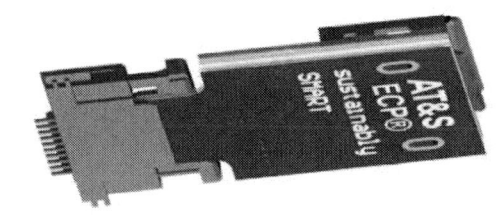

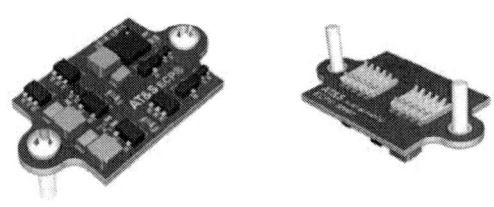

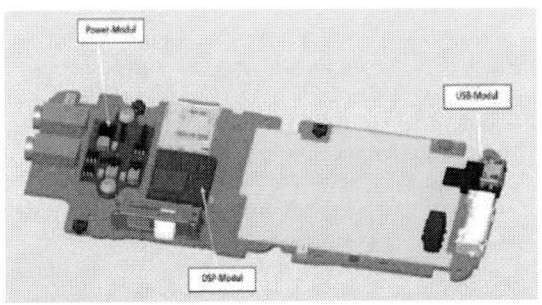

Figure 3: USB module with SMT components (top picture), power modules (top and bottom sides) and reconfigured device PCB with all 3 modules.

3. Results & Discusion

As mentioned above, within the Sustainably SMART project, a whole work package has been devoted for the development of highly dense modules with embedded components as a mean for system miniaturization through heterogeneous integration of components. In this paper, the authors would like to show the manufacturing results of the digital signal processing (DSP) module and the USB and power modules for the SPEECH device through embedding processes.

3.1 Manufacturing of the DSP Module

Analysis of the existing system architecture of the Digital Signal Processing (DSP) Unit of the Speech device has shown that the "new" embedded DSP functional module for the end user SPEECH will have 8 functional layers, of which 6 layers are in the core substrate. The 6 layer core has been designed by IZM-SIIT and was produced by AT&S. The biggest components are the DSP (10x10mm) and the FLASH (8.15 x 6.15mm). In order to achieve the smallest footprint of the module in the X&Y directions, these components have to be integrated on different layers. SPEECH has required to keep the FLASH component non-embedded on top of the whole module in order to remain interchangeable. Therefore, only the DSP and other accompanying electronic components were embedded. In order to minimize the outline, all 8 copper layers were used to connect the components in the module.

The 6-core layer was manufactured by AT&S and IZM performed all other process steps toward embedding. In specific, on the embedded layer (bottom of the 6-core laminate), the DSP chip, the 4-pad DSP crystal, the bus buffer, the voltage detector, the NPN transistor, the LC filter will be assembled and embedded. On the top of the 6-core layer, the Flash chip, the 2-pad clock crystal, capacitors and resistors will be assembled.

3.1.1 Embedding Technologies for the DSP Module

A number of processes should precede before the actual embedding of the bottom layer of the 6-core laminate.

(a) Solder mask application & silver metallisation

After the core was received by AT&S, IZM applied a dedicated layer of 15μm solder mask only on the bottom side of the core. The solder mask should be defined -only on the areas of assembly- and leave the copper open in the rest of the area because the subsequent lamination of the prepregs can work particularly well on rough copper and not on very smooth solder masks. Therefore, the solder mask should be applied only on the needed areas for assembly.

The TOP side of the 6-layer core is left unprocessed, because there will be no assembly and the rough copper should stay uncovered to ensure the good adhesion of the prepregs in the next steps of epoxy lamination for embedding.

(b) Resist application, photolithography & silver metallisation

The boards after solder mask application are processed in the clean room for dry film AM 140 lamination, illumination, and final development in order to get the definition of the pads that have to be deposited with silver.

A Lemmer equipment was used for electroless immersion silver that deposited 0.2μm silver on the copper pads. The silver coating ensured the good solderability and assembly of all pre-packaged components prior to embedding. Fig. 4 shows the bottom side of a 9"x12" board after silver metallisation, resist stripping and prior to component assembly.

Figure 4: Bottom side of the 6-core laminate (9"x12") after silver metallisation prior to assembly.

(c) Assembly of components & Underfilling

At the Bottom side of the core, the DSP and crystal components were placed using an ASM Siplace pick&place machine with an accuracy of +/- 20μm. Both crystals are positioned near the DSP to guarantee correct timing and lower external influences as well as keeping the copper traces short.

IZM can take advantage of economical large scale manufacturing, using its state-of-the-art facilities processing 9"x12" or 18"x24" panels. In this case, each panel (test board), 9"x12" in size, has incorporated 22 modules. Fig. 5 shows a closer look of the DSP and crystal components assembled on the bottom of the 6-core layer prior to embedding.

Figure 5: View of the bottom side of the 6-layer with DSP and crystal components to be assembled, prior to their embedding.

The components after assembly have to be underfilled in order to prevent them move during embedding where the actual lamination profile is reaching for short time over the melting point of the solder joints. The components after underfilling have been measured to find out the exact thickness in order to decide the thickness of the prepregs that will be used for the embedding of the bottom and the top layer of the 6-core laminate. The same underfilling practice has been followed also in embedding of pre-packaged components for highly miniaturized medical microsystems as shown in literature [6,7], where components could be shifted few microns from their original soldered position after lamination processes.

(d) Set-up processes for embedding & actual embedding

The two boards with the components assembled on the bottom side have to pass a brown oxide process to roughen the copper so as to ensure a good adhesion of the epoxy resin prepregs during lamination. Then, based on the component height measurements performed, a stack of prepreg layers should be cut based on the Gerber data and positioned with the cavities around the components at the bottom side of the core. For the embedding, 3 different prepreg type layers were processed with cavities around the components.

All the layers were positioned in the metallic plate stapel and were laminated in the Laufer Press under 30 bar pressure and a constant vaccum. After embedding, the 9"x12" boards are shown in Fig. 6. These boards will be further processed for microvias and through vias opening.

(e) Via opening & via metallisation processes

Subsequently, the boards were processed in order to open vias; namely, the laser microvias using a combination of UV/CO_2 laser and the through-vias

78

using mechanical drilling. The UV laser was used for quick removal of the copper layer and the CO_2 Laser for milder removal of the epoxy layers. The microvias were 100μm in diameter.

The complete board with laser microvias and mechanically drilled vias is shown in Fig. 6, and is then copper electroplated for the filling of the laser microvias and also a copper layer of 15μm on each via wall of the through-via.

Figure 6: Embedded board after laser microvias, through-vias and copper metallisation.

(f) Solder mask application and silver metallisation

The boards are subject to MecETCH for roughening the copper again in order to have a good adhesion of the resist. The pass in the MecETCH machine takes place in a velocity of 0.7 m/min. The AM140 film resist is applied by roll and vaccum lamination simultaneously. After the definition of the pads, solder mask film of 30μm was laminated and photolithographically processed to define the soldered areas of the top surface where the assembly of the SMT components will take place. Fig. 7 shows the final embedded DSP boards, after solder mask application and silver metallisation, ready for the assembly of the components on the top side.

(g) SMT component assembly on the embedded DSP board

As a last step of the creation of the SPEECH DSP board, the flash memory chip, the 2-pad crystal, capacitors and resistors will be assembled on the top. By jetting, solder paste is deposited on the silver pads and the components were placed using a FINE PLACER system. Subsequently, the assembled boards passed reflow processes. Fig. 7 shows the SMT components on the whole embedded DSP board before singulation of the DSP modules.

Figure 7: SMT components assembled on the DSP embedded board.

After the assemly of the SMT components, the DSP modules were singulated from the DSP board. The DSP modules were only 15mmx15mmx2.8mm, including also the SMT components. The thickness of the DSP embedded part was only 1.9 mm. Fig. 8 shows the DSP modules after singulation.

Figure 8: Final DSP modules after singulation from the embedded DSP board.

3.1.2 Analysis of Embedded DSP Modules

The embedded DSP modules were analysed by cross sectioning and X-Ray tomography. Fig. 9, shows a representative cross section of the DSP module on the top of the figure and another perspective of the module after X-Ray tomography at the bottom of the figure. In this specific cross section, the embedded pre-packaged DSP die and quartz crystal can be seen at the bottom of the 8-layer module. On the other side, the capacitors (C) and flash memory BGA component assembled as SMT components can be also seen. In addition, the mechanically drilled vias, the microvias, and all 8 copper layers of the DSP module are visible. The X-ray tomography picture presented at the bottom of the figure shows another slice view of the DSP module, delineating again in high definition the embedded DSP component with the flash and capacitors on top. The DSP module is 15mmx15mmx1.9 mm for the embedded part. Including the flash memory and passive on top, its thickness reaches 2.8mm. Some DSP modules have successfully passed moisture sensitivity level 3 (MSL3) tests, 3x lead-free reflows.

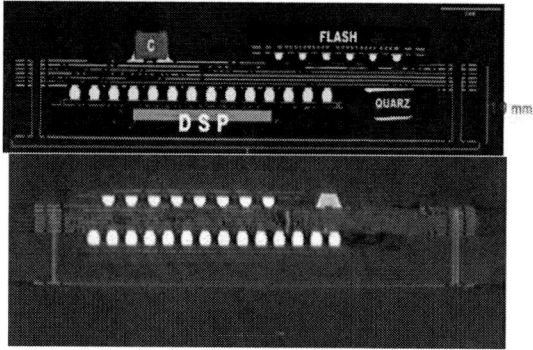

Figure 9: Cross section view (top) and X-Ray tomograph (bottom) of the DSP modules. Their size is 15mmx15mmx1.9mm (embedded part) and up to 2.8mm in height including the flash and capacitors as SMT components on top.

3.2 Manufacturing of the USB Module

The DVR USB module provides the primary functionality of data transfer and also is responsible for power supply for charging the battery. It is the main data /power interface of the DVR towards charging device & docking station. As in many electronic handheld devices the rigid connection between the mainboard and the USB port is one of the crucial points for devoice failure. Occurring failures are mainly driven by inaccurate pluging or un-plugging, abuse forces on plugged cables or drop of the tools. All these events cause stress peaks on the rather weak port-solder-board interface.

The demonstrator USB Module was designed to avoid the mentioned handling stresses. Additionally it offers the possibility to quickly exchange a potentially non-functional module. The basic concept for the DVR USB module is to utilize a standard ZIF connector that is rigidly soldered to a backbone board and to connect the actual USB module via that connector to the backbone board, while actual module and ZIF connector are decoupled by an intermediating flex segment.

The USB module features a 6-layer design (6 copper layers) with embedded passive components that are transferred inside the board to minimize the footprint as well as some embedded ferrites to improve signal quality. Additonally the embedding of the passives into the USB substrate causes a better protection of the interfaces against impact loads and other mechanicals stresses.

For dielectrics a midT$_g$ FR4 material R1566 form Panasonic Electric Works Europe was applied, a commonly used material for industrial and handheld applications. Figure 10 shows a top and bottom view of the USB module as well as the USB connectors and the passives assembled on its top.

Figure 10: 6-layer base USB module (top and bottom view). USB module with USB connector and passives assembled on its top.

3.3 Manufacturing of the Power Module

Main function of the Power module is to provide the needed voltage levels for the operation of the different DVR blocks, while combining a small form factor with a non-permanent interconnection to the backbone. A CAD drawing of the power module is shown in Figure 3. The power module is designed in a 4-layer stackup, again based on AT&S center core embedding technology. Material again is taken from the Panasonic R1566 fsamily. Figure 11 shows the top and bottom view of the manufactured power module.

Figure 11: 4-layer Power Module. Top side with passives assembled and bottom side with spring connectors assembled.

3.4 Functional Device PCB

The chosen end user device in this project is a professional, digital dictation recorder. The speech will be digitized in several, selectable file formats and is stored on a SC-Card. The device is powered by a Lithium-Polymer rechargeable battery. File transfer to the PC as well as charging can be done via a Micro-USB-Socket at the lower end of the device. Figure 12 shows the digital voice recorder (DVR).

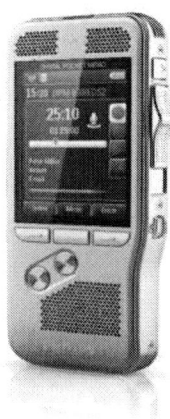

Figure 12: Speech digital dictation recorder.

The DSP module integrates vital functional components which would be otherwise assembled on larger area of the device PCB. In specific, the DSP (Digital Signal Processor) has the following functions: (a) to execute the differenct compression algorithms of the audio data (dss, MP3), (b) to generate the different file-container formats like Wave, DSS or MP3 and (c) to manage the data transfer to and the file system (FAT32) of the SD-Card, (d) to manage the USB-Interface.

A second integrated component is the Microcontroller, which performs the following functions: a) steers the full color LCD Display b) scans all control-elements of the UI (e.g. buttons,

light sensor, position sensor). A third integrated component is the Flash memory. It stores the Firmware of the device. It saves all settings of the device (e.g. file format, microphone characteristic, serial number, keywords). Figure 13 shows the DSP embedded module that co-integrates all these three components and some peripheral components needed. The connection between the Module and the DVR's mainboard is realized by reflow soldering, as shown in Figure 13.

Figure 13: DSP Module on mainboard.

The electronics require several different supply voltages which are derived either from the built in rechargeable battery or from USB-Power (when USB is connected to the device). Charging a lithium-polymer battery requires a certain charging algorithm which is realized with a special charging integrated circuit. These functions/components are condensed in the so called Power Module. The connection between the power module and the main board is realized via 20 gold plated spring contacts. The mechanical attachment is achieved via two screws, as shown in Figures 3 and 11.

The USB-Connection is done via a Micro-USB Socket which enables the USB communication between the device and the USB-Host (e.g. PC). To protect the interface against ESD (Electro Static Discharge) ESD and EMV protection components are placed close to the USB-Socket. These components are combined in the third module of the Project named USB-Module. The connections between the module and the mainboard are realized via a connector. Figure 10 shows the top and bottom side of the USB module.

The Production process can be optimized with respect to assembly processes if 1 mainboard panel contains 4 mainboards.

The DSP-Module will be reflow-soldered on to the mainboard after automatic pick & place process. The Power-Module as well as the USB-Module will be mounted manually on the mainboard when also other mechanical components (e.g. Main Slide Switch) will be assembled. Figure 14 shows all 3 Modules assembled on the mainboard. After the final assembly, the mainboard will be 100% automatically tested on dedicated test apparatus which is connected via several dozen test pins with the mainboard. Once the PCB has passed this step, the PCB is released and will transferred to the final assembly line. When the product is then completely assembled, a final 100% test will be done before the device will be packed.

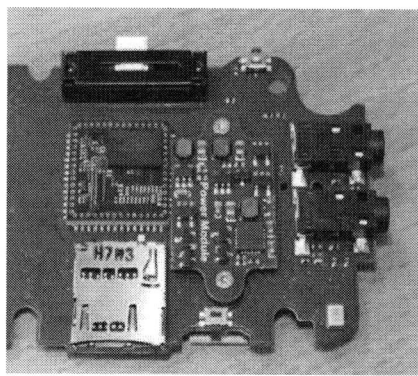

Figure 14: DSP and Power modules (top picture) and USB Module (bottom picture) assembled on the device DVR mainboard.

Conclusions

This paper has presented the development of embedding technologies for the heterogeneous integration of pre-packaged components toward module miniaturization. These developments are integrated in larger effort undertaken in the Sustainably SMART project toward holistic processes with large environmental impact. In this communication, the manufacturing embedding processes were fully described for the transformation of the existing digital signal processing unit of Speech device to a highly miniaturized embedded DSP module. At least 50% space reduction on the Speech backbone was achieved with the embedded DSP version. The DSP modules have been significantly miniaturized, manufactured in large scale formats of 9"x12" using

state-of-the-art PCB facilities and have a total size of 15mmx15mmx1.9mm with flash memory and passives on top of the module. In addition, this paper has shown the manufacturing of two other embedded USB and power modules with a high degree of component embedding, operational flexibility and robustness. A new configuration layout on the DVR mainboard has been achieved with the 3 Modules. The paper has also described in detail the current state of the DVR mainboard after the assembly of all new 3 modules by Speech and the production processes toward device build-up by Speech.

Acknowledgments

The research leading to these results has received funding from the European Union's Horizon 2020 research and innovation programme under grant agreement No 680604.

References

[1] A.Ostmann, A. Neumann, P. Sommer, H. Reichl, "Buried components in printed circuit boards", Advancing Microelectronics, May/June 2005, pp. 13-18.

[2] L.Boettcher, D.Manessis, A. Ostmann, S. Karaszkiewicz, H.Reichl, "Embedded Chip Packages –Technology and Applications", Proc. SMTA 2008, Orlando, FL, August 2008, ISBN 978-0-9789465-7-9, pp. 43 – 48.

[3] D.Manessis, L.Boettcher, A. Ostmann, R. Aschenbrenner and H. Reichl, "Chip embedding technology developments leading to the emergence of miniaturized system-in-packages", Proceedings in ECTC 2010, Las Vegas, NV, USA, June 1-4, 2010, ISBN: 978-1-4244-6412-8, ISSN: 0569-5503, pp. 803-810.

[4] L. Boettcher , D. Manessis , S. Karaszkiewicz, A. Ostmann, "Next Generation System in a Package Manufacturing ", Proceedings in SMTA 2010, Orlando, FL, USA, October 24-28, 2010, ISBN 978-0- 9840562-3-1, pp. 222-228.

[5] D. Manessis, S-F.Yen, A. Ostmann, R Aschenbrenner and H. Reichl, "Technical Understanding of Resin Coated Copper (RCC) lamination processes for realisation of reliable chip embedding technologies", ECTC 2007, Reno, NV, May 29- June 1, 2007, pp. 278-285.

[6] D.Manessis et al. "Manufacturing of Miniaturised Microsystems for Low Power Wireless Body-Area-Networks (BAN) Medical Applications-Technological Challenges and Achievements", in digital form, ESTC 2012, Amsterdam, NL.

[7] D.Manessis et al. "Large-scale manufacturing of embedded subsystems-in-substrates and a 3D-stacking approach for a miniaturised medical system integration", in digital form, EMPC 2013, Grenoble, FR.

Ultra-thin Capacitors in Silicon for 3D-Integration and Flexible Electronics

Mamta Pradhan, Saleh Ferwana, Christine Harendt, Harald Richter and Joachim N. Burghartz

Institut für Mikroelektronik Stuttgart, Allmandring 30a, 70569 Stuttgart

Phone: +49 711 21855 431, and E-mail: pradhan@ims-chips.de

Abstract

Current electronic systems and circuits possess approximately 80% of passive components. Thus, high performance integrated silicon-based passive components with high density are required to reduce the overall cost and bulkiness of the systems. This paper presents the development of high-density ultra-thin 3D capacitors in silicon, one of the important passive components exploited in approximately all the circuits for SiP, 3D-integration and flexible systems. The 3D structure is achieved by using deep reactive ion etching (DRIE) process. The substrate and the polysilicon act as the two electrode plates with silicon dioxide as the dielectric. Two designs with different pitches are investigated with successful back-grinding of the substrate to 50 µm thin and 22 µm ultra-thin singulated capacitors. Capacitive densities of 36.25 nF/mm² and 28.75 nF/mm² over a frequency bandwidth of 100 KHz are presented. The reduced thickness and flexible nature of these ultra-thin capacitors (<50 µm) makes them a suitable candidate for a wide range of applications.

Key words: ultra-thin capacitors, flexible electronics, passive components, hybrid system in foil (HySiF)

Introduction

Miniaturization based growth in microelectronics industry has led near to the exhaustion of Moore's law. Further exploration in this regard has given rise to "More than Moore" technologies [1]. Major effort has been put into producing smaller and smaller active devices in order to satisfy multiple industrial sectors say sensors in Internet of things (IoT), mobile health care, energy appliances, radio and television. These active devices are encircled by high number of passive components. Around 80 % of the highly integrated electronics systems or circuits are possessed by passives [2], yielding large foot print and high weight. Hence, high performance integrated silicon-based passive components with high density are required in order to reduce the overall cost, volume and weight of systems. Among these passives, capacitors play a major role, due to its requirement in coupling of AC signals, decoupling (bypass capacitors) and snubbing circuits [3]. They consume a lot of crucial space on the System on Chip (SoC) and System in Package (SiP) when used for buffering and coupling-decoupling purposes. The high-density silicon-based capacitors would reduce the overall area along with high integration and reliability. The applications of these capacitors vary according to the capacitive values from pico-farads to micro-farads for various functions such as energy storing devices [4], switch capacitive filters in electrocardiogram (ECG) [5], switched capacitor converters [6] etc.

Development of ultra-thin capacitors with high capacitive densities proposes lucrative integration in hybrid systems-in-foil (HySiF) for flexible electronics. Although flexible circuits have been in electrical/electronic applications since World War II [7], the bending capacity of these circuits is limited by the deformity of circuits containing surface mounted passive devices (SMD) on the foil. Flexible electronics is expanding a lot in the recent years due to rise in applications which need conformity over curved surfaces, majorly in smart and wearble electronics, medical implants etc.

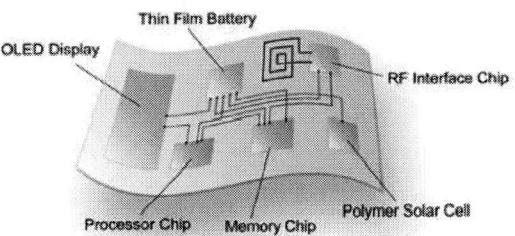

Figure 1: A schematic illustration of HySiF with various discrete electronic components such as display, battery, wireless communication etc., combined with large-area thin film, organic electronics and ultra-thin chips together in flexible foil [1]

Currently, the flexible electronic industry is widely occupied by thin film organic and inorganic semiconductors and 2D materials [8]. The large area organic electronics is cheaper due to high throughput roll-to-roll printing technology, but it comes with the cost of low mobility and low performance [9]. Devices with high mobility 2D semiconductors such as graphene and carbon nanotubes are also slower than silicon, owing to the channel length or the device technology [10]. Devices on silicon wafers, when thinned down to < 50 μm exhibit excellent physical bending abilities. They provide the desirable electrical characteristics in a particular bending state [11]. Thus, ultra-thin passive devices in silicon promote the overall development of silicon micro-electronics as well as HySiF (Figure 1) in the flexible world.

This paper presents high-density ultra-thin capacitors with a thickness of ~22 μm, by exploiting the high aspect ratio trenches in silicon using DRIE etching. Instead of using high k-dielectrics for increasing the capacitive density, this work has been carried out with a low k-dielectric of SiO$_2$. This encourages a feasible fabrication of the capacitors along with the better breakdown voltage for the same dielectric thickness [12].

Capacitor Design

This section describes the structure and design specifications of the fabricated capacitors. The high capacitance density is achieved by increasing the effective surface area of the capacitors in low foot print. Deep vertical cylindrical structures are fabricated in silicon to achieve the desired capacitance. The three dimensions which can affect the effective surface area are the diameter of the cylinder, the depth of the cylinder and the pitch of these cylindrical structures.

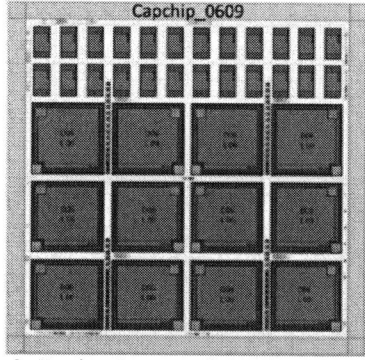

Figure 2: Design layout of the capacitors

Two capacitor designs namely D06L06 and D06L09 [13] are investigated in this work as presented in Figure 2. Both the designed capacitors are square blocks of 2mm x 2mm. Each block contains multitude of vertical cylinders etched in the silicon substrate with a diameter of 600 nm and depth of ~15 μm. The only variation in the two designs is the pitch of vertical cylinders. Design D06L06 has a pitch of 1.2 μm where as D06L09 has a pitch of 1.5 μm. Due to this, two different

capacitance densities are acheived. Due to a smaller pitch of D06L06 (denser in cylindrical trenches), a higher surface area, the capacitance value is expected to be higher than that of D06L09.

Process flow

The schematic of the process flow is depicted in Figure 3. A 150 mm p+ wafer of (10-20) mΩ-cm is thermally oxidised to a layer of 15 nm SiO$_2$ (a). Then, it is uniformly implanted with BF$_2^+$ resulting in p++ implanted surface layer (b).

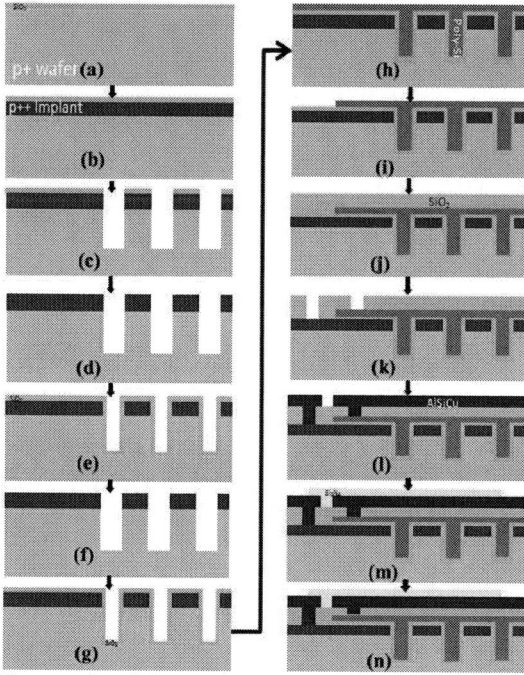

Figure 3: Schematic process flow for the fabrication of 3D capacitors in silicon

The lithography for the cylindrical structure pattern is performed on the wafer. It is followed by the DRIE etching of silicon, creating ~15 μm deep hollow cylinders (c). The scanning electron microscope (SEM) image of these deep cylindrical structures in silicon is presented in Figure (4, 5).

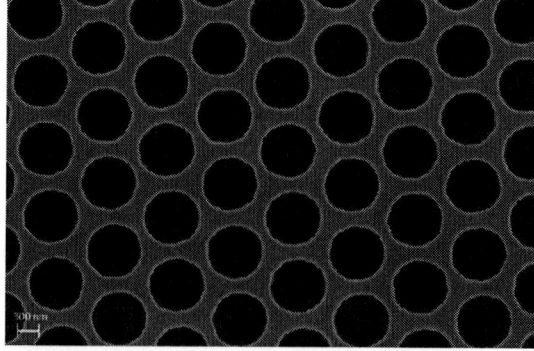

Figure 4: Top view of the cydrindrical structures in silicon

Figure 4 presents the top view of the capacitors, where as Figure 5 presents the side cross-section of the capacitor structures. It can be seen in Figure 5-b and Figure 5-c that the cylindrical structures are wide at the top and tapered to the bottom. This is useful in the uniform deposition of the polysilicon in the later layers.

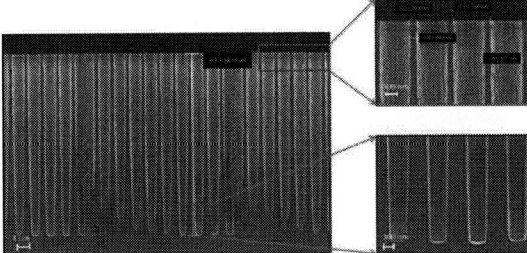

Figure 5: SEM cross-section of the capacitor substrate: (a) deep cylindrical trenches in the silicon (b) trench view near the top surface, (c) tapered trench view at the bottom

Then, the 15 nm of SiO_2 on the remaining wafer is wet etched (d). After this, a cleaning oxide of 50 nm is thermally oxidized on the wafer and wet etched (e, f). This oxide etching helps in making the inner surface of the cylinders smooth, which improves the breakdown voltage of the capacitors. The SEM image of the inner surface before and after the oxide etching is presented in Figure 6.

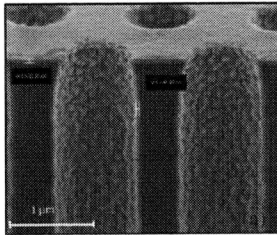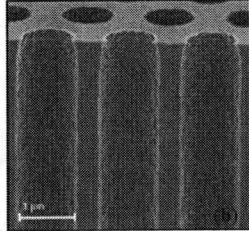

Figure 6: SEM cross-section of the cleaning oxide: (a) rough inner surface with the oxide, (b) smooth inner surface after wet etching

The dielectric layer of 15 nm SiO_2 is then grown using dry oxidation for a denser and high-quality dielectric (g). A 500 nm of in-situ doped polysilicon layer is deposited on the wafer which is patterned and etched away from the substrate contact regions (h, i). The SEM image of this polysilicon layer is shown in Figure 7. An islolation layer between the two terminals is then created by the deposition of 300 nm of SiO_2 (j) using a Plasma Enhanced Chemical Vapour Deposition (PECVD). Then, the vias are etched for the metal contacts to the polysilicon and the substrate (k). AlSiCu is then sputtered on the wafer and patterned (l), followed by the deposition of a passivation layer of plasma oxide and Si_3N_4 as protection layers, and finally the contact pads are patterned and opened (m). To have smooth side walls, the capacitors are singulated by trenching 70 μm deep into the silicon using DREI process, after analyzing the electrical response of the capacitors on the wafer. The 675 μm thick

wafer is then backgrinded to 400 μm (n) using coarse grinding and then finely back-thinned to 50 μm and 22 μm sequentially.

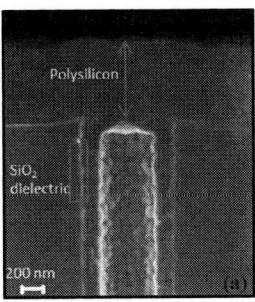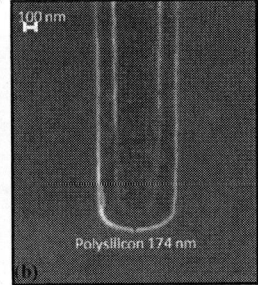

Figure 7: SEM cross-section of the deposited in-situ polysilicon: (a) cylindrical structures covered on the top, (b) layer of polysilicon in the inner surface of the cylinder

In the above process flow, the substrate acts as the first electrode and the second electrode is formed by the in-situ doped polysilicon. In another process, the second electrode is developed by using a $POCl_3$ diffused polysilicon instead of the in-situ doped one. The remaining process flow remains the same on both the wafers. Both the wafers are then measured for the electrical response of the capacitors.

Electrical Measurements and Results

Both types of above mentioned capacitors are measured for the capacitance value using a MFIA LCR meter from Zurich Instruments. During the capacitance measurements, the voltage is swept from -2.5 V to 2.5 V with an AC voltage of 300 mV at a frequency of 1 kHz. The impedance (Z) and phase (Θ) of the measurements are recorded and then calculated into capacitance (Cs) and resistance (Rs) using a series model (Cs-Rs). Owing to the semiconductor nature of the electrodes and the silicon dioxide dielectric, the C-V curve depicts the characteristics of a MOS capacitor at high frequencies.

As shown is Figure 8, the capacitance value of D06L06 is higher than that of D06L09. At 0 Volts and 1 kHz, the capacitance density of D06L06 and D06L09 with in-situ doped poly is observed to be 36.25 nF/mm^2 and 28.75 nF/mm^2 respectively. It is also observed that the capacitance value is higher for the capacitors with in-situ doped polysilicon as compared to the wafer with $POCl_3$ diffused polysilicon. The capacitance density of the capacitors with $POCl_3$ diffused polysilicon is 31.4 nF/mm^2 and 27.07 nF/mm^2 for the respective designs. After back-grinding to 50 μm, the capacitors are again measured on the supporting foil as well as after detachment from the foil. These thin capacitors are further back-grinded to 22 μm thus, creating ultra-thin capacitors in silicon which are flexible in nature. The consistency in the C-V measurements of both 50 μm thin and 22 μm ultra-thin capacitors for both the designs is presented in Figure 9. They present the same value of capacitance as observed on the 675 μm thick wafer.

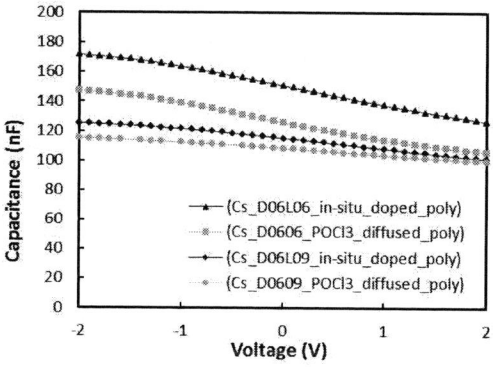

Figure 8: C-V curve of capacitors with in-situ doped polysilicon in comparision with POCl₃ diffused polysilicon

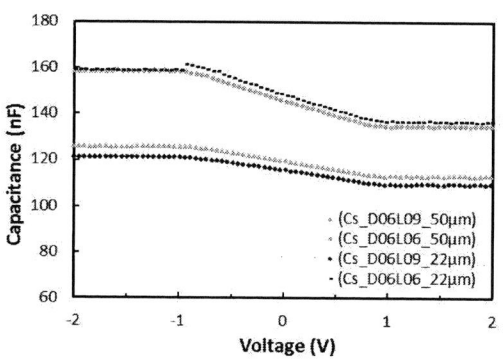

Figure 9: C-V curve of 50 μm thin capacitors compared to 22 μm ultra-thin capacitors

The capacitors are also measured with a frequency sweep of 10 Hz to 1 MHz at a constant voltage of 0 Volts with an alternating supply of 300 mV. For 50 μm thin capacitors, the bandwidth is observed to be of 20 kHz, where as from Figure 10, it can be seen that for 22 μm ultra-thin capacitors, the bandwidth is 100 kHz, thus depicting an improvement in the frequency range for ultra-thin capacitors. After the C-V and C-F measurements, the breakdown voltage of these capacitors is measured using Agilent E5270B Precision IV Analyzer. In order to measure the leakage current, a constant voltage is applied to the capacitors with a single step of 1 V for an average of a minute and the corresponding current through it is measured. The capacitors are discharged to the neutral state by applying 0 Volts for the same time duration. The next step voltages are applied to the capacitor in the same manner as described above

till the breakdown takes place. In Figure 11, it can be seen that the capacitor breaksdown at -7 V and 5.5 V.

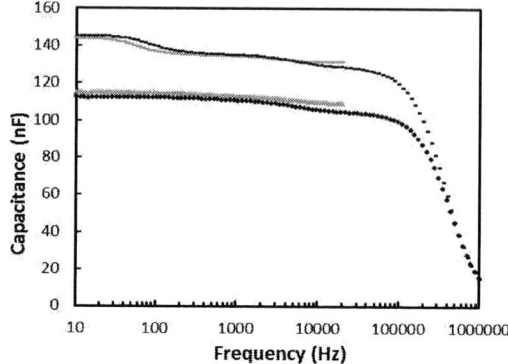

Figure 10: C-F measurements of 50 μm thin capacitors as compared to 22 μm ultra-thin capacitors

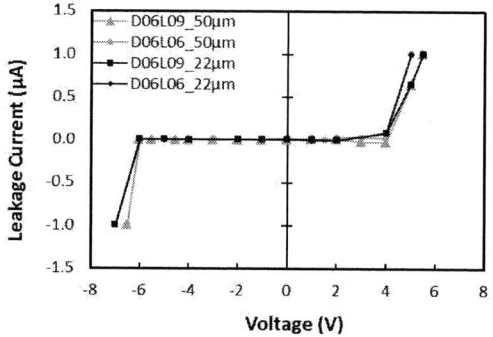

Figure 11: Leakage current through the capacitors at breakdown voltage

In order to measure the yield of these capacitors on the wafer, all the capacitors are measured using a wafer prober at 0 V and 1 kHz. It is observed that for design D06L06 the capacitance value increased gradually from anti-flat to the flat side of the wafer, where as for D06L09, the centre of the wafer has lowest value with the highest being at the flat side. This could be attributed to the unintentional extra etching of the cylindrical trenches on the flat side of the wafer. Capacitance variation of 13 % and 6.6 % wafer for D06L06 and D06L09 respectively is recorded on the whole wafer. Wafer yield of 100 % and 71% is measured for D06L06 and D06L09 respectively. The average value of capacitance of the 2x2 mm² capacitors on the whole wafer resulted in 143 nF and 115 nF for D06L06 and D06L09 respectively. The average value of the series resistance of these capacitors on the whole wafer resulted in 189 Ω for D06L06 and 230 Ω for D06L09. The measured curves are presented in Figure 12. The charging and

86

discharing time of these capacitors is very low due to small series resistance.

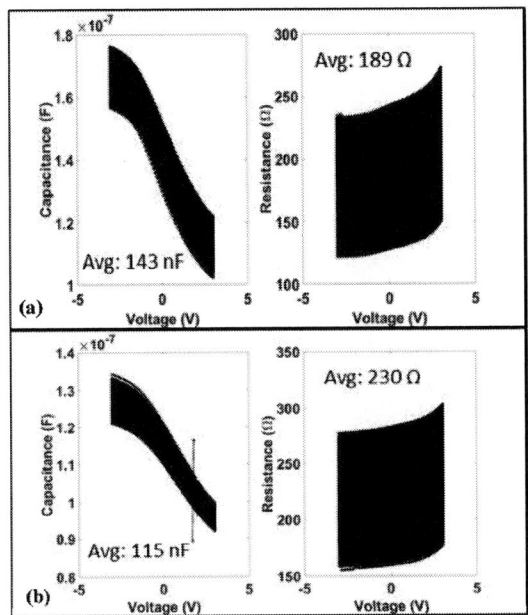

Figure 12: Variation of capacitance and resistance on the wafer for (a) D06L06 and (b) D06L09

Conclusion

Successful development and fabrication of the high density ultra-thin 3D capacitors in silicon has been presented in two designs.

The low series resistance depicts fast charging and discharging of the capacitors making it desirable for low power applications

Thus, these ultra thin capacitors are a suitable passive component for integrating in SiP, Package on Package (PoP) or 3D-IC, moreover due to the bendability; they are perfect candidate to be integrated in thin foils for realizing felexible electronic circuits. The broad bandwidth of the 22 μm ultra-thin capacitors broadens its application range in hybrid SiF.

Acknowledgements

The authors wish to acknowledge the contributions of members of IMS team, M. Jurisch, B. Leibold, D. Nikolaev, G. Heimpel, J. Tudarinow, U. Kohler N. Entenmann, S. Schmiel and D. Kühnle in the fabrication process.

References

[1] Joachim Burghartz, "Ultra-thin Chip Technology and Applications", Springer Publishing Company, Incorporated, 1st edition, Chapter 1, pp. 9-10, 2010.

[2] H. Johari and F. Ayazi, "High-density Embedded Deep Trench Capacitors in Silicon With Enhanced Breakdown Voltage", IEEE transactions on Components and Packaging Technology, December 4, pp. 808-815, 2009.

[3] R. Ramos, "Film Capacitors in Power Applications: Choices and Particular Characteristics Needed", IEEE Power Electronics Magazine, Vol. 5, pp. 45-50, March, 2018.

[4] K. R. Ramya and B. M. Madhu, "Design of actuation control unit with ultra capacitors as the embedded energy back up for electro-mechanical applications" International Conference on Energy, Communication, Data Analytics and Soft Computing (ICECDS), August 1-2, pp. 3575–3578, 2017.

[5] N. Singh and P. P. Bansod, "Switched-capacitor filter design for ECG application using 180nm CMOS technology", International Conference on Recent Innovations in Signal processing and Embedded Systems (RISE), October 27-29, pp. 439–443, 2017.

[6] D. Sivaraj and M. Arounassalame, "High gain quadratic boost switched capacitor converter for photovoltaic applications", IEEE International Conference on Power, Control, Signals and Instrumentation Engineering (ICPCSI), September 21-22, pp. 1234–1239, 2017.

[7] J. Fjelstad, "Flexible Circuit Technology", BR Publishing, 4th edition, Chapter 2, pp. 27-28, 2011.

[8] A. Nathan, A. Ahnood, M. T. Cole, S. Lee, Y. Suzuki, P. Hiralal, F. Bonaccorso, T. Hasan, L. Garcia-Gancedo, A. Dyadyusha, S. Haque, P. Andrew, S. Hofmann, J. Moultrie, D. Chu, A. J. Flewitt, A. C. Ferrari, M. J. Kelly, J. Robertson, G. A. J. Amaratunga, and W. I. Milne, "Flexible Electronics: The Next Ubiquitous Platform", Proceedings of the IEEE, May 13, pp. 1486–1517, 2012.

[9] J. N. Burghartz, W. Appel, C. Harendt, H. Rempp, H. Richter, and M. Zimmermann, "Ultra-thin chips and related applications, a new paradigm in silicon technology", In 2009 Proceedings of the European Solid State Device Research Conference, September 14-18, pp. 29–36, 2009.

[10] Shoubhik Gupta, William Taube Navaraj, Leandro Lorenzelli and Ravinder Dahiya, "Ultra-thin chips for high-performance flexible electronics", npj Flexible Electronics, March 14, pp 1-16, 2018.

[11] Ravinder S. Dahiya, Salvatore Gennaro, "Bendable Ultra-Thin Chips on Flexible Foils", IEEE Sensors Journal, Vol. 13, pp. 4030 – 4037, October 2013.

[12] P. Jain and E. J. Rymaszewski, "Embedded thin film capacitors-theoretical limits", IEEE Transactions on Advanced Packaging, Vol. 25, pp. 454–458, August 2002.

[13] Liyuan Cai, "Entwurf und Herstellung eines Kondensatorchips basierend auf der Chipfilm^{TM} -Technologie", Studienarbeit, INES, University of Stuttgart, April 2014.

Life-Cycle Assessment for Power Electronics Module Manufacturing

Benjamin Schellscheidt, Jessica Richter, Thomas Licht

Faculty of Electrical Engineering and Information Technology
Department of Microelectronics
HSD University of Applied Sciences Düsseldorf
Münsterstraße 156, 40476 Düsseldorf, Germany
+49 211 4351 3145   benjamin.schellscheidt@hs-duesseldorf.de

Abstract

It is often difficult to determine the environmental impact of industrial manufacturing processes, because various resources, chemicals and sources of energy are used. In addition, environmental impacts can depend in geographic scope and need to be evaluated depending on the manufacturing location. One possible solution to this problem is the ISO-standardized Life-Cycle Assessment technique, which allows determining the environmental impact of a product or service. It can be universally applied to products, production methods, disposal and recycling processes and services. All environmental impacts that can be allocated to the examined process are taken into account, from raw material production through manufacturing and shipping to recycling or deposition. Here, we show that we successfully applied this technique in a current, publicly funded research project "ReffiMaL", where an update to multiple steps in a manufacturing process for high power electronic switches is examined for its overall environmental impact. We were able to assign quantifiable data in all major impact categories (e.g. climate change, deforestation, ozone depletion) to all individual manufacturing steps. This enabled us to compare two processes with differing raw material and processing techniques, affirming no major increase in environmental impact. This is the first usage of a Life-Cycle Assessment for manufacturing change evaluation in the power semiconductor field that we are aware of.

Project "ReffiMaL" is funded by the German Federal Ministry of Education and Research.

Key words: Green Electronics, Green Manufacturing, Power Electronics

Green Electronics

With an increasing demand for environmentally friendly and sustainable products by consumers and customers in many industries [1], manufacturers are currently facing the challenge of assessing their products and production techniques for their environmental impact. This can often be a difficult task, because in most industries there is little precedent for a viable approach to both collecting the relevant production data and a sensible method to translate the data into a meaningful statement about environmental impact.

Life Cycle Assessment

A possible solution to this challenge is to determine the environmental impact by using Life Cycle Assessment (LCA). This technique is ISO-standardized [2] and offers the advantage of being universally applicable to products, production methods and services. In a Life Cycle Assessment, only environmental aspects are assessed, economic and social aspects are deliberately excluded.

A Life Cycle Assessment, when carried out according to the ISO standard, is divided into four main phases.

In the first phase, the assessment's goal and the functional unit is defined. The functional unit is a definition of the service that is provided by the assessed product or production method. By defining the functional unit through the provided service, alternative products with the same function become comparable to each other because they can be referenced to a common functional unit. Further, the environmental impact categories are chosen, for example deforestation and global warming

In the second phase, an inventory of all energy and material flows from and to the environment is made for the assessed system. This is called the Life Cycle Inventory (LCI). In a third phase, the LCI then allows to evaluate the contribution of each of its items to the defined environmental impact categories.

In a final phase, the findings of the prior investigations are interpreted. This interpretation should be consistent with the study's scope and intended audience to obtain usable results.

All of the above steps in the creation of a LCA are interdependent, and often a result or finding in a later step leads to a necessary revision of a previous step. This makes LCA an iterative approach, where every iteration brings one closer to a comprehensive and consistent overall assessment.

Extent of LCA in this Study

A full Life Cycle Assessment accounts for all interactions with the environment that can be allocated to the assessed product or service throughout its complete life cycle. This includes everything from gathering resources from the environment, processing these resources into products, waste, energy usage and all interactions with the environment through recycling or deposition of a product at the end of its lifetime. The LCA technique, however, can also be applied to studies of smaller extent. In this paper, an LCA was created for a range of manufacturing steps of a power electronics module. The steps included coating of the modules base plate with a solderable metal layer, application of a solder mask, manufacturing of the solder preform and the soldering process itself.

We chose these specific manufacturing steps for evaluation because they are involved in the creation of the connection between the modules baseplate and the ceramic insulator. This connection is the subject in a publicly funded project called ReffiMaL [3], which we describe in more detail below. We assessed the ecological impact of these steps using LCA methods both before and after the changes that came with the project. The two data sets then enabled us to compare ecological impacts between the standard and modified manufacturing processes and assess how the environmental burden changed with the changes in manufacturing.

Impact Categories and Data Sources

In any LCA study, measured ecological impacts are aggregated in impact categories to reduce the result data to a manageable yet descriptive amount. Impact categories are chosen according to the study's intended purpose and audience. We intend this LCA to give a general overview of the environmental impact. We therefore chose an equally broad range of impact categories.

Impact categories can be divided into three groups, each corresponding to a part of the ecosystem they are taking effect in. Categories for land are *land consumption of arable and forest areas*, *land consumption by soil sealing*, *land use*, and *terrestrial eutrophication*. Categories related to the water cycle and water bodies are *water consumption*, *aquatic eutrophication*, and *acidification*. Related to the atmosphere and air quality are the categories *summer smog*, *stratospheric ozone depletion* and *greenhouse effect* (*global warming*).

Each impact category is assigned an appropriate measure to quantify impacts. Land use is measured in area and time period of use, water in volume. The remaining categories are assigned a chemical substance causing the respective ecological impact. For the greenhouse effect, for example, this would be carbon dioxide (CO_2), the most famous greenhouse gas. All contributions of the investigated process to each category are then expressed in equivalent amounts of the chosen measure or chemical substance, allowing the aggregation of all impacts in a category into a single number. To give an example, methane (CH_4) causes a much higher greenhouse effect per mass than carbon dioxide [4] and must therefore be converted into an equivalent mass of carbon dioxide to be aggregated in the greenhouse effect impact category.

Table 1: Impact categories and representative measures and substances

Impact Category	Measure or substance
Land Consumption – Arable and Forest Areas	m^2a
Land Consumption – Soil Sealing	m^2a
Land Use	m^2
Water Consumption	L
Aquatic Eutrophication	Phosphate (PO_4)
Summer Smog	Ethylene (C_2H_4)
Stratospheric Ozone Depletion	Trichlorofluoromethane (CFC-11, CCl_3F)
Terrestrial Eutrophication	Phosphate (PO_4)
Greenhouse Effect	Carbon dioxide (CO_2)
Acidification	Sulfur dioxide(SO_2)

When performing an LCA, it is very important to have a reliable and accurate data source that provides information on ecological impacts for all substances and forms of energy used in the investigated process. Many commercial and openly accessible databases exist for this purpose, like agri-footprint, USDA LCA Commons Inventory Data, and ecoinvent. We chose to use "ProBas", a database maintained by the German Environment Agency [5]. Using a single, known-good data source for all materials and forms of energy ensures consistent data quality and comparability of results, and is recommended by the ISO standard [2].

Project ReffiMaL

The introduction of renewable energies lead to fundamental changes in the field of electrical energy production, distribution and use [6]. To meet the requirements for climate change, efficient use of resources and sustainable manufacturing, new and innovative solutions for power electronic modules and power semiconductors are needed.

Currently used material systems in power modules are well established and have a well studied, predictable reliability and life time [7]. With the industries focus on reliability and longevity, aspects like material efficiency and environmental effects have long been viewed as secondary. Further, the complexity of the system and the distribution of expertise over multiple manufacturers along the value-added chain have led to only very few innovations in recent times.

The publicly funded project "ReffiMaL" [3] addresses this problem by bringing together manufacturers along the value-added chain as project partners. The partners in this project are Infineon Technologies AG, where the module is assembled, DODUCO Solutions GmbH, who apply a solderable coating to the baseplates, PFARR Stanztechnik GmbH, who manufacture the solder preform, and HSD University of Applied Sciences Düsseldorf, who provide support in material screening and selection, and conduct the Life Cycle Assessment.

Power Electronics Module and Materials

Subject of the investigation was a power electronics module manufactured by Infineon Technologies AG.

Figure 1: Power electronics module, produced by Infineon Technologies AG

The module mainly consists of a baseplate, ceramic insulators, the actual semiconductor chip and electrical conductors. The connection between the baseplate and the ceramic insulator is established by soldering. Figure 2 depicts the resulting material stack for both the standard, well-established process and the modified, more resource efficient and environmentally friendly process.

The modified process features a thinner solderable coating of the base plate (red colored area), which is also selectively applied only to the required parts of the surface. The remaining, bare part of the base plate then naturally acts as a solder stop mask due to its non-wetting behavior, rendering a separate stop mask creation unnecessary. The baseplate itself (light blue colored) consists of a metal-matrix-composite material that incorporates silicon carbide particles in an aluminum matrix (AlSiC).

While in the standard process, the coating was applied in a galvanic process, the new approach in the modified process is to use physical vapor deposition (PVD), which involves less use of chemicals and a high internal rate of recycling for the used metals. This reduces the required amount of material to less than half of what was used in the standard process.

A second change in the modified process is the reduction of solder thickness (yellow colored area) and the incorporation of spacers into the solder preform to prevent excessive tilting of the ceramic insulator during soldering. This reduces the overall amount of material used and omits the need for an extra manufacturing step to apply spacers so the baseplate. By incorporating the spacers into the solder preform, material composition of the solder changes towards a higher content of copper, resulting in higher thermal conductivity [8].

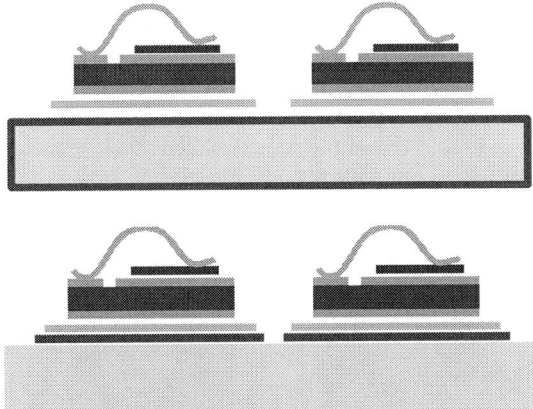

Figure 2: Cross section of material stack

Results

We were able to determine material flows and energy usage of all major manufacturing steps in the creation of the baseplates solder connection to the ceramic insulator, both for the standard and for the modified process. This allowed us to determine the quantity of ecological effect in each chosen environmental impact category and to compare the standard and modified process against each other. Relative changes of impact are listed in table 2.

Table 2: Environmental impact changes

Impact Category	Change from standard to modified Process
Land Consumption – Arable and Forest Areas	97%
Land Consumption – Soil Sealing	91%
Land Use	97%
Water Consumption	101%
Aquatic Eutrophication	86%
Summer Smog	77%
Stratospheric Ozone Depletion	95%
Terrestrial Eutrophication	96%
Greenhouse Effect	98%
Acidification	80%

A reduction in environmental impact in all categories was determined, with the exception of water consumption. The average impact per category of the modified process is 92% of the standard processes impacts, with values ranging from 77% to 101%.

We deliberately refrained from an analysis of the gathered data that determined environmental impact for single manufacturing steps or a single project partner. We also refrained from further condensing the calculated effects in each impact category into a single benchmark number, as this would require designating relative values of importance to the impact categories. This is not allowed according to the ISO standard [2].

Conclusion

The reduction of calculated environmental impact in all categories but one gives us high confidence that we can conclude an actual improvement in environmental protection from this study. The global challenge of climate change through emission of greenhouse gasses is not worsened through the modified process; we calculated a slight reduction in the category of global warming (98%).

Apart from the specific results in this study, it is noteworthy that today's databases for environmental impacts of resources and energies are sufficient to conduct a Life Cycle Assessment in the field of power semiconductors. We hope this encourages others to use Life Cycle Assessment as a method of determining environmental impact of manufacturing in the industry.

Although we did not calculate environmental impact for each project partner, the gathered data is sufficient to do so in future evaluations. It is possible to conduct an LCA to identify manufacturing processes with a prominent environmental impact, to then apply and control targeted measures to lessen the impact.

Acknowledgments

We would like to thank our partners in project ReffiMaL DODUCO Solutions GmbH, Infineon Technologies AG and PFARR Stanztechnik GmbH. We further thank the German Federal Ministry of Education and Research for funding the project through their funding initiative MatRessource [9].

References

[1] Flatters, Paul and Willmott, Michael, 2009: Understanding the Post-Recession Consumer. In: Harvard Business Review, July-August 2009 issue

[2] ISO 14040:2006 Environmental management – Life cycle assessment – Principles and framework
ISO 14044:2006 Environmental management – Life cycle assessment – Requirements and guidelines

[3] https://matressource.de/de/projekte/projekte-alphabetisch/reffimal/

[4] Forster, P., V. Ramaswamy, P. Artaxo, T. Berntsen, R. Betts, D.W. Fahey, J. Haywood, J. Lean, D.C. Lowe, G. Myhre, J. Nganga, R. Prinn, G. Raga, M. Schulz and R. Van Dorland, 2007: Changes in Atmospheric Constituents and in Radiative Forcing. In: Climate Change 2007: The Physical Science Basis. Contribution of Working Group I to the Fourth Assessment Report of the Intergovernmental Panel on Climate Change [Solomon, S., D. Qin, M. Manning, Z. Chen, M. Marquis, K.B. Averyt, M.Tignor and H.L. Miller (eds.)]. Cambridge University Press, Cambridge, United Kingdom and New York, NY, USA.

[5] Prozessorientierte Basisdaten für Umweltmanagementsysteme „ProBas", www.probas.umweltbundesamt.de

[6] Gaviano, Antonello, Weber, Karl and Dirmeier, Christian, 2012: Challenges and Integration of PV and Wind Energy Facilities from a Smart Grid Point of View. Energy Procedia. 25. 118–125. 10.1016/j.egypro.2012.07.016.

[7] Bayerer, Reinhold, Herrmann, Tobias, Licht, Thomas, Lutz, J. and Feller, Marco, 2008: Model for Power Cycling lifetime of IGBT Modules - various factors influencing lifetime.

[8] Chung, D.D.L., 2001: Materials for thermal conduction. In: Applied Thermal Engineering 21 (2001) 1593-1605

[9] https://www.matressource.de/de/

Embedding as a key Board-Level Technology for Modularization and Circular Design of Smart Mobile Products: Environmental Assessment

Karsten Schischke[1], Dionysios Manessis[1], Jakub Pawlikowski[1], Tobias Kupka[3], Thomas Krivec[3], Rainer Pamminger[5], Sebastian Glaser[5], Gerhard Podhradsky[4], Nils F. Nissen[1], Martin Schneider-Ramelow[1], Klaus-Dieter Lang[1,2]

[1] Fraunhofer Institute for Reliability and Microintegration IZM, Gustav-Meyer-Allee 25, 13355 Berlin, Germany, phone: +49.30.46403-156, karsten.schischke@izm.fraunhofer.de
[2] Technische Universität Berlin
[3] AT&S AG
[4] Speech Processing Solutions GmbH
[5] TU Wien

Abstract

Embedding technologies for heterogeneous integration of active and passive components are a promising approach for functional modules on the printed circuit board level. Modularity and embedding related miniaturization are key enablers for some upcoming trends in a Circular Economy: Better serviceability and reparability of devices, reduced material consumption, lower printed circuit board production impacts due to "outsourcing" of complex functionality to smaller modules, resilience of the product platform against component obsolescence for longer product life cycles, potentially increased reliability of embedded components. For a broader framing of the modularity discussion, this paper also gives an outlook on other modular approaches in the field of mobile devices and smartphones in particular. The embedding technology with its miniaturization potential might reduce the total environmental impacts of manufacturing printed circuit board assemblies, supporting Circular Economy strategies. The additional environmental footprint of modularising a smartphone in a way done by Fairphone increases the manufacturing-related greenhouse gas emissions by 4,6%. Embedding – targeting at a better serviceability, but not necessarily simplifying do-it-yourself repairs – only slightly increases the carbon footprint in absolute terms. In case of a digital voice recorder this additional carbon footprint is 0,66 kg CO_2-eq. compared to an add-on of 1,64 kg CO_2-eq. for Fairphone's modularization approach.

Key words: environmental Life Cycle Assessment, carbon footprint, component embedding

1. Introduction

Circular Economy gained a momentum in the recent past as a business concept for a more sustainable and resource-efficient economy. Keeping materials in the economy through extended product lifetimes, better servicing and repair, reuse and remanufacturing and ultimately better recycling are at the core of the circular economy. This is not only a matter of changed business models, but also design adaptations can foster circularity [1].

Modularity is one of the perceived sustainability strategies to achieve more circularity by design [2].

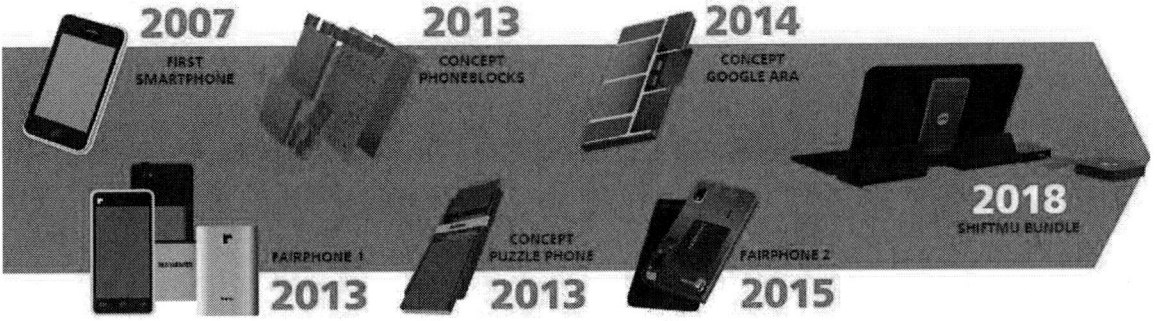

Figure 1: Modular smartphones (selection)

1.1 Modularisation of smart mobile devices

Since 2013 modular design approaches for smartphones in particular (see also [3, 4]) have been disclosed by various players (see Figure 1), the Fairphone 2 being the most prominent example of a modular smartphone. Most of these devices feature a kind of modularity where modules come with individual housings. Rather large connectors connect modules electrically and mechanically. Consumers are supposed to replace modules manually or with very limited use of tools.

Table 1: Environmental impact Fairphone 2 (raw materials acquisition and manufacturing)

Environmental indicator	unit	Fairphone 2	thereof: Modularity components (module housing, board-to-board connectors incl. PCB area)	
Global warming potential	kg CO_2-eq.	35,98	1,64	(4,6%)
Abiotic resource depletion, minerals	g Antimony-eq.	1,48	0,833	(56,4%)
Abiotic resource depletion, fossil	MJ	148,03	17,03	(11,5%)
Humantoxicity	kg DCB-eq.	8,35	0,358	(4,3%)
Ecotoxicity	kg DCB-eq.	0,107	0,0069	(6,5%)

Life cycle assessments show that this kind of modularity comes at additional environmental impacts. Typically the manufacturing of sub-housings, additional PCB board area for connectors and the gold-coated connectors as such increase environmental impacts of the device significantly (see Table 1) [5], and also the material consumption for such designs can be critical from an environmental perspective [6]. The carbon footprint of the device increases by 4,6% or 1,64 kg CO_2-eq. Proske et al. [5] have shown that in case of the Fairphone 2 this additional impact can be compensated easily by an extended lifetime through do-it-yourself repairs. This requires of course, that this DIY philosophy is embraced by the consumers and that the OEM continues to provide spare parts.

A less visable trend is found among the large market leaders in the smartphone business: Increasingly a kind of internal modularity is implemented, where numerous functional modules are connected with the mainboard through flex PCBs and board-to-board connectors. The iPhone X features 14 such connectors on the mainboard (Figure 2). This presumably facilitates professional repairs of the devices, thus being at least partly in support of the circular economy.

What is not found (yet) among mobile devices is modularity on the printed circuit board level, enabled by embedding technology. This approach is the novelty of the product concept presented here and in [7, 8]. Embedding as such can reduce the carbon footprint of a printed circuit board assembly as shown by Kupka et al. [9] for an exemplary design and AT&S' ECP® technology. It remains to be seen throughout this analysis, if this environmental benefit of embedding also materializes for modular product designs.

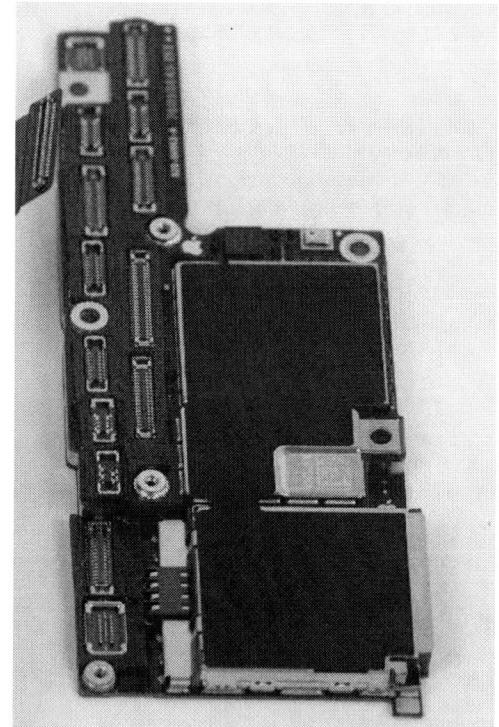

Figure 2: Mainboard connectors iPhone X

1.2 Policy context

EU ecodesign policies currently address product reparability, reusability, upgradeabilty, recyclability and other aspects summarized under the term "material efficiency". Regulation under the European Eco-design Directive [10] and the related standardisation mandate M/543 currently paves the way for product designs which feature circularity criteria. Embedding might be a key technology to comply with these upcoming requirements.

1.3 Contribution of embedding technology to circularity

As such, embedding technologies for heterogeneous integration of active and passive components are a promising approach for splitting a complex printed circuit board into individual modules. This has a four-fold circularity aspect:

(1) A modular design is a viable strategy to mitigate the risk of omponent obsolescence: In case of individual components being discontinued, a resdesign requires only a module redesign then, not a full board redesign, thus a better long-term compatibility of board and modules.

(2) Embedded components tend to be more reliable than surface mounted components, if thermal management is properly implemented.

(3) In case of a malfunction, repair case or if reuse and remanufacturing is intended, a hybrid system with modules assembled on a PCB allows for better replacement of individual functional blocks. A reversible interconnection technology facilitates such replacements.

(4) "Outsorcing" complex circuitries into embedded modules can facilitate the design of a less complex mainboard with fewer layers, thus less manufacturing effort.

This paper analyses the two latter aspects of circular design: How much better needs repairability / reuse to be for a better circularity? And: Can the "outsourcing" of complexity in smaller embedded modules help to reduce the overall system carbon footprint?

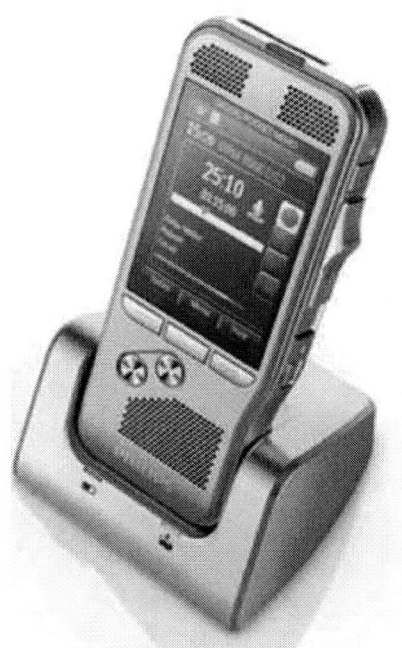

Figure 3: Digital Voice Recorder DPM 8000

2. Scope

The analysis refers to the development of a printed circuit board assembly for a digital voice recorder (Figure 3), split in a backbone and functional modules. Functional modules comprise a Digital Signal Processor Module, a power module and an USB module for Universal Serial Bus connectivity (Figure 4).

Figure 4: USB modules on panel (demonstrator production)

The embedding approach for the DSP module is based on the "face-down" integration of pre-packaged semiconductor components. The DSP module is comprised of 8 copper layers with 6 inner PCB layers and with in total 68 components embedded and surface mounted on top of the module [7].

All 3 modules can theoretically be assembled with a reversible interconnection technology, but momentarily for a demonstrator only the power module and the USB module feature a reversible interconnection technology. The USB module connects with a flex segment to a conventional board-to-board connector. The power module features a spring array on the bottom side and is held in place by two screws (see Figure 5).

Figure 5: Bottom side of power module with spring contacts

Figure 6 depicts a schematic view of spring assembly and materials.

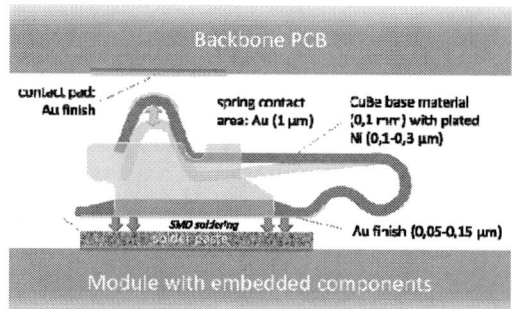

Figure 6: Spring contacts (schematics).

The backbone board, the carrier of these three modules, features the same size as the monolithic device before re-design. The backbone is a 6-layers board. As the DSP circuitry is now in the DSP module, the complexity of the backbone board is reduced and it might be possible to reduce the layer count to 4. In a first demonstrator the backbone board is still produced as a 6-layer board to demonstrate reliably the modularization approach as such, but a proper iterative re-design of the backbone considering audio quality and EMC has the potential to be realized as a 4-layers board. Although not yet demonstrated, such kind of design is taken into account in the assessment to show the evironmental potential of the overall approach. For further details of the modules, the embedding technology and the overall PCB assembly see Manessis et al. [7, 8].

The designs considered for the environmental comparison are the following:

(1) **Standard design**: 6 layer PCB with digital signal processing, USB connectivity, power management all integrated

(2) **6L backbone with 3 modules**: DSP, USB and power modules featuring embedding technologies, assembled with reversible and non-reversible interconnection technologies on the backbone.

(3) **4L backbone with 3 modules**: Same as above, but the backbone requires only 4 PCB layers (theoretical calculation only).

3. Methodology

The assessment of environmental impacts of the embedding approach follows the Life Cycle Assessment methodology applied to greenhouse gas emissions, based on ISO 14.067 [11].

3.1 Functional unit

The functional unit, i.e. the reference point to make technologies comparable, is one unit of a processed printed circuit board for a digital voice recorder, including surface-mounted modules – where applicable – to realise the same functionality.

3.2 System boundaries

This assessment focusses on those aspects, which make a significant difference in the life cycle of embedded modules versus standard PCB assemblies. System boundaries are cradle-to-gate, i.e. from mining of raw materials to manufacturing of the PCB, embedded modules and related interconnection technology.

All the electronics components are excluded from the assessment as these remain largely unchanged, regardless if certain functions are moved to modules or not. It is worthwhile noticing that the PCB(s) and modules (without components) are only a fraction of the carbon footprint of the whole device. Semiconductor components typically feature a higher environmental footprint than the PCB substrate. This is relevant when considering yield loss of a technology. As the design assessed in this paper is on the prototyping level no yield data is available and as such yield losses are not included in the assessment.

The assembly (soldering) of the DSP module is excluded from the assessment as well as this is an integrated step of the anyway required SMD soldering process.

As a worst case approach this analysis does not include the recycling of production waste. As shown by Kupka [12] recycling of industrial waste from PCB processing might result in an overall carbon footprint credit of roughly 5% for saved primary resources.

Table 2: System boundaries

Included in system boundaries
All raw materials and energy from cradle to final product for
• PCB processes
• Embedding processes
• Module connectors
Assembly of embedded components and SMD components assembled onto the modules
Assembly of USB and power module on backbone
Production infrastructure (overhead: pressurized air, water supply, HVAC, chemicals regeneration, wastewater treatment, light / electricity supply, server room, lab)

Excluded from system boundaries
Electrical and electronics components (SMD and embedded)
Assembly of SMD components and the DSP module
Transports supply chain
Use phase
End-of-life of the product (disposal, recycling)

3.3 Data sources and modelling

The analysis of the printed circuit board manufacturing process and AT&S' proprietary ECP® technology for component embedding is an adaptation of Kupka's assessment of the PCB production at AT&S [12]. Secondary data is derived largely from the Ecoinvent database 3.4. The chosen carbon intensity of the electricity supply is that of the EU28 (447 g CO_2-eq./kWh, reference year: 2013) [13]. The chosen electricity source is highly relevant for the overall impacts of PCB production as the differences in energy intensity vary broadly within Europe and even more on a global scale. Production at AT&S in Austria benefits from a high share of renewable energy in the electricity grid mix. The actual carbon footprint of producing the backbone and the modules with embedded components The embedding technology by Fraunhofer IZM are lab scale processes. Energy and material consumption of the embedding process is approximated with industrial data from PCB processes. Data for the electroless immersion silver process, which is part of the Fraunhofer embedding technology, is estimated based on data by Schneider-Ramelow et al. [14] and Sitek [15]. Data for connectors are derived from layout files, assessment of manufacturing processes for (larger) connectors in power applications [16], and internal re-modelling of metal forming and plating processes with the LCA software tool GaBi.

The number of units per panel is estimated, considering an optimized panel layout. The actual demonstrators are produced with less units per panel (see Figure 4) than what would be produced at large scale. The number of units per panel assumed for this analysis are listed in Table 3.

Table 3: Calculated units per panel

	Panel size	units per panel
Backbone PCB	21 x 24 in²	44
USB module	18 x 24 in²	800
Power module	18 x 24 in²	380
DSP module	18 x 24 in²	640

Backbone PCB, USB module and power module are modelled with an ENIG surface finish (electroless Nickel-Gold), although for the demonstrator the USB module is produced with an OSP finish (see Figure 4).

3.4 Environmental indicator

The chosen environmental indicator is the carbon footprint in CO_2-equivalents. This represents the Global Warming Potential of emissions throughout the product life cycle.

4. Results

4.1 Comparison of design alternatives

The carbon footprint of the standard 6-layer printed circuit board (before assembly) of the digital voice recorder is slightly above 1 kg CO_2-eq. The carbon footprint of the whole digital voice recorder is roughly 4,7 kg CO_2-eq. [9], but be aware that this figure has been calculated before the more granular PCB modelling data became available and system boundaries and background data used differs from the data used in this analysis.

The three modules featuring embedding technology add an additional carbon footprint of 0,65 kg CO_2-eq. Impact of connectors for the USB and power module are another 0,007 kg CO_2-eq. In case the reduced complexity of the backbone allows in the end for a reduction of PCB layers, the carbon footprint of the backbone goes down to 0,90 kg CO_2-eq. (see Table 4).

Table 4: Carbon footprint results (cradle-to-gate)

	Carbon footprint (kg CO_2-eq.)		
	Standard design	6L backbone with 3 modules	4L backbone with 3 modules
PCB / backbone	1,01	1,01	0,90
USB module	-	0,17	0,17
Power module	-	0,32	0,32
DSP module	-	0,16	0,16
Connectors	-	0,007	0,007
Totals	**1,01**	**1,67**	**1,57**
For comparison: digital voice recorder (1 unit)	4,69	-	-

The difference between the 6 and the 4 layers board is rather low as the environmental impacts are mainly driven by the epoxy / glass fiber / copper laminate substrate material (core and prepregs), and the gold finish processes, and less so with additional lamination, lithography and etching processes of additional layers. In particular, the ENIG process is the reason, why the impact of the power module is twice as much as of the DSP module and why the USB module despite the smaller size is of a similar carbon footprint as the DSP module.

4.2 Detailed results: DSP module

The individual carbon footprint results per process step for one DSP module are depicted in Figure 7, following the sequence of process steps. The overall carbon footprint of 0,164 kg CO_2-eq. is almost evenly split among the high density interconnect 6 layer substrate required for the routing, the actual embedding processes, and the aggregated overhead for both sub-processes. The initial laser drilling and the lay-up processes include the core and prepreg materials. The lamination press is a time consuming and energy intensive process. Other relevant processes in terms of carbon footprint are the numerous wet chemistry processes for copper plating, viafilling and immersion silver. However, immersion silver is a low impact alternative to the use of gold finishes [14].

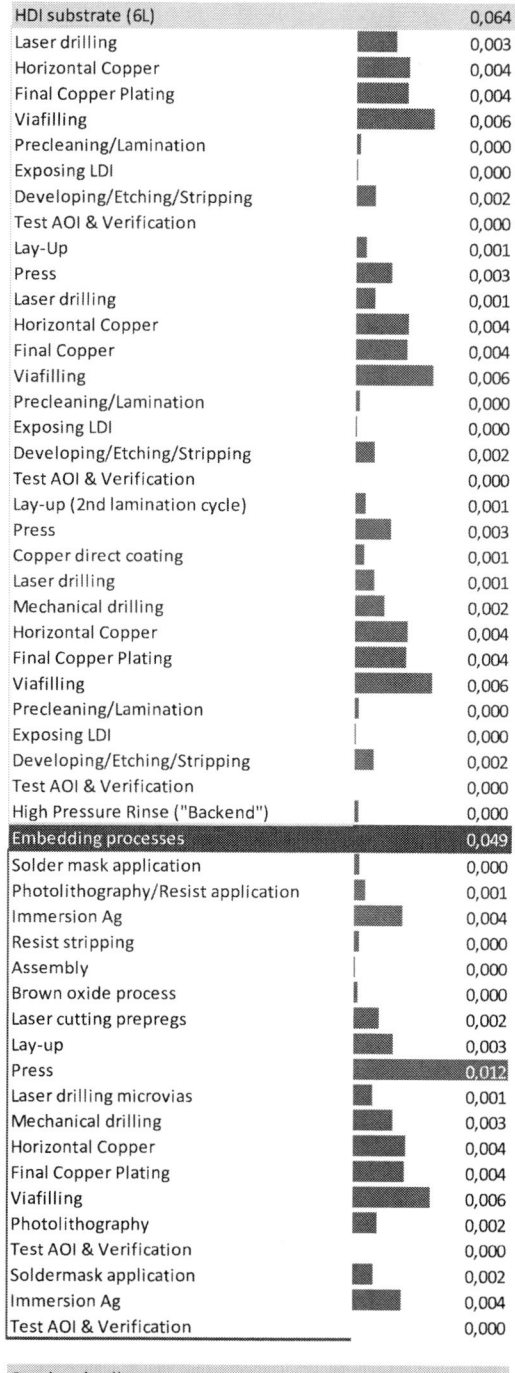

HDI substrate (6L)	0,064
Laser drilling	0,003
Horizontal Copper	0,004
Final Copper Plating	0,004
Viafilling	0,006
Precleaning/Lamination	0,000
Exposing LDI	0,000
Developing/Etching/Stripping	0,002
Test AOI & Verification	0,000
Lay-Up	0,001
Press	0,003
Laser drilling	0,001
Horizontal Copper	0,004
Final Copper	0,004
Viafilling	0,006
Precleaning/Lamination	0,000
Exposing LDI	0,000
Developing/Etching/Stripping	0,002
Test AOI & Verification	0,000
Lay-up (2nd lamination cycle)	0,001
Press	0,003
Copper direct coating	0,001
Laser drilling	0,001
Mechanical drilling	0,002
Horizontal Copper	0,004
Final Copper Plating	0,004
Viafilling	0,006
Precleaning/Lamination	0,000
Exposing LDI	0,000
Developing/Etching/Stripping	0,002
Test AOI & Verification	0,000
High Pressure Rinse ("Backend")	0,000
Embedding processes	0,049
Solder mask application	0,000
Photolithography/Resist application	0,001
Immersion Ag	0,004
Resist stripping	0,000
Assembly	0,000
Brown oxide process	0,000
Laser cutting prepregs	0,002
Lay-up	0,003
Press	0,012
Laser drilling microvias	0,001
Mechanical drilling	0,003
Horizontal Copper	0,004
Final Copper Plating	0,004
Viafilling	0,006
Photolithography	0,002
Test AOI & Verification	0,000
Soldermask application	0,002
Immersion Ag	0,004
Test AOI & Verification	0,000
Overhead - all processes	0,052

Figure 7: Carbon footprint (kg CO_2-eq.) of the DSP module per process step

5. Interpretation and discussion

Just designing certain functionalities into modules, but keeping the PCB backbone as a 6 layer structure results in a significantly increased carbon footprint (+ 66%). Even if the backbone can

be successfully re-designed into a 4 layer PCB the related reduction in carbon footprint is over-compensated by the additional greenhouse gas emissions of the embedding technology for the 3 modules. Reducing the number of modules does not change the picture: The DSP module alone comes with a higher carbon footprint than the delta between a 6 and a 4 layers backbone. Limiting the modules to the USB module and the power module and keeping the digital signal processor directly on the backbone PCB would mean again a higher complexity of the backbone and no chance to reduce the number of layers. Going the opposite direction and to modularize even more does not result in a better environmal assessmen either: In the early stages of the project a fourth module was under discussion, which could have been the audio part of the digital voice recorder. With the above correlation on the one hand of PCB layer number and carbon footprint and on the other hand embedding technologies and carbon footprint a potential further reduction to a 2 layers backbone still would not compensate module production. Theoretically, instead of reducing the number of PCB layers also a size reduction of the backbone could be a viable way to reduce the carbon footprint, but this relates to multiple other design issues and has not been implemented in this project. The functionality integrated in the DSP module corresponds to more than 3 cm² board area on the initial board, see Figure 8.

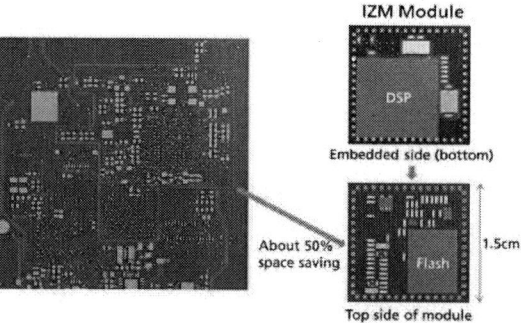

Figure 8: Theoretical design optimization through embedded DSP module concept [7].

A size reduction of the PCB backbone in the range of 3 cm² does not yet break even with the additional impacts of the 3 modules featuring embedding and the connectors, but would reduce the delta significantly.

The use of gold is a typical environmental hot-spot for small electronics products. The upstream processes of gold mining, extraction and refining are very energy intensive and related to harmful emissions. That is why in environmental Life Cycle Assessments gold frequently is an issue. For reliability reasons gold finishes are preferable. This argument is even more relevant for detachable components in modular designs where contact surfaces are supposed to provide good electrical contact even after several mating-unmating cycles.

It has to be stated as of now, that modularization also on the PCB level leads to a higher carbon footprint. As a consequence, this modularization approach only pays off, if it yields an extension of the average lifetime (through better reparability, reuse) across all units put on the market. The modularization related impact of embedding is roughly 0,66 kg CO_2-eq. and as such much lower than the carbon footprint of the modularization parts of the Fairphone 2 (1,64 kg CO_2-eq.). For the digital voice recorder this means that a lifetime extension of more than 15% across all units has to be achieved to compensate the additional impacts. With the right servicing model in place [17] this is likely to be achievable.

6. Conclusion abnd outlook

An environmental Life Cycle Assessment is a useful approach to verify the environmental effects of an intended design or technology change, giving indications of environmental hot-spots. Such an analysis can steer the development of a circular business model in the right, i.e. sustainable, direction.

Embedding technology is an enabler for modularization on the board level and as such can support circular economy strategies. Just as other modularization approaches module production results in an additional environmental impact compared to conventional designs. However, this additional impact is much lower for embedded modules than for other modular concepts. As such, embedding of active and passive components can be a sound strategy to reduce the carbon intensity of products, if it leads to extended lifetimes for parts and products by fostering reparability, reuse and remanufacturing.

The technology and carbon footprint effects are demonstrated on the example of a digital voice recorder. For other mobile applications, such as smartphones, the potential to achieve an overall favourable lifecycle impact reduction through modularization and embedding is higher: Such devices feature a significantly higher total carbon footprint than a digital voice recorder and consequently increasing lifetime of individual units and modules pays off much faster.

7. Acknowledgements

The project sustainablySMART has received funding from the European Union's Horizon 2020 research and innovation programme under grant agreement no. 680640.

References

[1] C.A. Bakker, M.C. den Hollander, E. van Hinte, Y. Zijlstra, "Products That Last - product design for circular business models", Delft, The Netherlands, November 6, 2014

[2] M. Proske, M. Jaeger-Erben, "Decreasing obsolescence with modular smartphones? – An interdisciplinary perspective on lifecycles", Journal of Cleaner Production, Vol. 223, Pages 57-66, March 2019

[3] S. Hankammer, R. Jiang, R. Kleer, M. Schymanietz, "From Phonebloks to Google Project Ara. A Case Study of the Application of Sustainable Mass Customization", Procedia CIRP, Vol. 51, pp. 72-78, 2016

[4] K. Schischke, M. Proske, N.F. Nissen, K.-D. Lang, "Modular Products: Smartphone Design from a Circular Economy Perspective", Proceedings of Electronics Goes Green 2016+, Berlin, Germany, September 2016

[5] M. Proske, M., C. Clemm, N. Richter, " Life Cycle Assessment of the Fairphone 2 - Final Report", Fraunhofer IZM, Berlin, November 2016

[6] K. Schischke, N.F. Nissen, K.-D. Lang, "Getting the Balance Right between Circular Design and the Footprint of Modularity Materials for Smart Mobile Devices", 2019 MRS Spring Meeting, Phoenix, AZ, USA, April 22 - 26, 2019

[7] D. Manessis, J. Pawlikowski, A. Ostmann, K. Schischke, R. Aschenbrenner, M. Schneider-Ramelow, T. Krivec, G. Podhradsky, K.-D. Lang: "Embedding technologies for heterogeneous integration of components in PCBs-an innovative modularisation approach with environmental impact", Proceedings of EMPC 2017 – 21st European Microelectronics Packaging Conference, Warsaw, Poland, September 10-13, 2017

[8] D. Manessis, K. Schischke, J. Pawlikowski, T. Krivec, G. Schulz, G. Podhradsky, R. Aschenbrenner, M. Schneider-Ramelow, A. Ostmann, K.-D. Lang: "Embedding technologies for the manufacturing of advanced miniaturised modules toward the realisation of compact and environmentally friendly electronic devices", Proceedings of EMPC 2019 – 22nd European Microelectronics Packaging Conference, Pisa, Italy, September 16-19, 2019

[9] T. Kupka, G. Schulz, T. Krivec, W. Wimmer, „"Modularization of Printed Circuit Boards through Embedding Technology and the Influence of highly integrated Modules on the Product Carbon Footprint of Electronic Systems", Proceedings of CARE Innovation 2018, Vienna, Austria, November 26-29, 2018

[10] Directive 2009/125/EC of the European Parliament and of the Council of 21 October 2009 establishing a framework for the setting of ecodesign requirements for energy-related products, OJ L 285, 31.10.2009, p. 10–35

[11] ISO/TS 14067:2018, "Greenhouse gases -- Carbon footprint of products -- Requirements

and guidelines for quantification", International Organization for Standardization

[12] T. Kupka, "Comparison of the environmental performance of standard printed circuit board production and embedded component packaging at AT&S, Leoben", master thesis at TU Wien, Vienna, Austria, May 17, 2018

[13] A. Moro, L. Lonza, "Electricity carbon intensity in European Member States: Impacts on GHG emissions of electric vehicles", Transportation Research Part D: Transport and Environment, Vol. 64, pp. 5-14, October 2018

[14] M. Schneider-Ramelow, J. Müller, J.-M. Göhre, H. Reichl, "Immersion Ag as an Alternative in "Green" COB Technology", Proceedings of Electronics Goes Green 2008+, Berlin, Germany, September 8-10, 2008

[15] J. Sitek et al., "Scientific Case Study Reports and Evaluations, ILCD datasets", Deliverable 4.2, project "LCA to go", September 17, 2013

[16] E. A. Olivetti, H. Duan, R. E. Kirchain, "Exploration of Carbon Footprint of Electrical Products – Guidance Document for Product Attribute to Impact Algorithm Methodology", MIT Materials Systems Laboratory, June 2013

[17] R. Pamminger, S. Glaser, W. Wimmer, G. Podhradsky, „Guideline Development to Design Modular Products that meet the Needs of Circular Economy", Proceedings of CARE Innovation 2018, Vienna, Austria, November 26-29, 2018

Recovery of valuable BGA components from used electronic mobile devices and their application in new electronic products.

Marek Koscielski[1], Janusz Sitek[1], Piotr Ciszewski[2], Piotr Dawidowicz[2], Aneta Arazna[1], Kamil Janeczek[1], Wojciech Steplewski[1], Gerhard Podhradsky[3], Roland Ambrosch[4]

1 Tele and Radio Research Institute, Ratuszowa 11 Street, 03-450 Warsaw, Poland

2 Semicon Sp z o.o., Zwolenska 43/43A Street, 04-761Warsaw, Poland

3 Speech Processing Solutions GmbH, Gutheil Schoder Gasse 8-12, 1100 Wien, Austria

4 Pro Automation GmbH, Franzosengraben 10, 1030 Vienna, Austria

* Corresponding Author: marek.koscielski@itr.org.pl

Abstract

A positive trend towards circular economy in many domains has been observed. This approach is based on reuse, repair activities that would lead to new functional device or a retrofitting that would lead to a new product. The standard period of use of mobile electronic devices is short and although the device as whole or partly might be still functional it usually is recycled as whole. It has been proven that recycling might not be the optimal way to source the materials that would be reused as there is possibility of the use of the components almost "as is" to bring them to second life. The presented article focuses on the ball grid array (BGA) chips that are often used in nowadays electronic devices. The sourced chips were retrofitted to a new design devices that has been tested.

Key words: BGA chip recovery, circular economy, reuse, repair, IMC (intermetallic compounds), disassembly, printed circuit boards, reliability tests.

Introduction

The pro-environmental movement of the European legislation is observed. One of the biggest changer up to now was ruling of the RoHS which lead to ban of lead in solder joints also other chemical compounds were restricted. Nowadays the modification and amendment of the chemicals is also coming into force named as RoHS2. The green oriented movements are also putting stress on the Circular Economy. This trends aims on the extension of products life, it can be achieved in several ways eg. recycling, repair or reuse.

As lots of mobile devices are now produced and they often exchanged to newer ones not due to functional drawbacks or malfunctions. The circular economy could be of help here to minimalize the negative effects of the growing waste of electronic equipment. The usage of some chips that are of higher value or components that consist of precious. The production of the new subassembly is not only the used material, but also natural resources, water, transport, energy etc. Re-use requires less of this type of investment, giving a big advantage to the environment. During the research the main goal was to recover the BGA chips. These types of components are very often used today, and their price can be very high. Their re-use has economic justification. This kind of housing is very often used in processor, memory chip or so called SiP (System in Package). As a donor device the digital voice recorder (DVR) was used, the assembled PCB can be seen in figure 1, were two BGA type chips are present. The operation of removal of the component may influence its structure, it is therefore necessary to tests many aspects of this process. Replacement of components is already used in the repair of various systems, but the use of this methodology on a mass scale requires careful analysis of all critical points.

Figure 1. disassembled DVR with marked chips to be recovered

In the article recovery and new device assembly processes are described. There was presented also the results of examination the new device in which recovered chips were used. The authors applied thermal stresses and examination of recovered chip before and after such tests. The X-Ray analysis, intermetallic compounds measurements as well as functional tests of new devices were performed.

As it was already mentioned a DVR was used as a donor of the chips. The device can be quite easily dismantled and was provided by one of the partners Speech Processing Solutions. The above mentioned procedures for desoldering and remanufacturing were followed and the chips ready for reuse were obtained.

Recovery of the chipsdesoldering

During the studies the desoldering station was used to recover the chips. Afterwards a procedure depicted on the figure 2 was followed.

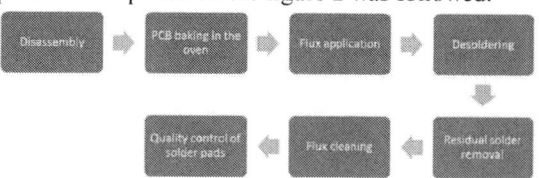

Figure 2. Scheme of recovery of the chip-desoldering process.

During the recovery of the chips that can be of a value here are the list of steps that have to be followed for a successful recovery. As PCBAs are embedded in a housing a process of disassembly has to be followed. This can lead to unscrewing of the fasteners opening of the lids. Those processes were carried manually as we performed the tests on a lab scale. As an alternative of such approach for a faster and a cheaper way might be usage of grinding and automation eg. collaborative robots. This was proven also to work well during the studies. The next steps that follow are quite typical for desoldering that is application of the flux that will lower the surface tension and enable to remove the chip more easily, followed be application of heat by IR or other type of heat transfer devices or ovens. As a next step removal of a residual solder has to follow and then also flux residues are removed. Subsequently quality control of the recovered chip would take place by optical inspection (AOI) and comparison of the recovered components pads and/or X-Ray inspection (AXI) that can assess the quality of the wire bonds. Quality control of recovered components is crucial and can be automated. At today's state of electronics, mass control stations using optical, X-ray and electro-functional techniques would allow the recovery of a significant number of valuable components.

Recovery of the chipsremanufacturing

If the desoldered chip has passed the quality control the next steps that follow are connected with the remanufacturing of the BGA chip. This would lead to reballing, which is illustrated in figure 3, that would enable to solder the chip back on to new device or repaired one.

Not all steps from the presented graph would be mandatory in a higher throughput eg. laser marking but might came useful to differentiate the samples after the process. The recovery starts with the application of a tacky flux on to which solder spheres are placed. To check if all spheres are in place and to check any misalignment, as that might lead to shorts during soldering, an AOI might be used. Also different techniques of application of solder balls have been taken into account and tested see [1]. Soldering takes place in a controlled reflow oven, or other dedicated device that is suitable to provide heat in a control manner.

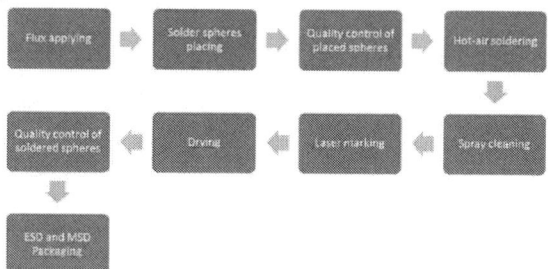

Figure 3. Scheme of recovery of the chip-remanufacturing process.

The subsequent step is cleaning of the flux residue and any other foreign debris. Next step is marking that can be achieved through laser or other techniques are also possible. Drying the chip is also crucial to prevent the components from popcorning defect, which manifests in delamination, then additional steps of inspection are foreseen and as a last step packaging.

Design of a new device

The design of a new device was planned and executed. The device was a mobile alarm clock with the possibility to play the song. The schematics are shown on figure 4.

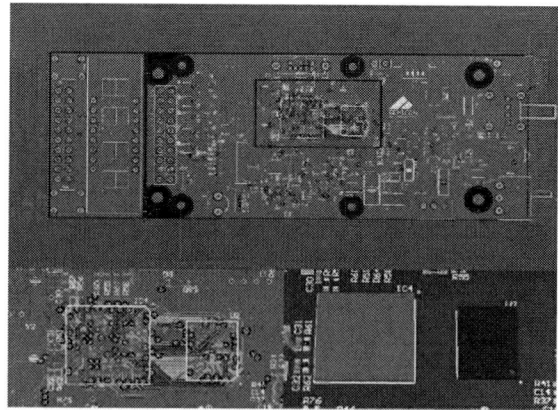

Figure 4. Layout of the produced PCBA

The assembly was made on a standard SMT line with the stencil printer, pick and place machine, and reflow oven at the end. The quality was checked according to the IPC standard, also the Xray images with additional laminography and computer tomography have been done. There were no problems observed during the production also the software had been written to the memory without issues. The assembled PCBA is presented in figure 5 showing a magnification on the chips that were laser marked with "Re" sign to notify the re-cycling of the chip.

Figure 5. Layout of the produced PCBA top and the marking of the recycled chips at the bottom.

Carried out tests

After the device was made, standard reliability tests were performed, which could indicate the weaknesses of the applied component recovery technology. It was agreed that the assembled units will undergo different climatic tests to check the quality of the whole new device and also the re manufactured chips. The units have gone a constant heat at 105°C for 250 h and 500 h. After such tests the Xray images from the recovered chips were made using X-ray control system type NIKON XT V 160.

Another crucial thing that had been examined was the thickness of the Intermetallic Compounds layer (IMC) before (T0) and after thermal stresses (TS 250 h, TS 500 h). The thickness was examined through execution of cross-section using Digital microscope KEYENCE VHX-5000. The joints before and after the climatic stresses were monitored. The sample was etched to enable the measurements of the IMC. IMC thickness was measurements for the three randomly selected solder balls in the first, sixth and eleventh plane of CPU solder balls. The comparison of thickness was performed.

Results of the carried out tests

All the samples were examined and had worked properly, only in one case of a sample after thermal shock test had issues with switch. It was examined that the problematic switch did not withstand the thermal conditions, and the temperature in that case could be lowered as according to the data sheet the maximum temperature should be lowered then 125°C.

The example X-Ray images of recovered components made immediately after new devices assembled are presented in the figure 6. In all cases no malfunction could be observed. Also the examination was repeated after the thermal test (see fig.7) and no cracks, bridging or another soldering anomalies were observed. Solder balls were uniform in size and shape and they were centered in the solder pads. There was observed no differences between quality of soldering before and after thermal tests.

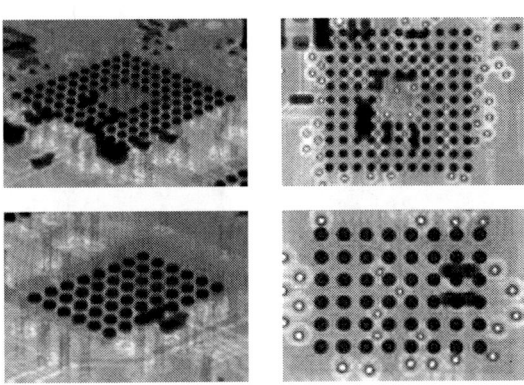

Figure 6. Exemplary Xray images top-down and tilt from CPU (top) and memory (bottom) before the tests.

Figure 7. Exemplary Xray images top-down and tilt from CPU (top) and memory (bottom) after 250h TS test.

Exemplary image from the IMC measurements are presented in figure 8. However, average values of IMC thicknesses are shown in the figure 9. We can observed typical phenomenon of increase of IMC thicknesses with aging time. This result is similar to that obtained in the article [11] and [12] which authors studied, inter alia, influence of aging time on IMC thickness growth. The thicker values were monitored in our work (4.05 µm after 500 h TS) were also quite typical that have been noted in the literature. The authors of the work [12] noticed maximum IMC thickness about 3.2 µm after aging by 500 h at 120 °C. However, the authors of the work [11] observed IMC with a thickness of 5.69 μm after aging 500 h at 150 °C. The differences in the thicknesses of IMC for the same time of aging result from the differences of aging temperature as well as materials of tested solder [9,11].

Figure 8. Exemplary image from the thickness of IMC measurements.

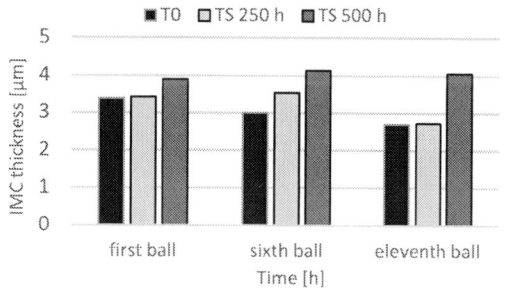

Figure 9. IMC thicknesses for different time of thermal stresses.

Summary

It has been shown that recovery of the valuable chips from mobile devices is feasible. It has been checked that some rigors have to be taken into account as in standard manufacturing process. The usage of circular economy approach might lead to social responsible and minimize the negative effects of obtaining raw materials. The designed devices were functional even after harsh climatic tests. There was observed no defects in the solder joints of tested components after thermal stress during functional tests and Xray inspection. The thickness of intermetallic compounds had comparable values to those contained in the literature. It have been noted an increase of IMC thicknesses with aging time growth. Recovered components, can be re-soldered after the regeneration process. With the current production and consumption of components, it allows a significant saving of natural resources and reduction of waste production.

Acknowledgements

The research leading to these results has received funding from the European Union's Horizon 2020 research and innovation programme under grant agreement No 680604.

References

[1] J. Sitek, M. Koscielski, A. Arazna, K. Janeczek and W. Stęplewski, "Investigations of BGA components' balls remanufacturing techniques for Circular Economy applications," *2018 7th Electronic System-Integration Technology Conference (ESTC)*, Dresden, Germany, 2018, pp. 1-6

[2] J. Sitek, M. Koscielski, A. Arazna, W. Steplewski, K. Janeczek, "Challenges of reuse and remanufacturing of modern chips in smart mobile devices", Electronics Goes Green 2016+, 2016, ISBN: 978-1-5090-5208-0, DOI: http://dx.doi.org/10.1109/EGG.2016.7829846.

[3] European Commission, „Closing the Loop - an EU Action Plan for the Circular Economy", Brussels, (2015)

[4] J, Korhonen, C. Nuur, A. Feldmann, S. Eshetu Birkie, "Circular economy as an essentially contested concept", Journal of Cleaner Production, Volume 175, 20 February 2018, Pages 544-552, https://doi.org/10.1016/j.jclepro.2017.12.111.

[6] I. Stobbe, H. Griese, H.Potter, H. Reichl, L. Stobbe, "Quality Assured Disassembly of Electronic Components for Reuse", 2002 IEEE International Electronics and the Environment Symposium;

[7] J. Sitek et al., "Investigations of temperature resistance of memory BGA components during multi-reflow processes for Circular Economy applications," 2017, 21st European Microelectronics and Packaging Conference (EMPC) & Exhibition, Warsaw, 2017, pp. 1-7. doi: 10.23919/EMPC.2017.8346853

[8] "sustainablySMART", Sustainable Smart Mobile Devices Lifecycles through Advanced Re-design, Reliability, and Re-use and Remanufacturing Technologies, Project ID: 680604, Available: https://cordis.europa.eu/project/rcn/198769_en.html

[9] J. Sitek et al., "The influence of materials and technological parameters on IMC growth speed and thickness at solder joints in SMT for special applications", 42nd International Microelectronics and Packaging IMAPS Poland 2018 Conference, September 23-26, 2018, Gliwice, Poland.

[10] J. Sitek et al., "Investigations of BGA components' balls remanufacturing techniques for Circular Economy applications,", 7th Electronics System-Integration Technology Conference, September 18-21, 2018, Dresden, Germany, 2018

[11] T. An, F. Qin, "Effect of intermetallic compound microstructure on the tensile behavior of Sn3.0Ag0.5Cu/Cu solder joint under various strain rates", Microelectronics Reliability,, vol. 54, no. 5, 2014. Pp. 932-938

[12] Z.L. Li, L.X. Cheng et al., "Effects of joint size and isothermal aging on interfacial IMC growth in Sn-3.0Ag-0.5Cu-0.1TiO2 solder joints" Journal of Alloys and Compounds, vol. 697, no. 3, 2017, pp. 104-113.

Challenges of SMT assembling on biodegradable PCB substrates

Attila Géczy, István Hajdu, László Gál, Csaba Norbert Barna, Márk Kovács, Gábor Harsányi

Budapest University of Technology and Economics, Dept. of Electronics Technology
Budapest, Egry J. u. 18. H-1111

gattila@ett.bme.hu

Abstract

With the increase of e-waste and the amount of produced commercial electronics, which can be considered as future electronic waste, the processing of this waste, and the question of possible alternative substrates raise alarming questions. Bio-based and biodegradable substrates emerge for substituting the conventional materials with environmentally friendly solutions. Our paper presents investigations into assembling with reflow soldering on printed circuit boards (PCBs) fabricated from bio-based and biodegradable substrates (PLA, CA), with the focus on surface mount technology (SMT) and the thermal load, which affects the substrates heavily. In the paper the solderability on such biodegradables is investigated from lead-free reflow temperatures, also from the aspect of reduced temperatures, with different heating factor values. It was found that the substrates are very susceptible to high thermal loads.

Key words: biodegradable, bio-based, PCB, substrate, reflow

Introduction

With the increase of e-waste and the amount of produced commercial electronics (which can be considered as possible future electronic junk), the processing of this waste, and the question of possible alternative substrates raise alarming questions. Bio-based and biodegradable substrates emerge for substituting the conventional materials with environmentally friendly solutions. While the topic is widely researched, is basics are still not established well in the literature or in actual production.

There are regulations worldwide, which aim to tune electronics manufacturing processes to be more friendly to the environment. The goals of such regulations point to reducing the hazardous substances (e.g. the RoHS directive, the widespread regulations on leaded solder alloys [1-2]), but the Flame-Retardant NEMA grade 4 (FR4) material is still the main substrate for the Printed Circuit Board industry. To find substitutes for common, epoxy-glass fiber based substrates, novel application of bio-based epoxies must be considered. Bio-based epoxies or different biodegradables are in the focus of development and research: the introduction of these substrates to electronics manufacturing could help to decrease the harmful impact of electronics assemblies. On the other hand, the application of bio-based epoxies, or biodegradable plastics are not straightforward: such substrates perform considerably weaker during regular SMT assembly from the aspect of thermal and mechanical properties.

To sum up recent advances in the field, it is important to show the advances of different research groups. Staat et al. investigated biodegradables from the aspect of mechanical and electrical parameters [3-7]. The team followed work recently by investigating the applicability of biodegradables (Polylactic acid, PLA; Cellulose acetate CA) generally as a PCB substrate, where the idea is to combine different biodegradables and additives for substrate applications. [8,9] Biodegradables were extensively investigated as a substrate with dielectric parameters by Ohki et al. [10, 11]. Mattana et al. also considered the use of PLA [12] for sensor applications [13]. Antennas for 2.45 GHz application were constructed from PLA [14], and CMOS research was also extended with such substrates [15]. Schramm et al. dealt with the aspect of the fabrication of such materials [16], Baecker thoroughly investigated the degradation [17]. Aspects of future application [18], electrochemical migration failure/reliability concerns on the surface of biodegradables [19], carbon-fiber reinforced, glucose- based composites [20] performance modeling [21] further sensor applications [22-24] and the possibility of technology adoption to robotics was also in the focus of researchers involved with biodegrables in electronics [25].

More recently plant based PCB were produced by Vijay Kumar Guna et al. [26]; Unda et al. and Adnan et al. also investigated biodegradable substrates for sensor applications [27] and solid state devices [28]; Sun investigated doped, conductive PLA for the application in microsystems [29]; Boussatour used PLA for microwave applications in their researches.[30]. Chen et al. presented UHF RFID platform fabricated on 3D-printed biodegradable palstics. [31] Zhang et al. worked with polycaprolactone (PCL) with conductive Fe fillers in order to present a micropatterned Fe-PCL interconnect. [32] Lee et al. investgated paper-substrate based organic field-effect transistors for implementations in degradable biomaterials. [33]

Figure 1: SMT assembled printed circuit board on PLA substrate and the deformation of the substrate after soldering. [9]

As the recent listing of literature presents, the topic is widely investigated. Also it can be said, that the PLA and the CA materials are in the main focus of the scientific community. Heat sensitivity and deformations along reflow soldering processes are also reported [9] – this aspect is investigated in the paper as a challenge of the biodegradable substrates.

Experimental

In our previous studies, we presented progress on subtractive PCB copper patterning, and investigated low temperature lead-free soldering on such boards, with a road to produce commercial electronics on biodegradables.

In the current paper we focus on the aspects of reflow soldering steps on Polylactic Acid (PLA), Cellulose Acetate (CA) PCBs. For the experiments we set up a special test board with subtractive copper patterns (~10 μm thickness) prepared for surface mounted resistors in the dimensions of 1206 down to 0402. The boards were positioned on sample holders to avoid excessive lateral deformation (melting) and any resulting contamination of the oven. The parameters of the used materials are presented in Table 1. The parameters reveal, that the use of such materials in general SMT assembling is problematic due to the heat sensitive nature of the substrates. The paper aims to highlight the problems arising at usual lead-free reflow processes.

For reflow, we used an Asscon Quicky Vapour Phase Soldering (VPS) oven [34] with so called "heat level" temperature control, where the heat transfer medium was Galden LS230 with the boiling point temperature of 230°C. Standard lead-free tin-silver-copper paste was used (SAC305). VPS method was

chosen due to its uniform heating. The profile was set to minimize thermal load on the boards.

Table 1: Used Material parameters.

CA - Material properties			
Mechanical/Thermal Properties	*Value*	*Dimension*	*Standard*
Density	1.29	g/cm³	ISO 1183
Tensile Strength	62.5	MPa	ISO 527
Melting Temperature	180-190	°C	ISO 3146-C
Glass Transition Temperature	104	°C	ISO 306

PLA - Material properties			
Mechanical/Thermal Properties	*Value*	*Dimension*	*Standard*
Density	1.55	g/cm³	ISO 1183
Tensile Strength	32	MPa	ISO 527
Melting Temperature	140-160	°C	ISO 3146-C
Glass Transition Temperature	105	°C	ISO 306

Afterwards a second approach was applied with an experimental VPS oven with Galden HT170, with the boiling point (maximum) temperature of ~170°C. For the soldering we have used 58Bi42Sn (138°C melting point) alloy. Three profiles were used, where the heating was set according to standard VPS heating (step function converting to exponential nature). The evaluated profiles (P1-P2-P3) had the approximate heating factor [35] of 390, 640 and 840 [s°C] respectively. The resulting problems are evaluated with optical analysis and destructive cross-section tests.

Results

Initial soldering results in high temperature VPS oven revealed excessive deformation of PCB with deformation observable on the traces, the pads, under and around the components. Figure 2 reveals serious deformation along the components from 1206 to 0402. The traces are deformed visibly, also the boards deformed heavily on the sample holder, which resulted in unacceptable results from aspect of applicability. While CA has higher melting point and glass transition temperature, heavy deformation and voiding was inspected in CA substrates as well. It is apparent that neither of the substrates are eligible for soldering in direct, lead-free reflow soldering temperatures. After the visual qualitative inspection, further quantitative inspection were omitted.

0603 resistors were evaluated on PLA substrates according to the deformations and further failure points, caused by the thermal stresses with the three reduced temperature profiles. On selected occasions it was found that the traces and the pads were lifted from the substrate. This was observed mainly during P2 and P3 profiles. This failure is presented in Figure 3.

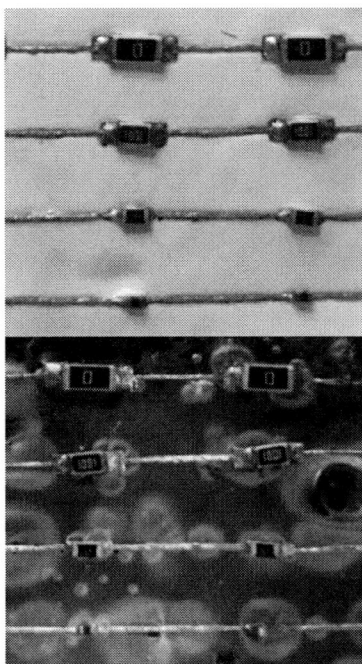

Figure 2: Serious deformation and damage of substrates on PLA and CA (top and bottom respectively) substrates after reflowing SM components (dimensions: 1206, 0805, 0603, 0402) with vapour phase soldering in a heat-level controlled VPS oven (parameters: Galden: LS230; Tmax=230°C, solder alloy: SAC305).

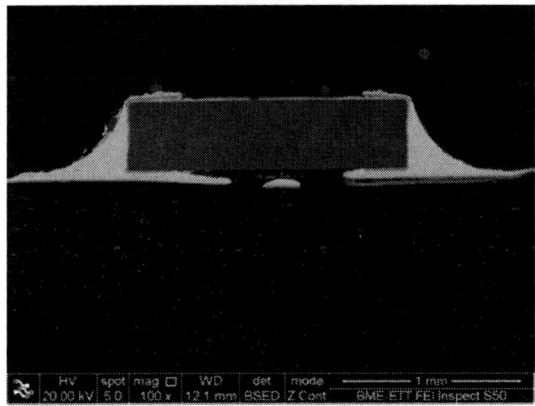

Figure 3: Typical delamination failure of top copper layer with P2 profile in VPS on PLA (Parameters: Galden: HT170; Tmax=170°C, solder alloy: Bi42Sn58)

Further failures were also investigated, where the copper adhesion had no problems. Figure 4. shows a typical case, where the substrate sags under the component, and the vertical deformation was a considerable 180,5 µm compared to the edges of the pads (practically the plane of the substrate).

The sagging was intercepted with the following method, ten components were analyzed profile to profile according to the distance of the bottom side of the component and the depth of the pit. The results are presented in Figures 5 and 6.

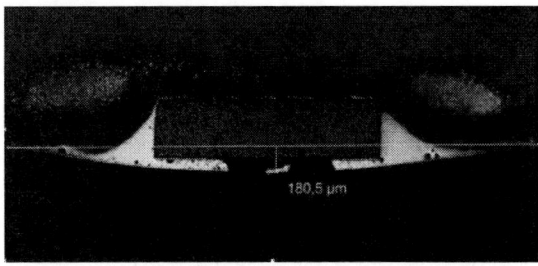

Figure 4: Sagging of substrate under 0603 SMD chip-size component after soldering in a vapour phase reflow oven with P2 profile on PLA (Parameters: Galden: HT170; Tmax=170°C, solder alloy: Bi42Sn58)

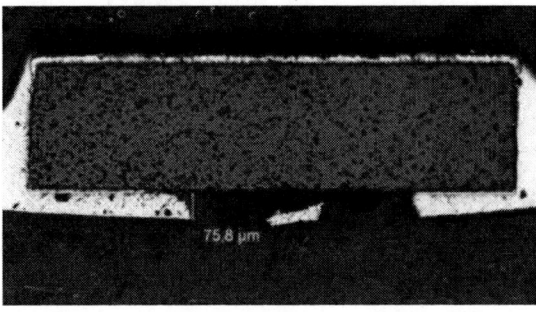

Figure 5: Sagging of substrate under 0603 SMD chip-size component after soldering in a vapour phase reflow oven with reduced temperatures on PLA (Parameters: Galden: HT170; Tmax=170°C, solder alloy: Bi42Sn58)

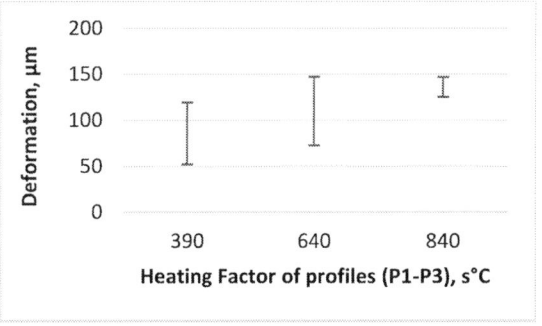

Figure 6: Deformation values: distance from the bottom of the component to the bottom of the sagged pad in the substrate according to thermal load on the assemblies.

It is apparent, that with the higher thermal load (P1 to P3) the sagging is more significant. It is interesting to see, that the min-max deviation is changing according to the increase of the load. For the P3 profile, the deviation is decreased, possibly due to the fact, that in the sample holder, the substrate reached a maximum of its possible deformation due to the component. The distances presented in Figure 6 are also containing the thickness of the copper layer; the thickness of the solder layer is depending on the sag of the substrate, but for further elaboration of the sagging, modelling of the phenomenon would be recommended. Without reinforcement, even fast profiles with reduced reflow temperatures cause problems on PLA substrates. It is interesting to see, that with small heating factor values, such as 390, 640 and 840 [s°C], very significant deformation was observed. Calculating with the melting temperature of the PLA, it is recommended for PLA based PCBs, to present maximum reflow temperatures around 140-150 °C. This aspect was successfully implemented in our previous paper [36], where the heating factor was limited to 460, 280, 252, 146 [s°C], respectively. These values are in line with more recent heating factor investigations [35], but considerably lower than usual reflow soldering values used in common reflow soldering, emphasizing serious limitations of the biodegradables.

Results

It was shown, that both PLA and CA is not recommended as PCB substrate for reflow soldering with usual lead-free temperatures, due to the excessive damage on the substrates. It was furthermore presented on the more sensitive (thus from the aspect of the problem, more representative) PLA, that even with reduced reflow temperatures, the possibility of copper delamination and sagging of the substrate underneath the components is observable. The sagging tends to a value, where the effect possibily saturates due to the fixing of the sample. For future work, the modelling of the sagging is suggested, which might be used to catalyze the investigation of reinforcement application enhancement of the substrates. Also additions in a PLA composite [9] might be a proper way to enhance the substrates, and ease the challenges presented in the paper.

Acknowledgements

The research reported in this paper was partially supported by the Higher Education Excellence Program of the Ministry of Human Capacities in the frame of the Biotechnology research area of Budapest University of Technology and Economics (BME-FIKP-BIO).

References

[1] Zs. Illyefalvi-Vitéz, J. Pinkola, G. Harsányi, Cs. Dominkovics, B. Illés, L. Tersztyánszky, Present Status of Transition to Pb-free Soldering, Proceedings of 28th IEEE-ISSE conference 2005, pp. 72-77.

[2] Illyefalvi-Vitéz, Krammer O, Pinkola J "Testing the Impact of Pb-free Soldering on Reliability" ESTC Dresden, Germany, pp. 468-472. (2006)

[3] Staat, A., Hohlfeld, T., Standau, T., Weimann, I., Bauer, R., Harre, K., Substrates based on renewable resources for printed circuit boards, IEEE ISSE (2014), pp. 50-53.

[4] Staat A, Vogt M, Harre K, Bauer R, Effect of Incorporation of Different Additives in sustainable Polymers on Selected Electrical and Mechanical Properties, IEEE ISSE, 2015. 21-25.

[5] Albrecht Georg Staat, Hardi Köhler, Maximilian Vogt, Reinhard Bauer, Kathrin Harre , Effect of Incorporation of Additive Combinations in Polylactic Acid on Selected Electrical Properties, European Polymer Federation Congress, Germany, 2015.

[6] Albrecht Georg Staat, Rico Mende, Rico Schumann, Kathrin Harre, Reinhard Bauer, Investigation of Wiring Boards Based on Biopolymer Substrates, 39th International Spring Seminar on Electronics Technology "Printed Electronics and Smart Textiles", At Pilsen, CZ, DOI: 10.1109/ISSE.2016.7563165 · [8] Lungen S, Klemm A, Wohlrabe H, Evaluation of the quality of SMDs according to vacuum vapour phase soldering, 38th IEEE ISSE, (2015.) pp. 218 – 222.

[7] Albrecht Staat; Kathrin Harre; Reinhard Bauer, Materials made of renewable resources in electrical engineering, 40th International Spring Seminar on Electronics Technology (ISSE), 2017. pp. 1-5.

[8] Henning, C., Schmid, A., Hecht, S., Harre, K. and Bauer, R. (2019) "Applicability of Different Bio-based Polymers for Wiring Boards", Periodica Polytechnica Electrical Engineering and Computer Science, 63(1), pp. 1-8. doi: https://doi.org/10.3311/PPee.13431.

[9] Carolin Henning, Anna Schmid, Sophia Hecht, Cindy Rückmar, Kathrin Harre, Reinhard Bauer, Usability of Bio-based Polymers for PCB, IEEE ISSE 2019. (Accepted, in press.)

[10] Ohki, Y., Hirai, N.; Electrical Conduction and Breakdown Properties of Several Biodegradable Polymers, IEEE Transactions on Dielectrics and Electrical Insulation, 2007 (14), 1559 – 1566.

[11] Hirai, N., Ishikawa, H., Ohki, Y.; Electrical Conduction Properties of Several Biodegradable Polymers, Annual Report Conference on Electrical Insulation and Dielectric Phenomena, 2007.

[12] Mattana G., Briand G., Marette A., Vásquez Q. A., F. de Rooija N., Polylactic acid as a biodegradable material for all-solution-processed organic electronic devices, Organic Electronics, Volume 17, February 2015, pp. 77–86.

[13] Vásquez Quintero, A., Frolet, N., Marki, D., Marette, A., Mattana, G., Briand, D., de Rooij N.F., Printing and encapsulation of electrical conductors on polylactic acid (PLA) for sensing applications, 2014 IEEE 27th International Conference on Micro Electro Mechanical Systems (MEMS), 2014 pp. 532 - 535.

[14] Haroon, Ullah, S., Flint, J.A., Electro-textile based wearable patch antenna on biodegradable polylactic acid (PLA) plastic substrate for 2.45 GHz, ISM band applications, 2014 International Conference on Emerging Technologies (ICET), 2014, pp. 158 – 163

[15] Hwang, S.-W., Song, J.-K., Huang, X., Cheng, H., Kang, S.-K., Kim, B. H., Kim, J.-H., Yu, S., Huang, Y. and Rogers, J. A. (2014), High-Performance Biodegradable/Transient Electronics on Biodegradable Polymers. Adv. Mater., 26: 3905–3911.

[16] Schramm, R. Reinhardt, A., Franke, J., Capability of Biopolymers in Electronics Manufacturing, IEEE ISSE (2012) pp. 345-349.

[17] Baecker Matthias, Schusser Sebastian, Leinhos Marcel, Poghossian Arshak, Schoening Michael J., Sensor system for the monitoring of degradation processes of biodegradable biopolymers, Sensors and Measuring Systems 2014; 17. ITG/GMA Symposium; 2014.

[18] Mihai Irimia-Vladu, "Green" electronics: biodegradable and biocompatible materials and devices for sustainable future, Chem. Soc. Rev., 2014, 43, pp. 588-610

[19] B. Medgyes, I. Hajdu, R. Berenyi, L. Gal, M. Ruszinko, G. Harsanyi, Electrochemical migration of silver on conventional and biodegradable substrates in microelectronics, 37th International Spring Seminar on Electronics Technology (ISSE), 2014, pp. 256 – 260.

[20] Niedermann P., Szebényi G., Toldy A.: Characterization of high glass transition temperature sugar-based epoxy resin composites with jute and carbon fibre reinforcement. Composites Science and Technology, 117, 62-68 (2015)

[21] A. Elkamel, L. Simon, E. Tsai, V. Vinayagamoorthy, I. Bagshaw, S. Al-Adwani, K. Mahdi, Modeling the Mechanical Properties of Biopolymers for Automotive Applications, Proceedings of the 2015 International Conference on Industrial Engineering and Operations Management Dubai, United Arab Emirates (UAE), March 3 – 5, 2015

[22] Ramendra K. Pal, Ahmed A. Farghaly, Congzhou Wang, Maryanne M. Collinson, Subhas C. Kundu, Vamsi K. Yadavalli, Conducting polymer-silk biocomposites for flexible and biodegradable electrochemical sensors, Biosensors and Bioelectronics, Volume 81, 15 July 2016, 294–302

[23] Srinivasulu Kanaparthi and Sushmee Badhulika, Solvent-free fabrication of a biodegradable all-carbon paper based field effect transistor for human motion detection through strain sensing, Green Chem., 2016, 24th March 2016, DOI: 10.1039/c6gc00368k

[24] Andrea Luvisi, Alessandra Panattoni, Alberto Materazzi, RFID temperature sensors for monitoring soil solarization with biodegradable films, Computers and Electronics in Agriculture, Volume 123, April 2016, Pages 135–141

[25] Jonathan Rossiter, Jonathan Winfield, Ioannis Ieropoulos, Here today, gone tomorrow: biodegradable soft robots, Electroactive Polymer Actuators and Devices (EAPAD), Proc. of SPIE Vol. 9798, 2016,

[26] Vijay Kumar Guna, Geethapriya Murugesan, Bhuvaneswari Hulikal Basavarajaiah, Manikandan Ilangovan, Sharon Olivera, Venkatesh Krishna, and Narendra Reddy, Plant-Based Completely Biodegradable Printed Circuit Boards, IEEE TRANSACTIONS ON ELECTRON DEVICES, VOL. 63, NO. 12, DECEMBER 2016, pp. 4893-4898.

[27] Kassan Unda, Ali Mohammadkhah, Kwang-Man Lee, Delbert E. Day, Matthew J. O'Keefe, Chang-Soo Kim, Sensor substrates based on biodegradable glass materials, IEEE SENSORS, 2016. pp. 1-3.

[28] Shihab Adnan; Kwang-Man Lee; Mohammad Tayeb Ghasr; Matthew J. O'Keefe; Delbert E. Day; Chang-Soo Kim, Water-Soluble Glass Substrate as a Platform for Biodegradable Solid-State Devices, IEEE Journal of the Electron Devices Society, 2016, Volume: 4, Issue: 6, pp. 490 - 494

[29] Zhumei Sun and Luis Fernando Velásquez-García, Monolithic FFF-Printed, Biocompatible, Biodegradable, Dielectric-Conductive Microsystems, JOURNAL OF MICROELECTROMECHANICAL SYSTEMS, VOL. 26, NO. 6, DECEMBER 2017, pp. 1356-1370

[30] G. Boussatour, P.-Y. Cresson, B. Genestie, N. Joly, and T. Lasri, Characterization of Biodegradable and Biosourced Polylactic Acid (PLA) Substrate in a Wide Frequency Range (0,5 - 26 GHz), IEEE MTT-S International Microwave Workshop Series on Advanced Materials and Processes (IMWS-AMP 2017), 20-22 September 2017, Pavia, Italy

[31] Chen, X., He, H., Ukkonen, L., & Virkki, J. (2018). 3D-Printed Eco-Friendly and Cost-Effective Wireless Platforms. 2018 7th Electronic System-Integration Technology Conference (ESTC). doi:10.1109/estc.2018.8546358

[32] Zhang, T., Tsang, M., Du, L., Kim, M., & Allen, M. G. (2019). Electrical Interconnects Fabricated From Biodegradable Conductive Polymer Composites. IEEE Transactions on Components, Packaging and Manufacturing Technology, 1–1. doi:10.1109/tcpmt.2019.2905154

[33] Lee, C.-J., Chang, Y.-C., Wang, L.-W., & Wang, Y.-H. (2019). Biodegradable Materials for Organic Field-effect Transistors on a Paper Substrate. IEEE Electron Device Letters, 1–1. doi:10.1109/led.2018.2890618

[34] Machine learning-based prediction of component self-alignment in vapour phase and infrared soldering, Oliver Krammer , Péter Martinek , Balazs Illes , László Jakab, Soldering & Surface Mount Technology, Volume: 31 Issue: 3, 2019

[35] Petr Veselý, Eva Horynová, Jiří Starý, David Bušek, Karel Dušek, Vít Zahradník, Martin Plaček, Pavel Mach, Martin Kučírek, Vladimír Ježek, Milan Dosedla, (2018) "Solder joint quality evaluation based on heating factor", Circuit World, Vol. 44. Issue: 1, pp.37-44, doi:10.1108/CW-10-2017-0059.

[36] A. Geczy, T. Garami, B. Kovacs, D. Nagy, L. Gal, M. Ruszinko, I. Hajdu, Soldering tests with biodegradable printed circuit boards, 2013. IEEE 19th International Symposium for Design and Technology in Electronic Packaging (SIITME); doi:10.1109/SIITME.2013.6743641

Flip Chip Assembly on Coreless Substrate Challenge with Die Bond Solution

James Su, Yu Po Wang, Mike Tsai, KC Tsai, David Lai

Siliconware Precision Industries Co. Ltd. (SPIL)
No. 123, Sec 3, Chung Shan Rd., Tantzu, Taichung, Taiwan

Tel: 886-4-25341525 ext. 5075; E-mail: Jamessu@spil.com.tw

Abstract

Flip Chip Packages are the popular packages in many different applications. Flip Chip can provide the ideal solution for low I/O to high I/O, high electrical performance demand in where high frequency, high speed are required. By using coreless substrate with embedded fine-trace substrate (ETS) technologies to achieve package miniaturization requirement.Comparing with conventional substrate, coreless substrate technologies eliminate the substrate core, and utilize substrate layers to interconnect the chip and the PCB board. It brings benefits not only with low Z-height, lightweight, but also offers short interconnection distance and good signal integrity. The ETS coreless technology is a promising solution for the next generation substrate. To approach these requirements, we would like to discuss the suitable process for laser assisted bonding (LAB), thermal compression bonding (TCB) combining with molding underfill (MUF) technology and comparing with traditional mass reflow (MR).

A Flip Chip Package and ETS coreless substrate combination are used for this paper experiment. The major challenge is thinner ETS coreless substrate will induce more warpage concern because of no rigid core material as supporting structure, thus ETS coreless substrate with less stiffness, is easier to bring handling issue during assembly manufacturing process. Thermal compression bonding could offer less stress to reduce CTE mismatch between silicon chip and coreless substrate. The study result of warpage comparison among TCB, LAB and mass reflow process, shows that the TCB could provide less warpage. Furthermore, TCB technology is popular used for thin core, coreless and advance silicon node product field. For LAB, it can provide the better throughput than TCB, and also provide more accuracy fine pitch bonding than MR.

Keywords- Laser-Assisted Bonding, thermal compression bonding, mass reflow, Flip Chip, Coreless Substrate

Introduction

Flip chip technologies have received growing attention for a mainstream interconnection solution in the semiconductor assembly field due to their ability to achieve increasing device functionality and package miniaturization. The flip chip technology is also related to next generation of high performance semiconductor package structures such as HBW-POP, System in Package (SiP) or some complex designs. Therefore, there are a lot of different demands for advanced flip chip packages and technologies from semicondustor industry.

The substrate cost is high and the manufacturing process of core shoud be the major cost for the substrate. To remove the core or make it thiner, the substrate with die attach will become more challenging. The conventional mass reflow (MR) process is the most popular used for die attaching in flip chip and surface mount technology (SMT) technology field because the soldering have high productivity and high self-alignment effect of solder joint. However, there is limitation of bump pitch with mass reflow (MR) process. The thermal compression bonding (TCB) process is really slow UPH for mass production compare with mass reflow (MR). So far, we need to provide a solution is under study for high volume production, precision bump join and reduce the die attaching stress. Laser assisted bonding (LAB) should be the good solution for these problems solving.

In this paper, the warpage is the always challenges for the flip chip die attached especially for the mass reflow (MR) with die bond on coreless substraste. Using laser assisted bond can prevent thermal budget and reduce the warpage effienticlly. Refer to Figure 1 for the substrate trend photos.

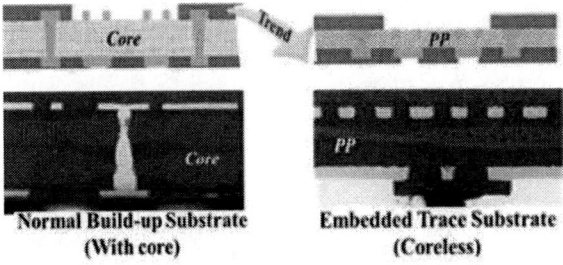

Figure 1: Normal Substrate Trend to ETS Coreless Substrate Photos

(Normal Build-up Substrate (With core) / Embedded Trace Substrate (Coreless))

Test Vehicle Decription

For the flip chip package and small form factor requirement, the test vehicle (TV) will do the comparsion between mass reflow (MR) and laser assisted bonding (LAB) for the package size 50 x 50 mm. The basic test vehicle (TV) information was shown in Table 1

Table 1: Mass Reflow vs. Laser Assisted Bonding

PKG	PKG size (mm)	Die			Sub.	
		Bump type	Size (mm)	THK (mil)	Bump pad	THK (mm)
FCBGA	50*50	Cu-pillar	25*30	31	SOP	1.5

Process Flow

First at all, the major assembly procedure and process flow sequences were defined in Figure 2. The fine pitch of flip chip die attach was used mass reflow process to join with coreless substrate. The detail process includes the flux dipping, die bond on OSP surface finished substrate, automated optical inspection, thermal reflow for solder melting and flux cleaning process; the process flow shows in Die Attach~ Reflow step of Figure 2.

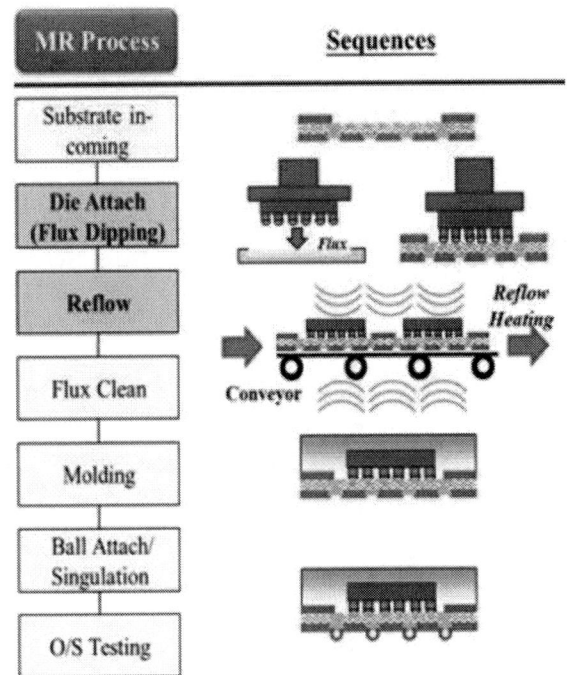

Figure 2: Mass Reflow-Die attach of process flow

After Molding process flow, the flip chip package is manufactured by molding process and the flux dipping, ball attach, thermal mass reflow (MR), Singulation process and open or short testing for each of package.

This standard thermal reflow process induced too much thermal leading for both die side of silicon and coreless substrate, and it makes the both of materials warped. The warpage is the major problem for all of advanced flip chip packages. The solder wicking for left side and right sides of coreless substrate shows in Figure 3.

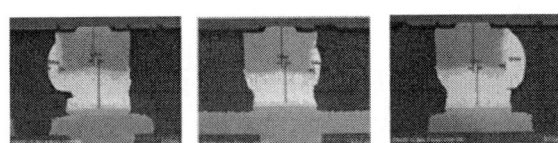

Figure 3 X-section image of Mass Reflow

Laser assisted bonding (LAB) offers the less time of laser energy than mass reflow (MR). The major process can do the die attach in 2 secs and it will prevent the die/ substrate mismatch and bump void.

The process flow sequences were defined in Figure 4. The fine pitch of flip chip die attach used the laser assisted bonding (LAB) process to join with coreless substrate, and detailed process includes the flux dipping, die put on OSP surface finished substrate, laser assisted bonding, automated optical inspection, for solder melting and flux cleaning process; the process flow shows in Die Attach step of Figure 4.

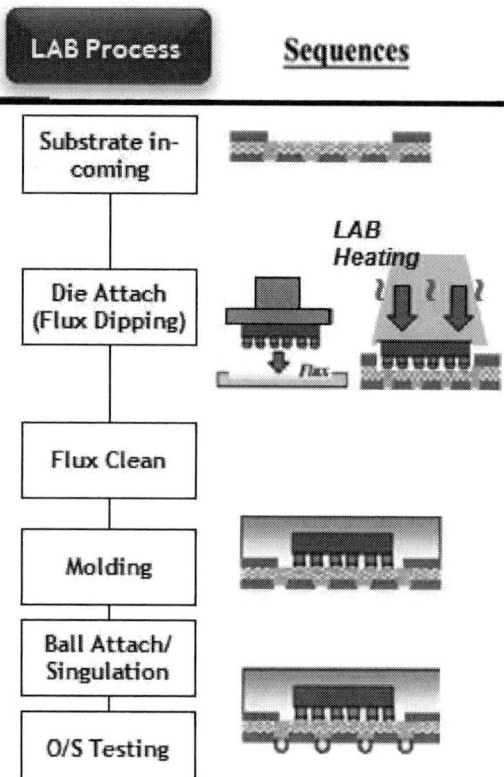

Figure 4: Laser Assisted Bonding -Die attach of process flow

From the step of molding process flow, it repeats the pervious laser assisted bonding (LAB) process for the package.

The laser assisted bonding machine had same high bonding accuracy to control the alignment to substrate side, and offer laser energy to silicon chip for solder melting and joined with OSP substrate together. The advantage for laser assisted bonding structure is less thermal budge than mass reflow because laser emission only work 1~2 seconds and most of thermal in bump with surface finish OSP substrate.; refer to Figure 5.

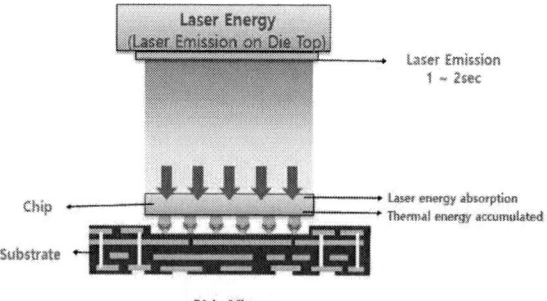

Figure 5: Working Principle of Laser Assisted Bonding

This laser assisted bonding process spent less time with thermal budget for die side of silicon with coreless substrate, and it prevents the both of materials less warpage. The warpage makes bump mismatch or bridge for all of flip chip advanced packages. The bump join of coreless substrate shows in Figure 6.

Figure 6: X-section image of Laser Assisted Bonding

Result & Discussion

This paper had demonstrated the test vehical (TV) flip chip into 50 x 50 mm packge size. By using new assembly process technology, the high integration flip chip package warpage only have 60% compared with conventional mass reflow. There is no nonwetting, no void and passes reliability tests which means laser assisted bonding (LAB) is workable for coreless substrate with Cu pillar bump compared with conventional mass reflow. The demonstrated photo as shown in Table 2.

Table 2: Overall result comparison between convenital mass reflow (MR) & laser assisted bonding (LAB)

	Output							
Process	PKG WPG	Wetting result			RT			
	Avg.	Non-wetting	Bump void	Avg. SOH	u-HTST 168hr	HTSL 1000hr	TCT 1000c	X-section
MR	1	0/10K	0/10K	58.5	Pass	Pass	Pass	
LAB	0.6X	0/10K	0/10K	71.4	Pass	Pass	Pass	

Comparing solder bump and Cu pillar bump for mass reflow (MR). As shown in Figure 7 for solder x-section photos shows substrate expanding is much less in laser assisted bonding (LAB) than mass reflow (MR) process which means non-wetting and bump bridge can be prevented effectivly.

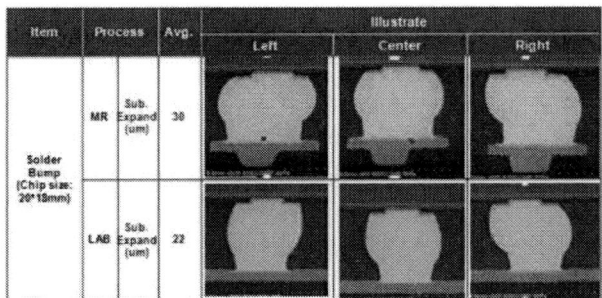

Figure 7: Solder Bump X-section image of mass reflow (MR) and laser assisted bonding (LAB) Bonding

There are two major reasons that the laser assisted bonding (LAB) can prevent a high risk of bump to trace shorting especially as need increases for finer bump pitch designs. The first one is the bump hight average higher than mass flow (MR) because of underfill gets much more gap to fill up easier without void. And the second is the better bump sharp, the sharp looks slender, and bump to trace get longer distance to prevent bump bridge issue. As shown in Figure 8.

Item	Process		Avg.	Illustrate
Cu-pillar bump (Chip size: 8*7mm)	MR	SOH (um)	52	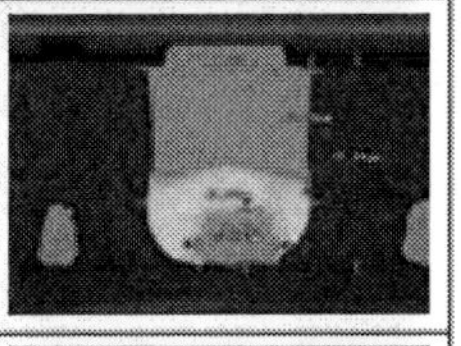
		Sub. Expand (um)	7.0	
	LAB	SOH (um)	59	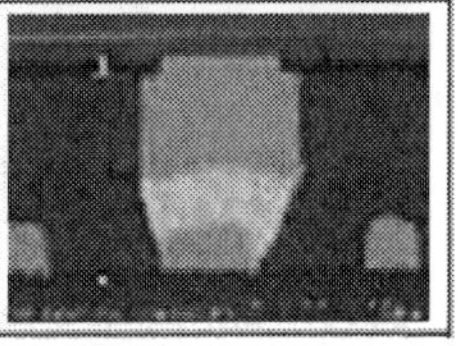
		Sub. Expand (um)	4.0	

Figure 8: X-section image of mass reflow (MR) and laser assisted bonding (LAB)

The last one is the substrate warpage preventing, the mass reflow (MR) process cacuse solder creep due to the substrate thermal leading. Refer to Figure 3.

By comparing with mass reflow, the laser assisted bonding (LAB) overall bump shape with warpage are better, and the stand off hight (SOH) is much higher. The reliability test items are follow standard JEDEC condition includes temperature cycling test, uHAST and high temperature storage in Table 2. Each reliability test are used O/S testing and SAT inspecation were then performed to confirm the quality performance and got the passed result for all reliability test items including MSL3, TCT1000, u-HAST168 and HTSL1000.

Conclusions

In this paper, we demonstrated the die attach process and technology. Based on these results, it is clear that the laser assisted bonding LAB chip attach methodology generates better reliability performance than conventional mass reflow process. Through review of these results, it shows that not only can LAB and mass reflow technologies achieve bump pitch reduction with a finer LW/LS substrate and escaped traces, but also device performance improvements. The laser assised bonding (LAB) can provide 40% warpage enhancements during the thermal assembly process for flip chip packages. Completed package level reliability test for following test conditions, passed MSL 3, TCT1000, u-HAST168 and HTS1000. This paper provided the higher throughput and accuracy bonding solution for further advanced fine pitch with coreless package requirement. Refer to the Table 3.

Table 3: Overall comparison for convenital mass reflow (MR), laser assisted bonding (LAB) & thermal compassion bonding (TCB)

	Method	DB + MR	DB + LAB	TCB
	Method	Non contact (air)	Non contact (IR light)	Contact
Heating	Object Chip	Yes	Yes	Yes
	Component	Yes	Yes	NA
	Sub	Yes	NA	NA
	Valid Area (mm)	☆☆☆	☆☆	☆
	Cycle Time	☆	☆☆☆	☆☆
	DB accuracy	☆	☆☆	☆☆☆
	Bump to trace distance	☆	☆☆	☆☆☆
	Z height control	NA	NA	Yes
	Thermal stress	☆☆	☆☆☆	☆☆☆
UPH	FC PKGs	☆☆☆	☆☆	☆

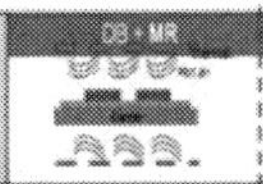

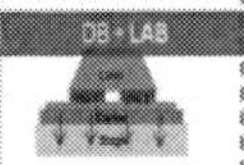

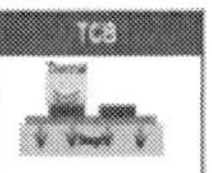

Acknowledgements

This research was supported by SPIL Research and Development Team. The authors would like to thank the R&D of SPIL under the leadership of SPIL Taiwan KC Tsai and David Lai. Their support was critical to the successful development of all die attach processes. Furthermore, the authors would like to thank all our co-developers including suppliers for strong support for the fine pitch of flip chip packaging die attach process development.

References

1. Mike Tsai, Albert Lan, Yan Han Yao, Meng Yueh Wu, Cheng Kai Chang, Roger Lo and Eason Chen, "Alternative Fine Pitch Solution of Low Cost and High Throughput Thermal Compression Bonding by using Capillary Underfill", in *Proc. 65th Electronic*

Components and Technol. Conf. (ECTC), 2015, pp.465-469.

2. Chih-Chun Hsu, Sheng-Mou Lin, Ying-Chih Chen, Wen-Zhou Wu, Che-Ya Chou, Charles Nan-Cheng Chen, Gavin Gao and Chen-Chao Wang, "3-D Antenna-in-Package Design with Metallic Coating Technique", in *Proc. 68th Electronic Components and Technol. Conf. (ECTC)*, 2018, pp.1472-1478.

3. Quan Qi and Carlton Hanna, "Flip Chip Solder Joint Pad Optimizations for a Connectivity SiP Application", in *Proc. 67th Electronic Components and Technol. Conf. (ECTC)*, 2017, pp.1045-1050.

4. Yanggyoo Jung, Dongsu Ryu, Minho Gim, Choonghoe Kim, Yunseok Song, Jinyoung Kim, Juhoon Yoon, Choonheung Lee, "Development of Next Generation Technology using Homogenized Laser-Assisted Bonding", in *Proc. 66th Electronic Components and Technol. Conf. (ECTC)*, 2016, pp.88-94.

5. MIL-STD-883, TEST METHOD STANDARD MICROCIRCUITS

6. JEDEC Solid State Technology Association. JESD22-A104C: Temperature Cycling;

Corrosion of Tin-Indium Solder during the Manufacturing Process of Biomedical Ultrasound Transducers

F. Bertocci[1], A. Grandoni[1]

[1]Global Transducer Technology, Esaote spa, Via di Caciolle 15, 50127, Firenze (Italy)

francesco.bertocci@esaote.com, andrea.grandoni@esaote.com

Abstract

The development of facile, reliable and low temperature soldering plays a crucial role when the physical and chemical characteristics of the materials are very critical. In case of piezoelectric material, i.e. high dielectric constant PZT (Lead Zirconate Titanate), the Curie temperature is in the range of (130-150)°C as for the Rhombohedral-to-tetragonal phase Temperature (T_{RT}) of PMN-PT (Lead Magnesium Niobate – Lead Titanate) that is usually below 100°C. The study aims to analyze the defects occurring during the manufacturing process of an Ultrasound (US) transducer for medical imaging with a low melting temperature Sn/In (52/48) solders. The effects and the root cause of fouling of the Sn/In solderings due to the galvanic corrosion have been detected by means of Scanning Electron Microscope (SEM) and Energy Dispersive X-ray (EDX) analysis. The paper has been focused on the mechanical dicing process of the US transducers and, in particular, on the blade cooling water that showed large quantities of metals, i.e. Tungsten (200mg/L). The cooling water quality explains the kinetics and the root cause of the galvanic corrosion. The residual metals of the mechanical dicing process in the cooling water activated the galvanic corrosion phenomenon on the US transducers due to the low redox potential of the indium by making prohibitive the weldability and the reworkability of the Sn/In solder.

Key words: lead-free solders, corrosion behavior, low melting temperature, tin-indium, Scanning Electron Microscopy (SEM), Energy Dispersive X-Ray (EDX) spectroscopy, ultrasound (US) transducer, biomedical imaging

I. INTRODUCTION

The use of solder alloys, i.e. tin-silver-copper (Sn–Ag–Cu, SAC), or the application of Electrically Conductive Adhesives (ECAs), became mandatory especially for the final revision version on the use of Hazardous Substances in the Electrical and Electronic Products, knows as China RoHS 2 (in force in 2016, 1 July). The traditional eutectic tin-lead solder (Sn/Pb, 63/37 – melting point at 183°C) solder joints have been phased out due to high toxicity of lead for the environment and for the human health [1][2]. The restriction on the usage of Pb has stimulated the research and the industry to invest substantial resources, in order to develop and to validate various Pb-free solders. Many alternative solders have been proposed thanks to good solderability, excellent performance, and high ductility [3]. However, many lead-free alternatives have a higher melting temperature than Sn/Pb alloy. The SAC solder is the most widely used lead-free solder material with a melting point greater than 217°C and the mainly application is devoted to Surface-Mount Technology (SMT) and electronic packaging [4][5]. The literature provided several drawbacks due the application of these lead-free solder alloys, i.e. electrochemical migration (ECM), intermetallic compounds or reduction of surface resistance [6][7]. Some electronics manufacturing processes require low temperature processing for thermal-sensitive materials and products, i.e. PZT [8] or PMN-PT [9]. In the past, the authors designed and characterized ICAs (Isotropic Conductive Adhesive) as Sn/Pb solder replacement in US transducers for medical imaging [10][11]. The application was very promising thanks to low curing temperature (down to 24°C), proved stability over time and low thermal stress thanks to the Coefficient of Thermal Expansion (CTE) that better match the specific substrate [12]. On other hand, the development of new sophisticated processes, the training of the operators and the non-reworkable feature of ICAs made this solderless technique quite binding to make effective use in a manufacturing process for US transducers. Electrical connections by means of Anisotropic Conductive Film (ACF) for US transducer array devoted to medical imaging applications have been developed [13], but the reliability over time and the flexibility soldering process and materials have to be validated [14][15]. A significant interest has been oriented towards the usage of eutectic tin-indium (Sn/In, 52/48 – melting point at 118°C) alloy. Indium is the softest metal with a very low melting point (157°C). The Tin-Indium alloy became highly attractive candidate as the lead-free solder not only for a melting temperature lower than Sn-Pb alloy, but also for good wettability, flexibility and adequate strength for the manufacturing process of the US transducers. In literature Sn/In has been applied in semiconductor devices for improving the heat transfer efficiency from chip to packaging [16]. Some drawbacks are pointed out on the effects of indium on the corrosion resistance in Sn-Ag-based solder, i.e. Sn-Ag-In [17][18].

The purpose of this paper is to detect and to analyze in detail the defects occurring during the manufacturing process of US transducers based on low melting temperature Sn/In solder. The next section provides the motivation for this study focusing on the description of the US transducer and on the manufacturing process operation that generates the defects. The third section illustrates the effects and the galvanic corrosion behavior of the Sn/In-based soldering by means of Scanning Electron Microscope and Energy Dispersive X-ray analysis. In this context, the mechanical dicing operation of the manufacturing process has been taken into account and, in particular, the cooling water quality. Discussion section explains the kinetics and the root cause of the fouling due to the galvanic corrosion phenomenon due to the low redox potential of the indium leading prohibitive the solderability and the reworkability of the Sn/In solder. The fifth section conclude the paper.

II. THE TECHNICAL PROBLEM

The authors considered the phased array ultrasound transducer for medical imaging devoted to the diagnostic of cardiac muscle (Figure 1).

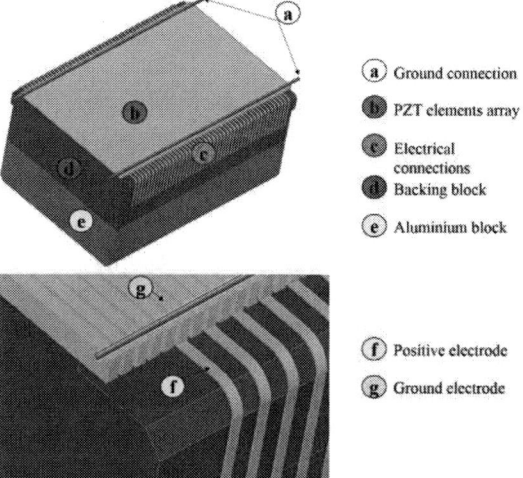

Figure 1: Perspective view of phased array transducer for diagnostic of cardiac muscle

The transducer includes the backing block (Figure 1d) made in rubber that is disposed on an aluminium block (Figure 1e) behind the piezoelectric (PZT) elements array (Figure 1b) [19]. The backing layer reduces the signals distortion by suppressing the unnecessary ultrasound waves propagation coming from the PZT elements. The

latter are formed starting from a PZT plate (23 mm of lenght and 14 mm in width) [11] that is divided in an array of elements by means of mechanical dicing operation. The transducer is diced into 128 piezoelectric elements that are arranged in line 64 on odd side, and 64 on even side. Each PZT element on positive side is connected to the corresponding electrical connection (Figure 1c). The matching layer (not showed in Figure 1) is disposed over the PZT elements array allows ultrasound signals, generated by PZT elements, to be efficiently transferred toward a target by matching sound impedances of the materials involved in the transducer. In fact, the matching layer has an intermediate value between the sound impedance of the piezoelectric element and the target. A thin conductive metal, i.e. gold, composes the electrodes of each piezoelectric element. The positive electrode (Figure 1f) connection provides the electrical signals and the opposite separated corresponding negative electrode (Figure 1g) act as the ground reference. The soldering of two metal wires along the two edges of the US transducer (Figure 1a) allows for the connection to the ground reference for each PZT element. An easy and robust way able to connect each PZT element to the ground reference by means of the ground wires consists in a pre-tinning operation along the edges of the PZT plate based on Sn/In solder before the mechanical dicing. The latter is one of the main process operations that affects the final quality of the US transducer. Important factors, i.e. feed rate and rotation speed of blade, and cooling water quality, have been studied and evaluated for reducing the performance degradation caused by mechanical dicing process [20][21]. It is important to emphasize that the fragility and the geometry of piezoelectric elements (width around 150 μm or less) do not allow for performing a precise and a compliant ground wire soldering without a pre-tinning; furthermore the lacking of a ground connection for one or more piezoelectric element lead the US transducer to a scrap. For this reason, the pre-tinning operation before the mechanical dicing becomes very important, in order to prepare and to perform in the best way the soldering operation of the ground wires connection.

The issue occurred during the process of mechanical dicing, in which this operation is performed for a batch size of 10 US transducers (total working time around 5 hours – around 30 minutes for each piece). The formation of fouling (in the paper also called encrustations) on the Sn/In-based pre-tinning and the existence of efflorescences on the aluminium block (Figure 2) have been detected. The encrustation of pre-tinning avoids the possibility to perform the ground wire soldering due to the very low wettability of the Sn/In-solder. The efflorescences on the aluminium block are a negative effect to be take into account for determining the behaviour of the defects and for the potential contamination that can invalidate the final quality of the US transducer.

Figure 2: Transducer with defects due the presence of encrustations and efflorescences

III. EXPERIMENTAL RESULTS

A. Characterization Measurements: Morphological and Microchemical Characteristics of the defects

The nature of the defects required analytical investigations, in order to understand the phenomenon leading to the efflorescences and encrustations with the aim to propose potential solutions for the manufacturing process of US transducer for medical imaging. For this reason, the sample (Figure 2) processed by means of dicing saw machine (DISCO DAD3350) has been analyzed through SEM (Philips XL20) equipped with EDX. This instrumentation allows for acquiring images with high resolution and for characterizing the chemical elements (semi-quantitative analysis) of the area under investigation. Figure 3 shows the characterization of a deposit removed by adhesive tape from the aluminium block surface of the transducer. In particular, the morphological analysis allowed to determine the structure of the efflorescence (small non-homogeneous particles), in which the greater grain has an irregular shape with diameter of 100 µm (Figure 3a). The microchemical characterization revealed that Tungsten (W) and Aluminium (Al) mainly compose the grain with a relative composition of around 39% and 10% respectively (Figure 3b). Therefore, the EDX analysis revealed that the efflorescences removed from the aluminium block consist essentially of tungsten and aluminium in oxided form, due to the presence of Oxigen (38%). Figure 4 shows the area of aluminium block of the transducer without the efflorescences. The morphological analysis by means of SEM revealed that the aluminium has been corroded and the effects are the dark fractures (red arrows). The portion of defective material by the corrosion consists of aluminium oxides. It appeared

evident that during the mechanical dicing operation a corrosion phenomenon has been activated.

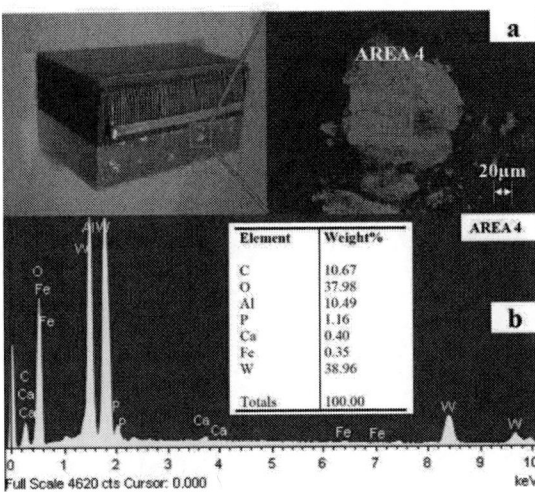

Figure 3: SEM (a) and EDX (b) analysis of the efflorescence deposit of the aluminium block surface

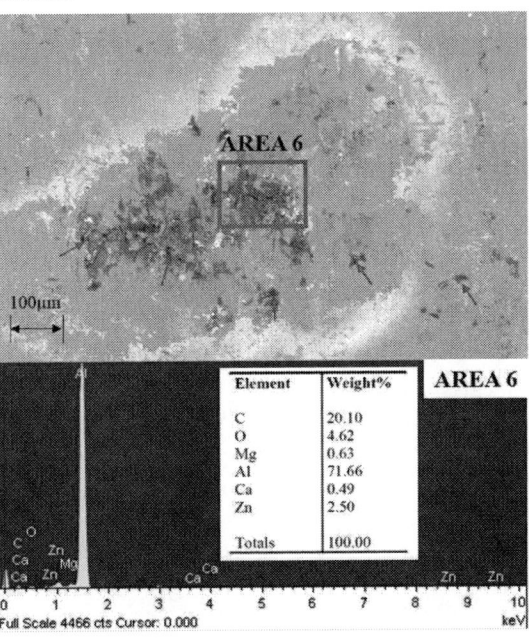

Figure 4: SEM and EDX analysis of the aluminium block area without the efflorescences deposit

The authors decided to prepare a sample of US transducer (Figure 5) with two different pre-tinnings on the PZT plate: one based on standard Sn/Pb alloy (left edge) and the second based on Sn/In alloy (right edge), in order to intercept and to investigate the morphological and microchemical characteristics of the defects. Therefore, the sample has been left in a continuous wet condition under the blade cooling water for 5 hours (the same working time for dicing 10 US transducers) without performing the mechanical dicing process, in order to eliminate potential variability source due to the generation of other defects, i.e. delaminations or micro-cracks [20][21].

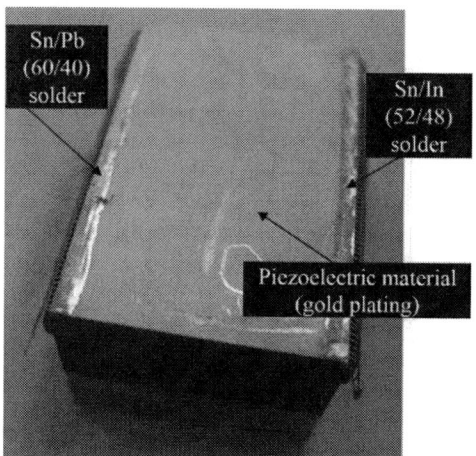

Figure 5: Transducer with two different pre-tinnings: Tin-Lead-based (left edge) and Tin-Indium-based (right edge) solderings

A series of SEM images in different points of the sample surface under analysis have been carried out (Figure 6).

The encrustations were cleary recognized on the gold surface of the PZT plate and the homogeneity of Sn/In solder was very poor (Figure 6a). In Figure 6b, instead, a perfect clean and homogeneous Sn/Pb solder has been observed.

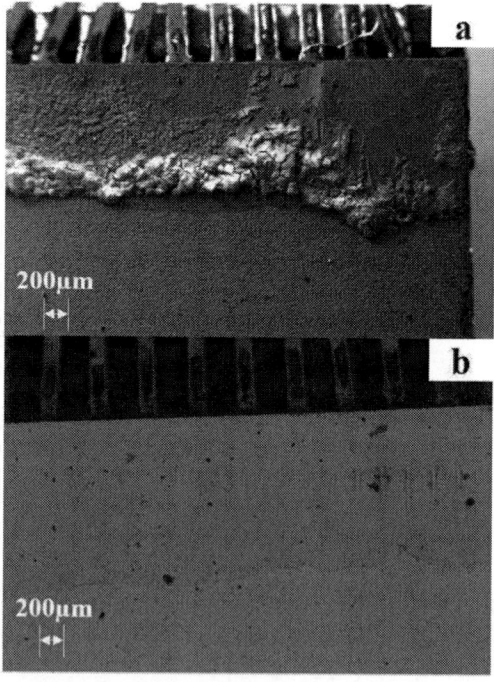

Figure 6: SEM analysis on top surface of US transducer with two different pre-tinnings. a) Tin-Indium-based soldering; b) Tin-Lead-based soldering

Three areas of the encrusted Sn/In-based soldering have been investigated (Figure 7) and the microchemical analysis for measuring the relative elementary composition in [%] are collected in Table 1. The analysis on AREA 1 clearly showed the presence of Tin (32%) and Indium (47%) with small amounts of Tungsten (around 2%). Oxygen indicates the presence of oxidized compounds.

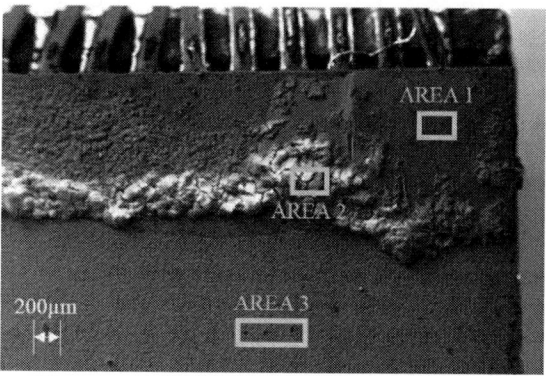

Figure 7: SEM inspection on three different areas of the encrusted Tin-Indium solder

The AREA 2 aims to investigate the chemical elements composition in the maximum concentration of the encrustation. The predominant composition of the encrusted solid is based on Indium (28%) and Tungsten (40%) in oxided form for the presence of Oxigen (28%). The Tin is completely absent. The encrustations on the Tin-Indium-based solderings are located close to the interface with gold of the PZT plate, rather than on the outer edges of the piece. By this preliminary consideration, the solderability of Sn/In-based pre-tinning becomes prohibitive due to the change of the eutectic alloy in the pre-tinning soldering.

Table 1: relative chemical composition based weight by means of EDX

Element	Weight [%]		
	AREA 1	**AREA 2**	**AREA 3**
C	-	-	5.59
O	18.23	31.79	7.03
Ni	-	-	3.18
In	47.33	28.24	-
Sn	31.96		-
W	2.48	39.96	1.98
Au	-	-	82.21
Totals	100.00	100.00	100.00

In order to double-check the EDX analysis and the total lack of Sn in the encrusted pre-tinning, the deposit have been removed from AREA 2 by means of an adhesive tape. Therefore, the encrustation has been characterized in detail (Figure 8). The analysis confirmed that the core of defect is based on Indium (45%) in oxided form due to the 47% of Oxygen and that the eutectic Sn/In solder changed the characterist due to the absence of tin. A

118

not negligible quantity of W (more than 6%) has been detected.

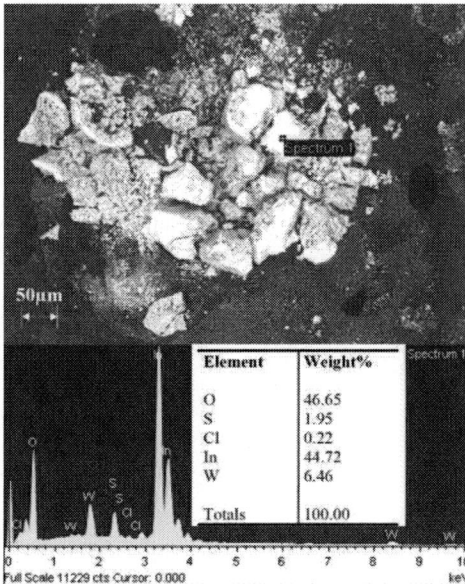

Element	Weight%
O	46.65
S	1.95
Cl	0.22
In	44.72
W	6.46
Totals	100.00

Figure 8: SEM and EDX characterization of encrustation removed from Sn/In-based pre-tinning

AREA 3 is composed by gold (82%), although small amount of Nickel (3%) and Tungsten (2%) have been found.

B. Morphological and Microchemical Characteristics of the blade cooling water

The previous analysis led the authors to analyze the blade cooling water that is in direct contact with the transducer during the mechanical dicing operation. The cooling liquid became the prime suspect for the generation of the defects. The cooling water is continuously recycled by an ad-hoc filters chain. For the best results of cut quality is mandatory to keep it very clean by filtering micro-grains that could generate potential damaging in the dicing process. Therefore, the chemical analysis and the characterization of the cooling water have been provided by selecting a liquid sample (Table 2). The color was oplascent with white particles in suspension. The pH was 7.8 and the conductivity @25°C was 578.0 µS/cm. The sulfates, bicarbonates, chlorides and sodium are the main molecules dispersed in the clear liquid. For the determination of the presence of heavy metals in the water sample, the method of acidification has been adopted. During the operation of acid addition, the liquid sample under analysis became a numb due to the formation of a precipitate. In order to understand the nature of the solids and of residual metals precipitated after the acidification operation, two chemical characterizations have been performed: the

first analysis has been addressed to the liquid, the second one to the precipitated particles.

Table 2: Chemical analysis of the anions and cations of the cooling water. Method 1 is UNI ES ISO 10304-1:2009, 2 is ARPAT CNR IRSA 2010, 3 is UNI ES ISO 14911:2001

	Parameter	Method	Results [mg/L]
Anions	Sulfates	1	67.6
	Nitrates	1	0.3
	Carbonates	2	<3
	Bicarbonates	2	116.0
	Chlorides	1	70.2
	Nitrites	1	<0.13
	Fluorides	1	0.1
	Bromides	1	<0.10
	Phosphates	1	2.1
Cations	Lithium	3	<0.10
	Sodium	3	89.5
	Potassium	3	2.3
	Ammonium	3	<0.50
	Calcium	3	23.2
	Magnesium	3	6.6

The results of the suspended and precipitated metals (signed by *) in the liquid sample are reported in Table 3.

Table 3: Chemical analysis of the liquid sample after the acidification and of the precipitated metals (signed by *). Method 4 is UNI ES ISO 11885:2009

Metal	Method	Results [µg/L]
Alluminium	4	61.0
Boron	4	63.8
Cobalt	4	15.1
Chrome *	4	72.3
Iron	4	197.4
Manganese	4	9.3
Nickel	4	246.6
Lead *	4	545.4
Copper	4	23.6
Selenium *	4	2763.0
Silicon *	4	1066.0
Tin	4	<5
Strontium	4	535.8
Tallium	4	<10
Titanium *	4	84.1
Vanadium	4	19.0
Zinc	4	396.0
Indium	4	<5
Gold	4	<5
Zirconium *	4	132.6
Tungsten	4	196,200.0

119

The presence of tungsten ($\approx$200mg/L) in suspension is very significant. In fact, W composes some materials of the US tranducer and, therefore, the filters chain cannot filter the particles completely. In the precipitated metals the presence of Lead, Zirconium and Titanium are not negligible. These elements compose the piezoelectric material and they are the processing residues of the PZT plates by means of the mechanical dicing operation.

IV. DISCUSSION

The root cause of the defects, i.e. encrustations, corrosion, on the Sn/In-based solderings and on the aluminium block has to be found in the reduction potential of the residual metals species produced during the dicing operation of the US transducer for realizing the PZT elements array (Table 4).

Table 4: Oxidation-reduction reactions for the main elements of the US transducer

	Reduction reaction	Reduction potential (V)
In	$In^{3+} + 3e^- \leftrightarrow In$	-0.34
	$In^{3+} + 1e^- \leftrightarrow In^{2+}$	-0.49
	$In^{3+} + 1e^- \leftrightarrow In^{2+}$	-0.40
	$In^{2+} + 1e^- \leftrightarrow In$	-0.14
	$In_2O_3 + 6H^+ + 6e^- \leftrightarrow 2In + 3H_2O$	-0.19
	$In(OH)_3 + 3H^+ + 3e^- \leftrightarrow In + 3H_2O$	-0.17
	$In(OH)_3 + 3e^- \leftrightarrow In + 3OH^-$	-1.00
	$In + H^+ + 1e^- \leftrightarrow InH$	-1.95
	$InOH^{2+} + H^+ + 3e^- \leftrightarrow In + H_2O$	-0.27
Pb	$Pb^{2+} + 2e^- \leftrightarrow Pb$	-0.13
	$Pb^{4+} + 2e^- \leftrightarrow Pb^{2+}$	+1.67
	$PbO + H_2O + 2e^- \leftrightarrow Pb + 2OH^-$	-0.58
	$PbO + 2H^+ + 2e^- \leftrightarrow Pb + H_2O$	+0.25
	$PbO_2 + 4H^+ + 2e^- \leftrightarrow Pb^{2+} + 2H_2O$	+1.46
	$Pb_3O_4 + 2H^+ + 2e^- \leftrightarrow 3PbO + H_2O$	+0.97
	$Pb(OH)_2 + 2H^+ + 2e^- \leftrightarrow Pb + 2H_2O$	+0.28
Al	$Al^{3+} + 3e^- \leftrightarrow Al$	-1.66
	$Al(OH)_3 + 3e^- \leftrightarrow Al + 3OH^-$	-2.30
	$Al_2O_3 + 6H^+ + 6e^- \leftrightarrow 2Al + 3OH^-$	-1.49
	$Al(OH)_3 + 3H^+ + 3e^- \leftrightarrow Al + 3H_2O$	-1.47
W	$WO_2 + 4H^+ + 4e^- \leftrightarrow W + 2H_2O$	-0.02
	$WO_3 + 6H^+ + 6e^- \leftrightarrow W + 3H_2O$	-0.09
	$2WO_3 + 2H^+ + 2e^- \leftrightarrow W_2O_5 + H_2O$	-0.03

The oxidation-reduction reactions analysis of the main metallic species of the US transducer (In, Pb, Al) and of the cooling water (W) explains how the Indium is extremely more vulnerable to oxidation due to a lower redox potential than lead. This evaluation allows for justifying the phenomenon of galvanic corrosion for the Sn/In alloy leading to a poor solderability of the pre-tinning after the mechanical dicing, while the Sn/Pb alloy maintains the same eutectic characteristic (higher reduction potential than Indium). Tungsten, in large quantities in the cooling water, is reduced in contact with Indium and Aluminium, leading to galvanic corrosion phenomenon. As Indium, the aluminium oxidises easily due to the lowest reduction potential. The quality of cooling water and the residual metals have a significant effect for the galvanic corrosive phenomenon and its kinetics. Furthermore, the weldability tests have been also performed providing a sacrifical anode by short-circuiting the US transducer with Mg (Magnesium)-based metallic wire before the mechanical dicing operation. The Mg has been chosen for the lower redox potential than Indium, but the benefit has been slight by considering the difficulty to implement in practice this solution. Therefore, this test did not justify a deepening of this strategy.

V. CONCLUSION

The replacement of Sn/Pb solder led to consider lead-free solders by respecting the stringent requirement of low temperature stress for the materials, i.e. PZT, involved in the construction of US transducer, in which the Curie temperature is less than (130-150)°C. The authors detected and analyzed in detail the effects and the root cause of the defects occurring during the manufacturing process of US transducer with solderings that are based on low melting temperature Sn/In (52/48) alloy. The morphological and the microchemical analysis by means SEM and EDX, respectively, proved that the Sn/In-based soldering is affected by the phenomenon of galvanic corrosion during the mechanical dicing process. The blade cooling water quality and the reduction potentials of the metals in the liquid explained the kinetics of the galvanic corrosion on Indium and Aluminium that compose the US transducer for medical imaging. In fact, indium is more vulnerable to oxidation due to a low reduction potential, in comparison to metals with higher reduction potential, i.e. Pb and W. The latter is in large quantities in the cooling water. Tungsten is one of the main elements of the US transducer materials and, therefore, it cannot be totally eliminated. Tests showed that after 5 hours the galvanic corrosion is rather advanced making poor the weldability and the reworkability of the Sn/In alloy. A mitigation plan will be oriented to reduce the exposition of the piece under water during the mechanical dicing process up to 1 hour (batch size equals to 2). A sealing based-polymer layer on the Sn/In-based pre-tinning can contribute for avoiding the cooling water contact and the growth of corrosion. Further development will focus on the evaluation of the water pH effect on the corrosive phenomenon kinetics by minimizing the galvanic corrosion over 5 hours.

ACKNOWLEDGEMENTS

The authors thank Francesca Gambineri (Laboratori Archa srl) for the contribution on the morphological and micro-chemical analysis.

REFERENCES

[1] Y. Shu, K. Rajathurai, F. Gao, Q. Cui, Z. Gu, "Synthesis and thermal properties of low melting temperature tin/indium (Sn/In) lead-free nanosolders and their melting behavior in a vapor flux", Journal of Alloys and Compounds, Vol. 626, pp. 391-400, 2015.

[2] The Restriction of Hazardous Substances in Electrical and Electronic Equipment (RoHS). Official Journal of the European Union: L37/19-L 37/23.

[3] L.F. Li, Y.K. Cheng, G.L. Xu, E.Z. Wang, Z.H. Zhang, H. Wang, "Effects of indium addition on properties and wettability of Sn-0.7Cu-0.2Ni lead-free solders", Journal of Materials and Design, Vol. 64, pp 15-20, 2014.

[4] S. Cheng, C.M. Huang, M. Pecht, "A review of lead-free solders for electronics applications", Journal of Microelectronics Reliability, Vol. 75, pp. 77-95, 2017.

[5] Z.W. Wang, X. Li, H. Chen, "Preparation and Properties of SMT Environmental Protection and High Efficient Lead-free Solder Paste", Proceedings of the 7th International Conference on Energy and Environmental Protection (ICEEP), Shenzen, China, July 14-15, pp. 1103-1107, 2018.

[6] L. Hua, C. Yang, "Corrosion behavior, whisker growth, and electrochemical migration of Sn-3.0Ag-0.5Cu solder doping with In and Zn in NaCl solution", Journal of Microelectronics Reliability, Vol. 51, No. 12, pp. 2274-2283, 2011.

[7] S.K. Lin, R.B. Chang, S.W. Chen, M.Y. Tsai, C.M. Hsu, "Effects of zinc on the interfacial reactions of tin-indium solder joints with copper", Journal of Material Science, Vol. 49, No. 10, pp. 3805-3815, 2014.

[8] Q.Q. Zhang, F.T. Djuth, Q.F. Zhou, C.H. Hu, J.H. Cha, K.K. Shung, "High frequency broadband PZT thick film ultrasonic transducers for medical imaging applications", Journal of Ultrasonics, Vol. 44, No. 1, pp. e711–e715, 2006.

[9] C.M. Wong, Y. Chen, H. Luo, J. Dai, K.H. Lam, H.L. Chan, "Development of a 20-MHz wide-bandwidth PMN-PT single crystal phased-array ultrasound transducer", Journal of Ultrasonics, Vol. 73, pp. 181-186, 2017.

[10] M. Catelani, V. L. Scarano, F. Bertocci, R. Berni, "Optimization of the Soldering Process With ECAs in Electronic Equipment: Characterization Measurement and Experimental Design", Journal of IEEE Transactions on Components, Packaging and Manufacturing Technology, Vol. 1, No. 10, pp. 1616-1626, 2011.

[11] M. Catelani, V.L. Scarano, F. Bertocci, "Implementation and Characterization of a Medical Ultrasound Phased Array Probe With New Pb-Free Soldering Materials", Journal of IEEE Transactions on Instrumentation and Measurement, Vol. 59, No. 10, pp. 2522-2529, 2010.

[12] M. Catelani, V.L. Scarano, F. Bertocci, R. Singuaroli, P. Palchetti, A. Grandoni, "Thermal stress on silver conductive adhesive solder joints: performance evaluation of medical ultrasound array transducer", Proceedings of the 2009 IEEE Instrumentation and Measurement Technology Conference, Singapore, Singapore, May 5-7, pp. 468-471, 2009.

[13] H. Nguyen, T. Eggen, K.E. Aasmundtveit, "Assembly of transducer array using anisotropic conductive film for medical imaging applications", Proceedings of the 2015 IEEE European Microelectronics Packaging Conference (EMPC), Friedrichshafen, Germany, September 14-16, pp. 1-6, 2015.

[14] L. Frisk, S. Lahokallio, J. Kiilunen, "Comparison of microvia HDI PCBs with ACF interconnections in accelerated life testing", Proocedings of the 2017 IEEE Eurpoean Microelectronics Packaging Conference (EMPC), Warsaw, Poland, September 10-13, pp. 1-6, 2017.

[15] C. Chung, G. Sim, S. Lee, K. Paik, "Effects of Conductive Particles on the Electrical Stability and Reliability of Anisotropic Conductive Film Chip-on-Board Interconnections", Journal of Transactions on Components, Packaging and Manufacturing Technology, Vol. 2, No. 3, pp. 359-366, 2012.

[16] T. Fałat et al., "Investigation on interaction between indium based thermal interface material and copper and silicon substrates", Proceedings of the 2013 IEEE Eurpoean Microelectronics Packaging Conference (EMPC), Grenoble, France, September 9-12, pp. 1-5, 2013.

[17] F. Rosalbino, E. Angelini, G. Zanicchi, R. Marazza, "Corrosion behaviour assessment of lead-free Sn-Ag-M (M = In, Bi, Cu) solder alloys", Journal of Materials Chemistry and Physics, Vol. 109, No. 2, pp. 386-391, 2008.

[18] R.M. Shalaby, "Effect of silver and indium addition on mechanical properties and indentaion creep behavior of rapidly solidified Bi-Sn based lead-free solder alloys", Journal of Materials Science & Engineering A, Vol. 560, pp. 86-95, 2013.

[19] L. Spicci, "FEM simulation for pulse echo performances of an ultrasound imaging linear probe", Proceedings of the 2012 Comsol Conference, Milan, Italy, October 10-12.

[20] M. Fuelg, G. Mackh, L. Frey, "Assessment of dicing induced damage and residual stress on the mechanical and electrical behavior of chips", Proceedings of the 2015 IEEE 65th Electronic Components and Technology Conference (ECTC), San Diego, CA, USA, May 26-29, pp. 214-219, 2015.

[21] S. McCann, Y. Sato, V. Sundaram, R.R. Tummala, S.K. Sitaraman. "Prevention of Cracking From RDL Stress and Dicing Defects in Glass Substrates", Journal of IEEE Transactions on Device and Materials Reliability, Vol. 16, No. 1, pp. 43-49, 2016.

Influence of copper wire material to corrosion resistant packages and systems for high temperature applications

Sandy Klengel[1], Robert Klengel[1], Jan Schischka[1], Tino Stephan[1], Matthias Petzold[1]
Motoki Eto[2], Noritoshi Araki[2], Takashi Yamada[2]

[1]Fraunhofer Institute for Microstructure of Materials IMWS, Halle, Germany
[2] Nippon Micrometal Corporation, 158-1 Sayamagahara, Iruma-City, Saitama 358-0032, Japan

Email: sandy.klengel@imws.fraunhofer.de
phone: +49 345 5589 125

Abstract

We present studies for a newly developed gold-palladium coated copper wire with focus to the automotive industry. This wire material was designed to prevent halide induced interface corrosion and also sulphur induced pitting corrosion. The base material was alloyed with special element to systematically adjust the intermetallics formed at the interface to the pad metallization and to define the microstructure of copper base material to prevent sulphur enhanced corrosion. For reliability testing the wire was bonded to a silicon die with aluminum metallization and a nickel-palladium-gold plated lead frame. The mold compound contains sulfur as well as low amount of chlorine (< 10wt ppm). The package was tested by high temperature storage at T=175°C for 2,000 hours in comparison to a standard gold-palladium coated wire and a bare copper wire. We show results of high resolution analyses (Scanning Electron Microscopy, Transmission Electron Microscopy, nanospot-EDS-mapping) for the characterization of the intermetallic compounds formed between bond wire and die metallization and also for the stitch bond contact itself. The running corrosion mechanism will be investigated in detail. The results show that the newly developed gold-palladium coated copper wire shows a much higher reliability compared to the standard wire materials.

Key words: APC wire, copper wire bonding, reliable packaging, automotive electronics packaging, high temperature electronics, materials for harsh environments

Introduction

Wire bonding is still the most common interconnect technology for semiconductors. During the last years many efforts have been continuing to increase the integration density and to minimize the packaging size. Copper wire was used to substitute gold wire materials. In the past years, bare copper, palladium coated (PCC) and gold-palladium coated (APC) copper wires were developed especially to meet the requirements in automotive electronics industry for harsh environments. Microelectronic packages often contain low amounts of contaminants caused by the encapsulation material or the semiconductor processing. Mold compounds include sulfuric adhesion promoters or chlorine in low ppm-range and pads on silicon dies for wire bonding can show small amounts of fluorine caused by the back-end processing. Especially the wire bonded contacts or wire bond materials behave very sensitive in presence of contaminants. The reaction between the contaminants and the wire bonded contacts lower the reliability of the complete package or systems significantly. Material development especially for the wire bond material could improve the reliability

of the systems. Residues on pad surfaces or low amounts of halides (like Cl in ppm-range) in the encapsulating material cannot be excluded in wire bonded packages so far. The APC wires show an improved behavior against halide induced corrosion but for them high temperature induced pitting corrosion induced by sulfur containing environments and mold compounds is limiting for reliability [1], [2]. In order to apply copper wire to automotive devices, the wire material has to meet the demands for long term reliability specific to automobiles including stable operation under harsh environment with high temperature and humidity. To meet these demands, a new type of APC wire has been developed and introduced to the market [2],[3].

In this paper, we present studies for the new APC wire "EX1R" with focus to the automotive industry. This wire material was developed to prevent halide induced interface corrosion and also sulfur induced pitting corrosion. The base material was alloyed with special element to systematically adjust the intermetallic compound (IMC) formed at the interface to the pad metallization and to define the microstructure of the copper base material to prevent sulfur enhanced corrosion. We show high resolution analyses results (Scanning Electron Microscopy SEM, Transmission Electron Microscopy TEM, nanospot-EDS-mapping for element identification) for the characterization of the intermetallic compounds formed between bond wire and die metallization of different wire types. Also pitting corrosion mechanisms at stitch bond contacts will be investigated in detail. The results show that the newly developed APC wire "EX1R" indicates a much higher reliability especially for harsh environments with regard to halide and sulfur enhanced corrosion.

Failure Mode "Halide induced corrosion of intermetallic copper compounds at ball bond contacts with additional influence of sulfur"

In copper-aluminum system different intermetallic compounds will be formed after wire bonding and developed during thermal ageing. In general copper rich and aluminum rich IMCs are known. Figure 1 shows the binary phase diagram of the Al/Cu system, indicating possible IMCs in the temperature range less than T=300 °C [4].

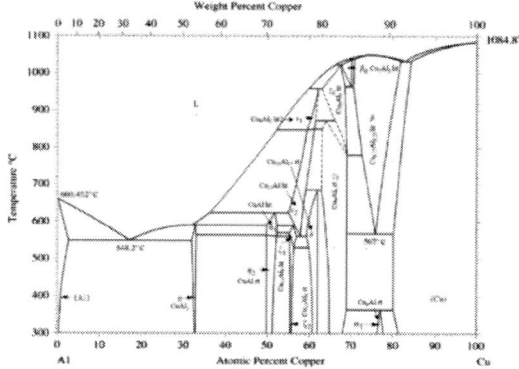

Figure 1: Binary phase diagram of system Cu-Al [4]

Figure 2 shows a scheme of the IMC type location for copper-aluminum wire bonded contacts.

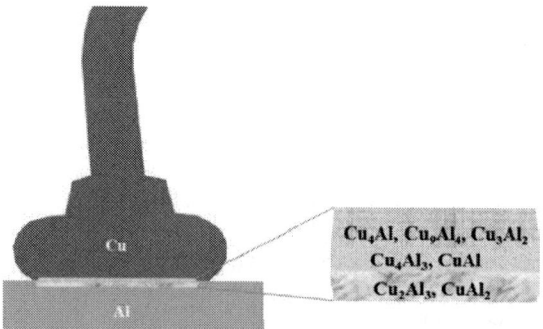

Figure 2: Location of the intermetallics for copper-aluminum wire bonded contacts

As reported many times before the copper-aluminum intermetallic compounds can be attacked

by halides leading to a limited reliability of the complete wire bonded system. This is most likely caused by a galvanic driven corrosion. This corrosion process starts due to the difference in electrochemical potential between Al and Cu in the presence of an electrolyte, i.e. humidity in combinations with ions like Cl- or other halides.

$$Cu_mAl_n + 3nCl^- = nAlCl_3 + mCu + 3ne^- \quad (1)$$

$$2AlCl_3 + 3H_2O = Al_2O_3 + 6HCl \quad (2)$$

During this corrosion reaction the copper rich IMC Cu_9Al_4 or Cu_3Al_2 is decomposed to amorphous aluminum oxide and dispersed Cu particles.

Several publication showed that the addition of palladium in copper-aluminum system would improve the corrosion resistance. With palladium coated copper the generation of single copper rich IMCs was inhibited and the IMC growth slowed down in general. It was also indicated that palladium layer could act as barrier layer for chlorine ions penetration and work as reaction partner with HCl forming $PdCl_2$ [5],[6],[7],[8]. I. Qin et al. [9] found that Pd substitutes Cu in copper rich Cu_9Al_4 intermetallic compound leading to a higher corrosion resistance. Also our investigation at bare copper, PCC and APC wires after HAST testing verify these results [10]. Figure 3 shows a scheme of IMC type location for APC and PCC bond wire contacts.

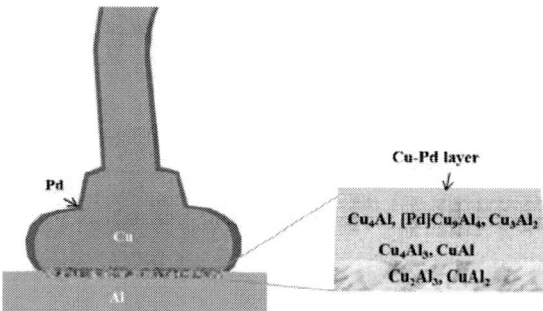

Figure 3: Location of the intermetallics for palladium coated copper-aluminum wire bonded contacts

However, due to the differences in electrochemical potential between Pd and Cu a galvanic cell could be formed and additional failure modes become present especially for HTS testing. If there is bulk copper exposed a crevice corrosion is able to start [7].
In HTS testing also the presence of sulfur from adhesion promoters has a significant influence to the halide induced interfacial corrosion behavior of bare copper, PCC and APC ball bond contacts. It was shown that the sulfur attacks the copper bulk of the ball bond contact by a pitting corrosion process or forms an $Al_2(SO_4)_3$ from the copper rich IMC as intermediate reaction product during halide induced

interface corrosion [11], see Figure 4. During these processes a gap is formed. This allows the contaminants/halides to penetrate faster into the interconnection area by driven capillary force [12],[13].

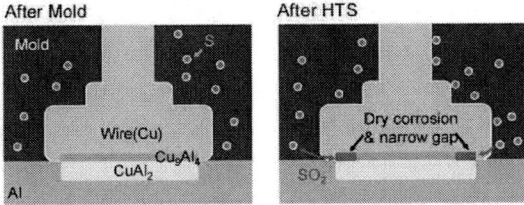

Figure 4: Assumed HTS failure mechanism for copper bonded ball bond contacts overlaying or in addition to the halide induced corrosion assumed by Saruwatari et al. [11].

Additionally reactions between sulfur and copper can take place forming a copper sulfide.

Failure Mode "Sulfur induced corrosion on stitch bond contacts for palladium coated wire types"

In HTS testing especially for temperatures T>150°C also the presence of sulfur from adhesion promoters has a significant influence to the reliability of PCC and APC wires and their stitch bonded contacts. It is assumed that at these temperatures the thiols are thermal decomposed in the mold compound leading to formation of H_2S [14]. If the 80nm thin Pd coating is damaged during stitch bonding processing a galvanic cell at the interface Pd/Cu is formed. By the influence of moisture, pitting corrosion will start, leading to a degradation of the copper bulk material. Lee et al. hypthesized that this mechanism can also take place under "dry" conditions for temperatures up to T=200°C [7]. Then additionally the influence of sulfur becomes prominent acting as catalyst. As reaction product a copper oxide or copper sulfide will be formed next to the reaction area [15]. Figure 5 shows a scheme for the sulfur induced pitting corrosion at stitch bonded contacts.

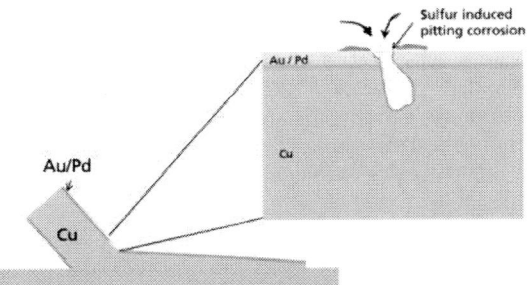

Figure 5: Scheme for reaction mechanism of sulfur induced corrosion on Pd coated copper wires

Krinke et al. stated in [16] that this process is dependent on time, temperature and the level of Pd

coating damage. As the stitch bond is the thinnest part with only several µm thickness the pitting area is big in comparison. Thus, this type of degradation can lead to completely corroded areas at stitch bond contacts resulting in "electrical opens".

Experimental

Samples and reliability testing

Within our studies, three types of test specimen were used for the different research objectives: a bare copper wire and a conventional APC wire serving as reference samples in comparison to the optimized APC wire "EX1R". All samples had a wire diameter of d=20µm. For reliability testing the wires were bonded to a silicon die with aluminum metallization (ball bond contact) and a nickel-palladium-gold plated lead frame (stitch bond contact) and finally encapsulated by a conventional mold compound (amount of Cl < 10wt-ppm). Afterwards the package was tested for 2,000 hrs by HTS at T=175°C.

Investigation methods

For microstructural analysis the test specimen after 2,000hrs HTS testing were prepared in cross section using a JEOL cross section polisher (JEOL GmbH, Freising, Germany). The cross section polisher uses an argon beam to provide high quality section surfaces and to obtain cleanly polished interface zone between the wire bond materials and substrate metallization. Following the preparation, high resolution microstructural analyses were performed by scanning electron microscopy (SEM) and energy-dispersive X-ray Spectroscopy (EDS) using a Zeiss Supra-55 VP (Carl Zeiss Microscopy GmbH, Oberkochen, Germany) with an EDAX-Trident System (AMETEK EDAX GmbH, Wiesbaden, Germany). For high resolution analyses of the running corrosion mechanism and intermetallic compounds, electron- transparent lamellas were prepared out of the interface between wire material and substrate metallization. Then, Transmission Electron Microscopy (TEM) was performed working with an FEI Tecnai G2 F20 (FEI Company, Eindhoven, The Netherlands) operated at U=200 keV for EDS line scans and with an FEI Titan[3] 80-300 (FEI Company, Eindhoven, The Netherlands) operated at U=300 keV for EDS mappings.

Results

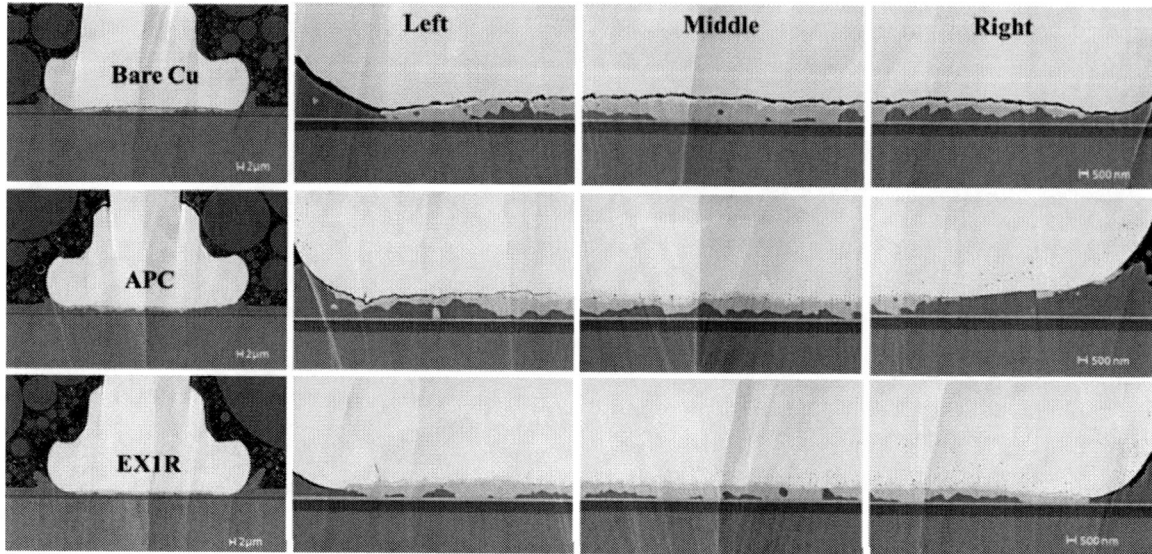

Figure 6: SEM results of the interface between bare Cu, conventional APC, EX1R wire and aluminum pad metallization after HAST. Bare copper shows a completely corroded interface while for the conventional APC wire faily corroded interface was found. The optimized APC wire "EX1R" shows only starting corrosion processes at the outer areas of the contact that stop after a few micrometer.

SEM analyses were performed at the interface between ball bond contact and silicon die metallization as well as of the stitch bond contact after high rate ion polishing. For the ball bond contact, the bare copper wire shows a nearly complete corroded interface and conventional APC wire also shows a fairly corroded interface. The optimized APC wire "EX1R" shows a starting corrosion at the outer areas of the ball bond contact. The corrosion stops after a few micrometer. Comparable TEM analyses of the interface area show that the optimized "EX1R" has a palladium modified intermetallic compound at these area that prevents the progression of corrosive attack. Additionally alloying element is enriched on top of the copper rich IMC preventing the chemical reactions taking place as described above.

The the stitch bond contacts show some attacked areas for the conventional APC wire and the optimized APC wire "EX1R". For the EX1R wire these areas are very small compared to the conventional APC wire. High resolution TEM analyses show that the chemical reaction between Pd and Cu is stopped, as the attacked area is passivated by an oxidizing reaction of the alloyed elements at the corrosion front.

A. Scanning Electron Microscopy after HTS (ball bond contacts)

Figure 6 shows the results of SEM analyzes comparing bare copper wire, conventional APC wire and optimized APC wire "EX1R". The bare copper wire has a completely corroded interface between the wire bond material and silicon die metallization, and the conventional APC wire also has a fairly

corroded interface. The IMC that is next to the aluminum metallization seems not attacked. The corrosion area is only 50-100 nm thick but results in a complete delamination of the ball bond contact. The interface of the optimized APC wire "EX1R" and the pad metallization shows only a starting corrosion at the outer areas. The corrosion stops after a few micrometer and the main interconnection area is free from any corrosion processes.

In addition, nano-scale voids are observed in the vicinity of bonded ball surface for all three wire types. These voids are caused by the flame-off process for FAB formation and are considered not affecting the corrosion process; previous experiment indicates that the positions of those voids do not change before and after HTS testing.

B. Transmission Electron Microscopy after HTS (EX1R ball bond contact)

Figure 7 shows a high resolution TEM image of the intermetallic compound formed at the interface between EX1R wire material and Aluminum pad metallization after HTS testing. The interconnection area shows two IMCs. One is located near to copper and one is located near to aluminum.

The nano-spot EDS mapping in Figure 8 illustrates, as described for HAST testing in [10], a protective layer of the alloyed element is formed on top of the copper rich intermetallic compound. Additionally in Figure 8 the implementation of Pd in the Cu-Al-IMC can be seen clearly. These modification prevent the chemical reactions taking place in molded packages with bare copper or PCC/APC wire materials. Compared to bare copper wire and conventional APC wire a higher reliability of "EX1R" wire is recognizable.

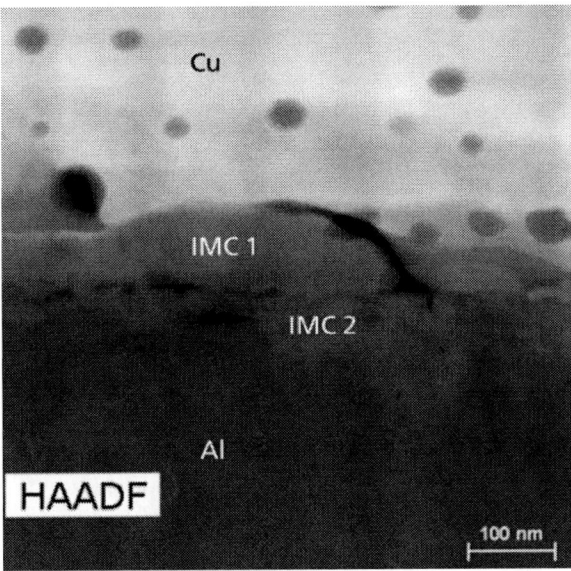

Figure 7: High resolution TEM image of interconnection area between EX1R wire and aluminum pad metallization. Two different intermetallic compounds are visible. No corrosion processes are detectable.

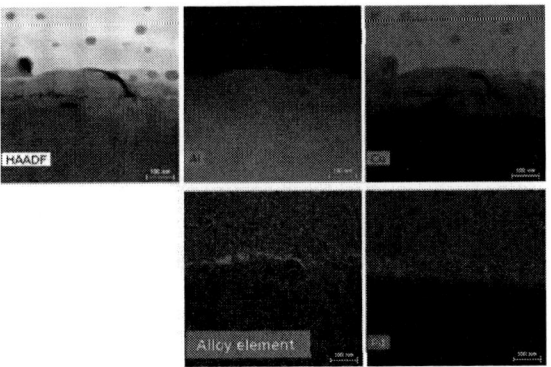

Figure 8: Nano-spot EDS mapping of the interface between EX1R wire and aluminum pad metallization showing the presence of palladium in the intermetallic compound and enrichment of the alloyed element between IMC and copper wire material.

C. Scanning Electron Microscopy after HTS (stitch bond contacts)

SEM analyses at cross sectioned stitch contacts show degradation of the copper bulk material especially for the conventional APC and the EX1R wire. The bare copper wire seems to be unaffected by any attack. Figur 10 shows an overview for all three wire types. There are voids/decomposed material of several μm detectable in the copper bulk for the conventional APC wire at the stitch contact. Also to wire itself is affected. The optimized APC wire "EX1R" shows only minor voids/decomposed material at the stitch bond contact. The wire material far from the bond contact does not show any defects.

D. Transmission Electron Microscopy: EX1R wire (stitch bond contact)

Additional TEM analyses were performed for the EX1R stitch bond contact to investigate the voids formed at the copper bulk material. Figure 9 shows this area right below the Pd coating. The decomposed material covers an area of about 300-400nm. Nanospot EDS analyses, see Figure 11, reveals the presence of sulfur in this area. The interface to copper bulk, that is not affected, is covered by an alloying element Also a clear oxgen signal is detectable. Thus, it can be determined that the corrosion process is stopped by a passivation reaction of this alloying element.

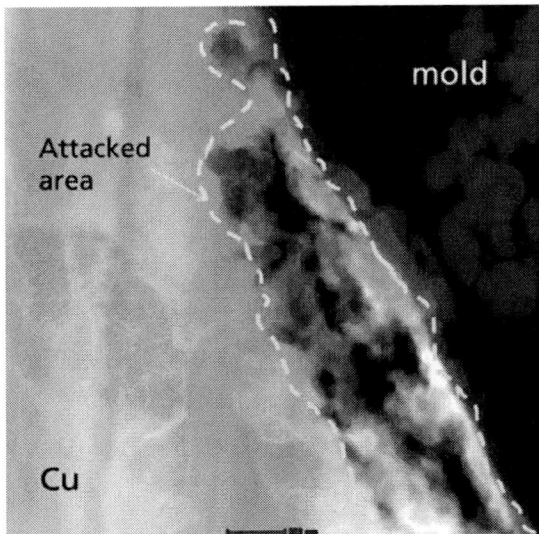

Figure 9: High resolution TEM image of the attacked area at the EX1R stitch bond contact.

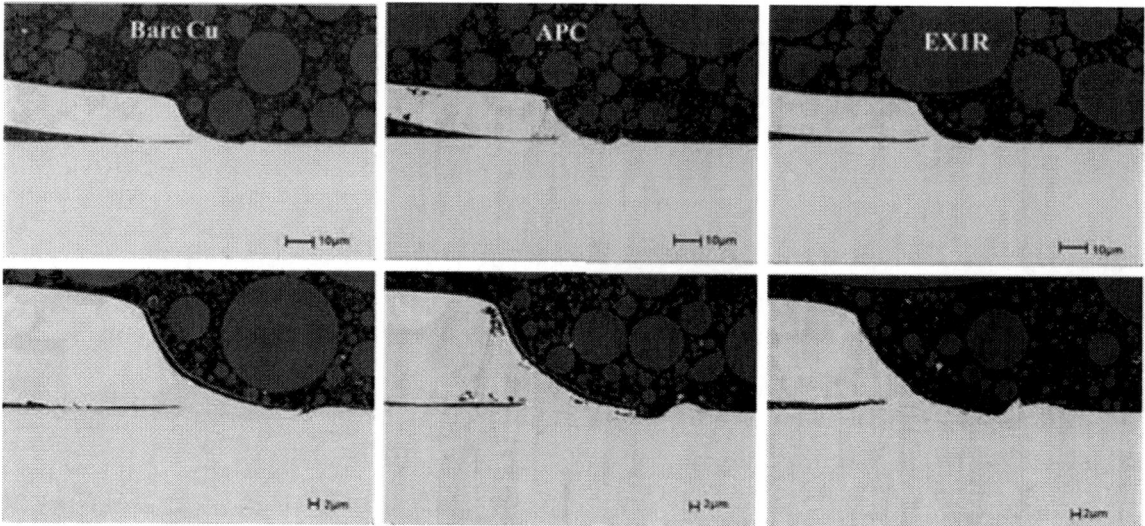

Figure 10: SEM results of the cross sectioned stitch contact for bare Cu, conventional APC and EX1R wire. Conventional APC wire shows voids and degraded areas in te copper bulk material. Optimized APC wire EX1R is affected less having only minor voids. Bare copper seems not to be affected by any degradation effect.

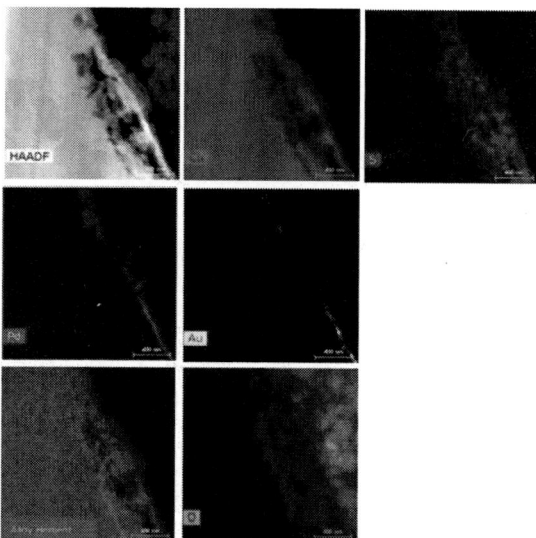

Figure 8: Nano-Spot EDS mapping showing the element distribution at the attacked area of Figure . The alloyed element is enriched at this area and prevent the corrosion reaction by a passivation reaction.

Summary and Conclusion

We presented a study of the new APC bond wire "EX1R" with potential to be applied for robust automotive applications. We showed results for its interconnection stability in HTS conditions in comparison to a bare copper and conventional APC wire. The wire material of "EX1R" was developed to prevent halide induced interface corrosion and also sulfur induced pitting corrosion. The base material was alloyed with special elements to systematically adjust the intermetallic formed at the interface to the pad metallization and to define the microstructure of the copper base material to prevent sulfur induced corrosion.

SEM analyses were done after HTS testing at the interface between wire bond contact and silicon die metallization after high rate ion polishing for all samples. Bare copper wire shows a nearly complete corroded interface and also at the interconnection of the conventional APC wire a fairly corrosive attack is detectable. The optimized APC wire "EX1R" shows a starting corrosion at the outer areas of the ball bond contact. But therefore, the corrosion stops after a few micrometer. The optimized "EX1R" shows a palladium modified intermetallic compound at these area that slows down the corrosive attack. Comparable results were published by März et al. for Au-Al IMC systems [15] and also for the EX1R wire after HAST testing [10]. For APC wire "EX1R" the alloying element is enriched on top of the copper rich intermetallic compound additionally preventing the chemical reaction taking place in the system Pd-Cu-S-Halide. Both effects lead to a high corrosion resistance of wire material and the bond contacts formed between wire material and pad/substrate metallization.

Additional SEM analyses at the stitch bond contacts for bare copper, conventional APC and optimized APC wire EX1R show degradation of the copper bulk material especially for the conventional APC and minor attack for the the EX1R wire. The bare copper wire seems to be unaffected by any attack. High resolution TEM analyses reveal the presence of sulfur in this area. Additionally an enrichment of alloyed element can be detected for the EX1R wire. The interface to the not affected copper bulk is covered by this element. Also a clear oxgen signal is detectable. Thus, it can be determined that the corrosion process at the optimized APC wire EX1R is stopped by a paasivation reaction of alloying element.

As halides and sulfur are common impurities in microelectronic packaging (e.g. chlorine in low ppm range in mold compounds, fluorine in low ppm range on aluminum pad metallization and sulfur in adhesion promoters in mold compounds) it is very important to use a modified wire material that will not react with these substances or limit their aggressive behaviour. The results show that the optimized APC wire "EX1R" is nearly inert to common corrosion processes in microelectronic packaging, indicating a much more higher reliability compared to common PCC, APC or bare copper wire types.

References

[1] Han, Mingchuan, et al. "Copper wire bond pad/IMC interfacial layer crack study during HTSL (high temperature storage life) test." 2016 IEEE 18th Electronics Packaging Technology Conference (EPTC). IEEE, 2016.

[2] M. Eto, T. Haibara, R. Oishi, T. Yamada, T. Uno, T. Oyamada "Thermal bond reliability of high reliability new palladium-coated copper wire," Electronic components and Technology Conference (ECTC), 2017, pp.1297-1302.

[3] M. Eto, T. Haibara, R. Oishi, T. Yamada, T. Uno, T. Oyamada "Newly developed high reliability palladium coated copper wire for automotive application," European Microelectronics Packaging Conference (EMPC), 2017.

[4] Kah, Paul, et al. "Factors influencing Al-Cu weld properties by intermetallic compound formation." International Journal of Mechanical and Materials Engineering 10.1 (2015): 10.

[5] Lim, Adeline BY, et al. "Evaluation of the corrosion performance of Cu–Al intermetallic compounds and the effect of Pd addition." Microelectronics Reliability 56 (2016): 155-161.

[6] Abe, Hidenori, et al. "Cu wire and Pd-Cu wire package reliability and molding compounds." 2012 IEEE 62nd Electronic Components and Technology Conference. IEEE, 2012.

[7] Lee, Chu-Chung Stephen, et al. "Copper versus palladium coated copper wire process and reliability differences." 2014 IEEE 64th Electronic Components and Technology Conference (ECTC). IEEE, 2014.

[8] Boettcher, Tim, et al. "On the intermetallic corrosion of Cu-Al wire bonds." 2010 12th Electronics Packaging Technology Conference. IEEE, 2010.

[9] Qin, Ivy, et al. "Molded reliability study for different Cu wire bonding configurations." 2013 IEEE 63rd Electronic Components and Technology Conference. IEEE, 2013.

[10] S. Klengel, et al. " A new reliable, corrosion resistant gold-palladium coated copper wire material." 2019 IEEE 69th Electronic Components and Technology Conference (ECTC). IEEE, 2019.

[11] Saruwatari, Tatsunori, et al. "Reliability improvement of Cu-wire bonded lead frame package for automotive applications." 2017 Pan Pacific Microelectronics Symposium (Pan Pacific). IEEE, 2017.

[12] Lee, Chu-Chung, et al. "Copper ball voids for Pd-Cu wires: Affecting factors and methods of controlling." 2016 IEEE 66th Electronic Components and Technology Conference (ECTC). IEEE, 2016.

[13] R. Rongen, et al. " Package Material Selection Criteria for High Temperature Automotive Applications." 2019 IEEE 69th Electronic Components and Technology Conference (ECTC). IEEE, 2019.

[14] C. J. Thompson, R. A. Meyer, J. S. Ball, "Thermal Decomposition of Sulfur Compounds. II. 1-Pentanethiol", J. Am. Chem. Soc., 1952, 74 (13), pp 3287–3289.

[15] Lee, Chu-Chung, et al. "Copper ball voids for Pd-Cu wires: Affecting factors and methods of controlling." 2016 IEEE 66th Electronic Components and Technology Conference (ECTC). IEEE, 2016.

[16] Krinke, Jörg C., et al. "High temperature degradation of palladium coated copper bond wires." Microelectronics Reliability 54.9-10 (2014): 1995-1999.

[17] März, B., et al. "Investigation of the palladium distribution in the intermetallic phase region of Au-Al wire bond interconnects." 2012 4th Electronic System-Integration Technology Conference. IEEE, 2012.

Investigation of the Microstructure of Mn-doped Tin-Silver-Copper Solder Alloys Solidified with Different Cooling Rates

Tamás Hurtony, Oliver Krammer, Balázs Illés, Péter Gordon

Budapest University of Technology and Economics, Department of Electronics Technology
H-1111, Egry József u. 18., Budapest, Hungary

+36 1 463 4268, hurtony@ett.bme.hu

Abstract

Due to the moderate price and the non-toxicity, manganese is considered as an ideal dopant for the SAC (SnAgCu) solder alloys. Manganese refines the grain of solder joints, yielding better thermomechanical properties. In present research, the microstructure of the manganese-doped alloys solidified with different technological parameters had been investigated. Sn/Ag0.3/Cu0.7 based solder alloy with three different Mn content (0.1, 0.4, 0.7% wt%) were reflowed on a copper substrate with tempered hot plate. They were solidified with different cooling rates from 0.3 to 4.5 K/s. Cross-sections have been prepared from the solder samples and the metallographic properties of the solder samples was investigated with optical and scanning electron microscopes. The characteristic features of the samples have been compared to conventionally used SAC305 (Sn/Ag3/Cu0.5) solder alloys, solidified with the same rates of cooling. Results showed that besides the grain refinement, the Mn content might also have effect on the evolution of intermetallic layer between the substrate and the solder alloy. The IMC grains of the layer were more elongated and more spalled grains had been observed close to the layer. However, independently of the cooling rate, the microstructure of the Mn containing solder alloys remained the same. This suggests that the macroscopic properties are also expected to be less sensitive to the cooling rate of the solidification.

Key words: lead-free soldering; manganese doping; cooling rate; selective electrochemical etching; scanning electron microscopy.

1. Introduction

The most common soldering technology for the automated, mass manufacturing of electronics devices is the reflow soldering technology [1] either by convection heating or by vapour phase [2,3].The restriction of hazardous substances urged the electronic industry to ban the conventional lead-based solder alloys in the application of reflow soldering and to introduce the lead-free substitution of these materials. The change over was not easy; the appropriate set of the reflow profile became harder due to the narrower technological window [4], and the first generation of lead-free solder alloys were lagging behind the conventional eutectic solders in both mechanical and reliability properties. Nowadays, the most commonly applied second-generation alloy is the Sn-Ag-Cu (SAC). Huge number of researches is investigating the properties of SAC alloys, and according to them, the process parameters and the mechanical behaviours of some SAC solders are more superior compared to lead-bearing alloys. However in certain applications, for example under extreme operation, e.g. overpressure [5] or extreme process environment [6], these improved solder alloys might not perform adequately; therefore the development of third generation lead-free solder alloys, even doped with

microalloys (additives with less than 0.2 wt.%), or with nanoparticles is still a current topic.

The improved properties are pushed to the limit by adding minor alloying components (tracing elements) and/or nanoparticles into the SAC base solder in order to achieve e.g. better thermomechanical or drop-shock performance. As mentioned, one approach can be adding nanoparticles to the SAC solder alloy. Ani et al. added TiO_2 and NiO nanoparticles, and they found that these additions can improve the wettability of the alloy, and can reduce the rate of growth of the intermetallic layer. The highest fillet height was observed by adding 0.01 wt% NiO [7], and the lowest intermetallic layer thickness was reached by adding 0.05 wt.% TiO_2 [8]. Thus, these reinforcing nanoparticles can enhance the quality and reliability properties of solder joints for ultra-fine package assembly.Due to the moderate price and the non-toxicity, the manganese is considered as an ideal dopant for the SAC solder alloys. Mn reported to have a grain refinement effect on the SAC solder alloys both in nanoparticle [9] and in micralloyed [10] form. At soldering temperature and below that the solubility of Mn in the Sn base solder alloy is relatively low; so even at very low concentration, Mn containing micro precipitation might appear inside the solder bulk which may act as a

130

crystallization centre for the solidification of the SAC alloy. As the consequence of finer microstructure, these alloys might be more resistant against intensive and dynamic external mechanical exposure, such as drop tests [11]. The addition of minor composing element might also give the possibility to reduce the silver content of the SAC solder which might yield both in reliability increase [12] and in significant cost reduction.

2. Materials and methods

The structural analyses and the measurement of intermetallic layer thickness was carried out for Sn/Ag0.3/Cu0.7 based solder alloys doped with 0.1, 0.4, 0.7 wt.% manganese and for a SAC305 (Sn/Ag3/Cu0.5) alloy as a reference. The solder bar had been pre-formed to solid solder wires with a diameter of 1 mm by cold forming wire process. The bumps were formed by cutting solder wires (ø1 mm) to an exact length of 2 mm. This was achieved by preparing laser marks into the wire at equidistant positions. The solder preform had been cleaned with sonic bath in isopropanol. The solder preform had been reflowed on a 0.2 mm thick pure copper sheet which was previously cut into 15x5 mm substrates. In order to obtain relatively same geometry of the reflowed solder balls, solder mask was applied to prevent uncontrolled wetting by the solder balls. Kapton® tape was adhered onto the surface of the substrate and an exactly 900 µm diameter round aperture was cut on the mask also with laser. Four apertures had been formed on every substrate for reflowing the balls made from the four different alloys on the same substrate. This resulted in solder balls with approximately 2 mm in diameter, and with a solder substrate interface of 1.2 mm length at the symmetrical axis of the balls (Fig. 1).

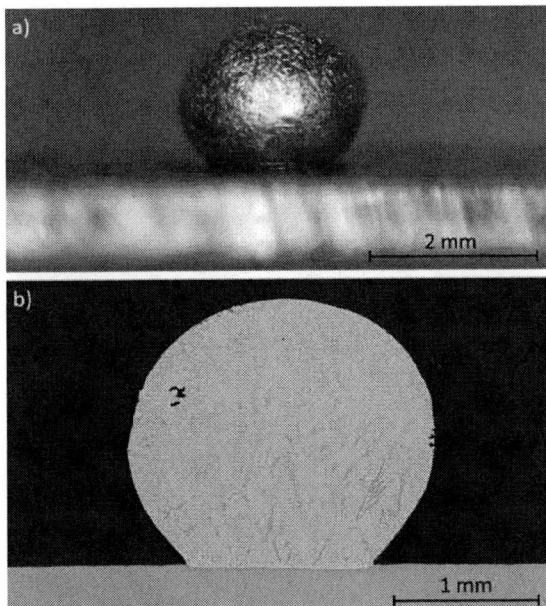

Figure 1: The formed solder ball: a) optical microscopy image; b) cross-section

The application of such geometry provided high enough solder volume for not limiting the dissolution of the copper atoms into the molten solder, while the spherical shape of solder bump ensured the uniform heat distribution all over the solder interface. Furthermore the morphology of the intermetallic layer is expected to be similar in all direction.

The solder balls had been reflowed on a custom made hot plate equipment. The heat was applied by a power resistor in order to have the minimal obtainable thermal inertia of the hot plate. Forced convectional cooling was applied to set the different cooling rates of the samples. A 3D printed nozzle was designed accommodating to the bottom of the power resistor to a fixed position (Fig. 2). The pressure of the blowing air was set by a regulator. Thermal interface material was applied between the hot plate and the substrate, and the temperature was measured on the hot plate by thermocouples.

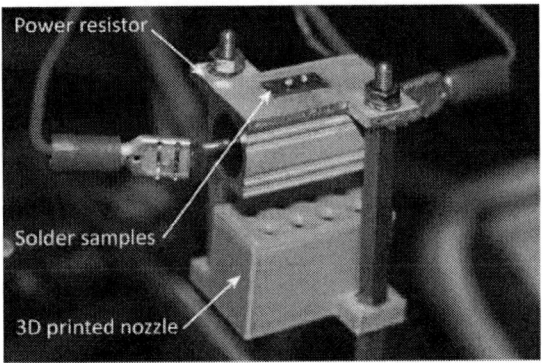

Figure 2: The custom made hot plate for preparing the solder ball samples with different cooling rate

The rates of cooling were set to 0.3, 2.5 and 4.5 K/s. The reflow profiles for cooling rates of are illustrated in Fig. 3.

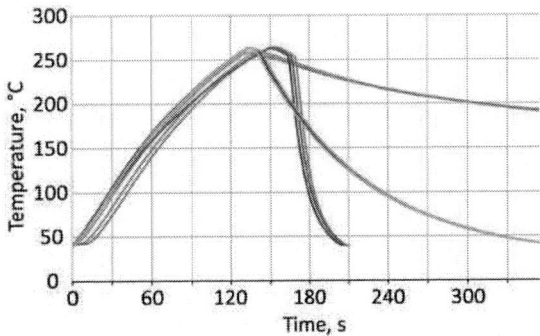

Figure 3: The applied reflow profiles

Once the peak temperature of the hotplate (260 °C) was reached, the current load of the resistor was turned off and the air stream was released by a valve. The time needed for reaching the peak temperatures was approximately 2 minutes in every case. After melting the solders, they were in liquid form for at least 15 seconds before applying the cooling. As it can be seen in Fig. 3., the temperature profiles of consecutive experiments were really similar to each other. The applied cooling rates of samples are reproducible. After the reflow soldering, the solder balls had been cross-sectioned. The microstructure of the solder joints and the thickness of the intermetallic layer inside them were analysed by scanning electron microscopy. Backscattered electron detector (BSE) was used for the structural analyses and EDX (energy-dispersive X-ray spectroscopy) was utilized for material composition measurements. The fine microstructure of the solder bulk had been revealed with a special selective electrochemical etching process [13].

3. Results

Due to the fact that no diffusion barrier layer was deposited onto the substrate significant copper dissolution was expected. The major part of the dissolved copper usually forms Cu_6Sn_5 intermetallic compound at the solder substrate interface but they can also crystalize inside the solder bulk. Higher amount of IMC was observed inside the solder bulk in the case of the slowly cooled samples (Fig. 4). This is because the total time above liquidus is higher, therefore higher amount of Cu atom can be dissolved into the solder bulk.

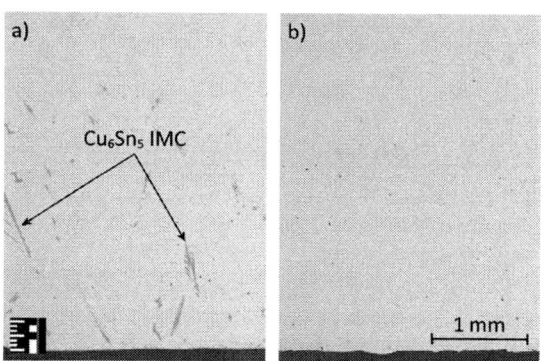

Figure 4: SEM micrographs of the SAC305 samples: a) solidified with 0.3 K/s; b) solidified with 2.5 K/s

The average thickness of the intermetallic layer was also slightly higher in the case of the slowly solidified samples (Fig. 5).

Figure 5: The intermetallic layers of the SAC305 samples: a) solidified with 0.3 K/s; b) solidified with 2.5 K/s

Due to the moderate cooling rate needle-like primary Cu_6Sn_5 IMCs had been formed inside the

solder bulk (Fig. 6). Similar trend was observable in the case of all Mn containing samples, but the needle-like crystals of primary Cu_6Sn_5 IMCs were not such long as it was observed in the case of the SAC305 sample. The morphology of the intermetallic layer of both Mn containing and SAC305 solder samples were relatively similar to each other. Continuous and scallop-type intermetallic layer was formed in the case of every sample. However, the intermetallic layer of the 0.1 wt.% Mn containing solder were slightly different from the others. Independently of the cooling rate, the IMC grains of the layer were more elongated, and more spalled grains had been observed close to the layer in this sample.

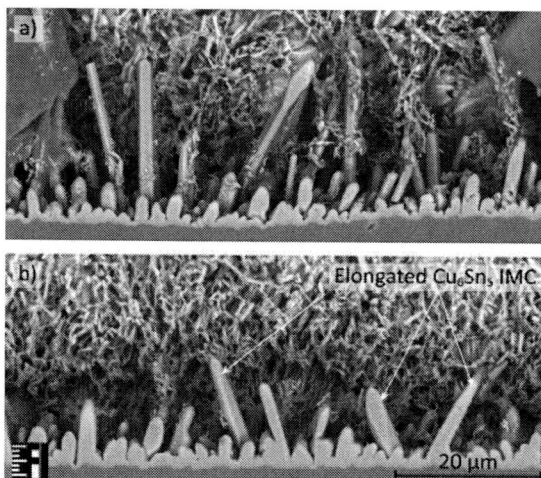

Figure 6: The intermetallic layer of the SAC-Mn01 solder samples: a) solidified with 0.3 K/s; b) solidified with 2.5 K/s

During the selective electrochemical etching process only the pure β-tin phases had been extracted from the cross-sectioning plane of the samples in a depth of approximately 50 μm. The dendritic arms of the tin phases had been extracted and the Ag_3Sn intermetallic compound remained in the interdendritic regions. They are forming a fibrous like network which strengthens the structure of the solder bulk. No plate-like primary Ag_3Sn IMC had been observed on any of the samples (Fig. 7).

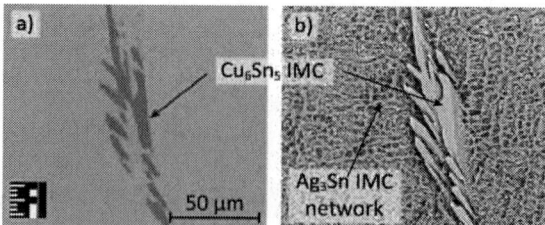

Figure 7: SEM micrographs of the SAC305 solder samples solidified with 0.3 K/s: a) before the selective etching; b) after the etching process

The manganese precipitates appears as granular features inside the solder bulk (Fig. 8). These features act like nucleation centers for the

132

Ag$_3$Sn IMCs, therefore they are usually forming around the Mn particles.

Figure 8: Material composition map of a SAC-Mn04 solder sample. Mn precipitates are observable surrounded by Ag$_3$Sn intermetallics.

Regardless of the cooling rate, the microstructure of the Mn containing solders was relatively similar to each other. The microstructure of the solder bulk containing any Mn was significantly different than that was observed in the case of SAC305 sample (Fig. 9). The Ag$_3$Sn IMCs are also rather forming needle-like structures, and the presence of them is not restricted to any interdendritic region. In fact no dendritic structures of tin phases had been recognized inside the solder bulk of Mn containing solder. The Mn precipitations are most likely preventing the formation of regular β-tin dendrites.

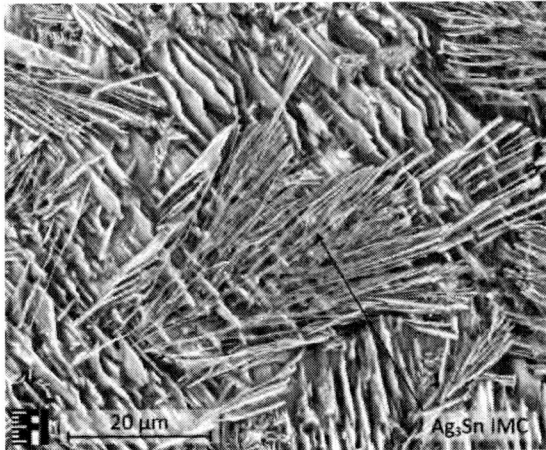

Figure 9: SEM micrographs of the SACMn01 samples solidified with 2.5 K/s

Comparing the microstructure of the Mn containing solder close and relatively far from the substrate solder interface it can be observed that the relative abundance of manganese precipitates is significantly higher close to the interface. Since the density of the Mn atoms is only slightly higher than the density of the Sn this cannot be explained by the gravitational forces. More likely this is caused by

the segregation of Mn precipitates during the solidification. Although the hot plate is cooled directly by the air stream, the solder balls are not solidifying from bottom to top. Because of the free convection they are also loosing heat from above, therefore the solidification of the balls are usually occurring from top to bottom and as the consequence of this the Mn are precipitates are segregating to the bottom of the solder.

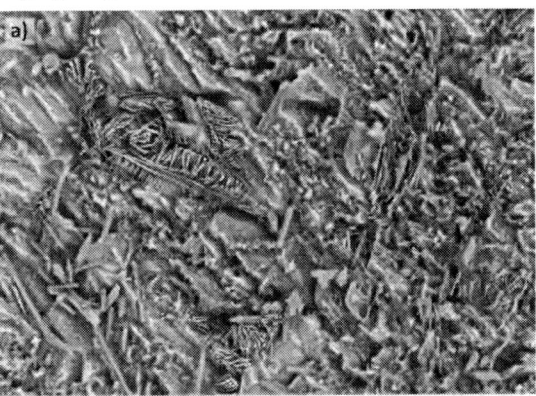

Figure 10: SEM micrographs of SAC-Mn07 solder samples solidified with 0.3 K/s cooling rate: a) close to the solder-substrate interface; b) far from it

4. Conclusions

The microstructure of Mn containing solder alloy solidified with different cooling rate had been observed and compared to commercial SAC305 solder alloy. It was found that even 0.1 weight percent of Mn content can significantly alter the microstructure. Due to the grain refinement effect of the Mn particles inside the solder bulk, no β-tin dendrites can be formed and the Ag3Sn IMCs are not only present inside the interdenritic regions, but they are structurally distributed inside the solder.

Independently of the cooling rate of the solder samples the microstructure of the Mn containing solder alloys remained stationary. This suggests that the macroscopic properties are also expected to be less sensitive to the cooling rate of the solidification. Significantly higher amount of Mn particles had been observed close to the substrate of the solder which might be in correlation with the directionality of the solidification of the solder samples. As the solder is solidifying from top to the bottom the Mn particles are being segregated from the solution, and their concentration become higher close to the substrate.

Acknowledgements

This research was supported by the National Research, Development and Innovation Office – NKFIH, FK 127970.

References

[1] P. Veselý, E. Horynová, J. Starý, D. Bušek, K. Dušek, V. Zahradník, M. Dosedla, "Solder joint quality evaluation based on heating factor", Circuit World, Vol. 44, No. 1, pp. 37-44, 2018.

[2] A. Géczy, "Investigating heat transfer coefficient differences on printed circuit boards during vapour phase reflow soldering", International Journal of Heat and Mass Transfer, Vol. 109, pp. 167-174, 2017.

[3] L. Livovsky, A. Pietrikova, "Measurement and regulation of saturated vapour height level in VPS chamber", Soldering & Surface Mount Technology, early cite, 2019. https://doi.org/10.1108/SSMT-10-2018-0040

[4] F. Steiner, V. Wirth, M. Hirman, "Relationship of Soldering Profile, Voids Formation and Strength of Soldered Joints", Proceedings of the 42nd International Spring Seminar on Electronics Technology (ISSE 2019), Wroclaw, Poland, May 15-19, 2019.

[5] W.Y.W. Yusoff, N. Ismail, N.S. Safee, A. Ismail, A. Jalar, M.A. Bakar, "Correlation of microstructural evolution and hardness properties of 99.0Sn-0.3Ag-0.7Cu (SAC0307) lead-free solder under blast wave condition", Soldering & Surface Mount Technology, Vol. 31, Issue 2, pp. 102-108, 2019.

[6] A. Geczy, M. Fejos, L. Tersztyánszky, "Investigating and compensating printed circuit board shrinkage induced failures during reflow soldering", Soldering & Surface Mount Technology, Vol. 27, Issue 2, pp. 61-68, 2015.

[7] F.C. Ani, A. Jalar, A.A. Saad, C.Y. Khor, M.A. Abas, Z. Bachok, N.K. Othman, "Characterization of SAC – x NiO nano-reinforced lead-free solder joint in an ultra-fine package assembly", Soldering & Surface Mount Technology, Vol. 31, Issue 2, pp.109-124, 2019.

[8] F.C. Ani, A. Jalar, A.A. Saad, C.Y. Khor, R. Ismail, Z. Bachok, M.A. Abas, N.K. Othman, "SAC–xTiO2 nano-reinforced lead-free solder joint characterizations in ultra-fine package assembly", Soldering & Surface Mount Technology, Vol. 30, Issue 1, pp. 1-13, 2018.

[9] Y. Tang, S. Luo, G. Li, Z. Yang, C. Hou, "Effects of Mn nanoparticle addition on wettability, microstructure and microhardness of low-Ag Sn-0.3Ag-0.7Cu-xMn(np) composite solders", Soldering & Surface Mount Technology, Vol. 30, Issue 3, pp. 153-163, 2018.

[10] L.W. Lin, J.M. Song, Y.S. Lai, Y.T. Chiu, N.C. Lee, J.Y. Uan, "Alloying modification of Sn–Ag–Cu solders by manganese and titanium", Microelectronics Reliability, Vol. 49, pp. 235-241, 2009.

[11] W. Liu, N.C. Lee, A. Porras, M. Ding, A. Gallagher, A. Huang, S. Chen, J.C.B. Lee, "Achieving high reliability low cost lead-free SAC solder joints via Mn or Ce doping", Electronic Components and Technology Conference, pp. 994-1007, 2009.

[12] B. Medgyes, E. Roman, A. Bohnert, S. Szurdán, X. Zhong and G. Harsányi, "Electrochemical Migration Investigations on SAC-Bi-xMn Solder Alloys", 2018 IEEE 24th International Symposium for Design and Technology in Electronic Packaging (SIITME), Iasi, pp. 80-212, 2018.

[13] T. Hurtony, A. Bonyár, P. Gordon, G. Harsányi, "Investigation of intermetallic compounds (IMCs) in electrochemically stripped solder joints with SEM", Microelectronics Reliability, Vol. 52, pp. 1138-1142, 2012.

TLPB Improves Solder Connections by On Chip Creation of Intermetallic Phase Precursors

Anna Steenmann, Jessica Richter, Benjamin Schellscheidt, Thomas Licht

Faculty of Electrical Engineering and Information Technology
Department of Microelectronics
HSD University of Applied Sciences Düsseldorf,
Münsterstraße 156, 40476 Düsseldorf, Germany
+49 211 4351 3564   anna.steenmann@hs-duesseldorf.de

Abstract

In this paper, we present a conceptual design of an on-chip solder stack to connect silicon devices faster and more reliable. Almost all electric devices rely on solder layers to provide electrical and mechanical connections between components. We improve the solder connection at industry standard solder parameters of 300°C and some minutes of solder time. The solder stack is implemented as a transient liquid phase bonding (TLPB) system to realize interconnections. The creation of extremely thin layers accelerates the formation of intermetallic phases (IMPs) in solder processes. By using this effect, a dramatic increase in the diffusion dynamic can be achieved. We created the solder stack with Ficks's law of diffusion using diffusion constants of copper and tin to deliberately form the phase Cu_3Sn.

Key words: copper-tin stack; interconnection technologies; intermetallic phases (IMPs); reliability; soldering; transient liquid phase bonding (TLPB)

Introduction

Today nearly every electric device relies on solder layers to provide electrical and mechanical connections between components. Solder connections are produced at process temperatures of 250°C to 450°C applied for some seconds or minutes with a typical solder thickness of about 50μm. The process time and temperature depend on component size and composition of materials. [1]

An ideal solder connection is composed of intermetallic phases (IMPs) at the interfaces between device and solder, and substrate and solder. Typically, a thin region of tin remains between the two IMP layers. IMPs of copper and tin are Cu_6Sn_5 and Cu_3Sn. The development of IMPs is decisive for a good mechanical connection because of their higher melting point and mechanical stability. Under mechanical and thermal stress the connection usually fails at its weakest point, the thin layer of tin. [2]

In todays and future applications, the reliability of solder connections is important for a long-lasting electrical device. Solder connections are expected to withstand high thermal and mechanical stress to achieve a long lifetime and high reliability. To achieve these requirements, connection technologies are subject to continuous research and improvement. Since solder connections often are the weakest link regarding the reliability of electric components [3] alternative joining technologies are becoming more popular, such as sintering, conductive adhesive or transient liquid phase bonding (TLPB).

In this paper we investigate a solder stack on silicon chips to connect the silicon device directly by TLPB. For this we use the diffusion of a high melting component in a second component that is liquid at process temperature. Ongoing diffusion leads to the formation of IMPs with a melting point above process temperature, thus solidifying the connection at a constant temperature. [5] By this isothermal solidification, the solder contact becomes more durable against mechanical and thermal load and can be used at high temperatures exceeding 300°C. [1]

Methods

We investigated a TLPB solder connection for a typical high-temperature application like power electronics. Especially in the automotive sector, power electronics depend on huge copper areas and thicknesses to provide good thermal conductivity. For this application, we bonded silicon chips on direct copper bonded substrates (DCB-substrates) with a specially designed stack of thin copper and tin layers. This supports the transformation of the solder connection to the more stable IMP Cu_3Sn.

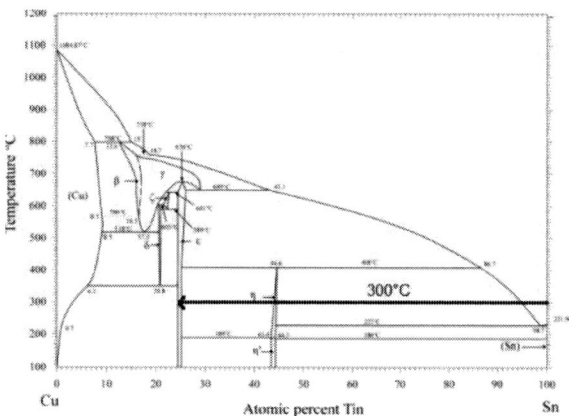

Figure 1: Copper-tin phase diagram [4]

Figure 1 shows a phase diagram of copper (Cu) and tin (Sn) where the isothermal solidification into IMPs at 300°C is shown by the arrow. Tin becomes liquid by heating up the system to 300°C. Over time copper diffuses into tin and increases the concentration of copper in tin. This process, represented by the arrow in the phase diagram, leads to the formation of solid IMPs lowering the amount of liquid tin. A micro section of the resulting bond layer is shown in figure 2.

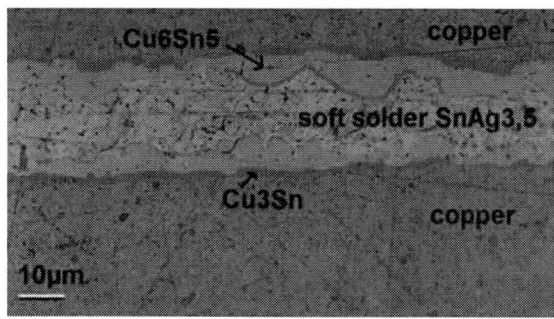

Figure 2: Micro section of a phase formation between copper and tin. The red line denotes the interface between Cu_6Sn_5 and soft solder

Figure 2 depicts a micro section of a copper-tin-copper connection with IMPs. Copper and tin form IMPs of two possible chemical compositions, Cu_6Sn_5 and Cu_3Sn. Cu_6Sn_5 is the faster building phase with a melting point of about 415°C. Cu_3Sn grows as a thin layer at the contact area, Cu_6Sn_5 as lumped accumulations adjacent to the Cu_3Sn phase. Over time copper diffuses into liquid tin and Cu_6Sn_5 to transform the whole connection into Cu_3Sn with a melting point of about 673°C. [5]

Calculation

For the construction of the solder layer directly onto a silicon chip we aimed for an atomic ratio of three parts copper and one part tin to build the Cu_3Sn phase. Required time of diffusion depends on soldering temperature and required diffusion length, which is affected by the layer thickness. Equation (1) demonstrates the relation of the parameter for solder connections. Therefore a reduced time of diffusion is possible by increasing the temperature T and decreasing the thickness of the diffusion length x_D. [6]

$$x_D^2 = 4 \cdot D(T) \cdot t \qquad (1)$$

To allow usage of standard industrial soldering equipment, we aimed for a process temperature of 300°C. With the frequency factor D_0 for copper $0,16 \cdot 10\text{-}4 m^2/s$, $7,7 \cdot 10\text{-}4 m^2/s$ for tin, and activation energy Q 199kJ/mol for copper and 107kJ/mol for tin we calculate the diffusion constant of copper and tin with (2) [5]. The diffusion constant of copper in tin at 300°C is calculated with (3). [6]

$$D(T) = D_0 \cdot e^{-(Q/kT)} \qquad (2)$$

$$D_{ab}(T) = D_a(T) \cdot w_b + D_b(T) \cdot w_a \qquad (3)$$

For a desired solder time of a few seconds, (1) yields a layer thickness of approximately 800nm. The maximum diffusion length of 800nm needs about 2 seconds at 300°C to transform to the Cu_3Sn phase.

The objective of this paper is the creation of a solder layer directly on silicon chips to connect them to a metalized ceramic board like a DCB. To ensure good wetting, a solder thickness of several micrometers is needed to compensate for DCB surface roughness. We built a solder stack of copper and tin that meets all aforementioned requirements of the solder connection. Figure 3 depicts the layer build-up of the copper-tin stack.

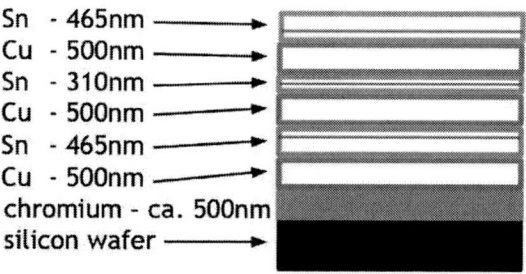

Figure 3: Copper-tin stack with chromium adhesion layer on silicon wafer

Experimental Results

Solder Stack

We deposited chromium as an adhesion layer via magnetron sputtering on a silicon chip followed by the copper-tin solder stack. The next copper layer (500nm) transforms 311nm tin into Cu_3Sn. Approximately half of the next 500nm copper layer diffuses into the lower tin layer and transforms the rest of the 465nm tin layer into Cu_3Sn phase. The other half of the copper diffused into the upper tin layer, transforms half of it into Cu_3Sn, and so on to

136

the final layer. The red line in tin layers illustrates the calculated diffusion length of copper into tin. The rest of the subsequent tin layer wetted the DCB and builds Cu₃Sn with DCB copper.

building of IMPs directly during the formation of the layers at room temperature. The phenomena we see here are very promising for a fast and reliable creation of IMPs using sputtering processes.

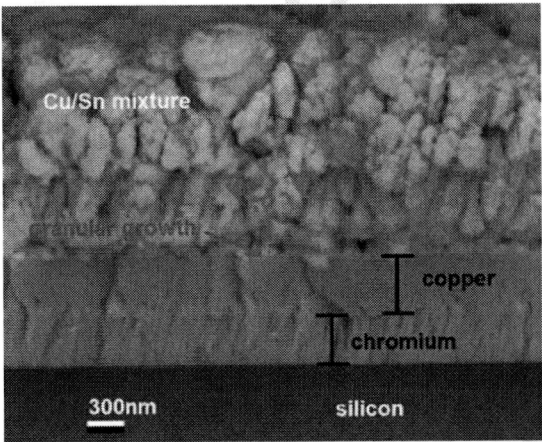

Figure 4: Side view of magnetron sputtered copper-tin stack (electron microscope image)

The magnetron sputtered multilayer of copper and tin is shown in figure 4. On the chromium layer a copper layer is followed by the first tin and next copper layer. We sputtered the layers consecutively without intermediate exposure to oxidizing atmosphere with a deposition rate of 0,5nm/s for copper and 2nm/s for tin. Table 1 shows all parameters for the sputter process.

Table 1: Parameters for the sputter process

material	Magnetron-sputter conditions			
	pressure range	temperature	growth rate	gas
copper	5·10-3 mbar to 5·10-1 mbar	no heating and no cooling	0,5nm/s	N₂/Ar
tin	5·10-3 mbar to 5·10-1 mbar	no heating and no cooling	2nm/s	N₂/Ar

When deposited by sputtering at room temperature, copper exhibits columnar growth patterns, whereas tin tends to form a layer of granular thickness. [7] Recognize the granular growth of the tin layer in figure 4. The first tin layer growth lumped on the columnar copper layer, leading to less well defined layers in the following stack. By using this effect a dramatic increase in the diffusion dynamic can be achieved.

The advantage of columnar copper and granular tin growth is a better mixture of copper and tin. Thereby the interface area increases and the required diffusion length decreases. The solder time for Cu₃Sn solidification is reduced.

The analysis of solder stacks before and after tempering shows no modification of the phase structure. This suggests the formation of IMPs before tempering or soldering. The columnar growth and mixture of copper and tin encouraged the

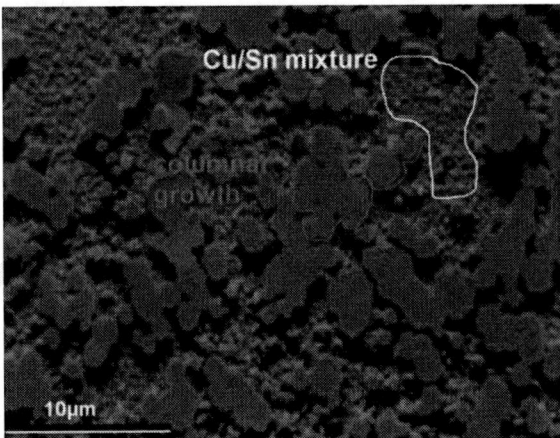

Figure 5: Top view of copper-tin stack on DCB with detached silicon chip (electron microscope image)

In a new series of test, we deposited an extra layer of tin for 200 seconds, which yields a layer height of 400nm, on the existing stack for enough tin to wet and contact the DCB surface. We then soldered the chip to the DCB in a formic acid atmosphere to reduce oxidation effects. Figure 5 shows the solder surface of the DCB with detached silicon chip.

After removing the silicon chip the solder stack is visible on the DCB. The copper columns exhibit a flat top with a copper-tin mixture between the gaps. This supports the assertion that copper and tin intermixed while sputtering and the required diffusion length decreased enough to form IMPs before soldering.

Diffusion Length

Another series of experiments show the behavior of copper diffusion into tin. To verify the calculated diffusion constant of copper into tin we evaluated the phase building at copper-tin interfaces. We studied the diffusion length at a composition of copper and soft solder as tin. The typically power electronic application has the advantage of more copper and soft solder that is permuted into IMPs in our experimental soldering processes (figure 2).

We varied soldering temperature and soldering time to determine the dependency of diffusion length. The composition is soldered for 10 minutes at 250°C, 300°C, and 400°C. Soldering time varied at 300°C for 10, 20, and 40 minutes. After soldering we cut and ground samples to evaluate the diffusion length via light microscopy.

The deviation of the calculated and evaluated diffusion length at 300°C soldering temperature is shown in figure 6.

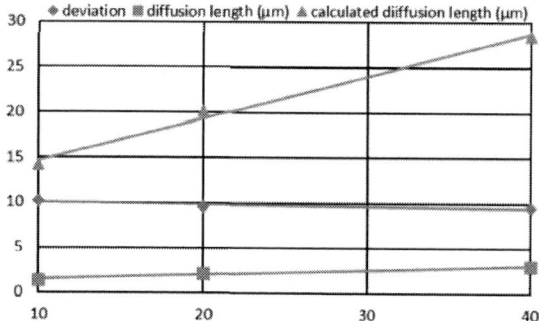

Figure 6: Deviation, calculated, and evaluated diffusion length at 300°C over various solder times from 10 to 40 minutes

The soldering time in figure 6 varied from 10 to 40 minutes. Both diffusion lengths increases and the deviation between evaluated and calculated diffusion length is nearly continuous factor 10. Due to diffusion behavior in IMPs is excluded in the used equations for calculation a deviation is expected. Ficks's law of diffusion is a model based on the assumption for pure metal diffusion. The case of diffusion in IMPs is excluded. This hypothesis is supported by the minor decrease of deviation between calculated and evaluated diffusion length. The more IMPs has been created, the lower is the deviation of diffusion lengths.

In addition to the soldering time we varied the soldering temperature from 250 to 400°C at 10 minutes soldering time (figure 7).

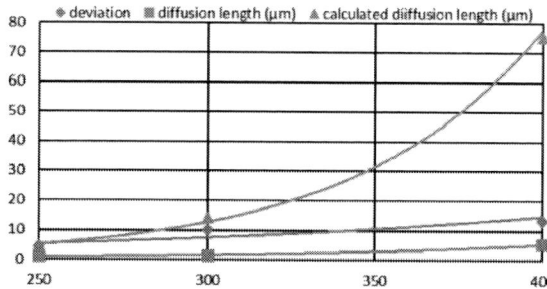

Figure 7: Deviation, calculated, and evaluated diffusion length at 10 minutes over various solder temperatures from 250 to 400°C

The calculated diffusion length of copper in tin increases exponential while the evaluation shows a less increasing diffusion length. The deviation between evaluated and calculated diffusion length increases over soldering temperature. This is an effect of the constructed diffusion constant at each temperature. The deviation factor is constant at linear time variation but diffusion length increase with the square of rising temperature. Consequently the factor of deviation increases due to higher temperatures.

Discussion

We designed a multilayer copper-tin-stack to connect silicon chips on DCBs by TLPB for a higher reliability in power electronic applications. To use the standard solder process temperature of about 300°C we deposited a thin layer stack directly on silicon chips. With this, the transformation to a solid solder connection is possible in seconds. Because of solid body and surface diffusion on every interface of copper and tin, IMPs solidify the solder connection at process temperature. The generated connection is capable of enduring higher mechanical and thermal load. It is usable at temperatures over 300°C without melting and with better mechanical stability. Low solder temperatures reduce thermal and mechanical stress and increase the reliability of electric devices and solder connections. [5]

In this paper we show the combination of columnar copper growth followed by granular growth of a tin layer. By a multilayer structure with six layers we achieved an extreme mixture utilizing the different growth mechanism. The diffusion model used here yields only an estimation of the actual diffusion mechanics. In our experiments, the formation of the IMP can be realized directly during sputtering of copper layer and following tin layer. Because of the good mixture, phase building is approximately completed after depositing the layer. The solder stack bonds only at the interface between copper of the DCB and subsequent tin layer with a large amount of IMPs for a durable solder connection. This process will be investigated in more detail in the future.

Another aspect in this paper is the evaluation of diffusion lengths from copper into tin, to design a new enhanced copper-tin stack. Evaluated diffusion length of copper into tin confirm Fick's equation of diffusion. The deviation of evaluated and calculated diffusion length is constant for desired soldering conditions and makes the assumption usable for further series of experiments.

With these results and minor modifications in a next series of tests it may be possible to create IMPs at every layer interface at process temperature and a more stable solder connection. The layers should mix less and the building of IMPs and solidification of the solder stack will take longer. With this it should be possible to prevent IMP formation before soldering.

With these findings, we will improve the solder stack to make connections more reliable by using advantages of TLPB and build IMPs while soldering for a durable solder connection. In general, every electric device performance can be improved by using TLPB for solder connections. The process is easily implemented in industrial soldering machines due to the standard process temperature and duration. It creates a connection with the advantage of a stable solder connection usable at temperatures over 300°C.

References

Book

[1] J. Wilde, Jahrbuch Mikroverbindungstechnik 2015/2016 p. 178 ff.– Performance – Vergleich von TLP-Bonding, Sintern, Kleben und Löten.

Conference publication

[2] D. Feil, T. Herberholz, M. Guyenot und M. Nowottnick, „Performancevergleich eines Diffusionslötprozesses auf Sn-Cu-Basis (HotPowCon) mit Weichlöten und Ag-Sinter" in Elektronische Baugruppen und Leiterplatten: EBL 2018: multifunktionale Aufbau- und Verbindungstechnik - Beherrschung der Vielfalt: Vorträge der 9. DVS/GMM-Tagung in Fellbach am 20. und 21. Februar 2018, S. 236–242.

Conference publication

[3] P. Jacob, „Reliability testing and analysis of IGBT power semiconductor modules" in IEE Colloquium on `IGBT Propulsion Drives', London, UK, Apr. 1995, S. 4.

Book

[4] T. B. Massalski, Hg., Binary alloy phase diagrams. Metals Park, Ohio: American Soc. for Metals, 1986.

Paper

[5] K. Bobzin, L. Bosse, E. Lugscheider, „3.2 Transient Liquid Phase Bonding" Springer-Verlag, 2005.

Paper

[6] D. Gupta, Hg., Diffusion Processes in Advanced Technological Materials. Berlin, Heidelberg: Berlin Heidelberg, 2005.

Paper

[7] B.A. Movchan, A.V. Demchishin; Structure and Properties of Thick Condensates of Nickel, Titanium, Tungsten, Aluminium Oxides, and Zirconium Dioxide in Vacuum
Fiz. Metal. Metalloved. 28: 653-60 (Oct 1969)

Design of Interfacial Chemistry for Inducing Low Temperature Sintering of Silver Micro-fillers within Epoxy-based Binders

Masahiro Inoue and Shiho Nakazawa

Gunma University, 29-1 Hon-chou, Ota 373-0057, Japan

masa-inoue@gunma-u.ac.jp

Abstract

Low temperature sintering of silver micro-particles was successfully promoted within an epoxy-based binder by controlling interfacial chemistry (with no nanoparticles and related materials) to obtain a low electrical resistivity (~1 x 10^{-5} Ωcm). Although the particles were rarely sintered in the powder compacts at 160-170 °C, they were apparently sintered within a binder containing certain mercapto compound at this temperature range. Analysis of curing behavior of the adhesive indicated that sintering of the particles can be induced during the gelation stage. The sintering behavior is quite different depending on molecular structure of the compound. This materials design concept will be useful to develop advanced electrically conductive adhesives.

Key words: electrically conductive adhesives, epoxy resin, silver, sintering, electrical resistivity

Introduction

Electrically conductive adhesives (ECAs) are often used as printing pastes and interconnection materials in the electronics packaging technology [1]. However, development mechanism of electrical conductive paths and microstructure in the adhesives remains unclear. According a physical model about inter-filler contacts, metallic fillers make conductive contacts through an insulative gap [1,2]. To enhance electrical and thermal conductivities of the adhesives, the inter-filler electrical resistance is required to be reduced.

Because materials design for forming direct bonding between fillers is effective to reduce the interfacial resistance, introduction of nano-fillers was examined to induce low temperature sintering of micro-fillers within the polymeric binder in the last decade [3,4]. In this case, metal nanoparticles or precursors of the nanoparticles are mixed into the adhesives in conjunction with micro-fillers. The micro-fillers can be directly connected within the polymeric binder by the low temperature sintering of nanoparticles during curing.

By contrast, the authors have been successfully achieved low temperature sintering of silver (Ag) micro-fillers within epoxy-based binder only by controlling binder chemistry [5]. In this materials design, surface diffusion on silver micro-fillers can be successfully enhanced although no silver nanoparticles or its precursors were mixed into the adhesives. The present work aims to analyze low temperature sintering behavior of silver micro-fillers within epoxy-based binders to discuss the interfacial phenomenon for enhancing surface diffusion and necking of the micro-fillers.

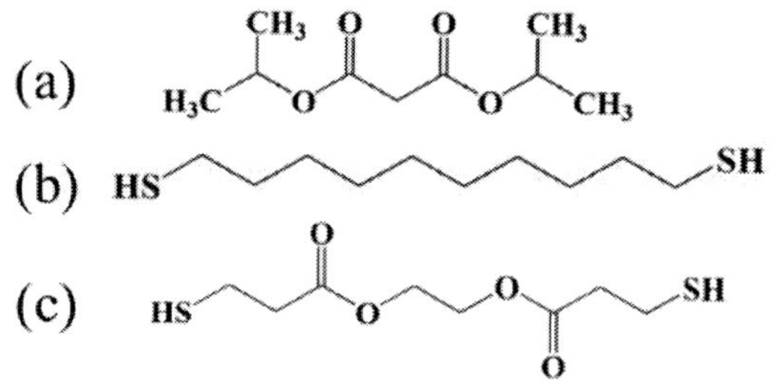

Figure 1: Organic compounds mixed into epoxy-based binders; (a) malonic acid diisopropyl ether, (b) 1,10-dimercaptodecane and (c) ethyeleneglycol bis(3-mercaptopropionate)

Experimental procedure

N,N-Diglycidyl-4-glycidyloxyaniline (Sigma-Aldrich, St. Louis, MO, USA) was used as a main component of binders. The epoxy resin was mixed with several components including malonic acid diisopropyl ether, 1,10-dimercaptodecane, and ethyeleneglycol bis(3-mercaptopropionate) (shown in Fig. 1), to obtain the epoxy-based binder. Composition of these components was fixed to 30mol% deficient from the stoichiometric composition in the binders. Ag water atomized particles (average particle size: 2.5 µm; Fukuda Metals and Foils Co., Ltd., Kyoto Japan) were subsequently mixed into the binders up to 82 wt% to obtain ECA pastes.

Then, the ECA pastes were printed on a glass substrate in dimensions of 76 mm x 3 mm x 30 µm, following by curing at 150-230 °C for 1 h in air.

140

Figure 2: Electrical resistivity of the adhesives with (a) malonic acid diisopropyl ether, (b) 1,10-dimercaptodecane and (c) ethyeleneglycol bis(3-mercaptopropionate), as a function of curing temperature

Electrical resistivity of the cured samples was measured using the four-probe method with a low-resistance meter (MCP-T610, Mitsubishi Analytic Co., Ltd.) at ambient temperature. Cross-sectional microstructure of the adhesives was observed with a scanning electron microscope (SEM) (Elionix, ERAX - 8900 M).

To analyze curing behavior of the adhesives, thermogravimetory-differential thermal analysis (TG-DTA) (Seiko Instruments Co., TG/DTA6200) was conducted. In addition, free-damped oscillation method (FDOM) with a rigid-body pendulum simultaneously with electrical resistance measurement [6] was also performed to examine relationship between curing behavior and electrical conductivity development. The condition of these measurements was fixed at 230 °C for 1 h (heating rate: 10 °C/min, cooling rate: 5 °C/min).

Results and discussion

Electrical resistivity of the adhesives

Figure 2 shows electrical resistivity of the ECAs as a function of curing temperature. The curing temperature dependence of electrical resistivity was quite different depending on the binder chemistry. When the binder containing malonic acid diisopropyl ether was used, the adhesive exhibit a very high electrical resistivity below 180 °C of curing temperature. Although electrical resistivity of the adhesive gradually decreased above 200 °C, relatively high resistivity (10^{-3} Ωcm) was obtained by curing even at 230 °C.

The adhesives composed of the epoxy-based binders containing mercapto compounds (1,10-dimercaptodecane and ethyeleneglycol bis(3-mercaptopropionate) exhibit very low electrical resistivity (~1 x 10^{-5} Ωcm) by curing at 230 °C. However, development behavior of electrical conductivity during curing was different depending on molecular structure of the mercapto compounds.

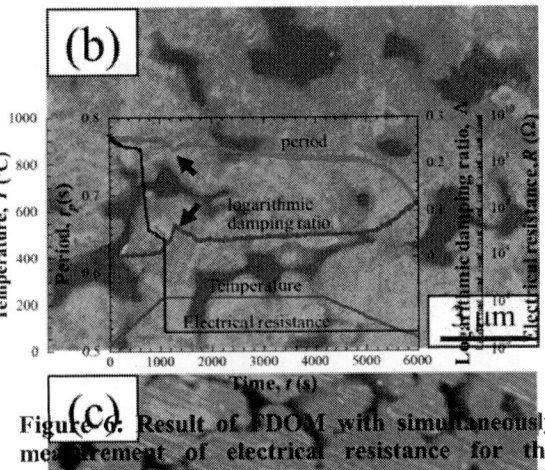

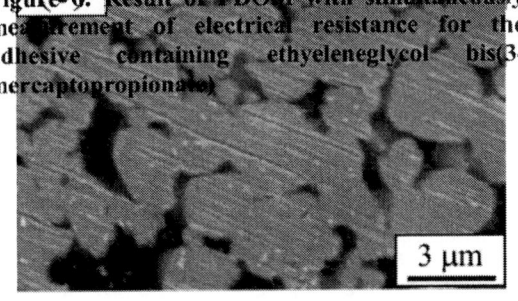

Figure 6: Result of FDOM with simultaneously measurement of electrical resistance for the adhesive containing ethyeleneglycol bis(3-mercaptopropionate)

Figure 3: Cross-sectional SEM image of the adhesives with (a) malonic acid diisopropyl ether, (b) 1,10-dimercaptodecane and (c) ethyeleneglycol bis(3-mercaptopropionate) cured at 230 °C for 1 h

Electrical resistivity of the adhesive with 1,10-dimercaptodecane was insufficiently decreased when it was cured below 200 °C. By contrast, electrical resistivity of the adhesive with ethyeleneglycol bis(3-mercaptopropionate) was significantly decreased between 160 and 180 °C. This implies the mercapto compounds play an important role in accelerating electrical conductivity development of the adhesives during curing.

Cross-sectional microstructure of the adhesives

Because electrical resistivity of the adhesives was different depending on the binder chemistry, their microstructure was observed using SEM. Figures 3 (a)-(c) show cross-sectional SEM images of the adhesives containing malonic acid diisopropyl ether, 1,10-dimercaptodecane, and ethyeleneglycol bis(3-mercaptopropionate) cured at 230 °C for 1 h.

In the adhesive with malonic acid diisopropyl ether, Ag micro-particles dispersed in the epoxy-based binder were rarely sintered (Fig. 3 (a)). Because the electrical contacts through a insulative gap were formed between the particles, the adhesive exhibited relatively high electrical resistivity.

By contrast, Ag micro-particles was apparently sintered in the adhesives with the mercapto compounds as shown in Fig. 3 (b) and (c). The particles formed conduction paths by directly

Figure 4: Cross-sectional SEM image of the adhesives with (a) 1,10-dimercaptodecane and (b) ethyeleneglycol bis(3-mercaptopropionate) cured at 160 °C for 1 h

bonding within the binder to significantly decrease electrical resistivity of the adhesives. Therefore, the mercapto compounds are effective to induce low temperature sintering of silver micro-particles within the epoxy-based binder.

Figure 2 also indicated that electrical conductivity development in the adhesives can be varied depending on molecular structure of the mercapto compounds. To investigate difference in microstructural development of the adhesives, microstructure of the samples cured at 160 °C for 1 h was observed as shown in Figs. 4 (a) and (b).

In the case of the adhesive with 1,10-dimercaptodecane, sintering of Ag micro-particles rarely occurred at this temperature (Fig. 4 (a)). By contrast, sintering of the particles had been already induced significantly in the adhesive with ethyeleneglycol bis(3-mercaptopropionate) (Fig. 4 (b)). The SEM observation clarified Ag micro-particles can be sintered within the epoxy-based binder even at ~160 °C.

Analysis of curing behavior of the adhesive

To examine sintering behavior of the particles, curing behavior of the adhesives was analyzed. Figure 5 shows TG-DTA curves of the adhesive with ethyeleneglycol bis(3-mercaptopropionate). The

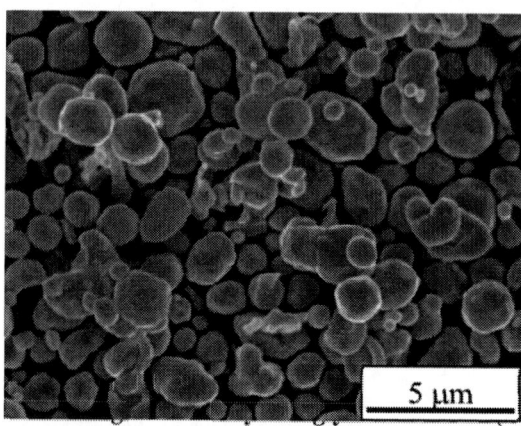

Figure ... surface of powder compact prepared using the Ag micro-particles after heated at 170 °C for 1 h

DTA curve indicates exothermic reaction that corresponds to curing reaction of the epoxy-based binder occurred at 160-230 °C. In this case, the mercapto compound can behave as curing agent.

In addition, significant weight loss was observed at the same temperature range in TG curve. This weight loss is thought to be caused by thermal decomposition of the mercapto compound.

Although curing reaction of the binder was found to occur at 160-230 °C by TG-DTA, information about polymer structure development was not obtained by the measurement. To discuss the polymer structure development, curing behavior of the adhesive was analyzed from the viewpoint of viscoelastic characterization. Variation in electrical resistance of the adhesive was also measure in addition to the viscoelastic characterization.

In particular, curing behavior of the adhesive was investigated by the FDOM simultaneously with electrical resistance measurement. In the FDOM measurement, period and logarithmic damping ratio of oscillation of the pendulum are proportion to $1/(E')^{1/2}$ and E", where E' and E" are storage modulus and loss modulus, respectively [6]. Because variations in viscoelastic properties of the adhesive are detected by the FDOM measurement, polymer structure development in the binder during curing can be analyzed.

Figure 6 shows result of the FDOM measurement simultaneously with electrical resistance measurement for the adhesive with bis(3-mercaptopropionate) under the same condition to the TG-DTA measurement shown in Fig. 5.

The logarithmic damping ratio initiated to increase at 160-170 °C in Fig. 6. Gelation of the binder was suggested to occur in this temperature range because exothermic reaction caused by curing reaction of the binder was detected at the same temperature range by DTA (Fig. 5). The period and logarithmic damping ratio subsequently varied at 200-230 °C (shown by arrows in Fig. 6). This means microstructural development that accompanied with increase in elastic modulus occurred during the gelation stage.

In the gelation stage, electrical resistance of the adhesive significantly decreased through two steps, 160-170 °C and 200-230 °C. This implies electrical conductivity development was promoted through the microstructural development during the gelation stage. This microstructural development corresponds to the sintering of Ag micro-particles within the binder.

After that, the period of oscillation of the pendulum gradually decreased during isothermal heating at 230 °C. This decrease in period means increase in storage modulus caused by cross-linking of binder molecules occurred in this stage.

Sintering of Ag micro-particles was clarified to be induced during the gelation stage. In this stage, significant weight loss caused by thermal decomposition was also detected by TG. The decomposition is thought to be suitable to induce sintering of Ag micro-particles.

Interfacial chemistry to promote low temperature sintering of Ag micro-particles

Low temperature sintering of metal particles is usually discussed from the viewpoint of driving force based on surface energy of the particles. To increase the driving force of sintering, use of nanoparticles is effective. Therefore, low temperature sintering of metal particles have been studied by using nanoparticles and related materials.

By contrast, the materials design described in this work provides another concept for promoting low temperature sintering of metal particles. Although Ag micro-particles have no extensive driving force for sintering, the particles were initiated to be sintered within the epoxy-based binder at ~160 °C. As a comparison test, powder compact of the Ag micro-particles was heated at 170 °C for 1 h. Figure 7 shows microstructure of the compact after heating. Because the compact was very brittle, its fracture surface was observed. The Ag micro-particles were not sintered during heating under this condition.

Therefore, sintering of the particles was apparently accelerated within the epoxy-based binder. The results obtained in this work implies that the mercapto compounds are key materials for promoting low temperature sintering of Ag micro-particles.

Surface diffusion of Ag is known to enhance by chemisorption of S and O atoms due to decrease in activation energy for surface diffusion [7]. Similar effect for surface diffusion enhancement was also reported in a mercapto compound adsorbed on Au. The authors propose that low temperature sintering of Ag micro-particles can be induced by using similar interface chemistry when appropriate chemical agents were mixed into the binder. Although the interfacial chemistry in the adhesives discussed in this work has been never clarified yet, several models were suggested by the density functional theory as shown in Fig. 8.

Conclusions

Low temperature sintering of Ag micro-

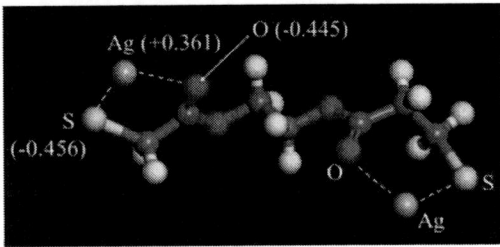

Figure 8: A model for interaction between a mercapto compound and silver atoms simulated by the density functional theory. Numbers indicated in this figure show atomic charge of each atom

particles was successfully achieved within an epoxy-based binder to obtain a low electrical resistivity (~1 x 10^{-5} Ωcm). A mercapto compound can effectively induce sintering even at ~160 °C. Interfacial chemistry is the key of this materials design. This concept will be useful to develop advanced ECAs.

Acknowledgements

The authors thank Center of Monodzukuri Research Organization (MRO), Ota, Japan for cooperation about microstructural characterization using scanning electron microscopy.

References

[1] J. E. Morris, "Conductive Adhesives for Electronics Packaging," J. Liu, Ed., Port Erin: Electrochemical Publications, pp.36, 1999.

[2] R. Holm, "Electric Contacts: Theory and Application," Springer-Verlag, Berlin, 1967.

[3] H. Jiang, K.-S. Moon, Y. Li, C. P. Wong, Ultra high conductivity of isotropic conductive adhesives, Proc. ECTC 2006, (2006), pp.485-490.

[4] K. Sasaki, N. Mizumura, Development of low-temperature sintered nano-silver pastes using metallo-organic (MO) technology and resin

reinforcing technology, Proc. Mate 2015, (2015), pp.239-244.

[5] M. Inoue, Y. Sakaniwa, Y. Tada, Binder chemistry for ultra-highly conductive pastes by low temperature sintering of silver micro-fillers, Proc. ECTC2016, (2016), pp.2068-2074.

[6] Y. Sakaniwa, M, Iida, Y. Tada, M. Inoue, "Conduction Path Development in Electrically Conductive Adhesives Composed of an Epoxy-based Binder," Proc. ICEP2015, pp. 252-257, 2015.

[7] P. A. Thiel, M. Shen, D.-J. Liu, J. W. Evans, Adsorbate-enhanced transport on metal surfaces: Oxygen and sulfur on coinage metals, J. Vac. Sci. Technol. A28, (2010), pp.1285-1298.

Copper Pillar Bumps on Acoustic Wave Components

J. Schober[1*], J. Berger[1], C. Eulenkamp[2], K. Nicolaus[3] and G. Feiertag[1]

[1] Department of Electrical Engineering and Information Technology, University of Applied Sciences Munich, Lothstrasse 64, 80335 Munich, Germany

[2] Department of Applied Sciences and Mechatronics, University of Applied Sciences Munich, Lothstrasse 34, 80335 Munich, Germany

[3] Advanced Packaging, RF360 Europe GmbH a Qualcomm – TDK Joint Venture, Anzinger Strasse 13, 81671 Munich, Germany

* Corresponding author email: jakob.schober@hm.edu

Abstract

In this study, we give an evolutional overview of packaging solutions for acoustic wave components, for the last two decades. Nowadays, acoustic wave components are essential for mobile data communication. Due to the need of a cavity above the micro-acoustic structures and steadily increasing requirements on electrical performance and size reduction, packaging has experienced a rapid progress. To achieve further miniaturization of acoustic wave components, we present a manufacturing method for the replacement of solder balls by copper pillar bumps as interconnects on one of the thinnest commercially available package, namely, TFAP™ (Thin Film Acoustic Package) of RF360. The round copper pillar bumps consist of copper, nickel and a tin cap. Spin coating and photolithography followed by electroplating is used to create copper pillar bumps on wafer level. Resist and seed layer are removed in a batch process. The tin is reflowed in formic acid atmosphere to form the solder cap. First copper pillars are cut with focused ion beam to examine the cross section with a scanning electron microscope.

Key words: Copper Pillar Bump (CPB), Surface Acoustic Wave (SAW), Bulk Acoustic Wave (BAW), Wafer Level Packaging (WLP), Radio Frequency (RF), Thin Film Acoustic Package (TFAP)

I - Introduction

Mobile data communication in the consumer and industrial sectors is already indispensable from everyday life. The global number of mobile broadband subscribers was 5.9 billion in the fourth quarter of 2018 and is still increasing [1]. According to Ericsson Mobility Report, the mobile data transfer volume rose nearly 88 percent between the fourth quarters of 2017 and 2018 [1]. The Internet of Things (IoT) is expected to additionally induce a significant increase of cellularly connected devices [1]. This trend will lead to an increasing demand of affordable and high performance RF-modules. RF-modules facilitate the selection of the required RF-frequency for the respective mobile data communication such as GPS (Global Positioning System), Bluetooth, Wi-Fi and LTE (Long-Term Evolution). Surface acoustic wave (SAW) filters and bulk acoustic wave (BAW) filters are commonly used filter technologies for wireless broadband communication of mobile devices. AWCs (acoustic wave components) together with passive components (e.g. inductors, capacitors) and semiconductor chips compose the RF-module on a substrate. SAW and BAW filters convert an electrical signal into a mechanical wave using piezoelectric materials. Depending on the design

only a certain frequency band is converted subsequently from the mechanical wave back to an electrical signal with low attenuation. The other frequencies experience high attenuation, resulting in the filter function. In [2] details of SAW, BAW and modules are provided. Mobile devices further advance and so do the requirements for RF-modules and their AWCs. Especially performance, costs and size are crucial criteria to compete on the global market. Nowadays, AWCs and their package are manufactured on wafer level, i.e. keeping production costs constant and simultaneously shrinking the AWCs, lead to lower production costs per component. In this study, we present one possibility to shrink the AWCs by miniaturizing the footprint of the interconnection to the substrate by replacing SBs (solder balls) with CPBs (copper pillar bumps). That shrinks the device and reduces manufacturing costs per device.

II – Package Evolution

Packaging plays an essential role for the electrical performance of AWCs. Above SAW and BAW structures, a cavity is needed to ensure the unaffected propagation and evolvement of the acoustic wave. The package protects against environmental influences like moisture or ionic attack, which can lead to corrosion of the SAW and

CSSP, DSSP and TFAP are products of Qualcomm Technologies, Inc. and / or it subsidiaries

BAW structures. Furthermore, mass altering contaminations or electrostatic discharge are prevented, which can cause device performance change and damage. Very thin protection layers can additionally protect the structures. [3]

Flip chip mounting to solder AWCs on a substrate is now state of the art. Then, the AWCs are typically over-molded to improve reliability. Thus, the package has to provide resistance against high pressure and temperature during the molding process. Furthermore, small size, competitive manufacturing costs, while still maintaining high electrical performance and reliability are the demands for new package solutions.

Already in the year 2000, EPCOS, which is now RF360 Europe, published the proprietary CSSP™ (Chip Sized SAW Package). Here, the SAW device is placed via flip chip mounting on a carrier substrate with SBs. A cavity manufactured by the proprietary PROTEC chip passivation over the SAW structure enables the use of an underfiller as shown in Figure 1. This offers enhanced opportunities to integrate passive components. [4]

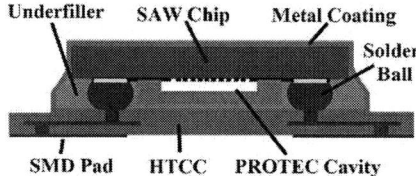

Figure 1: Schematic cross section of CSSP 1 SAW Package of EPCOS. [3]

In 2002 R. C. Ruby et al. presented an hermetically sealed wafer-scale package for thin-film bulk acoustic wave resonators (FBAR) by bonding two silicon wafers onto each other; one wafer carrying the FBAR structures and the other containing etched via holes and gaskets. Both wafers are aligned before bonding and the lid wafer is thinned after bonding. The package is gold-wire bonded to an IC package for testing. [5]

M. Goetz et al. developed an advanced SAW package for RF SAW devices in 2002. A silicon cap wafer is bonded via an adhesive to a piezoelectric wafer with SAW structures. The silicon cap wafer is etched via deep reactive ion etching to achieve cap separation and accessible bond pads for wire bonding. RF sputtering and chemical vapor deposition (CVD) ensure a hermetic coating. [6–8]

M. Franosch et al. developed a non-hermetic SU-8 polymer package for BAW filters. They present a two-step manufacturing process containing the so called "sacrificial layer method". First, a sacrificial layer, which is easily removable, is structured where the cavity is to be created. SU-8 is deposited on top, structured with holes. The sacrificial layer is removed by etching or dissolution through the holes of the SU-8. In the second step, a fresh layer SU-8 seals the holes creating the

encapsulated cavity. The polymer package withstands molding conditions. [9]

A novel approach to form a hermetic package for SAW devices was published by O. Kawachi et al. The SAW chip is flip chip mounted on a ceramic. Subsequently SnAg-solder (tin silver) is mechanically applied under heat on a batch of mounted SAW chips. The SnAg-solder seals the SAW chip. After half dicing a nickel coating is plated and the devices are singularized. [10]

To circumvent the issue of mechanical stress induced by different CTEs (coefficient of thermal expansion), J.-h Lim et al. bonded two lithium tantalate (LT) wafers to form a hermetically sealed SAW package. Sand blasting or laser drilling and wafer polishing are used to structure via holes. These are filled with copper by electroplating to create interconnects in the cap wafer. A Sn-Au solder wall seals the cavity after wafer bonding. Vias are interconnected with SnAu-solder (tin gold) as shown in Figure 2. A FEM (finite element method) numerical study by S. Gao et al. shows the thermo-mechanical behavior of this package regarding the bonding process. The calculations indicate the insensitivity of thermally induced stress generated by the package thickness. However, CTE and Young's modulus show partially significant influence. Furthermore, increasing bonding temperature leads to an increase in induced stress. [11–14]

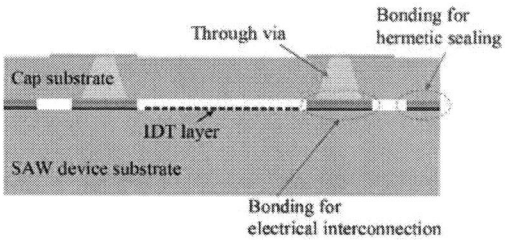

Figure 2: Schematic cross section of LT to LT bonded SAW package developed by J. Lim et al. IDT = interdigital transducer. [11]

G. Feiertag et al. published in 2007 the advance of EPCOS's commercially available package technologies for SAW components. Starting with the standard ceramic box package, where the SAW chip is adhesive-bonded and electrically connected through wire bonding before the ceramic box is sealed with a metal lid. Owing to performance limits caused by broad impedance variations of the bond wires and the massive space savings, the already stated CSSP package generation was introduced in 2000. A SAW chip is soldered on a high or low temperature cofired ceramic (HTCC or LTCC) interposer. The LTCC offers further size reduction of RF-modules by integrating inductances and capacitors. On HTCC substrates, coils can be processed in the package cavity [15]. Three CSSP generations were developed. The second CSSP

146

generation is called CSSP Plus (CSSP 2) and is shown in Figure 3. SB size reduction and creating the cavity with a polymer foil before hermetic metal deposition allows the removal of the underfiller and PROTEC process, leading to a further shrinking of the package size. [3]

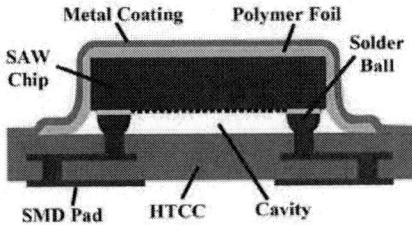

Figure 3: Schematic cross section of CSSP Plus (CSSP 2) SAW Package of EPCOS. [3]

The third CSSP generation is divided into a non-hermetic and a hermetic package. The cavity of the non-hermetic package is created by a polymer foil and glob top as illustrated in Figure 4. SAW structures are protected against corrosion by a thin inorganic passivation layer. [3]

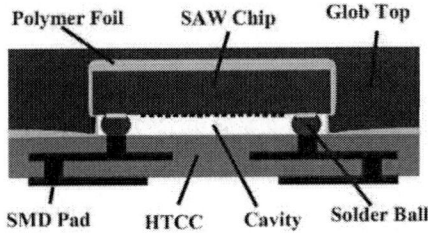

Figure 4: Schematic cross section of non-hermetic CSSP 3 SAW package of EPCOS. [3]

CSSP Plus and CSSP 3 withstand the molding process if the chip size is small. A larger chip and cavity size is necessary for SAW duplexers. To withstand the molding process even for large SAW chips, the hermetic CSSP 3 package was developed. A copper frame and copper pillars are electroplated onto the ceramic substrate. The SAW chip is flip chip mounted and soldered before a polymer foil is laminated on top. The polymer foil is opened at the copper frame followed by metal deposition to ensure hermeticism. The copper frame and pillars spread the strong forces during molding and avoid package failure. A schematic cross section is shown in Figure 5. [3]

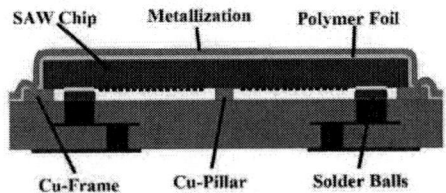

Figure 5: Schematic cross section of hermetic CSSP 3 SAW package of EPCOS. [3]

A thin film package was introduced by J. L. Pornin in 2007. A polymeric sacrificial layer is structured onto the BAW structure. Silicon dioxide (cap layer) is deposited and holes are structured via dry etching. After dissolution of the sacrificial layer another polymer is spin coated to seal the holes forming the cavity. An under bump metallization (UBM) is deposited after structuring the sealing layer. Lead free bumping, backside grinding, sawing, pick and place, flip chip soldering, underfiller application and over-molding on a substrate are done without damage to the BAW filter. In Figure 6 a cross sectional view of the cavity is shown. The hole in the cap layer is only slightly filled with the polymeric sealing layer. No sealing polymer flows into the cavity. [14]

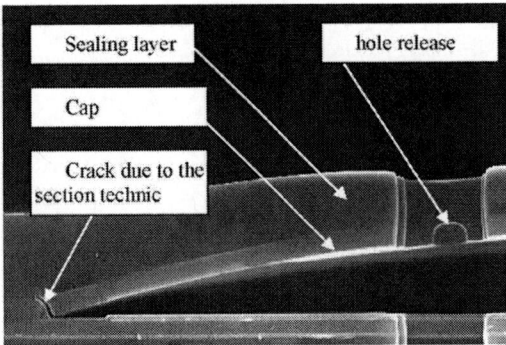

Figure 6: Cross sectional scanning electron microscope (SEM) image of thin film package. Hole release position in cap layer located with focused ion beam (FIB) cut. [14]

T. Fukano et al. presented a resin based wafer level package (WLP) for SAW filters as shown in Figure 7. The aluminum alloy SAW structure is made on LT. A resin based hollow structure is produced in the next step. Metal post terminals and solder ball grid array allow the electrode linkage. To ensure moldability, a reinforcement study was

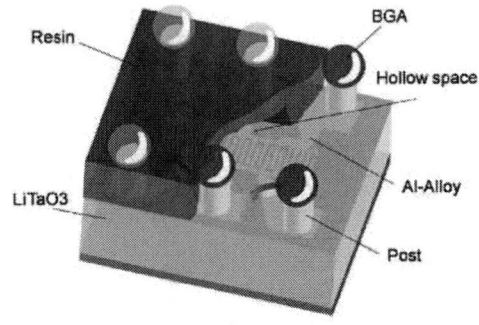

Figure 7: Schematic sketch of resin based SAW WLP package. [16]

performed. The reinforcement material with significant higher stiffness was deposited onto the package. The analysis showed that metals as reinforcement can be used despite the risk of parasitic induction, if the layers are distant enough from the terminal electrodes. [16,17]

A novel approach to further shrink the AWCs devices was published by K. Koh et al. in 2008. Two structured SAW devices were bonded face-to-face. Metal filled via holes enable electrical connection to the SAW electrodes. Even a 3D package stack of SAW devices was proposed, as shown in Figure 8. [18,19]

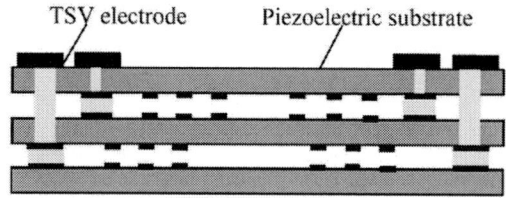

Figure 8: Three dimensional package proposal for SAW devices realized by TSV and face-to-face bonding. [19]

K. Sakinada et al. described a SAW WLP using a two-step resist process without a sacrificial layer. Therefore, firstly a resist is structured to form the sidewalls followed by a dry film lamination on top. A moldability study showed positive effects by hard bake, UV crosslinking and geometry improvements of the resist. Furthermore, remaining impurities in the package resist, which could corrode the SAW structures, were reduced. [20]

Imprinted Laminate WLP based on SU-8 was presented by J. H. Kuypers. Cavities are created in a SU-8 dry film by imprinting. This dry film is aligned and transfer-laminated subsequently to the device wafer. Lithography structures the openings for electrical connections. [21]

A true chip size package fabrication method for commercial SAW devices is described by C. Bauer et al. Adhesive wafer bonding of either two lithium tantalate or two lithium niobate wafers create the required cavity. Hence, the first wafer carries the SAW structures. Additionally adhesive polymer bars and frames are patterned via photolithography. Then, the second wafer is optically aligned and bonded onto the first one. Backgrinding singularizes the lids before a Ti (titan) adhesion and a Cu (copper) seed layer is sputtered. An electroplated redistribution layer composed of copper and nickel on top of the lid provides the electrical interconnection path. As the lid edges are very steep, photolithography requires spray coating followed by inclined exposure in a mask aligner [22]. SBs are deposited on a copper, nickel and gold UBM for interconnection to a substrate. The so called DSSP™ (Die Sized SAW Package) is completely manufactured on wafer level and withstands over-molding processes. [23] In Figure 9 schematic cross sections of DSSP 1 are shown. The first cross section describes the standard configuration. The remaining ones propose potential variations with integrated inductors. [24]

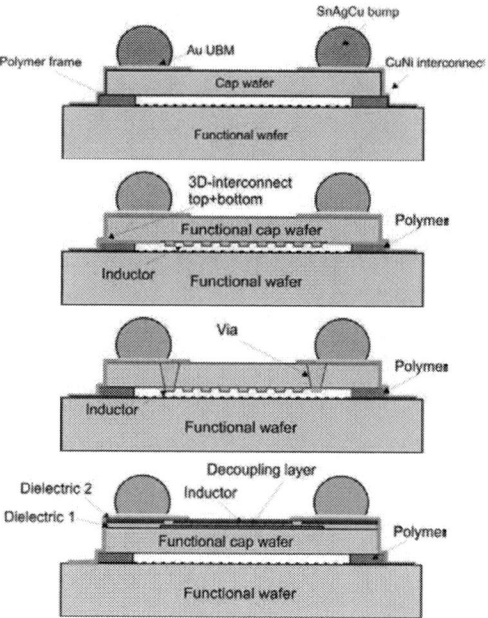

Figure 9: Schematic cross section of DSSP 1 WLP Package of EPCOS and variation proposals with inductor integration. [24]

Furthermore, a second commercial generation of the Die Size SAW Package, namely, DSSP 2 was introduced in 2013. DSSP 2 is a hermetically sealed WLP. This is achieved by replacing the adhesive by an electroplated frame and pillars. Bottom and top wafer are bonded together containing the frame made of either copper/copper or copper/tin/copper. As electrical paths cannot be routed under the metal frame, vias are drilled in the cap wafer and filled with copper. UBM and SBs placement is performed. In Figure 10 a schematic DSSP 2 sketch is shown with an optional dielectric decoupling layer between lid and coils. [24]

Both, CSSP and DSSP generations are perfectly suitable for SAW and BAW filters applications [25].

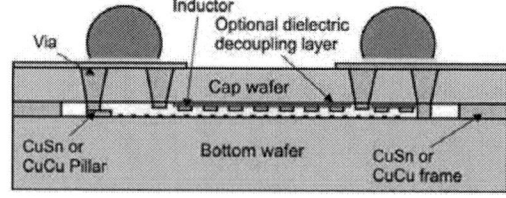

Figure 10: Schematic cross section of DSSP 2 WLP of EPCOS with integrated inductors. [24]

G. Fattinger et al. published a WLP for BAW devices creating the required cavity using the two-step lithography process. First, the sidewalls of the package are created around the BAW structure by an epoxy material; followed by a second layer of epoxy material to create the package roof. The package withstands over-molding. Instead of wire bonding or solder bumps on a UBM, in this study copper pillars are deposited to perform flip chip mounting on a substrate. Copper and tin are subsequently deposited to form the pillars. Pillar height controls the stand-off between package and substrate and proper under fill during over-molding is achieved. A schematic cross section is shown in Figure 11. [26]

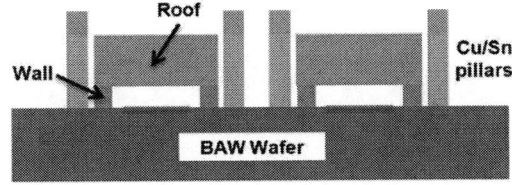

Figure 11: Schematic cross section of BAW WLP with copper pillar interconnections. [26]

R. Aigner et al. suggest eliminating the space needed by the copper pillars and probe pads on a BAW chip, as shown in Figure 11, by placing the copper pillars on top of the cavity as illustrated in Figure 12. [27]

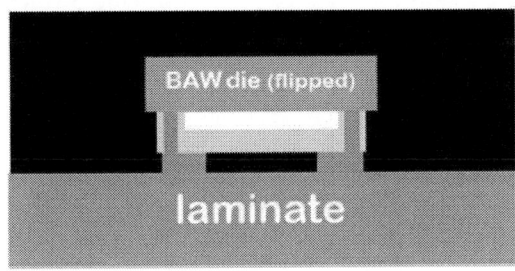

Figure 12: Schematic cross section of μBAW device in package with copper pillars as interconnection. [27]

T. Bauer et al. report the commercially available TFAP (Thin Film Acoustic Package) of EPCOS. Sacrificial layer method and sandwiching a soft between two stiff materials ensure excellent

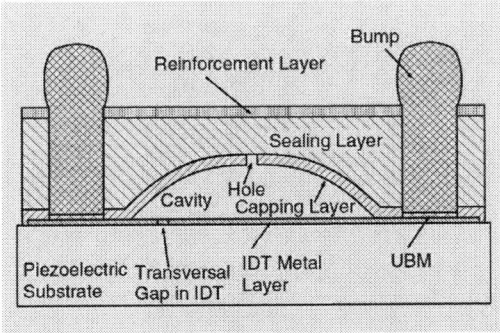

Figure 13: Simplified schematic cross section of wafer level TFAP for SAW and BAW devices of EPCOS. IDT = interdigital transducer. [25]

cavity stability. Therefore, a sacrificial resist layer is deposited on the SAW or BAW structure by photolithography followed by an edge rounding

process. The first dielectric layer is deposited on top and small release holes and openings for electrical connection are produced. The sacrificial layer is removed residue-free and without interaction to SAW and BAW structures. After that, a polymer is spin coated to seal the first dielectric layer. The third dielectric layer is called reinforcement and is also opened in the area of the electrical interconnections. Before solder bumping a UBM is created. In Figure 13 a simplified schematic cross section of TFAP is shown. [25]

A far as we know TFAP is one of the thinnest package solutions for direct RF-module integration. It fulfills the current challenging market requirements for SAW and BAW devices while still matching tightened performance specifications. In Figure 14 the package evolution and miniaturization of SAW duplexers on LT of RF360 (former EPCOS) are illustrated. [28]

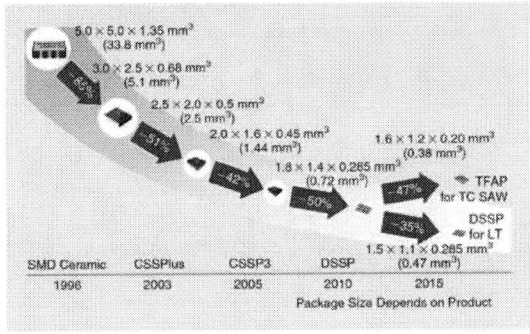

Figure 14: Miniaturization of SAW duplexers of RF360 (former EPCOS). Area times height. [25]

III – Copper Pillar Bumps on TFAP

Introduction

To save valuable chip area, the size of interconnects is constantly reduced. However, further miniaturization of SBs is limited, due to stand-off requirements for moldability and bridging between SBs. Especially if the package cavity and interconnects sit on the same side of the chip, the interconnects need to additionally compensate the package height to fulfill the stand-off requirements. This leads to even higher SB diameters. With CPBs it is possible to adjust the height in a wide range independent of the footprint (e.g. diameter) and thus stand-off can easily be set. Furthermore, a much finer pitch (distance between interconnects) is possible compared to SBs. Smaller footprint and smaller pitch allow chip miniaturization leading to more devices per wafer and new freedom for design. CPBs consist of a metal layer stack. The base is copper (Cu) with a solder cap on top. A nickel (Ni) diffusion barrier between Cu and solder is optional. Tin (Sn) or Sn alloys are used as solder material.

Materials and Methods

In this study CPBs were processed on TFAP. A multi-reticle design was created containing

149

several geometry variations. The designs include daisy chain and SAW filter devices. The package manufacturing was performed on wafer level as stated above. However, the process was customized for CPBs, thus stopped before UBM process for SBs.

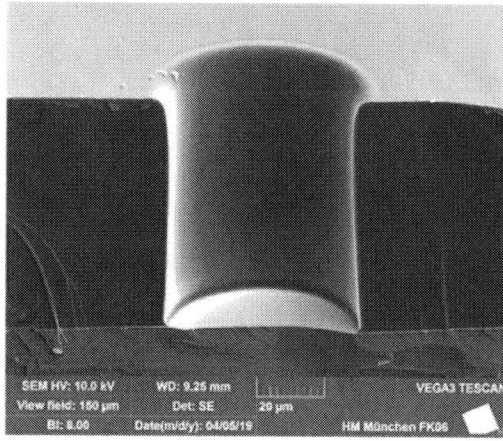

Figure 15: Cross sectional SEM image of blank wafer with Ti/Cu seed layer and structured resist.

Instead of the UBM, a thin Ti adhesion layer followed by a thin Cu seed layer was sputtered on the wafer to ensure electrical conductivity for electro plating. A photosensitive resist was then spin coated on the wafer. With a single coating process a uniform film thickness of around 85 μm was achieved. Edge bead removal (EBR) kept film thickness variation at the wafer edge to a minimum. The functional wafer was aligned to a glass mask in a standard mask aligner and exposed to UV-light. Afterwards development opened the resist and defined the CPB position. Figure 15 shows a cross sectional SEM image of a blank wafer (without package) and a sputtered Ti/Cu seed layer including the structured resist.

Cu, Ni and Sn were deposited via electroplating. Fast deposition rates were achieved,

Figure 16: Two SEM images of the same CPB on TFAP. Left: Before Ti/Cu seed layer etching. Right: After Ti/Cu seed layer etching.

still comply with good thickness uniformity. The resist was stripped in a standard batch tool without leaving any residues. Major challenges were the

Ti/Cu seed layer etching, due to identical material of seed layer and CPB base. We developed a wet batch process, which allows simultaneous etching of multiple wafers. Only a minor attack of the CPB base and the other materials occurred as shown in Figure 16.

The plated pillar was reflowed to create the solder cap. Therefore, the whole wafer was heated above the solder melting temperature in a reducing formic acid and nitrogen atmosphere. Tin oxides on

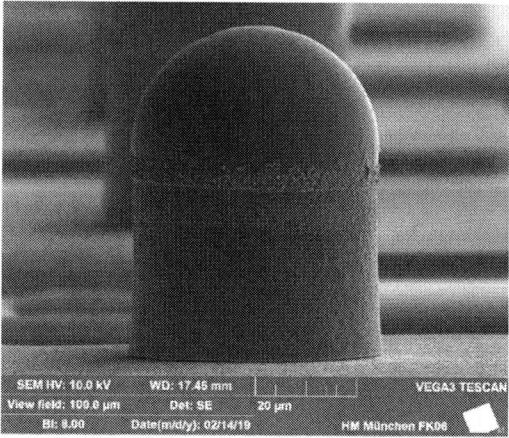

Figure 17: SEM image of completely processed CPB on TFAP.

the surface, which can prevent a round solder cap shape, are reduced by formic acid during reflow. The solder cap allows automated bump height inspection and easier chip placing. Figure 17 shows a completely processed CPB on TFAP.

<u>Analysis</u>

FIB cut analysis was performed to investigate interfaces within the CPB (Cu, Ni and Sn cap) and its substrate as shown in Figure 18. There are no

Figure 18: SEM image of CPB with FIB cut on TFAP. CPB stack of Cu, Ni, Sn. CPB is fully processed, including seed layer etch.

defects to be found in the Cu or Ni layer. Proper interfaces between Sn, Ni, Cu and the substrate can be observed. Only a slight undercut due to resist footing at the CPB base is visible.

Summary and Outlook

In this study, we give a rough overview of AWC package evolution in the last two decades. The main trend is the shrinkage of area and height of the components. Nowadays face-to-face bonding of two wafers with adhesive or metal frames and polymer based thin film packages are general standard. Flip chip mounting on interposers or PCBs and over-molding are state of the art. Driven by market demands package solutions need to be low priced, small in dimensions, suitable for direct RF-module integration, over-moldable and must still fulfill requirements as to excellent electrical performance and reliability. The replacement of SBs through CPBs promises further device miniaturization. Hence, we present the first step for replacement, a wafer level manufacturing method for CPBs on a piezoelectric TFAP chip, which already is one of the smallest package solutions.

Further investigations will be made regarding mechanical stability and reliability of CPBs compared to SBs on TFAP. Therefore, PCBs for flip chip placement and over-molding are already being manufactured. Influences of geometry modifications of the CPBs and their substrate will be examined as well. Failure modes provoked by induced environmental stress will be characterized. Thus, mechanical and FIB cross sections will be prepared and examined with SEM and EDX (energy dispersive X-ray spectroscopy). Computer tomography (CT) will be used as a non-destructive method.

Acknowledgements

The authors want to thank *Bayerisches Staatsministerium für Wirtschaft, Energie und Technologie* for their financial support.

At RF360, our thanks go to designers and production engineers for their technical support.

References

[1] Ericsson, "Ericsson Mobility Report Q4 Update February 2019,".

[2] A. Hagelauer, G. Fattinger, C. C. W. Ruppel et al., "Microwave Acoustic Wave Devices: Recent Advances on Architectures, Modeling, Materials, and Packaging", *IEEE Transactions on Microwave Theory and Techniques*, vol. 66, no. 10, pp. 4548–4562, 2018.

[3] G. Feiertag, H. Krüger, and C. Bauer, "Surface acoustic wave component packaging", in *European Microelectronics and Packaging Conference*.

[4] H. Meier, T. Baier, and G. Riha, "Miniaturization and advanced functionalities of SAW devices", in *2000 IEEE Ultrasonics*

Symposium. Proceedings. An International Symposium (Cat. No.00CH37121), pp. 395–401, IEEE, 22-25 Oct. 2000.

[5] R. C. Ruby, A. Barfknecht, C. Han et al., "High-Q FBAR filters in a wafer-level chip-scale package", in *2002 IEEE International Solid-State Circuits Conference. Digest of Technical Papers (Cat. No.02CH37315)*, pp. 184–458, IEEE, 3-7 Feb. 2002.

[6] M. Goetz and C. Jones, "Chip scale packaging techniques for RF SAW devices", in *27th Annual IEEE/SEMI International Electronics Manufacturing Technology Symposium*, pp. 63–66, IEEE, 17-18 July 2002.

[7] M. Goetz, C. Jones, J. Rao et al., "Advanced SAW packaging for modular integration", in *2nd Int. Symp. Acoustic Wave Devices for Future Mobile Communication Systems, 2004*, pp. 279–286.

[8] M. P. Goetz and C. E. Jones, "Modular integration of RF SAW filters", in *IEEE Ultrasonics Symposium, 2004*, pp. 1090–1093, IEEE, 23-27 Aug. 2004.

[9] M. Franosch, K.-G. Oppermann, A. Meckes et al., "Wafer-level-package for bulk acoustic wave (BAW) filters", in *2004 IEEE MTT-S International Microwave Symposium Digest (IEEE Cat. No.04CH37535)*, pp. 493–496, IEEE, June 6-11, 2004.

[10] O. Kawachi, K. Sakinada, Y. Kaneda et al., "P3N-2 Packaging of SAW Devices with Small, Low Profile and Hermetic Performance", in *2006 IEEE Ultrasonics Symposium*, pp. 2289–2292, IEEE, 02.10.2006 - 06.10.2006.

[11] J.-h. Lim, J.-s. Hwang, J. Kwon et al., "An Ultra Small SAW RF Filter using Wafer Level Packaging Technology", in *2006 IEEE Ultrasonics Symposium*, pp. 196–199, IEEE, 02.10.2006 - 06.10.2006.

[12] S. Gao, J. Hong, T. Kim et al., "Study of a Wafer Level Package (WLP) for Surface Acoustic Wave (SAW) Filter", in *2006 International Conference on Electronic Materials and Packaging*, pp. 1–4, IEEE, 11.12.2006 - 14.12.2006.

[13] T. H. Kim, W. K. Jeung, S. J. Yang et al., "Miniaturization and Optimization of RF SAW Filter Using Wafer Level Packaging Technology", in *2007 Proceedings 57th Electronic Components and Technology Conference*, pp. 574–579, IEEE, 29.05.2007 - 01.06.2007.

[14] J. L. Pornin, C. Gillot, G. Parat et al., "Wafer Level Thin Film Encapsulation for BAW RF MEMS", in *2007 Proceedings 57th Electronic Components and Technology Conference*, pp. 605–609, IEEE, 29.05.2007 - 01.06.2007.

[15] R. D. Koch, C. Bauer, J. E. Kiwitt et al., "Ultra low-profile self-matched HTCC SAW-

duplexer: manufacturing and performance challenges", in *2011 IEEE International Ultrasonics Symposium*, pp. 1829–1832, IEEE, 18.10.2011 - 21.10.2011.

[16] T. Fukano, Y. Okubo, J. Nishii et al., "0806 SAW Filters Using Wafer Level Packaging Technology", in *2007 IEEE Ultrasonics Symposium Proceedings*, pp. 68–71, IEEE, 28.10.2007 - 31.10.2007.

[17] T. Fukano, Y. Okubo, J. Nishii et al., "Wafer-level packaged SAW filters with resistance to transfer molding", in *2008 IEEE Ultrasonics Symposium*, pp. 108–111, IEEE, 02.11.2008 - 05.11.2008.

[18] K. Koh, T. Yamazaki, Y. Terao et al., "Study on SAW devices having face to face aligned packaged structure", in *2008 IEEE Ultrasonics Symposium*, pp. 1596–1599, IEEE, 02.11.2008 - 05.11.2008.

[19] K. Koh, H. Okitsui, and K. Hohkawa, "Fabrication of SAW devices with small package size using through substrate via technology", in *2009 IEEE International Ultrasonics Symposium*, pp. 2688–2691, IEEE, 20.09.2009 - 23.09.2009.

[20] K. Sakinada, A. Moriya, M. Kitajima et al., "A study of Wafer Level Packaging of SAW filter for module solution", in *2009 IEEE International Ultrasonics Symposium*, pp. 2692–2695, IEEE, 20.09.2009 - 23.09.2009.

[21] J. H. Kuypers, S. Tanaka, and M. Esashi, "Imprinted laminate wafer-level packaging for SAW ID-tags and SAW delay line sensors", *IEEE transactions on ultrasonics, ferroelectrics, and frequency control*, vol. 58, no. 2, pp. 406–413, 2011.

[22] B. L'huillier, M. Hornung, D. Tönnies, et al., eds., *Application of an angular exposure system to fabricate true-chip-size packages for SAW devices*, IEEE, 2011.

[23] C. Bauer, T. Heuser, V. Dragoi et al., "Adhesive Wafer Bonding Applied for Fabrication of True-Chip-Size Packages for SAW Devices", *ECS Transactions*, vol. 50, no. 7, pp. 363–370, 2013.

[24] F. M. Pitschi, C. Bauer, R. D. Koch et al., "Approaches to wafer level packaging for SAW components," in *2013 IEEE MTT-S International Microwave Symposium Digest (MTT)*, pp. 1–3, IEEE, 02.06.2013 - 07.06.2013.

[25] T. Bauer, C. Eggs, K. Wagner et al., "A Bright Outlook for Acoustic Filtering: A New Generation of Very Low-Profile SAW, TC SAW, and BAW Devices for Module Integration", *IEEE Microwave Magazine*, vol. 16, no. 7, pp. 73–81, 2015.

[26] G. Fattinger, P. Stokes, V. Potdar et al., "Miniaturization of BAW devices and the impact of wafer level packaging technology", in *2013 IEEE International Ultrasonics Symposium (IUS)*, pp. 228–231, IEEE, 21.07.2013 - 25.07.2013.

[27] R. Aigner, G. Fattinger, M. Schaefer et al., "BAW Filters for 5G Bands", in *2018 IEEE International Electron Devices Meeting (IEDM)*, 14.5.1-14.5.4, IEEE, 01.12.2018 - 05.12.2018.

[28] E. Schmidhammer, T. Metzger, and C. Hoffmann, "Multiplexers: A necessary extension for 4G/5G systems", in *2016 IEEE MTT-S International Microwave Symposium (IMS)*, pp. 1–4, IEEE, 22.05.2016 - 27.05.2016.

Stencil Printing and Flip-Chip Bonding for Assembly of Pixelated X-ray Detectors using PCB-type Interposer and Flexible Printed Circuit Boards

Andreas Schneider[1], Simon P. Cross[1], Matthew D. Wilson[1], Matthew C. Veale[1], Matthew Hart[1], Paul Seller[1], Marcello Borri[2], Dan Beckett[1], John D. Lipp[1]

UK Research and Innovation, Science and Technology Facilities Council; [1] Rutherford Appleton Laboratory, Technology, Harwell Campus, Didcot OX11 0QX, UK; [2] Daresbury Laboratory, SciTech Daresbury, Keckwick Lane, Daresbury, Warrington WA4 4AD, UK

+44-1235-44-6164 / andreas.schneider@stfc.ac.uk

Abstract

In recent years multi-layered Flexible Printed Circuit (FPC) boards have advanced considerably in complexity, allowing versatile designs for electronic packaging applications which require interconnect through-via [1]. This makes these boards also attractive for interposers (pitch-convertors) in X-ray detector applications where pixelated sensors are connected to readout Application-Specific Integrated Circuits (ASIC) in order to match different pixel-pitches between sensor and ASIC. Thin boards with total thickness of approx. 140μm are also available and desirable for low radiation absorption of detector packaging in high-energy particle experiments (e.g. CMOS imaging sensor for particle tracker [2,3]). These X-ray detector sensors and ASIC are based on a wide range of materials (Si, Cd(Zn)Te, GaAs). The mismatch in thermal expansion of these different materials is a major challenge. Bonding of sensors and ASIC to flexible boards increases the risk of distortion of the assembly, with potential delamination of components from the board. The paper describes the bonding processes. Stencil printing is used to form an array of dots from electrically conductive adhesive on the sensor, interposer or FPC. ASIC and interposer board are studded with a similar array of gold studs using a ball bonding technique. Subsequently flip-chip bonding is used to bond a studded component to an array of adhesive dots on the printed counterpart. Optimal bonding parameters are evaluated using shear testing and analysis of contact area. The flatness of boards is analysed. Bonded devices are tested for their electronic functionality as X-ray detectors.

Key words: X-Ray Detector, Interposer, Multi-Material Bonding, Gold Ball Studding, Stencil Printing, Flip-Chip

Introduction

In recent years the Science & Technology Facilities Council has developed and delivered world-leading X-ray imaging systems to synchrotrons and free electron lasers across the world [4]. In many of these applications detector modules are produced by flip-chip bonding the readout ASIC directly to a sensor. In these devices the pixel pitch of the ASIC and sensor are identical.

In the case of the Large Pixel Detector (LPD) system delivered to the European XFEL in 2017 [4,5], the pixel pitch of the silicon sensor (500μm) is different from that of the ASIC (250μm × 400μm xy-pitch). An LPD detector module consists of a single silicon sensor that is bonded to 8 individual LPD ASIC. Two essential fabrication methods are required for the assembly of a LPD module:

1) Based on a well-established process of silicon-based sensors, it is possible to produce devices with multiple metal layers allowing a redistribution of interconnects on the sensor surface.

2) The wire bonds that provide the connection between the individual ASIC and the rest of the data acquisition system sit behind the Si sensor to allow the tiling of individual modules to produce large detector areas. In order to achieve this, a 500μm-thick Si interposer is bonded to the ASICs to provide sufficient clearance for the wire bonds. Once these wire bonds have been made, the Si sensor is then bonded to the interposer.

While the use of these techniques has been very successful in the case of LPD, they do have a number of disadvantages, especially for newer sensor technologies that are based on compound semiconductor materials like Cd(Zn)Te and GaAs. In these detector materials it is currently not possible to produce multiple metal layers on the detector surface, meaning that either the sensor pitch must match that of the ASIC or the redistribution must happen at an interposer level. The existing silicon-based interposer technology is expensive. An alternative cheaper and more accessible solution that provides additional functionality (like the inclusion of surface mount components) is therefore desirable.

Motivation and Objectives

A recent development in Printed Circuit Board (PCB) technology provides a potential solution. Multi-layered, highly integrated, and complex Flexible Printed Circuit (FPC) boards which allow fan-out from a small pixel array on the bottom surface to a larger array at the top side of the board are commercially available now [1]. This sort of PCB-type interposer opens new possibilities for pitch-convertors and interconnects as alternative to the more difficult process methods in Through-Silicon Via (TSV) and Through-Glass Via (TGV) technology [6].

These PCB-type interposers comprise multiple layers of FR-4 and copper tracks. The copper tracks and via for the fan-out conductor layout [1] can be designed as flexibly as for standard PCB. In addition, existing technologies from the microelectronic packaging industry can be applied to attach Surface Mount Devices (SMD) to these boards. A further advantage of these boards is that individual layers are thin (35μm FR-4 with 5μm Cu conductor tracks) and allow the fabrication of FPC boards with low mass, which is also a requirement for particle tracking detectors as used in particle physics [2,3]. Further, depending on the design of the pitch-converter, it is possible to bond and connect sensor and ASIC without wire bonding and with the sensor covering the entire area of the assembly to enable close butting of modules on all four sides.

Flip-chip bonding is one of the essential techniques for the hybridization of ASIC and X-ray sensors and has been common practice for many years [7]. Usually the electrical contact between the pixelated sensor and the ASIC pixels is achieved using indium bump bonding or a bond with electrically conductive adhesive to gold ball studs on the ASIC readout chip [7–9]. The latter method has been successfully applied in the fabrication of the LPD assembly and it is also used for the spectroscopic imaging HEXITEC X-ray detector, which is another development of the Rutherford Appleton Laboratory (RAL) [10–12]. The HEXITEC detector system has been used in a diverse range of applications areas from materials science at large synchrotron facilities to the detection of illicit materials in the laboratory. One application area where the HEXITEC technology has prompted significant interest is that of medical imaging such as SPECT [13], which operates at high X-ray energies (>100keV) for large images with ~1mm resolution [13].

This paper describes the first attempt to bond a HEXITEC assembly with a PCB-interposer/pitch-convertor. This method is distinct from a TSV process for a similar HEXITEC detector system [14] in which through-via in the ASIC are used. As described in [14], this TSV method is a costly process at wafer level of the ASIC fabrication. A

PCB-interposer instead is more cost-effective and is fabricated independently from ASIC and sensor.

A difficulty of using PCB-type interposers is the large mismatch in thermal expansion between the PCB and other materials such as the Si ASIC and sensor materials such as Cd(Zn)Te and GaAs. The Coefficient of Thermal Expansion (CTE) for these materials ranges from 2.6 ppm/°C to ~45 ppm/°C [15], with Si having the lowest and the PCB material the highest among those materials. This is relevant because the flip-chip bonding process is carried out at elevated temperature in order to cure the adhesive. This heating process can lead to distortion of the assembly or delamination of the ASIC from the PCB-interposer, which is detrimental and may result in failure of the whole device. This is also important for scientific applications of the HEXITEC detector in which the system is subjected to vibration and temperature gradients [16]. The final packages must be robust enough to withstand these stresses. The mechanical stability of the assembled devices is established by shear testing. This paper therefore also investigates an underfill between the bonded areas which potentially gives more rigidity to the assembled detector and avoids delamination.

As an example for the detector hybridization process with a PCB-interposer, an assembly of a HEXITEC ASIC (80×80 pixels on 250μm pitch) to a GaAs sensor (80×80 pixels on a 500μm pitch) has been chosen and the bonding processes are described here with discussion of process modifications required for assembly.

Process and Inspection Methods

A standard hybrid detector such as the HEXITEC detector [10,11] consists of an ASIC bonded to a pixelated X-ray sensor made from a semiconductor with high atomic number. The 250μm-pitch pixels of the 80×80 channel readout array on the ASIC are studded using a ball-bonding technique. For the HEXITEC ASIC, studs with a diameter of 50μm and a height of ~30μm are deposited with an automatic thermosonic Palomar 8000 Ball Bonder. A matching array of 80×80 dots of electrically conductive epoxy adhesive loaded with silver particles is printed onto the pixel array of the sensor with a Reprint Mantis 23 stencil printer. The ASIC is bonded to the printed sensor after alignment by a thermal compression flip-chip process using a SET FC150. The ASIC is controlled and read out with I/O pads that protrude from one edge of the sensor. The pads are connected to a PCB and subsequent data acquisition system using Al-wire wedge bonding. The same equipment (apart from a wire wedge bonder) is used for building the HEXITEC hybrid detector with the PCB-type interposer/pitch-convertor.

For inspection, bonded devices are shear tested with the XYZtec Condor 100 bond tester using a shear gauge for 50kgf force (1kgf ≈ 9.8N).

High-resolution SLR-camera images of the sheared surface areas are analysed using ImageJ software in order to quantify regions of adhesive where the ASIC studs have been in contact with the interposer.

For improving the rigidity of the assembled detector with PCB-type interposer an underfill material is tested. The underfill is a 2-component adhesive which is initially mixed in a Synergy Devices Ltd. Speedmixer and subsequently applied through a syringe to the interface of the bonded ASIC/interposer assembly.

Selected samples are scanned with an optical profiler Cyber CT-100 for the analysis of the distortion of bonded samples.

PCB-Type Interposer

The layered PCB-type interposer is designed at RAL and fabricated by GS Swiss PCB. Its size is approximately 40mm×40mm. On the top side (side 'A') there is an array of 80×80 contact pads with a 500μm pitch and an additional row of bond pads for the sensor guard ring to which the sensor will be bonded. The 80×80 500μm-pitch pads are routed to a matching array of 80×80 pads on 250μm pitch on the bottom side (side 'B') in the centre of the PCB to which the ASIC is bonded. The I/O pads of the ASIC are also bonded to side 'B' of the PCB and routed to SMD components and a connector that will be soldered onto the side 'B' of the PCB as a final step of the assembly, together with a bias wire which is attached to the top cathode of the sensor.

For a HEXITEC detector with a PCB-type interposer as pitch-convertor, the same ASIC is used as for the standard hybrid HEXITEC detector, with the exception that the wire bond pads (readout I/O connections) are also studded.

Figure 1a: Sensor side of interposer / side 'A'.

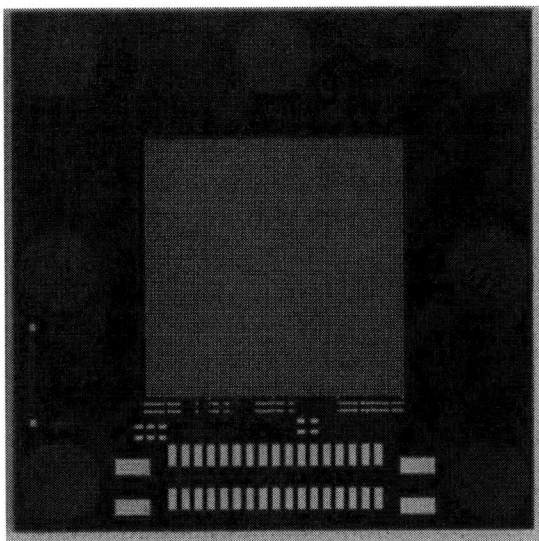

Figure 1b: ASIC side of interposer / side 'B'.

Fig. 1a+b show the sensor and the ASIC side of the PCB-interposer. Onto side 'A' a full-sized 40mm×40mm sensor or a smaller sensor with 500μm pitch can be bonded. For initial tests of the PCB-interposer, only a 20mm×20mm GaAs sensor from a standard 80×80 pixel HEXITEC hybrid detector is bonded to the centre of side 'A'. Even though in this case not all 6400 pixels of side 'A' are used and routed to the ASIC on the opposite side of the interposer, this initial experiment will give sufficient information about the viability of the bond processes for this type of interposer.

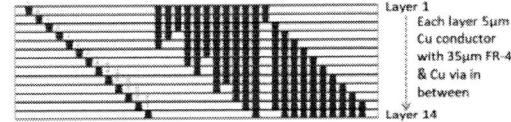

Figure 2: Sketch of interposer cross-section.

Fig. 2 indicates the basic layout design of the PCB-interposer in cross-section which consists of 14 layers FR-4 with Cu tracks and via connections between the layers. The surface finish of the contact pads is made of Electroless Nickel Electroless Palladium Immersion Gold (ENEPIG), which is ideal for gold ball studding. Due to the fabrication process of the interposer these contact pads are recessed by ~10μm below the top surface. Sufficiently high gold studs need to be applied.

Studding of ASIC and Interposer

For the PCB-interposer the same type of ASIC as for a standard HEXITEC detector is used and bonded to side 'B' of the interposer. The 250μm-pitch 80×80 pixel array of the ASIC is gold ball studded with the Palomar 8000 ball bonder using a 20μm gold wire as usually done for HEXITEC. Additionally, the I/O pads on 100μm pitch along one edge of the 80×80 pixel array are also studded with the same 20μm wire and same studding parameters. This and the following flip-chip bonding process permit the omission of the Al-wire wedge bond process step that is required for the standard HEXITEC hybrid detector.

Some semiconductor sensors made with materials such as Cd(Zn)Te are unsuitable for the ball-bonding process of gold studs because of their brittleness. Further, the GaAs sensor's pixel contacts are made from materials which are incompatible with gold studs. The alternative option here is to stud the 80×81 array of side 'A' of the PCB-interposer with a 500µm pitch. Additionally this has the advantage that the tall studs compensate for the 10µm recess of the PCB pixel contacts. Due to the layered fabrication process of the PCB solder resist, those pixel contacts are approx. 10µm below the top surface.

Since the pixel contacts of this array are 125µm in diameter, which is about 2½-times larger than those on the HEXITEC ASIC, slightly larger studs are formed on side 'A' using a 25µm gold wire which are easier to process for tall studs than studs formed with a 20µm wire. Because a larger wire is used and the PCB-interposer material (FR-4, Cu) is softer and more flexible than the Si of the ASIC, ball studding parameters such as studding force, bond time, ultrasound input have to be increased to achieve the required bond energy. Studs on this 500µm-pitch array have a diameter of ~78µm and total height of ~35µm and are ~28µm above the PCB surface.

Stencil Printing and Flip-Chip Bonding

As for the standard HEXITEC detector, an array of electrically conductive adhesive is printed onto the pixel array of the GaAs sensor followed by the flip-chip bonding of the sensor to the studded side 'A' of the interposer. In order to form a permanent connection between sensor and interposer, the adhesive is cured at 150°C in situ under bond pressure in the flip-chip bonding process. The resulting sensor/PCB-interposer bond is similar in rigidity and robustness to the standard HEXITEC hybrid detector without interposer.

In the second step of the hybridization (ASIC to PCB-interposer), the adhesive has to be printed onto the bond pads array of side 'B' of the interposer followed by flip-chip bonding of the studded ASIC to the side 'B' of the interposer. Flip-chip bond parameters such as bond force have to be adjusted rather than using the same parameter as in the first step of the hybridization (sensor to side 'A'). If the bond process for this second step is not optimized, only a weak bond between ASIC and interposer is formed and the ASIC is prone to delaminate from the PCB. Optimized stencil design, print and flip-chip bond parameters are essential to achieve an improved bond of the ASIC to the interposer.

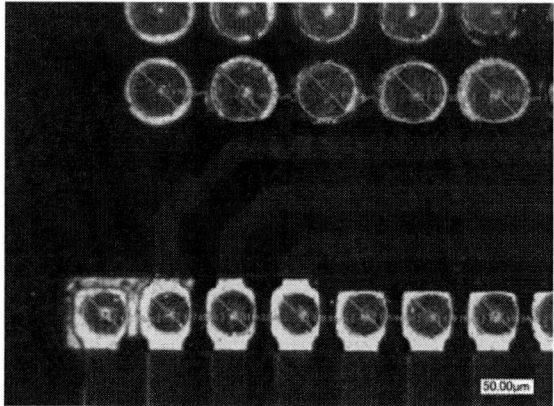

Figure 3: Adhesive dots printed onto the contact pattern of the interposer's side 'B'.

Unlike the uniform pixel size of the array on the sensor, the print pattern on side 'B' of the interposer must match different sizes of contact pads – those for the pixel array and those for the readout I/O pads. The dot size is determined not only by the aperture size in the stencil but also by the correct print gap between stencil and interposer.

Fig. 3 shows adhesive dots and their dimensions on different contact pads of side 'B' of the interposer. At the top of the image are dots of diameter ~190µm which occupy the pixel pads, whereas at the bottom of the image are the smaller dots (<140µm in diameter) on the readout pads that avoid shorting of these contacts. If the contact pads are not filled optimally with adhesive, then the contact volume to the studs of the ASIC is not sufficient to build a good bond during flip-chip bonding.

A further possible reason for the delamination of the ASIC might be slight deformation of the interposer because of the already adhered sensor on the opposite side. As a result the contact area between ASIC and interposer could be reduced because not all studs of the ASIC join with the adhesive. A higher bond force during flip-chip bonding/curing might well compensate for this deformation. To test this hypothesis, shear tests on several ASIC bonded to interposers are carried out.

Shear Test Results

In order to verify the influence of the bond force during flip-chip bonding on the bond strength between ASIC and interposer a variety of shear tests are carried out. At first several studded ASIC to PCB-type interposers are prepared in the same manner as described above using a range of bond forces (10kgf to 45kgf). After curing, the ASIC is sheared off the interposer and the required shear force is measured. Fig. 4 shows the relationship between bond force and shear force.

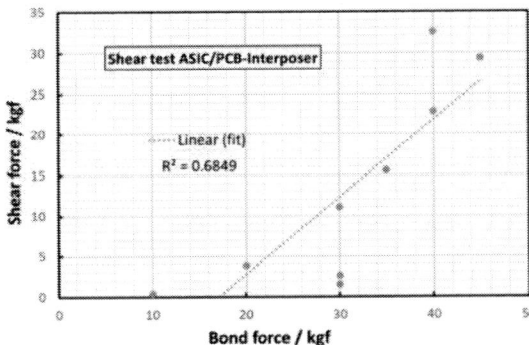

Figure 4: Relationship between shear force and flip-chip bond force.

The shear force increases with increasing flip-chip bond force in a linear tendency but with weak statistical correlation (dotted line in graph of Fig. 4). The strength of the bond between ASIC and interposer also depends on other factors such as uniformity of adhesive, stud and adhesive dot shape as well as adhesive age and pot life. Hence a large variation in shear force is observed even for the same bond force (30kgf and 40kgf). However, a higher bond force is certainly required for an improved bond of the ASIC. Guided by the linear fit of the data points, a flip-chip bond force of at least >17kgf is required for achieving any bond between ASIC and PCB-interposer. The ASIC of both samples which have been bonded at 40 kgf cracked during the shear test and parts of the ASIC remain on the interposer. Both samples have a shear strength of 23 kgf and 32.5 kgf, respectively. The contact area of the ASIC is investigated after shear testing.

Inspection of Contact Area

In order to inspect the contact area of ASIC bonds, a series of images of sheared ASIC are taken with a high-resolution SLR camera and then processed with ImageJ software. A typical processed image of an ASIC is shown in Fig. 5. The amount of silver-loaded epoxy on the studs of the sheared ASIC is evaluated and quantified with the ImageJ software. Initially the raw image of the ASIC is modified in order to remove obvious artifacts, and contrast / background colour are adjusted. That leaves just the representation of the adhesive in the processed image as lightly coloured circles (Fig. 5). This image indicates that almost no bond occurs in the centre (dark area) but studs at the edge of the ASIC show remaining adhesive on the studs (bright circles).

This image is converted into a binary image, which then is used with the particle analysis tool of ImageJ to evaluate the number of epoxy-coated studs.

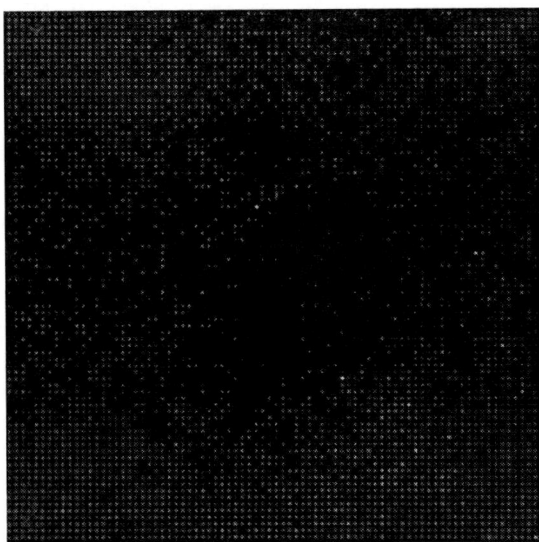

Figure 5: Processed image using ImageJ for an ASIC bonded with 45kgf.

Alternatively the area of the identified circles in the image is calculated as the percentage area of the whole image. For samples in which the ASIC is not fully sheared off, it is assumed that the bonded area under the remaining ASIC chip on the interposer is the same or better than the part of the ASIC which is analysed. The evaluated percentage area of the ASIC bond is plotted against the bond force for each sample (Fig. 6). Obviously the relationship between percentage area and bond force follows a similar linear trend as the shear force/bond force relationship. One of the samples in the data set (△) has been bonded at 40kgf bond force but only for a short period and subsequently cured ex situ in an oven instead of during the flip-chip bonding. The sample has an extremely small bond area (<1%), and fell off the interposer before a shear test could be carried out. All of the other samples in the data set have been bonded and cured in situ during the flip-chip bonding. This indicates that an in situ curing is also essential for a good flip-chip bond.

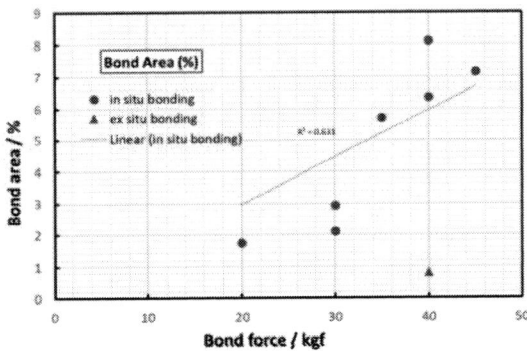

Figure 6: Relationship between calculated bond area of ASIC and applied bond force.

Underfill

In order to improve the bond strength between ASIC and interposer and add rigidity to the assembly, the application of an underfill after the flip-chip bonding is investigated. The underfill increases the contact area between the bonded components and also redistributes the stress and strain caused by the CTE mismatch [17]. A two-component underfill is mixed in a Speedmixer and applied to one side of the interface gap of the ASIC / interposer using a syringe. For the optimization of this process, studded ASIC are flip-chip bonded to glass slides so that the underfill in the gap can be observed after curing. The flip-chip process is the same as for an in situ bonded ASIC onto the interposer. Given that the underfill is drawn into the gap at the interface by capillary forces, only a very small amount of underfill (~0.05g) is required to cover the full area of the HEXITEC ASIC. When more underfill is injected, a large amount spills out on the opposite side from the application site (Tab. 1). Since some of the SMD components have to be soldered close to the edge of the ASIC on the interposer, a spill of underfill on these contacts needs to be avoided and the leakage minimized.

Underfill curing temperatures at room temperature (RT) and at slightly elevated temperatures (45°C, 60°C, 80°C) are investigated and further an additional sample was cured in a vacuum vessel at RT. Curing at RT requires at least one day whereas curing at 80°C is achieved in approx. 3h. However, air inclusion, voids, and bubbles are observed in all cases. Minimal voids and inclusions occur during curing at 60°C (Fig. 7). Curing in vacuum slightly increases the amount of bubbles compared with a sample cured at the same temperature under atmosphere.

Table 1: Leaked amount of underfill at opposite side from application.

Amount applied / g	Distance from edge where leaked underfill is observed / μm
0.05	<480
0.09	< 1280
0.14	>2850

Despite the better coverage of underfill at 60°C, the higher cure temperature may lead to further distortion of the PCB-interposer. The amount of underfill which diffuses into the gap and cures at RT seems to be sufficient to increase the contact area between ASIC and interposer for adequate strength.

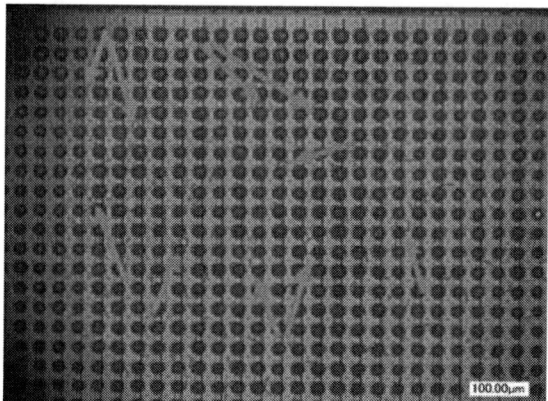

Figure 7: ASIC bonded to glass side and added underfill (cured at 60°C); arrows indicate areas with inclusions.

Complete Pitch-Converter with ASIC and Sensor

With all the optimization steps in mind, a pitch-convertor with a 20mm×20mm GaAs sensor and a HEXITEC ASIC bonded to the PCB-interposer has been produced using additional underfill. A photo of the whole unit (Fig. 8) shows the top of the interposer with the ASIC. The GaAs sensor can be seen in the mirror underneath the unit. The assembled device is inspected for bowing and distortion of the unit using an optical profiler.

Fig. 9a+b depicts the height profile of sensor and ASIC of this pitch-converter unit. The lighter colour in the height contour images indicates elevated areas and the darker colour lower areas of the profiled sensor and ASIC. The sensor is lower in the centre than at its edges. The ASIC appears to be higher in the centre than at its edges relative to the PCB surface (dark surrounding). The height difference from edge to centre for both ASIC and sensor is about 100μm to 150μm.

Figure 8: Assembled pitch-convertor unit with ASIC and GaAs X-ray sensor.

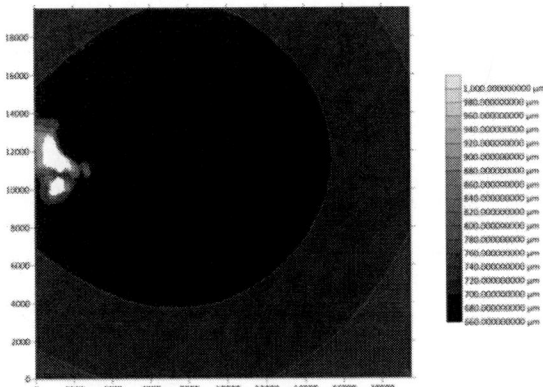

Figure 9a: Height contour of the sensor on the unit of Figure 8.

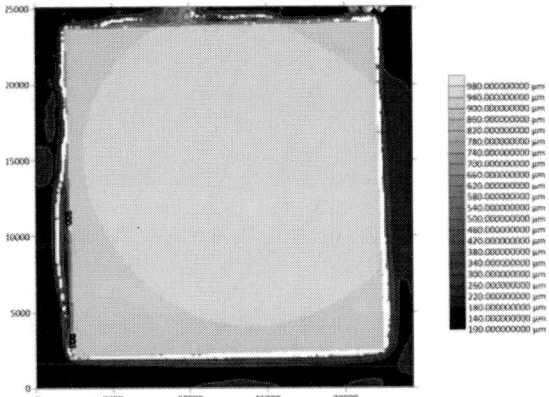

Figure 9b: Height contour of the ASIC on the unit of Figure 8.

The profiling confirms the findings of the bond area inspection (Fig. 5). Both indicate that the assembly is not fully flat and is not uniformly bonded across the interposer. The bowing of the assembly means that there are a number of poor connections between the I/O pads of the ASIC and the interposer. By gently manipulating the assembly manually it was possible to demonstrate connection between the ASIC, PCB interposer and GaAs sensor, although this was not stable enough to allow data to be collected from the system. Further development is required for a working device, especially regarding the flatness of the unit. The next step for future development will be bonding of a 40mm×40mm GaAs sensor to the interposer. Improvement in device rigidity and functionality will be expected.

Flexible Printed Circuit Boards

A layered PCB as described above can also be fabricated into thin FPC boards (approx. 140μm thick) which are also attractive for X-ray detector applications requiring minimal X-ray absorption in the electronic package (e.g. CMOS imaging sensors for particle tracker [2,3]). Here the flexibility of the FPC boards without any chip bonded to it complicates the bonding process. In order to improve the rigidity of FPC boards before bonding to sensor chips, temporary support substrates are tested. These are glass slides or Si wafers of the size of the FPC which are bonded temporarily to the FPC boards using thermal release tapes or water-soluble adhesive. A thermal release tape is selected that will release the FPC at higher temperature than the required flip-chip bond temperature so that several sensor chips can be bonded to the FPC without losing contact to the support substrate. The water-soluble adhesive will be tested if it does not change properties and adhesion during the bonding process.

The supports, especially glass slides, enable inspection of the quality of the temporary bond between slide and FPC board before the bonding process. It is difficult with both temporary adhesive layers to avoid air inclusions between FPC board and support structure (Fig. 10) due to the topography and flexibility of the chosen FPC. Consequently the opposite side of the FPC is still not totally flat which can adversely affect the quality of the flip-chip bonding. However, the glass and Si support substrates are sufficiently flat for the vacuum chucks of the flip-chip bonder.

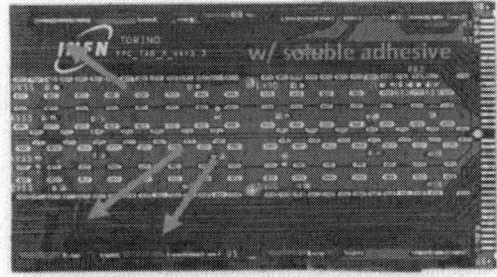

Figure 10: FPC bonded to glass slides as support substrate using water-soluble adhesive or thermal release tape; arrows indicate air inclusions.

No conclusive results for temporarily bonded support substrates have been achieved yet and further tests must be carried out to find out whether the release process will affect the bond between sensors and FPC.

Conclusion

Layered PCB with via and fan-out conductor tracks are used for building X-ray detectors using flip-chip bonding, stencil printing, and ball bond studding. In particular a PCB-interposer is used as demonstrator for a pitch-converter which reduces a pitch of the pixelated detector from 500μm to the 250μm pitch of the readout ASIC. This has several advantages, for example the bonding method does not require wire bonding and detector units are four-side buttable. The bowing and distortion of the whole unit is the main challenge because of the flexibility of the PCB-interposer and the mismatch in thermal expansion of the different materials bonded to the PCB. The paper describes how to overcome these issues by investigating the effect of flip-chip bond force on the bond strength of the ASIC to the PCB-interposer. In addition, underfill which is added in the gap between ASIC and interposer is examined.

Increasing the flip-chip bond force up to 45kgf increases the bond strength, as is shown in shear tests. However, inspection of the bond area on the ASIC reveals that more interconnect contacts are made at the edge of the ASIC than in the centre. This is corroborated by optical height profiling of a fully built unit which shows distortion and a bowing of the whole unit with the ASIC bulging upward away from the interposer. This distortion still impedes the full functionality of the unit as X-ray detector. Further development for improvement is required. This will also be interesting for future advances in the microelectronic packaging of CMOS imaging sensors for particle tracker using thin flexible FPC boards with low mass.

Acknowledgements

The authors would like to acknowledge STFC's CfI programme for funding of the project.

References

[1] GS Swiss PCB brochure https://www.swisspcb.ch/fileadmin/exceet/downloads/GS_Swiss/GS-CompanyBrochure_EN_2018.pdf .

[2] P. Martinengo, "The new Inner Tracking System of the ALICE experiment – CERN" https://www.phenix.bnl.gov/WWW/publish/mxliu/sPHENIX/Documents/2017-Feb-06-conferencefile-ITS_QM2017_v8_4_3.pdf, 2017.

[3] G. Aglieri et al., "Monolithic active pixel sensor development for the upgrade of the ALICE inner tracking system", Journal of Instrumentation Vol. 8, C12041, Dec., 2013.

[4] M.C. Veale, et al., "Characterisation of the high dynamic range Large Pixel Detector (LPD) and its use at X-ray free electron laser sources", J. Inst. Vol., No. 12, p. P12003, Dec., 2017.

[5] A. Koch, et al., "Performance of an LPD prototype detector at MHz frame rates under Synchrotron and FEL radiation", J. Inst., Vol. 8, p. C11001, Nov., 2013.

[6] Jihye Kim et al. "Electrical characteristics analysis and comparison between through silicon via(TSV) and through glass via(TGV)", Conf. IEEE EDAPS, Seoul, South Korea, 14 – 16 Dec. pp. 93-96, 2015.

[7] W. J. Greig, 'Integrated Circuit Packaging, Assembly and Interconnections', Springer, first edition, Boston, MA, Chapter 10 & 11, pp. 143-191, 2007.

[8] G. Harman, 'Wire Bonding in Microelectronics', McGraw Hill Professional, third edition, New York, Chapter 3 & 9, pp. 58-61 & 342-345, 2010.

[9] A. Schneider, et al., "Interconnect and bonding techniques for pixelated X-ray and gamma-ray detectors", J. Inst. Vol. 10, C02010, February, 2015.

[10] M. C. Veale et al., 'HEXITEC: A High-Energy X-ray Spectroscopic Imaging Detector for Synchrotron Applications', Synchrotron Radiation News, Vol. 31, No. 6, pp. 28-32, November, 2018.

[11] M. D. Wilson, et al. "A 10cm×10cm CdTe Spectroscopic Imaging Detector based on the HEXITEC ASIC", J. Inst., Vol. 10, P10011, October, 2015.

[12] L. Jones et al., 'HEXITEC ASIC – a pixelated readout chip for CZT detectors', Nucl. Instrum. Methods Phys. Res. A, Vol. 604, No. 1-2, pp. 34-37, June, 2009.

[13] W Scuffham, et al., "A CdTe detector for hyperspectral SPECT imaging", J. Inst. Vol. 7, P08027, August 2012.

[14] P. Seller, et al., "Through silicon via redistribution of I/O pads for 4-side butt-able imaging detectors," IEEE Nuclear Science Symposium and Medical Imaging Conference Record (NSS/MIC), Anaheim, CA, 27 Oct.-3 Nov., pp. 4142-4146, 2012.

[15] Isola, "High Performance Laminate and Prepreg / Typical Values", download of technical datasheet https://www.isola-group.com/products/all-printed-circuit-materials/370hr/ , © 2016.

[16] W. H. Baumgartner, et al., "The HEXITEC Hard X-Ray Pixelated CdTe Imager for Fast Solar Observations", SPIE High Energy Opt and IR Detect. for Astron. VII, Proc. Vol. 9915, p. 99151D, June, 2016.

[17] O. Rusanen, J Lenkkeri, "Thermal Stress Induced Failures in Adhesive Flip Chip Joints", Int. J. Microcircuit & Elect. Pack., Vol. 22, No. 4, pp. 363-369, Oct.-Dec., 1999.

High-energy X-ray Tomography for 3D Void Characterization in Au–Sn Solid-Liquid Interdiffusion (SLID) Bonds

Knut E. Aasmundtveit[1,2*], Kim Robert Tekseth[3], Dag W. Breiby[1,3], Hoang-Vu Nguyen[1]

[1] Dept. of Microsystems, University of South-Eastern Norway, 3184 Borre, Norway
[2] Dept. of Physics, Norwegian University of Science and Technology (NTNU), 7491 Trondheim, Norway
[3] PoreLab, Dept. of Physics, Norwegian University of Science and Technology (NTNU), 7491 Trondheim, Norway

Phone: +47-3100 9658, and E-mail Address: kaa@usn.no

Abstract

Au–Sn SLID bonding is a technique originally developed for harsh environment applications. The technology has recently shown promising results for ultrasound transducer fabrication. Characterizing the spatial and size distributions of voids is crucial for developing a fabrication process that satisfies acoustic requirements. This measurement is traditionally done by optical or electron cross-section microscopy that gives the void distribution in a randomly selected physically cut plane. X-ray micro computed tomography is a powerful tool for non-destructive three-dimensional imaging of void distributions, but is challenging to use in high-density materials like the ones used in ultrasound transducers.

We demonstrate that monochromatic, high-energy synchrotron X-ray tomography can give 3D images of such a challenging sample, resolving μm-sized voids in the bondline. The void distribution is highly non-uniform, implying that traditional cross-section microscopy would give different results depending on the plane of sectioning. Computed tomography allows the voids to be parametrized and treated statistically, revealing a wide distribution of void sizes, a tendency to form oblate voids with size-dependent orientation, as well as porous networks.

Key words: SLID bonding; X-ray Tomography; electro-acoustic devices; PZT; voids in bonding

I. Introduction

Solid-Liquid Interdiffusion (SLID) bonding is a technique based on a binary or ternary metal system that transforms to high-temperature stable intermetallic compounds (IMCs) at moderate bonding temperatures [1, 2]. Cu–Sn [3-5] and Au–Sn [6-8] are the most studied SLID systems, with processing temperatures above 232 °C and 278 °C, and with IMCs melting at about 700 °C and 500 °C, respectively. Initially developed for harsh environment applications, SLID bonding has proved to provide well-defined thin-layer metallurgical bonds, as well as excellent robustness [9] and possibility for fine-pitch bonding [3, 5]. The low to moderate bonding temperatures also enables the use of SLID bonding for temperature-sensitive materials such as poled piezoelectric materials, polymers and ferromagnets.

We have previously demonstrated Au–Sn and Au–In–Bi SLID bonding for ultrasound applications [10, 11]. The desired properties are thin-layer metallurgical bonds, with acoustic impedance matched to the bonding partners. For this application, the amount of voids in the bond must be minimized, as acoustic waves scatter at these voids. The standard method to characterize voids in the bond is through cross-section microscopy. We have previously compared such cross-sections with electrical impedance spectroscopy in order to investigate what level of voiding can be accepted for given acoustic requirements [12]. However, cross-section microscopy is a destructive technique, and it only gives 2D information from the chosen cross-section surface. Whether this cross-section surface is representative for the sample, remains an open question.

In this work, we have used X-ray tomography to access 3D information of the voids in the bonding layer of an ultrasound transducer dummy sample. The sample studied is a Lead Zirconate Titanate (PZT) die bonded to a Resonant Backing Layer (RBL) substrate using Au–Sn SLID bonding. The tomography data are compared with cross-section microscopy of samples manufactured with the same process.

The transducer sample is challenging to measure using X-ray computed tomography (CT) owing to the heavy elements (including lead and zirconium) giving strong absorption of the X-ray beam. For this reason, earlier attempts at using synchrotron CT with a modest beam energy of about 20 keV were unsuccessful. Similarly, despite the high maximum photon energy of home laboratory setups (e.g. Nikon XT H 225 ST, with photon energies ranging upto 225 keV), the polychromatic design of these instruments gives reduced resolution and weak contrast. For these reasons, CT setups based on monochromatic high-energy X-rays, as available at a few synchrotron beamlines around the world, are needed to image these mm-sized transducer samples in 3D. To the best of our knowledge, other imaging techniques cannot non-destructively resolve the internal μm-sized voids in the bonding layer. Scanning Acoustic Microscopy (SAM) is an alternative technique for non-destructive 3D imaging of voids, but the high acoustic impedance of PZT and RBL reflects the acoustic probing signal, limiting the signal-to-noise ratio and resolution obtained.

II. Experimental

A. Test vehicles

1) Materials

The test samples consisted of a PZT die bonded to a tungsten carbide-based RBL substrate using a eutectic Au–Sn preform. The dies had dimensions 5×5 mm^2 with Au coated on both sides. On the bonding side, the Au layer was electroplated to a thickness of 10-13 μm. The absolute roughness of the dies' bonding surface was in the range 1.2-1.5 μm. The substrates had dimensions slightly larger than the dies to facilitate the bonding process. The electroplated Au layer on the bonding side of the substrates had a thickness in the range 10-12 μm. The absolute roughness of the substrates' bonding surface was in the range 0.3-0.5 μm. No patterning was applied to the bonding surfaces of the dies and the substrates. The eutectic Au–Sn (Au 80 wt% - Sn 20 wt%) preform used in this work was supplied by Indium Corporation (USA). The preform had a thickness of about 25 μm, and was cut to dimensions suitable for bonding.

The Au thickness on the bonding side of the PZT dies and the RBL substrates, as well as the preform thickness, were selected to assure a surplus of Au in the bond-layer. The chosen thickness values of the different parts result in bonds with a layered structure of Au / Au–Sn (ζ phase) / Au, which has been reported to provide Au–Sn bonds with high mechanical strength and long-time reliability [9]. Note that the ζ phase is Au-rich, containing 8-16 at% / 5-10 wt% Sn.

2) Bonding process

Prior to bonding, PZT dies and RBL substrates were cleaned using an ultrasonic bath. The eutectic Au–Sn preform surfaces were cleaned on both sides with N$_2$ airflow to blow off dust particles (if present).

For each sample, a piece of Au–Sn preform was sandwiched between a die and a substrate. The substrate, the preform and the die were manually aligned on a flat ceramic heater. Next, the entire sample was kept fixed on the heater using a clamp, and was put inside a vacuum chamber. A bonding temperature profile was then applied under the control of a programmable PID controller (EZ-ZONE PM, Watlow, USA). The temperature profile applied included two main steps:

i) Heating a sample from room temperature to 250 °C and maintaining the temperature for 5 minutes to bake out any residual moisture and to assure a uniform temperature distribution in the bonding layers. Additionally, initial bonding is formed by solid-state diffusion, ensuring that the subsequent bonding will proceed with uniform growth of IMC layers [13].

ii) Heating the sample from 250 °C to the final bonding temperature of about 310 °C and maintaining the temperature for 15 minutes. After that, the sample is cooled down to room temperature.

B. Characterization

1) Synchrotron X-ray Computed Tomography

The X-ray measurement of the transducer sample of external dimensions $5 \times 0.5 \times 0.4$ mm^3 (diced from the originally bonded sample for optimal tomography resolution) was carried out at the high-energy CT beamline ID19 [14, 15] at the European Synchrotron Radiation Facility (ESRF) in Grenoble, France. The transducer sample was placed on a goniometer and illuminated with a parallel monochromatic X-ray beam with a photon energy of 68 keV. Approximately 2000 projections were acquired at equally spaced angles by rotating the sample about a vertical axis perpendicular to the beam. The measurement geometry used corresponds to an isotropic voxel size of 1.39 μm. The actual resolution was at least a factor 2-3 lower. The 3D reconstruction was obtained with the filtered back-projection reconstruction algorithm. The 3D volume was filtered using a wavelet-FFT filter [16] to reduce the presence of ring artifacts while retaining the structural information. Finally, segmentation using simple thresholding was used to separate voids from the sample phases. Statistics of the pore properties were obtained using Matlab®.

2) Cross-section microscopy

Open cross sections for microscopy of the bonded samples (PZT-RBL) were prepared using a conventional cross-sectioning process followed by Ar ion milling (IM4000, Hitachi High-Technologies, Japan). The bondline of the samples was inspected by optical microscopy (NEOPHOT 32, Jenoptik Jena GmbH, Germany), as well as scanning electron microscopy (SEM) and energy dispersive X-ray spectroscopy (EDX) (SU3500, Hitachi High-Technologies, Japan).

III. Representative Cross-sections – Tomography and Microscopy

A. X-ray Tomography

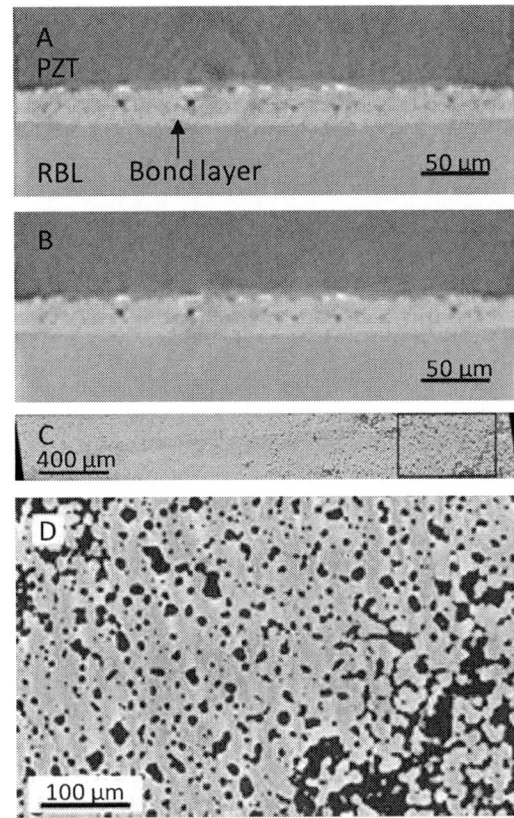

Figure 1: Reconstructed X-ray tomograms. A) Cross-section displaying the PZT die, RBL substrate and bond layer. B) Processed cross-section using wavelet-FFT filtering to remove ring artifacts. C) Reconstructed cross-section revealing the distribution of voids in the bond layer (cross-section normal to the one in A). The black wedges correspond to regions outside the field of view. D) Magnified cross-section of C, highlighting the segmented voids in dark/ green.

Figure 1 shows reconstructed X-ray tomograms, representing virtual cross-sections at different locations and for different orientations in the sample. It is clearly seen that the high photon energy in use (68 keV) gives a good material-specific contrast and signal-to-noise ratio, despite the high absorption in all three layers of the sandwich structure, namely PZT, Au/ Au–Sn, and RBL.

The reconstructed cross-section normal to the bond plane, cf. Figure 1 A), corresponds to a traditional cross-section micrograph. The bondline is clearly distinguished from the PZT and RBL layers, and voids can clearly be discerned in the bondline. The ring patterns are a common artefact in computed tomography. Figure 1 B) represents the same

reconstructed cross-section, at the same position as the one in Figure 1 A), after wavelet FFT filtering. Evidently, the ring artefacts are suppressed, without deterioration of the structural information. Suppression of the ring artefacts allows a more precise determination and analysis of the voids in the bond layer.

Figure 1 C-D) show reconstructed cross-sections in the bond layer, revealing information that is not readily available in traditional cross-section micrographs. The voids have a broad size distribution, and the void density is highly non-uniform. This implies that the information obtained through conventional cross-section micrographs would exhibit a large variation, depending on the position where the sample is cross-sectioned.

B. Cross-section microscopies

An optical micrograph of a cross-sectioned sample, manufactured with similar parameters as the one used for X-ray tomography, is given in Figure 2. Figure 2 A) has the same magnification as Figure 1 A-B, and the two techniques indeed give similar results in terms of bond layer shape and dimensions, as well as the spatial and size distributions of the voids. Figure 2 B) shows a magnified view, with a high-resolution objective ($NA = 0.9$). The optical micrograph can resolve smaller voids than our X-ray tomography setup allows, limited only by the diffraction limit of the microscope in use. Figure 2 B) resolves voids with dimensions down to ~0.3 µm. Note also the good material contrast obtained in the optical micrographs, easily distinguishing Au from Au–Sn intermetallics. Since all the materials in the sample are high-density materials, the overall material contrast is less pronounced in the X-ray tomography measurements.

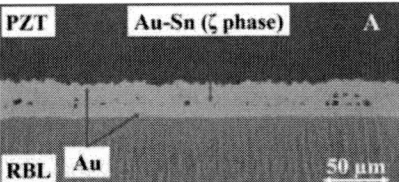

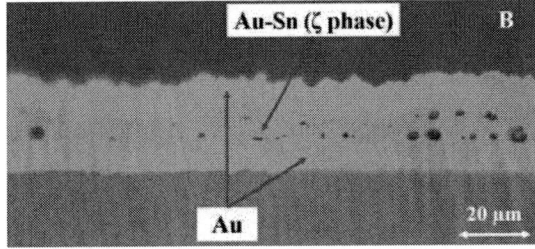

Figure 2: Optical micrograph of a cross-sectioned sample, being similar to the sample in Figure 1. A) Same scale as in Figure 1 A-B, showing comparable results. B) Magnified view. Small voids are observed in the bond layer, with sub-µm resolution. Reproduced from [12].

Scanning electron microscopy of cross-sections is able to resolve even smaller voids. Figure 3 shows a SEM micrograph of a Au–Sn SLID bonded model transducer. The image has a moderate magnification, still resolving voids < 0.1 μm. Note that the material contrast is much lower than in the optical micrograph, being primarily a channeling contrast differentiating crystal orientations of grains (exposing the difference between small-grain sputtered Au seed layer and large-grain Au electroplated layers). More details on contrast for studies of SLID bonds can be found in our previous publication [17].

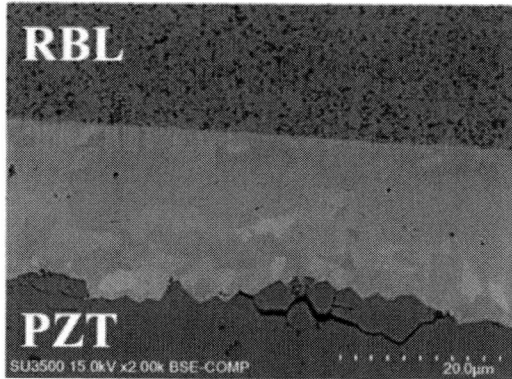

Figure 3: SEM micrograph of a cross-sectioned sample Au–Sn SLID bonded sample, with Au / Au–Sn / Au SLID bond between RBL and PZT.

IV.   3D Analysis of Void Content

The power of X-ray tomography lies in the ability to extract 3D microscopic images from a sample non-destructively, allowing detailed analysis of the internal structures in the sample.

A.   Void parameterization

The voids were labeled using a connectedness of 6, meaning that two neighbouring void voxels will be considered to belong to the same void only if they are face-to-face connected. Considering the experimental uncertainty relating to the resolution and the contrast, only voids consisting of more than 9 voxels were included in the analysis. With an isotropic voxel size of 1.4 μm, this corresponds to a minimum void size of 27 μm³ for inclusion in the statistical analysis. For consistency with the limited resolution, the few voids having only one voxel width along one dimension were discarded.

For a statistical analysis that reflects the dimension, shape and orientation of the voids, each void was parameterized as an ellipsoid, with three principal axes p_1, p_2, p_3 each defined as having the same second central moment as the void. The length and orientation of these three principal axes were used as the model parameters to describe each void. Note that p_1 is taken to be the longest principal axis

of the void, and p_3 the shortest – their orientation with respect to the sample geometry is defined through Euler angles yaw, pitch and roll.

B.   Number and volume of voids

When labeling voids using a connectedness of 6, a total of about 6300 voids were detected in the bonding layer volume. This number was reduced to ~3100 labeled voids by only considering those voids consisting of more than 9 voxels. These ~3100 voids correspond to an estimated void fraction of the bond layer of 4 %. The single largest void contributed approximately 10 % of the total pore volume, while 50 % of the total pore volume could be accounted for by the 54 largest voids.

Figure 4 shows the normalized cumulative void volume: The voids are sorted from the largest (#1) to the smallest, and for each void the volume is normalized towards the total volume of voids in the bond layer. It can easily be read from Figure 4 that the largest void indeed accounts for 10 % of the total void volume, and that the 10-11 largest voids contribute 40 % of the total volume.

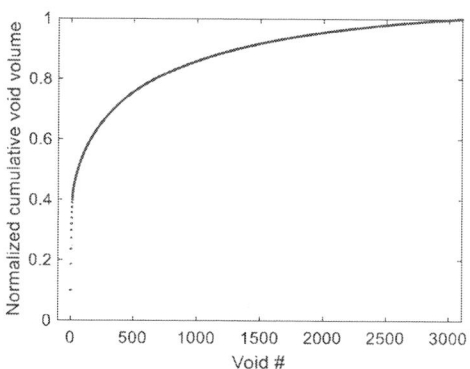

Figure 4: Normalized cumulative void volume.

C.   Principal axes – length and orientation

The principal axes lengths exhibit a large distribution: The mean principal axis lengths were 9 μm (p_1), 6 μm (p_2) and 4.5 μm (p_3), while the longest principal axes were 292 μm, 184 μm and 13 μm. The much lower maximum value for p_3 is compatible with the void size being restricted in one dimension by the actual bond layer thickness. Figure 5 gives the distribution of principal axes lengths, showing the data for axes of length < 25 μm.

Figure 6 shows the orientation distribution of p_1, for the 200 largest voids as well as for the remaining smaller voids. Void #200, selected to differentiate between large and small voids in Figure 6, has principal axis lengths 13.8 μm (p_1), 8.3 μm (p_2) and 7.5 μm (p_3). The yaw angle is uniformly distributed for both data sets, whereas the pitch and roll angles exhibit qualitatively different distributions for the large and the small voids.

The bond layer thickness imposes a constraint on the void extension, implying that large voids must have their long principal axes in-plane, as seen in the data. Notably, also the small voids are seen to exhibit a preferred orientation. The largest voids are oblate with $p_1 \sim p_2 \gg p_3$. The long axes are necessarily oriented in the bonding plane, confined by the ~15 µm thickness of the bonding layer. If looking at only the largest voids (selecting a larger void than #200 for defining large voids), this orientation distribution would be even narrower than the one presented in Figure 6. The smaller voids (#200 - #3000) also tend to be non-spherical and oriented, however with the long axis perpendicular to the bonding layer.

While the largest voids have a solidity of about 0.6, voids from about #2000 ($p_1 = 9.7$ µm, $p_2 = 5.3$ µm, $p_3 = 2.2$ µm) have a solidity near unity, meaning they are homogeneous objects.

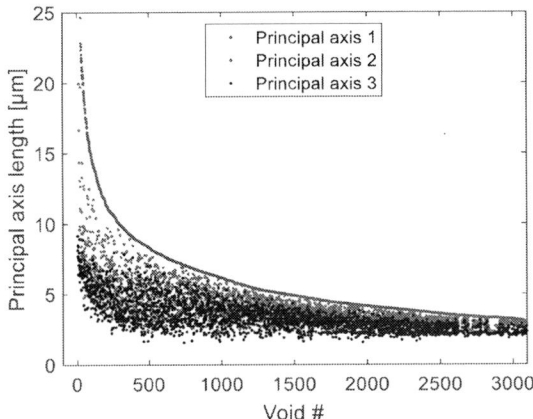

Figure 5: Principal axis lengths in descending order of the longest principal axis p_1.

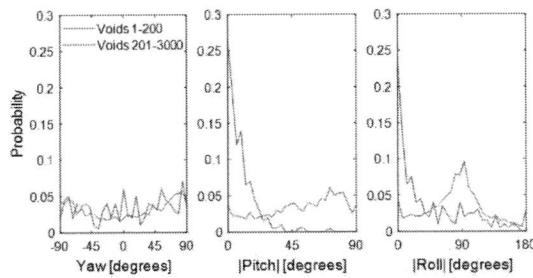

Figure 6: Orientation of the longest principal axis (p_1), for the 200 largest voids and for the remaining smaller voids. The yaw orientation distribution exhibits no preferred orientation. The pitch and roll angle distributions clearly signify that for the large voids, *both* the longest and second longest principal axes (p_1 and p_2) tend to be oriented parallel to the bonding layer. Also small voids are oriented, but with their longest principal axis (p_1) perpendicular to the bonding layer.

D. Larger voids vs smaller voids

From the distributions shown above, it is clear that the majority of voids has a moderate size and follows well-defined distributions where a statistical approach may be appropriate. The large voids are statistical outliers. Hence, it is appropriate to investigate these individually. Figure 7 shows orthogonal views of the largest void detected, false-colored red. This large void appears to be a connected network of smaller voids.

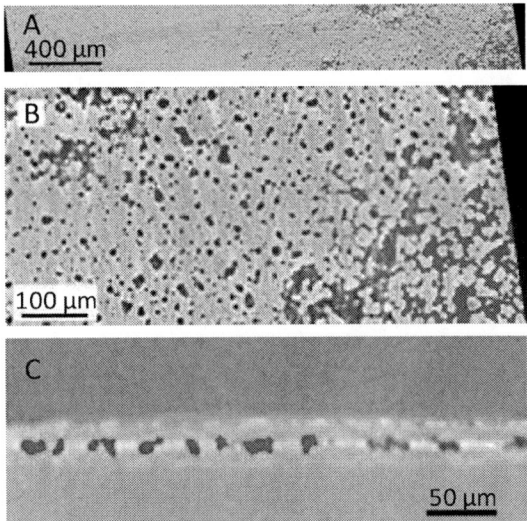

Figure 7: Reconstructed X-ray tomograms, showing the largest void in the sample, highlighted in red. A) Cross-section in the bond plane. The black wedges correspond to regions outside the field of view. B) Magnified cross-section in the bond plane. C) Cross-section normal to the bond plane.

For a statistical treatment of the moderately sized voids, we omitted the large voids accounting for about 50% of the total void volume, cf. Figure 4. This gave a cutoff at about 450 voxels from the analysis, corresponding to omitting the 66 largest voids from the analysis. The maximum principal axis lengths in the truncated data set were $p_{1,max} = 34$ µm, $p_{2,max} = 15$ µm, $p_{3,max} = 11$ µm. The mean values of the principal axis lengths do not change much compared with the data set including all voids, but the standard deviation is now a relevant number. The mean values and standard deviations for the principal axis lengths are $p_1 = (8 \pm 4)$ µm, $p_2 = (5.5 \pm 1.9)$ µm, $p_3 = (4.4 \pm 1.4)$ µm.

V. Discussion

The three-dimensional nature of tomography measurements gives access to vastly more information than what is obtainable from traditional cross-section microscopy. The tomography measurements and analysis reported in this paper, reveal a sample with a highly inhomogeneous void distribution, and a broad distribution of void sizes. This implies that cross-section microscopy would give qualitatively different results depending on the actual position where the sample is cross-sectioned. Comparing Figure 1 B and Figure 7 C gives a clear indication how different two reconstructed cross-sections from the same sample can be, where the only difference is the position of the reconstructed plane. The three-dimensional nature of the tomography data also allows a straightforward reconstruction of in-plane cross-sections, clearly imaging the spatial distribution of voids. Mapping such inhomogeneity through cross-section microscopy would require a large number of physically cut cross-sections. The destructive nature limits how many cross-sections can be made, and thus limits the amount of information that can possibly be retrieved.

Comparing Figure 1 B and Figure 2 A demonstrates that X-ray tomography and cross-section microscopy indeed give comparable results, if a corresponding reconstructed cross-section is selected from the tomography data. This comparison is therefore important, since it validates the accuracy of the information retrieved from the tomography data. The advantages of cross-section microscopies include the easy instrument access, higher resolution and better material contrast, as well as being a well-established and recognized technique for bond characterization.

Tomography reveals large voids and complex porous networks that would not be observable in a randomly cut cross-section, whereas cross-section microscopy reveals small voids that cannot be resolved with the X-ray tomography setup used. Our analysis of 3D void distribution in the bond layer can thus take into account only the voids that are larger than 3-4 µm. For ultrasound transducer applications, it is the larger voids that will most severely impede the acoustical performance. We have previously demonstrated by comparing cross-section microscopy with electrical impedance spectroscopy that a bond can have a distribution of smaller voids without sacrificing the acoustical performance [12]. The relevant length scale for what should be considered small or large voids, is the ultrasound wavelength for the application.

Uniformity, homogeneity and reproducibility are required for industrial manufacturing processes. For a mature process where a high uniformity and homogeneity are achieved, cross-section microscopy will remain a highly relevant characterization method. We have shown here that X-ray tomography gives unique information for process development, by providing 3D image data for an inhomogeneous sample.

VI. Conclusions

Synchrotron-based X-ray tomography gave non-destructive three-dimensional information about the void content in the bond layer of a Au–Sn SLID bonded ultrasound transducer dummy sample. Using high-energy monochromatic X-rays, as enabled by the selected synchrotron beamline, allows high-quality tomograms to be retrieved despite the high X-ray absorption of the sample being rich in high-density materials.

The tomography data exhibited a highly inhomogeneous void distribution, and a large variation in void sizes. Statistical analysis of void size, shape and orientation can be retrieved, information that is not obtainable by traditional cross-section microscopy. We have demonstrated the arbitrariness inherent in a cross-section analysis of an inhomogeneous sample, and propose X-ray tomography as a valuable tool in manufacturing process development.

Acknowledgements

The authors thank Zekija Ramic and Anh-Tuan Thai Nguyen, both USN, for Au electroplating of samples, and Tung Manh (USN) and Trym Eggen (GE Vingmed Ultrasound AS) for valuable discussions.

The present work was funded by the Research Council of Norway through the NANO2021 program (project number 235302; B-EAM – Advanced Assembly Technologies for Electro Acoustic Module used in Ultrasound Cardiovascular Applications). The Research Council of Norway is also acknowledged for the support to the Norwegian Micro- and Nano-Fabrication Facility, NorFab, project number 245963.

KRT and DWB thank the Research Council of Norway for funding through its Centres of Excellence scheme (project 262644) and FRINATEK (project 275182).

References

[1] K. E. Aasmundtveit, T.-T. Luu, H.-V. Nguyen, A. Larsson, and T. A. Tollefsen, "Intermetallic Bonding for High-Temperature Microelectronics and Microsystems: Solid-Liquid Interdiffusion Bonding," in *Intermetallic Compounds - Formation and Applications*, M. Aliofkhazraei, Ed., ed Open Access: Intech Open, 2018.

[2] L. Bernstein, "Semiconductor joining by solid-liquid-interdiffusion (SLID) process.," *Journal of the Electrochemical Society*, vol. 113, pp. 1282-88, 1966.

[3] H. Huebner, S. Penka, B. Barchmann, M. Eigner, W. Gruber, M. Nobis, *et al.*, "Microcontacts with sub-30 μm pitch for 3D chip-on-chip integration," *Microelectronic Engineering*, vol. 83, pp. 2155-2162, Nov-Dec 2006.

[4] T. T. Luu, A. Duan, K. E. Aasmundtveit, and N. Hoivik, "Optimized Cu-Sn wafer-level bonding using intermetallic phase characterization," *Journal of Electronic Materials*, vol. 42, pp. 3582-3592, 2013.

[5] C. Honrao, T. C. Huang, M. Kobayashi, V. Smet, P. M. Raj, and R. Tummala, "Accelerated SLID bonding using thin multi-layer copper-solder stack for fine-pitch interconnections," in *2014 IEEE 64th Electronic Components and Technology Conference (ECTC)*, 2014, pp. 1160-1165.

[6] T. A. Tollefsen, A. Larsson, O. M. Lovvik, and K. Aasmundtveit, "Au-Sn SLID Bonding-Properties and Possibilities," *Metallurgical and Materials Transactions B-Process Metallurgy and Materials Processing Science*, vol. 43, pp. 397-405, Apr 2012.

[7] W. R. Johnson, C. Q. Wang, Y. Liu, and J. D. Scofield, "Power Device Packaging Technologies for Extreme Environments," *IEEE Transactions on Electronics Packaging Manufacturing*, vol. 30, pp. 182-193, 2007.

[8] A. Rautiainen, H. B. Xu, E. Osterlund, J. Li, V. Vuorinen, and M. Paulasto-Krockel, "Microstructural Characterization and Mechanical Performance of Wafer-Level SLID Bonded Au-Sn and Cu-Sn Seal Rings for MEMS Encapsulation," *Journal of Electronic Materials*, vol. 44, pp. 4533-4548, Nov 2015.

[9] T. A. Tollefsen, A. Larsson, M. M. V. Taklo, A. Neels, X. Maeder, K. Hoydalsvik, *et al.*, "Au-Sn SLID Bonding: A Reliable HT Interconnect and Die Attach Technology," *Metallurgical and Materials Transactions B-Process Metallurgy and Materials Processing Science*, vol. 44, pp. 406-413, Apr 2013.

[10] T. Manh, H. V. Nguyen, D. Le-Anh, K. Aasmundtveit, L. Hoff, T. Eggen, *et al.*, "Au-Sn Solid-Liquid Interdiffusion (SLID) bonding for piezoelectric ultrasonic transducers," in *2016 IEEE International Ultrasonics Symposium (IUS)*, 2016, pp. 1-4.

[11] K. E. Aasmundtveit, T. Eggen, T. Manh, and H. V. Nguyen, "In-Bi low-temperature SLID bonding for piezoelectric materials," *Soldering & Surface Mount Technology*, vol. 30, pp. 100-105, 2018.

[12] H. V. Nguyen, M. Tung, T. Eggen, and K. E. Aasmundtveit, "Au-Sn Solid-Liquid Interdiffusion (SLID) bonding for mating surfaces with high roughness," in *2016 6th Electronic System-Integration Technology Conference (ESTC)*, 2016, pp. 1-6.

[13] H. Etschmaier, H. Torwesten, H. Eder, and P. Hadley, "Suppression of Interdiffusion in Copper/Tin Thin Films," *Journal of Materials Engineering and Performance*, vol. 21, pp. 1724-1727, 2012/08/01 2012.

[14] E. Boller, P. Tafforeau, A. Rack, V. Fernandez, L. Helfen, M. Rénier, *et al.*, "Synchrotron-tomography with micro, nano and high temporal resolution for industrial and academic use," presented at the 3rd International Conference on Tomography of Materials and Structures, Lund, Sweden, 2017.

[15] T. Weitkamp, P. Tafforeau, E. Boller, P. Cloetens, J. P. Valade, P. Bernard, *et al.*, "Status and evolution of the ESRF beamline ID19," *AIP Conference Proceedings*, vol. 1221, pp. 33-38, 2010.

[16] B. Munch, P. Trtik, F. Marone, and M. Stampanoni, "Stripe and ring artifact removal with combined wavelet - Fourier filtering," *Optics Express*, vol. 17, pp. 8567-8591, May 2009.

[17] K. E. Aasmundtveit, H. Jiang, T. A. Tollefsen, T. Luu, and H. Nguyen, "Phase Determination in SLID Bonding," in *2018 7th Electronic System-Integration Technology Conference (ESTC)*, 2018, pp. 1-6.

Hybrid Cu particle paste with surface-modified particles for high temperature electronics packaging

Sri Krishna Bhogaraju[1]. Omid Mokhtari[1], Jacopo Pascucci[1], Alexander Hanss[1], Maximilian Schmid[1], Fosca Conti[2], Gordon Elger[1]

[1]Institute of Innovative Mobility (IIMo), Technische Hochschule Ingolstadt, Ingolstadt, Germany

[2] Department of Chemical Sciences, University of Padova, Padova, Italy

+49-84193486424, srikrishna.bhogaraju@thi.de

Abstract

Die-attach bonding is a key process to realize high-temperature operation of power semiconductor devices. Ag sintering has been in the forefront of the research in the past decade as a suitable alternative to high temperature solders such as AuSn. However, the high cost of Ag and low electromigration resistance have been deterrants to large scale industrialization.

Cu is ~100 times cheaper than Ag and more abundantly available. It is easy to process and recycle and displays mechanical, thermal and electrical properties comparable with Ag.

In this contribution, the research is focused on developing Cu sintering as an alternative to Ag sintering. Different sintering pastes have been prepared in-house by combining Cu particles in micro and/or nano- scale with polyethylene glycol 600 (PEG600). The performances of the pastes have been analysed and evaluated under different bonding conditions.

Shear strengths of approximately 88MPa has been achieved while working with a combination of surface modified Cu-alloy particles and PEG600 as binder. Surface modifications on the Cu-alloy particles are obtained through a selective etching of the alloying element. Sintering was performed under a bonding pressure of 20MPa, at 275°C, for 30min, under N_2. In comparison with hybrid Cu particle paste under the same experimental conditions, the result is very promising and better by a factor 2.

Key words: Cu particle sintering, nanoporous Cu, oxidation protection, reducing binder, surface modifications

Introduction

The need for operating at higher temperatures as well as the advancements in miniaturization of microelectronics require urgently new materials and processes to tackle the challenges in the field of assembly and bonding technology. Die-attach bonding is a key process to realize interconnects for high temperature applications. With the increase in the usage of high power electronics devices, it is imperetive to find sustainable and reliable interconnects, which withstand operation temperatures above 150°C.

Transient liquid phase (TLP) bonding has been suggested as a suitable alternative for high temperature electronics applications. TLP offers the advantage of low processing temperatures while interconnects remain stable at relatively high operating temperature. However, brittle intermetallic compounds (IMC) and void formation are reported to be major drawbacks with TLP bonding, both of which have detrimental effects on the long term reliability of the interconnects[1]-[3].

Sintering, specifically silver sintering, has been reported to be a suitable alternative for high-temperature applications with shear strengths up to 75MPa while sintering under pressure[4][5]. However, sintered Ag particles have their own drawbacks, expecially because of high cost and low-electromigration resistance. Further, the composition of these pastes is mainly of nano particles, which increases the cost due to special processing and handling requirements.

Cu is ~100 times cheaper than Ag, particularly because it is more abundant and easily available. It has a lower coefficient of thermal expansion (16.5μm/mK) than Ag (18.9μm/mK) and has nearly the same thermal (401W/mK) conductivity as Ag (429W/mK). However, the higher melting point of Cu (1084°C) compared to Ag (961°C) translates to the requirements of slightly higher sintering temperature.

A challenge while working with Cu particles is the easy and fast oxidation under air. This is detrimental to the bondability, mechanical (shear strength) and thermal integrity of the joint [5]-[6].

The use of hydrogen or formic acid enriched N_2 atmosphere as reducing agents or vacuum during Cu sintering is an approach that has been popular with researchers to address the issue of oxidation of Cu particles [7]. However, H_2 brings the need for extensive safety precautions to be followed and formic acid leads to a corrosive environment, both of which are not conducive for large-scale industrialization. Further, the insufficient penetration of the reducing agent into the interconnect during sintering, necessitating an approach to include a reducing binder in the paste composition is reported[8]-[9].

Reports on the state of the art in copper sintering show the dependence on the reducing atmosphere of either formic acid or H_2. Further, in case of sintering under N_2, temperatures in excess of 350°C are reported. In all cases the pastes are mainly composed of nano scale spherical particles. [10]-[22]

In this contribution we present our recent development of innovative sintering pastes, which combine micro- and nano-scale Cu particles dispersed in a binder with reducing properties, thereby allowing for sintering under N_2 atmosphere. Average shear strengths of 88MPa are recorded while sintering under pressure with the new pastes.

Test chip specifications and assembly parameters of the sintering process

High power 1mm^2 light emitting diodes (LEDs) with Ni/Au metallization are used as standard test chips during the sintering processes. Stencil printing is adopted as the preferred paste application method. The printing of the paste was performed with a semi-automatic stencil printer (PBT, Go3v) with a motorized double blade squeegee and a stencil thickness of 75 μm. The squeegee speed was 13 mm/s, the squeegee pressure 20 N and the stencil separation 2.3 mm/s. The paste was printed on an Al-IMS substrate with ENIG metallization. Pastes were prepared in a Thinky planetary-rotary mixer.

Figure 1 shows the the schematic representation of the sintering process carried out in four main steps. In step one (Figure 1a), the sinter paste is printed onto the substrate using a stencil printer. In step two (Figure 1b), the substrate and the chip are heated to the sintering temperature of 275°C at a rate of 1K/s but the chip is not yet placed on the printed paste. In step three (Figure 1c), the chip is placed in the paste once the sintering temperature is reached. In step four (Figure 1d), sintering is carried out at 275°C under 60L/hr N_2 flow for 30 min with a bonding pressure of 20MPa. The sintering profile used in the experiments is reported in Figure 2.

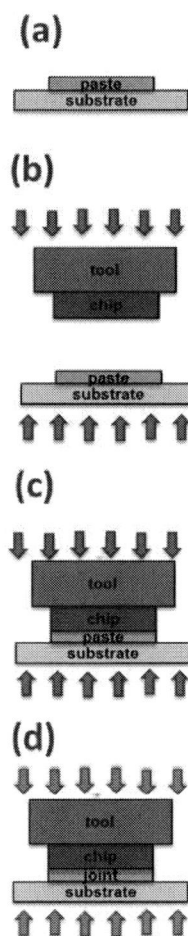

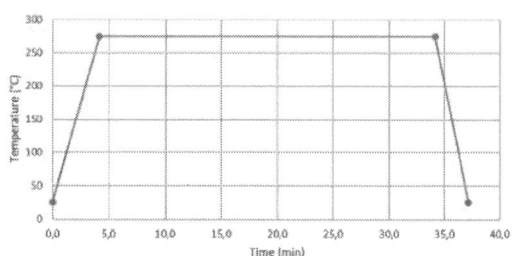

Figure 2 - Sintering profile for Cu sintering under bonding pressure of 20MPa at sintering temperature of 275°C for 30min with in-house developed Cu sinter pastes.

Composition and preparation of sintering pastes

For the purpose of this study, three different pastes were prepared, which were obtained by combining Cu particles in micro and nano scale with polyethylene glycol 600 (PEG600). The pastes are listed in Table I. Cu nano particles (80nm) were purchased from NanoGrafi Co. Ltd. Cu micro particles were procured from Osaka University. Cu/Zn-alloy micro particles were purchased from Zhangqiu Metallic Pigment Co. Ltd. PEG600 is a common binder and was used without any previous additional treatment.

Table I - Sintering paste composition

Paste	Composition
A	nano Cu + PEG600
B	micro Cu + nano Cu + PEG600
C	etched Cu/Zn-alloy particles + PEG600

The sintering pastes were prepared by mixing the reagents in a planetary-rotary mixer from Thinky and using the following profile: 4min of mixing at 1000rpm followed by 5min at 500rpm. In case of Cu/Zn-alloy particles, etching was carried out to dealloy the Zn. HCl was chosen as etchant as it creates soluble $ZnCl_2$ salts, thereby not leaving any unwanted residues. Further, the presence of chloride anions is reported to increase the copper surface diffusivity coefficient by five orders of magnitude compared to chloride free electrolytes [23].

Figure 1 - Schematic representation of the sintering process. Arrows indicate heating (red) or cooling (blue).

Analysis equipments and characterization methods

A Keyence VHX-900F 3D microscope was used to optically inspect the quality of the printing and the analysis of the fracture surface after shear testing of the sintered interconnects. The assemblies were prepared using a Fineplacer Pico pick and place equipment. Microstructure analysis was carried out on a LEO Leo Gemini 1530 scanning electron microscope (SEM). Shear testing of the sintered interconnects is conducted on a XYZ-Condor Sigma Lite shear tester with a shear speed of 250µm/s and a shear height of 25µm.

Results and discussions

Figure 3 shows the SEM analysis of the surface of the Cu/Zn-alloy particles before and after a 4 hrs etching process with 12 M HCl. The effects of the selective removal of zinc are evidents in Figure 3b.

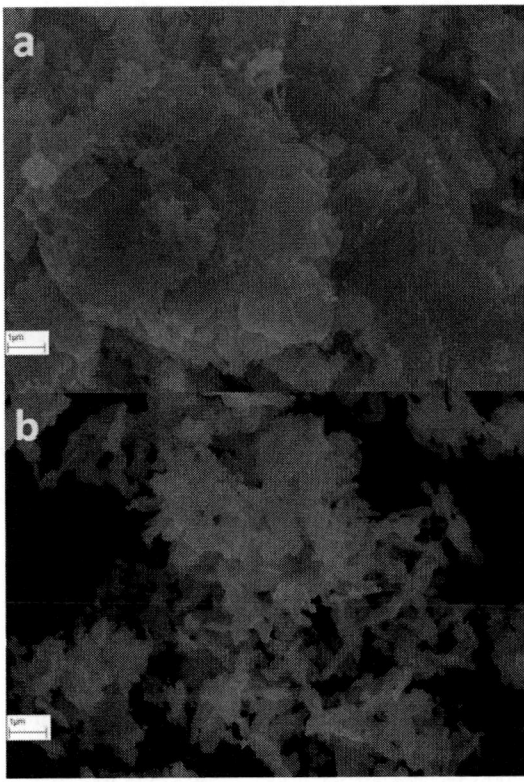

Figure 3 - (a) orignal Cu/Zn-alloy particles, (b) after etching for 4hrs in 12M HCl (1 μm scale) resulting in particles with increased surface area due to nano-structured modifications.

The surface modifications through the selective etching are aimed to create high energy free surfaces. Increased surface area leads to a higher specific surface energy γ (J/m²) distributed over the total surface area A (m²). This is highly beneficial for sintering, since the driving force for sintering is the reduction of the total surface energy following equation 1.

$$\Delta(\gamma A) = \Delta\gamma*A + \gamma*\Delta A \qquad (1)$$

Passing from the original Cu/Zn-alloy particles to the etched sample, an increment of at least 10% is observable in the surface area. The estimation is a preliminary result obtained analysing the SEM images (Figure 3).

Figure 4 shows the shear strength values for the three investigated sintering pastes. The results are 43 MPa, 48 MPa, and 88MPa for paste A, B, and C, respectively. Sintering pastes composed of surface modified μ-Cu/Zn alloy particles returned the higher average shear strength of 88MPa, which is approximately double in comparison to the other Cu-based particles pastes investigated in-house . Further, paste C can be sintered at temperature as low as 275°C and under a N_2 environment, which is an added advantage. The same paste under pressureless sintering conditions of 300°C for 60min returned an average shear strength of 30MPa.

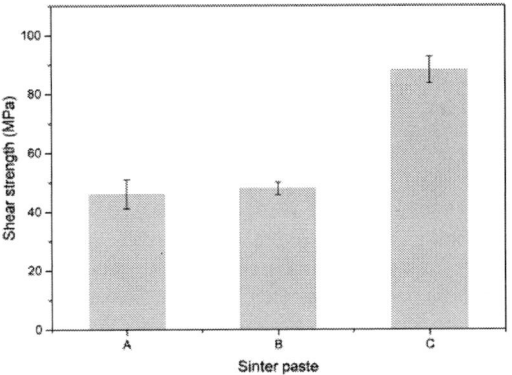

Figure 4 - Shear strength results for sinter pastes A, B and C used for sintering under a bonding pressure of 20 MPa, a temperature of 275°C and a bonding time of 30 min.

We attribute the improvement of shear strength to the etching of Zn. The selective etching of Zn from the Cu/Zn-alloy creates nano-structured modifications in the particles, which enhance the surface area. This increment leads to particles with high energy per unit volume and with increased curvature along the surfaces and edges. There effects provide optimal material characteristics for enhanced sintering, as described in eq. (1).

SEM analysis of the surface of the joint following shear tests after sintering with paste C under sintering conditions of 20MPa bonding pressure for 30min sintering time at 275°C sintering temperature under N_2 atmosphere shows a ductile fracture mode (Figure 5) .

Figure 5 - SEM analysis of the fracture surface after shear testing of the interconnect realized by sintering paste C under bonding conditions of 20MPa, 275°C, 30min.

To have a better understanding of the phenomena occurring in the etched paste C, additional analyses were carried out. This best performing paste is composed of surface modified Cu-based particles and a study of the impact of the sintering parameters was investigated. As can be seen in Figure 6, an increase in the bonding pressure from 0.1 to 10 MPa leads to a mechanically stable interconnect. Best results were obtained while sintering with a bonding force of 20MPa for 30min in N_2 atmosphere, as already shown in Figure 6 in comparison to paste A and B. At 0.1 and 10 MPa of bonding pressure the shear stress of paste C resulted to be 16 MPa and 38 MPa, respectively.

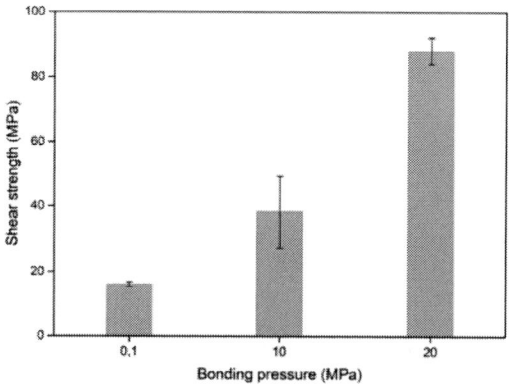

Figure 6 - Effect of bonding pressure on the shear strength of the sintered interconnect under sintering conditions of 275°C for 30min. Results refer to paste C.

In order to investigate the impact of the sintering temperature on the mechanical integrity of the sintered interconnect, bonding pressure (10MPa) and bonding time (30min) were held constant and the sintering temperature varied. An increase in temperature showed a marked improvement in the shear strength, as shown in Figure 7. In the 250-300 °C range, the shear strength pass from 38MPa to 63 MPa, with an improvement of 40%.

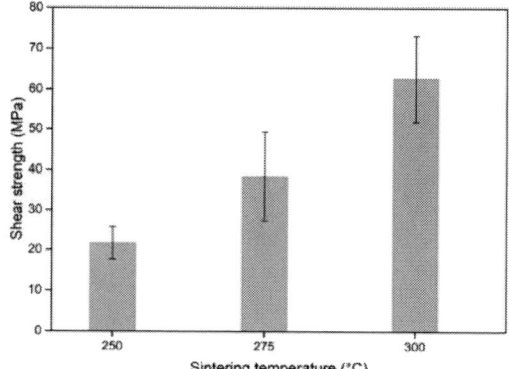

Figure 7 - Effect of sintering temperature on the shear strength of the sintered interconnect under bonding pressure of 10MPa and sintering time of 30min. Results refer to paste C.

The impact of the sintering time on the mechanical integrity of the interconnection was invesitigated. As plotted in Figure 8, the shear strength increased with increased bonding time while keeping sintering temperature and bonding pressure constant at 275°C and 20 MPa, respectively. The shear strength improves from 33 MPa at 10 min to 88 MPa at 30 min of sintering time. This effect can be well explained by the phenomenon of sintering which is mainly driven by surface diffusion followed by densification and grain growth.

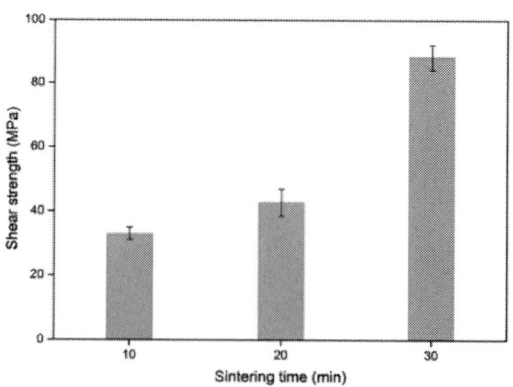

Figure 8 - Effect of sintering time on the shear strength of the sintered interconnect under bonding pressure of 20 MPa and temperature of 275 °C. Results refer to paste C.

The use of a bonding force is well known to enhance the efficiency of sintering. In case of Ag sintering, application of a bonding pressure while sintering seems to aid the particles sinter better together realizing dense interconnects, which offer good reliability. On the countrary, pressureless Ag sintered interconnects have shown a rapid degradation under high thermo-mechanical stress conditions[5]-[6].

The effect of the bonding force on sintering can be described by the Mackenzie-Shuttleworth model [23]. A representative equation is reported in eq. (2), where dp/dt is the densification rate and $(\gamma/r + P_{applied})$ is the driving force for densification. In this model, the densification rate is a function of six parameters, which can be described as in the following: $P_{applied}$ is the applied sintering pressure, γ is the surface energy, r is the particle radius, α is the geometrical constant, ρ is the density and η is the densification viscosity.

$$\frac{dp}{dt} = \frac{3}{2}\left(\frac{\gamma}{r} + P_{applied}\right)(1-\rho)\left(1 - \alpha\left(\frac{1}{\rho}-1\right)^{\frac{1}{3}}\ln\left(\frac{1}{1-\rho}\right)\right)1/\eta \quad (2)$$

A clear correlation as represented by eq. (2) is observed in our results, wherein the application of a bonding pressure is observed to enhance the shear strength, as shown in Figure 6 for the bonding pressure range between 0.1 and 20 MPa. The surface modifications realized on the particles of paste C increase the surface energy of the particles. Sinter paste composed of surface modified particles returns the best shear strength results as observed in comparison to paste A and B (Figure 4).

Figure 9 shows the interconnect realized under a bonding conditions of sintering at 275°C for 30min under N_2 atmopshere. While in case of sintered interconnect represented in Figure 9a a bonding pressure of 20MPa is applied during sintering, a bonding pressure of 1MPa is applied in case of sintered interconnect represented in Figure 9b. The analysis of the cross-sections of the interconnects further justifies the use of the Mackenzie-Shuttleworth model, wherein the application of the bonding pressure results in the realization of a dense interconnect.

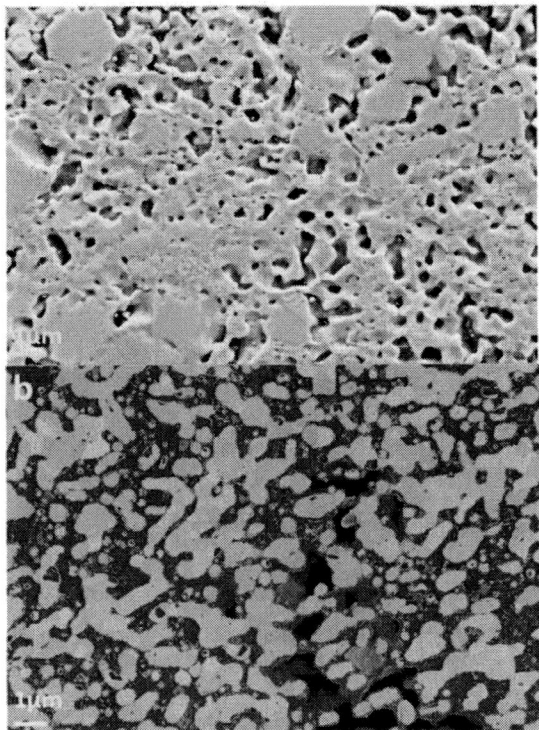

Figure 9 – SEM analyses (1 µm) of a sintered interconnect obtained using a bonding pressure of (a) 20MPa, and (b) 1MPa, while sintering at 275°C for 30min.

Conclusions

A novel Cu particle paste composed of surface modified micro Cu/Zn alloy particles mixed with PEG600 is developed. The surface modifications are realized through a low-cost wet chemical etching process where the alloying element, in this case Zn, is selectively etched from the composition. The paste can be sintered at temperatures as low as 275°C in a N_2 environment under the application of a bonding pressure. The surface modifications are observed to enhance sinterability between the particles as well as between particles and substrate. An average shear strength of 88MPa is obtained under sintering conditions of 275°C for 30min under N_2 atmosphere. The value is a factor 1.5 greater than the highest reported in literature [10]-[22] and a factor two greater than the comparable hybrid Cu particles pastes investigated in the present study.

Acknowledgements

The authors gratefully acknowledge Prof. Hiroshi Nishikawa from Osaka University for the micro-scale Cu particles and Simon Keim from Technische Hochschule Ingolstadt for the support with SEM analysis. This research was funded by the German Ministry of Education and Science within the project ProLoMa (Project number 13FH147PX6) under the program "Forschung an Fachhochschulen mit Unternehmen (FHprofUnt)".

References

[1] O. Mokhtari, "A review: Formation of voids in solder joint during the transient liquid phase bonding process-Causes and solutions", Microelectronics Reliability, 98, 95-105, 2019.

[2] O. Mokhtari, H. Nishikawa, "Transient liquid phase bonding of Sn–Bi solder with added Cu particles", Journal of Materials Science: Materials in Electronics, 27 (5), 4232-4244, 2016.

[3] O. Mokhtari, H. Nishikawa, "The shear strength of transient liquid phase bonded Sn–Bi solder joint with added Cu particles", Advanced Powder Technology, 27 (3), 1000-1005, 2016.

[4] S.K. Bhogaraju, A. Hanss, G. Elger, M. Schmid, F. Conti, "Evaluation of silver and copper sintering of first level interconnects for high power LEDs", Proceedings of the 2018 Electronic System-Integration Technology Conference (ESTC), Dresden, Germany, September 18-21, S. 1-8, 2018.

[5] A. Hanss, S.K. Bhogaraju, G. Elger, M. Schmid, F. Conti, "Process development and reliability of sintered high power chip size packages and flip chip LEDs", Proceedings of the 2018 Electronics Packaging and IMAPS All Asia Conference (ICEP-IAAC), Kuwana, Mie, Japan, April 17-21, S. 479-484, 2018.

[6] A. Hanss, S.K. Bhogaraju, G. Elger, M. Schmid, F. Conti, "Reliability of Sintered and Soldered High Power Chip Size Packages and Flip Chip LEDs", Proceedings of the 2018 IEEE 68th Electronic Components and Technology Conference (ECTC), San Diego, California, U.S.A, May 29-June 1, S. 2080-2088, 2018.

[7] F. Conti, A. Hanss, C. Fischer, G. Elger, "Thermogravimetric investigation on the interaction of formic acid with solder joint materials", New Journal of Chemistry 40, 10482-10487, 2016.

[8] S.K. Bhogaraju, O. Mokhtari, J. Pascucci, F. Conti, G. Elger, "Improved sinterability of particles to substrates by surface modifications on substrate metallizations," Proceedings of the International Conference and Exhibition on High Temperature Electronic Network

(HiTEN), IMAPS, Oxford, U.K., July 8-9, 2019, submitted for publication.

[9] S.K. Bhogaraju, O. Mokhtari, H. R. Kotadia, J. Pascucci, F. Conti, G. Elger, "A multi-pronged approach to low-pressure Cu sintering using surface-modified particles, substrate and chip metallization", Proceedings of the International symposium on Microelectronics, Boston, U.S:A., October 1-3, 2019, submitted for publication.

[10] X. Liu, H. Nishikawa, "Low-pressure Cu-Cu bonding using in-situ surface-modified microscale Cu particles for power device packaging", Scripta Materialia, 120, 2016.

[11] M. Yao, N-C. Lee, "Nano-Cu sintering paste for high power devices die attach applications", Proceedings of the 2018 IEEE 68th Electronic Components and Technology Conference (ECTC), Dresden, Germany, Setember 18-21, S. 557-563, 2018.

[12] J. Li, X. Yu, T. Shi, C. Cheng, J. Fan, S. Cheng, Z.Tang, "Low-Temperature and Low-Pressure Cu–Cu Bonding by Highly Sinterable Cu Nanoparticle Paste", Nanoscale research letters, 12. Jg., Nr. 1, S. 255, 2017.

[13] T. Yamakawa, "Influence of joining conditions on bonding strength of joints: efficacy of low-temperature bonding using Cu nanoparticle paste", Journal of electronic materials, 42. Jg., Nr. 6, S. 1260-1267, 2013.

[14] Y. Zuo, J. Shen, Y. Hu, R. Gao, "Improvement of oxidation resistance and bonding strength of Cu nanoparticles solder joints of Cu–Cu bonding by phosphating the nanoparticle", Journal of Materials Processing Technology, 253, 27-33, 2018.

[15] J. Kahler, N. Heuck, A. Wagner, A. Stranz, E. Peiner, A. Waag, "Sintering of copper particles for die attach", IEEE Transactions on Components, Packaging and Manufacturing Technology, 2. Jg., Nr. 10, S. 1587-1591, 2012.

[16] H. Nishikawa, T. Hirano, T. Takemoto, N. Terada, "Effects of joining conditions on joint strength of Cu/Cu joint using Cu nanoparticle paste", Open Surface Science Journal, 3, 60-64, 2011.

[17] H. Nakako, C. Sugama, Y. Kawana, M. Negishi, Y, Yanaka, D. Ishikawa, Y. Ejiri, "Sintering Cu Bonding Paste: Cycle Reliability and Applications", Proceedings of the International Exhibition and Conference for Power Electronics, Intelligent Motion, Renewable Energy and Energy Management, (PCIM Europe), VDE, Nurenberg, Germany, June 5-7, S. 1-6, 2018.

[18] Y. Gao, W. Li, C. Chen, H. Zhang, J. Jiu, C. Li, K. Suganuma, "Novel copper particle paste with self-reduction and self-protection characteristics for die attachment of power

semiconductor under a nitrogen atmosphere", Materials & Design, 160. Jg., S. 1265-1272, 2018.

[19] H. Nishikawa, T. Hirano, T. Takemoto, "Bonding process of Cu/Cu joint using Cu nanoparticle paste", Transactions of JWRI, 40. Jg., Nr. 2, S. 33-36, 2011.

[20] J. Liu, H. Chen, H. Ji, M. Li, "Highly conductive Cu–Cu joint formation by low-temperature sintering of formic acid-treated Cu nanoparticles", ACS applied materials & interfaces, 8. Jg., Nr. 48, S. 33289-33298, 2018.

[21] T. Fujimoto, T. Ogura, T, Sano, M. Takahashi, A. Hirose, "Joining of pure copper using Cu nanoparticles derived from CuO paste. Materials Transactions, 56(7), 992-996, 2015

[22] Y. Gao, H. Zhang, W. Li, J. Jiu, S. Nagao, T. Sugahara, K. Suganuma, "Die bonding performance using bimodal Cu particle paste under different sintering atmospheres", Journal of Electronic Materials, 46(7), 4575-4581, 2017.

[23] T. Egle, C. Barroo, N. Janvelyan, A. C. Baumgaertel, A. J. Akey, M. M. Biener, and J. Biener, "Multiscale Morphology of Nanoporous Copper Made from Intermetallic Phases", ACS applied materials & interfaces, 9(30), 25615-25622, 2017.

[24] J. K. Mackenzie, and R. Shuttleworth, "A phenomenological theory of sintering", Proceedings of the Physical Society, Section B, 62(12), 833-852, 1949.

Functional Printing of MMIC-Interconnects on LTCC Packages for sub-THz applications

Martin Ihle[1], Steffen Ziesche[1], Christian Zech[2], Benjamin Baumann[2]

[1]Fraunhofer Institute for Ceramic Technologies and Systems IKTS, Dresden, Germany

[2]Fraunhofer Institute for Applied Solid State Physics, Freiburg

martin.ihle@ikts.fraunhofer.de

Abstract— **In this paper, the authors presents about an Aerosol Jet printed interconnect technology between a GaAs MMIC and an LTCC packaging. Conductive silver nanoparticle and a dielectric polymer-based glob-top are utilized to fabricate die attach, dielectric ramp, and GCPW (Grounded Coplanar Waveguide) transmission line interconnect structures in order to interface a GaAs MMIC with an LTCC (Low Temperature Co-fired Ceramic) packaging substrate. The GCPW lines were printed using the Optomec Aerosol Jet 300 printer with silver nanoparticle ink from Paru. The printed minimal resolution of the interconnection was 34 μm with a gap of 23 μm. Simulation and measurement results of printed interconnects were evaluated in a frequency range between 1 and 225 GHz. The printed interconnects including the MMICs and GCPWs on LTCC have shown an insertion loss of ca. 2 dB at 140 GHz. The reflection coefficient stayed below -10 dB until 210 GHz.**

Keywords—LTCC, aerosol jet printing; high frequency interconnects; system-in-package

I. INTRODUCTION

The enormous progress of semiconductor technologies allows the realization of RF circuits with carrier frequencies in the high millimeter wave (mmw) range. Examples are full transceiver concepts realized in Gallium Arsenide (GaAs) [1] or silicon-germanium (SiGe) [2] technology. The D-Band around 110-180 GHz and H-Band from 180-320 GHz seem very attractive in that context as the antenna size is around or lower than 1 mm, especially for system-in-package (SiP) solutions [3]. Particular applications of above 200 GHz will offer prospective opportunities, because the natural attenuation of the atmosphere offers a unique and wideband window in this frequency range [4]. With the help of this high bandwidth, cost-effective, high-resolution radar sensors [5] or ultra-high bitrate communication lines could be realized. The described problems in the frequency range of 100 to 300 GHz could be solved by the usage of Low Temperature Co-fired Ceramics (LTCC) in combination with high-resolution Aerosol Jet printed transmission lines [6].

Typically wire-bonding and flip-chip bonding are the most popular technologies for interconnecting MMICs to the packaging substrates. In wire bonding the series reactance of the bond wire changes with increasing frequency, hence constraining the system to a narrow-band [7]. Therefore flip-chip interconnects suffer from reliability issues by the co-efficient of thermal expansion (CTE) mismatch and weak heat

transmission between the chip and package substrate [8]. Solutions with on-chip antennas have the drawback of a limited antenna efficiency and directivity [9]. Concluding, the well-known interconnection technologies are limited to realize miniaturized, narrow band solutions, which require extra design work to tailor to the specific application [10].

Aerosol Jet printing is a mask-less direct-writing technology, which can print sub-micro particle inks with a resolution down to 10 μm on planar or non-planar substrates [11]. A first study has shown compromising results of Aerosol Jet printed GCPWs (Grounded Coplanar Waveguide) on LCP in D-Band with a promising insertion loss of 0.35 dB/mm at 110 GHz [12]. Latest publications have been demonstrating the direct Aerosol Jet printing on GaAs-based microwave monolithic integrated circuits (MMICs) for on-chip interconnects [13]. Previously an Aerosol Jet direct written interconnect using ramps onto an attenuator MMIC achieving a loss of approximately 0.65 dB per interconnect at 20 GHz was demonstrated [14]. This paper extends the frequency bandwidth to 200 GHz.

This paper reports about a fast and low- cost Aerosol Jet printed interconnect technology between a GaAs demonstrator MMIC and an LTCC packaging. Conductive silver nanoparticle and a dielectric polymer-based glob-top are utilized to fabricate die attach, dielectric ramp, and GCPW transmission line interconnect structures in order to interface a GaAs MMIC with an LTCC packaging substrate. Advantageously this wideband interconnection doesn't require any gold pads on the LTCC substrate and works with cheaper silver metallization. The CPW lines were printed using the Optomec Aerosol Jet 300 printer. The printed minimal resolution of the interconnection was 34 μm with a gap of 23 μm. Simulation and measurement results of printed interconnects were evaluated in a frequency range between 1 and 225 GHz. This paper demonstrates a loss of ca. 2 dB at 140 GHz (correlated 0.8 dB/mm) for printed interconnects including the MMICs and GCPWs lines on LTCC. According to the current development situation the printed interconnection could efficiently be used until 200 GHz.

II. COPLANAR WAVEGUIDE CHARACTERIZATION ON LTCC

A. LTCC Manufacturing Process

Fig. 1 shows the presented manufacturing process for the LTCC RF-substrates in combination with Aerosol Jet printing.

At first a UV laser was used to drill 100 μm vias inside of the LTCC top layer. Subsequently the blind vias were filled with silver paste with the help of the remaining mylar foil and a screen printer. For Ferro A6M-E the 51 μm thick green tape, respectively (see Fig. 1, "LTCC Co-firing Process"). The following silver ground layer was printed on the second layer via screen printing. With additional supporting LTCC layers the final laminate was realized with an isostatic lamination process following the sintering process at maximum temperature of 850 °C. After sintering a lapping and polishing step could take place if necessary to decrease layer thickness and surface roughness. Finally a silver top metallization was fully printed except of openings with 200 μm space where the Aerosol Jet fine-line printed coplanar waveguides were placed (see Fig. 1, "LTCC Postfiring Process")

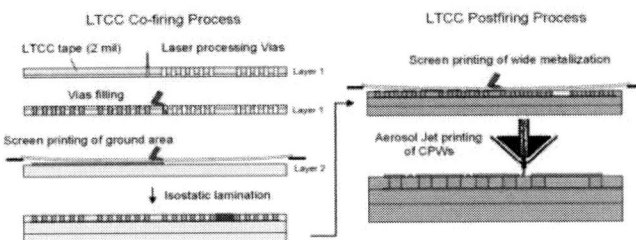

Fig 1: Process flow of LTCC-RF-Package manufacturing

B. Aerosol Jet Printing Process

Aerosol jet (AJ) direct-write printing generates a dense aerosol mist of material-laden ink micro droplets (1-5 μm droplet size) by pneumatic atomization process. The resulting continuous aerosol stream is focused in the printing head using a sheath gas and guided to the sample. The long standoff distance (1-5 mm) between printing nozzle and substrate with negligible change in printing diameter allows also the printing on 3D surfaces without adapting the height of the printing head. A nanoparticle silver ink was printed into the left openings to complete the fine-line GCPWs on LTCC (see Fig. 1, "LTCC Postfiring process"). The printed lines were sintered at 600 °C for 2 minutes and got a specific resistance of ca. 2.5 μΩ*cm and line thickness of 5-6 μm. Fig. 2 shows a detail view of a realized coplanar waveguide on unpolished Ferro A6M-E after printing 2(a) and sintering 2(b) which have a width of 34 μm, a gap of 23 μm between the ground and signal planes, and a thickness of 5 μm. The Aerosol Jet printing tolerance of the inner lines differed from ± 1-3 μm. Additionally, the surface roughness of the printed CPW structures is illustrated in Fig 2 (c). The Ra roughness (arithmetic average roughness) of the signal and ground planes is found to be about 1.28 μm. These values represent comparable mean roughness which affects RF skin effect losses.

C. Characterization

Since the AJ printing technology allows 3D and multilayer transmission lines from electronic components to printed circuit board lines, it is necessary to characterize the electromagnetic behavior of the Aerosol Jet printed interconnects. Therefore, grounded coplanar waveguide (GCPW) test

structures with different lengths were realized. To accurately de-embed the RF probes, calibration standards (Line, Short, Open, Thru) were printed as well. Line structures with different line lengths (0.25, 0.5, 0.75, 1, 1.25, 1.75, 2, and 5 mm) of ground-signal-ground (GSG) CPWs were printed on the targeted LTCC substrates. S-parameters have been measured in several frequency bands from 1 to 330 GHz. For 1 to 150 GHz, an Anritsu MS4647B VNA is used with 3739C broadband testset and Cascade Microtech 100 um pitch GSG probes. For G band (140 to 220 GHz), a Keysight N5245A PNA-X with VDI WR5.1 extension is used and for H band (220 to 330 GHz) a Keysight N5224A PNA with VDI WR3.4 extension and 60 um GSG probes.

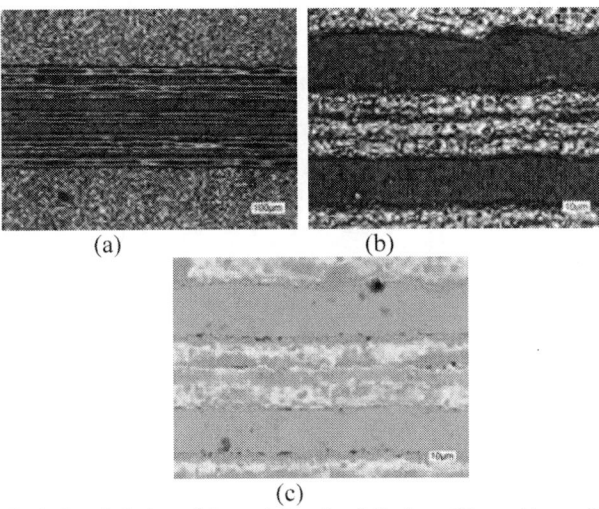

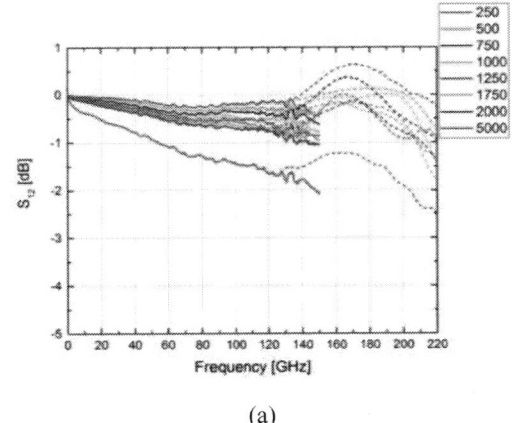

Fig 2: Detailed view of Aerosol Jet printed Coplanar Waveguides on LTCC surface unpolished Ferro A6M-E after printing (a) after sintering (b) and heat map view of the printed waveguides after sintering (c)

Furthermore, Fig. 3 shows the measurements of the transmission coefficient (S_{21}) for the printed CPW on Ferro A6M LTCC packaging after TRL calibration with respect to the line standard. The printed GCPWs on Ferro A6M show low attenuation of about 0.4 dB/mm at 140 GHz and 0.7 dB/mm at 240 GHz respectively.

(a)

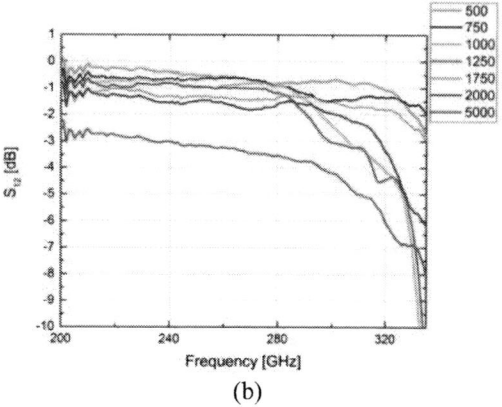

(b)

Fig 3: Measured S_{12}-parameter of lines with different line lengths for Ferro A6M-E up to 220 GHz (a) and from 200 to 330 GHz (b)

Based on these measurements, the relative effective permittivity for the LTCC material was also extracted. The values of the relative permittivity for Ferro A6M material are found to be 3.8 at 140 GHz and 4.0 at 240 GHz respectively

III. MMIC-INTERCONNECTS

A. Design

The goal of this printing strategy is to realize a simple transition from a GCPW transition line on the LTCC package to a transmission on the MMIC so that the band width limitation of the packaging is not defined by the interconnection.

The design of the 3D interconnects is composed of the following elements shown in Fig. 4: an MMIC die (from Fraunhofer IAF), LTCC packaging substrate, a die attach material, a dielectric ramp structure, and GCPW transmission lines. For this work, a transmission GaAs die with a line length of 1 mm and a thickness of 50 µm was used. A die-attach material is used to adhere the silicon die to the LTCC packaging substrate. In order to create a transition between the top of the die and the substrate below, ramp structures are also required to transition the 50 µm thickness of the die. These components are realized with a thermally conductive epoxy from EPO-TEK (H70E-4), low CTE and dispense jet-able 2 component adhesive. At last, silver 50 Ω transmission lines were precisely printed between the lines on the LTCC packaging and the MMIC pads using Aerosol-Jet printing. These conductive patterns are realized with Paru PG007-AJ, a silver nanoparticle-based ink utilizing an alcohol-based solution for low-temperature (< 250 °C) processing. After sintering, the printed silver patterns exhibit resistivity in the range of 5-6 µΩ*cm. All detailed dimension of the package and printed lines are listed in Fig. 4.

Aerosol jet printing brings some of the highest resolution and broadest ranges of possible materials among 3D direct printing tools, but is slower than many others like Ink- Jet printing, because just one nozzle is typically used for printing. So this technology may be best suited for small structures such as short line or small antennas or of interconnects as shown here.

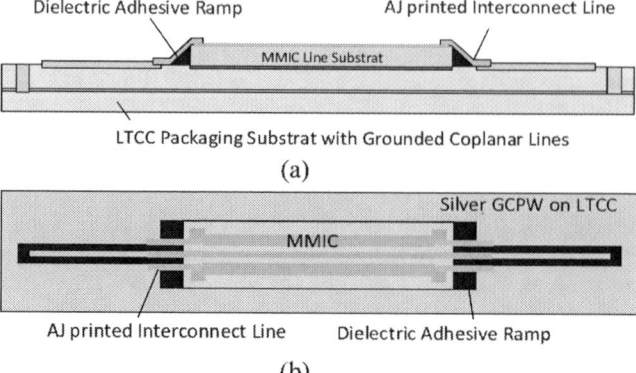

Fig 4: Cross-section (a) of proposed Aerosol Jet printed Interconnection to GaAs-MMIC and Top View (b); Dimension of the package and printed lines: Impedance 50 Ω, Line width 34 µm, Gap width 23 µm, Ground width 40 µm, Length adhesive ramp Interconnection line length: 2.5 mm, Transition length: 0.3 mm, Line on MMIC: 1 mm, Printed line on LTCC: 1.2 mm

Images of the printed 3D GCPW interconnect samples are shown in Fig. 5. The Paru PG007-AJ ink was printed with a pneumatic atomizer through a 150 µm nozzle. Three silver lines were AJ printed overlapping side by side to realize the defined interconnection line width of 34 µm and 23 µm gap. Furthermore three layers of the described layout were printed to achieve a conductor thickness of approximately 5 µm.

B. Measurement

Simulations were performed using ANSYS HFSS. As described in chapter II several measurement equipment was used to measure the printed Interconnection from 1 to 225 GHz. The measured and simulated S-parameters for the complete printed GCPWs-Interconnection are depicted in Fig. 6.

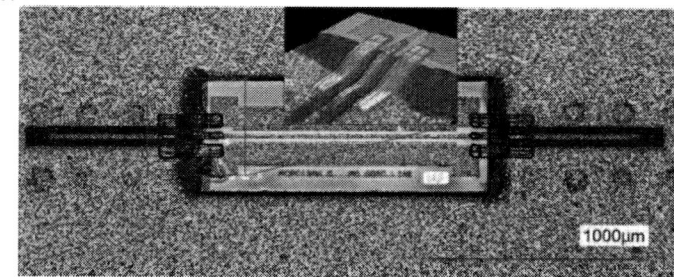

Fig 5: Aerosol Jet printed interconnect to GaAs-MMIC on LTCC with inset detail of printed interconnect (Impedance 50 Ω, Line width 34 µm, Gap width 23 µm)

These results, shown in Fig. 6, are the end to end measurement of the entire printed component, including the 1.2 mm long GCPW on LTCC and 0.3 mm the transition line up to the MMIC die on each side.

Until a frequency of 140 GHz the printed interconnect performs well and measurements correlate nearly with simulated data. At 140 GHz the measured loss is ca. 2 dB for the complete structure. The return loss stays below -10 dB up to 200 GHz. This correlates to a line loss 0.8 dB/mm at 140 GHz.

Above 140 GHz the difference between measurement and simulated data increases (see also Fig. 7). At 200 GHz the measured line loss is 3.2 dB.

The measured transmission loss of interconnect is depicted in detail in comparison with printed GCPW on Ferro A6M (Fig. 7). The difference between a printed 2.5 mm GCPW and a 2.5 mm interconnection line is ca. 1 dB. From a practical point of the view the printed interconnects are applicable until 200 GHz.

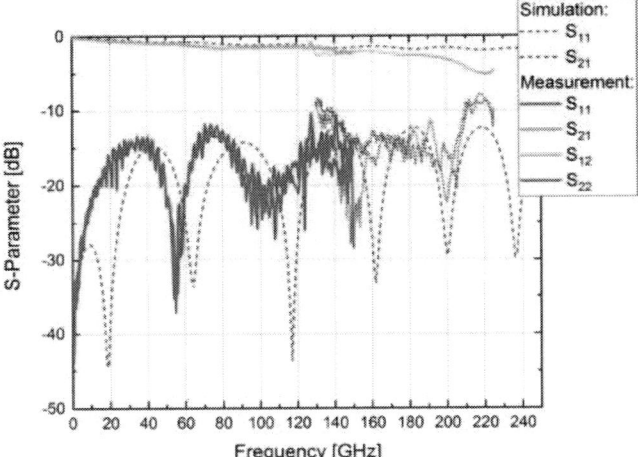

Figure 6: Simulated and measured S-parameters of the printed complete Interconnection lines with 2.5 mm length

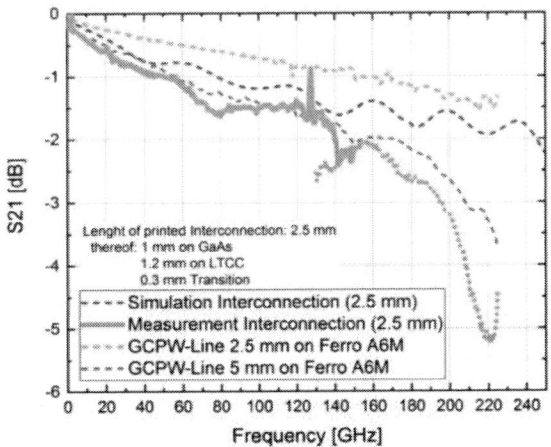

Figure 7: Detailed view of S21-Parameter of Interconnection in comparison with GCPW-Lines on LTCC

IV. CONCLUSION

In this paper, we presented a direct-writing technology technique to create wideband interconnects as an alternative approach to conventional bonding or flip chip processes especially for frequencies above 100 GHz. This novel Aerosol Jet printing technique does not need additional gold contacts on LTCC and cold be used with standard MMIC contacts. It

also can interconnect sensitive components like thinned GaAs, which are hard to handle with short bond wire contacts. Additionally this technique does not suffer from disadvantages like parasitic effects, discontinuity and a narrow band.

To the best of our knowledge this paper present for the first time a high frequency additively manufactured interconnect to a GaAs MMIC component above 110 GHz. The loss of the two interconnects and GCPW leading up to the MMIC device were measured to be 2 dB (correlated 0.8 dB/mm) at 140 GHz.

ACKNOWLEDGMENT

This work was supported by Federal Ministry for Education and Research (BMBF) within the framework of "Validating the Innovation Potential of Scientific Research – VIP+"

REFERENCES

[1] T. Bryllert, V. Drakinskiy, K. B. Cooper and J. Stake, "Integrated 200–240-GHz FMCW Radar Transceiver Module," in *IEEE Transactions on Microwave Theory and Techniques*, vol. 61, no. 10, pp. 3808-3815, Oct. 2013.

[2] K. Schmalz, W. Winkler, J. Borngraeber, W. Debski, B. Heinemann and J. C. Schyett. (2010, May). 122 GHz ISM-band transceiver concept and silicon ICs for low-cost receiver in SiGe BiCMOS. presented at IEEE MTT-S International Microwave Symposium

[3] H. Song, "Packages for Terahertz Electronics", Proceedings of the IEEE, Volume: 105, Issue: 6, June 2017

[4] H. Liebe, "MPM-An Atmospheric Millimeter-Wave Propagation Model", International Journal of Infrared and Millimeter Waves, Vol. 10, No. 6, 1989

[5] M. Charis et al, "Very High Resolution Radar at 300 GHz", Proceedings of the 11th European Radar Conference, 8-10 Oct 2014, Rome, Italy

[6] M. Ihle, S. Ziesche, C. Zech and B. Baumann, "Compact LTCC Packaging and Printing Technologies for Sub–THz Modules," *2018 7th Electronic System-Integration Technology Conference (ESTC)*, Dresden, 2018, pp. 1-4.

[7] Y. P. Zhang and D. Liu, "Antenna-on-chip and antenna-in-package solutions to highly integrated millimeter-wave devices for wireless communications," IEEE Trans. Antennas Propag., vol. 57, no. 10 PART 1, pp. 2830–2841, 2009.

[8] Z. Zhang and C. P. Wong, "Recent Advances in Flip-Chip Underfill: Materials , Process , and Reliability," vol. 27, no. 3, pp. 515–524, 2004.

[9] P. H. Park and S. S. Wong, "An on-chip dipole antenna for millimeter-wave transmitters," *2008 IEEE Radio Frequency Integrated Circuits Symposium*, Atlanta, GA, 2008, pp. 629-632.

[10] B. Zhang et al "Integration of a 140 GHz Packaged LTCC Grid Array Antenna With an InP Detector", IEEE Transactions on Components, Packaging and Manufacturing Technology, Vol. 5, No. 8, August 2015

[11] M. Ihle, U. Partsch, S. Mosch, A. Goldberg, "Aerosol Printing of High Resolution Films for LTCC Multilayer Components", Journal of Microelectronics and Electronic Packaging 9(3):133-137, July 2012

[12] Cai, Y. Chang, K. Wang, W. Khan, S. Pavlidis, J. Papapolymerou, "High Resolution Aerosol Jet Printing of D- Band Printed Transmission Lines on Flexible LCP Substrate", IEEE MTT-S International Microwave Symposium , 2013

[13] X. Lang, X. Lu, M. Yihong, D. Scherrer, T. Chung, E. Nguyen, R. Lai "Direct On-Chip 3-D Aerosol Jet Printing With High Reliability", IEEE Transactions on Components, Packaging and Manufacturing Technology, Vol. 7, No. 8, August 2017

[14] M. T. Craton, C. Oakley, J. D. Albrecht, P. Chahal and J. Papapolymerou, "Fully Additively Manufactured Broadband Low Loss High Frequency Interconnects," *2018 Asia-Pacific Microwave Conference (APMC)*, Kyoto, 2018, pp. 327-329

Characterization of Alternative Sinter Materials for Power Electronics

David Stenzel[1&3], Christian Schwarzer[2&3], Michael Schnepf[2&3], Ly May Chew[3], Thomas Blank[1], Jörg Franke[4] and Michael Kaloudis[2]

[1] Karlsruhe Institute of Technology, Karlsruhe, Germany

[2] Aschaffenburg University of Applied Sciences, Aschaffenburg, Germany

[3] Heraeus Deutschland GmbH & Co. KG, Hanau, Germany

[4] Institute for Factory Automation and Production Systems (FAPS),

Friedrich-Alexander-University Erlangen-Nuremberg, Nuremberg, Germany

david.stenzel@heraeus.com

Abstract

Silver Sintering for die attach has gained increasing interest in past years. Compared to known interconnect materials such as Adhesives and Solder, the unique superior thermal and electrical properties of sintered silver joints enable power electronics to operate at greater reliability and at higher temperature. Copper, touted as the alternative, possess similar properties to silver and being the same material as metalized substrate can reduce CTE mis-match. Up till now copper sintering attracts great interest as an alternative option for die attach material. This study investigates the influence of using copper as the non-silver additives for formation of silver sintered layers as well as the effects on material properties of the sintered joints for die attach applications. The different sinter materials are evaluated by tensile strength, Young´s modulus, die shear strength and electrical conductivity. Process-oriented test specimen were used to replicate the die attach layer formed during pressure sintering. This contribution demonstrates that the mechanical and electrical properties of the sintered layer are affected by non-silver particles. It was found that the used filler materials can lead to a decrease of mechanical and electrical properties in comparison to the pure silver interconnection. It was also found, that the copper paste obtained a relatively strong bonding and confirms the feasibility as an alternative die attach material.

Key words: silver sinter, filler material, copper sinter

Introduction

Within the last years the demand for power electronics packaging with higher efficiency, faster switching frequencies and smaller die size is increasing. Therefore the current trend is to replace the common silicon-based dies by wide band gap (WBG) semiconductor devices such as silicon carbide (SiC) or gallium nitride (GaN) [1, 2].
To take full advantage of WBG semiconductor properties, a suitable die attach material is necessary to establish a reliable interconnection between the semiconductor and carrier substrate. Solder materials are facing a challenge because they cannot fulfill the demand of supporting operating temperature above 150 °C. Silver sintering has become an attractive die attach technology over the past years due to its high melting point (961 °C) and its superior thermal and electrical properties [2].

The feasibility of silver sintering as die attach material for power electronic devices has been reported by numerous studies. Long term reliability tests demonstrate that silver sintered joints exhibits high reliability than lead-free solder joints [3–9].

Silver sinter joints can be obtained by sintering process at moderate temperatures (230-250 °C) and applying a mechanical pressure between 10-30 MPa with the longest time of 5 min [10, 11].
These days a wide range of silver sinter materials as well as sinter press equipment are commercially available for high volume production. Therefore, silver sintering is becoming more important for power electronics packaging industry.
Copper sinter material has recently attracted growing attention to be considered as an alternative die attach material due to its relative low metal costs and its similar material properties in comparison to silver material [13, 14]. The focus of previous researches is based on nano-scale copper particles which has been reported to form sinter joints at moderate temperatures by pressureless sintering [15, 16]. For micro-scale copper powder, it has been reported that pressure is required to apply during sintering process in order to generate a strong sinter joint [17].
In general, it is known, that for copper sintering, oxides from copper particle must be removed from the surface by reduction using hydrogen or other agents to create a strong sinter joint. Currently

copper sintering is facing many challenges such as copper oxidation and long-term reliability. However, our current study on micro-scale copper sintering shows promising results without applying additional reduction steps [18].

Our previous studies have illustrated that the silver sinter material properties can be modified by adding a small amount of non-silver particles and consequently improved the mechanical properties and reliability of the silver sintered joint 12].

The aim of this study is to prepare special specimen to allow process-oriented measurement. By use of this test specimen the influence of non-silver materials on the formation of sinter layer and the material properties of sintered joints for die attach in power applications is investigated.

Experimental

The experimental investigation of sinter material focuses on standard methods like Tensile Test and four-point probe method for measuring key properties of sintered joints (required for design and simulation of Power electronics) and destructive Die Shear tests.

Four different sinter pastes for die attach were evaluated in this study: microscale silver paste, microscale copper paste, silver pastes with copper flakes and copper powder. Each paste was used to prepare an appropriate specimen for every single test method which includes sinter strips, tensile test specimen and SiC dies attached on Si₃N₄-AMB substrates. The sinter strips were used to measure the electrical conductivity.

The mechanical properties such as tensile strength and Young´s modulus were determined by using special tensile test specimen.

The bonding strength of sinter joints were investigated by die shear test, which is a standard method for testing sintered or soldered interconnections. Furthermore, voids and delamination in the sinter joints were inspected by C-mode scanning acoustic microscopy (C-SAM) measurement. Scanning electron microscopy (SEM) was used to study the microstructure of Cu flakes and Cu powder as well as the fracture surface after tensile test.

Sinter materials

Four different sinter pastes were under investigation. All sinter pastes which were used in this study were provided by Heraeus. Both silver paste and copper paste consist of a micro scale flake-shaped powder with organic solvents, whereas, the other two pastes were prepared by replacing 10 wt% of silver particles in the standard silver paste by non-silver particles such as copper flakes and spherical copper particles. Table 1 summarizes the overview of all sinter pastes which were investigated in this study. A SEM image of each filler particles is shown in figure 1.

Table 1: Overview of tested sinter pastes

Paste	Filler	Abbreviation
Ag-paste	N/A	AG
Cu-paste	N/A	CU
Ag-paste	Cu flakes	AG-CF
Ag-paste	Cu powder	AG-CP

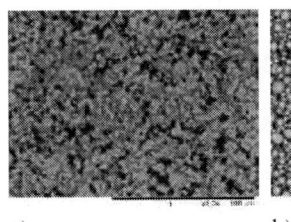 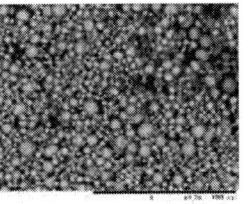

a)   b)

Figure 1: SEM images of filler materials. a) Cu flakes, b) Cu powder

Sinter process

Before sintering, the wet printed sinter pastes were dried in an oven to remove the organic solvents. The silver paste and all silver pastes with filler materials were dried at 140 °C for 10 min in air, whereas the copper paste was dried at 80 °C for 30 min in N₂ to avoid oxidization of the copper material.

After drying, Ag metallized SiC dies were attached onto the dried sinter paste at 130 °C with a placement force of 2000 g for 2 seconds. After die placement, all samples were sintered with a pressure of 10 MPa for 3 min. The sinter process of silver-based pastes and copper paste differ in their temperature and atmosphere as the sinter temperatures correspond approximately to a homologous temperature of 0.4. As a result, sinter process for silver-based pastes was performed at 230 °C in air whereas, copper paste was conducted at 300 °C in N₂ atmosphere to prevent oxidation of copper layer by using a sinter press [21].

Tensile strength and Young´s modulus

Tensile testing is performed to determine the mechanical properties such as tensile strength, Young´s modulus and elongation of sintered specimen. Tensile testing is a standardized method where specified specimens are stretched with a continuous strain rate until fracture. The test is based on the specifications of DIN EN ISO 6892-1. During the test, the applied force and the change in length of the samples is measured continuously.

The stress σ, strain ε and Young´s modulus are calculated by using Eq. (1), (2) and (3), respectively. Where F is the measured force and A is the nominal cross-section area.

$$\sigma = \frac{F}{A} \quad (1)$$

$$\varepsilon = \frac{\Delta L}{L_0} \quad (2)$$

Where ΔL is the change of length and L_0 is the initial length.

$$E = \frac{\sigma}{\varepsilon} \quad (3)$$

Process-oriented tensile test specimens were prepared by a specially developed sinter tool which is shown in figure 2. It enables the preparation of seven equal specimens at the same time with a homogenous pressure distribution which was confirmed by a pressure indicating film (figure 3).

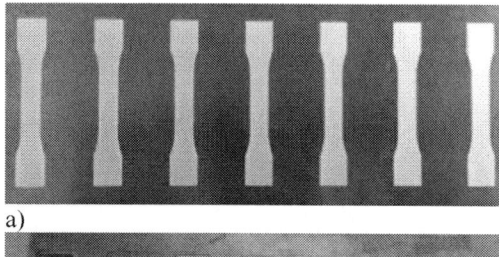

a)

b)

Figure 2: Sinter tool for tensile test specimen. a), b) Top tool with seven stamps for each specimen

The sinter pastes were printed on the bottom tool (figure 2 a) by using a stencil with a thickness of 150 µm and subsequently sintered with the process parameters described above. The layout of specimens is based on the standard DIN 50125 Type H for tensile test pieces and is shown in figure 4 a. The thickness of the samples was adjusted according to the dimension of a die attach interconnection.

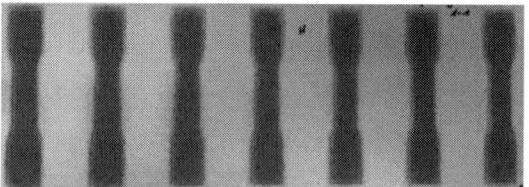

Figure 3: Homogenous pressure distribution of sinter tool

For test procedure, the specimens were clamped vertically into the mechanical specimen holder (see Figure 4 b) and were loaded with a constant test speed of 0.1 mm/min at room temperature until breakage. During the test, the applied force was measured by a 0.5 kN load cell and the change in length was measured by the distance of traverse. The tensile strength was determined from the maximal force and the nominal cross-section of the specimen.

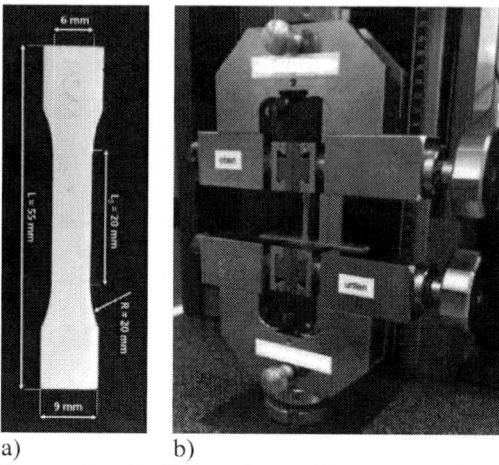

a)   b)

Figure 4: a) Sintered tensile test specimen, b) specimen holder

Electrical conductivity

The four-point probe method is used for measurement of sheet resistance of thin films or surfaces. Therefore, four needle-type electrodes are placed in-line on the surface of specimen (figure 5). A constant current I is flowing between the two outer needles. In consequence of conductivity of the material a potential difference U can be measured between the two inner needles which is used to calculate the sheet resistance $R_\square$ as shown in Eq. 4. [19]

$$R_\square = K * \frac{U}{I} \quad (4)$$

Where K is a correction factor regarding to the needle electrodes position and dimensions of specimen.

The specific resistance of a specimen can be determined by Eq. 5. Where d is the layer thickness.

$$\rho = d * R_\square \quad (5)$$

Finally, the electrical conductivity σ is defined by Eq. 6.

$$\sigma = \frac{1}{\rho} \quad (6)$$

For determination of the electrical conductivity the four-point probe method was performed by using appropriate test equipment of our laboratory which can be seen in figure 6 b. The measurement of electrical conductivity was used to evaluate the influence of sinter paste on the electrical properties of sinter joint. Therefore, sintered stripes with the dimensions shown in figure 6 a were printed on Al_2O_3 ceramic substrate by using a stencil with a thickness of 200 µm and have been sintered as described in section "Sinter process".

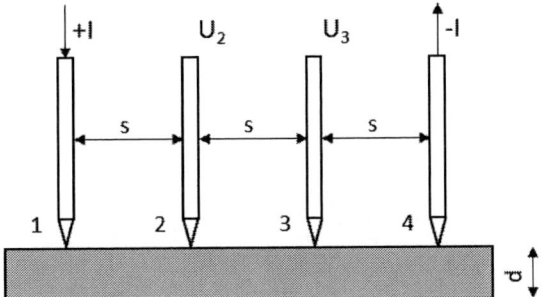

Figure 5: Schematic setup of four-point probe measurement

For each sinter paste, 5 sinter stripes were produced and the sheet resistance of every single specimen has been measured 5 time. After measuring the thickness of the individual specimen, by using a micrometer as well as confocal laser scanning microscopy, the electrical conductivity was calculated.

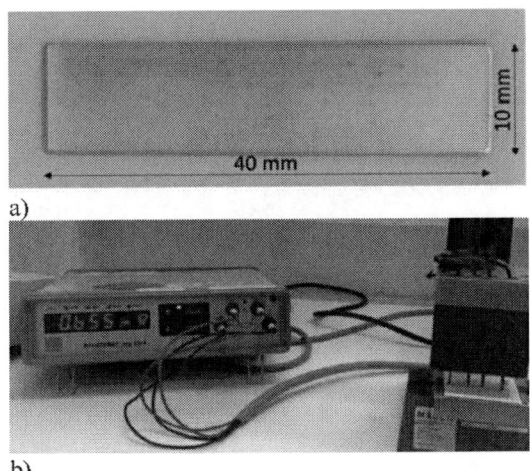

Figure 6: a) Sintered strip, b) four-point probe measurement equipment

Die shear strength

The bonding strength of sintered joints is quantified by a destructive die shear test. Die shear measurement is a standard test method to determine the shear strength of interconnection materials according to MIL-STD-883E method 2019.5. During the test, a force with a constant speed is applied on to a semiconductor chip until fracture which is attached to a substrate using sinter paste as bonding material. The shear strength σ is determined by the shear force F and the respective area A of semiconductor as indicated in Eq. 7.

$$\sigma = \frac{F}{A} \quad (7)$$

All sinter pastes were used as joining material to attach mechanical 4 x 7.3 mm² CPM2-1700-0045B silicon carbide power MOSFETs with a thickness of 380 µm from CREE/Wolfspeed on silver metallized Si_3N_4-active metal brazed (AMB) substrates.

The shear strength of sintered samples was quantified by a destructive hot die shear test (HDS) as shown in Figure 7 a. The test was performed at 260°C whereby the SiC dies were sheared with a shear height of 80 µm and a constant test speed of 300 µm/s. The shear strength was calculated from the maximum applied force and the respective area of semiconductor. After testing the fracture surface was visually inspected to define the failure criteria of die attach. For each sinter paste 10 samples were tested by using a shear tester for the test procedure. A schematic diagram of die shear test is shown in Figure 7 b.

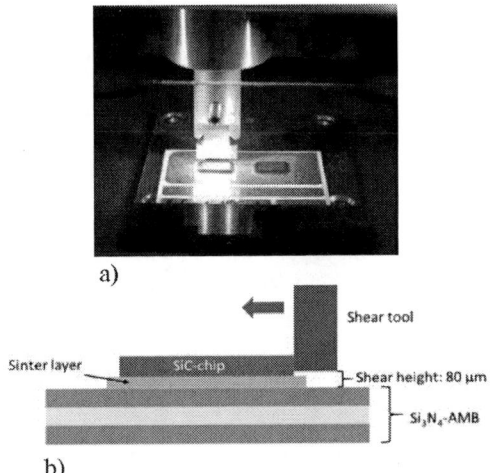

Figure 7: a) Shear tool with specimen on hotplate, b) schematic diagram of die shear test

Results

After completing the experimental test procedure, the influence of different sinter pastes on the mechanical properties of the sinter layer and the formation of the sinter joint for die attach were evaluated. The results of tensile test, four-point probe method and hot die shear test are summarized in table 2. In general, the results show that the silver paste AG has the highest values for tensile strength, electrical conductivity and initial shear strength compared to silver pastes consist of non-silver filler. Furthermore, the copper paste CU is slightly stronger and stiffer than AG but the values of electrical conductivity and initial shear strength are lower.

Tensile strength and Young's modulus

The evaluation of tensile test results shows, that paste CU has the highest tensile strength with an average of 68.4 MPa followed by paste AG with 61.7 MPa. All silver sinter pastes consisting non-silver particles show lower tensile strengths. The tensile strength is reduced by 25 % for paste AG-CP and by 50 % for paste AG-CF in comparison to paste AG.

A similar behavior is observed for Young's modulus. The copper paste CU has the highest Young's modulus with an average of 15 GPa which was expected due to the higher stiffness of bulk copper compared to silver. For silver paste AG a Young's modulus of 8.5 GPa was measured and the pastes contain filler materials show the same decrease of Young's modulus as for tensile strength. However, it is conspicuous that all values are relatively low in comparison to previous studies [7, 8]. The difference in results most likely due to different measurement methods used for elongation detection. Furthermore, the clamping of tensile test specimen may affect the tensile test measurement. It is required that the fragile specimen do not get stressed or damaged by handling before the measurement was realized.

Additionally, SEM images of fracture surface after tensile test were carried out for evaluation of sinter joint which are shown in figure 8. The silver paste AG (8 a) shows a homogenous sinter layer with a strong bonding between the primary flake shaped silver particles. In contrary, an anisotropic very dense sinter layer was observed for the copper paste CU (8 b). So, the initial structure of copper powder is still existing after sintering.

A very isotropic distribution of flake shaped Cu particles was found for the sinter joint of paste AG-CF (8 c). But it seems that the flakes prohibit the formation of a close silver matrix. The sinter joint of paste AG-CP (8 d) shows a homogenous strong sintered silver matrix including the added non-silver particles. But it can be seen, that there is no strong bonding between the silver matrix and the spherical copper particles.

In conclusion and regarding to the mechanical properties of sintered joint, the particle shape and especially the bonding between the silver matrix and filler particles have a significant influence.

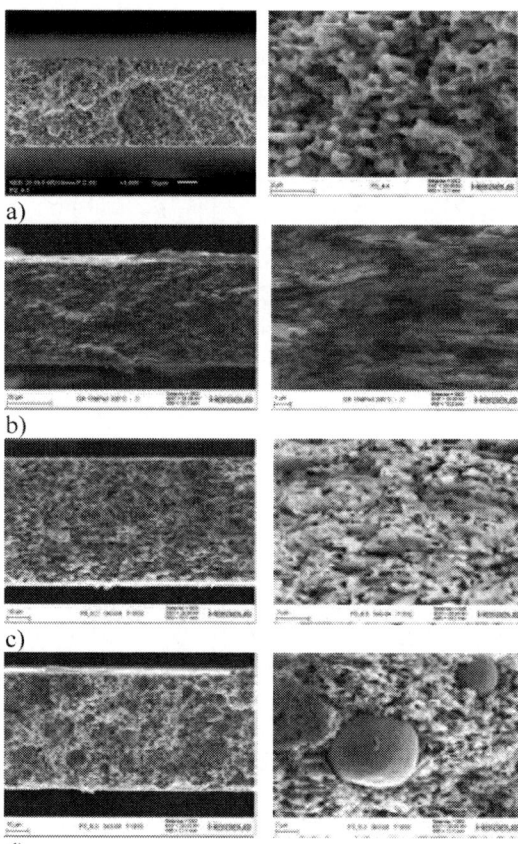

Figure 8: SEM images of fracture surface after tensile test. a) AG, b) CU, c) AG-CF, d) AG-CP

Electrical conductivity

The highest electrical conductivity was determined for the silver paste AG with an average of 30.1 x 10^6 S/m. The conductivity was reduced approximately by half in comparison to bulk silver. In general, all silver pastes with filler show lower values depending on the filler material and the particle shape. The electrical conductivity is reduced

Table 2: Results of tensile test, four-point probe method and initial hot die shear test

Sinter paste	Tensile strength [MPa]	Youngs modulus [GPa]	Electrical conductivity *10^6 [S/m]	Initial Shear strength [MPa]
AG	61.7 ± 3.9	8.5 ± 0.4	30.1 ± 1.0	49.8 ± 1.9
CU	68.4 ± 7.1	15.0 ± 0.4	17.6 ± 0.4	29.5 ± 1.5
AG-CF	33.1 ± 4.8	5.3 ± 0.2	12.3 ± 0.1	2.6 ± 0.8[1] | 13.7 ± 2.7[2]
AG-CP	46.4 ± 4.6	6.6 ± 0.3	18.9 ± 0.6	6.4 ± 2.3[1] | 27.5 ± 3.0[2]

Sinter temperature: [1]230 °C, [2]300 °C

by 35 % for paste AG-CP and by 60 % for paste AG-CF in comparison to paste AG.

This effect can be attributed to the bonding between the Ag-matrix and filler particle as well as to the electrical conductivity of filler material.

A low electrical conductivity was also determined for the copper paste CU with an average of 17.6×10^6 S/m. In comparison to bulk copper the conductivity of copper sintered layer is only about one third. This relatively low value can be referred to the porosity of the sinter layer, a possible thin oxide layer on the copper surface or due to organic residues left from the sinter paste.

By comparing the results of four-point probe method with the values of tensile strength a correlation between both was observed. According to this, the electrical conductivity is decreasing with a reduction of tensile strength, which implies that the measurement of electrical conductivity can be used as a reference value to evaluate the formation of sinter joint. However, the electrical conductivity of sinter and filler materials has also an impact on the electrical properties of sinter layer.

Die shear strength

The results of hot die shear test demonstrate, that there is no direct relationship between tensile strength and die shear strength. Contrary to the tensile strength the silver paste AG achieves the highest shear strength with an average of 50 MPa followed by the paste CU with an average of 30 MPa. In contrast, both pastes AG-CF and paste AG-CP have a very low die shear strength (< 10 MPa). An adhesion failure at the die was observed indicating that a very weak bonding formed between the sinter layer and the metallization of the SiC die.

For better comprehension of the observed failure mode an auger electron spectroscopy analysis of the die backside was carried out before and after sintering. The results show that a significant higher carbon content on the die backside after sintering was observed for both AG-CF and AG-CP pastes which is likely to be the reason for the relatively low die shear strength. After measuring the results of sintering at 230 °C an additional sintering test at 300 °C showed a steep increase of the bonding strength up to 13.7 MPa for the sinter paste with the flake shaped copper filler and even 27.5 MPa for the paste with added copper spheres. This indicates that for the interparticle bonding between copper and silver particles higher temperatures are required. It is assumed that at lower temperatures, even when diffusion of copper into silver, according to literature, is faster and relates in slightly more diffusion length, compared to silver-silver diffusion, the remaining oxides or coating on surfaces may impede the bonding of copper and silver particles [22].

The results of C-SAM measurement of initial samples sintered at 230 °C which are shown in figure 9 verify the results of HDS. The image of paste AG shows a good bonding of the sinter joint in terms of a dark coloration. In contrary, both pastes AG-CF and AG-CP show only a grey coloration with single bright areas which indicates a weaker bonding between sinter layer and SiC chip. The evaluation of copper joint was very difficult because of the same acoustic velocity of AMB substrate and sintered copper layer. Nevertheless, no bright areas could be detected which would indicate delaminations.

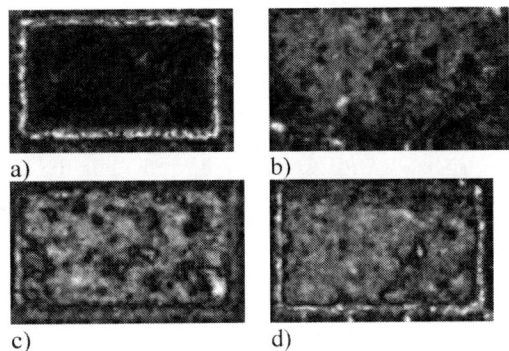

Figure 9: C-SAM images of sinter materials. a) AG, b) CU, c) AG-CF, d) AG-CP

Conclusion

This paper focused on the preparation of process-oriented test specimen to determine the tensile strength, the Young's modulus and the electrical conductivity of a sintered interconnection. The properties of a silver sinter paste as well as sinter material with copper filler material and a copper sinter paste was evaluated. The results of this study demonstrate that the mechanical and electrical properties of the sintered layer are influenced by non-silver filler materials. These findings correspond to earlier investigation.

It was found that replacing silver particles by copper fillers leads to a reduction in mechanical and electrical properties, in comparison to the pure silver paste by using a standard sinter process. Higher sintering temperatures led to significantly increased in shear strengths and allows the presumption that the diffusion-based interparticle bonding of copper fillers and silver particles requires temperatures beyond 230°C. It was also found, that the copper paste obtained a relatively strong bonding and confirms the feasibility as an alternative die attach material. Furthermore, the results of die shear test show low relation to tensile strength of sintered specimen.

In conclusion Ag particles of a sinter layer cannot be easily replaced by copper fillers in a standard process without deteriorating the mechanical and electrical properties of sintered joint.

It is expected, that the sintering parameters especially the temperature and the sinter pressure have a significant influence on the tensile strength and Young's modulus of sinter joint. Further studies will be conducted to investigate the process influence on the material properties of the sintered joints. Additional tensile tests of sintered specimen with different filler materials will be performed by using a greyscale correlation for strain determination to verify the previous test results.

Furthermore, the long-term reliability of sintered joints will be evaluated by performing a thermal shock test.

Acknowledgment

The authors would like to thank Dr. Michael Jörger, Dr. Stefan Gunst and Wolfgang Schmitt from Heraeus Deutschland GmbH & Co. KG for their superb scientific guidance and Habib Mustain from Wolfspeed, a Cree Company for providing the SiC devices used in this study.

References

[1] K. S. Siow, "Are Sintered Silver Joints Ready for Use as Interconnect Material in Microelectronic Packaging?," *Journal of Elec Materi*, vol. 43, no. 4, pp. 947–961, 2014.

[2] H. S. Chin, K. Y. Cheong, and A. B. Ismail, "A Review on Die Attach Materials for SiC-Based High-Temperature Power Devices," *Metall and Materi Trans B*, vol. 41, no. 4, pp. 824–832, 2010.

[3] K. S. Siow and Y. T. Lin, "Identifying the Development State of Sintered Silver (Ag) as a Bonding Material in the Microelectronic Packaging Via a Patent Landscape Study," *J. Electron. Packag*, vol. 138, no. 2, p. 20804, 2016.

[4] Luis Alberto Navarro Melchor, "Evaluation of Die-Attach Materials for High Temperature Power Electronics Applications and Analysis of the Ag Particles Sintering Solution," Dissertation, Electronic Engineering Department, Universität Autònoma de Barcelona (UAB), Barcelona, 2015.

[5] R. Dudek, R. Döring, S. Rzepka, C. Ehrhardt, M. Hutter, J. Rudzki, F. Osterwald, R. Eisele, S., "Investigations on Power Cycling Induced Fatigue Failure of IGBTs with Silver Sintered Interconnects," in *2015 European Microelectronics Packaging Conference (EMPC): Date: 14-16 Sept. 2015*, 2015.

[6] T. Herboth, "Gesinterte Silber-Verbindungsschichten unter thermomechanischer Beanspruchung," Dissertation, Albert-Ludwigs-Universität, Freiburg im Breisgau, 2015.

[7] C. Mertens, *Die Niedertemperatur-Verbindungstechnik der Leistungselektronik*. Düsseldorf: VDI-Verl., 2004.

[8] M. Knörr, *Verbinden von Leistungshalbleiterbauelementen durch Sintern von nanoskaligen Silberpartikeln*, 1st ed. Aachen: Shaker, 2011.

[9] S. Klaka, *Eine Niedertemperatur-Verbindungstechnik zum Aufbau von Leistungshalbleitermodulen*, 1st ed. Göttingen: Cuvillier, 1997.

[10] T. Krebs *et al.*, "A breakthrough in power electronics reliability — New die attach and wire bonding materials," in *2013 IEEE 63rd Electronic Components and Technology Conference*, Las Vegas, NV, USA, May. 2013 - May. 2013, pp. 1746–1752.

[11] Schmitt Wolfgang and Thomas Krebs, "Adjust the mechanical properties of sintered silver layers using additives," in *CIPS 2016 - 9th International Conference on Integrated Power Electronics Systems: 8-10 March 2016*, 2016.

[12] W. Schmitt and L. M. Chew, "Silver Sinter Paste for SiC Bonding with Improved Mechanical Properties," in *2017 IEEE 67th Electronic Components and Technology Conference (ECTC)*, Orlando, FL, USA, May. 2017 - Jun. 2017, pp. 1560–1565.

[13] Hideo Nakako, Dai Ishikawa, Chie Sugama, Yuki Kawana, Motohiro Negishi, Yoshinori Ejiri, "Highly Reliable Package Bonding with Copper Sintering Paste," Sep. 2017.

[14] Hideo Nakako, Dai Ishikawa, Chie Sugama, Yuki Kawana, Motohiro Negishi, Yoshinori Ejiri, "Sintering Copper Die-Bonding Paste Curable Under PressurelessConditions," in *PCIM Europe 2017 ; International Exhibition and Conference for Power Electronics, Intelligent Motion, Renewable Energy and Energy Management: 16-18 May 2017*, 2017?

[15] A. A. Zinn *et al.*, "A novel nanocopper-based advanced packaging material," in *2016 IEEE 18th Electronics Packaging Technology Conference (EPTC)*, Singapore, Nov. 2016 - Dec. 2016, pp. 1–6.

[16] B. H. Lee, M. Z. Ng, A. A. Zinn, and C. L. Gan, "Application of copper nanoparticles as die attachment for high power LED," in *2015 IEEE 17th Electronics Packaging and Technology Conference (EPTC)*, Singapore, Dec. 2015 - Dec. 2015, pp. 1–5.

[17] J. Kahler *et al.*, "Sintering of Copper Particles for Die Attach," *IEEE Trans. Compon., Packag. Manufact. Technol.*, vol. 2, no. 10, pp. 1587–1591, 2012.

[18] Christian Schwarzer, Ly May Chew, Michael Schnepf, Thomas Stoll, Jörg Franke and Michael Kaloudis, "Investigation of Copper Sinter Material for Die Attach," 2018.

[19] M. Gombotz, "Untersuchungen an dünnen Molybdän/Wolfram- und Chromschichten am computerunterstützten Vier-Spitzen-

Messplatz," Diplomarbeit, Institut für Industrielle Elektronik und Materialwissenschaften, Technische Universität Wien, Wien, 2002.

[20] Espec Corporation, Ed., "Liquid to Liquid Thermal Shock Test," [Online] Available: https://www.espec.co.jp/english/products/trust ee/test/liquid.html. Accessed on: May 21 2019.

[21] Christian Schwarzer, Wolfgang Schmitt, Michael Schnepf, Jörg Franke, Michael Kaloudis, "Investigation of a Sintering Process for Micro-Scale Copper Material," Apr. 2019. *Electronic in Harsh Environments Conference, Amsterdam, Netherlands*

[22] E. A. Brandes and G. B. Brook, "Smithells Metals Reference Book seventh edition," Reed Educational and Pmfessiond Publishing Ltd., 1992.

Heterogeneous Integration of Surface Mounted Devices for Bendable Electronics

Daniel Ernst, Nico Richter and Thomas Zerna

Centre of Microtechnical Manufacturing, Technische Universität Dresden, Dresden, Germany

+49 351 463 36941, daniel.ernst@tu-dresden.de

Abstract

Electronic devices becomes more and more mechanical flexible [1]. Additive manufacturing, e.g. printing of functional layers is one solution to build up such devices. However, for higher integration of function it is useful to integrate traditional solid components. Heterogeneous integration of solid electronic devices on flexible foil substrates enable high productive processes, which are established very well for traditional solid electronics.

The present work investigated in reliability of surface mounted devices. As interconnecting technology, a soldering process was used in the first step. Adhesive bonding technologies will be also investigate in the next step. With regard to rollable electronics, as already shown on CES2019 [3], an adapted test setup was designed to simulate comparable stress on the test specimen. In detail the mechanical stress also take effect to the solder junctions of the devices with this test setup.

A special test specimen was designed with a variation of electronic components with regard to the size (01005, 0201, 0402) and mounting orientation. The bending test was applied with different bending radii of 5 mm and 15 mm.

The Weibull analysis allows to estimate lifetime of those modules. As expected the smaller the device the higher the lifetime is. The same effect was observed for a bigger bending radius, which was also expected. But, with the evaluated results, it is now possible to estimate lifetime of applications in the user environment. [4]

Key words: bendable, heterogenous integration, open form

Introduction

Traditional electronic is mostly mechanically rigid. This includes electronic devices, such as common displays or mobile phones, electronic modules (e.g. mainboards) and components. On the other hand, especially flexible or bendable modules offer a wide range of new applications. In addition, organic electronics is becoming more prominent for special applications. However, full flexible electronics cannot ever reach the functionality and performance of rigid electronics. Beside that, the costs of suitable technologies are often higher than for rigid technologies. Therefore, a heterogenous integration may be a solution to solve these restrictrions. That also means, the use of rigid components allows using wellknown and comparable cheap technologies for flexible electronics. This results in local rigid areas, but the whole system is mostly flexible, also called global flexible.

One main challenge is the adaption of technogies for rigid electronics to be used for flexible electronics. The second challenge is to ensure the long-term reliability of those manufactured heterogenous devices. The requirements are quiet different. For rigid electronics, the thermal mismatch of the materials is typically the main impact factor on reliability issues. Flexible electronic devices have the potential to reduce this issue significantly, because of this flexibility. However, the flexibility behavior itself may be a problem for interconnecting and the rigid components especially with regard to mechanical loads, such as bending or twisting.

Figure 1 Example of heterogenous integration [5]

Analysis of Reliability Impact Factors and Design of Test

As already mentioned, for flexible electronics the requirements for a reliable system are different in comparison to rigid electronics. In detail the following mechanical properties has to be considered for reliability investigations [2]:
- Stretching,
- Bending and
- Twist and Bow.

In this work, the investigation focuses on cyclic bending of heterogeneous integrated modules. Different designs are possible to realize a bending load to a device or module. The main parameter is the radius of the bending. Depending on the later application test, setup has to consider the bending load at work life. A typical case is kind of a freeform bending as shown in figure 2. That means there is finally a not clearly defined bending radius. However, for the first step the influence of the real bending has to be investigated. The results of lifetime evaluation may be used for design optimization to reduce critical bending radii.

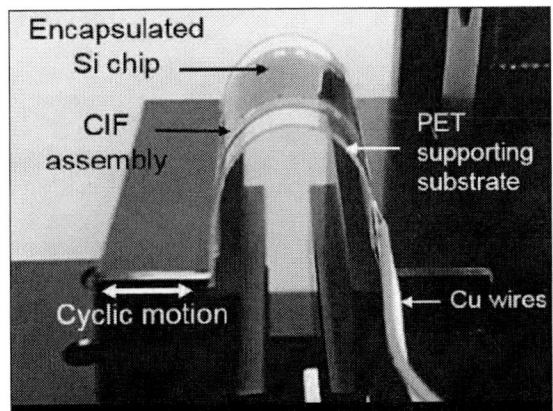

Figure 2 Example for dynamic bending tests [8]

During a bending test two areas of the substrate are of main interest. The inner layer of the substrate towards the center point of the bending will be compressed during bending. The outer layer will be stretched. The principal is shown in figure 4

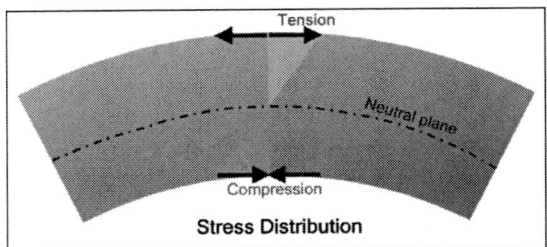

Figure 3 Bending principal [7]

A former developed test equipment, already described in [2], was modified and optimized. The design of the test setup allows a bending of test samples with a defined bending radius and moving speed. It is possible to vary those parameters as shown in table I. The controlling is realized by an Arduino microcontroller. In principal, this test setup can be used for any bendable electronics that meets the geometrical size restrictions. The final test setup is shown in figure 3.

TABLE I Test parameter

Parameter	Values
Bending radius	(5, 10, 15) mm
Movement speed	(1…50) mm/s
Maximum size of test sample	60 x 200 mm²

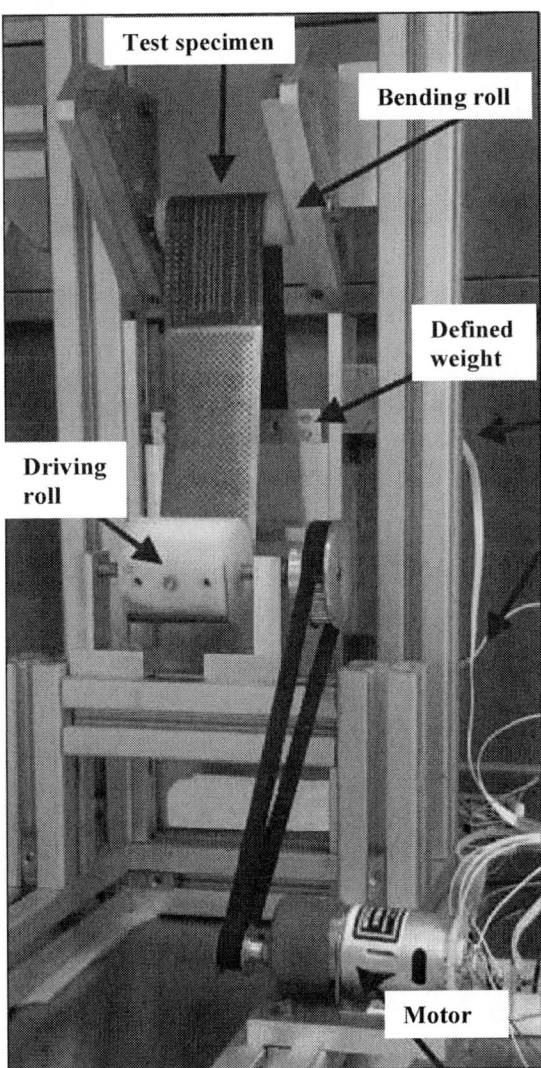

Figure 4 Test setup [4]

For the characterization of the reliability of SMD components mounted on flexible substrates, a test specimen was designed. Looking into a cross section of an electronic module we typically do have a stack of different materials. The common built up is (bottom to top):

- Substrate,
- Conductive layer,
- Interconnecting layer,
- Electronic components and
- Passivation.

The lowest mechanical stress is expected to happen in the so called neutral plane. According to experience, the interconnecting layer is the most critical layer with regard to reliability. That means the design has to consider the different thicknesses of the layers. Typical values are shown in table II.

TABLE II Typical thicknesses of materials

Layer	Material	Thickness
Substrate	PI, PEN, …	12…50 μm
Conductive layer	Cu, Ag, …	1…35 μm
Interconnecting layer	Solder joint, Adhesive	10 μm
Electronic components	SMD, bare dies	10…500 μm
Passivation	Epoxy, silicone, …	5…500 μm

In best case, the lowest mechanical stress induced by bending is in the interconnecting area. Hence, the substrate thickness should be as equal as possible compared to the thickness of the components.

However, typically flexible substrates are thinner as compononents, as shown in table II. The test samples in this work are designed with standard passive SMD components with different sizes (01005, 0201, 0402). Innolot IL89 from Heraeus was chosen as solder alloy because of its very good reliability properties. The substrate used was Polyimide (50 μm) with cladded copper (35 μm).

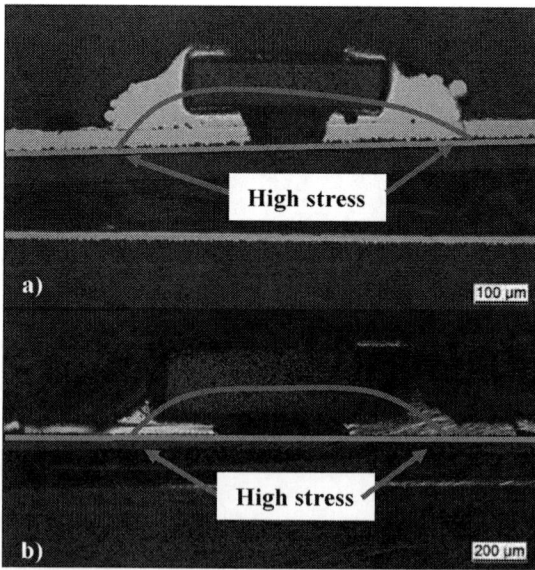

Figure 5 Cross section of SMD; a) chip size 01005, b) chip size 0402 (orange: substrate neutral plane, red: geometrical neutral plane)

In figure 5 the neutral planes are shown for 01005 and 0402 SMD components mounted on flex. The geometrical neutral plane is in the interconnection, which should be good. However, because of the stiffness of the components the resulting neutral plane of bending stress will end at the edges of the geometrical neutral planes, because there is only a low bending expected underneath the component itself (see figure 6). This results in high stress at these edges directly in the area of the solder joints and the copper structure.

Obviously, there is no suitable solution to avoid this stress in this area by the use of standardized substrates and components. Therefore, it is to expect, that most failures will occur in the solder joint after bending test. To reduce the stress an orthogonal orientation was also investigated on the test specimen. Three test specimen were produced. the bending radius was varied between 5 mm and 15 mm for testing. The moving speed was chosen between 20 mm/s and 30 mm/s.

Results after Bending Cycle Test

The test specimen were cycled up to 100,000 bending cycles. It has been expected that, the lower the bending radius is the higher is the mechanical load to the module.

The orientation of the SMD regarding to bending has also an influence on reliability. Furthermore the bending is also more critical the larger the components are. This has been expected because the area of bending is smaller in that case. This can also be seen by comparing the resulting gaps which occurs during bending at the edges of the SMD components, see figure 6.

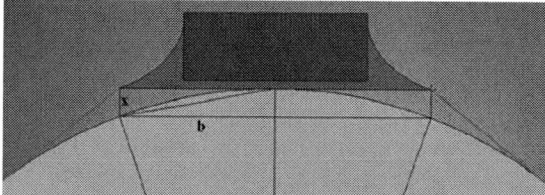

Figure 6 Principle cross section of bended module with mechanical rigid SMD component for gap description (x = gap) [4]

The gap is at least a decade higher for longitudinal oriented SMD in comparison to orthogonal oriented components. The resulting mechanical stress during bending is mainly absorbed in the substrate material for orthogonal SMD. Therefore, even after 100,000 cycles no failures were observed for those components, independent from the component size. Because of those results the following examinations considered mainly the longitudinal oriented SMD. The functionality of the SMD has been evaluated by measuring the resistance.

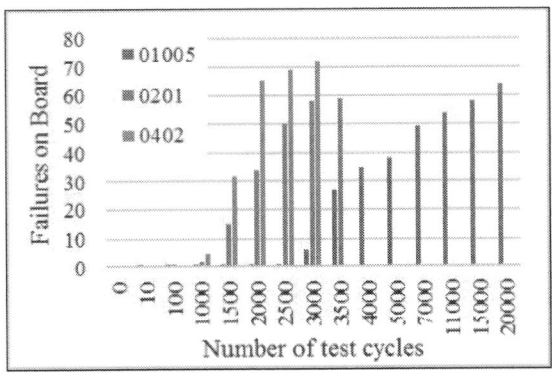

Figure 7 Failure rate of longitudinal oriented SMD after bending with 15 mm bending radius

With regard to the failure modes, two main types of failures can be observed. On one hand cracks in the solder joints were observed (see figure 8). On the other hand cracks of the copper pads or lines were observed (see figure 9).

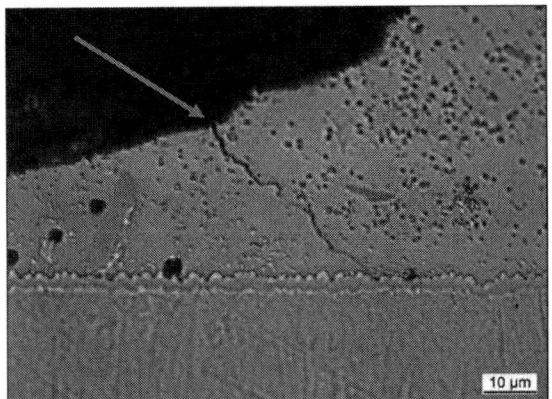

Figure 8 Crack in solder joint after bending

To evaluate the cracks in detail, metallographic preparation are necessary. Because of the high amount of failures, only samples were investigated and evaluated. As a result, the amount of the cracks in the copper is around twice higher as in the solder joint.

However, with that results it is possible to estimate the total lifetime with a Weibull analysis. Therefore, the results were clustered with regard to orientation, size of component and bending radius. The characteristic lifetime T_C can be calculated with the following equation:

$$T_C = \frac{(n-i)N_{pr} + \sum N_i}{i}$$

n — total amount of tested SMD
N_{pr} — total amount of bending cycles
N_i — failure cycle of single SMD
i — amount of failures

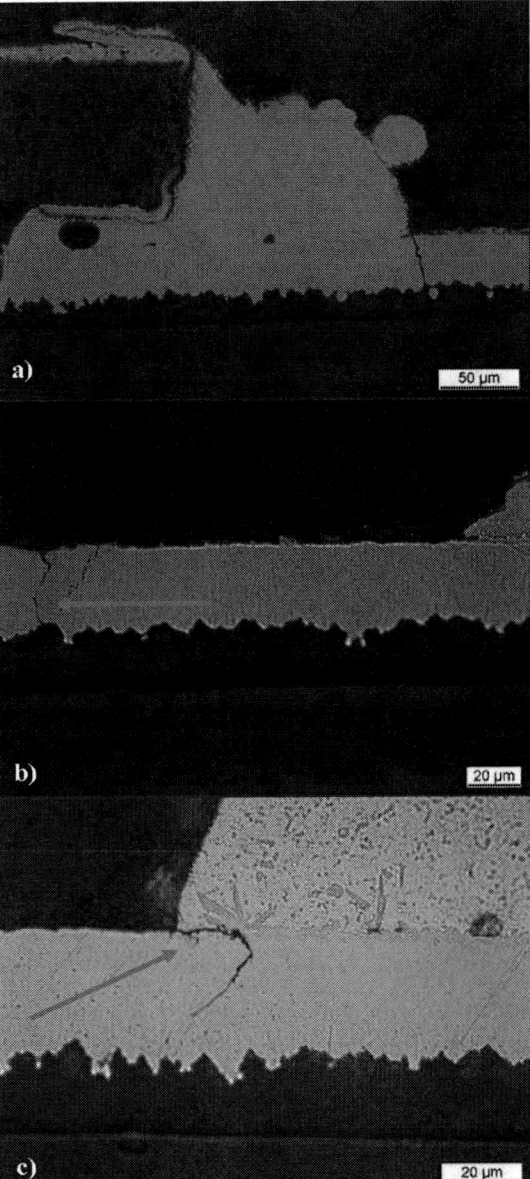

Figure 9 Crack in copper structure: a) at solder edge; b) outside interconnecting area; c) crack started in solder joint and spreading to copper

In table III the results are shown. Obviously, the lifetime can be increased significantly by the use of smaller SMD. Especially for 5 mm bending radius the lifetime doubles by halving the component size.

TABLE III Results of Weibull analysis

Bending radius [mm]	Chip size	Estimated lifetime T_C [cycles]
5	0402	310
5	0201	635
5	01005	1255
15	0402	2144
15	0201	2841
15	01005	8992

The analysis of impact factors shows also the highest impact by the bending radius with an F-value around 154 followed by chip size with an F-value of 64. The F-value describes the significance of an impact factor to the variance of the test results and is calculated as the ratio of the total mean square to the mean square of failures.

Conclusion and Outlook

The present work focused on evaluating the long-term reliability of hetereogenous integrated PCBs, which mainly means the combination of flexible substrates with mechanical rigid SMD components and technologies. For characterization of mechanical lifetime an existing bending test equipment was optimized and used for stressing the test specimen. The test specimen was designed to cover common used passive SMD components with chip sizes of 01005, 0201 and 0402. By the use of soldering also a common technology were chosen for interconnecting.

With the results of the bending tests an estimation of lifetimes was possible. But, with regard to the failure modes this lifetime may be increased by optimization of the test design. The highest amount of failure were detected in the copper structure near the solder joint. So, the standard design should be changed in that way, that the copper lines are connected orthogonal to the solder pads as shown in figure 10.

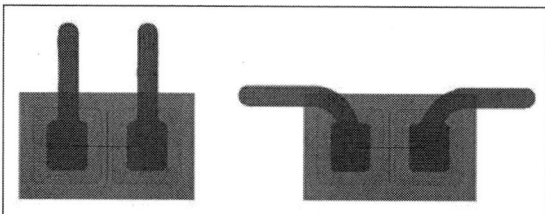

Figure 10 Optimized design rule for SMD mounting

This design rule is used for the following investigations. Furthermore, first FEM simulations also considering the new design for lifetime simulation. First results show no critical areas in the copper layer itself and lower stresses at the edge of the solder joints in comparison to the old design. [6]

Next step is to evaluate the new design and also alternative packaging technologies, such as adhesive bonding and embedding technologies. With the FEM simulation, a model-based approach shall enable faster results of long-term reliability, especially for investigations based on concrete applications.

References

[1] https://news.ihsmarkit.com (05.02.2019)

[2] D. Ernst, T. Zerna, K.-J. Wolter, "Influences of Organic Materials on Packaging Technologies and their Consideration for Lifetime Evaluation", 34th International Spring Seminar on Electronics Technology (ISSE), High Tatras, Slovakia, 2011

[3] https://www.cnet.com (05.02.2019)

[4] N. Richter, „Zuverlässigkeitsbewertung von oberflächenmontierten Bauelementen nach dynamischer Biegebelastung", diploma thesis, TU Dresden, Dresden, Germany, 2018

[5] https://www.allflexinc.com (07.06.2019)

[6] E. Dallmann, "Schadensanalyse und Schadenssimulation von oberflächenmontierten Bauelementen nach Biegebelastung", diploma thesis, TU Dresden, Dresden, Germany, 2019

[7] https://www.tutelman.com (23.07.2018)

[8] Ji-Hye, K. et al: "Effects of Anisotropic Conductive Films Gap Heights on the Bending Reliability of Chip-In-Flex Packages for Wearable Electronics Apllications" IEEE 67th Electronic Components and Technology Conference (ECTC), 30.05-02.06.2017, Orlando

Assembly of very fine pitches Infrared focal plane array with indium micro balls.

O. MAILLIART, S. RENET, F. BERGER, A. GUEUGNOT, S. BISOTTO, S. GOUT, L. MATHIEU, Y. GOIRAN and T. CHAIRA

CEA Grenoble, LETI, 38000 Grenoble, France

Abstract

Flip chip is a high-density and highly reliable interconnection technology, which is mandatory for the fabrication of high end heterogeneous imaging arrays. In the field of cooled infrared detectors, indium balls interconnections technology is well mastered and widely used down to 10μm pitch. Control of ultra-fine pitch (<10 μm) and high interconnections count flip-chip bonding technology represents a challenge on the roadmap of next generation devices. In the context of the common laboratory DEFIR with LYNRED (ex SOFRADIR), CEA-LETI has addressed the issue of assembling very small pitches focal-plane-arrays (FPA). First, we developed new wafer level processes and managed to produce Under Bump Metallization (UBM) and indium micro balls at pitches as fine as 7,5 μm and 5 μm. These balls arrays were used to develop bonding processes on mechanical and electrical test vehicles up to a resolution of 1280 x 1024 pixels. We investigated in details the influence of many aspects of the process such as the flatness difference that can be caught up, the alignment accuracy, the formation of Inter Metallic Compounds (IMC) at UBM/In interface. These new processes were successfully used to solder CMT (Cadmium Mercury Tellurium) TV range (640x512 pixels) detectors at 7,5 μm pitch onto ROICs specially designed to fulfil the project's needs. First electro-optical characterizations revealed a connection rate higher than 99% and it has been possible to acquire IR pictures. In addition, some ageing tests have been performed on these functional detectors and an electrical (with daisy chains) 2k² demonstrator at 5 μm pitch is to be achieved.

Key words: Very fine pitch interconnections, In balls, flip chip, hybridization

Introduction

Pitch reduction in Infra Red (IR) systems has been a subject of major concern over the past few years. A number of papers accurately describes the interest and challenges of the miniaturization of infrared sensor systems [1] [2] [3] [4] [5] and sub 10μm pitch IRFPA are now commercially available [6] [7] [8]

Besides, studies for improving the existing small pixel devices are continuing [9] and large format ultra-small pitch IRFPA have been demonstrated, as early as 2015 with the LW MCT 5μm pitch 1280 x 720 array from DRS [10].

The French Direction Générale de l'Armement (DGA) has supported the study that has led to the results presented in this paper. We focus on the fabrication of the interconnections onto the readout circuit at wafer level and the hybridization with the detector. The main demonstrator is a 7,5 μm pitch 640 x 512 MWIR MCT device hybridized to an analogical ROIC, by an indium bump process. In addition, we developed a 5μm pitch hybridization process on mechanical components.

Interconnections processing

Indium ball is the reference hybridization technology used in the field of cooled infra red sensors. This interconnection is composed of two elements (cf. Figure 1):

- An Under Bump Metallization pad (UBM) realised by deposition and ion beam etching (IBE) of three different metallic layers including a gold wetting layer.

- An Indium ball deposited by evaporation on a mask and lift off. The In ball is a solder that melt at 156°C and will assume electric conduction and mechanical strength between the MCT and the readout circuit.

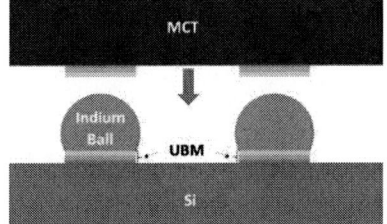

Figure 1: schematic representation of In balls on UBM during hybridization

We used a reference process initially developed for 15 μm pitch and transferred it directly at 7,5 μm and 5 μm pitch without any change. We met several issues due to the dimensional changes of the interconnections and carried out the developments presented hereunder.

UBM manufacturing

The reference process developed during the past few years leads to the creation of well-shaped and plane UBM at 15 μm pitch (cf. figure 2-a). It consists in depositing three successive metallic layers onto the wafer and creating the UBM by ion beam etching (IBE) throw a resin mask.

The reduction of the pitch from 15 μm to 7.5 μm and 5 μm results in an important reduction of the UBM diameter. The main issue we had to adress concerning the UBM manufacturing is what we called "etching wall defects". The huge reduction in the pad size results in the formation of a wall at the periphery of the UBM as shown in figure 2-b&c. These walls must be removed as they will prevent the solder from wetting well the surface of the UBM and therefore will be detrimental to the creation of a strong interconnection.

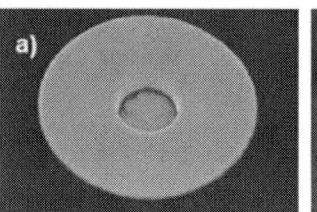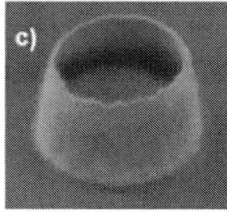

Figure 2: UBM obtained with reference process at 15μm (a), 7,5μm (b) and 5μm (c) pitches

Further investigations confirmed that these walls are formed during the UBM etching step (IBE). It appears that residues from etched materials deposit onto the resin as illustrated on figure 3-a and Figure 3b. Figure 3-c represent a UBM after the stripping of the resin with the standard process. It points out that the created etching wall is almost as high as the UBM.

Note that this kind of etching walls were also observed after etching for higner pitches (10μm and 15μm), but they were easily removed during strepping. We assume that the difficulty at smaller pitch comes from a less favourable form factor of the UBM (same thickness but smaller area).

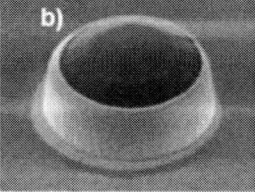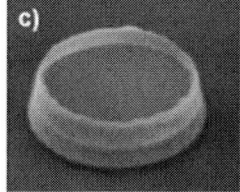

Figure 3: MEB pictures of a UBM after different steps: a) lithography, b) UBM IBE etching, c) resin stripping

A study of the etching process (resin type and thickness, current, tension, angle) has been carried out so as to define optimized parameters for 7,5 μm and 5 μm pitches. Explored conditions are synthetized in figure 4. It appears that wall formation is very sensitive to this parameters. Etching angle seams to be the key parameter in wall elimination, an increase is favourable to wall elimination (first column in figure 4). The beam will etch the side of the resin during the process and therefore will eliminate the deposited residues when formed. On the opposite, a smaller angle (then the beam is almost perpendicular to the wafer) contributes to thicken the wall (right column). Other parameters have lesser effect even if a moderate increase of intensity and tension is also beneficial.

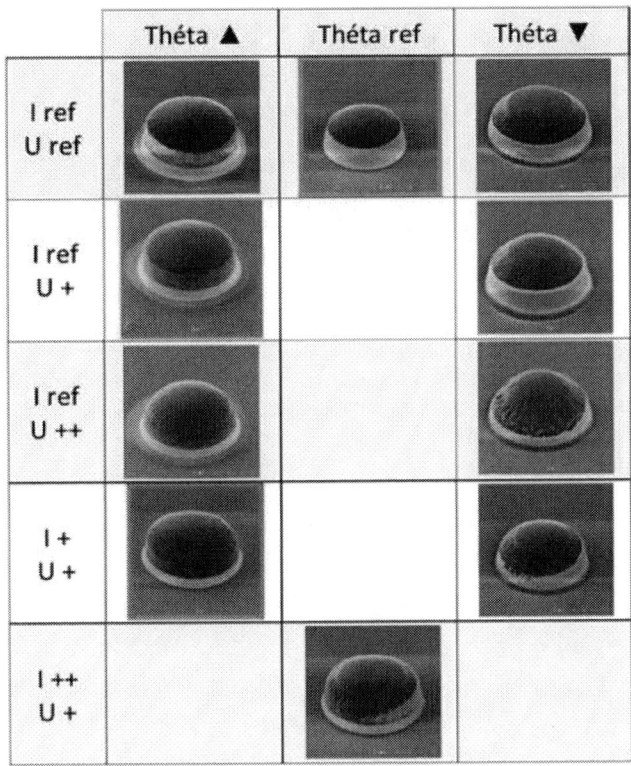

Figure 4: Overview of the influence of main etching parameters on the peripheral walls during etching of the UBM

The developed new UBM process is compatible with the production of UBM pads as fine as 1µm in diameter and presents the advantage of being common to 15 µm, 10 µm, 7,5 µm and 5 µm pitches.

In balls

We used a slightly modified version of the 10µm pitch process for indium deposition. Indium is deposited onto a patterned resin by evaporation under vacuum in an Alliance Concept EVA 760 equipment. The Indium bumps are obtained by lifting off the resin.

Hybridization

In this study, we first investigated hybridization of large Interconnection Network (IN) arrays to set the misalignment limit (mechanical component). The self alignment assembly is done by a reflow of the indium µballs interconnections on a Pick&Place SET FC300 machine with formic acid vapor (figure 5). The yield and reliability of that technology has been widely studied [1] [2] and CEA LETI has a great experience with this kind of processes down to 10µm pitch.

Figure 5 shows the morphology of the interconnection after assembly at a pitch of 7,5 µm. Si and MCT substrates have been well joined together and the alignment seems to be very satisfying. As we had forecast, these SEM observations show a great proportion of intermetallic compounds inside the interconnections what could have an impact on the reliability of the component as indium and IMC have very different mechanical properties. Thanks to these components, we will be able to carry out studies to evaluate the reliability of interconnections with high IMC proportions.

In addition, this results have been obtained with UBM which have been designed for 30µm and 15µm pitches. Therefore, if necessary, we can adapt the UBM, particularly the thickness and/or the chemistry of the different layers in order de reduce the IMC proportion in the interconnection.

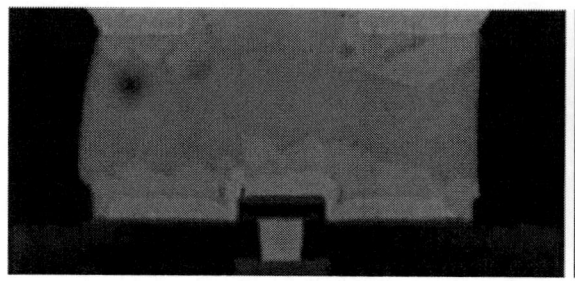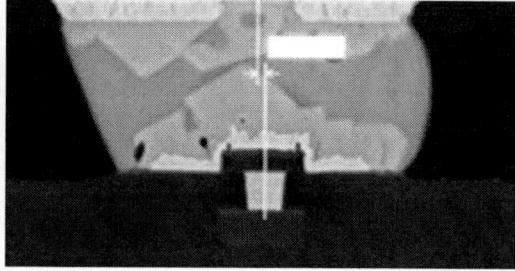

Figure 5: 7.5 µm pitch : cross section after hybrid 640 x 512 MCT (a) and 1280 x 1024 MCT (b)

Same kind of hybridizations have been successfully done at 5µm pitch (cf. figure 6) with the very same process . For these first experiments at very fine pitch the alignment is less accurate but still acceptable. The observed proportion of residual pure indium has been measured on several cross section samples and can be as low as 35% in volume. In some cases, a junction is created between IMC growing from the top and bottom chips what might drastically modify the mechanical behaviour of the interconnection. The alignment accuracy according to the format and the pitch has been quantified and is summarized in the table 1 below.

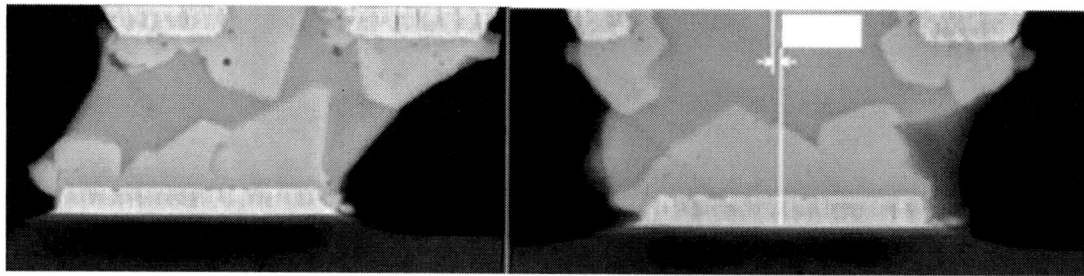

Figure 6: 5 µm pitch : cross section after hybridization 640 x 512 MCT (c) and 1280 x 1024 MCT (d)

pitch	640x512	1280x1024
7.5µm	<0.1µm	0.4-0.5µm
5µm	0.4 to 0.8µm	

Table 1: Misalignment for 7.5 µm and 5 µm hybridization process

Considering all these results on mechanical test vehicles, we were confident to realize a MCT demonstrator (whith a ROIC) in a 640 x 512 format and with a 7.5 µm pitch. We used the same process as above and obtained quite good results (comparable to the assembly on Interconnection Networks).

Figure 7 below shows a cross section of some indium interconnections of this functional component. The alignment accuracy is as good as for IN components and no shortcuts or open circuits have been observed.

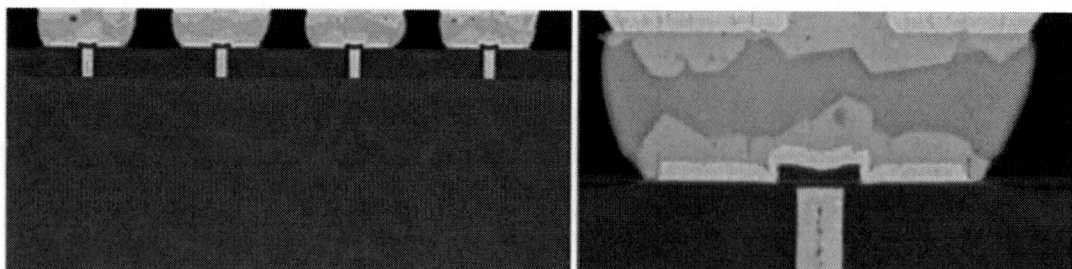

Figure 7: 7,5µm pitch : cross section after hybridization 640 x 512 MCT on a ROIC

Fifteen FPA have been hybridized, characterized, and tested under ageing or working conditions. Protocol and results are more detailed by Bisotto and Al. [11].

Electro-optical characterizations were particularly interesting for hybridization evaluation as it makes it possible to measure precisely the hybridization yield. The average hybridization yield measured after ageing (100 cycles between 93K and 293K, 15 minutes per cycle) is as high as 99.96%.

Images

Finaly, we used our 7,5µm pitch demonstrator to realize some indoor and outdoor images and videos.

Figure 8: outdoor images acquired with 7,5µm 640 x 512 demonstrator. Up: day. Down:night

Conclusion

In this work, we aimed to transfert a standard 10µm pitch hybridization process to 7,5µm and 5µm pitch without bringing major evolution.

We developed all the processes (interconnection production and hybridization) on mechanical test vehicles and managed to produce functional demonstrators at 7,5µm pitch.

We pointed out that the process is compatible with a 5µm pitch even if the alignment was less accurate than for 7,5 µm. A 2k² mechanical demonstrator at 5µm pitch is currently on the go.

Electro-optical characterizations revealed that the hybridation yield was as high as 99,96% after ageing.

This results are quite encouraging, and 7,5µm Full HD array seems to be reachable in the near future.

Acknowledgments

We acknowledge support by the French DGA, and thank more specifically JC Peyrard. This work has been done in thecontext of DEFIR the common laboratory of LYNRED and CEA-Leti.

References

[1] Driggers R G, Vollmerhausen R, Reynolds J P, Fanning J and Holst G C 2012 Infrared detector size: how low should you go? Opt. Eng. **51** 063202

[2] Holst G C and Driggers R G 2012 Small detectors in infrared system design *Opt. Eng.* **51** 096401

[3] A Rogalski et al 2016 Challenges of small-pixel infrared detectors : a review Rep. Prog. Phys. 79 046501

[4] O Gravrand, N Baier, A ferron, F Rochette, J Berthoz, L Rubaldo and R Cluzel 2014 MTF Issues in Small-Pixel-Pitch Planar Quantum IR Detectors journal of electronic materials vol 43, n°8

[5] Reibel Y, Pere-Laperne N, Augey T, Rubaldo L, Decaens G, Bourqui M L, Bisotto S, Gravrand O and Destefanis G 2014 Getting small, new 10 *µ*m pixel pitch cooled infrared products *Proc. SPIE* **9070** 907034

[6] SOFRADIR Daphnis 1280x720 MCT MWIR 10µm digital http://www.sofradir.com/product/daphnis-hd-mw/

[7] LEONARDO Superhawk 1280x1024 MCT MWIR 8µm https://www.leonardocompany.com/en/-/superhawk

[8] David Jeckells, R. Kennedy McEwen, Sudesh K. Bains, Martin Herbert, SELEX ES Infrared Ltd, 2016 Further developments of 8µm pitch MCT pixels at Selex ES Proc SPIE 9819 981953

[9] David Jeckells, R. Kennedy McEwen, Sudesh K. Bains, Martin Herbert, SELEX ES Infrared Ltd, 2016 Further developments

of 8μm pitch MCT pixels at Selex ES Proc SPIE 9819 981953

[10] J. M. Armstrong, M. R. Skokan, M. A. Kinch, J. D. Luttmer 2014 HDVIP five-micron pitch HgCdTe focal plane arrays Proc SPIE 9070 907033

[11] S.Bisotto, J.Abergel, B.Dupont, A.Ferron, O.Mailliart, JA. Nicolas, S.Renet, F.Rochette, JL. Santailler; 7.5 μm and 5 μm pitch IRFPA developments in MWIR at CEA-Leti

Sintering of oxide-free copper pastes for the attachment of SiC power devices

Luca Del Carro[1], Chunlei Liu[2], Fabio Koller[2], Alfred A. Zinn[3], Thomas Brunschwiler[1]

1: IBM Research – Zurich, Säumerstrasse 4, 8803 Rüschlikon, Switzerland

2: ABB Switzerland Ltd. Corporate Research, RD-S2 Segelhofdtrasse 1K, 5405 Baden-Dättwil, Switzerland

3: Kuprion Inc., Palo Alto, CA 94303, USA

Abstract

The introduction of wide-band-gap semiconductors such as silicon carbide (SiC) in power electronic devices has allowed operation temperatures beyond 300°C. However, these high operational temperatures are not suitable for traditional die-attach materials such as solder. Therefore, a bonding material that is stable beyond 300°C, with high electrical and thermal conductivities and low cost is essential. In this regard, copper (Cu) nanoparticle-based pastes are promising candidates due to their stability at high temperatures and elevated thermal and electrical conductivities. Furthermore, the recent introduction of oxide-free Cu pastes based on amine-passivated Cu nanoparticles enables these materials to be processed in inert atmosphere.

We report here on attaching SiC dies to direct-bonded copper ceramic substrates by sintering oxide-free Cu pastes. Applying bonding pressure during sintering improves the die-attach quality by reducing the percentage of residual voids and increasing shear strength. We succeeded in further reducing the percentage of voids by using a paste containing a solvent with a low boiling point and high vapor pressure. Furthermore, we developed a novel dual-layer process to deposit Cu paste that yields residual percentages of voids in the die attach as low as 5% and a shear strength of 29 MPa. Finally, the reliability of the die attach by oxide-free Cu paste was investigated with thermal shock cycling.

Key words: sintering, copper nanoparticles, copper paste, die attachment, power electronic packaging.

Introduction

Wide-band-gap semiconductors such as silicon carbide (SiC) have been introduced in power electronic devices to enable operation temperatures beyond 300°C [1, 2, 3]. These high operational temperatures are not suitable for traditional packaging materials to connect the device to the substrate, called the die attach. In fact, high-melting solders based on lead (Pb) have been restricted due to their environmental toxicity, whereas lead-free formulations re-melt above 250°C [4, 5, 6, 7]. Therefore, a bonding material that is stable above 300°C, has high electrical and thermal conductivities and low cost is essential.

In this context, pastes based on metal nanoparticles (NPs) dispersed in a carrier solvent have become an attractive alternative. Owing to the high area-to-volume ratio of NPs, these materials can be applied and sintered at temperatures below 300°C. After sintering, the area-to-volume ratio of the material drops, and the material stays solid until its melting point, which is often above 800°C. Among the various possible metals, copper (Cu) NP-based pastes are promising candidates due to their high electrical and thermal conductivities and moderate price compared to silver NP-based pastes.

However, Cu NPs readily oxidize when exposed to ambient atmosphere, forming oxide shells that limit the diffusivity of the Cu atoms, hence preventing sintering. Recently, a passivation of Cu NP surfaces with a mixture of amines has been reported to prevent oxidation [8, 9, 10], leading to an oxide-free Cu paste potentially able to replace state-of-the-art die-attach materials.

In this work, we attached SiC dies with an area of 25 mm^2 to ceramic substrates by sintering an oxide-free Cu paste. First, we examine the effects of bonding pressure and solvent content of the mixture on the percentage of residual voids in the die attach. Then, we develop a novel process of dual-layer deposition of Cu paste, which minimizes the percentage of voids and consequently improves mechanical performance. Finally, we investigate the reliability of the die attach using an oxide-free Cu paste and thermal shock cycling.

Materials & Methods

Copper pastes:

We used Cu NPs (Kuprion Inc., Palo Alto, USA) synthesized via a bottom-up approach in a non-aqueous glyme medium by reducing anhydrous copper(II) chloride with sodium borohydride in the presence of an amine mixture on hexylamine (boiling point 131°C) and dihexylamine (boiling point 195°C) [11]:

$$CuCl_2(s) + 2\ NaBH_4(s) \rightarrow Cu^0\ (s) + 2\ NaCl(s) + H_2(g) + 2\ BH_3(g)\ .$$

After reduction, Cu NPs were formed with a size distribution from 10 to 60 nm [12]. These NPs were passivated by the amine mixture to form an inverse micelle structure on the surface [8]. Passivation was performed in order to hinder the aggregation of the NPs and to protect them from oxidation. To the best of our knowledge, the amine mixture cannot reduce the copper oxide.

Finally, we added a carrier solvent to obtain a paste with a metal load of 95–97 wt-%. We used two paste formulations, one containing 1-decanol (decanol paste) and one containing 1-octanol (octanol paste), as carrier solvents, see Table I.

Table I: Cu paste formulations produced with the same batch of Cu NPs.

Cu paste	Carrier solvent	Solvent boiling point [°C]	Solvent vapor pressure @20°C [Pa]
Decanol	1-decanol $C_{10}H_{22}O$	233	1.134
Octanol	1-octanol $C_8H_{18}O$	188	8.7

Test samples:

Silicon carbide (SiC) chips with a titanium-nickel-silver (Ti-Ni-Ag) back metallization and a size of of 5×5 mm^2 were bonded to a direct-bonded copper (DBC) substrate containing a copper layer bonded to the ceramic substrate (alumina, Al_2O_3).

Bonding processes:

We used two different bonding processes, one without and one with a pre-drying step, see Figure 1a-b.

In the first case, see Figure 1a, copper paste was dispensed onto the copper metallization of the substrate, and a chip was placed by a flip-chip bonder (T-3000-FC3, Dr. Tresky AG, Switzerland) onto the paste with a pressure of 2 MPa for 10 s. Subsequently, the assembly was mounted in a spring system to apply bonding pressure during sintering, see Figure 1c. At complete deformation of the springs, a force of 73 N acts on the plate. Finally, the assembly was sintered in a reflow oven (ATV Technologie, Vaterstetten, Germany) in nitrogen atmosphere.

In the second case with a pre-drying step, a 40-μm layer of copper paste was initially stencil-printed onto the substrate and dried in the oven at 80°C for 30 min in a formic acid atmosphere (8 vol%), see Figure 1b. Next, fresh Cu paste was dispensed onto the dried layer and the chip was placed on the paste. Then, the assembly was mounted in the spring system and sintered in the reflow oven in nitrogen atmosphere.

Figure 1d depicts an optical image of samples assembled with these two bonding processes.

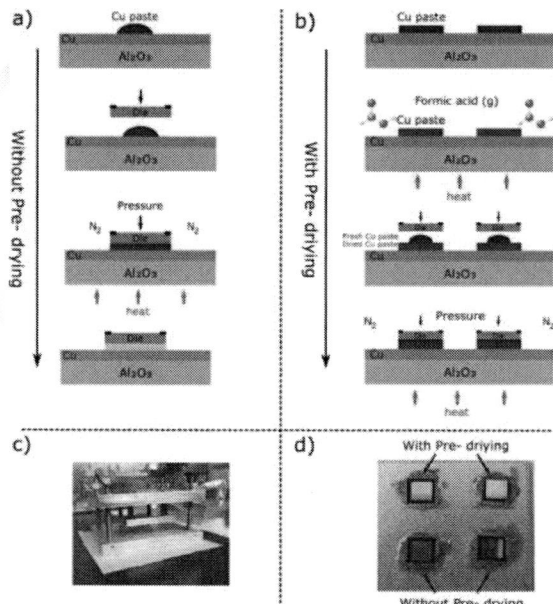

Figure 1: Schematic representation of the bonding processes a) without pre-drying step and b) with pre-drying step; c) spring system to apply bonding pressure during the sintering process; d) samples assembled without and with pre-drying step.

Characterization:

After the bonding process, we used scanning acoustic microscopy (SAM, Sonoscan Gen5) to assess the bond interface between the two materials. Ultra-sound waves are emitted by a transducer. The waves are reflected at interfaces between materials with different acoustic impedances. The time of flight of the reflected wave indicates the location of the interface within the sample. By scanning over the desired area, the microscope converted the signal into a grayscale image in which voids and cracks appeared brighter than bonded areas. We used a transducer with a frequency of 100 MHz and focal length of 0.5" for these experiments. The percentage of voids was evaluated from SAM images with image-analysis software (Fiji).

We then used a liquid-to-liquid thermal shock cycler (TSC) (Espec, TBS-51) to evaluate the thermal reliability of the bond by exposing the sample to temperature changes. The temperature profile of one cycle in this study was −40°C for

5 min and 150°C for 5 min with a 2-min transfer time between cycles.

We used a scanning electron microscope (SEM, VEGA, Tescan) to identify the microstructure of the bond interface and sintered copper. The shear strength was measured using a Nordson Dage 4000 with a 200 kg shear tester cartridge. The total strength was normalized on the die area of 25 mm^2.

We then performed a thermogravimetric analysis (TGA) with a Mettler Toledo TGA 851. For each analysis, about 15 mg of paste was placed in an aluminum capsule with a punched opening of 0.5 mm diameter in the lid. All measurements were performed by running a heating profile from 25 to 400°C at a rate of 20 K/min in nitrogen atmosphere.

Results & Discussion

A key indicator of die-attach quality is the percentage of residual voids after sintering, which must be as low as possible to guarantee high electrical, thermal and mechanical coupling between the substrate and the die. We identified the percentage of residual voids as a function of bonding pressure and the solvent content of the paste mixture, see Figure 2a. First, samples with decanol-based paste were sintered in a nitrogen atmosphere at 300°C for 10 min with a sintering pressure of 0 to 3 MPa. For increasing bonding pressure, scanning acoustic microscope (SAM) analysis revealed that the percentage of voids was reduced from 55.6 to 31.8%. This reduction leads to an increase of the bond shear strength from 1.4 to 22.5 MPa. Second, the percentage of voids was further reduced to 24.6% by switching from decanol to an octanol-based paste. We also investigated the thermal behavior of these paste mixtures using thermogravimetric analysis (TGA) to record the mass changes in the samples during sintering with respect to the initial mass measured at room temperature, see Figure 2b. Our analysis showed an onset of mass loss above 50°C in both samples. For the octanol sample, most of the mass loss took place between 125 and 185°C, leading to a total mass loss of 2.9% at 300°C. For the decanol sample, the loss was concentrated between 150 and 250°C, leveling off at 300°C with a total mass loss of 4.6%.

These results demonstrate that the residual voiding is greatly reduced by applying bonding pressure, thus improving the performance of the die attach. The pressure compacts the die-attach material during sintering by reducing the size of the voids and cracks formed by evaporating the carrier solvent [13, 14]. Furthermore, the bonding pressure increases the sintering driving force, thus enhancing the densifying mechanisms such as grain boundary diffusion and creep flow [15, 16, 17, 18, 19]. The pressure results in an unconstrained uniaxial stress on the die attach that mitigates the packing defects of the particles and increases their contact area by improving surface diffusion.

Our thermal analysis attributed the measured mass losses to the evaporation of the organic solvents used in the mixtures. The faster mass loss of the octanol paste compared to decanol is due to its higher vapor pressure and lower boiling temperature, see Table I. Moreover, due to its higher volatility, the octanol-based paste contains a smaller amount of solvent than the decanol-based paste. These differences impact the percentage of residual voids after sintering, as verified by SAM image analysis, suggesting that the octanol-based paste is preferential for die-attachment applications.

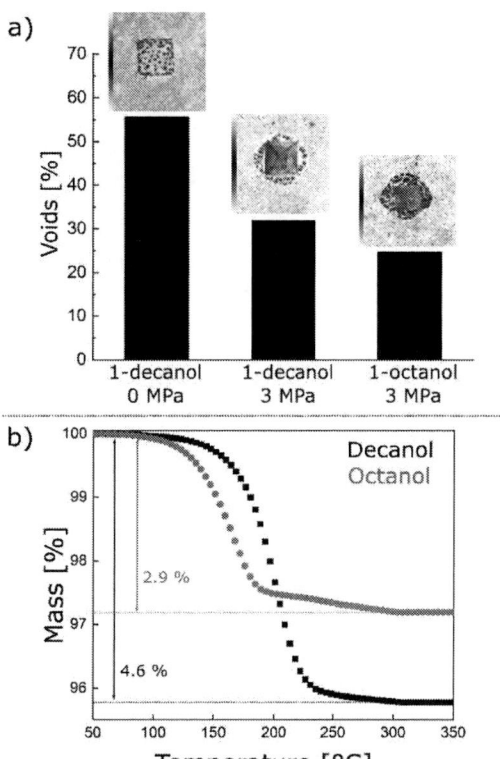

Figure 2: a) Percentage of residual voids as a function of the solvent content of the mixture and the bonding pressure. Representative scanning acoustic microscope (SAM) images of the samples are shown. b) Thermogravimetric analysis (TGA) of decanol and octanol-based copper pastes.

To reduce the percentage of voids further, we developed a process of dual-layer deposition of Cu paste for die attachment, see Figure 1b. Herein, a 40-µm-thick film of octanol paste is stencil-printed on the substrate and then dried at 80°C for 30 min in a formic acid-enriched nitrogen atmosphere. Subsequently, a fresh layer of octanol paste is dispensed onto the dried layer, and then the die is sintered in nitrogen atmosphere at a temperature of 300°C for 10 min with a bonding pressure of 0.7 MPa.

SAM analysis shows that the percentage of voids is reduced from 40.8 to 5% for samples bonded without and with the pre-drying step, respectively (Figure 3). By extending the dwell time from 10 to 60 minutes, the percentage of voids in those samples bonded without pre-drying drops from 40.8 to 26%, whereas no significant change was observed for the pre-dried samples. For the latter samples, a shear strength of 28 and 29.4 MPa was measured for dwell times of 10 and 60 min, respectively. Then, samples were processed at a lower sintering temperature of 220°C and a dwell time of 60 min, which yielded a percentage of residual voids above 55% and shear strength below 2 MPa for samples with and without the pre-drying step.

Our proposed process of dual-layer deposition significantly decreases the percentage of voids because the fresh Cu paste fills the cracks formed in the first layer during drying. The formic acid atmosphere used during the drying step creates a shell of copper formate on the nanoparticles that prevents oxidation [20]. During the bonding process, this passivation shell is thermally desorbed, enabling the diffusion of the Cu atoms, thus the sintering [21]. The elevated shear strength measured in the pre-dried samples indicates a good adhesion between the two layers [22]. For pre-dried samples, the dwell time does not affect the percentage of voids nor the shear strength, suggesting that the extent of sintering is not increased by a longer time at a given temperature [23]. On the other hand, the high residual voids in samples sintered at 220°C for 60 min implies that a higher temperature is necessary to achieve a strong adhesion between chip and substrate. These results suggest that temperature plays a predominant role with respect to the dwell time in the sintering-based die attachment.

To assess the reliability of the bond, we performed thermal shock cycling (TSC) tests on octanol-based samples bonded with and without the pre-drying step at different values of bonding pressure and dwell time, see Figure 4a. After 200 cycles, all the analyzed samples showed delamination below 5% of the total bonding area concentrated at the edges of the die attach.

After TSC, a cross section of the pre-dried sample sintered at 300°C for 60 min was prepared and imaged with a scanning electron microscope (SEM), see Figure 4b. The images show a good diffusion of the sintered copper at the chip and substrate metallization layers. Furthermore, no bond line between the two layers of copper paste was observed.

The concentration of delamination at the corners reveals the high strength of the bond between the chip and the substrate. This debonding is caused by the large coefficient of thermal expansion (CTE) mismatch between the sintered copper and the two joined parts, which results in thermo-mechanical stress during TSC.

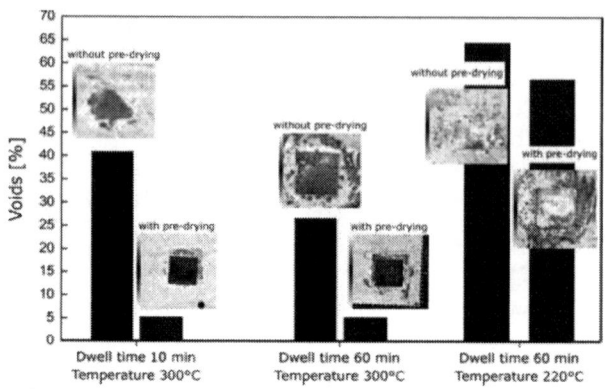

Figure 3: Percentage of voids after bonding process with and without pre-drying as a function of dwell time and maximum temperature. Representative SAM images of the samples are shown.

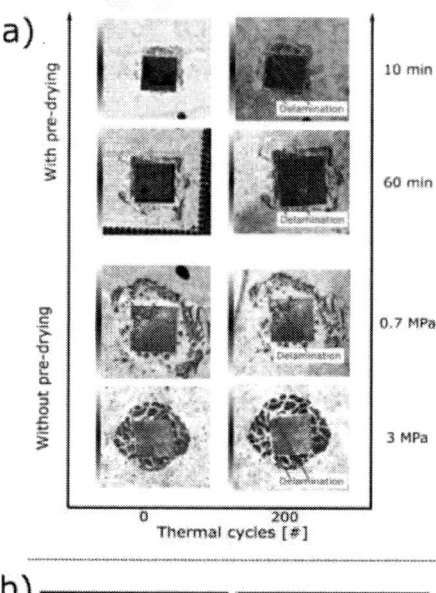

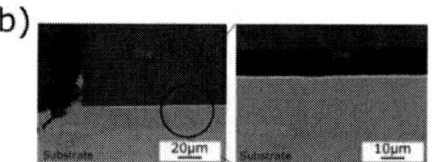

Figure 4: a) Scanning acoustic microscope (SAM) images of octanol-based samples before and after thermal shock cycling (TSC) as a function of the bonding process, bonding pressure and dwell time described here; b) scanning electron microscope (SEM) cross section image of a TSC-tested sample sintered at 300°C for 60 min with 0.7 MPa bonding pressure.

This stress is dissipated by cracks that form at the bonding interfaces [24]. The low delamination observed in all the samples can be attributed to the high porosity of sintered copper, which leads to a concentation of the stresses in the ligaments between the nanoparticles and hence to plastic deformation of the material with consequent stress dissipation [25].

The absence of a bond line between the two layers of copper paste testify to their successful merging during sintering.

Conclusion

This work reported the attachment of SiC dies to a direct-bonded copper ceramic substrate by sintering an oxide-free Cu paste at 300°C in inert atmosphere.

Applying bonding pressure during sintering improved the quality of the die attach by reducing the percentage of residual voids and increasing the shear strength. We reduced the percentage of voids further by using a carrier solvent with a low boiling point and high vapor pressure in the mixture.

Furthermore, we developed a novel dual-layer process to deposit Cu paste, leading to percentages of residual voids in the die attach as low as 5% and a shear strength of 29 MPa. The sintering temperature proved to play a greater role than the dwell time. These results were obtained by applying a bonding pressure of only 0.7 MPa, which makes the process attractive for mass production.

Finally, we investigated the reliability of die attachment with an oxide-free Cu paste by thermal shock cycling, achieving a delamination of only 5% at the edges of the die after 200 cycles.

Acknolewdgements

The authors thank Bruno Michel, André R. Studart and Heike Riel for their continuous support and Lilli-Marie Pavka for copyediting this manuscript.

References

[1] Y. Mikamura *et al.*, "Novel Designed SiC Devices for High Power and High Efficiency Systems," *IEEE Trans. Electron Devices*, vol. 62, no. 2, pp. 382–389, Feb. 2015.

[2] P. Neudeck, R. Okojie, and L. Chen, "High-temperature electronics-a role for wide bandgap semiconductors?," *Proc. IEEE*, vol. 90, no. 6, pp. 1065–1076, 2002.

[3] L. Lorenz, T. Erlbacher, and O. Hilt, "1 - Future technology trends," in *Wide Bandgap Power Semiconductor Packaging*, K. Suganuma, Ed. Woodhead Publishing, 2018, pp. 3–53.

[4] V. R. Manikam and K. Y. Cheong, "Die attach materials for high temperature applications: a review," *IEEE Trans. Components, Packag. Manuf. Technol.*, vol. 1, no. 4, pp. 457–478, 2011.

[5] K. Suganuma, "2 - Interconnection technologies," in *Wide Bandgap Power Semiconductor Packaging*, K. Suganuma, Ed. Woodhead Publishing, 2018, pp. 57–80.

[6] European Commission, *DIRECTIVE 2011/65/EU OF THE EUROPEAN PARLIAMENT AND OF THE COUNCIL of 8 June 2011 - ROHS*, vol. 54, no. 1 July. 2011, pp. 88–110.

[7] K. Suganuma, S. J. Kim, and K. S. Kim, "High-temperature lead-free solders: Properties and possibilities," *Jom*, vol. 61, no. 1, pp. 64–71, 2009.

[8] A. A. Zinn, R. M. Stoltenberg, J. Chang, Y. L. Tseng, and S. M. Clark, "NanoCopper as a soldering alternative: Solder-free assembly," *16th Int. Conf. Nanotechnol. - IEEE NANO 2016*, pp. 367–370, 2016.

[9] B. H. Lee, M. Z. Ng, A. A. Zinn, and C. L. Gan, "Evaluation of copper nanoparticles for low temperature bonded interconnections," *Proc. Int. Symp. Phys. Fail. Anal. Integr. Circuits, IPFA*, vol. 2015-Augus, pp. 102–106, 2015.

[10] S. Nagao *et al.*, "Thermal reliability of SiC device with Cu sintering die-attach processed at 250 ° C in N2 gas," 2017, no. HiTEN, pp. 193–196.

[11] A. A. Zinn, "Lead solder-free electronic," 2009.

[12] A. A. Zinn, "NanoCopper-Based Solder-Free Electronic Assembly Material," no. October, 2014.

[13] V. Raja Manikam and K. Yew Cheong, "Die Attach Materials for High Temperature Applications: A Review," *Ieee Trans. Components, Packag. Manuf. Technol.*, vol. 1, no. 4, pp. 457–478, 2011.

[14] M. Knoerr and A. Schletz, "Power semiconductor joining through sintering of silver nanoparticles: Evaluation of influence of parameters time, temperature and pressure on density, strength and reliability," *2010 6th Int. Conf. Integr. Power Electron. Syst.*, pp. 1–6, 2010.

[15] R. M. German, "Sintering theory and practice," *Solar-Terrestrial Phys.*, p. 568, 1996.

[16] R. Castro and K. van Benthem, *Sintering: mechanisms of convention nanodensification and field assisted processes*, vol. 35. Springer Science & Business Media, 2012.

[17] Z. Z. Fang and H. Wang, "Densification and grain growth during sintering of nanosized particles," *Ceram. Trans.*, vol. 209, pp. 389–400, 2010.

[18] Z. Z. Zhang and G. Q. Lu, "Pressure-assisted low-temperature sintering of silver paste as an alternative die-attach solution to solder reflow," *IEEE Trans. Electron. Packag. Manuf.*, vol. 25, no. 4, pp. 279–283, 2002.

[19] Z. A. Munir, U. Anselmi-Tamburini, and M.

Ohyanagi, "The effect of electric field and pressure on the synthesis and consolidation of materials: A review of the spark plasma sintering method," *J. Mater. Sci.*, vol. 41, no. 3, pp. 763–777, 2006.

[20] T. Yonezawa, H. Tsukamoto, Y. Yong, M. T. Nguyen, and M. Matsubara, "Low temperature sintering process of copper fine particles under nitrogen gas flow with Cu2+-alkanolamine metallacycle compounds for electrically conductive layer formation," *RSC Adv.*, vol. 6, no. 15, pp. 12048–12052, 2016.

[21] F. Conti, A. Hanss, C. Fischer, and G. Elger, "Thermogravimetric investigation on the interaction of formic acid with solder joint materials," *New J. Chem.*, vol. 40, no. 12, pp. 10482–10487, 2016.

[22] Y. Gao, C. Chen, S. Nagao, K. Suganuma, A. S. Bahman, and F. Ia, "Highly Reliable Package using Cu Particles Sinter Paste for Next Generation Power Devices," no. May, pp. 7–9, 2019.

[23] T. Wang, X. Chen, G. Q. Lu, and G. Y. Lei, "Low-temperature sintering with nano-silver paste in die-attached interconnection," *J. Electron. Mater.*, vol. 36, no. 10, pp. 1333–1340, 2007.

[24] J. G. Bai and G.-. Lu, "Thermomechanical Reliability of Low-Temperature Sintered Silver Die Attached SiC Power Device Assembly," *IEEE Trans. Device Mater. Reliab.*, vol. 6, no. 3, pp. 436–441, 2006.

[25] J. Hu, L. Yang, and M. W. Shin, "Meachanis and thermal effect of delamination in light-emitting diode packages," *Mater. Sci.*, no. September, pp. 28–30, 2005.

Aluminium Nitride based bio-MEMS for vascular graft monitoring

L. Natta[1,2], V. M. Mastronardi[1], F. Guido[1], L. Algieri[1,2], S. Puce[1,2], F. Pisano[1], F. Rizzi[1], R. Pulli[3], A. Qualtieri[1] and M. De Vittorio[1,2]

[1] Center for Biomolecular Nanotechnologies @UNILE, Istituto Italiano di Tecnologia, Arnesano (LE), 73010, Italy

[2] Università del Salento, Lecce, 73100, Italy

[3] Department of Vascular Surgery, University of Bari "Aldo Moro", Bari, 70121, Italy

Phone number: +39 08321816246 e-mail: lara.natta@iit.it

Abstract

Various techniques are able to monitor the hemodynamic conditions inside a prosthetic graft, thus providing useful information to optimize the patient's therapy but even on the graft status. However, all the current technologies affect the mechanical properties and functionality of the device. A proof-of-concept experiment is reported in this paper, which demonstrates the ability of an AlN-based sensor, integrated on the extraluminal surface of the prosthesis, to continuously monitor the inner blood flow, avoiding any interference with the graft structure. To this aim, mechanical properties and dimensions of the sensor are optimized in order to maximize its response to a variation of hemodynamics parameters. The very performing features of the sensor in terms of sensitivity, conformability and biocompatibility of the used materials, make the device able to successfully monitor the graft functionality, opening new perspectives for the development of an implantable sensorized vascular graft system.

Key words: Vascular graft, Aluminum Nitride, MEMS, Piezoelectric material, Flexible electronics

1. Introduction

A vascular graft is a synthetic tube used to bypass diseased vessels as, for example, in case of atherosclerosis or aneurisms. These devices are properly designed to be compliant with the real arterial structure in order to reduce the mechanical strain and blood turbulence after the implantation [1]. However, although all the recent technological improvement in the performance of these devices and even if after the surgical procedure follow up processes are defined to monitor their status, vascular graft can fail without any premonitory symptoms [2]. In particular, as the graft becomes occluded, its stiffness increases, inducing an hemodynamically compromised blood flow, but without generating evident clinical signs. This continuous and uncontrolled deterioration can finally lead to thrombosis or ischemia [2]. To overcome these issues, a continuous monitoring of the hemodynamics in to the graft is extremely important to extend its lifetime, and to provide an early warning about its failure [3]. In this respect new solutions based on different analysis techniques have been investigated [1, 3-5]. The first approach, exploits Micro-Electro-Mechanic System (MEMS) integrated in the inner part of the prosthesis in direct contact with the blood flow [3, 5]. But traditional rigid sensor are inappropriate for chronic vascular graft monitoring, due to their low compliance with the soft surface of the device, and because can locally generates turbulences in the blood stream, increasing the graft occlusion rate [1]. More recently, a new class of flexible sensors were developed to be directly integrated on the external graft surface in order to follow the wall expansion due to the pulsatile blood flow [1]. A possible approach in this sense, is the use of a conductive PDMS-based strain gauge wrapped around the extraluminal surface of the prosthesis, but the dimensions of the sensor still affect the mechanical properties of the graft. Considering the wide application of piezoelectric flexible sensors in the cardiovascular field [4, 6, 7], an innovative approach to continuously monitor a vascular graft could exploit these class of materials. Indeed, by virtue of its peculiar property of instantaneously generating an output voltage as a consequence of a rapid variation of the dynamic pressure, piezoelectric based sensors are very appealing to monitor the blood flow in an artificial vessel. One demonstration in the literature reports on a sensor unit based on two polyvinylidene difluoride (PVDF) sensors (28 μm) encapsulated in Mylar® foil and wrapped around the graft [4]. The sensor is able to monitor hemodynamic variations and the presence of small occlusions in the graft, but its dimensions still affects the mechanical behavior of the graft structure. In this paper we report on an ultrathin, lightweight and flexible smart patch, based on piezoelectric Aluminum Nitride (AlN), integrated on the extraluminal surface of prosthesis, for the continuous monitoring of the vascular graft. In particular, the AlN is inherently biocompatible and non-toxic [8], with high electromechanical coupling [9] and intrinsic high resistivity [10]. The mechanical properties of the sensors are optimized in order to not modify the structure of the vascular graft and its functionality, but ensuring a very good responsivity

2. Materials and Method

The fabrication process developed for the proposed AlN-based piezoelectric sensor relies on the fabrication of a multilayered structure. The sensing element is embedded in a polymeric matrix making the whole system able to conformally adhere to the extraluminal surface of the vascular graft [11]. Polyimide (PI2555®, HD Microsystem) was chosen as structural polymeric layer for the deposition and patterning of the sensitive element, because of its desirable structural and mechanical properties, and for the possibility to strictly control the thickness and smoothness of the deposited film. The heterostructure is composed by:

- Aluminum nitride-interlayer (AlN-IL, 120 nm), that allows the adhesion of the entire structure to the polymeric substrate and promote the texture orientation of the other layers
- Molybdenum bottom electrode (Mo, 200 nm)
- Piezoelectric aluminum nitride (AlN, 1 μm)
- Molybdenum top electrode (Mo, 200 nm)

Fig. 1 reports the main steps of the fabrication protocol. The first step consists in the deposition of a poly(methyl methacrylate) (PMMA, 950PMMA® MicroChem) sacrificial layer on a rigid silicon substrate that allows the detachment of the freestanding structure at the end of the fabrication process. The polyimide (PI) is then spun on this layer with a total thickness of 4 μm, and cured in two steps (130 °C for 60 minutes on hot plate and with a ramp up to 250 °C for 90 minutes in oven).

On this polymeric substrate the piezoelectric stack is deposited by means of a reactive DC magnetron sputtering and then patterned with standard surface micromachining techniques. The AlN layers are deposited at room temperature by using a high-purity Al target (99.9995%) with a 1:1 gas mixture ratio (N2:Ar) at a working pressure of $2.8 \times 10\text{-}3$ mbar, under a DC-pulsed power of 750 W and 1000 W for AlN-IL and piezoelectric AlN, respectively. Mo layers are sputtered using a pure Mo target (99.95%) in Ar atmosphere under DC power of 200 W and a working pressure of $5 \times 10\text{-}3$ mbar. In particular, the AlN-IL/Mo films are patterned in rectangular geometries by using an inductively coupled plasma (ICP) etching system with defined gas mixtures of BCl_3 and Ar. Then exploiting the same etching process both the AlN piezoelectric thin film and Mo top electrode (deposited in a second deposition run) are patterned.

Finally, 4μm-thick layer of PI is conformally deposited to mechanically ad electrically insulate the entire structure and consequently, the etching of the PMMA sacrificial layer allows the detachment of the freestanding structure. Exploiting a chemical surface functionalization process an additional layer of polydimethylsiloxane (PDMS) is then added (about 15 μm) to the structure in order to increase the overall flexibility of the system, obtaining a perfect adhesion with the vascular graft. To guarantee electrical connections, vias are opened using O2 plasma, through the top polyimide coating. The total thickness of the final device, including the PDMS sheet, is about 20 μm, with an active area extension of 4 mm x 5 mm.

3. Results and Discussion

The developed flexible piezoelectric smart patch is conformally attached to the extraluminal surface of the vascular graft, exploiting a PVA-based glue. In this way the cyclic expansion and contraction of the graft walls, allows the sensor to generate a proportional output voltage (V_{piezo}) . This is acquired to extract information about the hemodynamic conditions inside the artificial vessel. The responsivity of the sensor to a pressure variation

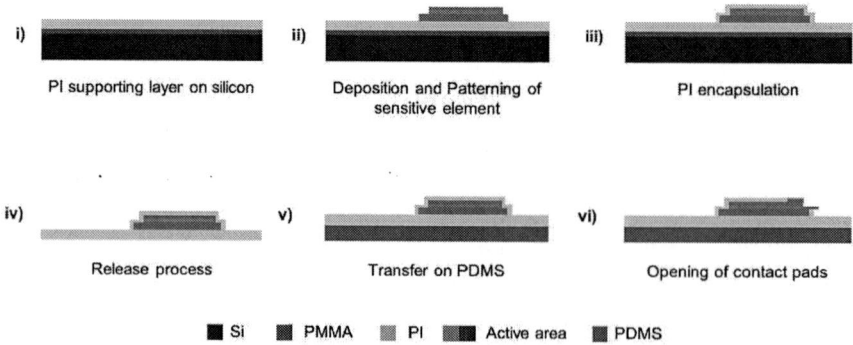

Fig. 1: Fabrication protocol for the AlN-based piezoelectric sensor

was evaluated in a physiological range between 5 kPa and 50 kPa, that represents the pressure range for blood vessels pulses [12]. In particular, the sensor was calibrated by applying this pressure range on a PDMS supporting membrane on which the sensor was attached. The generated open circuit voltage was then measured with an oscilloscope. The results reported in Fig. 2a demonstrated that the resulting output voltage generated by the piezoelectric patch is directly correlated with the applied pressure on the membrane. Indeed, due to the progressive increasing of the applied pressure in the vessel corresponds to a linear increase of the electrical output generated by the piezoelectric material, with a calculated sensitivity of 0.012 V Pa^{-1} m^{-2}.

Moreover, for the described application, a very high flexibility of the sensor is mandatory in order to not modify or damaging the correct functionality of the implanted device. In this perspective, the Young Modulus of the whole sensor structure was calculated by a dynamic mechanical analyzer (DMA, TA Q800), with a uniaxial tensile test. The considered rectangular sample had a measured thickness of 17 μm, a width of 9.57 mm and length of about 17 mm, and it was subjected to a controlled tension between 0 up to 1 N. The Young modulus was then calculated as the slope of the linear part of the obtained stress-strain curve (dotted line in Fig. 2b). The extrapolated value of 95 MPa is more than five times lower than the values reported for a typical vascular graft (E_{graft} =0.5 GPa), as desired [13].

Moreover, in order to simulate and quantify the deformation induced by the piezoelectric smart patch on the graft, due to its attachment on the external surface, a finite element model (FEM) of the sensorized vascular graft was designed. The piezoelectric patch structure was defined, following the real sensor configuration, with thepreviously described sensitive element, embedded between two polyimide layers on a PDMS structural layer, with a thickness of about 20 μm. To provide the model with realistic features, the measured mechanical properties of the stack (previously reported) were used in the FEM description of all the smart-patch layer materials' data. The considered vascular graft model was described as a 1 mm-thick polytetrafluoroethylene (PTFE) ring, recovering its material features from literature [14].

The piezoelectric sensor was integrated on the external surface of the vascular graft. The FEM analysis were set up exploiting structural mechanics, describing the behavior of the patch using the "plate theory" [15], and the bending rigidity of the patch was described by the equation (1).

$$D = \frac{2Eh^3}{3(1 - v^2)} \qquad (1)$$

where E is the Young's modulus, h the patch thickness and v is the Poisson's ratio.

The FEM results reported that there is a radial displacement of the graft surface of less than 0.02 nm when the sensor is attached on its external surface (Fig. 2c). This evidence underlines that the implant, as was designed, does not affect the whole

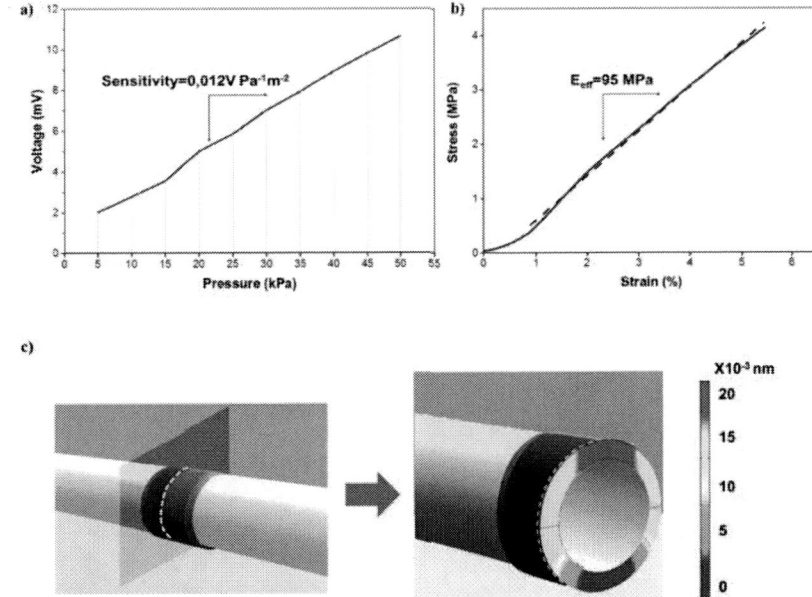

Fig. 2: a) Sensor response related to the application of a physiological pressure range between 5 and 50 kPa. b) Stress-strain curve, and the measured effective Young modulus of the piezoelectric sensor. c) Radial displacement on the transverse cross section of the vascular graft due to the integration of the flexible sensor on the external surface.

correct configuration of the vascular graft, preserving its functionality.

The sensor-graft system was then tested in order to verify its capability to provide information about the conditions of the artificial vessel. The most common way to assess the hemodynamic conditions of a natural or artificial vessel is the evaluation of the pulse wave signal (Fig. 3 yellow inset). It is characterized by the repetition of three main elements: the systolic peak, the dicrotic notch and the diastolic peak and it can provide important information about the blood pressure and different clinical information.

According to these considerations, and in order to mimic the arterial behavior both in physiological or pathological conditions, a proper measurement set up was designed (Fig. 3). The sensorized graft was integrated in a pulsatile fluid cycle in which a pump pushes water (set to a flow of about 350 ml/min [16]) through a solenoid valve with variable flow. The valve opening percentage and relative time interval were controlled in order to create a cycling routine that induce the repetition of the previously mentioned parameters (systolic and diastolic peaks and dicrotic notch). A modification of the features of these fundamental factors are significant marks of an undesired variation in the patient health status [17].

The piezoelectric sensor output was measured, by the use of an oscilloscope. In a first analysis the sensor, attached on the vascular prosthesis, was used to evaluate a flow rate variation inside the graft, induced by modifying the peak-to-peak time interval. This parameter, closely related to the R-R interval of the electrocardiogram (ECG), can define the duration of a complete heart cycle. Moreover, its variation can influence the usefulness of some hemodynamic indexes [18]. The pulse frequency of the flow was varied in an interval between 55 and 85 bpm, and the sensor was able to detect the imposed variation showing a great stability of the generated signal (Fig. 4a).

A variation of two important hemodynamics parameter was induced: the augmentation index (AI) and the time interval between systolic and diastolic peaks (ΔT_{DVP}). The first one is defined as the ratio between the diastolic (P_2) and the systolic (P_1) peak (equation (2)) and provide important information about the mechanical properties of the vessel, allowing the recognition of modifications in the graft wall compliance due to the formation of a plaque inside the vessel [18-20]. The second one (equation (3)), strictly related to the AI, is associated to the transit time of pressure wave from the central cardiovascular system to the periphery[21, 22].

$$AI(\%) = P_2/P_1 *100 \qquad (2)$$

$$\Delta T_{DVP} = t_{P_2} - t_{P_1} \qquad (3)$$

In order to simulate the variation of the augmentation index, that represents an index of the vessel elasticity [23], a change in the P_2 / P_1 ratio was imposed. The P_2 amplitude was varied in order to modify the AI, starting from a value of about 0.45, compatible with a safe vessel, until 0.80, typical of a highly stiff one. To do so, the opening percentage of the solenoid valve corresponding to the diastolic peak (P_2) was varied. The plots reported in Fig. 4b demonstrate the ability of the sensor to identify the modification in the pulse wave profile related to the formation of plaques in the artificial vessel. Indeed, the voltage generated by the sensor, proportionally increases with the increase the pressure applied to the graft wall by the water flow due to the variation in the solenoid valve opening percentage , establishing that the sensor, is able to detect the deviation from the physiological value of the Augmentation Index. A similar approach was then used to induce a variation of the transit time of pressure wave (ΔT_{DVP}) by modifying the opening time of the outlet channel of the solenoid valve between P_1 and P_2.

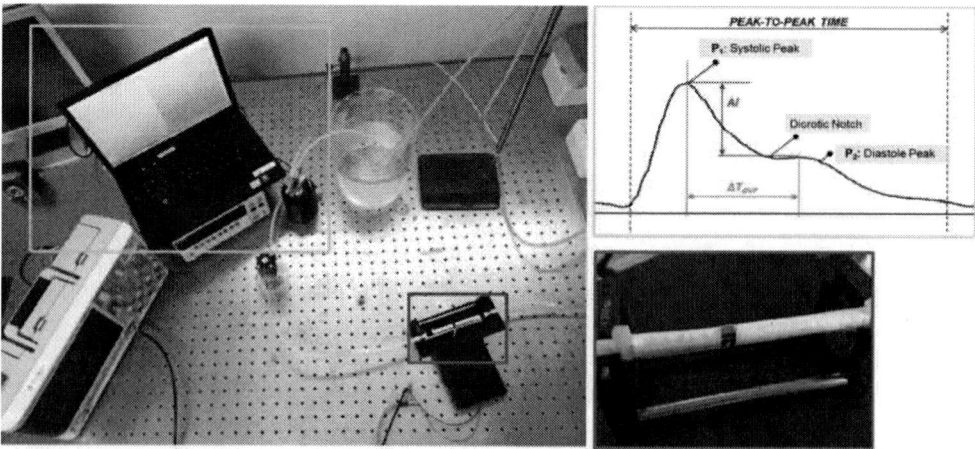

Fig. 3: Measurement set up. In the Yellow inset the pulse wave signal with its parameters, and reporting the peak-to-peak time, the Augmentation Index (AI) and the ΔT_{DVP}. In the red inset a particular of the sensorized graft system

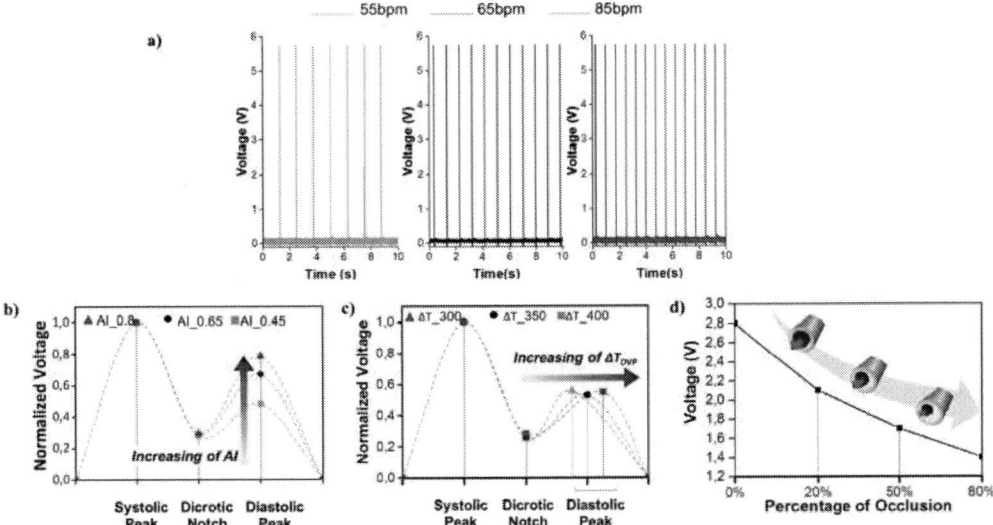

Fig. 5: In vitro test, for the evaluation of the device capability to continuously monitoring the graft status. a) Voltage response to a variation of the peak-to peak time. b) Voltage generated by the piezoelectric sensor due to the induced variation of the Augmentation Index. c) Voltage response of the sensor due to a variation of the time interval between the systolic and diastolic peak. d) Sensor output due to the different percentage of occlusion into the vascular graft.

Then, from the analysis reported in Fig. 4c, it is shown that the sensor can detect even small variations of the time interval between P_1 and P_2, following the imposed time step of about 50 ms of the solenoid valve.

Finally, as the re-occlusion of the graft is one of the most common failure causes for a vascular implant [24], different levels of occlusions were introduced in the flow system. The initial diameter of the vessel was reduced by the use of a surgical forceps, between 20% and 80%. The results reported in the Fig. 4d demonstrate that the sensor is able to recognize the flow patterns in the vessel related to the presence of the stenosis, congruently varying the generated output voltage with the reduction of the vessel diameter. In particular a flattening of the trends was observed properly differentiating the different degrees of induced occlusion.

4. Conclusion

The reported results demonstrate that the AlN-based piezoelectric sensor is able to provide important information about the hemodynamics behavior of a vascular graft when coupled with its external surface. The thin and flexible structure of the device, its really sensitive and predictable response together with its intrinsic biocompatibility make the system well suited for the proposed application. It offers an accurate measurement of the graft wall motion due to the variation of the blood flow condition inside it. The generated data could then be useful to alert the patient and the physician as a consequence of an early change in graft function, opening new perspective for a real-time vascular graft surveillance.

References

[1] S.J.A. Majerus, J. Dunning, J.A. Potkay, K.M. Bogie, Flexible, structured MWCNT/PDMS sensor for chronic vascular access monitoring, 2016 IEEE SENSORS, 2016, pp. 1-3.
[2] N. Singh, A.N. Sidawy, K.J. DeZee, R.F. Neville, C. Akbari, W. Henderson, Factors associated with early failure of infrainguinal lower extremity arterial bypass, Journal of Vascular Surgery 47(3) (2008) 556-561.
[3] J.H. Cheong, S.S.Y. Ng, X. Liu, R. Xue, H.J. Lim, P.B. Khannur, K.L. Chan, A.A. Lee, K. Kang, L.S. Lim, C. He, P. Singh, W. Park, M. Je, An Inductively Powered Implantable Blood Flow Sensor Microsystem for Vascular Grafts, IEEE Transactions on Biomedical Engineering 59(9) (2012) 2466-2475.
[4] R.F. Neville, S.K. Gupta, D.J. Kuraguntla, Initial in vitro and in vivo evaluation of a self-monitoring prosthetic bypass graft, Journal of Vascular Surgery 65(6) (2017) 1793-1801.
[5] M. Je, J.H. Cheong, C.K. Ho, S.S.Y. Ng, R. Xue, H. Cha, X. Liu, W. Park, L.S. Lim, C. He, K. Cheng, X. Zou, Z. Chen, L. Yao, S.J. Cheng, P. Li, L. Liu, M. Cheng, Z. Duan, R. Rajkumar, Y. Zheng, W.L. Goh, Y. Guo, G. Dawe, Wireless sensor microsystems for emerging biomedical applications (Invited), 2015 IEEE International Symposium on

Radio-Frequency Integration Technology (RFIT), 2015, pp. 139-141.

[6] D.Y. Park, D.J. Joe, D.H. Kim, H. Park, J.H. Han, C.K. Jeong, H. Park, J.G. Park, B. Joung, K.J. Lee, Self-Powered Real-Time Arterial Pulse Monitoring Using Ultrathin Epidermal Piezoelectric Sensors, Advanced Materials 29(37) (2017) 1702308.

[7] T. Sekine, R. Sugano, T. Tashiro, J. Sato, Y. Takeda, H. Matsui, D. Kumaki, F. Domingues Dos Santos, A. Miyabo, S. Tokito, Fully Printed Wearable Vital Sensor for Human Pulse Rate Monitoring using Ferroelectric Polymer, Scientific Reports 8(1) (2018) 4442.

[8] J. Nathan, K. Lynette, M. Alan, Flexible-CMOS and biocompatible piezoelectric AlN material for MEMS applications, Smart Materials and Structures 22(11) (2013) 115033.

[9] C. Giordano, I. Ingrosso, M. Todaro, G. Maruccio, S. De Guido, R. Cingolani, A. Passaseo, M. De Vittorio, AlN on polysilicon piezoelectric cantilevers for sensors/actuators, Microelectronic Engineering 86(4-6) (2009) 1204-1207.

[10] S. Trolier-McKinstry, P. Muralt, Thin film piezoelectrics for MEMS, Journal of Electroceramics 12(1-2) (2004) 7-17.

[11] L. Natta, V.M. Mastronardi, F. Guido, L. Algieri, S. Puce, F. Pisano, F. Rizzi, R. Pulli, A. Qualtieri, M. De Vittorio, Soft and flexible piezoelectric smart patch for vascular graft monitoring based on Aluminum Nitride thin film, Scientific Reports 9(1) (2019) 8392.

[12] T.Q. Trung, N.-E. Lee, Flexible and Stretchable Physical Sensor Integrated Platforms for Wearable Human-Activity Monitoringand Personal Healthcare, Advanced Materials 28(22) (2016) 4338-4372.

[13] R.Y. Kannan, H.J. Salacinski, P.E. Butler, G. Hamilton, A.M. Seifalian, Current status of prosthetic bypass grafts: A review, Journal of Biomedical Materials Research Part B: Applied Biomaterials 74B(1) (2005) 570-581.

[14] C.S. Jørgensen, W.P. Paaske, Physical and mechanical properties of ePTFE stretch vascular grafts determined by time-resolved scanning acoustic microscopy, European Journal of Vascular and Endovascular Surgery 15(5) (1998) 416-422.

[15] J.N. Reddy, Theory and analysis of elastic plates and shells, CRC press2006.

[16] P. Lewis, J. Psaila, W. Davies, K. McCarty, J. Woodcock, Measurement of volume flow in the human common femoral artery using a duplex ultrasound system, Ultrasound in Medicine and Biology 12(10) (1986) 777-784.

[17] H.-T. Wu, H.-K. Wu, C.-L. Wang, Y.-L. Yang, W.-H. Wu, T.-H. Tsai, H.-H. Chang, Modeling the Pulse Signal by Wave-Shape Function and Analyzing by Synchrosqueezing Transform, PLoS ONE 11(6) (2016) e0157135.

[18] I.B. Wilkinson, H. MacCallum, L. Flint, J.R. Cockcroft, D.E. Newby, D.J. Webb, The influence of heart rate on augmentation index and central arterial pressure in humans, The Journal of Physiology 525(1) (2004) 263-270.

[19] M. Koobatian, C. Koenigsknecht, S. Row, S. Andreadis, D. Swartz, Surgical Technique for the Implantation of Tissue Engineered Vascular Grafts and Subsequent In Vivo Monitoring, Journal of visualized experiments : JoVE (98) (2015) e52354-e52354.

[20] C. Dagdeviren, B.D. Yang, Y. Su, P.L. Tran, P. Joe, E. Anderson, J. Xia, V. Doraiswamy, B. Dehdashti, X. Feng, Conformal piezoelectric energy harvesting and storage from motions of the heart, lung, and diaphragm, Proceedings of the National Academy of Sciences 111(5) (2014) 1927-1932.

[21] M. Elgendi, On the Analysis of Fingertip Photoplethysmogram Signals, Current Cardiology Reviews 8(1) (2012) 14-25.

[22] C. Dagdeviren, Y. Su, P. Joe, R. Yona, Y. Liu, Y.-S. Kim, Y. Huang, A.R. Damadoran, J. Xia, L.W. Martin, Conformable amplified lead zirconate titanate sensors with enhanced piezoelectric response for cutaneous pressure monitoring, Nature communications 5 (2014) 4496.

[23] W.W. Nichols, Clinical measurement of arterial stiffness obtained from noninvasive pressure waveforms, American Journal of Hypertension 18(S1) (2005) 3S-10S.

[24] S. Pashneh-Tala, S. MacNeil, F. Claeyssens, The Tissue-Engineered Vascular Graft—Past, Present, and Future, Tissue Engineering Part B: Reviews 22(1) (2015) 68-100.

Active cooling using fluid channels in a silicon-ceramic composite substrate

M. Fischer, T. Werthes, C. Kleinholz and J. Müller

Institute of Micro- and Nanotechnologies MacroNano®,

Technische Universität Ilmenau, Gustav-Kirchhoff-Str. 7, Ilmenau, Germany

Phone: + 49 3677 693413, michael.fischer@tu-ilmenau.de

Abstract

The silicon-ceramic composite substrate (SiCer) allows the combination of MEMS and LTCC technologies in one wafer substrate. The expansion coefficient of the specially developed LTCC is adapted to that of the silicon over a wide temperature range, allowing cofiring of silicon and LTCC for achieving a strong bond interface. The combined very different substrate properties greatly increase the technological potential for the design of new sensor applications. In addition to the implementation of micromechanical components in the silicon layer with simultaneous wiring and electronic integration in the LTCC layer, it is also possible to realize fluid channels in both layers and connect them to each other. The bond connection between silicon and LTCC is considered hermetic above a certain bond frame width. In the first part of this paper, the technology for the generation of channels in the SiCer substrate is presented and a measurement method for determining the gas tightness or the leak rate of the bond connection is shown. In the second part, the technology for the setup of an active fluidic cooling system (water based) in a SiCer substrate is presented. The test module is fluidically and electrically connected in a special test bench. The silicon area in the middle of the module, exposed by DRIE (deep reactive ion etching), contains the cooling channels as well as a mounted thermal test die simulating a lossy component and to measure the temperature on the silicon die.

Key words: SiCer, silicon-ceramic composite substrate, fluid channels, active cooling

Introduction

The requirements on a carrier system for the cooling of e.g. electronic and sensor components are continuously increasing over the last few years due to steadily increasing power density. Therefore, the selection of the circuit carrier material has a huge impact on the heat management. Due to its higher thermal conductivity, a glass-ceramic circuit carrier shows a significantly better performance than an organic circuit carrier. If the glass-ceramic substrate is combined with silicon, a further significant improvement in heat dissipation will be expected, since the thermal conductivity of silicon is at least 100 to 150 times higher than that of the glass-ceramic. A SiCer substrate combines both substrate classes in an optimal way and thus opens up new possibilities in the design and manufacture of active cooling solutions.

The objective of the present work is the investigation and evaluation of an integrated fluidic cooling concept in SiCer with an emphasis on miniaturization. For this purpose, the sealing bond frame widths (Fig. 1) need to be minimized. Design rules of the bond frame width are based on the results of leak test measurements. The fluid actively cools a separated silicon area with a mounted heater chip, demonstrating the use of the composite substrate for higher power dissipation (Fig. 1).

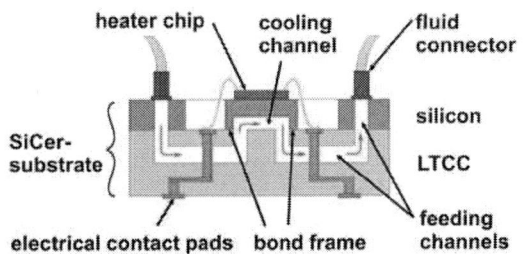

Figure 1: Schematic structure for the demonstration of SiCer as an integrated fluid system.

Manufacturing concept of SiCer with channels

Basically, the composite substrate is built up from a pre-processed LTCC multilayer and a thin, also pre-processed silicon wafer [1], but the number and arrangement of the individual LTCC and silicon layers can also vary. The expansion coefficient as well as the powder morphology and material composition of the LTCC (BCT, bondable ceramic tape) are adapted to silicon. A typical bond between silicon and LTCC is achieved by deposition of a wetting layer and pressure-assisted sintering (3.27 kPa, 900 °C). In order to join the pre-processed BCT multilayer in the unsintered state with the silicon layer, isostatic lamination (21 MPa,

82°C) is used. If cavities are involved (e.g. chambers and channels), they need to be supported during lamination to prevent them from collapsing. A common method is to fill the cavities with carbon inlays, which can be used both in the LTCC layer and in the silicon layer. After pre-processing of both substrate layers (see Fig. 2), they are stacked and laminated with high alignment accuracy using an optical stacking system (Fraunhofer IOF). During the subsequent sintering process, the carbon is completely removed in an extended burn-out step (combustion). The LTCC in this quasi-monolithic SiCer substrate shrinks only in z direction ("zero-shrinkage") due to the constraint sintering condition. The substrate is then post-processed in wafer format. Silicon cut out areas and fluidic connection holes to the LTCC channels are finally generated by means of Deep Reactive Ion Etching (DRIE) (see last step in Fig. 2). The etching process automatically stops at the interface to the LTCC.

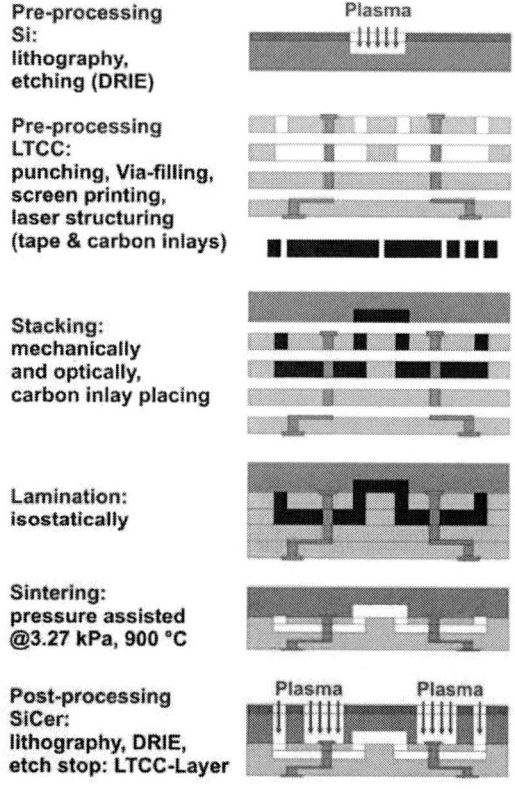

Figure 2: Process sequence for fabricating fluid channels in a SiCer substrate.

Gas tightness of the SiCer bond

The gas tightness of the SiCer bond as a function of the bond area (bond frame width) is an essential criterion in the design of fluid channels. The smaller the bond area can be designed, the larger the area of the cooling channel or the number of cooling channels per area can be implemented.

The tightness of the SiCer bonding was determined by means of a helium leak detector (vacuum method, Smart Test, Pfeifer Vacuum GmbH). With the vacuum method, the test object is evacuated by a vacuum pump [2]. If there is a leak channel in the object to be tested (SiCer bond frame), a constant gas stream flows from the outer side with the pressure p1 in the direction of the evacuated side with the pressure p2, where p2 < p1 applies. A mass spectrometer serves as a leak detector to determine the test gas flow. Helium is used as the test gas because it can be detected with high sensitivity due to its low molar mass. SiCer test chips were set up for the measurement, which were connected to the vacuum flange of the leak detector using a clamping device (see Fig. 3). In the clamping device, the SiCer test chip is held between two O-rings. The SiCer test chip has a ring-shaped bond frame and is assembled in different widths (100, 250, 500 and 1000μm).

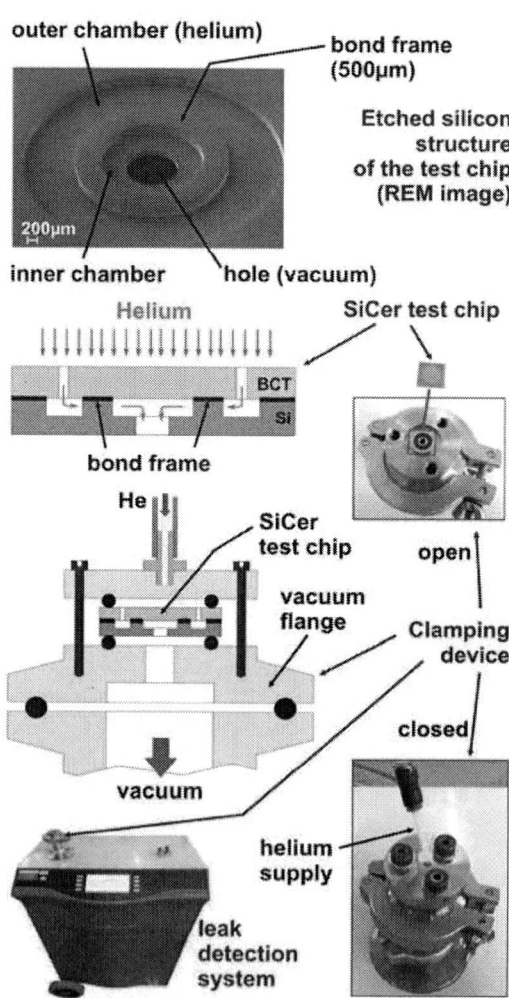

Figure 3: Arrangement for determining the leakage rate (tightness) of a SiCer bond frame.

212

The technological realization of these test specimens was carried out with the process sequence described in the previous section. The etched silicon structure with the circular chamber shape is shown in the upper part of Fig. 3.

Measurement results tightness

Figure 4 shows the number of tested dense (blue) and leaky (orange) SiCer test chips as a function of the bond frame width. (with the bond frame width of 100 µm only 9 samples were available for the measurement due to manufacturing defects) During the measurements, all test objects with a limit leak rate $< 1 \cdot 10^{-8}$ (mbar·l)/s were considered leaky [3]. There is a clear dependence of hermeticity on the bond frame width. The probability of a leak that runs through the whole bond frame interface decreases with increasing bond frame width. While the yield of a 100 µm bond frame width is 66.67 %, a bond frame width of 250 µm leads to a yield of 91.67 %. The test specimens with bond frame widths of 500 µm and 1000 µm are 100 % tight.

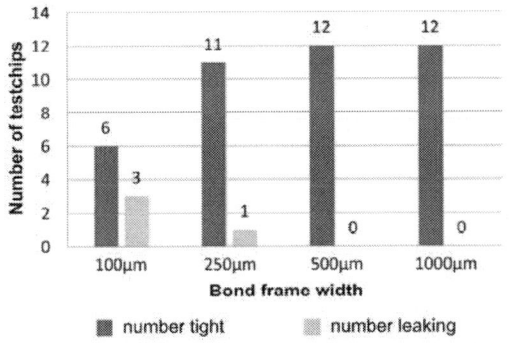

Figure 4: Number of tight SiCer test chips as a function of their bond frame width.

SiCer system with active cooling

Channels in LTCC for use in active cooling of e.g. power electronics have been investigated and described thoroughly [4], [5]. In order to increase the effect of heat transfer, thermal vias extending into the LTCC channel were introduced [6]. However, due to the relatively low thermal conductivity of classical LTCC materials (1-4 W/m K) and the limited thermal via density, the heat transfer is always limited. An alternative material with higher thermal conductivity is silicon (150 W/m K). Channels can also be created in silicon and used for more effective active cooling [7]. With SiCer it is possible to combine both material classes. This opens up new possibilities in the design and construction of microfluidic cooling systems (e.g. prevent lateral heat flow).

A substrate structure with a silicon thickness of 380 µm and a LTCC thickness (sintered) of 500 µm (equivalent to 4 layers with a sintered thickness of 125 µm) was selected for the realization

of an active cooling system in SiCer. The chip land pad size (free etched silicon area) is 5x5 mm² and the silicon cooling channel width is 300 µm (Fig. 5) and height is 250 µm respectively. The minimum bond frame width between channel segments is 250 µm. On the one hand, this choice was made to achieve a sufficiently long cooling channel (the wider the bond frame, the shorter the channel at constant area). On the other hand, it is to be expected that a 100 % liquid tightness will be achieved with this bond frame width. The resulting channel length in the silicon is 27.4 mm. The feed channel geometries in the LTCC layer result on the one hand from the foil thickness of the LTCC and on the other hand from the necessary module size for the fluidic test setup (channel height sintered: approx. 125 µm, channel width: 1 mm, channel length for inlet and outlet: 2 x 23.0 mm).

The technological implementation of the SiCer cooler module was again carried out with the process sequence shown in Figure 2. After separating the individual modules by wafer sawing, a heater chip (PST1, Delphi) was mounted on the etched silicon area using a thermally conducting silver adhesive (DM6030Hk, Diemat, 60 W/m K).

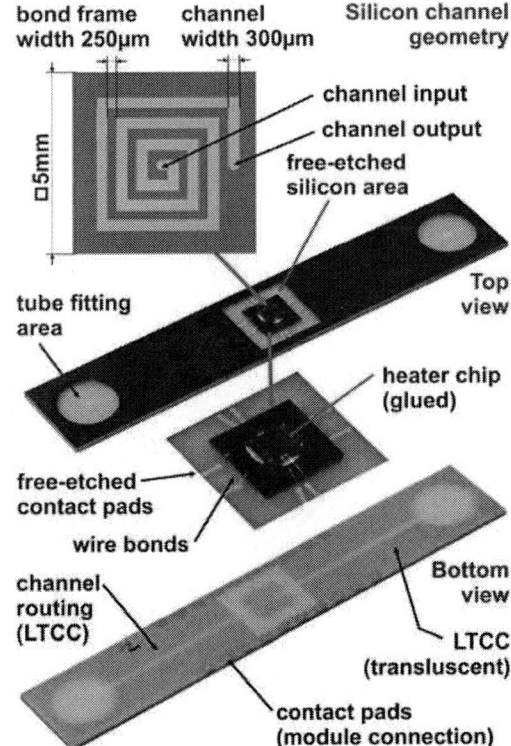

Figure 5: Module structure details of the SiCer cooling system.

During the silicon etching process, also the gold bond pads at the interface between the LTCC and silicon are exposed. They are used to electrically contact the heater chip to the LTCC by wire bonding.

Fluidic and electrical connections and wiring of the SiCer module

The electrical and fluidic connection to the SiCer cooler module is made by a flexible fluidic test bench. The electrical contact to the heater chip is realized via a block of spring contact pins from the bottom side of the module. The fluidic connection is made by means of adjustable and specially manufactured connection blocks made of PEEK (polyetheretherketone) and matching, surface-sealing screw connections of HPLC technology (high pressure liquid cromatography). At the same time, the module is mechanically clamped by the screw connection (see Fig. 6).

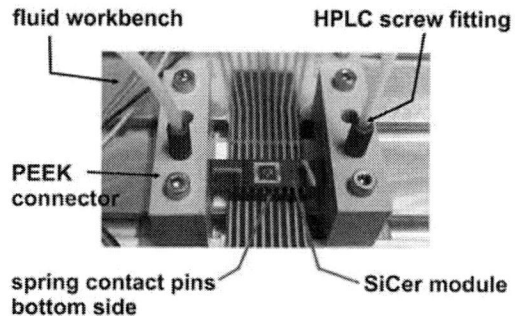

Figure 6: Electric and fluidic SiCer module connection.

A piston pump (Teledyne Isco, 1000D) was used to investigate the thermo-fluidic behaviour of the SiCer cooler module (see Fig. 7). With this set-up it is possible to realize a precise constant volume flow and thus directly set a defined flow rate.

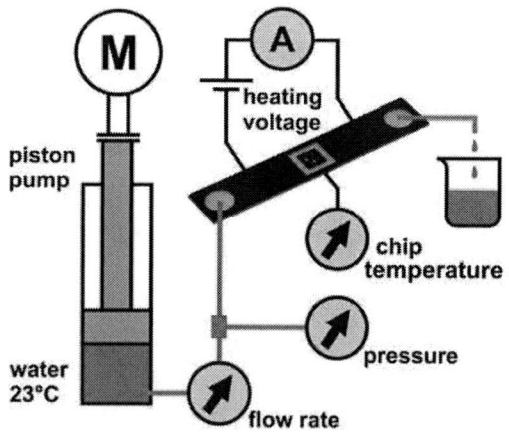

Figure 7: Electric and fluidic wiring of the SiCer module.

Furthermore, a pressure gauge was inserted between the pump and the SiCer cooler to determine the pressure drop across the module. The connections between all components were realized by Teflon tubes and tube fittings. The thermal test chip PST1

(Delphi) has a thin-film resistor as heat source and a diode network for temperature measurement. Since the forward voltage of a diode decreases with increasing temperature, the temperature can be detected at a constant diode current.

Investigation of thermo-fluidic behavior

All measurements were made at room temperature with deionized water as cooling medium (23°C).
First, the cooling effect was investigated as a function of the flow rate. The chip temperature of the SiCer module was set to 80°C (U=4.2 V, I=0.15A at the heating resistor). The module was then driven at flow rates from 0.25 ml/min to 2 ml/min in 0.25 ml/min steps. The respective flow rate was held constant for 5 min before the next higher flow rate was set. The curve in Fig. 8 shows a clear cooling effect with increasing flow rate. With a flow rate of 2 ml/min, a temperature reduction of more than 50 °C can be achieved on the chip of the SiCer module.

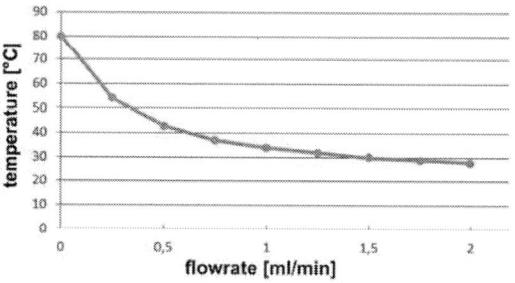

Figure 8: Cooling behaviour of the test setup as a function of the flow rate.

Another important parameter is the pressure rise as a function of the flow rate in the system. The characteristic curve in Fig. 9 shows an approximately linear pressure increase with increasing flow rate in the described operating mode.

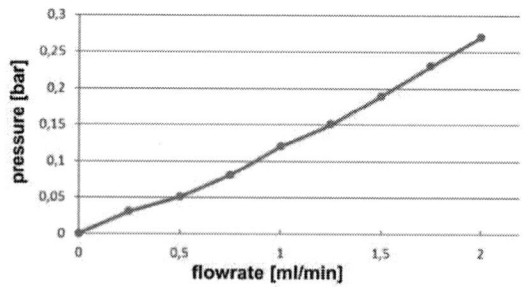

Figure 9: Pressure generation in the test setup as a function of the flow rate.

Even though the system was not tested up to burst pressure, the proven compressive strength of the SiCer-module can be considered as very good.

Figure 10 shows the cooling gradient as a function of the flow rate. The higher the flow rate, the faster is the temperature drop on the heater chip. It can be established that after about 175 sec no further reduction of the temperature in the system can be achieved at all investigated flow rates.

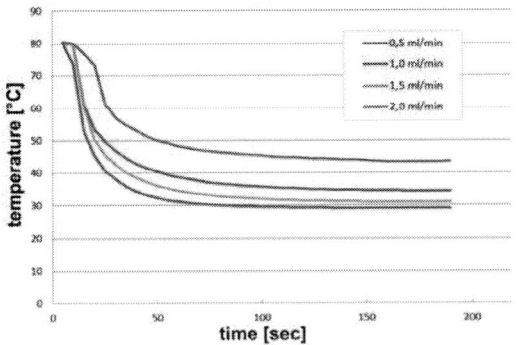

Figure 10: Cooling time and final temperatures at different flow rates.

Conclusion

The thermo-fluidic behavior of the silicon-ceramic composite substrate (SiCer) as well as the technology concept for the fabrication of cavities (channels) at the bond interface was investigated using an active cooling system. The gas tightness of the bond connection of the SiCer substrate was determined as a function of the minimum bond frame width (250 μm) and a cooling channel structure was dimensioned based on this. For the manufactured SiCer cooling module an electro-fluidic test bench was developed and built which was used to characterize the thermo-fluidic behavior. For example, the cooling system allows a chip temperature of 80 °C to be reduced by more than 50 °C within 3 minutes with a flow rate of 2 ml/min.

Acknowledgement

The authors gratefully acknowledge the financial support by the DFG, German Research Foundation, FOR 1522 MUSIK (MU 3171/1-1).

References

[1] M. Fischer et al, "Radio Frequency Microelectromechanical System-Platform Based on Silicon-Ceramic Composite Substrates", Journal of Microelectronics and Electronic Packaging, Vol. 12, No. 1, pp. 37-42, March, 2015.

[2] W. Große-Bley, "Lecksuchtechniken", in K. Jousten, Wutz Handbuch Vakuumtechnik, 8. Auflage, Wiesbaden, Springer Wiesbaden, 2004.

[3] MIL-STD-883, 2014, Test Method Standard Microcircuits – Seal, Rev. J, Method 1014.14.

[4] D-w. Hu et al, "Investigation of Cooling Performance of Micro-Channel Structure Embedded in LTCC Substrate for 3D Micro-System", IEEE 11th International Conference on Solid-State and Integrated Circuit Technology, 2012.

[5] A. Pietrikova et al, "Simulation of cooling efficiency via miniaturised channels in multilayer LTCC for power electronics", Journal of Electrical Engineering, Vol. 68, 2017, S. 132-137.

[6] L-Y. Zhang; Y-F. Zhang, "Simulation on heat transfer of microchannels and thermal vias for high power electronic packages", 15th International Conference on Electronic Packaging Technology, 2014, S. 508-510.

[7] G. Xia et al, "Experimental and numerical study of fluid flow and heat transfer characteristics in microchannel heat sink with complex structure", Energy Conversion and Management, 105, 2015, S. 848-857.

Monolithic Ceramic IR-Emitter in Multilayer Technology

Adrian Goldberg*, Birgit Manhica*, Steffen Ziesche*

Fraunhofer IKTS*, Winterbergstraße 28, 01277, Dresden

+49 351 25537783, Adrian.goldberg@ikt.fraunhofer.de

Abstract

This paper presents the research results arose from the cooperation project for the development of a monolithic Infrared (IR) emitter in ceramic multilayer technology for requirements in a gas analyzer as IR radiation source. Currently available IR emitters on the market, such as micro lamps or silicon based IR-Emitter have a limited range of wavelengths (e.g. when using glass or sapphire windows), low long-term stability, often early failures in open operation, low radiated power (due to poor emission levels or low radiator temperatures < 500 °C) and poor modulation ability. Therefore, the aim of the cooperation is to develop an economically viable solution of a high-temperature and long-term stable ceramic IR-emitter as infrared radiation source at high temperatures (≥650 °C) and pulsation frequenzies up to 5 Hz.

Key words: LTCC technology; IR-Emitter; multilayer ceramic; high temperature, printed LTCC, sacrificial layer

1. Introduction

Infrared (IR) analysis is a widely used method for the characterization of gases, liquids and solids. It is based on the analysis of characteristic absorption bands (vibrational states) of the molecules, whereby the middle and far IR (MIR, FIR) regions are characterized by particularly characteristic spectral signatures and are therefore called fingerprint regions. In addition to a suitable detection unit and the measuring section (sample chamber), an IR radiation source (emitter) is always required as an integral part of the analysis technology. Figure 1 shows the typical structure of a simple gas analyzer. The selectivity to a particular substance is here e.g. achieved by an integrated into the detector narrow-band filter, however, the source emits broadband. The previously available IR emitters on the market, such as micro lamps and silicon based heaters have a limited range of wavelengths (e.g. when using glass or sapphire windows), low long-term stability, often early failures in open operation, low radiated power (due to poor emission levels or low radiator temperatures < 500 °C) and poor modulation ability.

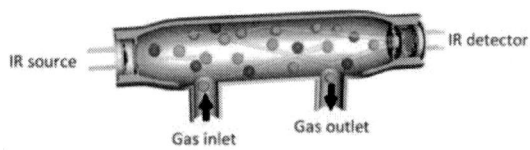

Figure 1: Basic structure of a typical NDIR gas analyzer with IR-emitter and detector

Therefore, the goal in the research project is to realize a monolithic ceramic solution of an IR emitter with ceramic embedded heater structures using a technological approach which allows on the one hand for a very small thermal mass of the emitter and therefore good modulation ability and on the other hand, can operate at higher temperatures (> 650 ° C).

2. Technical innovation and goals of development

The basic idea of the desired solution is to develop a new generation of high-temperature-stable, monolithic, miniaturized and cost-reduced ceramic IR emitters in combination with the already established mass production suitable LTCC multilayer technology and new layer deposition processes (screen-printable LTCC paste, sacrificial materials and aerosol printing technology). Figure 2 shows the concept of a monolithic ceramic IR-emitter with thin ceramic membrane (with integrated heater) embedded in a ceramic package with cavities and infrared window based on Si-ARC technology.

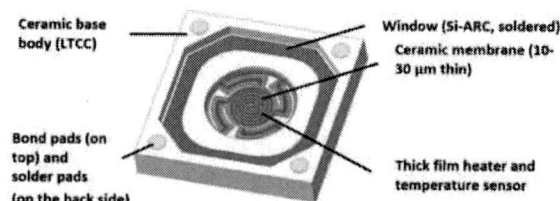

Figure 2: Concept of a monolithic ceramic IR-Emitter

Such a ceramic IR-emitter should fulfill the following requirements:

- Spotlight area 1 to 5 millimeters (scalable, technology potential < 20 mm²)
- Thermal-source up to 650 °C, pulsatility: 5 Hz
- Emissivity: ≥ 0,8, spectral range: 2…14 µm

- Electrical power consumption: ≤ 1 W, voltage lower 10 V
- Hermetical sealed and long-term stable

3. State of the art LTCC technology and new layer deposition technics

The fabrication of multilayer components follows the process chain shown in Figure 2. After cutting the unsintered individual foils the electrical connections from layer to layer (vias) and possibly required openings for cavities were punched. This is followed by metallization of the vias by stencil printing process and screen printing of the individual foils with the required functional structures (printed conductors, resistors). Subsequent process steps are the stacking of the individual tapes and their compression (lamination) at elevated temperature (70-90 °C). After the cofiring process (the firing process of all materials contained in the multilayer), the monolithic ceramic component have the required functional properties.

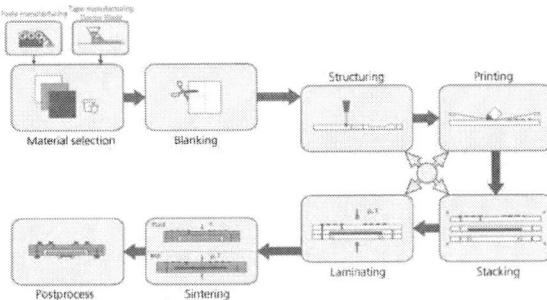

Figure 3: multilayer process chain

In addition to electrical functions, non-electrical functions like channels and chambers for the passage or mixing of fluidic media [1], [2] or membranes or mechanical springs [3], [4], [5], [6] for the pressure and force measurement can be formed.

Common to all such applications is the formation of the required non-electrical functional elements with the available dielectric base tapes. Since these are only available in foil thicknesses of about 50-250µm and is the limit for the manufacturing resolution in z-direction. The solution that goes beyond the prior art (Table 1) consists in the combined application of the sacrificial paste technique with the screen or stencil printing of the previously processed tapes form base materials (LTCC). Using this technology self-supporting microstructures can be produced. The achievable layer thicknesses of 10 - 20µm reduces the possible steps in the z-direction and allow higher structure resolution even in the lateral dimensions.

The previous multilayer technology allows the combination of differently structured single layers. With the described novel paste or printing technology additive structures can also be generated and positioned with high precision, which was

previously not possible. This new development is patented by Fraunhofer IKTS [7].

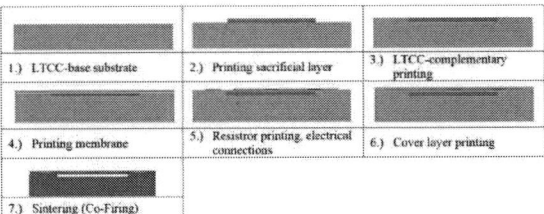

Figure 4: schmetic process chain for LTCC based IR-Emitter

4. Experiments/ Calculations/ Simulations

Figure 5 shows the the layout in cross section of the developed ceramic IR-Emitter. A target thickness of the membrane of 10 to 30 microns was achieved. In the membrane, a resistor as heater are to be embedded and connected with vias through the entire structure to the outside contact pads. Cavities with sacrificial paste technology in the range 250 µm thickness are built in order to allow the membrane to bend during the heating time and protect the ceramic membrane body against overheating. As a result, a high temperature operation with very low thermal mass und a pulsatility upt to 5 Hz can be achieved.

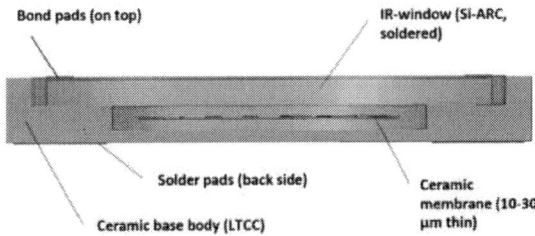

Figure 5: IR-Emitter layout cross-section

To build such a ceramic IR-emitter the individual technology steps has to be developed and optimized with the help of test structures. The following steps show the development of the individual technologies up to the complete sensor.

A. Development of a screen-printable sacrificial and LTCC paste

For a spatially resolved deposition of sacrificial pastes, different paste formulations with different binder variants were tested. The layer deposition of the membrane layers was also carried out with a LTCC paste. This ensures optimal compatibility. After investigation the composition of sacrificial and LTCC-paste for producing a double cavity could be found. Crucial for a residual-free sintering of the sacrificial paste is a channel access from the outside to the inner cavities. So oxygen easily reach the graphite for burnout during the sintering process. Figure xx shows a screen-printed membrane with a thickness of about 20 microns with integrated openings and embedded heater.

B.   Deposition of functional layers

The standard thick-film materials like ruthenium oxide, silver, gold, platinum and mixed compositions are screen-printed to a thickness of 10-15 µm. With a total thickness of a screen printed ceramic membrane as small as 20 microns, such thicknesses are unsuitable for implementation. In this development project, commercial screen printeable and aerosol printable metal paste were tested for its property to achieve layer thicknesses of about 5 µm. Figure 6 shows the working principle of an aerosol jet printer.

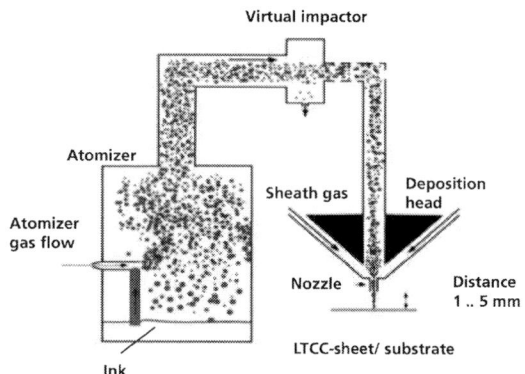

Figure 6: Aerosol jet printing principle

To determine the working princible of infrared heater on a thin printed LTCC membrane, different aerosol jet printable heater materials were deposited and the resistance and heating rate was measured. With aerosol jet gold, silver, platinum inks were printed on their suitability as a heater and examined. Figure 7 shows an example of a screen-printed LTCC membrane with an aerosol jet printed gold heater. In addition, very thin printable thick film pastes were tested. In this case, a co-fire platinum paste developed specifically for IKTS and a commercial ruthenium oxide paste were tested for their suitability. With the measured resistance of the different heater materials in comparison the thickness, the projekt partner simulate different heater and membrane geometries and calculate the necessary heater length for the best adaption to the electronics.

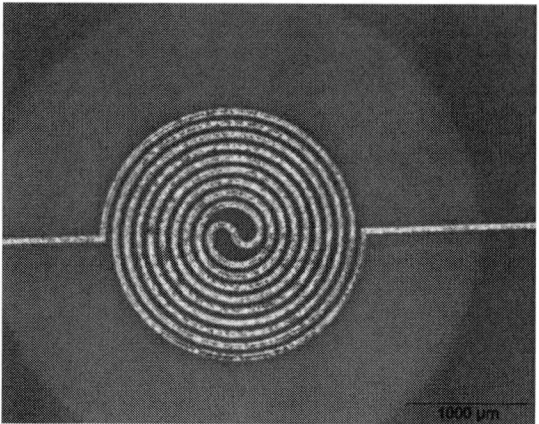

Figure 7: Screen printed LTCC-membrane (thickness ~20 µm) and aerosol jet printed gold heater (thickness: ~5 µm) on top

B.   Membrane structuring by laser

Figure 8 shows a membrane simulated with an aerosol-printed heater made of platinum. To reduce the thermal mass and the best possible thermal decoupling to the body, the unneeded parts of the membrane were reduced to the bare minimum and constructed a suspended only four retaining webs membrane.

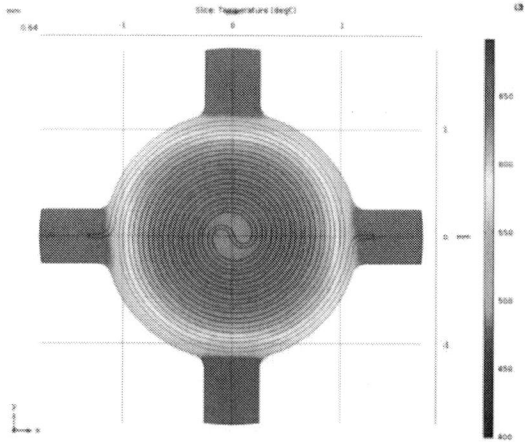

Figure 8: Simulated LTCC membrane with only four webs for best thermal and mechanical decoupling

For the testing of different layouts of such membrane structures, a manufacturing process has been developed. In the process, a cavity is filled with a sacrificial paste, so that a smooth plane is formed on which an LTCC layer is screen-printed. After drying the layers, the deposition of the heater can be done with the aid of screnn printing or by aerosol jet. The layers should be as thin as possible. This is followed by covering the heater layers with another layer of LTCC and the common sintering process. During the sintering process of the ceramic, the printed sacrificial layer burns out and leaves behind a self-supporting, functionalized membrane which, for example, can form an IR-emitter (Figure 7). This is followed by the structuring of the membrane by means of a laser with the respective mechanical, thermal decoupling. Figure 9 shows a structured membrane with inetgrated silver heater.

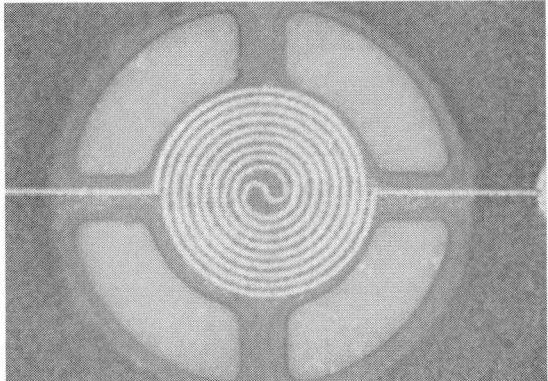

Figure 9: Structured LTCC membrane by laser with integrated silver heater

C. Layout and manufacturing process of the monolithic ceramic IR-Emitter in Multilayer Technology

The thickness of the radiator is about 1.3 mm with a cavity depth of 250 microns below and above the membrane. To accommodate the SI-ARC window is a 500 micron deep cavity with a solderable ring for fixing the window in the ceramic body. As a result, a hermetic encapsulation is achieved. To control the process and as a safety aspect, an additional temperature sensor is provided inside the ceramic housing. It is screen-printed with a platinum paste as a PT-100 sensor and integrated into the setup. Figure xx shows the integrated temperature sensor in the overall IR emitter concept.

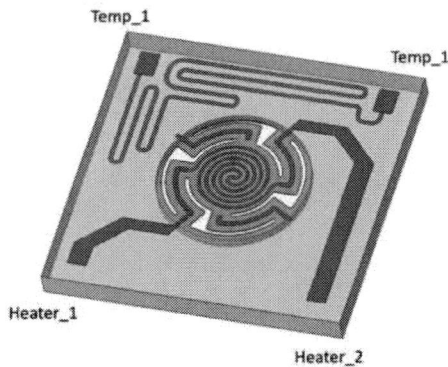

Figure 10: IR-emitter layout with integrated temperature sensor and heater element

The manufactoring sequence of the individual sensor layers follows the scheme as you can see in figure 3 and starts like a typical multilayer ceramic panel process. Thereby, 49 ceramic IR-emitter are produced simultaneously in a 4 inch format. In contrast to the standard multilayer technology, the cavities are built up in the additive process using the screen-printable LTCC paste and sacrificial materials. The principle sensor design of differential sensor you can see in the figure 2

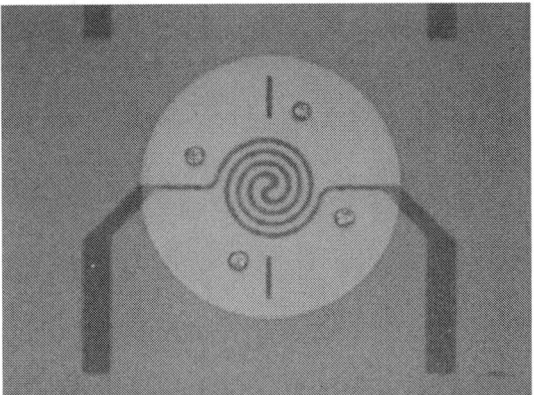

Figure 11: Printed LTCC-membrane with heater element and with sacrificial layer filled cavity

With the new developed manufacturing process, the channel are structured by laser and the cavities then filled with sacrificial material by stencil printing. The filled cavity is then covered with another screen printed LTCC layer for membrane formation with a thickness of about 25 micrometer (in the drying state). With screen printing the termination (Ag, AgPd) and the heater elements (Pt, RuO2) were formed. The heater elements based on silver and gold were printed with aerosol jet and is shown in Figure 11. The thickness of the aerosol-printed heaters is approximately 5 μm. The thickness of the screen printed heater elements is approximately 8 μm.

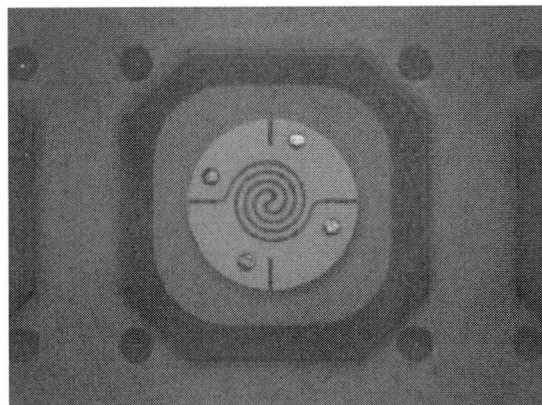

Figure 12: Isostatatic lamination of all layers

This is followed by stacking and isostatic lamination of layers with soldering pads for Si-ARC window and layer with cavity and the contacts already printed on the top side and the vias up to the screen-printed terminations on layer 2 at 70 ° C. and 200 bar. Figure 12 shows a ready isostatic-laminated structure.

Figure 13: Sintered LTCC-IR-emitter in panel process

After lamination of the individual layers follows the sintering process with an adapted profile for burnout the sacrificial materials and sintering of the LTCC. The burnout of the organics in the tape takes place at 375 ° C for 60 minutes. In the heating phase up to the peak temperature the combustion of the sacrificial materials is carried out at about 700 ° C with the followed sintering of the LTCC. The full sintering is completed at 850 ° C with a 15 minute hold time. A sintered IR-emitter is shown in Figure 13. The shrinkage rate is about 13 percent. After the sintering process the different possible membrane geometries were cutted by laser (UV-Laser with 20 Watt power output) and the last process is a singularization process by means of a wafer saw out of the LTCC panel. In Figure 14 is a ready lasered membrane geomtrie shown.

Figure 14: laser structured membrane with only four webs

4.   Results and Discussion

After the sintering process and sawing of the ceramic IR-emitter, all samples were measured (resitance of heaters and temperature sensors) and the functionality tested. In order to be able to evaluate the production parameters a 3D scan of the geometries and an optical inspection was done. Initial measurements of the temperature sensors gave an average resistance of 35 ohms and about 30-50 ohms for a heater made of screen printed platinum. An example of a measured surface of a manufactured ceramic IR emitter is shown in Figure 15. Two levels of cavities are visible, with the first level representing the plane for the Si-ARC window and the second level the heater membrane. In principle, very smooth ceramic membranes with integrated heaters have been achieved. The maximum warping of the membranes was 8 µm

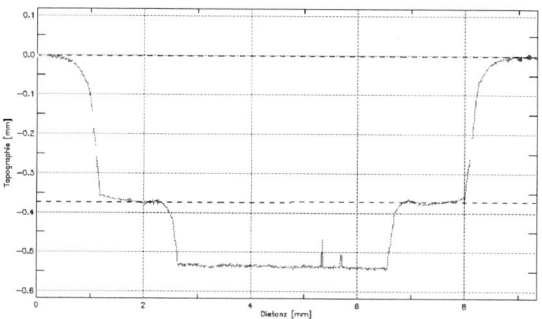

Figure 15: Measured surface profile of a manufactured ceramic IR-emitter

Based on the test structures as shown in Figure 7, temperature cycle tests were carried out to see if the different materials are suitable in principle. With the help of an infrared camera, the heating rates were determined as a function of the heating power. Figure 16 shows the results in comparison. In principle, measured temperatures are only to be considered as comparative values, since an exact emission coefficient was not known. However, the measurements show that all materials up to 550 ° C seem quite sustainable. An exact determination of the properties such as emission coefficient, wavelength spectrum and pulsation capability will be carried out in the next few weeks by the project partner with the first available ceramic IR emitter.

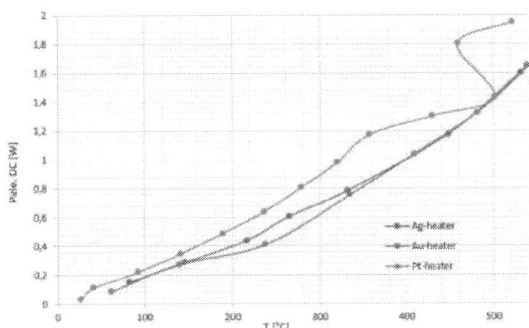

Figure 16: Comparison of heater materials on membrane test- structures up to 550 ° C

[7]   Patent, WO2017/021291 A1, "Methode for production a component from ceramic materials, U. Partsch,   A   Goldberg,   29.07.2016

Acknowledgements

The work carried out here was supported by the Saxon development bank (SAB).

References

[1]   Walter   Smetana,   Bruno   Balluch, GüntherStangl,   Sigrid   Lüftl,   Sabine   Seidler: "Processing procedures for the realization of fine structured channel arrays and bridging elements by LTCC-Technology", Microelectronics Reliability 49 (2009) 592–599

[2]   Nuria Ibanez-Garcia, Cynthia S. Martinez-Cisneros, Francisco Valdes, Julian Alonso: "Green-tape ceramics. New technological approach for integrating   electronics   and   fluidics   in microsystems", Trends in Analytical Chemistry, Vol. 27, No. 1, 2008.

[3]   Lenz, C.; Ziesche, S.; Partsch, U.; Neubert, H.: "Low Temperature Cofired Ceramics (LTCC)-Based   Miniaturized   Load   Cells",   Proc.   35th International   Spring   Seminar   on   Electronics Technology, ISSE, Bad Aussee (2012)

[4]   Karol Malecha, Thomas Maeder, Caroline Jacq: "Fabrication of membranes and microchannels in   low-temperature   co-fired   ceramic   (LTCC) substrate using novel water-based sacrificial carbon pastes", Journal of the European Ceramic Society 32 (2012) 3277–3286.

[5]   Partsch, U.; Gebhardt, S.; Arndt, D.; Georgi, H.; Neubert, H.; Fleischer, D. and Gruchow, M.:   „LTCC-Based   Sensors   for   Mechanical Quantities", CICMT – Ceramic Interconnect and Ceramic Microsystems Technology, Denver 2009, Proceeding

[6]   Uwe Partsch, Christian Lenz, Steffen Ziesche, Carolin Lohrberg, Holger Neubert, Thomas Maeder: „LTCC-Based   Sensors   for   Mechanical Quantities",   MIDEM2012,   48th   International Conferenceon   Microelectronics,   Devices   and Materialswith   the   Workshop   onCeramic Microsystems, September 19th - 21th, 2012, Otočec, Slovenia, Proceedings.

Electrical Modeling and Analysis of Through-Silicon-Via Crosstalk Based on Scalable Physical Lumped Circuit Model for 3D Packaging

Xiangyang Shi, Rong Zeng, Liming Lv

Institute of Electronic Engineering, China Academy of Engineering Physics. Mianyang, China

shixy08@foxmail.com

Abstract

In this paper, we investigate the transportation mode of TSV structure for subsequent modeling G-S TSVs. Next, a scalable physical lumped circuit model of ground-signal TSVs is established. The scalable G-S TSVs model is developed with closed-form analytic RLGC equations derived from physical structure, and can be changed with structure parameters in design process accordingly. The validity of this model is verified with full-wave FEM-based simulator up to 20 GHz. Based on this accurate physical lumped circuit model, we investigate crosstalk of TSV array in both frequency domain and time domain. The method of modeling and analysis of TSV array crosstalk proposed in this paper can effectively reduce simulation time and computing resource over finite-element-based 3-D full-wave simulation.

Key words: Through-Silicon-Via, physical lumped circuit model, crosstalk, 3D packaging

Introduction

3D packaging using through-silicon-via (TSV) has become popular owing to increased integration density, reduced transmission delay, and high bandwidth [1]. However, one of the most notable problems of TSV is crosstalk noise, which arises due to unwanted jam among TSVs. This crosstalk will increase the noise margin undermining signal reliability and increasing the bit error rate in signal transmission. Studies show that the crosstalk problem is not negligible in TSVs because the crosstalk coupling noise may lead to transmission errors [2-3]. For this reason, analysis of TSVs crosstalk is crucial to the design of TSV-based three-dimensional integration and has caught the attention of many researchers.

Traditionally, the characteristic of TSVs crosstalk is analyzed by full-wave finite element method (FEM) [4]. Full-wave FEM-based simulator solves electromagnetic equations of elements by given boundary conditions. Therefore, this solution can accurately character TSVs crosstalk. However, the problem of FEM is the increase in computation time as structural complexity increases. Furthermore, TSVs crosstalk varies with structure parameters of TSVs (e.g. diameter and height), made it complex to evaluate the characteristic of TSVs crosstalk. Hence, it is necessary to search a fast and accurate method to analyze TSVs crosstalk varied with structure parameters. Modeling and analysis TSVs crosstalk based on lumped circuit model of TSV can efficiently reduce computation time. It is noteworthy that electronic parameters in such reported model are extracted from electromagnetic solver (such as ANSYS Q3D Extractor) [5], which increased

workload for different structure parameters. Therefore, a convenience and accurate lumped circuit model varied with TSV structure parameters becomes a key method to analyze TSVs crosstalk efficiently.

In this paper, we analyze TSVs crosstalk based on scalable physical lumped circuit model of ground-signal (G-S) TSVs. The scalable G-S TSVs model is developed with closed-form analytic RLGC equations derived from physical structure, and can be changed with structure parameters in design process accordingly. The validity of this model is verified with full-wave FEM-based simulator. Based on this accurate physical lumped circuit model, we investigate crosstalk of TSV array in both frequency domain and time domain.

The rest of the paper is organized as follows. In section II, a scalable physical lumped circuit model of G-S TSVs is established, and is verified with full-wave FEM-based simulator up to 20 GHz. Based on this model, the crosstalk of multi-TSVs is analyzed both in frequency domain and in time domain in section III. The conclusion is summarized in Section IV.

Scalable Physical Lumped Circuit Model of G-S TSVs

In this section, a scalable physical lumped circuit model is established. This model is developed with closed-form analytic RLGC equations based on compact model. The equations are derived with physical structure. Thus, the model represents the physical meaning of G-S TSVs with closed-form equations.

Fig. 1 shows the typical structure of G-S TSVs and its geometric parameters. Both signal and ground TSVs have similar structure. Since the signal propagates along with metal-insulator-semiconductor (MIS) structure as shown in Fig. 1, it allows the existence three transmission modes, i.e. dielectric quasi-TEM mode, slow-wave mode and skin-effect mode [6] . The transmission mode is primarily depended on signal frequency and Silicon resistivity. Therefore, it is necessary to investigate the transportation mode of MIS TSV structure before establishing G-S TSVs model.

The transmission mode divided in frequency domain can be obtain from four frequencies, namely dielectric relaxation frequency of silicon (f_e), relaxation frequency of the interfacial polarization (f_s), characteristic frequency for skin-effect in silicon (f_δ) and characteristic frequency of slow-wave mode (f_0). They can be written as:

$$f_e = \frac{1}{2\pi\varepsilon_0\varepsilon_{Si}\rho_{Si}} \tag{1}$$

$$f_s = \frac{1}{2\pi\varepsilon_0\varepsilon_{SiO2}\rho_{Si}}\frac{t_{ox}}{t_{si}} \tag{2}$$

$$f_\delta = \frac{\rho_{Si}}{\pi\mu_0\left(t_{si}\right)^2} \tag{3}$$

$$f_o = \frac{1}{\left(\frac{1}{f_s}+\frac{2}{3f_\delta}\right)} \tag{4}$$

respectively, where ε_0, ε_{Si}, ε_{SiO2} and μ_0 are permittivity of free space, relative permittivity of silicon, relative permeability of SiO$_2$ and permeability of free space respectively. ρ_{Si} represents the resistivity of silicon. t_{ox} is the thickness of SiO$_2$ and t_{si} is the distance between two TSVs defined in Fig. 1.

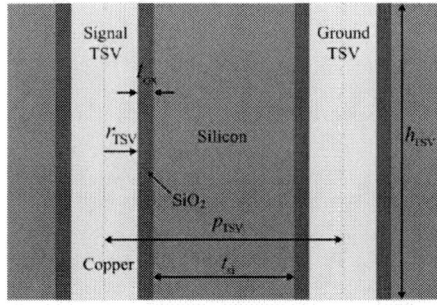

Figure. 1: Structure of G-S TSV and geometric parameters

Table. 1. Properties of fundamental modes

Signal frequency	Transmission mode
$f\leq0.3f_0$	Slow-wave
$f\geq1.5f_e$	Quasi-TEM
$f\geq4f_\delta$	Skin-effect

The frequency ranges of three transmission modes [7] are listed in Table. 1. Assuming that t_{ox} and t_{si} are fixed as 0.5 μm and 66 μm respectively, the resistivity-frequency domain chart is shown in Fig. 2. For low resistivity silicon (LRS) with 0.1

Ω·m resistivity, the frequency range of dielectric quasi-TEM mode is above 20 GHz. A signal with dielectric quasi-TEM mode region transmits through MIS structure without an effect of Si substrate resistivity ideally. Whereas in slow-wave mode and transition region (frequency less than 20 GHz), silicon substrate is regarded as semiconductor neither dielectric nor metallic. Consequently, the physical lumped circuit model of G-S TSVs differs in different transportation mode.

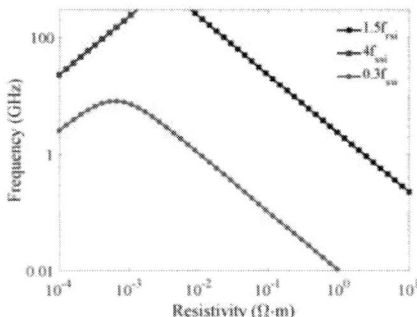

Figure. 2: Resistivity-frequency domain chart

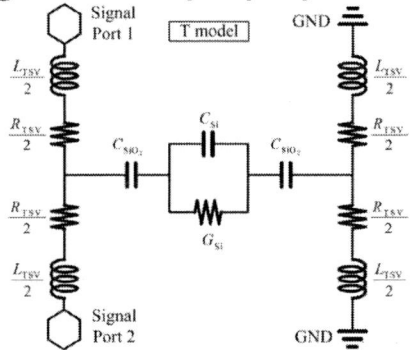

Figure. 3: Scalable physical lumped circuit model of G-S TSVs

Based on the previous analysis, a scalable physical lumped circuit model of G-S TSVs with T-shape is established in Fig. 3. The model is applied to frequency range of slow-wave mode and transition region. Therefore silicon substrate is equivalent as the parallel connection of capacitance (C_{si}) and conductance (G_{si}) between G-S TSVs as shown in Fig. 3. C_{si} and G_{si} are modeled as shown in:

$$C_{Si} = \frac{\pi\varepsilon_0\varepsilon_{si}h_{TSV}}{\cosh^{-1}\left(\frac{p_{TSV}}{2r_{TSV}}\right)} \tag{5}$$

$$G_{si} = \frac{\pi\sigma_{si}h_{TSV}}{\cosh^{-1}\left(\frac{p_{TSV}}{2r_{TSV}}\right)} \tag{6}$$

where h_{TSV}, p_{TSV}, r_{TSV} and σ_{Si} are height of TSV, pitch of G-S TSVs, radius of TSV and conductivity of silicon separately. The SiO$_2$ layer surrounding TSV is necessary for electrically isolating TSV and silicon substrate. Hence the SiO$_2$ layer is equivalent as insulator capacitance C_{SiO2} as shown in Fig. 3. It can be derived from the coaxial-cable capacitance model:

$$C_{SiO2} = \frac{2\pi\varepsilon_0\varepsilon_{ox}h_{TSV}}{\ln\left(\frac{r_{TSV}+t_{ox}}{r_{TSV}}\right)} \quad (7)$$

Since high frequency current flows close to the surface of conductor due to the skin effect, the resistance of TSV RTSV is modeled with direct current resistance R_{DC} and alternating current resistance R_{AC}. The analytic equation of TSV resistances is shown as follows:

$$R_{TSV} = \sqrt{R_{DC}^2 + R_{AC}^2} \quad (8)$$

$$R_{DC} = \frac{\rho_{TSV}h_{TSV}}{\pi r_{TSV}^2} \quad (9)$$

$$R_{AC} = \frac{\rho_{TSV}h_{TSV}}{\pi\left(2r_{TSV}\delta(f)-\delta(f)^2\right)} \quad (10)$$

where ρ_{TSV} represents the resistivity of TSV and $\delta(f)$ is the skin depth defined as:

$$\delta(f) = \frac{1}{\sqrt{\pi f \mu_0 \mu_{TSV}\sigma_{TSV}}} \quad (11)$$

where f, μ_{TSV} and σ_{TSV} are signal frequency, relative permittivity and conductivity of TSV. Besides, TSV also acts as inductance with signal frequency increasing. Hence the TSV inductance L_{TSV} is necessary for modeling. Since the height of TSV is finite, L_{TSV} is modified as:

$$L_{TSV} = \alpha\frac{\mu_0\mu_{TSV}h_{TSV}}{2\pi}\left[\ln\left(\frac{2h_{TSV}}{r_{TSV}}\right)-1\right] \quad (12)$$

where α is correction factor.

To validate the lumped circuit model established in this paper, the scattering parameters (S-parameters) from lumped circuit model is compared with that from FEM simulator. The structural parameters are fixed as r_{TSV} of 15 μm, h_{TSV} of 120 μm, p_{TSV} of 100 μm and t_{SiO2} of 0.5 μm. The comparison of S-parameters is shown in Fig. 4. The S-parameters from lumped circuit model and FEM simulator are well matched up to 20 GHz. Hence the validity of the lumped circuit model is validated and this model can be used for analysis of TSVs crosstalk.

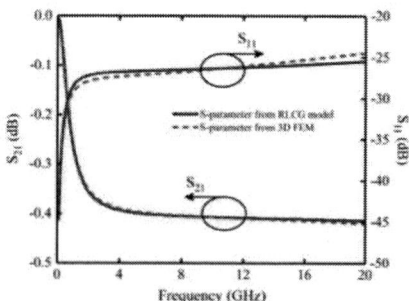

Figure. 4: Comparison of S-parameters from RLCG model and 3D FEM simulator

Crosstalk Analysis of TSVs

Crosstalk in TSV networks is a critical issue in 3-D integration and may have significant impacts on signal integrity, especially for high density TSV arrays on a lossy silicon substrate. This section mainly analyzes the crosstalk of multi-TSVs both in frequency domain and in time domain based on the model established in previous section.

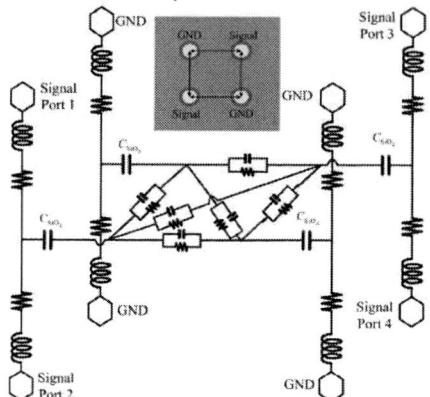

Fig. 5. 2 × 2 TSV array structure and equivalent circuit model

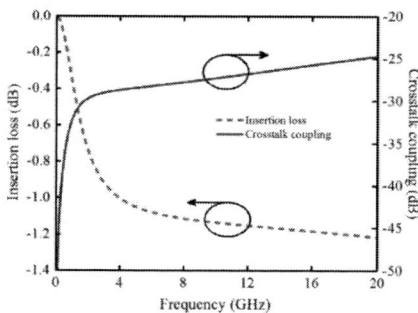

Figure. 6: The crosstalk of TSV array in frequency domain

To specify the method of modeling and analysis of TSVs crosstalk, a 2 × 2 TSV array structure as shown in Fig. 5 is chosen as the research object. The structure includes two high-speed G-S TSVs, and the pitch between signal TSV and ground TSV is 100 μm. Two group of G-S TSVs are coupled by silicon substrate, which can be equivalent as the parallel connection of capacitance and conductance as analyzed in section II. Thus, the equivalent circuit model of TSV array is described as in Fig. 5. The crosstalk of TSV array can be analyzed based on this equivalent circuit model.

Fig. 6 shows the frequency domain crosstalk simulation of TSV array. The insertion loss increases compared with signal G-S TSVs since the signal is jammed by other G-S TSVs. The crosstalk between two group of G-S TSVs increases as signal frequency increasing.

The crosstalk between two TSVs in time domain is analyzed with step signal input and

224

periodic signal input as shown in Fig. 7 and Fig. 8. The TSVs are terminated at both ends by a 50 Ω resistance. The step signal has a 1.8 V voltage step and rise time varies from 50 ps to 200 ps, which correspond to the knee frequency of 20 GHz to 5 GHz. From Fig. 7, the maximum crosstalk voltage decreases from 35 mV to 19 mV with the rise time increasing from 50 ps to 200 ps. The magnitude of periodic signal is 1.8 V, and the duty ratio is 20%. The rise time and fall time both are 100 ps. From Fig. 8, the maximum crosstalk voltage is 35 mV.

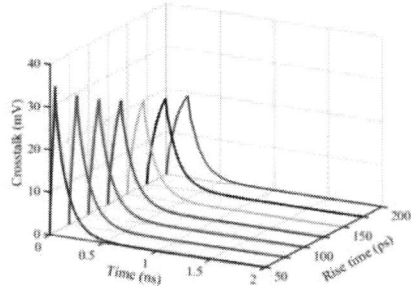

Figure. 7: The crosstalk of TSV array using step signal input in time domain

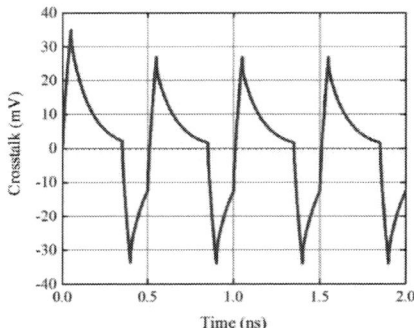

Figure. 8: The crosstalk of TSV array using using periodic signal input in time domain

Conclusion

In this paper, the transportation mode of MIS TSV structure is investigated. A scalable physical lumped circuit model of G-S TSVs is established. This model is developed with closed-form analytic RLGC equations based on compact model. The accuracy of the lumped model is verified with full-wave FEM-based simulator up to 20 GHz. Based on the G-S model, we analyze the crosstalk of TSV array in both frequency and time domain. The method of modeling and analysis of TSV crosstalk proposed in this paper can effectively reduce computation time for 3D packaging.

Acknowledgements

This project is supported by CAEP Foundation. (Grant No. PY2019059)

References

[1]. D. H. Kim, S. Mukhopadhyay, S. K. Lim, "TSV-Aware Interconnect Distribution Models for Prediction of Delay and Power Consumption of 3-D Stacked ICs," IEEE Transactions on Computer-Aided Design of Integrated Circuits and Systems, vol. 33, pp.1384-1395, 2014.

[2]. S. M. S. Harb, W. Eisenstadt, "A study of characterizing crosstalk effects in 3-D vias," Circuits & Systems (LASCAS), 2017 IEEE 8th Latin American Symposium on. IEEE, 2017, pp.1-4.

[3]. A. E .Engin, S. R. Narasimhan, "Modeling of crosstalk in through silicon vias," IEEE Trans. Electromagn. Compat, vol. 55, pp.149-158, 2013.

[4]. D. Kostka, T. Song, S. K. Lim, "3D IC-package-board co-analysis using 3D EM simulation for mobile applications," Electronic Components and Technology Conference (ECTC), 2013 IEEE 63rd. IEEE, 2013, pp.2113-2120.

[5]. L. Zhao, L. Yang, X. Sun, et al, "Research of methodologies to enlarge the isolation in 3D interconnection," Electronic Packaging Technology & High Density Packaging, 2009, International Conference on. IEEE, 2009, pp.331-334.

[6]. I. Ndip, B. Curran, K. Lobbicke, et al, "High-frequency modeling of TSVs for 3-D chip integration and silicon interposers considering skin-effect, dielectric quasi-TEM and slow-wave modes," IEEE transactions on components, packaging and manufacturing technology, vol. 1, pp. 1627-1641, 2011.

[7]. H. Hasegawa, M. Furukawa, H. Yanai, "Properties of microstrip line on Si-SiO2 system," IEEE Trans. Microwave Theory Tech, vol. 19, pp.869-881, 1971.

New Encapsulation Concepts for Medical Ultrasound Probes – A Heat Transfer Simulation Study

Nu Bich Duyen Do[1*], Erik Andreassen[1,2], Stephen Edwardsen[3], Anders Lifjeld[3], Hoang-Vu Nguyen[1], Knut E. Aasmundtveit[1], and Kristin Imenes[1]

[1]Department of Microsystems, University of South-Eastern Norway, Horten, Norway
[2]SINTEF, Oslo, Norway
[3]GE Vingmed Ultrasound AS, Horten, Norway

*Phone: (+47) 3100 9852, and E-mail Address: Nu.B.Do@usn.no

Abstract

Thermal management is important for medical ultrasound probes to maintain optimal performance, reliability, and lifetime of the devices, as well as to avoid heat-induced damage to the patients' tissue. This paper presents heat transfer simulations of the scan head of a trans-esophageal echocardiography (TEE) ultrasound probe, which operates temporarily inside the human esophagus for cardiac imaging. The current encapsulation design of the scan head requires manual assembly of several prefabricated parts to ensure functionalities such as heat spreading, electromagnetic interference (EMI) shielding, electrical isolation and biocompatibility. New encapsulation concepts that provide a more efficient manufacturing process while maintaining the multi-functional performance are desirable. The objective of this study is to screen encapsulation designs and materials which can simplify the encapsulation of the scan head. The main output to consider from the simulations is the maximum surface temperature of the scan head, which must be below 43 °C to ensure thermal safety for patients. Two encapsulation concepts are analyzed: single-layer encapsulation and double-layer encapsulation. The simulation results show that a double-layer encapsulation, such as a polymer-coated metal encapsulation or a metallized polymer encapsulation, can fulfill the requirements regarding heat transfer, EMI shielding, electrical isolation and biocompatibility.

Key words: trans-esophageal echocardiography, ultrasound scan head, heat transfer simulations, encapsulation

I. Introduction

Ultrasound imaging plays an important role in the diagnosis of cardiovascular diseases. Imaging from the inside of the esophagus, known as trans-esophageal echocardiography (TEE), has the capability of giving high quality images due to the proximity of the esophagus to the heart. The TEE probe is passed down the esophagus and provides 3D images of the heart in real time [1].

The scan head of a TEE probe, referred to as 'scan head' in this paper, contains the ultrasound transducer and driving electronics. The starting point in our study is a scan head with an encapsulation which requires manual assembly of several prefabricated parts, as illustrated in Figure 1. It is therefore desirable to reduce the number of prefabricated parts and process steps to simplify the scan head assembly. Developing new encapsulation designs, as well as utilizing new functional materials, are potential approaches to reach a more efficient manufacturing. New encapsulation concepts must also maintain high image quality and medical safety for the patients.

Thermal management in a TEE scan head is important with respect to patient safety, since the transducer and the electronics in the scan head generate heat during the operation. Hence, heat distribution and heat transfer in the TEE scan head must be well-controlled to avoid hot spots on the scan head surface. According to the standard IEC 60601-2-37 of requirements for the basic safety and essential performance of ultrasonic medical diagnostic equipment, the temperature limit of medical devices in direct contact with the human body should be lower than 43 °C to avoid thermal damage to biological tissues [2]. Thermal management is also critical to the performance, lifetime and reliability of the scan head.

Electrical isolation, electromagnetic interference (EMI) shielding, and biocompatibility are other vital concerns when it comes to patient safety and selection of a new encapsulation concept. Encapsulation materials used in the TEE scan head must ensure electrical isolation while surfaces in contact with the human body needs to be biocompatible. In addition, the encapsulation materials should provide EMI shielding to protect the device from signal interference and ensure proper operation of the device [2].

In this study, potential encapsulation designs and materials, with respect to heat dissipation, are evaluated by means of numerical simulations of heat transfer. EMI shielding, biocompatibility, and electrical isolation are also parameters to consider when selecting new designs and materials for TEE scan head encapsulation.

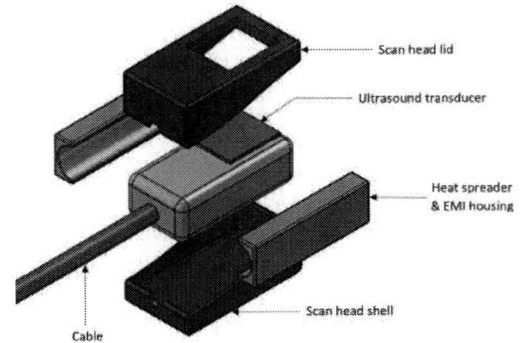

Figure 1: Simplified encapsulation design of a TEE scan head. The 'heat spreader & EMI housing' parts spread the heat from the 'ultrasound transducer' and provide EMI shielding for the scan head. The 'scan head lid' and the 'scan head shell' parts define the outer shape of the scan head, and these are in contact with biological tissue.

II.   Heat Transfer Simulation Model

An idealized 3D model of the TEE scan head is simulated in COMSOL Multiphysics 5.3a. This model includes a heat source, a heat sink and an outer encapsulation. The heat source represents electronic components generating heat within the scan head, such as the ultrasound transducer and the driving electronics. The simplified heat source in this study is modelled as a silicon (Si) solid, since Si is the material of the component generating the most heat within the scan head. The heat source is in contact with an aluminum (Al) heat sink. In the model, the heat source generates a power of 1 W. This power value, being higher than the typical value for such a scan head, is selected in order to have a certain safety margin. The encapsulation covers the entire scan head, except the heat source's surface, which is the ultrasound lens, as shown in Figure 2. This encapsulation and the exterior surface of the heat source are in direct contact with the esophageal medium. The remaining volume inside the scan head, such as cables, other electronic components, organic substrates, is modelled as air for simplifying the simulations, thus representing a worst-case scenario.

In this study, steady-state thermal conduction is considered the main mechanism of heat transport. The thermal contact between domains in the scan head is assumed to be perfect. The heat transfer can be described by the thermal diffusion equation, which is derived from the principles of conservation of energy and Fourier's law [3].

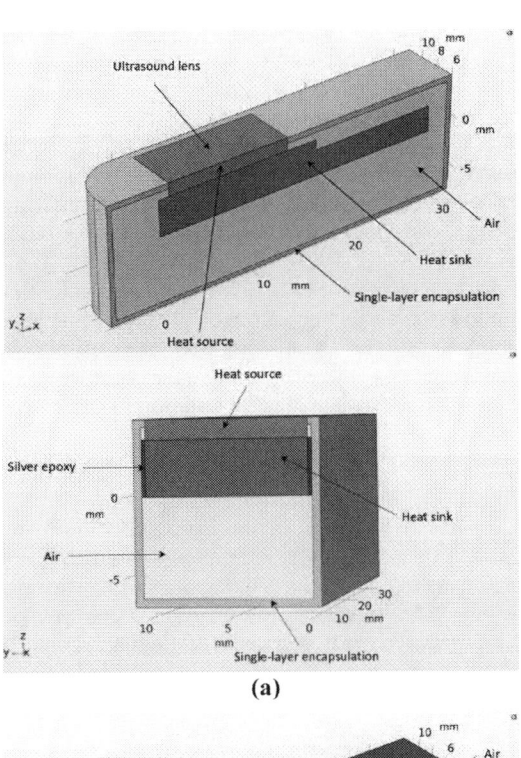

(a)

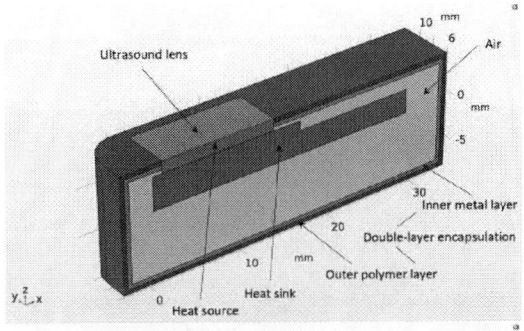

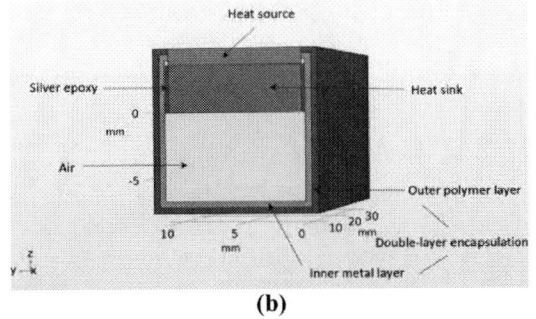

(b)

Figure 2: Cross-sections of the scan head model consisting of a Si heat source, an Al heat sink, and (a) a single-layer encapsulation, or (b) a double-layer encapsulation with an inner metal layer and an outer polymer layer. The heat sink is in contact with the inner surface of the encapsulation along its two long sides via silver-filled epoxy. The remaining volume inside the scan head contains air. The ultrasound lens on the surface of the heat source is a window for transmission and reception of ultrasound signals.

The steady-state heat equation is:

$$k\nabla^2 T + q = 0 \qquad (1)$$

where k is the thermal conductivity (W/(m·K)) of the solid material; ∇^2 is the Laplace operator; T is the scalar temperature field (K); q is the rate at which energy is generated per unit volume of the medium (W/m^3).

A convective heat flux boundary condition is applied to the outer surface of the scan head, representing the heat flux to the esophagus, maintained at human body temperature. The expression for this boundary condition [3] is:

$$q_0 = h_c(T_{ext} - T) \qquad (2)$$

where h_c is the heat transfer coefficient (W/(m^2·K)); T_{ext} is the temperature of the external medium (K), (37 °C in our case); T is the temperature of the scan head surface (K); q_0 is the convective heat flux (W/m^2). In our simulations, the heat transfer coefficient, h_c, was set to 400 (W/(m^2·K)). This value was calculated from esophageal heat transfer measurements on a pig. The 3D model was meshed using free tetrahedral elements with a minimum element size of 5 μm in order to describe the heat flow adequately in the thinnest layers in this study.

In this study, the main constraint is to keep the maximum surface temperature of the scan head below 43 °C to ensure thermal safety for patients. Another constraint is the outer dimensions of the scan head, which should be similar to, or smaller than, those of the existing scan head. This sets limits for the thickness of the encapsulation. Material properties of the heat source (Si), the heat sink (Al) and air were taken from COMSOL's database. Typical data for encapsulation materials were used. In further studies, encapsulation materials will be selected based on simulation results and material databases (e.g. MatWeb, NIST).

III. Results and Discussions

The objective of the heat transfer simulations in this paper is to identify appropriate encapsulation designs and materials that are able to satisfy the thermal safety requirement for the TEE ultrasound scan head as well as to simplify the assembly of the scan head. The simplification of the assembly process can be related to:

- Reducing the number of prefabricated parts to reduce the number of process steps, or
- Selecting materials and corresponding packaging techniques suitable for automatic processes.

1. Heat transfer of TEE scan head with single-layer encapsulation

In a single-layer encapsulation design, the scan head is encapsulated by a single layer made of a single homogeneous material or a composite. This layer can be applied in an automatic process, such as a molding process with a thermoplastic or a thermosetting polymer. Such molding processes are well-developed packaging techniques suitable for automatic encapsulation of electronic components [4]. Hence, the molding processes can be employed for encapsulating the scan head in automated steps or for manufacturing prefabricated parts where any components or modules in the scan head may not endure the pressure and temperature of certain molding techniques.

All types of materials, such as polymers, metals, ceramics, composites, are considered to find a material appropriate for single-layer encapsulation design. Thermal conductivity k is the only parameter of the material needed for steady-state heat conduction. Therefore, the value of k was varied in a wide range (0.1 – 500) W/(m·K) to simulate heat transfer for different material categories. Thermal conductivity of polymers and their composites are often in the range of (0.1 – 3) W/(m·K), while common ceramics, metals, and alloys have higher thermal conductivity [3], [4].

In heat transfer simulations for single-layer encapsulation, the thickness of encapsulation layer was kept at 0.5 mm or 1 mm, while the thermal conductivity was varied from 0.1 to 500 W/(m·K). A thickness value of 1 mm was chosen as the upper value since this is the encapsulation thickness of the current scan head. The thickness of 0.5 mm was based on the typical lower limits for some of the relevant encapsulation processes, such as injection molding of polymers and die casting of metals, to facilitate the manufacturing of the encapsulation layer [5], [6].

An example of a heat transfer simulation of the scan head encapsulated by a single-layer encapsulation is shown in Figure 3. The encapsulation layer is Cu with a thickness of 0.5 mm, and a thermal conductivity of 400 W/(m·K). The surface temperature of the scan head in Figure 3 varies from 38 °C to 39.2 °C. The maximum surface temperature occurs at the opening surface of the heat source, which is not encapsulated (see Figure 2).

Figure 4 shows the effect of the thermal conductivity on the maximum surface temperature for two encapsulation thicknesses of 0.5 mm and 1 mm. For both thickness values, the maximum surface temperature decreases with increasing thermal conductivity. Increasing the thermal conductivity from 0.1 to 50 W/(m·K) leads to a large drop in the maximum surface temperature. When increasing thermal conductivity to values higher than 50 W/(m·K), the reduction in maximum surface temperature is relatively small. Figure 4 also indicates that when the thermal conductivity of the encapsulation material is higher than 20 W/(m·K) for 0.5 mm thickness, or higher than 10 W/(m·K) for 1 mm thickness, the maximum surface temperature of the scan head will satisfy the requirement that no hot spots on the surface are hotter than 43 °C.

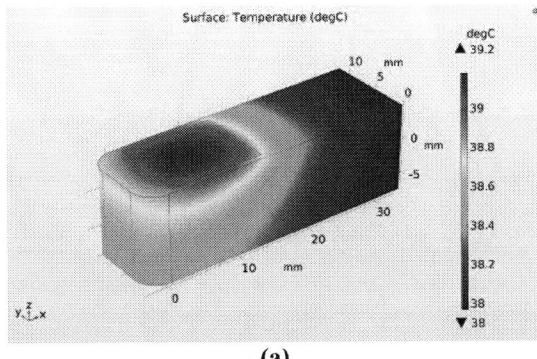

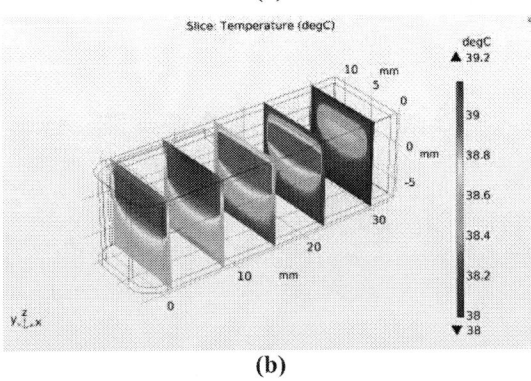

Figure 3: (a) Surface temperature, and (b) bulk temperature of the scan head having a 0.5 mm single-layer encapsulation with k = 400 W/(m·K)

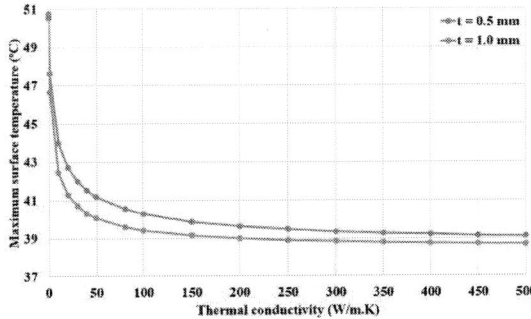

Figure 4: Effect of the thermal conductivity of the encapsulation material on the maximum surface temperature of the scan head. Simulation results for single-layer encapsulation with a thickness of 0.5 mm and 1 mm.

From a heat transfer point of view, a single-layer encapsulation can provide good thermal management for the scan head when the encapsulation material has a thermal conductivity higher than 20 W/(m·K) and a thickness in the range of (0.5 – 1.0) mm. However, other factors, such as EMI shielding, electrical isolation, biocompatibility, processing, have to be considered when selecting the encapsulation material for the TEE scan head.

To the best of our knowledge, there is no single material which can satisfy all the requirements. Polymers are good candidates for electrical isolation and biocompatibility, as well as being suitable for automatic encapsulation processes. However, they have low thermal conductivity and they are transparent to EM waves. Ceramics can provide good thermal dissipation and electrical isolation, but the EMI shielding performance of ceramics is poor. Metals is perfect for thermal dissipation and EMI shielding, but they do not provide electrical isolation.

On the other hand, there may be composite materials (e.g. polymer-based composites) with thermal conductivity higher than 20 W/(m·K) which also can satisfy the contradictory requirements, such as electrical isolation and EMI shielding. Potential composite suitable for the single-layer encapsulation designs will be investigated further.

2. Heat transfer of TEE scan head with double-layer encapsulation

Metals can satisfy most of the requirements, such as heat transfer, EMI shielding, and biocompatibility. Since the scan head is in direct contact with the patient's esophagus, the outer encapsulation layer must provide electrical isolation. This can be achieved by covering the outer surface of a metal layer with a polymer layer. In this paper, two double-layer encapsulation designs are considered.

- Polymer-coated metal encapsulation: a double-layer design in which a thin coating of a biocompatible polymer covers the outer surface of prefabricated metal parts
- Metallized polymer encapsulation: a double-layer design in which a thin metal layer is deposited on the inner surface of prefabricated parts made of a biocompatible polymer.

2.1. Selection of materials for double-layer encapsulation designs

For double-layer encapsulation designs, many metals can be used, since electrical isolation and biocompatibility are handled by the polymer layer. Some common metals used for electronics packaging are Ag (k = 428 W/(m·K)), Cu (k = 400 W/(m·K)), Au (k = 318 W/(m·K)), Al (k = 236 W/(m·K)) [3], [4]. Cu is the metal chosen in this paper since it has high thermal conductivity and low price. In addition, there are many well-developed methods for processing Cu [4]–[6].

Polymer materials for polymer-coated metal encapsulations must be able to provide thin and durable layers. There are several techniques for applying smooth, uniform, thin coatings of a polymer on the outer surface of a metal part. Spraying or dip coating of polytetrafluoroethylene (PTFE), a biocompatible polymer, on Cu parts prefabricated by die casting [7]–[9] is an example.

Regarding the metallized polymer encapsulations (i.e. a metal layer being thinner than the polymer layer), the polymer part(s) can be prefabricated by injection molding, and then a thin metal coating can be applied to the inner surface of the polymer parts. Electroless plating followed by electroplating can be used for applying the metal coating, providing metal layer with thickness from nm to mm [8]. The thickness of the metal coating must satisfy the EMI shielding requirement (see Sect. 2.3). Such a double-layer encapsulation could be a Cu-metallized part made by injection molding in an appropriate polymer material such as polyethersulfone (PES) [6], [7].

In fact, the thermal conductivity of most (unfilled, biocompatible) polymers are in the range of 0.1 – 0.3 W/(m·K). Hence PTFE and PES can be used as representative polymers.

2.2. Heat transfer of TEE scan head with polymer-coated metal encapsulation

An encapsulation consisting of Cu parts coated with PTFE is considered. The latter will be in direct contact with the human esophagus. The Cu part may be fabricated by die casting, and the PTFE coating can be applied with an established coating method.

In our simulations, the Cu thickness was 0.5 mm and 0.9 mm, while the PTFE thickness was varied from 10 μm to 100 μm. The Cu thickness of 0.5 mm was chosen based on a typical lower limit for thicknesses achieved by die casting and the required mechanical integrity of the metal part [5]. The Cu thickness of 0.9 mm Cu was chosen as an upper value, so that the total thickness of the double-layer encapsulation does not exceed the encapsulation thickness of 1 mm of the current scan head. The PTFE thickness range was based on capabilities of available coating methods [8], [10].

An example of a heat transfer simulation of a scan head having polymer-coated metal encapsulation is presented in Figure 5. In this case, the Cu inner layer is 0.5 mm thick, while the PTFE outer layer is 100 μm thick. Figure 5 shows that the surface temperature of the scan head varies from 37.6 °C to 39.4 °C. The maximum surface temperature occurs on the outer surface of the heat source.

Figure 6 shows the effect of the thickness of the PTFE coating on the maximum surface temperature for two Cu thickness values 0.5 mm and 0.9 mm. For both Cu thicknesses, the effect of PTFE thickness ranging from 10 μm to 100 μm is negligible. Furthermore, maximum temperatures obtained with 0.5 mm and 0.9 mm Cu are quite close. Any combination of Cu having thickness in the range of (0.5 – 0.9) mm with PTFE having thickness in the range of (10 – 100) μm will give a maximum surface temperature of about 39 °C, which satisfies the thermal safety requirement (max. 43 °C).

From a heat transfer point of view, a polymer-coated metal encapsulation can provide good thermal management for the scan head. However, the thickness of the polymer coating should be considered carefully. First, the coating

must be uniform, so that there are no areas with coating significantly thinner than the nominal value. Secondly, the thinnest coatings in the range simulated may be prone to damage when the scan head is used, such as scratches exposing the metal. This may be detrimental for the electrical isolation or the biocompatibility. The electrical surface resistance of such polymer-coated metal encapsulations should be measured in order to identify the most appropriate polymer coating thickness and coating process. In addition, measurements of EMI shielding effectiveness and biocompatibility should also be conducted.

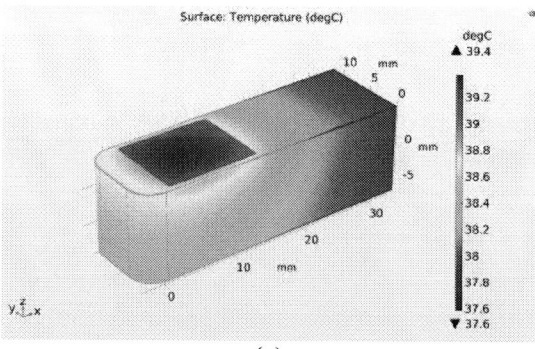

(a)

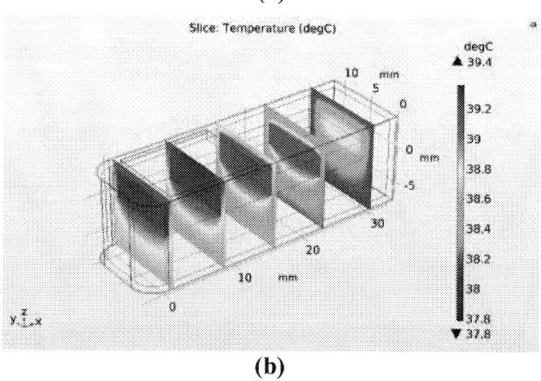

(b)

Figure 5: (a) Surface temperature, and (b) bulk temperature of a scan head with polymer-coated metal encapsulation. The Cu inner layer is 0.5 mm thick, the PTFE outer layer is 100 μm thick.

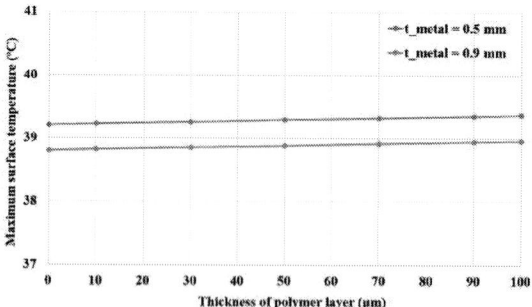

Figure 6: Effect of polymer layer thickness on the maximum surface temperature of the scan head encapsulated by polymer-coated metal structure having metal thickness of 0.5 mm or 0.9 mm.

230

2.3. Heat transfer of TEE scan head with metallized polymer encapsulation

In this type of double-layer encapsulation, we have considered a PES part with the inside metallized with a thin layer of Cu. The polymer part(s) could be prefabricated by injection molding, and the Cu layer could be deposited by conventional metal plating techniques, such as electroplating or electroless plating.

The simulations were applied for two PES thickness values (0.5 mm and 0.9 mm), and the Cu thickness vas varied from 10 µm to 100 µm. The PES thickness of 0.5 mm was chosen due to the lower limit of injection molding [6], [8], and the consideration of the mechanical integrity of the polymer part. The PES thickness of 0.9 mm was chosen as an upper value, so that the total thickness of the double-layer encapsulation does not exceed the encapsulation thickness of 1 mm of the current scan head. The Cu thickness range was selected based on the capabilities of plating techniques, and the requirements for EMI shielding [8], [11].

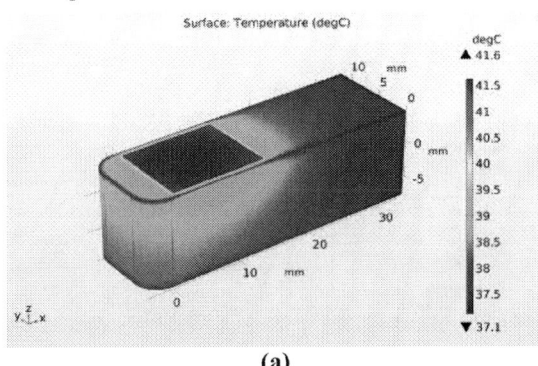

(a)

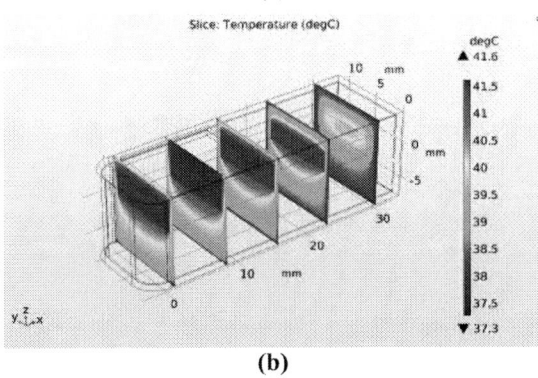

(b)

Figure 7: (a) Surface temperature, and (b) bulk temperature of a TEE scan head with metallized polymer encapsulation. The outer PTFE thickness is 0.5 mm while the inner Cu thickness is 100 µm.

An example of a heat transfer simulation of a scan head encapsulated as mentioned above is shown in Figure 7. This double-layer encapsulation design has a 100 µm Cu inner layer and a 0.5 mm PES outer layer. The surface temperature of the scan head in this case varies from 37.1 °C to 41.6 °C. The maximum surface temperature occurs at the outer surface of the heat source. Note that the maximum surface temperature (41.6 °C) is higher than that of the polymer-coated metal encapsulation (around 39.4 °C, Sect. 2.2). In the latter case, the heat spreading metal layer was thicker while the thermally insulating polymer layer was thinner.

Figure 8 shows the effect of Cu thickness on the maximum surface temperature, for PES thicknesses of 0.5 mm and 0.9 mm. For both PES thicknesses, the maximum surface temperature decreases when increasing the thickness of Cu layer from 10 µm to 100 µm. Figure 8 also indicates that the thermal safety limit of 43 °C is reached when the Cu layer thickness is above a certain value, e.g. above 50 µm when the PES thickness is 0.5 mm.

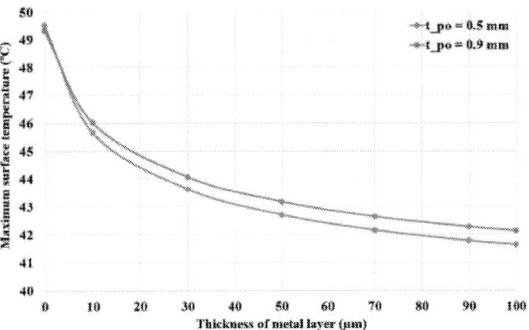

Figure 8: Effect of the metal layer thickness on the maximum surface temperature of a scan head with metallized polymer encapsulation. Results for Cu metal layers and two PES polymer thicknesses: 0.5 mm and 0.9 mm.

From a heat transfer point of view, a metallized polymer encapsulation can provide adequate thermal dissipation when the Cu layer thickness is above a certain value. The Cu thickness must also provide sufficient EMI shielding.

The EMI shielding effectiveness (EMISE) of PES–Cu encapsulations with different thickness combinations (i.e. 0.5 – 0.9 mm PES and 10 – 100 µm Cu) was calculated based on ref. [11]. The results show that the proposed metallized polymer encapsulations can provide adequate EMISE. In all cases considered, the EMISE is higher than 100 dB in the frequency range of (0.03 – 3) GHz, which is a common frequency range for testing EMISE of medical devices used in professional healthcare facility environment [12], [13]. These values are above the shielding requirements for medical devices classified in the same group as the TEE scan head (group 1, class B). For example, the peak limit for radiated disturbance of these devices, at a measurement distance of 3 m, in the frequency range (1 – 3) GHz is 70 dB [13]. Nevertheless, EMISE measurements of such encapsulations must be conducted to verify the theoretical results and to select the most suitable metal layer thickness.

IV. Conclusions

The heat transfer of a TEE scan head was simulated. The target of the simulations was to identify promising encapsulation concepts which can ensure the safety for patients (maximum scan head surface temperature below 43 °C), and simplify the assembly process of the scan head. Two encapsulation concepts were analyzed, single-layer encapsulation and double-layer encapsulation.

Our simulation results show that a single-layer encapsulation, with appropriate thermal conductivity and thickness, can provide sufficient heat transfer for the scan head. However, further analysis is required to figure out specific material(s) which can comply with other functional requirements. The materials should have thermal conductivity higher than 20 W/(m·K), while satisfying requirements such as EMI shielding and electrical isolation.

On the other hand, a double-layer encapsulation, such as a polymer-coated metal encapsulation or a metallized polymer encapsulation is able to provide the multi-functional performance required for this scan head. For a polymer-coated metal encapsulation, any combination of inner metal layer thickness in the range of (0.5 – 0.9) mm and outer polymer layer thickness in the range of (10 – 100) μm will satisfy the thermal requirement. For a metallized polymer encapsulation, the inner metal layer must have a certain thickness in order to satisfy the thermal requirement. As an example, a Cu layer must be thicker than 50 μm when the thickness of a typical polymer material is 0.5 mm.

In further work, prototypes will be fabricated, and experimental measurements will be performed in order to verify the proposed encapsulation concepts. Both manufacturability and thermal management will be considered.

Acknowledgements

This work was funded by the Research Council of Norway through the BIA program (grant number: 269618; *Mechanical miniaturization in interventional medical instruments*), and by the Norwegian Micro- and Nano-Fabrication Facility (NorFab, project number: 245963/F50). The authors gratefully acknowledge Prof. Tao Dong, Zhongyuan Shi, and Thao Vo at the University of South-Eastern Norway for valuable discussions.

References

[1] E. C. Pua, S. F. Idriss, P. D. Wolf, and S. W. Smith, "Real-time 3D transesophageal echocardiography," *Ultrason. Imaging*, vol. 26, no. 4, pp. 217–232, 2004.

[2] IEC, *IEC 60601-2-37 - Particular requirements for the safety of ultrasonic medical diagnostic and monitoring equipment.* 2004.

[3] F. P. Incropera, D. P. DeWitt, T. L. Bergman, and A. S. Lavine, *Fundamentals of Heat and Mass Transfer 6th Edition.* 2007.

[4] R. Tummala, *Fundamentals of Microsystems Packaging.* McGraw-Hill Education, 2001.

[5] B. Andresen, "Die Casting Engineering - A Hydraulic, Thermal and Mechanical Process," *Engineering*, 2004.

[6] E. A. Campo and E. A. Campo, "The Complete Part Design Handbook," in *The Complete Part Design Handbook*, 2012.

[7] A. J. T. Teo, A. Mishra, I. Park, Y. J. Kim, W. T. Park, and Y. J. Yoon, "Polymeric Biomaterials for Medical Implants and Devices," *ACS Biomaterials Science and Engineering.* 2016.

[8] A. A. R. Elshabini-Riad, *Thin film technology handbook.* 1998.

[9] Dupont Teflon and N. & I. Coatings, "Applying Teflon Coatings."

[10] T. Schneller, R. Waser, M. Kosec, and D. Payne, "Chemical solution deposition of functional oxide thin films," *Chem. Solut. Depos. Funct. Oxide Thin Film.*, vol. 9783211993, pp. 1–796, 2013.

[11] C. R. Paul, *Introduction to Electromagnetic Compatibility: Second Edition.* 2006.

[12] IEC, *IEC 60601-1-2 - Medical electrical equipment - Part 1-2: General requirements for basic safety and essential performance - Collateral Standard: Electromagnetic distrubances - Requirements and tests.* 2014.

[13] CISPR, *EN 55011:2007 - Industrial, scientific and medical equipment - radio-frequency disturbance characteristics - Limits and methods of measurements.* 2007.

Thermomechanical Properties of Fan-Out Wafer Level Package Fabricated with Various Epoxy Mold Compound

Haksan Jeong, Kwang-Ho Jung, Choong-Jae Lee, Kyung Deuk Min, and Seung-Boo Jung[*]

School of Advanced Materials Science and Engineering, Sungkyunkwan University, Suwon-si, Korea

Corresponding author: Tel. +82 31 290 7359; Fax +82 31 290 7371; E-mail sbjung@skku.edu

Abstract

Fan-out package has advantages such as higher I/O density, higher electrical performance, ultra-thin and low power consumption. However, Fan-out package have some issues caused by warpage induced by difference of coefficient of thermal expansion between epoxy mold compound (EMC) and Si die. Therefore, the warpage behavior of the FOWLP component molded with three kinds of EMCs were investigated by using shadow moiré method and finite element method analysis. The environmental reliability of fan-out package component was evaluated by thermal shock test and temperature-humidity test. The warpage behavior of fan-out package component changed from convex to concave with increasing the temperature. The warpage of fan-out package component molded by EMC with lower CTE was reduced because of smaller CTE mismatch between chip and EMC at room temperature. However, the warpage of FOWLP component molded by EMC with lower modulus was reduced the warpage even though CTE of EMC was higher than others due to stress-strain relaxation at 260 °C. The electrical resistance of fan-out package component increased with increasing aging time and cycles during temperature-humidity test and thermal shock test.

Key words: Fan-out package, Epoxy mold compound, Warpage, Temperature-humidity test, Thermal shock test,

Introduction

The trends in electronic devices for 5G, internet of things (IoT) and wearable applications has demanded the miniaturization, lightening and higher funcstionality [1,2]. According to these demands, the 2.5D and 3D packaging technologies such as package-on-package (PoP), interposer, wafer level chip scale packaging (WLCSP) and so on have been developed fast [3,4]. Among the advanced packaging technologies, the fan-out package such as fan-out wafer level package (FO-WLP) and fan-out panel level package (FO-PLP) has many advantages such as a higher I/O density, thin, higher electrical performance, low power consumption, low cost and so on. [5,6]. So, fan-out package is adopted as advanced package technology in companies. [7,8]

Structural characteristics of fan-out package are fan-out region and substrate-less structure. The molded area of fan-out package is larger than the area of of Si die, so, package area of fan-out package is larger than that of Si die. Therfore, I/O density of fan-out package can increase. Furthermore, fan-out package has redistribution layer (RDL) which is thinner than printed circuit board (PCB). So, fan-out package has better heat dissipation and thinner thickness compared to flip-chip ball grid array (FC-BGA). However, fan-out package has higher warpage because of thinner substrate-less structure.

The warpage was generated by a coefficient of thermal expansion (CTE) mismatch between different materials. The warpage affects the fine pitch process and the yield of the fan-out package. And the warpage could affect the reliability of the device too [9-11].

We fabricated the fan-out package component with three kinds of EMCs and evaluated the warpage of a fan-out package component by shadow moiré method and finite element method (FEM) analysis. The electrical resistance of the daisy chain was measured under thermal shock tests and the constant temperature-humidity test.

Experimental procedure

Figure 1 shows the process flow of fan-out package component. The fan-out package component was 8 x 8 x 0.075 mm³ Si die encapsulated inside a 14 x 14 x 0.3 mm³ package with 15 μm RDL. First of all, carrier is laminated with adhesive, as shown in figure 1 (a). The bare Si picked and place on the carrier. The Si die was molded by three kinds of 300 um EMCs in thickness. The three kinds of EMC were identified as EMC1, EMC2 and EMC3. The molded die encapsulated by EMC was heated at 200 °C for 2 h and then very slowly cooled to room temperature. Then molded

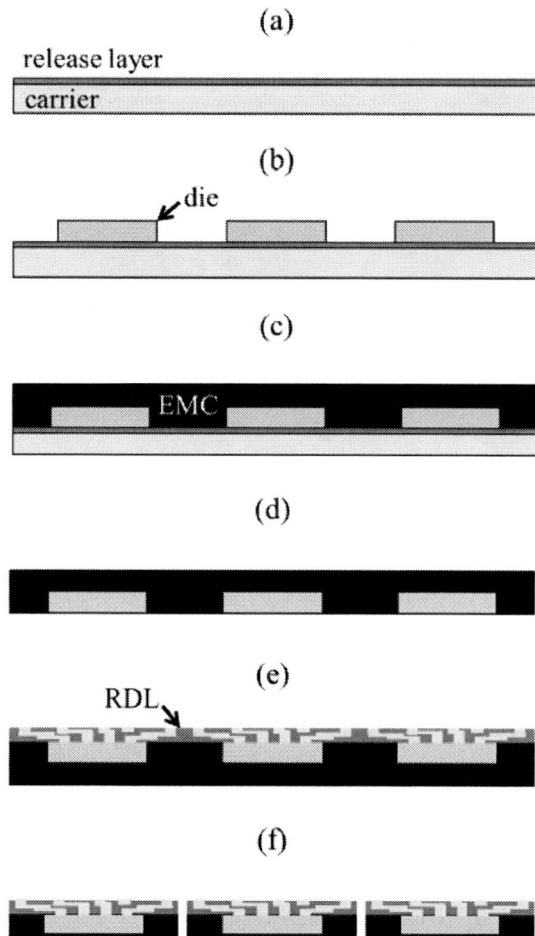

(a)

release layer

carrier

(b)

die

(c)

EMC

(d)

(e)

RDL

(f)

Figure 1. Schematic diagram fan-out package (a) laminate carrier (b) pick and place the Si die (c) EMC molding process (d) release and peel tape (e) RDL process and (f) singulation

chips were detached from the carrier wafer, as shown in figure 1 (d). The conventional photo lithography and electroplating processes were used for RDL fabrication. Patterning of positive type photosensitive polyimide (PSPI) is performed using photolithography with positive image masks. The Ti/Cu seed layer was sputtered using a sputter system (SNTEK, Republic of Korea). After patterning the photoresist, 3 µm Cu RDL was electroplated on the wafer. The photoresist and seed layer were removed. The 2nd ~ 5th RDL were fabricated by the same process as the 1st RDL. After the RDL fabrication, wafers were singulated, as shown in figure 1 (f).

The modulus, CTE and glass temperature were measured from 25 °C to 260 °C using dynamic mechanical analyzer (DMA8000, PerkinElmer, USA) and thermo mechanical analysis (TMA, Seiko Exstar 6000, Seiko Inst., Japan), respectively.

In compliance with JESD22-B112A, the warpage behavior for fan-out package component

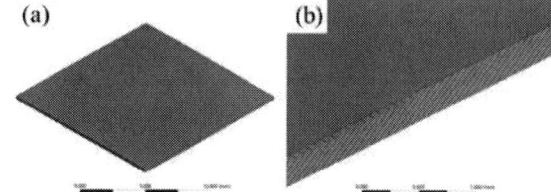

(a)   (b)

Figure 2. (a) FEM analysis model and (b) cross-sectional plane of model of fan-out package component

was measured from 20 °C to 260 °C, using the shadow moiré equipment (PS200, Akrometrix, USA). Also, finite element method analysis was carried out to compare stress and strain distribution for warpage of fan out package component. The figure 2 shows the FEM analysis model of fan-out package component. The electrical resistance of FOWLP component was measured during thermal shock test and temperature-humidity test to evaluate the bonding property between metal and dielectric layers and the effect of humidity on the reliability of RDL pattern. In compliance with JESD22-A104D, the thermal shock test was performed in the temperature ranging from -40 °C to 125 °C using a thermal shock test chamber (TSA-101S, ESPEC, Japan). The dwell time was 15 minutes. The temperature-humidity test was carried out at 85 °C and 85% relative-humidity (RH) using temperature-humidity test chamber (PL-1KPH, ESPEC CORP., Japan) for 1000 h. The fan out package components were observed using a field emission scanning electron microscope (FE-SEM, TSM-7500f, JEOL, Japan).

Results and discussion

Figure 3 shows the cross-sectional micrographs of multi-layers of fan-out package component. The fan-out package component composed of the EMC mold, Si chip and RDL. The total thickness of each RDL layer was designed to be 3 µm. RDL was well formed on the fan-out region without any defects after RDL process.

Figure 4 shows the DMA and TMA results of three kinds of EMCs. And table 1 shows the CTE, modulus and Tg of various EMCs. The CTE of EMC2 was lower than that of other EMCs. The CTE at 25 °C and 260 °C of EMC2 was 6.2 ppm/°C and 63.48 ppm/°C, respectively. However, Tg of EMC2 was lower than that of other EMCs and Tg of EMC3

(a) EMC   (b) EMC

die

PI layers   PI layers

50 µm   50 µm

Figure 3. SEM micrographs of the fabricated multi-layer consisting of Cu and PI layers on the (a) Si die and (b) EMC mold

234

(a)

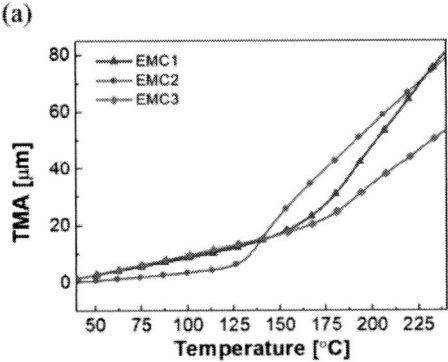

(b)

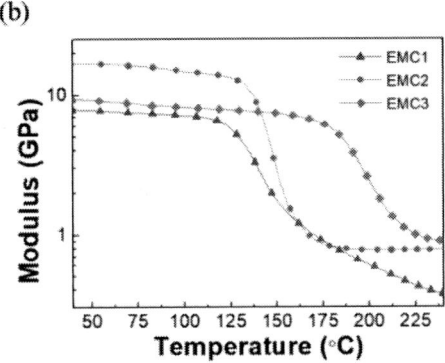

Figure 4. (a) TMA and (b) DMA behavior of various EMC

was higher than EMC1. So, thermal expansion of EMC3 from 25 °C to 260 °C was the shortest among various EMC, as shown in figure 3 (a). The modulus of EMC1 at 25 °C and 260 °C was lower than that of EMC2 and EMC3.

Figure 5 shows the warpage behavior of fan-out package component. The shape of component was changed from convex warpage to concave warpage with increasing temperature from 25 °C to 260 °C. At 25 °C, the warpage of fan out package component molded by EMC3 was lowest. However, the warpage of fan out package component molded by EMC1 was lower than that by other EMCs.

The figure 6 shows the FEM analysis of warpage behavior for fan-out package component with various EMCs at 25 °C and 260 °C. The PSPI was used in the FEM model of RDL whose thickness was 15 μm. The CTE and modulus of EMC obtained from the DMA and TMA results used in figure 4. The CTE and modulus of RDL used in this study were 49 ppm/°C and 2.2 GPa, respectively. [12]. We assumed stress-free temperature was the

Table 1. CTE, modulus and Tg of various EMCs

	CTE at 25 ℃ (ppm/℃)	Modulus at 25 ℃ (GPa)	T_g	CTE at 260 ℃ (ppm/℃)	Modulus at 260 ℃ (GPa)
EMC1	13.54	8.25	151 ℃	86.45	0.27
EMC2	6.2	16.7	130 ℃	63.48	0.78
EMC3	13.12	9.5	160 ℃	46.46	0.86

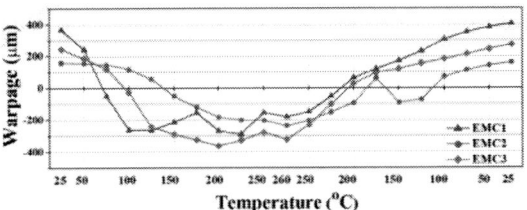

Figure 5. Warpage behavior of fan-out package component molded by various EMCs

post curing temperature which is 200 °C because the shrinkage of EMC during post curing was ignored. The results of shadow moiré method and FEM results showed a tendency to match. The warpage shape of FEM results at 25 °C and 260 °C was convex warpage and concave warpage, respectively. The simulated warpage of the fan-out package component molded by EMC2 at 25 °C was 125 μm which is lowest value among that by various EMCs. The simulated warpage of the component molded with EMC1 was 332 μm.

The thermomechanical stress of fan-out package component was induced by CTE mismatch between EMC and Si die [13]. So, warpage of fan-out package increased with CTE mismatch between EMC and Si increased. However, the warpage of fan-out package was decreased with modulus of EMC was decreased because of lowering the stress relaxation. Therefore, the effect of the material properties of EMC on warpage was analyzed by FEM. Figure 7 and 8 shows the contour plots of von Mises stress and equivalent elastic strain distribution of EMC for the fan-out package component with various EMCs. As shown in Figures 7, the maximum von Mises stress and maximum elastic strain were concentrated on the interface between corner of the Si die and EMC. At 25 °C, the maximum von Mises stresses of the fan-out package

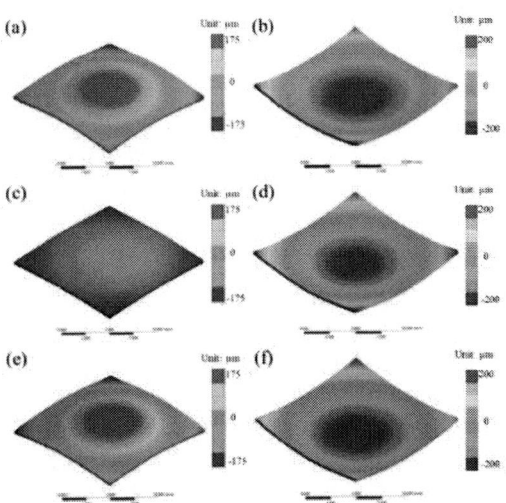

Figure 6. FEM analysis of warpage behavior for fan-out package component with various EMCs at (a,c,e) 25 °C and (b,d,f) 260 °C : (a,b) EMC1, (c,d) EMC2, and (e,f) EMC3

235

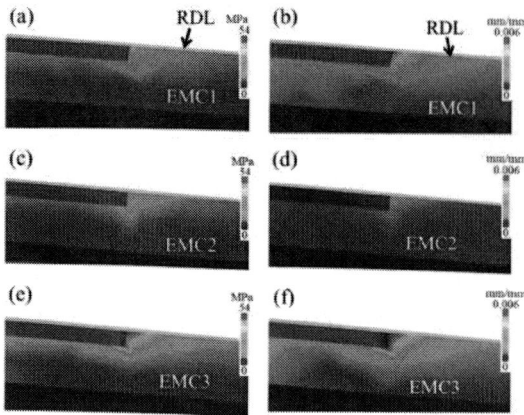

Figure 7. Contour plots of (a,c,e) von-Mises stress and (b,d,f) elastic strain for fan-out package component with (a,b) EMC1, (c,d) EMC2, and (e,f) EMC3 at 25 °C

component molded by EMC1, EMC2 and EMC3 was 38.29 Mpa, 28.18 Mpa and 53.19 MPa, respectively. As shown in figure 7, the maximum elastic stain of the fan-out package component molded by EMC1, EMC2 and EMC3 was the 0.00505 mm/mm, 0.00175 mm/mm and 0.00584 mm/mm, respectively. The CTE of EMC2 was smallest so thermomechanical stress induced by CTE mismatch was lowest. Therefore, the maximum von Mises stress of the fan-out package component molded with EMC2 was lowest. As shown in figure 8, maximum von Mises stresses of the fan-out package component molded by EMC1, EMC2 and EMC3 was 2.70 Mpa, 4.36 Mpa and 4.61 MPa, respectively. The maximum elastic stain of the fan-out package component molded by EMC1, EMC2 and EMC3 was the 0.00597 mm/mm, 0.00605 mm/mm and 0.00630 mm/mm, respectively. The simulated von Mises stresses, elastic stain of EMC1

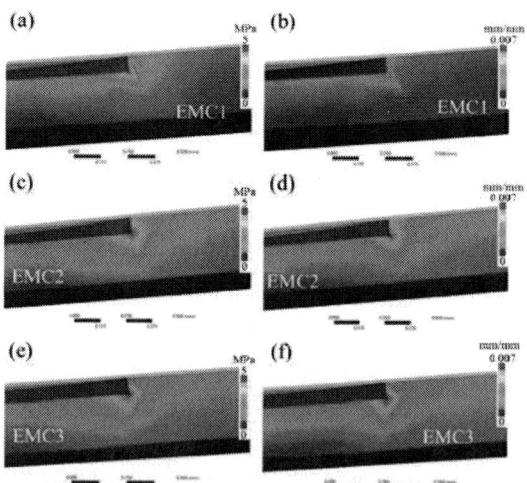

Figure 8. Contour plots of (a,c,e) von-Mises stress and (b,d,f) elastic strain for fan-out package component with (a,b) EMC1, (c,d) EMC2, and (e,f) EMC3 at 260 °C

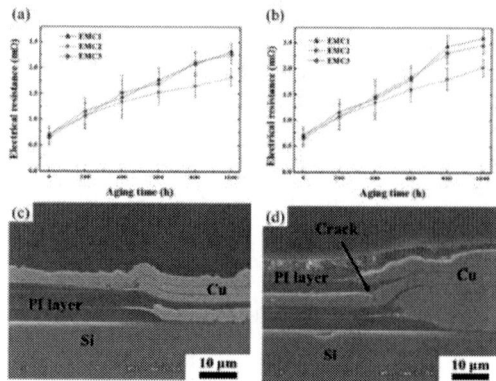

Figure 9. Electrical resistance of fan-out package component with various EMCs during (a) thermal shock test and (b) temperature-humidity test. Cross-sectional SEM micrograph of fan-out package component after (c) temperature humidity test and (d) thermal shock test

was smallest due to the lower stress relaxation of the EMC. At 260 °C, warpage of the package has more influence on warpage of the package than the CTE of EMC.

Figure 9 shows the electrical resistance and cross-sectional SEM micrographs of fan-out package component molded with various EMCs after temperature humidity test and thermal shock test. The electrical resistance before environmental reliability test was about 0.7 mΩ. The electrical resistance increased during environmental test at all fan-out package component. The electrical resistances of fan-out package component molded with EMC1, EMC2 and EMC3 after temperature humidity test for 1000 h was 2.32 mΩ, 2.20 mΩ and 2.26 mΩ, respectively. The electrical resistance at all component increased with increasing aging time. However, the electrical resistance did not significantly different after temperature-humidity test. After 1000 thermal shock cycles, the electrical resistances of fan-out package component molded with EMC1, EMC2 and EMC3 was 2.03 mΩ, 2.6 mΩ and 2.46 mΩ, respectively. the electrical resistances of fan-out package component molded with EMC1 was lower than those with EMC2 and EMC3. After temperature-humidity test. The discontinuous Cu electrode and defects were not observed after temperature-humidity tests as shown in figure 9 (c). However, crack was observed on the Cu electrode after thermal shock test as shown in figure 9 (d).

Conclusion

The warpage behavior of the FOWLP component molded with three kinds of EMCs were investigated by using shadow moiré method and finite element method analysis. Warpage of the fan-out package component was estimated from 25 °C and 260 °C. The warpage shape of fan-oout package component was chaged from convex to concave with increasing temperature. The warpage was affected by CTE and modulus of EMC. The warpage of fan-out package component molded by EMC with lower CTE was reduced because of smaller CTE mismatch between chip and EMC at room temperature. However, the warpage of fan-out package component molded by EMC with lower modulus was reduced the warpage even though CTE of EMC was higher than others due to stress-strain relaxation at 260 °C. The electrical resistance of FOWLP component increased with increasing aging time and cycles during temperature-humidity test and TS test. The warpage was smaller for lower EMC CTEs and lower EMC modulus values.

Acknowledgements

This research was supported by Basic Science Research Program through the National Research Foundation of Korea(NRF) funded by the Ministry of Education(2018R1D1A1B07047138).

This work was supported by the Technology Innovation Program (or Industrial Strategic Technology Development Program) (20003390, Development of Metal-Silicon Nitride Ceramics Core Part Bonding Materials for Eco-Friendly Electric Vehicles) funded By the Ministry of Trade, Industry & Energy (MOTIE, Korea)

References

[1] J.W. Kim, Y.C. Lee, S.S. Ha, S.B. Jung, J. Mater. Sci.-Mater. Electron."Failure behaviors of BGA solder joints under various loading conditions of high-speed shear test" J. Mater. Sci.-Mater. Electron. 20 (2009) 17

[2] A. Yoshida, J. Taniguchi, K. Murata, M. Kada, Y. Yamamoto, Y. Takagi, T. Notomi, A. Fujita, "A Study on Package Stacking Process for Packge-on-Packge (PoP)", Proceedings of the Electronic Components and Technology Conference (ECTC), USA, May 30 – June 2, 2006

[3] Y.L. Tzeng, N. Kao, E. Chen, J.Y. Lai, Y.P. Wang, C.S. Hsiao, "Warpage and Stress Characteristic Analyses on Packag-on-Package (PoP) Structure", Proceedings of the Electronics Packaging Technology Conference (ECTC), USA May 29 - Junge 1, 2007

[4] W. Lin, M.W. Lee, "PoP/CSP Warpage Evaluation and Viscoelastic Modeling", Proceedings of the Electronic Components and Technology Conference (ECTC), USA, May 27-30, 2008

[5] C.H. Liu, L.Y. Chen, C.L. Lu, H.C. Chen, C.Y. Chen and S.C. Chang, "Wafer Form Warpage Characterization Based on Composite Factors Including Passivation Films, Re-Distribution Layers, Epoxy Molding Compound Utilized in Innovative Fan-Out Package" Proceedings of the Electronic Components and Technology Conference (ECTC), Lake Buena Vista, FL, May 30- June 2 pp 847-852. 2017

[6] Chih-Hsun Hsu,Wen-Yang Li,Chi-Jen Chen,Yih-Jenn Jiang,Jui-Feng Tai, Chang-Fu Lin, C. Key Chung "Construction of FO-MCM with C4 bumps built first using Chip last assembly" Proceedings of the 2019 Electronic Components and Technology Conference (ECTC), Las vegas, USA, May 28-31 2019

[7] V. S. Rao, C. T. Chong, , S. W. Ho, M. Zhi, C.S. Choong, P. S. Lim, D. Ismael, and Y. Y. Liang, "Development of High Density Fan Out Wafer Level Package (HD FOWLP) with Multi-layer Fine Pitch RDL for Mobile Applications" 2016 IEEE 66th Electronic Components and Technology Conference (ECTC), pp. 1522-1529, 2016.,

[8] X. Hua, H. Xu, Z. L. Zhang, D. Chen, K. H. Tan, C. M. Lai, J. H. Lau, M. Li, M. Li, E. Kuah, N. Fan, W. Kai and K. Cheung "Development of chip-first and die-up fan-out wafer-level packaging", 2017 IEEE 19th Electronics Packaging Technology Conference (EPTC), pp. S23-1–6, December 2017.

[9] M. Takekoshi, K. Nishido, Y. Okada, N. Suzuki, T. Nonaka, "Warpage suppression during FO-WLP fabrication process", Proceedings of the Electronic Components and Technology Conference (ECTC), USA May 30 - June 2, 2017

[10] C. Palesko, A. Palesko, E.J. Vardaman, "Cost and Yield Analysis of Multi-Die Packaging using 2.5D Technology Compared to Fan-out Wafer Level Packaging", Proceedings of the Electronics System-Integration Technology Conference (ESTC), Finland, September 16-18, 2014

[11] S.S. Deng, S.J. Hwang and H.H. Lee, "Warage Prediction and Experiments of Fan-Out Waferlevel Package During Encapsulation Process" IEEE Trans. on compon. Package. and Manuf. Technol. 3, 3, pp452-458, 2013

[12] F.X. Che, D. Ho, M.Z. Ding, X. Zhang, "Modeling and Design Solutions to Overcome Warpage Challenge for Fan-Out Wafer Level Packaging (FO-WLP) Technology", Proceedings of the Electronics Packaging Technology Conference (EPTC), Singapore, December 2-4, 2015

[13] P. Sun, V.C.K. Leung, B. Xie, V.W. Ma, D.X.Q. Shi, "Warpage Reduction of Package-on-Package (PoP) Module by Material Selection & Process Optimization", Proceedings of the International Conference on Electronic Packaging Technology & High Density Packaging (ICEPT-HDP), China, July 28-31, 2008

Packaging Challenges and Reliability Performance of Compound Semiconductor Focal Plane Arrays

Stoyan Stoyanov[1], Chris Bailey[1], Rhys Waite[2], Christopher Hicks[2] and Terry Golding[3]

[1] Computational Mechanics and Reliability Group, University of Greenwich, London, UK
[2] Microchip Technology Inc., Caldicot, UK
[3] Amethyst Research Inc., Ardmore, OK, USA

Phone: +44 (0)20 8331 8520   E-mail: s.stoyanov@gre.ac.uk

Abstract

The development of new high-performance Focal Plane Arrays (FPAs) for imaging systems is driven by advances in photodetector material growth and processing, readout integrated circuits and IR detector chip hybridisation/packaging. The hybridisation of the IR detector chip and the readout integrated circuit (ROIC) through flip-chip bonding is a key packaging challenge for pixel arrays with very small indium bumps and 10-30 µm pitch sizes. This paper details the development and use of finite element models that can be used to assess and optimise the compression bonding process, and can enable insights into the impact of chip misalignment on the resulting flip-chip quality and the bonding equipment placement accuracy requirements for a given FPA specification. In addition, the fatigue performance of the indium interconnects of different fine pitch FPAs is evaluated and compared. The modelling results point that high quality interconnects and robust, defects-free assembly require micrometre placement accuracy. It is also possible that indium joints of higher resolution, larger size FPAs accumulate less damage under cryogenic temperature cycling compared to less dense, smaller in size, focal plane arrays.

Key words: Finite element modelling, focal plane array, compression bonding, flip-chip assembly, reliability, cryogenic temperature, compound semiconductors.

Introduction

Focal plane arrays (FPAs) are image sensing devices featuring typically a two-dimensional rectangular array of light-sensing pixels. They have been traditionally developed for military applications such as target acquisition, night vision, and weapon and missile seekers. In the past few decades, there has been substantial research and development in the area of IR detectors, particularly in relation to new detector materials and in the design and fabrication of very fine pitch, high resolution pixel array structures [1-3]. At present, FPAs are increasingly designed, fabricated and used in a wide range of non-military applications, from medical diagnostics and industrial process control to security and surveillance to astronomy and civil space applications [4-7]. There is a strong and emerging interest to expand the use of FPAs into the commercial market.

The operating principle of FPAs involves the pixels detection of incoming infrared radiation and conversion of photons into respective electrical signals in order to form an image. The focal plane array comprises of two parts: (1) the IR detector array and (2) the readout integrated circuit. The detector pixel array is the infrared-sensing element of the focal plane array and the readout integrated circuit is the signal processing component.

The IR detector and FPA technology developments are driven by demands for improved sensitivity and resolution of the imaging systems, needs for high-yield low-cost fabrication capabilities, and continuing advances in the sensing material compositions and their processing [8,9]. One of the main hurdles that is yet to be overcome is the establishment of supply chains for FPAs production and use within a wider commercial market.

The FPA packaging is also known as hybridisation. It is among the key challenges for the industry. This is the process of the physical assembly through flip-chip bonding and electrical integration of the IR detector chip and the ROIC. Indium solder is the industry standard for FPA interconnection material. The hybridisation concept/process directly impacts yield, reliability and reproducibility, and dictates the ability and level of success to which high resolution and small pitch size pixel arrays can be handled. Higher stand-off height and higher uniformity of the indium bumps have been pursued in FPA fabrication, and as a result manufacturing techniques for indium bumping of IR detectors and ROIC chips have been extensively researched and developed [10].

Advanced FPAs feature high density pixel/interconnect arrays utilising very small size indium bumps and chip contact pads which make the assembly of the detector chip onto the readout integrated and hence requires advanced, high precision placement equipment. The reason for this is that achieving high quality interconnection between the detector pixel matrix and the readout chip is a critical requirement for the proper functional performance of the imaging device.

In this paper, a computational approach utilising finite element models is developed and used to provide critical assessments of the technical challenges and requirements associated with the alignment accuracy and the compression bonding flip-chip process for the packaging of high-resolution, large pixel array imaging IR detector chips. The bonding process models are used for quantitative evaluation of the effects of inaccurate placement of the detector chip onto the ROIC, and enabled insights into the resulting indium joint characteristics such as shape and stand-off height and the risks for poor interconnection quality.

In addition, the thermo-mechanical fatigue damage of the indium solder interconnects in assembled FPAs under thermal cycling at cryogenic temperatures (300K to 77K) is also assessed. Most IR FPA are integrated with a cryocooler to provide operating temperature of the device below 150K. This is required in order to reduce the thermally-induced noise and the noise-to-signal ratio and thus improve dramatically the FPA spatial resolution and sensitivity characterises. As liquid nitrogen is the most common cooling solution, operating temperature of 77K defines the thermal load considered in this investigation. The modelling study aimed to generate new insights into the reliability performance trends of FPAs when respective resolution/pixel matrix size increase and I/O pitch sizes decrease.

FPA Specifications: Geometry and Materials

Geometric Data

IR imaging sensors with pixel arrays of 320x256, 640x512 and 1280x1024 are investigated. The respective FPAs feature pitch size in the range 12 μm to 30 μm, indium interconnect volumes from 0.896E-6 mm^3 to1.4E-6 mm^3, and contact pad sizes in the range of 8-10 μm. An outline of a typical IR detector chip/ pixel array structure, with details for the mesa pixel and pad layout are presented in Fig. 1. In addition, this figure shows also a schematic of the pre-bonded flip-chip stack of the detector and the ROIC chips. Larger in size non-pixel peripheral joints (supported by respective pad layouts) complete the FPA assembly interconnection requirements.

The key specifications of the three investigated high resolution FPAs are provided with Table 1. For example, the 1280x1024 IR detector chip has pitch size 12 μm and mesa pixel size 9 x 9 μm, with a 3 μm gap between two adjacent pixels.

The contact pads and the mesa pixels have sub-micron thickness. The contact pads on the ROIC and the detector chips have square shape with dimensions as reported in Table 1.

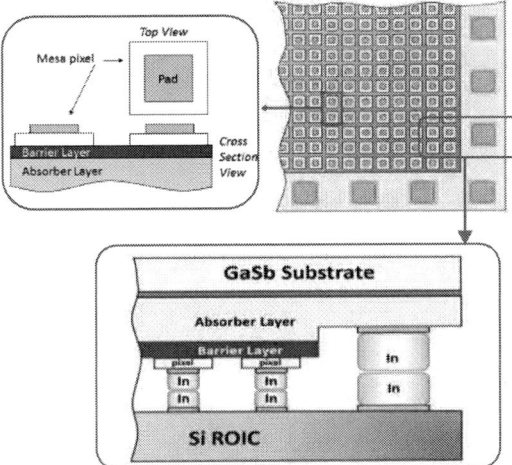

Figure 1: IR detector pixel array outline and schematic of the pre-bonded flip-chip assembly (IR detector and ROIC chips).

Table 1: FPA specifications.

FPA Assembly	FPA #1	FPA #2	FPA #3
Pixel Array	320x256	640x512	1280x1024
No. of pixels	81,920	327,680	1,310,720
Pixel Pitch (μm)	30	20	12
Pixel size (μm)	25x25	17x17	9x9
Contact pads (μm)	10x10	10x10	8x8
ROIC Chip Reference	FLIR ISC9809	FLIR ISC0402	FLIR ISC1308

As detailed in Table 1, commercial off-the-shelf ROIC chips from the infrared and imaging systems company FLIR have matching specifications for the three detector pixel array format resolutions. For example, the readout integrated circuit for the assembly of the 1280x1024 IR detector chip is the FLIR ISC1308 ROIC [11]. This ROIC has a 1280x1024 contact pads matching layout at 12 μm pitch size.

The IR detector chips are Group III-V compound semiconductor materials. The gallium antimonide (GaSb) substrate of the detector chip is a pre-assembly feature only and commonly designed to have thickness comparable with the thickness of the ROIC. Once the FPA hybrid stack is formed and the IR chip is bonded onto the ROIC, the GaSb substrate is removed. Thus, the photo-sensing chip in a final FPA is only few microns tick, comprising the absorber layer, barrier layer and the mesa pixels only as detailed in Fig. 1. A solder joint is assumed to be obtained by two pre-bumped indium deposits – on the IR detector chip and the ROIC pads respectively. For the discussed assemblies in this work, the indium deposits are equally split volume-wise between the opposing pads of the flip-chip structure (see Fig 1, bottom).

Material Data

The numerical (finite element) modes for the analysis of the FPA assembly bonding processes and the thermo-mechanical FPA reliability performance under cryogenic temperature cycling loads require also material data in addition to the geometric and topology data of the flip-chip hybridised chips. In the models, FPA materials, excluding the indium solder, are considered as having elastic behaviour. Temperature dependent data, where available, is included in the respective models. The required material properties for the ROIC chip (Si), IR detector chip substrate (GaSb), contact pads (Au) and interconnect material (In) include Young's modulus, Poisson's ratio and coefficient of thermal expansion (CTE). These are summarized in Table 2. The utilized linear coefficient of thermal expansion data for silicon and indium is sourced from reference [12]. The absorber layer, pixel and barrier layer materials, as illustrated in Fig. 1, are all Group III-V compound semiconductor materials with values for Young's modulus, Poisson's ratio and CTE in the range of 62 MPa, 0.32 and 5.5-5.9 ppm/K respectively.

The material behaviour of the indium solder in particular is very important for the utilized simulation models. The indium visco-plastic behaviour is modelled using the Anand inelastic strain rate model [13].

Table 2: Material data used in models.

Material	Young's Modulus (GPa)	Poisson's Ratio	CTE (ppm·K^{-1})
Indium (In)	20.54 @ 77K 16.24 @ 187K 12.70 @ 300K	0.433 @ 77K 0.441 @ 187K 0.449 @ 300K	Ref [12]
Silicon (Si)	168.0	0.28	
Gold (Au)	77.2	0.42	14.4
Gallium antimonide (GaSb)	76.0	0.30	1.4 @ 77K 4.5 @ 140K 6.0 @ 200K 6.5 @ 300K

Compression Bonding Modelling of FPAs

Compression bonding is the most common hybridisation technique for of the IR detectors because of the indium's ductile and malleable material behaviour. The ability to model the compression bonding of FPAs can be very beneficial. It can help the assessment of the effects of chip misalignment on the resulting feasibility and quality of indium bump formation, and can be also used to provide model-based capability for process optimisation. These benefits are demonstrated in this section of the paper.

Modelling Approach

The modelling approach is developed to enable the assessment of the impact which applied compression loads on the IR detector chip have on the indium joint formation process. Finite element method and ANSYS simulation software [14] are employed to develop two-dimensional finite element model of the FPA assembly at local level, representing one pixel unit spatial domain and the formation of a single joint within the full assembly array.

Figure 2 details the finite element model of the FPA for simulation of the compression bonding condition and the application of the respective model boundary conditions (BC): (1) the applied pressure (i.e. bonding force) and (2) the structural constraints (displacement degrees-of-freedom, DOF) along with the contact boundary for the indium-to-indium joint formation. The contact boundary is defined over the regions where the expected (i.e. IR detector chip indium bump to ROIC indium bump contact boundary) or potential (indium to vicinity of the pad area) contact may take place during bonding. ANSYS contact pair (element types TARGE169 and CONTA172) modelling capability is exploited, with "no separation" condition for the behaviour of the contact surface and friction coefficient 0.2.

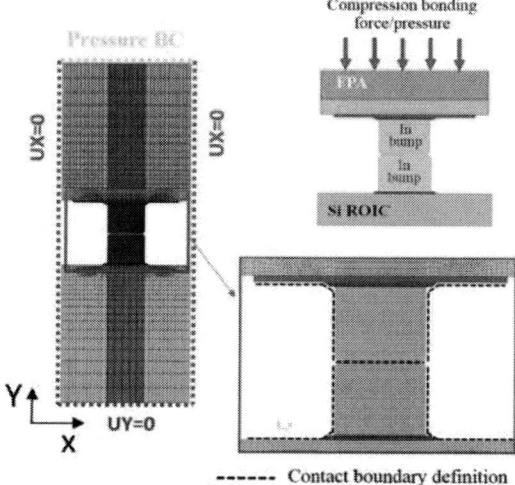

------ Contact boundary definition

Figure 2: Finite element model for compression bonding simulation of focal plane arrays.

Non-linear transient simulations are undertaken under the conditions for large deformation. The formation of the single indium joint is a result of a diffusion process (not simulated here) at the established contact interface of the two indium deposits and under sustained bonding force that is applied. The developed model assumes that there is no lateral movement of the flipped chip during bonding. With this study, the modelling results are obtained under the assumption for bonding process undertaken at room temperature.

The applied compression bonding load profile consists of the phases. The pressure is first ramped linearly from 0 to the load profile peak value (P_{max}), and then, in the second phase, the pressure is maintained constant at that level for the remaining duration of the load application. The reported results are obtained with a load profile where the ratio of pressure ramp up time to constant pressure time is 1:5. The applied pressure (or force) peak value and the ramp up/dwell durations of the load application are process parameters that need to be uniquely optimised for a given FPA so that the shape and the bonding quality of the resulting indium joint at the end of the applied load profile meet specification requirements.

Modelling Results

An accurate placement of the IR detector chip onto the ROIC chip so that respective contact pads on both sides for the array of indium joints match is the key requirement to enable the assembly by means of applying a compression force. With the developed process model, it is possible to evaluate how different placement accuracy may impact the formation of the indium joints. The accuracy of placement of the IR detector chip can be defined with the value of Δp (see Fig. 3, top) which measures the offset between two corresponding pads (on the ROIC and IR detector sides). Here, two different levels of misalignment are assessed; referred to as Cases A and B (see Fig. 3, bottom):

- Case A: Exact positioning ($\Delta p = 0 \mu m$)
- Case B: Moderate misalignment ($\Delta p = 3 \mu m$).

Below, the modelling predictions are demonstrated only in the case of bonding the 320x256 FPA. Analysis of misalignment effects for any other FPA, with different pixel resolution and geometric specification, and using the same process model can be performed in a very similar way.

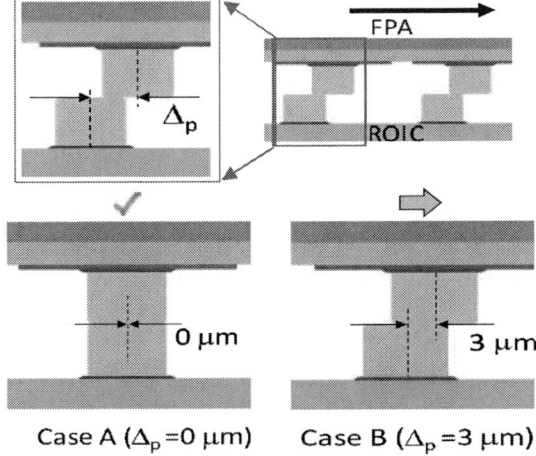

Figure 3: Simulated misalignment cases for the compression bonding of the 320x256 FPA.

Figure 4 details the shape deformation of the indium joint assuming the placement accuracy scenarios (Cases A and B). The last row details the deformed shape of the final joint achieved at the end of the applied bonding profile. It should be noted that due to the inelastic (visco-plastic) behaviour of indium, the deformation of the formed joint takes place throughout the entire application of the bonding pressure/force, including during the phase of the profile where the applied pressure is kept constant. The final shape of the formed joint is achieved at the end of the compression profile when the pressure from the bonding force is removed.

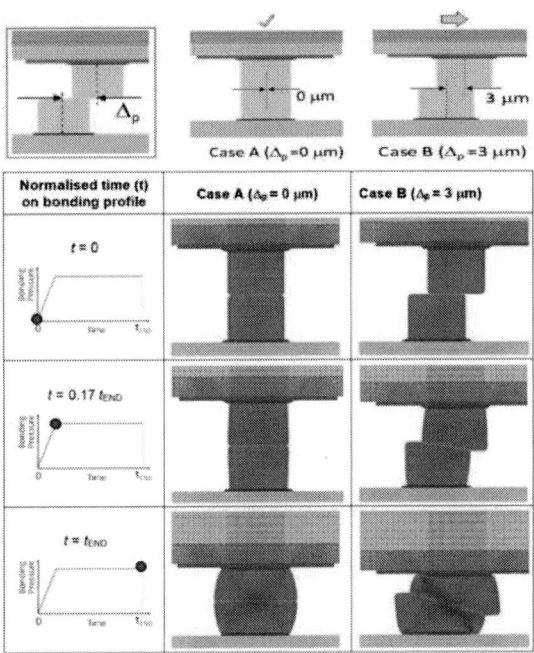

Figure 4: Simulations results from compression bonding analysis of the effect of placement accuracy on indium joint formation for 320x256 FPA.

The main conclusion is that misalignment of even few microns may be difficult to accept and tolerate at such small dimensions of the formed interconnects. In this instance, a misalignment of 3 μm may still result in the formation of a joint but, as evident from the modelling results, the indium joint shape is compromised quality-wise and the interconnect comes with a very uncertain (variable) and significantly lower stand-off height. Presence of such attributes mean that the FPA performance and expected reliability will be compromised. Although for this specific FPA assembly there is no risk of bridging between adjacent joints, misalignment is also problematic in the context of implementing a robust control of the bonding force magnitude needed to avoid excessive indium bump collapse and avoid the risk of joint formation outside the area of the contact pad.

It is clear that the formation of joints is highly sensitive not just to the placement accuracy of the IR detector chip onto the ROIC chip but also to the bonding profile parameters: (1) the maximum pressure/binding force, and (2) the load ramp and dwell time durations. Because the process model is capable of assessing the magnitude of the joint deformation during bonding, it can be used to support the bonding process optimization for given equipment and alignment accuracy.

An example of such model use can be given by assessing the compression load magnitude changes required in the case of the three FPAs discussed in this work. This is because the FPAs have different pixel array density and additional geometric variations. Assuming perfectly aligned flip-chips for the bonding of the 320x256, 640x512 and 1280x1024 FPAs, respective process models were employed and used to identify the required bonding pressure condition that can achieve a predefined target stand-off height value of 9.5μm for the indium joints. For the analysed compression bonding profile with duration indicated by the normalised time value t_{END} and ratio of load ramp up time to dwell time at peak pressure 1:5, the optimal maximum pressure value (P_{max}) for the compression bonding has scaled across the three focal plane arrays as follows:

- 320x256 FPA: $P_{max} = P_{nom}$;
- 640x512 FPA: $P_{max} = 1.5\ P_{nom}$;
- 1280x1024 FPA: $P_{max} = 2.0\ P_{nom}$.

These results show that compression bonding set up is not trivial and changes substantially as IR detector arrays change specification. The process of identifying the optimal process parameters of the compression bonding profile can be greatly aided by adopting the discussed process models.

Reliability Modelling of FPAs

The thermo-mechanical modelling of the thermal fatigue damage in indium joints of FPAs under cryogenic temperature cycling load conditions can aid the qualitative analysis for the expected reliability performance for different high resolution focal plane arrays. Here, such demonstration is detailed for the three FPA's being investigated. As no failure data for indium joints of FPA structures and respectively no life-time model are available, at present the reported model predicted damage induced in the indium joints of the FPA is not correlated to corresponding cycles to failure.

Thermo-mechanical modelling

A thermo-mechanical fatigue model for each of the three high resolution FPAs is constructed as a full three-dimensional (3D) slice model. A model captures explicitly all assembly details in a cross-sectional slice along the diagonal line of the FPA. Note that the GaSb substrate is only a pre-assembly feature of the detector array and therefore once the

assembly (hybridization) is completed, the substrate is removed and is not part of the final FPA. Due to existing symmetry along the diagonal line of the FPA, the slice model includes the domain from the central (neutral) point of the assembly to the corner of the ROIC chip, through the whole assembly thickness.

The slice modelling approach is superior to two-dimensional models and is an appropriate, accurate enough, modelling strategy when full 3D model is not practical (as in this case due to extremely large pixel arrays). Figure 5 illustrates the slice model developed for the 320x256 FPA, with similar models also developed for the FPAs with resolutions 640x512 and 1280x1024 pixels. The figure provides also a detailed view on the finite element mesh and the geometric representation at the level of a single solder joint in the assembly. These models assume no misalignment of the flipped IR detector chip and thus perfectly formed shape-wise indium solder joints.

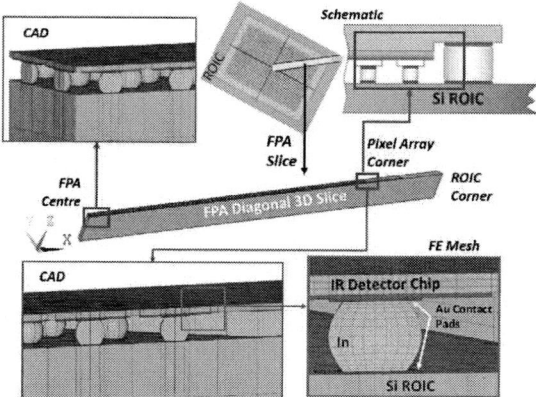

Figure 5: 3D slice model illustration developed for the fatigue analysis of indium joints in 320x256 FPA.

Non-linear transient thermo-mechanical analysis that simulates the responses of the FPAs under applied (imposed isothermal) temperature cycling load is undertaken using ANSYS simulation software. The domain is meshed with element type SOLID185. The range of the temperature cycle is 293K to 77K over 1 hour, with 15 minutes ramp/dwell times. Mesh size of the developed 3D slice models for the three FPAs is in the range 330,000 to 575,000 mesh elements. Standard for a 3D diagonal slice model structural boundary conditions are applied: (1) for the displacement degree-of-freedom (DOF) in perpendicular direction to the slice we have symmetry on the "front" face of the slice and coupled degree-of-freedom (out-of-slice plane) for the nodes on the "back" side of the slice, (2) all DOF fixed at a point along the central line of the assembly, and (3) symmetry at the face of the slice defining half-diagonal symmetry.

The low cycle thermal fatigue, being driven by the accumulation of permanent deformation (i.e.

inelastic strain) experienced by the solder joints, can be evaluated with the simulation models by predicting the plastic work per unit volume in the solder joints. The fatigue damage indicator is therefore defined as the accumulated visco-plastic strain energy density accumulated in the indium joints over one temperature cycle. The damage indicator, denoted as ΔW_{ave}, is calculated as an average of the accumulated visco-plastic strain energy density (over one temperature cycle) in an interfacial layer of the indium joint with the contact pad where the highest damage is predicted, and hence where the crack is expected to initiate and propagate. This calculation is undertaken on the most stressed joint, expected and confirmed to be the corner one in the array for the studied FPAs, with layer thickness for the averaging result 0.5 μm. Such averaging approach is best practice as point results can be influenced by finite element mesh sizes and stress singularity effects.

Modelling results for thermal fatigue damage in indium joints

The thermo-mechanical simulation results show that with all three FPA structures the most critical (most damaged) joint in the array is the one located at the corner. The maximum concentration of the accumulated inelastic energy density is found at the interface of the joint with the contact pad on the ROIC side of the package, and at the pad region facing towards the centre of the assembly (inwards). The simulation results for the predicted thermal fatigue damage are illustrated in Fig. 6.

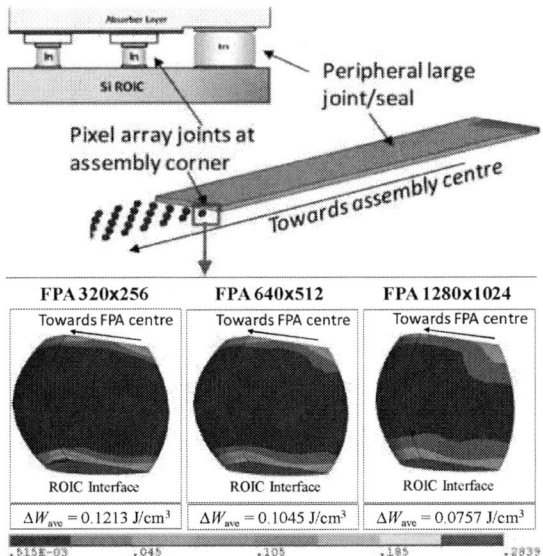

Figure 6: Simulation results for the thermal fatigue damage of indium solder joints in the form of plastic work (J/cm³) accumulated over one temperature cycle. The results detail the state of the most critical corner solder joint of the studied high-resolution FPA assemblies.

Based on the predictions obtained with the finite element models, the damage parameter (ΔW_{ave}) for the critical indium joint is found to be:

- 14% lower with the 640x512 FPA compared to the 320x256 FPA assembly;
- 38% lower with the 1280x1024 FPA compared to the 320x256 FPA assembly.

Figure 7 provides a graph (point result where the peak plastic work is found) that details how the damage accumulates over the temperature cycle. The damage is predominantly induced as the temperature load is ramped, up or down, while much smaller accumulation of damage takes place during the dwell periods of the cycle.

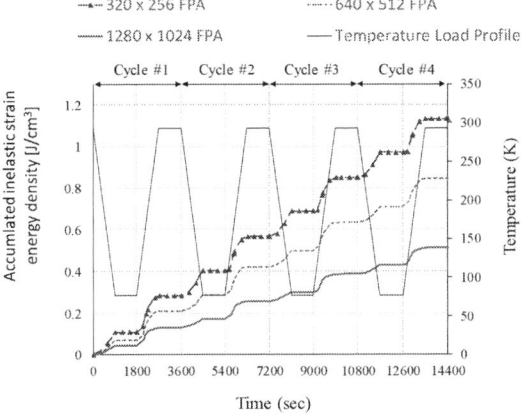

Figure 7: Accumulation of damage in indium joints during cryogenic temperature cycling – maximum inelastic energy density point result associated with the critical corner indium joint of the FPA assemblies.

These thermo-mechanical results for indium solder damage show that as the pixel array increases its resolution and density, and despite the fact that the size of the assembly gets larger (and thus the corner joint gets further away from the assembly's neutral point), the damage in the indium joints in the array, including the one at the corner, is getting smaller. The above finding is important and is not trivial. It can be only confirmed either through the presented modelling approach or by undertaking actual experimental tests of thermal cycling on the three FPA assemblies.

The reason for the finding that more complex, high density pixel array assemblies may potentially be more reliable in the context of the thermal fatigue reliability of the indium solder interconnects is this: the shear strain impact from the in-plane linear CTE mismatch between the ROIC and the detector chip is accommodated better as the array size increases and this causes the shear strain per joint to get smaller. This can be seen as an improved ROIC - IR detector chip compliance that can be potentially featured by larger pixel arrays.

It should be noted that the 1280x1024 FPA joints have smaller interfacial area with the contact pad and hence in real terms a crack will need to propagate over smaller distance to cause the failure. While this is an implication for the actual time-to-failure, the damage prediction itself will still imply having the slowest rate of crack initiation and propagation with the high risk FPA.

The large peripheral indium interconnect is found to play, in addition to the electrical signalling functional role, an important structural role. Simulation results not detailed in this paper reveal that the larger peripheral joints reduce the value of the predicted thermal fatigue damage, ΔW_{ave}, in the corner solder join of the array by as much as 75-80%. The damage in the large peripheral joint itself should not be of a concern as the damage is at lower or similar level as for the corner array joint but the length of the crack to fail the joint is greater by factor of 50 (8-10 µm for the array joint vs >500 µm for the peripheral joint).

Conclusions

Finite element models for the compression bonding process of flip-chip hybridisation of high resolution IR detector chips with readout integrated circuits and for the thermo-mechanical reliability performance of the resulting FPA fine-pitch indium solder joints have been developed and demonstrated. The models enable critical capabilities for the optimisation of compression bonding process parameters and assessing the flip-chip alignment requirements as well as supporting FPA design-for-reliability activities.

The main findings from the undertaken studies in relation to the IR detector and the ROIC chip hybridisation are as follows:

- Misalignment of the IR detector chip onto the ROIC in the flip-chip bonding process has major impact on the quality of the resulting joints in terms of shape, out-of-pad contact and most importantly the expected stand-off height. This can affect the long-term performance and reliability of the focal plane array.
- Uncertainty in placement accuracy makes the optimal compression bonding parameter setup problematic as different levels of misalignment result in different characteristics of the indium joints under same compression bonding conditions. This makes practically impossible to adopt a robust assembly condition that can assure high quality of the resulting FPAs.
- Assembly equipment capable of providing accuracy of chip placement <3µm and better is a key requirement.

In terms of the fatigue reliability of the indium interconnects of FPAs under temperature cryogenic cycling loads, the modelling study enabled to conclude:

- The most damaged (corner) joints in higher resolution FPAs may accumulate less damage compared to less dense focal plane arrays. This can happen despite the fact that the higher resolution FPAs may be larger in size and thus have the corner joints at larger distance from the neutral (FPA centre) point of the assembly.
- Although the above result was observed in this work, it needs to be model-confirmed in each instance of a given FPA structure. This is because the shear force distribution depends on both the total number of interconnects and the actual physical land size of the matrix array (respectively the pitch size).

Acknowledgements

This work has been funded by Innovate-UK through the "Technically high element alignment (THEIA)" project, project number 103439.

References

[1] P. Martyniuk *et al.*, "New concepts in infrared photodetector designs", Applied Physics Reviews, Vol. 1, Issue 4, pp. 1-35, 2014.

[2] R. Zhao, W. Ma, S. Wang, X. Yu, Y. Feng and Y. Zhao, "Design and fabrication of a sandwich framed focal plane array for uncooled infrared imaging", Proc. 18th International Conference on Solid-State Sensors, Actuators and Microsystems, Anchorage, AK, USA, June 21-25, pp. 1318-1321, 2015.

[3] Y. Ou *et al.*, "Design, fabrication, and characterization of a 240×240 MEMS uncooled infrared focal plane array with 42-µm pitch pixels", Journal of Microelectromechanical Systems, Vol. 22 , No. 2, pp. 452-461, 2013.

[4] C. Downs and T.E. Vandervelde, "Progress in infrared photodetectors since 2000", Sensors, Vol. 13, pp. 5054-5098, 2013.

[5] A. Rogalski, "Infrared detectors for the future", Acta Physica Polonica, A, Vol.116, No. 3, pp. 389-406, 2009.

[6] Antoni Rogalski, "Infrared Detectors", Taylor & Francis, Boca Raton, 2nd edition, 2011.

[7] T. Sprafke and J.W. Beletic, "High-performance infrared Focal Plane Arrays for space applications", Optics and Photonics News, Vol. 19, Issue 6, pp. 22-27, 2008.

[8] A. Karim and J.Y. Andersson, "Infrared detectors: Advances, challenges and new technologies", IOP Conference Series: Materials Science and Engineering, Vol. 51, 01200, pp. 1-8, 2013.

[9] A. Rogalski, "Next decade in infrared detectors", Proc. SPIE Vol. 10433, Electro-Optical and Infrared Systems: Technology and Applications XIV, 104330L, 2017.

[10] J. Jiang *et al.*, "Fabrication of indium bumps for hybrid infrared focal plane array applications",

Infrared Physics & Technology, Vol. 45, Issue 2, pp. 143-151, 2004.

[11] FLIR, https://www.flir.co.uk/products/isc9809/

[12] X. Cheng, C. Liu and V.V. Silberschmidt, "Numerical analysis of thermo-mechanical behavior of indium micro joint at cryogenic temperatures", Computational Materials Science, Vol. 52, pp. 274-281, 2012.

[13] R.W. Chang and F.P. McCluskey, "Constitutive relations of indium in extreme temperature electronic packaging based on Anand model", Journal of Electronic Materials, Vol. 38, No. 9, pp. 1855-1859, 2009.

[14] ANSYS, www.ansys.com

Evaluation Method of Electronic Performance Margins for Sip Based on Field-Circuit and Multiple Physical Cooperative Simulation

Zhaorong Li[1], Gang Xu[2]*, Xuejiao Huang, Liming Lv, Rong Zeng and Wei Zhong

Institute of Electronic Engineering, China Academy of Engineer Physics, Mianyang, China

[1]lizhro@foxmail.com, [2]xugangthu@163.com

Abstract

Owing to advantages, such as multi-function and high integration, etc, system in package(SiP) is more and more popular in the electronic industry. But when designing a SiP product, it also put forward higher requirements to design for reliability, environmental adaptability, for example. Among them, the primary consideration is the more accurate margin analysis of electrical performance in different working environments. Electrical performance evaluation is no longer a simple link simulation. It needs to consider the impact of internal and external working environment including the electromagnetic(EM), thermal and mechanical field for the system which all need to be accurately considered in the electrical performance margins evaluation for the SiP. Based on the method of field-circuit and multi-physical cooperative simulation, an accurate method for evaluating the electrical performance margin of SiP is proposed this paper. In this evaluation method, we can bring many key influencing factors, such as the structure, IC & components, the material parameter fluctuation, the working environment condition, and so on. Using this collaborative simulation, we can establish different accuracy of the evaluation model of the electrical performance margin, which is suitable for the design requirements of different stages of the system. In order to validate this evaluation method, we have designed and manufactured a simplified verification model for multi-physical field simulation from a practical radar SiP product.

Key words: SiP, electrical performance margins evaluation, multi-physical Coupling Simulation, Field-Circuit Cooperative Simulation

Introduction

With the rapid development of modern electronic technology, electronic products are developing towards multifunction, high performance and miniaturization. To that end, the most famous system integration concepts: system in package(SiP) and system on chip(SoC) are just put forward. Comparing with SoC, SiP technology has more advantages [1]. And in recent years, System in Package (SiP) has been more and more popular in the electronic industry because of the advantages of multi-function, high integration and high performance. While SiP's high density and complex hybrid integration make performance improvement, but also brings new problems, such as more difficult process manufacturing and difficult debugging [2]. So, when designing SiP a product, potential reliability problems, for example, environmental adaptability, require us to find out as accurately as possible in advance. Among them, the primary consideration is margins evaluation for the performance of the system in different environments. Nowadays for the electrical performance evaluation of SiP, simple link simulation can not meet the requirements of accurate evaluation. It is necessary to consider the influence of complex working environment from internal and external EM, thermal and mechanical fields on the electrical performance

of the system, so as to achieve the purpose of high accuracy margin evaluation, provide high reliable guidance for the design of SiP and improve the reliability of the product.

This paper, the evaluation method of electric performance for the SiP design is firstly explained in detail. The main idea is multi-physical and field-circuit cooperative simulation. Then by using this evaluation method, we analyzed the power gain and the isolation of stray signal in three different temperature working environment. At last, the margins were evaluated, and this method was verified at the same time.

Multi-Physical Field Coupling Simulation

In the process of working, the performance of electronic system depends not only on its chips and components, but also on EM interference, thermal effect, stress and deformation and so on. With the increasing integration, especially in SiP, multi-physical fields have a significant impact on the electrical performance. Based on the workbench platform of ANSYS, the multi-physical field coupling simulation of EM-thermal-mechanical can be concluded together. So that the impact of the thermal distribution, stress and deformation in different external environments for the SiP's EM characteristics could be took in performance analysis. As shown in Figure 1, it can be used as an input file for the analysis of system performance margin, and improve the accuracy of evaluation.

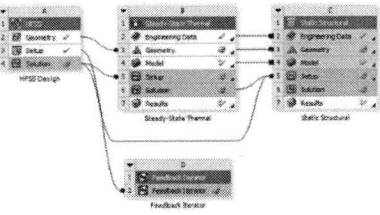

Fig. 1. Multi-physical field coupling simulation on workbench platform

Field-Circuit Cooperative Simulation

The method of field-circuit cooperative simulation analysis is based on SNP model [3], as shown in figure 2. By way of combining EM filed and system link simulation softwares, the EM characteristic of the structure and the IC performance could be analyzed together to ger a more accurate evaluation for the system performance. The key model, as shown in Figure 2 SNP model is a scattering parameter model, which is widely used in many electronic applications. It has strong generality and practicability. Therefore, this model is used as the data model of field-circuit cooperative simulation and analysis method in this paper.

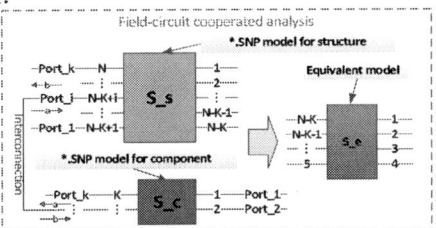

Fig.2. Field-Circuit Cooperative Simulation

Evaluation Method of Electrical Performance Margins for SiP

Based on the method of multi-physical field and field-circuit cooperative simulation, a method for evaluating the electrical performance margins of SiP is proposed, as shown in Figure 3. In this evaluation method, we can bring many key influencing factors, such as the structure, IC & components, the material parameter fluctuation, the working environment condition, and so on to consider together. But as regarding the level of the complexity and the requirement to the simulation work, we divided different evaluation level.

In the stage of system scheme design, on the ADS platform, the primary evaluation of the three-temperature(-55, 25, 85) electrical performance can be realized by using the behavior-level circuit model and the three-temperature test data of ICs and components. Owing to this evaluation only considers the effect of temperature on the performance for the Ics and components, so that it could be used as the pre-simulation performance verification.

In the stage of system board design, by method of field-circuit cooperative simulation, the factors of structure, such as EM loss, coupling and so on can considerd in the evaluation of electrical performance. This could be regarded as the secondary level evaluation. At this level, the transmission loss optimization and EMC analysis for the SiP would be finished.

After the design of the board and the package structure for the SiP was completed, the working state of SiP under three temperatures can be simulated based on the method of multi-physical field. For this deign stage, the environmental adaptability anlysis is necessary and important. In the single thermal simulation, the main heat sources are ICs and components, and the simulation condition is just the equivalent working environment. But if through multi-physical simulation, the heat from transmission loss on metal and dielectric will be regarded as another kind of heat source for thermal analysis. As a result that this thermal distribution could be calculated more accurately than the single thermal analysis. According to the thermal distribution, the temperature of the key Ics and components can be able to read out, therefore the real performance of those ICs and components could be calculated from the temperature's measured performance data by means of interpolation. At this stage, we considerd the influence of temperature, that also the great influence factor for a SiP product.

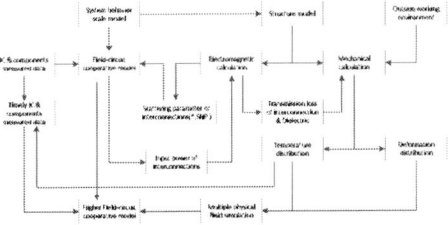

Fig. 3.The evaluation procedure of SiP electronic performance margins

When coming to mechanical analysis stage, the thermal-stress and the loading condition will be analyzed for the package structure. At different ambient temperature and working state, those influence factors could be shown in the deformation and stress distribution. Thus the deformation can be considered into the electromagnetic characteristic by the multi-physics method. This deformation-electromagnetic analysis is very important for high RF system especially. Because even the deformation is so small compared with the structure, but there is some influence to the centre frequency, which is very important for resonator such for example the filter. For this stage, the job is very heavy, but the efficiency maybe not so nice, so we do not suggest do this work for every SiP.

All in all, based on the method of field-circuit and multi-physical cooperative simulation, the electric performance of the SiP could be evaluated accurately. But the SiP's high integrity and complexity result in high difficulty, and there are so many factors to be considered. In a word, we need flexible use this evaluation method to analysis the performance according to the real demand at different design stage.

as well as structural deformation on the electrical performance of SIP by be considered according to the demand, and the evaluation of the three-temperature electrical performance margin of SiP can be realized. At this stage, there are many factors to be considered, the structure is complex, the workload is large, and the result precision is high. As the final step of system post-simulation verification. So that different accuracy level margins evaluation of the electric performance for the SiP would come out.

Experimental verification module

In order to verify the feasibility and accuracy of this evaluation method based on multi-physical field and field-circuit cooperative simulation, it was applied in practical projects, and an experimental verification module was designed to carry out further verification.

We all know that for a complete SiP product, its structure has tens thousands of complex structure, and hundreds of various components, as a result that the multi-physical field simulation will be very difficult. And a complete SiP product always has a high cost. Those mean that a complete is not suitable for extensive experimental comparison and verification. Therefore, based on the principle of retaining the basic functions and structural characteristics of a microwave SiP, a multi-physical field verification module was designed as shown in Fig. 4. Its main features are as follows showing. In the circuit function, it contains radio frequency functions such as frequency conversion, amplification, embedded filter, power heating and so on. In the packaging structure, it has BGA stacking, SMT table sticking, bonding wire, tin welding, connector, etc. It not only retains the basic characteristics of SiP module, but also simplifies the structure and cost as much as possible, so that has the ability of experimental verification.

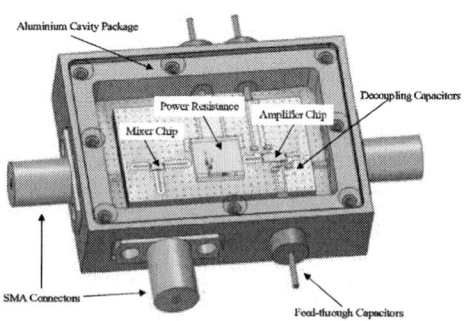

Fig. 4. 3D structure of Experimental verification module

In this experiment module, mixer chip is reserved to simulate the effect of signal intermodulation caused by frequency conversion in SiP. High-power devices such as power amplifiers are simulated by power resistance to generate high heat, which leads to the rise of system temperature and affects electrical performance. Retained amplifier to simulate the influence of non-linear devices on signal. Embedded cross-toe filter, as shown in Figure 5, simulates thermal stress deformation how to produce influence on passive structure performance.

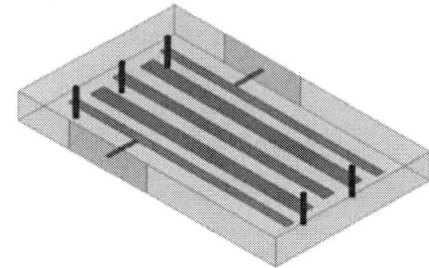

Fig. 5. 3D structure view of the embedded cross-toe filter

Primary Evaluation with IC performance

In the initial stage of SiP design, combining the performance of the chips and the basic function units, with the system circuit link, we realized the primary evaluation of the electrical performance of the system. As shown in Fig. 6, a primary evaluation model for the electrical performance of the module is established. The simulation results are shown in Fig. 7. The figure (a) is circuit link gain and the left one is the output signal frequency spectrum. It can be seen that the circuit link gain of the module is 9.29-9.39 dB, the in-band flatness is about 0.1 dB, and the stop-band rejection for the noise is upto 41.2 dBc.

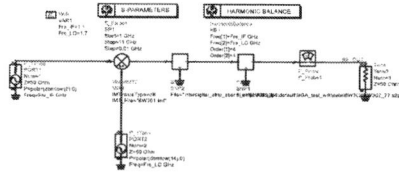

Fig. 6. System link with IC performance

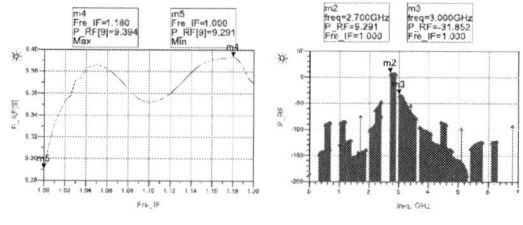

(a) Gain   (b) frequency spectrum

Fig. 7. The electric performance evaluation on (a) and (b) of the system

248

EM Field Simulation

After circuit design, board and package design, it comes to time to analysis the EM characteristic. Here we mainly analyzed design transmission loss and the EMC. The simulation results are shown in Fig. 8. The transmission characteristics of the embedded cross-toe filter and the interconnection structure are shown in the (a) figure. The EM field distribution of the module is shown in the (b) figure, the input power is come from the system circuit simulation, in order to has a better EM field distribution to analysis the EMC. According to the results, we can say that the loss and filter both have good performance, and the EM field also meet the design demand.

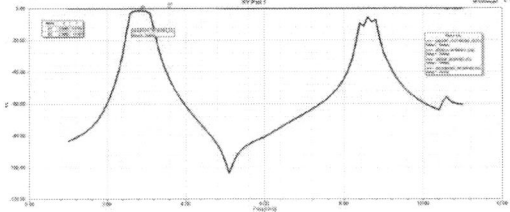

(a) Transmission characteristics

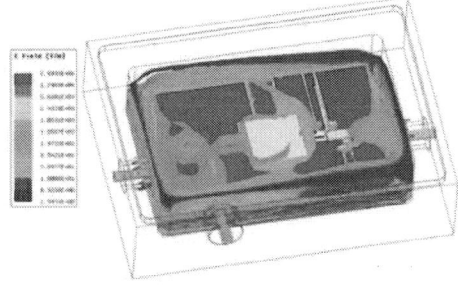

(b) Electric field distribution

Fig. 8. The (a) and (b) of the verification module

Thermal Distribution Simulation

Through the simulation works above, we think the system meet function basically, but we don't know how about the performance in the real working state. This is the work we did at next staps. When the module works, the power resistor heats up 5W and the amplifier power is 0.35W. Through thermal simulation analysis, the thermal distribution of the module at three temperatures environment (-55 ,+25 ,+85) were all obtained as shown in figure 9, and the temperature of the chips and components were extracted as shown in table 1. Based on the linear interpolation method, the three-temperature performance test data of the amplifier are extended, and the working performance of the module at room temperature is obtained. As shown in Figure 10, it can be seen that the signal gain of the amplifier decreases with the increase of temperature. At 77 C, the gain is affected by temperature and decreases to 16.7dB. It can be seen that the influence of temperature on the electrical performance of the

system is obvious, and the consideration of temperature must be strengthened in the design process.

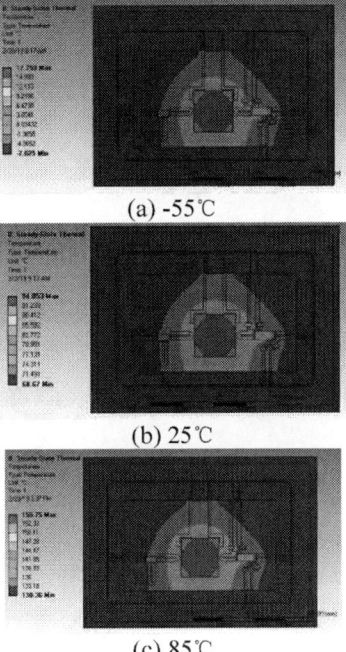

(a) -55℃

(b) 25℃

(c) 85℃

Fig. 9. Thermal Distribution Simulation on the temperature of (a)(b)(c)

Tab. 1 The temperature of the key components

IC & Component	-55(℃)	25(℃)	85(℃)
Mixer	-4	74	133
Amplifier	-1	77	136
power resistance	17	94	155

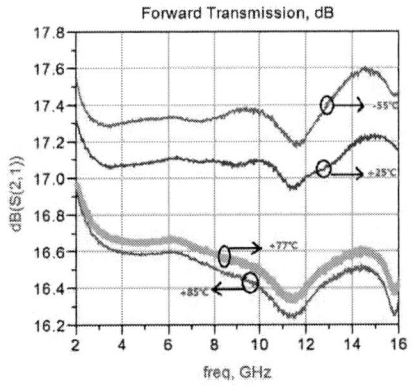

Fig. 10. Gain of the amplifier decreases with the increase of temperature

Advanced Evaluation with multi-physical field

On the basis of multi-physical field simulation, the EM characteristics of the module structure are obtained, and the *.SNP file is output. The thermal distribution of each working state is obtained too, the temperature of each chip and component under the corresponding state were extracted, and the corresponding performance were obtained by interpolation calculation method. Based on field-circuit cooperative simulation, the accurate evaluation model of electrical performance was established, as shown in Figure 11., and the simulation results are shown in Figure 12(a)(b)(c).

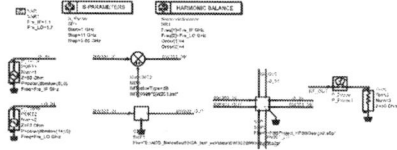

Fig. 11 System circuit link with multi-physical characteristics

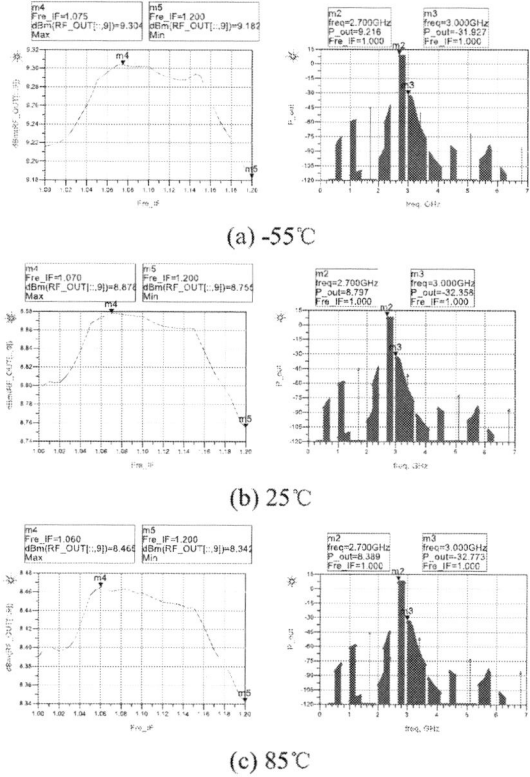

(a) -55℃

(b) 25℃

(c) 85℃

Fig. 12 The Gain and frequency spectrum evaluation at three temperature for the system

Margins evaluation for the Gain and frequency spectrum

Comparing the simulation results of the module in Fig. 12 at three temperatures, it is found that the output signal power decreases with the increase of temperature. Especially when at 85 ℃, the temperature of the amplifier chip reaches 136 ℃. And for its thermal resistance is 47 ℃/W, therefore, the junction temperature of the amplifier chip is 152.45 ℃, which is lower than the limit temperature of 175 ℃. It can be seen that this module can work normally at three temperatures working states. And we can observed the margins that the circuit link

gain is about 8.34-9.30 dB and the stop-band rejection for the noise is greater 41.1 dBc. Compared with the primary evaluation and accurate evaluation of this module, the self-heating effect of the module causes the chips temperature to rise and then its performance to change. So relatively speaking, the latter is closer to the actual work situation and more accurate.

Conclusion

Based on the method of multi-physical field and field-circuit cooperative simulation, the real working state of the system can be simulated. The influence of the working environment, such as EM characteristics, thermal and mechanical, could be considered in the electric performance analysis for the SiP. The more influence factors are taken into account, the more fidelity for the real situation, but also the more job need to be down. Therefore, we should use different accurate level evaluation method to obtain the margins of the performance according the actual requirement

References

[1] R.R Tummala, "Moore's law meets its match (system-on-package)", Spectrum, Vol. 43, No. 6, pp. 44-49, 2006.

[2] J Priest, M Ahmad, L Li, et al, "Feasibility Study of a SiP for High Performance and Reliability Product Application." High Density Microsystem Design and Packaging and Component Failure Analysis, 2005.

[3] L Wang, L Wang, X Li, et al, " Multi physical field simulation of medium voltage switchgear and optimal design", 2016 27th International Symposium on Discharges and Electrical Insulation in Vacuum (ISDEIV), 2016.

[4] Q Cheng, Z Zhang，N Xie, and etc. "Multi-Physical Fields Coupling Simulation and Performances Analysis of a Novel Heated-Tip Injector", Journal of Thermophysics and Heat Transfer, Vol. 30, No. 3, pp. 1-12, 2016.

[5] D. Pan, L. Li, and M. Wang, "Modeling and Optimization of Air-core Monopole Linear Motor Based on Multi-physical Fields", IEEE Transactions on Industrial Electronics (2018):1-1.

[6] Z Li, L Lv, X Huang, et al. "progressive optimization method for RF SiP system", journal of microwaves, Vol. 34, No. 1, pp. 29-35, 2018.

Assessment of FOWLP process dependent wafer warpage using parametric FE study

Ghanshyam Gadhiya[1,2], Birgit Brämer[1], Sven Rzepka[1], Thomas Otto[1]

[1]Fraunhofer ENAS, Micro Materials Center, [2]Chemnitz University of Technology

Technologie Campus 3, D-09126 Chemnitz, Germany

ghanshyam.gadhiya@enas.fraunhofer.de

Abstract

The paper presents the steps for the parametric finite element model creation of the wafer, material characterization of the adhesive tape, analytical and finite element study of the wafer warpage considering the bifurcation, gravity effect on the wafer bow assessing the warpage of mold/Si bilayered structure under thermal loading. The analytical results are compared to finite element analyses (FEA) considering the linear and nonlinear deflection. Consequently, the FEA approach has been used to study the deformation of 12" reconstituted wafers in their FOWLP fabrication process. By changing the temperature, the deformation of the wafer shows a bifurcation point, at which the warpage changes between the spherical and cylindrical shapes. The bifurcation region has been analyzed for the relevant range of overmold thicknesses in order to provide the guidance to optimum wafer and process designs that avoid the excessive warpage. For different wafer structures, the study determines also the effects of the gravitational force on the wafer bow as well as its influence in combination with the thermal mismatch. Finally, the FOWLP process induced warpage has been demonstrated by FEA incorporating the geometrical nonlinearity, gravity and ground support by means of contact elements.

Key words: Gravity effect, bifurcation, DMA measurement, wafer warpage, parametric FE modeling

Introduction

System-in-package (SiP) solutions based on the fan-out-wafer-level-packaging (FOWLP) technologies are getting significant attention for the heterogeneous system integration offering the shortest time-to-market, high reliability, functional safety and low costs. One of the challenges is the excessive wafer warpage, while going through various fabrication process steps such as molding, mold curing, carrier debonding, redistribution layer processes and grinding. The wafer having different components with different material properties warps when it is subjected to the temperature change during these processes because all the materials involved have their specific and diverse properties regarding the thermal expansion and stiffness. The wafer warpage study is crucial to keep the warpage within defined range to complete subsequent processes for FOWLP packaging successfully. Unfortunately, the physical assessments are costly and particularly time consuming. The virtual prototyping based on the numerical models offers an attractive alternative allowing the warpage analysis by varying the material and geometry parameters like die or mold thickness, die occupation ratio, Young's modulus etc.

At first, the paper introduces an approach to create fully parametric finite element models using ANSYS[TM] [1] for the reconstituted wafer with many components in a single package like multiple dies, vias and integrated passives, which can cover full portfolio of current and future SiP products based on FOWLP. Using this approach, various FE models for different wafer structures are created. Second, the material characterization was performed using the Dynamic Mechanical Analysis (DMA) tool to obtain the viscoelastic material properties of the adhesive tape. The wafer deformation behavior is then studied using analytical solutions provided by different authors for simple two layers plate system. Next, the analytically calculated wafer deformations for epoxy molding compound (EMC)/Si structure are compared with the numerical simulation, which shows very good agreement. It is then used to understand realistic system such as the reconstituted wafer having mold and dies. Simulating the dependence of the wafer warpage on the overmold thickness, a bifurcational behavior is found. The shape of the warpage changes from the spherical to cylindrical when the wafer is cooled from 150°C to 25°C. The bifurcation initiation points have been identified for a range of overmold thicknesses, which can further guide to design optimum wafer maps and process flows that avoid excessive wafer warpage.

The effect of constant gravitational force on the EMC wafer warpage is investigated at different temperatures and the influence of gravity combined with the thermal mismatch is simulated for different wafer structures including i) only EMC wafer, ii) mold/Si wafer and iii) reconstituted wafer. Finally, the total FOWLP process induced warpage is analyzed by FE analysis considering the geometrical nonlinearity, gravity and ground support modelled by contact elements.

Parametric Finite Element Modeling

The process steps for creating the parametric FE model have been outlined in Figure 1 with corresponding example models, which starts by defining the SiP geometry parameters, material properties, element types and real constants. The next parametric area modeling step has been adopted from the fully parametric modular system of models for SiP products based on FOWLP [2]. This step provides flexibility to create FE model for any combination of components like multiple dies, vias and integrated passives in the package. The intermediate steps create wafer areas, dies or package area matrix according to the given wafer map and assigns the real constants to the areas according to the material properties to be assigned. The last step generates the areas mesh, extrudes areas to create volumes and assigns material properties to volume elements according to real constant numbers. Same strategies can be used to create parametric FE model for the panel by replacing the circular wafer areas with the rectangular panel areas.

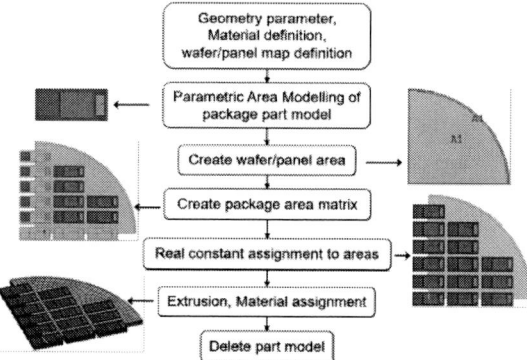

Figure 1: Parametric FE modeling workflow for wafer.

In this work, all the simulations are performed with the quarter 3D FE models meshed using 20 node solid elements with multilayered shell option. The results are verified using the quadratic solid elements without shell for some cases. The symmetric boundary constraints are applied on the symmetry planes due to the quarter symmetry of the wafer. All the materials are assumed isotropic.

Material Characterizations

It is very important to have material properties measured from the actual sample for the accuracy of the finite element analysis. Therefore,

the viscoelastic properties of the adhesive tape is characterized by DMA. In this measurement, the frequency sweep is used to extract the material response under the sinusoidal loading at different frequencies for a certain temperature range.

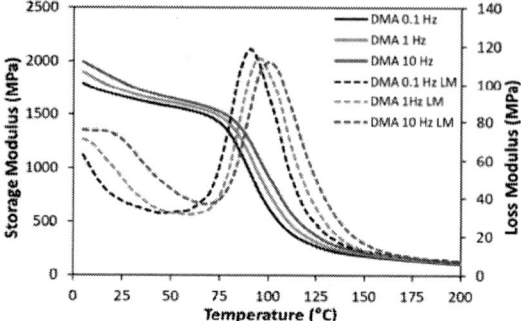

Figure 2: DMA measurement data for the adhesive tape (solid lines represent storage modulus).

The storage modulus and the loss modulus extracted from the measurement are shown in Figure 2 for the adhesive tape for the temperature between 5°C and 200°C. The elastic modulus at room temperature is approximately 2 GPa and the glass transition temperature is 96°C. Similarly, the measured material properties for the epoxy molding compound are given in [3] which has the glass transition temperature of 150°C. Finally, the Prony series parameters calculated from the master curve for both the materials are used in the FE simulations.

Analytical Approaches for Bilayer Wafer bow

A simple bilayer structure is considered in order to understand the deformation behavior of the system under the thermal loading condition. Numerous authors have demonstrated the analytical solution considering the linear deformation of the multilayer structure [4-7].

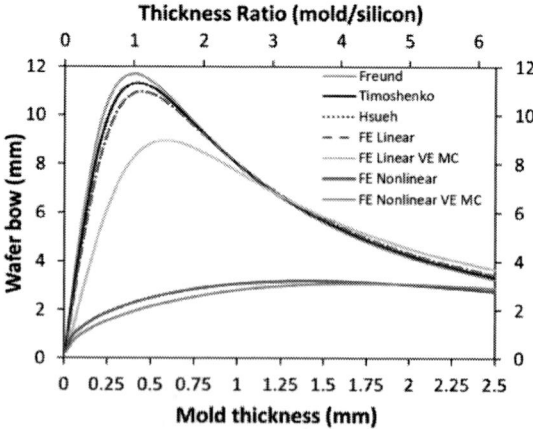

Figure 3: Comparison between the analytical and simulated wafer deformation for different mold thicknesses on 0.4 mm thick Si layer.

The quarter 3D FE model of 300 mm diameter wafer made of mold compound and silicon layer is modeled. The wafer is cooled down from 150°C reference temperature to room temperature in 2 minutes. Different cases have been simulated with the constant 0.4 mm silicon layer thickness, varying mold compound thickness from 10 μm to 2.5 mm with linear elastic and viscoelastic properties of mold compound. The FE results are compared with the analytical solutions from Timoshenko [4], Freund et. al [5] and Hsueh [6] as shown in the Figure 3. For all the cases, the wafer warps symmetrically in a spherical shape due to the complete symmetry in geometry, material properties and thermal loading. Considering the linear deflection only, the maximum wafer bow calculated at the edge of the wafer shows similar trend of an increase initially for the mold/Si thickness ratio < 1, interestingly reaching its maximum near ratio = 1 and then decreasing for the higher ratios. The highest difference of about 0.4 mm between the three analytically calculated warpages is found at ratio = 1. The best match is found for the calculated wafer bow between the FE and analytical solution from Hsueh. When we consider the viscoelastic properties of the mold compound, the maximum deformation is 9.24 mm, which is not at the ratio = 1 but shifted to the higher ratio.

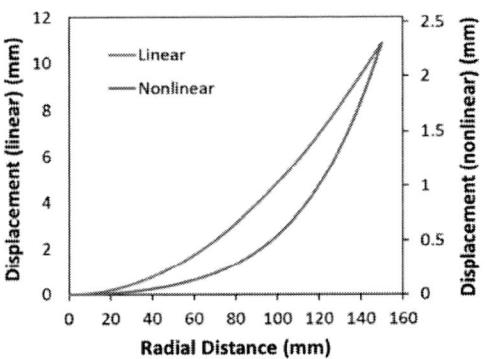

Figure 4: Comparison between the uniform and nonuniform displacement over wafer radial distance for ratio = 1.

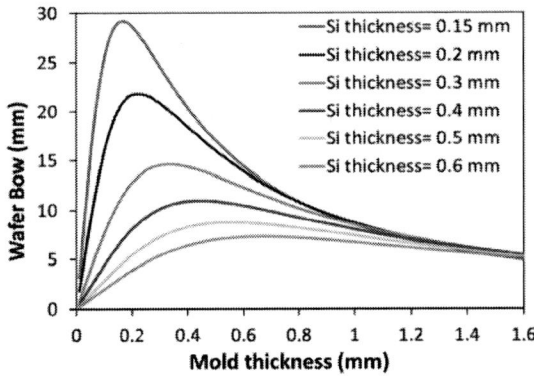

Figure 5: Analytically calculated wafer bow as function of mold and Si thickness.

Considering the geometric nonlinearity, the simulated maximum wafer bow is 3 mm at 1.8 mm

mold thickness for viscoelastic EMC. It is three times smaller than the results with the linear deflection assumption. This indicates that the 12'' diameter thin wafer shows highly nonlinear behavior, which must be incorporated in the finite element analysis. This is also visible in the form of non-uniformity in the displacement (or curvature) over the radial distance for the nonlinear case as compared in Figure 4 with the uniform displacement for the linear case. The difference between the wafer deflections is small for the EMC with linear elastic and viscoelastic properties.

The maximum warpage also depends on the silicon layer thickness and the applied temperature loading. Because of the excellent match between the simulated and analytically calculated wafer bow using analytical equation from Hsueh, the effect of Si thickness has been presented in Figure 5 only using analytical solution [6]. For thin Si layer, the deformation is higher as compared to thick Si. Also, the highest warpage is near the mold/Si thickness ratio = 1 for all the cases.

FOWLP based Reconstituted Wafer Warpage Simulation

After studying the bilayer plate structure, the typical FOWLP based reconstituted wafer has been modelled as shown in Figure 6 which has about 1900 silicon dies (size = 0.4×4.85×4.85 mm). Again, the wafer deformation is simulated for different overmold thicknesses considering the linear and nonlinear deflection.

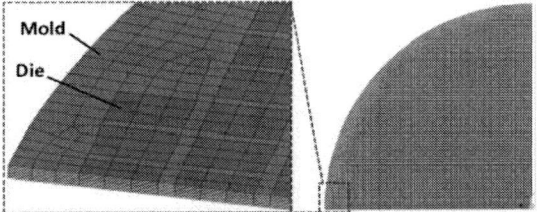

Figure 6: Quarter FE model of the reconstituted wafer.

For the linear deflection, it is possible to calculate the wafer bow analytically using the effective material properties for mold and Si as shown by various authors [7-9]. The simulated results show similar trend like bilayer plate structure (Figure 7) having the maximum bow of 7.36 mm at overmold/die thickness ratio = 1.25.

Considering the geometrical nonlinearity, the simulated wafer bow shows the bifurcation behavior over certain range of overmold thickness e.g 0.1 to 1 mm as represented in the Figure 7. It means that the wafer warps in a symmetric spherical shape upto 0.1 mm and after 1 mm thick overmold. For the spherical shape deformation, the out-of-plane displacement in one direction at the wafer edge is same as the other direction, $w_x = w_y$. For the other thicknesses, the wafer bifurcates and deforms asymmetrically exhibiting a cylindrical shape, which means that $w_x \neq w_y$. The maximum and minimum wafer edge displacements are shown with the red and blue lines for the asymmetric warpage. Interestingly, the maximum displacement w_{max} follows the trend similar to the linear deflection case upto certain overmold thicknesses. If we assume symmetric deformation in the bifurcation region and connect the black lines by drawing the trend line in Figure 7, it will show similar behavior like shown in Figure 3 for nonlinear case.

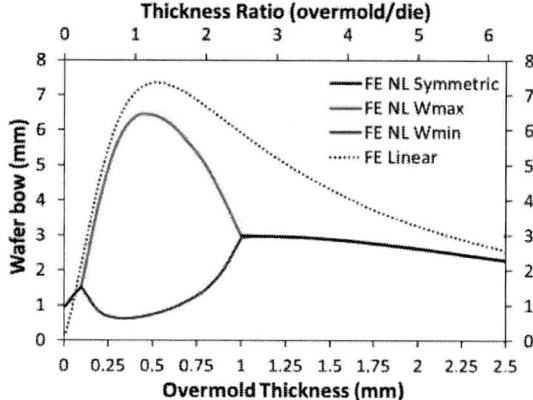

Figure 7: Simulated wafer bow as a function of overmold thickness considering the linear and nonlinear deflection (Viscoelastic EMC).

The area under the red and blue line is called bifurcation region where the wafer tends to show sudden change in the shape after bifurcation point if the overmold thickness is within this range. As seen in the Figure 5, the strong dependence of the warpage on the Si thickness, the bifurcation region could also shift left or right in the x-axis accordingly. The area under the bifurcation region could also vary depending on various factors e.g die thickness, overmold thickness, silicon occupation ratio, material properties and expected process temperature changes. If there are more layers like, redistribution layers than they have to be considered also for the study of bifurcation region. In order to prevent excessive wafer warpage, it is useful for the wafer map designer to know about this region for optimizing these factors.

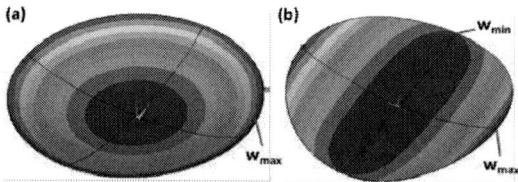

Figure 8: (a) Spherical shape wafer deformation ($w_{max} = 3$ mm) for overmold/die thickness ratio = 2.5 and (b) cylindrical deformation of reconstituted wafer for thickness ratio = 1.25 showing $w_{max} = 6.43$ mm and $w_{min} = 0.73$ mm.

The reconstituted wafer warped in the spherical and cylindrical shapes for nonlinear case

are illustrated in Figure 8 for different overmold/die thickness ratios. For the ratio of 2.5, the wafer warpage becomes symmetrical again having nearly 3 mm wafer bow. For the ratio of 1.25, the wafer edge displacement is 6.43 mm in one direction and 0.73 mm in another.

The evolution of the wafer edge displacement over the temperature change is depicted in Figure 9 for different overmold thicknesses. As explained in the cited publications [11-14], it shows three regions: linear stable spherical ($w_x = w_y$), nonlinear stable spherical ($w_x = w_y$) and after bifurcation stable cylindrical ($w_x \neq w_y$). After the bifurcation point, the spherical shape is possible but it is unstable. Despite that, the simulation shows the stable spherical shape for some overmold thicknesses. Therefore, the small perturbation force is applied on one point at the edge of the wafer to initiate the bifurcation.

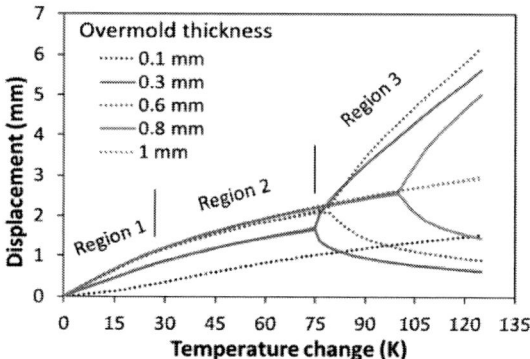

Figure 9: Wafer edge displacement as a function of temperature change from 150°C to 25°C for different overmold thicknesses showing bifurcation behaviour.

It is evident that the bifurcation starting temperature is dependent on the geometric design parameters like the thickness of the overmold. In the example analyzed, bifurcation starts when the temperature change exceeds 117 K, 87 K, 71 K, 75 K, 100 K, and 120 K for overmold thicknesses of 0.1 mm, 0.2 mm, 0.3 mm, 0.6 mm, 0.8 mm, and 1 mm, respectively. That means, there is a critical overmold thickness between 0.3 to 0.6 mm, at which the bifurcation starts at the temperature change of 71 to 75 K already, while it takes 50 K more for other thicknesses. Similar trends exist for the material properties like the magnitude of glass transition temperature and process conditions like duration at certain temperatures. Hence, the risk of excessive warpage caused by the cylindrical deformation can safely be prevented by choosing an optimum geometric, material, and process design which can be found upfront by the numerical simulations.

Mold Compound Wafer Deformation due to the Gravitational Force

The 1.5 mm thick 300 mm diameter wafer made up of only mold compound material is modelled to study the effect of gravitational force on the wafer warpage at different temperatures. The wafer edge at the periphery is fixed in the out-of-plane direction. The constant gravitational force is applied keeping also the temperature constant.

The simulated wafer warpage results are shown in the Figure 10. For -50°C and 0°C, the wafer deformation is very small within -200 µm. As the temperature increases, the wafer warps more for given duration. For example, it reaches the displacement of -0.7 mm and -1.9 mm in 60 minutes at temperature of 100°C and 125°C.

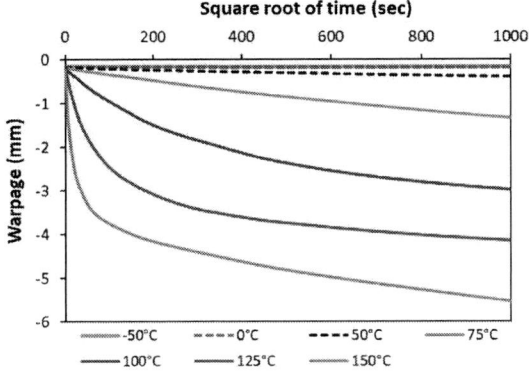

Figure 10: Wafer warpage due to the gravitational force at different temperatures.

Due to the higher viscoelastic relaxation of mold compound at higher temperature (150°C) near the glass transition temperature, the wafer bows initially upto -3.5 mm within the first hour. Then it deforms with the constant rate and reaches -5.5 mm after 11.5 days. This study is useful to understand the wafer deformation when it is stored in the shelf for a longer period or when the wafer deformation changes from warped state to flat during the heating on the vacuum chuck.

Gravity effect on the wafer deformation

Four different cases have been simulated to understand the effect of gravity on the wafer warpage for the assumed temperature profile as given in Figure 11.

- **Case 1:** Only mold compound wafer (thickness = 0.48 mm)

As we have seen in the previous section, the wafer deforms very much during the first dwell period of 4 minutes at 150°C. During the cooling to room temperature and heating phase, the wafer stays at the same warpage and again it deforms more at similar rate during the second dwell time reaching more than -3 mm.

- **Case 2:** 0.1 mm thick EMC and 0.38 mm thick Si bilayer wafer structure

Similar to case 1, the wafer deforms during first 2 minutes because of the viscoelastic relaxation of mold compound but at lower rate reaching only -0.33 mm due to the stiff Si layer. During the cool

down, the wafer warps more due to the thermal mismatch between both the layers. Again, the wafer recovers from -1.12 mm to -0.32 mm during the heating phase. The change in the wafer deformation is minor for further dwell period and the trend is similar for the next cycle.

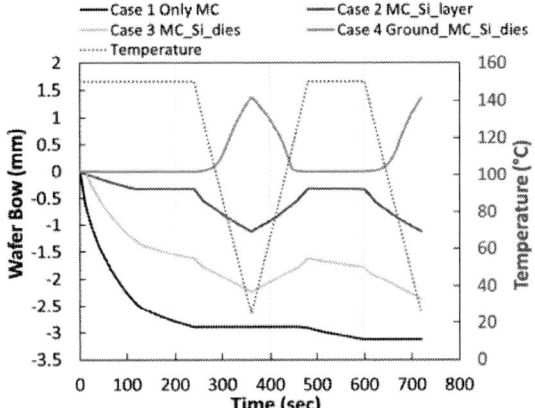

Figure 11: The effect of gravity on the deformation of 4 different types of wafer structures.

- **Case 3:** FOWLP based reconstituted wafer (Figure 6) with 0.38 mm thick dies

It is now obvious to have the average wafer deformation between the previous 2 cases because this case is combination of both the cases having the silicon occupation ratio accordingly. It shows less relaxation of EMC during the dwell time compared to case 1 due to the presence of dies reaching -1.6 mm, which is still large compared to case 2. For the cooling and heating phase, the wafer deforms according to the thermal mismatch similar to case 2 but shows less recovery in comparison. During the second dwell time, the EMC relaxation increases the warpage by 0.15 mm. The next cooling further increases the displacement to -2.35 mm, which is higher compared to the previous cycle.

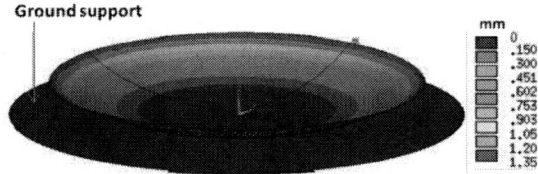

Figure 12: The wafer deformation for case 4 at room temperature considering the ground support.

- **Case 4:** same wafer as the case 3 but simulated with the ground support

Previous three cases are simulated considering the free deformation of the wafer in out-of-plane direction, which is not the case in reality. Normally, the wafer is placed on or held by vacuum chuck during the FOWLP processes, which supports the wafer and does not allow it to deform freely like in the aforementioned cases. Since it is important to include this support in the FE analysis as well, contact elements have been used to provide a rigid frictionless ground support as explained in [14]. The wafer is held on the ground by the gravity without any out-of-plane constraints. Now, the wafer stays flat on the ground at high temperature during the dwell time. During the cooling, the thermal mismatch forces the wafer to deform which turn out to be of 1.36 mm in the opposite direction (Figure 12) compared to 3 cases. After heating, the wafer becomes flat again.

FOWLP process induced wafer warpage simulation

Finally, the previous studies are used to investigate the FOWLP process induced wafer warpage using finite element analysis. The assumed temperature profile is plotted in Figure 13, which has the tape annealing, molding, mold curing and carrier debonding (removal of steel and tape) steps. During these processes, the wafer is hold on the vacuum chuck and flipped for debonding step, which changes the gravitational force direction. The element birth and death technique is used to replicate the processes like inclusion and removal of the material similar to the molding and debonding process. Three sets of boundary conditions are applied in the simulation:

(1) without gravity and without ground support (wafer is fixed in its center only),

(2) with gravity but still without ground support (wafer is fixed in its center only), and

(3) with gravity and with ground support using contact elements.

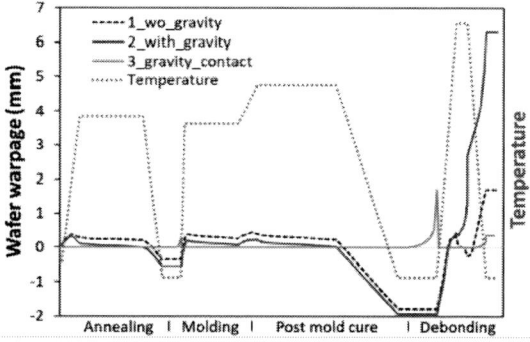

Figure 13: The evolution of process induced wafer warpage (temperature not scaled).

It should be noted that all the other constraints are applied and removed as required according to the process steps. The viscoelastic properties are used for mold compound [3] and adhesive tape (Figure 2). The geometric nonlinearity is taken into account in all the sets. The wafer warpage results are summarized in Figure 13 which are extracted at the wafer edge for the first two sets. Regarding the wafer warpage orientation, the positive warpage values represent the convex shape while the negative values indicate the concave shape

of the wafer bow for set 1 and set 2. With the change in the gravitational force direction for the debonding step, the warpage orientation also changes. For set 3, it has always positive warpage in the concave shape.

The wafer shows small deformation before the mold cure step due to the thermal mismatch between the steel, tape and dies. For the first three steps, the small effect of gravity is visible between set 1 and set 2. The wafer warps mainly after the post mold cure and debonding steps. Here, the large differences can be seen between the results of the three sets of boundary conditions. First, the effect of gravity is substantial at higher temperature due to the viscoelastic relaxation of EMC. After debonding, the resulting warpage is increased from 1.7 mm (set 1) to more than 6 mm (set 2) when adding this effect to the simulation. Hence, the gravity must not be neglected. Second, the ground support is an essential effect as well. Without considering the ground support, the warpage magnitudes grow too high. In reality, the fabrication process becomes critical and needs corrective measures when the warpage is beyond 2 mm. In the example chosen for this analysis, no warpage problem was experienced. However, the simulation using the set 2 boundary conditions would overestimate the warpage.

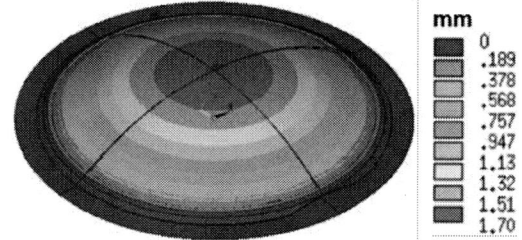

Figure 14: The wafer deformation at room temperature after post mold cure (set 3). The wafer lift up from the center using the ground support at the edges.

It has become evident that the only set 3 is able to replicate the actual processes realistically. Hence, the analysis had to apply these conditions and to include the contact elements. This makes the simulation more complex, computationally expensive, and frequently causes fails due to divergence. Applying the contact stabilization measures to avoid the divergencies needs to be done with care in order not to cause overdamping, which might artificially lower the warpage magnitudes. In this example, the resultant wafer warpage is 1.7 mm after post mold cure (Figure 14) and 0.36 mm after debonding. They are both below 2 mm threshold, above which corrective actions would have been necessary.

Conclusion

The analytical and FE study of the mold/Si bilayered structure under thermal loading reveals that there is a large difference in the wafer deformation for the linear and nonlinear deflection.

Thin wafers show a highly nonlinear deformation, which must be incorporate in the FE analysis.

The gravitational force as well as the ground support has significant effect on the wafer warpage. Both effects need to be considered in the numerical models.

For the typical range of overmold thicknesses, the FOWLP wafer may show a bifurcation effect in the warpage resulting in an excessive wafer bow where the shape changes from spherical to cylindrical. The onset of this effect depends on the chosen geometry, material, and process conditions, i.e., it can be avoided by an optimized configuration that can be determined systematically upfront by simulation.

Finally, the FOWLP process induced warpage is replicated by FE analysis incorporating the geometrical nonlinearity, gravity and ground support by means of contact elements.

Acknowledgements

Authors would like to thank Mr. Remi Pantou for the material characterization. This work was performed within the project EuroPAT-MASiP, which has received funding from the ECSEL JU under grant agreement No. 737497. The ECSEL JU is supported by the European Union's Horizon 2020 research and innovation programme and by national programmes of Austria, Finland, France, Germany, Hungary, Ireland, Netherlands, Portugal, and Sweden. Part of this work was also performed under the Federal Cluster of Excellence EXC 1075 "MERGE Technologies for Multifunctional Lightweight Structures" supported by German Research Foundation (DFG).

References

[1] ANSYS™ Multiphysics, version 19, User's manual, Ansys Inc., 2019.

[2] G. Gadhiya, B. Brämer and S. Rzepka, "Automated Virtual Prototyping for Fastest Time-to-Market of New System in Package Solutions", Electronic System-Integration Technology Conference (ESTC), Dresden, Germany, pp. 1-7, 2018.

[3] G. Gadhiya, B. Brämer, S. Rzepka, T. Otto and H. Kuisma, "The Creation of a Validated Scheme for the Automated Optimization of Systems in Package Designs", Smart System-Integration (SSI), Barcelona, Spain, 2019.

[4] S. P. Timoshenko, "Analysis of Bi-Metal Thermostats", J. Opt. Soc. Am., 11, pp. 233–255, 1925.

[5] L. B. Freund, J. A. Floro and E. Chason, "Extensions of the Stoney formula for substrate curvature to configurations with thin substrates or large deformations", Applied Physics Letters, 74, 1987-1989.

[6] C. H. Hsueh, "Modeling of elastic deformation of multilayers due to residual stresses and external bending", J. Appl. Phys., vol. 91, pp. 9652, 2002.

[7] E. Suhir, "Predicted thermally induced stresses in, and the bow of, a circular substrate/thin-film structure," Journal of Applied Physics, vol. 88, pp. 2363-2370, 2000.

[8] T. Lin et al., "Warpage simulation and experiment for panel level fan-out package", 2016 IEEE CPMT Symposium Japan (ICSJ), Kyoto, pp. 129-131, 2016.

[9] F.X. Che, D. Ho, M.Z. Ding, D.R. MinWoo, "Study on process induced wafer level warpage of fan-out wafer level packaging," 2016 IEEE 66th Electronic Components and Technology Conference (ECTC), pp. 1879–1885, 2016.

[10] C. Zhu, P. Guo and Z. Dai, "Investigation on wafer warpage evolution and wafer asymmetric deformation in fan-out wafer level packaging processes," 2017 18th International Conference on Electronic Packaging Technology (ICEPT), Harbin, pp. 664-668, 2017.

[11] M. Finot, S. Suresh, "Small and large deformation of thick and thin-film multi-layers: Effects of layer geometry, plasticity and compositional gradients," Journal of the Mechanics and Physics of Solids, Volume 44, Issue 5, Pages 683-721, 1996.

[12] J. Schicker, W. A. Khan, T. Arnold and C. Hirschi, "Stress-warping relation in thin film coated wafers", Modelling and Simulation in Materials Science and Engineering, 25. 025005, 2017.

[13] L. M. Dunn, Y. Zhang and M. V. Bright, "Deformation and structural stability of layered plate microstructures subjected to thermal loading", Microelectromechanical Systems, Journal of 11. 372 - 384.

[14] D. Shin and J. J. Lee, "Analysis of Asymmetric Warpage of Thin Wafers on Flat Plate Considering Bifurcation and Gravitational Force", IEEE Transactions on Components, Packaging and Manufacturing Technology, vol. 4, no. 2, pp. 248-258, Feb. 2014.

Development of mechanically compliant flip chip interconnect using single metal coated polymer spheres

Daniel N. Wright, Branson D. Belle, and Joachim S. Graff

SINTEF AS, Gaustadalleen 23 C, 0373 Oslo, Norway

Daniel.nilsen.wright@sintef.no

Abstract

Most available fine pitch interconnects, like micro bumps and copper pillars, are not particularly compliant whereas available compliant interconnects, like plastic core solder balls, are not fine pitch. Using Ag-plated polymer spheres (MPS) in conjunction with a nano-Ag conductive ink has the potential to achieve mechanically compliant flip chip interconnects since the structural integrity is maintained by the flexible polymer core while the electrical conductivity is maintained by the Ag plated shell. Additionally, the low processing temperature means that it is relevant for systems that require low temperatures or that are very sensitive to thermo-mechanical stress, like MEMS sensors.

Previous work has shown that a major challenge in the proposed process was the confinement of the conductive ink onto the Au pad. This work focuses on finding an oleophobic coating that can be patterned to confine the ink on the contact pads. Two materials were tested, a fluoroacrylate additive for photoresists and a fluoropolymer that needed to be patterned separately. The latter showed superior oleophobicity and was therefore chosen. Patterning by positive and negative photoresist was tested. Using positive photoresist as a masking layer for reactive ion etching proved incompatible with the desired output. The use of negative photoresist with a lift-off technique showed potential, but needs to be optimized. Using reactive ion etching through a stencil mask showed the best results.

Key words: Metal coated polymer spheres, MPS, flip chip, low temperature

Introduction

Most available fine pitch interconnects, like micro bumps and copper pillars, are not particularly compliant whereas available compliant interconnects, like plastic core solder balls, are not fine pitch. Using Ag-plated polymer spheres (MPS) in conjunction with a nano-Ag conductive ink has the potential to achieve mechanically compliant flip chip interconnects since the structural integrity is maintained by the flexible polymer core while the electrical conductivity is maintained by the Ag plated shell, as illustrated in Figure 1. Additionally, the low processing temperature means that it is relevant for systems that require low temperatures or that are very sensitive to thermo-mechanical stress, like MEMS sensors.

Previous work on establishing such an interconnect focused on using ink jet printing to deposit nano-Ag ink onto the surface. The main issue with this approach was to confine the ink on the electrical pad, as the surface energy of the surrounding passivation layer was usually lower than the gold surface of the pad and hence the ink would run out [1].

This work takes a new approach by applying and patterning an oleophobic coating to the passivation layer. The work concentrates on finding the appropriate coating, measuring contact angles and patterning the oleophobic coating

Experimental

Material selection

Two oleophobic materials were identified for testing our approach:
- Cytonix FluorAcryl 7298
- CYTOP

Cytonix FluorAcryl 7298 is a perfluoropolyether (PFPE) acrylate used as an additive in other UV curable coatings to reduce surface tension and improve water, oil and stain resistance. For this work it was mixed into both a positive (AZ 4533, Merck Materials) and negative photoresist (Ma-N 1440, Microresist Technology) at different mix ratios. An approximate amount of Cytonix was added into a UV-opaque bottle on a scale and then the appropriate amount of photoresist was added to reach the desired concentration. The mixture was stirred on a magnetic stirrer for at least 5 minutes.

CYTOP in an amourphous fluoropolymer designed for water and oil repellency for transparent surfaces. For this work, CYTOP type M was diluted in the appropriate solvent at 1% and 5% concentration for initial testing. Subsequent development of patterning processing used 1% concentration.

To ensure adhesion to the passivation layer, a silane coupler (aminopropyltrimetoksysilane) was mixed in pure ethanol at 0.05%. Since the silane coupler degrades quickly, a new batch was made within 3 hours of each experiment.

The nano-Ag ink used was Silverjet DGP 40LT-15C from ANP. This ink is designed for ink jet printing and is based on triethylene glycol momoethyl ether (TGME) which has a low vapour pressure and thus dries very slowly at room temperature, which is a vital characteristic for this process, as is relies attaching MPS to the ink while the ink is wet.

Sample preparation

The photoresist with Cytonix additive mixtures were spun on quartered Cz-Si (100) wafer at 3000 rpm for 30 s followed by a 30 s bake at 100 °C on a hot plate.

To ensure a representative process for the samples with negative photoresist, areas of these samples were exposed to UV using a Heidelberg UV writer with a dwell time of 60 µs. The samples were then developed in Microposit MF319 developer for 60 s.

The samples with positive photoresist were developed in Microposit MF19 for 60 s after which they received a post bake at 100 °C for 60 s on a hot plate.

For Cytop, the Cz-Si wafers first received the silane coupler spun at 1500 rpm for 30 s, followed by drying at 80°C for 1 minute. Thereafter, the Cytop mixtures were spun at 1000 rpm for 30 s, targeting thicknesses of 0.5 µm and 1.0 µm for 1% and 5% concentrations, respectively. After spinning, the samples were heat treated on a hot plate with the following profile; 30 min at 50°C, 45 min at 80°C and 30 min at 180 – 240 °C.

Contact angle measurements

Contact angle measurements were carried out using a manual contact angle setup. The Si wafer was placed on the holder and drops of Ag-ink were applied using a micro-pipette set at 2 µl. The actual volume varied slightly between the measurements due to the pipette tip being plastic. After application of the drop, the reference line was adjusted to intersect with both edges of the drop. The measuring line was then aligned at both edges of the drop. Each material was tested with at least 8-10 drops.

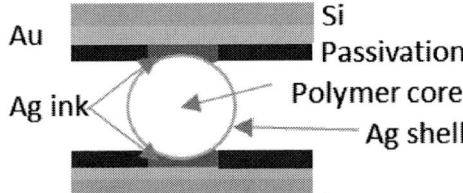

Figure 1: Illustration of the MPS interconnect. Nano-Ag ink supplies electrical and mechanical contact between Au electrical pad and MPS.

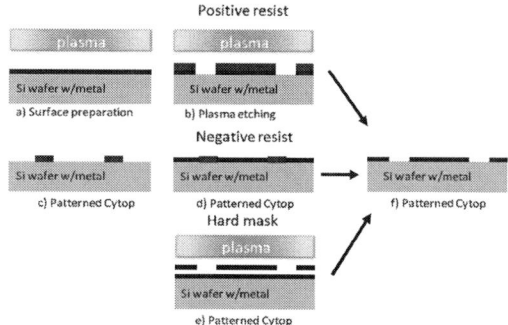

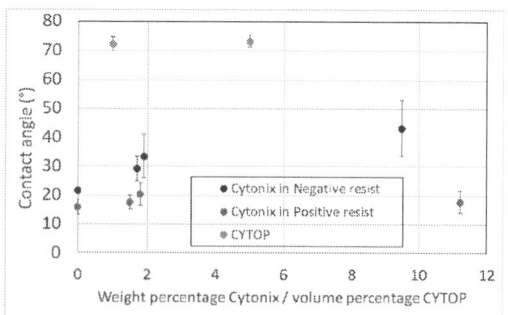

Figure 2: The three different approaches to patterning the Cytop coating.

Patterning Cytop

Three approaches to patterning the Cytop coated samples are illustrated in Figure 2. The first approach used positive photoresist to define an etching mask. Since the Cytop is oleophobic a surface treatment in an O_2 plasma (a) was needed to lower the surface energy in order for the photoresist to adhere to the Cytop. The photoresist was then patterned, developed and used as a mask for plasma etching (b). The second approach was to apply negative photoresist prior to Cytop application and do a lift-off process (c, d). The last approach was using a hard mask for patterned plasma etching of the Cytop (e). The specific parameters used for each approach is covered in the next section.

Results and discussion

Contact angles

The results of the contact angle measurements can be seen in Figure 3. For Cytonix in positive photoresist, the contact angle did not increase much outside the margin of error. For the negative photoresist, the addition of Cytonix increased the contact angle from $22° \pm 1°$ to $43° \pm 10°$. The high uncertainty on the latter measurement was due to some unevenness in the photoresist. Although the Cytnoix did improve the contact angle, it did not render the photoresist sufficiently oleophobic to repel the Ag ink when the sample was tilted to let the ink run off. Additionally, it was found that the TGME in the

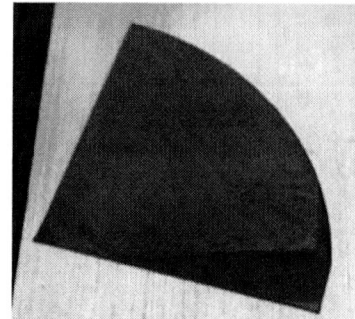

Figure 4: Positive photoresist adhering to the Cytop coating after surface treatment of the Cytop.

Ag-ink dissolved both the positive and negative photoresists. These issues are both vitally important to the process and thus further testing of the Cytonix was abandoned.

The contact angle of the Cytop was basically identical for both 1% and 5% concentrations, at $72° \pm 2°$. More important, however, is the fact that when tilted, the ink ran off the sample with minimal residue left on the surface, making it very compliant to the intended process.

Patterning Cytop with positive photoresist

For patterning the Cytop with positive photoresist, the photoresist had to be applied onto the Cytop. Depositing any coating on a surface that is designed to be oleophobic is counterintuitive. Contact measurements revealed that the photoresist had a contact angle towards the Cytop of $69° \pm 3°$. The photoresist also failed to adhere to the Cytop when spun. To increase the surface energy of the Cytop it was plasma treated (O_2, 50 sccm, 500 W, 0.65 Torr, 1 minute). The contact angle was measured to $67° \pm 3°$, which is a minimal change within the margin of error, but it resulted in the photoresist adhering to the surface when spun, as seen in Figure 4. However, after the photoresist was stripped, the contact angle towards the Ag ink was only $6° \pm 1°$ and thus not compatible with the intended process. The supplier of Cytop recommended a heat treatment (80 °C for 1 hr

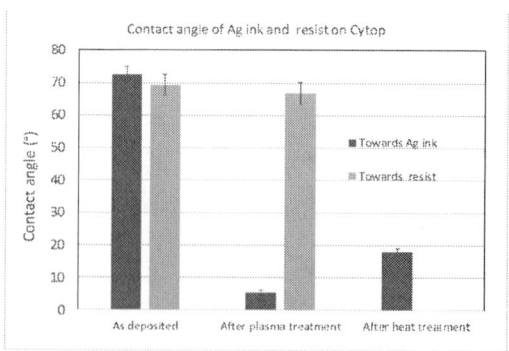

Figure 3: Contact angle results for Cytop and Cytonix in positive and negative photoresist.

Figure 5: Contact angle results for Ag ink and photoresist on Cytop after different treatments.

followed by 180 °C for 1 hr) to revitalize the oleophobic properties. However, this only increased the contact angle to $18° \pm 1°$, while also leaving Ag ink residue when the sample was tilted. These results are summarized in Figure 5 Hence, it was concluded that using positive photoresist to pattern the Cytop was not compatible with the intended process.

Patterning with negative photoresist

An obvious advantage of using a lift-off approach with a negative photoresist is that the photoresist is applied before the Cytop, hence the adhesion of the photoresist to the surface is not a problem. However, one challenge is that the coupling agent needed for Cytop's adhesion to any passivation layer is diluted in ethanol, which is an organic solvent that can dissolve the photoresist. A short test confirmed this.

A design of experiment was set up to find parameters that would render the photoresist capable of withstanding the brief exposure to ethanol while still being able to strip the photoresist after use. Six samples of Au coated Si dies were first spun with photoresist at 3000 rpm for 30 s, targeting a photoresist thickness of about 2 µm. The photoresist was then soft baked at 100 °C for 2 min and subsequently exposed in a UV writer in a pattern containing 40 µm diameter contact holes. The photoresist was then developed for 60 s, rinsed and dried in N_2, after which the samples received a post bake for 1 minute at different temperatures.

In the first round of tests, the UV dose in the writer was varied together with the post bake temperature. After treatment, the samples were placed in the spin coater, ethanol was applied and then the sample was spun at 1500 rpm for 30 s. Exposure at the higher ranges tested were chosen for further work on optimising the post bake temperature. Samples were thereafter spun with Cytop followed by a heat treatment for 1 hour at 50 °C and then 80 °C. The photoresist dots did not seem affected by the Cytop. Subsequently the photoresist was attempted removed in acetone.

Table 1 shows micrographs summarising the findings of the above process. After the post bake the 40 µm dots are fairly similar, but the samples baked at 200 °C have a slightly more defined edge. After ethanol spin, the samples baked at 180 °C and 190 °C both showed some photoresist residue close to the dot, while the sample baked at 200 °C did not.

On samples that were post baked at 180 °C the photoresist was largely removed after 1 minute in acetone. There were a few dots where there seemed to be residual photoresist.

On samples post baked at 190 °C the photoresist seemed removed but the hue suggested a slight residue residing on the dots. This was confirmed by scraping the surface of the pad using a probe needle.

Samples post baked at 200 °C had most of the photoresist intact after exposure to acetone. Several trials with long acetone exposures, and in combination with ultra sound, failed to remove the photoresist completely.

Similar tests were done using dimethyl sulfoxide (DMSO) to strip the photoresist. DMSO completely removed photoresist that had been post baked at 190 °C and 200 °C, which was a promising result. After application of Cytop, however, the photoresist appeared impossible to remove completely. On closer inspection by SEM, seen in Figure 6, the residues observed in the optical microscope was largely remains of the Cytop coating. This suggests that the Cytop coating had created protective cap over the photoresist dots. In some cases the Cytop remained almost fully intact in a punctured state. In other cases the Cytop was ripped but largely intact. Hence, the integrity of the Cytop coating is undermining the ability to

Table 1: Micrographs of 40 µm photoresist dots after development, baking, a brief exposure to ethanol and photoresist removal in acetone.

T (°C)	Before bake	After bake	After ethanol spin	After Cytop application and photoresist removal in acetone	
180					
190					
200					

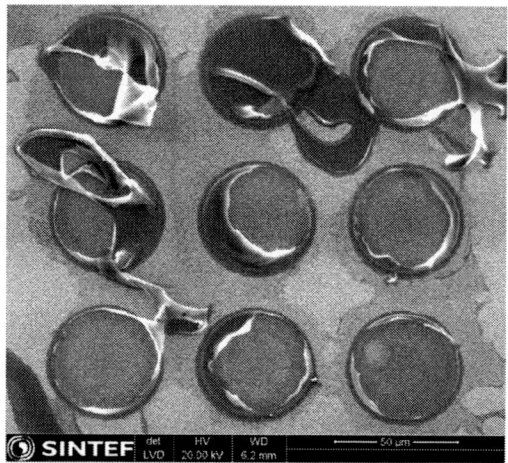

Figure 6: Remains of Cytop coating after attempted liftoff using DMSO to remove photoresist.

Figure 7: Si sample with stencil mask attached face down for plasma opening of Cytop coating.

remove it using a lift-off process. One possible solution, not studied in this work, is to apply thinner coatings of Cytop that will enable full removal.

Patterning with stencil mask

For these tests Si samples were prepared with Cytop as described in the experimental section. A Si stencil mask from the HyperConnect project [ref] was laid face down onto the sample and attached with tape, as seen in Figure 7. The stencil mask was fabricated from 400 μm thick Si wafer with 300 μm recesses in areas with 40 μm diameter through holes. The samples were thereafter exposed to a 5 minute O_2 plasma etch at 300 W. Nano-Ag ink was thereafter spun onto the sample at 400 rpm.

By comparing the left and centre image in Figure 8, one can clearly see that the pattern transfer from the stencil mask onto the Cytop coating was largely successful. The size of the etched holes, however, are highly dependent on the hole's proximity to the recess wall and hence the stencil mask design will have to be design optimised for the transfer pattern. Within the large recess openings most holes were within 2 μm of the

targeted diameter, as can be seen in the right most image in Figure 8. White light interferometry (WLI) was used to analyse the ink volume of eight nano-Ag ink droplets in one of the arrays in Figure 8. The results can be seen in Table 2. The large variation in droplet volume was mainly due to the variations in droplet height, not diameter. The cause of this large variation is not clear, but there are probably many factors that can influence the result and that have to be understood and optimised for a more even droplet calumet to be achieved.

Table 2: *results* of droplet analysis using WLI

Parameter	Value
Droplet volume (μm3)	307 ± 179
Height (μm)	0.6 ± 0.3
Diameter (μm)	34 ± 2

MPS application and analysis

After a successful pattern transfer using plasma etching and a stencil mask, MPS was applied onto the samples by drizzling from a meshed canister, resulting in very random scattering of MPS. The sample was the dried at at 80 °C for 5 minutes and sintered at 150 °C for 30 minutes. Some results are shown in Figure 9. The image on the left shows at least four individual MPS fairly well centered in the nano-Ag ink droplet. However, as the MPS have been applied in

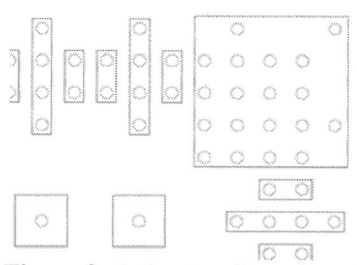 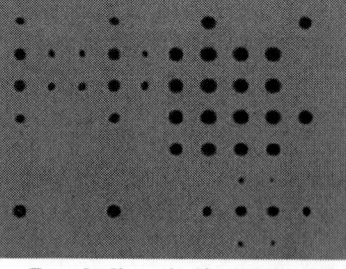 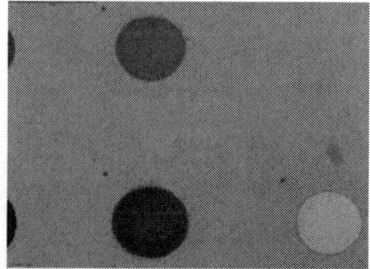

Figure 8: Left: stencil mask pattern. Purple lines indicate the recess lines while green circles are the through holes. Middle: Nano-Ag droplets on sample surface. The pattern transfer is good, but the size of the etched hole openings depends largely on the proximity to the recess walls. Right: Three hole openings in the Cytop. The color difference in the holes on the left probably indicate difference in Nano-Ag ink drop volume. The right most hole has not managed to attract nano-Ag ink.

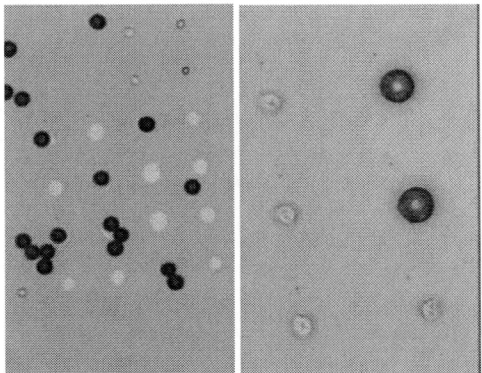

Figure 10: MPS attached to Si sample by patterned nano-Ag ink. Left: at least four MPS that are individually centered on an ink droplet. Right: Two MPS more or less perfectly positioned on the ink droplets.

a random fashion, there are also agglomerations of MPS on pads. One surprising result, especially considering the hydrophobic nature of the Cytop coating, is the residing MPS outside the of the opened holes in the coating. One would expect that the adhesion to the Cytop was sufficiently weak to allow these MPS to be removed in the air flow. Hence, for further development of the process the air flow would need to be optimised in order to remove superfluous MPS wile retaining the intended attached MPS.

The image to the right shows two MPS perfectly located according to the droplet pattern, showing that this process has potential for accurately placing MPS as individual contacts.

Figure 10 shows an SEM of the neck formed by sintered nano-Ag ink. It clearly shows that the ink has been drawn up onto the MPS surface. As both the ink and MPS surface are Ag, elemental analysis could not be used to distinguish between the ink and the metal layer. It is fair to assume that the rough surface of the MPS is pivotal for generating the capillary forces needed to draw the ink onto the MPS.

One potential challenge of this interconnect as depicted in Figure 10 its mechanical strength. A wider neck would be desirable and that would demand a larger volume of ink residing on the contact openings before MPS application. However, one can envisage a process where a die that has received MPS on each contact pad is flipped and dipped into a thin reservoir of nano-Ag ink for attachment to a substrate die. In the process of dipping the rough surface of the MPS will most likely drawn the nano-Ag ink up onto the already formed neck. This will of course depend on the depth of the dipping reservoir. In this process the oleophobic properties of the Cytop coating is advantageous once again as it would limit the nano-Ag ink from protruding farther than the contact opening edge.

Figure 9: SEM image of the neck between the substrate and the MPS formed by sintered nano-Ag ink.

The fabrication process was repeated for a sample with an Au coating on Si for measurement of electrical resistance. A probing needle was carefully positioned on the top of three MPS and the resistance measured between this and the Au surface was 6-7 Ω per MPS. If using a dipping and flip chip process as mentioned above, the increase in nano-Ag ink would most certainly decrease this resistance.

Conclusion

This paper describes the development of a fine pitch and mechanically compliant interconnect comprising a single metal coated polymer sphere (MPS). The work has focused on finding a process for confining nano-Ag ink on contact pads, as previous work showed that the area surrounding the contact pads often has lower surface energy than the Au pad. A flouroacrylate additive for photoresists was found to have insufficient oleophobic properties for the process. However, a fluoro-polymer coating was found to have excellent oleophobic properties. Methods for patterning the fluoro-polymer were tested. Using a negative photoresist for a lift off technique showed promising results, but needs to be optimised to allow full removal of the Cytop coating. The most reliable method was reactive ion etching through a stencil mask. Single MPS on contact pads were measured to have a resistance of 6-7 Ω.

Acknowledgement

This work has been made possible by funding from the Norwegian Research Council and the MUPIA project (European Union's HORIZON 2020 Programme, Clean Sky II, Grant no. 785337)

We would also like to that our colleague Terje Didriksen for preparation of silane coupling agents.

References

[1] D. N. Wright, M. M. V. Taklo, A. B. Vardøy, and H. Kristiansen, "Metal coated polymer spheres for compliant fine pitch ball grid arrays," in *2014 International 3D Systems Integration Conference (3DIC)*, 2014, pp. 1–7.

[2] "Final Report Summary - HYPERCONNECT (Functional joining of dissimilar materials using directed self-assembly of nanoparticles by capillary-bridging)," Aug. 2016.

A novel DAF Solo Mount Method for SDBG Process

Yusuke Fumita[1], Shinya Takyu[1], Atsushi Uemichi[1], Kunio Yomogida[1], Tsuyoshi Tazawa[2], Yoshinobu Ozaki[2]

[1]LINTEC Corporation, 2-1-2 Koraku, Bunkyo-ku, Tokyo, 112-0004 Japan

[2] Hitachi Chemical Co., Ltd., 7-7 Shinkawasaki, Saiwai-ku, Kawasaki-shi, Kanagawa, 212-0032 Japan

+81-50-9015-0705, y-fumita@post.lintec.co.jp

Abstract

SDBG (Stealth Dicing Before Grinding) process is known as an effective process for thin chips. In SDBG process, DAF (Die Attach Film) is used for 3D chips stacking package. Conventionally, DAF and wafer are mounted on a ring frame at the same time by a pressing roller. However, the pressing force tend to be unequal to the whole wafer, which causes uneven kerf width after BG tape peeling. In this study, we inspected a novel DAF mount method called "DAF solo mount method", and this method gives more balanced pressing force to DAF. Thus, this method can keep even kerf width after BG tape peeling compared to the conventional method.

Key words: SDBG, DBG, DAF solo mount method, CV, even kerf width

1. Introducton

As the digital network information society is making rapid progress, there is a strong demand for high functionality, while minimization of mobile devices are required. Since mobile devices that support high-level information networks are getting portable, they are required to become smaller and lighter. Thus, the miniaturization of semiconductor devices is necessary [1], [2]. 3D chip stacking packages using DAF (Die Attach Film) have been developed to achieve these goals. With this, thinning chips is an important aspect for the miniaturization, and SDBG (Stealth Dicing Before Grinding) process has been very successful. This process is known to be a very effecticve process [3], [4], [5]. Figure 1 explains this process. 1) BG (Back Grinding) tape is laminated to the pattern side of a wafer. 2) Stealth dicing is done from back side of the wafer. This makes the modified layers inside Si wafer and the crack reaches the pattern side. 3) Wafer under goes back grinding process (chips are singulated and modified layers gone). 4) DAF is mounted on a ring frame and BG tape peeled. 5) DAF is expanded for singulation. In SDBG process, DAF expanding singulation is one of most difficult step. Conventionally, DAF and wafer are mounted on a ring frame at the same time by a pressing roller. However the pressing force of the roller to the DAF tends to be unequal to the whole wafer, and this causes uneven kerf width after BG tape peel (figure 2). In this condition, kerf width tends to become unequal after DAF expansion. This has a bad influence in the next step which is chip pick up. Pick up failure is a very time-consuming error mode [6]. This implies that equal pressing force to the DAF

layer is quite important. In this study, we inspected a novel DAF mount method called "DAF solo mount method," and this new method is applied to SDBG process to improve the balance of pressing force to DAF. Process of this method is as follows (figure 3). 1) DAF is mounted to the ring frame. 2) Wafer is then mounted to the DAF with ring frame. 3) BG tape is peeled off. This method enables equal pressure force to be applied to the DAF, since primarily only DAF is mounted to the ringframe. With this condition, the kerf width after DAF expanding singulation is expected to be even. The equalness of kerf width after DAF expand is the key for reduction of pick up error. This also means UPH (Unit Per Hour) will increase. For this reason this paper will compare the CV (coefficient of variation) of the kerf width with the conventional method to experiment the effectiveness of the new DAF solo mount method.

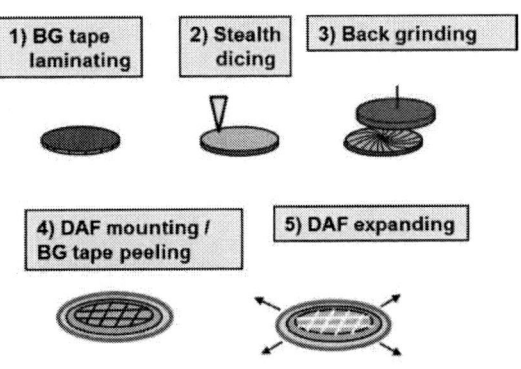

Figure 1: SDBG process flow

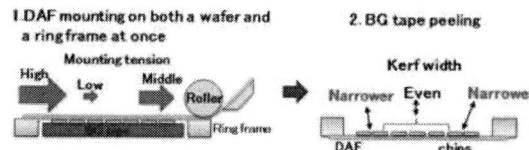

Figure 2: Conventional method of DAF mounting

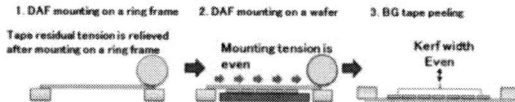

Figure 3: DAF solo mount method

2. Experiment

Sample SDBG wafer will be prepared to compare the CV of the conventional method and DAF solo mount method.

2.1 Materials

Wafers are φ300mm (12inch) Si mirror and thickness is 775um. BG tape is E-3100UN (LINTEC Corporation). DAF-A of Hitachi Chemical Co., Ltd. is used. These tapes are both commercial tapes.

2.2 BG tape lamination

BG tape laminator is RAD-3520F/12 (LINTEC Corporation). The laminating speed is 20 mm/sec., laminating pressure 0.1MPa, and wafer table temperature is 50℃.

2.3 Stealth dicing

Stealth dicer is DFL7361 (DISCO Corporation). Laser beam is irradiated from the back side of Si wafer. This makes two modified layers which generates a crack that extends to the surface of the wafer. SD (Stealth Dicing) conditons are as below (table 1).

Table 1: SD conditions

	Speed	Power	Defocus	CP
Pass 1	700mm/s	1.7W	−168um	−6um
Pass 2	700mm/s	1.7W	−146um	−6um

2.4 Back grinding

Grinder is DGP 8761 (DISCO Corporation). Z1 wheel is #340 of GF13-SD340-BE065-50, the grinding wheel revolution speed is 3200 rpm, the grinding feed speed is 1 um/sec. Z2 wheel is #6000 of GF13-SD6000-V696-50, the revolution speed is 3000 rpm, the feed speed is 0.3 um/sec. Z3 DP (Dry Polish) is DPW-018 DP-F05, the polishing load is 200N. Final thickness of the chips are 30um but still with BG tape support.

2.5 DAF mounting

The singulated wafer supported by the BG tape is mounted on DAF by two methods. One is conventional method and the other is DAF solo mount method. DAF mounter is RAD-2510F/12 (LINTEC Corporation) with mounting speed 10 mm/sec., temperature of wafer table and press roller is 50℃, pressure of the roller is 0.3MPa, for the conventional method. DAF, wafer, and ring frame are mounted at the same time. As for the new method, DAF is mounted on ring frame without wafer. Difference in conditions are faster mounting speed, 20 mm/sec., which provide higher tensile strength to the DAF.

2.6 BG tape peeling

After the wafer is mount, heat seal peeling tape is attached to the edge of BG tape for peeling. Temperture of the heater head is 210℃, attachment time is 3sec. Next, the peeling unit moves in three phases for careful BG tape peel. First is 1mm/sec. (from start to 5mm distance), next, 2mm/sec. (5-40mm), then 4mm/sec. (40-320mm).

2.7 DAF expanding singulation and heat shrink

DAF is expanded on the cool expansion stage of DDS2300 (DISCO Corporation.) for DAF singulation. Next, the outer periphery of the DAF is shrunk by heat to maintain the wide kerf width. Process condition is below (table 2 and figure 4).

Table 2: DDS2300 conditions

Cool expanding		Heat shrinking	
Cooler tempature	0℃	Table type	Porous
Cooling time	120s	Ascending amount	7mm
Ascending amount	12mm	Ascending speed	1mm/s
Ascending speed	100mm/s	Heater tempature	220℃
Waiting time	0s	Heating distance	20mm
		Rotation speed	5deg/s
		Cooling time	10s

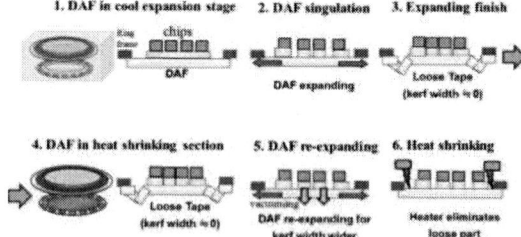

Figure 4: Process of DAF expansion

2.8 Measuring the kerf width

The kerf width after DAF expansion and heat shrink were then measured by image capturing. Measurement interval is about 3.5mm and each MD (Machine Direction) and TD (Transverse Direction) were measured (figure 5). Do note that MD kerf extends in the same direction as the DAF mounting direction.

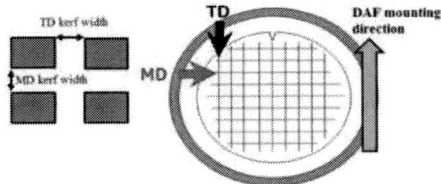

Figure 5: Relation of MD and TD

3. Results of kerf width measurement

3.1 Evenness of kerf width before DAF expansion

SD crack is nearly 0um just after DAF mount and kerf width literally canntot be measured before DAF expansion. However having the hypothesis that mounting method impacts the chip kerf, we decided to use the DBG (Dicing Before Grinding) method to confirm the kerf with just after mounting. DBG is an effective process (figure 6) since kerf width of blade cut can be measured.

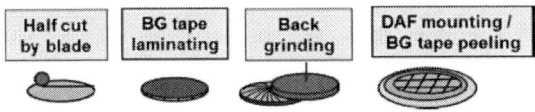

Figure 6: DBG process flow

Wafer size is Si mirror φ300mm (12inch), chip size is 10mm*10mm, wafer thickness is 30 um, DAF-A is mounted by two methods, conventional and the DAF solo mount method. Kerf width of MD and TD was measured after BG tape peel. The result is stated below (table 3).

Table 3: CV of kerf width before DAF expansion

Conventional method	MD	TD	DAF solo mout method	MD	TD
measuring point	1383	1382	measuring point	1384	1381
Avg /um	31.0	26.8	Avg /um	28.4	27.0
SD /um	2.44	1.64	SD /um	1.99	1.61
CV	0.079	0.061	CV	0.070	0.060

Avg: Average SD: Standrad Deviation
CV: Coefficient variation (CV= SD / Avg)

Coefficient of variation (CV) is a measure of relative variability and low CV reveals even kerf width. DAF solo mount method in table 3 shows kerf width of MD improvement of CV value from 0.079 to 0.070. The distribution of the kerf width for both methods can be compared visually in figure 7.

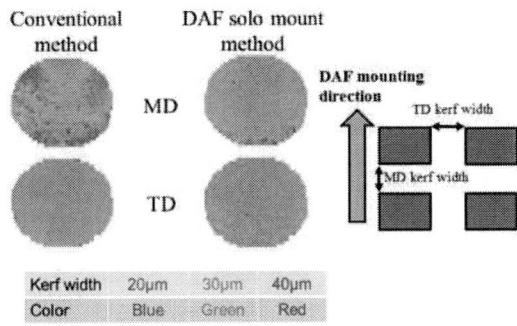

Kerf width	20µm	30µm	40µm
Color	Blue	Green	Red

Figure 7: Distribution of kerf width

3.2 CV of kerf width after DAF expansion

In this evaluation, kerf width of SDBG wafer after DAF expansion is measured. Wafer size is Si mirror φ300mm (12inch), chip size is 10mm*10mm, wafer thickness is 30 um, DAF-A is mounted by two methods, conventional and DAF solo mount method. The result is below (table 4).

Table 4: CV of kerf width after DAF expansion

Conventional method	MD	TD	DAF solo mout method	MD	TD
measuring point	2003	2008	measuring point	2008	2008
Avg /um	25.2	25.7	Avg /um	22.6	24.8
SD /um	4.65	4.19	SD /um	4.23	4.59
CV	0.185	0.163	CV	0.187	0.185

As shown above, the CV of DAF solo mount method was higher than conventional method. Contrary to expectations, evenness of kerf width after expanding became worse by DAF solo mount method.

3.3 Improvement of kerf width after DAF expansion

Generally, edge of wafer is difficult to expand and kerf width of wafer edge tends to be narrow. In this evaluation, we measured the kerf width of wafer edge after DAF expansion. Figure 8 shows the measuring point in yellow and table 5 shows the result of this evalution.

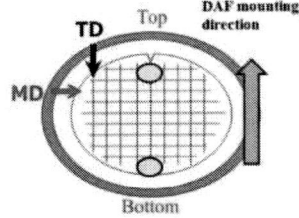

Figure 8: Measuring point of wafer edge

Table 5: Kerf width of top and bottom of wafer

Conventional method	MD	TD	DAF solo mout method	MD	TD
Top/um	25.5	32.7	Top/um	30.2	34.5
Bottom/um	13.6	24.0	Bottom/um	15.6	25.8

Table 5 shows that kerf width of the DAF solo mount method is wider than the convetional method. It indicates that the DAF solo mount method can improve the kerf width of wafer top and bottom after expansion. However, as shown in table 4, kerf width of the wafer as a whole is affected by DAF expanding process. Both DAF mounting and expanding conditions highly relate to the CV of kerf width.

4. Discussions

DAF solo mount method shows even kerf width before DAF expansion compared to conventional method (table 6).

Table 6: CV before and after DAF expansion

Before DAF expansion

Conventional method	MD	TD	DAF solo mout method	MD	TD
CV	0.079	0.061	CV	0.070	0.060

After DAF expannsion

Conventional method	MD	TD	DAF solo mout method	MD	TD
CV	0.185	0.163	CV	0.187	0.185

One of the factors for keeping kerf width even after expansion is DAF mount condition. In the DAF solo mount method, DAF tensile strength is higher than the conventional method during DAF mounting. This is because, as shown in figure 9, DAF mounting with insufficient tension can cause air bubbles when the wafer is next mounted to the DAF.

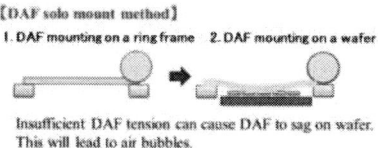

Insufficient DAF tension can cause DAF to sag on wafer.
This will lead to air bubbles.

Figure 9: Possible failure case of DAF mounting

However, by mounting DAF with strong tension, this can cause the DAF to shrink after expansion as a trade off effect (figure 10).

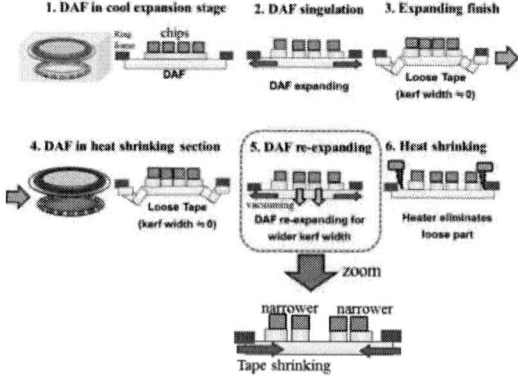

DAF mount strong tension leads to tape shrink

Figure 10: Tape shrinking after expanding

DAF-A used in the experiment was a type that was easy to shrink, and kerf width had a tendency to narrow down after DAF re-expansion. Since the properties of DAF affect kerf width after expansion, we conducted an additional evaluation by choosing a different type of DAF. Details of the evaluation is below. Wafer is Si mirror φ300mm (12 inch), chip size is 8mm*12mm, wafer thickness is 15 um, and DAF-B (Hitachi Chemical Co. Ltd.) is mounted in two methods using the same mounting conditions as DAF-A.The results of MD and TD CVs are below (table 7).

Table 7: CV of kerf width after DAF-B expansion

Conventional method	MD	TD	DAF solo mout method	MD	TD
measuring point	1648	2479	measuring point	1648	2462
Avg /um	50.8	44.3	Avg /um	49.5	50.8
SD /um	5.35	5.68	SD /um	4.61	5.30
CV	0.105	0.128	CV	0.093	0.104

Table 7 shows that the CV of the DAF solo mount method is clearly lower than the conventional method. This implies that the DAF solo mount method can provide evenness of kerf width after DAF expansion using DAF-B. We can understand that physical property of DAF relates to the kerf width after DAF expansion. However even with DAF-A, we anticipate that the evenness of kerf width can be kept wide by lowering the tensile strengh during DAF mount in the DAF solo mount method. We will continue to try and find the best mounting conditon.

5. Conclusions

Experimenting the DAF solo mount method, lead us to understand that this method can provide a balanced pressing force to the DAF in SDBG process. This method can keep even kerf width before DAF expansion. However sustaining even kerf width of the wafer during DAF expansion was much more difficult than expected. The tape properties of DAF and mounting conditions alone proved to be influencing factors. Yet, we were successful to find that the top and bottom of wafer, which is often the location with narrowed kerf width, showed improvement with the proposed DAF solo mount method.

Acknowledgements

The authors would like to thank Tomonori Minegishi and Keiichi Hatakeyama of Hitachi Chemical Co., Ltd. for their contribution, Kensuke Tamura, Tadatomo Yamada and Shino Moritani of LINTEC Corporation for their support in this work. This work was supported by consortium JOINT (Jisso Open Innovation Network of Tops).

References

[1] S.Takyu, T. Kurosawa, N. Shimizu and S. Harada, "Novel Wafer Dicing and Chip Thinning Technologies Realizing High Chip Strength," Proc. 56 th Electronic Components and Technology Conference (ECTC), San Diego, CA, May 2006, pp 1623-1627.

[2] S. Takyu, J. Sagara and K. Kurosawa, "A Study on ChipThinning Process for Ultara Thin Memory Devices, Proc. 58th Electronic Components and Technology Conference, Orlando, FL, May 2008, pp. 1511-1516.

[3] N. Suzuki, Y. Kondo, K. Atsumi, N. Uchiyama, T Ohba, "High Throuhput and Improved Edge Straightness for Memory Applications using Stealth Dicing", IEEE 68th Electronic Components and Technology Conference (ECTC), 2018, pp. 2174-2179.

[4] S. Takyu and T. Kurosawa, "Chip thinning technologies for chip stacking packages," Internathinal Conference on Electronics Packaging , pp. 64-68, April 2013.

[5] K. Aizawa, J. Maeda, Y. Hasegawa and S. Takyu, "A Study on Back Grinding Tape for Ultara-thin Chip Fabrication" 20th International Conference on Electronic Materials and Packaging, Hong Kong, PP.1-4, Decmber 2018.

[6] S. Takyu, T. Kurosawa, and A. Tomono, "Novel DAF (Die Attach Film) separation technologies for ultra-thin chip," IEEE CMPT Symposium Japan, pp. 111-114, May 2012.

Increasing the productivity of the novel atmospheric pressure sputtering technology for 3D chip interconnection

J. Bickel[1], H.-D. Ngo[1,2], M. Schneider Ramelow[2], K.-D. Lang[2,3]

[1]University of Applied Sciences, Berlin, Germany
[2]Fraunhofer IZM, Berlin, Germany
[3]Technische Universität Berlin, Berlin, Germany

+49 30 50193671, jan.bickel@htw-berlin.de

Abstract

Three-dimensional printing and rapid prototyping are becoming more and more important in the field of research and low volume production. There are many systems that can print a wide variety of materials. In the field of thin-film technology, as required for MEMS applications, only a few approaches are available. One of the promising technologies in this field is sputtering at atmospheric pressure with miniaturized plasma sources. This paper introduces a new anode nozzle geometry that increases the productivity of this novel technology substantial. It shows that three-dimensional thin-film metal structures can be printed more economically.

Key words: atmospheric pressure sputtering, harsh environment, socket, flow analyses, plasma

Field of application

In the field of high-temperature (HT) sensors, not only the temperature resistance of the actual sensor semiconductor material but also the temperature resistance of the associated assembly and interconnection technology (ICT) are critical for the trouble-free functionality of the entire system. As a rule, rare and expensive metals such as gold, platinum and palladium are used in the HT-ICT. These can still be used at chip level in a very targeted manner and with minimal consumption. This means that the cost pressure from the use of such materials is very low here. In the field of ICT, however, this procedure has so far been very difficult or even impossible.

This could be changed with the technology of partial sputtering at atmospheric pressure (PAPLD). The biggest obstacle here at present is the lateral enlargement of the structures in a deposition process and the process gas consumed in the process, which has so far greatly limited the economic use of this technology. In this paper the optimization of the outlet nozzle is investigated to minimize these disadvantages.

Working principal and process parameter of PAPLD before optimization

This work uses the advantage of the sputtering technique at atmospheric pressure. [1] [2]
This technique can be used to realize thin metal layer onto any topological surfaces. Figure 1 illustrates the working principle of the plasma source for deposition.

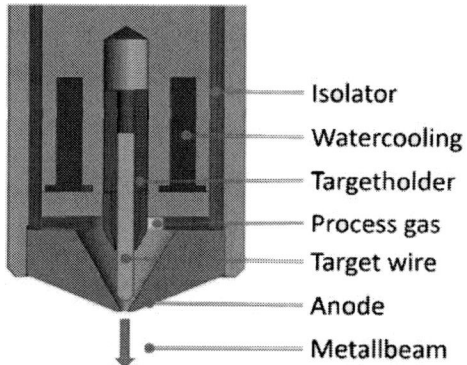

Figure 1: Working principle of the plasma source for 3D deposition

For a sputter process a capacitive coupled, direct current, helium plasma is generated around a target wire made from the metal that is to be sputtered. The wire is held by a target holder, which is used to adjust the distance between the target and the anode, to transport the heat away from the target wire and to channelize the process gas into the source. Therefore, a gas flow of 1000 sccm is used to transport the metal particles from the wire onto the substrate surface and prevents the backflow of air into the plasma. Meanwhile, instead of using a controlled arc-plasma, a glow discharge plasma is used, so there is no plasma jetting out of the source, which allows the process area and deposit material to remain at a lower temperature than the conventional arc-jet-sources [3]. Different target materials have been tested. The best deposition results were achieved with gold, platinum and palladium. The deposited particle size is between

10 nm and 50 nm. The sputter rate is controlled by the plasma power, which can be regulated approximately between 2 and 85 W for a stable process and depend on the heat-carrying capacity of the target wire. The plasma U-I-characteristic is similar a conventional low pressure plasma characteristic. The compactness of the layer is with approximately 20% at standard conditions very low due to the low process temperature and the diffuse trajectory of the sputtered particle. Therefore, the energy of the layer forming species is low in comparison with low pressure sputter technologies. In order to obtain compact layers, therefore it is necessary to supply the missing energy by means of heat. By heating the substrate during the process it is possible to increase the compactness of the layer to about 90%. The profile of a static sputtered layer normally shows a Gaussian distribution in the cross section. With the smallest standard anode/nozzle the profile can be seen in Figure 2. The minimum profile width that can be produced is just over 3.5 mm. The layer growth rate is 32 nm/s.

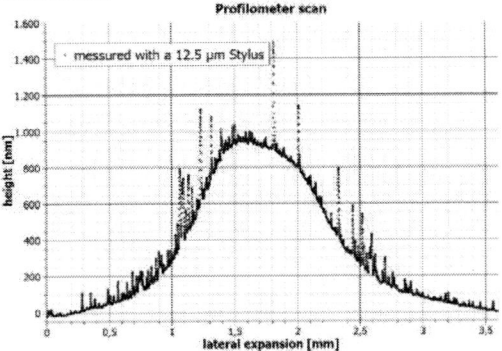

Figure 2: Tactile profile scan of a 30 s deposition with the standard anode

Boundary parameters for optimization

A key point of the sputter technologies in general is the mean free path (MFP). It can be calculating with formula (1) were (k_B) is the Boltzmann constant and (d) is the kinematic diameter of the process gas atoms and depend on the pressure (p) and the temperature (ϑ). Because of the difficulty of determine d it also can be calculated with formula (2) by using the viscosity (η) and the molar mass (M) of the process gas as the factor for the atomic cross-section. For this technology the pressure is given by the actual barometric pressure at the time of processing and it is for standard conditions 101.3 kPa. As the process gas is Helium, all parameters are well known. For standard conditions, the MFP is calculating as formula (1) at 120 nm and as formula (2) at 200 nm. For the highest possible process gas temperatures of 600 °C the MFP increases to 350 nm respectively 700 nm

$$\lambda = \frac{k_B \cdot \vartheta}{\sqrt{2} \cdot \pi \cdot d^2 \cdot p} \qquad (1)$$

$$\lambda = \frac{\eta}{p} \sqrt{\frac{\pi \cdot k_B \cdot \vartheta}{2 \cdot M}} \qquad (2)$$

It can be assumed that the distance between the electrodes is a critical factor for a working sputtering process. But it can be observed, that the glow discharge plasma burns around the whole target wire. If the distance between the target and the anode is too large, a filamentous plasma channel is formed, which over lines the distance to the area of the glow discharge plasma. Within this channel, a voltage drop of up to approximately 40 V can occur. For this reason, the electrical field loses energy, which is needed for the acceleration of the projectiles ions. It is assumed that, for an optimal process, the maximum electric field strength must refer to the length in the mean free path distance, since the energy available to the process is that of the working voltage. Therefore, the distance between the electrodes should still be minimal as possible at the top of the target wire.

Fluidic Optimization

In order to make this technology usable for micro structuring processes, the profile diameter of a static deposition rate (SDR) must first be reduced. By using smaller nozzle openings, the sputtered target material can generally be focused on a smaller area. However, care must be taken to ensure that the sputtered target material does not land on the nozzle itself, it should be transported out of the source as efficiently as possible. For this, the distance between target and anode must be as large as possible. In the area where the process gas/metal mixture is to leave the anode, a contact between metal and anode is unavoidable. Therefore, the flow velocity in this area must be as high as possible and the turbulence of the process gas must be as low as possible in order to minimize the residence time of the sputtered metal particles in the vicinity of the nozzle. For a stable plasma, however, it is necessary that the distance between the electrodes is not too large. A compromise must therefore be found. In order to find this, the finite element method (FEM) was used in this work. It is assumed that the highest sputtering rate is to be found at the location of the strongest electric field. Concentrating the sputtering process on the front part of the target has the advantage that the metal particles only have to travel a shorter distance.

Finite element method

A CFX simulation was used to investigate the flow processes. This was used to model and simulate the smallest standard anode. For the verification of the model the simulation results were compared with real measurements. For this work the qualitative measuring device of a Schlieren camera was used [4] [5] [6]. Accordingly, the results should rather be understood as qualitative analysis. Against this background, a transient simulation was carried out to ensure better comparable results. The critical point of the flow simulation is the moment when the laminar flow changes into a turbulent flow. There is no general method for the transition range of both flow types in one model. For simulations near the walls, the shear stress transport (SST) method is generally better suited than the standard k-e approaches or other Reynolds averaged Navier-Stokes (RANS) methods. Even better, however, is the Detached Eddy Simulation (DES) method, especially with a detailed mesh near the wall [7]. For free flow simulations the LES method has proven itself, but only with increased computational effort [8].

In the standard process a 1000 sccm helium is used. This provides a more laminar flow pattern. The best results were achieved with an LES simulation. With this simulation even smaller fluctuations in the transient range along the laminar flow could be correctly simulated. The left column of Figure 3 shows a comparison of a simulation and a real Schlieren-photography. The simulation shows the helium concentration in linear scale from 0% to 10%. Since the Schlieren camera shows the smallest difference in the refractive index of a gas, a resolution in the ppm range can be assumed. In the turbulent range, the LES method has poorly mapped the drifting apart of the process gas flow. The SST method has proven to be a better variant. Although the single micro vortices that form at the boundary layer between the process gas flow and the surrounding air are no longer mapped in detail, However, the distribution of the process gas is very good with the SST method. The right column of Figure 3 shows a comparison of a SST simulation and a real photography.

Laminar 1 slm LES   Turbulent 2.5 slm SST

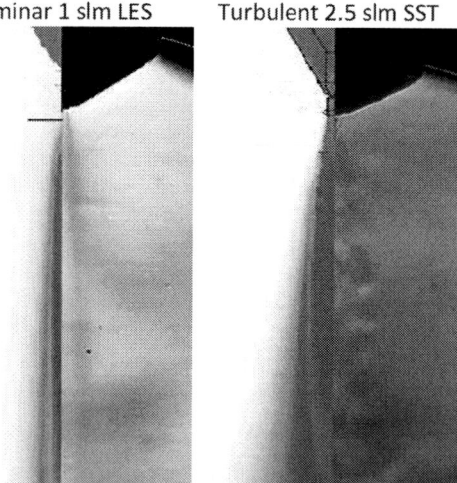

Figure 3: Comparison of simulation with reality. Each time on the left side is the simulation of the Helium distribution and each time on the right side is the Schlieren photography.

In order to determine when turbulent flow components predominate, the vortex viscosity could be identified as a sufficient criterion. In areas where the vortex viscosity exceeds the natural dynamic

viscosity of the process gas, the SST method provides more accurate results of helium distribution. The individual vortices themselves are not shown. However, these are not interesting either, as it is only a matter of verifying the model and the turbulent flows should not occur in the actual operational area.

After the verification of the free flow model and the simulation method, the process gas flow during the coating of a flat surface was simulated with a LES simulation. Again, the helium concentration was used as the parameter to be studied. The helium still has a 100% concentration within the source and only outside the source is it reduced by mixing with the surrounding air. It is therefore assumed that the density of the sputtered metal particles and thus the distribution during layer growth is qualitatively very similar to the helium distribution.

In Figure 4, the helium distribution pattern was between 80% and 100% and was scaled to the size of a static deposition profile. It is easy to see that the qualitative distributions are very similar. However, it must be noted that the size of the deposition cannot be simulated directly this way, only the pattern of the distribution can be predicted.

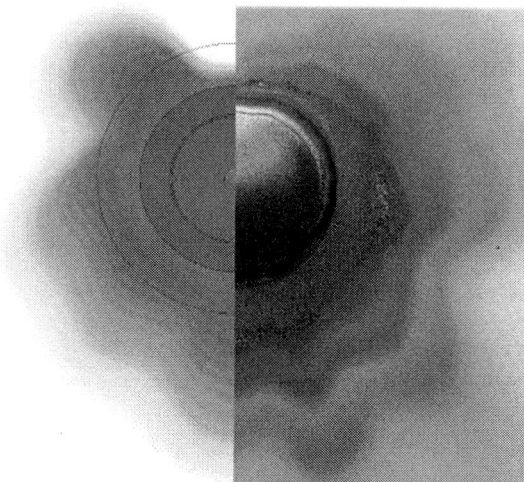

Figure 4: Comparison between a real deposition right and a scaled simulation of the helium distribution left. Processed with a standard nozzle with 0.5 mm nozzle opening and 1000 sccm flow.

Modell details

The buoyancy of the helium was taken into consideration. In order to obtain as realistic results as possible, the source was not simulated exactly at right angles to the earth's gravitational pull, but tilted by 5 degrees. Also for this reason a low side wind at the beginning of 0.1 m/s was added to evaluate the stability of the forming process gas flow. Furthermore, diffusion was also taken into consideration, since it is not clear in the steady-state to what extent diffusion exceeds convection. For the diffusion of helium in air, the value of 1.5 for kinematic diffusivity was used. Within the source, the wall roughness was set at 20 μm. However, a thermal consideration was omitted due to the high complexity and inaccuracy in the plasma area. The assumption was made that the thermal influences have a subordinate role for the process gas flow. This thesis was verified by a Schlieren image with plasma switched on and off and confirmed as long as the plasma burns stably. When an arc occurs, however, the surrounding helium heats up very rapidly and very strongly and expands, resulting in an increased gas flow from the source. However, this event is rare and undesirable and relatively unimportant for the normal process. The mesh was created with tetrahedral elements and particularly fine in the curvatures and bottlenecks. The number of nodes in all models was kept above 3 million and the transitions of the mesh element sizes were realized very slowly. In order to reduce the computing effort, only half of the model was simulated. Less than half cannot be realized, otherwise the influence of lift and crosswind cannot be taken into consideration. Furthermore, it was investigated to what extent the internal complexity of the sputter source has to be reproduced exactly. Or to what extent simplifications in mode geometry can be made in favor of a simpler mesh.

Optimization of the nozzle

With the verified model a new anode was designed. It is assumed that the sputtered material is carried only a few orders of magnitude above the MFP far through the actual sputtering process. With a maximum MFP of 700 nm, no kinetic energy from the sputtering process will be present after about 10 µm due to the particle collision. However, this length is sufficient for the process gas stream to carry

the metal particles from this point. An increased flow velocity to reduce the fluidic boundary layer is therefore not necessary. Since the nozzle directs the process gas within the source, care must be taken in the design that no unnecessary turbulence occurs, which leads to the metal particles being carried somewhere else in the source.

There are two basic approaches to a nozzle shape (see figure 5 below). One is the hole nozzle, where the nozzle wall is at a maximum distance from the target wire to the tip of the target. Because of this, the process gas flow is very slow inside the source, so there is no turbulence and the electric field is at its highest at the tip of the target. The disadvantage of this nozzle is that it is only suitable for coating planar substrates, as the outer edge of the nozzle contacts the substrate during tilting. The other is the cone anode, where the nozzle walls continuously converge towards the target. This nozzle is suitable for coating three-dimensional geometries and to a certain extent it can also be driven into recesses.

In preliminary tests, the opening diameter of the nozzle was initially fixed at 400 µm. One of these preliminary tests was shown in Figure 4. By means of a flow simulation, the process gas flow at which no more turbulent flow occurs was found at 250 sccm. The flow should be as high as possible so that on the one hand the residence time of the nanoparticles in the source is as short as possible, on the other hand it is assumed that the kinetic energy of the particles leads to better layer growth if this is as high as possible. Furthermore, it is assumed that a low particle concentration in the process gas counteracts

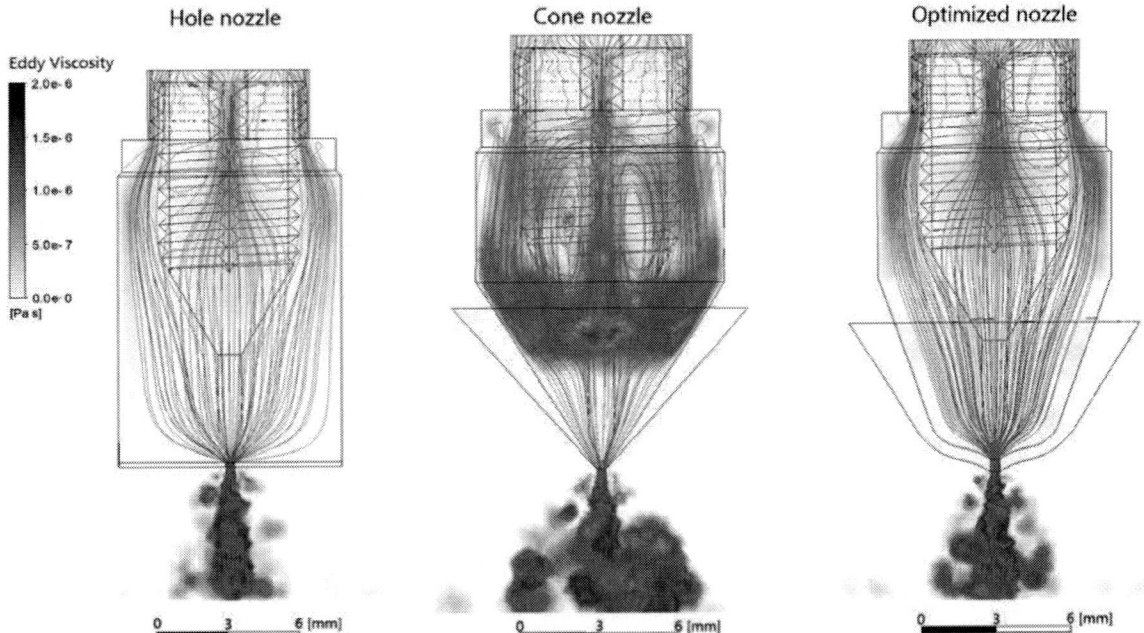

Figure 5: Three different nozzle geometries in comparison. Visible are the flow lines and the eddy viscosity. On the very left a primitive hole nozzle, in the middle a primitive cone nozzle, on the very right the optimized nozzle. All anodes were simulated with 250 sccm helium flow and have an opening diameter of 400 µm.

a conglomeration of the particles before they hit the substrate. Again, the eddy viscosity was used as the parameter for the analysis. At a flow rate of 1000 sccm it is at a maximum of 80 µPa s. These simulations could again be verified experimentally by Schlieren photography. The simulations of the flows within the source for these simple basic forms are very left and in the middle in Figure 5. It can be clearly seen the close to vortex-free flow at the hole nozzle. On the contrary, a large vortex occurs at the cone nozzle. The centre of the vortex is located above the target holder, but the spurs of the vortex continuously transport sputtered material from the target to the target holder. The sputtered material is thus increasingly deposited again at the transition from the target to the holder. This behaviour can also be observed in reality. However, it also exists in an attenuated form at the hole anode. Due to the tapered walls, an additional impulse is generated at the conical nozzle, which drives the gas flow apart after leaving the source, which can also be seen in the comparison of the simulations in Figure 5. Furthermore, it was recognized that guiding the process gas flow within the narrow point in the nozzle has a centring effect on the process gas flow. However, target material is also deposited particularly strongly in this guidance. The deposition of the sputtered target material on a wall can be seen in connection with the strength of the turbulences. In a theoretically perfect laminar flow of an aerosol of nanoparticles, only the outermost particles would come into contact with the wall and thus have the possibility to deposit there. The particles flying in areas farther away from the wall cannot interact with the wall. It is only through the turbulence that the nanoparticles can come close to the wall and thus participate in the formation of layers. For this reason, the analysis of the degree of turbulence is very interesting for the growth process of the layer and exactly this is described by the eddy viscosity.

From this knowledge a combination of both forms was developed, which combines the advantages of both nozzles and eliminates or at least reduces the disadvantages. The simulation of the flow of this optimized anode can be seen on the right in Figure 5. There is a vortex-free current within the entire source and the micro vortices leading to the increase in eddy viscosity are minimal in the area of the target and only slightly stronger than the hole anode. Due to a small bow in the area of the outlet, the velocity gradient was related to a longer distance and thus attenuated, which leads to less turbulence. To reduce the drift of the gas flow, the bottleneck of the nozzle was slightly extended. The eddy viscosity is thus outside the source between the hole anode and the cone anode.

With this new anode, a static deposition rate (SDR) of up to 350 nm/s can be achieved, so that the SDR is 11 times higher than with the previous 800 µm anode. If the masses deposited on the substrate are compared by integrating the Gaussian distribution, the result is 5 mg/h for the old 800 µm nozzle and 7.4 mg/h for the new 400 µm nozzle. The efficiency of the nozzle has therefore increased by 50 % despite the smaller diameter of the opening.

In order to better predict the deposition profile in its expansion, the eddy viscosity was examined on a planar substrate and compared with a deposition. From the previously described concept of particle deposition, layer growth can only take place in areas where the eddy viscosity has a significant size. Figure 6 shows a superimposition of a simulation and a real measurement. The new optimized 400 µm nozzle with a flow of 250 sccm helium at a distance of 200 µm above the substrate was investigated.

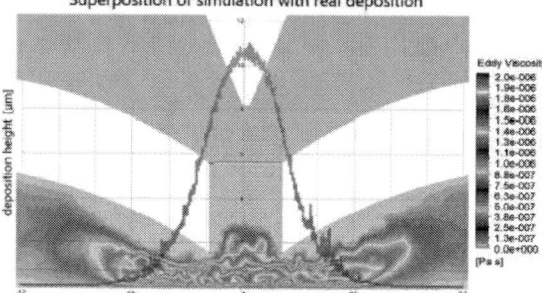

Figure 6: Cross section of a superposition of a simulationn of vortex viscosity and a real measured deposition.

Conclusion

By simulating the process gas flow, it was possible to develop a nozzle geometry with which significantly higher SDR can be achieved and much finer structures can be produced. Only a quarter of the previous process gas flow is required at a 50% higher yield at the same time, which considerably reduces the process costs of such a sputter tool. The simulation model also allowed the process parameter range to be narrowed down and the requirements for a fully automatic coating system to be defined. Furthermore, it is possible to analyse the coating behaviour of any nozzle and surface geometry in detail.

Application perspective

The aim was to model the deposition process in order to know the distance and tolerances required for a process before being able to test it with a concrete system. This data can then be used to define the exact requirements for a system and thus save development costs. For this purpose, three dimensional depositions are investigated to achieve this aim.

Acknowledgements

This work was created within the framework of an AIF-funded ZIM project. We would like to thank also the BMWi for the financial support.

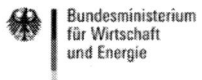 

We would also like to thank Aurion GmbH and Beaplas GmbH for their support in researching this new technology.

References

[1]. **GmbH, Beaplas.** *DE 20 2014 105 706 U1* Deutschland, 2014.

[2]. **Jan Bickel, Et al.** Sputtern an Atmosphärendruck. *Galvanotechnik.* 11 01, 2015, pp. 2268 - 2273.

[3]. **Andreas Schütze, Et al.** The Atmospheric-Pressure Plasma Jet: A Review. *IEEE TRANSACTIONS ON PLASMA SCIENCE.* 12 6, 1998, pp. 1685 - 1694.

[4]. **Hair, L. M.** *Test data from solid propellant plume aerodynamics test program in Ames 6 x 6 foot supersonic wind tunnel (shuttle test FA7) (Ames test 033-66).* s.l. : NASA United States , 1975. NASA-CR-150346, RTR-016-2

[5]. **Gerhard Emig, Elias Klemm.** *Technische Chemie Einführung in die chemische Reaktionstechnik.* Berlin : Springer Berlin, 1975. 978-3-540-23452-4 (ISBN).

[6]. **A., Schöne.** *Durchflußmessung. In: Meßtechnik.* s.l. : Springer, Berlin, Heidelberg, 1997. ISBN: 978-3-540-60095-4.

[7]. **F. R. Menter, M. Kuntz, R. Langtry.** *Ten years of industrial experience with the SST turbulence model.* Germany : Begell House, Inc., 2003.

[8]. **Tkatchenko, I., Kornev, N., Jahnke, S. et al.** *Performances of LES and RANS Models for Simulation of Complex Flows in a Coaxial Jet Mixer.* s.l. : Springer Science + Business Media B.V. , 2006. ISSN: 1386-6184.

[9]. **Polzin, J.** *Strömungsuntersuchungen an einem ebenen Diffusor.* s.l. : Springer-Verlag, 1940. ISSN: 0020-1154.

[10]. **Gerasimov, Dr Aleksey.** Turbulence Modelling LES & DES. [Online] 2006. http://www.ae.metu.edu.tr/seminar/Turbulenc e_Seminar/UGM2006_Turbulence_LES_DE S.pdf.

New Assembly Technology for VCSEL arrays comprising ultra-thin diodes

Rainer Dohle, Senior Member, IEEE, Thomas Friedrich, Jing Guo, and Jörg Goßler

Micro Systems Engineering GmbH, Schlegelweg 17, 95180 Berg (Oberfranken), Germany

Phone+49 9293 78 717, rainer.dohle@mst.com

Abstract

Objective of this paper is the development of an assembly technology for VCSELs 360 micron by 360 micron in size and a thickness of 72 micron with high positioning accuracy for automated production. Due to special customer requirements, a final thickness of the conductive glue of about 50 micron is necessary. For highest reliability, gold wire bonding of the top side contact of the VCSEL to a silicon substrate with gold metallization and printing gold-based conductive glue have been used. One VCSEL array contains 512 diodes. Special design features of the VCSEL secure that the relatively high thermal resistance of the cured conductive glue layer does not impair the optical properties or the life time of the VCSELs. Die bonding has been carried out with a Datacon EVO machine. The singulated laser chips were picked from a blue tape. For the interconnection of the top side contact we used gold wire and a Stand-Off Stich Bond process with special loop geometry. With an optimized screen printing process we did get a reproducible height of the conductive glue of about 50 micron with low standard deviation, as required, and no shorts at the silicon substrate. The position tolerance of the VCSELs meets the customer requirements. Gold wire bonding delivered excellent results, meeting the customer specification. For the first prototypes we did get a total yield of 99.9%. The optical spectrum of the VCSELs and other measurements indicate that the assembly processes do not harm the diodes or their optical properties.

Key words: VCSEL arrays, assembly technique, SSB, high reliability

Introduction

VCSEL arrays have a very broad application potential [1]. Objective of this paper is the development of an assembly technology for VCSELs 360 micron by 360 micron in size and a thickness of 72 micron with high positioning accuracy for automated production with high total yield, using gold-based conductive glue (because silver migration is a concern) securing high yield and extremely high reliability. Due to special customer requirements, a final thickness of the conductive glue of about 50 micron is necessary. For highest reliability, gold wire bonding of the top side contact of the VCSEL to a silicon substrate with gold metallization has been a customer requirement. That means a monometallic contact system has been used. One VCSEL array contains 512 laser diodes.

Methodology

Due to the boundary conditions, required by the customer, and the high reliability application, not all assembly techniques could be considered. We started with an automated dispensing operation of the conductive glue, but we could not get the required amount of conductive glue with sufficiently low standard deviation. After this experiment, we stamped the conductive glue automatically, using special 0.35mm and 0.47mm stamping tools with standoff pin, see fig. 1:

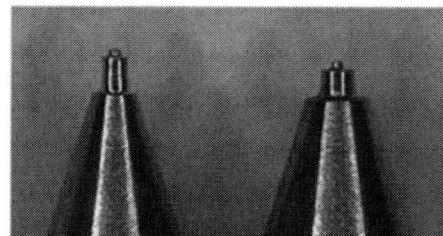

Figure 1: Stamping tools used for experiments

This approach was not conducive either. Therefore we started experiments with screen printing of the conductive glue. After first promising results we ordered a special screen with special surface coating and optimized the mesh size and the angle of the screen wires. This yielded to a very productive manufacturing technique meeting the requirements mentioned above. Special design features of the VCSEL secure that the relatively high thermal resistance of the cured conductive glue layer does not impair the optical properties or the life time of the VCSELs.

Die bonding has been carried out with a Datacon EVO machine, offering single part traceability. The singulated laser chips were picked from a blue tape.

For the interconnection of the top side contact we used gold wire with 25 micron diameter, using Stand-Off Stich Bond process with special loop geometry, required by the customer, see fig. 2:

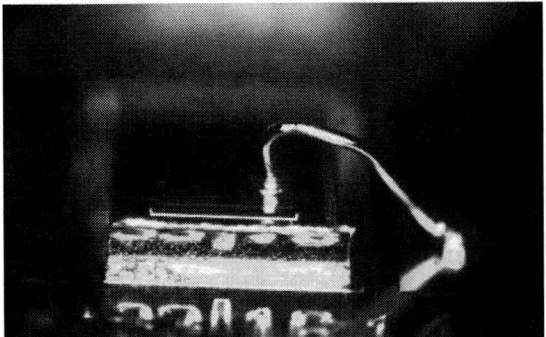

Figure 2: Wire bonded VCSEL

VCSEL Array Design

The VCSEL array consists of 512 laser diodes (semiconductor thickness 10 μm, backside gold metallization thickness 62 μm) as given above.

Stencil Printing

The required thickness of the conductive glue of about 50 micron presupposed the development of a non-standard printing process. We used a special screen VA 200-0.036D-22.5° with MS50µm film and surface coating.

Die Bonding

For the die bonding of those thin III-V-semiconductor chips special, unique process parameters and processing tools were developed.

In [2] are key properties for die pickup of ultra thin silicon dies from a wafer foil described: die strength (or fracture strength), bending stress during pickup, edge peel force and bulk peel force. The more fragile III-V semiconductor material with lower die strength makes the die pickup from the wafer foil even more difficult. The peeling of the dies from the wafer tape during die bonding is a very critical process [2], because die stress levels during peeling are significantly high. The thinner the die is, the higher the stress [2]. The wafer dicing method has not only an impact on die strength, but also the adhesion of the die curbs on the wafer foil [2]. Prior to optimization of our die bonding process we observed die cracking of the III-V semiconductor material, fig. 3 and fig. 4:

Figure 3: VCSEL damaged during die pickup

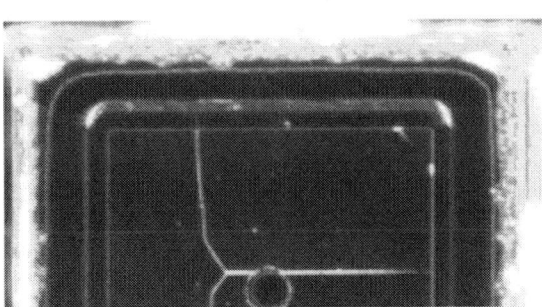

Figure 4: VCSEL damaged during die pickup

In order to prevent cracking of the fragile semiconductor material, we reduced the speed of the moving ejector part (needle). This facilitates the activation of the peeling process from the wafer foil. The higher the peel energy and the thinner the die the lower the process speed [2]. According to [2] the amount of the peel force has an impact on the maximum bending stress of the die and this force increases in most cases with speed. More theoretical background can be found in [2] - [5].

Due to the reversible elongation of the wafer foil (dicing tape 10003R from Ultron) during pick-up it is not possible for the vision system of the die bonding machine to identify the next die properly. We solved this problem by implementing three runs of the wafer map, picking every third die. After a while, the elongation of the wafer foil disappears, and during the next run the die can be identified and picked successfully.

Wire Bonding

Stand-Off Stitch bonding (SSB) involves the placement of a ball bump at one end of the wire interconnect, then placing a wire with another ball at the other end of the interconnect and stitching off the wire on the previous placed ball bump. Whether the need is due to poorly bondable materials, non-flat bonding surfaces, odd packaging situations, or just the need for high reliability; the integrity of a wire bond interconnect can usually be greatly improved through the proper use of Auxiliary Wires. Auxiliary Wires are defined as Security Wires, Security Bumps, or Stand-Off Stitch (also known as Stitch on Bump). The old stand-by Security Wire has been an asset for several decades, however, this is being replaced by Security Bumps which require a smaller second bond termination area. Further, Stand-Off Stitch (SOS or SSB) has many more applications and also have many side benefits that could be incorporated into a circuit design for better wire strength properties, fewer interconnects (die to die bonding), and lower loops. During SSB is bump first bonded to the substrate. After that a ball is ball-wedge bonded with the wedge on top of the first bump. This results in a near homogeneous stitch bond interconnect to the bump with an inherent improvement in stitch bond pull strength. Another use for SSB is Reverse Bonding (Stitch bond on bump on die bond pad) often resulting in a lower loop profile than standard forward wire loop and the loop is stronger because the wire hasn't been work annealed above the ball (in the heat affected zone). The SSB process allows a finer control of bump height and flatness and a reduced overall height (non-compressed about three times of the wire diameter). A major impediment to the implementation of SSB is the retraining of visual inspectors and the approval of quality department. [6] - [8]

Stand-Off Stitch bonding was carried out at a temperature of 80°C. The ball is pressed to the bonding pad on the die with sufficient force to cause plastic deformation and atomic interdiffusion of the wire and the underlying metallization, which ensure the intimate contact between the two metal surfaces and form the first bond. The loop geometry was optimized to keep enough distance to the edge of the die. A nondestructive pull-test was carried out on some sample to ensure that the wires are stable connected.

Geometrical Results and Yield

Die bonding results: Figure 5 shows a VCSEL array after die bonding:

Figure 5: VCSEL array with 512 laser diodes after die bonding onto a Silicon submount 20mm by 30mm in size

Figure 6 shows a detailed view with six VCSELs after die bonding:

Figure 6: Detail of the VCSEL array after die bonding

The die placement accuracy is ±10µm@3σ, the die rotation <5°. Figure 7 shows a detailed view with one VCSEL after die bonding:

Figure 7: One VCSEL after die bonding with prober mark

Wire bonding results: We obtained a distance between ball and VCSEL edge between 6 and 10 micron, a distance between ball and light emitting window of the VCSEL of about 60 micron, a ball diameter at the VCSEL of about 70 micron, and a loop height of 240 to 260 micron.

Figure 8 and 9 show wire bonded VCSELs, fig. 10 shows a detail of the SSB bond. Figure 11 shows a cross section of a wire bonded VCSEL under high magnification:

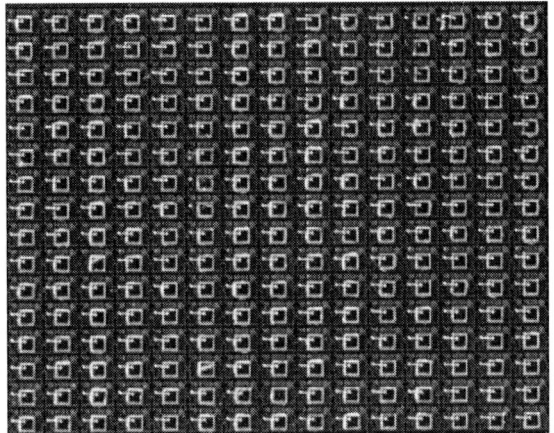

Figure 8: VCSEL array after wire bonding

Figure 9: Wire bonded VCSEL

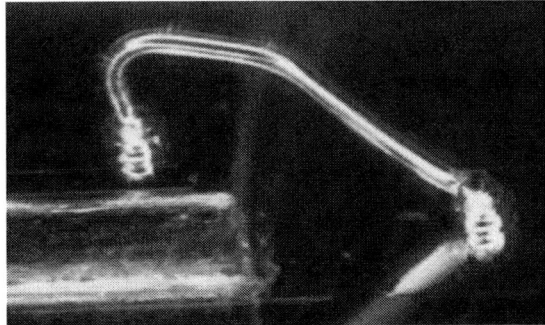

Figure 10: Detail of a SSB bond

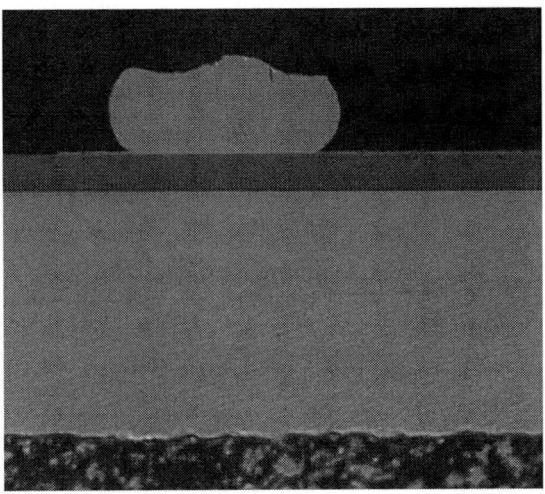

Figure 11: Cross section of a wire bonded VCSEL (detail)

Measurement Results

Figure 12 shows the U-I plot of one of the laser diodes after assembly at different temperatures.

Figure 13 shows the optical output power of one of the laser diodes after assembly at different temperatures.

Figure 14 shows optical output power and voltage of one VCSEL in dependence of the drive current.

Figure 15 shows the optical spectrum of one of the laser diodes after assembly. The shape of this optical spectrum demonstrates that the assembly and curing processes did not harm the VCSEL. This finding has been confirmed by life time experiments.

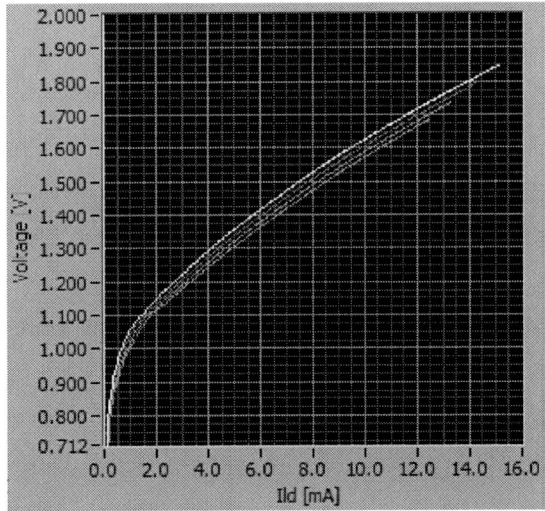

Figure 12: U-I plot of one VCSEL in dependence of the temperature (white: 10°C, red 20°C, green 30°C, blue 40°C)

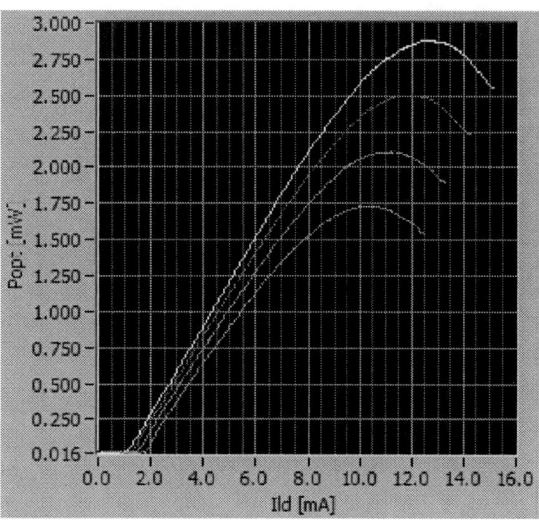

Figure 13 (a): Optical output power of one VCSEL vs. drive current in dependence of the temperature (white: 10°C, red 20°C, green 30°C, blue 40°C)

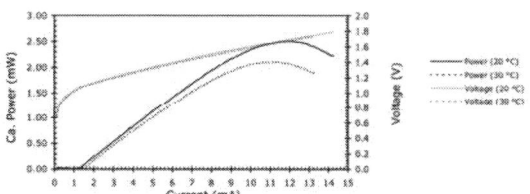

Figure 14: Optical output power and voltage of one VCSEL vs. drive current

The most common emission wavelengths of VCSELs are in the range of 750–980 nm (often around 850 nm), as obtained with the GaAs/AlGaAs material system [9]. However, we used VCSELs with longer wavelengths of about 1.5 μm (as required for, e.g., gas sensing) obtained with devices based on indium phosphide (InAlGaAsP on InP).

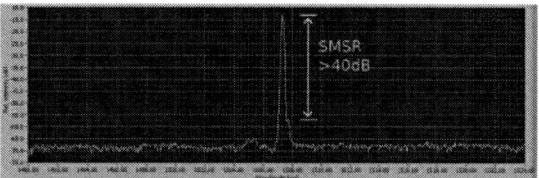

Figure 15: Relative intensity vs. wavelength at T=20°C and I=12mA. The side mode suppression ratio SMSR is larger than 40dB

The ground mode of this VCSEL is 1507.2nm, the side peak at 1507.5nm is from a polarization mode, and the peak in the underground noise at 1505nm is from the next higher transversal laser mode.

The short resonator of a VCSEL makes it easy to achieve single-frequency operation, even combined with some wavelength tunability [10].

Figure 16 shows the tuning characteristic of one VCSEL at 20°C and 30°C.

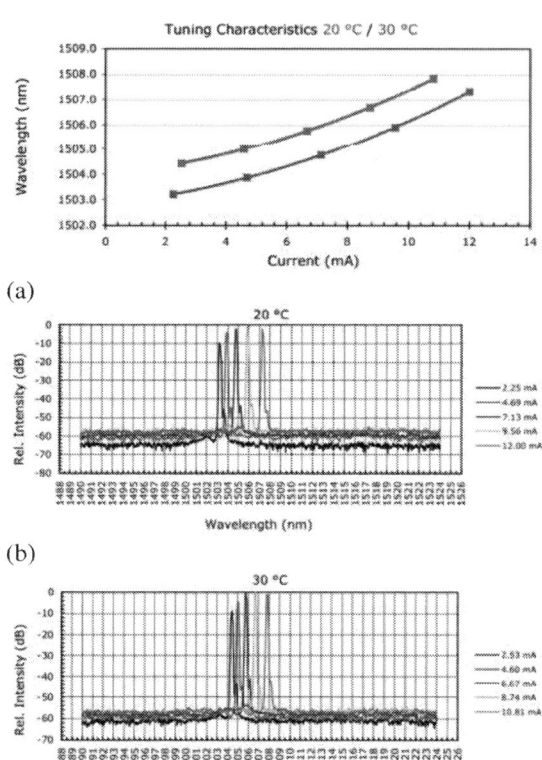

(a)

(b)

(c)

Figure 16: Tuning characteristic of one VCSEL at 20°C and 30°C at different currents (a), wavelength at different drive currents at 20°C (b), wavelength at different drive currents at 30°C (c)

Discussion

The die bonding yield depends on the needle speed during die pickup and the needle geometry. Carefully optimization has been necessary.

During the wire bonding, the bonding force leads work hardening, while the ultrasonic energy softens the wire. Due to the strong ultrasonic energy and vibration, the die has to be attached very firmly to the substrate otherwise this ultrasonic energy and force could influence the connection of the die and lift the die up together with the bonding tool. This lifted die problem could be solved by slightly reducing the ultrasonic energy and force, but lower pull strength between the wire and the die should also be considered at the same time, that means carefully optimization of the wire bond process has been necessary as well.

Summary

With the optimized screen printing process mentioned above we did get a reproducible height of the conductive glue of about 50 micron with low standard deviation, as required and no shorts at the silicon substrate. The position tolerance of the VCSELs in x and y direction has been about ±10 micron and theta less than ±5 degrees, meeting the customer requirements.

Gold wire SSB bonding (fig. 17) delivered excellent results, meeting the customer specification.

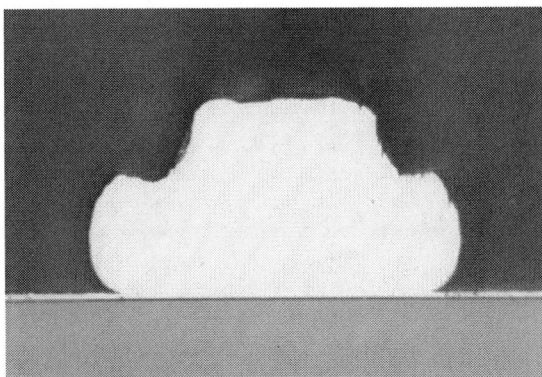

Figure 17: Cross section of a SSB bond at the substrate side

For the first prototypes we did get a total yield of 99.9%. The optical spectrum of the VCSELs and other measurements indicate that the assembly processes do not harm the diodes or their optical properties. At the conference we will present results from the serial production too.

Outlook

As mentioned in chapter "Die Bonding" already, has the wafer dicing method not only an impact on die strength, but also the adhesion of the die curbs on the wafer foil [2]. Therefore we will work on the improvement of the die singulation processes including investigation of other dicing tapes. This will allow us to reduce die bonding time and improve the die bonding yield further.

We will manufacture VCSEL arrays with 2048 laser diodes in the future as well.

Conclusions

We developed an assembly technology for VCSEL arrays using conducting glue and stand-off stitch wire bonding. Due to the monometallic contact system high reliability and long life time can be secured. The electro-optical and burn-in data furnished proof of the quality of the engineered assembly technology.

Acknowledgements

We acknowledge Simon Tätzner and Thomas Kätzel for valuable contributions to this paper. We would like to thank Thomas Dünne and Maximilian Wallrodt for part of the photos and Bernd Burger for cross sectioning and micrographs.

References

[1] R. Michalzik (ed.), "VCSELs. Fundamentals, Technology and Applications of Vertical-Cavity Surface-Emitting Lasers", ISBN 978-3-642-24986-0, Springer, 2013.

[2] S. Behler, W. Teng, A. Podpod, "Key Properties for Successful Ultra-Thin Die Pickup", Proceedings of the IEEE 67th Electronic Components and Technology Conference (ECTC), pp. 95-101, 2017.

[3] K. Kendall, "Thin-film peeling - the elastic term", J. Phys. D: Appl. Phys., Vol. 8, Issue 13, p. 1449-1452, 1975.

[4] D.A. van den Ende, H.J. van de Wiel, R.H.L. Kusters, A. Sridhar, J.F.M. Schrama, M. Cauwe, J. van den Brand, "Mechanical and electrical properties of ultra-thin chips and flexible electronics assemblies during bending", Microelectronics Reliability, Vol. 54, p. 2860-2870, 2014.

[5] S. Behler, "Dynamic Simulation of Die Pickup from Wafer Tape by a Multi-Disc Ejector using Peel-Energy to Peel-Velocity coupling", 17th Electronics Packaging Technology Conference (EPTC), p. 1-5, 2015.

[6] R.C. Garcia, "Improving Stitch Bond on Hybrid Thick Film Substrate using Stand-Off Stitch Wire Bond Technique", International Symposium on Microelectronics (IMAPS): Fall 2015, Vol. 2015, No. 1, pp. 425-429, 2015.

[7] H. Xu, A. Shah, B. Milton, I. Qin, "Wire Bonding Advances for Multi-Chip and System in Package Devices", International Symposium on Microelectronics (IMAPS): Fall 2018, Vol. 2018, pp. 583-588, 2018.

[8] D.J. Rasmussen and R. Thomsen, "Improved Bond Reliability Through the use of Auxiliary Wires (Security Bonds and Stand-Off Stitch)", 43rd International Symposium on Microelectronics (IMAPS): Fall 2010, Vol. 2010, pp. 474-478, 2010.

[9] J.L. Jewell, J.P. Harbison, A. Scherer, Y.H. Lee, L.T. Florez, "Vertical-cavity surface-emitting lasers", IEEE Journal of Quantum Electronics, Vol. 27, Issue 6, p. 1332- 1346, 1991.

[10] C.J. Chang-Hasnain, "Tunable VCSEL", IEEE Journal of Selected Topics in Quantum Electronics, Vol. 6, Issue 6, p. 978-987, 2000.

Evaluation of adaptive processes for the embedding of bare dies in IC substrates

Ruben Kahle[1], Friedrich-Leonhard Schein[2] and Andreas Ostmann[1]

1: Fraunhofer Institute for Reliability and Microintegration (IZM), 13355 Berlin
2: Technical University of Berlin, 13355 Berlin

Phone: +49 30 46403-680, E-Mail: ruben.kahle@izm.fraunhofer.de

Abstract

Highly integrated, advanced multi-chip packaging solutions combine application, logic and computing dies with memory or components for power management in a single package. A solution to achieve low fabrication costs is the close embedding of thin dies in IC Substrates based on large formats (600 x 600 mm²), known from PCB fabrication. In a consortium of partners from industry and research advanced technologies for Panel Level Packaging (PLP) are developed. Here, dies are symmetrically embedded under low stress into pre-manufactured IC substrates. The Embedding in Cores with Cavities (EiCC) targets towards low cost and thin packages (< 150 μm) with multiple, heterogeneous components. The biggest disadvantage is the potentially low yield due to low assembly accuracy and process tolerances during the embedding process. This paper presents recent results to optimize the yield of the EiCC process chain. We assemble two 6x6 mm², 100 μm thin dies with 25 μm high Cu pillars face down on a temporary adhesive foil with two assembly concepts, varying assembly throughput and accuracy. After embedding the stack in Ajinomoto Build-Up Film (ABF), laser drilled vias and a semi-additive Process (SAP) with 10 μm lines and space with a copper thickness of 5 μm acts as electrical routing between the daisy chain structured dies. Based on practical work we compare the known status of precision focussed manufacturing against a rule-based system that acquires data with a Coordinate Measurement Machine (CMM), rearranges fabrication plans and forwards data along the process chain.

Key words: bare dies, packaging, embedding, adaptive process

Introduction

Consumers ask for AI supported functions within smartphones or automotive applications to enhance the user experience. Therefore, advanced packages combine application and memory dies with supportive components for power management in a single thin (< 200μm) and reliable package. A promising variant of multi-chip fan out packages is the embedded device. The technique utilizes organic, glass fiber reinforced, premanufactured substrates.

Symmetrical embedding of bare dies into a core with cavities (EiCC) was demonstrated by Younggun et al. from University of Fukuoka demonstrated in 2015 [1]. It was shown that EiCC generates low residual stress on the dies during embedding and soldering, the fabrication substrates exhibited very low warpage and first reliability tests passed JEDEC 1 [1]. Although the fabrication of fine line substrate with copper line/space > 10 μm is a high volume technology, the embedding of bare dies with fine pitches (< 100μm) into IC substrates is not.

The great challenges are misalignment of the components due to relatively low assembly accuracy on high throughput assembly tools (3σ > 25μm), embedding induced repositioning and routing layout

dependend stretching/shrinking of organic substrates during fabrication.

To combine the potential low cost of PCB manufacturing with cost intensive fine pitch bare dies, the production tolerances can be costly optimized but the time to market is long and failure costs . In this paper a solution to enter the market faster with new embedding products in EiCC technology is presented. We use a process chain that acquires the actual position of embedded dies, redesign the electrical routing and feed-forward the information along a R&D manufacturing line.

Principle of adaptive processes

The basic principle of the adaptive approach is a feed-forward loop with actual data for each unique package on the substrate. Along a PCB process chain the following manufacturing element (e.g. laser-direct-imaging, laser-drilling station) can fabricate the routing of even misaligned dies due to a layout correction. For this purpose, previously measured coordinates of the embedded components in relation to substrate coordinate system are utilized.

A flexible mask-less lithography system, a routing system and a packaging method were demonstrated in the late 1980s to handle the die position errors [2, 3, 4]. Also, compression molded multi die packages, fabricated as reconfigured wafers and panels from Deca Technologies Inc. (ASE Group) use adaptive processes to solve die-shift issue during encapsulation [5].

In the following section, we present an embedding test vehicle with two bare dies in an organic substrate to demonstrate adaptive processes.

Testvehicle

The test vehicle is a 150μm thin module with two 6 x 6mm² dies each 100μm thick. Both dies compromise 576 pcs. of 25μm high Cu pillars with a diameter of 60μm. On the die surface daisy chain structures between the Cu connections allow electrical measurements (see Fig.1). The pitch of the outer three pad rows is 100μm. Laser vias with a diameter of 20μm in combination with contact pads of 60μm enable ±20μm tolerance via to pad. The 18 x 12mm² large package is a fan out design. It features die-to-die connections with a copper lines/spaces of 10μm. The semi-additive built up copper is 5μm thick. As a metal seed layer Ti/Cu was deposited 100/200nm thick on 25μm layers of nanofilled ABF GX-T61.

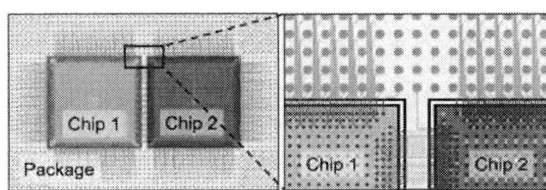

Figure 1: package design overview and bridge details between the dies

To lower the material costs we first used a smaller test panel of 305 x 227mm² with 120 dual chip packages (see Fig. 2).

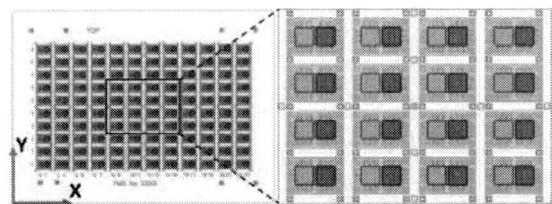

Figure 2: substrate design overview and detail

We placed registration elements in the free space of 3mm distance around the packages to sequential measure geometrical changes of the substrate along the process chain and enable global plus local alignment during assembly.

Fabrication of EiCC packages

The EiCC concept is based on pre-manufactured substrates that incorporate electrical connections and vias on multiple layers but most important through cavities. In this paper, Hitachi Chemicals core material was chosen (MCL-E-770G). It has a Tg >260 °C, a CTE xy of 4-6 ppm/K and a CTE z of 8-13 ppm/K.

The core was structured (print and etch) and UV laser milled to fabricate the cavities. The cavity was 80μm bigger than the die outline. To compensate relaxation of the core during the etching process, shrinkage factors where iteratively evaluated to minimize the initial geometric failures. Figure 3 depicts a simplified EiCC process.

We evaluated the adaptive approach in comparison to a process chain withot adaption where we focussed on accuracy in crucial process steps (here referred to as a „classic" process chain).

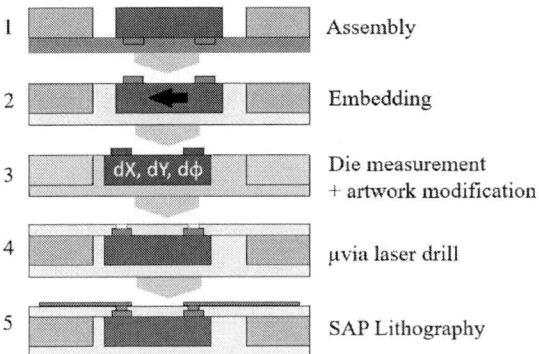

Figure 3: simplified adaptive process flow

1. Assembly

The substrates were prepared with a single sided adhesive foil. The foil was vacuum laminated on top of the substrate frontside with a lamination tool from Dynachem (VA 7224-HP7). Then, test dies with Cu pillar contacts were attached face down into the cavity onto the single sided adhesive foil. To assemble the dies, a Datacon EVO 2200 and a SiPlace CA3 were used to show the concept comparison, slow but accurate (cph < 500, $3\sigma < 10\mu m$) in contrast to fast and less accurate x, y, φ position (cph > 2500, $3\sigma > 25\mu m$), cph = chips per hour.

2. Embedding

A single layer ABF GX-T61 was laminated towards the backside of the silicon test dies. A vacuum lamination tool from Dynachem (VA 7224-HP7) pressed the ABF layer 90s at 100°C stage temperature with 0.6 Nm/mm² into the gap between die and cavity and evenly on the bottom side. According to [1] the temporary stack was treated with and without a prebake of 120 °C for 30min until epoxy gelation. In the next step, the adhesive tape was debonded and a second identical ABF layer laminated on the frontside of the dies. Finally the stack of substrate, dies and symmetric laminated ABF was cured in a PCB lamination press Lauffer (RMV7) with settings 1 Nm/mm², under a vacuum of 1 mbar, at a temperature of 150 °C for 1h. The process results in a planar dielectric layer over the chips with bumps inside

3. Measurement of embedded dies

To measure the coordinates and rotation of the embedded dies we peeled the PET covering foil off the ABF after lamination and placed the substrates with vacuum fixation in a CMM capable panel inspection tool (Rudolph Firefly S1200). During recipe creation, we tested brightfield, darkfield and and the built-in laser flurescence illumination. Using flurescence illumination, the embedded copper structures on the die show a clear contrast through the 25 µm thick ABF GX-T61 layer (see Fig. 4).

Comparing the optical focusability of the glossy ABF layer, after peeling off the PET, versus roughened chemical/plasma treated ABF layer, the glossy surface showed more stable results. To measure bump coordinates and to calculate the die center and rotation, we used a circular search algorithms for corner-oriented bumps.

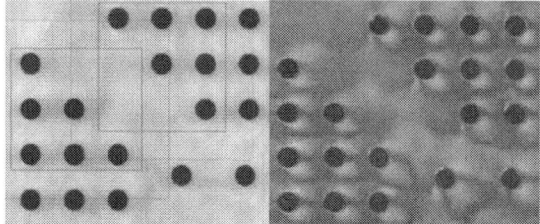

Figure 4: 60 µm Cu pillars under ABF in fluorescence light (left) and RGB microscope view in dark field (right).

Data of the die target coordinates, and measured x- and y-shift and the rotation were saved in a machine-readable ASCII format.

4. Artwork manipulation and export

The yield of single die packages can be optimized by using methods of translation and rotation of the digital artwork. Mask based steppers can solve this task in a feed forward loop combined with an internal or external CMM. To connect two or multiple embedded dies with each other it is not sufficient to move, rotate and scale the artwork only, because the angle and the direction of the die (re)positioning exceed the pad tolerances of 20µm.

To solve this task we evaluated the following method: We predefined cutting areas (blue) around the die that allow artwork manipulation. An in-house developed software tool was used to calculate flexible, rubberband alike connection traces (red), and relocatable plus fixed elements (green) as depicted in figure 5. The laser vias moving with the die connections are highlighted light green. Here, the die was rotated 0.15 °, and shifted -25 µm in x- and y-direction.

To recalculate the artwork, the CMM- and Gerber data of all CAM layers were imported and all elements recalculated using a multithread optimized program code. Complete substrate fabrication data were exported as Gerber-type for lithography and

Excollon for laser drilling. In addition, coordinates of measured registration marks on the substrates are included in the export to inhibit non-linear scaling when using drill registration marks in the PCB Core with higher production tolerances.

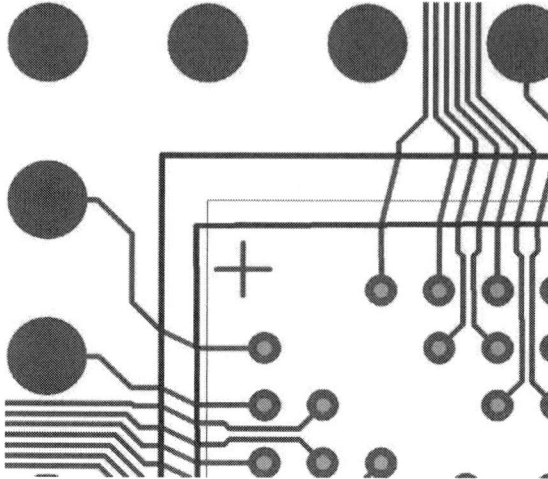

Figure 5: rubber band alike CAM rerouting

5. Laser drill

To fabricate micro vias through the ABF a Schmoll PicoDrill with 355 nm laser source was used in punching mode (single 16ps long pulses). Panel specific drill data was fed into the machine using Excollon data with actual data of the global alignment marks and the panel was registered using four alignment marks.

6. RDL Processing

The redistribution layer (RDL) is formed in a semi-additive process. At first a metal barrier and seed metal is deposited by sputtering Ti (100nm) and Cu (200nm). Then a 7µm thin dry film photo resist (Hitachi RY5107UT) is applied in a vacuum applicator. Laser Direct Imaging (LDI, Orbotech Paragon Ultra 200) exposes the RDL pattern to the resist.

The use of LDI and recalculated artwork enables substrate and package unique images by acquiring only four global fiducials. Then the resist is developed, copper structures are plated 5µm into the resist. Finally, the resist is stripped, and barrier metal and Cu seed layer are back-etched.

Results and Discussion

Embedding: According to [1] the mean failure repositioning vector |dXdY| of embedded dies between assembly and the embedding process is lower if the described preheat is applied. The preheat results in a gelation of ABF and inhbits a remelting. In our test group of 8 Panels with 960 dies the average repositioning was 23µm (σ =13µm). The vector diagram is depicted in Figure 6.

Using a fast cycle embedding press sheme without the preheating resulted in a chaotic movement of dies with a significant higher mean repositioning failure vector - only bounded by the geometry of the cavity. Nevertheless, we combined a larger cavity (see fig. 8) with a high-speed assembly machine and and a short cycle press program to show the benefits of the adaptive approach.

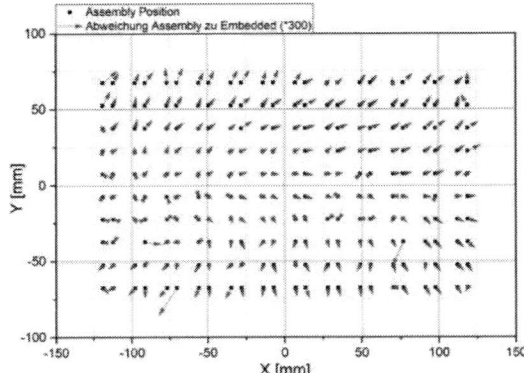

Figure 6: average repositioning with preheat [1]

Isolating position-defined defects, the accuracy-focused processes only had a yield of approx. 13% of all packages. In most of the cases both, the laser drills and the copper RDL were not aligned with the repositioned embedded die. Despite the fact, that we used local registration marks near the package, accurate assembly machine ($3\sigma < 10\mu m$) and an embedding optimized lamination sheme, only intensive engineering of the embedding process itself with complex CTE mismatch, material flows and potential non-linear scaling of the substrate might solve the problem.

In contrast, the adaptive process chain only utilizes four global alignment marks, a fast and less accurate assembly machine and a quick lamination cycle. As a result the adaptive approach compensated the repositioning and an overall first pass yield of 62% in an R&D environment with only 8 panels was achieved. The majority of the defects (24%) could be solved with a laser machine recalibration. Figure 7 depicts the rubber band alike RDL of an embedded die with $100\mu m$ pitch and figure 8 the well aligned interconnects.

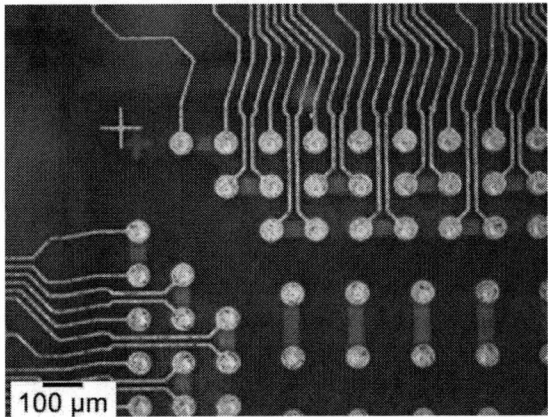

Figure 7: rubber band alike, adapted RDL of an embedded die in EiCC technique

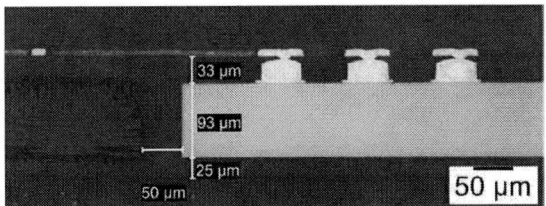

Figure 8: cross section of the test vehicle with 100 µm pitch, adapted RDL and laserdrill positions

Conclusion

Comparing the total process time of the accuracy-focused approach and the adaptive approach we conclude, that high troughput assembly (cph>2500) and short cycle embedding work best in combination with adaptive processes. In sum, we achieved a 50% shorter total production time for one panel with adaptive processes compared to the classic approach. Key elements of an adaptive process for embedded dies, ready to support further miniaturization [6] are:

- Fast ($t_{measure} < 60s$) and accurate ($3\sigma < 2\mu m$) measurement process of embedded devices with an external CMM that does not slow down lithography tools.
- High performance recalculation tools ($t_{calc} < 30s$) to manipulate, translate and export CAM data into fabrication and machine specific export working data.
- Efficient ripping algorithms of direct imaging systems ($t_{RIP} < 60s$) for large scale format up to 600 x 600mm².

Acknowledgements

This work was financially supported by the German Ministry of Education and Research within the project "PEKOS" (No. 16ES0718K).

References

[1] Younggun H., "Process Feasibility and Reliability Performance of Fine Pitch Si Bare Chip Embedded in Through Cavity of Substrate Core", IEEE Transacons on components, packaging and manufacturing technology, Vol 05, NO. 4, April, 2015

[2] Eichelberger C., „Adaptive lithography system to provide high density interconnect". US Patent US4835704A, 1986.

[3] Eichelberger C., „Multichip integrated circuit packaging configuration and method". US Patent US4783695, 1986.

[4] Haller T., „Locally orientation specific routing system". US Patent US5258920A, 1989.

[5] Bishop C., "Adaptive Patterning Design Methodologies", IEEE 66th Electronic Components and Technology Conference, 2016

[6] Ostmann A. , "High Density Interconnect Processes for Panel Level Packaging", 7th ESTC 2018, Dresden, Germany, Sept. 18-21, DOI:10.1109/ESTC.2018.854643

High throughput two-stage bonding technique for advanced wafer level packaging

S. Altenbockum[1], B. Oberhofer[2], T. Sonoda[3], and V. Rangelov[1]

[1]ATV Technologie GmbH, Vaterstetten, Germany

Phone: +49 8106 30500, silvia.altenbockum@atv-tech.de

[2]Panasonic Industry Europe GmbH, Ottobrunn, Germany

[3]Panasonic Smart Factory Solutions Co., Ltd, Fukuoka, Japan

Abstract

Thermo-Sonic Flip Chip Bonding (TS-FCB) is an already established technology for chip on wafer (CoW) applications in the industry. As a very fast and reliable interconnection process [1] it has been chosen for this project to temporarily tack small and medium sized ICs on the substrate. Its individual chip placement allows high yields per substrate. Metallic interconnections with high reliability were formed and characterized using a thermo-compression gang bonding oven with a capability to process substrates with topography [2]. Both techniques together combine high productivity with flexible, reliable, and well known fluxless process providing a new solution to drive the integration towards future needs.

Key words: Gang bonding, thermo-sonic bonding, flipchip, vacuum reflow soldering

Introduction

The continuous need for package miniaturization, higher performance and cost reduction has driven the advances in packaging technologies in the past decade. 3D wafer level integration [3], which is already well established in high-volume manufacturing and also newer techniques as fan-out wafer level packaging (FOWLP) allow heterogeneous integration in the upcoming packages to force future trends like Internet of Things (IoT), Industry 4.0, Electro Mobility with autonomous driving, green energy using high power and dense sensor integration to a wide range of products and infrastructure for "smarter" environment and life [4, 5].

In the present work, a two-stage approach has been evaluated by separating the chip placement and the permanent gang bonding of those. Existing thermo-compression processes as wafer-to-wafer, chip-to-wafer, and chip-to-chip bonding either show disadvantages in yield or need long process time making the process economically inefficient.

With the new idea of combining the two steps of thermo-sonic (figure 1) and thermo-compression bonding processes, a very fast chip-to-chip or chip-to-wafer alignment and tacking in the first step with a permanent final gang bonding in the second step is realized. With this setup, the loss of throughput is partially compensated by batch processing in a dedicated vacuum reflow oven with the capability of applying an isostatic pressure over the entire working area. Furthermore, there is a significant yield improvement, while the selection of only "good dies" can be realized. The presented approach shows higher flexibility in comparison to other similar methods. It enables the integration of dies with completely different technological origin, different sizes and heights.

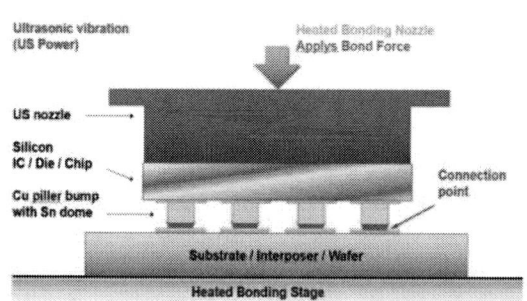

Figure 1: Schematic drawing of Thermo-Sonic Flip Chip Bonder

Thermo-Sonic Flip Chip Bonding

The thermo-sonic flipchip bonding (TS-FCB) process was used for temporary bonding. TS-FCB is a widely adapted technique in advanced packaging for automotive, LED and mobile devices. It provides high speed chip alignment and quick bonding without using flux [6]. To prove and demonstrate the process flow, an existing design of a chip substrate with outer dimensions of 10x10 mm² has been utilized. That size was considered as an applicable

286

base for future developments of mass production solutions for a chip on wafer (CoW) process with 6, 8 or even 12 inch wafer substrates.

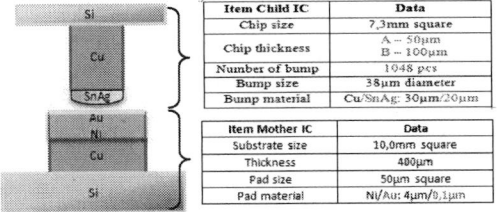

Item Child IC	Data
Chip size	7,3mm square
Chip thickness	A – 50µm B – 100µm
Number of bump	1048 pcs
Bump size	38µm diameter
Bump material	Cu/SnAg: 30µm/20µm

Item Mother IC	Data
Substrate size	10,0mm square
Thickness	400µm
Pad size	50µm square
Pad material	Ni/Au: 4µm/0,1µm

Figure 2: Test Vehicle with mother (substrate) and child IC

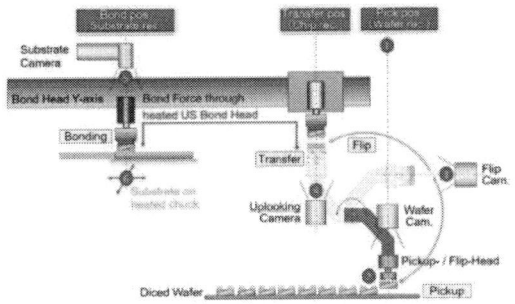

Figure 3: Schematic drawing of the TS-FC Bonding

(1) The wafer camera detects the chip at the pickup position.
(2) Pick up of the singulated ICs from dicing tape.
(3) Rotation to 90° to present the picked chip to the Flip camera.
(4) Rotation continues to 180° and the pick tool presents the chip for handover to the TSB bond head.
(5) The substrate camera defines the target position for the child IC on top of the mother substrate.
(6) The bond head moves with the child to the bonding position, detects touch down and starts US power while and ramping bond force.

For demonstration of fine pitch interconnection, a test vehicle (figure 2) consisting of a Si dummy chip (child) with 7.3 mm edge length on each side and a square Si substrate (mother) with 10 mm size were used. The child contained 1048 copper pillars, each with 38 µm in diameter and topped with a thin, pre-melted SnAg solder cap. The 50 µm square bond pad of the mother ICs were Ni-Au plated. This configuration was used with two different kinds of child ICs having thicknesses of 50 and 100 µm. The module provides a test arrangement containing 27 Daisy chains, which are used to measure the electrical resistance. The thermo-sonic bonding was performed at elevated temperatures below the melting point of the solder. The mother IC (substrate) stays heated at 200°C on a vacuum chuck and the child IC to be mounted on the mother, was heated to same temperature by the bondhead. The heatup of the child happened during the short handling period. Temperatures for mother and child have been selected equally in order to keep the mechanical stress as low as possible for the module after cool down to room temperature. The ultrasonic bond parameters were chosen in order to ensure a temporary bonding that allows a substrate transport to the following gang bonding process without the risk of displacement. The total process time for pre-attach was 300 ms per chip and has potential to be shortened. The applied ultrasonic power was set in a range of 0.2 to 0.4 W with a parallel ramped bond force up to 25 N after the chip contact. Figure 3 describes the mode of operation of the TS-FCB system [6].

A plasma pre-cleaning step before and after thermo-sonic bonding can be omitted increasing the cost efficiency. The subsequent batch step is processed under a reducing gas atmosphere removing the oxides on the surfaces of the units child- and the bottom side mother-IC and increase their solderability.

Thermo-Compression Bonding

For permanent bonding, an established equipment for vacuum reflow soldering with a dedicated press unit was used (figure 4).

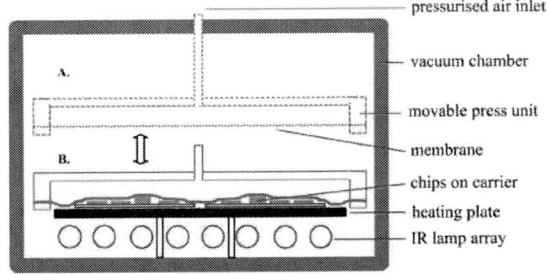

Figure 4: Schematic drawing of the integrated press unit in the bonding oven

Direct infrared heating ensured fast and flexible thermal profiles. Load and unload of the samples occurred at temperatures lower than 50 °C. Vapors of formic acid were used as a reducing agent in a controlled atmosphere environment. The duration of the pre-cleaning step during gang bonding was varied depending on the oxidation state of the bumps. To promote reliable bonding over the entire substrate or wafer, at least moderate mechanical pressure is required. The pressing unit is implemented within the vacuum chamber and consists of a flexible membrane clamped gas-tightly on a metal holder. By controlling the gas flow, the membrane expands and applies an isostatic pressure of up to 0.5 MPa over the entire working area of currently Ø160 mm. The flexible membrane based on an elastomer material can easily follow complex topographies and compensate component tolerances and warpage of large devices [2].

The cross-sections in figure 5 visualize the bonding process prior to and after solder reflow. A certain amount of pressure must be applied to the joining surfaces to keep these in contact with each other during the heating steps. The reducing atmosphere in the chamber provides good wetting on the landing pads and copper pillars' sidewalls. Vacuum is applied in order to remove voids in the solder layer. The isostatic pressure provides a uniform distribution of the distance between copper pillars and landing pads. Thus, the amount of solder forming the pillar cap can be reduced. This eliminates the risk of squeeze out and the interconnection density increases. Furthermore, the current setup can be used for transient liquid phase bonding (TLPB), also known as solid-liquid interdiffusion (SLID) process. As TLPB is based on diffusive processes, a very thin metallic interlayer to be melted is used. A good contact of the joining surfaces over the entire working area becomes more critical and ensures forming of intermetallic phases with higher melting point [7].

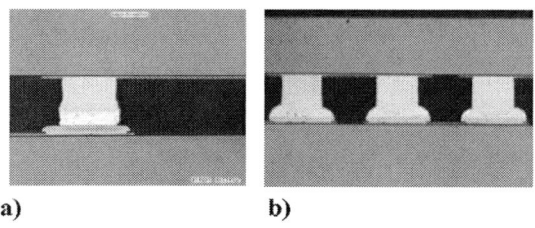

a)   **b)**

Figure 5: Cross-section of Cu-SnAg-Au interconnection a) after TS-FCB and b) after TS-FCB and gang bonding in a vacuum/formic acid environment

Process evaluation

Bonded samples were analyzed after the TS-FCB with various methods as described in the following chapter and these results were used to judge and compare the thermo-compression vacuum reflow process afterwards by doing same analytics again.

The placement accuracy was validated using IR microscopy, which allows a look through the polished silicon, where the structural circuit design of the chip and the substrate can be seen in parallel. Misalignment and shift can be quickly identified and measured. In figure 6, an example of a sample bonded with optimal process parameters is shown. The initial alignment accuracy achieved by the flipchip bonder could be maintained within a range

required for mass production during the gang bonding step.

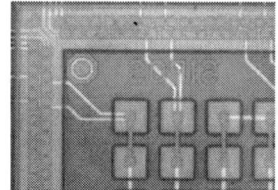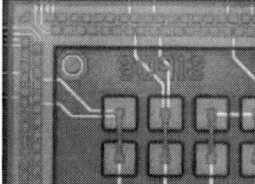

Figure 6: IR microscopy before (left) and after (right) permanent thermo-compression gang bonding

Further nondestructive tests like measurements of tilt and gap before and after final thermo-compression gang bonding have been done as well as electrical measurements. The original tilt after the first step of the process shows improvement after the second thermo-compression gang bonding (TCB) oven step. It results in a typical parallelism of about less than 2 µm over the total area of the test chip with an edge length of 7.3 mm. The constant gap over the whole area underneath the chip has a positive influence to the lifetime of the final modules caused by less thermal stress due to a homogenous and uniform thermal spread. In addition, the stable bond line thickness acts as a good base for stacking more dies on top of the lowest within a 3D package.

Electrical measurement of a Daisy chain setup on the test vehicle was done to verify the contact of all bumps. The electrical resistance through the Daisy chains correlates with the quality among the closed contacts of all bumps.

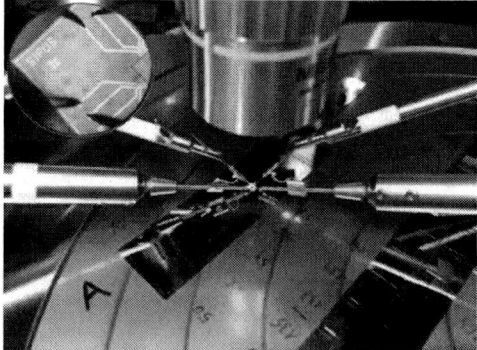

Figure 7: Electrodes for resistance measurement

For these measurements, the 4-wire method (figure 7) was applied to get a precise result of the resistance of the involved contacts. The resistance has been measured after the first step, which reflected the status after thermo-sonic bonding, and once more after the final second step of thermo-compression gang bonding. Nearly 50% of the samples failed after TS-FCB. As chips were bonded at high temperature under ambient atmosphere, oxidation of base metal surfaces Sn and Cu used for the interconnecting bump occurred and degraded some of the electrical contact materials. Other causes for open contacts from the perspective of bonding can be a non-parallelly adjusted bond tool versus bond stage, or poor thickness uniformity of the copper pillars or the solder cap on it respectively. The amount of the samples passed the electrical tests increased to 99.8% after the final thermo-compression gang bonding step. The measured resistance was in average 26.5 Ω over 320 IOs.

Additionally, samples were visually analyzed by Scanning Acoustic Microscopy (SAM) in order to detect bond defects such as cracks, voids or contaminations and reveal electrical open contact points [8]. The sample shown in figure 8 was bonded under optimal process parameters. There were no voids and delamination regions observed between the attached top chip (child IC) and the Si interposer (mother IC). All copper pillars could be completely connected to the pad.

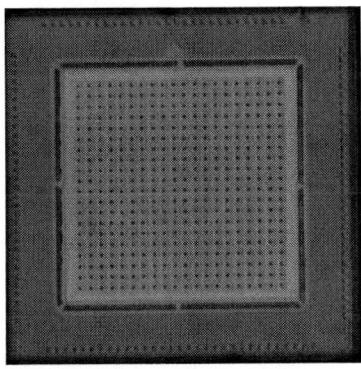

Figure 8: High resolution scanning acoustic microscope image of bonded sample after TS-FCB and gang bonding

The growth of the intermetallic compounds is influenced by the variation of the process parameters. Voids and joint quality was studied with preparation of cross-sections (figure 5). Tin has been transformed into a very thin intermetallic layer of Cu_3Sn at the interface Cu/Sn as well as in the intermetallic phase of $(Cu, Au)_6 Sn_5$ [9]. Due to the homogeneous pressure distribution, the surface unevenness is compensated, thus achieving a better joint quality.

For the analysis of the microstructure inside the solder joint, a scanning electron microscopy (SEM) with an energy dispersive x-ray spectroscopy (EDX) was used. EDX was operated at photon energies between 5 and 20 keV. This method quantifies the composition of the elements in the intermetallic phases. A sample of a copper pillar shown in figure 9 was bonded at 270 °C and pressure of 0.4 MPa in sequently reducing atmosphere and vacuum. The EDX color mapping

illustrates a uniform conversion of the copper pillar solder cap into intermetallic compounds over the entire joint area.

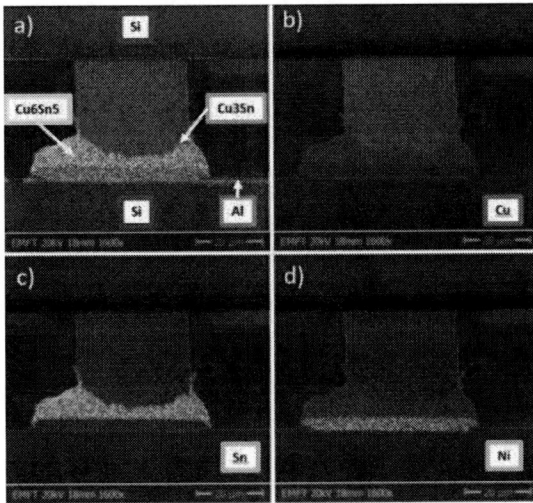

Figure 9: (a) EDX color mapping of Cu-SnAg-Au interconnection after TS-FCB and gang bonding. Single elements: (b) Cu, (c) Sn and (d) Ni

In order to find the optimum process window for the gang bonding, different parameters as peak temperature and duration, formic acid pre-cleaning time, and applied pressure were varied. The shear strength of the resulting joints was measured as a quality criterion. The minimum required shear force for the used test vehicle was defined with 10.48kgf which equals 10gf/bump.

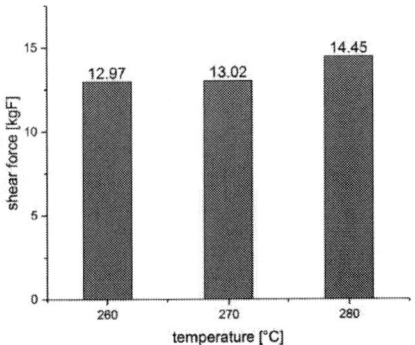

Figure 10: Shear force vs. temperature

The graph in figure 10 shows the shear force in relation to different peak temperatures between 260°C and 280°C at the same process conditions. The melting temperature of SnAg solder is 226°C. The slight increase in strength is most probably caused by the progressive formation of intermetallic compounds, whose growth is accelerated at higher temperatures.

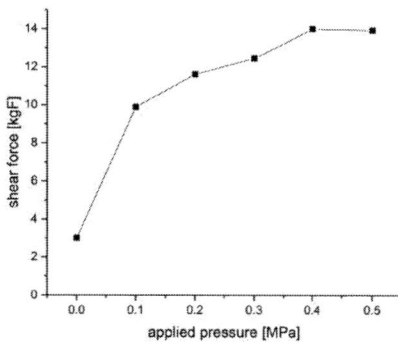

Figure 11: Shear force vs. applied pressure

Series of tests shown in figure 11 was carried out in order to investigate possible influence of the applied pressure on the resulting shear force. Peak temperature for the whole series was set to 270 °C. Applying no pressure on the top of the module during the gang bonding process resulted in poor joint strength. In order to achieve sufficient shear strength, a pressure above 0.1 MPa is required. By further increase of the pressure up to 0.4 MPa, the distance of the surfaces to be joined is lowered. The amount of remaining solder underneath the copper pillars decreases in contrast to that on the sidewalls. The ratio of the formed intermetallic compounds becomes higher. As a consequence the measured shear force increases linearly between 0.1 and 0.4 MPa. For values higher than 0.4 MPa a saturation level seems to be reached. With this experiment, the need for a pressure application and the enhancement by increasing it is clearly shown.

In figure 12, the shear force is plotted as function of the duration of the peak and the pre-cleaning step. Extending the process time during the liquidus state of the solder resulted in steadily increasing shear values. This can be explained by the progressive formation of intermetallic phases. The relationship to the long-term reliability of the bonds was not a scope of this work.

Initially, the shear force rises with increasing the pre-cleaning time under reducing atmosphere. For the current setup, it seems that 5 minutes are sufficient for removing the oxides from the joining surfaces. After this time a degree of saturation is reached. During the experiments, it was observed that the efficiency of the formic acid pre-cleaning plays a very significant role for the bonding quality. Further increase of the shear force in the range of 20 kgf seems achievable.

In further experiments using the same parameters, child ICs of different heights A and B were placed in close distances to each other and processed in the thermo-compression bonder as a batch. The shear tests showed comparable results to the child ICs even with different heights bonded within one run. No significant influence of the package height difference on the pressure distribution over the entire working area of the heating plate could be observed. In addition, the warpage across the child ICs which remained after thermo-sonic flipchip bonding could be reduced during the reflow in the membrane oven significantly from 8 to 1 µm. Thus, the reliability of the electrical contacts in the modules was significantly improved. A reduction and homogenization of the gap size can positively influence a following underfill process and the package reliability.

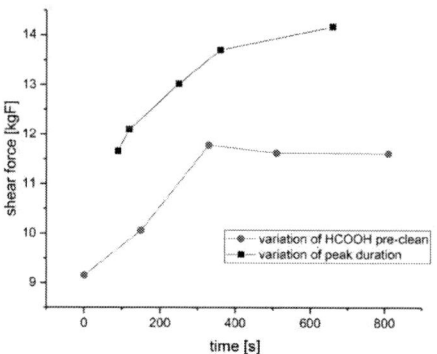

Figure 12: Shear force vs. pre-clean & peak duration

Conclusion

A new two-stage approach for CoC and CoW bonding was evaluated. A thermo-sonic flipchip bonder was used for fast and precise alignment and temporary bonding in the first stage. In a second stage, permanent thermo-compression gang bonding was performed in a vacuum reflow oven with a thermo-compression capability. For all experiments, an existing proven test vehicle has been selected. By optimizing the process parameters, the yield was significantly increased while the required additional time for permanent bonding was kept at levels ensuring high throughput at moderate costs.

The application of isostatic pressure can be utilized within a wide variety of electronic package designs without the need of height compensation tooling. However, maintaining of the alignment accuracy is a key factor that needs to be further investigated as a subject of a future study.

From metallurgical point of view, the Cu-SnAg-Au/Ni interconnect system in combination with the membrane oven has shown a promising outlook for other material systems such as Cu-Sn-Cu. The feasibility for reducing the thickness of the solder cap and the transition to a TLPB process forming high melting intermetallic compounds has great potential for high power applications. Other benefits of avoiding gold are naturally found at reductions in material costs and skipping the process step of forming golden stud bumps in the assembly flow.

Acknowledgements

The authors would like to thank Mrs. Denise Linke and Mr. Ramon Linke of Fraunhofer EMFT for the analytical support with preparing cross-sections, SEM imaging, and electrical measurements. Thanks also to Mrs. Helena Zhang of Micro-Power Scientific for her support with the SAM analysis.

References

[1] R. Windemuth and T. Ishikawa, "New flipchip technology", 2009 European Microelectronics and Packaging Conference (EMPC), Rimini, Italy, June 15-18, pp. 1-6, 2009

[2] V. Rangelov, S. Altenbockum, J. Kleff, C. Weber, H. Oppermann, and K.-D. Lang, "Development and Evaluation of Equipment Enhancements for Transient Liquid Phase Bonding (TLPB) and Sintering", 2017 Int. Conf. of Electrical and Electronic Technologies for Automotive (Automotive 2017), Torino, Italy, June 15-16, IMAPS Vol. 2017, pp. 1-5, 2017.

[3] G. Pauzenberg et al., "Thin Wafer Handling and Chip to Wafer Stacking Technologies", 2013 International Microsystems, Packaging, Assambly and Circuits Technology Conference (IMPACT), Taipei, Taiwan, October 22-25, pp. 59-62, 2013.

[4] T. Braun et al., "Trends in Fan-out wafer and panel level packaging," 2017 International Conference on Electronics Packaging (ICEP), Yamagata, Japan, April 19-22, pp. 325-327, 2017.

[5] M. J. Wolf, P. Ramm, A. Klumpp and H. Reichl, "Technologies for 3D wafer level heterogeneous integration," 2008 Symposium on Design, Test, Integration and Packaging of MEMS/MOEMS (DTIP), Nice, France, April 9-11, pp. 123-126, 2008.

[6] T. Ishikawa and T. Kojio, "Thermo-Sonic Flip Chip Methods on Copper or Solder Interconnect Structure", Micro Electronics Symposium (MES), Big Island, Hawaii, USA, pp. 99-102, 2016.

[7] W.D. Mac Donald and T.W. Eagar, "Transient Liquid Phase Bonding", Annu. Rev. of Mat. Sci., Vol. 22, pp. 23-46, August, 1992.

[8] S. Yamatsu et al., "3 D Stacking Process with Thermo-Sonic Bonding using Non-Conductive Film", 2018 Electronic Components and Technology Conference (ECTC), San Diego, CA, USA, May-June 29-01, pp. 2049-2054, 2018.

[9] H.T. Chen, C.Q. Wang, C. Yan, M.Y. Li and Y. Huang, "Cross-Interaction of Interfacial Reactions in Ni (Au/ni/Cu) – SnAg-Cu Solder Joints during Reflow Soldering and Thermal Aging", Journal of Electric Materials, Vol. 36, No. 1, pp. 26-32, August, 2007.

Effect of overmolding process on the integrity of electronic circuits

Mona Bakr [1], Yibo Su [2], Frederick Bossuyt [1], Jan Vanfleteren [1]

[1] *Center for Microsystems Technology, IMEC and Ghent University, Belgium*
mona.bakr@ugent.be
Tel.: +32 9 264 6606

[2] *Brightlands Materials Center, The Netherlands*

Abstract

Traditional injection molding processes have been widely used in the plastic processing industry. It is the major processing technique for converting thermoplastic polymers into complicated 3D parts with the aid of heat and pressure. Next generation of electronic circuits used in different application areas such as automotive, home appliances and medical devices will embed various electronic functionalities in plastic products. In this study, over-molding injection molding (OVM) of electronic components will be examined to insert novel performance in polymer materials. This low-cost manufacturing process offers potential benefits such as, reduction in processing time, higher freedom of design and less energy used when compared to the conventional injection molding method. This paper aims to evaluate the performance of this process and propose a series of alternative solutions to optimize the adhesion between and integration of electronics and engineering plastics. A number of methods are used to optimize the process so that the electronic circuits are not damaged during the over-molding, moreover to test the reliability of the system in order to control the continuity of connections between the electronic circuit foils and the electronic components after the OVM process. Correspondingly, we have performed specific tests for this purpose varying in some conditions: the type of injected plastic used, over-molding parameters (temperature, pressure and injection time), electronic circuit design, type of assembled electronic components, type of foils used, and the effect of using underfill material below the electronic component. From these tests, first conclusions were made. We have also studied adhesion between the foil and the over-molding material. In this case, various types of engineering plastics have been tested; polypropylene (PP), 30% weight percentage glass fiber filled polypropylene (GF-PP), Polyamide-6 (PA6) and 50% weight percentage glass fiber filled polyamide-6 (GF-PA6). It was proved that throughout the wide range of tested materials, (PA6) over-molded samples showed a better adhesion on the copper-polyimide foils than the rest. These plastics were over-molded on two types of polyimide (PI)/Copper (Cu) tracks foils with and without an adhesive layer between PI and Cu. It was obviously clear that the foils with an adhesive layer between PI and Cu had more delamination in the Cu tracks than the foils without an adhesive layer. Furthermore, it was shown that the presence of an underfill material has an effect on the system as the foils that had an underfill material below their components successfully had a better connection than the foils without an underfill material. Finally, experiments were executed using the two-probe method as an electrical measurement and microscope investigation as the visual inspection.

Keywords: Over-molding process, Flexible foils, Electronic circuits, Engineering plastics, Reliability

Introduction

Flexible electronic circuits are regularly an alternative for replacing the rigid printed circuit boards (PCBs) in different application areas such as automotive industry, medical devices, and home appliances. They require novel integration methods of electronic functions into products with some benefits such as decreased weight, higher precision, lower costs, reduced operation time and flexible custom design when compared to currently available electronics manufacturing and packaging methods [1]. Thermoset epoxies are the most widely used materials in electronic packaging and printed circuit boards, however, thermoplastics polymers offer superior properties; depending upon their chemistry

they can be very much like rubber, or as strong as aluminum. They are organic melt processable materials. This generally means that they are heated, formed then cooled in their final shape. They are environmentally sustainable and have precision in molding capability. In general, the combination of lightweight, high strength, and low processing costs make thermoplastics well suited to many applications. The most common methods of processing thermoplastics are injection molding, extrusion, and thermoforming. Injection molding is a manufacturing process where melted polymer is forced into a mold cavity under pressure. A mold cavity is essentially a copy of the part being produced. The cavity is filled with plastic, and the plastic changes phase to a solid product.

Typically, injection pressures range from 5000 to 30,000 psi. Because of the high pressures involved, the mold must be clamped shut during injection and cooling. The injection molding process is capable of producing large numbers of parts to very high levels of precision [2]. Injection molded thermoplastics are already used, for instance, in MEMS packaging as well as in 3D electronic circuits (Molded Interconnect Devices) [3]. One of the most common integration approaches is a system-in-foil application which has been known as a way to realize electronic systems. In this concept, a flexible polymer substrate is used as a base substrate where electronic components are assembled and then by using different manufacturing methods they can be encapsulated and formed to a product [4]. One possible way to directly integrate various functionalities into plastic products is the use of a conventional injection molding process to over-mold flexible electronic circuits [5]. Over-molding is a process of adding an additional layer of material over an already existing object. This process is regularly used to manufacture parts, sub-sections of parts, and for prototype development. Typically, the substrate material will be bonded and mechanically interlocked with other materials . This material is placed into the injection molding tool. Then the over-mold material is shot into or around the substrate which is in our case, a flexible foil with assembled electronic components. When the over-mold materials solidify, the two materials become joined together as one single product. Over-molding varies according to the materials' choice, so if the substrate is metal and the over-molding is plastic, any type of thermoplastic can be used, but in the case of over-molding a plastic part with another plastic, then there can be some compatibility issues, which will be discussed in the following sections.

Over-molded integrated foils process flow

The concept of over-molding integration is based on a combination of flexible electronic substrates, electronic components assembly, film forming, injection molding, and in-mold processes. The idea is to assemble SMD components on flexible substrates or circuit boards, and use the former as an insert in a conventional injection molding machine [6]. In such a way a required encapsulation process is needed for the desired application. By using flexible plastic foils and cost-effective printing methods together with over-molding process, there is a great potential for building a manufacturing technology platform that can provide products in various applications fields with complex shapes at low cost. This work focuses to implement test vehicles where electronic components are assembled on flexible copper-polyimide foils [7]. The process flow for flexible electronic foils is explained in detail :

1. Clean the copper surface & micro etch copper.
2. Laminate dry film photoresist onto the substrate using dry film laminator. UV illuminate the photoresist.
3. Develop the photoresist.
4. Etch samples and use stripping to remove the resist from the patterned copper.
5. Apply an OSP to protect the patterned copper from oxidation.
6. The flex circuit is laser cut, thus cutting it into two parts and cut an opening for the polymer to flow during over-molding process.
7. Finally, components are assembled using a conventional lead-free solder. The whole process is shown in figure 1

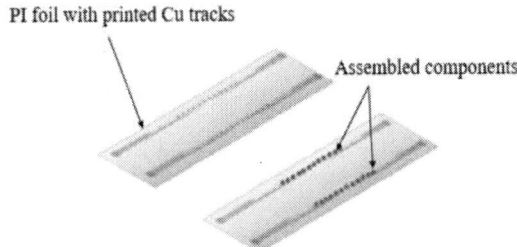

Figure 1: Integrated flexible foils process

The result is a flex foil with an assembled electronic circuit. This circuit is now prepared to an over-molding process, illustrated in fig.2.

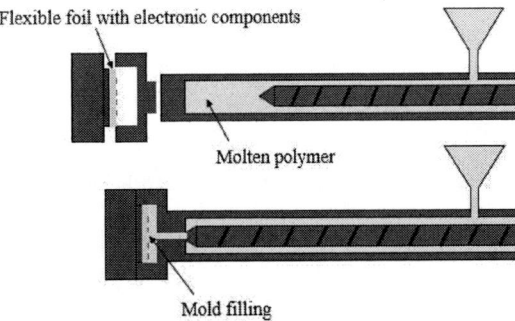

Figure 2: Schematic over-molding process

Before over-molding the desired polymer, the over-molded material (flexible foil with assembled components) should be dried in a convection oven to remove moisture, which can lead to expanding gas bubbles during the process. During over-molding as shown in fig. 2, the printed flexible foil in fig.1 is clamped in the mold. The over-molding polymer is heated beyond its melting temperature. After the desired temperature is achieved, the mold is closed on the foil. Once the polymer is melted the pressure moves the injection screw to push the material towards the mold. The polymer starts to flow into cavity with the integrated foil, thus achieving the desired product shape in a very short time (less than a second). Finally when the temperature is sufficiently low and the over-molding polymer has solidified the mold opens again and the over-molded foil is ejected from the machine. The result is a dogbone shaped plastic with integrated resistors fig. 3.

Figure 3: Over-molded sample with integrated resistors

Adhesion between engineering thermoplastics and flexible foils

Integration of flexible electronic circuits into engineering thermoplastics by over-molding technique can be realized in two approaches as schematically shown in fig.4: namely (1) encapsulating the flexible electronic circuits into engineering thermoplastic structure with undercut mold design (top), or (2) utilizing the adhesion between flexible electronic circuit foils and engineering thermoplastics to realize the structural integrity (bottom).

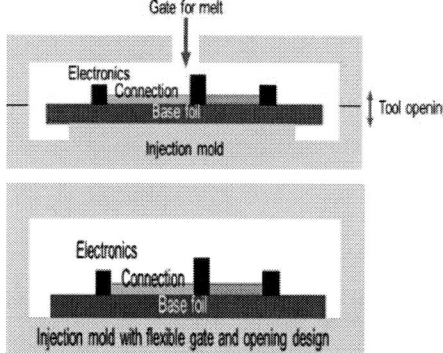

Figure 4: Mold design for encapsulating flexible electronic circuits (top) and mold design for adhering flexible electronic circuits with engineering thermoplastics (bottom)

The encapsulated flexible electronics can show a well-performed initial structural integrity with a wide range of base foil-engineering thermoplastic material combinations. However, if the adhesion between foil and thermoplastic material is not optimized, an undesired separation between base foil and engineering thermoplastic in long-term application may happen, which is inevitable due to the significant coefficient of thermal expansion (CTE) difference between typical foil material (neat polymer, CTE ~ 10^{-5}) and typical engineering thermoplastic (fiber filled polymer, CTE ~ 10^{-6}). The separation can either formulate wrinkles and bumps of flexible foil showing deteriorated aesthetic appearance or influence the joining between

electronics and base foil/connection resulting damaged functionality of the electronics. In comparing with encapsulation design, the adhering design offers more flexibility in designing the mold. In addition, the optimized adhesion between foil and engineering thermoplastic can result in a much-improved long-term structural integrity. Therefore, it is necessary to evaluate the adhesion performance between foils and engineering thermoplastics to screen the optimized material combination for over-molding. Fracture energy based peel test is used to evaluate adhesion between various types of foils and engineering thermoplastics. Since such test, as critically assessed, is capable to characterize fracture energy between dissimilar materials, which is independent of the geometry of test specimens, and mechanical property of materials [8]. In this study, the base foil (PI foil with Cu meanders) will be over-molded with different engineering plastics into the form of peel test specimens in order to check the best adhesion performance. The engineering thermoplastics used in this study are all commercialized materials, which are widely applied in automotive, and electronics industry. Four types of engineering thermoplastics are involved in this study Polypropylene (PP), 30% weight percentage glass fiber filled polypropylene (GF-PP), Polyamide-6 (PA6) and 50% weight percentage glass fiber filled polyamide-6 (GF-PA6). The injection molding parameters are in accordance with the recommendation from manufactures of engineering plastics, within the boundary, a relatively higher temperature is applied to promote adhesion. Injection molding melt temperature of 240 °C and mold temperature of 65°C is used for PP and GF-PP (with melting point 160 °C), injection molding melt temperature of 270 °C and mold temperature of 80°C are employed for PA6 and GF-PA6 (with melting point 220 °C). As it was mentioned previously , in our case the PI foil with copper tracks is placed into the mold and the injection process starts where the polymer , with opposite equal flow directions, covers the electronic components (resistors) as shown in fig.5.

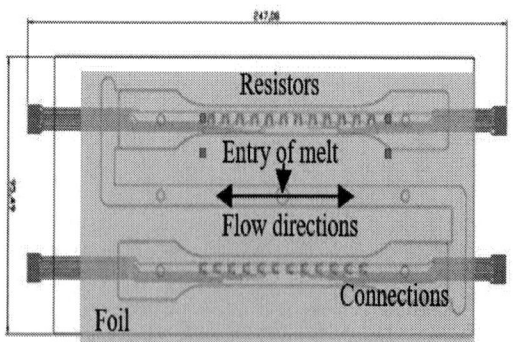

Figure 5: Schematic mold design

294

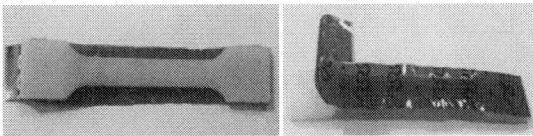

Figure 6: A trimmed over-molded flexible electronic circuit for peel test

The component will be further trimmed into the peel test specimen as shown in fig.6.The over-molded flexible electronics foil with PP is shown in fig.7. It can be observed that bonding between PP and foil is not completed. This is an expected observation, since the adhesion between dissimilar materials strongly relies on the formation of chemical bonds, and PP is lacking reactive chemical function groups.

Figure 7: Over-molded flexible electronics foil with PP

Figure 8 shows the over-molded flexible electronics foil with GF-PP, which shows a comparable result with fig. 8.

Figure 8: Over-molded flexible electronics foil with GF-PP.

Strong adhesion is observed between the foil and over-molded PA6 as shown in fig.9.This may be attributed to the high reactive C=O bond in PA6 which promote chemical bonding with PI on the foil surface.

Figure 9: Over-molded flexible electronics foil with PA6.

Figure 10 shows the over-molded flexible electronics foil with GF-PA6. The adhesion between UPISEL-N foil and GF-PA6 is also weak with a large area of separation. The weak bonding between foil and GF-PA6 is not attributed to interfacial thermal residual strain energy, which is indeed lower in foil/GF-PA6 system than foil/PA6 system. Since the CTE of GF-PA6 (10 ppm/°C) is more comparable to UPISEL-N foil (18 ppm/°C) than the PA6 (90 ppm/°C), this results in a much lower interfacial thermal residual strain energy tending to separate the interface. Anticipation on explaining this difference between PA6 and GF-PA6 is the effect of short glass fibers which may migrate to the interfacial region and partly embedded in foil during the over-molding process. As schematically illustrated in fig.11, since the adhesion between foil and glass fiber can hardly be realized, multiple microcracks may be initiated at the interfacial region, these cracks can propagate under a relatively low loading such as interfacial thermal residual stress.

Figure 10: Over-molded flexible electronics foil with GF-PA6.

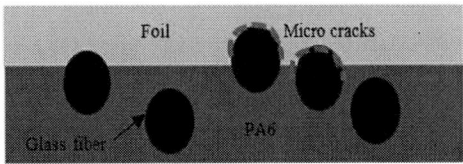

Figure 11: Schematically view of the interfacial region of the foil/GF-PA6 system

As a summary of this material screening test, a strong bonding is performed in the foil-PA6 system while the other material combinations hardly realize measurable adhesion. Therefore, PA6 material is employed in the following functionality test for its best adhesion performance.

Test Vehicle Description

To check the performance of the whole mechanism, experiments on the integrated foils during the over-molding process have been performed. A test vehicle realized on samples of PI-Cu foils are used to allow electrical testing of the embedded components by using contact pads on the flex foil. Each test sample has a total length of 247.08 mm and a width of 95.49 mm. Test structures had 24 pieces of SMD (0 ohm) resistors in 0402 packages. All SMD components were assembled using lead-free solder. Fig.12 shows the sample used in the test vehicle.

Figure 12: Sample before (left) and after (right) OVM

Test samples were studied in several conditions; without underfill, with underfill as well as with both underfill and glob top application. These conditions were compared in order to make an overall comparison for the test vehicles. Moreover, the over-molding process tested for two different cycles to check the integrity of the samples in different temperature and pressure conditions. Table1 includes the conditions of the test vehicle. Furthermore, all test vehicles performed on two types of polyimide (PI)/Copper (Cu) tracks foils with and without an adhesive layer between PI and Cu.

Table 1: Injection molding process parameters

Parameter	Cycle 1	Cycle 2
Material Temperature	250 °C	270 °C
Mold Temperature	50 °C	80 °C
Pressure	562 bar	451 bar
Injection time	0.22 sec	0.46 sec
Holding time	25 sec	25 sec
Cooling time	25 sec	25 sec

Electrical measurements before and after OVM

The measurement of the test samples is done pre and post over-molding by using a two-point resistance measurement system. Which means that a multimeter is connected by two probes to the contacting pads on the foil as shown in figure 13.

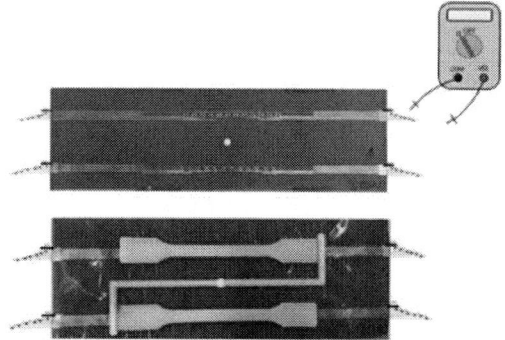

Figure 13: Two-probe measurement

Each test sample has four grounds marked in black like in fig.13 while the rest are connection lines numbered according to the resistor location. In this case, we have 24 resistance readings (one reading for each of the individual resistors) for each sample. Our aim is to check whether the components are still functional or not after over-molding process, thus the resistance values should not change after over-molding.

Results and discussion

It was shown that in figure 15 (cycle 2) the sample exposed to high temperature environment gave better results in adhesion and in the connection performance. Because higher temperature means

reduced viscosity of the liquid polymer, less injection pressure and accordingly less mechanical stress on the components. Also the resistance values were almost the same and did not get higher while in fig. 14 (cycle 1) the resistance values after overmolding became higher due to the higher mechanical stress on the components. Moreover, all components were still functional after over-molding process. Samples were visually inspected and no sample damages, component damages or displacement or other evident defects were observed.

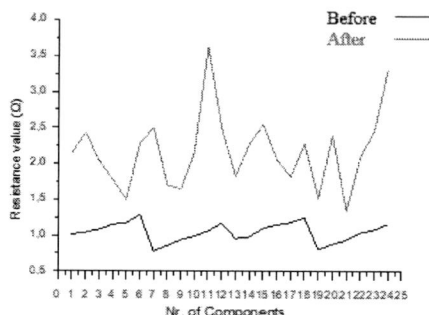

Figure 14: Measured resistance values before and after OVM at low temperature (cycle 1)

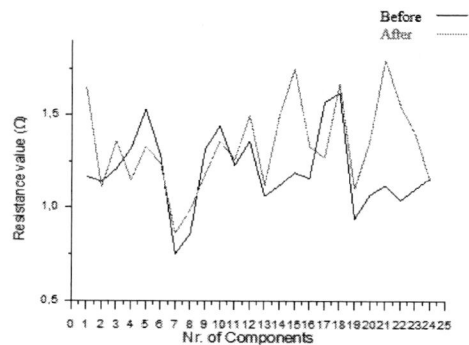

Figure 15: Measured resistance values before and after OVM at high temperature (cycle 2)

It was shown that UPISEL-N (without adhesive layer) polyimide foil is more compatible with over-molding process conditions. As shown in fig.16, the PI foil with adhesive layer showed delamination in the Cu tracks that lead to an open circuit in the connection post over-molding and sometimes the sample was completely destroyed during the over-molding process. For these reasons, all tests were performed on UPISEL-N foil.

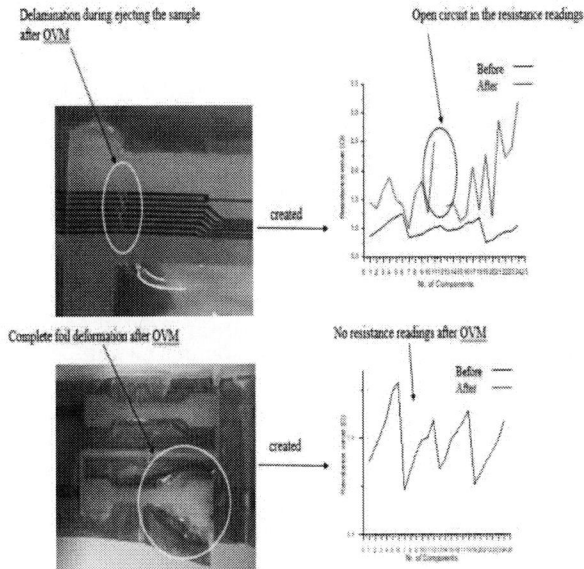

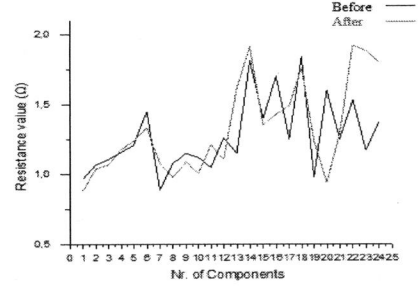

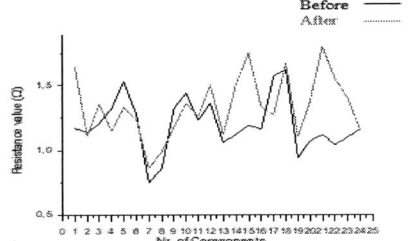

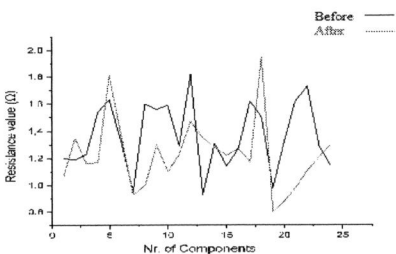

Figure 16: Failures in PI foil with an adhesive layer

It was shown that the presence of an underfill material below the components has a contribution to the measurements. In fig.17, some samples were over-molded without an underfill material, resulting in high resistance values than pre-over-molding readings. But, when adding underfill material as shown in fig.18, The resistance values become more stable compared to the readings in fig.17 after over-molding.

Figure 18: OVM with underfill material at 270 °C

Furthermore, the influence of glob top material was studied as well. Some components were surrounded with epoxy-based material and their resistance values are shown in fig.19, dried at 100 °C for 15 minutes resulting in slightly lower resistance values compared to fig.16 which was for components with underfill material below them with no glob top.

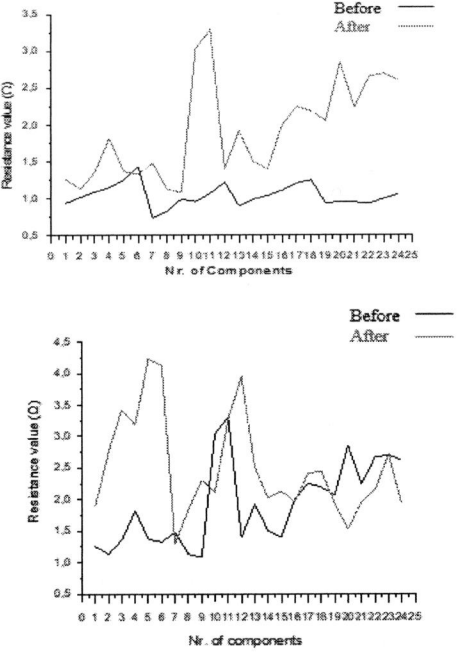

Figure 17: OVM without underfill material at 250 °C

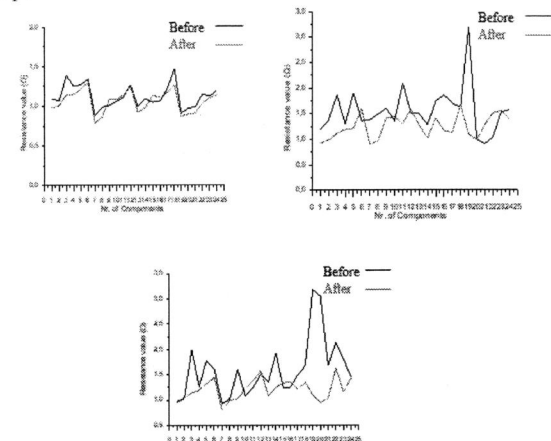

Figure 19: OVM of samples with glob top and underfill application at 270 °C

Simulation of OVM process

In order to understand the effect of injection molding process condition on the functionality of electronic components, the entire OVM process is simulated by Moldex3D software. Figure 20 shows the model established in Moldex3D environment, which excludes copper connection due to the feature of thin layer, small area and high thermal conductivity.

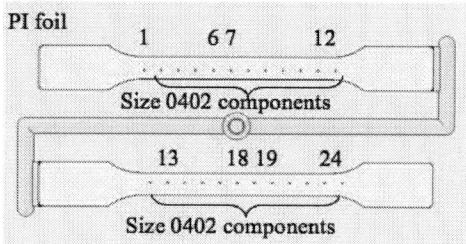

Figure 20: Model of simulation

The previously mentioned OVM cycle 2 is simulated and the temperature history of selected components from figure 20 is shown in figure 21. The simulation shows that the temperature of components drops from 250°C to 220 °C in 5 seconds, which finally reaches 120 °C in the end of cycle. Therefore, the soldering (melting point of 260 °C) under OVM cycle 2 (injection temperature of 270 °C) is not melted thus the functionality of resistors is possible to be retained.

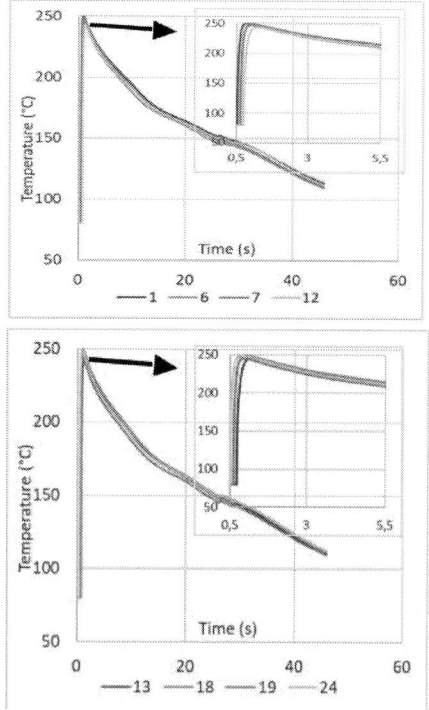

Figure 21: Temperature history of selected components

Conclusions

The technology presented in this paper proves that over-molding integration could be a feasible technology enabling an ideal integration of electrical features into 3D plastic products. It also shows that electrical functionality can be integrated efficiently inside the thermoplastic polymer. It also proves that the number of the embedded SMD components can be relatively high and still functional under pressure and temperature of the over-molding process.Underfill material showed better rsults, at least for the component and designs used in this study. However, the change in resistance was still within the measurement error of the two probe measurement method. Performing the over-molding in high temperature condition is essential for better adhesion results. PA6 showed the best adhesion compared to other materials discussed in this paper. Application of glob top is not yet confirmed whether or not it had a major influence but further experiments will be performed to clarify this point. Furthermore, different electronics packages such as capacitive touch, sensors, micro controllers and LEDs will also be a topic for further study.

Acknowledgments

This work is financed through the Flexlines project within the Interreg V-programme Flanders-The Netherlands, a cross-border cooperation programme with financial support from the European Regional Development Fund, and co-financed by the Province of East-Flanders, Belgium.

References

[1] Gonzalez et al, "Design and implementation of flexible and stretchable systems", Microelectronics Reliability, Vol.51, No.6, pp.1069-1076, March 2011.

[2] Sina Ebnesajjad, "Fluoroplastics", William Andrew Publishers, second edition, Volume 2, pp. 151-193, 2003.

[3] K. Gilleo, D. Jones, G. Pham-Van-Diep, "Thermoplastic Injection Molding: New Packages and 3D Circuits", Proceedings of the 10 ECWC Conference at IPC Printed Circuits Expo, 2005.

[4] M. Koyuncu, "Systems Integration: From Embedded Components to Integrated Flexible Systems", Proceedings of the Plastic Electronic Conference, Dresden, Germany, October 19- 21st, 2010.

[5] T. Alajoki et al.,"Hybrid in-mould integration for novel electrical and optical features in 3D plastic products," Proceedings of the 4th Electronic System-Integration Technology Conference, Amsterdam, Netherlands, pp. 1-6, 2012.

[6] N.J.Teh et al., "Embedding of Electronics within Thermoplastics Polymers by Injection Moulding", Proceedings of the IEEE/CPMT International Electronics Manufacturing Technology Symposium, pp. 10-18, 2000.

[7] B. Plovie et al.,"One-time deformable thermoplastic devices based on flexible circuit board technology." Proceedings of the 11th International Microsystems, Packaging, Assembly and Circuits Technology Conference (IMPACT), Taipei, pp.125-128,2016.

[8] Y. Su, M. deRooij, W. Grouve, L. Warnet, "Characterisation of metal–thermoplastic composite hybrid joints by means of a mandrel peel test", Composites Part B: Engineering, Vol. 95, pp. 293-300, June, 2016.

Process Flow and Cost Modelling for Fan-Out Panel Level Packaging

Mathilde Billaud, Hannes Zedel, Lutz Stobbe, Tanja Braun, Nils Nissen, Klaus-Dieter Lang

Fraunhofer Institute for Reliability and Microintegration IZM, Gustav-Meyer-Allee 25, 13355 Berlin, Germany
Technische Universität Berlin

Phone +49 30 46403-195, Mathilde.billaud@izm.fraunhofer.de

Abstract

In response to the emerging trends in packaging technologies, detailed cost analysis models are required to assess the economical and environmental feasibility of different technology options, business scenarios, design options, and process configurations. As established models do not provide a sufficient level of granularity both in representing realistic industrial processes and detailing the cost allocation, a new Multi Level Level Cost Model was developed for comparative cost assessments. The new strategy follows a bottom-up approach and allocates cost-relevant process data to every piece of equipment to allow differentiated cost analyses and identification of optimization strategies. The model is best suited for relative comparisons and enables realistic estimates of real packaging costs if the data input requirements are met adequately. The flexibility and efficacy of the model is demonstrated in this paper with an exemplary Chip First Fan-Out Panel Level Packaging process. The analysis investigates 1) the cost evolution in relation to the number of RDLs, 2) a preliminary estimate of yield requirements to compete with Wafer Level Packaging, 3) an analysis of substrate and package design choices with regard to substrate area utilization and its economic and environmental impact.

Key words: Cost model, cost analysis, panel level packaging, Fan-out, chip first

Introduction

As the range of packaging technologies is extending to meet the needs of miniaturization and heterogeneous integration, the alternative to produce packages on a larger substrate becomes also more concrete at the industrial level. The first high-volume production of Fan-Out Panel Level Packaging (FO-PLP) was launched in 2018 by SEMCO for the Samsung Galaxy Watch, taking advantage of a small form factor and high integration capability, as well as a production on a large and efficient substrate area [1]. In the past years, several actors invested into Fan-out (FO) on panel (e.g SEMCO, PTI and Nepes) to take benefits of their panel processing capabilities to build a FO line. Simultaneously, FO specialists (e.g ASE and JCET) are moving towards panel to reduce their production costs [2].

As part of the ongoing development of the panel level packaging process at Fraunhofer IZM, many materials and technological options are currently under investigation. In order to evaluate the economical feasibility of these devlopments, a multi-level cost model was developed.

This article outlines the methodological approach and illustrates the cost model with first calculation results based on a specific packaging technology (chip first FO-PLP).

State of the art of cost modelling for packaging technologies

There are only two companies offering cost analyses for Fan-Out packaging technologies: SavanSys Solutions LLC (published papers [3][4])

and System Plus Consulting (reverse cost analysis for a specific FO-WLP product to purchase [5]). The SavanSys methodology is based on a generic process flows, built on "activities" such as, die placement, reconstituated wafer, redistribution layer (RDL) or underbump metallization (UBM). The individual process steps and related parameters are not described in detail. By partially aggregating costs of different machines, the costs are then evaluated at the activity level, not at the process step level related to specific equipment, which may limit the precise identification of relevant bottlenecks. System Plus Consulting study is based on reverse engineering of a specific product and technology. They reconstituated a quite detailed process flow for the InFO Fan-Out wafer level packaging technology from TSMC to elaborate the cost analysis. The total costs and the cost per step are also disclosed. However, the aim of this analysis has been to estimate a price for one mature technology produced on wafer. This study cannot be utilized to compare different process technologies and assess their impact on costs.

The above mentioned analyses are mainly focused on wafer level packaging, mature technologies, and existing productions. Only one paper addresses costs on panel leve packaging [3]. More generally, these cost analyses are targeting OSATs, rather than materials and equipment suppliers. Generic process models are usually applied, without recognizing relevant technical parameters related to the specific process and its variations.

Current market interests are focused on assessing the economical feasibility of new manufacturing technologies in comparison to established practices and deriving recommendations on process developments as well as optimization strategies. Due to the limitations in features and precision of cost allocation as well as a lack of transparency of the methodology, the available models and literature data were considered insufficient to meet specific R&D requirements. Therefor, a new cost modelling approach was developed.

The main objectives of our Multi-Level-Cost Model for advanced packaging technologies are:

- Sufficiently accurate cost prediction for advanced and new packaging manufacturing process (e.g. FO-PLP)
- High data granularity and detailed cost allocation in order to allow differentiated comparisons of e.g. different technology options (e.g. chip-first or chip last, lithography technology, panel design, etc.)
- Adaptive to different products and their package design (e.g. number of dies, dimensions, number of RDLs, etc.)
- Adaptive to a specific business case (e.g. production location, production volume, production time frame, etc.)
- The Multi-Level approach enables to analyze costs from highest aggregation (complete process) down to specific process steps and technology options (e.g. invidual equipment and material options or configurations).

Multi- Level Cost Model Description

The cost model methodology is a based on a bottom-up approach in three major steps:

1. Digitalized model of the process flow: The model essentially comprises the sequence of the individual process steps and assigns the appropriate equipment to each step. Each piece of equipment is associated to the following parameters: equipment invest, floor space requirement including clean room class, processing duration per substrate (machine and operator), power consumption, and possibility of parallel processing. The model also includes material type and quantity, and costs per substrate, as well as the possible reusability. All data are collected in data bases.
2. Scaling to a specific product scenario: Once the technology and exact process sequence is established, the process model is scaled to a specific product, by taking into account the substrate and package size, the number and size of chips/SMD components in the package, as well as the number of required RDLs and balling.
3. Scaling to a specific production scenario: The fabrication is then scaled to a definite

business case with a chosen volume production or production time frame, production location and conditions, productivity/maintenance time, labor costs. Facilities costs, related to the chosen location, are taken into account (electricity, clean room CAPEX and OPEX, and rent).

The final results of this cost modelling approach are then highly detailed and specific to a process flow, to a product design, and production scenario. The implementation of the cost model allows for quick evaluations of cost impacts and comparative assessments of various scenarios.

Application to Chip first FO-PLP Process Flow

For WL or PL Fan-Out processes, two approaches are encountered and already available in mass production: "mold first" and "RDL first" [6]. Each strategy offers opportunities and drawbacks, but in this paper, we will focus on the mold first face-down approach (dies are placed face-down on the carrier). Figure 1 introduces the scheme of the final package analyzed by the cost model. In this configuration, the dies have Al pads. The packaging process involves three redistribution layers (RDLs) with photosensitive dielectric, Ni-Au UBM and SnAgCu balls.

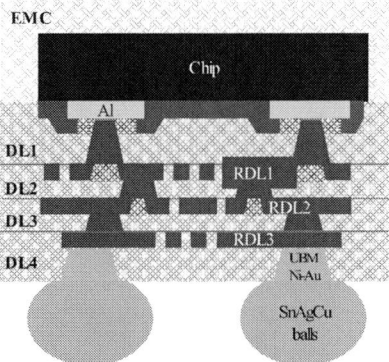

Figure 1: Scheme of the package design considered for this paper: Chip-first face-down approach, aluminium pads on the chip, 3 RDLs with photosensitive dielectric (DL), Ni/Au underbump metallization (UBM) and SnAgCu balls.

The detailed process flow is described in Figure 2. The packaging process flow starts with the component placement on the carrier and ends with package singulation. It is structured in four main modules (with their related challenges): assembly (die shift compensation, speed, and accuracy), molding (warpage, thickness, die aspect ratio, and die count on panel), RDL (e.g L/S reduction) and UBM/ballings [6]. The electrical characterization is not included in this model.

The bottom-up approach of this cost model enables to consider and adjust relevant technical parameters related to every process step in order to evaluate its impact on total costs. Table 1 sums up the main technical parameters influencing the package costs for the different modules.

Table 1: Main parameters influencing productivity and costs for the different technical modules of FO-PLP.

Module	Main parameters influencing productivity and costs
Assembly	Chip counts per package Package size Placement speed
Panel Embedding	Carrier type and reutilization rate Mold type and thickness Dies aspect ratios and counts
Redistribution layers on	Type of pads on chip (Al/Cu) Laser or photo structuring
EMC	RDL count and L/S dimension

Limitations

The bottom-up approach requires substantial understanding of the manufacturing process and its technology details. A complete generic process that consists of nonspecific modules can not be customized.

Data collection of at least the majority of relevant parameters is therefore compulsory but remains a challenge, since cost quotations from existing fabs including equipment, facility, and other indirect costs are mostly unavailable. Consequently, cost data from fab conditions are complemented by data from lab conditions (for equipments/materials prices and parameters), as well as by literature reviews, and expert knowledge (for facilities and labor costs). Process quality (yield) data do not exist at this level of development and cannot be properly estimated from the technology demonstration.Although the above-mentioned limitations are influencing the absolute cost estimates to some extent, the cost modeling results show realistic and relevant cost trends in relative comparisons. They are capable to support technology selection, optimization strategies, and decision making. Furthermore, they identify bottlenecks in early stages of manufacturing process development, as well as indicate feasible technology options and allow to estimate the return of investment for specific business scenarios.

Exemplary results

Description of the product assumptions

The following results are based on specific parameters for package/design and volume production, compiled in Table 2, unless otherwise stated. Due to lack of data on yield, a theoretical yield of 100% is considered in this comparative study.

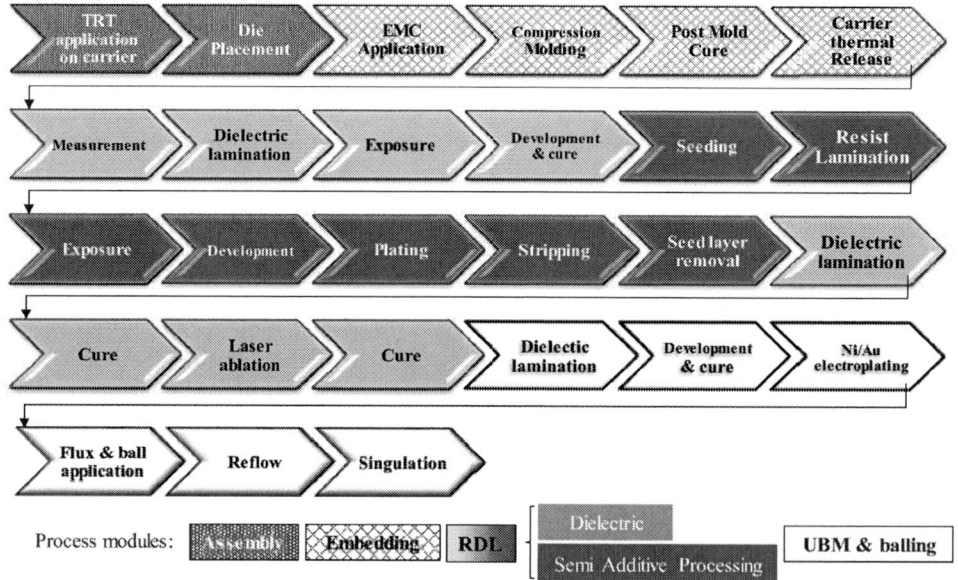

Figure 2: Description of the process flow considered for this paper: Chip first face-down approach with Al pads and photosensitive dielectric. Only one RDL is described here. TRT, EMC and UBM stand for respectively Thermal Release Tape, Epoxy Molding Compound and Underbump Metallization.

Table 2: Example of values considered in this study for the package/panel design and the production

Parameters	Value considered for this study
Package size	16 x 14 mm²
Panel size	457 x 610 mm² (Fraunhofer IZM full panel size)
RDL & L/S	3 RDLs & 10/10µm
Packages per panel	1,147
Volume production	50 M potential good packages per year
Production location	Taiwan

Distribution of Costs

Figure 3 represents the distribution of annual costs according to the main four aggregated items: equipments, materials, facilities and labor. The distribution is displayed for a process flow with only one RDL and with three RDLs. In both cases, materials represent the highest cost impact. By adding more RDLs to the process flow, additional equipments are required to achieve the same production volume, leading to an increased share of equipment costs. Since labor cost is related to the number of machines, the share of labor costs is also increasing. The smallest contribution is facility costs, including clean room CAPEX depreciation and OPEX, clean room electricity, floor space and light cost. The facility contribution might be underestimated because of data gaps on indirect costs, such as water system and automated material handling system. The annual cost investment is raised by 70% for three RDLs compared to one RDL (not shown on this picture).

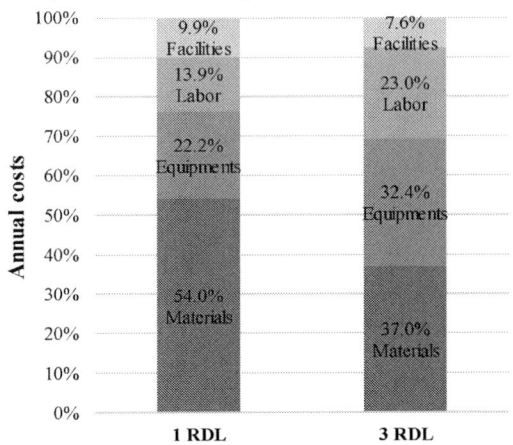

Figure 3: Distribution of costs for the annual production of 50M packages on panel (16x14mm²) according to the number of RDL in the process flow. The total annual cost is raised by 70% with three RDLs in comparison with one RDL.

Comparison of Yield between Wafer and Panel Geometry

The cost model analysis points out the beneficial effects of a panel geometry over a wafer geometry. Figure 4 shows the minimum yield that should be achieved on panel to be competitive with wafer level packaging (WLP), as function of chip costs. For a first evaluation of the geometry comparison, the theoretical yield on wafer is fixed at 100% and the same process flow is applied to both substrates.

Because packaging costs are relatively small compared to chip costs, the production on panel does not enable a substantially lower yield to be competitive with WLP despite the extremly increased number of packages per process step. Especially for high-value chips, PLP yield should be close to WLP yield to sustain a cost advantage over WLP.

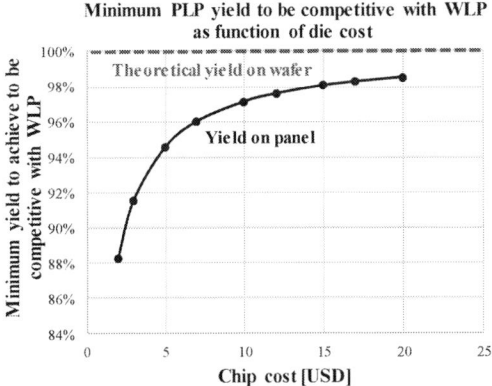

Figure 4: Comparison of a theoretical yield on wafer (100%) and the yield required on panel to reach the same total costs as on wafer.

Optimization of Package and Panel Design

The main advantage of panel over wafer is its high flexibility to a variety of package sizes. A direct key performance indicator (KPI) is a higher area utilization (AU) of the substrate for most of the package configurations. AU describes the proportion of the substrate area that is covered by packages. Maximizing AU has two major advantages:
- Time advantage: by having more packages per substrate, the number of substrates processed to reach a defined production volume is reduced
- Cost advantage: by reducing the amount of materials needed per package, since many materials are applied to the whole substrate and are subsequently wasted on non-package area (substrate edge exclusion area and package-to-package area).

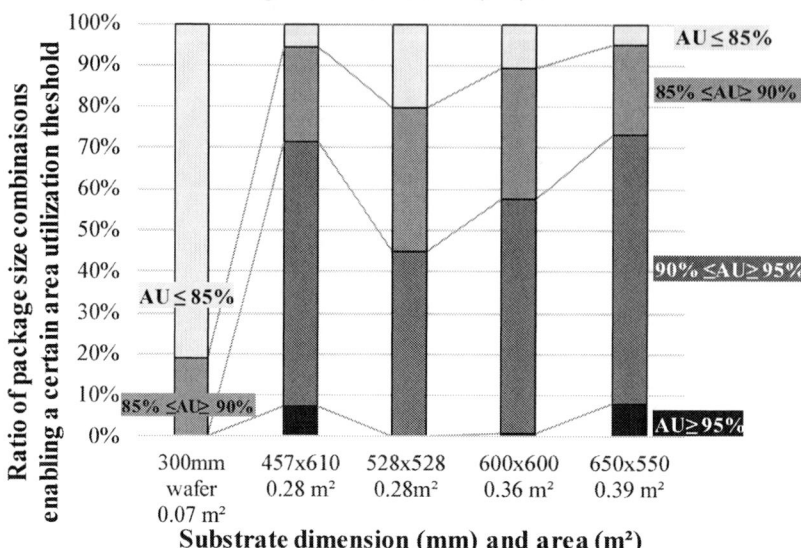

Figure 5: Comparison of the ratio of package size dimension combinations allowing a certain AU rate on various substrates (package edge lengths and widths varying from 1 to 50mm). Rectangular substrate shapes outperform squared and round shapes by providing higher package design flexibility.

The AU depends on the substrate and package sizes, the package-to-package distance, and the edge exclusion. For all package size combinations, with edge length and width between 1 and 50 mm (with a step of 1mm) and aspect ratio limited to three, the AU is calculated by the following formula:

$$Area\ utilization\ = \frac{package\ area\ \times\ packages\ count}{total\ substrate\ area}$$

Figure 5 shows the share of package combinations that allow a certain AU threshold (85, 90, or 95%), on wafer and on different panel sizes. The IZM Panel (457x610mm) enables a high AU (≥90%) for 70% of package size combinations. On the contrary, the 300mm wafer leads to an AU between 85 and 88% but only for small packages (<11x11mm²). For larger packages than 11x11mm², the AU remains under 85%. The panel achieves a much higher AU for a wider range of package size combinations, and especially for large packages. Furthermore, rectangular panels outperform squared panels for the AU of rectangular packages, because those can be placed either horizontally or vertically to maximize AU. Therefor, rectangular panels provide greater flexibility in package design

as more edge length combinations allow for high AU.

Impact on Waste Generation (sustainability)

The unutilized substrate area results in a materials overconsumption and waste generation, at each process step. To illustrate the order of magnitude of this material loss, the Figure 6 compares the amount of Epoxy Molding compound (EMC) loss for the production of 50M packages on a 457x610mm panel and on a 300mm wafer for different package sizes.

For this production volume, around 76,000 panels (457x610) or 350,000 wafers (300mm) with 20x20mm packages are required. The AU is 94.7% and 82.1%, producing 402kg and 1,725kg of EMC waste on unused area, respectively on panel and wafer. In total, the amount of wasted EMC can be reduced by 77% by producing on panel instead of wafer. The EMC thickness can be reduced but is always limited by the chip thickness and handling issues. Furthermore, EMC is one of the most expensive materials in the analyzed FOPLP process. A higher utilization of the substrate area, thanks to panel, is a simple and cost-effective solution to significantly reduce material wastes and thus production costs, as well as environmental impact

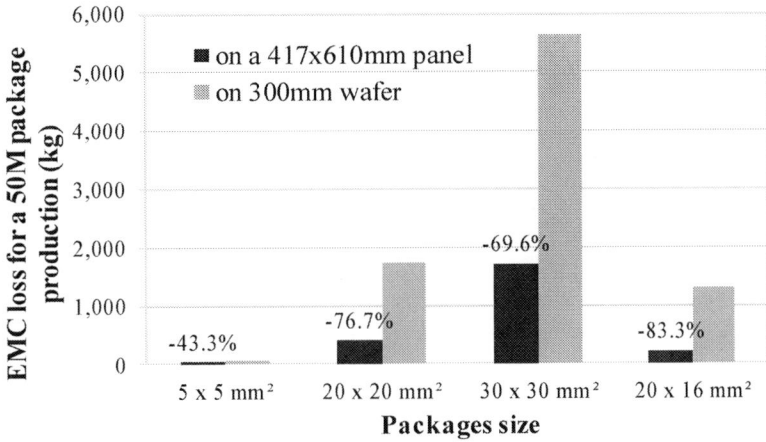

Figure 6: Weight of EMC (in kg) lost because of unused substrate area during the production of 50M packages (different sizes) on wafer (light grey) and on panel (dark grey).

Conclusion

In order to compare different packaging technologies, product designs and business cases, a new Multi Level Cost Model was developed. The model follows a complex bottom-up approach that represents realistically the industrial processes in both wafer and panel level packaging and achieves greater granularity of the results than established models. The detailed allocation of process data to derive differentiated costs per process step requires substantial data input to derive realistic conclusions, while missing data can be interpolated from generic data sets to some extent. The more data of the related process is available, the more meaningful are the assessments of cost optimization strategies and environmental impacts of different process steps. This mode offers to assess the economical and environmental sustainability of a new technology or manufacturing in relative comparison to other technologies. By using the model as a planning tool and incorporating sustainability targets at early stages of the technology development & implementation, both economical and environmental goals can be met simultaneously and more confidently. The high granularity of the cost analysis leads to detailed analyses of process parameter configurations, technology options (e.g. comparing number of RDLs as in Figure 3), and design choice interactions as well as highly transparent results.

The analyses revealed several key criteria for economical and environmentally friendly productions, such as the substrate aspect ratio and size optimization in order to maximize the area utilization rate (Figure 5). For packages larger than 11x11mm², wafer offer a significantly low utilization area compare to panel. In the case of the EMC deposition, a relatively expensive material in the mentioned process, the reduction of waste and costs associated to the production on panel and to a higher utilization rate was demonstrated (figure 6), emphasizing the environmental and economic relevance. Even though yield data is not yet available, a chip-cost dependent yield target could be estimated to make FO-PLP competitive against FO-WLP (Figure 4). Those analyses demonstrate the high flexibility of the new model and its applicability to all process related cost considerations. The efficacy primarily relies on process-related data availability, a complete understanding of the process flow and the relevant interactions between various parameters.

Future Work

Depending on customer requirements, the features of the model can be extended to e.g. account for splitting substrates mid-process, incorporating preliminary yield data once available, extend internal databases to allow for more and better generic interpolations of missing data, developing interfaces with e.g. design tools, etc.

References

[1] J. Kim *et al.*, "Fan-Out Panel Level Package with Fine Pitch Pattern," Proc. - Electron. Components Technol. Conf., vol. 2018-May, pp. 52–57, 2018.

[2] Yole Développement, "Fan-Out Packaging 2019 Technologies & Market Trends," 2019.

[3] A. Lujan, "Cost Comparison of Fan-out Wafer Level Packaging and Flip Chip Packaging," Addit. Conf. (Device Packag. HiTEC, HiTEN, CICMT), vol. 2017, no. DPC, pp. 1–37, 2017.

[4] A. P. Lujan, Ed., "A cost analysis of RDL-first and mold-first fan-out wafer level packaging," in 2016 International Conference on Electronics Packaging (ICEP), 2016, pp. 237–242.

[5] System plus consulting, "Apple A10 with TSMC's inFO packaging iPhone 7 Plus application processor," 2016.

[6] T. Braun *et al.*, "Panel Level Packaging - A View Along the Process Chain," *Proc. -* Electron. Components Technol. Conf., vol. 2018-May, pp. 70–78, 2018.

Hybrid Systems-in-Foil (HySiF) – Low Stress CFP Process for Biomedical Application

U. Passlack*[1], B. Albrecht[1], C. Harendt[1] and J.N. Burghartz[1,2]

[1]Institut für Mikroelektronik Stuttgart (IMS CHIPS), Allmandring 30 a, 70569 Stuttgart, Germany

[2]Institut für Nano-and Mikroelektronische Systeme (INES), Pfaffenwaldring 47, 70569 Stuttgart, Germany

* Corresponding Author: Email: passlack@ims-chips.de, Phone: +49 711 21855 448

Abstract

Flexible Hybrid Systems-in-Foil (HySiF) allow the combination of large-area thin-film electronics with ultra-thin, high performance silicon chips. Monitoring and determining the long-term reliability, and robustness, especially for biomedical application, is crucial. Biocompatible materials such as polyimides and noble metals are required, in order to prevent from side effects and to reduce the risk of rejection reactions. Based on the previously published data from our group, the scope of this work is to present the ongoing improvements of the process using a low stress polyamic acid based polyimide. The advantage over other organic materials is its low thermal expansion coefficient (CTE), which is close to that of silicon, resulting in reduced stress in the system, particularly, during the manufacturing process. However, unlike benzocyclobutene based polymers, polyimide PI2611 shows poor planarizing behavior, leading to more inhomogeneous substrate layers and thus, affecting process steps, such as, photolithography, silicon chip embedding, metal etching, via opening and, direct laser writing, especially in terms of accuracy. To address those issues, we have investigated the homogeneity of multiple polymer layers using both spectroscopic reflectometry and surface profilometry. Based on a fabricated HySiF demonstrator, the crucial process steps are discussed and the reliability of the system is verified by thorough measurements under biological conditions.

Key words: Chip-Film Patch (CFP), Hybrid Systems-in-Foil (HySiF), low CTE polyimide, benzocyclobutene (BCB)

I. Introduction

Currently, the utilization of flexible systems encompasses the field of consumer electronics and Internet of Thinks (IoT), and attracts an immense interest in the health and medical sectors. The demand for implementing more and more sensing components on a miniaturized flexible system with the benefit of high performance silicon chips has been considerably increased, which leads to the concept of Chip-Film Patch (CFP) technology [1-4]. CFP can act either as a foil interposer, where CMOS chips are embedded in a foil patch with one or multiple interconnect layers, fabricated with conventional IC processing technologies or as a stand-alone foil-system for printed electronic and organic compounds that are manufactured and interconnected directly on the foil surface [5]. Due to the mechanical and electrical properties, a variety of polyimides are used to build up flexible stand-alone systems. Particularly, aromatic polyimides have various useful characteristics and low dielectric constants which make them very suitable as dielectrics and passivation as well as flexible substrate films [6-7].

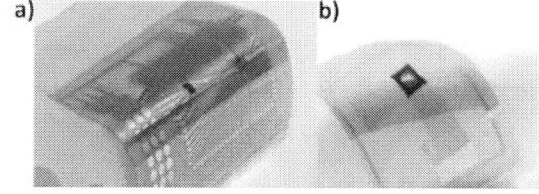

Figure 1: Photograph images of CFPs. a) Photograph of a flexible multichip demonstrator [2] and b) of a flexible stand-alone system, generated and characterized in this work.

To ensure flexibility as well as biocompatibility of the CFPs we use the polyimide PI2611 and benzocyclobutene BCB 4024-40 as the base materials. In this publication, we present preliminary results of thin-film measurements based on those polymers and, furthermore, demonstrating the behavior and the reliability of our CFPs under long-term stress in physiological environment.

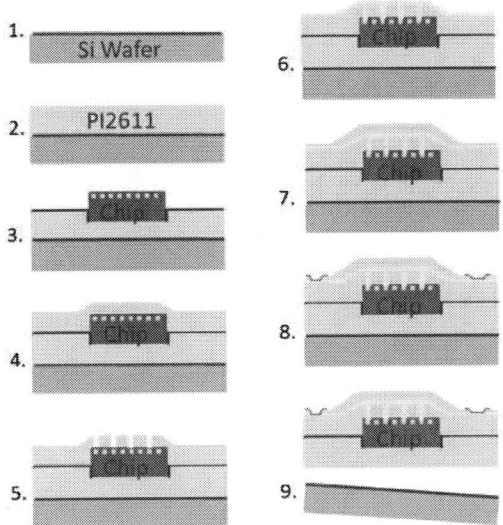

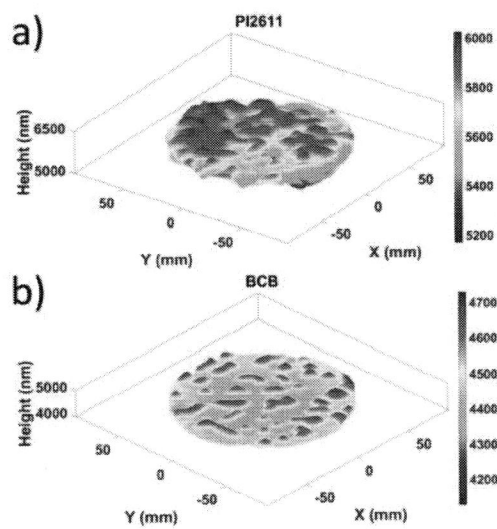

Figure 2: Schematic process flow of generating a CFP. 1) Surface preparation to increase the temporary adhesion of PI; 2) successive coating of PI followed by lithography and cavity etching; 3) spin-coating of a thin BCB layer (red colored) followed by the manual placement of the chip; 4) and 5) coating of PI with subsequent plasma etching to open the chip pads; 6) AlSiCu sputtering and structuring (grey colored); 7) single coating step of PI; 8) fan-out pad opening and evaporation of a noble metal (green colored) followed by 9) releasing the CFP.

Figure 3: A representative 3-D graph, computed in Matlab, to visualize a) a single coated PI2611 layer (min=5.2 μm, max=6.0 μm) and b) a single coated BCB layer (min=4.1 μm, max=4.7 μm), onto a 150 mm silicon substrate.

II. Process flow of the Chip-Film Patch

To investigate the film thickness of polyimide and BCB, we have fabricated flexible sensors in foil (Figure 1 b), using CMOS compatible Chip-Film Patch technology. An overview about the major processing steps is given in Figure 2.

The process starts with the temporary adhesion of the polyimide PI2611 onto a 150 mm silicon substrate using a spin-coating process to assemble a 15 μm thick layer system (1 and 2). To integrate ultra-thin chips, the PI is photolithographically structured by a laser direct writer, followed by an RIE etching process. The back-thinned CMOS chips (20-30 μm thickness) are subsequently placed into those cavities on an 800 nm thick coated BCB layer using a manual die bonder tool, followed by a curing process to finalize the bonding. In order to reduce intrinsic thermal mismatch between the silicon chips and the polymers, the chips are covered gradually with spin-coated polyimide to form a symmetrical stack (3 and 4). Micro vias are etched with RIE directly on the chip pads using either a hard mask (500 nm AlSiCu) or a soft mask (30 μm photoresist).

Once the micro via opening are verified optically and electrically, the connection from chip to foil level is formed by AlSiCu alloy metallization and structuring using a dry metal etching process (5 and 6). For the completion of the chip stack, and to ensure reliable interconnects, polyimide is spin-coated on top (7), followed by fan-out pad opening using a thick photoresist layer as a soft mask and plasma etching. Finally, a lift-off process is performed to cover the fan-out pads with an evaporated noble metal (8), which ensures biocompatibility, and in addition, preventing the CFPs from corrosion. The flexible foil assembly was finally characterized and the CFP is released with help of a laser cutter from the temporary carrier (9).

III. Study and monitoring of multiple polymer layers

Both polymers are directly spin-coated on flat silicon substrates with a subsequent curing process. After imidization reaction of polyimide and after ring opening and polymerization of BCB, the film thickness was determined by profilometry and spectroscopic reflectometry. Regarding the spectroscopic reflectometry, the optical response of the layer structure is measured and the resulted film thickness is inferred from the best fit of measured and modeled data.

Table 1. Compilation of viscosity as well as film thickness determination of PI2611 and BCB after single coating step.

Parameter	1st layer PI2611	1st layer PI2611 (optimized process)	1st layer BCB
viscosity (st)	125	125	3.5
min (μm)	5.2	3.8	4.1
max (μm)	6.0	4.1	4.7
mean (μm)	5.7	4.0	4.5
std (μm)	0.16	0.05	0.09
thickness deviation in percent (%)	14	7.5	13

Table 2. Overview of measured polyimide film thickness after subsequent multi-layer coating.

Thickness	1st layer PI2611	2nd layer PI2611	3rd layer PI2611
min (μm)	4.1	9.1	13.3
max (μm)	5.5	10.7	15.9
mean (μm)	4.9	10.0	14.7
std (μm)	0.19	0.28	0.36

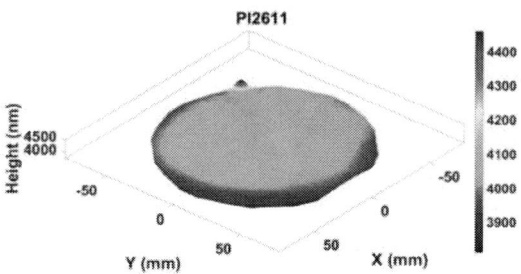

Figure 4: Uniformity profile of a polyimide coated flat substrate after optimization of the spin-coating process. In addition, the obtained, visible particle leads in an increased film thickness. A uniformity profile of 300 nm thickness deviation is determined.

Due to the anisotropic behavior of the polyimide [6], an adapted model is used. Furthermore, in order to get a reliable evaluation of the polymer uniformity, the 2-D grid, normally used for standard thickness measurement of photoresist or silicon oxide/nitride, was extended, by narrowing the distance from one data point to the next one (6 mm distance between each data point). The obtained data are analyzed in Matlab and interpolated 3-D images of the film thickness of both polymers are computed (Figure 3). The higher viscosity of polyimide [8] is leading very likely to a difference in uniformity compared to BCB (Table 1). A uniformity profile of 800 nm thickness deviation is achieved. On flat silicon substrate, BCB shows a more uniform behavior compared to polyimide (Figure 3 b). In fact, polyimide coating especially when using PI2611, is more challenging in order to achieve homogenous spin-coated profiles (Figure 3 a). Moreover, the determination of consistent thin-film measurements of either polyimide or BCB is restricted by the detection limit of the spectroscopic reflectometer. Thus, in order to verify the upper threshold of the reflectometer, substrates are coated with three polyimide layers and film thicknesses are determined after each coating step and summarized in Table 2. The trend of thickness deviation from the first measured PI layer (1.4 μm thickness deviation) to the third one (2.6 μm thickness deviation), is gradually increasing thus a solid upper limit of 15 μm polyimide thickness has been proposed. Beside material properties, uniformity is highly dependent on process parameters such as the spin-coating and curing. In order to evaluate the effect of modified spin speed and spin time, silicon substrates are coated with polyimide PI2611, and film thickness as well as uniformity is investigated. Notably, after parameter adjustment, coating uniformity could be improved. A more homogenous polyimide profile of around 300 nm thickness deviation is accomplished and visualized in Figure 4 and Table 1.

In this study, we have shown the feasibility of measuring multiple polymer layers on wafer level by a reflectometer with reliable results up to 15 μm film thickness. Nevertheless, the most challenging part is to achieve homogenous polymer coatings on structured substrates, particularly, on top of embedded ultra-thin chips. For this reason, the same measurement technique is used to investigate polyimide film thickness and homogeneity after chip embedding and subsequent spin-coating.

The 2-D grid for the travelling path has been extended, thus, helping to monitor minor changes on top of the chip surface more precisely, due to more accurate interpolation of data points (Figure 5).

In general, determination of polyimide thickness on chip surfaces, especially, on chip pads, is important in order to prevent over etching which could lead to exposed chip edges and consequently chip damage during the micro-via opening process. Analyzing the film thickness after one time polyimide coating above a chip, a thickness deviation of around 1.8 μm was obtained and visualized in Figure 5. Compared to an unstructured PI coated silicon substrate (Table 1, thickness deviation of 800 nm); a uniformity profile of 26.5% percent deviation over the chip was obtained.

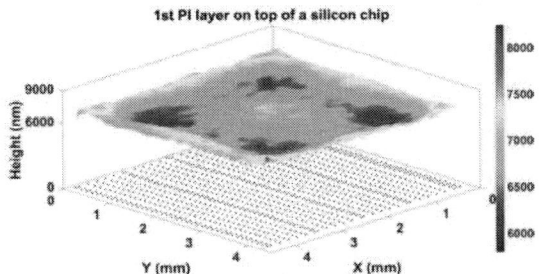

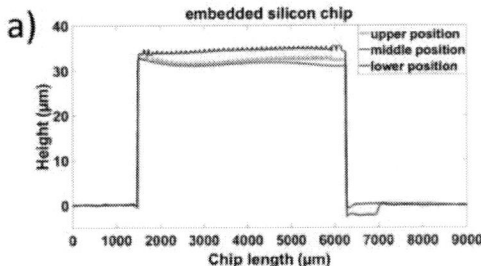

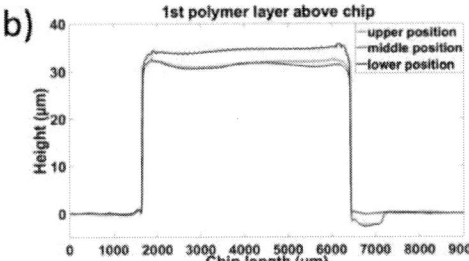

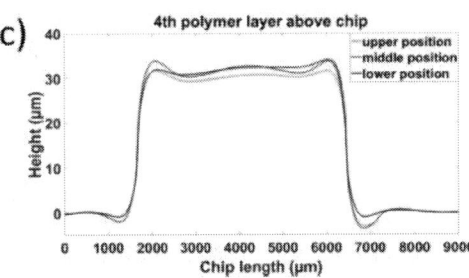

Figure 5: Topographic visualization of a polyimide layer, coated on a silicon chip (chip size 4.6 x 4.6 mm². A 3-D image of the chip surface was computed in Matlab (min= 5700 nm, max= 7500 nm, mean= 6827 nm, std= 352 nm). The 2-D grid illustrates the travelling path for data point measurement (110 µm distance between each measured point).

Considering the challenge to overcome a height difference of around 30 µm during the spin-coating process, the anomaly, visible in the coating profile is very likely due to the chip itself. This is in line with the plotted 3-D graph in Figure 5, where the thickness distribution over the chip surface has shown an increased topography, especially at the chip edges compared to the chip center.

Considering the manual embedding process of ultra-thin silicon chips, the warpage of chips due to thermal mismatch may also result in more inhomogeneity during polyimide coating and will be propagating with each additional polyimide layer.

To verify the spectroscopic results of polymer uniformity, height measurements of polymer layers are performed by the use of a surface profilometer.
Three different chip positions, relative to the wafer flat were measured: on an upper position above chip pads, a middle position, and on a lower position. The results are graphically depicted in Figure 6.
The height profile of polyimide is measured relative to the embedded chip, thus representing the first coated layer on the chip surface (Figure 6 b). In total, four polymer layers were coated above a chip, three polyimide layers and a final BCB layer. Compared to the measured height profile of the embedded silicon chip (Figure 6 a), the first polymer layer shows a slightly rise in the topography profile. By the spin-coating of each additional polymer layer, the difference in height topography is propagated thus leading in an inhomogeneous surface profile. An enhanced effect on the chip edges is visible. This profile is shown in Figure 6 c, and reflects especially at the chip edges a higher topography compared to the first polymer layer.

Figure 6: Height profile measurements of a) embedded ultra-thin silicon chip. b) First coated polymer layer above the chip and c) the fourth polymer layer above three preceded polymer layers on the chip. The measurements are performed at three different positions over the chip.

The ongoing work will focus on developing the processes of both, polyimide and BCB coating, mainly above integrated chips, thus helping to improve the uniformity on chip surfaces, facilitating the subsequent process steps.

IV. Reliability and long-term performance of the CFP

In order to assess the long-term performance of the test substrates, the CFPs are placed in a physiological solution at 37 °C, and the DC impedance has been recorded for 800 hours, in regular intervals.
After 400 hours the impedance values abruptly rise toward MΩ (Figure 7). Those contact problems were likely caused by delamination of the metal polyimide interface.

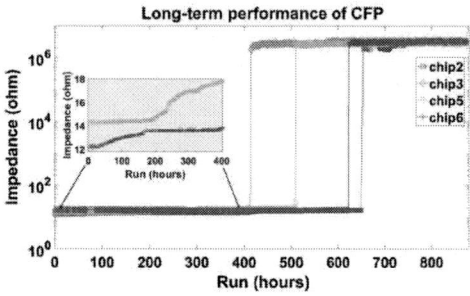

Figure 7: Trend of the measured impedance vs. time during the long- term performance of four independent chips. The close-up (from 0-400 hours), of two representative interconnects are chosen to visualize the smooth rising behavior of electrical signals.

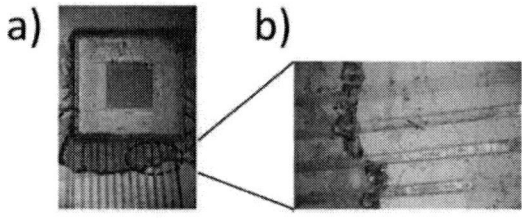

Figure 8: Photographs of a Chip-Film Patch. a) Close- up of an embedded ultra-thin chip in poly-imide foil, after soaking test in saline solution and b) zoom of corroded conductive paths near the chip.

Since polymer materials such as the chosen polyimide PI2611 are weak barriers against diffusion of gases, ions and water. Moisture can diffuse a lot easier through defects from the insulating material to reach the interface. The condensation of moisture in cavities at the interface, leading to a diffusion path through pores of the second layer and will be propagated. The effect of delamination is illustrated in Figure 8, showing images after the investigation of the long-term stability of CFPs. This confirms that delamination appeared at the metal-polyimide interface in striking distance of the embedded silicon chip, supposed to be a critical area for potential fracture and delamination reactions.

For the first time, we have verified the reliability and stability of the Chip-Film Patches in biological environment. To increase the functionality of the CFPs for the use in biomedical application, the optimization of our manufacturing process such as the combination and alternation of different coating materials, will be further investigated.

V.   Conclusion

We have provided an appropriate solution to monitor and improve the polyimide system, thus helping to control the process flow, which leads to an enhanced yield of functional flexible devices. To evaluate polymer uniformity on both flat and structural substrate, we have introduced a fast and non-invasive optical method. The preliminary results of the long-term performance of the Chip-Film Patches, in physiological environment will help us to improve our process, to generate optimized, functional devices, for the use in biomedical application.

Acknowledgements

We would like to thank all our coworker in the IMS clean room. Special thanks go to Wilfried Beller, Angela Schneider, Andreas Müller, Elisabeth Penteker, Saleh Ferwana and Shuo Wang. This work is supported by the German BMBF project FlexMax (16ES0775).

References

[1] J. Burghartz, "Ultra-Thin Chip Technology and Applications", Springer, New York, USA, 2011.

[2] B. Albrecht et al., "Multi-Chip Patch in low stress polymer foils based on an adaptive layout for flexible sensor systems", Electronics System-Integration Technology Conference (ESTC), Dresden, Germany, September, pp. 1-5, 2018.

[3] C. Harendt et al., "Hybrid systems in foil (HySiF) exploiting ultra-thin flexible chips", Solid-State Electron, Vol. 113, pp. 101-108, November, 2015.

[4] G. Alavi et al., "Adaptive Layout Technique for Microhybrid Integration of Chip-Film Patch", IEEE Transactions on Components, Packaging and Manufacturing Technology (CMPT), Vol. 8, pp. 802-810, May, 2018.

[5] M.-U. Hassan et al., "Combining organic and printed electronics in hybrid system in foil (HySiF) based smart skin for robotic applications", Proceedings of the European Microelectronic Packaging Conference (EMPC), Friedrichshafen, Germany, September, pp. 1–6, 2015.

[6] S.-Y. Park et al., "Anisotropic Thermal Characteristics of Graphene-embedded Polyimide Composite Sheets", Polymers & Polymer Composites, Vol. 24, No. 5, pp. 315-322, 2016.

[7] M. Kotera et al., "Microstructures of BPDA-PPD polyimide thin films with different thickness", Polymer, Vol. 54, pp. 2435-2439, 2013.

[8] HD Microsystems, Technical information, Sept. 2008, see www.hdmicrosystems.com

Dispensing Improvements with Drop on Demand Technology

Eric Cadalen*, Thomas Séon**, Christophe Josserand***, David Manteigas*

MBDA France, 1 avenue Réaumur 92358 Le Plessis Robinson cedex France

**Institut D'Alembert, UMR 7190, CNRS & Sorbonne Université, Paris, France*

***Laboratoire d'Hydrodynamique (LadHyX), UMR7646 CNRS-Ecole Polytechnique, 91128 Palaiseau CEDEX, France*

Abstract

Building on existing technologies in microelectronics dedicated to packaging and encapsulation, liquid polymer dispensing is a leading technology that is widely deployed. In this field, developments correlate with the application requirements and improvements are segmented in specific applications, such as conformal coating and capillary underfilling.

In the context of the developments, heterogeneity is a constant driver. On the one hand, dot diameter or line width reduction is a constant requirement for advanced microelectronic applications. On the other hand, equipment has to be compatible with existing or increasing dispensing areas. The consequence is a need for more accuracy on a larger area and smaller dispensing pattern units for higher panel volume. Drop on Demand (DoD) dispensing is a common answer to these challenging requirements.

The first building blocks of DoD for coatings were laid with inkjet printing more than 40 years ago, with optimal physical parameters corresponding to low-viscosity coating materials. Today's coatings are quite similar to inks and can be introduced in DoD dispensing equipment.

Starting from the results acquired with optimized coatings dispensed only with a dedicated pneumatic actuator, the experiments are extended to new materials and equipment technologies. The results are supported by high-speed camera measurements. The data acquired in these experiments lead to multiple process windows considering the predicted strong interaction between equipment and materials.

Key words: Drop on demand, conformal coating, piezoelectric and pneumatic actuator

Introduction

The processes used today in electronics, and in particular in fluid dispensing, use off-the-shelf equipment and materials. Drop on Demand (DoD) fluid deposition equipment shows potential of interest among the many solutions available.

Expected improvement is a better selectivity, integration density or process flexibility in the face of heterogeneity of the needs. But DoD implementation is a new set up with and equilibrium to be achieved between equipment and material or process parameters.

In the case of DoD, problems of satellite ejection going as far as atomization of the material may arise. Problems of droplet rebound, floating or spattering on impact must also be anticipated and avoided.

In an approach combining the properties of the fluids and the characteristics of the equipment, a process window can be defined using two dimensionless numbers, the Ohnesorge number and the Weber number. This process window must be defined by identification and control of the parameters of materials and equipment initially selected off-the-shelf.

Interest in DoD for heterogeneous fluids

The electronics industry uses fluid dispensing for a wide diversity of materials. These materials must be defined by their physical properties so that they are within the DoD process window. The dynamic viscosity of the materials used enables rankings to be determined and pre-selection by linking applications to the properties of the material.

Given that the dynamic viscosity of water at room temperature is 1 mPa.s, fluids with low viscosity from 1 to 25 mPa.s are typically found in the first category. These inks are used by ejection of unit droplets in inkjet printing to define a drawing or a diagram representing electrical interconnections with variable integration densities.

The range from 25 to 100 mPa.s includes coatings for varnishing or 'painting' surfaces by applying a thin membrane of the material, generally by spraying. The particular feature of the process is the use of a solvent lowering the viscosity and evaporating after deposition. Specified material thicknesses between 25 and 75 microns are obtained in the case of the conformal coatings used for ruggedized printed circuit boards. The higher viscosity range from 100 to 500 mPa.s includes encapsulants such as the underfill resins under components or integrated circuits. The viscosity can be adjusted by temperature modulation to accelerate the capillary effect. Encapsulants intended for filling or structural bonding can have viscosities of up to 10,000 mPa.s at room temperature.

The viscosity can increase still further to substantially exceed 10,000 mPa.s at room temperature, but use of such materials is often facilitated by increasing the temperature and thus lowering the viscosity. Above 100,000 mPa.s, the fluids become 'pasty' and their flow is controlled to form dams, for example, thanks to their thixotropic behavior.

Equipment heterogeneity

The available fluid deposition means are thus going to be combined with the materials pre-defined by their viscosity. The schematic in Figure 1 shows a heterogeneity of materials and equipment that can be exploited to define specific applications over an extended viscosity range. The equipment used for each viscosity range is defined in response to two recurring constraints: increasing integration density and increasing surface areas to be covered.

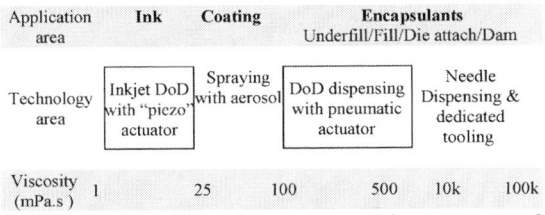

Figure 1: Typical equipment and applications per viscosity range

Drop on Demand (DoD) dispensing systems, based on selective ejection of droplets, which first emerged with pneumatic actuators, are dedicated in particular to underfill resins with viscosities between 100 and 500 mPa.s. The principle is to eject a droplet through a nozzle by a needle acting as a piston. The pneumatically-actuated needle has a tip optimized for the fluids.

Contactless DoD dispensing also avoids the proximity necessary between the surface to be coated and the needle. It is potentially a better response to the needs of emerging applications with 3D displacements and dispensing, or simply to the heterogeneities encountered between components of very different shapes and volumes with 2.5D displacements and dispensing. This makes it possible to make deposits in zones inaccessible to a needle in very small spaces between two components of different heights, for example.

Selectivity requirements

DoD is a breakthrough in terms of selectivity compared with previous processes, as the basic pattern is a droplet in contrast to 'dispensing' systems where the basic pattern is a bead and possibly a wide strip. DoD favors precise selective deposition and avoids the otherwise inevitable use of temporary masking.

For example, DoD dedicated to highly-integrated microelectronics now benefits from developments initiated more than 40 years ago with inkjet printing and optimal physical parameters corresponding to low-viscosity materials. The results include dedicated inks for marking with bubble jet printers, and piezoelectric actuators for desktop printers. Inks with viscosities from 5 to 25 mPa.s dedicated to DoD equipment can achieve basic droplet volumes between 2 and 150 picolitres.

Opportunities from green electronics

In parallel, 'low-viscosity' materials are evolving, as a consequence of legislative restrictions relating to the use of chemicals that pose risks for the user. Products initially based on organic solvents have been reformulated applying water-based chemistry. These restrictions on use also bring benefits in diversity of implementations. Changes are observed in the customary modes of thermal polymerization by temperature increase, UV polymerization by ultraviolet radiation, or moisture uptake.

The implementation processes are changing, with new 'dual' polymerization modes coupling two modes, for example ultraviolet radiation followed by moisture uptake. There are also 'hybrid' polymerization modes leaving the choice between polymerization by heating or by ultraviolet radiation. Each mode is going to facilitate implementation; in the case of dual mode, a zone shaded from UV radiation could be polymerized by moisture uptake.

In environmental terms, DoD technology promises lower consumption of materials, given the intrinsic selectivity of the process. Consumption can be limited substantially by reducing the size of the basic unit and dispensing only where necessary, in order to the meet the need for 'green electronics'.

DoD experiments

When the available materials and the equipment that can be used with them are examined, it is clear that interactions must be taken into account.

The materials include coatings with high viscosities from 100 to 500 mPa.s, and fluids with low viscosities from 25 to 100 mPa.s, which are more similar to inks. The low-viscosity coatings would only be compatible with systems using piezoelectric actuators identical to those used in office printers, although this remains to be confirmed.

The equipment includes systems with piezoelectric actuators operating at a maximum frequency close to 3000 Hz, with a nozzle diameter of 100 µm, and systems with pneumatic actuators operating on a cycle based on a maximum frequency of 160 Hz, with a nozzle diameter of 75 µm.

313

Process set up on low-viscosity fluids

An initial experimental approach is necessary in order to confirm the compatibility of low-viscosity fluids with a piezoelectric actuator (Figure 2). This approach was applied for the two low-viscosity coatings A and C in Table 1. When these coatings are used with a pneumatic actuator, substantial spattering of satellites on the impacted surface is observed [1]. A high-speed or stroboscopic camera is used to take nozzle exit photographs.

Fluid physical indicators	Coating low viscosity A	Coating high viscosity B	Coating low viscosity C
Viscosity (mPa.s)	50	359	91
Measured surface tension (mN/m)	23.5	21.7	21.6

Table 1: Fluid physical indicator

The low-viscosity coatings are observed to be well suited to a piezoelectric actuator with ejection at room temperature (Figure 2 (a) and (b)) showing a tail or satellites.

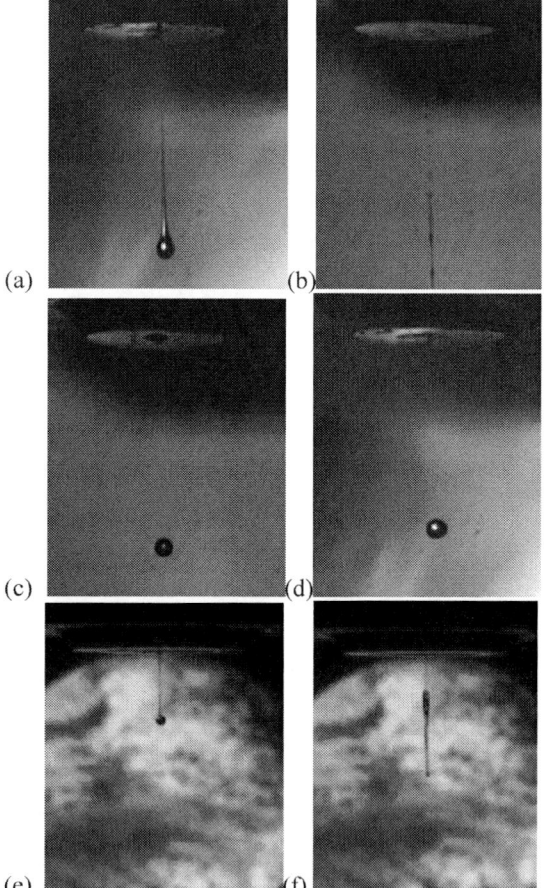

Figure 2: Droplet ejection for low viscosity fluid

   (a)   coating A with piezo actuator, room temperature
   (b)   coating C with piezo actuator, room temperature
   (c)   coating A with piezo actuator, 60 C
   (d)   coating C with piezo actuator, 70 °C
   (e)   coating C with pneumatic actuator, room temperature
   (f)   coating C with pneumatic actuator, 60 °C

This form is rapidly improved by increasing the temperature (Figure 2 (c) and (d)) to form a perfect sphere, resulting in ejection controlled before impact.

Low-viscosity coating ejection by pneumatic actuator results in a nozzle exit form that cannot be controlled. Despite initial attempts to optimize the temperatures or equipment parameters such as cycle time or ejected volume, the result after impact shows substantial satellite deposition. This can moreover be predicted on ejection in the photographs (Figure 2 (e) and (f)).

This first approach implemented with low-viscosity coatings demonstrates compatibility with the piezoelectric systems.

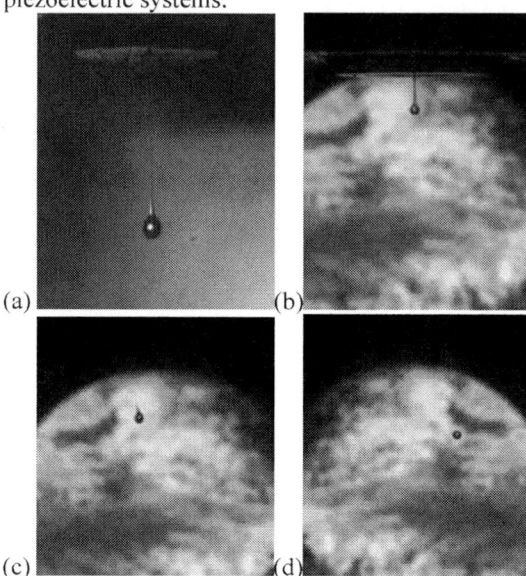

Figure 3: Droplet ejection for high viscosity fluid B

   (a)   coating B with piezo actuator, 75 °C
   (b)   coating B with pneumatic actuator, room temperature
   (c)   coating B with pneumatic actuator, room temperature, distance 1 cm from nozzle
   (d)   coating B with pneumatic actuator, room temperature, 2 cm from nozzle

Process set up on high-viscosity fluids

To fully confirm the suggested interaction between DoD materials and equipment, the use of high-viscosity coatings in piezoelectric systems should be tested. The results can be compared with the results obtained with the pneumatic system, which are satisfactory even at room temperature.

Initial industrial feedback on DoD deposition of conformal coatings has already been reported [1]. The tests on high-viscosity coating B led to the determination of a minimum temperature of 75 °C, which gives the result shown in Figure 3 (a). However, the form obtained is not spherical, as it is at this temperature for the low-viscosity fluids. Ejection of a high-viscosity coating is optimized more easily with a pneumatic actuator. Furthermore, the pairing of this actuator with a high-viscosity fluid enables the distance between nozzle and substrate to be increased, allowing time to form a spherical droplet. This result supports the possibility of working in 2.5 D with a working distance from at least 1 cm (Figure 3 (c)) to 2 cm (Figure 3 (d)).

The photographs taken can be used to quantify and compare the droplet sizes obtained. The droplets obtained by the piezoelectric systems are two- to three-fold smaller than those produced by the pneumatic systems, despite the former having larger nozzles. Two potential processes are thus identified, given in Table 2 and determined essentially by the material to be used.

An initial DoD approach for three coatings provides guidance for industrial choices based on a physical parameter, viscosity, given in Table 1. However, these experimental results do not yet enable prediction of behavior on ejection nor on impact. Even if satellites are reduced or eliminated on droplet ejection, the behavior on impact is likely to cause other satellites by splashing. Conversely, the droplet might bounce or float at the surface.

Industrial process	DoD by pneumatic actuator	DoD by piezoelectric actuator
Low-viscosity material	PROCESS 1 UNSUITABLE Satellites, substantial splashing	PROCESS 2 CONCEIVABLE Very high integration density Compatibility with many existing coatings
High-viscosity material	PROCESS 3 CONCEIVABLE High integration density Need to formulate suitable new coatings	PROCESS 4 UNSUITABLE Large temperature increase reducing the material to low viscosity

Table 2: Status of coating deposition procedures by DoD

Other physical parameters must be taken into account in order to understand the results obtained, for example the capillary forces, quantified by the surface energy of the fluid. Even though gravitational forces can be neglected [1], viscous forces, capillary forces and inertial forces are also to be taken into account in the customary dimensionless numbers: Reynolds number, Weber number and Ohnesorge number.

Technological building blocks

As a first step, the Newtonian behavior or "continuous flow" property will allow reference to a large volume of existing data on this type of fluid [2][3] in order to characterize the impact. In the case of Newtonian post-impact behavior, spreading and wetting is expected, considering that all the kinetic energy is transferred by viscous friction on the surface.

The viscosity measurements given in Table 3 are insufficient to characterize the rheological behavior of the fluids shown in Figure 4.

The first information to be obtained from Figure 4 is the Newtonian behavior of the three fluids, indicated by the fact that the shear stress is

proportional to the shear rate and will not be time-dependent.

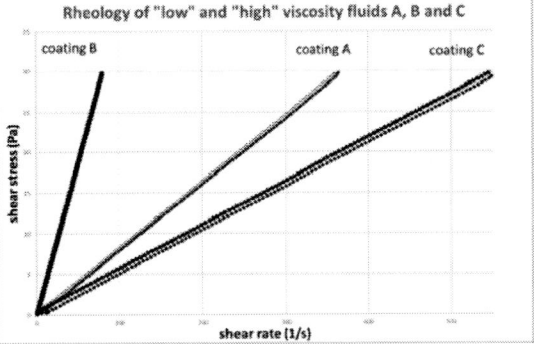

Figure 4: Coating viscometer measurements

Ratio of physical parameters

The impact of the droplet is characterized using two dimensionless numbers. The Reynolds number characterizes the part played by the viscosity, distinguishing between laminar flow at low Reynolds number and turbulent flow at high Reynolds number. It is defined as:

$$Re = \frac{\rho \, v \, d}{\eta} \qquad \text{Formula 1} \qquad \text{where:}$$

ρ: density (kg/m3)
d: characteristic linear dimension, typically drop diameter (m)
v: velocity of the fluid with respect to the object (m/s)
η : dynamic viscosity (Pa.s)

The surface tension comes into play in the second dimensionless number, the Weber number, given by:

$$We = \frac{\rho \, v^2 \, d}{\sigma} \qquad \text{Formula 3}$$

σ: surface tension (N/m)

The surface tension predominates at low Weber numbers, while droplet inertia is predominant at high Weber numbers.

The two numbers are used to model the impact. For example, at high viscosity [3], the maximum diameter on impact is described by the following function:

$$\frac{Dmax}{Do} \propto Re^{1/5} \qquad \text{Formula 2}$$

A function involving both numbers (We and Re) is used to model the impacts at higher speeds. A major issue for our applications is splashing during impact. Splashing depends in a complex manner on the impact parameters, and its control cannot be based only on experimental observations [4]: splashing may occur more easily with high-viscosity fluids than with low-viscosity fluids.

Empirically, a splashing threshold [3] has been obtained:

$$K = We\sqrt{Re} \qquad \text{Formula 4}$$

If the threshold value exceeds a critical value (3000 for impacts on solid substrates), droplet impact-related splashing may occur.

A combination of these two numbers is pertinent for the impact. The Ohnesorge number

(Oh) given by formula 5 is a ratio between the Weber number and the Reynolds number. It is used to compare the viscous forces with the capillary forces. The Ohnesorge number characterizes the liquid at the scale of the droplet: in particular, sprays form at low Oh values, while at high Oh values the liquid is not deformable and remains blocked at the nozzle exit:

$$Oh = \frac{\eta}{\sqrt{(\rho \sigma d)}} = \frac{\sqrt{We}}{Re} \qquad \text{Formula 5}$$

On the basis of this information, the dimensionless numbers can be linked with the physics of the materials (density, viscosity and surface tension) and the industrial equipment (ejection speed and diameter). The values obtained are given in Table 3 below.

coating	A	A	B	C	C
process	1	2	3	1	2
viscosity (mPas)	55	55	359	91	91
surface tension (mN/m)	23.5	23.5	21.7	24.6	24.6
density (g/cm3)	1.05	1.05	1.06	1.09	1.09
max. fall speed measured	10*	2	3.60	10	2.00
nozzle diameter (µm)	75	100	75	75	100
Re	14.32	3.82	0.80	8.98	2.40
We	335	18	48	332	18
Oh	1.28	1.11	8.64	2.03	1.76
splashing threshold K	1268	35	43	996	28

Table 3: Data characterizing the fluids
(*assumption, not measured)

Progress can consequently be made in understanding the behavior of the three coatings.

K is high for the low-viscosity coatings with pneumatic actuator, but remains below the defined splashing threshold value of 3000. The fall speeds are determined, with an assumed value for coating A, and contribute to the calculation of K. The post-impact satellites observed for low-viscosity fluids A and C ejected by pneumatic actuators are consequently formed on ejection before impact. In contrast, the satellites observed for coating B are linked with inadequate nozzle height. Figure 5 shows satellites formed at an impact distance corresponding to an ejection intermediate between those of images (c) and (d) in Figure 3. This distance is too short for the viscous forces to pull the droplet into a spherical form before impact.

Moreover, the fall speed is measured and shows high variability between processes. In contrast, it is constant with the pneumatic process if a single droplet is ejected. The measured speeds are also substantially (up to ten-fold) lower than the assumed values used previously [1].

DoD process window

On the basis of these experimental data, the DoD process 'window' can be obtained [6] by plotting the Weber number as a function of the Ohnesorge number for each coating. The results are

shown in Figure 6. The coatings that are outside the DoD printing area effectively show defects, but the most interesting and obvious observation is that materials A and C fall within the DoD printing area if a piezoelectric actuator is used (and set up).

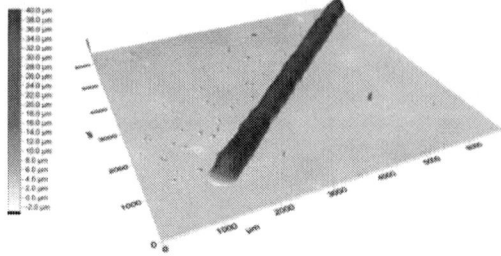

Figure 5: Satellites observed for coating B with an incorrect process setting

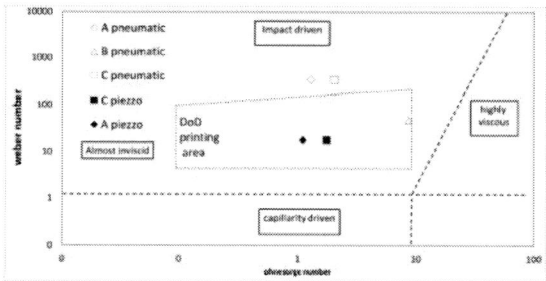

Figure 6: DoD ejection process window

The process adjustment and setting options are consequently as shown in Figure 7. Within the coordinate system formed by the Ohnesorge number as a function of the Weber number, representing the "DoD ejection process window", selection of material or equipment options is going to define the operating point.

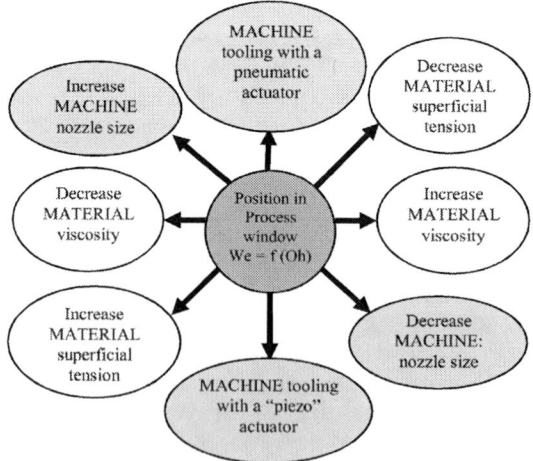

Figure 7: DoD process window adjustment

Ejection is controlled before impact. Further work is necessary to confirm wetting and droplet spreading without splashing on impact. Defects to be eliminated are spatter, rebound or any floating of the droplet at the impact surface. Rebound is linked with the air film that is formed between the droplet and

316

the surface [5], independently of its characteristics. By analogy with this study on various polar or non-polar fluids, a high value (between 50 and 100) of the Weber number avoids rebound and, all the more so, hovering on the air film. Comparison with the data in Table 3 shows that wetting must take place with impact control without rebound for coating B.

Measurement of impact parameters

This is effectively what is observed in images of droplet impact on a surface. The first step of the impact is a "viscous" behavior in less than 100 ms without any spattering or floating. Then a capillary flow of the drop will occur in the scale of a second before equilibrium. This "visco capillary" behavior is summarized by the curve in Figure 8 of the ratio of post-impact drop diameter to ejection diameter as a function of time.

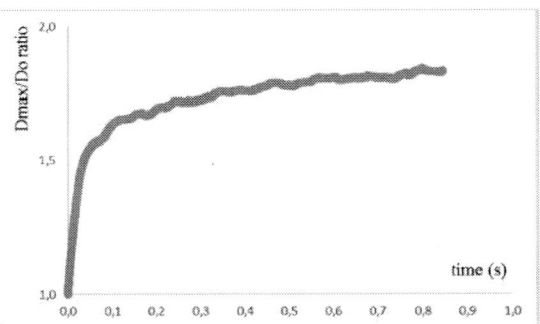

Figure 8: Dmax/Do ratio as a function of time

Experimental validation

Now that droplet ejection and impact are characterized, the next step is to control droplet volume by an experimental plan with a factor screening objective. The post-impact droplet can be defined by a truncated sphere model. The patterns will then be formed. A series of droplets with constant spacing results in the formation of a regular bead. A set of lines sufficiently close together results in a surface thanks to the visco-capillary behavior observed with coating B, for example.

A cause and effect diagram can be used to pre-select the potential influencing factors. The diagram in Figure 9 gives the root causes likely to have an effect on this basic volume.

The fixed factors are controllable constants in this complex process, and there are consequently many of them. The focus will be on reducing them in the case of high-viscosity fluid B dispensed by pneumatic actuator. This imposes, for example, a nozzle diameter of 75 µm with a DoD cycle of 160 Hz.

Nozzle temperature (factor A) is typically one of the noise factors, and has been added in a range representative of the conditions applicable to the process from 25 to 35 °C.

Similarly, the nozzle height is reduced deliberately to below 10 mm, which is sure to cause defects; efforts will be made to minimize them by

looking for key factors. The process could then be controlled by setting heights greater than 10 mm. The height between the dispensing nozzle and the impact surface may become a noise factor in the case of surfaces with high degrees of heterogeneity, and will be set considering variations of 2 to 8 mm in factor B.

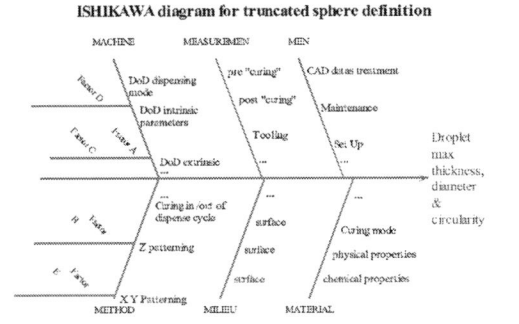

Figure 9: Cause and effect diagram

Factor D is identified as a noise factor linked with the dispensed droplet volume to fix. However, However, the potential volume variations are likely to modify repeatability or reproducibility of dots or regularity of lines. This volume is set by adjusting the needle stroke by 5 to 10 increments.

For controlling the process, there are many variable factors accessible without taking into account surface quality (flatness, roughness, etc.). The most obvious variables for getting from a dot to a controlled line are the spacing or overlap between dots to be matched with the displacement speed. The two factors selected are consequently factor E representing the head displacement speed from 20 to 40 mm/s and factor C representing the droplet spacing pitch from 300 to 500 µm.

The experiment plan is based on a factorial plan with two levels of 16 tests giving a resolution V enabling the suspected interactions to be studied, in particular between factors C and E. The other interactions may be taken into account, given that all the interactions are de-aliased from the five main factors.

The measurements are made by profilometry using a confocal detector in white light. For the sake of simplicity, the measurements are made after curing of the basic patterns (dots, lines, etc.) produced. The target for more-or-less complex patterns is a final thickness between 25 and 75 µm.

Characterization of the basic pattern

The droplet heights measured on an organic surface are distributed around the specified lower limit of 25 µm, enabling final height adjustments after overlaying of the patterns. Conversely, the droplet height does not depend on the needle stroke, a counter-intuitive result. This modeling consists mainly of error. The precision of the model is not sufficient to keep these results as significant, even though the modeling is.

Source	Term df	Error df	F-value	p-value	
Whole-plot	1	Not defined			
a-temperature	1	Not defined			
Subplot	4	10.00	12.26	0.0007	significant
B- nozzle height variation	1	10.00	12.71	0.0051	
D- needle stroke	1	10.00	20.25	0.0011	
E- DoD head speed	1	10.00	5.24	0.0451	
BE	1	10.00	10.84	0.0081	

Table 4: Anova of drop diameter

Conversely, the calculation of the volume and the diameter of the unit droplet is fully explained. This is illustrated by the analysis of variance (Table 4) resulting from fitting the stepwise regression. The diameter modeled as a Taylor series rank depends essentially on the actuator setting for droplet unit volume and on the tool height, which must be increased. Based on the results obtained in ejection images, this avoids spatter and the droplet volume is controlled, enabling sphere formation before impact.

These results are confirmed when the circularity of the pattern, i.e. the maximum diameter minus the minimum diameter, is taken into account. The same tendency (refer to the analysis of variance in Table 5) is observed on the height, which must be increased.

Source	Term df	Error df	F-value	p-value	
Whole-plot	1	Not defined			
a-temperature	1	Not defined			
Subplot	3	11.00	18.95	0.0001	significant
B- nozzle height variation	1	11.00	3.52	0.0874	
D--needle stroke	1	11.00	40.96	< 0.0001	
BD	1	11.00	12.36	0.0048	

Table 5: Anova of circularity parameter

The lack of precision of the drop height modeling can also now be explained: when the circularity value deteriorates, i.e. increases, the heights are modified, as can be seen for two examples in Figure 10 that illustrate the interaction between factors B and D. The other important factor in height control is observed visually on the 16 test specimens produced: the circularity of the pattern varies widely between tests and, conversely, the results are explained in the analysis of variance by a regression with a fit of more than 75% according to the terms of Table 5.

The setting obtained experimentally is also consistent with the ejection measurements: the best

circularity results are obtained for a maximum variation of 8 mm shown on the 3D graph of Figure 10 (b).

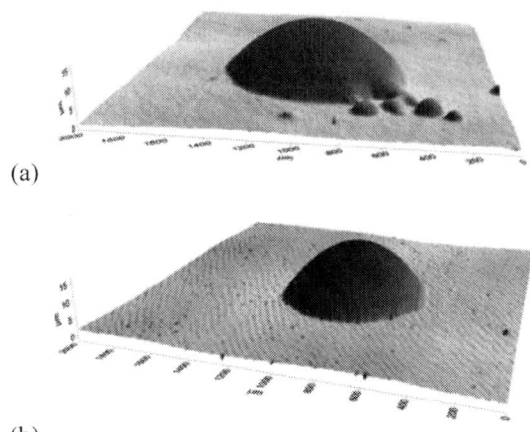

(a)

(b)

Figure 10: Interaction on circularity
(a) Nozzle height 2 mm, stroke 5 incr
(b) Nozzle height 8 mm, stroke 10 incr

Characterization of a line

For deposition, on a given surface, of solder mask coating characteristic of a printed circuit, the width of a basic line is easier to model and results in a regression fitting the model at more than 99.8%, which simply shows the influence of all the factors and six interactions in the screening objective.

Source	Term df	Error df	F-value	p-value
Whole-plot		Not defined		
a-temperature		Not defined		
Subplot	10	4.00	1613.84	< 0.0001
B-nozzle height variation	1	4.00	70.04	0.0011
C-repeating distance	1	4.00	8793.83	< 0.0001
D-needle stroke	1	4.00	1861.84	< 0.0001
E-DoD head speed	1	4.00	229.84	0.0001
aB	1	4.00	168.10	0.0002
aC	1	4.00	961.43	< 0.0001
aD	1	4.00	550.42	< 0.0001
aE	1	4.00	227.76	0.0001
BE	1	4.00	760.49	< 0.0001
CD	1	4.00	2514.64	< 0.0001

Table 6: Anova of line width

The factor with the greatest influence in the model appears in Table 6: the spacing between drops (repeating distance).

Given the large number of interactions influencing line width, this distance must consequently be a process adjustment variable for a given material, knowing that this parameter also influences line height. This result must be considered alongside the results obtained [6] where the droplet repeating distance is defined to obtain a balance between the effects of bulging and of gaps.

318

Conclusion

The first extractions of the DOE by screening performed on coatings dispensed by DoD show three improvements validated by appropriate video imaging.

The first result, counter-intuitive, is the hoped-for benefit for the highly heterogeneous 2.5D applications, enabling the nozzle to be raised at least 2 cm above the substrate and to 'remotely' reach zones inaccessible to other processes.

The second result is related to the optimization of the integration density by reducing the droplet unit volume of 15 nl deposited initially with a pneumatic actuator to 150 pl with a piezoelectric actuator.

The third advantage of the piezoelectric actuator is the optimization of the selectivity, illustrated by deposition speeds at frequencies at least 10-fold higher by combining them with compatible materials and much smaller unit volumes.

These results may be refined by RSM (Response Surface Methodology) or by extension to other potentially suitable materials such as the underfill resins.

Acknowledgements

The authors would like to acknowledge the members of the laboratory team at MBDA Le Plessis Robinson and the Institut d'Alembert at UPMC for their support during equipment and camera setup.

References

[1] Eric Cadalen, Olivier Maire, David Manteigas Development of selective conformal coating process based on advanced packaging for harsh environments Imaps EMPC 2017

[2] Brian L. Scheller and Douglas W. Bousfield "Newtonian drop impact with a solid surface" AIChE Journal June 1995 Vol.41, No 6

[3] C. Josserand, S.T. Thoroddsen "Drop impact on a solid surface" Annu. Rev. Fluid Mech. 2016. 48:365-91

[4] Xu L, Zhang WW, Nagel SR. 2005. Drop splashing on a dry smooth surface. Phys. Rev. Lett. 94:184505

[5] De Ruiter J, Lagraauw R, van den Ende D, Mugele."Wettability-independent bouncing on flat surfaces mediated by thin air films". Nat. Phys. 11:48–53

[6] Brian Derby "Inkjet printing of functional and structural materials: fluid property requirements, feature stability, and resolution" Annu. Rev. Mater. Res. 2010. 40:395-41

Thermosonic wedge-wedge bonding using dosed tool heating

Andreas Unger, Matthias Hunstig, Michael Brökelmann, and Hans J. Hesse

Hesse GmbH, Lise-Meitner-Str. 5, 33104 Paderborn, Germany

+49 5251 1560-681, andreas.unger@hesse-mechatronics.com

Abstract

This paper presents a novel thermosonic heavy wire and ribbon bonding process using a laser-heated bonding tool. The proposed thermosonic bonding process uses ultrasonic and thermal energy in a user-definable composition to achieve optimal connections. Through the substitution of vibration energy by thermal energy, ultrasonic vibration amplitude and/or normal force can be reduced, thus lowering the stress to the substrate. This improvement is most significant on substrates which react sensitively during purely ultrasonic bonding like battery caps, copper alloys, coated caps/clips or which are fragile like dies or sensors. Alternatively, the additional power can be used to reduce bonding time and therefore increase throughput and efficiency, or to produce stronger connections within the same bonding time. The temperature distribution at and around the bonding interface is investigated. During the process, it is necessary to ensure appropriate absorption of laser radiation at the tool tip. For this reason, a special tool with integrated optical fibre and optimized laser absorption was developed. Using a pyrometer and a closed-loop controller, the tool tip temperature is precisely controlled in real time. The temperature effect on bond strength is investigated; at a given bond duration and ultrasonic power, tool heating always increased bond strength significantly. Alternatively, the process time required for a desired bond strength could be reduced. Dosed tool heating using laser power is thus a very promising advancement in wedge bonding.

Key words: thermosonic bonding, heavy wire bonding, tool heating, laser heating

Introduction

Ultrasonic wedge-wedge heavy wire bonding is a well-established industrial process for connections in power electronics [1, pp. 33-38]. It is also increasingly used in other applications such as cell connections in automotive battery packs [2, 3].

Ultrasonic wire bonding is a "cold" process; instead of melting, the metals are bonded by interdiffusion and formation of intermetallic compounds induced by the ultrasonic vibration [1, p. 24]. As such, the process was originally done at room temperature.

Today, the vast majority of wedge-wedge wire bonds using Al or Cu wire are produced at room temperature. For thin Au wire applications in both wedge-wedge and ball-wedge technology it is common to heat the interface and the whole device to about 125 to 220 °C [1, p. 36]. Such a process, also increasingly used for Cu wire ball-wedge bonding [4], is known as thermosonic wire bonding because it uses both thermal and ultrasonic energy.

Effects of Heat in Thermosonic Wire Bonding

The main motivation for thermosonic bonding is to provide some of the activation energy required to initiate interdiffusion of the bonding materials in the form of thermal energy instead of kinetic ultrasonic vibration energy [1, pp. 33-36]. In some Au wire applications, especially where normal force and/or vibration amplitude are limited by the substrate sensitivity, this is necessary to obtain reliable bonds. Bonding at elevated temperature has several additional effects which can be positive also when processing other materials:

- Diffusion, a major driver of the bonding process, is accelerated at higher temperature. This can potentially reduce the required process time.

- Increasing temperature reduces wire material strength and reduces or eliminates strain hardening [5]. Thus, the same wire deformation can be obtained with less normal force. This is particularly advantageous on sensitive substrates such as dies or sensors, where too high a normal force is known to contribute to cratering [1, p. 256].

- Increased ultrasonic vibration has a similar effect on the stress-strain curve as increased temperature [6]. Thus, the same process potentially requires less vibration if run at an increased temperature. This, too, is particularly advantageous on sensitive substrates.

- The effects of substituting vibration energy by thermal energy can also be exploited to process wires which would require normal forces or vibration amplitudes beyond the substrate limits at room temperature. Examples for such systems are heavy copper wire bonds on sensitive silicon chips without special metallization [7, 8], but also ball bonds on thin microelectronic structures [9].

- Some material combinations, such as the aforementioned Au on Au, or Al on Ag [1, p. 37], can only be reliably bonded at increased temperatures.

Conventional Heating in Wire Bonding

Despite these potential advantages, thermosonic bonding is rarely used besides ball-wedge and Au wedge-wedge applications, mostly due to the limitations of available heating technology.

In current applications, heat is usually supplied through the workholder which clamps the package containing the substrate to be bonded [1, p. 6] Even if a pre-heating station is used, it takes a certain time before the substrate has reached its target temperature and the bonding process can begin. The whole package is heated, thus it must be designed to withstand the increased temperature regarding material limits and thermomechanical stress. In many applications, such as battery packs, this is not possible. It may additionally be necessary to prevent oxidation using an inert gas atmosphere around the process as it is done in Cu ball-wedge bonding [4].

Oxidation and thermal stress to the package can be avoided by not heating the whole package, but only the process zone.

In one of the first descriptions of a thermosonic process, Coucoulas [5] resistively heated the bonding tool using a special laboratory setup to obtain tool temperatures above 400 °C for bonding Al and Cu wires to thin films. While the experiments were successful, such a setup never appeared in commercial machines, probably due to the rather bulky setup and the inefficient heating system. A more compact bonding tool with integrated resistive heating has been proposed by Cho [10]. There are some commercial solutions for heating bonding tools by thermal radiation [11] and bonding tools of suitable material could also be heated by induction. In any case, at least a major part of the bonding tool is heated due to the necessary extent of the required parts and the good thermal conductivity of typical wedge bonding tool materials. For the same reason, heating takes some time and the maximum temperature is limited, as the ultrasound transducer must not exceed its maximum operation temperature.

Laser Heating in Wire Bonding

In ideal thermosonic bonding, only the process area is heated and only as much energy is supplied as is required to keep bond pad and wire at the required temperature level during the bonding process. The ideal heat source for such a process delivers high power directly to the process zone.

If the heating power is too low, parts must be heated well before the actual bonding process, using their heat capacity and thermal conduction to provide the required heating power to the process. If the

heating acts far from the process zone, one must rely on long distance thermal conduction to heat the process zone. Both cases result in a lot of wasted energy and undesired heating of parts that shall remain cool, such as the ultrasound transducer.

Lasers can provide high power to a small area and thus are a promising energy source for thermosonic bonding. Some researchers have therefore investigated laser heated wire bonding in the past:

Savu et al. [12] bonded heavy Au wire on Cu substrate, heating the wire with a near infrared (IR-A) laser before and during ultrasound vibration. They did not investigate the resulting bond strength. Liu et al. [13] use an IR-A (808 nm) laser to heat bond pads on a MEMS locally and only shortly before they are contacted by the Au wire ball. Schneider et al. [14] use an IR-A (1080 nm) laser to heat heavy Cu wire before applying ultrasonic vibration to bond the wire to Cu plates. Both teams of researchers observe an increase in bond strength with increased laser power.

However, such processes suffer from the high reflectance of typical bond pad and wire materials such as Ag, Al, Au, Cu. Only a small percentage of infrared laser energy, typically less than 5 % [15], is converted into heat in the process zone. The rest is reflected in an uncontrolled way, causing undesired heating of other parts and posing a potential hazard. The laser power must be much higher than the heat flow required by the process.

Uncontrolled laser reflection into the environment can be greatly reduced, albeit not completely avoided, by guiding the laser beam inside the bonding tool, for example using a fibre as shown in Figure 1, and only activating the laser when the bonding tool touches the wire. But this does not solve the issue of low absorption, most laser energy is still reflected back into the tool or optical fibre, possible damaging the fibre or laser. For the abovementioned disadvantages, no laser heating process is yet available in a commercial wire bonding machine.

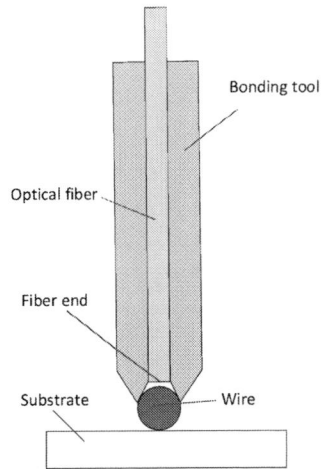

Figure 1: Schematic bonding tool design for direct wire heating with open tip

Novel Laser Heated Tool Approach

To overcome the already mentioned problems of low absorption rate of common lasers on standard wire and substrate materials, a bonding tool with integrated optical fibre and optimized laser absorption for indirect wire heating was developed. The laser radiation is used to heat the tool tip during or immediately before the bonding process to heat wire and interface by thermal conduction and thus enable a thermosonic bond process. This tool offers significant advantages compared to a tool with open tip as shown in Figure 1. Previous studies have shown that standard bonding tool material (tungsten carbide in cobalt matrix) has an absorption rate of approximately 70 % at IR-A wavelengths, much more than the less than 5 % of standard wire materials [15]. This dramatically increases the share of the laser power usable in the process, enabling the use of laser sources of relatively low power.

Mechanical and thermal long-time tests with various tool variants were carried out. The operating principle and the developed bonding tool are shown in Figure 2. central hole in the tool is used to place the end of the fibre centrally inside the tool, pointing towards the tip. The section towards the tool tip increases the effective area on which the laser beam is absorbed after leaving the fibre. This reduces the intensity of the incident radiation and the resulting local thermal stress for the tip. Another important benefit is the decreased resulting reflection to the laser system because of the beam trap character of the geometry. If energy is dissipated at the end of the fibre, the resulting heat can damage the fibre and also impairs accurate temperature measurements through the fibre as described below.

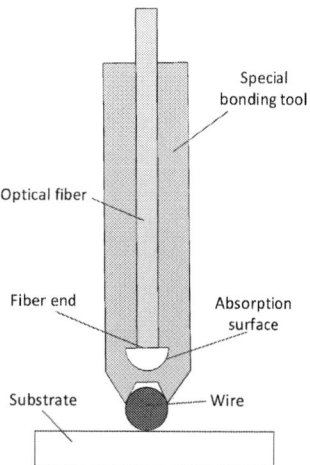

Figure 2: Schematic bonding tool design for indirect wire heating with laser absorption surface

Finite Element (FE) Heating Simulation

Based on previous investigations, the process temperature required to achieve the desired effects in bond formation (time reduction, normal force reduction, substitution of mechanical vibration power,

general bondability) was expected to be 150-200 °C, which is within the typical range for conventional thermosonic processes [1, p. 36]. In order to determine the required tool temperature for heating the interface to this temperature, a 3D finite element (FE) model of the contact partners bonding tool, wire and substrate/DBC was created using ANSYS. As shown in Figure 3, the copper wire with a diameter of 500 µm is placed under a standard wire bonding tool (tungsten carbide in cobalt matrix) on a 300 µm thick copper layer of a DBC. In order to produce the (thermal) contact conditions of a normal wire bond, the wire portion under the tool was deformed to model the deformation occurring in an ultrasonic bonding process in the so-called pre-deformation phase due to the applied normal force.

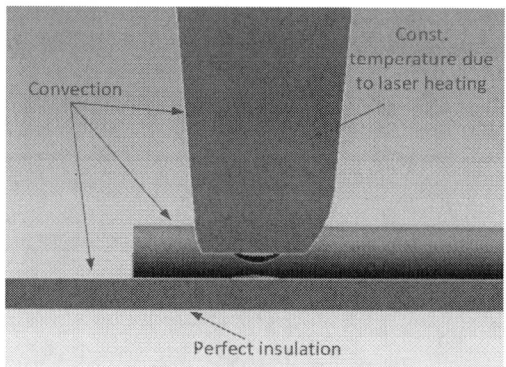

Figure 3: Finite element model for thermal analysis

The following boundary conditions were defined in the thermal simulation in order to obtain valid results: The temperature is assumed to be constant in the tool tip due to closed-loop temperature control (see paragraph 'Tool Temperature Control'). To consider the natural convection of the cooling air, a convective load is placed on the surfaces of the tool, wire and substrate with a convection coefficient of 5 W/m² and an ambient temperature of 22 °C. For the simulation, the ceramic underneath the DCB copper layer is assumed to be a perfect insulator. Wire length and substrate extent were chosen so large that their outer regions remained very close to ambient temperature.

Thermal conductivities are assumed as 400 W/(m·K) for copper and 70 W/(m·K) [16] for the tool. The simulation results are also affected by the contact settings. In ANSYS, the resistance to solid/solid thermal conduction per unit area at the interface is included as a user supplied real constant value of thermal contact conductance (*TCC*) in W/m²K. The relationship between *TCC*, heat flux and temperature is given by:

$$q = TCC \, (T_{\text{hot}} - T_{\text{cold}}) \qquad (1)$$

where q is the heat flux per unit area (in W/m²), T_{hot} (in K) is the local temperature on the hot surface and

T_{cold} (in K) is the local temperature on the cold surface [17]. The current understanding of the Thermal Contact Conductance under ultrasonic bonding conditions is limited. Thus, the values of *TCC* are difficult to predict. For this reason, the values of *TCC* for the contact tool/wire and wire/substrate are set to 35×10^6 W/(m²K) which corresponds to a thermally bonded contact of two solids. Due to the high thermal conductivity of copper and the relatively low conductivity of the tool material, the heat flow into the wire is more than twice as high as the heat flow into the upper part of the tool.

Figure 4 shows the FE-calculated temperature distribution in the interface after linearly ramping the tool temperature from 22 °C to 400 °C within 35 ms and then keeping it at this temperature for another 300 ms. The resulting temperature in the interface rises to 148 °C. The heat affected area in the substrate is very small compared to a conventional substrate heater. This approach thus enables local heat application to the interface during the bonding process without affecting the surrounding material very much.

In addition, simulations were performed for different tool temperatures between 100 °C and 500 °C to evaluate the effect of the tool temperature on the resulting interface temperature. The results can be seen in Figure 5. As expected, the interface temperature shows a typical bounded growth course. A steady interface temperature is not achievable for typical bond times. This can be explained by the high thermal conductivity of copper and the high thermal flux into DBC and wire.

A stationary analysis was also performed to calculate the temperature gradient in the bonding tool and the transducer. The result is presented in Figure 6. It shows that heat conduction from the heated tool tip only slightly increases the transducer temperature. Thus, heating-induced temperature effects on the transducer are very limited. This is especially important to ensure a stable bonding process over time.

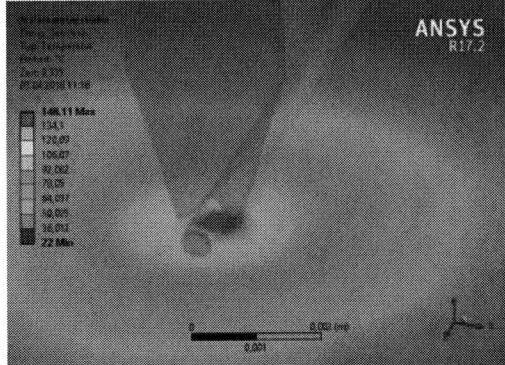

Figure 4: Transient temperature distribution in the interface of wire and DCB, snapshot for 400 °C tool temperature after 335 ms

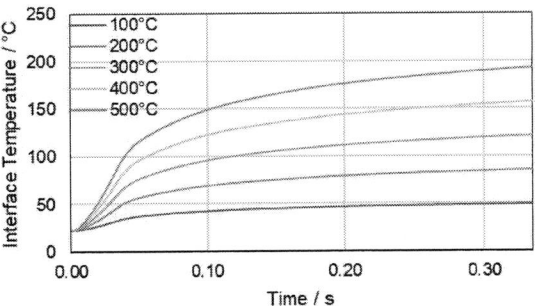

Figure 5: Temperature course in the bonding interface for different tool temperatures (22 °C before 0 s, specified temperature from 35 ms, linear ramp in between)

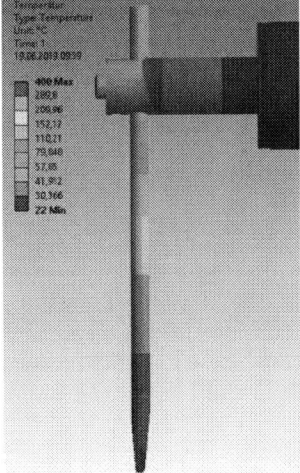

Figure 6: Steady state temperature distribution within bonding tool (slim rod) and transducer tip for a constant tool tip temperature of 400 °C

Tool Temperature Control

To provide the required energy in the real thermosonic process, a near infrared laser source has been integrated into the wire bonding setup via an optical fibre, cp. Figure 7. The temperature is controlled by pyrometry, with the pyrometer projected coaxially into the laser beam by a beam splitter to combine temperature measurement and laser heating in one optical fibre. This fibre ends in a heavy wire bonding tool for copper wires with 500 μm diameter. The pyrometer is connected to an external controller to measure and control the tool tip temperature simultaneously. Controlling the temperature ensures a consistent quality of bond connections. A Hesse Mechatronics Bondjet BJ959 automatic bonding machine equipped with a standard wire bondhead is used for bonding tests on rolled copper plates.

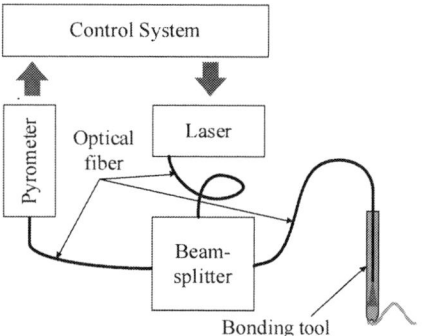

Figure 7: Schematic of the optical design for implementing a process control

For validation of the temperature control, a target temperature profile was defined for the tool without contact and without wire which started at 22 °C, jumped to 500 °C for 14 seconds, and returned to 22 °C. The resulting temperature of the tool tip measured using the pyrometer and simultaneously using a thermal camera. As the results presented in Figure 8 shown, the set target temperature is reached quickly and accurately. At the beginning, rapid warm-up is required, thus the controller quickly increases the laser power, cp. Figure 9. After reaching the target temperature, the laser power required to hold this temperature decreases significantly over time, as less heat is conducted into the already heated rest of system. Once the target temperature drops, the laser is switched off and the temperature slowly falls towards the 22 °C target, as there is no active cooling system. As a reference also shown in Figure 8, the temperature was measured with a thermal camera. Using similar emission coefficients, the data from pyrometer and thermal camera show good agreement.

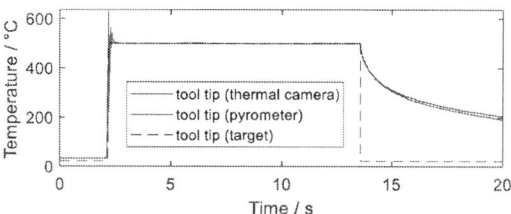

Figure 8: Set and measured temperature during temperature control experiment (free tool tip)

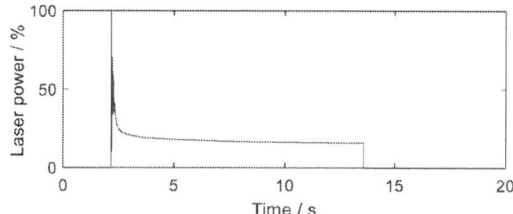

Figure 9: Laser power, set by temperate controller, during temperature control experiment (free tool tip)

Figure 10 shows the results of a similar investigation during a bonding process. Each plateau indicates a bonding phase using a constant tool temperature of 450 °C, ultrasound is applied for 335 ms. Between the heating phases, when the tool moves to the next bond location, the temperature drops to an undefined value. As the next bond starts, the temperature quickly rises to 450 °C again. Figure 11 shows a side view of all relevant components during this process, observed using a thermal camera. The results confirm that the high power of the laser directly and quickly heats up the tool tip which then heats wire and substrate.

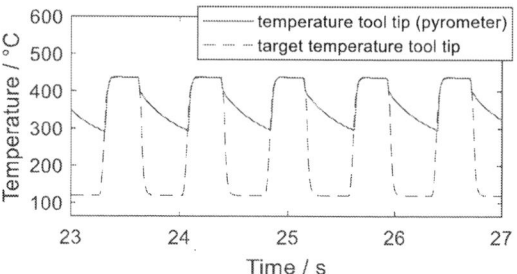

Figure 10: Resulting temperature plateaus during the wire bonding process

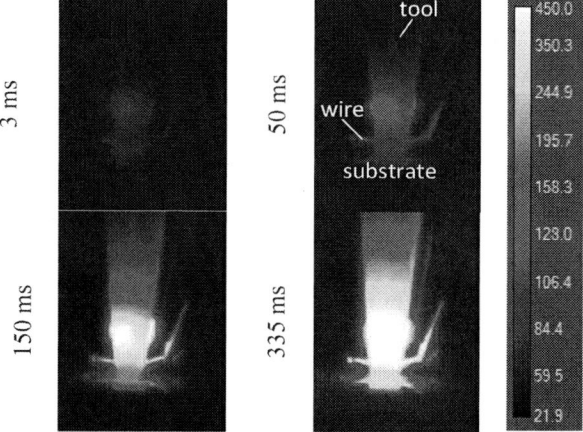

Figure 11: Temporal development of a tool heating process during a source bond (thermal camera images)

Wire Bonding Results with Tool Heating

For evaluation purposes, the mechanical strengths of 500 µm copper wire bonds created using the novel thermosonic process are investigated. Bonding tests showed a very positive influence of the added thermal energy. Figure 12 shows this clearly. The target shear force is approximately 6000 cN. The ultrasonic transducer voltage U_S required to achieve this bond strength in a "cold" ultrasonic process is defined as $U_{S0} = 100\%$. At $U_S = 50\%$, no bond connections were observed. However, if the tool tip is heated to a temperature of 450 °C, a shear force of about 4000 cN is reached already at this low voltage.

With such a heated tool, the target strength is already reached at $U_S = 70\ \%$, doubling the shear force of 3000 cN obtained in the corresponding cold process. At higher ultrasonic power, the gain is naturally reduced, still is approx. 33 % at $U_S = 100\ \%$, where the thermosonic process produces a shear force of 8000 cN. The associated bonds show excellent properties with a full-surface intimate welding, so that the shear test tool sheared through the wire material and the joining zone remained intact ('nugget', Figure 12, top right).

Further tests have proven that instead of reducing ultrasonic vibration power, heating the tool can also be used to reduce the process time while maintaining the same desired bond strength.

T=22°C (without Laser)   T= 450°C (with laser)

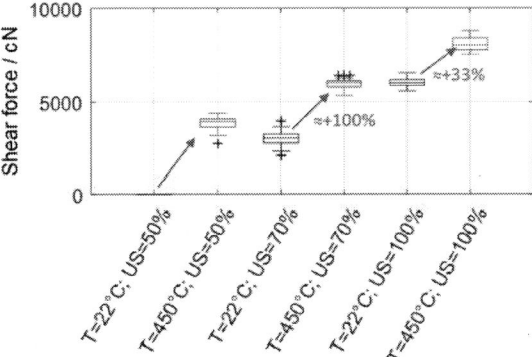

Figure 12: Shear strength of source bonds (500 μm Cu) produced using different ultrasonic voltage U_S and tool tip temperature T

Conclusions and Outlook

Thermosonic bonding, combining ultrasonic and thermal energy to form a bond, has several advantages over cold ultrasonic bonding. But it can currently only be used industrially by heating the whole package through the workholder, which excludes many otherwise promising uses. Lasers can provide high power to a small area and thus are very promising for the use in thermosonic bonding with focused local and temporal heating. The low absorption rate of most bonding materials is a critical drawback for direct laser heating of the bond. This drawback is overcome by heating the tool tip instead, which provides very high and constant thermal energy dissipation thanks to its high absorption (material property) and the special design of the absorption structure (geometry and surface).

In the presented setup, a bonding tool with integrated fibre was used. The fibre provides laser energy to the tool tip and is concurrently used for temperature measurement by pyrometry. The tool tip can be heated up to 500 °C, target temperatures up to this limit are reached precisely by a closed-loop laser power controller. Using a FE model, the effect of a heated tool tip on surrounding materials was analysed. The results show that the temperature of the local interface reaches the desired 150-200 °C in a very short time. This was validated using a high-speed thermal camera.

Investigation of the temperature effect on bond strength shows a significant benefit of laser assisted thermosonic bonding over conventional ultrasonic bonding. At a given bond duration and low ultrasonic power, bonds of significant strength could be formed using the thermosonic process, while no bond formed in the cold process. At larger ultrasonic power, the bond strength increased significantly compared to the cold process.

Thermosonic wedge-wedge bonding using a laser-heated bonding tool is thus a very promising advancement for enabling wire bonding applications on sensitive substrates requiring low normal force or low ultrasonic vibration as well as for reducing process time in any wire bonding process. While not investigated in this contribution, it is also expected to enable connections not feasible in a cold ultrasonic process or increase bondability like in gold bonding.

Further investigations with aluminium and copper wire will be performed in the future to verify the results for industrial applications like semiconductor modules with dies and for challenging processes like battery bonding. Additionally, deep parameter studies will be conducted, e. g. with shorter process times, as well as extensive endurance tests.

Acknowledgements

The contents of this paper are results of the joint research and development project "Inno-WeBo", conducted in cooperation with CeramOptec GmbH and Laser Zentrum Hannover e. V. and funded in the framework of "Zentrales Innovationsprogramm Mittelstand" (ZIM) by the German Federal Ministry for Economic Affairs and Energy due to a resolution of the German Bundestag.

References

[1] G. Harman, "Wire Bonding in Microelectronics", 3rd ed., McGraw-Hill, New York, 2010

[2] C. Ruoff, "A closer look at wire bonding", Charged – Electric Vehicles Magazine, No. 24, pp. 18-27, Mar/Apr, 2016, https://chargedevs.com/features/a-closer-look-at-wire-bonding/

[3] A. Das, D. Li, D. Williams, D. Greenwood, "Joining Technologies for Automotive Battery Systems Manufacturing". World Electr. Veh. J., Vol. 9, 22, July, 2018

[4] Z. W. Zhong, "Wire bonding using copper wire", Microelectronics International 26 (1), pp. 10-16, 2009

[5] A. Coucoulas, "Hot Work Ultrasonic Bonding - A Method of Facilitating Metal Flow By Restoration Processes" Proc. 20 IEEE Electronic Components Conf., pp. 549-556, 1970.

[6] B. Langenecker, "Effects of Ultrasound on Deformation Characteristics of Metals", IEEE Trans. Son. Ultrason. 13 (1), pp. 1-8, 1966

[7] K. Guth, D. Siepe, J. Görlich, H. Torwesten, R. Roth, F. Hille, F. Umbach, "New assembly and interconnects beyond sintering methods", Proc. PCIM, pp. 232-237, Nürnberg, Germany, 2010.

[8] J. Rudzki, F. Osterwald, M. Becker, R. Eisele, M. Poech, "Power Modules with Increased Power Density and Reliability Using Cu Wire Bonds on Sintered Metal Buffer Layers", 8th Int. Conf. Integrated Power Electronics Systems (CIPS), pp. 451-455, Nürnberg, Germany, 2014

[9] G. Singh, A. S. M. A. Haseeb, "Effect of Pedestal Temperature on Bonding Strength and Deformation Characteristics for 5N Copper Wire Bonding", Journal of Electronic Materials, 2016

[10] J. W. Cho, "Wire bonding wedge tool with electric heater", Patent US 5 958 270, 1999.

[11] DeWeyl Tool Company, "Heated Capillary Holder", https://www.deweyl.com/assets/pdfs/heater_tools.pdf, accessed 2019-06-05

[12] Savu, I. D.; Savu, S. V.; Sebes, G.: "Preheating and heat addition by LASER beam in hybrid LASER-ultrasonic welding", J Therm Anal Calorim Vol. 111, pp. 1221–1226, 2013

[13] Liu, Y; Sun, L.: "Laser-heating wire bonding on MEMS packaging", AIP Advances, Vol. 4, Issue 3, ID 031312, 2014

[14] F. Schneider, Y. Long, H. Ohrdes, J. Twiefel, M. Brökelmann, M. Hunstig, A. Venkatesh, J. Hermsdorf, S. Kaierle, L. Overmeyer, "Effect of laser assistance in ultrasonic copper wire bonding", Lasers in Manufacturing Conference, Munich, Germany, June 26-29, 2017

[15] U. Dürr: "Reproduzierbares Laserschweißen von Kupferwerkstoffen", Metall – Fachzeitschrift für Metallurgie, Vol. 62, No. 10, pp. 647-650, 2008 (in German)

[16] Goodfellow, "Tungsten Carbide/Cobalt (WC 94/Co 6) Material Information", http://www.goodfellow.com/E/Tungsten-Carbide-Cobalt-Ceramic.html, accessed 2018-08-08

[17] M. K. Thompson, J. Thompson, "Considerations for Predicting Thermal Contact Resistance in ANSYS", Proceedings of the 17th Korea ANSYS User's Conference, 2007

Effect of bonding temperature on shear strength of joints using micro-sized Ag particles for high temperature packaging technology

Hiroshi Nishikawa*, Myong-Hoon Roh**, Akira Fujita** and Nobuo Kamada**

*Joining and Welding Research Institute, Osaka University, 11-1 Mihogaoka, Ibaraki, Osaka 567-0047, Japan

+81-6-6879-8691, nisikawa@jwri.osaka-u.ac.jp

** KAKEN TECH Co. Ltd., 901 Shimofutamatacho, Higashiomi, Shiga 527-0065, Japan

Abstract

As a die attach process for power modules, we focus on a sintering process using metal particles, which can be operated at a low temperature. However, some drawbacks of this technology still remain. Recently, the bonding process using the micro-sized Ag particle paste has been studied for joint reliability and cost-effectiveness, but it is limited in terms of high applied pressure or relatively low joint strength. In this study, a low pressure of 0.4 MPa was applied during bonding by using micro-sized Ag particles for various bonding temperatures. The micro-sized Ag paste used in this study was composed of chestnut-burr-like and spherical particles with micro-size. Under low applied pressure, the joint with a high shear strength over 50 MPa was achieved when the bonding temperature and time were 260°C and 600 s, respectively. The microstructure, especially the Ag sintered area ratio, was closely related to the shear strength of the joints.

Key words: Micro-sized Ag particles, sintering process, shear strength, bonding temperature,

Introduction

As a recent trend, the silicon carbide (SiC) is of particular interest for semiconductor device. Compared with conventional silicon (Si) devices, SiC devices can operate with significant lower power loss and higher operating temperature, which contributes to miniaturization and higher performance of power devices. The next generation power devices are need to be operated at high temperature and have extensive use in a considerable variety of industries such as the automotive, aerospace and energy production industries. To assemble these power devices, there is a significantly increasing demand on the high temperature packaging and interconnection technologies. Now the European Restriction of Hazardous Substances (RoHS) directive currently exempts the use of high-lead-containing solders for high-temperature soldering. Although Pb-based solder has been used to join Si chips to base materials made of copper, a strong drive exists to find good Pb-free alternatives for high-temperature joining process. Research is needed to establish and characterize a new high-temperature joining process such as die attach process. Several materials and joining processes have been proposed as alternatives to high-lead-containing solders. For example, Au-, Zn- and Bi-based alloys have been investigated for use as lead-free solder, but their widespread use is unlikely because of their inferior properties and high cost [1-8].

A joining process using nanoparticles have been proposed as a solder alternative [9-11]. The sintering behavior of nanoparticles has attracted significant interest because it is well known that nanoparticles of metals have lower sintering and melting temperatures than the bulk metal. When the nanoparticles are heated, sintering takes place between them, producing a bulk-state bonded layer. However, there are some drawbacks of joining process using nanoparticle pastes; for example, it is difficult to produce suitable nanoparticle pastes for the process conditions and the residual organic materials after the joining process can induce unexpectedly large voids in the joint area [12, 13]. To avoid such problems, in our research group, we focus on the joint using an only micro-sized Ag particle paste. In our previous research, a joining process using a chestnut-burr-like micro-sized Ag particle paste was investigated as a replacement for high-lead-containing solders for high-power devices [14]. Under a nitrogen atmosphere, the good shear strengths of the oxygen-free Cu and ENIG disc joints could be obtained by heating at 300 °C for 600 s under a joining pressure of 10 MPa. In this study, a low joining pressure of 0.4 MPa was applied during bonding by using micro-sized Ag particles for various bonding times and bonding temperatures..

Experimental

The mixed Ag paste composed of micro-sized Ag chestnut-burr-like (CBL) and Ag spherical particles was used as a bonding material. This mixed Ag particle paste was provided by Kaken Tech Co.

Fig. 1 SEM images of micro-sized Ag particles. (a) chestnut-burr-like particles, (b) spherical particles

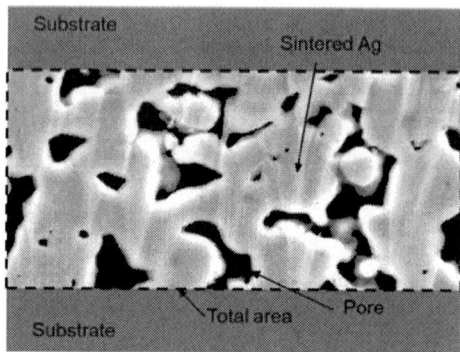

Fig. 2 Definition of sintered Ag area and total area.

Fig. 3 SEM images after heating of mixed Ag particle paste at 300 °C for 60 min in N_2.

Ltd. Fig. 1 shows CBL and spherical particles. The mixed paste was mainly composed of a glycol ether-based solvent and two kinds of Ag particles. The micro-sized Ag CBL particles had an average diameter of approximately 3 μm, and the Ag spherical particles had a diameter of approximately 1 μm. The weight ratio of CBL and spherical particles was 1:1, and the metal content of the Ag mixed paste was 86 mass%.

An electroless nickel immersion gold (ENIG) finished Cu disk was used as a bonding sample. The larger ENIG disk sample had a diameter of 10 mm and a thickness of 5 mm, and the smaller ENIG disk sample had a diameter of 3 mm and a thickness of 2 mm. Before bonding, ENIG Cu disks were cleaned with an ethanol solution for 300 s. The mixed paste was printed with a diameter of 5 mm and a thickness of 150 μm using a stainless steel metal mask on the larger ENIG disk and the small one was placed onto the paste. The pre-heating process was conducted at 130 °C for 300 s to evaporate the solvent and the bonding was performed at temperatures ranging from 200 to 260 °C. The holding time was varied from 0 s to 600 s in nitrogen atmosphere under a low-pressure of 0.4 MPa.

The shear strength of the joint was measured at a shear rate of 0.017 mm/s and a shear tool height of 200 μm from the surface of the bottom sample. The cross-sections of the joint were observed by field emission scanning electron microscopy (FE-SEM, SU-70, Hitachi Ltd.). The Ag area ratio was used as a measure of sinterability of the joint layer, instead of the relative density. The Ag area ratio is defined as the ratio of the sintered Ag area to the total area of the joint observed in the cross-section as shown in Fig. 2.

Results and discussion

In order to evaluate the sinterability of the mixed Ag particles, the paste was heated at 300 °C for 60 min in nitrogen atmosphere. The morphology after heating is shown in Fig. 3. Well connected particles were observed and it indicated that the mixed Ag particles nearby could be sintered in themselves due to large surface area of CBL particles and the smaller particles were helpful for sintering.

Figure 4 shows the average shear strength of the joints using the mixed Ag paste as a function of bonding time. The bonding temperatures was 260 °C under an applied pressure of 0.4 MPa. Even 0 s holding time at the bonding temperature of 260 °C, the average shear strength was 22.3 MPa, which is similar to that of high-lead-containing solders. Then, the shear strength rapidly improved to 36 MPa as the sintering time increased to 60 s. Finally, the joint with a high shear strength over 50 MPa was achieved when the bonding time was 600 s although applied pressure was 0.4 MPa.

Figure 5 shows the average shear strength of the joints using the mixed Ag paste for 600 s depending on the bonding temperature. The average shear strength of the joints bonded at 200 °C was about 20 MPa, which is similar to that of Pb-5Sn solder. When the bonding was conducted at 260 °C

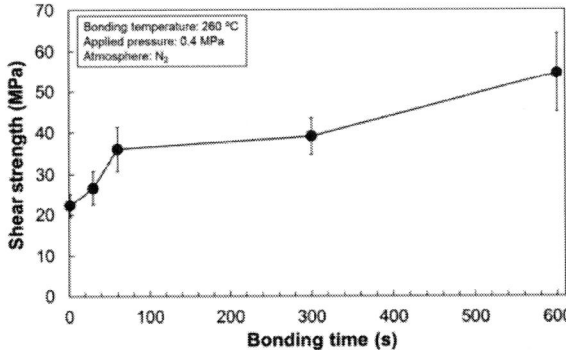

Fig. 4 Effect of bonding time on shear strength of joints using mixed Ag paste.

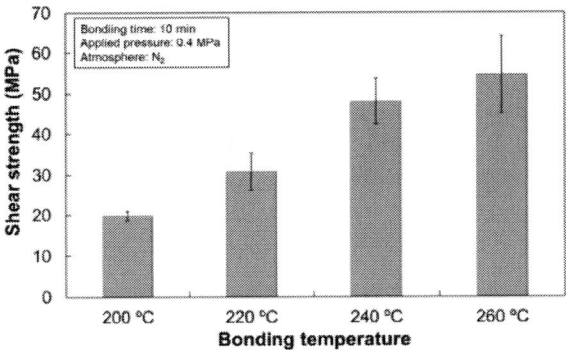

Fig. 5 Effect of bonding temperature on shear strength of joints using mixed Ag paste.

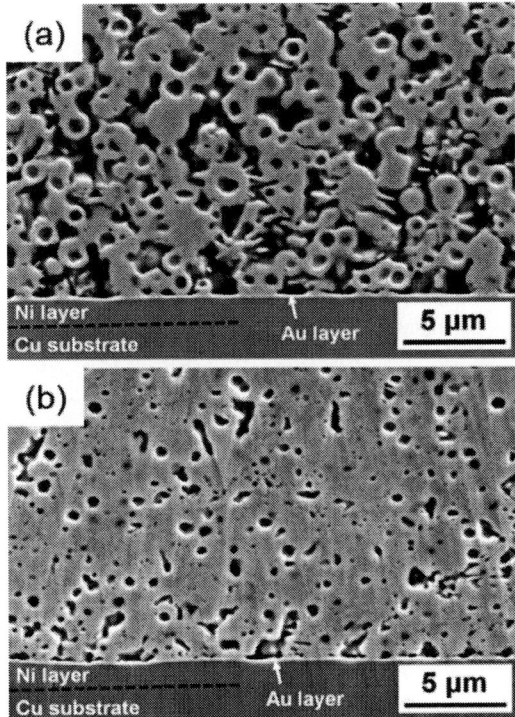

Fig. 6 shows SEM images of typical microstructure of joint using mixed Ag paste for 600 s under applied pressure of 0.4 MPa. (a) 200 °C (b) 240 °C

for 600 s, the shear strength of the joint reached 54.6 MPa. These results indicate that the low applied pressure could serve the low-temperature bonding process with good joint strength, even though the micro-sized Ag particle paste was used.

Figure 6 shows SEM images of the typical microstructure of the joint using the mixed Ag paste under applied pressure of 0.4 MPa. The bonding time was 600 s. Figure 6 (a) shows the typical microstructure of the joint at the bonding temperature of 200 °C and Figure 6 (b) shows the typical microstructure of the joint at the bonding temperature of 240 °C. Ag sintered bonding layer was densified as the bonding temperature increased. Generally, the yield strength of metallic materials is lost when heated, meaning that plastic flow easily occurs at lower stresses at elevated temperatures [15]. Reduction of the yield strength of Ag as a function of temperature was also reported by Carreker [16]. Thus, the density of sintered Ag may be improved as the bonding temperature increased when the pressure was constant. And the shear strength of the joint increased as the bonding temperature incrased because of the denser structure of the sintered bonding layer.

Conclusions

The mixed Ag paste composed of micro-sized Ag chestnut-burr-like (CBL) and Ag spherical particles was applied for the low-temperature

bonding process under applied pressure of 0.4 MPa. The paste was heated at 300 °C for 60 min in nitrogen atmosphere. Well connected particles were observed. When the mixed Ag paste was used, a bonding with the high average shear strength of 54.6 MPa was achieved at the sintering temperature of 260 °C although applied pressure was 0.4 MPa. The microstructure, especially the Ag sintered area ratio, was closely related to the shear strength of the joints.

References

[1] V. Chidambaram, J. Hattel and J. Hald, "Design of lead-free candidate alloys for high-temperature soldering based on the Au-Sn system," Materials and Design, vol. 31 pp. 4638-4645, 2010.

[2] Y.C. Liu, J.W.R. Teo, S.K. Tung and K.H. Lam, "High-temperature creep and hardness of eutectic 80Au/20Sn solder," Journal of Alloys and Compounds, vol. 448, pp. 340-343, 2008.

[3] V. Chidambaram, J. Hald, and J. Hattel, "Development of Au-Ge based candidate alloys as an alternative to high-lead content solders," Journal of Alloys and Compounds, vol. 490, pp. 170-179, 2010.

[4] M. Rettenmayr, P. Lambracht, B. Kempf and C. Tschudin, "Zn-Al based alloys as Pb-free solders for die attach," Journal of Electronic Materials, vol. 31, pp. 278-285, 2002.

[5] N. Kang, H.S. Na, S.J. Kim and C.Y. Kang, "Alloy design of Zn-Al-Cu solder for ultra high temperature," Journal of Alloys and Compounds, vol. 467, pp. 246-250, 2009.

[6] Y. Yamada, Y. Takaku, Y. Yagi. Y. Nishibe, I. Ohnuma, Y. Sutou, R. Kainuma and K. Ishida, "Pb-free high temperature solders for power device packaging," Microelectronics Reliability, vol. 46, pp. 1932-1937, 2006.

[7] T. Takashashi, S. Komatsu, H. Nishikawa and T. Takemoto, "Improvement of high-temperature performance of Zn-Sn solder joint," Journal of Electronic Materials, vol. 39, pp. 1241-1247, 2010.

[8] J.M. Song, H.Y. Chuang and Z.M. Wu, "Interfacial reactions between Bi-Ag high-temperature solders and metallic substrates," Journal of Electronic Materials, vol. 35, pp. 1041-1049, 2006.

[9] E. Ide, A. Angata, A. Hirose and K. F. Kobayashi, "Metal-metal bonding using Ag metallo-organic nanoparticles," Acta Materialia, vol. 53, pp. 2358-2393, 2005.
[10]T.G. Lei, J.N. Calata and G-Q. Lu, "Low-temperature sintering of nanoscale silver paste for attaching large-are chips," IEEE Transactions on Components and Packaging Technology, vol. 33, pp. 98-104, 2010.

[11]H. Nishikawa, T. Hirano, T. Takemoto and N. Terada, "Effect of joining conditions on joint strength of Cu/Cu joint using Cu nanoparticle paste," The Open Surface Science Journal, vol. 3, pp. 60-64, 2011.

[12]J. Yan, G. Zou, A. Wu, J. Ren, A. Hu and J.N. Zhou, "Improvement of bondability by depressing the inhomogeneous distribution of nanoparticles in a sintering bonding process with silver nanoparticles," Journal of Electronic Materials, vol. 41, pp. 1924-1930, 2012.

[13]H. Yu, L. Li and Y. Zhang, "Silver nanoparticle-based thermal interface materials with ultra-low thermal resistance for power electronics applications," Scripta Materialia, vol. 66, pp. 931-934, 2012.

[14]H. Nishikawa, X. Liu, X. Wang, A. Fujita, N. Kamada and M. Saito, "Microscale Ag particle paste for sintered joints in high-power devices, Materials Letters," vol. 161, pp. 231-233, 2015.

[15]R. M. German, "Sintering: From Empirical Observations to Scientific Principles," ELSEVIER, USA, pp. 306-309, 2014.

[16]R. P. Carreker, Jr., "Tensile deformation of silver as a function of temperatur, strain rate, and grain size," JOM, vol. 9, pp. 112-115, 1957.

Opto-electronics flip-chip bonding Automation and *in-situ* quality monitoring.

A. GRIFFART, P.METZGER

SET Corporation

131, Impasse Barteudet
74490 Saint-Jeoire - France

☎ +33(0)450 358 392

AGriffart@set-sas.fr

We are witnessing the growth and needs for solutions allowing the assembly of optical components, together on their sub-mounts. The solder quality, the yield and the throughput are often more important than the ultimate accuracy that the equipment can provide.
We foresee a technological limitation: It is not efficient to work with fiducials in this field, it is always better to align the (sub)device, using one or several of its active features. In this paper, we aim to show how a smart teaching of the vision software allows reliable and accurate recognition. We also emphasis that a high-quality machine produces in-situ and real time information that can prevent destructive control in many cases. Those two topics are explored with post-bonding sub-micron accuracy need mindset, and we show how this target allows to understand better the production yield topic.

Key words: Flip-chip, High accuracy, Opto-electronic, Automatic process, Monitoring.

Smart teaching for automatic recognition

Reduction of interconnexion pitch, smaller die dimensions and improvement in die performances require better bonding accuracy. New technological challenges appear, in particular for alignment accuracy and in advanced packaging.

Arising of the need

Nowadays, the demand for a submicronic post-bond accuracy is more and more frequent. It is already required for Focal Plane Arrays (FPAs). To achieve such high accuracy, several criteria must be reached together *[5]*:

- Accurate machine

- Fine fabrication

- Fine vision tools

- Stable clean room

After reaching these prerequisites and proving the feasibility, the question is to lower, as much as possible, the cost without compromising the final results.

The answer is partly on the fabrication side and mainly on the vision tools, teaching of references and alignment method employed, as it is easier to adjust one or few fabrication steps or re-teach a new alignment method, rather than changing clean rooms or equipment.

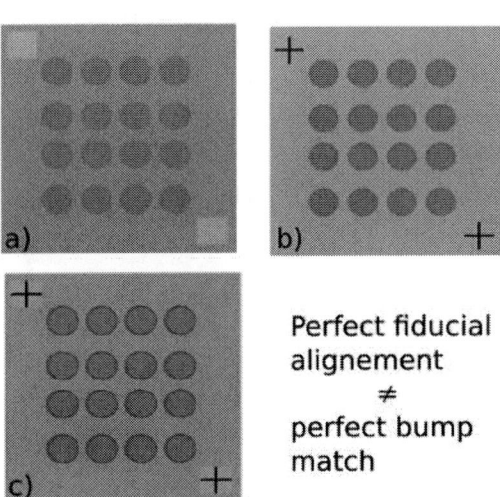

Figure 1: Alignment on fiducial vs alignment on active features.

This simple schematic represents an FPA a), and its asic b), that need to be bonded one onto the other. On the c) picture, we see an overlapped image of the alignment result.

This drawing allows to understand that, in some cases, discrepancies may occur between alignment marks (lithography) and bumps (fabrication). We notice the offset between the upper and the lower bumps. In this case, it is better to perform the alignment on the bumps themselves rather than on the squares and crosses. It is less costly than adjusting the lithography.

Smart teaching need for opto-electronic devices.

With the growth of photonic and optoelectronic, the packaging machines must handle smaller, denser and more demanding components.

It is more frequent to have optical emitters which must be bonded face down, to inject light directly into integrated waveguides. Therefore, the active alignment method becomes much less useful, as passive alignment of well-chosen areas give very good results.

The **Figure 2** gathers drawings explaining how optical flip-chip assemblies require this alignment on active features for the best results. In a) we show the optical emitter in red. And it's fiducial in black.

In b) we represent the grating coupler that will inject light to be guided in the device.

When looking at the a) drawing we see that the center position of the emitter and its fiducial does not match. Under we show a simple light mode which need to overlap with the grating so signal is efficiently transmitted.

In c) we show that if we align on the fiducial the mode will not overlap perfectly the grating and some signal losses are expected.

In d) we show that an alignment on active features, i.e the grating and the emitter, we have perfect matching of the mode and the grating coupler.

This example enlightens the new needs in opto-electronic flip-chip bonding. Those two devices need to be bonded in such a way that the light is coupled directly from the emitter into integrated grating.

Here, the light coupling ratio is directly linked to the post bonding accuracy. There is no possibility of tuning after soldering. As high bandwidth is more and more often requested for communications (in particular for 5G applications), devices have to integrate more and more optical components.

Even with fiducials of good quality, any deviation on fabrication will result in yield loss. This yield loss can be prevented by recognizing directly the "active feature" on the samples.

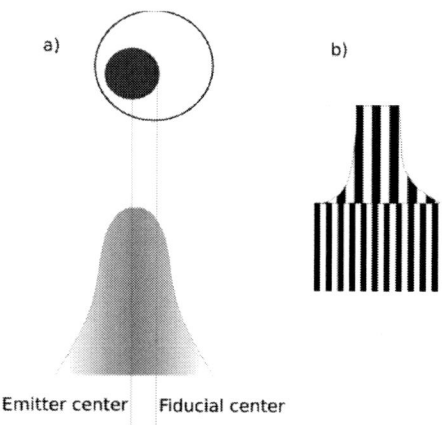

Emitter center   Fiducial center

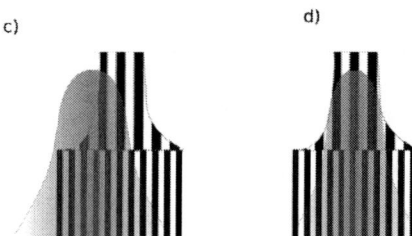

Figure 2: Example of opto-electronic flip-chip application with direct light coupling needing smart teaching.

EMPC-2019 A.GRIFFART

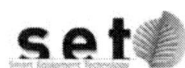

How it can be performed

As said above, the teaching is a critical step where the operator needs to evaluate which features will be the alignment reference and to choose a recognition method.

Basic guidelines for features selection could be to have a unique shape within the field of view and to have it reproducible among a components batch. Naturally, a sharp feature with a good contrast warrants better alignment results.

Usually, when preparing a full auto process for a batch, some deviation among the samples is observed. Therefore, the choice of the best suited feature on the samples and the best method is crucial.

This **Figure 3** presents a basic example to work around deviation of the chosen feature. Shapes are squircles with different order, the order value is given under each plot. We use the COGNEX® recognition system integrated in our machines, and the images were produced with Scilab®. In a) we see the image we use to teach the software, it is a squircle with an order of 1 000 000, that is considered as square.

Picture b) and c) are then recognized with a fitting function that was used to teach the square. Pictures d), e), f) are recognized with an edge finding function used to do another teaching of the a) image. The goal is to find the limit of squircle recognition when the software expects a square, for two different methods.

The first method allows to recognize squircle with an order down to 50, then the result is no more reproducible and fails dramatically for orders under 30. The second method allows to have fine recognition for any squircle order. This simple example based on plotted images shows how fluent production and high accuracy can be reached even with some deviation through a smart teaching.

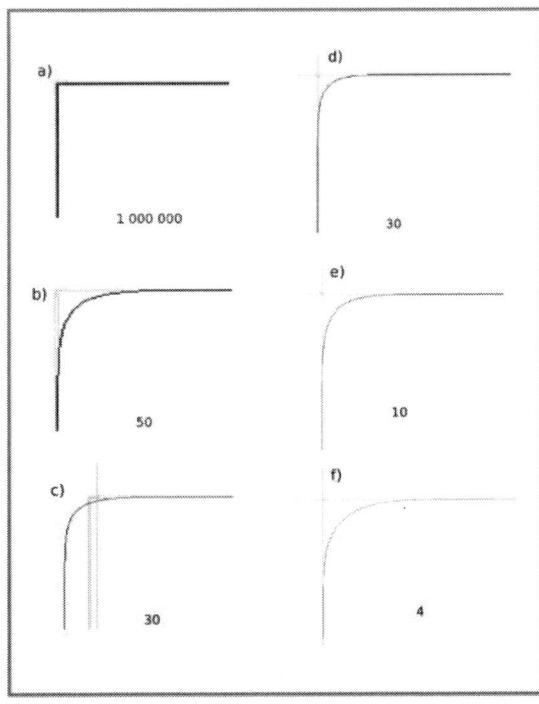

Figure 3: Results of testing two different vision tool methods when recognizing different squircles after teaching a square.

Process monitoring as *in-situ* quality control

Either in development or in production, having a way to ensure the bonding quality without increasing the total cycle time or destroy the components is an asset.

We often see researchers or production people struggling to interpret daisy chain measurement, having to wait for cross section imagery or other types of tests. For example, for reflow processes, it is common to have a visual inspection of the solder in order to evaluate both the eutectic and the wetting. Here, we propose a simple quality assessment based on the machine sensors and actuators. Indeed, to achieve high quality and high accuracy flip-chip assemblies, the machine must monitor itself thru several parameters with closed-loop systems. This gives a chance to produce *in-situ* information on the process and to extract the data during the bonding process.

It can be used both as a development help and as a production assessment.

Reflow solder quality control

We present an example of how this process recording allows to witness invisible changes on the material. This is based on production samples bonded with an Au/Sn, 80/20 (% weight) solder. In the two following graphs, we only plot the force and the position of the vertical motion bonding arm. Both bindings are done with the same force and temperature parameters. The only difference comes from the components. The sampling time is 0.1 second.

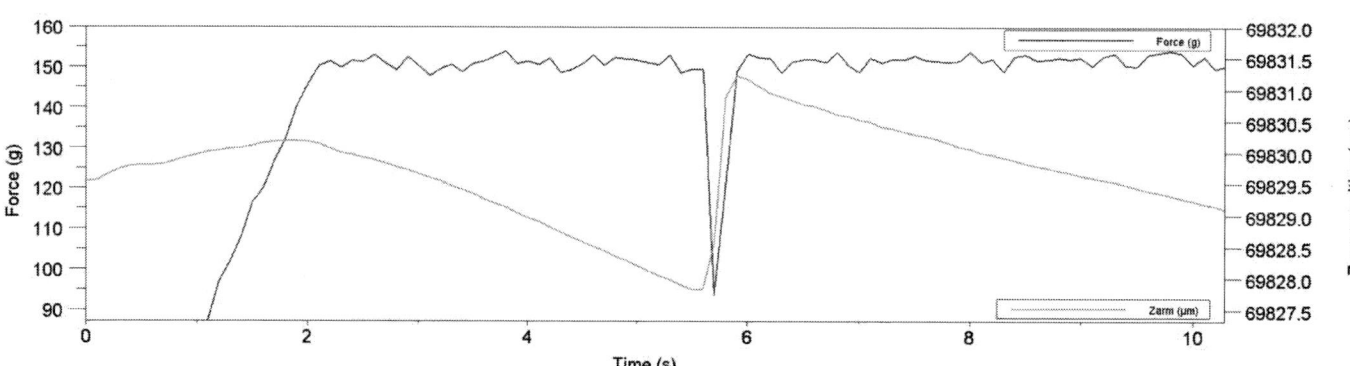

Figure 4: Plot of the reference behavior.

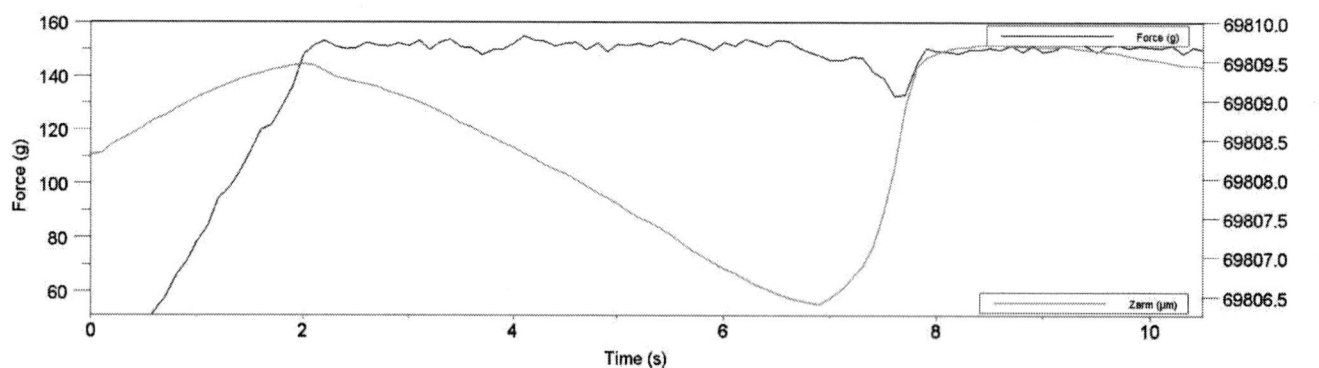

Figure 5: Plot of an oxidized solder behavior.

EMPC-2019 A.GRIFFART

On the **Figure 4** monitoring graph, we see the proper behavior of the solder pad. This correlation between force drop and arm motion is always the same: drop force duration, minimum force, Z displacement during force drop.

This graph resumes the melting behavior of the solder. When the temperature reaches the liquidus, there is no more mechanical resistance and the force drops (blue curve, units on the left vertical axis), in parallel the arm re-adjusts its position to match the required force (green curve, units on the right vertical axis). On the left and right part of this drop, we see the motion of the arm compensating the thermal expansion of components under temperature ramp.

By reading the graph, we determine that the solder height has been reduced by 3 μm during the reflow. Even with a force control process, the reliability of the samples and the machine allow to reproduce this value.

The **Figure 5** graph represents the same process as in **Figure 4** with the same samples. Before going for bonding, the solder was left 3 minutes under alignment temperature without the protective forming gas (H_2 5% in N_2).

The behavior of a diffused/oxidized solder is no more the same. Looking at the drop, we see that it did not occur in the same way. Comparing with the reference behavior, we can expect deviation on the results, and therefore less performance on this device (higher solder resistivity, voids,etc…).

Moreover, we can check for the melting temperature of the solder looking at the process monitoring data:

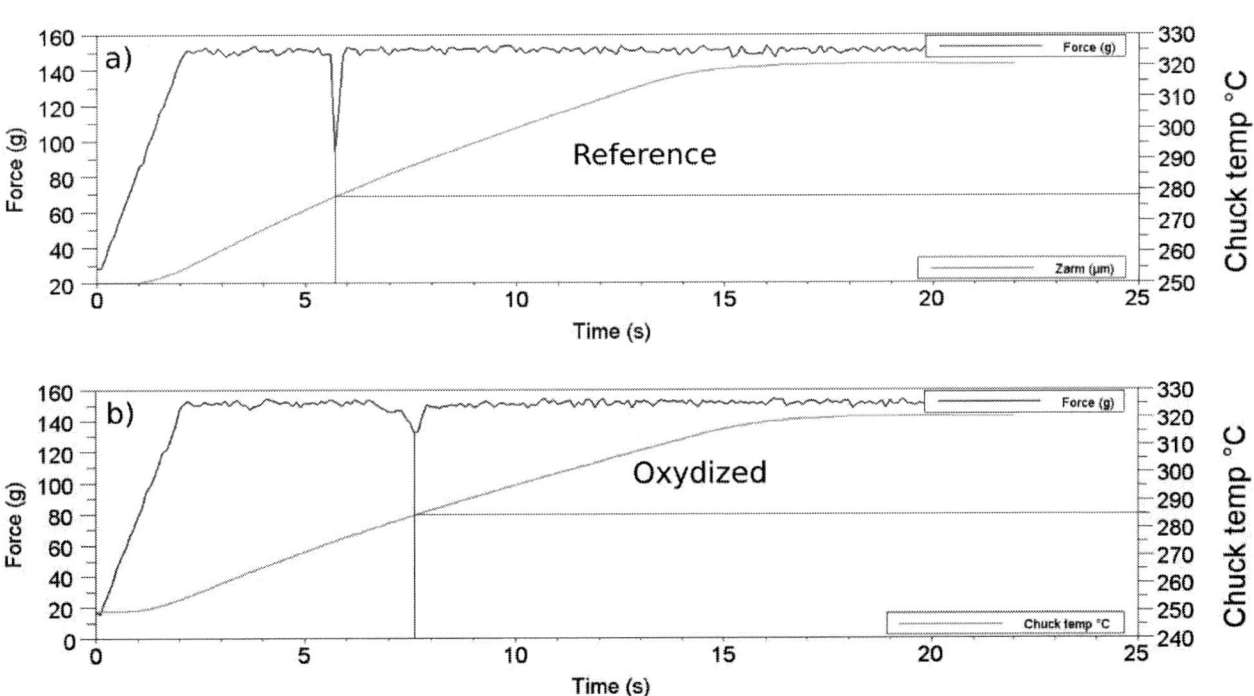

Figure 6: Comparison of melting temperature for the reference and oxidized solder.

The **Figure 6** is a plot of the force and chuck temperature over time. The blue curve (units on the left vertical axis) correspond to the force, and the green curve (units on the right vertical axis) is the chuck temperature.

We can relate the melting temperature (measured by the machine sensors) when pointing the minimum of the force.

We can check the corresponding chuck temperature (for this process, the arm and chuck temperature ramps were the same).

The a) upper graph indicate 278°C when the b) lower graph gives 284.9°C.

This kind of test can be used to check the solder quality among a batch and to allow the machine user to have information on the machine temperature calibration state.

We monitor the melting time vs heating chuck/arm temperature. Experimentally, with the flip-chip bonder, we match the theoretical melting point of the alloy.

This simple example allows to understand how this process monitoring is useful for both production and development. Deviation on the material will result in deviation on the graphs.

Position controlled processes

Depending on the application, it can be much better to bond with a Z control, instead of a force control. In other words, maintaining a controlled gap between chip and substrate, instead of controlling the bonding force, can give better bonding quality results, as avoiding any short-cut between adjacent bumps which are reflowed during the process.

Depending on the bump size, material and surface nature, the effect of surface tension can lead to different wetting degrees. Bump shaping for image sensors or laser bar soldering aims opposite result in term of solder wetting. So, it is crucial to be able to tune this effect only by the machine parameters.

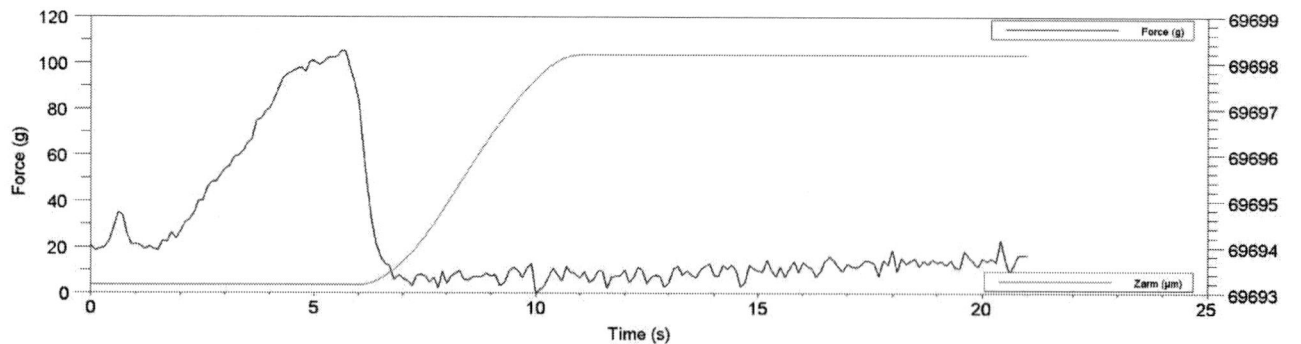

Figure 7: Plot of a Z-position process

This **Figure 7** shows the monitoring of the force and the position during the bonding. For this process, we program the machine to press down 5 μm after the thermal ramp is done, from the contact position. The force drop comes from the same physical effect: the bumps are melting at this moment.

As we drive the arm motion and not require any force control, this is only a drop after the melting.

Conclusion

In this paper, we have overviewed two important topics for the flip-chip activities: How automatic alignment need smart teaching to ensure best results with the lowest cost. And how a flip-chip bonder not only produce assemblies but also *in-situ* information on the components.

Approaches used in image sensor production can be transposed for optoelectronic application. This allows to reduce dramatically the dependence of the yield regarding some fabrication deviation. Also, given the growth of optical component integration, recognizing active features at the alignment step is a promising approach to ensure the bandwidth evolution of telecom devices. This method already found usage for 5G devices production.

High quality flip-chip bonders not only perform the assembly process but also produce valuables data thanks to the process monitoring. Factories and foundries nowadays use and need improvement on "4.0 production" concept. SET flip chip bonders already integrate all these functions.

References

[1] C. C. Lee and R. W. Chuang, "Fluxless non-eutectic joints fabricated using gold-tin multilayer composite," *IEEE Trans. Components Packag. Technol.*, vol. 26, no. 2, pp. 416–422, 2003.

[2] C. C. An, M. H. Wu, Y. W. Huang, T. H. Chen, C. H. Chao, and W. Y. Yeh, "Study on flip chip assembly of high-density micro-LED array," *Proc. Tech. Pap. - Int. Microsystems, Packag. Assem. Circuits Technol. Conf. IMPACT*, pp. 336–338, 2011.

[3] J. W. Yoon, H. S. Chun, J. M. Koo, and S. B. Jung, "Au-Sn flip-chip solder bump for microelectronic and optoelectronic applications," *Microsyst. Technol.*, vol. 13, no. 11–12, pp. 1463–1469, 2007.

[4] P. Metzger, D. Ph, J. Macheda, M. D. Stead, and K. A. Cooper, "DEVELOPMENT DONE ON DEVICE BONDER TO ADDRESS 3D REQUIREMENTS IN A PRODUCTION ENVIRONMENT," pp. 3–7, 2009.

[5] C. Avrillier and P. Metzger, "Flip-chip bonding: How to meet high accuracy requirements?," *EMPC 2017 - 21st Eur. Microelectron. Packag. Conf. Exhib.*, vol. 2018-January, no. September, pp. 1–6

Annexe:

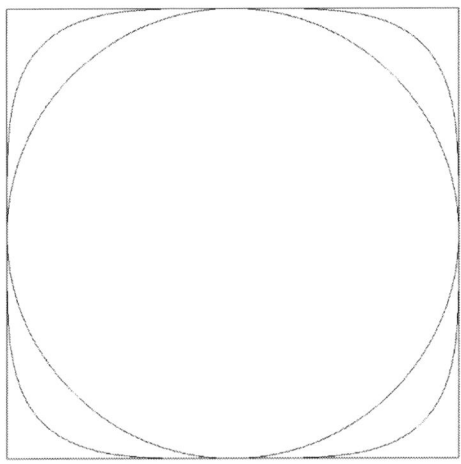

Figure 8 Squircle concept

This Figure 7 intend to explain simply the squircle concept. From the inner to the outer of the figure, we have: a circle, a squircle and a square.

It is easy to see that the squircle is in between the circle and the square.

From the circle definition equation:

$$X^2 + Y^2 = r^2$$

The squircle is defined as

$$X^n + Y^n = r^n$$

With *n=4* in the Figure 7 plot.

Then if we increase the order *n*, the shape tends to the square.

Flexible Electronics Printing by Electroplating

By Dr. Igor Kadija

ECSI Fibrotools, Inc. 7617 Astoria Place Raleigh, NC 27612 USA

1(201) 786-3122 ikadija@fibrotools.com

Abstract

In many electronics' industrial applications deposition of metal on isolated electronic structures is necessary to deliver functional products. Examples are Flexible Electronics, PWB surface finishing and IC manufacturing. The typical state-of-the-art approach by the industry is to apply electroless or immersion processing or, by application of metal containing pastes, both burdened with inherent problems. Flexible Electronics, having no common current collector, are predominantly produced by inkjet or screen printing of metal and carbon pastes over a flexible substrate. Having inferior conductivity compared to pure metals to achieve adequate performance, the pastes are significantly thicker and bulkier than the pure metal. It is desirable to have an efficient way for electroplating patterned isolated electronic structures, if possible without the seed layer. This calls for establishing electrical contacts with all isolated structures without interfering with the electric field access to all structures that need electroplating. Contact Electroplating Technology (CET), US Patent #10,184,189 [17], has been demonstrated with isolated patterned structures of various substrates including Flexible Electronics. Following a high-speed xeroxing or patterning of isolated electronic structures with a layer of negligible thickness and resistance as high as several kiloohms cm, the Electrochemical Printing Technology (EPT) enables upgrading of the same by deposition of pure metal of desired thickness up to 10 microns. Completely isolated electronic interconnects with spacing as low as 25 microns were uniformly electroplated within seconds with Ag, Cu, Ni and Au.

Keywords: flexible, electronics, electroplating, printing

Technology

The state-of-the-art production of flexible electronics is complex. It includes solvents, chemicals and high temperature treatments of inks or pastes and polymer structures to achieve functional products. Typically, inferior properties of the metal pastes, such as low conductivity and excessive thickness, limit the progress of the industry, particularly in the field of high-density flexible electronics. To optimize the process, recently new approaches were demonstrated in Flexible Electronics, such as sintering of metal nanoparticles, such as silver. For example, U.S. Pat. #9,343,233[12], describes a method of depositing a suspension of silver nanoparticles followed by sintering at 170-180° C. for 30 to 40 minutes. Again, the process remains complex and time consuming and includes pushing the substrate performance to the limits. More recently, efforts have been made to minimize chemical and thermal processing of the pastes. For example, ink jet printing technique is used for depositing nanoparticles of Cu, followed by drying, precuring at 100C and fusing with photo processing; Li, Yunjun et al. US Patent #10,231,344[24], The process still involves chemical and thermal

reactions/processing while delivering an inferior product. While much progress has been made in the fields of electroless and immersion plating, both processes are spontaneous and extremely slow, approximately 10-50 times slower then electroplating, ultimately delivering an inferior product.

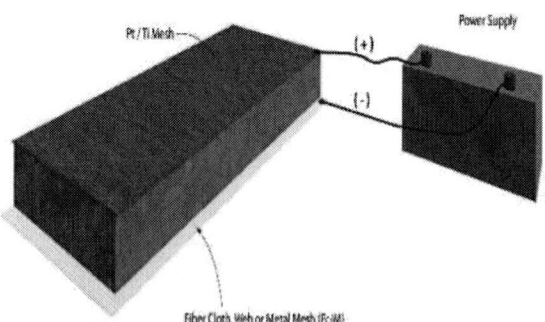

Figure 1. Schematics of the CET concept

For several decades' efforts to electroplate isolated features have been made in the past, including the author's and others [7,8,10,14-16].

None of these resulted in an efficient and industrially applicable solution. More recently a novel and potentially practical approach [17,18] is being made by ECSI Fibrotools. By applying the Contact Electroplating Technology (CET), (Figures 1 and 2) with up to several million electrical contacts per square inch, each isolated structure is simultaneously contacted and cathodically charged by the main power supply via the CET's working electrode.

The CET application in EPT addresses issues specific to flexible manufacturing. In flexible electronic manufacturing, low grade metal pastes are used in order to overcome the absence of peripheral electrical contact. This comes at the expense of line resolution capability which stands in the way of flexible electronic evolution. A 20-30 micron paste line is necessitated where a 3-micron thick pure metal line would suffice (10:1).

For EPT implementation, the substrate to be used as a precursor to electroplating must be patterned by Xeroxing, screen printing, ink-jet patterning, laser ablation or any other patterning technique. As long as these provide sufficient adhesion and have a resistance of few Kiloohms-cm or less, the patterns can be electroplated by EPT. The CET working electrode is part of a sandwich structure which consists of the counter electrode (perforated or expanded Pt/Ti mesh); an open porosity, chemically stable polymer-structure sponge that is wettable and the CET electrode itself. It is a highly conductive Fiber Cloth, Web or Metal mesh (FC/M) with the number of contacts points ranging from app. 200 psi to 5,000,000 psi plus, depending on specific pattern requirement. With EPT each isolated pattern is simultaneously charged by the main power supply via the CET electrode. Electroplating occurs directly on each patterned feature enabling well-controlled metal electroplating.

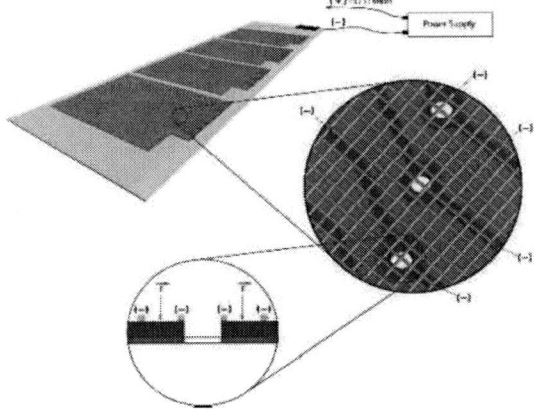

Figure 2. Schematics of the electrical contact nd the electric field distribution.

In order to provide enough supply of electrolyte and a satisfactory statistical distribution of contacts across the pattern, the sandwich can be reciprocating over the substrate. Reciprocation ensures that electrolyte is continuously exchanged at the interface. At the same time the reciprocation can be made either by constant contact combined with pumping action through the sponge or, can be repeatedly detached across the pattern thus allowing for a more even distribution of deposit.

While processing the patterned substrates, metal deposition takes place over the FC/M electrode as well as over the pattern on the substrate. For most practical situations, the area of the pattern to be electroplated with metal is significantly smaller than the area of the FC/M which, by design, is the same or larger than the area of the flexible substrate. Hence the portion of the current utilized in depositing the metal over the pattern is typically a small fraction of the total current. Under these circumstances, several options are available for managing deposit distribution and maintaining optimal operating capability: (1) Reverse pulsing, (2) Flipping the sandwich and, (3) Roller plating combined with Reverse pulsing technique.

(1) Reverse pulsing

Two variations of RP can be applied; a) Constant contact RP synchronized with reciprocating time and frequency and, b) RP application synchronized with "off" contact with the substrate.

With either option of electroplating, cathodic portion of the pulse is performed with the current density defined by the optimal conditions for the metal deposition. For a constant contact a), anodic portion of the pulse is designed for striping (or at least maintaining) the deposit on the minimum level.

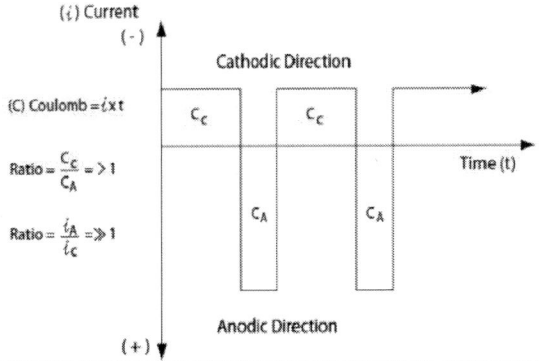

Figure 3. Schematic presentation of the reverse pulsing technique needed for maintaining constant contact in CET operation.

The technique is known in the art of plating

particularly in PWB processing. It has been practiced for many decades to remove excess plating at the excess metal deposit over the FC/M. The anodic pulse is typically designed with a much higher current density for a shorter time to reduce the excess metal growth from the edges of the through holes and at the sharp corners of the pattern. Figure 4. Shows the electric field "edge effect" and the potential use of the same to maintain process continuity. As a parallel to this condition for PWBs one can consider the composite structure of the FC/M/substrate as being an electroplated substrate with sections prone to "edge effect". While in constant contact with the substrate the above RP mechanism allows electroplating of PWB while removing most of deposit from the FC/M electrode.

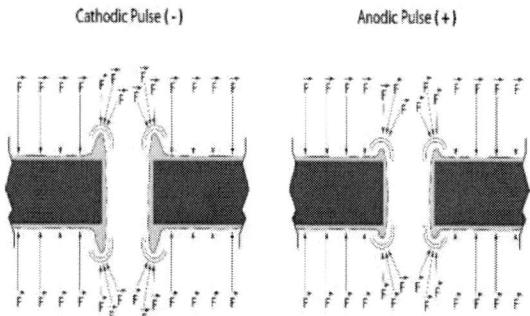

Figure 4. Anodic pulse preference for a higher electric field results in the removal of excess metal.

For option b), while in contact with the PWB substrate FC/M is charged as a cathode thus enabling electroplating of the substrate and FC/M. While detached from the PWB it is anodically polarized to remove extra metal deposit from FC/M only.

(2) Reversing the sandwich

Two identical FC/M electrodes can be employed. While operating as a cathode without any reverse pulsing, the working FC/M may reach the point of saturation with the metal. By flipping the sandwich and reversing the polarity, the same electrode can be used as a counter electrode and provide supply of metal ions by anodic dissolution. At the same time the previous counter-electrode can be utilized as cathode.

(3) Roller plating option

Roller plating option, Figure 6., with RP implementation leads to a practical EPT version for flexible electronics conveyorized processing. A series of CET sandwiches wrapped up in a roller enable RP method application for a conveyorized

system for Flexible electronics production. Due to the rolling action the level of immersion in the electrolyte can be maintained on a level not exceeding the thickness of the sandwich structure.

Regarding the points of contacts effects on uniformity, it was difficult to identify specific section of the surface that can be attributed to the contact. Using Olympus LEXT microscope, we have conducted a series of tests and 3-dimensional microscopic analysis with various IPC substrates. The evenly distributed random surface roughness of the order of several hundred nanometers made it impossible to attribute the specific points to the contacts made during electroplating. Nevertheless, no functional effect or correlation regarding the contact points could be determined.

CET Design Options

Reverse pulsing option (1), including both the continuous contact a). or reciprocating contact b), is an ideal combination for batch processing of items such as Direct Bond Copper (DBC) metallization, Ball Grid Arrays (BGA) and others. The items can be laid in a tray in large numbers to be electroplated in a batch production process.

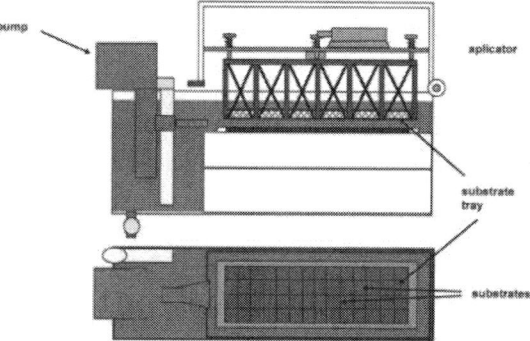

Figure 5. CET implementation for single or multiple small items batch processing.

EPT processing can be implemented in batch processing to enable electroplating single or multiple smaller area substrate. It can be modified for conveyorized processing by synchronizing the conveyor motion with EPT system actions. This option can be of interest to independent and captive R&D and Academic institutions involved in technological advancements.

Conveyorized processing is a viable option with all three versions. Option (1) can be synchronized to enable conveyor stepping motion of flexible substrate during "OFF" times of electroplating. So too, the flipping motion (2) of the EPT sandwich can be synchronized with conveyor stepping motion. Roller Plating Option

(3) is synchronized with Roller motion to enable electroplating as per required specs for any product.

Figure 6. shows the conveyorized EPT roller schematics for flexible electronics application. While the conveyor is moving the substrate through the processing chamber, the sequence of several rollers enables continuous electroplating of metal.

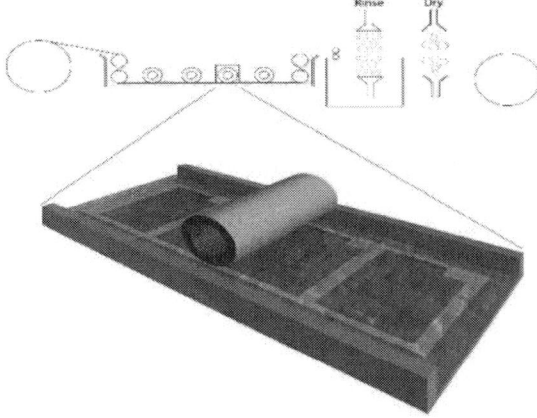

Figure 6. CET Conveyorized processing for PWB and/or Flexible Electronics.

EPT Examples

With the goal of demonstration of Flexible Electronics with EPT process, we have tested and have produced Ni, Cu and Ag electroplated samples of flexible substrates patterned with various techniques. One of the goal was to identify the electroplating uniformity and coverage of isolated patterns.

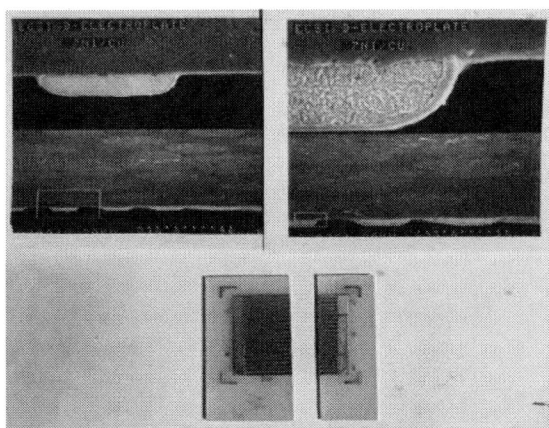

Figure 7. A cross-sectional image of etched standard pattern IPC-24 section shows uniformity and completeness of isolated Cu line encapsulation with 7-micron Ni deposited with CET technique, 200x and 700x.

Figure 7. shows an example of cross sections of a CET electroplated IPC-24 PWB standard etching test pattern. 50 microns thick, 200 microns wide Cu line was electroplated with 7-micron Ni.

Figure 8. shows PWB standard Cu etching test pattern electroplated with Ni. The photo shows uniform Ni coverage across the substrate containing the minimum of 50-micron isolated line/space pattern.

Figure 8a. 50 micron PWB isolated standard etching patterns CET electroplated with Ni.

A detailed 3-dimensional analysis of the pattern was carried out with the Olympus LEXT microscope. Having the capability to identify Z direction with selectivity down to 6 nanometers, this microscope shows deformations on the surface even with substrates of near mirror appearance. Figure 8b. shows summarized reading including Sa average roughness of 182 nm for Ni plated interconnects.

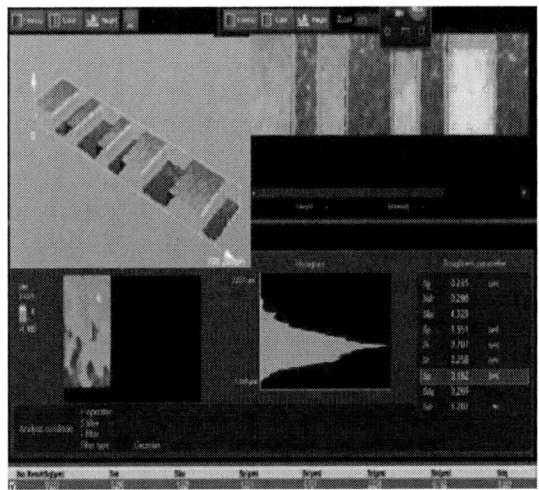

Figure 8b. Olympus LEXT Summary of the 3-dimensional analysis of Ni plated test pattern.

Figure 8c. Black and white 3-dimensional photo of the pattern section. The photo shows the Ni coverage of the interconnects reaching over the footing of the lines.

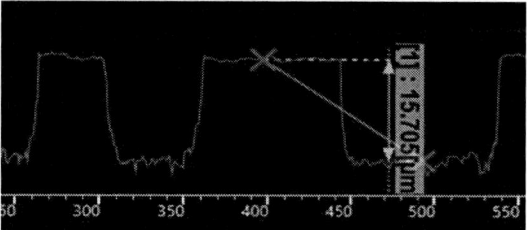

Figure 8d. Profile of the selected pattern indicating the height of 15.705 microns as expected for 1/2 oz Cu laminate.

In flexible electronics the resistors can be produced by applying carbon paste. We have used these patterns to demonstrate the conductivity upgrade by using EPT technology with silver electroplating. Figure 9. shows the results of using carbon patterns over milar produced as Flexible electronics resistors after electroplating with EPT method with silver. After 2 times 4 seconds EPT application a 5 mm long, 1.5 mm wide pattern having 350 ohms resistance was reduced to 0.2 ohms.

Figure 9. Sample of double sided carbon resistor conversion to one side silver wire with EPT processing

Another example of the implementation of the EPT technology is presented in an R&D sample of electrostatic carbon test images in Figure 10. A series of standard test patterns of 4 microns thick carbon patterns was electroplated with silver.

Uniformity in coverage across the various isolated interdigitated samples was achieved down to a 25-micron resolution.

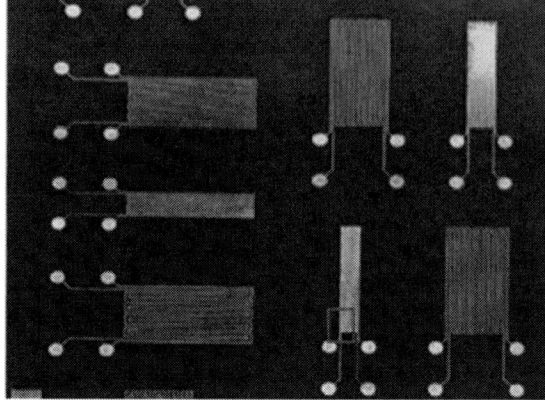

Figure 10a. Interdigitated PWB standard carbon test patterns after CET plating.

Figure 10b. Nominally 25-micron spacing CET electroplating of standard carbon patterns. Scale bars represent 100 microns.

EPT electroplating of flexible electronics devices was demonstrated on Laser Formed Electronic Device Patterns. Figure 11 shows the silver-plated high-density pattern formed by Laser Ablation of 100 nm Cu/milar laminate. The example shows the optimization from 100 ohms resistance over 5 mm distance and 2 mm wide line down to 0.01 ohms in 10 seconds after EPT plating.

342

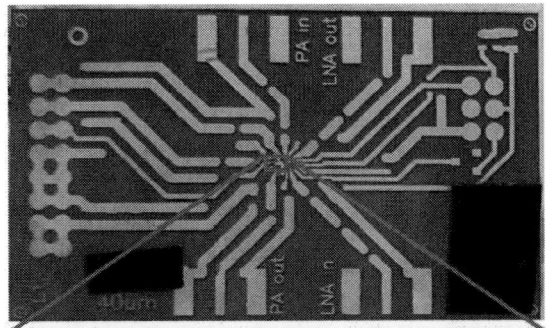
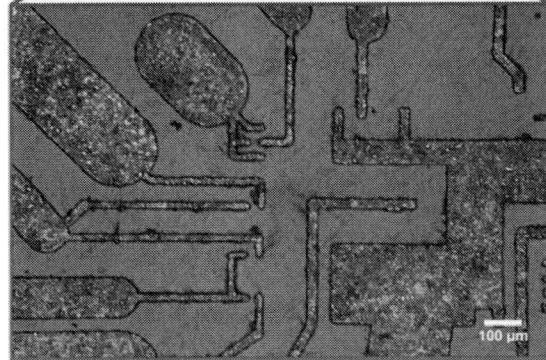

Figure 11. 40-micron resolution Laser Ablated sample of Cu/Milar laminate electroplated with silver brought resistance from 100 ohms to 0.01 ohms in approximately 10 seconds

Conclusions

The electroplating process is fundamentally the most desirable expedient way to deposit high quality metal on microscopic and nano size features. It was demonstrated that EPT enables well-controlled and verifiable quality of Flexible Electronic electroforming. The process is simple and can be applied to a wide range of electronic circuits and device applications.

References

[1] Bladon, J. J., (1996) A palladium sulfide catalyst for electrolytic plating. *J. Electrochem. Soc.*, 143, 1206-1210.
Journal Article

[2] J.O'M. Bockris and A.K.N. Reddy, *Modern Electrochemistry*, Plenum Press, 1970.
Book

[3] Boyd et. al., Dec. 2006, Plating Solutions for Electrochemical or Chemical Deposition of Copper Interconnect and Method therefor, US Patent 7,147,767.
Patent

[4] Chi-Chao Wan, A Review of the Technology Development of Direct Metallization, Proc. Natl. Sci. Counc. ROC(A) Vol. 23, No. 3, 1999, pp. 365-368.
Book

[5] Hanna, F. et al., Controlling factors affecting the stability and rate of electroless Copper plating, *Materials letters*, Vol. 58, 1-2, Jan.

2004 pp. 104-109.
Journal

[6] Hwang, et. al. Nov. 2016, Semiconductor package and method of forming the same, US Patent 9,484,292,
Patent

[7] Kadija, I., Method and Apparatus for Manufacturing Interconnects with Fine Lines and Spaces, US Patent 5,024,735.
Patent

[8] Kadija, I., Method and Apparatus for Manufacturing Interconnects with Fine Lines and Spaces, US Patent 5,114,558.
Patent

[9] Polakovic, F. and A. M. Piano, Process for preparing the through holes of a printed wiring board for electroplating, US Patent, 4,622.108.
Patent

[10] Schroeder R. et. al., June 2000, Method and Device for Electrochemical Treatment with Treatment Liquid of an Item to be Treated, US Patent 6,071,400,
Patent

[11] Suzuki et. al., Sept. 2004, Method of Manufacturing Printed Wiring Boards, US Patent 6,796,027,
Patent

[12] Tentzeris, et. al., May 2016, Additively deposited electronic components and methods for producing same, US Patent 9,343,233.
Patent

[13] Yang et. al. July 2005, Method of Plating Metal Layers over isolated pads on semiconductor package substrate, US Patent 6,916,685.
Patent

[14] Yang et. al., July 2012, Method for Fabricating Back end of the Line Structures with Liner and Seed Material, US Patent 8,232,195.
Patent

[15] Zaban et. al. June 2015, Nickel- Cobalt alloys as current collectors and conductive interconnects and deposition thereof on transparent conductive oxides, US Patent 9,064,985,
Patent

[16] Zhang, T. et. al., A Laser Printing based Approach for Printed Electronics, Applied Physics Letters 108, 103501 (2016).
Article

[17] Kadija, I., Apparatus and Method of Contact Electroplating of Isolated Structures, US Pat. 10,184,189, February 22, 2019.
Patent

[18] Kadija, I., Electrochemical Printing Technique (EPT) for Flexible Electronics, 2017 TechConnect World Innovation Conference, Washington, May 15-17, 2017.
Article from conference proceedings

[19] Goy CB1, et. al, Electrical characterization of conductive textile materials and its evaluation as electrodes for venous occlusion plethysmography. 2013 Aug;37(6):359-67, Elsevier

Journal Article

[20] The Characterization of Conductive Textile Materials Intended for Radio Frequency Application, IEEE Antennas and Propagation Magazine 49(3):28 - 40 July 2007.

Journal Article

[21] Peng, S. et al., The sheet resistance of graphene under contact and its effect on the derived specific contact resistivity, Carbon, Vol. 82, February 2015, Pages 500-505

Journal Article

[22] Rigosi, A., et al. Electrical Stabilization of Surface Resistivity in Epitaxial Graphene Systems by Amorphous Boron Nitride Encapsulation, ACS AuthorChoice.

Book

[23] Klapproth, K., Fast, Cheap, Molten Metal 3D Printing, New Materials & Applications, Nov. 2017

Book

[24] Li et al., Metallic Inc, March 2019, US Patent 10,231,344

Patent

Graphene-coated copper nanoparticles for thermal conductivity enhancement in water-based nanofluid

A. Hafid Zehri[1], Andreas Nylander[1], Lilei Ye[2], Johan Liu[1,3]

[1] Electronics Materials and Systems Laboratory, Department of Microtechnology and Nanoscience (MC2), Chalmers University of Technology, Kemivägen 9, Se-412 96, Gothenburg, Sweden

[2] SHT Smart High Tech AB, Aschebergsgatan 46, Se-426 69 Gothenburg, Sweden

[3] SMIT Center, School of Mechanical Engineering and Automation
Shanghai University, No 20, Chengzhong Road
Shanghai, Box 808, 201800, China
Email: johan.liu@chalmers.se

Abstract

The integration of metallic nanoparticles (NPs) in nanofluids was found to enhance the thermal properties of the mixture and affect the rheological properties of the base liquid. However, due to their size and electrochemical properties, the added metallic nanoparticles have a limited contribution to the thermal transport and their stability hinders further development of such an approach in thermal management. We investigated in this work the effect of the presence of graphene as a coating layer of on copper nanoparticles dispersed in water as a water-based graphene coated copper nanofluid. Electronics microscopy was deployed to investigate the presence and the number of layers of graphene around the metallic nanoparticles. The observed particles were found to have a spherical morphology with a full coating of several layers. The elemental characterization of the NPs showed the presence of graphitic structure confirming the nature of the coating. The thermal properties of the fluid were estimated versus loading fraction of graphene coated nanoparticles and temperature using a hot disk method. An increase of up to 17% was recorded at a concentration of 0.1 w.% at 45deg C. Dynamic Light Scattering and zeta potential were used to investigate the electrochemical properties of the produced nanoparticles. The particles were found to present weak surface charges corresponding to a zeta potential of 6mV that promoted the segregation of the NPs. The rheological properties of the resulted fluids were investigated using viscometer. The NFs were found to have a Newtonian behaviour.

Keywords: Graphene coated copper nanoparticles, nanofluids, thermal conductivity and rheology of nanofluids.

I. Introduction

Temperature control plays a critical role in electronics systems for the products life cycles and their performances. For the last decades, fluid-based solutions for thermal management provided a promising alternative in heat dissipation. From forced liquid cooling to liquid evaporation cooling, traditional approaches for heat removal resulted in low and controlled working temperatures[1]–[3]. However, standard fluids such as water, ethylene glycol and oil offer limited properties that will soon no longer be compatible with the high energy densities of the electronics development trend. The concept of nanofluids (NFs), on the other hand, quickly attracted increasing attention with the promise to improve the effective thermophysical properties of the fluids.

NFs, which are mixtures of base fluids (i.e. water, ethylene glycol and oil…etc.) with dispersed metallic and nonmetallic nanoparticles (NPs), showed in most of the research cases a considerable increase in the thermal properties of the overall fluid [4]. The benchmark exercise from the International Nanofluid Property reported the unanimity effect of increasing thermal properties of fluids based metallic and nonmetallic dispersions [5]. The results on aluminium oxide NFs thermal conductivity of water, ethylene glycol and oil NFs showed an increase between 2% and 32%, and 1% and 41% and 5% and 39%, respectively, depending on concentration, particle size and method of preparation in their respective fluids [6].

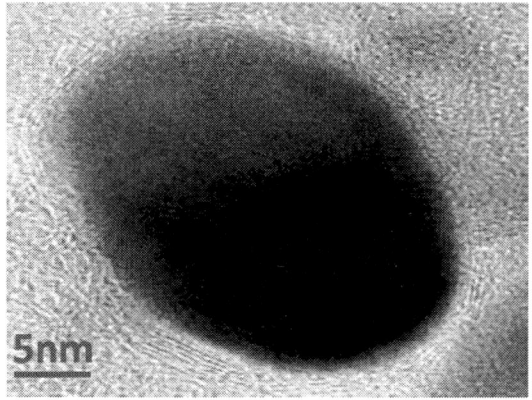

Figure 2: TEM observation of graphene coated copper nanoparticle. A coating of 4nm is seen covering the particle.

Dispersed silver NPs in water was also found to positively affect the thermal properties with an increase of 11% of the heat transfer coefficient at only 0.039 vol.% and result in an increase of 18% in the thermal conductivity at a concentration of 0.3vol% at 25° C [7]. In ethylene glycol, the same filler resulted in an increase up to 23% at 0.1vol% and 50° C and highlighted the effect of the temperature and the shape of particles in result to the aging process for up to 5 hours [8]. In the case of copper (Cu) NFs, the shape of the nanofiller was investigate by comparing nanocubes, nanosphere and short and long nanowires. The

result reported on the percolation mechanism where thermal conductivity was enhanced with a higher aspect ratio of the fillers to reach an increase of 40% at only 0.25vol% comparatively to 4.2% increase in the case of Cu nanocubes [9]. Graphene, as a promising material for electronics cooling, has also been investigated as NFs for heat managements. Thanks to its thermophysical

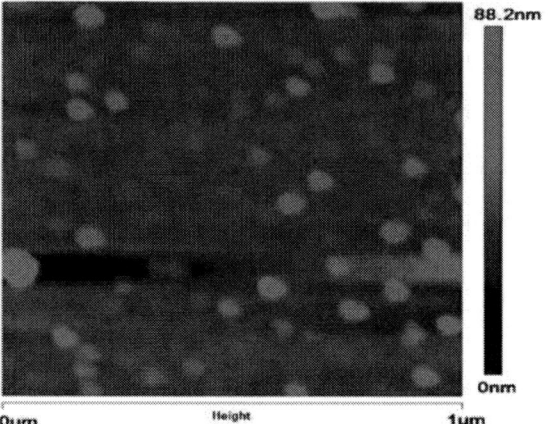

Figure 1: 1umx1um SPM topology mapping of coated nanoparticles. Data analysis showed a maximum of 87nm particles with a mean diameter of 57nm.

[10] and chemical properties [11], graphene offers high carrier mobility, thermal conductivity, stability and corrosion resistance due to its sp2 hybridization. It was found that the addition of graphene nanofiller to DI water increased the thermal conductivity up to 31.83% and the heat transfer coefficient up to 35.6% at 0.1%. Graphene nanoplatelets dispersed in DI water was investigated in pin-fin heat sinks and resulted in a lower thermal resistance with an increase of 23% in the convective heat transfer coefficient [12]. In heat pipe, graphene nanosheets in DI water were reported to enhance the heat transfer coefficient, the thermal conductivity, and the thermal efficiency with 62.3%, 27.6%, 93% respectively [13]. A higher improvement was measured in ethylene glycol where dispersed graphene resulted in an increase of up to 86% in thermal conductivity at 5.0 vol.% [14]. On the

346

other hand, graphene oxides nanoparticles were also dispersed in fluids and the resulted nanofluids were found to exhibit better stability due to the surface functionalization with oxygen groups. It was reported that the functionalized graphene reached an increase of 61% at 5vol% when dispersed in ethylene glycol. In water, an increase of 181% in the heat transfer coefficient of the base fluid was recorded at a concentration as low as 0.2 vol.% [15]. In this paper, we investigate the effect of adding graphene coated nanoparticles as a filler in deionized (DI) water and we report on the thermal and viscosity measurements. Electron microscopy was used to investigate the morphology and the quality of the coating. Elemental characterization of the NPs was provided by X-ray photoelectron spectroscopy (XPS) to investigate the surface of the coatings. Zeta potential measurements were to use to measure the surface charges. Hot disk measurements were used to measure the thermal conductivity of the nanofluids. Viscometer analysis was run to study the rheological properties of the nanofluids.

II. Materials and Methods

II.1 Sample preparation

Graphene-coated NPs were developed in collaboration with SHT Smart High Tech AB, Sweden. Concentrations of 0.1 w.%, 0.05 w.%, 0.02 w.%, and 0.01 w.% of graphene coated copper nanoparticles were prepared. The nanoparticles were dispersed to their respective concentrations into DI water and sonicated for 10min before each measurement. The solutions were later diluted further for analysis in TEM, AFM and zeta potential.

II.2 Characterization

In order to investigate the morphology and the quality of the coating around the NPs, Transmission Electron Microscope (TEM) FEI Tecnai T20 equipped with LaB6 filament was used combined to Scanning Probe Microscope (SPM) Bruker Dimension ICON in tapping mode

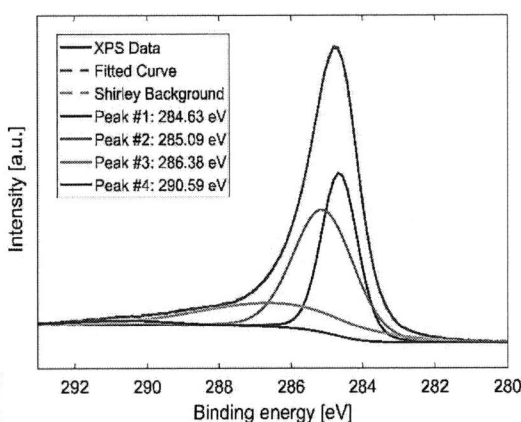

Figure 3: Deconvolution of the C1s peak from XPS analysis. Three main peaks are created in addition to a minor peak,

using an RFESP cantilever for revealing the surface topology. Few droplets of low concentration nanofluid samples were drier on top of silicon wafer prior to the analysis. Later on, the nanoparticles were centrifuged and dried on a hot plate. The powder was then collected and pressed before being clamped to a sample holder and loaded into a PHI 5000C XPS to confirm the presence of graphene in the coating around the NPs. The stability of the dispersion was assessed through the measurement of the zeta potential with DLS Nano S90 Malvern. Low concentrations of DI water mixed with graphene-coated were prepared prior to the analysis. The thermal properties of the water-based graphene coated NFs were evaluated using Hot disk TPS 2500 by varying the concentration from 0.01 w.% to 0.1 w.% and the temperatures from 25deg C to 45 deg C. Kepton sensor with 2mm radius (design 7577) was used with the characterization time of 1s to avoid the contribution from the convection. Finally, the dynamic viscosity of the NFs was measured with a dynamic shear rheometers (DSR) SmartPave 92 equipped with a flat rotating disk.

III. Results and discussion

The graphene coated copper nanoparticles were found to exhibit a spheroidal to spherical morphology. All the observed particles showed a nanometer thick coating around them. The number of layers around the particles reached 15 layers at most. Figure 1 shows a spheroidal copper nanoparticle of 73nm under TEM. The particle is shown exhibiting a 4nm thick coating surrounding it that corresponds to 11 layers of graphene.

The nanoparticles were analyzed with SPM in order to reveal their surface topology. A scan resolution of 512 × 512 was achieved for a surface of 1μm × 1 μm. Data acquisition was then followed by an image processing procedure. First, a three-point flattening was applied, to correct the tilt resulting from the tip-substrate interaction. A 2D surface mapping of the surface topography of the sample is shown in Figure 2. The image processing was followed later with data analysis by performing a particle analysis using Nanoscope analysis 1.5.

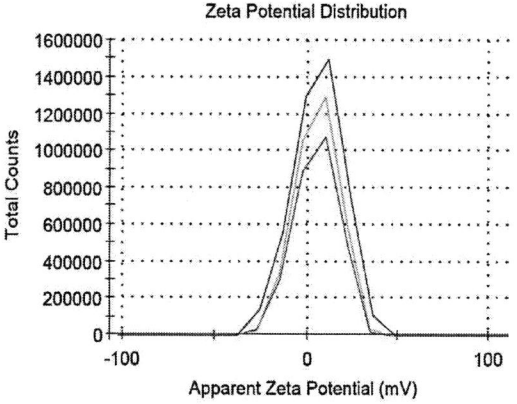

Figure 4: Zeta potential measurement. Three samples with low concentration graphene coated copper nanoparticles are shown to have an average zeta potential of 6mV.

The grains were marked using a height threshold of 0.11nm. The selected grains allowed later to determine the size distribution of the NPS. Results showed a size distribution with an average size of 57.5nm and a maximum of 87.27nm.

In order to confirm the origin of coating, the nanoparticles were centrifuged and dried at

80deg C on a hot plate and then pressed to small platelet on carbon adhesive before being loaded into the XPS instrument. The collected signal is generated from a 10um spot and offers a good representation of the powders. The results from the surface analysis of the powder are shown in figure 3. The deconvolution of the C1s peak can be seen in the same figure. Three main and one minor component could be extracted. The main peaks including Peak #1, #2 and #3 are located at 284.63, 285.09 and 286.38 eV respectively. the minor peak (i.e. Peak #4) is centred around 290.59eV. Comparing these data to XPS references, the location of these peaks in the spectrum could be matched and identified [16]-[17]. Peak #1 corresponds to the sp2 bonded carbon as the main lattice of the graphene structure. Peak #2 is interpreted as the sp3 peak that represents amorphous carbon or eventual damages to the graphene structure, while peak #3 could be matched with C-O-C bond. Finally, the minor peak at 290eV is attributed to C1s satellite. The slit shifts in values comparatively to the references are explained by the use the neutralizer and calibration of the instrument.

The zeta potential measurements of the graphene-coated nanoparticles dispersed in water are shown in Figure 4. The sample that was measured three times was found to have an average potential of 6.22mV that correspond to relatively low surface charges. The particles are therefore prompt to segregation. For this study, no surfactant was employed in order to insulate the effect of Nanoparticles. Before each measurement, the fluids were sonicated under 2W for 10min.

Figure 5 depicts the thermal conductivity measurements of graphene coated nanoparticles in water at temperatures varying from 25 to 45 deg C with 5 degrees gradients and with concentrations of 0.1w.%, 0.05w.%, 0.02w% and 0.01w.%. Tests on pure water were initially made in order to verify the accuracy of the instrument.

The addition of the NPs appears to have a positive effect on the heat transfer of the NFs.

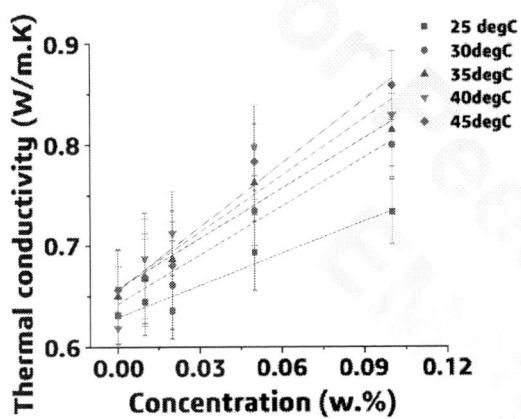

Figure 5: Thermal conductivity measurements of the different concentration of NFs at temperature of 25, 30, 35, 40 and 45 deg C.

At 25deg C, the thermal conductivities measured were found corresponding to the thermal conductivity of water and minor deviations were recorded for samples with 0.1w.% and 0.05w.%. However, an increase is seen at a higher temperature where the thermal conductivities of the NFs samples increased with the increase of concentration. At 30deg C, the thermal conductivity increased between 3% and 9% when the concentration NPs increased up to 0.1w.%. At 35deg C, the increase varied between 2% and reached 11% increase at 0.1 w.% At 40 deg C, the results showed almost the same improvement with a maximum of 15% increase in the thermal conductivity of 0.05 w.% NF. The thermal conductivities at 45deg C reached the highest increase that reached 17% increase.

The viscosity of the nanofluids plays a major role in its thermal conductivity that can positively contribute or hinder the heat transfer. As a Newtonian fluid, the viscosity of water does not change with the shear rate. However, the addition of NPs to the fluids might strongly affect the base fluid rheology. Figure 6 shows the results of measurement of the viscosity of water and the other nanofluids. The results confirm the nature

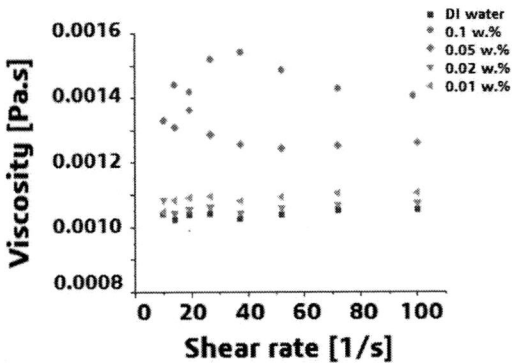

Figure 6: Dynamic viscosity measurements of DI water and different concentrations of NFs.

of the Newtonian behaviour of water where its viscosity remains relatively unchanged with the applied shear rates. For the NFs, it can be seen that the increase of concentration results in an increase in the viscosity of the fluid in all the samples. In the case of 0.1w% and 0.05 w.% NFs, the values of the measured viscosities displayed a different behaviour with an increase in viscosity at low shear rates. The later behaviour was found to be rather a consequence of aggregation of the NPs that occurs faster in the case of high concentration since the particles are more easily attracted to each other.

IV. Conclusion

This work aimed at investigating the effect of graphene coated NPs in water on the thermal and rheological properties of the base fluid. Results from the TEM confirmed the presence of several layers of coating surrounding the NPs, while SPM analysis showed a size distribution in the nanometer scale. Furthermore, the elemental analysis of the surface of the NPs confirmed the presence of the graphitic SP2 bonded structure centered around 284.8eV. The presence of graphene coated NPs within the liquid was later on found to have a positive effect on the thermal conductivity of the NF with an increase of up to 17% at 45 deg C with a concentration of 0.1 w.%. Finally, the rheological behaviour of the NFs was evaluated. The measurement of the dynamic viscosities showed a Newtonian behaviour of the NFs with and increased viscosity as the concentration of the NPs in the fluid increased. The low contribution of the NPs to thermal transfer and the deviation from the Newtonian behaviour at low shear rates is explained at this step by the weak surface charges measured with a low zeta potential. As a consequence, the NPs were prompt to segregation.

V. Acknowledgement

The authors acknowledge the financial support from the Swedish Board for Strategic Research (SSF) with the contract GMT14-0045, from Formas with the contract No: FR-2017/0009, from the Swedish National Science Foundation with the contract No: 621-2007-4660, from the Vinnova Siografen program as well as from the Production Area of Advance at Chalmers University of Technology, Sweden. We greatly acknowledge also the financial support from EU Horizon 2020 program "Smartherm" and "Nanosmart".

VI. References

[1] Y. F. Maydanik, S. V. Vershinin, M. A. Korukov, and J. M. Ochterbeck, "Miniature loop heat pipes-a promising means for cooling electronics," *IEEE Trans. Components Packag. Technol.*, vol. 28, no. 2, pp. 290–296, Jun. 2005.

[2] B. A. . Abu-Hijleh, "Enhanced solar still performance using water film cooling of the glass cover," *Desalination*, vol. 107, no. 3, pp. 235–244, Dec. 1996.

[3] M. S. Sehmbey, L. C. Chow, M. R. Pais, and T. Mahefkey, "High heat flux spray cooling of electronics," *Proceedings*, vol. 324, pp. 903–909, 2008.

[4] R. Saidur, K. Y. Leong, and H. A. Mohammad, "A review on applications and challenges of nanofluids," *Renew. Sustain. Energy Rev.*, vol. 15, no. 3, pp. 1646–1668, Apr. 2011.

[5] J. Buongiorno *et al.*, "A benchmark study on the thermal conductivity of nanofluids," *J. Appl. Phys*, vol. 106, p.

94312, 2009.

[6] V. Sridhara and L. N. Satapathy, "Al2O3-based nanofluids: a review.," *Nanoscale Res. Lett.*, vol. 6, no. 1, p. 456, Jul. 2011.

[7] R. L. Fragelli, L. E. de A. Sanchez, R. R. Ingraci Neto, and V. L. Scalon, "Refrigeration capacity of silver nanofluids under electrohydrodynamic effect oriented to heat removal in machining process," *Exp. Therm. Fluid Sci.*, vol. 96, pp. 11–19, Sep. 2018.

[8] T. Khamliche, S. Khamlich, T. B. Doyle, D. Makinde, and M. Maaza, "Thermal conductivity enhancement of nano-silver particles dispersed ethylene glycol based nanofluids," *Mater. Res. Express*, vol. 5, p. 35020, 2018.

[9] S. Bhanushali, N. N. Jason, P. Ghosh, A. Ganesh, G. P. Simon, and W. Cheng, "Enhanced Thermal Conductivity of Copper Nanofluids: The Effect of Filler Geometry," *ACS Appl. Mater. Interfaces*, vol. 9, no. 22, pp. 18925–18935, Jun. 2017.

[10] E. Pop, V. Varshney, and A. K. Roy, "Thermal properties of graphene: Fundamentals and applications," *MRS Bull.*, vol. 37, no. 12, pp. 1273–1281, Dec. 2012.

[11] E. Mousset, Z. Wang, J. Hammaker, and O. Lefebvre, "Physico-chemical properties of pristine graphene and its performance as electrode material for electro-Fenton treatment of wastewater," *Electrochim. Acta*, vol. 214, pp. 217–230, Oct. 2016.

[12] H. M. Ali and W. Arshad, "Effect of channel angle of pin-fin heat sink on heat transfer performance using water based graphene nanoplatelets nanofluids," *Int. J. Heat Mass Transf.*, vol. 106, pp. 465–472, Mar. 2017.

[13] T. Tharayil, L. G. Asirvatham, V. Ravindran, and S. Wongwises, "Thermal performance of miniature loop heat pipe with graphene–water nanofluid," *Int. J.*

Heat Mass Transf., vol. 93, pp. 957–968, Feb. 2016.

[14] W. Yu, H. Xie, X. Wang, and X. Wang, "Significant thermal conductivity enhancement for nanofluids containing graphene nanosheets," *Phys. Lett. A*, vol. 375, no. 10, pp. 1323–1328, Mar. 2011.

[15] M. Salem, T. Meakhail, M. Bassily, and S. Torii, "Thermal Transport Phenomena of Graphene Oxide Nanofluids in Turbulent Pipe Flow."

[16] J.-M. Feng and Y.-J. Dai, "Water-assisted growth of graphene on carbon nanotubes by the chemical vapor deposition method," *Nanoscale*, vol. 5, no. 10, p. 4422, May 2013.

[17] A. Nylander; Y. Fu, "Covalent Anchoring of Carbon Nanotube-Based Thermal Interface Materials Using Epoxy-Silane Monolayers," IEEE Transactions on Components, Packaging and Manufacturing Technology 2019 Volume: 9, Issue: 3

Anisotropic Composite Core Material for Inductor-based Fully Integrated Voltage Regulator

Suiying Ye[1], Arvind Sridhar[1], Luca Del Carro[1], Paul McCloskey[2], Ansar Masood[2], and Thomas Brunschwiler[1]

1: IBM Research – Zurich, 8803 Rüschlikon, Switzerland
+41 77-922-70-52, yeahss.study@gmail.com
2: Tyndall National Institute, Cork, Ireland

Abstract

Miniaturization of efficient integrated voltage regulators is critical in order to fully benefit from dynamic voltage and frequency scaling. A major challenge to this is the high parasitic loss of inductor cores at switching frequencies in the MHz range. In this study, we report on a novel soft magnetic core material with anisotropic magnetic properties. This composite core is fabricated using magnetic field directed assembly of magnetic particles in an epoxy matrix. Various experimental conditions were explored, and subsequent image analytics of the resulting samples was used to optimize the process parameters, in order to yield the desired morphology. Magnetic measurements revealed improved magnetization, lower coercivity and higher permeability along the percolation axis of the magnetic particles when compared to isotropic composites with the same particle loading. In addition, the permeability of the anisotropic composites remained close to constant up to 100 MHz. Finally, a funicular "necking" structure with nano-sized particles assembled around contact points of micro-sized particles in the composite was achieved via a bi-modal particle distribution, resulting in further improvements in magnetic properties. Our anisotropic composite core materials are attractive candidates to establish highly efficient power inductors.

Key words: system integration and embedded passives, integrated voltage regulator (IVR), magnetic core materials, composite materials, particle assembly, neck formation

Introduction

Dynamic voltage and frequency scaling (DVFS) is a technique to improve the energy efficiency of integrated circuits for specific computational workloads, leading to longer battery recharging intervals for mobile devices or reduced energy footprint for data centers. The full potential of DVFS can only be reached with the use of integrated voltage regulators (IVRs), which can supply the optimal voltage for every processor core on a microprocessor chip. IVRs can provide a fast dynamic response, with voltage switching speeds in the order of nanoseconds/V, as opposed to the microseconds/V of conventional off-chip DC-DC converters [1] - [4]. A major challenge in the miniaturization of efficient IVRs is the size of the power inductors while maintaining low parasitic losses at switching frequencies in the order of MHz [5]. To address this challenge, the exploration of novel soft magnetic core materials, which can significantly reduce the inductor size and increase the inductance density, has recently gained prominence.

Ferromagnetic cores applied in IVR inductors may result in high losses arising from magnetic hysteresis and eddy currents. Alloys, such as Co-Zr-Ta and Ni-Fe exhibit high permeability and high saturation flux density [6] - [10]. However, their low electrical resistivity results in prohibitively large eddy current losses, limiting their application to switching frequencies of less than ~30 kHz. Different strategies have been utilized to mitigate eddy currents, such as multilayered thin-film inductors [11] or granular/composite ferromagnetic materials with ferromagnetic particles embedded in dielectric matrix [12][13].

In this paper, we present a novel composite core material with anisotropic magnetic properties. This material is fabricated via magnetic field directed assembly of ferromagnetic microparticles in a dielectric matrix material. The directed assembly is achieved via the application of an external magnetic field during the curing of the matrix material. The physical mechanisms behind the directed assembly method will be described. We will demonstrate the alignment of magnetic particles into chains by an external magnetic field. Image processing will be used to explore the effects of varying process parameters. As a result, the proposed material exhibits a close to constant magnetic permeability over a large frequency range up to 100 MHz, similar to the particle-based composites previously studied in the literature. In addition, the proposed composite material also reduces eddy current losses due to its anisotropic morphology. Finally, a funicular "necking" structure was achieved by the addition of nanoparticles, enhancing the magnetic contact between the microparticles and hence, increase the effective permeability of the composite.

Composite Magnetic Material Fabrication

Figure 1 illustrates a desired pattern of the composite magnetic material, where particles form percolating chains along an external magnetic field $\vec{B}$. The addition of nanoparticles results in a funicular "necking" structure around contact points of micron-sized particles. This anisotropic arrangement is believed to feature comparable permeability as previous studied particle-based composites, while mitigating the eddy current loss due to the dielectric insulation of the particle chains in the direction perpendicular to the magnetic flux.

In our fabrication process, an electromagnet with a small air gap was chosen as the external magnetic source for the particle assembly. This configuration results in a straight and uniform magnetic field inside the small air gap. Samples were placed in the middle of the air gap, as illustrated in Figure 2a. The equivalent magnetic circuit representing the electromagnet device is shown in Figure 2b. The overall reluctance can be written as:

$$\mathcal{R} = \mathcal{R}_c + \mathcal{R}_g = \frac{l_c}{\mu_0 \mu_{rc} A_c} + \frac{l_g}{\mu_0 A_c},$$

where l_c is the perimeter of the magnetic core of the electromagnet, l_g is the length of the air gap and A_c is the cross-section of the core [cm^2]. Using Ampere's Law, the magnetic flux ϕ can be derived as:

$$\phi = \frac{Ni}{\mathcal{R}_c + \mathcal{R}_g} = BA_c,$$

Hence, the magnetic flux density B is proportional to the current i.

For the purpose of visualization, samples with magnetic microparticles with a one-order of magnitude larger diameter than finally targeted were fabricated. The particles were made from coprecipitation of iron oxide nanoparticles within a monodisperse polystyrene network. These particles (PS-Mag-135 and PS-Mag-10) were purchased from *Microparticles GmbH*, with a size of 135 µm and 10.31 µm, and high monodispersity with coefficient of variation (CV) = 1.6% and 2.4%, respectively. These magnetic particles were received as dispersed in water. They were first dried in an oven before further processing. For the actual inductor composite core materials, Ni-Fe alloy particles with max. 45 µm (Ni81/Fe19, at. %, refer as NiFe-m) and 40-100 nm (Ni80/Fe20, at. %, refer as NiFe-n) were purchased from *GoodFellow* and *US Research Nanomaterials, Inc.*, respectively. Both alloy particles were received in powder form. EpoFix resin and hardener, purchased from *Struers GmbH* with curing time of 12 h at room temperature served as matrix material. All materials were used as received without further purification.

Both isotropic composites under no external magnetic field and anisotropic composites assembled with the application of the external magnetic field

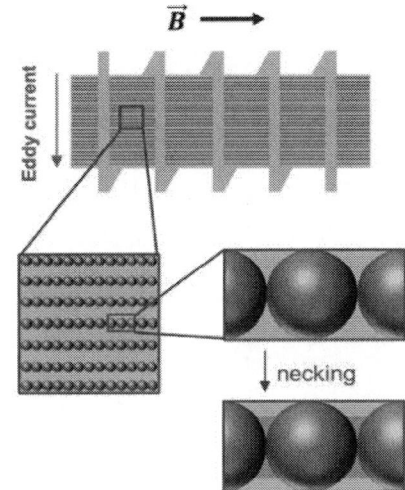

Figure 1: Schematic of particle arrangement in anisotropic magnetic cores of embedded inductors

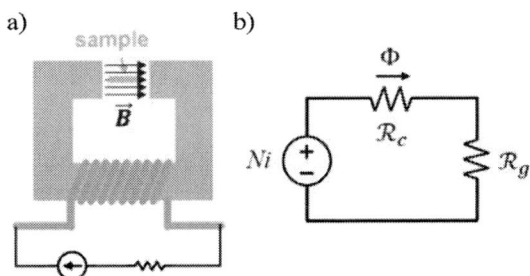

Figure 2: a) Schematic of an electromagnet with an air gap and a sample in the center, b) and its analogous magnetic circuit.

were fabricated. In a typical process, epoxy resin and hardener were weighed according to the desired ratio and stirred in a sample mold. Then a pre-calculated amount of magnetic particles was mixed into the sample mold, followed by moderate stirring. Next, the prepared mixture was either cured directly or while placed in the middle of the air gap of the abovementioned electromagnet.

Optical Microscope and Scanning Electron Microscope (SEM) images were used to study the fabricated composites. Magnetic property characterization including permeability (*PMM 9G; Ryowa Electronics, Japan*) and hysteresis loop (*MESA 200 HF; SHB Instruments*) measurements were carried out. These measurements required thin and squared samples (3×3×0.4 mm or 10×10×0.4 mm), which were cut from our fabricated samples using a dicing saw.

$\vec{B} \longrightarrow$

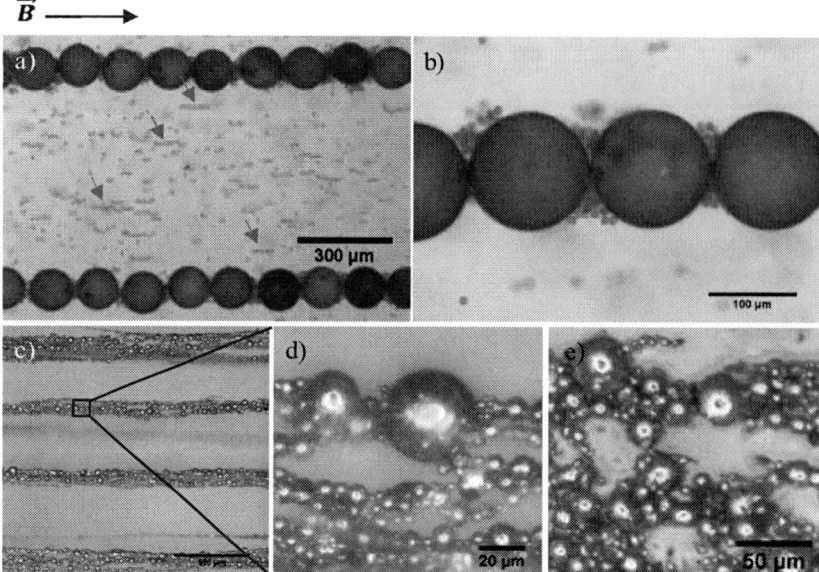

Figure 3: Optical images of a, b) mixed PS-Mag-135 (0.5 wt.%) and PS-Mag-10 (0.05 wt%.) after exposure to 40 mT for 1 h. c) 1 wt. % of NiFe-m after exposure to 40 mT for 3 h. d) A zoom-in region of c) as illustrated by black square. e) Mixed NiFe-m (silver appearance) and NiFe-n (black appearance) after exposure to 40 mT for 3 h.

Results and Discussions

1. Magnetic Field Directed Particle Assembly

When exposed to the external magnetic field, particles assembled to form chains along the direction of the field, with point-contacts between each other (Figure 3a, 3b). Particles were aligned during the liquid phase of the epoxy resin and then the particles were locked in place by the curing of the epoxy resin. The epoxy resin is non-magnetic and considered having negligible effects on the interaction between magnetic particles and the external magnetic field. By comparing the insitu-images taken during particle assembly under the external magnetic field (not shown here), with the optical images made after curing of the epoxy resin (Figure 3), one can conclude that the epoxy resin is capable of locking the particle arrangement once it is cured without affecting the patterns, even after the removal of the external magnetic field. In addition, branching of the particle-chains and chains of multiple particle widths (Figure 3c, d and e) occurred in all samples no matter which kind of particles were used. The mechanism will be discussed in detail below.

When magnetic particles of two different sizes were mixed in the composite, it was found that not only did the large particles align into chains under the influence of the magnetic field, but also most of the small particles agglomerated around the point-contacts of large particles. A funicular "necking" structure appeared as desired (Figure 3b). This phenomenon is attributed to the magnetic field permeating through the large particles forming high-

density flux lines near the point contacts, thereby driving the smaller particles towards these locations. The accumulation of smaller particles in this region thus forms "necking" structures. On the other hand, the small particles which are sufficiently far away from the large particles were not attracted by the field gradient but formed independent chains of their own due to the external magnetic field, similar to the large particles (indicated by the red arrows in Figure 3a). Similar effects could be observed in samples with NiFe particles (Figure 3e), where the shining silver particles are NiFe-m and the black fillings in-between are NiFe-n. The necking of small particles in-between larger particles has the potential to enhance the magnetic permeability of the composites, by reducing the effective reluctance to magnetic flux.

The physics behind the magnetic field-directed assembly has been reviewed in [14]–[18]. Typically, a particle acts as a single magnetic dipole

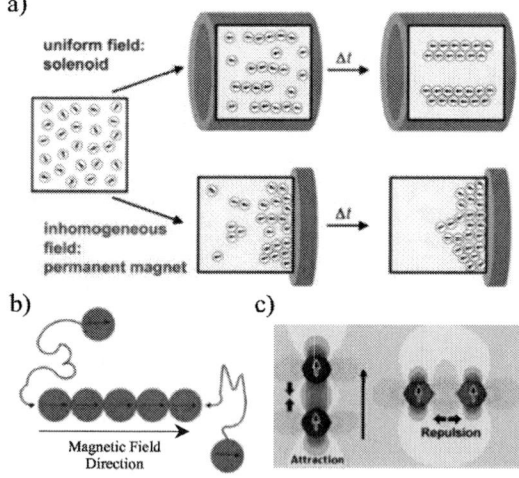

Figure 4: a) Schematic comparing magnetic field directed assembly of particles in uniform magnetic fields, such as a solenoid, and field gradients (i.e. inhomogeneous field) such as from a permanent magnet. b) Schematic illustrating the magnetization of particles as magnetic moments and their assembly leading to a chain-like structure. c) Illustration of particle interaction under external magnetic field.

under a magnetic field $\vec{H}$, with magnetic moment: $m = MV = \chi H$, where M is the magnetization and V is the volume of the particle. The interaction energy of a magnetic dipole with the external magnetic field, and the force (called packing force) exerted on the particle (dipole) can be expressed as: $U_m = m \cdot H$, $F_p = \nabla(m \cdot H)$. This force moves particles to regions of highest magnetic field strength, creating a particle concentration gradient. Two possible particle assemblies resulting from external magnetic fields are illustrated in Figure 4a: 1) a uniform magnetic field magnetizes a particle and generates torques to rotate and align its magnetic moment parallel to the field; 2) A magnetic field gradient, instead, exerts a force on the moment and attracts the particles onto the surface. A uniform field drives particle to assemble uniform chains. On the other hand, a field gradient pulls particles toward the region of the strongest field by strong attractive forces, as predicted mathematically.

The magnetized particle behaves like a tiny magnet resulting in a highly inhomogeneous magnetic field. This field will magnetize nearby particles as well as exert forces on them. Assembly behaviors are mainly due to this dipole-dipole interaction with a magnetic field created by a dipole 1 acting on dipole 2, as

$$H_1 = \frac{3(m \cdot r)r - m}{d^3},$$

where r is the unit vector parallel to the line pointing from the center of dipole 1 to that of dipole 2 and d is the center-center distance. Such a magnetic field from particle 1 acting on particle 2 is an inhomogeneous field with a magnetic field gradient and can thus enable the movement of particle 2. As a result, particle chains are formed, as illustrated on Figure 4b [19]. Here, the interaction energy of these two dipoles and the force exerted on dipole 2 can be written as

$$U_2 = mH_1 = \frac{(3\cos^2 \alpha - 1)m^2}{d^3},$$
$$F_2 = \nabla(mH_1) = \frac{3(1 - 3\cos^2 \alpha)m^2}{d^4},$$

where α $(0° - 90°)$ is the angle between the direction of the external magnetic field and the line connecting the center of the two particles (dipoles). In addition, it is clear from the equations that, the interaction of two particles depends on their configuration, with a critical angle 54.09°. This means that, attraction occurs when $0° \leq \alpha < 54.09°$, and particles repel each other when $54.09° < \alpha \leq 90°$, as shown in Figure 4c [20].

Moreover, such motion of particles only occurs when Brownian motion from thermal fluctuations can be overcome by the magnetization energy, which typically calculated as $E = -\frac{1}{2\mu_0}\chi V B^2$, where B is the applied magnetic field strength. It is worth to note that chains formed in a homogeneous field tend to arrange side by side repelling each other, as the case of the uniform field in Figure 4a. This is due to a minimization of the repulsive energy between magnetized particles.

On the other hand, a more complicated case results when a large non-uniform local field is created by nearby particles causing the particle at the origin to change its magnetization direction. Such forces give features to non-linear assemblies, branching of from particle chains, or into chains with a width of multiple particles (Figure 3c) [18].

2. Effects of Different Assembly Conditions

In order to optimize the fabrication process, we fabricated samples under different experimental conditions. The parameters explored include particle concentrations, magnetic flux densities and exposure time to the external magnetic field. Preliminary studies by image processing performed on optical micrographs of cured samples are described below.

Firstly, the directionality of the formed particle-chains was studied to detect if they were aligned nicely along the external magnetic field. This was implemented using a plugin in *Fiji ImageJ* called "Directionality", where Fourier component analysis was applied to measure angles of segments, and a local gradient orientation method was used to build histograms with regards to different directions [21]. This resulted in a measure of directionality for individual chains from the image and allowed to plot histograms indicating the amount of structures/lines oriented in a given direction (angle). Figure 5 shows an example of the computed histograms from samples fabricated under different magnetic flux densities, as well as the corresponding standard deviation (STD) in the insert. In the histogram, angle 0° stands for a particle-chain aligned along the external magnetic field, while angles $\neq 0°$ refer to the angle between a particle-chain and the field. A higher and narrower peak in the distribution indicates that a higher fraction of particle-chains was aligned along the field. Samples fabricated under a magnetic flux density of

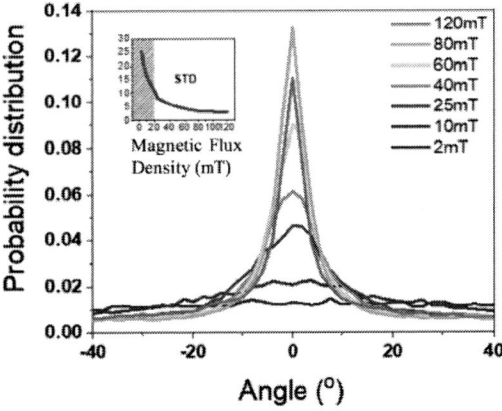

Figure 5: Directionality histogram and standard deviation of its Gaussian fit (inset) from samples of 0.5 wt. % PS-Mag-135 in epoxy fabricated under the external field with an exposure time of 60 min for different magnetic flux densities.

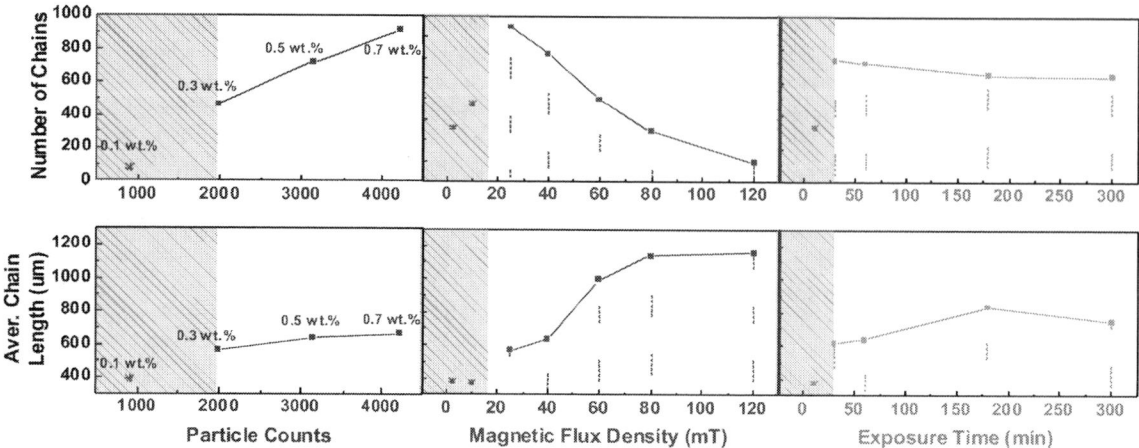

Figure 6: Data interpretation of number of chains and average chain length of PS-Mag-135 in epoxy under different experimental conditions. Grey areas stand for results with no remarkable particle-chains.

less than 25 mT show no significant peaks in the histogram and a high STD, indicating randomly distributed particles without the formation of chains, which is also revealed from optical images (not shown here). On the other hand, samples fabricated with a magnetic flux density larger than 25 mT show a clear peak in the histogram as well as a low STD. Above 60mT the improvement in the STD starts leveling off. A slight decrease of the STDs with regards to increasing magnetic flux density indicates better alignment of particles into straight chains. Nevertheless, such improvement is not significant. Similar results were also found in samples with different concentrations and exposure times to the external magnetic field. In summary, a clear alignment (i.e. a distinguishable peak as well as low and stable STD <8°) was found for concentrations >0.3 wt. % and an exposure time of 30 min.

Chains (defined as three contacted particles and more) were detected using a plugin called "Ridge Detection" [22]. Following further quantitative analyses, the number of chains and average chain length under different experimental conditions were calculated, and the results are shown in Figure 6. Consistent with the directionality analysis above, the chain analyses also reveal the same threshold of minimum concentration, magnetic flux density and exposure time (i.e. 0.3 wt.%, 25 mT and 30 min respectively), that is needed to form significant number and of chains lengths. Both parameters increase almost linearly with regards to the increasing particle concentrations. However, at high concentrations large lateral clusters of particles were formed, which can be attributed to interactions between neighboring chains as described previously. These samples were not considered for further analyses. The data under different magnetic flux densities indicates higher magnetic flux densities facilitate assembly of particles into long chains and due to same particle loading, the number of chains is thus reduced. At extremely high magnetic flux

densities, an accumulation of particles at the 2 extremities of the sample was found. This is the result of fringing fields at the air gap-metal core interface, whose effects increase for higher and higher flux densities, namely, a magnetic gradient. Hence, a

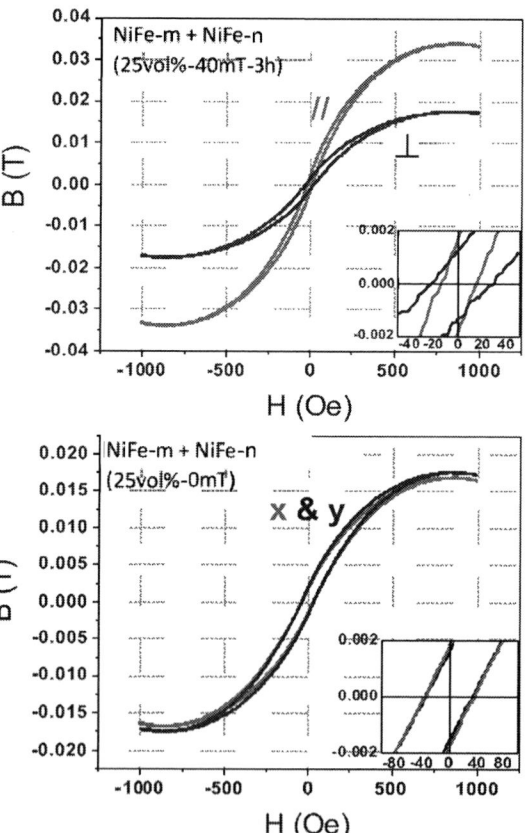

Figure 7: Hysteresis loops of anisotropic (top, exposed to 40 mT for 3 h) and isotropic (bottom, cured directly without exposure to the external magnetic field) composites fabricated with 25 vol% of NiFe-m in epoxy and 1/10 of NiFe-n by weight.

magnetic field around $40-60$ mT is recommended to be necessary and sufficient to assemble particles into long chains. An exposure time of 30 mins was found to be sufficient for the alignment of most particles during curing. There is only negligible improvement beyond this time. In all our further fabrication processes, a $60-180$ min exposure time was preferred.

3.   Magnetic Properties of the Composites

To test anisotropic effects, measurements on both axes (parallel and perpendicular to the external magnetic field) were performed. Due to the low concentration of magnetic particles in our composites and the limitation of the characterization equipment, samples were not saturated during hysteresis characterizations. However, remarkable differences between anisotropic and isotropic composites can be observed, as shown on Figure 7. In the B-H loop of the anisotropic composite, the magnetization curve of the axis parallel to the external magnetic field shows significantly higher slope compared to the one with the perpendicular axis, as well as a lower coercivity (~15 Oe comparing to ~29 Oe, // and $\perp$ axis respectively). On the other hand, no significant differences occur between measurements on the two axes in the B-H loop of the isotopic composite. Hence, anisotropic properties were demonstrated in anisotropic composites fabricated under exposure to the external magnetic field. The percolating direction parallel to the field results in faster magnetization and lower coercivity, which is important to mitigate core losses. These results reveal that better magnetic properties can be achieved from the magnetic field assisted fabrication.

Regarding to permeability measurements, one should note that since the samples were not saturated due to low magnetic loadings, results can only be interpreted relatively and the permeability value (which is low) is not worthy to be noted. Nevertheless, permeability measurements further demonstrated anisotropic properties on composites fabricated under exposure to an external magnetic field, with a significant higher initial permeability on the parallel axis (Figure 8a). On the other hand, negligible differences are shown on the isotropic composite (Figure 8b). In addition, the measured permeabilities are close to constant up to 100 MHz with a slight drop, which is a preferred property for high frequency applications.

In the comparison of samples with and without nanoparticle "necking" (Fig. 8c and d, respectively), larger difference between the parallel and perpendicular axis is found in the bi-modal sample, demonstrating the enhancement of the "necking" structure, which can be attributed to the improved contacts between microparticles resulting in an increase in magnetic flux in the sample.

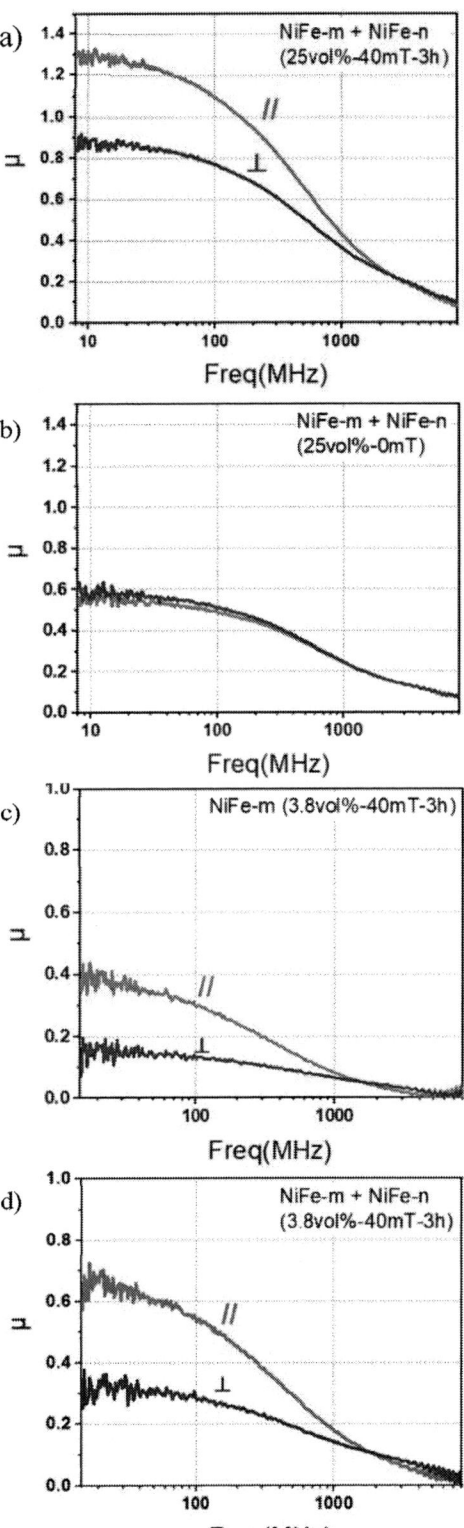

Figure 8: Permeability of anisotropic (a) and isotropic (b) composites with 25 vol% NiFe-m in epoxy and 1/10 NiFe-n by weight. c & d) are permeability of anisotropic composites with 3.8 vol% NiFe-m in epoxy with and without mixing 1/10 NiFe-n by weight.

Conclusion

We discussed the fabrication of both isotropic and anisotropic magnetic composite core materials, by using magnetic field-assisted particle assembly. Micron-sized particles arranged into chains while nano-sized particles accumulated at contact points of the larger particles, forming funicular "necking" structures. Image processing studies on the resulting patterns from different experimental conditions indicated the thresholds to trigger particle-chain assembly for various particle concentrations, magnetic flux densities and exposure times to the magnetic field. Moderate conditions (0.3 – 0.7 wt.%, 40 – 60 mT, 30 – 180 min) are sufficient to result in the desired chain patterns. The fabricated anisotropic composites exhibit different magnetic properties between axis parallel and perpendicular to the external magnetic field, with improved magnetization, lower coercivity and higher permeability measured on the parallel axis. Also, the magnetic permeability was further improved by the "necking" of the bi-modal particle distribution. These improved composites are attractive candidates for high-efficiency inductor-based fully integrated voltage regulators.

Acknowledgements

We gratefully acknowledge Andreas Bischof for technical assistance and Bruno Michel, Walter Riess and André R. Studart for continuous support. This work has been partly supported by the European Commission (H2020 program) within the project GaNonCMOS (Project No. H2020-ICT-721107).

References

[1] D. Nenni, "Reducing SoC Power Consumption using Integrated Voltage Regulators," *SemiWiki.com*, 2011. [Online]. Available: http://www.semiwiki.com/forum/content/694-reducing-soc-power-consumption-using-integrated-voltage-regulators.html.

[2] STIJN EYERMAN and LIEVEN EECKHOUT, "Fine-Grained DVFS Using On-Chip Regulators," *ACM Trans. Archit. Code Optim.*, vol. 8, no. 1, 2011.

[3] S. Mueller *et al.*, "Design of High Efficiency Integrated Voltage Regulators with Embedded Magnetic Core Inductors," *Proc. - Electron. Components Technol. Conf.*, vol. 2016-Augus, pp. 566–573, 2016.

[4] D. Hou, F. C. Lee, and Q. Li, "Very High Frequency IVR for Small Portable Electronics with High-Current Multiphase 3-D Integrated Magnetics," *IEEE Trans. Power Electron.*, vol. 32, no. 11, pp. 8705–8717, 2017.

[5] C. Devries and S. Kummerl, "Powerful solutions come in small packages Innovative SiP power modules simplify."

[6] H. L. Chan, K. W. E. Cheng, T. K. Cheung, and C. K. Cheung, "Study on Magnetic Materials Used in Power Transformer and Inductor," *2006 2nd Int. Conf. Power Electron. Syst. Appl.*, no. April, pp. 165–169, 2006.

[7] G. Herzer, "Modern soft magnets: Amorphous and nanocrystalline materials," *Acta Mater.*, vol. 61, no. 3, pp. 718–734, 2013.

[8] M. Schneider, R. Castagnetti, M. G. Allen, and H. Baltes, "Integrated flux concentrator improves CMOS magnetotransistors," *Proc. IEEE Micro Electro Mech. Syst. 1995*, p. 151, 1995.

[9] W. P. Taylor, O. Brand, and M. G. Allen, "Fully integrated magnetically actuated micromachined relays," *J. Microelectromechanical Syst.*, vol. 7, no. 2, pp. 181–191, 1998.

[10] D. S. Gardner, G. Schrom, F. Paillet, B. Jamieson, T. Karnik, and S. Borkar, "Review of on-chip inductor structures with magnetic films," *IEEE Trans. Magn.*, vol. 45, no. 10, pp. 4760–4766, 2009.

[11] R. Anthony, C. Ómathúna, and J. F. Rohan, "Magnetic Multilayer Fabrication Technology with Selective Activation of SU-8 Films," *J. Phys. Conf. Ser.*, vol. 757, no. 1, 2016.

[12] C. R. Sullivan, J. Qiu, D. V. Harburg, and C. G. Levey, "Batch fabrication of radial anisotropy toroidal inductors," *3D-PEIM 2016 - 2016 Int. Symp. 3D Power Electron. Integr. Manuf.*, 2016.

[13] M. L. F. Bellaredj, S. Mueller, A. K. Davis, Y. Mano, P. A. Kohl, and M. Swaminathan, "Fabrication, characterization and comparison of composite magnetic materials for high efficiency integrated voltage regulators with embedded magnetic core micro-inductors," *J. Phys. D. Appl. Phys.*, vol. 50, no. 45, p. 455001, Nov. 2017.

[14] G. Helgesen, A. T. Skjeltorp, P. M. Mors, R. Botet, and R. Jullien, "Aggregation of magnetic microspheres: Experiments and simulations," *Phys. Rev. Lett.*, vol. 61, no. 15, pp. 1736–1739, 1988.

[15] H. WANG, Y. YU, Y. SUN, and Q. CHEN, "Magnetic Nanochains: a Review," *Nano*, vol. 06, no. 01, pp. 1–17, 2011.

[16] T. M. Crawford, S. Carolina, and U. States, "Self-Assembled Magnetic Materials", *Comprehensive Supramolecular Chemistry II*, vol. 9, pp. 257-272, 2017.

[17] L. Hu, R. Zhang, and Q. Chen, "Synthesis and assembly of nanomaterials under magnetic fields," *Nanoscale*, vol. 6, no. 23, pp. 14064–

14105, 2014.

[18] J. B. Tracy and T. M. Crawford, "Magnetic field-directed self-assembly of magnetic nanoparticles," *MRS Bull.*, vol. 38, no. 11, pp. 915–920, 2013.

[19] J. WANG, S. CAO, S. XIA, and N. GAN, " The Assembly of Chain-Like Ni Arrays under Magnetic Field and Its Enhanced Properties ," *Int. J. Nanosci.*, vol. 09, no. 05, pp. 543–547, 2011.

[20] M. Wang, L. He, and Y. Yin, "Magnetic field guided colloidal assembly," *Mater. Today*, vol. 16, no. 4, pp. 110–116, 2013.

[21] J.-Y. Tinevez, "Directionality - ImageJ," 2010. [Online]. Available: http://imagej.net/Directionality.

[22] T. Wagner and M. Hiner, "Ridge Detection - Imagej," *Imagej.net*, 2016. [Online]. Available: http://imagej.net/Ridge_Detection.

Implementation of a prototype embedded system for in-car multipoint temperature measuring

Madalin Vasile Moise[1], Paul Mugur Svasta[1] and Alin Gheorghita Mazare[2]

[1] Center for Electronics Technology and Interconnection Techniques, Polytechnic University of Bucharest, Romania

[2] Faculty of Electronics, Comunications and Computers, University of Pitesti, Romania

+40751755058, madalin.moise@cetti.ro, paul.svasta@cetti.ro, alinmazare@yahoo.com

Abstract

The core of this paper is an explicit, complete design and implementation of a custom prototyping embedded system PCB capable of multipoint temperature measurement, hazard detection and other functionalities given by the possibility of attaching additional hardware. The additional hardware are modular printed circuit boards (PCB) that are stackable onto the main prototype PCB in order to increase its functionality. In this way, a GPS and GSM board module together with a relay controlled speaker are connected to the prototype system. They are used to transform the in-car multipoint temperature measuring system into an IoT device accessible everywhere, capable of detecting movement, temperatures, fire and generate alarms in a hazard situation. It is designed to be used in automotive industry but also in home or industry automatization. Thanks to a reduced complexity and having a simple interface, people which are new to electronics and programming can use this prototype to put their own ideas in practice fast and easy. The embedded system is developed on an ATmega 2560 microcontroller and the temperature detection is achieved using AMG8834 sensor which has a two dimensional area with 8x8 (64) infrared pixels providing 64 individual temperature readings. The experimental results were obtained by developing the module and using it to monitor an environment through GSM connection.

Key words: Embedded, PCB, microcontroller, Remote temperature monitoring, GSM modem, GPS module

Introduction

A stifling hot car isn't only uncomfortable, it's also dangerous for people's health. Overexposure to high air temperatures can cause heat illness such as heatstroke and hyperthermia, which could leads to a possible death.

Every year in the United States, an average of 37 children die after being left in hot cars, according to researchers of a new study published online in the journal Temperature [1].

Vehicles heat up rapidly, with the majority of the temperature rise occurring within the first 15 to 30 minutes [2]; interior temperatures rise very rapidly and were found to reach the critical temperature of 40°C in about 8 minutes on a typical clear summer day. A peak cabin temperature of 75.1°C can be reach while some surfaces of the interior can reach 80°C [3]; also in the event of a fire caused by a short circuit inside the vehicle, the temperature will only increase locally and it will take some time until it is detected by a car sensor.

The time passed until the fire is detected it's crucial, influencing resulting damage. By using the multipoint temperature sensor, we measure temperature simultaneously in multiple points with

higher speed and any detected danger can be countered faster.

The system has the ability to send alarms to outside world via GSM, send accurate GPS location, it can also receive commands, turns it into a system whose operation can be easily expanded with future improvements.

There are so many separated modules which incorporates GPS, GSM, or temperature monitoring on the market today, this system advantage over them is that it incorporate all in one with the capability to measure in multiple points the temperature of a surface (human body, car parts), not the temperature of surrounding air as the existing modules on the market do.

The proposed system mainly targets autonomous vehicles, as they become more and more advanced they will act as a driver, picking childs from school, transporting injured persons as an ambulance and many other activities. As there is not a human supervisor in the car the need for an alert in case of fire arise. It is designed for advanced tracking and monitoring and can be used in any vehicle to track it's location in an event of a fire, it comprises GSM modem (SIM800L), GPS module (PMB-648), microcontroller (ATMEGA2560), and temperature sensor (AMG8834). The microcontroller program is achieved using C/C++ language, with AT Commands to the GSM modem and serial to the GPS module.

The GSM modem and GPS module are interfaced with microcontroller ATMEGA2560 to alert the driver or the vehicle owner via short message (SMS) in the event of an emergency, when the reading temperature inside the car with AMG8834 increases.

The system additionally provides access to the measurement and troubleshooting points of different electronic components of the project; a much easier understanding of infrared temperature sensors; illustration of concepts used in modern electronics, such as intelligent temperature sensor, C/C++ programming, interrupt interfaces and I2C interface.

Overview of the platform

The proposed IoT platform is developed in three stages. Stage one is the development of a High Precision Infrared heat sensor which uses microelectromechanical systems (MEMS) technology capable of mapping a surface heat and making the user aware of any danger.

The second stage is the development of a prototype board based on a micrcontroller that is energy efficient, quick responding and easy to understand, allowing for further development and the third stage is designing an ergonomically case to incorporate everything in.

High Precision Infrared heat sensor which uses MEMS technology

The proposed approach will be controlled with the support of ATmega2560 which will be placed conveniently inside the vehicle. The SIM800L GSM/GPRS and PMB-648 GPS modules communicate with the microcontroller via serial interface while AMG8834 matrix temperature sensor use I2C for communication. While most of the temperature sensors used inside the car are measuring the ambient temperature, our concept detects the heat generated by the human body or by objects as infrared rays, resulting in a better accuracy in temperature monitoring and movement detection as the sensor distinguishes the heat between the air and human body or objects.

AMG8834 sensor uses infrared (IR) radiation versus conduction heat transfer, which provides a unique solution that allows for new levels of performance and reliability to be achived in many constrained applications.

Fire involves a set of dangers, the first and most obvious are burns, also others such as smoke inhalation poisoning. This project tries to create an economic device that helps to prevent fires and their early detection in order to save lives and economic goods.

In there is an increase of temperature in one or more points inside the car an alarm will be raised and a message containing information about the location and temperature will be sent to a predefined list of contacts via GSM connectivity. The flames are detected by processing the data collected by the MEMS sensor and analyse the constant change of each pixel value in consecutive frames.

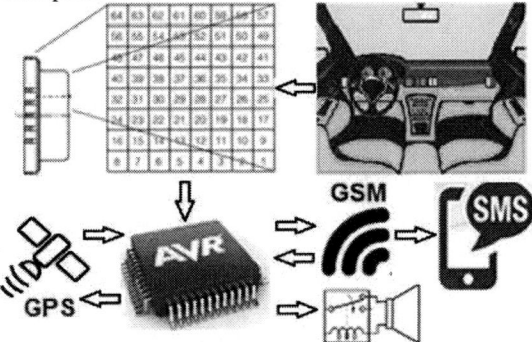

Figure 1: System block diagram

Figure 1 illustrates the proposed IoT system architecture. It consists of GSM modem (SIM800L), GPS module (PMB-648), microcontroller (ATmega2560), High Precision Infrared Array Sensor based onF Advanced MEMS Technology (AMG8834) and a relay driven speaker. The main components of the system can be described as follows.

MEMS Temperature sensor AMG8834

The AMG8834 is a Grid-Eye infrared array sensor in 14 pin SMD module, it detects the heat (infrared rays) of the human body and other objects, from -20°C to 100°C with ±3°C accuracy for up to 10 times per second. Each pixel have a field of view of 7.5° and the range in up to 7 meter making it the perfect choice to be used inside the car. It allows forming the thermal image of surrounding objects in the form of an 8x8 matrix and send it to the microcontroller. Each cell of the matrix is associated with the temperature of the object that has fallen into the visual field of a particular sensitive cell.

The AMG8834 module is a hybrid assembly placed in a cermet case. It combines an optical lens, a ceramic base, an infrared detector in the form of an 8x8 MEMS array, a thermistor and a digital-to-analog integrated chip, which is needed for processing signals from the array, control, and communication with a microcontroller.

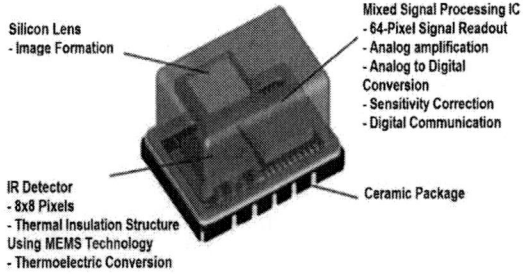

Figure 2: AMG8834 features

The principle of operation of the AMG8834 is as follows: thermal infrared rays from surrounding objects are directed by a lens to a sensitive matrix of 8 x 8 sensors. The viewing angle of each sensor is approximately 7.5 °, the total viewing angle of AMG88 is 60°. The flux of IR radiation, being absorbed by the sensors of the matrix, is converted into thermo-emission. Further, the signals are fed to the amplifier using the multiplexer, after this, data is digitized and the thermal image is generated using digital signal processing methods. After processing, thermal image is available to the user. The communication with AMG8834 is made using the high-speed I2C bus. The frequency of reading can vary from 1 to 10 frames/s.

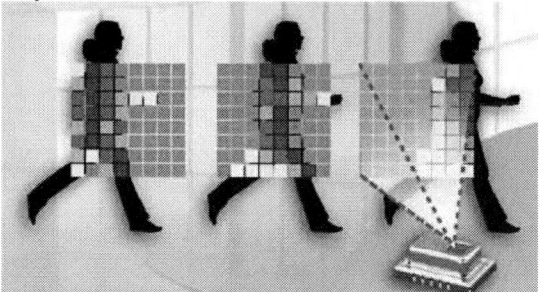

Figure 3: Principle of operation of the AMG8834

GSM Module SIM800L

SIM800L is a miniature cellular module which allows for GPRS transmission, sending, receiving SMS and making and receiving voice calls and supports SIMCOM enhanced AT commands.

Low cost, small footprint and quad band frequency support make this module the perfect solution for any project that requires long range connectivity. After connecting power the module boots up, searches for cellular network and login automatically.
Specification:
  - Power consumption:
    • Sleep mode < 2.0mA
    • Idle mode < 7.0mA
    • GSM transmission (avg): 350 mA
    • GSM transmission (peek): 2000mA
  -Supply voltage: 3.8V - 4.2V
  -Supported frequencies: 2G Quad Band (850 / 950 / 1800 /1900 MHz).

GPS Module PMB-648

The GPS module has built-in SiRFstarIII chipsets receivers which give unparalleled GPS performance and precision, 20 parallel satellite-tracking channels for fast acquisition and reacquisition, a low power consumption and ultra mini size only 32x32mm with a built-in antenna. It is designed for Car Navigation, Marine Navigation, Fleet Management, AVL and Location-Based Services, Auto Pilot, Personal Navigation or touring

devices, Tracking devices/systems and Mapping devices application. The communication with the microcontroller is made via serial port TTL at 4800 bps (Optional 9600, 19300, 38400 bps).

ATmega2560

The platform microcontroller is an ATmega2560 microprocessor which is one of the most popular Atmel's AVR microprocessor. The Atmel ATmega2560 is a low power CMOS 8 bit microcontroller based on the AVR enhanced RISC architecture. By executing powerful instructions in a single clock cycle, the ATmega2560 achieves throughputs approaching 1MIPS per MHz allowing the system designer to optimise power consumption versus processing speed.

This is programmed in C and much of the programming involves direct access to the processor's registers. A separate hardware programmer is needed to communicate between the programming environment and the processor. The microcontroller has a maximum operating frequency of 16MHz, 256KBytes of flash, 4Kbytes of EEPROM, 8Kbytes of SRAM, and a variety of analog and digital peripherals.

Platform Developing

We have designed the platform as open-source hardware/software allowing the developers to build prototypes much cheaper and faster customizing it in order to meet their needs.

The printed circuit board (PCB) is capable of multiple functionalities by attaching additional hardware. The role is to reduce the complexity of work in new systems and to facilitate immediate results. We designed the PCB so that a user unfamiliar with the field can easily understand them and quickly develop an application exponentially cheaper and faster process.

For building the platform, we used both existing modules on the market (e.g. temperature sensor) but also products that we had optimized, by doing so a later developer will be able to extend the range of the platform usage by easily adding new or optimized modules.

Platform is designed to make easy access to ATmega2560 inputs and outputs, to have a low energy consumption and to have troubleshooting, programming, communication and debugging interface through an ICSP (In Circuit Serial Programming) interface. We have also considered ease of attaching new modules by including connectors for each microcontroller pin on the PCB.

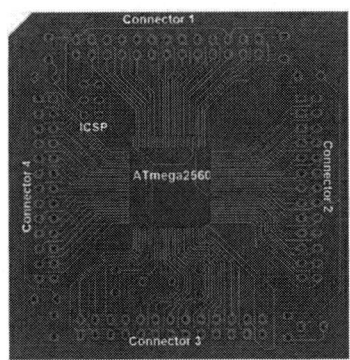

Figure 7: Platform Printed Circuit Board (PCB)

We have also designed an ergonomically case to incorporate everything in one place.

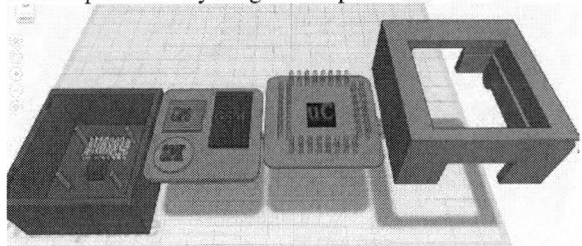

Figure 8: Overview of the stackable modules and case

It is designed in such a way that everything is stacked one on top of the other like a sandwich while still it allows access to measurement and future adding stacked modules.

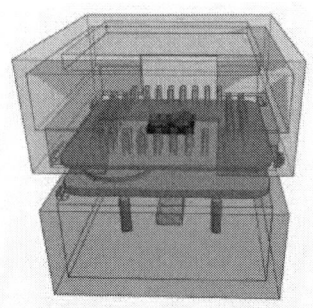

Figure 9: Case enclosure

Results

If there is a short circuit inside the dash board, the temperature increase will be just locally, the sensor will detect it faster than a thermometer that is measuring the temperature of the surrounding air. The sensor can also be used to detect motion inside the car which can be helpful if a pet is forgotten inside the car or if a burglar want to steal the car.

In order to test the sample application, we have set up a test where we heat up a part of surface at two different temperatures and measure its temperature, if a high temperature is read the microcontroller will send a text message containing the temperature information and GPS location. The first measurement is perform at ambiental temperature, the second one is after we've slightly heat up one zone and the third after applying a higher temperature on the same surface.

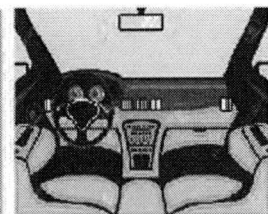

Figure 10: Ambiental temperature read

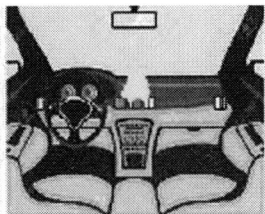

Figure 11: Temperature read after slightly heat up

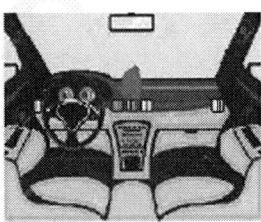

Figure 12: Temperature read after high heat is applied

Conclusions

Infrared thermal sensing and imaging sensor make it possible to measure and map surface temperature and thermal distribution passively and no intrusively. Thermal infrared detectors are distinguished by the advantages of a wide wavelength response, no requirement for cooling, high-temperature stability, high signal-to-noise ratio and low cost. We've designed the systems so that a user unfamiliar with the field can easily understand them and quickly develop an application.

References

[1] Evaluating the impact of solar radiation on pediatric heat balance within enclosed, hot vehicles, abstract, Jennifer K. Vanos, Riane Middel, Michelle N. Polletty, Nancy J. Selover, 23 may 2018, Taylor & Francis online.

[2] McLaren, Null, Quinn. Heat Stress from Enclosed Vehicles: Moderate Ambient Temperatures Cause Significant Temperature Rise in Enclosed Vehicles. Pediatrics Vol. 116 No. 1. July 2005.

[3] Temperatures in Cars Survey, Russell Manning, John Ewing, February 2009

PrintPOWER – Paste systems for multifunctional copper power modules

K. Reinhardt, S. Körner, and U. Partsch

Fraunhofer Institute for Ceramic Technologies and Systems IKTS, Winterbergstraße 28, 01277 Dresden

+49 351 2553 7837, kathrin.reinhardt@ikts.fraunhofer.de

Abstract

The aim of the present work was the development of paste systems for use in the high temperature and low temperature range, which can be used for the construction of copper-based power electronic components. In doing so, a feasibility was to be elaborated which shows the combination and functionality of all paste systems, so that in future individual power modules can be set up. For this purpose, the following single-paste systems were developed or further developed: copper thick film pastes, glass insulations, silver polymer pastes and Al₂O₃ polymer insulation. To verify the compatibility of the individual developed paste systems, a printed multi-layer module was set up and the layers tested for functionality. It could be shown that the developed paste systems can be combined with each other and can be sintered or hardened under nitrogen atmosphere without suffering a loss of function. Accordingly, it is possible in the future to build application-related power modules with the previously developed paste systems.

Key words: Thick-film power modules, copper multilayer circuits, high temperature pastes, low temperature pastes, rheology, 3D dispensing

Introduction

Power electronic systems are the backbone of the energy transition and environmentally friendly individual electro mobility. Modules in which power semiconductors and other components are integrated form the heart of such systems. Copper thick films combine the advantages of DCB substrates, which are dominant in the module sector, and the silver thick-film technology used for highly reliable signal processing circuits. The focus of the work was the development of copper-compatible paste systems (conductive and dielectric pastes) and their combination to create multilayer module structures on ceramic substrates. The printed modules on the multifunctional substrates should be able to contain, in addition to the power semiconductors, various other active and passive components. This integrated functionality at the module level, let these modules count to the class of the "Intelligent Power Modules" (IPM).

Experimental

A. Materials and Paste Preparation

For the development of high-temperature (HT) and low-temperature (LT) suspensions, the materials listed in Table 1 were used as inorganic main components. For the HT copper metallizations the inorganic powders were mixed together in a roller bench for 25 min. The mixture was ground together in a dissolver with a solution of polyacrylate in texanol. The pre-dispersed suspensions were finally homogenized using a three-roll mill (EXAKT 120E). The HT glass isolation as well as the LT suspensions (Ag metallization and Al₂O₃ isolation) were comparable formulated. The inorganic powders were dispersed in polymer solutions (HT isolation: polyacrylate/texanol solution; LT suspensions: PVB/Terpineol solutions) in a SpeedMixer™ (DAC 150 FVZ Hauschild & Co. KG) and after this also homogenized using a three-roll mill.

Table 1: Investigated suspensions depending on the inorganic main component.

Type of suspension	Inorganic main component	FESEM image
HT metallization	Copper	
HT isolation	Glass	---
LT metallization	Silver	
LT isolation	Alumina	

B. Printing Tests

All HT suspensions and the LT insulation suspension were screen printed. The printing tests were carried out on a screen printer Microtronic II from EKRA. The characterization layout for the HT Cu suspensions is shown in figure 1a) and for the HT glass and LT polymer suspensions in figure 1b).

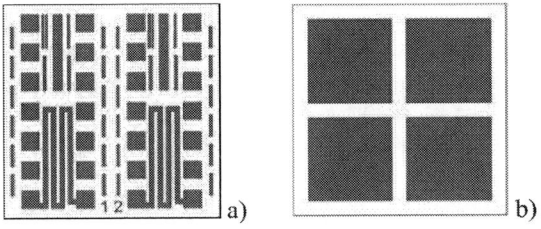

Figure 1: Screen-Printing test layouts for a) HT Cu suspensions, b) HT glass and LT Al₂O₃ insulations.

The LT silver suspensions were deposited on a dispenser from Asymtek, Axiom 1000 X-1010.

For the demonstrator a test layout shown in figure 2 was used, where the feasibility of the combination of all pastes for using in a power module were tested.

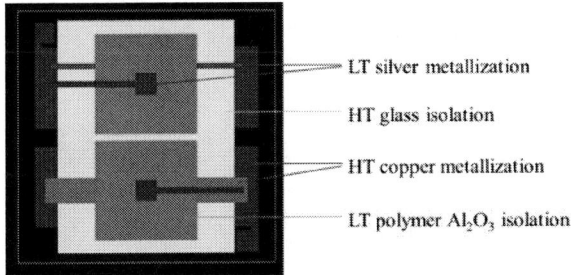

Figure 2: Demonstrator test layout for printed power electronics modules.

C. Electrical and mechanical characterization:

The following investigations were carried out on the deposited layers:

• Electrical (resistance measurements) and
• Mechanical characterization (wire-peel, pull stud and tesaband test)
fired and hardened layers (including aging tests such as thermal aging (150 °C, 100 h).

Results and Discussions

A. HT copper metallization

The main aim for HT copper metallization was the further development of the copper paste system with regard to achieve higher aspect ratios for higher conductivity. For this different surfactants were tested which should improve the printing resolution. Therefore, it is necessary to use additives, which burn out under nitrogen atmosphere

without any residue. The results of the following three copper pastes are presented:

• Cu paste A: without any surfactant,
• Cu paste B: with surfactant I (softener),
• Cu paste C: with surfactant II (surface additive).

All copper layers were fired at 955 °C for 15 minutes under nitrogen atmosphere in a belt furnace. An overview of the printing qualities and the reached aspect ratios (AR = Height / Line width) is compiled in figure 3. The addition of additives allows an increase in the layer thickness and a reduction in the line width. It can be seen that the Cu paste C with the surface additive achieves the highest aspect ratio (0.21), whereas the omissions of additives results in a very low aspect ratio (Cu paste A, AR = 0.05). This shows that the addition of additives can be used specifically for setting geometric line properties. Direct control is possible. However, an improvement in the aspect ratio results in a deterioration of the surface quality, so that for the Cu paste C a light screen structure can be observed.

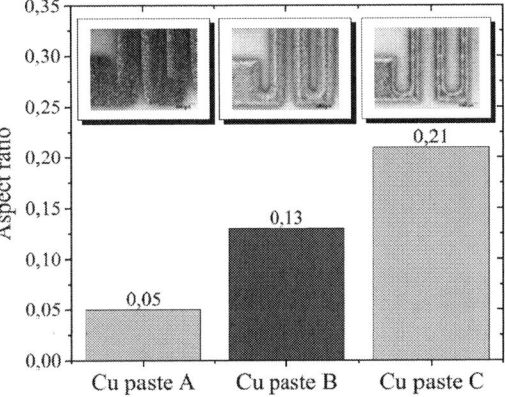

Figure 3: Aspect ratios of the examined HT Cu metallizations with and without surfactant addition on Al₂O₃ substrates, including microscopic pictures.

The Cu pastes were also evaluated for their electrical conductivity (Figure 4). By adding additives, the electrical conductivity could be increased. Thus, an increase of 430 kS/cm for Cu paste A to 451.4 kS/cm was achieved for Cu paste C, which corresponds to 77.8 % of the Cu bulk resistance. This increase can be explained on the one hand with the improvement of the aspect ratio of the printed layers, on the other hand, the addition of additive allows an increase in the solids content. Thus, for Cu paste A only a maximum solids content of 85 % was achieved, whereas Cu paste B contains 88 % solids and Cu paste C 89 %. The reduction in conductivity despite increasing the solids content for Cu paste B may possibly be explained by incomplete burnout of the additive during the nitrogen-firing step. Carbon residues can negatively affect the sintering behavior and increase the electrical resistance of the layer.

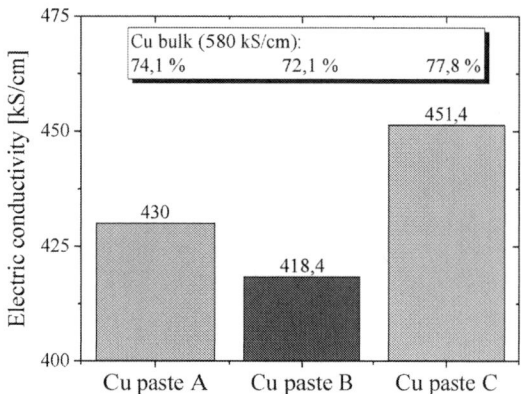

Figure 4: Electric conductivity of the examined HT Cu metallizations with and without surfactant addition on Al₂O₃ substrates.

The mechanical adhesion of the layers was characterized by wire peel test. In this case, the adhesion of the sintered layer to the substrate is not measured directly, but the adhesion of a soldered connection to a sintered layer is measured. Figure 5 compares the results for the three HT Cu suspensions. The measurements being carried out initially after firing and after an aging of 100 h at 150 °C. It can be observed that the adhesive strength increases with increasing solids content. For the Cu paste C a maximum adhesion of 30.8 N/4mm² are achieved. With the exception of Cu paste B, the characterized pastes show no reduction in adhesive strength after aging at 150 °C. for 100 h. Again, the decrease in adhesion of Cu paste B after the aging step can be explained with possible carbon residues within the layer, which could reduce the layer properties. This indicates that the surfactants can probably influence the microstructure on the surface of the Cu layers (connection to the solder) or at the interface to the Al₂O₃ substrate. So the adhesion can be specific adjust.

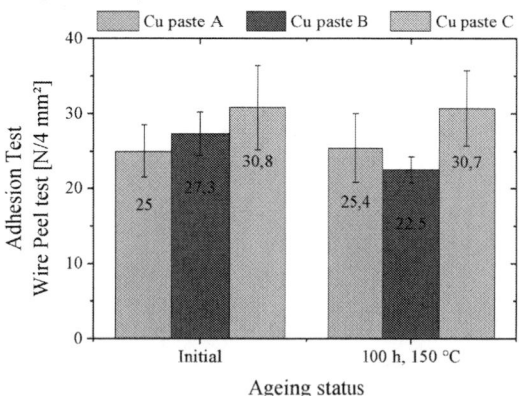

Figure 5: Wire-peel test results of the examined HT Cu metallizations with and without surfactant addition on Al₂O₃ substrates.

In conclusion can be said that varying the composition of copper pastes, in particular with rheological additives, it was possible to specifically set different layer geometries, different conductivities and different adhesion behaviors.

B. HT glass isolation

For the isolation of the HT Cu layers two different glasses were selected, one commercial (glass I) and one IKTS glass (glass II). First, they were tested on whether they can be sintered under a nitrogen atmosphere, so that forms a firm, defect-free layer and no reactions with the copper layers occur. Table 2 compares the sintered layers of the HT glass insulation suspensions of glass I and II as a function of the sintering end temperature on non-printed Al₂O₃ substrates and printed HT Cu layers (Cu paste A).

Table 2: Overview of screen printing qualities of HT glass insulations (glass I & II) on non-printed Al₂O₃ substrates and printed HT Cu layers.

Type of suspension	850 °C, 15 min, N₂	955 °C, 15 min, N₂
Glass I on Al₂O₃ blank		
Glass I on Cu paste A		
Glass II on Al₂O₃ blank		
Glass II on Cu paste A		

The glass I suspension shows glazed sintered layers on Al₂O₃ (row 2) irrespective of the sintering end temperature, which have a light gray color. The gray coloration can be attributed to small carbon residues within the glass layer, which probably occur during the nitrogen burn-out of the polymer solution. Nethertheless it do not prevent the sintering behavior of the glass layer. Whitish stains within the glass layers may possibly indicate dewetting or poor leveling behavior of the suspension on the substrate surface. If the glass I suspension is applied to the HT Cu layer and sintered (row 3), interactions occur in the form of bubble formation between the glass insulation and the Cu layer as the sintering end temperature increases. A maximum sintering temperature of 850 °C is therefore possible for the glass I suspension on HT Cu layers.

The glass II suspension has a matt glass surface on Al_2O_3 (row 4) for a final sintering temperature of 850 °C., which forms a glassy transparent layer only at a sintering temperature of 955 °C. This shows that for the sintering of the glass II a temperature of 955 °C is necessary to form a dense insulation layer. Furthermore a strong surface structure can be established, which can be explained here due to the lower viscosity of the glass suspension (due to the chosen polymer solution). A comparable image is observed for the printing of the glass II suspension onto the HT Cu layer (row 5). Again, only a glassy, transparent layer is formed at a sintering temperature of 955 °C. Furthermore, no significant interactions of the glass insulation with the Cu layer can be occur.

The HT glass insulation layers were tested for their mechanical adhesion on Al_2O_3 and on the HT Cu metallizations. The pull stud test was carried out, in which the adhesion forces between the layers are measured directly by means of an epoxy stamped aluminum stud. For the test, the following samples were selected, which showed the best sintering results:

- Glass I: 850 °C, 15 min, N_2
- Glass II: 955 °C, 15 min, N_2

Figure 6 compares the results of the pull stud test of the selected samples. The glass insulation layers on blank Al_2O_3 show maximum adhesion values between 465 N/5.73mm² and 523 N/5.73mm², without any ageing cycle. The highest adhesion to Al_2O_3 can be observed for the glass I insulation, which was sintered at 850 °C. It was also possible to observe predominantly ceramic fracture for all layers on Al_2O_3 as the fracture pattern, that means the weakest point in this layer composite is the Al_2O_3 ceramic and high adhesive properties of the glass insulations can be assumed.

Comparing the results presented above with the adhesion tests of the glass insulations on HT Cu layers, the adhesion values for all glass insulations are reduced, but to varying degrees depending on the respective glass insulation. For the glass I insulations a significantly greater decrease in adhesive strength is occurred, which decreases by more than 65% (162 N/5.73mm²). The glass II insulation has a 34% percent reduction, but has the highest adhesive properties on HT Cu layers (306 N/5.73mm²) compared to all other glass insulations. The fracture pattern of the glass insulation on Cu layers is not like for the pure Al_2O_3 substrates predominantly ceramic fracture, but the aluminum stamp triggers in the peel test above all the glass layer, the Cu layer still adhered to the substrate. This could be determined for all examined samples.

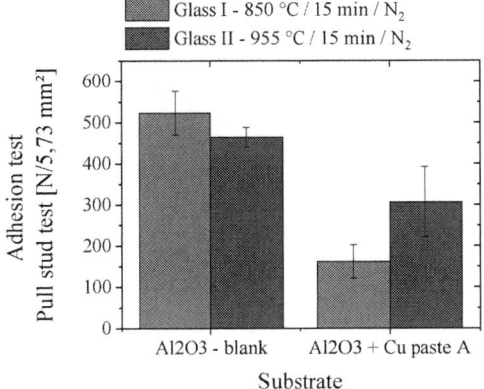

Figure 6: Pull stud test results of the investigated HT glass insulations on non-printed Al_2O_3 substrates and printed HT Cu layers.

In conclusion, it can be said that a direct relationship to the sintering end temperature cannot be found with the samples selected here, but a clear difference with respect to the underlying layer of the glass insulation.

C. LT silver metallization

To replace wire bonding in later power electronic components, polymer based pastes will be evaluated. These can be deposited in addition to the screen printing, inter alia, with 3D substrate compatible deposition methods such as dispense jetting. For the LT Ag metallizations one silver powder was tested which was mixed in two different PVB polymer solutions. A short-chain (Ag paste A) and a longer-chain polymer (Ag paste B) was selected. The silver pastes were deposited by dispense jetting on Al_2O_3 substrates with prefabricated contact pads. This was followed by a variation of the curing temperature and curing time for both pastes and a subsequent electrical characterization.

Figure 7 shows the dependence of the sheet resistance on the curing conditions of the LT Ag pastes A and B. As the thermal input increases, the resistance of the functional layers decreases. The Ag paste B layers tend to have a lower resistance (8.2 mOhm/sq @ 280°C for 30 min) than the Ag paste A layers (9.2 mOhm/sq @ 280°C for 30 min) - this is a result of the slightly higher silver content in the hardened layers of the Ag paste A. With regard to a further lowering of the sheet resistance, current work focuses on a variation of the silver powder and an optimization of the binder system in order to improve curing conditions to lower temperatures and times. Sheet resistance less than 6 mOhm/sq can be observed for curing conditions of 200 °C for 10 minutes.

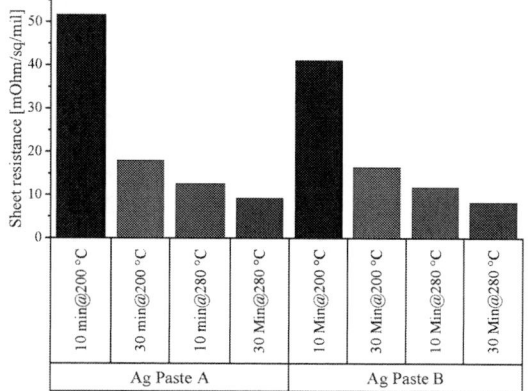

Figure 7: Sheet resistance of LT Ag pastes A and B depending on the curing conditions.

D. LT polymer Al_2O_3 isolation

The functionality of the insulation layers was demonstrated by measuring the insulating resistance and breakdown voltage. Figure 8 shows the insulation resistance (columns) and the associated breakdown voltage (symbols) of the polymer-based Al_2O_3 insulation paste as a function of the extent of the aging of the suspension (black - initial measurement after layer hardening, red - aging for 500 h at 150 °C, blue - aging for 500 h at 85 °C and 85 % relative humidity). The test structures were first built up by screen printing. As electrodes for the power supply, the silver polymer pastes developed in the project were used. The characteristic values of the initial measurement for the insulation resistance are in the range of 4.1E12 ohms - the pertinent breakdown voltage is 1000 V. The aging of the samples at 150 °C for 500 h causes the insulation resistance to increase again to 4E13 ohms - compared to the initial ones the breakdown voltage remains constant at 1000 V. The increase in the insulation resistance can either be an indication that the layers were initially not completely cured or that the polymer absorbs humidity, which was removed by the long removal time at 150 °C and so on the insulation resistance increases. An indication of the second possibility shows the aging at 85 °C and 85 % relative humidity. After a storage time of 500 h, the insulation resistance drops to 2.2E10 ohms and the breakdown voltage drops to 488 V (without sample 3, with sample 3 to 570 V). Whether this process is reversible or can be reduced by adjusting the parameters during initial hardening remains to be investigated.

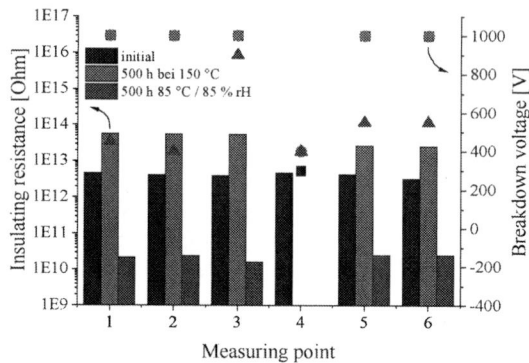

Figure 8: Insulation resistance (columns) and associated breakdown voltage (symbols) of the polymer-based Al_2O_3 insulation paste depending on the aging condition: black - initial measurement after layer hardening, red - aging for 500 h at 150 °C, blue - aging for 500 h at 85 °C and 85 % relative humidity.

E. Demonstrator

For demonstration purposes and to verify the compatibility of the developed paste systems, a multi-layer module was printed and various functional tests were performed. For the demonstrator setup, the following suspensions were selected:

- HT Cu metallization: Cu paste A (without surfactants)
- HT glass insulation:
  Glass paste I, sintering end temp.: 850 °C
  Glass paste II, sinter end temp.: 955 °C
- LT Ag metallization: Ag paste B
- LT polymer Al_2O_3 isolation.

The structure was based on the layout in figure 2. The printing order was chosen as follows:

→ Print and firing first HT Cu metallization

→ Print and firing HT glass insulation

→ Print and firing second HT Cu metallization

→ Print and curing first LT Ag metallization

→ Print and curing LT polymer Al_2O_3 insulation

→ Print and curing second LT Ag metallization.

It could be shown that the developed paste systems can be combined with each other and can be sintered or hardened under nitrogen atmosphere without suffering a loss of function (see figure 9). Accordingly, it is possible in the future to build application-oriented power modules with the previously developed paste systems.

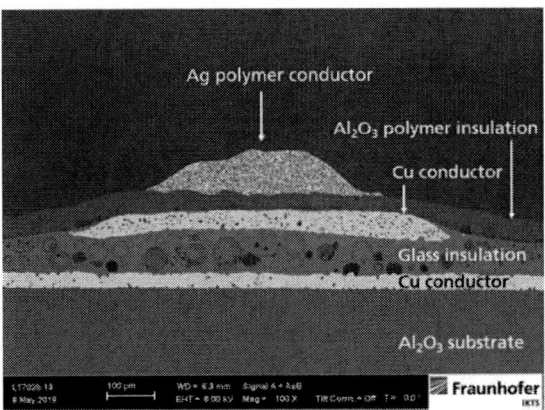

Figure 9: Cross-section of a printed multi-layer module demonstrator.

The characterization of the demonstrator was based on the measurement of sheet resistances between the individual metallizations. Figure 10 shows the demonstrator for the printed power electronics module with numbered measuring points for resistance determination. These are sintered copper pads (blue; Cu paste A), a sintered glass insulation (beige; glass I or II), a sintered copper conductor (red; Cu paste A) and a hardened silver conductor (gray; Ag paste B) and polymer-based insulation layer (green; Al2O3 paste). The sheet resistances of the HT Cu metallization (red interconnect structure Figure 10) and the LT Ag metallization (gray interconnect structure Figure 10) are shown in Figure 11. The resistors have not been normalized to a layer thickness.

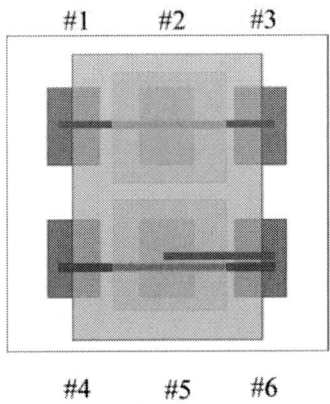

Figure 10: Demonstrator layout for printed power electronics modules with numbered measuring points for describing electrical resistance measurements.

The sheet resistance of the silver polymer paste B is 8.6 mOhm/sq (substrate Cu paste A / glass I) or 6.0 mOhm/sq (substrate Cu paste A / glass I, red columns) respectively lower than comparable curing conditions on test substrates (Figure 7). However, the hardening of the LT silver conductive path from # 1 to # 3 takes place in triplicate: initially after printing, a second time after the insulation layer

has been printed, and finally after the last print of the conductor from # 6 to # 5. Since the sheet resistance decreases with increasing thermal input of the paste, this difference can be explained. The difference between the two substrates is explained by the surface quality of the fired-glass insulations and the associated layer geometry of the silver guides. The sheet resistance between # 4 and # 5 is expected to be between that of the copper and the silver paste. The sheet resistances of the HT-Cu interconnects are 1.9 mOhm/sq (substrate Cu paste A / glass I) or 1.7 mOhm/sq (substrate Cu paste A / glass II, blue columns) in a resistance range that corresponds to the preliminary investigations.

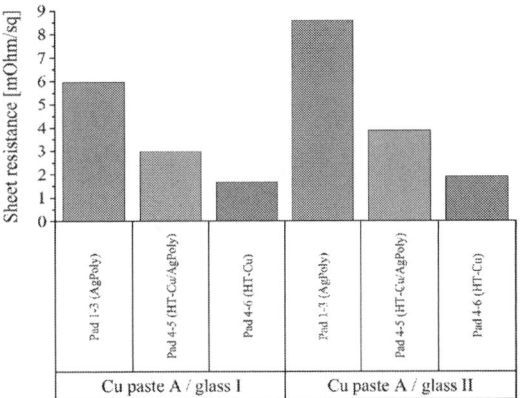

Figure 11: Sheet resistance of substrates Cu paste A / glass I and Cu paste A / glass II. red - sheet resistance Ag paste B from # 1 to # 3; green - sheet resistance Cu paste A / Ag paste B from # 4 to # 5; blue - sheet resistance Cu paste A from # 4 to # 6.

Conclusion

By using the developed copper thick film pastes and adapted insulation pastes for inert firing or curing, several functions can be embedded in power module structures. Furthermore, new contacting options were evaluated via 3D-printed silver polymer suspensions, which could replace established technologies, such as bonding or Siemens Planar Interconnect Technology (SiPLIT), and are more cost-effectively and with high reliability.

As part of these investigations, the following single paste-systems have been developed:

- Copper thick film pastes for thick print and fine line applications: By varying the composition of copper pastes, in particular with rheological additives, it was possible to specifically set different layer geometries, which enable both thick printing and fine line printing.
- Glass insulations for nitrogen atmospheres: Two glasses were evaluated which sinter densely and transparently at temperatures of up to 955 °C and have only small carbon residues within the layer. This makes it possible to use the developed glass pastes as dielectric layers, but

also as covering layers. Furthermore, the development of glass insulation for sintering temperatures $\geq$ 850 °C allows a multilayer structure with HT-Cu systems, so that robust circuits can be integrated in several levels.

- Silver polymer pastes: For the low temperature range, a silver metallization has been developed, which has very good conducting properties and can be deposited by means of dispensing printing processes. This will allow future use of new 3D contacting options that can replace established technologies such as bonding or the SiPLIT method more cost-effectively and with higher reliability.

- Al_2O_3 polymer insulation: In addition, an Al_2O_3 polymer insulation was developed, which has good insulation properties and can complete the assembly of a power multilayer module.

Metalization impact on heat transfer through sintered nanosilver based thermal joints

Krzysztof Stojek[1], Jan Felba[1], Przemysław Matkowski[1], Damian Nowak[1], Mateusz Kanus[1] and Andrzej Mościcki[2]

1) Wrocław University of Science and Technology, Faculty of Microsystem Electronics and Photonics, Wrocław, Poland
2) Amepox Microelectronics Co. LTD, Łódź, Poland

+48 71 320 49 76, krzysztof.stojek@pwr.edu.pl

Abstract

Research aim was to develop thermal analysis method of thermal joints made by sintered silver nanoparticles in real structures. Each layer in sandwich-like structures in typical microelectronics structures introduces thermal resistance of layer and simultaneously thermal resistance of contacts between neigbouring layers. Each thermal resistance affects heat transfer. The idea was to find if reducing number of layers improves heat transfer efficiency. Thermal analysis were performed with using infrared camera (thermographic method), where main advantage is that measurements are contactless. Typical thermal analysis methods for strongly conductive layers (like sintered silver) are insufficient,because temperature difference might be at level 0.1 K and using variable signals may solve such problem.

Key words: thermography, silver sintering, heat transfer, thermal joints, silver nanoparticles

Introduction

The miniaturization in microelectronics and faster operation of devices like transistors on processors surface, LEDs and other elements considered as power electronics cause increased heat generation during work. Heat dissipation becomes crucial in modern electronics, especially if considering second level of microelectronics packaging. Both, industrial and scientific societies are interested in developing materials, which role is to efficiently transfer heat from heat sources (usually semiconductor chip) to heat receivers (e.g. DBC-substreate). Of course, two additional requirements should be fulfilled: mechanical joining of an electronic element to base structure and its electrical connection. Mostly, the reflow soldering technology is used as all these targets are met; joining with the use of an electrically/thermally conductive adhesive is used to a much lesser degree. Both these techniques are strongly limited in case of high temperature and high-power electronics. As an alternative, in electronic packaging, a low-temperature joining technique (LTJT) by silver sintering can be used. The joining temperatures of LTJT and SnAg soldering are similar (usually below 300 °C); however, the maximum operating temperature, electrical conductivity, thermal conductivity, density and coefficient of thermal expansion (CTE) for the LTJT in comparison with soldering are much higher. The mechanical strength of joins usually are not so high, but applicable in electronic packaging [1, 2].

As standard, metallized surfaces are joined. There are many technological factors determining the quality of the joints [3], however usually two parameters are distinguished the process; the sintering temperature and external pressure. The typical sintering temperature and pressure are about 300 °C with 10-20 MPa pressure assistance. During the research used sintering process was called low temperature and low pressure because the values ot them were 230 °C and 0.5 MPa, respectively.

As standard, metallized surfaces are joined. However, when tested joining materials have an addition of epoxy resin it is possible to manufacture thermal joints for non-metalized semiconductor's surfaces [4]. Such advantage gave us opportunity to determine if metalization on surfaces affects heat transfer or not. It is important if we consider multilayer matelization. In such case every material is introducing thermal resistance of its own and also thermal resistance of contact between neigbouring layers, which affects heat transfer most significantly. For heat transfer, the lower thermal resistance, the higher heat transfer, than the lower themperature difference between elements and at the end the heat source temperature decrese. In such manner, removing metalization may lead to removing layer thermal resistance and also contacts thermal resistance, which should improve heat transfer.

The idea is to remove metalizations and bond to non-metallized semiconductor, which could reduce manufacturing cost by process simplification. So the clue of the reaserch was to determine the metalization impact on heat transfer through silver based thermal joints.

Used Materials

Within research laboratory developed material for sintering was done by Amepox Microelectronics LTD. The composition was 95 wt.% of silver in ratio 40:60 silver microflakes to silver nanoparticles nanoparticles. Epoxy resin was 5 wt.%. Sintering was done for silver layer formation and epoxy resin addition was responsible for mechanical strength.

Silver sintering

Research for application of silver sintering process in thermal joints manufacturing was researched in order to determine proper materials composition and process parameters. The number of factors affects both, thermal and mechanical properties, like material composition, amount of liquid fraction - viscosity, amount of epoxy resin, surface metalization types, surface area, material of joined elements, roughness of joined surfaces, dimensions and shape of particles, deposition type, thickness of deposisted layer, pre-drying, pressure, drying and sintering time in proper temepratures, atmosphere and cooling, etc [3]. During the research most of the factors were constant, the only variable was the presence of metalization. Sintering was done in 230 °C with assistance pressure 0.5 MPa in time 60 minutes in air conditions.

Samples

The samples were done by two bare-die self made silicon chips. Their geometry and size result from previous experiments enabling minimization of measurement error [5]. The first chip was threaten as a controllable heat source and second as a heat receiver – substrate. The area of the chip was 5 mm x 5 mm and their thickness was 3000 μm. Samples after sintering was attached to printed circuit board with using electrically conductive adhesive to enable supply of electrical power to the system as a controllable heat source (fig. 1).

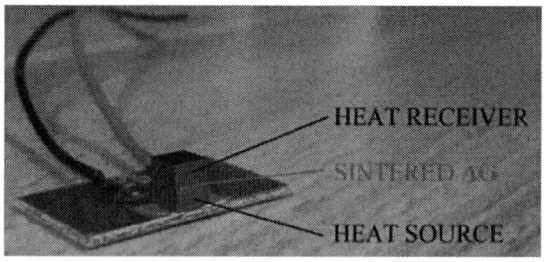

Figure 1: The sample.

Thin film of silver was fabricated by magnetron sputtering from circular target onto silicon substrates using the Pfeiffer Classic 570 vacuum system (Wetzlar, Germany) with a WMK-100 magnetron source. It was fabricated under an Ar pressure of 5.8×10^{-3} mbar and applied constant power of 320 W with discharge current 0.2 A. The average thickness of silver thin film was 0.25 μm.

Temperature Measurements

Temperature measurements was done with using Flir A40m infrared (IR) camera. Such attitude gave us possibility to perform contactless analysis, so there was no temperature transfer disturbance. Additionally, temperature distribution was analysed. Before measurements, the camera was calibrated and analysed surfaces was coved with thin layer of material with defined emmisivity. In such approach the thermal measurements error was on level ± 2 °C or ± 2 % and were taken higher value from selected both. Despite the temperature measurement error, the more important factor is thermal sensitivity which describe the lowest distinguishable value between two neighbouring areas and such value was at the level 80 mK, so the detection difference at level 0.1 K (or °C) was possible.

Typical temperature analysis based on analysis in steady state conditions, where electrical power is supplied to the system, the conditions equilibrium appears and temperature stabilization is present – steary-state analysis (fig. 2) [6].

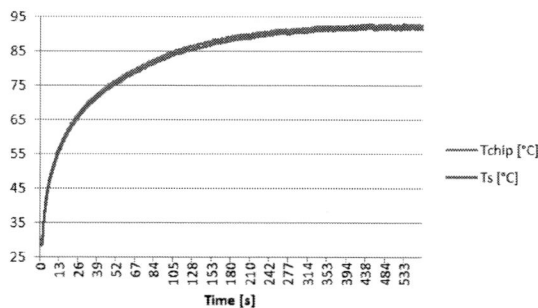

Figure 2: Steady-state temperature analysis [6].

Than the temperature on heat source and one heat receiver is measured and their difference in ratio to supplied measured power lead to thermal resistance determination. The temperature are performed in common plane for heat source (T_{CHIP}) and heat receiver (T_S), so it is possible to measure it simultaneously with incorporation the same error.

During our research the variable signals were used. Is was suspected, that it is possible to supply power to the system with using sinus signals. We used function generator and signal amplifier to supply power on level, that was transferred to temepratures near operation temepratures. In such

372

manner, the movie was recorded and was analyzed in time domain (fig. 3 – exemplar thermogram). Analized temperatures were mean temperature values for areas – red elipses 1 (heat source temperature) and 2 (heat receiver temperature). At the surface of heat receiver there is visible temperature disturbance, that was in such case surface covering and such thermograph was presented to show the impact of emmisivity on temperature measurements. During the research the layer was deposited correctly.

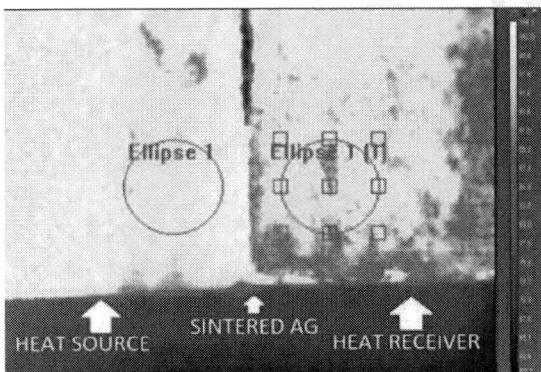

Figure 3: Examplar thermogram.

Results

As mentioned before, the samples were supplied with variable signals. Aproach based on putting amplified signals to the system and measure the temperature response for different frequencies of half of the sinus signal (fig. 4).

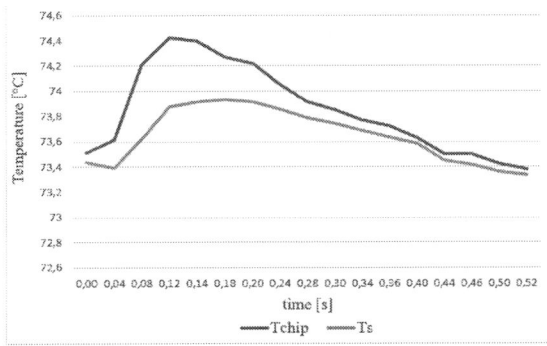

Figure 4: Amplified sinus supply temperature response on heat source and heat receiver.

Shown characteristics represents one cycle of heating and cooling by half of the sinus signal. Second half was cutted by using diode in the circuit. Presented characteristics was obtained with using 2 Hz. According to the result is was possible to determine maximal temperatures on both, heat source and heat receiver and in such case it was 74.4 and 73.9 °C respectivelly. The temperature difference was 0.5 °C. These temperatures were shifted in time obviously, which may also be analized but it is not the topic of presented paper.

Used IR camera have limitation determined by operating speed which is maximal 50 Hz. So to avoid signal aliasing maximal measurement frequency is 25 Hz. So the measurements were performed from 0.1 Hz to 25 Hz in selected values.

Such analysis was done in several frequencies: 0.1, 0.2, 0.3, 0,4. 0.5, 0.6, 0.7, 0.8, 0.9, 1, 2, 3, 4, 5, 6, 7, 8, 9, 10, 15, 20 and 25 Hz. Than the temperature difference was calculated to obtain statistics. In such case the conditions were reproduced and from structural point of view the metalization was the variable only. Mean value of temperature diffierence, as well as standard deviation and median was calculated (fig. 5).

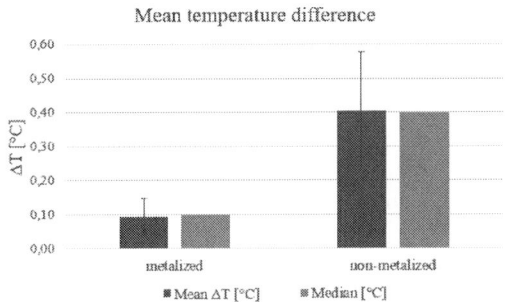

Figure 5: Statistical analysis of metalization impact on temperature difference via sintered silver joint.

Lower temperature difference was achieved for metalized surface of thermal joint. Nevertheless, the temperature difference between metalized and non-metalized strucures are on the level 0.3 °C.

Conclusions

Different approach for thermal analysis of silver based thermal joints were presented. With using half of the amplified sinus signal to heat the samples the temperature response was analysed. The lower temperature difference was achieved for metalized surfaces of joined elements, but the difference is small. This may lead to neglect using metalizations for joint in order to simplify process and cost reduction. Further reaserch will consider using pulse and square signals in analogy to presented approach. Temperature difference appears according to conductivity of thermal joint. Radiation is used for IR measurements. The convection in such manner becomes energy loss and migdy be unwanted effect. To solve this issue putting samples in vacuum chamber and performing radiation measurements through transparent for IR windows should be considered.

Acknowledgements

Special thanks to Waldemar Klausek, student at Faculty of Microsystem Electronics and Photonics of Wrocław University of Science and Technology.

The works were performed by Statutory Grant no 0401/0137/18, Wrocław University of Science and Technology.

References

[1] G. Bai, „Low-temperature sintering of nanoscale silver paste for semiconductor device interconnection", Ph.D. dissertation, Virginia Tech, Blacksburg, VA, Oct. 2005.

[2] K.S. Siow (*Editor*), "Die-Attach Materials for High Temperature Applications in Microelectronics Packaging", Springer 2019.

[3] J. Felba, „Technological aspects of silver particle sintering for electronic packaging", Circuit World, 2018, Vol. 44, No 1, pp.2-15, 2018.

[4] K. Stojek, J. Felba, T. Fałat, M. Kiliszkiewicz, A. Mościcki „The materials and technology parameters influenced on the mechanical properties of low temperature sintered silver joints", EMPC Conference, Warsaw 2017, DOI: 10.23919/EMPC.2017.8346924.

[5] K. Stojek, D. Nowak, J. Felba, A. Mościcki, T. Fałat, A. Surmiak, „Heat transfer efficiency measurements with using thermography for low-temperature and low-pressure sintered silver joints", ESTC Conference, Dresden 2018, DOI: 10.1109/ESTC.2018.8546351.

[6] K. Stojek, T. Fałat, J. Felba, P.K. Matkowski, W. Macherzyński, A. Mościcki, "Thermal joints based on sintered silver micro-and nano-sized particles". ISSE, Pilsen 2016, DOI: 10.1109/ISSE.2016.7563172.

Reworkable Anisotropic Conductive Adhesive for Assembly of Medical Devices

Hoang-Vu Nguyen[1*], Helge Kristiansen[2], Anders Lifjeld[3], Kristin Imenes[1], Knut E. Aasmundtveit[1]

[1] University of South-Eastern Norway, 3184 Borre, Norway
[2] Conpart AS, 2013 Skjetten, Norway
[3] GE Vingmed Ultrasound AS, 3191 Horten, Norway

* Phone: +47-3100 9658. E-mail address: hoang.v.nguyen@usn.no

Abstract

Reworkable anisotropic conductive adhesives (ACAs) are of interest when the material is used for the assembly of electronic modules with high value, such as in medical ultrasound probes. Commercially available ACAs are generally difficult or even impossible to rework due to the common use of thermosetting adhesive matrix. ACA material with competitive performance compared to conventional ACAs and with reworkability at modest conditions is developed in this work. Adhesive matrices comprising a blend of a thermosetting epoxy and a thermoplastic polymer are selected because it has shown potential to ensure good electrical and mechanical integrity whilst still allowing reworkablity for ACA interconnects. This paper presents the findings of favourable mixing ratios between an epoxy system compatible with ACA applications and a high-performance thermoplastic polymer that offer good mechanical strength combined with reworkability. Die shear strength at varying temperatures relevant for production/storage (23 °C), operation of medical ultrasound probes (50 °C) and rework of ACA bonds (190 °C) is studied. Complete removal of adhesive remaining on bonding surfaces, for successful rework, is verified. The results show a high die shear strength at both 23 °C and 50 °C for adhesive formulations comprising up to 67 wt% of thermoplastic polymer, being comparable to the shear strength obtained for common lead-free solders and conventional ACAs. The die shear strength at 190 °C drops dramatically, and agrees very well with the results from the rework evaluation.

Key words: anisotropic conductive adhesive, reworkable, repairable, medical devices

Introduction

Anisotropic conductive adhesives (ACAs) have emerged as an alternative to soldering for flip-chip interconnection of electronic devices to a variety of substrates. An ACA comprises a non-conductive adhesive matrix filled with a low concentration (below the percolation threshold) of mono-disperse conductive particles [1]. The particles are either solid metal particles or metal-coated polymer spheres. The mechanical strength of ACA interconnects comes from the adhesive matrix, while the anisotropic electrical conduction is established when the conductive particles are trapped between the mating bumps-pads on a chip and a substrate (Figure 1). ACAs can be either in paste-form or in film-form, being called anisotropic conductive paste (ACP) and anisotropic conductive film (ACF), respectively. ACF is the most popular used form of ACAs, due to the ability to better control the volume of material, the density and the distribution of conductive particles within the sample, as well as the ease of handling. Compared to solder technology, the ACAs offer major benefits

such as lower processing temperature, fewer processing steps, and finer interconnection pitch.

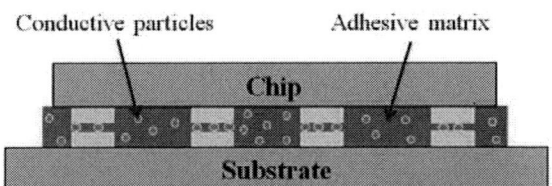

Figure 1: An illustration of an ACA assembly

The ACAs have widely been used as a standard interconnection technology in the assembly of flat-panel displays where soldering is not applicable for bonding chips to glass substrates, because of low pitch and metallization available [2]. The technology has also shown high potential for use in bonding chips to silicon/ceramic substrates, chips to organic substrates such as flexible and rigid printed circuit boards (flex and FR4), flex to flex, flex to FR4 [1, 2]. The main reason is the ACA technology can provide electrical connection, mechanical strength and sealing/underfilling in one quick bonding step. Furthermore, the technology is capable of providing high interconnect yield and reliable interconnects [3-6].

The present work is motivated through the assembly of medical ultrasound probes in which electronic modules are bonded using an ACF. Since the cost of these modules are high, it is a huge advantage if ACF bonds are reworkable. That means one is able to separate mating modules from each other, remove the residual adhesive on bonding surfaces, and reuse / re-bond functional module(s), either during production or when repairing returned products. Commercially available ACFs are generally not reworkable due to the common use of thermosetting adhesive matrix [2]. A limited number of commercially available ACFs are claimed to be reworkable/reparable. However, a successful rework is still challenging, particularly the process of removing the residual adhesive on the bonding surfaces [7]. ACA materials with competitive performance compared to conventional ACAs and with proper reworkability at modest conditions are investigated in this work. Enabling the reworkability of the ACAs demands modifications of its adhesive matrix. Adhesive matrices comprising a blend of a thermosetting epoxy and a thermoplastic polymer are of interest because it has shown potential to enable the reworkablity whilst still ensuring good electrical and mechanical performance as well as the reliability of adhesive-based interconnects [1, 8].

This paper presents the findings of favourable mixing ratios between an epoxy system compatible with ACA applications and a thermoplastic polymer that result in adhesive blends with good mechanical performance while enhancing the reworkability of the adhesive. Test samples include silicon-based dies and substrates being bonded using adhesive blends with varying ratios of the epoxy and the thermoplastic. Die shear strength at varying temperatures relevant for production/storage, operation of ultrasound probes and rework of ACA bonds is studied. The ability to fully remove the adhesive remaining on bonding surfaces, for successful rework, is also verified.

Experimental

Preparation of adhesive blends and adhesive films

The thermosetting epoxy system comprises a low-viscosity bisphenol F resin, a curing agent and a curing catalyst. The mixing ratio between the components of the epoxy system is selected to ensure the compatibility with the ACA applications in terms of viscosity, curing temperature and curing time. The thermoplastic polymer used is a polysulphone (PSU) with low molecular weight to promote resin flow during bonding. The PSU materials are amorphous polymers with excellent thermal stability (glass transition temperature T_g ranging from 180 to 280 °C), good mechanical properties and resistance to chemicals such as dilute acids, alkalis, electrolytes [9, 10]. Since PSU is

supplied as pellets, 1-Methyl-2-pyrrolidinone (NMP) was used to dissolve the PSU. NMP is selected thanks to its good solvency properties to PSU. In addtition, this choice is acceptable because NMP is also used as a solvent for a wide range of polymers, and as a cleaning agent in a variety of industrial applications such as electronics, automotives, chemicals, pharmaceuticals [11]. Conductive particles were not included in the adhesive formulations in this work. This is supposed to have insigificant impact on the mechanical performance of the adhesive due to the low concentration of the particles in ACA formulations.

A blend of epoxy and PSU was prepared by adding epoxy resin, curing agent and curing catalyst to the NMP-solution of PSU. A uniform paste without air bubbles was obtained, using a SpeedMixer DAC 150.1 FVZ-K from FlackTek Inc. The paste was then coated onto a glass substrate in form of stripes by using Kapton tape (thickness of about 70 µm) as mask. The stripes were dried at room temperature for 2 days. After drying, adhesive films could be peeled off the glass substrate, and were used similarly as ACF.

In the present work, adhesive formulations comprising 25 – 67 wt% of PSU were prepared. The adhesive films after drying have a thickness in the range of 30 to 45 µm, depending on the formulations.

Preparation of samples for die shear tests

Test samples consisting of a bare silicon die bonded to a bare silicon substrate were prepared for destructive die shear tests. The dies and the substrates were washed using acetone and isopropanol, and then dried with nitrogen gas. The assembly process was similar to a typical ACA/ACF bonding process. An adhesive film was cut to an appropriate size, and was subsequently applied on the substrate. For samples being bonded using only epoxy, a drop of the epoxy was deposited on the substrate. The die was assembled on the adhesive using a flip-chip bonder FinePlacer Pico from Finetech GmbH. A bondline thickness of 20 µm between the die and the substrate was controlled by means of spacers. The bonding was performed by applying a sufficient force at 190 °C for 5 minutes. The bonding time was selected somewhat longer than needed in order to ensure a high curing degree of the adhesive.

Die shear tests at varying temperatures & visual inspection

An F&K DELVOTEC Bond Tester 5600 was used to conduct the destructive die shear tests. A custom-built sample holder, which could clamp the sample and keep it at a specified temperature, was machined in aluminum. The sample holder was mounted on the workstage of the bond tester, being compatible with the conventional shear test procedure. Figure 2 illustrates the setup for the die shear testing trials. The sample holder is heated by a Watlow Firerod cartridge heater with feedback from a thermocouple installed inside the holder. The actual temperature in the bondline during the die shear tests was checked by measuring the temperature on top of the substrate and on top of the die. The precision in temperature control of our test samples was found to be within +/- 1 – 5 °C, depending on the test temperature. The die shear tests were performed at room temperature (about 23 °C), 50 °C and 190 °C, representing temperatures relevant for production/storage, operation of medical ultrasound probes, and rework of bonded electronic modules, respectively.

Visual inspection was performed on all samples subjected to the destructive die shear tests to analyze fracture surface, and to measure bonding area. The die shear strength of each test sample was determined based on the measured maximum force and the bonding area.

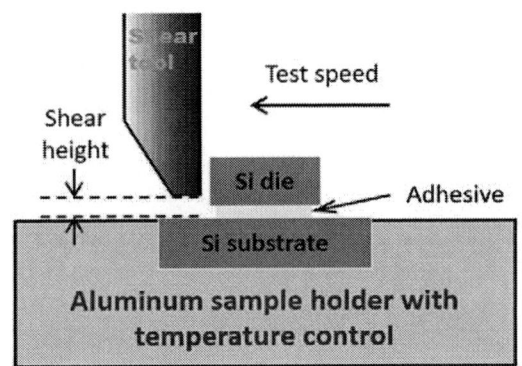

Figure 2: Illustration of a test specimen in the custom-built die shear test fixture

Evaluation of reworkability

The rework process includes two main steps; separating the die from the substrate of a bonded sample, and removing the adhesive remaining on bonding surface of the die and the substrate. In this work, test samples with a bonding area relevant for real applications (e.g. from 25 to 36 mm²) were subjected to the rework evaluation. A sample was heated up to 190 °C on a hot plate and kept for about 1 minute. The die was then separated from the substrate by using tweezers to apply a twisting force between them. After separation, the residual adhesive on the bonding surfaces was removed by rubbing a cotton swab soaked in the solvent NMP at room temperature, followed by applying a light force on the residue at about 50 – 60 °C using a wooden spatula.

Results

The die shear strength at room temperature (23 °C) of varying adhesive formulations is shown in Figure 3. Adhesive formulations comprising 25 –

67 wt% of PSU exhibit comparable die shear strength. Compared to bare epoxy, the die shear strength of adhesive formulations with PSU is about 30 % lower. The visual inspection of samples tested at room temperature showed mainly cohesive fracture for samples based on adhesive formulations with PSU whereas cohesive-adhesive fracture was observed for bare epoxy samples.

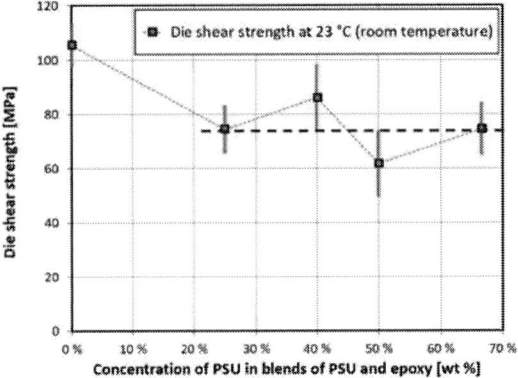

Figure 3: Die shear strength at room temperature of adhesive formulations of interest. Each datapoint is the average value of at least 6 repeated tests.

The die shear strength at 50 °C of the adhesive formulations comprising 25 wt%, 50 wt% and 67 wt% of PSU is shown in Figure 4. The die shear strength retains high values (over 50 MPa) at 50 °C. Adhesive formulations comprising 50 – 67 wt% of PSU exhibit comparable die shear strength which is somewhat lower than that of the formulation with 25 wt% of PSU. The visual inspection showed predominantly cohesive fracture for all samples tested at 50 °C.

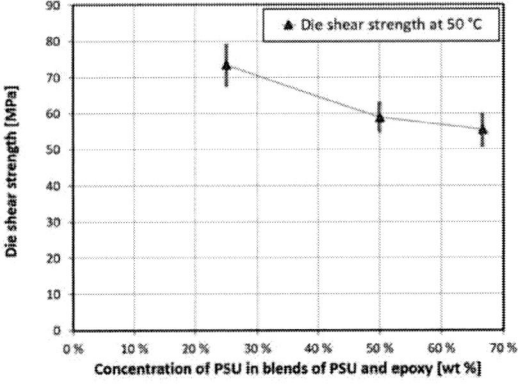

Figure 4: Die shear strength at 50 °C of adhesive formulations of interest. Each datapoint is the average value of at least 6 repeated tests.

Figure 5 shows the die shear strength at 190 °C of varying adhesive formulations. While epoxy still maintains a die shear strength over 4 MPa, all adhesive formulations with PSU exhibit a considerably lower die shear strength, being about 1 MPa and below. The higher the concentration of PSU in the adhesive formulations is, the lower the die shear strength is. The failure mechanisms of samples tested at 190 °C are similar to that of samples tested at room temperature; mainly cohesive fracture for samples based on adhesive formulations with PSU, and cohesive-adhesive fracture for bare epoxy samples. Figure 6 shows typical fracture surfaces of test samples bonded with bare epoxy and with an adhesive formulation comprising PSU. These samples were sheared at 190 °C.

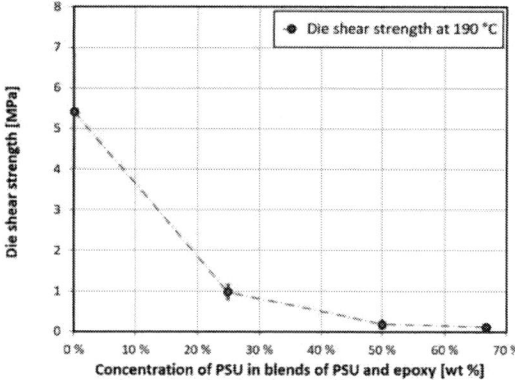

Figure 5: Die shear strength at 190 °C of adhesive formulations of interest. Each datapoint is the average value of at least 5 repeated tests.

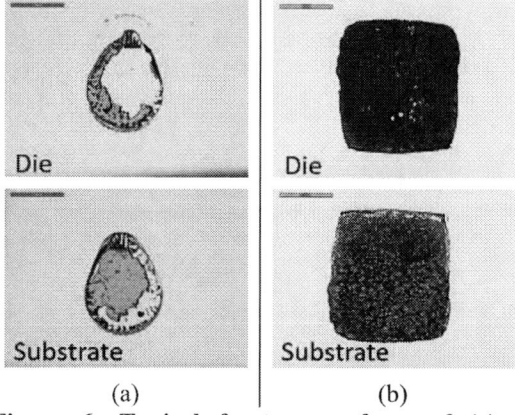

(a)   (b)

Figure 6: Typical fracture surfaces of (a) a sample bonded with bare epoxy, and (b) a sample bonded with an adhesive formulation comprising PSU. The samples were sheared at 190 °C.

Table 1 shows the results from the rework trials. The reworkability is enabled when PSU is included in formulations with a concentration from 25 wt%. However, adhesive formulations comprising more than 40 wt% of PSU allow a successful rework with less effort, compared to those with less concentration of PSU. Figure 7 shows typical result from a rework trial for test samples based on adhesive formulations comprising PSU. After separating the die from the substrate of a bonded sample, the bonding area (in the range of 25

– 36 mm^2) of both the die and the substrate is covered with adhesive (Figure 7-a). After the process of removing the residual adhesive, clean bonding surfaces are obtained (Figure 7-b), indicating a successful rework.

Table 1: Results from the evaluation of reworkability

Adhesive formulations	Possibility to separate a die from a substrate (*)	Removal of residual adhesive on bonding surface	
		Possibility	Difficulty
Bare epoxy	No	No (**)	Very hard
25 wt% PSU	No	Yes (**)	Hard
40 wt% PSU	Yes	Yes	Moderate
50 wt% PSU	Yes	Yes	Easy
67 wt% PSU	Yes	Yes	Easy

* Samples with a bonding area from 25 to 36 mm^2
** Trials using samples with a considerably smaller bonding area

(a) Before removal of residual adhesive

(b) After removal of residual adhesive

Figure 7: Typical result from a rework trial for samples based on adhesive formulations comprising PSU

Discussion

Figure 8 shows the comparison of die shear strength measured at room temperature (23 °C), 50 °C and 190 °C for adhesive formulations with varying concentrations of PSU. The temperature ranging from 23 to 50 °C has a minor effect on the mechanical die shear strength of the adhesives tested in this work. The die shear strength in such temperature range retains a high value (over 50 MPa), being comparable to the shear strength measured at room temperature for common lead-free solders (30 – 65 MPa [12-14]) and conventional ACF used in Chip-on-Glass applications (40 – 110 MPa [15, 16]). When the temperature is increased to 190 °C, the die shear strength of all adhesive formulations drops dramatically, compared to the corresponding values at room temperature and 50 °C. This implies the possibility to remove a die from a substrate of a bonded sample by applying a low force in certain conditions.

The remarkable drop in die shear strength at 190 °C of all adhesive formulations is attributed to the reduction of the modulus of polymer materials at

378

temperatures within and above its T_g range. The test temperature of 190 °C is within the T_g range of PSU (180 – 280 °C [9, 10]) and is above the T_g of cured epoxy, which is typically lower than the curing temperature applied (190 °C in this work). At such temperature, the modulus of both PSU and epoxy is reduced considerably [9]. However, the modulus of PSU used in this work, which is an amorphous polymer and has low molecular weight, is probably reduced more than that of the cured epoxy, which is highly cross-linked. This explains the dramatic drop in die shear strength at 190 °C for all adhesive formulations, as well as the reason for considerably lower die shear strength of the formulations with PSU compared to that of the bare epoxy (Figure 5).

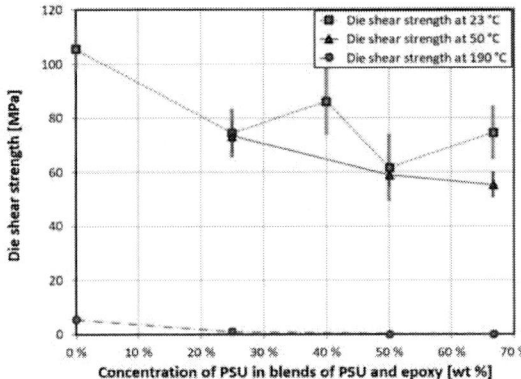

Figure 8: Die shear strength at varying temperatures of adhesive formulations with varying concentrations of PSU

The temperatures selected for die shear tests in this work are relevant for the production/storage (23 °C), the operation of medical ultrasound probes (50 °C) as well as the rework of bonded electronic modules inside the probes (190 °C). The high die shear strength in the temperature range of 23 – 50 °C as well as the remarkably low die shear strength at 190 °C indicate the high potential for using adhesive formulations comprising PSU as the matrix for the ACA under development. Such an ACA, even with a high concentration of PSU, is supposed to provide sufficient mechanical and electrical performance as well as integrity for use in medical applications whilst facilitating the reworkability.

The failure mode of all adhesive formulations comprising PSU is similar, being predominantly cohesive fracture. The failure mode is the same even with varying shear test temperatures. That means the adhesion of adhesive formulations with PSU to the surface of silicon dies and silicon substrates is comparable. This indicates a well dispersion of PSU and epoxy in the adhesive films manufactured in this work.

The results from the rework evaluation (Table 1) and from the die shear testing trials at 190 °C, the rework temperature (Figure 5), agree very well with each other. The lower the die shear

strength at the rework temperature is, the easiser the rework process is. A die shear strength below 1 MPa at the rework temperature is an indication of satisfactory reworkability.

Further environmental tests relevant for medical applications such as temperature cycling, temperature aging, humidity testing are crucial to select the adhesive formulations that can provide good mechanical integrity as well as good electrical performance, whilst still allowing an easy rework. Electrical characterization for the ACA under development is also planned.

Conclusions

Adhesive formulations comprising a thermosetting bisphenol F epoxy system and a thermoplastic polysulphone (PSU) are developed for use as an adhesive matrix in a reworkable anisotropic conductive adhesive (ACA). Die shear strength of adhesive formulations comprising 0 – 67 wt% of PSU was characterized at varying temperatures. The test temperatures were 23 °C, 50 °C and 190 °C, being relevant for production/storage, operation of medical ultrasound probes (the target application for the ACA under development), and rework of ACA bonds, respectively. The reworkability of these adhesive formulations was also evaluated in practice.

High die shear strength at 23 °C and 50 °C, being over 50 MPa, was attained for adhesive formulations comprising up to 67 wt% of PSU. The die shear strength at rework temperature (190 °C) drops dramatically, and agrees very well with the results from the rework evaluation. The lower the die shear strength at the rework temperature is, the easiser the rework process is. A die shear strength below 1 MPa at the rework temperature is an indication of satisfactory reworkability. The results from the present paper confirm the feasibility of using blends of epoxy and PSU as the adhesive matrix for the reworkable ACA under development.

Acknowledgements

The present work was funded by the Research Council of Norway through the BIA program (project number: 269618/O20; *MMIMI – Mechanical miniaturization in interventional medical instruments*). The Research Council of Norway is also acknowledged for the support to the Norwegian Micro- and Nano-Fabrication Facility (NorFab, project number: 245963/F50).

The authors would like to thank Birgitte Kasin Hønsvall, Zekija Ramic and Anh-Tuan Thai Nguyen, all at University of South-Eastern Norway (USN), for their support in the experimental work. We would also like to thank Erik Andrcassen at USN and SINTEF for fruitful discussions during the work.

References

[1] J. E. Morris and J. Liu, "Electrically Conductive Adhesives: A Research Status Review," in *Micro- and Opto-Electronic Materials and Structures: Physics, Mechanics, Design, Reliability, Packaging*, vol. 2, E. Suhir, Y. C. Lee, and C. P. Wong, Eds. U.S.A.: Springer US, 2007, pp. B527-B570.

[2] P. J. Opdahl, "Anisotropic Conductive Adhesives," in *Handbook of Visual Display Technology*, J. Chen, W. Cranton, and M. Fihn, Eds. 2 ed. Switzerland: Springer International Publishing, 2016, pp. 1533-1541.

[3] A. Larsson, F. Oldervoll, T. A. T. Seip, H.-V. Nguyen, H. Kristiansen, and Ø. Sløgedal, "Anisotropic Conductive Film for Flip-Chip Interconnection of A High I/O Silicon Based Finger Print Sensor," in *The 22nd Micromechanics and Micro systems Europe Workshop*, Tønsberg, Norway, 2011, pp. 186-189.

[4] H.-V. Nguyen, T. Eggen, B. Sten-Nilsen, K. Imenes, and K. E. Aasmundtveit, "Assembly of Multiple Chips on Flexible Substrate Using Anisotropic Conductive Film for Medical Imaging Applications," in *The 64th Electronic Components and Technology Conference*, Orlando, Florida, USA, 2014, pp. 498-503: IEEE.

[5] P. Palm, J. Maattanen, Y. De Maquille, A. Picault, J. Vanfleteren, and B. Vandecasteele, "Comparison of different flex materials in high density flip chip on flex applications," *Microelectronics Reliability*, vol. 43, no. 3, pp. 445-451, 2003.

[6] M. J. Yim *et al.*, "Highly Reliable Flip-Chip-on-Flex Package Using Multilayered Anisotropic Conductive Film," (in English), *Journal of Electronic Materials*, vol. 33, no. 1, pp. 76-82, 2004.

[7] Z. W. Zhong, "Various Adhesives for Flip Chips," *Journal of Electronic Packaging*, vol. 127, no. 1, pp. 29-32, 2005.

[8] K. Moon, C. Rockett, C. Kretz, W. F. Burgoyne, and C. P. Wong, "Improvement of adhesion and electrical properties of reworkable thermoplastic conductive adhesives," *Journal of Adhesion Science and Technology*, vol. 17, no. 13, pp. 1785-1799, 2003.

[9] J. M. G. Cowie and V. Arrighi, *Polymers: Chemistry and Physics of Modern Materials*, 3rd ed. CRC Press - Taylor and Francis Group, 2007.

[10] V. R. Sastri, *Plastics in Medical Devices - Properties, Requirements, and Applications*. William Andrew, 2010, p. 352.

[11] Mitsubishi-Chemical. N-Methyl-2-Pyrrolidone (NMP) [Online]. Available: https://www.m-chemical.co.jp/en/products/departments/mcc/c4/product/1201005_7922.html

[12] T. Siewert, S. Liu, D. R. Smith, and J. C. Madeni. Properties of Lead-Free Solders [Online]. Available: https://www.msed.nist.gov/solder/NIST_LeadfreeSolder_v4.pdf

[13] Y. A. Su, L. B. Tan, V. B. C. Tan, and T. Y. Tee, "Rate-Dependent Properties of Sn-Ag-Cu Based Lead Free Solder Joints," presented at the 11th Electronics Packaging Technology Conference, Singapore, 9-11 Dec. 2009, 2009.

[14] Y. Tian, C. Wang, S. Yang, P. Lin, and L. Liang, "Shear Fracture Behavior of Sn3.0Ag0.5Cu Solder joints on Cu Pads with Different Solder Volumes," presented at the International Conference on Electronic Packaging Technology & High Density Packaging (ICEPT-HDP), Shanghai, China, 28-31 July 2008, 2008.

[15] J. H. Zhang, Y. C. Chan, M. O. Alam, and S. Fu, "Contact Resistance and Adhesion Performance of ACF Interconnections to Aluminum Metallization," *Microelectronics Reliability*, vol. 43, no. 8, pp. 1303-1310, 2003.

[16] K.-C. Chen, H.-T. Li, C.-W. Hsu, and C.-P. Yang, "Properties and Reliability Test of Anisotropic Conductive Film in Chip on Glass Package," in *The 1st Electronics Systemintegration Technology Conference*, Dresden, Germany, 2006, vol. 1, pp. 51-55.

Advanced Interconnection technology with Laser Assisted Bonding Process for PET Substrate

Yong-Sung EOM*, Ki-Seok JANG, Jiho JOO, and Kwang-Seong CHOI

Electronics and Telecommunications Research Institute, 138 Gajeongno, Yuseong-gu, Daejeon, 305-700, Korea

+82 42 860 5547, *yseom@etri.re.kr

Abstract

As a state-of-the-art interconnection technology, laser assisted bonding process has been applied to the LED bonding process with PET substrates. In conventional reflow processes, when PET substrate is exposed to the 150°C environment, the PET substrate is easily deformed due to the low glass transition temperature of 70°C. Therefore, it is necessary to carry out the semiconductor bonding process without changing the dimension of PET substrate during the bonding process. In this study, laser assisted bonding process was carried out using solvent free epoxy based solder paste which was newly developed. LED device (2mm x 2mm x 1mm) was bonded to a 0.2mm thickness PET substrate with a laser of 980nm wavelength. The absorption, transmission and reflection of the laser were examined on the PET substrate. The power and time of the laser were precisely tuned to optimize device and substrate temperatures during the laser reflow process and prevent deformation of the PET substrate. Temperature of LED device and the PET substrate were measured by IR camera and thermocouple during laser bonding process. The laser bonding process using the newly developed solvent free epoxy based solder paste was successfully performed on LED device and PET substrate without any damage. Laser reflow bonding process are expected to be extensively used in semiconductor packaging in the near future.

Key words: laser, interconnection, bonding, reflow, pet

Introduction

With the increasing electrical interconnection density for semiconductor packaging, flipchip bonding technology has been developed as a thermo compression bonding process using conventional reflow process. Solder is an excellent interconnection material widely used in the flipchip bonding process based on thermo-compression technology in the semiconductor packaging industry [1-4].

However, this conventional interconnection process has been limited by the following two reasons. First, the process time of thermo-compression is long and UPH (unit per hour) is low, the other is the dimensional instability during the interconnection process due to the difference in CTE (coefficient of thermal expansion) between the device and the substrate. Because of these two reasons, laser assisted bonding (LAB) is carefully adopted in the semiconductor packaging industry [6-13].

In this study, we used solventless hybrid underfill composed of epoxy resin and solder powder in 980nm wavelength LAB process as shown in Figure 1 [14]. PET based substrate and LED device were used in the LAB process. In the LAB process, the LED device was completely interconnected without any deformation of the PET substrate, and LED worked well with the given electrical current.

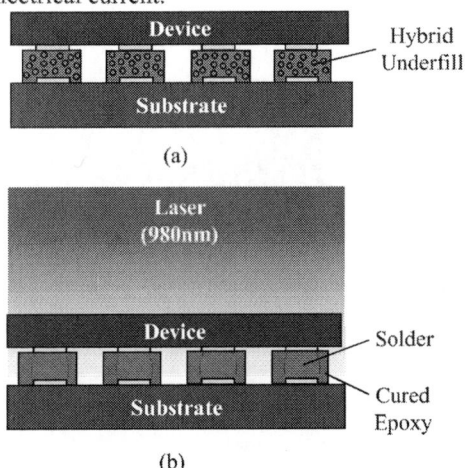

(a)

(b)

Figure 1: Schematic of LAB process

Materials and Experiment

For the LAB process, a hybrid underfill composed of solventless resin and solder powder was prepared. The chemical components of solventless resin are epoxy as a base resin, reductant to remove oxide on solder, curing agent and some addictives. The solder powder used in this study is Sn/58Bi type 4 with a melting temperature of 138°C. The viscosity of the hybrid underfill with a solder content of 88 wt.% was 150,000 cPs at 10rpm in an ambient temperature of 24°C measured by BrookField viscometer. LED (model: 603-B23, Everlight Co.) and PET substrate used in this study are shown in Figure 2. Dimension and applied current of LED were 2mm x 2mm x 1mm and 20mA, respectively. The thickness of PET substrate was 0.35mm.

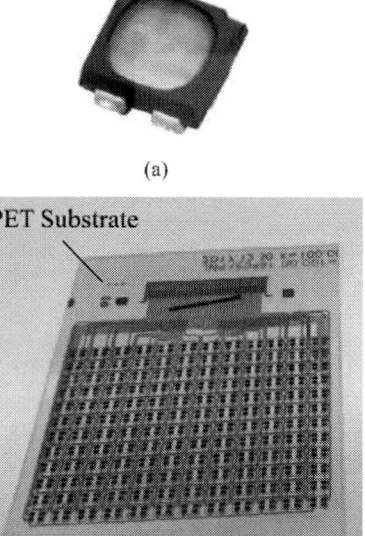

(a)

PET Substrate

(b)

Figure 2: Pictures of: (a) LED device, (b) PET substrate

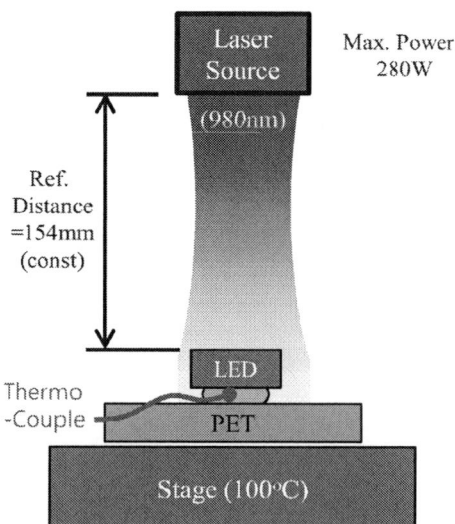

Figure 3: Schmatic of LAB process to measure process temperature between LED and PET substrate.

Figure 3 shows set-up to measure process temperature during LAB process. Maximum laser power, wavelength of laser, laser applied area and

distance between laser source and LED sample were 280W, 980nm, 15mm x 15mm and 154mm, respectively. In order to minimize the laser power in the LAB process and to prevent PET deformation, the temperature of the stage was maintauined at 100°C, which is lower than the temperature range of PET deformation. A UV/VIS/NIR spectrometer (Lambda 750, PerkinElmer) was used to measure laser absorption, transmittance and reflectance of PET substrate at the wavelength of 980nm used in the LAB process of this study. Thermocouple was placed between the LED device and the PET substrate to measure the temperature of the hybrid underfill during the LAB process as shown in Figure 4. The processing temperature was also measured withy an IR camera during the actual LAB process. After LAB process, solder interconnect cross-section between the LED pad and the metal pad of the PET substrate was observed. A 20mA current was applied to the packaged sample for evaluation of the LAB process.

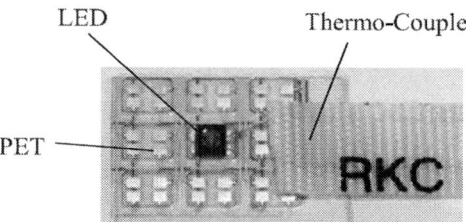

Figure 4: Picture of sample to measure the processing temperature by a thermocouple placed between the LED and the PET substrate during the LAB process.

Results and Discussion

The PET substrate used in this study was exposed in a 180°C oven to evaluate dimensional stability in processing temperature. As shown in Figure 5, it was clearly observed that the conventional surface mounting technology (SMT) reflow process was impossible because the PET substrate was completely deformed.

PET Substrate

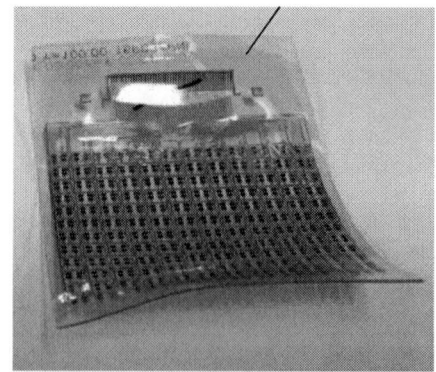

Figure 5: Picture of the PET substrate after heat treatment in 180°C oven.

In order to predict the temperature of the PET substrate in the LAB process, the laser absorption, transmittance and reflectance of the PET substrate were measured by an IR spectrometer as shown in Figure 6. The laser absorption, transmittance and reflectance at the wavelength of 980nm used in this study were 2.0%, 88.6% and 9.4%, respectively. Therefore, it can be easily judged that the PET substrate will not be deformed in the LAB process because only 2.0% of the energy exposed from the laser source was converted to the temperature of the PET substrate.

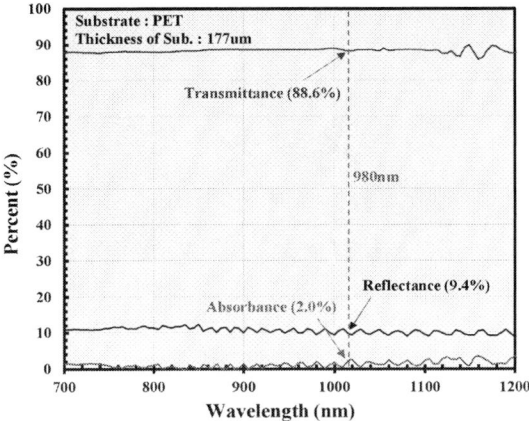

Figure 6: Measurement result of laser absorption, transmittance and reflectance as a function of wavelength.

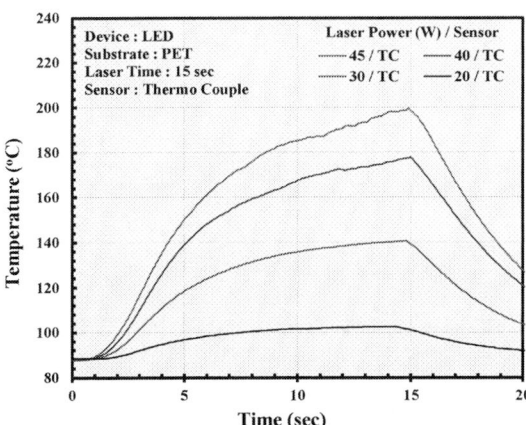

Figure 7: Measured temperatures between LED device and PET substrate by the thermos-couples according to laser power.

The temperatures between the LED device and the PET substrate shown in Fig. 4 were measured by thermocouples according to several laser powers. The laser of 980nm wavelength was exposed to sample for 15 seconds with the power of 20W, 30W, 40W and 45W. The temperatures were increased with increasing time as shown in Figure 7. The peak temperatures of 20W, 30W, 40W and 45W were 102°C, 140°C, 178°C and 200°C,

respectively. The laser power was determined to be 40W to melt solder of the hybrid underfill in the LAB process because the melting temperature of solder in the hybrid underfill is 138°C.

Figure 8: Picture of sample to detect temperatures of spot 1 and 2 by IR camera.

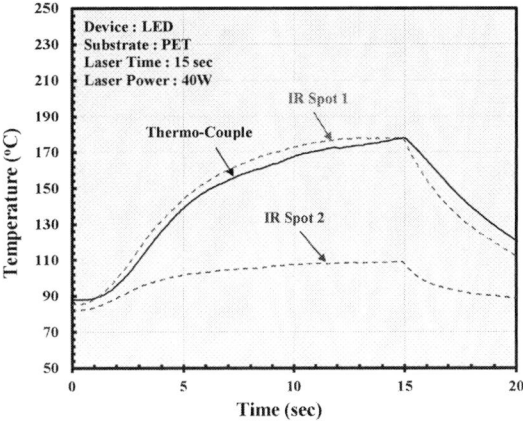

Figure 9: Comparison of temperatures measured by contact thermocouple and non-contact IR camera.

Figure 8 shows the locations to measure temperature by IR camera while the thermo-couple is placed between LED device and PET substrtate. IR Spots 1 and 2 are the locations of the surface of LED device and PET substrate, respectively. We compared the temperatures measured by a contact thermos-couple (placed between LED device and the PET substrate) and a non-contact IR camera as shown in Figure 9. There is a good agreement between the temperatures measured by a non-contact IR camera (spot 1) and contact thermocouple, and the temperature of the PET substrate of spot 2 by IR camera is very low because only 2.0% of the laser energy is absorbed by the PET substrate. The maximum temperature of spot 2 is 110°C and does not cause thermal deformation of PET substrate while the temperature of spot 1 is 176°C, which is higher than the melting temperature of solder. Thus, electrical interconnection using the hybrid underfill has become possible through the LAB process without deforming the PET substrate.

The LAB process with a laser power of 40W was performed for 15 seconds for electrical interconnection between the LED device and the PET substrate using hybrid underfill. Figure 10

383

shows an image that captures the temperature of IR camera. The red and yellow colored areas indicate the LED device and the PET substrate, respectively. As expected, a rapid temperature rise was observed only in the LED device area.

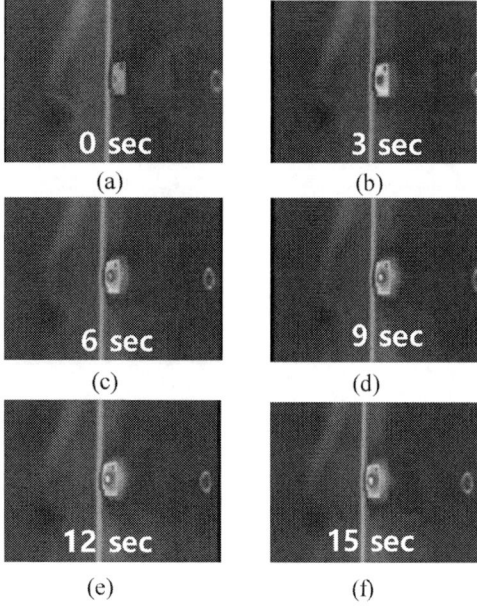

(a)   (b)

(c)   (d)

(e)   (f)

Figure 10: Captured images of temperatures by IR camera in LAB process.

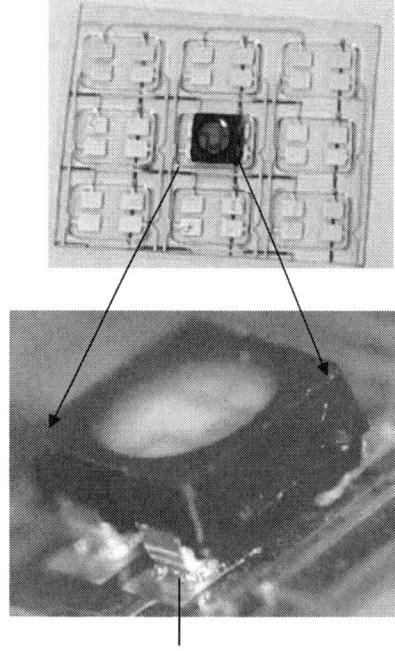

Solder Interconnection

Figure 11: Picture of packaged LED device on PET substrate.

Figure 11 shows LED device packaged on the PET substrate after the LAB process. It was clearly observed that the LED device was completely interconnected by solder wetting in the LAB process. To observe the solder interconnection between the LED device and the PET substrate, the cross-section of the metal pad of the LED device was obserced as shown in Figure 12. The matal pad of the LED device was completely covered with solder for 15 seconds of LAB processing time, and a small void was observed between pads of the LED device and the PET substrate. For high reliability, this void can be eliminated by optimization of the hybrid underfill and the LAB process condition.

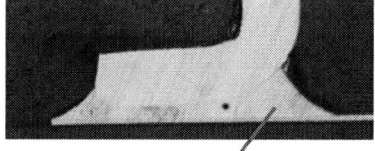

Solder Interconnection

Figure 12: Cross-section of solder interconnection between the LED device and the PET substrate.

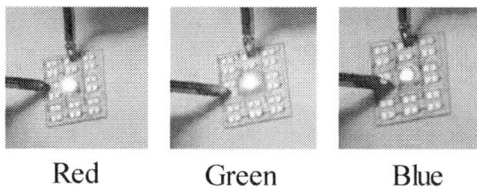

Red   Green   Blue

Figure 13: Picture the LED device packaged on the PET substrate with 20mA of current after the LAB process.

A current of 20mA was applied to the LED device packaged on the PET substrate after the LAB process. Red, green and blue light were perfectly observed, and the hybrid underfill material and LAB process successfully demonstrated interconnection process.

Conclusion

An advance interconnection technology with laser assisted bonding (LAB) process and the hybrid underfill was introduced in this study. The wavelength of the laser used was 980nm, the hybrid underfill material was consisted of epoxy resin and solder (type 4) powder. A PET substrate with severe dimensional defomation at 180°C was used to evaluate the LAB process. The hybrid underfill as the interconnection material was screenprinted onto the PET substrate, and the LED device was placed on the hybrid underfill. The laser output power, time and exposure area of the LAB process were 40W, 15 seconds and 15mm x 15mm, respectively. The maximum temperature was measured at 176°C on the surface of the LED device, no dimensional deformation of the PET device was observed after the LAB process. A cross section of solder interconnection between the metal pad of the LED device and the PET substrate was observed, and the solder was completely covered on the metal pad. In conclusion, the LAB process and the hybrid underfill has proven to be excellent interconnet technologies that can be highly productive in the semiconductor industry.

Acknowledgements

This work was supported by the Ministry of Trade, Industry & Energy (MOTIE, Korea) under Industrial Technology Innovation Program. No.10082367, "Eco-friendly Interconnection Paste and Process with Laser Bonding for Flexible LED Module". This work was also supported by the Korea Institute of Energy Technology Evaluation and Planning (KETEP) and the Ministry of Trade, Industry & Energy (MOTIE) of the Republic of Korea (No. 20183010014310 and 20000352). The authors would like to thank Jihae Son, Iseul Jung and for their support in the sample preparation and measurements.

References

[1] K. Honda, A. Nagai, M. Satou, S. Hagiwara, S. Tuchida, H. Abe, "NCF for Pre-Applied Process in Higher Density Electronic Package Including 3D-Package," Proc. IEEE Electronic Components and Technology Conf. (ECTC 12), 2012, pp. 385-392.

[2] T. Nonaka, Y. Kobayashi, N. Asahi, S. Niizeki and K. Fujimaru, "High Throughput Thermal Compression NCF Bonding," Proc. IEEE Electronic Components and Technology Conf. (ECTC 14), 2014, pp.913-918.

[3] K. Matsumura, M. Tomikawa, Y. Sakabe and Y. Shiba, "New Non Conductive Film for high productivity process," IEEE CPMT Symposium Japan (ICSJ 15), 2015, pp.19-20.

[4] S.-G. Ahn, H.-K. Kim, D.-W. Kim, D. Hiner, K.-S. Kim, T.-K. Hwang, M.-J. Lee, D.-B. Kang, and J.-H. Yoon, "Wafer Level Multi-Chip Gang Bonding Using TCNCF," Proc. IEEE Electronic Components and Technology Conf. (ECTC 16), 2016, pp. 122-127.

[5] D.-S. Ryu, "Advanced Interconnect with Laser Assisted Bonding," Proc. Semicon Taiwan, 2015.

[6] Y.-Y. Jung, D.-S. Ryu, M.-H. Gim, C.-H. Kim, Y.-S. Song, J.-.Y Kim, J.-H. Yoon, C.-H. Lee, "Development of Next Generation Flip Chip Interconnection Technology using Homogenized Laser-Assisted Bonding," Proc. IEEE Electronic Components and Technology Conf. (ECTC 16), 2016, pp. 392-400.

[7] T. Nakamura, F. Shafiq, T. Otani, O. Watanabe, T. Maeda, Y. Hagiwara, K. Honjo, T. Owada, D. Mori, O. Egashira, T. Nagamatsu, "Improvement of C2W collective bonding reliability and UPH through innovations in machine, materials and methods," Proc. IEEE Electronic Components and Technology Conf. (ECTC 17), 2017, pp. 108-115.

[8] K.-S. Choi, W. A. Braganca Junior, K.-S. Jang, H.-C. Bae, and Y.-S. Eom, "Development of Stacking Process for 3D TSV (Through Silicon Via) Structure using Laser," Proc. International Symposium on Microelectronics (IMAPS 17), 2017, pp. 67-71.

[9] L. Del Carro, T. Brunschwiler, M. Kossatz, L. Schnackenberg, M. Fettke, I. Clark, "Laser sintering of dip-based all-copper interconnects," Proc. IEEE Electronic Components and Technology Conf. (ECTC 18), 2018, pp. 279-286.

[10] L.A. Wentlent, M. Genanu, T. Alghoul, "Effects of Laser Selective Reflow on Solder Joint Microstructure and Reliability, " Proc. IEEE Electronic Components and Technology Conf. (ECTC 18), 2018, pp. 425-433.

[11] K.-S. Choi, W. A. Braganca Junior, l.s. Jeong, K.-S. Jang, S.-H. Moon, H.-C. Bae, Y.-S. Eom, M.-K. Cho, and S.- I. Chang, "Interconnection Process using Laser and Hybrid Underfill for LED Array Module on PET Substrate," Proc. IEEE Electronic Components and Technology Conf. (ECTC 18), 2018, pp. 1561-1567.

[12] Wagno Alves Braganca Junior, Yong-Sung Eom, Keon-Soo Jang, Seok Hwan Moon, Hyun-Cheol Bae, and Kwang-Seong Choi, "Collective Laser-assisted Bonding Process for 3D TSV Integration with NCP, ETRI J. Accepted, 2019.

[13] K.-S. Choi, Y.-S. Eom, S.-H. Moon, J.-H. Joo, L.-S. Jeong, "Enhanced Performance of Laser-Assisted Bonding with Compression (LABC) Compared with Thermal Compression Bonding (TCB) Technology", Proc. IEEE Electronic Components and Technology Conf. (ECTC 19), 2019, will be published.

[14] Y.-S. Eom et al., "Characterization of Fluxing and Hybrid Underfills with Micro-encapsulated Catalyst for Long Pot Life", ETRI J., Vol. 36, no. 3, pp. 343-351, 2014.

Advanced Application Capabilities of Thick Printed Copper Technology

J. Hlina[1], J. Reboun[1], J. Johan[2], and A. Hamacek[1]

[1]Department of Technologies and Measurement / RICE, Faculty of Electrical Engineering, University of West Bohemia, Pilsen, Czech Republic
[2]Elceram a.s., Hradec Kralove, Czech Republic

+420377634528, hlina@ket.zcu.cz

Abstract

This paper is focused on the advanced capabilities of Thick Printed Copper (TPC) technology used for power electronic substrate manufacturing. TPC technology is based on screen-printing of special copper pastes on ceramic substrates and its firing in a nitrogen atmosphere. This technology can be used as an alternative solution to conventional metallization techniques such as DBC or AMB and enables wide-range interconnection capabilities which cannot be realized with these conventional techniques. TPC interconnection capabilities include multilayer structures, copper plated vias and integrated passive components. Parameters and reliability of above-mentioned TPC structures after accelerated aging and thermal shock cycling are described in this paper.

Keywords: Thick printed copper, TPC, power electronic substrates, alumina, multilayer structures

Introduction

Power electronic puts a high demand on electronic components and substrates and with the development of different smart power modules, the complex interconnections are required. TPC technology offers wide-range interconnection capabilities which cannot be realized by conventional technologies such as DBC (Direct Bonded Copper) or AMB (Active Metal Brazing) with reasonable costs and technological demands [1, 2].

TPC technology is based on sequential screen-printing of copper paste on Al_2O_3 eventually AlN substrates and its subsequent drying and firing in an inert atmosphere [3, 4]. A two-paste system including adhesion and build-up paste is used for manufacturing of TPC substrates [5]. Adhesion paste creates a reliable connection with the ceramic substrate and build-up paste enables increasing the thickness up to 300 μm [6]. Additive deposition technology brings the advantage of high patterns resolution, material savings and manufacturing capability of substrates with complex interconnections containing multilayer circuits, copper plated vias or direct integrated passive components [7, 8]. Due to these benefits, TPC technology is suitable for the realization of advanced smart power modules and other special power electronic devices.

Advanced TPC substrates

Our current research is focused on multilayer TPC structures, copper plated vias and printed resistors compatible with TPC technology. Above-mentioned technological possibilities enable the realization of advanced power electronic substrates with complex interconnections for power modules. Examples of these modules are shown in Figure 1 and Figure 2.

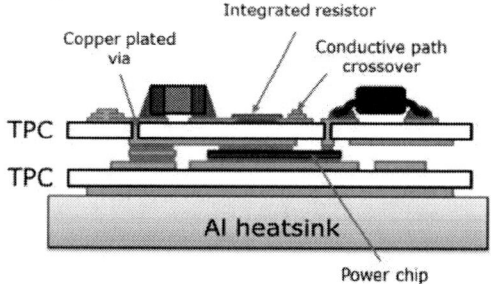

Figure 1: Sandwich topology with integrated drivers, sensors, resistors and copper plated vias.

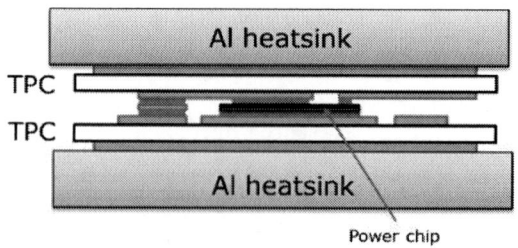

Figure 2: Sandwich topology with double-sided cooling.

Power module in Figure 1 is based on sandwich topology (two TPC substrates and power chips between them) and contains also control circuits and drivers on the top of the TPC substrate.

The driver part includes multilayer TPC structures (conductive path crossovers), integrated resistors and sensors. Power and driver circuits are connected by copper plated vias.

Power module in Figure 2 is also based on sandwich structure and compared to the power module in Figure 1 has double-sided cooling and does not contain control circuits.

A special test pattern with Multilayer TPC structures and copper plated vias (Figure 3) were designed and manufactured. This pattern was created by copper films printed by paste Heraeus C7403 and nitrogen fireable dielectric paste Heraeus IP9319D and contains structures for testing of electrical and mechanical parameters and also structures to verify print capabilities (overprints of the dielectric film, print resolution etc.). Copper plated vias with a different diameter in alumina substrates are also included in the test pattern.

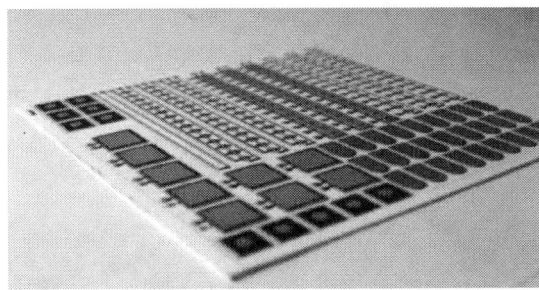

Figure 3: Realized special test pattern with multilayer TPC structures and copper plated vias.

Multilayer TPC structures

Multilayer TPC test structures were used for measurement of the dielectric film resistivity, its dielectric constant and dielectric strength, study of dielectric inner structure, measurement of copper traces resistance as well as the measurement of copper adhesion on the dielectric film. The reliability of multilayer TPC structures was verified by accelerated aging at elevated temperatures (dry heat test at 155 °C for 1000 hours according to standard IEC 60068-2-2) and thermal shock cycling (1000 cycles from -40 °C to 125 °C according to standard IEC 60068-2-14). After these tests, all the electrical and mechanical parameters of TPC test structures were measured again to study the changes.

Cross-sections of multilayer test specimens before and after thermal shock cycling observed by SEM microscope Phenom ProX are shown in Figure 4. It is obvious that dielectric film is porous. However, despite the porous structure of the dielectric film, multilayer TPC structures show sufficient electrical parameters.

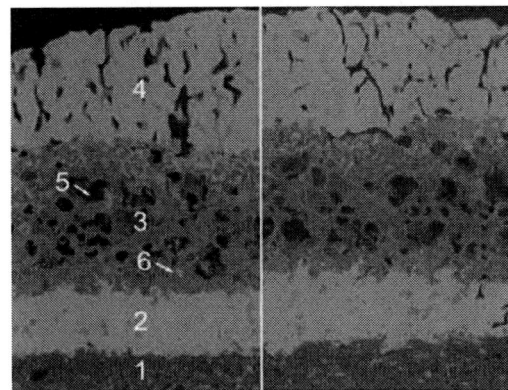

Figure 4: Cross-section of multilayer TPC structure before (left) and after thermal shock cycling (right) – 1: Al₂O₃ substrate, 2: copper film 1, 3: dielectric film, 4: copper film 2, 5: pore, 6: pore filled with glass phase.

Table 1: Parameters of multilayer TPC structures

Parameter	Before tests	After thermal aging	After thermal cycling
Capacity at 100 kHz (pF)	43.56	44.27	43.75
Dielectric constant at 100 kHz (-)	5.08	5.16	5.10
Dielectric loss at 100 kHz (-)	0.005	0.003	0.006
Resistivity (10^{12} Ω.m)	2.89	20.97	125.23
Impedance at 100 kHz (kΩ)	36.36	35.84	36.16
Phase angle at 100 kHz (°)	-89.71	-89.83	-89.66
Breakdown voltage (kV)	0.578	0.653	0.617
Dielectric strength (kV/mm)	8.75	9.90	9.34
Adhesion force (N/mm²)	22.33	12.00	12.76

Parameters of multilayer TPC structures are described in Table 1. Dielectric strength was measured according to standard IEC 60243-1 at AC voltage with a frequency of 50 Hz and a voltage rate of 50 V/s. Adhesion of the second copper film was measured by Pull off test (ASTM D4541). Description of the test specimen and test method are shown in Figure 5. Test specimens for adhesion measurement contain copper film on the bottom side of the alumina substrate (5 mm by 5 mm square pattern). The pattern on copper film 2 (circle pattern with a diameter of 3 mm) was dimensioned with a smaller area than the pattern on copper film 1. Therefore, the separation during Pull off test does not occur on the bottom side of the substrate. Brass rods were soldered with SnPb solder alloy to both sides of test specimens and specimens were placed to test device. The separation was always observed between the dielectric film and the first copper film.

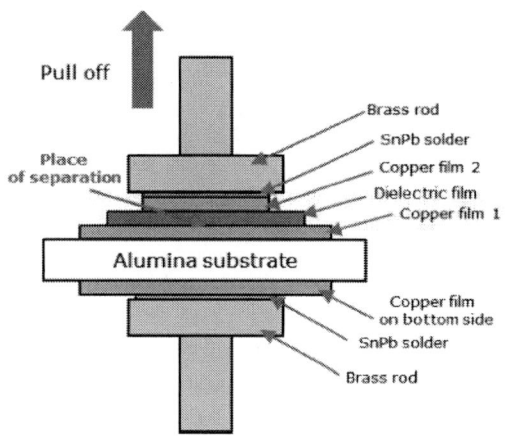

Figure 5: Description of test specimens and test method of adhesion measurement.

Accelerated aging and thermal shock cycling did not cause changes in dielectric film structure (Figure 4), deterioration of electrical parameters (Table 1) and any delamination of copper or dielectric films. On the contrary, some electrical parameters such as resistivity and dielectric strength increased. This phenomenon is probably caused by recrystallization of the dielectric film during accelerated aging and thermal cycling. However, accelerated aging and thermal cycling caused significant decreasing of adhesion of second copper film (43-46 %).

The standard manufacturing process of multilayer TPC structures is based on separately firing of each film. These structures can be also realized using co-firing of copper and dielectric films and co-fired structures show similar parameters as separately fired films. Using of co-firing can significantly reduce production costs. Co-firing of multilayer TPC structures was described in detail in [8].

Multilayer TPC structures can be used mainly for conductive path crossovers in complex interconnections where the connection of components is not possible in a single layer.

Copper plated vias

Copper plated vias with a different diameter (from 100 µm to 500 µm with a step of 50 µm) were included in the special test pattern. Holes in the ceramic substrate were drilled by CO_2 laser filled with copper paste Heraeus C7403 using screen-printing. Filling of holes has been supported by vacuum applied from the bottom part of the substrate which sucking the paste from the top part of holes. The substrate with copper plated vias was dried and fired and then copper plated vias were overprinted on both sides with copper film and fired again. This method brought the problem with cracks

on the interface of copper film and copper plated vias and also the uneven filling of holes (Figure 6).

Therefore, another method of realization of copper plated vias was tested. This method is based on the firing of copper paste filled holes and overprinted copper films in one step. This method brings a uniform filling of copper plated vias without any defects (Figure 7).

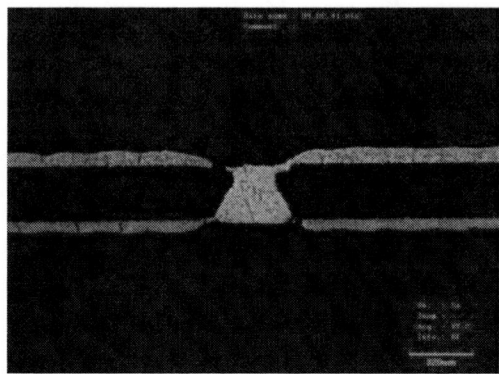

Figure 6: Cross-section of copper plated via with defects.

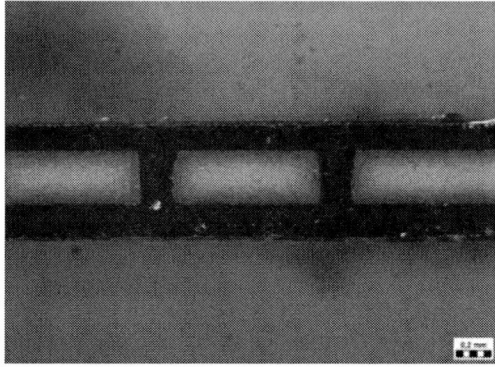

Figure 7: Cross-section of copper plated vias with uniform filling without any defects.

Realization of copper plated vias with a higher diameter than 500 µm was also tested. It has been verified that it is possible to realize functional copper plated vias with a diameter up to 700 µm. On the contrary, the realization of copper plated vias with a lower diameter than 300 µm is technologically demanding.

The resistance of copper plated vias was measured by four-wire resistance method and for all diameters was less than 10 mΩ. Resistance was also measured after accelerated aging and thermal shock cycling and the resistance of copper plated vias has not been changed. Accelerated aging and thermal shock cycling did not cause any delamination, defects and loss of functionality of copper plated vias.

Copper plated vias can be used for connection of copper films through the ceramic substrate or, because they are completely filled with copper, they can be also used as thermal vias for improvement the heat transfer through the ceramic substrate.

Printed resistors

TPC technology also enables the realization of directly integrated printed resistors. The main benefits of these resistors are better cooling efficiency and lower thickness compared to standard wire-wound resistors. The structure of the printed resistor with TPC terminals is shown in Figure 8. It consists of the resistive film (standard resistive thick film paste for firing in oxygen) protected with glass overglaze. Revealed end-parts of the resistive film are covered with copper film. Resistors with TPC terminal can be also realized in the discrete form (usually with the copper film on the bottom side for easy mounting to heatsink).

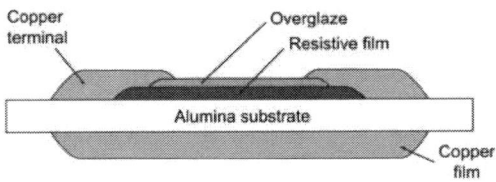

Figure 8: Structure of printed resistor with TPC terminals.

Specimen of the printed resistor is shown in Figure 9. It contains two square resistive structure (resistive paste Heraeus R8921 with a sheet resistance of 100 Ω/square). Realized resistors have nominal resistance value 390 Ω (nominal resistance values scattering after firing is ±10 %) and TCR (Temperature Coefficient of Resistance) ±120 ppm/°C. Nominal resistance values and TCR values are stable after accelerated aging by dry heat test and thermal shock cycling.

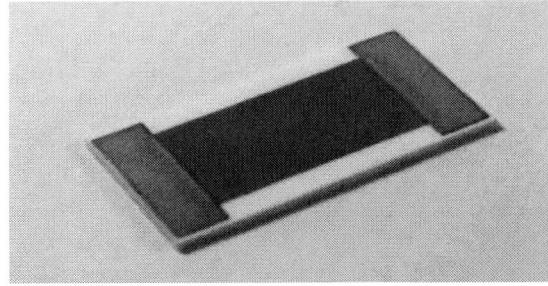

Figure 8: Printed resistor with TPC terminals.

Realized printed resistors show sufficient electrical parameters and stability. However, using this process is complicated to achieve low nominal resistance values which are necessary for the realization of integrated shunt and current sensing resistors which are very demanding in power electronic circuits. This problem can be solved by using resistive materials based on constantan.

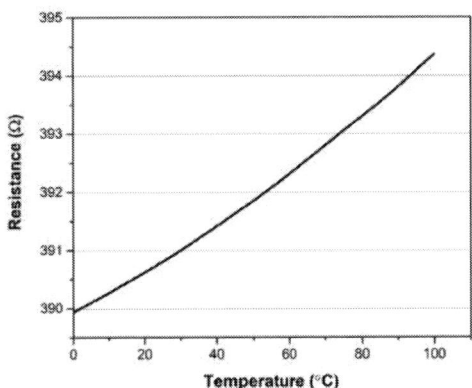

Figure 9: Temperature characteristic of printed resistor with TPC terminals.

Conclusion

It was verified that TPC technology is very suitable for the direct realization of complex interconnection structures such as conductive path crossings and copper plated vias as well as for the integration of printed passive components. Realized multilayer structures show sufficient electrical parameters for multilayer circuits. The dielectric strength is 8.8 kV/mm, resistivity is 2.9×10^{12} Ω.m and dielectric constant at 100 kHz is 5.1. It was also verified that copper plated vias can be realized with the diameter from 0.1 mm to 0.7 mm with resistance less than 10 mΩ. The best copper plated vias filling was achieved for the hole diameters from 0.3 to 0.7 mm All tested samples of multilayer structures, plated vias and printed resistors were fully functional after thermal aging and shock cycling without any signs of delamination or structural defects.

Acknowledgements

This research has been supported by the Ministry of Education, Youth and Sports of the Czech Republic under the project RICE – NETESIS, No. LO1607 and under the project ITI, No. CZ.02.1.01/0.0/0.0/18_069/0009855.

References

[1] A. Miric, P. Dietrich, "Inorganic substrates for power electronics applications", Heraeus Deutschland, 2013.

[2] S. Grohman, T. Smolinsky, J. Williams, C. Toy, "New improvements in thermal management: Thick print copper thick film as a replacement for Direct Bond Copper", Proceedings of the 18th European Microelectronics and Packaging Conference (EMPC), pp. 1-8, 2011.

[3] F. Rotman, D. Navarro, S. Mellul, "Optimised nitrogen-based atmosphere for copper thick film manufacture part 1: monitoring of oxygen doping in nitrogen", Microelectronics International, Vol. 8, No. 2, pp. 5-15, 1991.

[4] T. Blank, B. Levrer, T. Maurer, M. Meiser, M. Bruns, M. Weber, "Copper thick.film substrates for power electronics applications", Proceedings of the Electronic System-Integration Technology Conference (ESTC), pp. 4-9, 2014.

[5] J. Reboun, J. Hlina, P. Totzauer, A. Hamacek, "Effect of copper- and silver-based films on alumina substrate electrical properties", Ceramics International, Vol. 44, No. 4, pp. 3497-3500, February, 2018.

[6] P. Gundel, M. Bawohl, M. Challingsworth, M. Choisi, V. Garcia, M. Gaul, K. Kersken, Ch. Modes, I. Nikolaidis, R. Persons, J. Reitz, C. Shahbazi, "Thick printed copper as highly reliable substrate technology for power electronics", Proceedings of PCIM Europe 2015, pp. 19-21, 2015.

[7] J. Reboun, K. Hromadka, V. Hermansky, J. Johan, "Properties of power electronic substrates based on thick printed copper technology", Microelectronics Engineering, Vol. 167, pp. 58-62, January, 2017.

[8] J. Hlina, J. Reboun, V. Hermansky, M. Simonovsky, J. Johan, A. Hamacek, "Study of co-fired multilayer structures based on Thick Printed Copper technology", Materials Letters, Vol. 238, pp. 313-316, March, 2019.

Study on the Embedded Flexible Hybrid Stack Package using Polymer Elastic Bump Interconnection Method

Jae Hak Lee, Yong Jin Kim, Seung Man Kim and Jun-Yeob Song

Dept. of Ultra-Precision Machines and Systems, Advanced Manufacturing Systems Research Division
Korea Institute of Machinery and Materials (KIMM)
Daejeon, Republic of Korea

jaehak76@kimm.re.kr

Abstract

In this study, polymer elastic bump (PEB) bonding process is suggested newly, which could reduce stress concentration remarkably near bump and chip face by soft polymer bump material with electro-conductive metal line pattern and usage of non-conductive adhesive film (NCF) and also lower bonding temperature during face-down interconnect process between flexible substrate and ultra-thin chip. Elastic deformation and restoration behavior of PEB bump during bonding process are investigated using FEM analysis to calculate bump-to-pad contact area related to electrical contact resistance and verified them with respect to bonding force and temperature variation using micro tribometer. Through experiments, we find optimal bonding conditions and effect of bonding parameters such as bonding force, time and temperature. As bonding force increases up to 0.5N/bump, soft PEB is deformed elastically more and more. As result of that, contact area between PEB and metal pad is larger and kelvin resistance is lower below 4.75mΩ. The PEB exhibited nearly 100% elastic deformation and restoration at room temperature. However, as the temperature increased, the plasticity increased and plastic deformation gradually increased. In order to solve the Au cap metal line crack problem due to excessive deformation of PEB, spiral type and spoke type PEB are developed. In order to achieve high integration and smaller form factor of flexible semiconductor packages, a flexible semiconductor package of 2 stack was developed. The bending fatigue test of the flexible semiconductor package using PEB bumps was carried out and no change in daisy chain resistance occurred at 500,000 cycles at a bending radius of 10 mm.

Key words: flexible package, ultra-thin chip, polymer elastic bump, interconnection, stack

Introduction

Electronics on thin substrate such as flexible substrate will create a revolution in the electronics industry, enabling ultra-light and ultra-thin, flexible, easy-to-wear electronic products. System in Foil (SiF) package could realize flexible, multi-functional and higher performance package using flexible substrate and ultra-thin heterogeneous chip integration. Of course, many components are rigid, but thin chips by thinning down to 50 μm or less offer the opportunity for a manufacturer to create very flat, lightweight and flexible systems.

In this study, polymer elastic bump (PEB) bonding process is suggested newly, which could reduce stress concentration remarkably near bump and chip face by soft polymer bump material with electro-conductive metal line pattern and usage of non-conductive adhesive film (NCF) and also lower bonding temperature during face-down interconnect process between flexible substrate and ultra-thin chip. Compared to conventional bonding process by

metal bumps, it improves flexibility of package because PEB is softer than metal bump and NCF adhesive film keeps mechanical contact between PEB and pad. In addition, the conventional metal bump NCF face-down bonding has a problem that the bump connection part is open-circuited due to adhesive shrinkage, but the PEB face-down connection part maintains close contact between the bump and the pad during the adhesive shrinkage due to the elastic deformation and restoration behavior of the PEB.

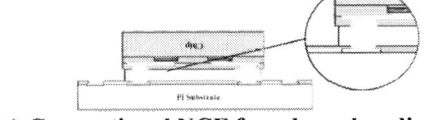

(a) Conventional NCF face-down bonding using metal bump

(b) NCF face-down bonding using PEB

Figure 1: Overview of PEB face-down interconnection process

In order to achieve high integration and smaller form factor of flexible semiconductor packages, a flexible semiconductor package of 2 stack was developed, that ultra-thin chip is embedded in flexible substrate using face-down interconnect of PEB and the second chip is attached with face-up and interconnected using a wire bonding technique.

Simulation of PEB Face-Down Bonding

The behavior of the bump according to the load applied to the bump was analyzed. The metal bump has a very small deformation amount of 0.002 μm at a load of 0.5 N / Bump due to the high elastic modulus, whereas the polymer elastic bump is deformed more than 2 μm. In addition, in the case of polymer elastic bumps, the bump shape is a smooth hemispherical shape, and deformation easily occurs at the bump end as the load increases.

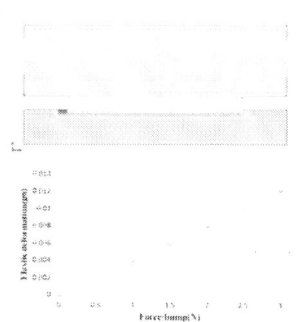

(a) Metal bump deformation

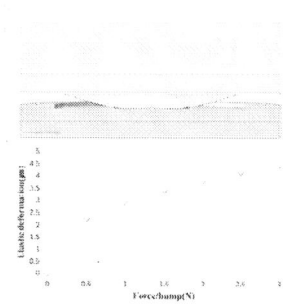

(b)PEB deformation

Figure 2: Analysis results of elastic deformation of metal bump and PEB w.r.t. increase of bonding load

The contact area between the polymer elastic bump and the substrate pad according to the load applied to the bump was analyzed. As the load increased, the contact area rapidly increased due to the elastic deformation of the polymer elastic bump, so that a uniform contact area and a low contact resistance could be maintained.

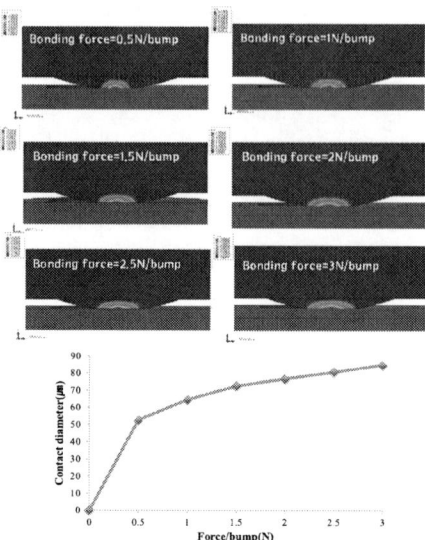

Figure 3: Analysis results of contact area (diameter) change between polymer elastic bump and pad according to bonding load

In order to solve the Au cap metal line crack problem due to excessive deformation of PEB, spiral cap and spoke cap PEB are developed. For the proposed spoke cap PEB and spiral cap PEB, the stress distribution was compared with the metal cap PEB at a load of 2.5 N / Bump and temperature of 25 °C and 225 °C. As the temperature increases to 225 °C, the stress is increased. As the bump deformation increases and the deformation of the polymer bump increases, the stress concentrates on the central part of the metal line formed on the bump. If cracks occur, it is predicted that cracks will occur first in the center and propagate to the outer part. Also, it can be seen that the bump is deformed and the stress is relatively large at the outer periphery where the slope changes rapidly. The spoke cap and spiral cap have less stress than the metal cap bump covering the entire polymer bump because the metal line on the polymer bump is patterned so that it can be easily deformed, which is believed to be caused by the relaxation of the stress.

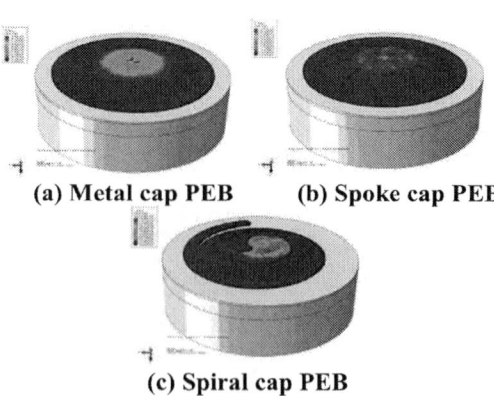

(a) Metal cap PEB   **(b) Spoke cap PEB**

(c) Spiral cap PEB

Figure 4: Von-Mises Stress of PEB during bump deformation at 25°C and 2.5N/Bump

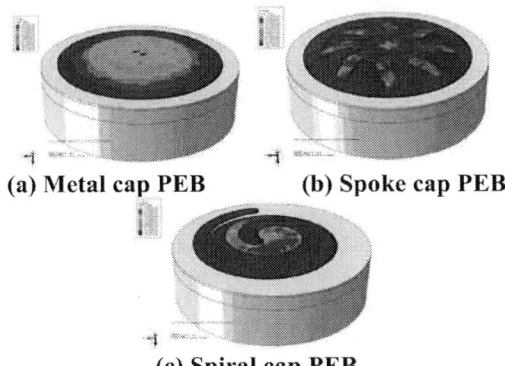

(a) Metal cap PEB   (b) Spoke cap PEB

(c) Spiral cap PEB

Figure 5: Von-Mises Stress of PEB during bump deformation at 225°C and 2.5N/bump

In order to analyze the deformation of PEB bumps according to temperature, elastic deformation behavior was analyzed under the load conditions of 2.5 N / Bump, 25 °C and 225 °C as described below. At room temperature, the modulus of elasticity of the polymer bump decreased to 348.85 MPa as the temperature increased to 225 °C. As a result, the bump deformation was 3.73 μm at room temperature, but the deformation increased to 8.73 μm at 225 °C. And it can be predicted that even if there is a bump height deviation, it is possible to easily perform bump height compensation using a small load.

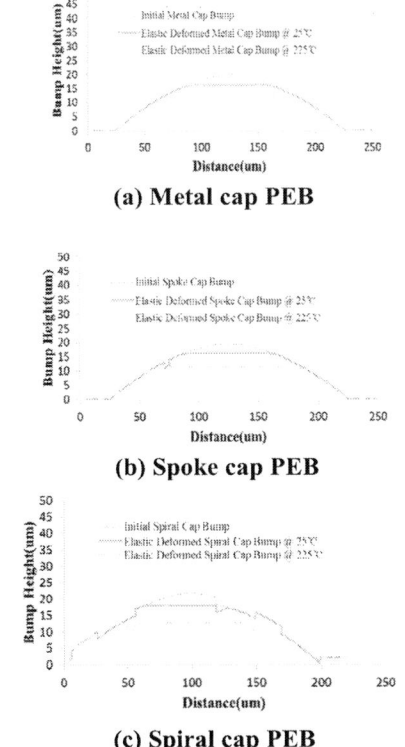

(a) Metal cap PEB

(b) Spoke cap PEB

(c) Spiral cap PEB

Figure 6: Elastic deformation height of PEB w.r.t. temperature

Experimental Results and Discussions

Fine PEB fabrication process of 50 μm diameter is developed using thermal reflow process and uniformity of bump diameter and height are evaluated that it is below 1um variation. In order to improve the Au metal cap cracking phenomenon due to excessive deformation of PEB, the following 200 μm diameter spiral cap type and spoke cap type PEB bump forming technology have been developed for robust PEB development. Even if the PEB is excessively deformed, the stress is low on the Au metal line.

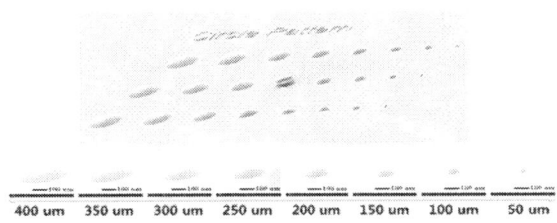

400 um   350 um   300 um   250 um   200 um   150 um   100 um   50 um

(a) PEB bank after thermal reflow process

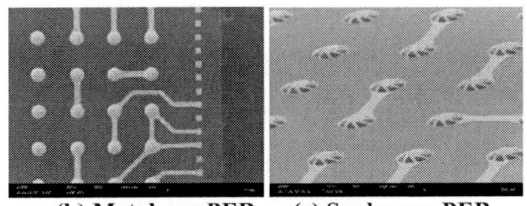

(b) Metal cap PEB   (c) Spoke cap PEB

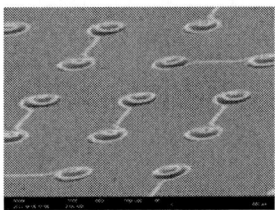

(d) Spiral cap PEB

Figure 7: Fabricated Polymer elastic bump (PEB) shape by thermal reflow process

The contact area change of PEB is investigated in real time as the load changes from 0 to 2.5 N/bump. The contact area increases as the load increases and the contact area decreases as the load decreases. Similar to the results of the PEB bump analysis, when the load is 1 N/bump, the contact diameter is about 60 μm. As the material of the bump is polymer, elastic deformation and restoration are close to 100 % at room temperature. However, as the temperature increases, the plasticity increases and the plastic deformation gradually increases.

393

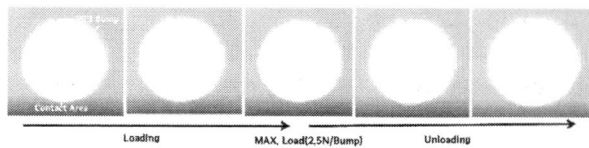

Loading   MAX, Load(2.5N/Bump)   Unloading

Figure 8: Contact area variation of PEB w.r.t . load

Through experiments, we find optimal bonding conditions and effect of bonding parameters such as bonding force, time and temperature. For the analysis of parameters and optimization of PEB bump face-down bonding process conditions, we made glass PEB bump chip with easy to observe the void trap and connection as below. The chip size was 5mm x 6mm, the bump diameter was 200μm, and the bump height was 20μm, which was the same as the thin PEB bump flexible chip to be fabricated. The process optimization is performed according to the process parameters of bonding load and temperature.

The Fig.9 is the result of measuring the contact resistance of PEB bump with bonding temperature fixed at 200 ℃ as bonding load increases. As the load increases at the same temperature condition, the elastic deformation and the visco-plastic deformation of soft PEB bump increase so that the contact area increases. The contact resistance was reduced and the uniformity characteristic of the contact resistance was also improved by this. The bump contact resistance is 2mΩ when the connection load is 0.8N/bump or less. It is confirmed that the contact resistance is lower than general ACF bonding. Even when the connection load is small as 0.4N/bump, the polymer bump easily deforms and the result of that contact resistance is low below 3.5 mΩ.

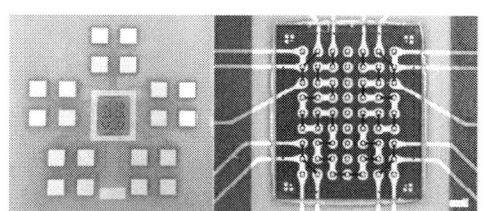

(a) Glass PEB NCF face-down bonding specimens with daisy chain & Kelvin pattern

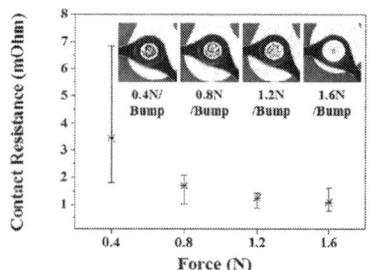

(b) Measured Kelvin resistance variation w.r.t. bonding force

Figure 9: Bump Kelvin resistance variation w.r.t. bonding force

Ultra-thin chip is fabricated through wafer bumping and RDL process, edge trimming, wafer thinning, thin chip transfer on polyimide film which makes it easy to handle ultra-thin chip and protect from die cracking by edge chipping because back-side polyimide film of chip suppresses crack opening and propagation and finally die singulation using dicing blade step cut, which prevents from bonding tool contamination caused by NCF adhesive creeping on back-side of ultra-thin chip during bonding process.

In order to achieve high integration and smaller form factor of flexible semiconductor packages, a flexible semiconductor package of 2 stack was developed as shown in figure 10 (a). In the first layer, a flexible ultra-thin chip of 20 μm thickness is embedded with face-down interconnect and the second chip is attached with face-up and interconnected using a wire bonding technique. After that, the package was molded using wire encapsulation and flexible silicon molding materials. The bending fatigue test of the flexible semiconductor package using PEB bumps was carried out and no change in daisy chain resistance occurred at 500,000 cycles at a bending radius of 10 mm.

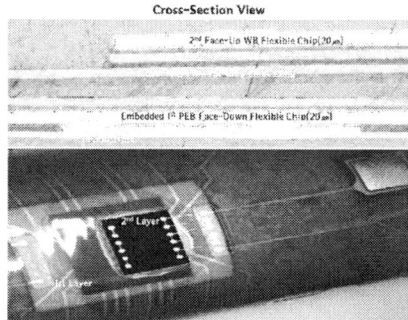

(a) Embedded flexible hybrid package

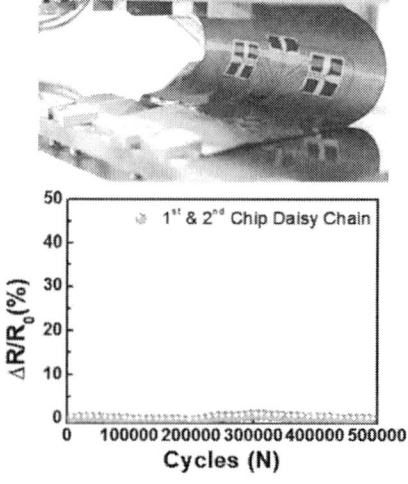

(b) Daisy chain resistance variation w.r.t. bending fatigue cycles

Figure 10: Bending Fatigue Test Results of embedded flexible hybrid stack package

Conclusions

In this study, polymer elastic bump (PEB) face-down bonding process is suggested newly, which could reduce stress concentration remarkably near bump and chip face by soft polymer bump material with electro-conductive metal line pattern and usage of non-conductive adhesive film (NCF), which is more flexible and has lower contact resistance than conventional ACF bonding, without crack and damage of ultra-thin chip. And also, the elastic deformation and restoration of PEB can solve the bump co-planarity problem. In order to achieve high integration and smaller form factor of flexible semiconductor packages, a flexible semiconductor package of 2 stack was developed. The bending fatigue test of the 2 stack flexible semiconductor package using PEB bumps was carried out and no change in daisy chain resistance occurred at 500,000 cycles at a bending radius of 10 mm.

Acknowledgements

This research was supported by "Development of Interconnection System and Process for Flexible Three Dimensional Heterogeneous Devices" funded by MOTIE (Ministry of Trade, Industry and Energy) and "Development of Nano-Based Manufacturing Key Technologies for Next-Generation Transparent Display with Flexibility" funded by nst (National Research Council of Science & Technology) in Korea.

References

[1] A. Dietzel, J. van den Brand, J. Vanfleteren, W. Christiaens, E. Bosman, and J. De Baets, "Ultra-thin Chip Technology and Applications", 1st ed., edited by J. N. Burghartz (Springer Verlag, New York, 2011), pp. 141–157.

[2] S. C. Kim, and Y. H. Kim, "Review paper: Flip chip bonding with anisotropic conductive film (ACF) and nonconductive adhesive (NCA)", Current Applied Physics, vol. 13, pp. S14-S25, 2013.

[3] N. Sawasaki, T. Ishihara, M. Mouri, Y. Murase, S. Masui, and H. Nakamoto, "Front-end Device Technology for Human Centric IoT", FUJITSU Sci. Tech. J, vol. 52, no. 4, pp. 61-67, 2016

Digital Core Supply in Automotive FC-BGA Package: a Smart Modularity Solution for Design and Validation

Cristina Somma, Aurora Sanna, Giovanni Graziosi

STMicroelectronics, via C. Olivetti 2, Agrate Brianza, Italy

cristina.somma@st.com; aurora.sanna@st.com; giovanni.graziosi@st.com

Abstract

The growing complexity of automotive digital applications requires a constant evolution of IC and package technologies. Thanks to its flexibility in supporting high signals density, TE-FCBGA (Thermal Enhanced Flip Chip Ball Grid Array) is the current package solution for high-end digital devices. An emerging criticality for this platform is the increasing static current consumption together with decreasing voltage levels, in particular for digital core supply interconnections. This work describes an innovative methodology for power supply optimization at package level, exploiting the concept of design modularity. The first step is to identify an overall layout strategy, working in a co-design context thanks to a close cooperation between package and silicon development engineers. The die to package connection of core supply is typically localized in the central bump-out area and benefits of intrinsic symmetry and uniformity, which are key features for current distribution robustness. This allows to define the ideal package interconnection as a matrix of small equal portions (modules) and to work on a single module that becomes the basic brick of the final layout. The optimization of the single module ensures a more efficient design flow and allows an early estimation of the overall electrical performance, which is obtained by simulation and represents a pre-layout strategy validation. The second step is the full design implementation: at this stage, the uniformity of modular structure can be locally affected by different layout constraints. Post-layout electrical simulation is hence needed to evaluate the impact of non-idealities and quantify the mismatch against early prediction: this drives a layout refinement in the finalization phase. The target is the best trade-off between performance and technology complexity, resulting in package costs optimization.

Key words: Design Modularity, FC-BGA, Power and Ground matrix, DC IR drop analysis

Introduction

TE-FCBGA (Thermal Enhanced Flip Chip Ball Grid Array) is the reference package technology for High-End digital devices. The main advantage related to this package platform is its intrinsic flexibility that perfectly matches the high interfaces integration requirements. The evolution of IC devices and package substrates results in new challenges for the digital core supply interconnections assessment.

The increasing static current consumption combined with decreasing voltage level is an emerging critical aspect for this package technology that has to be addressed and analyzed carefully to ensure the target performances. This work starts from a high density matrix of bumps and balls commonly connected to digital core supply and ground, and a uniform current distribution. This methodology proposes a new co-design and co-analysis solution based on an elementary cell to estimate and optimize the core digital supply DC voltage drop of the full package layout.

Central Connection Matrix Description and Verification Methodology

In a digital integrated circuit (IC), the highest power consumption is typically associated to the core power supply. At package development level, a design priority is reserved to the interconnection of this supply, which typically has a dedicated area below the central bumps matrix.

On this area, FC die bumps and BGA balls have an optimized connection: direct, short and robust. Red rectangle shown on Figure 1 indicates the typical power supply and ground nets location on FC-BGA.

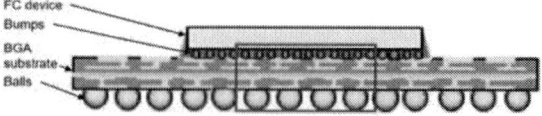

Figure 1: FC-BGA package cross-section: power supply and ground location delimited by red rectangle

IC and BGA central matrix connectivity can have different configurations based on bumps and balls position constraints but, usually they have anyway a common approach defined during design activity: uniform and regular distribution of power supply bumps or balls that alternates ground bumps or balls. On Figure 2, a detail of possible IC power bumps (red) and ground bumps (green) assignment that is propagated within central matrix.

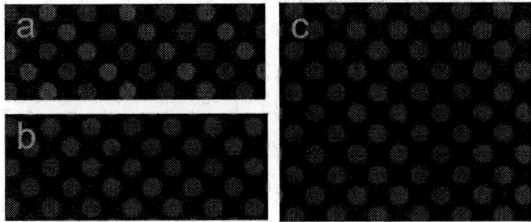

Figure 2: Detail of possible FC bumps connectivity distributions; power supply and ground sequence as diagonal 1 row shown on case a, diagonal 2 rows on case b and linear 3 rows on case c

A very important parameter for a good core supply design is the interconnection uniformity in terms of DC behavior (i.e. voltage drop generated by the average current flowing through package nets). In a practical case, at die level, a large area is dedicated to core power grid, that supplies blocks having different functionalities and switching activities. This means that the power consumption over the whole area is generally not homogeneous. At package level, consequently, the amount of current flowing through each bump can be not the same. Despite this, a robust package design strategy should start from the simplified assumption of a uniform current distribution on all die pads. In fact, local effects might be still unknown at the first stages of package design, and they may also change depending on the different operating conditions. Given this assumption, a robust package layout must guarantee a uniform connectivity from balls to bumps, avoiding any kind of current bottleneck that might generate an excessive voltage drop on some die locations. When a more realistic current distribution is applied, this strategy prevents the worst case of a layout weakness combined with a current peak in the same area. Since the connection matrix is made by vias connected together by plane shapes on the different substrate layers, if the parallel paths density is high enough and the uniformity is guaranteed, this will allow a homogeneous current distribution.

The simplest way to check the layout performance is to run a distributed DC IR drop simulation. The common technique, supported by different commercial tools for package electrical verification, starts from creating an ideal DC voltage source to represent the Voltage Regulator Module (VRM), that is supposed to be on the application PCB: the ideal source is directly applied between power and ground balls, thus neglecting any non-ideal effect due to PCB interconnections. The second step is to define a DC current source between each power bump and the ground bumps, to represent the absorption from each die connection. Simulation allows to verify the voltage drop and current density maps over the whole layout structure. If the same current absorption is applied to each bump, as consequence of the Ohm's law it can be easily concluded that bumps showing a higher voltage drop are connected to balls through a higher resistive path. The current density map highlights the layout bottlenecks that cause higher resistances, in order to drive layout improvements.

Modular Structure and Pre-Layout Strategy Validation

The described analysis requires a complete package design to be performed, so it is widely used as post-layout verification technique. The relevant question is how to apply the same methodology for a pre-layout estimation of the connection matrix performance, in order to validate a design strategy.

The answer lies in the modularity concept, combined with the linearity of the Ohm's law that regulates the DC electrical behavior. If the connection matrix is perfectly uniform in terms of vias and planes structure, it can be imagined as the combination of a certain number of elementary "bricks" (modules). Once the minimum brick is defined and designed, it is possible to perform a DC IR drop analysis on it, before the full layout implementation. If the absorbed current is the same on each bump, the voltage drop inside the module will be a realistic representation of the voltage drop in the full connection.

The minimum module identification comes from a reduction to lowest terms of the balls-to-bumps ratio inside the central matrix (Figure 3).

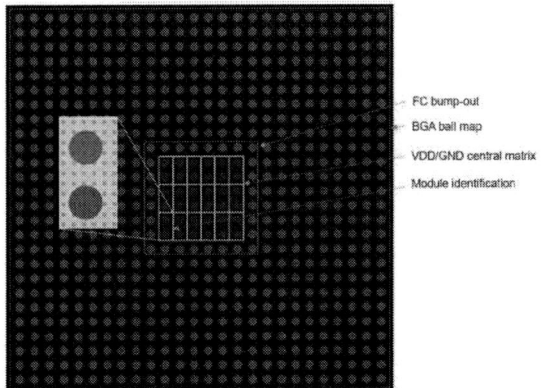

Figure 3: Basic module identification based on central matrix grid

Fundamental steps are the following:
- Based on die size, the total area of the central matrix is identified

- Ball pitch and ball pad dimensions are considered to determine the number of balls in the area; similarly, bump pitch and pad dimensions define the number of bumps
- Considering the target of a good symmetry between power and ground paths, 50% of the balls is assigned to power net and 50% is assigned to ground net. The same process is valid for bumps
- Balls-to-bumps ratio is calculated and reduced to lowest terms, considering that at least 1 power ball and 1 ground ball are needed
- The minimum balls-to-bumps ratio defines the number of balls and bumps in the module and, consequently, the module area

A direct implication of the described procedure is that different ball pitch and bump pitch options can be evaluated with the module technique, simply changing the brick dimensions based on the minimum balls-to-bumps ratio. Table 1 shows how numbers are defined for varying ball pitch from 0.65 mm to 0.8 mm and bumps number from 1800 to 1600, in the same area. Note that a ball pitch variation does not imply a bump pitch change: the example shows the combination of two independent parameters. Higher pitch means lower balls or bumps density and, consequently, higher current density. However, this effect can be balanced by an increased number of vias or other design optimization strategies, so the final voltage drop is not necessarily worsened. Comparative simulation quantifies the impact.

	0.65mm Ball-Pitch 1800 bumps	0.8mm Ball-Pitch 1600 bumps
CENTRAL MATRIX AREA	6mm x 6mm	6mm x 6mm
NUMBER OF BALLS	50 VDD BALLS 50 GND BALLS	32 VDD BALLS 32 GND BALLS
NUMBER OF BUMPS	900 VDD BUMPS 900 GND BUMPS	800 VDD BUMPS 800 GND BUMPS
BALLS TO BUMPS RATIO	50:900 = 1:18	32:800 = 1:25
MINIMUM MODULE STRUCTURE	1 VDD ball, 18 VDD bump 1 GND ball, 18 GND bumps vias/planes connection from balls to bumps	1 VDD ball, 25 VDD bump 1 GND ball, 25 GND bumps vias/planes connection from balls to bumps
NUMBER OF MODULES IN THE FULL MATRIX	50	32
TOTAL CURRENT ABSORBED BY DEVICE	10A	10A
CURRENT PER BUMP	11.11mA (10A/900bumps)	12.5mA/bump (10A/800bumps)
CURRENT IN THE MODULE	~200mA (11.11mA x 18 bumps)	312.5mA (12.5mA x 25 bumps)

Table 1: Module dimensioning for varying balls and bumps density

Once the module area is defined, bumps and balls in the brick must be connected by a matrix of vias, connected together by plane shapes on each substrate layer. The number of vias and the shapes structure are fundamental parts of the design strategy, and depend on technology constraints such as layers stack-up, vias dimensions and minimum spacing rules. Starting from bumps count, a balanced bumps-via-ball ratio has to be fine-tuned in order to

avoid layout bottlenecks or expensive over-design approaches. Also the distribution of vias, power supply and ground layers sequence and objects proximity have a relevant importance on module creation.

Finally, to perform a DC IR drop analysis, the current per bump information is needed. In the example with 0.65 ball pitch and 900 power bumps, if the device is supplied at 1V and its average power consumption is 10W, the total average current will be 10A. This corresponds to 11.11mA/bump (10A / 900 bumps, as reported in Table 1). The same uniform distribution is applied to the return current flowing through the 900 GND bumps. This means having ~200mA in a module that uses 18 bumps both for VDD and GND (Table 1). The module behavior when 200mA are applied should reflect the behavior of the full matrix when 10A are applied.

The module example with 1:18 balls-to-bumps ratio is shown in Figure 4.

Figure 4: Module with 1 VDD ball, 18 VDD bumps (red); 1 GND ball, 18 GND bumps (blue)

The package level target, in this case, is to guarantee a voltage drop < 5 mV (0.5% of the nominal supply voltage). Simulation provides the voltage distribution in power and ground nets (VDD and GND), as reported in Figure 5: for VDD, maximum voltage corresponds to the ideal value and it is located close to the ball, minimum voltage corresponds to the maximum package drop, and it is located in the bumps area; the opposite concept is valid for GND: minimum voltage corresponds to the ideal value (0V), so the maximum voltage is observed on bumps area and represents the maximum drop.

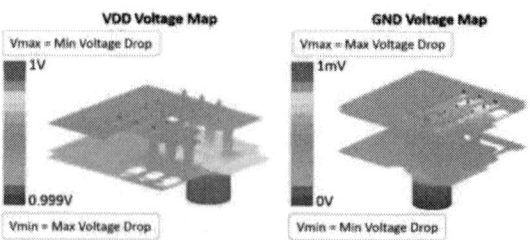

Figure 5: Voltage distribution in the module: VDD and GND connections

The sum of average VDD drop and average GND drop represents the average voltage drop in the module, that corresponds to ~2mV in the example. This number is perfectly aligned with the target. When the module will be repeated to build the full connection matrix, a similar value of average drop is expected if the starting assumptions are correct.

Besides the average drop, the analysis also shows the voltage difference among bumps, which is an indicator of the connection uniformity. Figure 6 shows that this difference in the module is one order of magnitude lower than the average drop (0.2 mV - 0.3 mV). This is proof of a homogeneous connection resistance from ball to bumps. A slight difference between power and ground behavior is expected and depends on the alignment between ball and bumps, that might be not perfectly symmetric for the two nets, also impacting on alignment of the whole vias matrix.

Figure 7 finally shows the current density map for both VDD and GND nets in the module: this allows to verify that there are no relevant bottlenecks in the structure, and confirms the good uniformity of the connection. Current density distribution and bumps level voltage difference are strictly related: if a critical voltage drop is identified on some specific bumps, some bottlenecks are most probably affecting the current flow in that area.

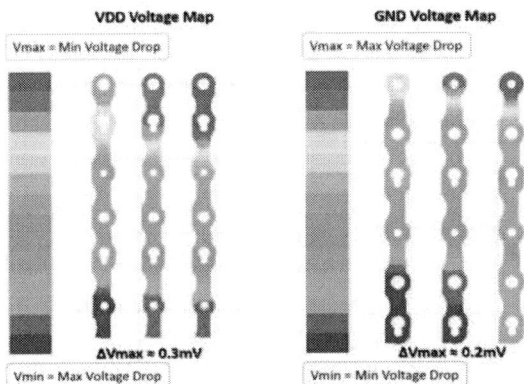

Figure 6: Voltage map on VDD and GND bumps area

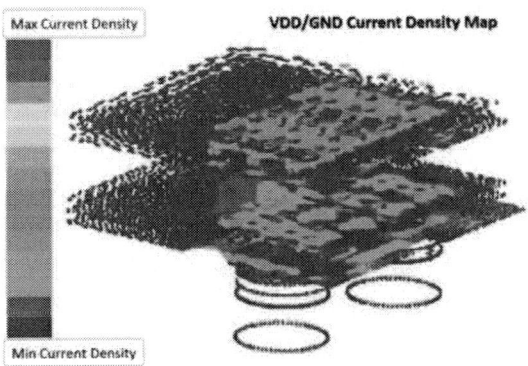

Figure 7: Module current density map on VDD and GND

Potential differences may occur from module to full layout analysis, due to some effects that are neglected in this pre-layout study: positive impact of large planes, negative impact of design non-idealities that break the matrix uniformity, behavior of peripheral connection paths lying outside the central matrix. All these elements will be discussed in the next paragraphs.

From Module to Full Matrix and Layout Optimization

Once the basic module has been optimized and validated following all the above techniques, the phase of module implementation on substrate layout can start.

Figure 8 shows on the left picture, the central matrix of a FC bump-out and a BGA balls-map on the right that, usually, are connected to digital core supply and ground nets. The difference on bumps or balls colors indicates power supply and core ground nets. The blue box in the picture on the right represents the die central matrix area position and the green box is the die boundary.

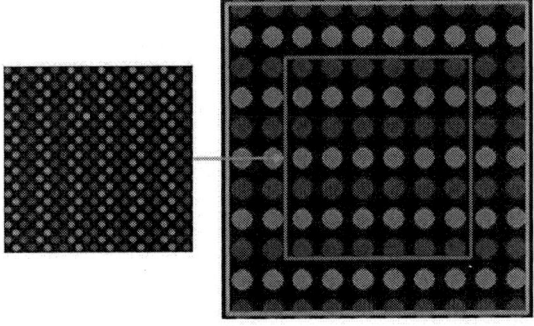

Figure 8: Digital core supply and ground nets: FC central bumps matrix (picture on the left) and BGA central ball matrix (picture on the right)

The implementation of module on full layout consists in a replication of the basic brick inside this area. As detailed before, this structure is composed of micro-vias, metal traces, core buried vias and small portions of planes that connect a group of bumps to some balls. These elementary cells are copied and placed several times, each adjacent to the other to fill the small equal portions of the grid previously identified. During the matrix copy/paste operation, not only the structure instantaneously connects each group of bumps-ball but, also all modules are attached together through planes boundary and traces as highlighted with red line on Figure 9. Wide power and ground planes help the current redistribution that mitigates the effect of absorption variability among different die areas. This positive effect is neglected in the

399

module-based pre-layout analysis that is, in this sense, pessimistic.

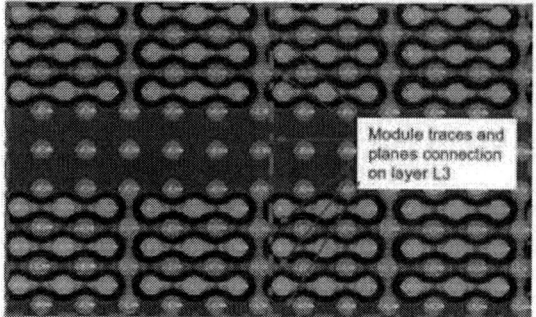

Figure 9: Inner layer detail of the connection of same module replicated twice

The ideal implementation results in a complex matrix of high bumps count that connects a proportional number of BGA balls (Figure 10).

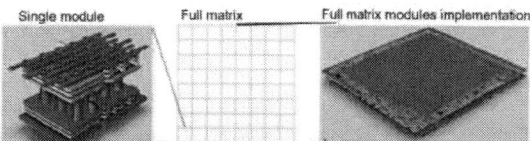

Figure 10: Full grid implementation

However, the module full implementation on a complex digital product introduces some non-idealities that locally affect the full matrix layout. Physical, spacing and electrical constraints on the matrix boundary or on specific locations, and some irregularity on bumps or balls connectivity impact the digital supply or ground net layout of module in that specific area. Figure 11 shows single module layout of an inner layer, the local critical effect seen on that layer, caused by the presence of some signals close to matrix edge and finally, the local layout enhancement that compensates the degradation effect implementing a selective reinforcement of power supply connection.

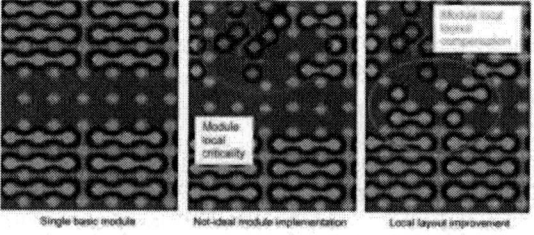

Figure 11: Example of non-ideal effect and the local module optimization

The full matrix DC IR drop simulation checks the effect of non-idealities at intermediate layout development level, in order to correct potential criticalities before finalization. Figure 12 shows an example: looking at the voltage difference among VDD bumps on the top metal layer, some higher drop locations are observed. Investigating in the layout structure, a discontinuity is recognized in the connection below those bumps.

Thanks to some local layout optimizations, DC voltage drop can be contained. The main actions applied for layout refinement are: increase the number of vias or shift some of them close to critical area, reduction of planes bottlenecks and creation of parallel connections in the identified critical path. Figure 12 also shows the final result after layout optimization: the difference among bumps is now minimized, and this also reduces the average voltage drop. The improvement level is in the order of 25%.

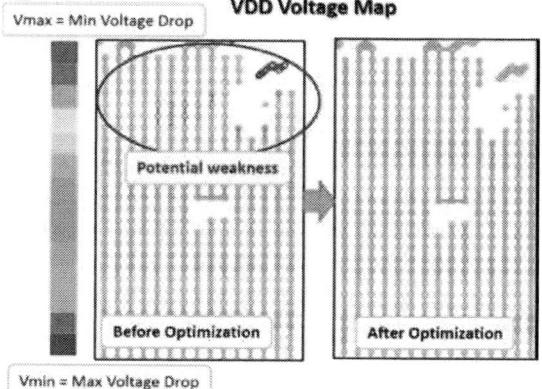

Figure 12: Effect of layout optimization on voltage drop uniformity among power bumps

The importance of layout optimization is furthermore reinforced by the fact that current will be not uniformly distributed, but some die portions will absorb more current than the others. For this reason, all the area that shows a potential criticality has to be improved as much as possible to avoid that, exactly in that weak region, an increase of current causes an excessive voltage drop.

In addition, also the target to have a perfect symmetry between power and ground balls or bumps count can be subjected to a slight overbalancing caused by peripheral bumps or balls connectivity. This non-ideality will not have a relevant weight on the final result but the local layout improvements must be accurate to mitigate any possible not-optimum effect.

After this optimization phase, FCBGA layout will be completed, finalized and validated.

Full Layout Final Validation

After all the local optimizations have been applied, the final step is to evaluate the performance of the full layout. Figure 13 shows the voltage drop map that includes the peripheral connections.

Table 2 shows that the module is very well representing the behavior of the full layout in terms of average voltage drop. Voltage difference among bumps is higher in the full layout case, but this is expected due to boundary effects: bumps that deviate more significantly from the average are, in fact, located on the matrix edge. However, reported numbers are more than acceptable and demonstrate the benefit of all the implemented improvements.

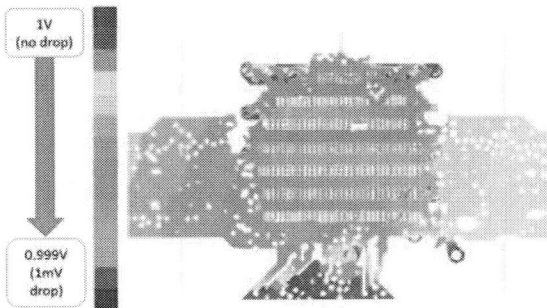

Figure 13: Full layout voltage drop map on VDD

	Pre-Layout (Module)	Full Layout
Average voltage drop from balls to die (VDD drop + GND drop)	2mV	1.98mV
Maximum voltage variation among VDD bumps	0.3mV	0.5mV
Maximum voltage variation among GND bumps	0.2mV	0.55mV

Table 2: Voltage drop values in pre-layout and full-layout analysis

Further Development: Vias Optimization for Cost Reduction

The virtual validation described in the previous paragraph shows a full alignment with electrical targets. From a performance viewpoint, package design is now ready for manufacturing.

Actually, another interesting output of DC IR drop simulation should be reviewed: the map of current flowing in all VDD and GND vias.

Figure 14 shows how this map can highlight the vias carrying a negligible current amount. If these vias are removed from design, no impact on voltage drop is expected. This helps to prevent overdesign and to potentially reduce manufacturing costs by shortening the time required for drilling process. For a safe approach, it is recommended to run this analysis not only under the uniform die current assumption, but also by stressing the structure with unbalanced die current distribution (based on realistic use case). This prevents to create weak paths in potentially critical areas. In addition, the possible degradation of dynamic behavior should

be considered before removing vias. Dedicated AC simulations are available for this check.

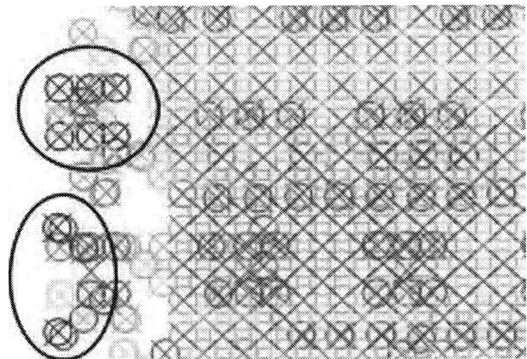

Figure 14: Map of current flowing in vias: blue color indicates a negligible value

Conclusions

The paper has described a methodology to validate power supply design strategy at package level, thanks to a pre-layout design and analysis based on the modularity concept. The alignment between pre-layout and post-layout results demonstrates the method effectiveness. The importance of electrical simulation to support all the steps of package design flow has been underlined and demonstrated through a dedicated example. As further development, the presented analysis can be used not only for pure performance optimization, but also to find the best trade-off between performance and manufacturing cost.

References

[1] G. Graziosi, C. Somma, A. Sanna, "BGA package design techniques for electrical performance awareness", MiNaPAD Forum, 2017

[2] G. Graziosi, P. Joubert Doriol, Y. Villavicencio, C. Forzan, M. Rotigni, D. Pandini, "Advanced modeling techniques for system-level power integrity and emc analysis", European Microelectronics and Packaging Conference, 2009, Page(s): 1 – 6

[3] J. Socha , S. Chitwood, B. Brim, "PCB power integrity flow — efficiently designing-in large pin-count, smaller pitch devices", 2017 IEEE International Symposium on Electromagnetic Compatibility & Signal/Power Integrity (EMCSI)

[4] R. Sjiarel, "Power integrity simulation of power delivery network system", 2015 IEEE 19th Workshop on Signal and Power Integrity (SPI) Pages: 1-5

Transfer Molding Simulation to Predict Filling Flaws and Optimize Package Design

M. Rovitto and A. Cannavacciuolo

STMicroelectronics, via C. Olivetti, 2, 20864 Agrate Brianza (MB), Italy

Phone: +390396037115, Email: marco.rovitto@st.com

Abstract

The process of microchip encapsulation by epoxy resin injection is simulated by employing moldflow modeling. In this work, the use of modeling allows to reproduce the unbalanced flow behavior of the resin within the mold cavity during the filling stage which leads to the formation of voids in the molded package. Simulation results are compared to experimental findings, in which the molding process is interrupted at designed points in time. Comparison shows good agreement between numerical outputs and experimental data and demonstrates the effectiveness of the model. After modeling is validated, the numerical methodology is applied to optimized package design in order to minimize the risk of issues caused by the filling stage. As a result, geometric variations show a significant effect on filling front progression avoiding the generation of structural defects during molding process.

Key words: package, molding, resin, moldflow, void, weld line

Introduction

Nowadays, the dominant technique for microchip encapsulation using epoxy molding compound (EMC) is transfer molding. The process gives proof of its high reliability level, robustness, and accuracy. Although these benefits, improper selection of process conditions, leadframe layout or mold chase design, as well as the complexity of EMC material properties during molding could raise the risk of fabrication issues. For instance, the unbalanced flow behavior of the EMC within the mold cavity during the filling stage can lead to the formation of voids, weld lines, and die pad tilting which may induce to package flaws and, subsequently, production loss [1]. The understanding of the process of EMC flow into the cavity is therefore essential to establish its impact on package quality and predict the actual defects in the molded package. For this purpose, integrated computer-aided engineering (CAE) technology gives a comprehensive solution for simulating the complicated physical phenomena inherent in transfer molding process for chip encapsulation [2]. In particular, CAE simulations help to determine the optimal package design by choosing the right material characteristics and process conditions in order to avoid the generation of filling risks.

In this work, molding process simulation is employed to reproduce the transient process of filling the mold cavity with EMC material. The filling analysis is performed on a standard quad-flat package (QFP) mold chase. In a typical QFP mold chase, the EMC encompasses the entire bottom side

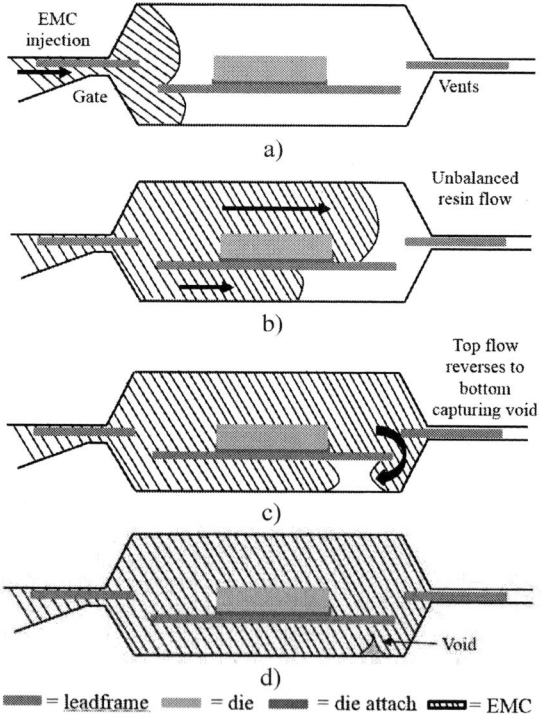

Figure 1: Schematic description of the void generation during transfer molding process into the QFP mold chase. a) EMC enters the gate insert. b) Unbalanced flow behavior of the EMC within the mold cavity during the filling stage. c) Top flow is faster than the bottom flow and reverses at the bottom side. d) As a consequence, void is formed.

of the cavity, where the gate insert is located, as depicted in Fig. 1a. In this specific situation, the resin flows into the cavity and differences in melt front progress between the top and the bottom of the mold cavity are evident (see Fig. 1b). Since the top cavity is filled faster than the cavity of the gate insert, the top melt front tends to flow reversal and entraps air at the bottom of the QFP cavity, as shown in Fig. 1c. When the resin flow compresses air at the bottom side, a weld line forms at that location representing a potential spot of weaker package structure (Fig. 1d). Simulation analysis is able to evidence unbalanced resin flow behavior. Further, filling results obtained from simulations are compared with experimental trials, well-known as short shots, where the filling progress is interrupted at different points in time. Once the CAE simulation is validated, a rigorous numerical investigation is conducted to investigate which is the main factor, such as leadframe design, material properties, and process conditions, which significantly effects the melt flow within the QFP cavity. In this way, the CAE simulation methodology is applied in virtual package design optimization in order to minimize the risk of issues caused by the filling stage. Controlling manufacturing failures results in either assembly yield gain and package quality improvement.

Geometry and Material Models

CAE modeling for microchip encapsulation lies in the difficulty in the generation of a suitable mesh for moldflow analysis as well as in the selection of resin material models.

The first challenge could be overcome by adopting an efficient grid generation procedure [2]. The three-dimensional (3D) outline is imported from a CAD file, where leadframe, die, wires, mold case, and runner are defined. Then, a two-dimensional (2D) triangular surface mesh is generated by considering the geometrical details of the 3D design. The extrusion of the 2D triangular mesh in the thickness direction allows to mesh the whole 3D domain from an automatic generation of prismatic grid. In this way, solid mesh for moldflow analysis is obtained, as depicted in Fig. 2. The local details of each component are showed and simulation boundary conditions are defined. Due to symmetry, cull is modeled as half-cylinder. The entrance of the resin into the mold chase is defined at the bottom of the cull, which is the location where the pellet is inserted into the plunger.

Once solid mesh and boundary conditions for modeling analysis are defined, the simulation flow procedure requires the selection of the EMC material models which better describe the chip encapsulation process. The filling process in mold cavity is usually represented as a transient issue with moving resin front. It can be mathematically described by the conservation equations of the 3D transient non-isothermal process, as presented in details in [2, 3].

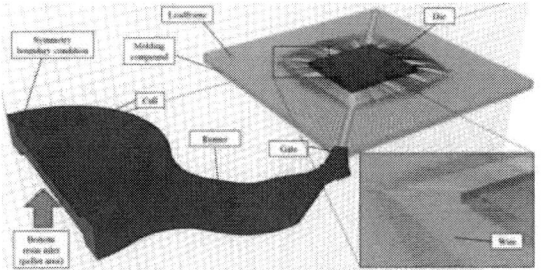

Figure 2: Profile view of the solid mesh layout employed in the moldflow analysis. All package component materials, cull, runner, and gate insert are listed. The zoomed-in detail view of the structure depicts wires. Further, imposed boundary conditions for the simulation are presented on the left side.

Then, the choice of material models needs to take into account rheological and curing kinetic behavior of the EMC. For this purpose, Cross-Castro-Macosko model is widely used to describe the viscosity change of EMC with temperature and conversion rate obtained from experimental rheological data [1]. Further, the curing reaction of epoxy resins is typically well reproduced by the model proposed by Kamal because of its accurate prediction of the EMC curing kinetics within the experimental data [4].

Regarding other components involved in the solid mesh design in Fig. 2, liner elastic material properties of copper are chosen for leadframe and wires, while silicon material properties are selected for die. Non-linear behavior of these materials is not relevant for the purpose of the analysis. Non-linearity could be important for advanced studies, such as investigation on wire sweeping and paddle shift [5].

Simulation Settings

This work focuses on the modeling of the filling stage of the molding process for microchip encapsulation. The simulation settings of resin filling into the mold chase cavity used for this specific test case are here described by following realistic molding machine information and process parameters. The filling procedure is made of two steps, which are controlled by plunger velocity, temperature, and pressure in the molding machine. These steps are called pre-heating and transfer molding process.

First, the EMC pellet is placed into the plunger cavity and pre-heated at 175°C for 5 seconds [6]. Thermoset might slightly react during pre-heating. With a higher initial conversion, the curing proceeds faster and the viscosity builds up accordingly. This step is reproduced in the model as resin inlet boundary condition, as described in the previous paragraph and depicted in Fig. 2. After preheating phase, the transfer of the EMC begins.

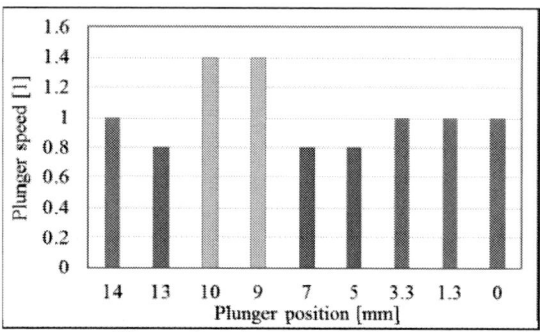

Figure 3: Injection profile used in the simulation for molding process.

The EMC temperature is 175°C. The time needed to fill the cavity is 5.6 seconds. The plunger presses the polymer pellet into the mold chase cavity by following the injection profile shown in Fig. 3. Typically, injection velocity slows down towards the end of the filling stage in order to avoid wire sweep issues [1]. During this phase, molding machine maintains a transfer pressure of about 10MPa.

Results and Discussion

Once design and simulation parameters are set up, unbalanced resin flow behavior within the QFP mold chase cavity can be investigated by employing molding process simulation. In parallel, moldflow modeling must be validated by comparison with short shot investigation. Simulation results are therefore compared with experimental short shots taken at different points in time.

Fig. 4 shows filling simulation results in terms of melt front progress at different plunger positions. Fig. 4a depicts the filling progress up to 8mm of plunger movement evaluated at the top and the bottom side of the cavity. The characteristic waviness of the melt front obtained with simulations is in good agreement with wave front behavior observed in short shot data at this plunger position. This demonstrates the effectiveness of the model. The simulation result exhibits an unbalanced resin flow behavior between the top and the bottom of the cavity. The top flow has much less friction resistance and runs ahead with respect to the bottom flow. Since the bottom melt front runs slower than the top front, die pad is not homogeneously filled, as depicted in Fig. 4b. As a consequence, the fast top melt front tends to flow reversal and entraps air at the bottom of the QFP cavity leading to weld line formation and, subsequently, nucleation of a void. In general, weld line is the line formed during the mold filling process by two different melt fronts joining together with sharp angle and indicates potential spots of weaker structure. The darker the weld line, the weaker the structure. Weld-lines look like cracks on the appearance of plastic parts. The local mechanical strength in the weld-line area could be drastically lower. It could be one of the most significant problems for structural applications due

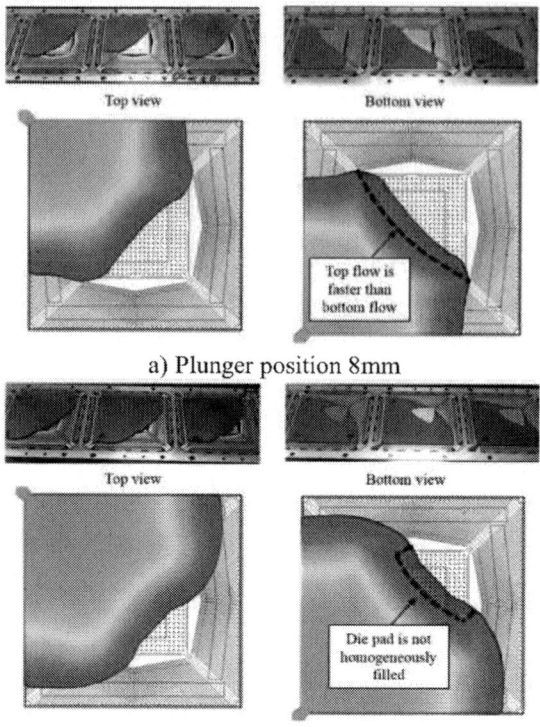

a) Plunger position 8mm

b) Plunger position 6.5mm

Figure 4: Top view of the simulation short shots taken at a) 8mm and b) 6.5mm of plunger movement. For each short shot, comparison with experimental front progress is presented.

to potential failure in the weld-line areas. Through CAE simulation, part geometry and weld-lines strength can be transferred into the stress solver. In particular, weld line strength is associated to the weld line meeting angle. It is the meeting angle between two converging melt fronts, ranging from 0 to 180 degree. If the welding angle is 180 degree, then the two melt fronts can be considered as one. If the welding angle is 0 degree, then two melt fronts converge head-on. Generally, the smaller the meeting angle is, the weaker the strength of the part becomes and the more critical the weld line will be after the part is ejected. By following this approach, Fig. 5 shows the distribution of meeting angle of melt fronts on weld lines. The zoomed-in area in Fig. 5 reveals that the most critical weld line is observed at the bottom area, where the top resin front flows reversal and meets the bottom melt front with low convergence. In fact, the weld line meeting angle is approximately 10 degree at this location. It means that the local mechanical strength around the area is significantly reduced due to the presence of the weld line which results in high probability of void generation.

Once the site of critical weld line formation is evaluated in the structure, molding process simulation becomes necessary in the development of new leadframe package design in order to lower the

404

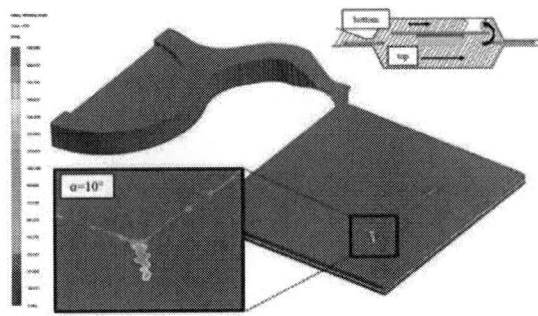

Figure 5: Profile view of the weld line meeting angle distribution evaluated in the standard QFP package. The minimum value is located at the package bottom side, where the top resin front meets the bottom melt front.

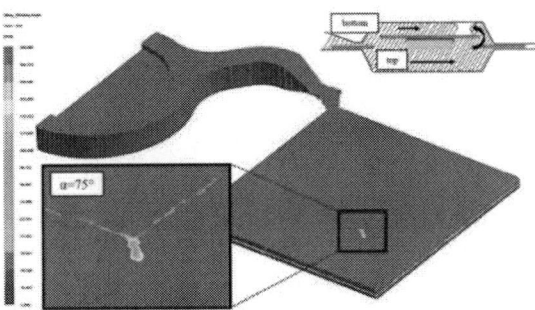

Figure 6: Profile view of the weld line meeting angle distribution evaluated in the optimized QFP package design. Higher meeting angle at the weld-line area is observed. This reduces the risk of generation of void at this location.

unbalanced melt flow within the QFP cavity and, consequently, the risk of void generation. For this purpose, the moldflow analysis is conducted on a different 3D solid mesh layout which include modification in frame die pad design. The same boundary conditions, material models, and simulation settings presented before are employed in the investigation. By analyzing the weld line meeting angle distribution in the structure depicted in Fig. 6, leadframe geometric changes result in higher meeting angle at the weld-line area with respect to the value obtained for the original design. Since two melt fronts meet with good convergence, the investigation on an optimized leadframe design shows a reduction of uneven resin flow behavior during the filling stage. The result leads to lower risk of critical weld line formation and, subsequently, crack generation compared to the original test case. Design optimization becomes therefore crucial for minimization of the risk of issues caused by the filling stage resulting consequently in packaging yield gain and quality improvement.

Conclusion

In this work, the investigation of the process of microchip encapsulation within a QFP mold cavity by EMC injection is carried out with the help of numerical modeling. Moldflow simulation is employed to reproduce the incomplete filling behavior due to the unbalanced resin flow between the top and the bottom of the cavity. Furthermore, simulation results show that the most critical weld line formation is located at the package bottom. The local mechanical strength reduces at this location leading to void generation. The effectiveness of the simulation results is confirmed by comparison of the melt front progress into the cavity with experimental short shots taken at different time steps. Once the model is validated, CAE methodology is employed to analyze the moldflow behavior with different package designs. The assessment is performed by investigating the weld line meeting angle distribution within the QFP cavity. Simulation result shows package design optimization allows for higher meeting angle compared to the value obtained with the standard design. In this way, the risk of issues caused by the filling stage is strongly reduced leading, subsequently, to package quality improvement.

Acknowledgements

The authors would like to thank Mr. Goran Liu and Mr. Cedric Liu from CoreTech System Co., Ltd. (Moldex3D) for their support on modeling and simulation activity and for the fruitful discussions about this study. Further, a special acknowledge goes to colleagues from STMicroelectronics Back-End Manufacturing and Technology R&D in Agrate Brianza (Italy) for their contribution to the project.

References

[1] M. Roellig, S. Meyer, M. Thiele, and S. Rzepka, K.-J. Wolter, "Modeling, Simulation and Calibration of the Chip Encapsulation Molding Process", Proc. EuroSimE, Delft, Netherlands, April 26-29, pp. 1-7, 2009.

[2] R.-Y. Chang, W.-H. Yang, S.-J. Hwang, and F. Su, "Three-Dimensional Modeling of Mold Filling in Microelectronics Encapsulation Process", IEEE Transactions on Components and Packaging Technologies, Vol. 27, No. 1, pp. 200-209, March, 2004.

[3] S. Moon, Z. Li, S. Gokhale, and J. Wang, "3-D Numerical Simulation and Validation of Underfill Flow of Flip-Chips", *in* IEEE Transactions on Components, Packaging and Manufacturing Technology, Vol. 1, No. 10, pp. 1517-1522, October, 2011.

[4] M. R. Kamal and M. R. Ryan, "Injection and Compression Molding Fundamentals", A. I. Isayev (Ed.), first edition, Marcel Dekker, Inc., New York, Chapter 4, pp. 329-376, 1987.

[5] Y.-Y. Chou, H.-S. Chiu, and W. Yang, "Three-Dimensional CAE of Wire Sweep and Paddle Shift in Microchip Encapsulation", Proc. IMPACT, Taipei, Taiwan, October 21-23, pp. 35-38, 2009.

[6] P. Gromala, J. Duerr, M. Dressler, K. M. B. Jansen, M. Hawryluk, and J. de Vreugd, "Comprehensive Material Characterization and Method of Its Validation by Means of FEM Simulation", Proc. EuroSimE, Linz, Austria, April 18-20, pp. 1/8-8/8, 2011.

New approach to apply 1,2,3-benzotriazole as a capping layer on UBMs for 3D TCB stacking and investigation of oxidation protection and solder wetting

Adil Shehzad[1,3], Jaber Derakhshandeh[1], Bernhard Wunderle[3], Lin Hou[1,2], Samuel Suhard[1], Kenneth June Rebibis[1], Andy Miller[1], Gerald Beyer[1] and Eric Beyne[1]

1) imec, Kapeldreef 75, 3001 Leuven, Belgium
2) Dept. Material Engineering, KU Leuven, Leuven, Belgium
3) Faculty Electrical engineering and and information technology (TU Chemnitz), Chemnitz, 09111, Germany

Phone: +32 (0)16 281567, E-mail: adil_shehzad19@yahoo.com

Abstract

In this paper, the oxidation protection and solder wettability is investigated after coating an organic compound 1,2,3-benzotriazole (BTAH, $C_6H_5N_3$) or BZT on copper UBMs. Since BZT serves as a very strong corrosion inhibitor for copper and its alloys, therefore the investigation is specifically focused on oxidation prevention during shelf life and also during bonding of copper UBMs. Different characterization and experimental analysis are performed involving water contact angle CA, XPS, FT-IR to identify the best coating recipe for the organic film in terms of film stability. Afterwards, stacking is performed using 3D (three dimensional) test vehicle, containing Cu based UBMs which faces the TCB (thermal compression bonding) process for fine pitch interconnection. Then wettability and performance of stacked coupons are analyzed by SEM cross-section and electrical resistance measurements.

Keywords: Benzotriazole, Corrosion inhibitor, Oxidation protection, Wettability, 3D Interconnects, Capping layer

Introduction

High density interconnects have fundamental role in enabling the miniaturization trend following the Moore's Law [1]. So, to address the scaling down limit and reduce the power consumption for future products, advanced 3D package is an evolving technology that can offer a fine pitch interconnection with high bandwidth and low cost [1, 2]. 3D interconnects mostly involve TSVs (Through Silicon Vias) and inter-block connections that can be provided by ball grid array, micro-bump or bump-less Cu-Cu (copper to copper) direct bonding between the stacked chips [3].

In this work Sn (Tin) solder capped micro bumps with 20 μm pitch Cu UBMs were used rather than Cu-Cu direct bonding to allow more efficient bonding process comprising fast thermo-compression bonding (TCB). TCB process provides good analogy for fine pitch interconnections involving (chip to chip) and (chip to wafer) arrangement where the bump pitches are smaller around sub 50 μm [4, 5]. Due to the fact that Cu as a UBM material oxidizes very fast either during shelf time or bonding, the focus of investigation in this study is oxidation prevention and wetting improvement on Cu surface [6]. When Cu undergoes oxidation, it allows 3-4 nm oxide layer present on surface which serves as entrapment layer leads to poor wetting by solder. Hence, no reaction

takes place between solder and Cu to create inter metallic compound (IMC). In turns the joint formation without any IMC might have weak electrical and mechanical connection which can make the device non-reliable [6, 7]. Therefore, the idea is to achieve an oxide free surface prior to TCB. For this purpose, surface treatment or passivation layer is required which could prevent the oxide growth over the surface. Thus, to attain the desired purpose an organic compound called BZT (Benzotriazole) is proposed.

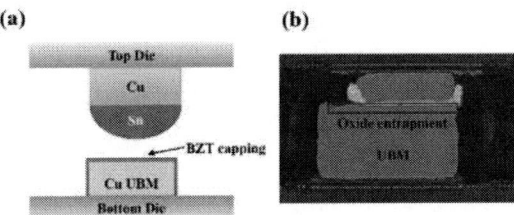

Figure 1: (a) BZT capping layer on UBM (b) X-section UBM oxidation [6]

Benzotriazole as Capping Layer

Normally besides expensive Au (gold) capping layer for UBM materials, SAM (Self-assembled monolayers) provide the best cost-effective option to be used as capping layers but in this work BZT is applied as passivation layer for Cu UBMs. The main reason of its proposal in this application is because of its strong corrosion inhibition effect on metal surfaces, specially Cu and its alloys [8]. BZT as a corrosion inhibitor material has a strong background since it has been used in industries for metal surface treatments. It is well known that the most common chemical mechanism which initiates and later facilitates corrosion is oxidation in ambient environment. And due to the reason BZT provides very active corrosion protection action, it is believed that it has the ability to inhibit surface oxidation on metal surface [8]. So, based on this information idea was proposed to choose BZT as a capping material also providing a cheaper solution in comparison to Au layer.

Benzotriazole BTAH or BZT is a heterocyclic organic compound with molar mass of 119.124 g/mol. It can also be written as (1,2,3-benzotriazole) where 1,2,3 refers to tri-azole network containing three nitrogen atoms [8]. This compound in simple form involves benzene ring which is attached to tri-azole network as show in Fig 2. Since this compound has a benzene ring attached that's why it belongs to the class of aromatic compound represented by chemical formula $C_6H_5N_3$ [9].

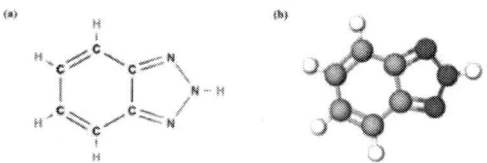

Figure 2: (a) Structure of BZT (b) 3D model [9, 22]

Proposed Layer Formation Mechanisms

In this work study belongs to the application of BZT specifically on Cu surface. It is very important to discuss the basic layer formation mechanism through which an invisible barrier film forms over Cu surfaces. BZT forms a protective layer comprises of complex between Cu and BTAH when a Cu is allowed to immerse in a solution containing BZT [8]. Different mechanisms were proposed regarding molecular adsorption or complex formation for BZT. Firstly, Cotton proposed the arrangement of surface complex of Cu–BTAH when Cu is dipped in a BTAH solution forming soluble Cu ion [10]. Later Cotton and Scholes claimed that a polymeric complex of 50 Angstrom thickness is formed as BTA by substitution of H atoms through N-H bond using a lone pair of electrons (free electron) from respective nitrogen atoms [11]. Another proposed theory was linear film formation on Cu(I) cuprous oxide surface which was much larger in thickness up to several thousand angstroms and it is due to dominant adsorption process in comparison to Cu(II) cupric oxide surface because of Cu(I) ions transportation from copper metal towards the film [12].

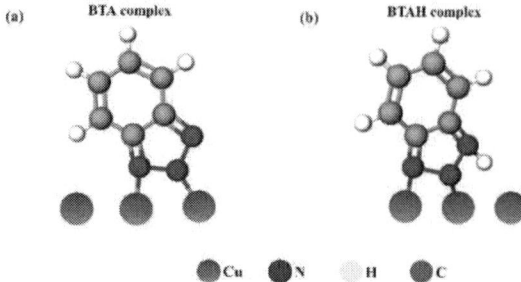

Figure 3: BZT complex formation model on Cu surface [22]

According to the suggestions from Morito and Suetaka, Cu protection is dependent upon Cu(I) BTAH complex that stays parallel to the Cu surface and not because of chemisorption of BTAH film [13]. However, it was concluded by Mansfield that BTAH kind of complex only chemisorbs on a copper oxides surfaces which enables the growth of films more than monolayer and in that case Cu(I) is highly suitable for the adsorption process instead of Cu(II) [14]. As proposed by Roberts, for the first BTAH layer which grows on the Cu_2O substrate, in this case Cu(I) ions could be connected by two bonds to the two oxygen atoms in the Cu(I) lattice, and the other two attached to the deprotonated BTAH molecules after removal of proton, (hydrogen cation H^+) as shown in below equation [15].

$$Cu(I) + BTAH \rightarrow Cu(I)BTA + H^+$$

Following the equation right explanation is the adsorption of BTAH in its molecular state enabling the first chemisorbed layer in the form of Cu(I) BTA complex aligning parallel to Cu surface. This Cu(I) surface complex was revealed through TOF-SIMS (Time of flight secondary ion mass spectrometry) [16]. Another key aspect is adsorption behavior of BZT over the metal surface. Previous study has shown that Langmuir adsorption isotherm fitted best for BTAH adsorption on Cu metal represented by following relation [9]:

$$\frac{C}{\theta} = \frac{1}{K_{ads}} + C$$

Where C is concentration of benzotriazole adsorbed on Cu surface. θ is fractional surface coverage by BTAH molecule and K_{ads} is adsorption equilibrium constant [9]. This relation endorse that increase in the concentration of BZT increases the surface coverage as well as the film stability [9]. All this information collected was helpful in defining the coating recipe in this work.

Design of Experiment

Experimental design was carried out based on the type of wafers and structures used for experimental analysis as shown in the table 1.

Table 1: Types of wafers used for experiments

Wafers	Features	Purpose
Blanket Cu	Metallic Cu	CA, XPS, FTIR
Wafers with damascene process	Cu Pads 20µm	AFM
Wafers with Cu bump	Test vehicle Cu UBM	Stacking, SEM Electrical test

As already shown in table that different wafers and structures were utilized to carry out the experimental analysis and characterization after coating with BZT. Blanket Cu samples were used to perform CA, XPS and FT-IR analysis. PTCW (Packageable test chip version W) test vehicle with damascene Cu layer was used specifically for AFM measurements. At the end stacking was done on 3DEM (3D electromigration) test vehicle with Cu bumps using TCB process and then characterized for electrical testing and cross section SEM analysis.

Sample Preparation

Sample preparation was done mainly using immersion coating because it is evident that BZT forms much stable complexes in stationary environment. Initially different concentrations of BZT were tested on blanket copper surface starting from 1mM, 10mM, 25mM, 40mM, 55mM and 100mM in order to define a best coating recipe in terms of surface coverage, film stability in ambient atmosphere as well as thermal stability. Same coating procedure was followed to coat all types of wafers and test structures to ensure the comparable results.

Coating process involve sequential steps. Firstly, cleaning with IPA, then dipping in commercial MS6020 solution (contains Citric acid) to remove surface oxides of Cu followed by rinsing step using deionized water. Afterwards allowing the Cu samples to immerse in solution containing BZT. To dissolve BZT, IPA solvent was used on supplier's recommendation providing 53.5% solubility. In the end two step rinsing was performed to remove unnecessary residuals, undissolved solute particles left inside the BZT solution as well as to minimize and control the film thickness specially during second step rinsing. After rinsing, samples were taken out and dried out gently with nitrogen flow. Whole process is illustrated in Fig 4.

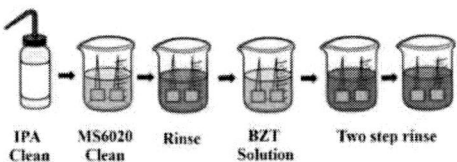

Figure 4: Sample preparation procedure

Results and Discussion

Water Contact Angle

Contact angle measurement was performed using sessile water drop to validate the BZT adsorption on Cu surface after coating. It was mainly to look how the wetting property is changing after BZT treatment. Contact angle was measured before and after coating i.e. on polished Cu surface and on BZT passivated surface. On blanket Cu the measured CA was higher nearer towards hydrophobic response, either close to 90° or > 90° and this is because of possible surface oxides [9]. After BZT passivation, wetting property was changing more towards hydrophilicity for all the concentrations tested. Low CA is due to the possible breakage of Cu oxide film, increasing the surface roughness and as a result water drop is expanding over surface [9]. Higher concentrations i.e. 40mM and 55mM showed promising decrease in CA around 60° to 66° initially after coating.

CA was measured on regular basis to test the film stability over time in ambient air. The results showed that high concentrations were more stable with 55mM showing film stability up to 5 days with good surface coverage. Below is shown the average and standard deviation plot for 25mM, 40mM and 55mM derived from measured values between day1 and day5. It can be seen that the values with lower concentrations are deviating more showing larger variation in CA over time. It means that films with lower concentrations are not much stable and this might be due to difference in surface coverage. While 55mM has shown good stability over time with just a difference of 2° in consecutive prepared samples as shown in Fig 5.

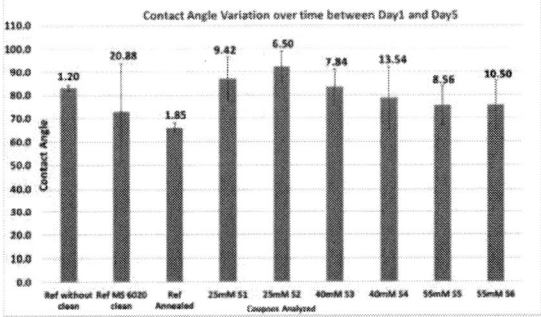

Figure 5: Contact angle variation w.r.t various concentrations

Contact angle measurement is just a quick method to determine the wetting property and behavior of surface, but it does not tell about the oxidation state. Therefore, through this experiment it cannot be revealed that if the surface is getting oxidized changing the wetting behavior of surface over time. Therefore, to verify the oxidation state of Cu surface XPS surface analysis was executed.

X-ray Photoelectron Spectroscopic Analysis

As in this work focus of investigation was oxidation protection of BZT thus to validate if BZT surface treatment on Cu surface is preventing the surface oxidation of Cu, XPS study was performed. For this purpose, surface spectra were analyzed before and after BZT passivation on Cu surface. Initially based on elemental analysis survey scans were recorded and the resultant peaks were translated into percentage atomic concentrations to describe in terms of quantitative analysis. The percentage atomic concentration for Carbon, Nitrogen, Oxygen and metallic Cu of all the samples analyzed in this XPS study are shown as bar chart in Fig 6.

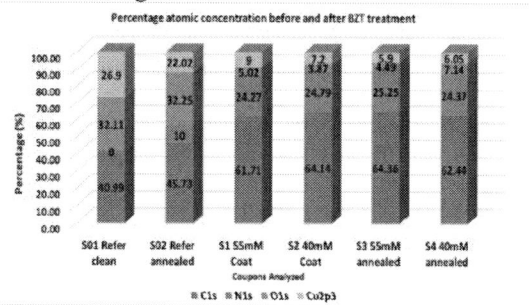

Figure 6: Percentage atomic concentration XPS study

Concentration regarding each element particularly elements associated with BZT like carbon and nitrogen along with oxygen bonded to metallic Cu were calculated. It was found that concentration of oxygen and metallic Cu were higher in reference samples prior to coating. In this study comparatively higher concentrations of BZT i.e. 40mM and 55Mm were examined. Thermal stability was also checked before and after coating annealed at 100°C for 1hr. In the reference sample annealed at 100°C the nitrogen was detected with an approximate concentration of around 10% which was added possibly from the tool. In case of all BZT treated samples a significant decrease in amount of oxygen was observed as well as increase in nitrogen and carbon content. Specially for 55mM the amount of oxygen is more or less same around 5% even after annealing, which suggest that 55mM has higher film stability against oxidation in ambient atmosphere relative to 40mM. The concentration of metallic Cu has also reduced after BZT passivation as demonstrated in Fig 6.

Hence, based on the results acquired from XPS experimental data it is evident that there is a stable passivation of BZT layer on Cu surface and at higher concentration film provides good stability against oxidation and prevents the oxide growth in ambient air when the Cu surface gets treated with BZT.

XPS Depth Profile Analysis

Analysis of the depth profile provides very useful mean to identify chemical species as well as the quantity present at the surface of the film or near to the surface [17]. The idea was to determine the film formation over surface based on chemical composition of BZT after coating on Cu surface. Samples were prepared with 55mM concentration which was identified as the best concentration as authenticated by CA and XPS spectral analysis regarding film stability. Coupons were analyzed for the depth profile before and after coating. From Fig 7, it can be seen in reference sample that a noticeable signal from metallic Cu is present along with weak oxygen single from surface. After BZT treatment a strong shift in the signals from different elements was observed. These signals from various components belong mainly to BZT layer involving Carbon and Nitrogen, providing a reliable evidence of BZT layer formation over the Cu surface.

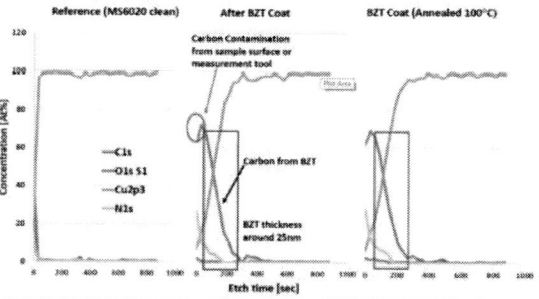

Figure 7: Depth profile before and after coating XPS study

Thermal stability was also monitored after annealing BZT coated sample at 100°C for 1hr. Film has shown a good thermal stability with just a slight shift in oxygen signal while strong signals from Carbon and Nitrogen are still present giving an indication of BZT film on Cu substrate even after annealing. Therefore, both spectral and depth profile analysis have shown a stable passivation of BZT on Cu surface limiting the oxygen content and preventing the oxide growth on Cu surface.

FT-IR analysis

IR study was carried to validate the BZT layer formation on Cu surface based on associated functional groups and chemical bonds. BZT is an organic compound with polymeric nature and ATR technique provides a very useful mean of qualitative analysis of polymeric materials [18]. As ATR possess higher sensitivity to chemical variations that's why ATR (Attenuated total reflection) sampling technique was used to perform experiments [18]. In this work VariGATR™ a single reflection horizontal ATR accessory is used with Germanium ATR crystal which possesses spectral range from (5000 to 650 cm-1).

Experiments were executed on samples before and after coating BZT on Cu surfaces involving three different concentrations i.e. 25mM 40mM and 55mM. Thermal stability was also analyzed for all concentrations after annealing samples at 100°C for 1hr similarly as done in XPS analysis. In reference samples without coating broad peaks were observed in fingerprint region is mainly from surface and it involves the associated bonds of Cu with either oxygen or OH groups. After the surface treatment with BZT, very prominent peaks in the fingerprint area can be seen as shown in Fig 8. Fingerprint region in IR spectra mainly belongs to the compound after film formation. The peaks determined from IR spectra were identified on the basis of absorption spectrum. As benzotriazole is an aromatic compound in which benzene ring is attached to three nitrogen atoms where C, N and H are the main constituents so mainly the bonds from carbon, nitrogen and hydrogen are expected [9].

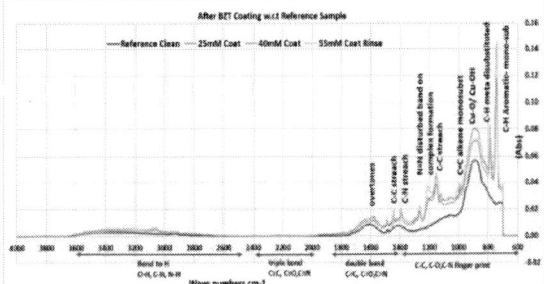

Figure 8: IR spectrum acquired after BZT passivation on Cu substrate

The blue curve at the bottom showing the reference sample without coating, while after coating with BZT the sharp peaks can be visualized. So, through analysis of detected peaks mainly in finger print region, the strongest absorption for aromatic compounds can be observed between (650-900cm^{-1}) which is due to (C-H) out of plane vibrations of aromatic ring, also called aromatic peaks from C-H bonds, as revealed in BZT coated samples. The other sharp bands between (1380-1450cm^{-1}) can be seen for (C-N) and (C-C) due to stretching vibrations [19]. A region with IR peaks between (1150cm^{-1} - 1210cm^{-1}) was noticed because of disturbed band mainly from (N=N) stretching during Cu-BTA complex formation [19]. From IR spectra it is understood that there is successful BZT complex formation over the metal surface.

To check thermal stability, samples were again analyzed using IR after annealing and it was found that most of the peaks were detected in the same region of absorption spectrum and they were not shifted much along X-axis as demonstrated in Fig 9. Little change in the shape of some peaks was observed upon heating at 100°C and it might be due the bonds that become dense result in changing the chemical environment. Regarding intensities along Y axis they are truly absolute values. Variation in the intensities with respect to reference sample is due to the difference in applied pressure on the measured samples during measurement, since ATR spectra is sensitive to the resulting pressure [20].

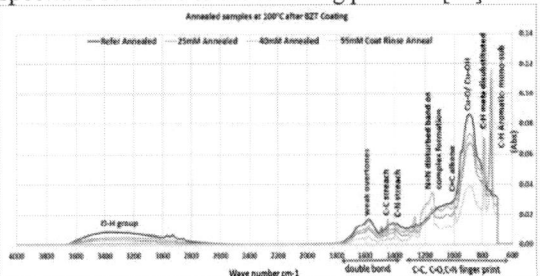

Figure 9: IR spectrum after annealing at 100°C for 1 hr after BZT coat

IR study has also confirmed a surface passivation of BZT on Cu with good thermal stability as discovered by obtained IR data.

AFM Analysis

AFM analysis was performed over PTCW dies with 20um pitch Cu pads embedded in the Si₃N₄ Substrate. The idea was to validate BZT layer formation on Cu pads and also to measure the BZT layer roughness and then to interpret layer thickness after BZT coating.

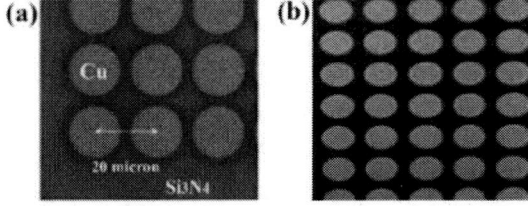

Figure 10: PTCW dies Top view (a) AFM Image (b) SEM Image

Samples in this study were prepared with shorter and longer immersion time to see the impact on layer thickness when Cu surface is exposed to BZT solution. Results have shown that after BZT coating, layer thickness increased with increase in immersion time. 2 hr exposure time has provided the BZT layer with thickness around (12-15nm) and with 18 hr dipping time, layer thickness has increased to (42-45nm) as shown in Fig 11. Roughness of surface also increases after passivation determined by Rq and Ra, the parameters used to define degree of roughness where "Ra" is mean roughness and "Rq" is root mean square average.

411

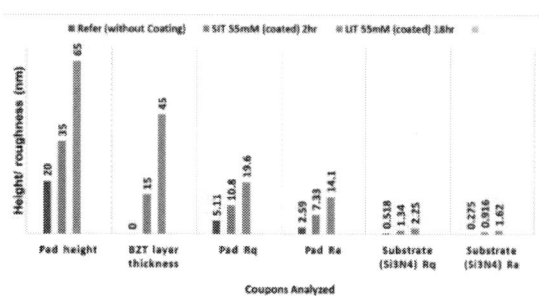

Figure 11: Pad height and measured roughness before and after coating

AFM data has also shown a successful BZT passivation on Cu surface.

Description of Test vehicle

In this work stacking was performed on 3DEM test vehicle. These chips were developed with imec's 65nm node BEOL standard process [21]. Bottom chips are in size of 7.2x7.2 mm² carrying 20 μm pitch Cu bumps while top chips were 5.2x5.2 mm² in size. The focus of stacking in this work was mainly CuSn system but two different types of top wafers were utilized. Firstly, top wafers with CuSn bumps with 5μm/5μm bump height were used with No flow underfill (NUF) [21]. Later top wafers with CuNiSn with bump height of 5 μm/0.2 μm/7 μm were applied with wafer level underfill (WLUF) [21]. The bottom chips containing Cu bumps were same on which BZT capping layer was applied. The diameter of top bumps was 7.5 μm while for lower bumps it was 12.5 μm to allow extra misalignment margin during TCB stacking. Stacking was done on coupons of 9 dies. Each die involves four electromigration structures and two daisy chains. On rough estimation each chip stacked involves 43848 bump joints (including dummy and functional bumps) whereas daisy chain contains around 800 bump connections.

3D Stacking Using TCB Process

After applying BZT coating on bottom chips involving Cu bumps, stacking was done with top chips using fast TCB process. Bonding force used for stacking was 20 N. As discussed, earlier TCB bonding is believed to be the finest method for fine pitch interconnection. In this work TCB process was carried out mainly in four process steps [21].

1. Picking top die at standby temperature
2. Heating up to higher temperature and dwell for 3 seconds
3. Heating up to peak temperature and dwell for 3 seconds. At this point at certain interface temperature, Sn is melted and allowed to react with Cu.
4. Cooling

Later in this work stacking was also done using different TCB process with a different bonding profile involving longer dwell time. In this case both bottom and top chucks were heated at elevated temperatures intended to reach the interface temperature enabling bonding.

Electrical Testing and Cross Section analysis

The focus of this work was to analyze the wettability between Cu UBM and Sn solder after applying BZT as a capping layer on bottom UBMs. Stacking was performed using different conditions i.e. with and without underfill materials, different immersion times and coating procedures to investigate the wettability. For this purpose, test vehicle was coated using 55mM concentration. After stacking, electrical resistance of daisy chains was measured. As a reference sample, test chip with gold capping layer was stacked to ensure a better comparison in terms of electrical data and wetting. The electrical yield for 20 μm pitch daisy chains along with X-section of all respective samples are shown and discussed below. Y-axis is cdf (cumulative distribution function) percentage and x-axis is the resistance values in Ohm.

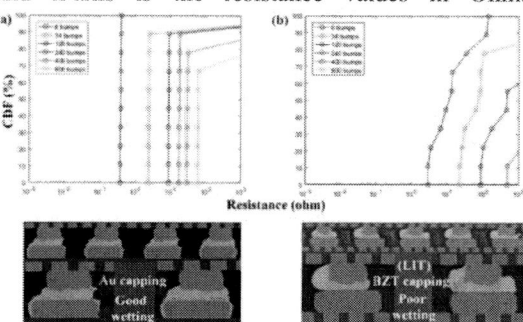

Fig 12: (a) Au capping (b) BZT capping (LIT)

As shown in Fig 12. firstly the comparison was made with a thick Au layer providing good wetting with CuSn system leading to IMC formation during TCB process while in case of organic BZT layer with (Longer immersion time – 18 hrs) and an expected thickness of (42–45nm) the wetting was not good to achieve IMC. While the electrical yield was very poor around 10% in evaluation to Au layer which was 90% for daisy chains. Therefore to further improve the results more experiments were carried out and are explained below.

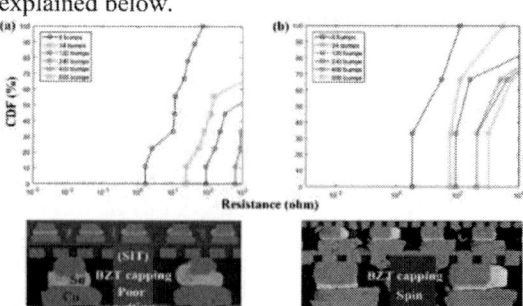

Figure 13: (a) BZT capping (SIT) (b) Spin coated

412

Therefore, in case of thick organic layer not be able to decompose during fast TCB process, idea was to minimize layer thickness as well as roughness. For this reason, shorter immersion time (SIT) and spin coating methods were tested to minimize layer roughness and to control thickness of layer but still wetting was too poor to achieve IMC in both cases. The expected BZT layer thickness using SIT was around (10-15nm). Regarding electrical yield, for spin coated sample it was better up to 30% but not so promising. To improve the wettability with BZT layer, in next step bottom coupons were coated using different solvent (methanol) providing higher solubility for BZT and higher pH in comparison to the IPA solvent and to look for the impact on wetting. Data obtained is illustrated in Fig 14.

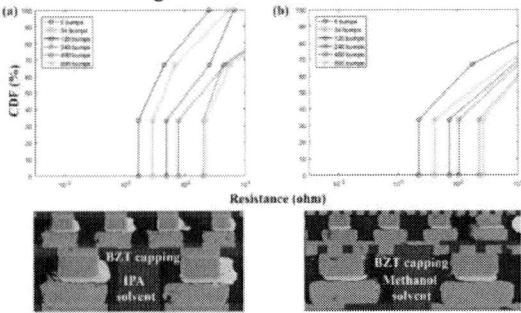

Figure 14: BZT layer (a) IPA solvent (b) Methanol solvent

Different solvent was tested because it is established from previous study that stability of BZT film increases upon increase in the pH. As at higher pH the deprotonation of BZT molecule increases and thus give rise to BTA⁻ form instead of BTAH and complex exists as Cu(I)-BTA which can increase oxidation protection [9]. The idea was if the layer thickness is to reduce further using much shorter immersion time and spin coating, stability of film must not get effected. Although electrical yield has improved up to 30% yet there is no IMC formation to create a reliable solder joint because of poor wetting. Therefore, bumps with BZT capping have shown bad yield involving higher resistance with wider distribution compared to Au capping which has shown a narrow distribution for electrical yield. However, stacking was continued for wetting improvement with BZT as a capping layer using optimized coating procedures. Stacking was executed with and without underfills to look for wetting behavior during bonding as shown in Fig 15.

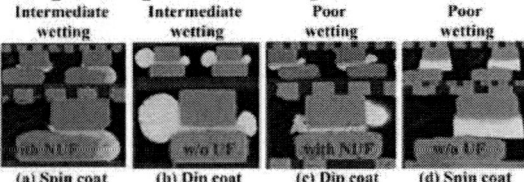

Figure 15: Wetting investigation with and without UF

The UF materials used in this work were without fluxing agents as the aim was to look for wetting purely from BZT. From Fig 15 (a) and (b), wetting is not perfect but intermediate wetting can be seen with (b) involving better alignment and Sn squeezed out where minor IMC is formed while in case for sample (c) and (d) wetting was poor. From these tests looks like BZT is not decomposed when using fact 6s TCB profile. Further optimization of TCB profile and BZT coating is needed to obtain complete wetting.

Conclusion

In this paper two main aspects were reported i.e. oxidation protection action of BZT in ambient atmosphere and wetting ability of BZT film in 3D stacks when using it as a capping on Cu UBMs. Various experimental studies were implemented to identify the best coating recipe for 3D stacking application. Regarding oxidation prevention of Cu, XPS study has shown a significant reduction in surface oxide after BZT treatment preventing Cu oxide growth in ambient air along with good thermal stability. CA, FT-IR and AFM data have also shown a stable passivation of BZT layer on Cu. Wetting investigation after 3D stacking have shown reasonable progress in terms of electrical data but not really productive to achieve IMC formation. Initially with fast TCB process thinner organic layers around (3-4nm), ideally monolayers are preferred like SAM layers rather than a thicker layer to enable good wetting. Even then benzotriazole can be successfully introduced to enhance the shelf life of wafers due to its oxidation protection behavior and also in conventional mass reflow bonding where the bonding is carried out with reflow process involving bonding time in minutes range.

Acknowledgements

The authors are really grateful and would like to acknowledge the efforts and support of all the partners especially group members of 3D program as well as Imec's clean room facility and supporting staff for their assistance.

References

[1] M. Gerber, C. Beddingfield, O'Connor S, M. Yoo, M. Lee, D. Kang, S. Park, C. Zwenger, R. Darveaux, R. Lanzone, K. Park. "Next generation fine pitch Cu Pillar technology Enabling next generation silicon nodes" In 61st Electronic Components and Technology Conference (ECTC), Chicago, May 31 pp. 612-618, 2011.

[2] Lim, Dau Fatt, Jun Wei, Chee Mang Ng, and Chuan Seng Tan. "Low temperature bump-less Cu-Cu bonding enhancement with self assembled monolayer (SAM) passivation for 3-D integration." In 60th Electronic Components

and Technology Conference (ECTC), pp. 1364-1369, 2010.

[3] K. Sakuma, P. S. Andry, C. K. Tsang, S. L. Wright, B. Dang, C. S. Patel, B. C. Webb, "3D chip-stacking technology with through-silicon vias and low-volume lead-free interconnections." In IBM Journal of Research and Development Vol. 52, no. 6, pp. 611-622, 2008.

[4] D. Hiner, D. W Kim, A. SeokGeun, K. Kim, H. Kim, M. Lee, D. Kang. "Multi-die chip on wafer thermo-compression bonding using non-conductive film." In 2015 IEEE 65th Electronic Components and Technology Conference (ECTC), pp. 17-21. IEEE, 2015.

[5] D. Vos, Joeri, L. Bogaerts, T. Buisson, C. Gerets, G. Jamieson, K. Vandersmissen, A. La Manna, and E. Beyne. "Key elements for sub-50μm pitch micro bump processes." In 2013 IEEE 63rd Electronic Components and Technology Conference, pp. 1122-1126. IEEE, 2013.

[6] J. Derakhshandeh, I. D. Preter, K. Vandersmissen, D. Dictus, L. Di Piazza, L. Hou, S. Guerrieri, "Cobalt UBM for fine pitch microbump applications in 3DIC." In 2015 IEEE International Interconnect Technology Conference and 2015 IEEE Materials for Advanced Metallization Conference (IITC/MAM), pp. 221-224. IEEE, 2015.

[7] P. Keil, D. Lützenkirchen-Hecht, R. Frahm. "Investigation of room temperature oxidation of Cu in air by Yoneda-XAFS." Proceedings of AIP conference, Vol. 882, no. 1, pp. 490-492. AIP, 2007.

[8] M. Finšgar, M. Ingrid. "Inhibition of copper corrosion by 1, 2, 3-benzotriazole: a review." Corrosion science Vol. 52, no. 9, pp. 2737-2749, 2010.

[9] P. F. Khan, V. Shanthi, R. K. Babu, S. Muralidharan, R. C. Barik. "Effect of benzotriazole on corrosion inhibition of copper under flow conditions." Journal of Environmental Chemical Engineering Vol. 3, no. 1, pp. 10-19, 2015.

[10] I. Dugdale, and J. B. Cotton. "An electrochemical investigation on the prevention of staining of copper by benzotriazole." Journal of Corrosion science Vol. 3, no. 2, pp. 69-74, 1963.

[11] J. B. Cotton, I. R. Scholes. "Benzotriazole and related compounds as corrosion inhibitors for copper." Journal of British Corrosion Vol. 2, no. 1, pp. 1-5, 1967.

[12] K. F Khaled, A. A. Mohamed, N. A. Al-Mobarak. "On the corrosion inhibition and adsorption behaviour of some benzotriazole derivatives during copper corrosion in nitric acid solutions: a combined experimental and theoretical study." Journal of applied electrochemistry Vol. 40, no. 3, pp. 601-613, 2010.

[13] N. Morito, W. Suetaka. "In situ ultra-violet high-sensitivity reflection studies of the corrosion inhibition of Cu by Benzotriazole." Journal of Japan Inst. Metals Vol. 37, no. 2, pp. 216-221, 1973.

[14] F. Mansfeld, T. SMITH, E. P. Parry. "Benzotriazole as corrosion inhibitor for copper." Journal of corrosion Vol. 27, no. 7, pp. 289-294, 1971.

[15] R. F. Roberts, "X-ray photoelectron spectroscopic characterization of copper oxide surfaces treated with benzotriazole." Journal of Electron Spectroscopy and related phenomena Vol. 4, no. 4, pp. 273-291, 1974.

[16] V. Brusic, M. A. Frisch, B. N. Eldridge, F. P. Novak, F. B. Kaufman, B. M. Rush, and G. S. Frankel. "Copper corrosion with and without inhibitors." Journal of the electrochemical society Vol. 138, no. 8, pp. 2253-2259, 1991.

[17] N. C. Erickson, S. N. Raman, J. S. Hammond, R. J. Holmes. "Depth profiling organic light-emitting devices by gas-cluster ion beam sputtering and X-ray photoelectron spectroscopy." Journal of Organic Electronics Vol. 15, no. 11, pp. 2988-2992, 2014.

[18] K. R. Kirov, H. E. Assender. "Quantitative ATR-IR analysis of anisotropic polymer films: Extraction of Optical Constants" Journal of Macromolecules Vol. 37, no. 3, pp. 894-904, 2004.

[19] M. E. Biggin, A. G. Andrew. "Infrared studies of benzotriazole on copper electrode surfaces: role of chloride in promoting reversibility." Journal of Electrochemical Society Vol. 148, no. 5, pp. 339-347, 2001.

[20] F. Friedrich, G. W. Peter. "Contact pressure effects on vibrational bands of kaolinite during infrared spectroscopic measurements in a diamond attenuated total reflection cell." Journal of Applied spectroscopy Vol. 64, no. 5, pp. 500-506, 2010.

[21] T. Wang, J. D. Messemaeker, V. Cherman, K. J. Rebibis et al. "Thermal compression bonding of 20 μm pitch micro bumps with pre-applied underfill-Process and reliability." In 2015 IEEE 17th Electronics Packaging and Technology Conference (EPTC), pp. 1-5, IEEE, 2015.

[22] T. Saavedra, M. Escobar, C. A. Ocayo, F. Tielens, and Santos, "1, 2, 3-Benzotriazole derivatives adsorption on Cu (1 1 1) surface: A DFT study." Journal of Chemical Physics Letters Vol. 689, pp.128-134, 2017

Investigation of mechanical and microstructural properties of a new, corrosion resistant gold-palladium coated copper bond wire

Georg Lorenz[1], Falk Naumann[1] Robert Klengel[1], Sandy Klengel[1], Matthias Petzold[1]
Motoki Eto[2], Noritoshi Araki[2], Takashi Yamada[2]

[1]Fraunhofer Institute for Microstructure of Materials IMWS, Halle, Germany

[2] Nippon Micrometal Corporation, 158-1 Sayamagahara, Iruma-City, Saitama 358-0032, Japan

Email: georg.lorenz@imws.fraunhofer.de

Phone: +49 345 5589 188

Abstract

In this study we present a newly developed gold-palladium coated copper (EX1R) wire with focus on automotive industry. By using an advanced material composition, halide induced interface corrosion and also sulphur induced pitting corrosion can be prevented sucessfully. Therefore the copper base material was alloyed with further elements to systematically adjust the intermetallics formed at the interface to the pad metallization and to define the microstructure of the copper base material. To show the improved reliability, in a first step three copper bond wires(bare copper wire; palladium coated copperwire (APC) and the new EX1R wire) were bonded to a silicon die with aluminum metallization and a nickel-palladium-gold plated lead frame and encapsulated with chlorine containing mold compound. In a second step, these packages were stressed using a highly accelerated stress test (HAST) and the degradations were analyzed using scanning electron microscopy (SEM). To further compare the new bond wire type with conventional bond wire materials a microstructural analysis of the grain structure using electron backscatter diffraction (EBSD) was carried out on the three copper sample wires. Additionally, the mechanical properties of the samples were investigated by tensile and nanoindentation experiments with the objective to predict interactions between the plastic deformation behavior and chip damage risks during the bond process. The results show that the newly developed gold-palladium coated copper wire shows a much higher reliability compared to the standard wire materials. Furthermore it could be shown that plastic deformation behavior of the new wire type is very similar to that of conventional palladium coated copper wires.

Key words: APC wire, wire bonding, corrosion resistance, reliable packaging, high temperature electronics, harsh environments

Introduction

Due to demanded cost reductions and to withstand harsh environmental conditions copper wires are introduced to a greater extend in automotive electronics. These wires often show a higher reliability at elevated temperatures but also are highly prone to different corrosion mechanisms [1] (e.g. halide induced corrosion in the contact interface between copper and aluminum or to temperature induced pitting corrosion due to sulphur). Palladium coated wires are able to prevent corrosion effects by protecting the bare copper from oxidation. In ball bonding processes a homogeneous distribution of the palladium coating after the ball flame off process is very important to maintain these desired properties [2]. The distribution of the palladium coating after ball formation can be influenced by additional alloys or flame off process parameters.

In this study a new gold-palladium coated copper wire alloyed with additional elements is analyzed. It has shown a very homogeneous palladium distribution on the Free Air Balls (FAB) and therefore also tremendous improvements in highly accelerated stress tests (HAST) by preventing corrosion effects is.

Focus of this investigation is the combined microstructural analysis with mechanical tests in order to analyze effects to the bonding properties as they are both closely linked to the bondability of bond wire materials. Three different wire types were chosen for the analyses (one bare copper bond wire, one conventional palladium coated copper wire (APC) and the new palladium-gold coated copper wire with additional alloying (EX1R). The microstructures of all three wires were analyzed on cross sections of FABs and the connected wire necks. Further, hardness and modulus of all samples were investigated and afterwards correlated to the microstructural results. In order to complete the mechanical characterization of the new bond wires and to compare these results to the material properties of standard copper bond materials tensile tests on all bond wires were performed.

Halide induced corrosion mechanism of intermetallic copper compounds (IMC)

In copper-aluminum systems different intermetallic compounds will be formed after wire bonding and developed during thermal ageing. In general copper rich and aluminum rich IMCs can be found in literature [1]. Figure 1 shows a scheme of the location of the IMCs for bare copper wire bonded contacts on aluminum pad metallization. As reported numerously in literature (e.g.[3]), the copper-aluminum intermetallic compounds can be altered by halides leading to a limited reliability of the complete wire bonded system.

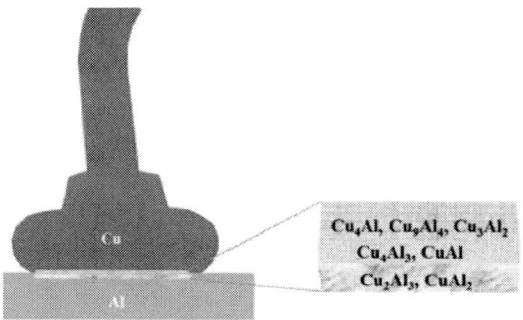

Figure 1. Location of the intermetallics for bare copper wire bond contacts on aluminum pad metallization

Lim et al. [4] investigated the electrical impedance behavior and the corrosion currents of different Cu-Al-IMCs in chlorine rich solutions. They showed that corrosion currents of the IMC shift to higher values with increasing aluminum content indicating that the aluminum rich IMC are less resistant against chloride attack.

Theoretical calculations done by H. Abe et al. [5] were used to predict possible IMC creation and reactivity between chlorine ions and single intermetallic compounds. By use of chemical model simulation techniques, generation of copper rich and copper poor IMC was indicated, while copper rich IMC is most likely corroded by chlorine ions during HAST testing.

Boettcher et al. [6] explained that effect by the self-passivation of aluminum. They concluded that the copper rich IMCs have a weak self-passivation oxide, compared to aluminum rich IMCs. Thus, the copper rich IMCs should be more prone for corrosive attack by chlorine. The findings

in [5] and [6] correlate with effects seen in molded packages during accelerated stress testing.

The simulation of Abe et al. [5] showed also that the addition of palladium in copper-aluminum system would improve the corrosion resistance. With palladium coated copper the generation of single copper rich IMCs was inhibited and the IMC growth slowed down in general. The simulation also indicated that a palladium layer could act as barrier layer for chlorine ion penetration and work as reaction partner with HCl forming $PdCl_2$.

Similar results were also published by C. Lee et al. [7] showing that interfacial crack line propagation and IMC growth is slowed down by the use of Pd-coated Cu wire compared to bare Cu wire. They assumed that Pd agglomerates are formed at the interface between copper wire and copper rich IMC. I. Qin et al. [8] found that Pd substitutes Cu in copper rich Cu_9Al_4 intermetallic compound leading to a higher corrosion resistance. Figure 2 shows the proposed IMC system for palladium coated copper wires on aluminum pad metallization.

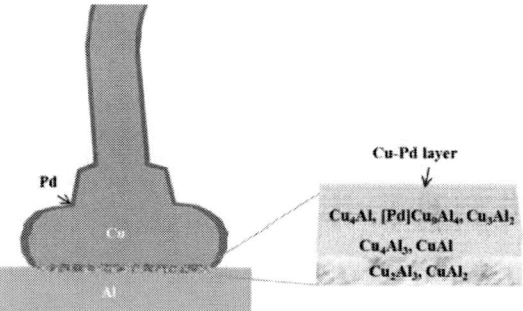

Figure 2. Location of the intermetallics of palladium coated copper wire bond contacts on aluminum pad metallization

Reliability testing of bonded sample wires

Within this study, three types of test specimen were used for different research objectives: a bare copper wire and a conventional APC wire serving as reference samples in comparison to the optimized APC wire "EX1R". All samples had a wire diameter of d=20 μm. For reliability testing the wires were bonded to a silicon die with aluminum metallization (ball bond contact) and a nickel-palladium-gold plated lead frame (stitch bond contact) and finally encapsulated by a chlorine containing (concentration of 21 ppm) mold compound. Afterwards the package was tested for t=192 h in a Highly Accelerated Stress Test (HAST) at T=130°C and H=85% rH with a voltage applied of U=4 V bias. After the tests, cross sections of the stressed bond wires were prepared and analyzed using SEM (see Figure 3).

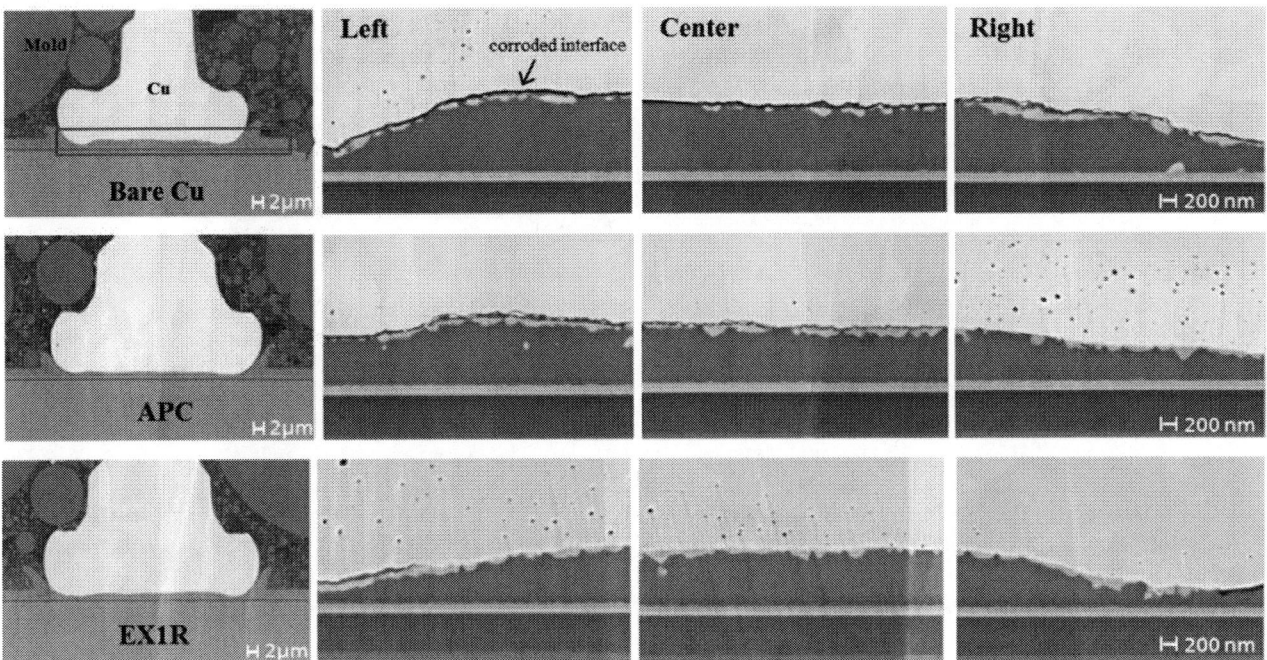

Figure 3. SEM results of the interface between bare Cu, conventional APC and EX1R wire on an aluminum pad metallization after HAST. Bare copper shows a completely corroded interface while for the conventional APC wire fairly corroded interface was found. The optimized APC wire "EX1R" shows only starting corrosion processes at the outer areas of the contact that stop after a few micrometer.

The bare copper wires show a nearly completely corroded interface. Conventional APC wires show a slightly improved but still fairly corroded interface. The optimized APC wire EX1R shows a starting corrosion at the outer areas of the ball bond contact. The corrosion stops after a few micrometers. Additional TEM analyses [9] of the interface area show Cu_2Al_3 IMC at the pad metallization side for bare copper and also for the optimized "EX1R" wire. These IMCs were not attacked by the corrosion process. In the corroded area of the copper wire Cu_9Al_4 residues are detectable. The optimized "EX1R" shows a palladium modified Cu_9Al_4 intermetallic compound at these areas wich prevents the corrosive attack. The reason for this is an additional alloying element enriching the top of the copper rich IMC that works as a diffusion barrier and ion catcher. Compared with the reactions taking place at the interface of the bare copper wire and the conventional APC wire a significantly higher reliability of "EX1R" wire is recognizable.

Microstructural characterization of grain structure using EBSD

To implement these new, high reliable wire bond materials into existing manufacturing processes by replacing other wire bond materials, elasto-plastic material properties play a key role regarding the risk of chip damage during bonding.

Especially brittle low k structures underneath of the bond contact can react very sensitive to the mechanical loads resulting from the bonding process. Therefore the microstructure and the deformation behavior of the three wire types were investigated as follows.

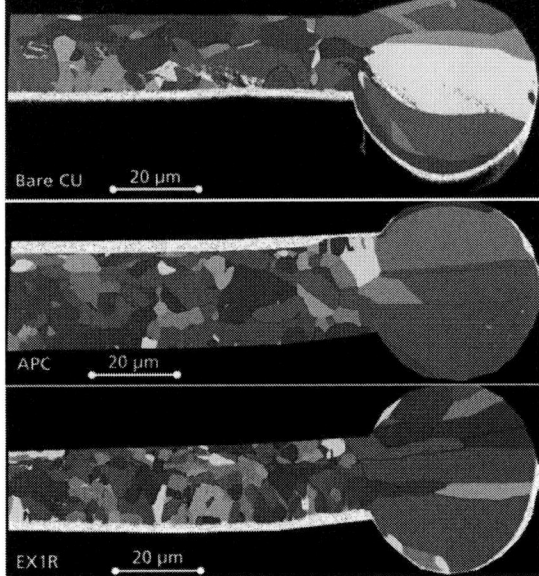

Figure 4. Grain structure of the different wires with Free Air Ball (top: bare copper; center: conventional APC; bottom: EX1R wire)

In a first step the grain structure of the three wire types was analyzed using EBSD techniques. Therefore partly processed specimens consisting of the Free Air Ball (FAB) and a wire end were produced using equal flame off parameters. All samples had a wire diameter of d=20µm. Afterwards cross sections have been prepared by mechanical grinding with a final ion polishing step. The grain distribution of the samples is exemplarily shown for each wire type in Figure 4.

Comparing the wire regions on the left side, the EX1R wire shows a slightly smaller grain size compared to the other two wires. This indicates a slightly harder material behavior as grain size correlates to the material hardness according to the (Hall–Petch relation) that is described in the nanoindentation section. For the FAB region, the grains show an increased size due to the flame off process. This should correlate to a softer material behavior (compared to the unaffected wire) and will especially be important for the ball bonding process step to avoid damage of the semiconductor dies.

Mechanical characterization using tensile testing

In order to get a general overview of the mechanical properties of the unprocessed wire material, tensile test were performed according to DIN EN ISO 6892-1 with the objective to obtain stress-strain diagrams for each wire. For each wire type, at least 8 samples have been tested. A strain rate of 1,5%/min was set together with a sample length of 100mm. All tests were performed at room temperature using a universal testing machine Zwick 1445.

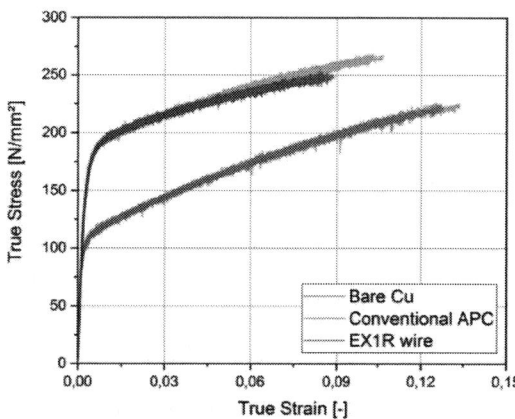

Figure 5. Average modulus trend over the process wire cross section

Evaluating the tensile test results (see Figure 5), no significant difference in the initial yield strength between the APC and EX1R wires can be seen. The strain hardening of the conventional APC wire is slightly higher compared to the EX1R wire. The bare copper wire shows significantly low initial yield strength compared to the other two wires. It

also can be seen that the strain hardening for the bare copper wire is the highest of all wires. The bare copper wire further shows the highest fracture strain of all wires, followed by the conventional APC and EX1R wire.

Investigation of mechanical properties using indentation testing on cross sections

As shown before, the grain structure and hence the mechanical properties of the relavant FAB change due the ball flame off process. Therefore cross sections of the bond wire types with the processed Free Air Ball have been prepared using mechanical preparation followed by ion polishing. The samples where placed into a KLA G200 nanoindenter for indentation testing.

In these tests a Berkovich-shaped indenter was used to determine indentation hardness and modulus of the FAB region. For all experiments the test machine was set to a continuous stiffness mode (csm) method where elastic modulus and hardness can be determined as a function of the indentation depth. The analysis of depth-dependent information allows a better estimation of possible preparation influences or other affecting factors (e.g. oxide layers on top of the sample surface). 15 indents on 4 samples for each processed wire type were placed along the wire axis covering the FAB and reaching to the heat affected zone (HAZ) of the wire (see Figure 6). The maximum indentation depth for all indents was set to 200 nm allowing a distance from indent to indent of 4 µm.

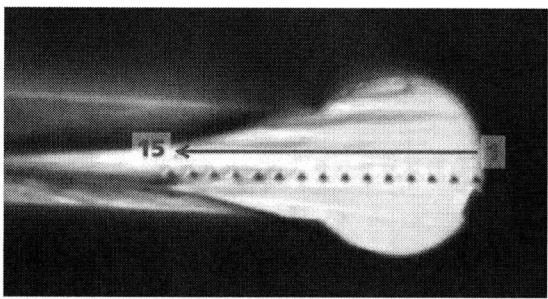

Figure 6. Exemplary shown: locations of indents along wire cross sections

Hardness and modulus of the indentation measurements were evaluated differently. The modulus was calculated between 10 and 20 nm indentation depth as the stiffness is highly affected by the underlying embedding material and decreases with higher indentation depths. In contrast hardness was evaluated at depths between 150 and 180 nm as for its calculation the effect of tip rounding at very low indentation depths has a significant impact that is reduced at higher indentation depths. Further the lower stiffness that affects the modulus at higher indentation depths is reduced when calculating the hardness as the elastic part is getting smaller with increasing indentation depths.

Comparing the modulus of the three wire types it can bee seen, that there are almost no differences between the wire (see Figure 7).

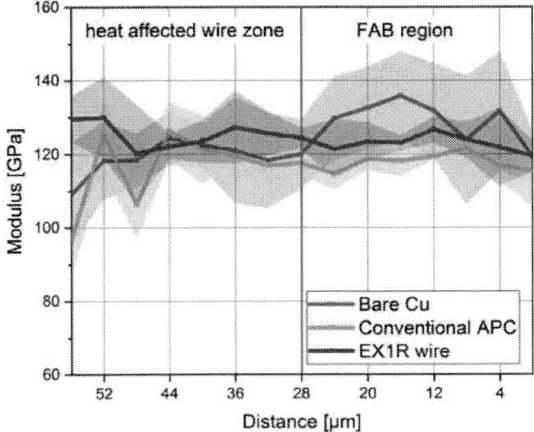

Figure 7. Average modulus trend over the processed wire cross section

Also between the FAB area and the heat affected wire zone no significant variations are found. This behavior is expectable as the grain size should not have a stron influence to the elastic properties. The values determined also show a good agreement with Young's moduli found in the literature for copper bulk material at room temperature.

In contrast, differences can be found in the hardness values when comparing the HAZ areas of the three different wire types (see Figure 8).

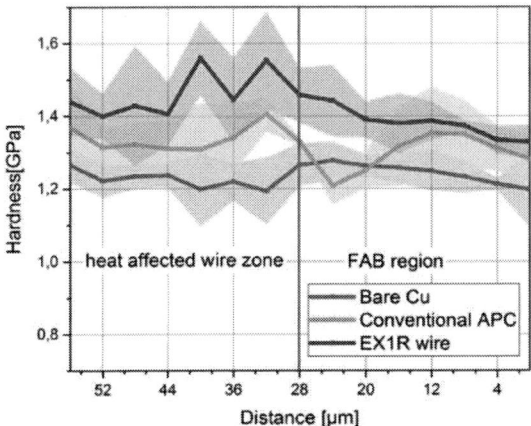

Figure 8. Average modulus trend over the processed wire cross section

For the FAB region these differences are reduced. This can be explained by the grain growth caused by the flame off process of the FAB. The relation between the grain size and hardness is described by the well known Hall-Petch relationship (see equation 1).

$$R_\varepsilon = \sigma_{i\varepsilon} + k_\varepsilon \cdot D_m^{-\frac{1}{2}} \qquad (1)$$

Hereby R_ε describes the yield strength of the material that can be qualitatively correlated to the hardness. D_m describes the mean grain size and $\sigma_{i\varepsilon}$ and k_ε are material specific constants.

Variations between single test series for both hardness and modulus values result from the local test method affecting only very small material volumes. At this small indentation depths, surface effects, grain boundarys and the anisotropic behavior of single grains can strongly influence the measurements.

Differences found in the tensile test between bare copper material and the APC and EX1R wire can qualitatively also be found in the nanoindentation experiments of the HAZ eventhough they are not that pronounced. These smaller differences might be explained by different reasons. First, both types of experiments adress different test voluminas. Therefore the Nanoindentation experiments are influenced for example by local grain orientation and effects of the surface topology. Further the grain structure in the heat affected zone might be changed differently by the FAB forming process. Due to this smaller differences in hardness compared to the diffences measured in the tensile tests can be explained. Last but not least, the different test directions of the two experiments might also influence results. The indentation experiments were carried out perpendicular to the wire axis where the tensile tests had a test direction parallel to the wire axis. In further analyses it is planned to also observe the grain structure on cross sections perpendicular to the wire axis. In the FAB region (wich is relevant for the ball bonding process) these differences are further reduced as expected from the grain structure analysis explained by the Hall-Petch relation as described before.

Summary and Conclusion

In this study a new APC bond wire "EX1R" with potential to be applied for robust automotive applications was analyzed with regards to his mechanical bahavior. The wire material of "EX1R" was developed to prevent halide induced interface corrosion and also sulfur induced pitting corrosion and showed drastically improved stability in HAST conditions compared to bare copper and conventional APC wires. To analyze the mechanical behavior which is relevant for the bondability of the wires microstructural and micromechanical tests have been carried out. Tensile testing of the wires showed a very similar mechanical behavior of the new EX1R wire compared to conventional APC wires. In the nanoindentation experiments and grain structure analysis using EBSD it was shown, that the flame off process to create the FAB during bonding increases the grain size and reduces the differences in hardness between the three tested wire types. Therefore it can be concluded, that the new material does not only show an improved stability in HAST conditions but also show a very similar mechanical defomation behavior compared to a standard APC bond wire. Therefore no additional chip damage risks, when bondig over active, areas are expected when introducing the new wire material into production processes.

References

[1] Kah, Paul, et al. "Factors influencing Al-Cu weld properties by intermetallic compound formation." *International Journal of Mechanical and Materials Engineering* 10.1 (2015): 10.

[2] Hang, C. J., et al. "Growth behavior of Cu/Al intermetallic compounds and cracks in copper ball bonds during isothermal aging". *Microelectronics reliability*, 2008, 48. Jg., Nr. 3, S. 416-424.

[3] Liu, Hai, et al. "Reliability of copper wire bonding in humidity environment." *2011 IEEE 13th Electronics Packaging Technology Conference.* IEEE, 2011.

[4] Lim, Adeline BY, et al. "Evaluation of the corrosion performance of Cu–Al intermetallic compounds and the effect of Pd addition." *Microelectronics Reliability* 56 (2016): 155-161.

[5] Abe, Hidenori, et al. "Cu wire and Pd-Cu wire package reliability and molding compounds." *2012 IEEE 62nd Electronic Components and Technology Conference.* IEEE, 2012.

[6] Boettcher, Tim, et al. "On the intermetallic corrosion of Cu-Al wire bonds." *2010 12th Electronics Packaging Technology Conference.* IEEE, 2010.

[7] Lee, Chu-Chung Stephen, et al. "Copper versus palladium coated copper wire process and reliability differences." *2014 IEEE 64th Electronic Components and Technology Conference (ECTC).* IEEE, 2014.

[8] Qin, Ivy, et al. "Molded reliability study for different Cu wire bonding configurations." *2013 IEEE 63rd Electronic Components and Technology Conference (ECTC).* IEEE, 2013.

[9] Klengel, Sandy, et al. "A new reliable, corrosion resistant gold-palladium coated copper wire material" *2019 IEEE 69th Electronic Components and Technology Conference (ECTC).* IEEE, 2019.

High spatial resolution measurements of thermo-mechanical stress effects in flip-chip packages

V. Cherman, M. Lofrano, M. Gonzalez, G. Van der Plas, K.J. Rebibis and E. Beyne

imec, Kapeldreef 75, 3001 Leuven, Belgium

vladimir.cherman@imec.be

Abstract

This work focuses on the experimental evaluation of thermo-mechanical stress induced in flip-chip BGA packages. The proprietary Si chip with integrated piezoresistive stress sensors is used to electrically measure chip package interaction (CPI) effects with high spatial resolution of 5µm. The electrical measurements are supported by finite element modeling which consecutively simulates different assembly steps such as flip-chip, application of underfill (UF), molding, and uses the experimentally determined thermo-mechanical properties of the packaging materials.

Key words: Chip Package Interaction, CPI, fcSCP packaging, Thermo-Mechanical stress

Introduction

Performance and reliability of microelectronics components to great extent is determined by the packaging assembly technology and thermo-mechanical interactions between the chip and the package [1]. The scaling trends of the recent years, particularly, the introduction of low-k dielectric materials into the back-end-of-line (BEOL) have significantly amplified the problem as lowering the dielectric constant in the material has a side effect of reducing its mechanical strength [2-3]. The safe reliability margins of the packaging technology are narrowing and thus it is becoming important to get more understanding about mechanical stress levels induced during the package assembly process. It can enable the prediction of failure modes in the package and in the back-end interconnects and their possible mitigation. Moreover, the package induced mechanical stress in Si chip may further modify the residual stress built up during the wafer processing [4] and create an additional shift in electrical performance of front end of line (FEOL) devices through their piezoresistive response [5]. Demands are increasing for accurate evaluation of package-induced mechanical stresses using the experimental methods and the thermo-mechanical modeling. The widely used experimental technique for evaluation of mechanical stresses induced in Si chip after packaging or after the reliability qualification is based on electrical measurements of integrated stress sensors such as field effect transistors (FET) or doped Si resistors, both exploiting the enhanced piezoresistivity of crystalline silicon [5, 6]. Rosettes of Si resistors [7] or van der Pauw designs [8] are theoretically and experimentally investigated and widely used for evaluation of CPI effects. In [9] the pseudo-hall sensor was introduced where two additional channel contacts allow accurate measurements of in-plane shear or a combination of two normal in-plane components of stress depending on sensor orientation [10]. Many applications of such sensors are limited by wire-bonded packages where two normal in-plane components of stress (or only their combinations) can be assessed while out-of-plane stresses are considered as negligibly small [11]. Moreover, in [12] it was shown that the rosettes of resistors fabricated on (111) Si wafer can be used to determine all six components of mechanical stress, thus they can be used for CPI study of more advanced packaging technologies such as flip-chip packages, chip-scale packages (CSP) *et al.* Dimensions of Si-based piezo resistors are typically in the order of hundreds micrometers which makes it difficult to use them for capturing stress gradients with high spatial resolution, for example local variations of stress caused by a flip-chip pillar or even a micro-bump and a through silicone via (TSV) in 3D stacked ICs [13, 14]. FETs, in this respect, are better candidates for stress sensing applications where high spatial resolution is required. Application of FETs for evaluation of local variation of stress induced by flip-chip pillar was reported in [15] and by microbumps in 3D stacked integrated circuits (3DSICs) in [16].

In this work, a test chip with calibrated FETs as stress sensors is used to assess CPI induced mechanical stress under copper pillars in flip-chip chip scale (fcCSP) packages. Finite element simulations are used to support and interpret the observed phenomena.

Test chip and stress sensors

The developed CPI test chip contains stress sensing elements which are designed as n-type and p-type FETs with channel dimensions of 3.8µm x 4.0µm arranged in the regular arrays with the pitch of 5µm as it is shown in Figure 1. The test chip is built on a planar 65nm CMOS platform utilizing (100) plane Si wafer with the flat (notch) parallel to [010]. Similar FET arrays are designed where the channels of the transistors are either parallel to the notch, i.e. [010] or perpendicular to it, i.e. [100].

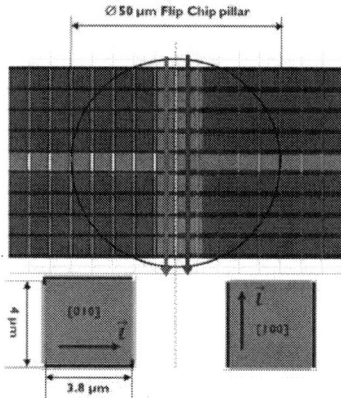

Figure 1. Arrangement of two n-FET arrays (8x9 FETs per array) with orthogonal channel orientations below a 50µm in diameter flip-chip pillar. Sensors along blue and red lines are measured and their response is used for CPI study. (Arrays of p-FETs have similar arrangement)

As it is shown in Figure 1, two FET arrays with orthogonal channel directions are arranged one next to the other and centered below the 50µm in diameter flip-chip pillar. One cell of the test chip contains four flip-chip pillars placed with the pitch of 120µm (Figure 2). Two of these pillars are used for CPI experiments in combination with, respectively, p-type and n-type FET arrays underneath. The cells are further arranged in 4.32mm square dies (as in Figure 3), defining the minimal chip size which can be used for the CPI evaluation of the packages.

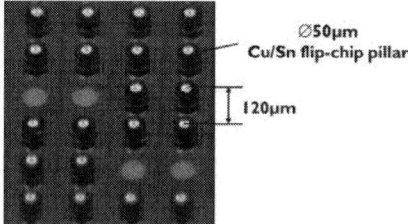

Figure 2. Image of the cells with and without plated flip-chip pillars.

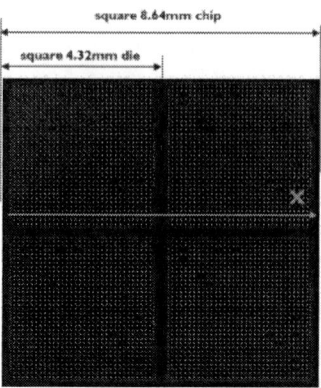

Figure 3. Photograph of the test chip consisting of four square 4.32mm dies. 12 cells in every die do not have pillars and are used for reference measurements.

Set of 12 cells in the die do not contain pillars and can be used for the reference CPI measurements. The maximal chip size is not strictly defined (limited) and can be theoretically scaled up to the size of the 300mm wafer with the step of 4.32mm. Within the die, all FETs are connected to the external IOs (peripheral flip-chip pillars) through the multiple pass-gates and can be selected by the digital serial code. To account for the high series parasitic impedance of the digital pass gates, the source and drain terminals of the stress sensing transistors have force and sense contacts. In the current study, square chips of 8.64mm in size (as in Figure 3) are used to evaluate the CPI effects of molded and bare-die fcCSP packages assembled using the thermo-compression bonding flip-chip process with capillary underfill (TCB-CUF) with the external package dimensions of 14x14mm² (as in Figure 4).

Figure 4. Assembled bare-die and molded fcCSP packages included in the CPI study.

Experimental details and discussions

Prior to CPI study, the FETs were calibrated, that is the currents of n- and p-type transistors at the pre-defined biasing conditions were measured as functions of in-plane and out-of-plane mechanical stresses. Four-point bending setup (as in [13]) was used to induce known uni-axial in-plane stress and to determine in-plane longitudinal (π_{11}) and transverse (π_{12}) equivalent piezoresistive coefficients. To determine the out-of-plane calibration coefficient (π_{13}), nanoindenter was used to apply know vertical compressive force on the flip-chip pillar plated on top of the FET arrays (as in [17]). It should be noted that vertical pressure applied to the flip-chip pillar induces both out-of-plane stress and the in-plane stresses at the level of the Si die. Numerical modeling was used to simulate the calibration procedure (illustrated in Figure 5) and to take it into account when calculating the π_{13} coefficients.

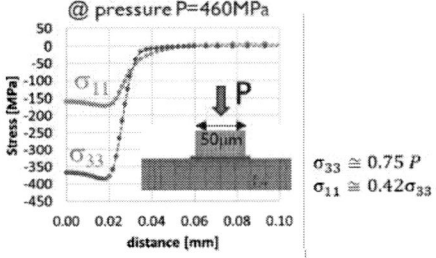

Figure 5. Simulated distribution of mechanical stress in Si below flip-chip pillar induced by external vertical pressure.

The experimentally determined calibration coefficients are summarized in the table below:

Table 1. Calibrated equivalent piezoresistive coeffitients of integrated stress sensors.

	π_{11} [ppm/MPa]	π_{12} [ppm/MPa]	π_{13} [ppm/MPa]
n-FET	300	180	-560
p-FET	-60	-210	263

The calibration was done on the set of samples taken from the separate wafers and it was shown that the lot-to-lot, wafer-to-wafer and transistor-to-transistor variations in combination with the measurement errors can contribute up to ±10% in the accuracy of the measured piezoresistive coefficients. Thus, the reported in Table 1 calibration coefficients can be applied to any sensor fabricated in the same technology from the same maskset. However, the sample-to-sample variations of the initial I-V characteristics of FETs are comparable with the expected shift of their performance due to mechanical stress, which may have significant effect on the accuracy of the CPI results. To mitigate this problem, all FETs included in the CPI study are measured at the wafer-level at reference (zero) stress condition and at fixed temperature of 25°C. In addition, the test chip is equipped with the programmable die identification module and temperature sensors (diodes). This allows to trace every individual stress sensor between the reference wafer-level test and the final CPI measurement of the packaged test chip and do temperature compensation. To further minimize the measurement inaccuracy and account for sample-to-sample variations unavoidable during package

assembly, the set of 10-50 identical packages is included in CPI study while median response of the stress sensors is used for CPI conclusions. As it was stated in [12], calculating all three components of stress (σ_{11}, σ_{22} and σ_{33}) remains challenging task while difference of two in-plane components can be easily and accurately evaluated by comparing currents of sensors with two orthogonal channel orientations located one next to the other as, for example, indicated by red and blue arrows in Figure 1:

$$\sigma_{11} - \sigma_{22} = \frac{\Delta I_0 - \Delta I_{90}}{\pi_{11} - \pi_{12}}$$

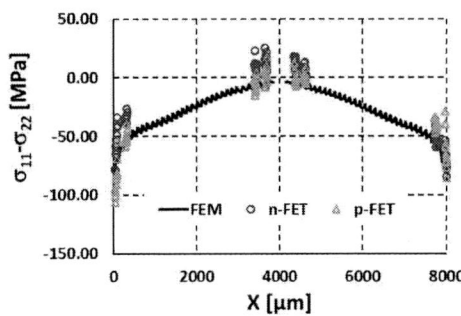

Figure 6. Difference of two in-plane stress components measured by n- and p-type stress sensors and simulated numerically.

Difference of two in-plane components of stress calculated based on the independent measurements of n-type and p-type stress sensors below flip-chip pillars located at the edges and in the middle of the package (along green line in Figure 3) is compared with the one simulated using finite element modeling (FEM) in Figure 6. The magnitude of $\sigma_{11}-\sigma_{22}$ is close to zero at the middle of the package due to its square symmetry and is increasing to 100MPa at the edges.

To account for the global thermo-mechanical stress induced by the mismatch in coefficients of thermal expansion (CTE) between dissimilar packaging materials and for the local distributions of stress around flip-chip pillars, a multilevel sub-modeling technique [18] was used for CPI simulations. The FE model was further strengthened by the correct thermo-mechanical material properties measured using nanoindentation and optical [19] techniques. The observed in Figure 6 mismatch between simulation and experimental results is less than ±15MPa which confirms the accuracy of the FE model and that of electrical measurements. Three components of stress (σ_{11}, σ_{22} and σ_{33}) determined using the validated FE model are shown in Figure 7 as functions of the coordinate (X) indicated as the green line in Figure 3.

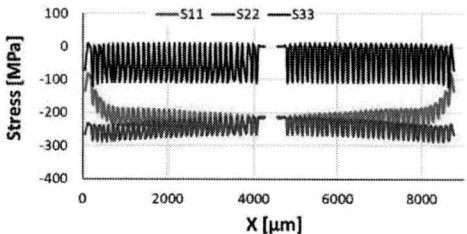

Figure 7. Results of the FE simulations for three components of stress (σ_{11}, σ_{22} and σ_{33}) along green line in Figure 3.

Simulations show that on the global (chip-level) scale, two in-plane components of stress are compressive with the amplitude of -200MPa in the middle of the package. At the vertical left and right edges of the die, σ_{11} component of stress decreases to -100MPa. Vertical component of stress is around zero at the global scale. The observed oscillations of all components of stress have periodicity of 120µm, which corresponds to the pitch of flip-chip pillars. These oscillations represent local stress variations induced by the pillars. Simulations predict that the local vertical stress induced in Si by the pillar is -100MPa compressive.

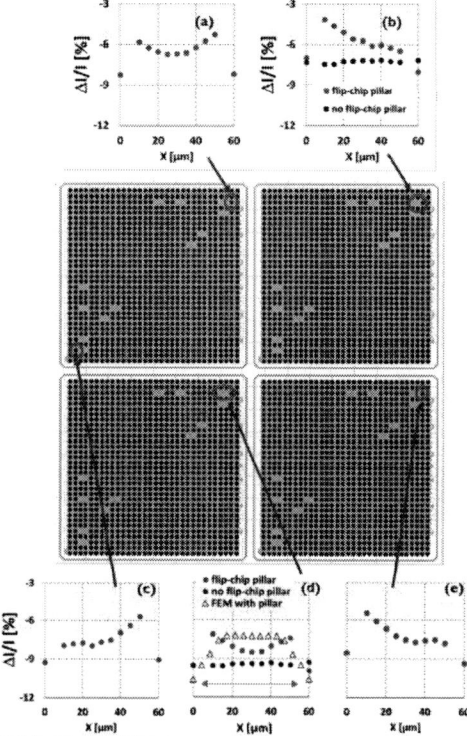

Figure 8. Electrical response of n-type stress sensors measured at different locations on the packaged (molded) test chip. (Red arrow indicates the footprint of 50µm in diameter flip-chip pillar)

Pillars also increase two in-plane components of stress by 50MPa through the same mechanism which was observed during vertical calibration (Figure 5).

To experimentally study local effects of flip-chip pillars on stress, change in electrical response of n-type FETs below the pillars (as illustrated by blue and red lines in Figure 1) located in different positions on the test chip is measured and shown in Figure 8. In the middle of the chip, two sets of sensors are measured representing, respectively, cases with and without plated pillars (as in Figure 2). When no pillar is present above the sensors, their response is flat but shifted down by 8-9% (black dots in Figure 8d). Most of this shift is the result of the global in-plane compressive stress induced in the die due to CTE mismatch between the laminate substrate and Si, while the contribution of the vertical tensile stress induced due to shrinkage of the underfill is not significant. Sensors measured in the close location on the die which has flip-chip pillar (blue dots in Figure 8d) show significant variance along the pillar middle line where the response of the affected sensors (below the pillar footprint) is increased in average by 3%. This increase can be attributed to the vertical pressure induced by the pillar towards silicon, where vertical stress can be estimated as -84MPa compressive using piezoresistive coefficients of n-FETs from Table 1 and equations from Figure 5. FE simulations predict similar response of the stress sensors (black tringles in Figure 8d) although the experimentally observed complex (convex-concave) shape of sensors response below the pillar cannot be fully demonstrated based on the simulation results. Vertical pressure applied by the flip-chip pillar towards the Si die can further be rationalized by a simple model shown in Figure 9.

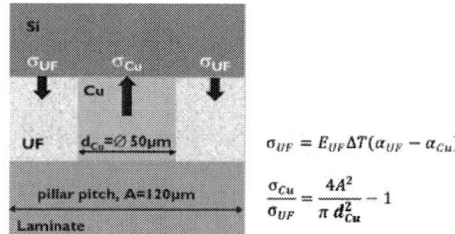

Figure 9. Simplified model of the thermo-mechanical system consisting of copper flip-chip pillar and surrounding underfill.

In this model, pressure induced by the relative shrinkage of the underfill (σ_{UF}) is proportional to its Young's modulus and to the temperature increase during curing of the UF. The reaction pressure of the pillar towards Si die (σ_{Cu}) is proportional to the pillar area density. Considering typical material properties and curing conditions ($E_{UF} \cong 10\text{GPa}$, $\alpha_{UF} \cong 30\text{ppm/K}$, $\alpha_{Cu} \cong 16\text{ppm/K}$,

$T_{cure} = 150°C$), these pressures can be estimated as $\sigma_{UF} \cong 18MPa$ and $\sigma_{Cu} \cong 114MPa$. The last one can be translated into vertical compressive stress induced by the pillar in Si using equation from Figure 5 as $\sigma_{33} \cong -85MPa$, which is in a good agreement with the experimental predictions.

To study the effect of overmolding on stress induced in Si, local response of the sensors below pillars located in the middle of bare-die and molded packages is compared in Figure 10. It can be seen, that the overmolding results in increase of the sensors response by 1-1.5%. Assuming that this process affects only global in-plane stresses, its contribution can be estimated as additional 20-35MPa of compressive stress induced in the Si die being already flip-chip assembled to the substrate.

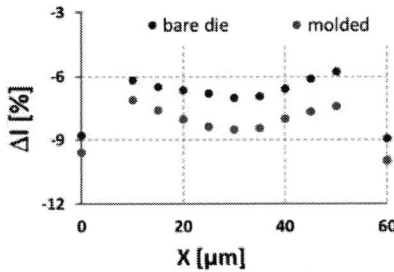

Figure 10. Response of stress sensors below pillars located in the middle of bare-die and molded packages (as in Figure 4).

Analysis of sensors below pillars located at the edges of the die (Figures 8c and 8e) indicates that the pillar-induced stress distribution breaks its local symmetry when measured far from the middle of the die. Similar behavior is registered for the sensors measured along the top horizontal edge of the die (Figures 8a and 8b), that is, symmetrical sensor response under the pillar in the middle of the top edge and asymmetrical response below the corner pillar. The observed phenomenon is explained by the CTE mismatch between laminate substrate and Si which induces bending moment on the pillar in the direction of the center of the package. At the level of the Si die, i.e. where the pillar has an anchoring point, this bending moment is translated into in-plane and vertical compressive stress on one side of the pillar and vertical tensile stress on the other side. This scenario is demonstrated in Figure 11, where the fraction of the FE model close to the die edge is shown in combination with the compensated response of stress sensors. The compensation is done by subtracting the global stress contribution measured at no pillar location (black dots in Figure 8b) from the response of the sensors affected by flip-chip pillars.

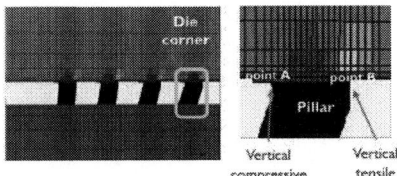

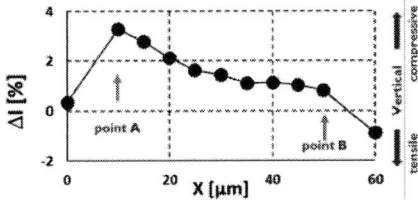

Figure 11. Simulated distribution of vertical stress in Si die above flip-chip pillar (top) and (bottom) compensated response of piezoresistive stress sensors above pillars located at the corner of the packaged die (as in Figure 3b).

While vertical compressive stress has no direct impact on reliability of the soft BEOL below the flip-chip pillar, tensile vertical force at its outer edge can be a potential source of cracks and delaminations in the backend layers as it was experimentally demonstrated by BABSI tests in [18].

Conclusions

The proprietary Si test chip with integrated piezoresistive stress sensors is used to electrically measure CPI effects. Dense arrangement of the sensors allows assessment of mechanical stress with spatial resolution of better than 10μm. This enables to separate global stress induced in the IC package due to the CTE mismatch between Si and the laminate substrate from the local stress modulation caused by the "pushing" action of the flip-chip pillar against Si surface. Electrical measurements are validated and rationalized by the finite element modeling which consecutively simulates different assembly steps and uses the experimentally determined thermo-mechanical properties of the packaging materials.

It is demonstrated that high spatial resolution stress measurements can help in better understanding of failure modes in advanced microelectronic packages with dens pillar interconnects.

Acknowledgements

The authors would like to thank core partners of the imec Industrial Affiliation Program (IIAP) on 3D Integration Technology.

References

[1] K. IKEGAMI, "Mechanical Problems in the Production Process of Semiconductor Devices,"

JSME Int. journal. Ser. 1, Solid Mech. strength Mater., vol. 33, no. 1, pp. 1–12, 2017.

[2] K. Chanda and V. Mahadev, "A fabless company's perspective on large die Chip Package Interaction (CPI) challenges," in IEEE International Reliability Physics Symposium Proceedings, 2014, no. 001, pp. 1–4.

[3] Liu, C. S. et al. CPI challenges in advanced Si technology nodes. Proc. 2013 IEEE Int. Interconnect Technol. Conf. IITC 2013 8–10 (2013).

[4] Hu, S. M. Stress-related problems in silicon technology. J. Appl. Phys. 70, (1991).

[5] Smith, C. Piezoresistance effect in germanium and silicon. Phys. Rev. 919, (1954).

[6] Kanda, Y. A graphical representation of the piezoresistance coefficients in silicon. IEEE Trans. Electron Devices 29, 64–70 (1982).

[7] Suhling, J. C. & Jaeger, R. C. Silicon piezoresistive stress sensors and their application in electronic packaging. IEEE Sens. J. 1, 14–30 (2001).

[8] Mian, A., Suhling, J. & Jaeger, R. The van der Pauw stress sensor. Sensors Journal, IEEE 6, 340–356 (2006).

[9] Doelle, M., Ruther, P. & Paul, O. A novel stress sensor based on the transverse pseudo-hall effect of MOSFETs. in The 16th Annual International Conference on Micro Electro Mechanical Systems, 2003. MEMS-03 Kyoto. IEEE, Kyoto, Japan 490–493 (2003).

[10] Doelle, M., Peters, C., Ruther, P. & Paul, O. Piezo-FET stress-sensor arrays for wire-bonding characterization. J. Microelectromechanical Syst. 15, 120–130 (2006).

[11] Schreier-Alt, T., Chmiel, G., Ansorge, F. & Lang, K. D. Piezoresistive stress sensor for inline monitoring during assembly and packaging of QFN. in Proceedings - Electronic Components and Technology Conference 2126.

[12] Roberts, J. et al. Measurement of die stress distributions in flip chip CBGA packaging. 2010 12th IEEE Intersoc. Conf. Therm. Thermomechanical Phenom. Electron. Syst. ITherm 2010 1–13 (2010).

[13] Ivankovic, A. et al. FET arrays as CPI sensors for 3D stacking and packaging characterization. 2012 IEEE Int. Reliab. Phys. Symp. 2E.3.1-2E.3.9 (2012).

[14] Perry, D. et al. An efficient array structure to characterize the impact of through silicon vias on FET devices. 118–122 (2011).

[15] Cherman, V. et al. Evaluation of mechanical stress induced during IC packaging. Proc. - Electron. Components Technol. Conf. 2018-May, 2168–2173 (2018).

[16] Cherman, V. et al. 3D stacking induced mechanical stress effects. in Proceedings - Electronic Components and Technology Conference 309–315 (2014).

[17] Vanstreels, K. et al. Advanced experimental back-end-of-line (BEOL) stability test: Measurements and simulations. Microelectron. Eng. 137, 54–58 (2015).

[18] Lofrano, M., Vandevelde, B. & Gonzalez, M. A multilevel sub-modeling approach to evaluate 3D IC packaging induced stress on hybrid interconnect structures. Microelectron. Eng. 120, 85–89 (2014).

[19] Salahouelhadj, A. & Gonzalez, M. CTE measurements for 3D package substrates using Digital Image Correlation. 2016 17th Int. Conf. Therm. Mech. Multi-Physics Simul. Exp. Microelectron. Microsystems, EuroSimE 2016 1–6 (2016).

ALD Coatings to Mitigate Against Tin Whiskers and Upgrade the Environmental Durability of Electronic Circuit Boards

Terho Kutilainen*, Marko Pudas**, Mark A. Ashworth***, Geoffrey D. Wilcox***, Jussi Hokka ****

*Oy Poltronic Ab, Finland,

**Picosun Oy, Finland,

***Materials Degradation Centre, Department of Materials, Loughborough University, UK,

****ESTEC, ESA, The Netherlands

Abstract

Picosun is undertaking a research programme, funded by the European Space Agency, in collaboration with Oy Poltronic Ab and Loughborough University, to evaluate conformal coatings made with atomic layer deposition (ALD) as a method of mitigating the growth of tin whiskers and for corrosion protection. ALD is a mature, key enabling technology in modern IC industries, and it is also utilised for corrosion protection in various special applications. As such, ALD is an interesting option for the protection of assembled electronic circuit boards and modules when extremely thin, practically massless, reworkable, transparent and inert coatings are needed, especially in specific high-end applications. The results show that the ALD process can significantly reduce tin whisker growth and thus offers considerable potential as a reworkable whisker mitigation strategy. The latest output of these, started in 2015 (Phase 1) and 2018 (Phase 2), are presented.

Key words: corrosion, tin whiskers, atomic layer deposition, ALD.

Introduction

Electronics production is going lead free and thus the availability of lead (Pb) containing electrical, electronic, and electromechanical (EEE) parts, materials and processes is decreasing, forcing space electronics companies to look for lead free alternatives despite the RoHS exemptions allowed for space applications. The possibility to upgrade industrial graded components is interesting due to the limited availability of space-qualified parts. These were the main drivers behind this study. The primary goals of the project were to evaluate atomic layer deposition (ALD) conformal coating as a method for mitigating the growth of tin whiskers and to develop a conformal coating method for component boards for space electronics.

Tin Whiskers

Tin whiskers can cause reliability issues for electronic components and assemblies. One and probably the most common failure mechanism is electrical shorting. Whiskers can grow, for example, between adjacent conductors and cause a transient short circuit provided enough current driving capacity is present to burn the whisker open, or the short may be permanent and cause failure in performance or heat up the device. Whiskers can also break loose and cause reliability issues, especially in devices with very small electrical clearances such as micromechanical, microelectronic or optical devices. The JEDEC/IEC standard classifies a feature as a whisker if it possesses an aspect ratio (length/width) greater than 2 and as being of significance if its length is greater than 10 μm. A few different whisker morphologies are shown as examples in Fig. 1. [1, 2, 3]

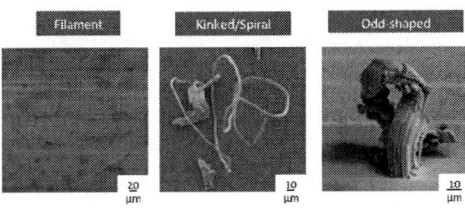

Figure 1: Examples of whisker morphologies

Upgrading of components

Commercial-off-the-shelf components (COTS) have limited capability to withstand harsh environmental conditions and reliability issues are more likely to occur compared to space level electrical, electronic, and electromechanical parts. Environmental protection using ALD encapsulation could offer improved reliability performance of plastic components. We chose an internally connected daisy chain QFP100 package for this specific task along with some other test structures for other purposes. In this paper we present the general structure of the designed test board and explain the environmental tests performed.

Atomic Layer Deposition

ALD uses sequential, self-limiting and surface-controlled gas phase chemical reactions to achieve control of film growth in the nanometre/sub-nanometre thickness regime. The ALD film is dense, crack-, defect- and pinhole-free and its thickness and structural and chemical characteristics can be precisely controlled on atomic scale. ALD process is digitally repeatable and it can be performed at relatively low temperatures. This gives the possibility to construct not only single material layers but also doped, mixed, or graded layers and nanolaminates, whereas low process temperature allows coating of also sensitive materials such as plastics and polymers.

Tin Whisker Test Method

Tin whisker test coupons comprised copper plates electroplated using a tin-copper alloy. The Sn-Cu alloy deposit used in this study is known to grow whiskers in a short time, which makes it possible to observe whisker growth and identify differences between coatings within the limited durations of the project phases, each of which was less than two years. The coupons have been electroplated in separate batches for each specific ALD coating. In Phase 1 coupons were prepared in the UK and ALD coated in Finland after a relatively short, but varying, transportation time. In Phase 2 plating was performed in Finland in order to minimise the time between electroplating and coating.

The Cu coupons have dimensions of 20 mm x 40 mm x 0,4mm of which the electroplated area is 20 mm x 20 mm. The whisker densities were evaluated for selected samples from each batch using optical microscopy. The whisker growth was evaluated over 20 (Phase 1) or 25 (Phase 2) frames per sample using a 20x objective lens, equivalent to an area/frame of 0.57 mm^2). The approximate locations of the frames on the test coupon are shown as blue dots in Fig. 2. Filaments and large eruptions were counted at three-month intervals. Some selected batches from Phase 1 have been examined in Phase 2, corresponding to a total storage time of some four years.

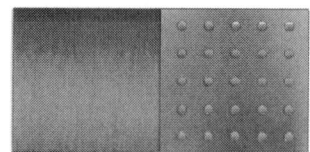

Figure 2: Tin whisker test coupon

Printed Circuit Board Test Vehicle

For the environmental tests and the rework trials we have used a board designed in-house for it and for some other purposes such as functional demonstration. The board and the layout are shown in Fig. 3. The numbers explained below show locations of the test structures.

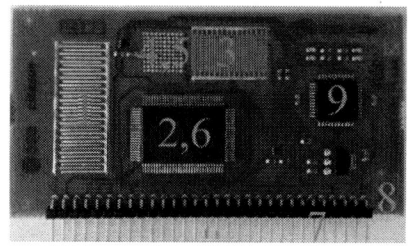

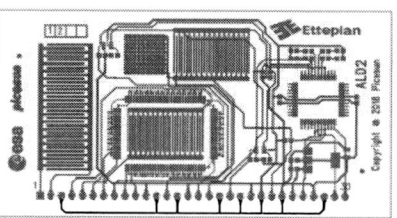

Figure 3: Test vehicle for environmental and rework tests

The test board consists of the following main parts and functions, of which only some are used for the work and the results presented in this paper:

1. SIR-test structure for corrosion testing, based on IPC PCB-B-25A, plated with a commercial SAC solder.
2. SIR-test structure, with 0,3 mm lines and spaces, not soldered. This is placed under a QFP package in order to study penetration of ALD coating under a low stand-off component.
3. SIR-test structure, similar to 2, solder plated. This structure is designed for demonstration purposes and it is connected to the functional part of the test board.
4. Test pattern located on the bottom layer on the component side, which, together with a similar pattern 5 on the top layer and the pregpreg between the layers, forms a capacitor that can be used for insulation resistance and absorbed humidity measurements.
5. Test pattern on the top layer is designed for tin whiskers analysis.

6. The QFP100 plastic package contains a daisy chain that is used for corrosion testing. The daisy chains are connected by microvias and conductors in inner layer of the board enabling analysis of the board reliability.

7. The output connector pins are tin electroplated and can thus be used for tin whisker observation in addition to electrical connections from and to the board.

8. The through holes for the row connector together with some additional wiring on the board can be used for conductive anodic filament (CAF) analysis.

9. The microcontroller in a QFN44 package, LEDs and some additional components form a functional demonstrator. Two I2C bus temperature detectors are used to monitor the temperature of the board and the microcontroller measures the impedance change in SIR 3.

Tin Whisker Test Results

In Phase 2, testing was continued of the most promising ALD coatings from Phase 1 and additional new ALD coatings were developed. The evaluated coating alternatives are referred to as individually numbered batches. The whisker density results after storage at room temperature for ~1 and ~ 3 years for the selected batches 1, 3, 4, 5 and 13 are summarised in Fig. 4. Batch 13 is a reference batch in which the samples have no coating but have gone through a similar thermal process to the coated samples. These samples are henceforth referred to as "thermal samples". The unprocessed reference coupons accompanied every coating batch (denoted later as B1, B3, etc.) except batch 1. Therefore, the chart (a) has no the data for B1. We can reasonably assume the reference values for this batch to be in the same range as those for the all other batches. The lower chart has the same scale on the y-axis to clearly demonstrate the considerable difference in whisker growth between the ALD processed coupons and the references. From the graph it is apparent that the whisker densities are very significantly reduced for the ALD processed samples compared with uncoated control samples.

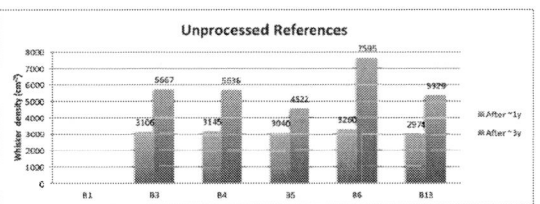

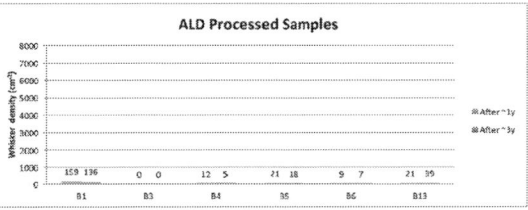

Figure 4: Whisker Densities of the Unprocessed (a) and the ALD Processed Samples (b)

The filament growth observed for batches 3, 4, 5 and 6 was on average less than ~20 whiskers cm^{-2} after ~1 year. No filament whiskers or large eruptions were recorded for any of the samples from batch 3. In comparison, the whisker density measured for the unprocessed reference samples was ~ 3000 cm^{-2} after ~1 year. For all batches, the whisker density of the unprocessed control samples has increased significantly from ~ 3000 whiskers cm^{-2} after ~1 year to > 5000 whiskers cm^{-2} after ~ 3 years.

An important observation is that the thermal samples also seem to suppress whisker growth. However, after 3 years, whisker counting does appear to indicate some increase in whisker densities for these samples, whereas the ALD coated samples showed no increase in whisker density compared with that measured after 1 year storage. SEM analysis of an ALD coated sample from B1 showed no evidence of whisker growth penetrating through the coating. All the observed whiskers were seen to be coated with ALD. Fig. 5 presents the same data as the chart (b) in Fig. 4 but using a different scale for improved clarity.

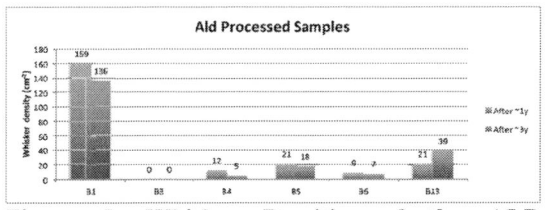

Figure 5: Whisker Densities of the ALD Processed Samples

A progressive reduction in the number of filaments and large eruptions was observed as the thickness of the ALD coating increased. The whisker densities measured for samples with 0.9 nm and 4.0 nm thick coatings are somewhat lower than the other data points. This is most likely due to these samples having a shorter delay between electroplating and ALD coating, i.e. less whisker growth was able to occur prior to coating. Fig. 6 shows the whisker density values measured for certain thicknesses and the delay time in arbitrary units supporting the assumption. This was the reason for arranging the electroplating and coating processes in the same location for Phase 2.

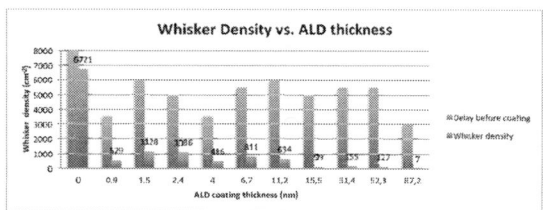

Figure 6: Whisker Densities vs. ALD Coating Thickness

Based on this result and observations on all manufactured batches having thicknesses from 87 nm up to 500 nm, we can assume that the coating thickness reaches the optimal performance at ~100 nm, depending on the exact ALD material formulation. The thermal samples without coating show a comparable reduction in whisker growth. On the other hand, the thermal references show some slight increase in whisker density over 3 years storage, unlike the ALD coated samples. Very probably this has occurred as the thermal samples have been exposed to the same thermal processing conditions and duration as the coated samples. This remains an object for further investigations.

In Phase 2 of the work several different ALD coatings were deposited immediately after electroplating of the SnCu. The most recent tin whisker growth data, up to 180 days from sample manufacture, shows without exception two things: (1) Tin whiskers grow rapidly and within 1 month all the reference samples have whisker densities in excess of 1000 whiskers/cm². (2) All samples coated with ALD or thermally processed have no tin whisker growth at the time of the analysis.

Corrosion protection

A surface insulation resistance (SIR) test structure in accordance with IPC PCB-B-25A (Fig. 3, pattern 1), a more dense finger pattern under a QFP100 package and two parallel daisy chains connected alternatingly in the board microvias and in the QFP100 package were used in the mixed flowing gas test (MFG) and the high temperature high humidity test undertaken by Trelic Oy, Finland. The MFG test was based on the EIA-364-TP65A monitored with copper coupons in accordance with the ASTM standard B810-01a. After the MFG, the boards were connected to a bias voltage and stored at 85°C and 85% relative humidity (RH) for 1000 hours according to JEDEC JESD22-A101D. The bias field strength was set to 25V/mm for the SIR

patterns and 50V/mm for the daisy chains. After these tests a voltage step-stress test was made increasing the electric field gradually up to 1400V/mm. All coated boards survived the tests without visible corrosion or failures. The uncoated reference boards showed some corrosion but no severe damage probably due to the high quality of the printed circuit board.

Figure 7: Boards in the MFG chamber

Rework

An ALD layer is in the range of 10^3 times thinner than a solder layer. Therefore, it was not expected to have any significant influence on the rework of solder joints. The thickness of the native oxide [4] on solder is comparable to the layer thicknesses created by the ALD process. Thus, rework of an ALD coated solder joint should not be an issue. This was a premise supported by the initial tests performed on the test boards used in the first phase and more recently confirmed in the second phase by documented rework trials made at RUAG Space Finland Oy Ab. Two components, a SOT223 package and a QFP44 package, were replaced manually without prior removal of the coating (Fig. 9). The soldering quality of the reworked parts complies with the requirements set for rework according to ECSS-Q-ST-70-38 C rev.1. After repair, the boards were recoated using ALD at Picosun.

Figure 9: Components removed and reassembled

Future outlook and conclusions

Based on these investigations and the monitoring of whisker growth over a period of almost 4 years we believe that ALD processing has potential to function as a highly effective whisker mitigation strategy. The ALD coating process provides both degassing and encapsulation for the components and substrate because the process takes place in vacuum at elevated temperatures. The ALD coated test boards have performed well in the harsh environmental tests and they may be easily reworked when required.

ALD provides practically massless coatings with superior properties, which makes the process a potential candidate to protect and upgrade electronics assemblies and components for space applications and others having similar requirements.

Further studies are in progress to both optimise the properties of the applied ALD coating and to clarify some of the observations.

Acknowledgements

This programme was funded by The European Space Agency under the European Component Initiative (ECI) Phase 4 and Strategic Initiative (StrIn) programmes. The authors also acknowledge use of facilities within the Loughborough Materials Characterisation Centre.

References

[1] Chien-Ming Huanga, Daniel Nunezb, James Coburnc, Michael Pecht. "Risk of tin whiskers in the nuclear industry". Microelectronics Reliability.Volume 81, February 2018, pp. 22-30

[2] NASA, Tin Whisker (and Other Metal Whisker) Homepage, https://nepp.nasa.gov/whisker/

[3] JEDEC. JESD22-A121A: Test Method for Measuring Whisker Growth on Tin and Tin alloy Surface Finishes. 2008.

[4] Giles Humpston, David M. Jacobson. Principles of Soldering, ASM International. 2004.

Comparison of QFN Chips Glued by ACA and NCA Adhesives on the Flexible Substrate

Martin Hirman, Frantisek Steiner, Jiri Navratil and Ales Hamacek

Department of Technologies and Measurements / Regional Inovation Centre of Electrical engineering
Faculty of Electrical Engineering, University of West Bohemia, Univerzitni 8, 306 14 Pilsen, Czech Republic

Contacts: hirmanm@ket.zcu.cz, +420377634546

Abstract

The paper deals with the QFN components assembly to the PET substrate with silver conductive pattern by anisotropically conductive adhesive (ACA) or non-conductive adhesive (NCA) with UV curing. Motivation for the experiment was to investigate and compare conductive joints prepared by ACA and NCA and make comparison of their reliability during the accelerated climatic ageing. For the NCA adhesive, the electrical conductivity was achieved by mechanical contact of the QFN component pads and substrate pattern pads. Therefore, the downforce on the QFN chip during the UV curing was important just as the adhesive's shrinkage rate. The assembly of the QFN components by non-conductive adhesive had similar or better properties in comparison with anisotropically conductive adhesive (ACA) joints and therefore non-conductive adhesives could be recommended for some special applications.

Key words: Non-conductive adhesive, anisotropically conductive adhesive, UV curing, PET substrate, conductive connection

Introduction

Currently, the popularity of flexible electronics is growing. Modern flexible substrates have a different properties than the common rigid substrates and it is not possible to use the standard connection methods (e.g. soldering, bonding) in all cases. The most of these substrates (e.g. PET, PEN, biodegradable ...) have lower thermal resistance and need flexible conductive joints. The soldering is not suitable due to the high temperature during the process and minimal flexibility of solder alloy [1,2].

The electrically conductive adhesives (ECA) are a good choice for solving the problems with low flexibility of joint and high temperature of the process [3–5], but they also have some disadvantages. The reliability of ECA joints is worse than soldered joints. The duration of the curing process is regularly longer than the duration of the soldering process. The cost of the joints made by ECA is also higher in comparison with soldering technology. Even so, the gluing of components to flexible substrates is necessary for some applications (e.g. wearable or printed electronics) [6].

Second choice is to use anisotropic conductive adhesive (ACA) where the curing process can be very fast (for UV curable ACA). Anisotropic conductive adhesives have similar advantages and disadvantages as ECAs but ACAs can be used for components with fine pitch package.

The disadvantage of ACAs is their worse electrical properties than the ECA joints.

Another known but neglected option is to use the non-conductive adhesive (NCA). The use of NCA for the electrical connection of lower level packaging (chips in the flip chip technology) was researched in 2000s [7,8], but using of NCA to connect higher level packaging components with flexible substrates has not been examined thoroughly. The use of UV curable NCAs for electrical connection of component with flexible substrate is very easy to realize and can be fast cured by UV light. The electrical connection is created by direct contact of component pads with conductive pattern on flexible substrate.

In our previous research, some tests with NCA were realized but direct comparison of conductive joints made by UV curable ACA and NCA have not been realized yet. Motivation for the experiment was to investigate and compare conductive joints prepared by ACA and NCA and also make comparison of their reliability during the accelerated climatic ageing.

Theory and Experiment

The testing of reliability can consist of different tests, e.g. bend testing [9], climatic ageing [10], vibration testing, etc. In our case, the climatic ageing was chosen for the experiment. The procedure of the realized experiment can be seen in Figure 1 as a process diagram.

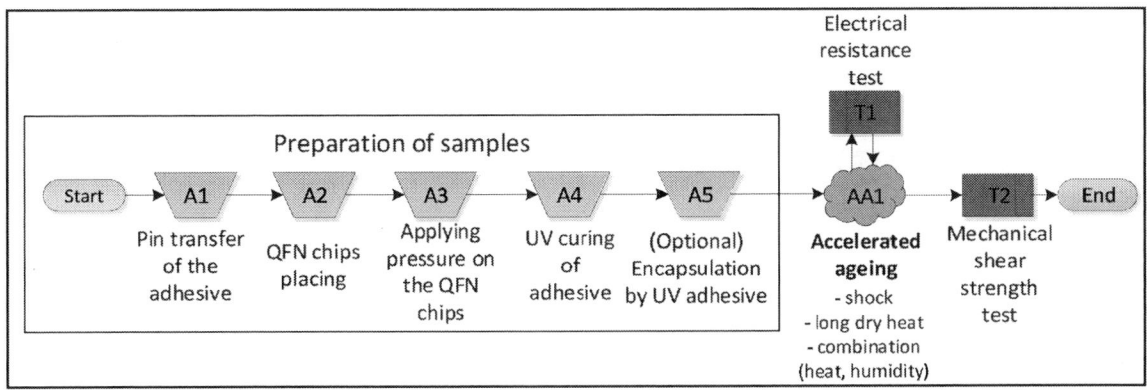

Figure 1: Procedure of performed experiment.

The flexible PET foil (DuPont Melinex) with screen-printed silver conductive pattern was used in the experiment. Two types of silver conductive pattern with dissimilar angles of QFN assembling (45°, 90°) were used, see Figure 2. The QFN chips (daisy chain = neighbouring pads are connected) were mounted by two adhesive types. One part of QFN chips was glued by UV curable ACA with application of pressure onto the components during the curing. Another part of QFN chips was glued by UV curable and flexible NCA with application of pressure onto the components during the curing. In case of the NCA, the pressure pushed all the adhesive from the space between screen printed pads and QFN chip pads out, so good mechanical-electrical contact was achieved while the adhesive is affixing the chip on the substrate.

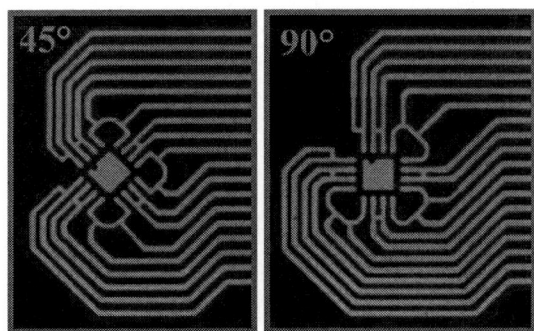

Figure 2: Patterns used for the experiment.

The ACA Elecolit 3063 and NCA Loctite AA3926 were used in the experiment. The Elecolit 3063 is described in datasheet as one component anisotropic conductive adhesive with good adhesivon to PET, Kapton, Mylar and other flexible substrates from Panacol company. This adhesive is consisted of acrylate resin with 3% of gold-coated particles with 10 µm size. The adhesive was analyzed by scanning electron microscope (SEM), where it has been proved that particles are nickel balls coated by gold, see Figure 3. The Loctite AA3926 is acrylic-based light-cured flexible non-conductive adhesive with wide randge of operation temperatures (–40 °C to +150 °C) from Henkel company.

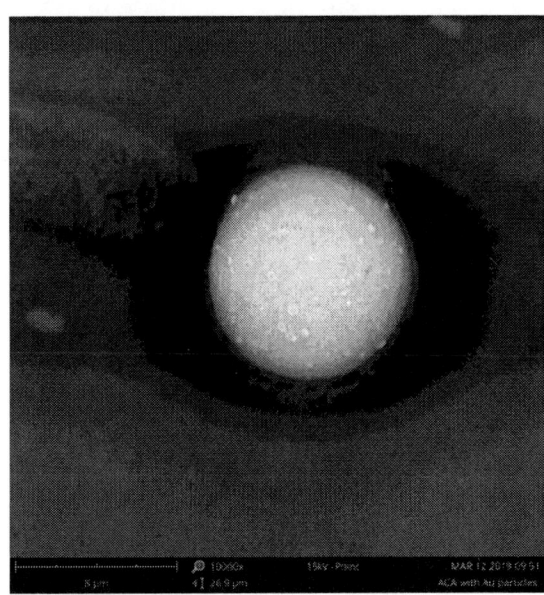

Figure 3: Detail of nickel particle with gold coating from SEM.

The electrical resistance and the mechanical shear strength of prepared samples were measured before and after the accelerated climatic ageing tests. The samples were submitted to three types of accelerated climatic ageing tests, see Table 1. The first type was realised in the shock chamber (85 °C / -20 °C / 1000 cycles). The second type was realized by a combination of tests consisting of a Cold test (-20 °C / 16 hours) according to the standard IEC 60068-2-1; a Dry heat test (85 °C / 16 hours) according to the standard IEC 60068-2-2 and a Damp heat, cyclic test (55 °C / 93 %RH / 12 hours and 25 °C / 97,5 %RH / 12 hours) according to the standard IEC 60068-2-30. The third type was a long Dry heat test (85°C / 1000 hours) according to the standard IEC 60068-2-2.

Table 1: The accelerated ageing tests performed in the experiment

Test 1 – Temperature shock	Test 2 – Combination	Test 3- Long dry heat
85°C / -20°C / 8 minutes / 1000 cycles	Cold (-20°C / 16 hours) Dry heat (85°C / 16 hours) Damp heat, cyclic (55°C / 93%RH / 12 hours and 25°C / 97,5%RH / 12 hours)	85°C / 1000 hours

The electrical resistance of glued joints was measured by four-point probe method. Each sample was connected to the Keithley 2701 device via the conector. The electrical resistance measured for each sample consisted of resistance of two silver conductive patterns (R1 and R5) and the resistance of two glued joints (R2 and R4) plus the resistance of connection inside the QFN (daisy chain), see Figure 4. The electrical resistance of connection inside QFN (R3) was a few mΩ. The electrical resistance of silver conductive patterns (R1 plus R5) was between 500 and 800 mΩ (650 mΩ in average). The electrical resistance of one joint is approximately the values from Figure 8 to Figure 11, which have to be decreased by 650 mΩ (approximately) and divided by 2 to have one joint resistance. The electrical resistance was measured before the ageing, during the ageing (at set intervals) and after the ageing.

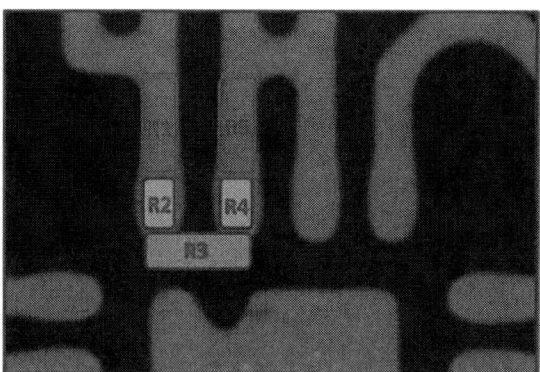

Figure 4: Theoretical structure of measured electrical resistance

The mechanical properties were studied by mechanical shear strength test. This test was realized before the ageing (first set of samples) and after the ageing (second set of samples). The test was realized by the device LabTest 3.030. The principal of the measurement can be seen in Figure 5. The hexagonal thorn pushes by force onto the component until the disruption of the joint appears. The flexible substrates had to be attached to the rigid PCB before testing because the samples (due their flexibility) could not be tested without reinforcement. The test was done according to the IEC 62137-1-2 standard.

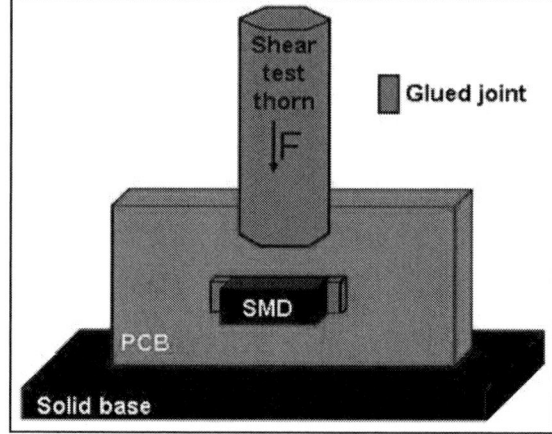

Figure 5: The principle of shear strength test

The ACA or NCA surrounds the QFN component from all sides. It follows that it is not possible to push onto the QFN side by thorn. The one side of adhesive above the QFN component was carefully removed by scalpel before the shear strength test (see Figure 6). The removing of the part of the adhesive caused reduction of mechanical shear strength but the removing was performed for all samples. This means that the comparison of measured values is still possible. The shear strength of the joint was not measured directly because the surface under load is not known. The maximal force required to shear off glued component from the substrate was recorded when the mechanical shear strength test was performed.

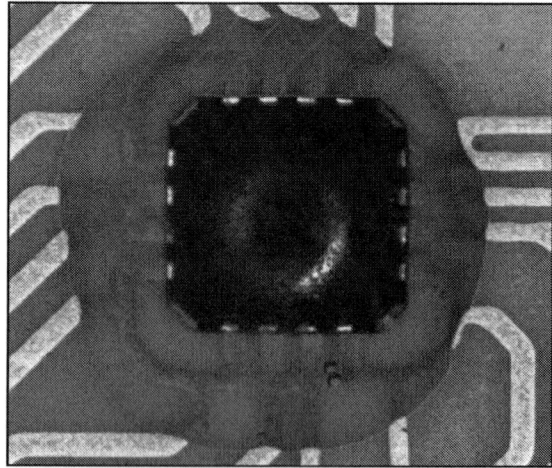

Figure 6: Encapsulated QFN component with red color marked of the surface for removing by scalpel before shear strength testing

434

Results and discussion

All measured values were statistically analyzed and are presented in the bar chart or boxplot diagrams in Figure 8 – 11.

The results of mechanical shear strength test (Figure 7) show that the ageing is not so significant factor for NCA adhesive joinst or encapsulated ACA joints. For the ACA joints without encapsulation, the long time ageing (temperature and shock) decrease the mechanical shear strength significantly. The significant difference for dry heat testing of QFN 45° encapsulated for NCA is probably a remote value.

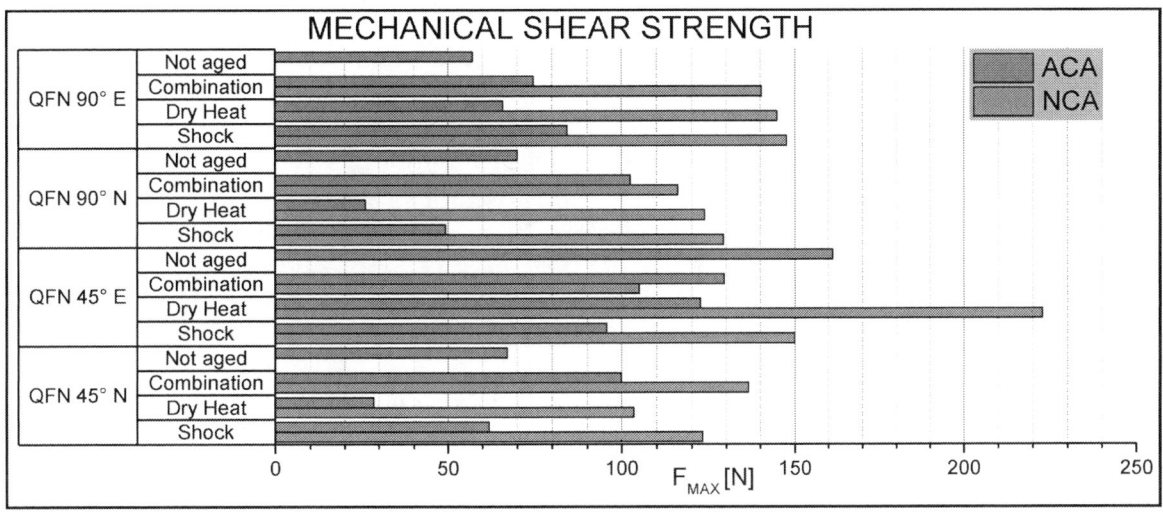

Figure 7: Bar chart of mechanical shear strength after the ageing.

The results of electrical resistance after combined (cold, dry and damp heat) ageing (Figure 8) show minimum changes for ACA joints. These results also show minimum changes for NCA joints after cold ageing and dry heat ageing, but significant increasing of resistance after damp heat cycling test is evident. The differences between the sample types were minimal.

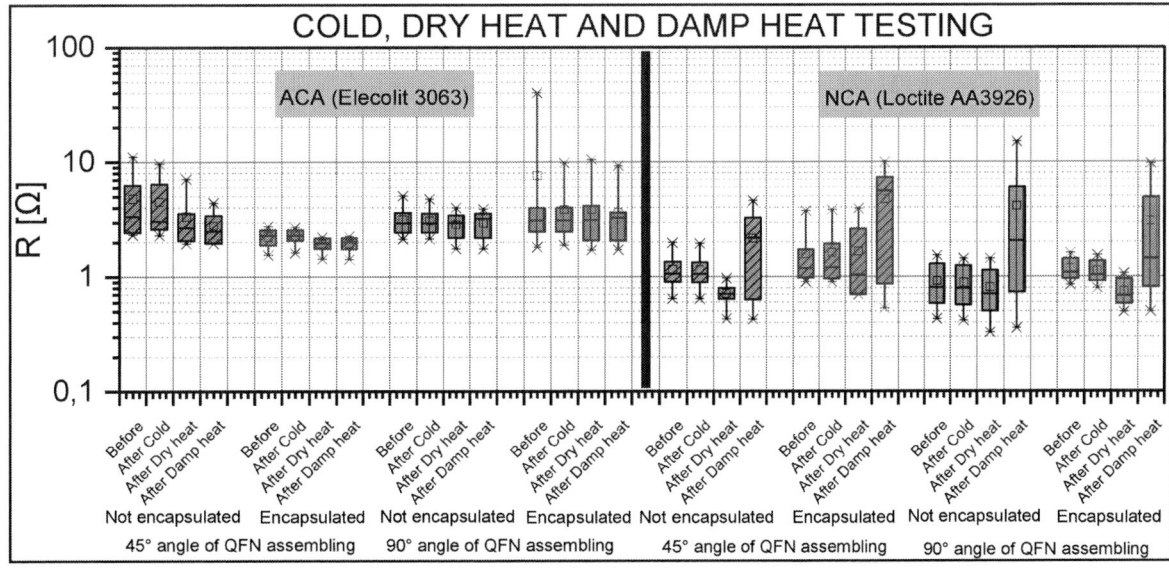

Figure 8: Boxplot of electrical resistance from combined ageing.

The results of joints electrical resistance after shock ageing (Figure 9) show little improving of resistance for ACA joints. On the other hand, some samples of ACA joints without encapsulation were destroyed during the test (i.e. samples had infinite resistance after the test), it follows that the boxplots for the ACA samples without encapsulation are not so relevant. The resistance of encapsulated ACA joints is better and all samples endured the test. The results of NCA joints show that NCA joints have better results in shock chamber without encapsulation, where total destruction of joint was not observed. The encapsulation of NCA joints has no significant effect on the electrical resistance during the ageing.

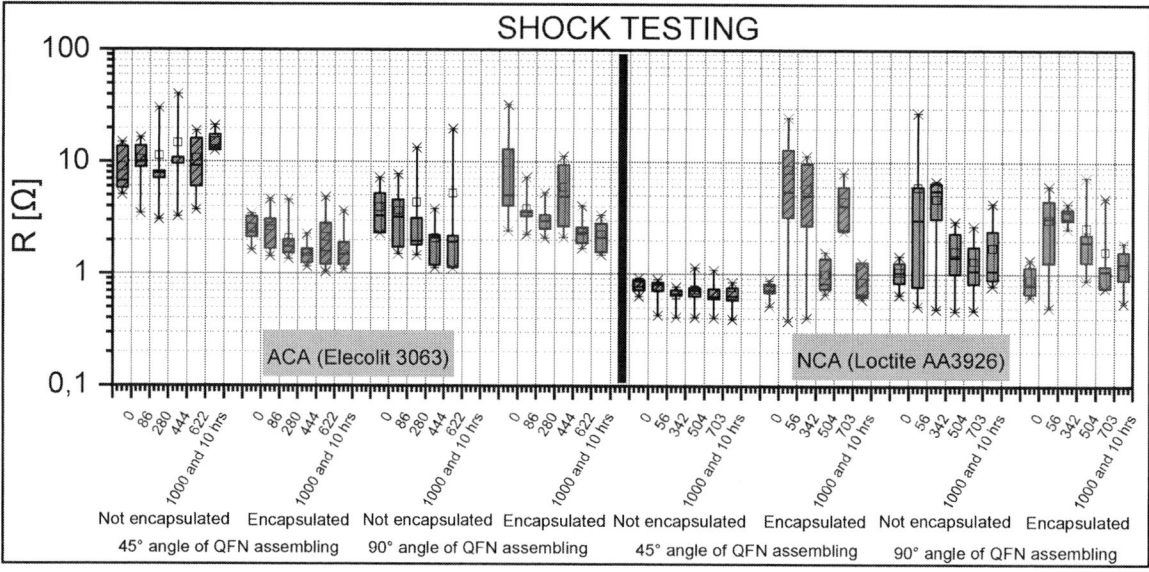

Figure 9: Boxplot of electrical resistance from shock ageing.

The boxplot diagrams of long dry heat ageing (Figure 10, 11) show that influence of dry heat ageing on ACA joints is not so significant for 45°, but if we consider the destroyed samples (with infinite resistance) the dry heat ageing is destructive for the ACA joints. The encapsulation of these samples improves the results but only slightly. It follows that ACA samples are not reliable after long dry heat ageing. For NCA joints the results show minimal effect of dry heat ageing on the electrical resistance. The samples with encapsulation seem to be a little better but if we consider the destroyed samples the joints without encapsulation are better but not significantly.

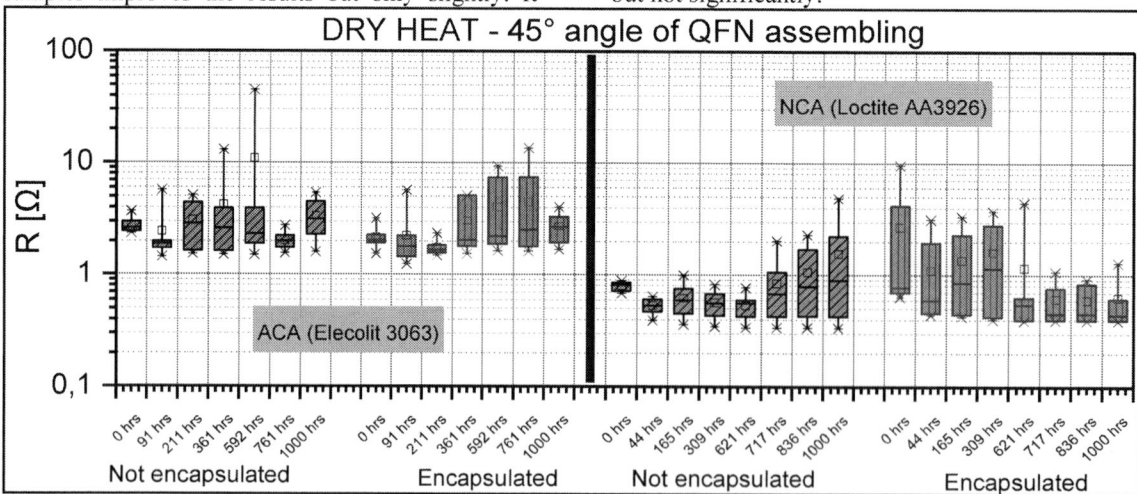

Figure 10: Boxplot of electrical resistance from long dry heat ageing for QFN components (45° angle).

436

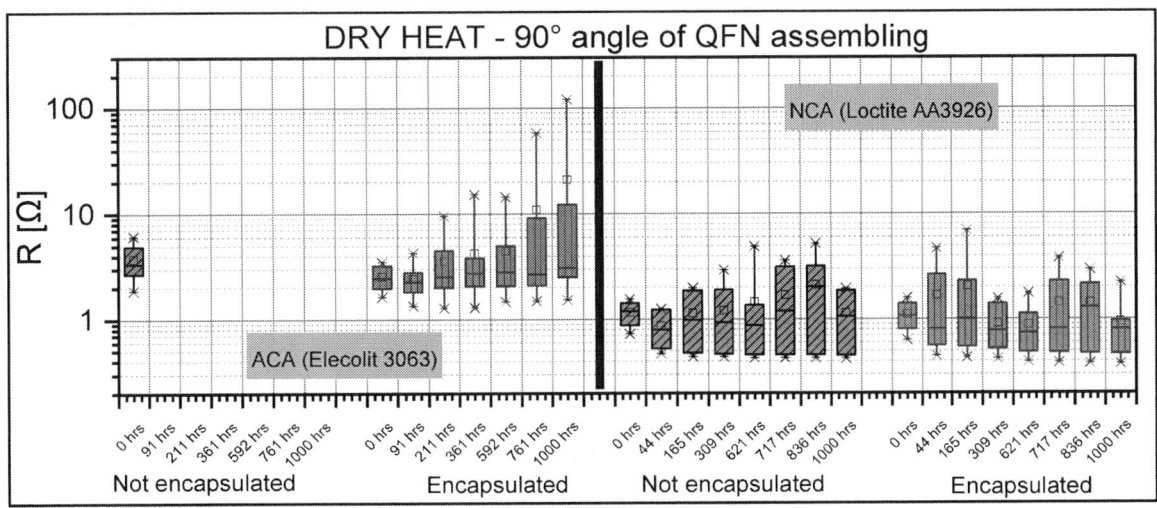

Figure 11: Boxplot of electrical resistance from long dry heat ageing for QFN components (90° angle).

Conclusions

The experiment provided proof that using of non-conductive UV curable adhesive to electrical connection of QFN chips onto the PET foil with printed silver conductive pattern is possible and the results could be better than with standard anisotropic conductive adhesive.

The results of the experiments showed that the mechanical shear strength of samples with ACA and NCA seems to be sufficient but for NCA is significantly better.

For ACA adhesive, the electrical resistance of samples is stable during the combination testing. The results of shock testing showed that some ACA joints without encapsulation were destroyed during the test, so this joint type is not stable for temperature shocks. The samples with encapsulation endure the test was a little increased at the end of the test but the samples were stable. The resistance of encapsulated ACA joints was in applicable limits during the whole test. The long dry heat testing of ACA samples provided proof that many samples were destroyed during the test, it follows the ACA sample are very unstable during the long dry heat ageing.

For NCA adhesive, the combination test caused a small increase of the electrical resistance but the samples are still applicable. The results of the shock testing showed that the electrical resistance grows during the test, but the resistance before and after the test is almost similar. All the time, the resistance of samples in the shock chamber is still in applicable limits. The resistance of NCA samples during long dry heat test is stable. The samples with encapsulation had a few destructions so the samples without encapsulation seem to be a little better.

All NCA results were similar or better than the ACA results, it follows that using of non-conductive adhesive to electrical and mechanical connection of component to the flexible substrate can be recommended for some special applications. The results also show that the properties of joints glued by UV curable non-conductive adhesive are similar or better than the properties of joints glued by isotropically electrically conductive adhesive tested in our previous research [11].

Acknowledgements

This research has been supported by the Ministry of Education, Youth and Sports of the Czech Republic under the RICE – New Technologies and Concepts for Smart Industrial Systems, project No. LO1607 and by the Student Grant Agency of the University of West Bohemia in Pilsen, grant No. SGS-2018-016 "Technology and Materials Systems in Electrical Engineering".

References

[1]   C. Mette, E. Stammen, K. Dilger, Int. J. Adhes. Adhes. 67 (2016) 49–53.

[2]   B.-S. Yim, J. Il Lee, B.H. Lee, Y.-E. Shin, J.-M. Kim, Microelectron. Reliab. 54 (2014) 2944–2950.

[3]   J.J. Licari, D.W. Swanson, Adhesive Technology for Electronic Applications - Materials, Processing, Reliability, 2., Elsevier Inc., 2011.

[4]   S. Ebnesajjad, A.H. Landrock, Adhesive Technology Handbook, 3., Elsevier Inc., 2015.

[5]   Y. Li, D. Lu, C.P. Wong, Electrical Conductive Adhesives with Nanotechnologies, Springer US, Boston, MA, 2010.

[6]   C. Sapsanis, U. Buttner, H. Omran, Y. Belmabkhout, O. Shekhah, M. Eddaoudi, K.N. Salama, in: 2016 IEEE 59th Int. Midwest Symp. Circuits Syst., IEEE, 2016, pp. 1–4.

[7]   F. Ferrando, J.-F. Zeberli, P. Clot, J.-M. Chenuz, in: 4th Int. Conf. Adhes. Join. Coat.

Technol. Electron. Manuf., n.d., pp. 205–211.

[8] C. Cheng-Li, A. Jong-Ning, L. Qing-An, L. Shi-Jie, H. Guo-Shing, in: Int. Conf. Electron. Mater. Packag., Taipei, 2008, pp. 208–211.

[9] P. Osypiuk, A. Dziedzic, W. Stęplewski, Solder. Surf. Mt. Technol. 28 (2016) 33–38.

[10] W. Steplewski, A. Dziedzic, J. Borecki, T. Serzysko, Circuit World 43 (2017) 19–26.

[11] M. Hirman, J. Navratil, F. Steiner, T. Dzugan, A. Hamacek, in: 2018 41st Int. Spring Semin. Electron. Technol., IEEE, 2018, pp. 1–6.

Tilting Behavior and Phase Formation of
Sn-Cu Composite Solder for Large Area Baseplate Solder Joints

Jessica Richter, Anna Steenmann, Benjamin Schellscheidt, Thomas Licht

Faculty of Electrical Engineering and Information Technology
Department of Microelectronics
HSD University of Applied Sciences Düsseldorf, Germany
+49 211 4351 3539   jessica.richter@hs-duesseldorf.de

Ralph Mädler

PFARR STANZTECHNIK GmbH Buttlar, Germany
ralph.maedler@pfarr.de

Abstract

This work is part of a publicly funded project called ReffiMaL (resource efficient material solutions for power electronics), which aims to reduce solder thickness significantly, while maintaining reliability. In this paper we present findings of the investigation of two types of composite solder materials, to determine the more favorable one regarding tilt, phase formation and reliability. Both preform types consist of the same amount of Sn and Cu, but in different three-dimensional arrangements. The layouts of these composite materials are designed in order to prevent the tilt of substrates during soldering process. The presented layouts consist of a high amount of pure copper in order to replace the more expensive metals Sn and Ag and simultaneously providing a lower thermal resistance of the joint compared with its monolithic solder materials opponents.

Key words: interconnection technologies; soldering; bond line tilt; composite solder; intermetallic phases (IMPs); reliability

Introduction

The state of the art technique of bonding a substrate to the baseplate in the build-up of high power IGBT modules is the usage of large area lead-free solder joints. For this purpose, preform-based solder materials are placed between the bonding partners before reflow soldering process. [1] Novel bonding techniques like "silver sintering" show promising results concerning durability and thermal properties but are used yet little due to material costs as well as disadvantages in the bonding process (e.g. extended bonding time under certain pressure) and remain an exception. [2]

A lot of research has been done in the past to understand the lifetime affecting parameters of solder-joints in order to increase the durability and thereby the reliability of the whole module stack. Besides the metallurgical side of influencing parameters (chemical composition of the solder material in the interaction with bonding process variables) [3;4;5;6], the geometric dimensioning of the solder joint is crucial for a long lifetime [1; 7; 8].

In case of baseplate solder joints it is desirable to create a nearly homogenous thickness of the so called bondline (BL) which is represented by the distance between substrate and baseplate after soldering process. In application conditions (alternating heating and cooling of the components

by dissipation of energy in the die) thermomechanical stresses are provoked and concentrate in the edges of the joining layer due to the global mismatch of the coefficients of thermal expansion (CTE) of the different components of the stack. Tilted or warped shapes of solder layers have to be prevented, otherwise the introduced stresses will fatigue the thinner sides of the tilted solder layer more rapidly, leading to premature failure. The difficulties to achieve a durable solder joint geometry can be found in the complex geometrical shape of the gap, which is formed between baseplate and substrate. Commonly used baseplates and substrates have a pre-bent geometry which is moreover changing during reflow soldering process due to the different CTEs. Solder in the liquid state is able to adapt to all these shape changes on the other hand providing little interior resistance to outer forces which can cause an undisered change in shape. To stabilize the shape of the solder layer different kinds of spacer-concepts can be found in the packaging of integrated circuits or power systems. For baseplates consisting of the metal matrix composite AlSiC, Al-wire bonds can be bonded onto the surface of the baseplate in needed spots, acting as spacers. These will not undergo a solid-liquid phase transition during soldering process and define a minimum of the so called bond line thickness (BLT) between baseplate and substrate of the stack. Baseplates consisting of

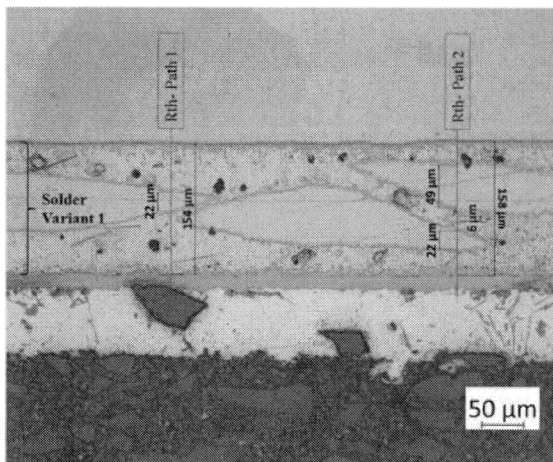

Figure 1: Micro section of a soldered composite solder Variant 1

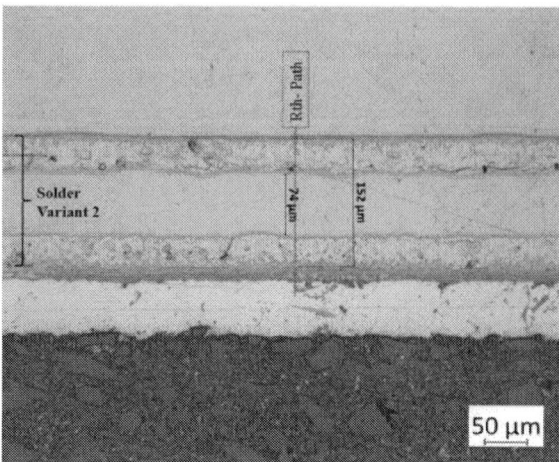

Figure 2: Micro section of a soldered composite solder Variant 2

copper can be supplied with stamped "bumps" which fulfill the same purpose [1; 9]

Another method is to place metallic meshes or high melting particles on top of the solder material before soldering. Due to their higher density, they start to sink into the liquid solder providing a minimum distance between baseplate and substrate due to their height. [10; 11; 12] Furthermore, there are solutions intending to have the spacer-elements integrated inside the solder materials before the soldering process, resulting in an increased wetting behavior. Known layouts of these preforms consist of either round particles being integrated in the solder material during casting process [13] or metallic meshes being incorporated in the solder [14].

In this paper two different layouts of preform based composite solder materials produced by roll bonding are used to join an AlSiC-Baseplate with two different kinds of substrates (Al₂O₃(DCB), AlN(AMB)) for the later assembly of a power semiconductor device. We designed the layouts of these composite materials in order to prevent tilt of substrates during soldering and thereby increasing the reliability of the joints. The presented layouts consist of a high amount of pure copper to replace the more expensive metals Sn and Ag and simultaneously provide a lower thermal resistance of the joint compared with its monolithic solder material alternatives, which is of great importance for the durability of the solder layer as well.

Experimental

Two variants of solder specimens were produced by cold roll bonding of SnAg3.5 and Cu strips (Cu-HCP) and were later die punched to rectangular preformed shapes. The overall thickness of the specimens was controlled to be about 150 μm. The first variant of the composite solders has a layout that is comparable with metal matrix composites. In the present case, the matrix is consisting of the solder material SnAg3.5 and copper particles as the second material being distributed inhomogeneously therein, having no connection to the surface sides of the preforms which are later connected to the baseplates and substrates (Figure 1). The second variant of the used composite solder specimen has a sandwich-type layout consisting of a copper core with a double sided coating of 35 μm thick layers of SnAg3.5 (Figure 2). Both layouts consist of nearly the same amount of copper.

The composite solder specimens were used in the construction of the stack by following schematic (Figure 3). Four substrates were soldered on each baseplate always having a preform of the variant 1 on the position A and D and a preform of the variant 2 on the position B and C. The corresponding substrates for position A and B were usual AlN(AMB)s with a Ni-metallization and usual Al₂O₃(DCB)s for the positions C and D. The baseplates consisted of the metal matrix composite AlSiC having a galvanized surface consisting of Ni. The stacks were reflow-soldered in a vacuum soldering furnace under N₂/H₂ atmosphere.

We varied the used soldering profiles in their peak-temperatures (ϑ_{Peak}) and holding times on peak-temperature (t_{Hold}) creating an experimental matrix according to:

$$\vartheta_{Peak} \cdot t_{Hold} = (300°C, 350°C, 400°C) \cdot \begin{pmatrix} 10min \\ 20min \\ 40min \end{pmatrix}$$

Figure 3: A soldered baseplate with marked cutting sections

Four cross section polishes per soldered specimen were made after cutting four samples out of each soldered substrate at four different places as marked in Figure 3 using a precision cutting machine.

Light optical microscopy was used to determine the BLT and to measure the intermetallic phases (IMP) size. The IMP composition was evaluated qualitatively using energy dispersive X-ray analyses (EDX).

To evaluate the bond line tilt of the soldered substrates, we analysed the difference in solder thickness between the four cut pieces. Sample 1 is the reference of every measured specimen.

Results and Discussion

The following three sections describe our findings regarding tilting behavior, thermal resistance and phase formation.

Tilting behavior

To determine the tilt of the specimen, we measure the distance between baseplate and substrate near the outer end of the bond line. To calculate the average, we determine the difference between specimen 1 and 2, 1 and 3 and 1 and 4. According to this, we can calculate the tilting in three directions. The results of the |ΔBLT| measurement are shown in Table 1.

Table 1: Average of the measured bond line tilt sorted by solder temperature and time

Temperture [°C] Time [min]	|ΔBLT| [µm]
300	**5,03**
10	2,17
20	6,01
40	6,92
350	**11,28**
10	18,17
20	9,92
40	5,75
400	**12,24**
10	10,45
20	12,59
40	13,68

The specimens that we soldered at 300°C achieved the best results over all combinations of substrate and solder.

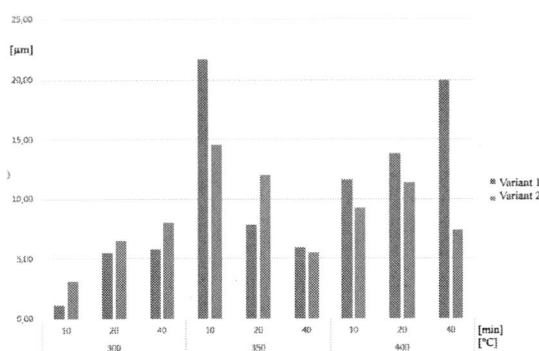

Figure 4: Comparison of the |ΔBLT| solder variants resolved after solder temperature and time

In Figure 4, the results of the tilting measurement of the two solder variants are shown. The left bars of each pair (blue) in the diagram show the composite solder variant 1, the right bars show variant 2 (orange). In general, the tilting of the substrate increases for higher soldering temperatures. One possible explanation is related to the increasing fluidity of the solder at higher temperatures. The less viscous a fluid gets, the less resistance it provides against shape changing outer forces. Furthermore, the higher the mismatch in coefficient of thermal expansion between the mating materials, and the higher the temperature change is, the higher the amount of shape change (elongation, bending etc.) will be. For the soldering profile with 300°C peak temperature and 10 min holding time the lowest tilt was measured. In average, solder variant 2 leads to less tilting compared to variant 1. Variant 1 provides better results at the lowest peak temperature of 300°C. A cautious explanation would be that the stiff middle Cu layer of variant 2 is balancing the liquid solder columns above and below it independently from their state of viscosity, whereas the inhomogeneous distributed Cu particles of solder composite variant 1 increase the viscosity of the solder for lower temperatures leading to less tilt. At elevated temperatures, the fluidity of the solder variant 1 increases and the Cu particles become more mobile, resulting in a higher tilt. This is also because the height of Cu particles and with that the minimum distance between the baseplate and substrate are distributed inhomogeneously for variant 1 whereas solder variant 2 provides this minimum distance by its homogenous thick, continuous Cu layer.

Table 1 also shows that on average every combination leads to tilting behavior under 15µm. We expect this to result in a homogeneous thermal conductivity of the system.

Phase formation

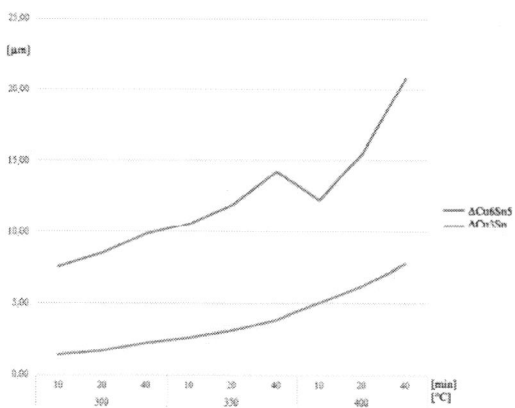

Figure 5: Cu-Sn-phase formation resolved after solder temperature and time

The results of the investigation of the phase formation shown in Figure 5 illustrate the exponential phase growth of Cu-Sn main phases. The phase growth can be described by the diffusion equation (1). Therefore a reduced time of diffusion is possible by increasing the temperature T and decreasing the thickness of the diffusion length x_D. [16]

$$x_D^2 = 4 \cdot D(T) \cdot t \qquad (3)$$

With the frequency factor D_0 for copper $0,16 \cdot 10^{-4} m^2/s$, $7,7 \cdot 10^{-4} m^2/s$ for Sn, and activation energy Q 199kJ/mol for Cu and 107kJ/mol for Sn we calculate the diffusion constant of copper and tin at 300°C with (4) [17].

$$D(T) = D_0 \cdot e^{-(Q/kT)} \qquad (4)$$

$$D_{ab}(T) = D_a(T) \cdot w_b + D_b(T) \cdot w_a \qquad (5)$$

For a solder time of 10 minutes at a solder temperature of 300°C, (5) yields a layer thickness of approximately $14,18µm$ Cu_3Sn.

The calculated result exceeds the measured result by a factor of 10. The deviation is caused by the used equation that relate to liquids, not to IMP [18].

In solder variant 1 the phases between the Cu particles can easily reach each other from opposing sides and fuse at higher temperatures or after longer soldering time. The large amount of Cu in these smaller Sn areas also leads to faster phase growth of Cu_3Sn. At 400°C the phase growth of Cu_6Sn_5 exceed 13µm after only 10 minutes soldering time. The two main phases of Cu and Sn have an influence on the thermal conductivity of the soldered system. The thermal conductivities are shown in Table 2.

Thermal resistance

The influence of tilting and phase formation on the thermal conductivity of the system is described by a simple addition of thermal resistances of the different layers. If the thermal spread is not taken into account, the thermal resistance along one virtual path through the stack can be represented by the following equation

$$R_{th} = d/(\lambda \cdot A). \qquad (1)$$

d is representing the thickness of each material layer of the stack, A is the constant spatial expansion of about $20cm^2$ in the calculated system and λ the thermal conductivity of the used material.

The thermal resistance of the solder is directly proportional to the thickness of the solder. A bond line with a nominal thickness of 150µm and a tilt of ±50µm leads to a solders contribution to the systems thermal resistance of about ±30% due to the varying thickness of the solder layer. Therefore, the cooling performance of the system would be dramatically impacted by a tilt of the bond line. The solder layers, especially between the ceramics and the base plate, contribute about 15% for the module's overall thermal resistance. The solder layer is usually based on more than 90% tin. The thermal performance of the solder joint is dominated by its tin content (thermal conductivity about: 65W/mK[1]). Other materials inside the module system have much higher thermal conductivity values. This makes the solder layer's properties one of the critical parameters for the thermal performance. A well-controlled solder thickness and low tilt between the mating surfaces is a good strategy to keep the influence of the solder layer at a minimum.

---

[1] Measured with Laser Flash Analysis according to ASTM E-1461, DIN 30905 and DIN EN 821

If we calculate the R_{th} for the solder variant 1 alongside a path (path 1 in Figure 1) where only one comparably thin Cu-particle is crossed a maximum on R_{th} can be found for the solder material. Here the BLT is about 154µm, the copper particle has a thickness of about 22µm. 5µm of Cu_6Sn_5 could be measured and 1µm of Cu_3Sn. The remaining 126µm are representing the solder material. This results in a computational R_{th} of 1.1 mK/W. Whereas the R_{th} if it is calculated alongside path 2, is resulting in its lowest value. Here the solder material has an overall height of only 59µm, copper a thickness of 49µm +6µm +22µm giving 77µm in total. Furthermore, 19µm of Cu_6Sn_5 phase could be measured and 3µm of Cu_3Sn. Exemplary, the R_{th} for the preform variant 1 is varying between 0.85 mK/W and 1.1 mK/W, based on the shown cross section polish in Figure 1.

For the preform variant 2, an R_{th} of 0.73 mK/W can be calculated alongside the path shown in Figure 2.

Compared to a monolithic homogeneous solder layer of 150µm, the thermal resistance of the soldered system is lowered by about 7% to 25% for composite variant 1 and by about 35% for composite variant 2.

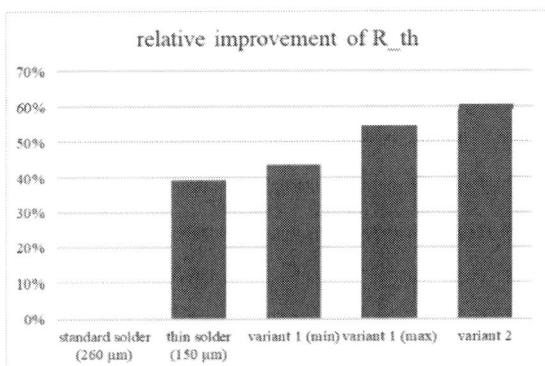

Figure 6: relative improvement of R_{th} of solders contribution to the systems thermal resistance

Figure 6 shows the relative improvement of the thermal resistance of the system in relation to a 260µm monolithic solder. The thermal conductivities of the materials are shown in Table 2. The thermal conductivity of the soldered system improved about 39% only by reducing solder thickness to 150µm. The three dimensional structure of variant 1 leads to a larger variability of R_{th} in the model.

Table 2: thermal conductivities

Material	thermal conductivity λ [W/m·K]
SnAg3,5[1]	65
Cu [15]	401
Cu_6Sn_5 [15]	34,1
Cu_3Sn [15]	70,4

Summary and Conclusion

To conclude, both composite solder variants show that they are able to prevent excessive tilt. Exemplary the thermal resistances for these materials in soldered state were determined. It could be shown that compared to their monolithic solder alternatives the included copper is helpful for an improved tilting behavior, resulting in a higher thermal conductivity and thus a better cooling performance. Minimizing the BLT variation helps to dissipate the heat created in a power electronic system in use. For the soldering profiles with a peak temperature of about 300°C the least tilt was measured, which is also a favorable process parameter concerning the phase growth kinetics.

Furthermore, the thermal conductivity of the system increases significantly by decreasing the solder thickness. For a solder thickness of 150 µm the tilting behavior could be optimized by means of Cu particles respectively the Cu core. The increased Copper content also has a positive influence on the thermal conductivity.

At solder variant 1 the phase growth in the small Sn areas between the Cu particles is increased in contrast to the particles surface regions facing towards larger Sn areas. This leads to a larger amount of Cu_3Sn in this area and therefore to a better thermal conductivity in this area. However, the used theoretical model to calculate R_{th} shows better results at solder variant 2. At higher solder temperature, these effect might have a larger impact on the thermal conductivity of the system. That could be a point at future studies.

Regarding the project's focus on resource and material efficient solutions, the two composite solder layouts show a possibility of how roll bonded material can be an effective alternative to monolithic material.

In the future, the reliability of these composite layouts has to be tested and has to be compared with their monolithic solder alternatives.

Acknowledgment

We would like to thank our other partners in project ReffiMaL DODUCO Solutions GmbH and Infineon Technologies AG. We further thank the German Federal Ministry of Education and Research (BMBF) for funding initiative MatRessource [19].

[1] Guth, K.; Mahnke P.: Improving the thermal reliability of large area solder joints in IGBT power modules. In: 4th International Conference on Integrated Power Systems. 7-9 June 2006. Naples, Italy

[2] Syed-Khaja, A. et al.: Process Optimization in Transient Liquid Phase Soldering (TLPS) for an Efficient and Economical Production of High Temperature Power electronics. In: CIPS 2016, 9th International Conference on Integrated

Power Electronics Systems, 8-10 March 2016, Nuremberg, Germany

[3] Liang, J. et al: Effects of Load and Thermal Conditions on Pb-Free solder Joint Reliability. In: Journal of ELECTRONIC MATERIALS, Vol 33., No. 12, 2014

[4] Li, Y. et al.: Root Cause Investigation of Lead-Free Solder Joint Interfacial Failures After Multiple Reflows. In: Journal of ELECTRONIC MATERIALS, Vol. 46, No. 3, 2017

[5] Yang, W. et al.: The Effect of Soldering Process Variables on the Microstructure and Mechanical Properties of Eutectic Sn-Ag/Cu Solder Joints. In: Journal of ELECTRONIC MATERIALS, Vol. 24, No. 10, 1995

[6] Sigelko, J. et al.: Effect of Cooling Rate on Microstructure and Mechanical Properties of Eutectic Sn-Ag Solder Joints with and without Intentionally Incorporated Cu6Sn5 Reinforcments. In: Journal of ELECTRONIC MATERIALS, Vol. 28, No. 11, 1999

[7] Lee, H.-T.; Huang, K.-C.: Effect of Solder-Joint Geometry on the Low-Cycle Fatigue Behavior of Sn-xAg-0.7Cu. In: Journal of ELECTRONIC MATERIALS, Vol. 45, No. 12, 2016

[8] Zimprich, P. et al.: Mechanical Size Effects in Miniaturized Lead-Free Solder Joints. In: Journal of ELECTRONIC MATERIALS, Vol. 37, No. 1, 2008

[9] Hayashi, K.; Izuta, G.: Improvement of Fatigue Life of Solder Joints by Thickness Control of Solder with Wire Bump Technique. In: ECTC conference proceedings, 2002

[10] Beerwerth, W.; Albrecht, G.: Verfahren zum Herstellen einer vorgegebenen Lotschichtstärke bei der Fertigung von Halbleiterbauelementen. Patent No.: DE 000002228703 A. Date of Patent: Jan. 10, 1974

[11] Mohl, N.: Verfahren zur Herstellung eines weitergebildeten Metallformkörpers und Verfahren zur Herstellung einer Leistungshalbleitereinrichtung mit einer Lotverbindung diesen weitergebildeten Metallformkörper verwendend. Patent No.: DE102013110812B3. Date of Patent: Oct. 09, 2014

[12] Keser, Helmut, Birr: Verfahren zum blasenfreien Verbinden eines großflächigen Halbleiter-Bauelements mit einem als Substrat dienenden Bauteils mittels Löten. Patent No. DE3442537 A1. Date of Patent: May 22, 1986

[13] Ushima, M.; Ishii, T.: Process for Producing a Solder Preform Having High-Melting Metal Particels Dispersed Therein. Patent No.: US 8,790,472 B2. Date of Patent: Jul. 29, 2014

[14] Booth, J.; Vijay, K.: Novel Technique to Reduce Substrate Tilt & Improve Bondlie Control between AlN substrate and AlSiC Baseplate in IGBT Modules. In: PCIM Europe conference proceedings, 2016

[15] Daniel, C.: „Diplomarbeit: Periodische Mikro-/Nano-Strukturierung verzinnter Materialoberflächen für elektrische Kontaktanwendungen und deren tribologisches Verhalten": April 2002

[16] D. Gupta, Hg., Diffusion Processes in Advanced Technological Materials. Berlin, Heidelberg: Springer-Verlag Berlin Heidelberg, 2005.

[17] K. Bobzin, L. Bosse, E. Lugscheider, „3.2 Transient Liquid Phase Bonding"

[18] Steenmann, A; Richter, J; Schellscheidt, B; Licht, T: TLPB imroved solder connections by on chip creation of intermetallic phase precursers. EMPC Pisa 2019

[19] https://www.matressource.de/de/

Simulation of temperature profiles in reflow ovens for soldering area array components

A. Yuile*, S. Wiese

Saarland University, Chair of Microintegration and Reliability, C 6 3 Campus, Saarbruecken, Germany

Corresponding author: adam.yuile@uni-saarland.de, +49 681 302 71827

Abstract

This paper presents computational fluid dynamics (CFD) simulations in the field of electronics manufacturing technology of area array components, namely a BGA, where the calculation of temperatures within the solder joint during a reflow soldering process are focused on. The CFD models were generated in the ANSYS Workbench and compute the flow field for the air, solder and solid structures such that the interaction between the air and solder are captured. The CFD models yielded the temporal progression of the temperature field and were furthermore capable of capturing the melting and solidification of the solder. The influence of the latent heat of the solder as a material property, enabled by the melting/solidification model, was studied in order to quantify its influence and, secondly, to verify that the latent heat was having the intended influence on the CFD model.

Key words: CFD, reflow soldering, latent heat, solidification

Introduction

Soldering failures in reflow ovens can often be attributed to inhomogeneous temperature distributions, owing to poor compatibility between the materials used [1], which are often difficult to anticipate with the fostering and influence of intermetallic compounds being difficult to determine [1], [2].

However, particularly with regards to the former, the CFD simulation approach can be used to optimise reflow soldering processes in the electronics manufacturing industry. Some of the main soldering defects frequently observed in practice include bridging, open joints, voiding, part cracking, poor wetting and skewing [3], many of which stem from not being able to maintain temperature uniformity to enable concurrent component soldering on differing positions on the electronic package.

Reflow oven profiles have been further complicated owing to two main factors. Firstly, the adoption of lead-free solder has typically enforced a smaller operating range of flow temperatures [4]. Furthermore, the required reflow temperatures for lead-free solders are typically higher than their leaded counterparts, hence increasing the risk of thermal damage [5]. Secondly, the miniaturisation of electronic packages and the increasing complexity of PCB assemblies in general has further highlighted the need to address the underlying issues [4]. However, it is hoped that CFD can contribute towards alleviating such issues by providing detailed insights into what is taking place.

In this study the local temperature profiles of solder on a BGA mounted chip on a PCB resolved through CFD are shown. A previous paper within the same research group [6] looked at the local solder temperature profiles in a reflow oven and this work intends to then extend this work by specifically studying the cool down phase of the solder and board.

CFD Models

Fluid flows are governed by various conservation equations, namely conservation of mass, momentum and energy and for the CFD approach these governing partial differential equations are approximated through algebraic relations and, using computers, are solved numerically through iterative means. This CFD approach allows for the flow field to be studied in a virtual flow laboratory and offers insights which are typically difficult to ascertain experimentally [7].

The ANSYS Workbench was used in order to generate the CFD model which was then solved iteratively in ANSYS Fluent. The final temperature field from [6] for the solid structures was used with the surrounding air initialised to atmospheric temperature, corresponding to the drawer being opened in a batch reflow oven to allow for a relatively rapid cool down.

The simulations represent a batched reflow oven for laboratory use where the reflowed components on the PCB remain static in space and do not translate within. This is in line with the SF02 reflow oven upon which the model was based, as per Figure 1. A conveyorised reflow oven is more of a classical reflow oven and these types of reflow ovens have also been analysed using CFD, for example in [8].

Figure 1: C.I.F FT02 reflow oven with drawer open

In Figure 2 a 3-D model is shown of the metallisation layers and the 5x5 BGA matrix upon which the chip is mounted onto the PCB. In Fluent these solid structures are housed within an air fluid volume that corresponded to the size of the FT02 reflow oven.

A predominantly hexahedral cut-cell mesh of approximately 300,000 cells was generated using the ANSYS Workbench and was capable of modelling the solidification of the solder. The cut-cell mesh approach may need to be revisited in the future should compatibility issues with the volume of fluid (VOF) method arise, a method which allows for simulating the interaction between fluid phases and surface tension.

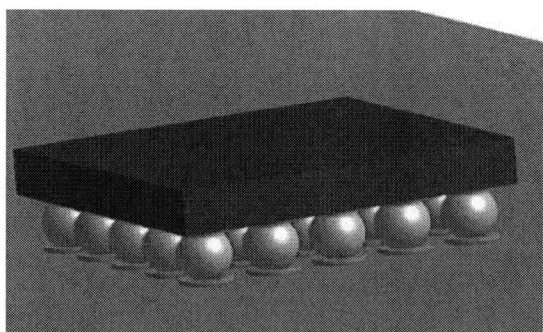

Figure 2: 3-D render of PCB, metallisation layers and chip mounted on 5x5 BGA

In Figure 3 an engineering drawing of the chip, BGA, PCB and metallisation layers from the 3-D model of Figure 2 is presented with all of the dimensions given in mm and details regarding the BGA's pitch etc. given in detail A. The metallisation thicknesses were chosen to be 35 μm on both the top and bottom sides. This represents a relatively thick

metallisation, especially with respect to the size of the solder balls, but this was deliberately chosen to minimise the aspect ratio of the cells in this area.

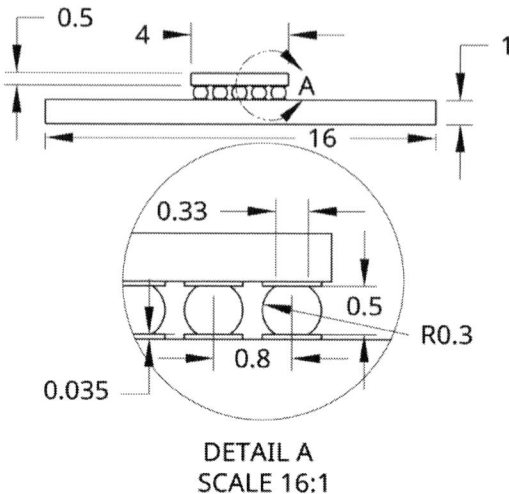

DETAIL A
SCALE 16:1

Figure 3: Engineering drawing of chip mounted on PCB through a 5x5 BGA matrix with dimensions in mm, including detailed view (DETAIL A) of BGA geometry

The reflow profile used in reflow ovens depends on the type of solder which is being reflowed. The reflow profile for lead-free solders with the FT02 oven is shown in Table 1.

Table 1: Reflow profile for SAC 3807 solder paste

Temperature 1	Duration 1	Temperature 2	Duration 2
140°C	60s	265°C	105s

This implied that the components to be soldered would first be pre-heated to 140°C for less than five minutes. Thereafter, the PCB is brought above the melting temperature of the particular solder for a maximum of one minute.

In [6] the flow field was initialised at 140°C and a radiator boundary condition used to heat the oven to Temperature 2 where effectively a steady state temperature field was achieved. This was then used as the initial condition for the cool-down of the PCB, albeit the surrounding air was re-initialised to atmospheric temperature, somewhat representing the case where the drawer opens in the batch reflow oven.

As in [6] the unsteady, pressure-coupled solver with gravity effects included and the ideal gas model used for air to effectively incorporate the buoyancy effects caused by the temperature-induced variations in density throughout the domain as well as the k-ε turbulence model were used for this analysis. A time-step of 0.01 s was chosen and found to provide numerically stable results with convergence achieved for each time step.

The thermal properties for the materials used in the domain were as follows in Table 2, where IC = chip and material properties similar to FR-4 were assigned for the PCB.

Table 2: Thermal material properties

	Cu	PCB (FR-4)	Solder	IC
ρ [kg/m³]	8978	1850	7000	2500
Cp [J/kg.K]	381	600	230	710
k [W/m.K]	387.6	0.29	63.2	100

The viscosity of the solder paste is known to be a function of the shear stress [9], which is another way of saying that the viscosity behaviour of the solder paste is non-Newtonian in nature, in addition to being a strong function of temperature and state. In this case it was assumed to be 1000 Pa.s, which is a representative viscosity for solder pastes observed at low shear rates.

In order to simulate the solidification of the solder during cool down the melting/solidification model was used in Fluent, where the solidus and liquidus temperatures of the solder were taken to be 217 and 220°C. The result of using the melting/solidification model being that the liquid fraction of the solder is modelled and can therefore be traced.

The melting/solidification model makes use of an enthalpy-porosity formulation where the liquid-solid zone (solder transiting between liquid and solid state) is treated as a porous region with porosity that is equal to the liquid fraction, with correspondingly appropriate sink terms added to the momentum and turbulence equations to account for non-fluid like behaviour [10].

When using the melting/solidification model the energy equation is also modified with the inclusion of latent heat [10], where the latent heat of the solder was specified to be 58.5 kJ/kg.

Results

Figure 4 shows the temporal progression of temperature within the domain from the initial flow field at 0.1, 1.0, 3.0 and 10.0 seconds respectively along a plane which slices through the 3rd column of the 5x5 BGA matrix, where a ghost outline of the components, as per those seen in Figure 2, is superimposed on the contours. Here one can observe that the temperature drops fairly rapidly in the PCB, with the heat lost to the surroundings, but that the solder balls, copper metallisations and chip exhibit more thermal inertia and hold on to their respective heat for longer.

By 10.0s of solution time the average temperature of the components is approximately 100°C lower than at the commencement of cool down and the temperature distribution within the components is further along the path of averaging out as they approach the surrounding atmospheric temperature.

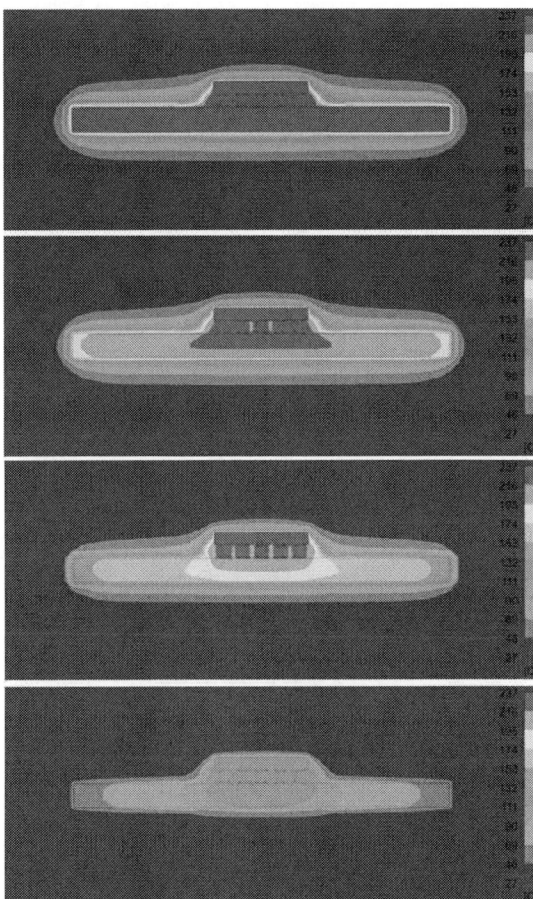

Figure 4: Temporal progression of temperature [°C], top 0.1s, 1.0s, 3.0s and 10.0s (bottom)

In Figure 5 the temporal progression of the solder liquid fraction, again at 0.1, 1.0, 3.0 and 10.0 seconds, shows the manner in which the solder solidifies. At 0.1 s the liquid fraction is uniformly 1, which is to say that the solder is entirely in its molten state.

It is observed that in this case the solder solidifies from the chip to the PCB side, first seen to initialise at 1.0 s where the solder liquid fraction first falls below 1. It's worth pointing out however that no copper traces outwith the metallisation layers have been modelled in the PCB which, given the strong disparity relative to the thermal properties of the PCB, would be bound to have a significant influence on the heat distribution.

At 3.0 s the solder balls are approximately half way through the solidification process, being as they are at a liquid fraction of 0.5 respectively and then after 10.0 s all of the solder balls are solidified, with a uniform liquid fraction of 0.

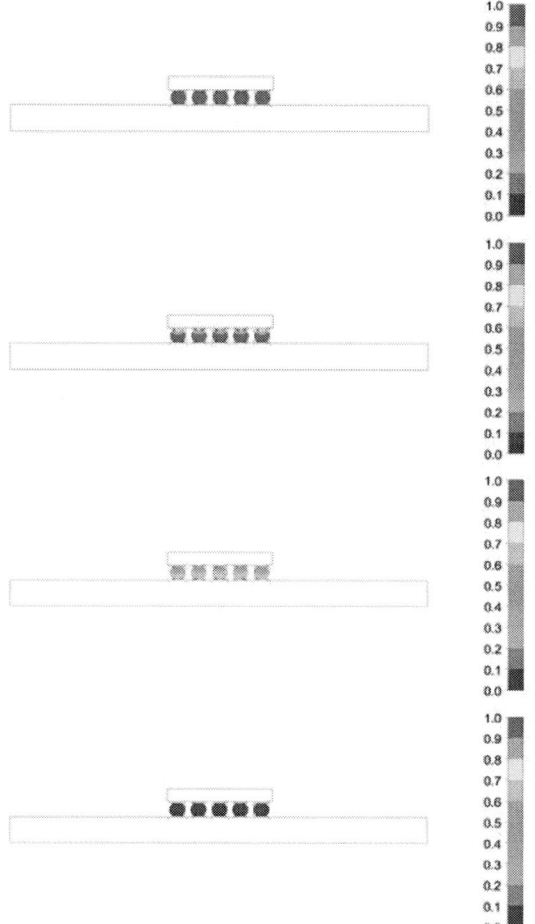

Figure 5: Temporal progression of solder liquid fraction (0 = solidified solder, 1 = molten solder), top 0.1s, 1.0s, 3.0s and 10.0s (bottom)

In order to demonstrate the influence of the solder latent heat, and indeed to verify that this parameter was having the desired influence on the transient progression of the simulations as expected, three different cases were studied. These cases were without latent heat (0 kJ/kg), the latent heat of 58.5 kJ/kg which was used for the main contour plots presented in Figure 4 and Figure 5 and finally double (117 kJ/kg) the original solder latent heat.

These three different configurations have been presented in Figure 6 for the central solder ball in the 5x5 BGA matrix (3rd row, 3rd column) where one can observe the plateau regions caused by the latent heat in the temperature traces, and likewise the relative absence thereof where no latent heat was modelled. The cooling rate appears to not be affected by the magnitude of the latent heat, but rather the latency time, as is to be expected.

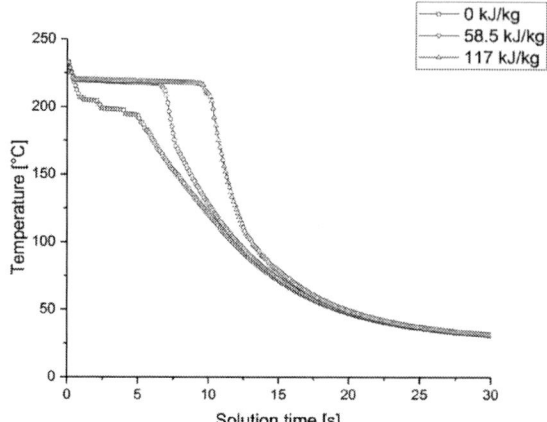

Figure 6: Time-temperature history of central solder ball for different solder latent heats, 0, 58.5 and 117 kJ/kg

It is quite typical in mechanical simulations of such area array components that the latent heat is insufficiently accounted for, but Figure 6 clearly demonstrates that the effect of the latent heat is not insignificant, as also stated for example in [1]. Simulation methods which capture the latent heat, with respect to the cooling phase in reflow soldering, should therefore be used where possible to feed representative input parameters to mechanical reliability assessments.

Conclusions

CFD has been shown to be a viable method for simulating local solder temperature profiles of area array components inside a reflow oven. It has also been demonstrated that the solidification of a BGA matrix can be captured using the CFD method, with the potential that more complicated models could ascertain the negative influence of for example large, proximal thermal masses on the reflow soldering process. Hence, it can be said that CFD could also help in making optimised design choices for both oven designs and PCB designs because of the insights offered through this virtual laboratory technique.

It was shown that modelling the latent heat carries a significant impact on the cooling after reflow soldering and methods, such as this CFD approach, should be used where possible to make improved assessments of reliability.

In terms of future work it would be interesting to investigate ways of modelling chip movement on the solder, relative to the PCB, and using VOF models to better capture the interaction between the solder and the air, in addition to increasing the complexity with copper traces and large proximal thermal masses.

References

[1] K. Dušek, A. Rudajevová, and M. Plaček, 'Influence of latent heat released from solder joints on the reflow temperature profile', *J.*

Mater. Sci. Mater. Electron., vol. 27, no. 1, pp. 543–549, 2016.

[2] A. C. Ferreira, S. F. C. F. Teixeira, R. F. Oliveira, N. J. Rodrigues, J. C. F. Teixeira, and D. Soares, 'CFD Modeling the Cooling Stage of Reflow Soldering Process', presented at the ASME 2016 International Mechanical Engineering Congress and Exposition, 2016, p. V002T02A024-V002T02A024.

[3] C.-S. Lau, M. Z. Abdullah, and F. Che Ani, 'Computational fluid dynamic and thermal analysis for BGA assembly during forced convection reflow soldering process', *Solder. Surf. Mt. Technol.*, vol. 24, no. 2, pp. 77–91, 2012.

[4] C.-S. Lau, M. Z. Abdullah, and F. Che Ani, 'Three-dimensional thermal investigations at board level in a reflow oven using thermal-coupling method', *Solder. Surf. Mt. Technol.*, vol. 24, no. 3, pp. 167–182, 2012.

[5] H. Yu and J. Kivilahti, 'CFD modelling of the flow field inside a reflow oven', *Solder. Surf. Mt. Technol.*, vol. 14, no. 1, pp. 38–44, 2002.

[6] A. Yuile and S. Wiese, 'Calculation of local solder temperature profiles in reflow ovens', in *2018 7th Electronic System-Integration Technology Conference (ESTC)*, 2018, pp. 1–5.

[7] V. Thakur, S. Mallik, and V. Vuppala, 'CFD simulation of solder paste flow and deformation behaviours during stencil printing process.', *Int. J. Recent Adv. Mech. Eng.*, 2015.

[8] I. Belov, M. Lindgren, P. Leisner, F. Bergner, and R. Bornoff, 'CFD aided reflow oven profiling for PCB preheating in a soldering process', in *Thermal, Mechanical and Multi-Physics Simulation Experiments in Microelectronics and Micro-Systems, 2007. EuroSime 2007. International Conference on*, 2007, pp. 1–8.

[9] F. V. Barbosa *et al.*, 'Rheology characterization of solder paste', in *ASME 2017 International Mechanical Engineering Congress and Exposition*, 2017, p. V007T09A035–V007T09A035.

[10] A. Fluent, 'ANSYS FLUENT theory guide: Version 13.0', *Ansys Inc Canonsburg*, 2010.

A Finite Element Modelling Approach of Test Setups for Multilayer Ceramic Capacitors

Vlad Serea, Steffen Wiese

Saarland University, 66123 Germany

+4968130271826, vlad.serea@uni-saarland.de

Abstract

This paper presents results of FEM simulations using ANSYS Mechanical on different bending fixtures for the testing of MLCC capacitors. The main purpose of this analysis is to compare the strain distribution in a three point bending fixture with the four point mechanism in order to better understand the underlying physical properties that lead to capacitor breakage. The simulations have been backed up by experimental data.

Key words: FEM, MLCC, Reliability

1. Introduction

Multilayer Ceramic Capacitors (MLCCs) represent major components in electronic systems and are widely used in several application fields. During component integration they have to sustain several external factors such as processing, soldering and board bending during assembly. Furthermore, in some applications, capacitors have to withstand extreme conditions, such as bending, large pressures or temperature variations. This is one of the main reasons of reliability tests, which have gained more interest in recent years, especially flexural tests.

The reliability of MLCCs strongly depends on the manufacturing process since it can produce flaws in the ceramic such as pores or micro cracks. These irregularities generate stress concentrations once tensile stress is applied and can lead to crack propagation through the ceramic material and (eventually) component failure [1].

When capacitors are mounted on boards, the elastic properties of the PCB change due to the increase in stiffness produced by the MLCC. This leads to an uneven strain distribution between edges of the beam and the middle. During three point bending, the maximum bending moment is in the middle right under the ram and decreases towards the supports. With a four point bending mechanism, the moment is constant between the two rams offering a much more regular stress distribution. This paper intends to offer a visualization of the strains throughout samples in different bending fixtures for a better grasp of the advantages and disadvantages of both test setups.

2. Experimental input

The FEM Model was built analog to the studied MLCCs, which is presented in Figure 1.

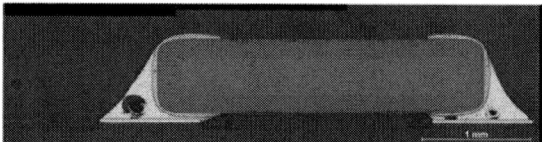

Figure 1: Cross section of the MLCC before bending, showing the alternating electrode layers throughout the ceramic body as well as the metallization and the solder shape.

The body is made of BaTiO3 which is a brittle ceramic and has alternating layers of electrodes made out of copper. The terminations are made out of nickel and the solder used is SAC305. The material properties used in the analysis according to the references can be seen in the Table 1.

Table 1: Material properties used for the FEM analysis

Material	Young's Modulus [GPa]	CTE [ppm/K]	Yield Stress [MPa]
PCB	23	14.4	-
Ceramic	90	$6.17*10^{-6}$	-
Copper	110	16.7	57
Nickel	207	13.3	148
Electrode	110	11.01	74
Solder	32	25.2	48

Previous papers in this study had the supports at a 40 mm distance according to [2] so that MLCC failure can be observed at lower displacements. The three point bending testing standard [3] uses a ram of 320mm radius to displace the boards by 2 mm when the fixed supports are 90 mm away. This was insufficient to generate any component breakage. For larger displacements, the test would be analog to the four point fixture since the ram will only push the board with its edges. This is the main reason a 90 mm distance between the supports was chosen in

450

both simulations and the ram used to displace the board in the three point fixture had a radius of 140 mm. This allows for an uniform contact between the board and ram for displacements of up to 7 mm [4]. Schematics are presented in the figures below:

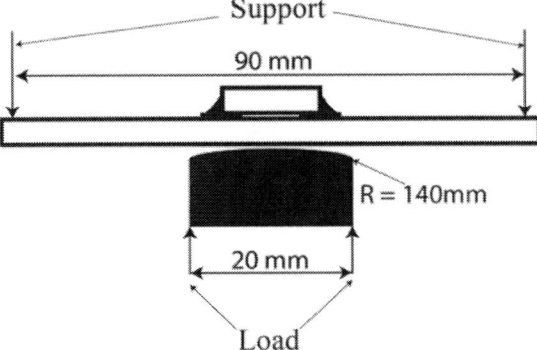

Figure 2: Schematic of the three point bending fixture and ram dimensions

For the four point fixture the board was displaced using to rolls of 3 mm radius placed 45 mm apart analog to the experiment.

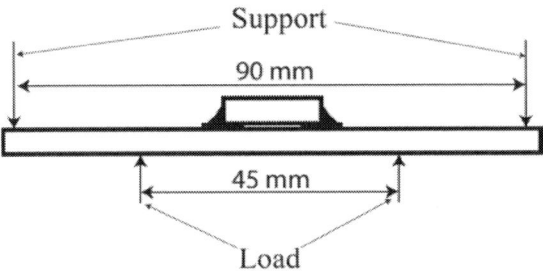

Figure 3: Schematic of the four point bending fixture with load positions

The mesh of the capacitor has been done with a body sizing of 40μm in order to increase accuracy and have a fine mesh between the plates and the body, since the plates are only 3μm thick. This body sizing was a good compromise between accuracy and computation time. The mesh in the area of the capacitor and copper pads can be seen in Figure 4:

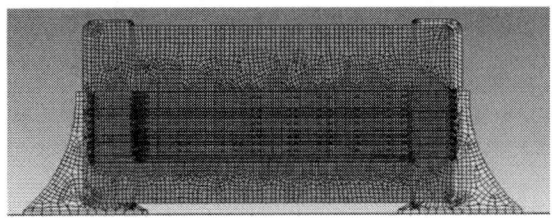

Figure 4: MLCC Mesh

Previous simulations done in this study used the PCB edges as a fixed support while the load is applied uniformly on a section of the PCB bottom. In the experiment the rams have a small contact area at the beginning that increases with the displacement. For a better comparison, circular rams have been constructed in order to displace the board.

The contacts have been defined as frictional using the Newton-Rhapson formulation and the friction coefficient between the board and supports has a value of 0.2 [5].

The displacement was 4 mm over 4 seconds in both cases. It has been broken down into more substeps for the solution to converge, so that each 1s time step was made up of 50 smaller substeps. Large deformations as well as non-linear effects are turned on.

3. Results

The strain distribution normal to the y axis, simulated for the three point bending fixture is presented in the figures below. It can be noticed that the strain underneath the capacitor is relatively small due to the increased stiffness in that region, as seen in Figure 5:

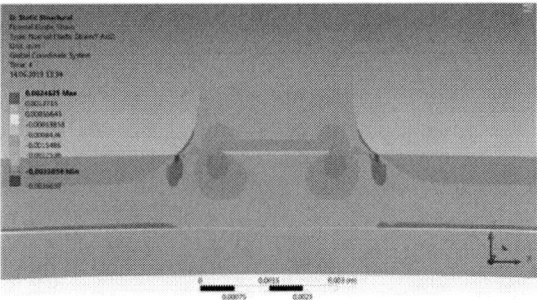

Figure 5: Front view of the elastic strain around the capacitor, normal to the Y axis with the three point bending fixture

Furthermore, with the three point fixture the strain along the PCB is also not constant, being almost zero near the supports and increasing toward the capacitor and decreasing again underneath it.

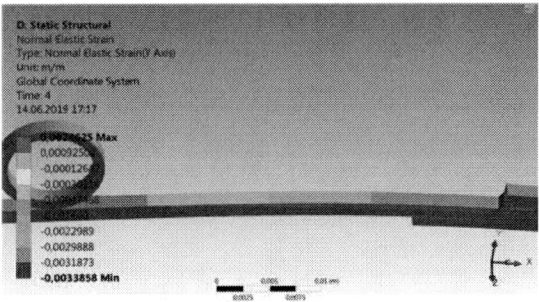

Figure 6: Side view of the elastic strain between the support (left) and capacitor (right), normal to the X axis with the three point bending fixture

The four point bending mechanism shows similar results underneath the capacitor, although the strain on the solder-PCB interface is lower, as seen in Figure 7.

451

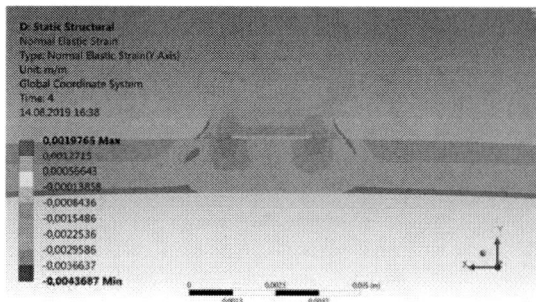

Figure 7: Front view of the elastic strain around the capacitor, normal to the Y axis with the four point bending fixture

Furthermore, in this case the strain on the upper side of the PCB is more constant in the area between the displacement rolls and the capacitor.

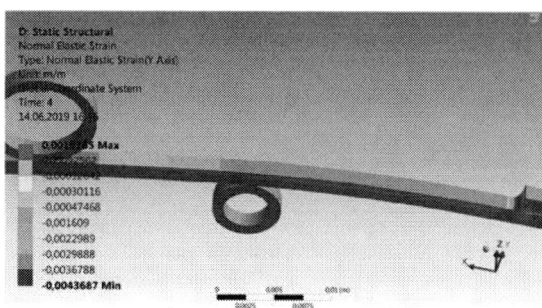

Figure 8: Side view of the elastic strain between the supports and capacitor, normal to the Y axis with the four point bending fixture

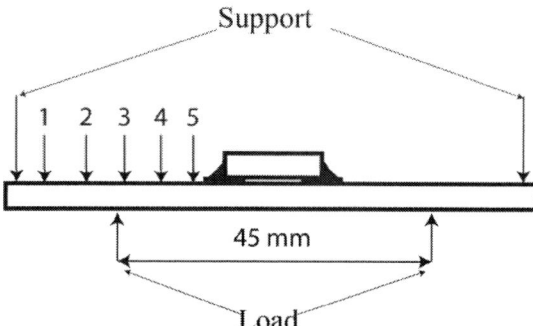

Figure 9: Position of the five probes along the board at 1 mm, 15 mm, 25 mm, 35 mm and 40 mm respectively

Probes have also been set along the topside of the PCB at specific distances between the support and the capacitor to investigate the relationship between displacement and strain. The results can be seen in Figure 9 and 10.

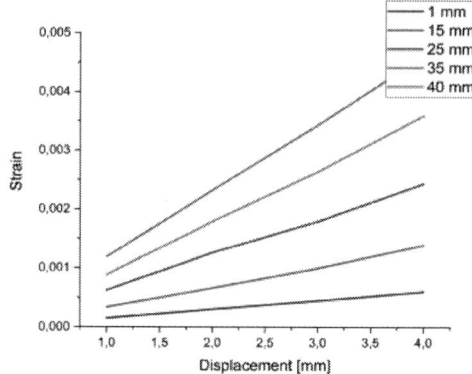

Figure 10: Displacement vs strain of three point fixture

We notice that the strain increases linearly with displacement for both bending mechanisms. As excpected, the values near the supports are almost 0 and increase when getting closer to the capacitor in the three point fixture. With the four point mechanisms, all values after the displacement roll are similar.

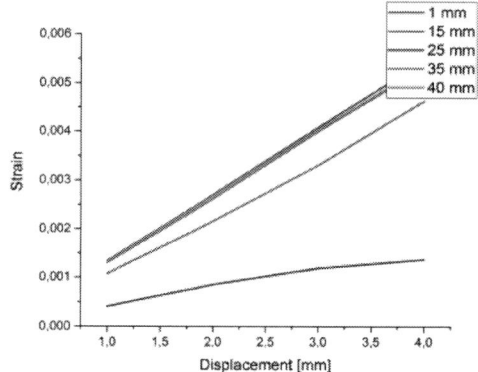

Figure 11: Displacement vs strain of four point fixture

4. Conclusions

The simulations done in this study show different strain distributions when using three and four point bending fixtures, respectively. Due to the soldered capacitor, the stiffness of the board increases in that region which minimizes the strains. It has also been shown that the strain on the board is more uniform when using the four point ram.

References

[1] Al Ahmar, Joseph, Erik Wiss and Steffen Wiese. "A crack propagation analysis of multilayer ceramic capacitors." 7th Electronic System-Integration Technology Conference (ESTC), Dresden, Germany, September 18-21, pp. 1-4, 2018.

[2] Al Ahmar, Joseph and Steffen Wiese. "Flex Cracking of Multilayer Ceramic Capacitors:

Experiments on Fracture Propagation." 2015 European Microelectronics Packaging Conference (EMPC), Friedrichshafen, Germany, September 14-16, pp. 1-4, 2015.

[3] Industries Association.AEC-Q200-005 REV A, „Board Flex Test", June 1, 2010

[4] EIA Proposed Draft PN-2271, Rev. G, 1993, Electronic Industries Association.

[5] Dumitru, Georgiana Ionela, Georgiana Chisiu and Andrei Tudor, "Trybological Characteristics of Printed Circuit Boards Determinate Through Micro-scratch Test", UPB Scientific Bulletin, Series D, Vol. 76, Iss. 4, 2014

Manufacturing of bio-compatible & degradable devices using inkjet technology for transient electronics

Kalyan Y. Mitra[1], Andreas Willert[1], Adam Sossalla[2], Michael Hoffmann[3] and Ralf Zichner[1]

[1]Fraunhofer Institute for Electronic Nano Systems ENAS, Printed Functionalities, Chemnitz, Germany

[2]Fraunhofer Institute for Biomedical Engineering, Sulzbach/Saar, Germany

[3]Fraunhofer Institute for Organic Electronics, Electron Beam and Plasma Technology, Dresden, Germany

Phone: +4937145001458, and E-mail: kalyanyoti.mitra@enas.fraunhofer.de

Abstract

The inkjet technology due to its μm-scale accuracy, up-scalability, efficient processing and industrial relevance, is widely accepted as a smart digital tool for manufacturing μ-electronic devices on rigid and flexible substrates. Within this research, the inkjet technology is implemented to manufacture bio-compatible and bio-degradable conductive electrodes, contacts for electrical signal transmission/stimulation and development of the multi-layered devices e.g. capacitors and thin-film-transistors (TFTs) along with suitable barrier layer characteristics, for implants in the medical applications. To accomplish such competitive goals, it is essential to select the most optimal functional materials, which would firstly fulfill the electrical needs of the device and secondly support processibility using the inkjet technology. The functional materials such as poly(3,4-ethylenedioxythiophene)-poly(styrenesulfonate), natural semiconductors and shellac are here utilized in form of commercially available and self-formulated inks, that are addressed carefully to deposit fundamental layers for printing capacitors and TFTs. The focus of this work is to decide on solution-processable device architectures e.g. BGBC TFTs, followed by the development and optimization of the deposition parameters for the specific materials tuned to defined layer thicknesses and concentrations. This supports in acquiring the desired electrical characteristics of conductivity, capacitance, charge transportation etc., and thus the device performance. The results show that the manufactured devices are achieved successfully on the bio-degradable substrates, processed entirely under 60 °C and ambient conditions. The electrical characteristics of the devices show direct dependency to the physical dimensions of the printed features, by exhibiting certain performance merits i.e. $210 \pm 50 \ pF/mm^2$ and $10 \ \mu A$ channel current.

Key words: Inkjet technology, bio-compatible, bio-degradable, conductor, capacitor and TFT

Introduction

The field of transient electronic is extremely innovative, where the electronic components are utilized for a defined electrical purpose over a certain time interval and then after it disintegrates within the operating environment. The most important goal is to know the defined device performance and their longevity, along with their physical degradability and compatibility, to the biological environment. The main motivation of this research work is to compliment this objective, by implementing a digital printing technology like inkjet for the deposition of the bio-degradable and -compatible (bd-bc) materials, to develop the functional layers in form of conductors, dielectrics, semiconductors and validate the fabricated passive and active devices.

There has been an extensive ongoing research regarding this topic during the past 5 years, which is recognized by the steadily emerging numerous scientific publications. The focus within these publications has been to search the suitable materials along their electronic and processing compatibilities for fabricating the individual components and devices. The development of these components and devices facilitate implementation into various bio-medical implants and analytical applications. Among the major publications, materials e.g. poly(lactic acid), poly(propylene) based synthetic paper, cellulose nanofibril, polycaprolactone, Ecoflex® foils from BASF, poly(glycolic acid), poly(L-lactide-co-glycolide), shellac, gelatin, caramelized glucose, parylene, poly(2-hydroxyethylmethacrylate), silk fibroin, polyvinyl alcohol, poly(1,8-octanediol-co-citrate) etc. are widely used as substrates; along with shellac, chicken egg albumen, deoxyribonucleic acid, nucleobases, silicon dioxide, guanine, adenine, magnesium oxide, cellulose nanofibril etc. as the dielectrics; gold, poly(3,4-ethylenedioxythiophene) poly(styrenesulfonate) PEDOT:PSS, magnesium, zinc etc. as conductors; and lastly zinc oxide, indigo, ß-carotene, parylene diimide, Indanthrene yellow G,

Indanthrene brilliant orange RF etc. as the bio-degradable semiconductors. These materials are typically deposited using conventional techniques to develop the electronically functional layers and devices. Some of these common techniques are the chemical vapor deposition, spin coating, drop casting, doctor blading, thermal evaporation, sublimation, electron beam, sputtering etc. The manufactured devices in broad terms have a limited scope for scalability and versatility, due to the deposition constraints. Some of the manufactured components and applications reported in literature are the resistors, capacitors, diodes, thin-film-transistors, sensors, actuators, antennas, inter-connects, inverters, volatile memory etc. over the fields of biological implants, chemical sensing and analysis. On the other side, these mentioned deposition techniques generate enormous material wastage because their patterning basis is subtractive. In contrast, the inkjet printing technology is digitally controlled (additive process) and endorses a precise deposition of the inks over various substrates and geometrical scales. Therefore, the inkjet technology is considered as a high potential manufacturing tool for the bd-bc absed μ-electronics. One of the other drawbacks, where the inkjet technology marks a positive statement, is the development of the entire electronic components and devices on totally independent bd-bc substrate platforms. [1] – [14]

In constrast to the state-of-the-art literature, here in this research, the authors attempt to validate the inkjet printing technology (up-scalable and industrial process) as a suitable tool to manufacture electronically functional layers and devices using bd-bc materials in sequential steps. The other objective is to evaluate the electronic properties and their suitability, towards the bio-medical implants for a single-shot response application. And, thus the main activites in this work include the demonstration of suitable bio-comptible materials e.g. PEDOT:PSS and to evaluate its electrical properties with respect to the pattern geometries i.e. layer thickness, pattern width & length and the deposition parameters. Once evaluated, then the suitable patterns can be designed for developing e.g. capacitors. The natural bd-bc material shellac was investigated, for which suitable inkjettable ink were formulated, for the development of the capacitors. The degradability of the shellac is explained by the works supported by Schaeffer et al. and Barnes et al. [13] - [16] Next to this, the validation of the bd-bc semiconductors is also envisaged using the inkjet printing technology. The merit of this research is set to develop a complete process chain i.e. utilized to print the capacitors and TFTs devices using the solution-based ink system and on the other hand accomplish the curing within 60 °C, which definitely widen the future scope due to the thermal compatibility with the various bd-bc substrates.

Methods and materials

For realizing the printed devices and their elemental components, a laboratory (lab) relevant inkjet printer DMP-2831 from Fujifilm Dimatix Inc. along with the DMC printheads (10 pL drop volume) were utilized. The inkjet system is capable to deposit printable inks in μm-scale accuracy and printing resolution ranging from 100 to 5080 dpi. The focus of this research is to investigate on the implementation of bio-degradable and -compatible materials (incl. substrates) that could be utilized for developing the electrical components e.g. conductors, dielectrics, semiconductors and devices e.g. capacitors & TFTs. Therefore, short-listing was performed to choose the correct materials for suitable applications, for the development of electronics (shown in Table 1). The Table 1 describes various functional material - inks and substrates, denoted in the category of reference materials denoted by "R" and investigation materials denoted by "In". The state of the art for the flexible printed electronics include implementation of PEN film, which is mechanically stable over high process temperatures (to accomplish curing, sintering and annealing). Here, for the development of reference electronic components and devices, PEN is used as well. The PEN exhibits T_g of about 180 °C. On the otherhand, bio-degradable and -compatible (bd-bc) substrates e.g. PLA & ORMOCER® are sensitive to high temperatures and thus are restricted to maximum 60°C. An exposure of higher temperature to these substrates lead to mechanical instability and physical/chemical degradation. In particular, the bd-bc substrate ORMOCER® is of high interest due to its chemical composition and degradation properties which are suitable for surgical implanatations. The Fraunhofer ISC manufactures the substrate in three steps: 1) dissolving the combinations of resins in appropriate solvents, 2) coating of the resin solution over glass and finally 3) pressing the cured resin film against with flexible PET substrate, to bring the film to a planar state with thickness between 80 - 125 μm.

Table 1: Shows list of functional materials used for the realization of the μ-electronics.

Material	Provider & details	Function
Polyethylen enaphthalate - PEN	125 μm thick Teonex® Q65HA, Teijin film Ltd.	"R" substrate
Poly(lactic acid) - PLA	50 μm thick Natavia NTSS50 film	"In" substrate
Degradable resins combined	80 - 125 μm thick, bioEXXVIII film from ORMOCER®	"In" substrate
Silver - Ag	Nano-particle based SI-J20X Ag ink from Agfa	"R" conductive ink
Magnesium - Mg	Commercial tape	"In" conductor

Poly(3,4-ethylenedioxythiophene)-poly(styrene sulfonate) - PEDOT:PSS	Clevious F-HC solar from Heraeus	"In" conductive ink
Cross-linked Polyvinylphenol - cPVP	Sigma Aldrich	"R" dielectric ink
Shellac	Sigma Aldrich	"In" dielectric ink
ß-carotene & Indigo	Commercially available	"In" semi-conductor ink

Next to this, "R" material inks were used for the development of the conductive electrodes and dielectrics. Here, two Ag nano-particles based inks were utilized for the fabrication of Gate (G), Source (S) and Drain (D) electrodes and cPVP was used as the dielectric for the development of the state of the inkjet printed TFTs. [17] - [18] In contrast, PEDOT:PSS ink was used as bio-compatible material to develop capacitors along with bd-bc shellac dielectric. Additionally, Mg based tape was also tested for the development of current conducting tracks. The PEDOT:PSS ink was purchased commercially from Heraeus. The shellac dielectric ink instead was formulated specially to comply with the inkjet printing process and to acquire the optimal electrical performance with respect to the physical layer geometries. Similarly, the bd-bc semiconductors like ß-carotene and Indigo were also purchased commercially and inks were formulated specially dedicated to the inkjet printing process. The lab based self-ink formulation of the bd-bc dielectric and semiconductors were done, due to the lack of current product specific commercialization. For each ink, a dedicated jetting waveform was optimized to achieve the best printing quality. Furthermore, sets of suitable digital pattern layouts were created to perform the investigations related to the development of conductive tracks shown in Figure 1 (a), capacitors in Figure 1 (b) and TFTs in Figure 1 (c). Figure 1 (a) shows group of patterns that represent conductive tracks with different track widths and lengths. As indicated in the image, the track width ranges from 0.5 – 2.5 mm with step of 0.5 mm, whereas the track length was varied from 5 - 20 mm with step of 5 mm. The focus of the investigation was to evaluate the function of electrical resistance vs. printed geometries. For keeping consistency, the PEDOT:PSS ink was deposited using only 15 µm drop space (1693 dpi) with 2 layers, along with constant substrate temperature of 30 °C. After the printing process, the patterns were cured at 60 °C for 30 mins in an oven.

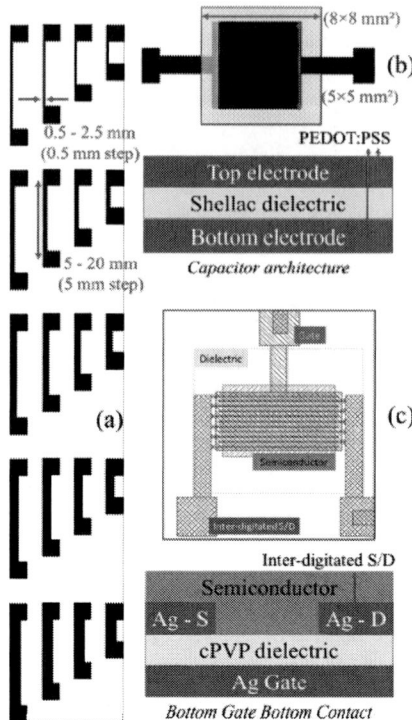

Figure 1: Images describing the device architecture & digital layouts with defined geometries for printing the (a) conductive tracks, (b) capacitors and (c) TFTs.

To execute the investigations related to the development of the capacitors, a basic conductive-insulator-conductive architecture was implemented, but with the exception of the electrodes printed using the PEDOT:PSS i.e "In" conductive ink. The dimension of the top and bottom electrode was kept identical, with the overlapping active area of approx. 5 × 5 mm². The dimension of the dielectric layer was modified to 8 × 8 mm², to avoid any interruption between the electrodes (shown in Figure 1 b). This finally leads to the fabrication of single capacitors. Next to this, the reliability of the inkjet printing process was also evaluated by replicated five devices in a row with same geometries. The parameters for depositing the PEDOT:PSS ink was kept exactly the same as it was mentioned for developing the conductive tracks. For fabricating the capacitors, the shellac "In" dielectric ink was utilized. Here, the investigation focus was put forward to optimize the dielectric thickness and morphology, obtained by changing number of layer 1 & 2 along with 5 µm & 10 µm drop spaces and substrate temperatures of 30 °C & 60 °C, respectively. Lastly, the evaluation of a bd-bc semiconductors was investigated by using state of the art device stack based on Bottom Gate Bottom Contact (BGBC) TFT stack architecture on PEN "R" substrate. [17] - [18] The architecture and material stack is shown in Figure 1 (c). A total of 60 TFTs were designed in a single array having channel width/length (W/L) ratios of 15 (0.75 mm W/ 50 µm L), 30 (1.5 mm W/ 50 µm L), 120 (6 mm W/ 50 µm

L), 180 (9 mm W/ 50 μm L), 240 (12 mm W/ 50 μm L) and 360 (18 mm W/ 50 μm L), respectively. As indicated, the design of the TFTs are based on the inter-digitated S/D electrodes, where the L is kept constant and W is varied from 0.75 - 18 mm, by multiplying S/D and channels. The semiconductors ß-carotene and Indigo were deposited such that it covers the area constituting of the interface of S/D and cPVP dielectric. Here, the deposition of the G & S/D electrodes were accomplished using Ag ink from Agfa with 25 μm (1016 dpi) & 23 μm (1104 dpi) drop spaces with 35 °C substrate temperature. The printed Ag electrodes were afterward sintered at 130 °C for 30 mins in oven. In contrast, the cPVP dielectric was self-formulated in lab and optimized for the inkjet printing process. The cPVP dielectric was deposited with 2 layers and 20 μm drop space (1270 dpi) with 30 °C substrate temperature. In this section, the focus is primary shifted to the validation of the semiconductors that could support device functionality and the deposition technology.

Results and discussions

In Figure 2 (a), a photograph is shown that represents the inkjet printed PEDOT:PSS electrodes. The printed electrode layers were found to be relatively smooth, when printed on either of the substrates. The ink's spreading over all the substrates was found to be moderate, leading to a tolerance <10 %. When the relationship between the pattern geometry to the electrical resistance of the conductive track is considered, then it can be said that the dependency is direct. In Figure 2 (b), the relation between the pattern geometry and electrical performance of the conductive track is shown.

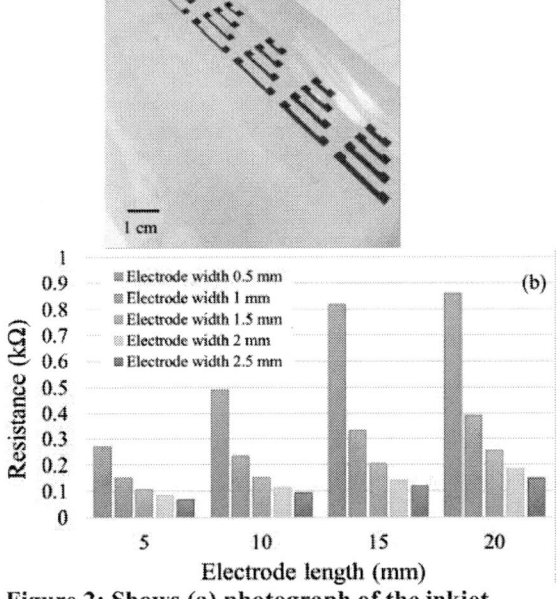

Figure 2: Shows (a) photograph of the inkjet printed PEDOT:PSS layers on PEN containing

different geometries and (b) graph representing their electrical resistances.

Additionally, the sheet resistance of the printed PEDOT:PSS layers were measured and found to be 35 ± 5 Ω/□. As the polymer-based conductive electrodes demonstrate the desired electrical characteristics, they can potentially be used as conductors in the μ-electronics dedicated to medical applications. Similar patterns were printed on PLA and Ormocere® substrates, to develop bio-compatible conductive tracks for implementing the developed conductors with solid-state or printed devices which are based on the multi-layer stack. A Mg tape was used for developing the conductive electrode tracks. At the present commercial stage, Mg cannot be considered as ready to be processed ink material for the inkjet printing technology. It is know that the degradation of Mg is much faster than other materials, therefore an another idea will be to generate a double conductive layer system, where the PEDOT:PSS is used as the protection barrier for the bottom Mg electrode. Such an implementation, where the PEDOT:PSS is used as a protection layer on top of the Mg tape, and in contrast where the PEDOT:PSS is utilized solely as an independent layer are shown in Figure 3 (a) & (b) respectively. The photograph simply demonstrates the developed electrodes entirely on the Ormocere® substrate.

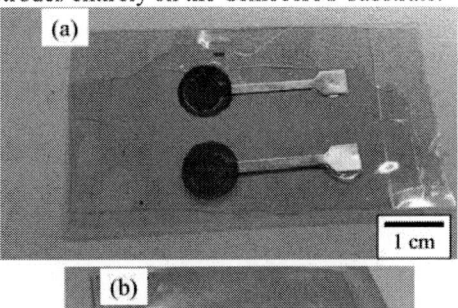

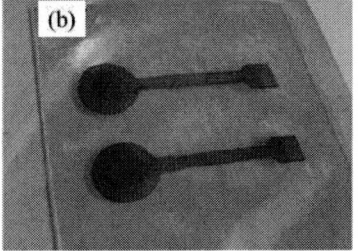

Figure 3: Photograph of printed PEDOT:PSS layer on bd-bc Ormocere® substrate for the implementation as (a) barrier layer for the Mg electrode or (b) as an independent conductor.

Next set of investigations were done on the development of capacitors. As indicated in the experimental section, to develop devices that are bd-bc, PEDOT:PSS was used for depositing the top and bottom electrodes and shellac as the dielectric material. The shellac ink formulation was adapted (concentration & solvent) to achieve the most optimal jetting properties. The best print quality was obtained at 7 wt. % shellac concentration and the ink formulated in ethylene glycol solvent. Although, the jetting of the ink was found favourable, the ink exhibited restriction towards spreading and drying. The generic problems e.g. the layer discontinuity with high irregularity can be seen from the microscopic images in Figure 4 (a) - (d), where the ink deposition volume was increased with differing the no. of layers from 1 - 2 (10 µm drop space) with 30 °C and 60 °C substrate temperatures.

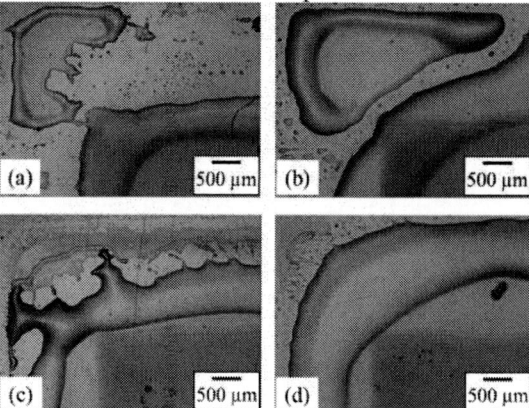

Figure 4: Microscopic images of the shellac dielectric layers printed on the PEDOT:PSS electrodes by varying the no. of layers & substrate temperatures (a) 1 layer & 60 °C, (b) 2 layers & 60 °C, (c) 1 layer & 30 °C and (d) 2 layers & 30 °C.

The microscopic images show that the best comprise is found when the dielectric is deposited with 2 layers and the substrate temperature is set at 30 °C. This phenomenon indicates that a slower drying process of the ink layer helps in the counter-acting the ink's high surface tension. The dielectric layer based on shellac was post-processed in two steps: 1) drying at 30 °C for 2 hrs until the solvent evaporate totally and 2) final curing at 60 °C for 2 hrs. The layer thickness after the entire curing process was measured to be 2.5 ± 0.5 µm (2 layers & 10 µm drop space). This 0.5 µm tolerance in the layer topology can further be explained directly, while the layer under-goes the phenomenon of coffee ring effect (CRE), spcially denoted by the occurance of thicker boundaries at the outer edge of the printed pattern. The CRE was found to be intensive, when more ink material was dosed into the confined area of the printed layer e.g. using 2 layers & 5 µm drop space. On the other hand, the dielectric layer obtained from 5 µm drop space with 1 & 2 layers was found to possess several challenges. One of them was to obtain a fully dried and cured layer, which is interpreted due to the high ink deposition volume and slow curing rate, triggered mainly by the low temperature and longer duration. Even if, the dielectric was cured completely, the measured layer thickness was found to be 4.5 µm (5 µm drop space & 1 layer), which is extremely high and not suitable for developing any kind of electronic passive or active devices.

The electrical resistance of the printed shellac dielectric (2 layers & 10 µm drop space) was measured across the electrodes and was found ~500 MΩ. This high resistance generate as a result of the thick dielectric layer. Next to this, the capacitance of the printed devices was measured and found to be 210 ± 50 pF/mm², for the dielectric processed by the mentioned deposition parameters. The measured capacitance and layer thickness were considered to re-calcultae the dielectric constant, which was found to be 3. Additionally, the fabrication reliability/yield of the printed capacitors vs. the layer thickness was also evaluated. The outcome concerning the fabrication yield is shown in Figure 5 (a), whereas a photograph of the inkjet printed capacitors using bio-degradable and -compatible materials is shown in Figure 5 (b).

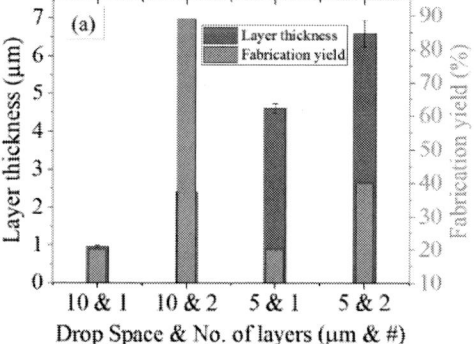

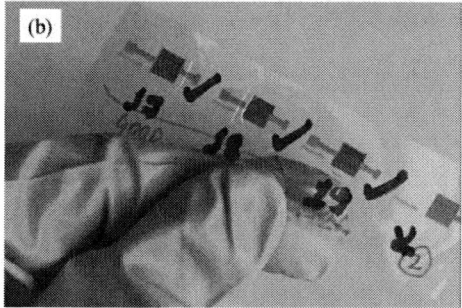

Figure 5: Shows (a) the dependency of the measured printed shellac thickness with the fabrication yield for the varied print parameters i.e. drop spaces & no. of layers and (b) a photograph containing the capacitors developed using the optimal deposition parameters.

The statistics for the fabrication yield was acquired using five devices for different dielectric thickness (acquired by different drop spaces & no. of layers). Firstly, it can be seen that the fabrication yield for the printed capacitors are not in conjunction with the layer thickness. It cannot be confirmed that as the shellac layer thickness increases, the fabrication yield for the printed capacitors also increase. In constrast, it is summarized that as the dielectric thickness increases more than 2.5 µm (i.e. with 5 µm drop space with 1 & 2 layers), then the dielectric is partially cured, leading to the electrical short-circuits and other device failures. Therefore, the fabrication yield is limited to the range 30 - 40 %. On the contrary, as the dielectric thickness is reduced below 2 μm (10 μm drop space & 1 layer), although the dielectric is entirely cured, but there occur some problems such as pin-holes and layer toplogical defects that lead to device failures. Therefore, the most reliable deposition parameter is 10 μm drop space with 2 layers, resulting into a fabrication yield of ~90 %. The obtained results regarding capacitors can be considered as very good, when the novelty of the research work is focused entirely on the exploitation of natural materials (inkjettable) utilized for the bd-bc electronics.

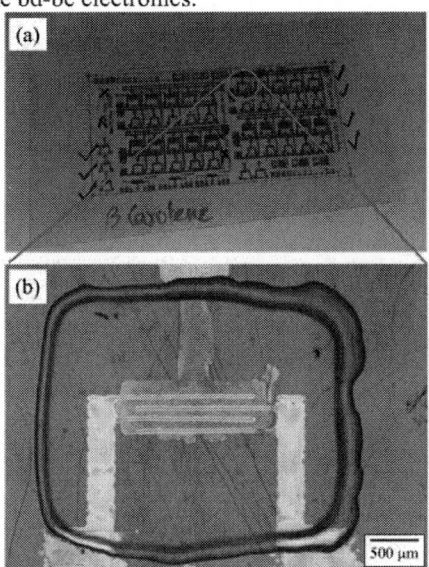

Figure 6: Illustrates (a) photograph of a printed TFTs array developed on PEN foil using standard materials and (b) a microscopic image of TFT without semiconductor.

Once the deposition of the fundamental layer stack for developing the capacitors were acquired, the main focus was shifted towards developing TFTs. For realizing the printing of TFTs, the very basic BGBC device architecture was selected. [17] – [18] The devices were developed with the deposition of Ag based G & inter-digitated S/D electrodes and PVP dielectric. Finally, the state-of-the-art semiconductor (SC) was here replaced by the deposition of natural indigo & ß-carotene based SCs. Appropriate ink formulations were developed using suitable solvents and concentrations for favourable jetting behavior for the SCs, through the printhead. A microscopic image of the printed TFT with standard TFT stack (Ag G, PVP dielectric and Ag S/D) without the semiconductor can be seen in Figure 6 (b). The exemplified TFT contains W/L ratio (18 mm/50 μm) of 360. Whereas, the photograph of an entire TFT array can be seen in Figure 6 (a). Here, the size of the TFTs are differed from 15 (0.75 mm/50 μm), 30 (1.5 mm/50 μm), 120 (6 mm/50 μm), 180 (9 mm/50 μm), 240 (12 mm/50 μm) and finally to 360 (18 mm/50 μm), over different rows and columns. The objective of the

research was to validate the functionality of the natural bd-bc SCs, that could potentially be used for fabricating the bd-bc TFTs.

For both the bd-bc SCs, the deposition process was conducted successfully on the standard TFT stacks. The entire fabrication of the standard TFT stacks (without SC) were accomplished with an over all post-treatment temperature of 150 °C with 2 hrs and thus PEN substrate was utilized. But for conducting the investigations related to the validation of the natural bd-bc SCs, the final curing temperature was restricted to 60 °C for 30 mins. Once the TFTs were finalized with the deposition of all the functional layers, they were electrically characterized. The TFTs were measured for their output characteristic curves to evaluate the correlation between the current vs. voltage, in relation to the operating conditions. The Figure 7 (a) & (b) shows the graphs corresponding to the TFT output characteristics when the TFT's W/L ratio is exemplified for 360.

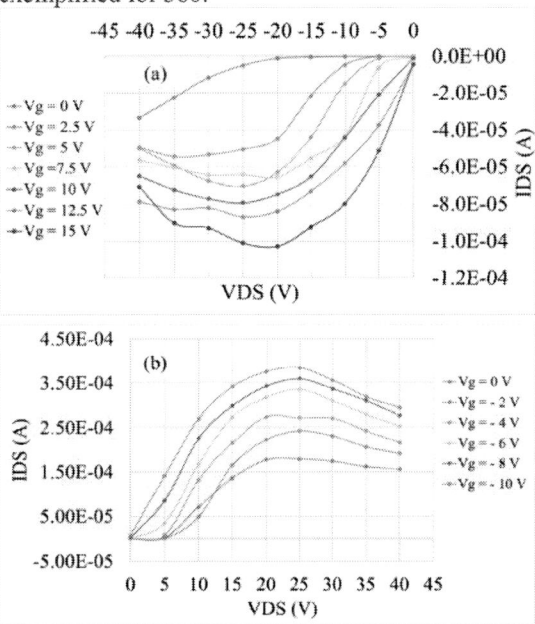

Figure 7: Show the output characteristic curves of the printed devices with W/L ratio of 360 developed using standard TFT stack along with (a) ß-carotene and (b) indigo based bd-bc SCs.

During the phase of the electrical characterization, V_{GS} was varied from 0 V to 15 V and V_{DS} was varied between 0 V to 45 V. It can be seen from both the graphs that the two SCs differ from each other drastically, the ß-carotene SC show p-type characteristics, whereas the indigo shows n-type characteristics. In both the cases, there is a linear relation between the applied voltage and obtained I_{SD}. In contrast to this, a definite channel I_{SD} saturation was not completely seen for the ß-carotene SC, while for the indigo SC a noticeable saturation in I_{SD} can be seen >25 V. The maximum driving I_{SD} in case of the TFTs with ß-carotene SC was found to range between 50 μA to 80 μA, where in case of indigo the I_{SD} ranges up to about 0.3 mA. These TFT output characteristics are considered as comparable to the literature and the basic SC functionality is hence validated. Although, it is also worth to say that the performance of the TFTs are not entirely ideal, but this could be improved by replacing the Ag based G and S/D electrodes by the ones from PEDOT:PSS.

Conclusion

It can finally be summarized that the fabrication of the electronic components using bd-bc materials are possible using the inkjet printing technology. The validation of such objective is challenging but is feasible by the right choice of materials and their stacking. The results from this investigation show that conductors, capacitors and TFTs can be developed with restricted electrical performances e.g. 35 ± 5 $\Omega/\square$ sheet resistance, capacitance of 210 ± 50 pF/mm^2 and driving current of 80 µA to 0.3 mA with linear and saturation properties for the TFTs, all within the post-process temperature of 60 °C. This leads to the future scope for investigation regarding the development of special TFTs with the implementation of all bd-bc materials including the substrate, so that an entire system is developed for the medical implantations, aiming at fabrication of transient electronics with predetermined functional durability.

Acknowledgements

The authors would like to present their humble gratitude to the internal Fraunhofer Gesellschaft (FhG) funding source for granting the project MAVO "bioElektron" (2016 - 2019) and the active participation of consortium partners FhG Institute for Organic Electronics, Electron Beam and Plasma Technology (FEP), FhG Institute for Biomedical Engineering (IBMT), FhG Institute for Silicate Research (ISC), and Fraunhofer Project Group Materials Recycling and Resource Strategies (IWKS).

References

[1] Y. H. Jung, T.-H. Chang, H. Zhang, C. Yao, Q. Zheng, V. W. Yang, H. Mi, M. Kim, S. J. Cho, D.-W. Park, H. Jiang, J. Lee, Y. Qiu, W. Zhou, Z. Cai, S. Gong, Z. Ma, "High-performance green flexible electronics based on biodegradable cellulose nanofibril paper", Nature Communications, Vol. 6, No. 7170, pp. 1-11, May, 2015.

[2] S.-W. Hwang, C. H. Lee, H. Cheng, J.-W. Jeong, S.-K. Kang, J.-H. Kim, J. Shin, J. Yang, Z. Liu, G. A. Ameer, Y. Huang, J. A. Rogers, "Biodegradable Elastomers and Silicon Nanomembranes/Nanoribbons for Stretchable, Transient Electronics, and Biosensors", Nano Letters, Vol. 15, pp. 2801-2808, February, 2015.

[3] M. Irimia-Vladu, P. A. Troshi, M. Reisinger, L. Shmygleva, Y. Kanbur, G. Schwabegger, M. Bodea, R. Schwödiauer, A. Mumyatov, J. W. Fergus, V. F. Razumov, H. Sitter, N. S. Sariciftci, S. Bauer, "Biocompatible and Biodegradable Materials for Organic Field-Effect Transistors", Advanced Functional Materials, Vol. 20, pp. 4069-4076, 2010.

[4] S.-W. Hwang, G. Park, H. Cheng, J.-K. Song, S.-K. Kang, L. Yin, J.-H. Kim, F. G. Omenetto, Y. Huang, K.-M. Lee, J. A. Rogers, "Materials for High-Performance Biodegradable Semiconductor Devices", Vol. 26, pp. 1992-2000, 2014.

[5] M. Irimia-Vladu, N. S. Sariciftci, S. Bauer, "Exotic materials for bio-organic electronics", Journal of Materials Chemistry, Vol. 21, pp. 1350-1361, October, 2011.

[6] K. Bajer, R. Malinowski, D. Bajer, S. Richert, "Properties of poly(lactic acid)/Ecoflex rigid foil sheets applied in thermoforming process", Polimery, Vol. 55, No. 7-8, pp. 591-593, 2010.

[7] M. Irimia-Vladu, "Green electronics: biodegradable and biocompatible materials and devices for sustainable future", Chemical Society Reviews, Vol. 43, No. 2, pp. 489-736, January, 2014.

[8] M. Irimia-Vladu, E. D. Głowacki, P. A. Troshin, G. Schwabegger, L. Leonat, D. K. Susarova, O. Krystal, M. Ullah, Y. Kanbur, M. A. Bodea, V. F. Razumov, H. Sitter, S. Bauer, N. S. Sariciftci, "Indigo - A Natural Pigment for High Performance Ambipolar Organic Field Effect Transistors and Circuits", Advanced Materials, Vol. 24, pp. 375-380, 2012.

[9] M. Irimia-Vladu, E. D. Głowacki, G. Voss, S. Bauer, N. S. Sariciftci, "Green and biodegradable electronics", Materials Today Review, Vol. 15, No. 7-8, pp. 340-346, July-August, 2012.

[10] M. Irimia-Vladu, E. D. Głowacki, G. Schwabegger, L. Leonat, H. Z. Akpinar, H. Sitter, S. Bauer, N. S. Sariciftci, "Natural resin shellac as a substrate and a dielectric layer for organic field-effect transistors", RSC Green Chemistry Communication, Vol. 15, pp. 1473-1476, March, 2013.

[11] V. R. Feig, H. Tran, Z. Bao, "Biodegradable Polymeric Materials in Degradable Electronic Devices", ACS Central Science, Vol. 4, pp. 337-348, February, 2018.

[12] S.-J. Kim, D.-B. Jeon, J.-H. Park, M.-K. Ryu, J.-H. Yang, C.-S. Hwang, G.-H. Kim, S.-M. Yoon, "Nonvolatile Memory Thin-Film Transistors Using Biodegradable Chicken Albumen Gate Insulator and Oxide Semiconductor Channel on Eco-Friendly Paper Substrate", ACS Applied Materials & Interfaces, Vol. 7, pp. 4869-4874, February, 2015.

[13] S. Ghoshal, M. A. Khan, R. A. Khan, F. Gul-E-Noor, A. M. S. Chowdhury, "Study on the Thermo-Mechanical and Biodegradable Properties of Shellac Films Grafted with Acrylic Monomers by Gamma Radiation", Journal of Polymer Environment, Vol. 18, pp. 216–223, April, 2010.

[14] L. M. Bellan, M. Pearsall, D. M. Cropek, R. Langer, "A 3D Interconnected Microchannel Network Formed in Gelatin by Sacrifi cial Shellac Microfi bers", Advanced Materials, Vol. 24, pp. 5187-5191, April, 2012.

[15] C. E. Barnes, "Chemical nature of Shellac", Industrial and Engineering Chemistry, Vol. 30, No. 4, pp. 449-451, April, 1938.

[16] B. B. Schaeffer, WM. Howlett Gardner, "Nature and Constitute of Shellac", Industrial and Engineering Chemistry, Vol. 30, No. 3, pp. 333-336, March, 1988.

[17] K. Y. Mitra, E. Sowade, C. Martínez-Domingo, E. Ramon, J. Carrabina, H. L. Gomes, R. R. Baumann, "Potential up-scaling of inkjet-printed devices for logical circuits in flexible electronics", AIP Conferce Proceeding, Vol. 1646, pp. 106-114, February, 2015.

[18] E. Sowade, K. Y. Mitra, E. Ramon, C. Martinez-Domingo, F. Villani, F. Loffredo, H. L. Gomes, R. R. Baumann, "Up-scaling of the manufacturing of all-inkjet-printed organic thin-film transistors: Device performance and manufacturing yield of transistor arrays", Organic Electronics, Vol. 30, pp. 237-246, 2016.

Substrate to Baseplate Attach: A Novel Solder Solution with an Embedded Metal Matrix for Enhanced Reliability

Karthik Vijay

Indium Corporation, 7 Newmarket Court, Kingston, Milton Keynes, UK

+44 1908 580400, kvijay@indium.com

Abstract

A standard solder preform is the typical interconnect solution between a DBC/substrate and the baseplate. In an IGBT power module with increasing power densities, it is becoming clear that a uniform bondline control between the power device and substrate has a direct link to mechanical reliability. This is especially true in automotive power modules (48V mild hybrid, HEV/EV platforms), industrial (inverters for welding, renewable energy), RF communication (5G and beyond). Current techniques to achieve a uniform bondline include: stamped bumps or stitched wire bonds on the substrate, that provide a standoff and preventing tilt. This paper looks at a novel technique that reinforces a solder preform with an embedded metal matrix that determines the desired standoff height for bondline control.

This study looked at bondline coplanarity, reliability and the process cost of ownership for a substrate (Cu & AlN) soldered to a baseplate (Cu & AlSiC) with (a) Standard solder preform with no bondline control; (b) Standard solder preform + baseplate with stamped bumps; (c) Solder preform + stitched wire bonds on the substrate; (d) Embedded metal matrix solder preform. The assembly was temperature cycled for 2 regimes; (a) -50/+150C; 1000 cycles; (b) -40/+125C; 1000 cycles. Samples were analysed by scanning acoustic microscopy to check for delamination.

Keywords: Bondline, Coplanarity, 5G, IGBT, Power module, Solder Preform, Metal Matrix

Introduction

Large area solder joints in RF and automotive multi-chip power devices experience fatigue caused by the periodic straining of the interconnection layers during thermal excursions as the device is operational. These stresses lead to delamination and cracks within the solder layer after many thermal cycles which increase the module to heat-sink thermal resistance and ultimately lead to early device failure [2].

Cracking and solder layer delamination occurs earlier in inhomogeneous solder joints due to stress concentration at thinner areas of the joint, Figure 1 shows how crack length within the solder joint increases greatly with solder layers thinner than 200µm illustrating how tilted samples where part of the joint is <200µm is more susceptible to cracking and delamination.

The advent of spacer technology allows control of the solder joint thickness for a given solder volume by reducing substrate tilt to achieve a homogenous solder layer as Figure 2 demonstrates.

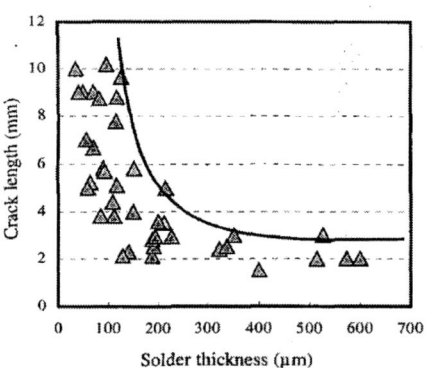

Figure 1: Correlation between solder joint thickness and induced crack length after thermal cycling [1]

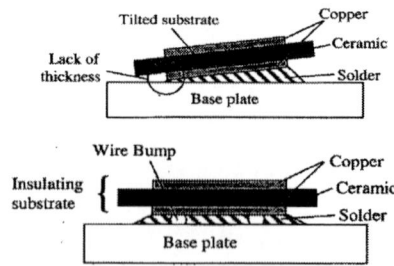

Figure 2: Substrate tilt example (top) and solution using wire bonds to achieve bondline uniformity

This is most commonly done in power modules by stitch bonding aluminium wire of a desired diameter to an AlSiC baseplate or for copper baseplate modules, copper 'bumps' can be stamped in baseplate, increasing the overall cost of the part (Figure 3).

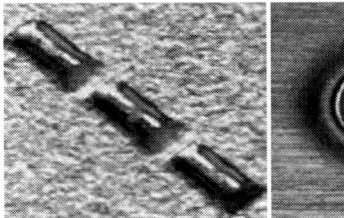

Figure 3: Traditional bondline control methods, aluminium wirebonds on AlSiC baseplate (left) & stamped 'bump' in copper baseplate (right)

The use of spacers in large area solder joints increases the joint lifetime by allowing for homogenous delamination, this occurs at a much slower rate than inhomogeneous delamination caused by substrate tilt [1-2]. This technology is well documented and employed today in power module assembly but this technique results in a high cost of ownership due to extra process steps and capital equipment costs.

An alternative solution to achieving a homogenous solder layer is proposed by using a solder preform engineered with an embedded metal matrix. In the same way as the traditional wire-bond method, when the solder melts during reflow the metal matrix remains intact and serves to maintain a uniform bondline thickness. As well as the obvious reduction in manufacturing time by the advent of a drop in solution to achieving bondline

Sample Preparation

To evaluate the lifetime of the embedded metal matrix preform, sample modules were made and evaluated against the traditional wire-bond method for achieving a homogenous solder layer;

these samples were also compared to reference modules with no bondline control; four of each variant was tested. The samples were then temperature cycled with a ΔT of 200K and cracking and delamination of the solder layer was monitored by scanning acoustic microscopy every 200 cycles.

Module assembly consisted of soldering ceramic AlN substrates (with Cu metalisation) to AlSiC baseplates using (a)200μm SnSb solder preforms with no bondline control as a reference; (b) Samples with the embedded metal matrix solder preform consisted of a 200μm matrix within the SnSb alloy; (c) samples with aluminium stitch bonds used 200μm diameter wire creating a standoff on the baseplate and (d) The Cu baseplate with stamped bumps had a manufactured 200um standoff, was assembled with a standard preforms and compared to a SnSb μm solder preform with an embedded metal matrix 200 thick.

SnSb alloy was selected over other candidate alloys, such as SnAg. SnSb is reported to perform better during thermal cycling due to a solid solution strengthening characteristic of Sb when alloyed with Sn.

Testing

Height was determined as the mean height variation across the top of the substrate at four points; the maximum deflection was also measured. Laser surface profiling showed that the embedded metal matrix solder preform had the least bondline variation. The sample with the embedded metal matrix shows the least co-planarity deviation (smallest ΔZ) at 52.5μm and a maximum deflection of ~60μm (Figure 4), followed by the wire-bonded sample at 56.5μm with a maximum deflection at ~70μm (Figure 5) and finally the sample without bondline control at 67.5μm and a maximum deflection of ~90μm (Figure 6).

Figure 4: Sample with embedded metal matrix; Mean co-planarity deviation = 52.5μm, Max Deflection = 60μm

Figure 5: Sample with Aluminium Stitch bonds; Mean co-planarity deviation = 56.5μm, Max Deflection = 70μm

Figure 6: Sample with no bondline control; Mean co-planarity deviation = 67.5μm Max Deflection = 90μm

Thermal Cycling

Samples were thermal cycled from -50°C to 150°C under the following conditions (Figure 7).

tdwell = 1 hour   Ts(max) = 150°C

ttransition = 30 seconds   Ts(min) = -50°C

ΔT = 200K

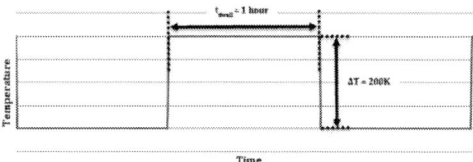

Figure 7: Representative temperature profile for thermal cycling test conditions

Upon thermal cycling, the samples with the SnSb solder preform with no bondline control exhibited delamination between substrate and baseplate at 600 cycles (Figure 8). For the samples that used the stitched wirebonds, cracking and delamination were observed at 800 cycles (Figure 9).

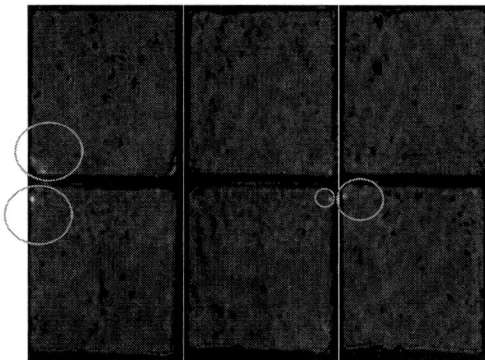

Figure 8: Samples made with no bondline control - SAM gate highlighting cracks in the solder layer at 600 thermal cycles

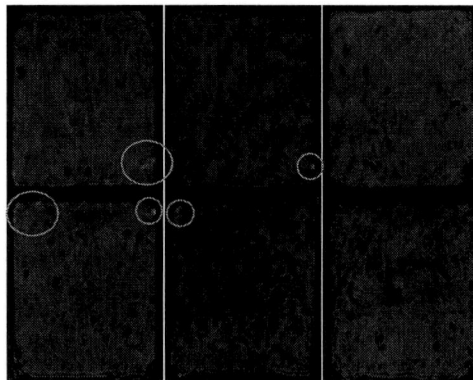

Figure 9: samples with Al wirebonds; SAM gate highlighting cracks in the Solder layer at 800 cycles

The embedded metal matrix solder preform did not show any delamination at 1000 cycle. The test was extended to 2000 cycles, and the embedded metal matrix solder preform did not exhibit delamination or evidence of thermal fatigue (Figure 10) [3].

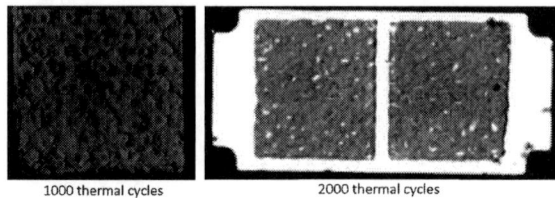

1000 thermal cycles   2000 thermal cycles

Figure 10: Samples made with reinforced matrix No cracks in the solder layer @ 1000 and 2000 thermal cycles

To compare bumped baseplate technique to the embedded matrix solder preform, (i) Cu DBC was soldered to a bumped Cu baseplate with a standard solder preform; (ii) Cu DBC was soldered to a standard Cu baseplate with the embedded metal matrix solder preform. The assemblies were subject to -40/+125C thermal cycling with a requirement of 1000 cycles.

The embedded metal matrix solder preform (Figure 11) showed no delamination even beyond 1000 cycles, whereas the baseplate with stamped bumps showed delamination at 800 cycles (Figure 12).

Figure 11: Embedded Metal Matrix showing no delamination

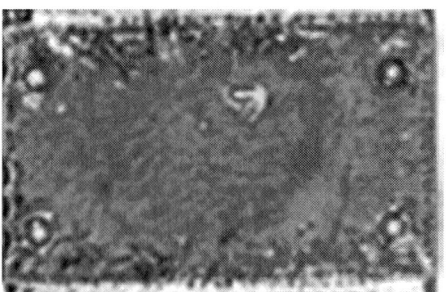

Figure 12: Substrate with stamped bump – showing some delamination

FEM Analysis

To evaluate the damage of creep strain, FEM (Finite Element Analysis) simulation models evaluated the maximum creep strain for two types of solder joints. One model studied the solder joint with no bondline control. The second model evaluated the solder joint with the embedded metal matrix. The two models were subject to isothermal storage at 125°C for10 hours [4].

Figure 13 compares the creep strains in the solder joint between the wirebond and embedded matrix techniques. After 10 hours isothermal storage at 125°C, the solder joint with the embedded matrix shows reduced creep strain compared to the stitched wirebond solder joint.

solder preform, there was equivalent wetting and voiding compared to a standard preform. The embedded matrix solder preform is thus a drop-in assembly solution.

The embedded metal matrix solder preform achieve the lowest cost of ownership as it eliminated the additional process steps and costs associated with the wirebond stitch and trim technique (process steps of wirebond stitch & trim, cost of dedicated wirebonders, maintenance, fixtures.), as well as costly bumped baseplates.

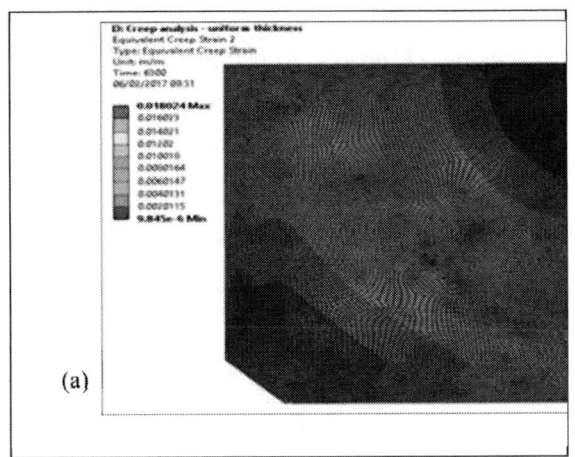 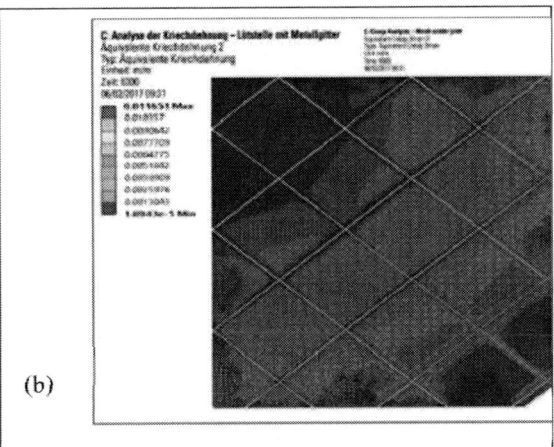

Figure 13: Creep response in solder joint with (a) wirebond control technique; (b) solder joint with embedded matrix

Summary

The SnSb solder preform with the embedded metal matrix consistently enhanced thermal cycling reliability compared to (a) a standard preform with no bondline control; (b) a standard solder preform assembled with the stitched wirebonds on the baseplate technique for bondline control; (c) a standard solder preform assembled with the bumped baseplate technique for bondline control

The increased reliability of the embedded metal matrix solder can be attributed to (i) the embedded metal matrix reinforcing the solder, thus enhancing creep resistance and significantly improving thermal cycling performance (ii) enhanced bondline control with the embedded matrix spread across the entire solder area, instead of being confined to only four corners.

In addition, for the same assembly process parameters used for the embedded metal matrix

References

Articles from conference proceedings

[1] K. Hayashi & G. Izuta, "Improvement of Fatigue Life of Solder Joints by Thickness Control of Solder with Wire Bump Technique", ECTC 2002

[2] K. Guth & P. Mahnke "Improving the thermal reliability of large area solder joints in IGBT power modules", Integrated Power Systems (CIPS), 2006

[3] L. Mills & K. Vijay "InFORMS vs the Trimmed Wirebond Technique to Achieve Uniform Bondline Control Between Substrate and Baseplate", PCIM Europe 2015

Realization of Embedded Passives using an additive Covalent bonded metallization approach

Sarthak Acharya, Shailesh Singh Chouhan, Jerker Delsing

EISLAB, Department of Computer science, Electrical & Space engineering,
Luleå university of technology, Luleå Sweden

{acharya.sarthak, shailesh.chouhan, jerker.delsing}@ltu.se

Abstract

Miniaturization is the call of the day. Electronics shrinking and scaling technology is the priority of all manufacturers. PCBA Industry is working towards the elimination of solder joints, reduction in use of discrete and bulky components, lowering of assemble span, minimized latency etc. Embedded passive technology is playing a significant role in this roadmap by providing better signal performance, reduced parasitic and cross-talk. In this work, the primary focus is to develop a cost-efficient and flexible fabrication methodology that will be suitable for bulk production. A sequential build up (SBU) procedure is adopted with an additive lithography process to realize the passives with minimum possible feature size (< 10 μm). A low cost insulating material, promising grafting solution and Laser assisted writing machine with optimized fabrication parameters are the highlights of this production method. A Computer Aided Design (CAD) software i.e. clewin is used during this process to pattern the mask for the entire process. Covalent bonded metallization (CBM) is the key process for the adhesion of copper layer on the desired site of the pattern. In the CBM process, a polymer surface is modified by grafting. The position of the surface modification is optically defined using a laser lithography system. Such surface modified samples are, then treated in an electroless copper process. Resulting in copper metallization only at the locations with a CBM modified surface. The verification of the copper deposition on the substrate is investigated using a high-resolution microscope followed by scanning electron microscopy (SEM). The confirmation of passive formation has been checked using kethley's source (electrical two-probe measurement). The first-order measured results showed the capacitance formed in the range of 0.3-8 pF. Further concrete measurements using standard methods are undergoing. One of the key advantage of this proposed process is its easiness and feasibility of at room temperature.

Key words: CBM, additive process, embedded passives, grafting material, Urethane coat, LASER patterning, Electroless Copper Plating, SBU process, PCB.

I. Introduction

Advanced electronics packaging is trending towards more compact architecture by embedding both active and passive components into multi-layered Printed Circuit Boards (PCBs). This embedded technology has increased the packaging density as well as the fabrication yield. The use of active components has reached at a saturation limit, instead embedded passives were adopted by the industries in order to avoid the solder-joint related issues such as strain induction and reliability [1]. Even discrete passives occupy extensive area on board. Therefore, passives with laminated foil or thin-film embedded passives implemented using different fabrication techniques are in use since two decades. Looking towards the latest market trend system-minitiaturtation a dedicated manufacturing process with the consideration of cost effectiveness, sustainable environment and reliability issues is needed. Achieving high Fabrication yield is one of the concerned topics for the PCB designers since the beginning. But keeping in align with the

Industrial fabrication protocol which includes checking of layer stacking, panel size control, impedance control calculation and many more. One of such implementations to reduce the real estate surface on the board, buried capacitance approach was preferred over the placement of passive components [2]. A good way to increase the yield is to embed the components as much as possible with smart integration technologies [4].

Heterogeneous integration for embedding active and passive components in multi-layered substrates is in trend using approaches like Laser processing, flip-chip bonding, electroless or electro-plating of Cu and embedded chip module in recent years for various applications [3]. One of such integrations of pre-packaged components with module miniaturization for speech signal processing using the face-down approach was implemented with 50% space reduction [4]. Few such fabrications of dielectric polymer composite on PCB were embedded non-linear passive components against Electrostatic Discharge (ESD) events [5]. A low pressure bonding method was also used to build passives for miniaturized Wi-Fi and Bluetooth modules. Nevertheless, one of the key ambiguities of these processes was adhesion of metallic layer [6], since a non-uniform growth of the plasma-enhanced process resulted in poor metallization.

Embedded structures were also in use for RF applications. Such embedded passives were made on organic substrates by Laser drilling at electrode Cu pads followed by electroless Cu plating [7]. Inductors and transformers were also made by micro-patterning of the Cu layer on substrates to enhance both the power density and thermal characteristics. However, AC-signal losses, Electromagnetic Compatibility issues and fabrication difficulties were the some of the main concerning factors [8-9]. Hence, fabrication of capacitors is the first choice by the PCB manufacturers and could be a great approach in designing of next generation folded electronics with flexible substrates [10].

System on Package (SoP) approach for heterogeneous electronics packaging is beneficial in the terms of greater performance, cost-saving, feature size reduction and less area penalty. One of such packaging techniques is named as Pocket Embedded Packaging (PEP) with aluminum substrate used for MMIC application [11]. Embedded substrates are also convenient in developing a low profile Fan out Package on Package [PoP] that supports 3D integration of the logic devices [12]. When looking towards the cost-effective approaches the preferred substrate by the industries is FR4 because of the properties like compatibility with electroless Cu plating, high yield factor and minimal Coefficient of Thermal Expansion (CTE) [13]. Therefore, for the proposed implementation the FR-4 substrate has been used.The proposed method is based on Covalent Bonded Metallization (CBM) technology that is functional up to micro level and can even possible to extend at a nano-scale in the near future. CBM

process is capable to fabricate embedded passives; metallic interconnects with the lowest feature size and multi-layer embedded die module. Thus, it achieves a tighter and robust heterogeneous integration.

Production methodologies emphasizing on embedded passives have ramped up in a wider scale for industrial product development. This paper proposesan additive Laser assisted production method to realize embedded passives on FR-4 subtract. In this approach, the CBM technique followed by electroless Cu plating was used to grow the micro-patterned metal tracks. Section II includes the details of various experimental steps.. Microscopy was used to investigate the structures; and the measurement of capacitances has s beenere done and the results are presented in Section III. Finally, the conclusion of the work is in Section IV.

II. Experimental Methodology

The proposed fabrication process is discussed in this section. During the production, the Sequential Build Up (SBU) technique was adopted due to the additive nature of the process. Starting with the cleaning of FR-4 substrate using ultrasonic cleaner, then the first layer of urethane as an insulating material was coated using spin coater. A proprietary grafting material provided by Cuptronics™ [14] was then coated as a second layer on top of the insulating material. In the next stage, Laser writing machine, LW405B by Microtech pvt. Ltd [15] was used for Laser assisted engraving. Finally, an electroless copper bath provided by JKEM pvt., [16] was used to grow Cu metal. This deposition of Cu metal is due to CBM phenomenon that is appeared because of activation of grafting material using Laser. A schematic view of the workflow is shown in Figure 1. Few key steps of this production technique are discussed in details in the following subsections.

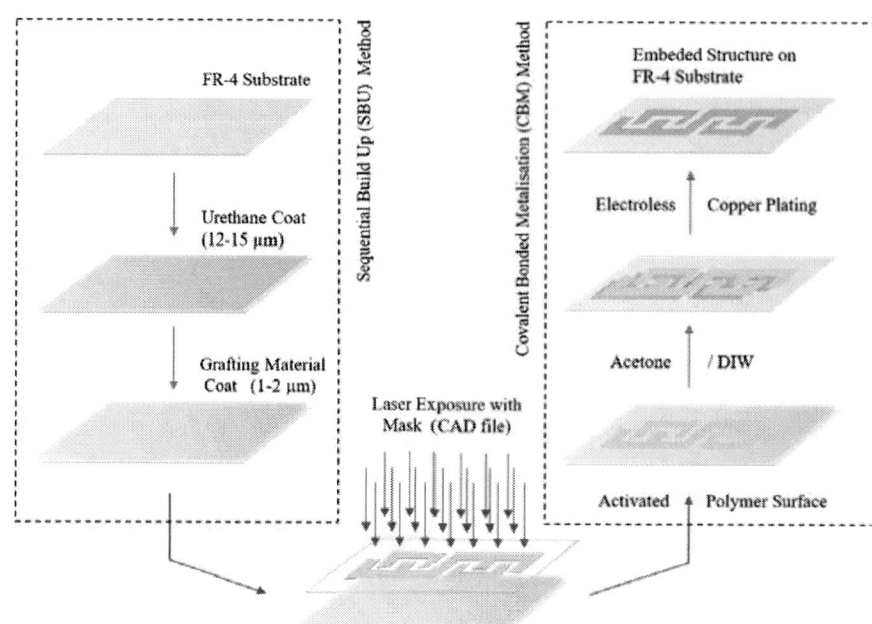

Figure 1: Schematic diagram of the SBU-CBM production

A. Laser Assisted Lithography

In this experiment, GaN solid-state Laser source with the wavelength of 375 nm was used for the lithography purpose. To perform lithography, first the desired mask was designed using the CAD software CLEWIN®. Next the mask was exported to lithography machine (LW045B) for the Laser exposure. The substrate with the urethane and grafting material coat was then placed inside the Laser machine to engrave the patterns on it. Few parameters associated with the Laser source needed optimization during this micro-patterning, which are listed in section III.

B. Covalent Bonded Metalisation (CBM)

CBM is a full additive technology based on Laser- polymerization for the robust adhesion of metals into polymer surface. Here, urethane is the polymer and the grafting solution containing monomers, was used to modify the surface during the UV-Laser exposure. The boundaries of the activation areas were confined using the desired CAD masks and rest of the area were remaining unaffected. During the exposure, bond break-and-form process between the grafting solution and polymers happens, which resulted in growing nucleation sites for the metallization. After this, the substrate is treated with electroless Cu bath to grow the metal layers on the activated areas. As the covalent bond between the polymer and the copper is formed by the surface alteration hence, the process is termed as Covalent Bonded Metalisation in the literature [17].

C. Electroless Copper Bath

For this experiment, a commercially available electroless Copper bath, PEC 660, by JKEM was used. PEC 660 comprises of four different baths i.e. activator bath (I), catalyst bath (II), reducer bath (III) and copper plating bath (IV). Each sample was processed in these baths subsequently. Proper Cu growth needs optimization of both exposure time and concentration of the bath. A typical growth rate of the Cu in the bath is is 0.04-0.05 μ/min.

III. Results and Discussions

In this section, the results related to the above production method are shown. After fabrication, the optical characterizations were done to investigate the feasibility of the process and the growth of metals on the substrate. A number of samples were prepared to optimize the parameters for a proper activation of the polymers in order to increase the rate of metalisation. Second, the electrical characterisation was performed to determine the value of capacitance of the fabricated embedded passive. The Process-controlled parameters, microscopic images and electrical measurement data were shown in the subsections below.

A. Process Parameters related to Laser and Copper Bath

There are two important steps in this fabrication that control the formation of structures on the substrate which are Laser exposure and electroless Cu bath.

Table 1: Process parameter of Electroless Cu Bath for perfect metalization

	Type	Name	Concentration	Time (minutes)	Temperature (°C)
			Electroless Copper Bath		
I	Activator	Precup-128	50 ml in 200 DeIonised Water (DIW)	1	25
II	Catalyst	Catcup-208	50 ml in 200 DIW	5	40
III	Reducer	Boric Acid+ACS Reducer-2074	2.1 gm in 100 ml DIW+1 ml of ACS-2074	5	25
IV	Cu Plating	PEC-660 (A/M/B)	9.25 ml of PEC-A+6.25 ml of PEC-M+10 ml of PEC-B in 100 ml of DIW	5-8	25

Former has three controlling parameters namely power gain, No. of repetition (exposure time) and D-step (rate of scan). The later also has three parameters, bath time, concentration and temperature of the bath. In this experiment the power, Number of repetition and rate of scan used were 7.0 mW, 6 (total14min 36sec) and 160µ/sec respectively. The electroless bath also required optimized bath duration, concentration and temperature of each bath to obtain a proper Cu growth. It should be noted that the Laser and bath parameters vary according to the selected polymers and grafting materials.

The optimized parameters used in this method for PEC 660 A are listed in Table 1

B. Optical Characterization

Two different optical characterizations were done to ensure the formation of the patterns and also to check the given feature size possible through the CBM technique. First, a high-resolution optical microscope (by FLIR pvt.) was used to examine the structures. Second, a Scanning Electron Microscope (SEM: JSM-IT300 by JEOL Inc.) was used to check and confirm the selected feature size. The images from optical microscope, for perfect growth with optimized parameters are shown in Figure 2. The SEM of the same sample with feature size is shown in Figure 3. It can be seen in Figure.4 that the capability of the proposed Laser based SBU-CBM method is to go to up to a feature size of 2.8 µm. Alteration in the parameters mentioned above resulted in improper activation of polymers and thus poor Metalisation, are shown in Figure 5. Another embedded passive architecture using this proposed production method is also shown in Figure 6. This is basically a Complex micro-patterned architecture to realize embedded-resistor.

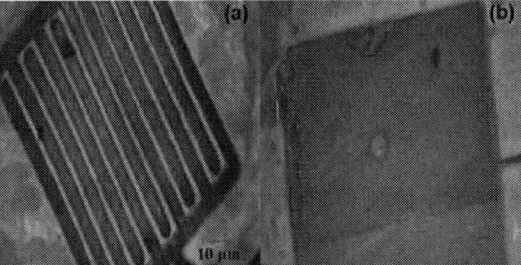

Figure 2: Optical microscopic image of the perfect growth of Cu using SBU-CBM method: (a) embedded passives fabricated on FR-4 with the finger size of 20µm and a gap of 9-10 µm resulting in 0.3-8 pF capacitance (b) Formation of the Cu pad

Figure 3: SEM images: (a) shows the embedded capacitor formation using SBU-CBM method (b) shows isolation between the fingers of the fabricated embedded capacitor which is in range of 6.5-6.8 µm

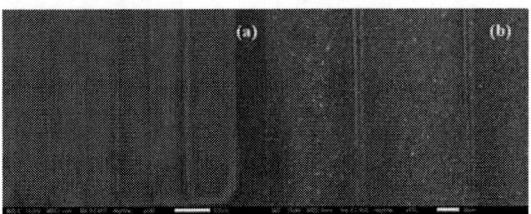

Figure 4: SEM images: (a) and (b) showthe minimum possible feature size of 2.8 µm that has been achieved using the proposed SBU-CBM fabrication technique.

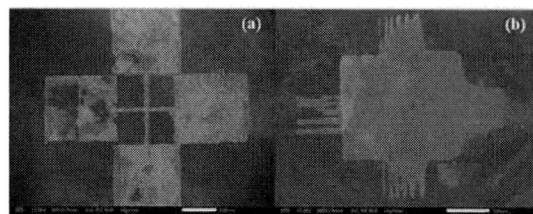

Figure 5: SEM image: (a) shows an improper selection of laser output power (b) shows an over-metalisation due to improper electroless plating parameters.

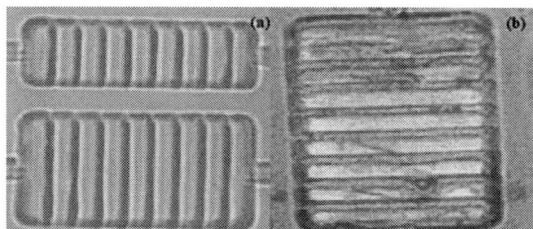

Figure 6: (a) and (b) Complex micro-patterned architecture realized using SBU-CBM method

C. Electrical Characterization

After, the confirmation of formation of embedded passives using the optical methods electrical characterization was performed to quantify the capacitance of the fabricated structures. The measurement was done using L-C-R meter (LCR Bridge HM8118 by HAMEG instruments) with a setup showed in Figure 7. To understand the reproducibility of proposed process, various samples with different mask (12different CAD files) were prepared with the final recipe to observe the formation of capacitances. The values of capacitance obtained using this fabrication method is in the range of 0.3 to 8 pF.

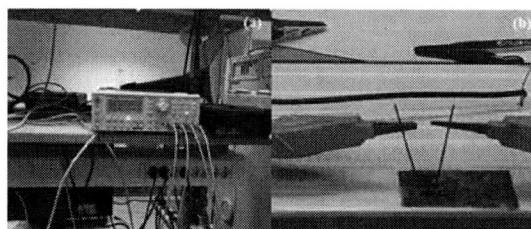

Figure 7: (a) LCR Bridge HM8118 instrument is used to quantify the capacitance of the embedded structures (b) shows the probing setup used for the measurement.

IV. Conclusion

This paper has focused on developing a new cost-efficient fabrication method for realization of embedded passives. Due to the fact that discrete passive components involve issues such as area penalty in PCB assembly, solder-joints reliability etc., new embedded passive technologies were preferred in manufacturing processes. Different structures were fabricated using the Covalent Bonded Metalisation (CBM) technique on FR-4 substrate. Optical characterizations were done using optical microscopy and SEM. A feature size of 2.8 µm was observed so far. An electrical measurement was also done to confirm and quantify the embedded capacitor Using this production methodology, further structures can be fabricated with reduced feature size along with the reliability and electrical characterizations to ensure the concreteness of this SBU-CBM method.

Acknowledgements

The authors would like to acknowledge the European Project: Productive 4.0 for the financial support. The work was done in the Clean Room, EISLAB, in the Department of Computer Science, Electrical and Space Engineering, LTU. The authors also thankful to JKEM pvt. and Cuptronics TM, Stockholm to provide the proprietary grafting material.

References

[1] Balmont M, Majek IB, Ousten Y. Ultra-thin actives for embedded components: halfway between thin film technology and embedded Surface Mounted Device. In2018 7th Electronic System-Integration Technology Conference (ESTC), 2018.

[2] Khan, Z. Improving fabrication yields. Printed Circuit Design and Fab, 27(1), 27-28, 2010.

[3] Zhang M, Shang J, Chen L. Embedding chip into substrates with cavities for hetergeneous integration. In 16th International Conference on Electronic Packaging Technology (ICEPT), pp. 1043-1046, 2015.

[4] Manessis, D., Pawlikowski, J., Ostmann, A., Schischke, K., Aschenbrenner, R., Schneider-Ramelow, M., Krivec, T., Podhradsky, G. and Lang, K.D., Embedding technologies for heterogeneous integration of components in PCBs-an innovative modularisation approach with environmental impact. In 21st European Microelectronics and Packaging Conference (EMPC) & Exhibition, pp. 1-8, 2017.

[5] Ghosh D. Embedded nonlinear passive components on flexible substrates for microelectronics applications. Journal of Materials Science: Materials in Electronics. 1;28(15):11550-6, 2017.

[6] Park SH, Ryu JI, Kim JC, Kang NK, Park JC, Kim YH. Fabrication and characterization of embedded active and passive device for wireless application. In Proceedings 60th Electronic Components and Technology Conference (ECTC), pp. 2035-2041, 2010.

[7] Kim H. Passive device embedded substrate for application of RF module. Circuit World. 3;42(2):84-8, 2016.

[8] Dou Y, Ouyang Z, Thummala P, Andersen MA. PCB embedded inductor for high-frequency ZVS SEPIC converter. In2018 IEEE Applied Power Electronics Conference and Exposition (APEC) pp. 98-104, 2018.

[9] Pascal Y, Petit M, Labrousse D, Costa F. Study of a Topology of Low-Loss Magnetic Component for PCB-Embedding. In2018 7th Electronic System-Integration Technology Conference (ESTC), pp. 1-7, 2018.

[10] Sterman Y, Demaine ED, Oxman N. PCB origami: A material-based design approach to computer-aided foldable electronic devices. Journal of Mechanical Design. 1;135(11):114502, 2013.

[11] Kim KM, Yook JM, Yeo SK, Kwon YS. Embedded IC technology for compact packaging inside aluminum substrate (pocket embedded packaging). In2007 European Microwave Conference, pp. 1125-1128, 2007.

[12] Jung BY, Ho DS, Sorono DV, Lim SP, Chen Z, Yong H, Lin B, Chong CT. Development of low profile Fan out PoP solution with embedded passive. In IEEE 16th Electronics Packaging Technology Conference (EPTC), pp. 597-600, 2014.

[13] C. Hong and M. Lee, "A novel FR-4 material for embedded substrate," 2011 6th International Microsystems, Packaging, Assembly and Circuits Technology Conference (IMPACT), Taipei, pp. 177-178, 2011.

[14] http://cuptronic.com/applications/plating-on-plastic/. (accessed on: June 6, 2019)

[15] Manual available on demand from Microtech pvt. Ltd., Palermo, Italy.
http://www.microtechweb.com/3d/3d.htm
(accessed on: June 6, 2019)

[16] http://jkem.se/perfekto/ (accessed on: June 6, 2019)

[17] James D. Rancourt, James B. Hollenhead & Larry T. Taylor (1993) Chemistry of the Interface Between Aluminum and Polyester Films, The Journal of Adhesion, 40:2-4, 267-285,

Ultra-thin polymer spray coating for advanced adhesive bonding applications

B. Matuskova*, J. Gasiorowski, J. Rimböck, T. Zenger, M. Eibelhuber, T. Uhrmann

EV Group, DI Erich Thallner Strasse 1, 4782 St. Florian am Inn, Austria

+43 7712 5311 0, B.Matuskova@evgroup.com

Abstract

Adhesive bonding using an ultra-thin polymer layer is getting more attention for more than Moore applications. However some challenges of ultra-thin layer coating remain and are subject to further research. For this purpose diluted BCB adhesive layers were evaluated to create bonding layers down to 30 nm adhesive thickness. BCB is a well-known and thus favourable candidate for this evaluation due to it's physical properties, such as low curing temperature, high degree of planarization, high optical clarity, along with good compatibility with various metallization systems.

Key words: ultra-thin coating, spray coating, BCB, adhesive bonding, heterogenenous integration

Introduction

Thin layer deposition on large area substrates was found to be a key process for various technological applications. Thin organic layers depositied on various substrates such as Si/SiOx wafers [1], glass and or metal/plastic [2] are core processes applied in the current device developments. The application of thin organic layers in optoelectronics [3], bio/photo/electro chemistry, organic and hybrid electronics [4] are in the focus of the current research. More than Moore as well as heterogenous integration are driving system integration, where die to substrate and die to wafer integration play a major role. Very promising use of ultra-thin polymer layers can be foreseen for adhesive bonding applications due to minimizing the possible impact of the adhesive layer to the die functionality. Compared to oher bonding techniques the main advantages of adhesive bonding include low bonding temperature, CMOS compatibility and high bonding strength while providing high flexibility and facilitated process integration. This technology is used for many applications such as packaging, microfluidic devices [5], hybrid integration of CMOS circuits and MEMS/ MOMS devices [6] or heterogeneous integration including silicon photonics [7].

In order to obtain homogeneous and uniform layers deposited on large area substrates different approaches were proposed. Spin coating, blade coating (also known as slot-die coating) as well as spray coating were recognized as most interesting and providing the highest technological viability. Spin coating is a well established technique providing very smooth layers with well control on the layer thickness. However, it is mainly applicable for round shaped substrates and has limitations for homogenous covering on non-flat, structured surfaces [8] as topography hinders uniform and flat adhesive layers. On the other hand blade (slot-die) coating was developed to deposit homogenous layers on large area substrates with various topographies [9]. Nontheless, this approach does not apply for thin layer deposition. Additionally both mentioned techniques demand increased material consumption and result in significant material losses. Along those lines spray coating technique turns out to be advantegeous since it allows overcoming remaining drawbacks of spin and/or blade coating processes but more importantly provides uniform coatings on non-planar surfaces.

Typically spray coating can be used to deposit uniform and homogenous layers on substrates with no limitation in shape, size and surface topography [10]. Though, usual application of spray coating with layer thickness below 100nm has shown various challenges [11]. Combining ultra-thin film thickness and flat surface requirement, especially inside cavities or vias, upsurges the process control complexity.

In this paper we elaborate EVG Omnispray coating technology with a clear focus of film thicknesses well below 100nm for heterogeneous integration as typically needed in photonic integration as well as in silicon photonics. This method facilitates precise ultrasonic droplet size control and therefore allows obtaining optimized and homogenous planar ultra-thin layers of benzocyclobutene (BCB) material (as reference) [12] with thicknesses down to 30 nm. Moreover, this technology was effectively applied to perform homogeneous coating inside cavities or vias.

Experimental

To enable the process flow the EVG100 series was used as it can combine spin and spray coating as well as other process steps like develop, bake and chill on dedicated modules. In this way individual process flows can be flexibly addressed with a very high degree of freedom in terms of material choice. The typical material range includes positive and negative resists, polyimides, double-sided coating of thin-resist layers, high-viscosity resists, and edge protection coatings. Ultra-thin resist layers are processed in a single process step using heated chuck. Due to precise recipe-controlled dispense, sprayed resist is not accumulated in the corners of the cavities. Moreover, spray-coating method saves resist material significantly, as there is no spin-off step included.

Because of its dielectric and optical clarity properties, Benzocyclobutene (BCB) from Dow Chemicals (marketed as Cyclotene 3022) was used as a reference material. With the purpose to obtain ultra-thin adhesive layers the material was diluted using solvents to adjust the viscosity of the sprayed material. Single side polished (100) Si/SiOx wafers (SiMat) were used.

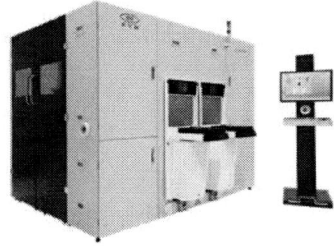

Figure 1 EVG 150 fully automated coating system

In order to evaluate the impact of substrate surface on the layer formation, EVG810 plasma activation tool was used for the surface treatment prior coating. The treatment was done under Oxygen atmosphere (0.3mbar) in dual frequency mode where top electrode was 150W and bottom 100W. Plasma activation substrate treatment was performed for 30s.

Variable Angle Spectroscopic Ellipsometry (VASE) was employed to characterize layer thickness and optical roughness. Woollam M-2000 rotating compensator ellipsometer with working in wide spectral range (190 – 1687 nm) was used. The ellipsometric measurements were performed at three different angles of incidence (60, 65, and 70°) in various points using an automated stage. The measured angle dependent ψ and Δ functions were modelled with CompleteEASE software to obtain information about layer thickness and optical roughness. The metrology details as well as modeling information can be found in the referenced paper [13].

Bruker Fast Scan Dimension Atomic Force Microscope (AFM) working in the scan assist mode (similar to tapping mode) and JSM-7500F Field Emission Scanning Electron Microscope (SEM) from JEOL were used to additionally analyze layer thickness and surface roughness.

Results and discussion

In order to perform ultra-thin layer spray coating process, at first the substrate surface properties impact on the thin film formation was analyzed. The deposition of organic molecules onto organic/inorganic substrates always requires specific cleanliness and/or surface energy (*e.g.* hydrophobic or hydrophilic behavior). The properties can be modified by applying *e.g.* (i) a thin interlayer (surface priming) or more advanced (ii) plasma treatment. The surface effect on thin layer formation was evaluated by depositing a 115 nm layer of BCB (reference thickness) on Si/SiOx with and without plasma treatment, respectively. The deposited layers were tested using VASE technique to extract information about the layer thickness and optical roughness values distribution over the coated wafer. The results are presented in Fig. 2 and Fig 3.

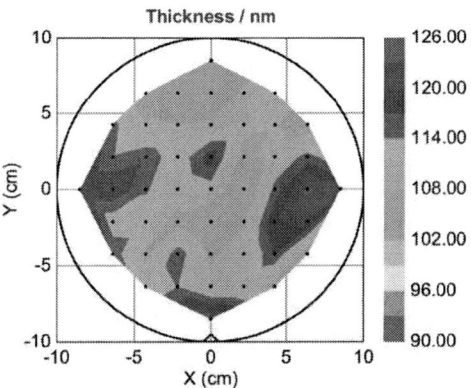

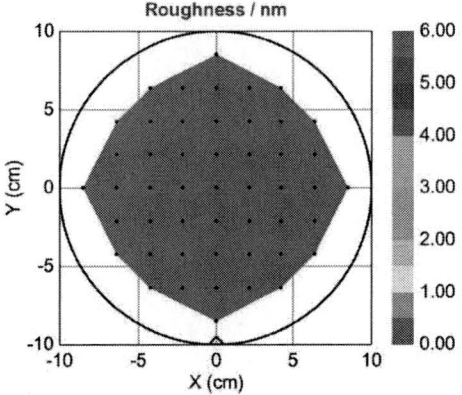

Figure 2 VASE results of BCB coated on blank Si/ SiOx wafer without plasma treatment presenting thickness (above) and roughness (below)

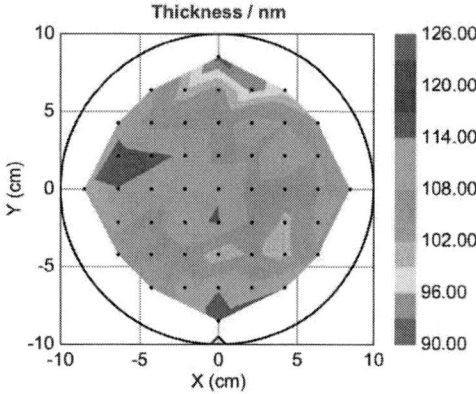

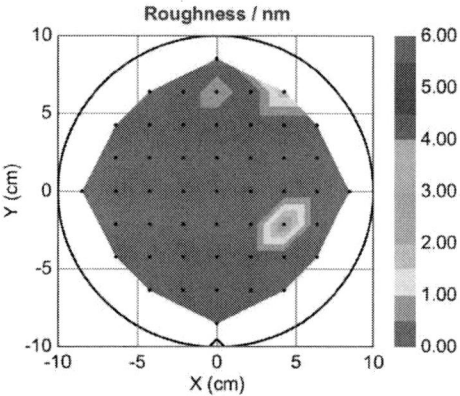

Figure 3 VASE results of BCB coated on plasma treated Si/SiOx wafer presenting thickness (above) and roughness (below)

Additionally, measurements to determine microscopic roughness have been performed on same samples as well using AFM on randomly selected area 2.5x2.5 μm^2. Fig. 4 compares both samples: blank and plasma treated Si/SiOx. Results show that microscopic roughness without plasma treatment is 4.76 nm and with plasma treatment is 4.02 nm.

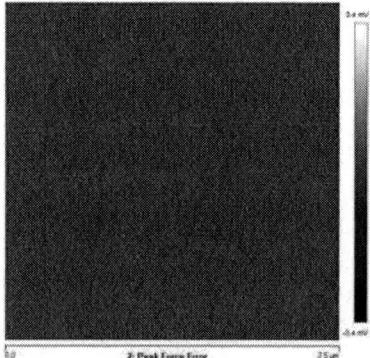

Figure 4A 4 Microscopic roughness of BCB coated on blank Si/SiOx wafer using AFM

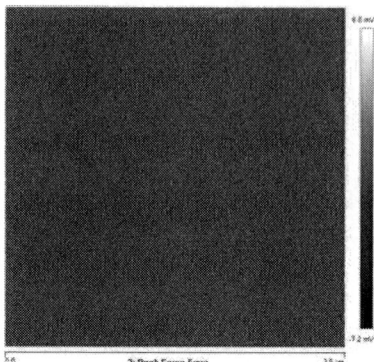

Figure 4B Microscopic roughness of BCB coated on plasma treated Si/SiOx wafer using AFM

The analysis of VASE results reveals that the surface treatment does not significantly influence the layer formation and homogeneity across the wafer. In the case of blank Si/SiOx wafers the BCB mean layer thickness was 111 nm, while for plasma treated substrates the mean thickness value was around 107 nm. The thickness uniformity variation within the wafer was found to be for all measured wafers around ±10%. This result proves the advantage of spray coating method application since it eliminates the need of substrate pre-treatment which is for some applications a key issue. When analyzing optical roughness evaluated using VASE for blank and plasma treated substrates, the roughness was found to be below 1 nm in both cases. The observed defects on the plasma treated substrate are most likely related to particles caused by the additional handling as the plasma treatement was done *ex-situ* on a semi automated tool.

Furthermore, the comparison between VASE optical roughness and AFM roughness is presented. Certain discrepancy between both measurement results can be observed. Namely, the VASE results show optical roughness lower than 1 nm, while from AFM roughness values are above 4 nm.

In order to evaluate the process scalability new samples with 30 nm and 115 nm (as reference) were prepared. The optimized process described above was employed again to obtain thinner layers. For further tests only plasma treated Si/SiOx wafers were used due to the substrate cleanliness reliability and identical substrate surface conditions for each evaluated substrate. Prepared samples were at first analyzed using VASE and thickness maps are presented in Fig. 5A for 30 nm layer thickness and in Fig. 5B for 115 nm layer thickness.

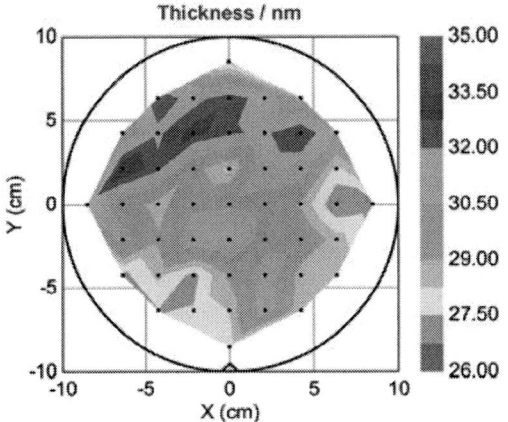

Figure 5A VASE map evaluation of 30nm layer thickness

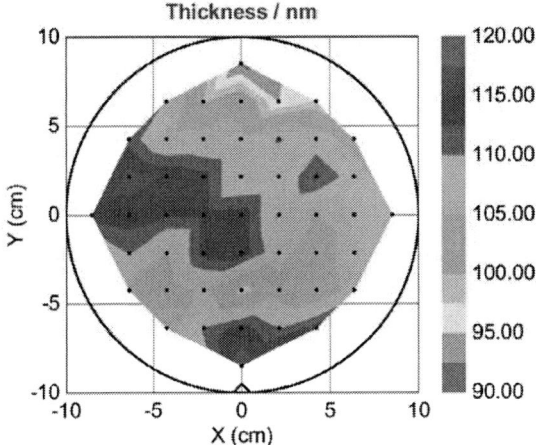

Figure 5B VASE map evaluation of 115 nm layer thickness

When analyzing the data very good correlation between the specified and measured results can be observed. In both cases a good homogeneity (±10%) of the prepared layers can be seen. The VASE data were additionally compared with thickness measurements done using AFM and SEM. The summarized results are shown in **Table 1**.

Table 1 Mean layer thickness (LT) and roughness comparison using three evaluation methods: VASE, AFM and SEM

	VASE		AFM		SEM
Target LT [nm]	Mean LT [nm]	Rough-ness [nm]	Mean LT [nm]	Rough-ness [nm]	Mean LT [nm]
30	30	0.1	25	6.9	31
115	108	0.1	110	4.0	111

For both BCB layers, SEM results were found to be in a very good agreement with VASE results additionally confirming the properness of the model in the VASE data anaylsis. The layer thicknesses determined by AFM are significantly different (lower values) and at the same time the roughness value determined by AFM is significantly higher to those obtained from VASE. In this case summing the values of layer thickness and roughness from AFM measurement the total value is exactly in the range of those measured by VASE and SEM. The observed difference of the roughness between VASE and AFM can be explained by the metrology operation principle which is related to the scanning area. The AFM determines the roughness in the microscale area (2.5x2.5um) while the VASE measurement area is in the milimiter scale. Another possible explanation of the observed roughness value mismatch could be the character of the roughness. This discrepancy can be explained looking on the theoretical background of both methods. The VASE results are obtained by analyzing specific response using mathematical model. In this model roughness is determined by effective medium approximation (EMA) algorithm, which is a mixture of complex dielectric functions of air voids and the substrate material with the 50:50 ratio. In this manner can be certain surface roughness calculated and thus directly summed up in to the layer thickness. On the other hand the AFM as a mechanical method is completely independent from the roughness character.

Finally, the process repeatability was analyzed. In this test, 12 reference samples with 115 nm target thickness were prepared and analyzed. The samples were processed and analyzed within 6 runs in the last 13 months (2 samples / run). The mean thickness values extracted from VASE are plotted as a function of the sample number in Fig. 6.

Generally the process is very stable as the wafer-to-wafer uniformity for all 12 samples is less then 5%. In production mode, the variation should be even better, since analysis shows that the variation within a run is much smaller than the run-to-run variation. This is most likely related to manual mixing preparation of the resist performed for each run freshly in small volumes.

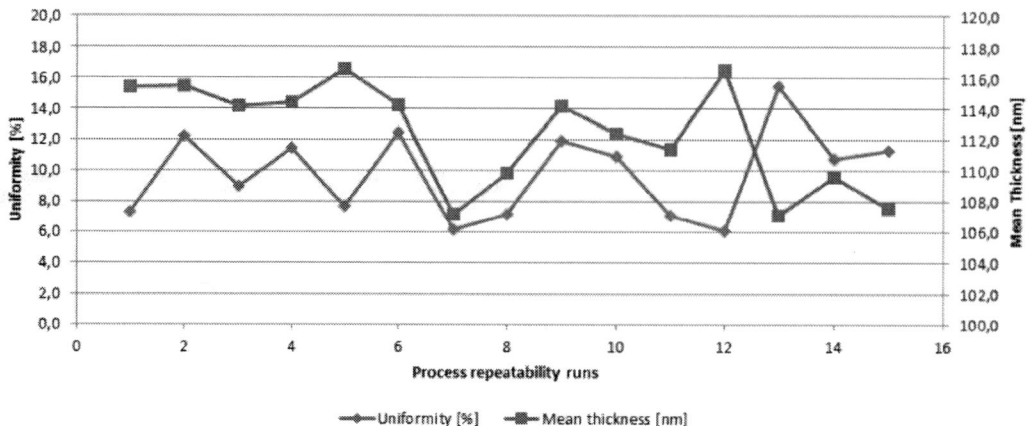

Figure 6 Process repeatability plot

Conclusions

In this study the feasibility of preparing ultra-thin layers (30 nm) using EVG´s Omnispray technology is presented. The multiple metrology tools (VASE, AFM, SEM) revealed that by using Omnispray technology homogeneous and repeatable thin and ultra-thin layers can be achieved. The application of the BCB material on Si/SiOx wafers makes this technology more advanteous as compared to other coating techniques used by high volume manufacturing semiconductor industry. Additionally, the achieved process stability determined by thickness measurement from samples prepared on different days within 13 months is remarkable. The observed value stability is in the same range as for μm layers which are state-of-the-art for current spray coating applications. The presented down-scaling process opens new opportunities for spray coating process development and applicability in current and new industry sectors.

References

[1] M. Bednorz et al, "Silicon/organic hybrid heterojunction infrared photodetector operating in the telecom regime", Organic Electronics, 14 (2013), 1344

[2] M. Kaltenbrunner et al, "Ultrathin and lightweight organic solar cells with high flexibility", Nature communications 3, Number 770 (2012)

[3] P. Meredith et al, "Electronic and optoelectronic materials and devices inspired by nature", Reports on Progress in Physics 76 (2013), 034501

[4] Mihai Irimia-Vladu, " 'Green' electronics: biodegradable and biocompatible materials and devices for sustainable future", Chem Soc Rev 43 (2014), 588

[5] A. Shores et al., "Thermoplastic films for adhesive bonding: hybrid microcircuit substrates", Proc. Electronic Components, Conf. (Houston, USA) pp 891–5

[6] A. B. Frazier et al., "Low temperature IC-compatible wafer-to-wafer bonding with embedded micro channels for integrated sensing systems", Proc. Midwest Symp. on Circuits and Systems 1996 (Rio de Janeiro, Brazil) pp 505–8

[7] G. Roelkens et al., "Ultra-thin BenzoCycloButene (BCB) bonding of III-V dies onto an SOI substrate", http://www.photonics.intec.ugent.be/download/pub_1868.pdf

[8] M. Eibelhuber et al "Combined Thick Resist Processing and Topography Patterning for Advanced Metal Plating", Proceedings EPTC 2018

[9] V. Zardetto et al, "Formulations and processing of nanocrystalline TiO2 films for the different requirements of plastic, metal and glass dye solar cell applications", Nanotechnology 24 (2013), 255401

[10] K. Gay et al "Advanced MEMS and packaging: resist and thin film adhesive processing solutions for lift-off processing", Proceedings IWLPC 2015

[11] N.P. Pham et al., Y, "Spray coating of photoresist for pattern transfer on high topography surfaces," J. Micromech. Microeng. 15 (2005) 691–697

[12] Chad Brubaker et al., *State of the Art Processing Schemes for BCB*, Proceedings IWLPC 2007

[13] J. Gasiorowski et al "Dielectric Function of Undoped and Doped Poly [2-methoxy-5-(3',7'-dimethyloctyloxy)-1,4-phenylene-vinylene] by Ellipsometry in a Wide Spectral Range", The Journal of Physical Chemistry C 117 (2013), 2201

Temperature evaluation of solder joints for adjusting reflow profiles

S. Wiese [1], M. Mueller [2], D. Čáp [2], D. Barth [1], A.Yuile [1], S. Schindler [3],

and I. Panchenko [2,4]

1) Saarland University, Chair of Microintegration and Reliability, Saarbrucken, Germany

2) Technische Universität Dresden, Institute of Electronic Packaging Technology, Dresden, Germany

3) Fraunhofer Center for Silicon Photovoltaics CSP, Halle (Saale), Germany

4) Fraunhofer Institute for Reliability and Microintegration, Moritzburg, Germany

Phone: +49 681 302 71820, E-mail: s.wiese@mx.uni-saarland.de

Abstract

In order to adjust the temperature profile of a reflow oven to a specific soldering task, thermocouples are placed on selected places on the printed circuit board. Since the thermocouples have a very small thermal mass and correspondingly short response times, compared to other types of temperature sensors, they are able to record the dynamic changes in temperature that occur during a reflow process. The goal is to monitor the rather rapid temperature changes of small components (e.g. CR, CC, LED) and the slower temperature changes at larger components (e.g. electrolytic capacitors, BGA). That way the temperature profile is optimised to securely solder all components within a limited total thermal load, to prevent damage of the components. In contrast to the established methodology to adjust temperature profiles the investigation presented in this paper focuses on measuring the solder temperature during the reflow process. A laboratory reflow oven is used, which is tailored for soldering lead-free solder prototypes and offers four channels for simultaneous temperature measurements through an integrated USB port, which also enables programmable control of the reflow profiles. In this work the temperatures of the solder are measured directly by making use of T-type thermocouples. Although having the comparative disadvantage of being less linear than their K-type counterparts, they have the advantage of being able to integrate better with solder, hence enabling a deeper study of specific solder temperatures, as opposed to merely the bulk component temperatures offered by K-type thermocouples.

Key words: lead-free solders, reflow soldering, reflow profile, solder process optimization, T-type thermocouples, temperature monitoring

Introduction

Soldering is the most important process for the assembly of electronic devices from individual components, e.g. ICs, diodes, resistors, and capacitors. The advantage of soldering compared to other joining technologies, such as gluing and welding consists mainly of two major aspects: (I) Molten solder provides an extraordinary selectivity in wetting on metallic and non metallic surfaces, which enables an autopositioning of components. Errors introduced by printing offsets (solder paste) and placement offsets (components) are reduced by strong wetting forces, which act between interconnect metallizations. (II) Soldered interconnects provide an exceptional functionality concerning electrical, thermal and mechanical properties of the joint. Moreover the reliability figures of soldered interconnects are superior to adhesive joints due to their better corrosion resistance and thermo-mechanical fatigue toughness.

Despite its preferable technological properties and joint functionality soldering is a complex joining process, which could lead to poor joining results, if the underlying physics are not sufficiently considered. From the physics standpoint soldering can be subdivided in five elementary mechanisms:

- melting of solder material
- wetting of joining surfaces
- dissolution of metallization material into the molten solder
- formation of intermetallic layers between metallisation and solder
- solidification of solder

In order to find a suitable path through these individual elementary mechanisms the soldering process needs to be carried out in a very controlled manner. Therefore manufacturers of solder paste give very precise instructions for adjusting soldering profiles. An example of such profile is given in figure 1.

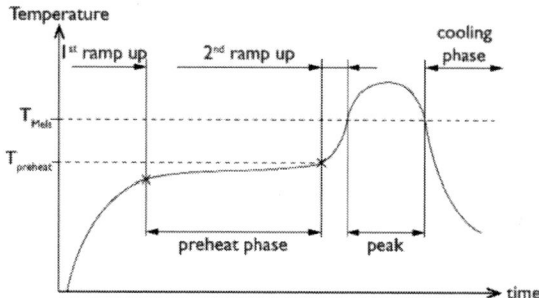

Figure 1: Soldering profile for a standard reflow process used for 2nd level interconnects in surface mount technology

The profile consists of various sections that address specific elementary processes. The first section is the preheat phase, which aims to lift the entire assembly to a higher temperature without damaging the organic base material of the printed circuit boad (PCB). Therefore a temperature is chosen, which is slightly above 150 °C. After the preheat phase the temperature is ramped up quickly above the melting point of the solder to enable the reflow of the solder paste. In order to keep the thermal damage to sensitive materials in the assembly as little as possible the ramp up phase turns immediately into the cooling phase after the peak temperature was reached. Due to the dynamic nature of the soldering peak the temperature will diverge between different locations of the assembly, which results in an earlier solder paste reflow at small components (e.g. chip capacitors) and a retarded solder paste reflow at bulky components (e.g. BGAs). Therefore an optimization of the solder reflow profile needs to be undertaken to provide robustness for the soldering process while keeping the thermal damage little [1,2].

Temperature monitoring

In order to optimize a specific reflow profile continuous temperature recordings were carried out during the soldering process, such as to monitor the temperature differences between small and bulky components. Usually a set of thermocouple wires is mounted at the locations of interest on the assembly and connected to a battery powered datalogger instrumentation. This setup is then moved through the reflow oven together with the assembly under test. Compared to other types of temperature sensors thin wired thermocouples have very low thermal mass, which allows to monitor the dynamic temperature profile accurately over time. Moreover they provide a very economical alternative, because after usage the wire can be cut from the tip and be welded to form a new thermocouple tip. In contrast many other sensors can be used only one time, when they were fixed to the assembly to provide a good thermal contact [1,2].

Thermocouple temperature sensing is based on a voltage that is generated by a temperature difference across the length of a wire. Since the generated voltage relates to the temperature in a material dependent ratio, a pair of two different metals or metallic alloys can be used as a temperature sensor. The most commonly used pair is the Ni-Cr/Ni-Al combination, also known as the K-type thermocouple. The K-type has a good sensitivity of approx. 41 µV/K and it impresses through its superior linearity compared to other material combinations. Therefore, in the time of analogue signal conditioners, when linearization was a challenge to the circuit design, K-type thermocouples became the prefered choice for precise temperature measurements. Most of the dataloggers that are used to optimize reflow profiles are designed for this thermocouple type, and K-type thermocouples are the standard in the electronic manufacturing industry [1-3]. However, there is another material combination that shows similar properties in the targeted temperature range (25 °C …300 °C). The Cu/Cu-Ni combination, also known as the T-type thermocouple, which has a sensitivity of approx. 43 µV/K at room temperature. However its sensitivity is strongly temperature dependent [3,4].

Figure 2 depicts the sensitivity of both thermocouples gained from a laboratory measurement using a Voltcraft TC-150 Calibrator generating the test temperatures, two Rigol DM3061 DMMs to measure the thermocouples voltage, and a two channel Beha Amprobe TMD-56 temperature-datalogger to provide cold junction compensation for the DMM inputs. The results of the calibration test show a fairly constant value of approx. 41 µV/K for the sensitivity of the K-type thermocouple, while the same value for the T-type thermocouple rises nearly linearly from 40 µV/K at 10 °C to a value of 58 µV/K at 365 °C.

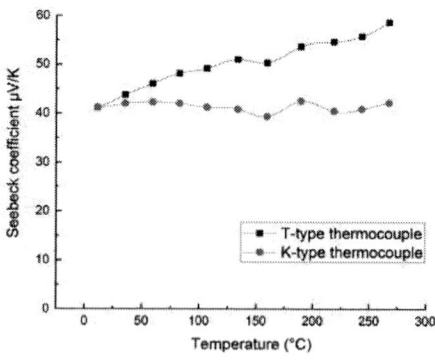

Figure 2: Experimental determination of the sensitivity of a K-type thermocouple (black line), which shows a fairly constant value of approx. 41 µV/K, and a T-type thermocouple (red line), which increases with temperature from 40 µV/K at 10 °C to a value of 58 µV/K at 365 °C.

Experiments

One of the benefits of T-type thermocouples is the improved wetting behavior of Sn-based solder materials (which is good for Cu and sufficient for Cu-Ni) compared to wetting on K-type materials Ni-Al or Ni-Cr. The goal of experimental investigations is to figure out, if T-type thermocouples are not only feasible to measure the board temperature but also the solder temperature during the joining process.

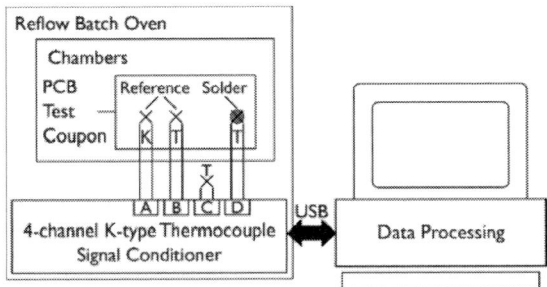

Figure 3: Schematic of the usage of the four channel temperature recorder applied within the laboratory batch reflow oven (LPKF Protoflow S) for evaluating the feasibility of T-type thermocouples to register the solder temperature during a reflow process.

In order to set up a simple pilot experiment a standard laboratory batch reflow oven (LPKF Protoflow S) was prepared for a reflow profile recording using T-type thermocouples. The laboratory batch oven is equipped with a four channel temperature recorder designed for K-type thermocouples.

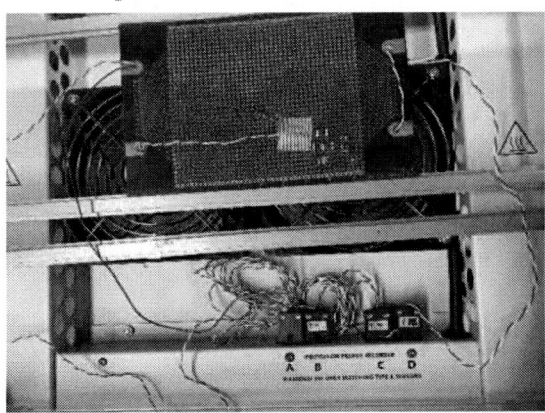

Figure 4: Photograph of the experimental setup for evaluating the feasibility of T-type thermocouples to measure the board and solder temperature during the joining process. Channel A is connected to a K-type thermocouple and Channels B to D are connected to T-type thermocouples. Channels A&B serve as reference board temperature measurements. Channel C measures the temperature at the input of the temperature controller. Channel D records the solder temperature.

The drawing in figure 3 illustrates the use of the temperature recorder and the photograph in figure 4 shows the placement of the thermocouples in the laboratory batch oven for the experiment. Channel A was connected to a K-type thermocouple, which was mounted directly on the board in order to record a reference temperature for the reflow process. Channel B was set up similarly to Channel A by the use of a T-type thermocouple. Channel C was connected to a T-type thermocouple that was placed in the plug & socket area of temperature recorder. Channel D was connected to a T-type thermocouple. The two wire tips are connected by a solder depot. This enables to directly measure the temperature in the solder during the reflow process. Figure 5 depicts schematically the connection of the two thermocouple wires by a solder depot.

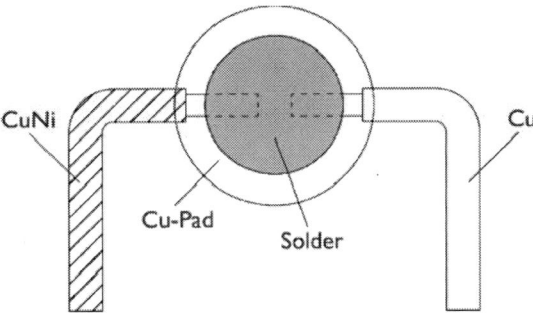

Figure 5: Schematic (top view) of the thermocouple tip connection of the two wires by a solder depot.

The 'solder-connection' of the two thermocouple wires can be realized in various ways. One possibility is to insert the two wires from the opposite sites into the solder depot (see figure 6).

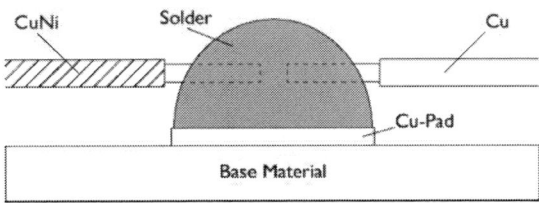

Figure 6: Schematic (side view) of thermocouple tip connection of the two wires by a solder depot on a Cu-pad of a PCB.

Another possibility is to use the Cu-pad as one wire tip [5]. The solder depot is dispensed onto the Cu-pad, while the CuNi-wire is inserted into the solder (see Figure 7).

The results of the measurement are depicted in figure 8. The thermocouple assignment refers to the schematic shown in figure 3. The black line refers to the K-type thermocouple (Channel A) and the red line refers to the T-type thermocouple (Channel B) measuring the board temperature as a reference. The blue line refers to the T-type thermocouple (Channel D), which is using the arrangement shown in figures 5 and 6 to measure the solder temperature during the reflow.

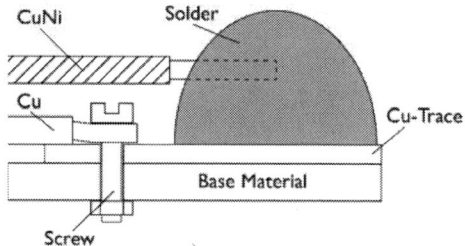

Figure 7: Schematic (side view) of a thermocouple tip 'solder- connection' using the Cu-pad as one wire tip [5]

The temperature measurement made by the T-type thermocouples was corrected by adjusting the values from the channel B measurement with that of channel A. That way a correct temperature value could be adjusted for T-type thermocouple measurement of the solder temperature (channel D). The comparison of the temperature values gained by channel D and those gained by channel B show the difference between an on board temperature measurement and the direct determination of the solder temperature.

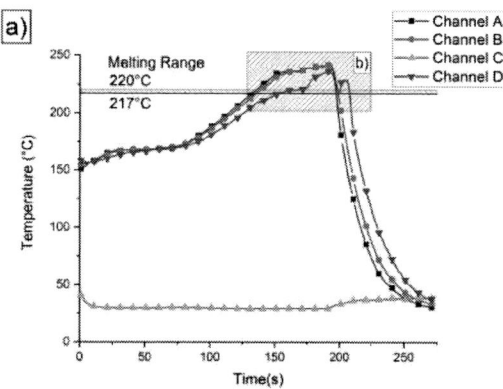

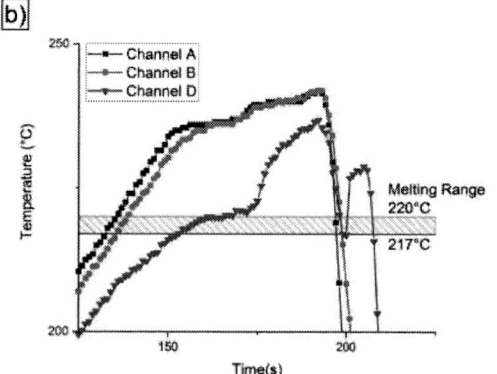

Figure 8: Results of the temperature measurements on a PCB in the batch reflow oven: a) complete graph, b) detail enlargement

Before approaching the peak phase of the reflow profile (see figure 1) the channel D thermocouple lags slightly behind the channel B thermocouple due to its higher thermal mass at the tip resulting from the relatively large solder depot. However by the entry of the peak phase the difference between the two recorded temperatures increases. While the channel B recording rises proportional to the oven temperature, the channel D temperature seems to be frozen at a specific value (220 °C), which equals the upper temperature of the melting range of the solder alloy (Chip Quik T3: SnAg3.0Cu0.5, $T_M = 217 ... 220$ °C). After a period of nearly 15 s also the recordings of channel D rise towards the peak temperature. An inverse behavior can be observed in the cooling phase. While the channel B recording declines exponentially, the channel D decline is interrupted at the melting temperature (217 °C), where the temperature rises shortly before it continues with the exponential decay. Both observations can be related to the latent heat that is absorbed or released, when the solder goes through solid-liquid phase transition.

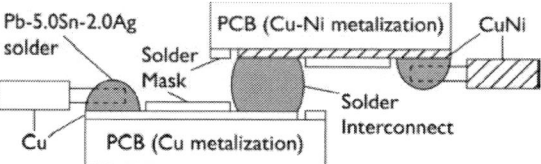

Figure 9: Schematic of a BGA ball acting as T-type thermocouple tip between a Cu-Ni and a Cu pad

Although the assemblies depicted in figure 6 and figure 7 are capable to measure the solder temperature during the reflow process, it is difficult to adapt those to true measurements on real components. Therefore a different measurement setup was developed, which is shown in figure 9. It is based on two identical PCBs with different metallization matching that of the T-type thermocouple. While the bottom PCB has a standard Cu metallization, the upper side features the necessary Cu-Ni. The Cu-Ni specimen is shown in detail in figure 10a. It is manufactured by standard PCB processes which are described in detail in [6]. Basicaly, a Cu-Ni foil (Cu-45Ni, thickness 5 μm) is laminated to a metallization free base material (FR4). The desired Cu-Ni structures are then etched with a $CuCl_2$ solution (with increased HCl and H_2O_2 content) and finally covered with a structured solder mask.

For the subsequent assembly process the five solder pads (⌀ 600 μm) in the center of the Cu-Ni specimen, shown in figure 10a are manually bumped with SnAg3.0Cu0.5 solder balls (⌀ 800 μm). The bumped Cu-Ni specimen is then soldered face to face to an identical Cu specimen as shown in figure 10b. The thermocouple extension wires, necessary for the connection to the measurement device (TC-08 Thermocouple Datalogger from Pico Technology), are hand soldered to the pads at the PCBs edges (see figure 10b) using a Pb-5.0Sn-2.0Ag solder that does not melt during reflow (melting point above 296 °C).

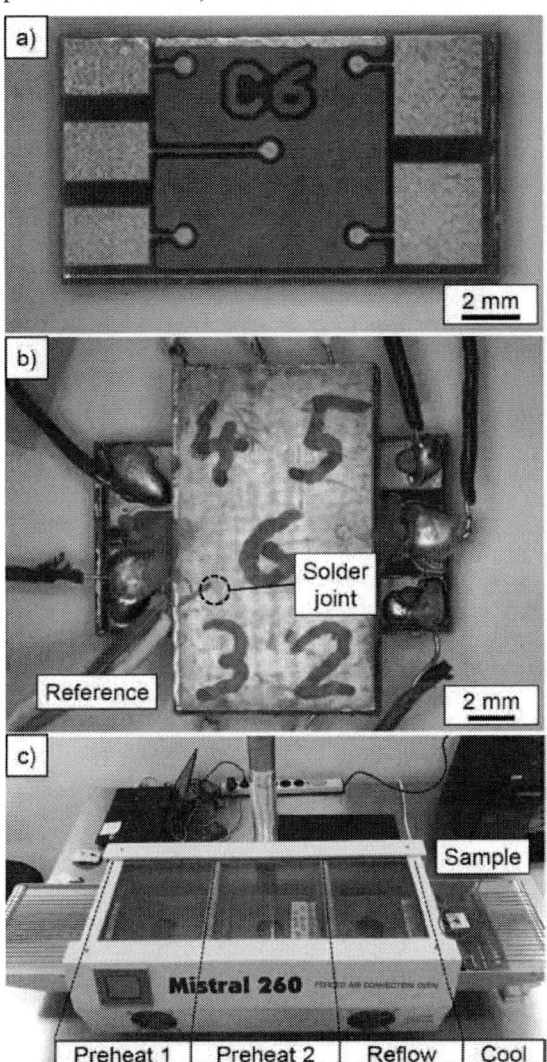

Figure 10: Experimental setup for the reflow experiment on BGA solder balls: a) structured Cu-Ni PCB specimen; b) assembled sample with measurement positions; c) Mistral 260 reflow oven during the experiment with the sample in the cooling zone

The TC-08 datalogger can handle different types of thermocouples and provides a cold junction compensation. It uses a multiplexer to enable an eight channel thermocouple input and one channel for the automatic internal cold junction compensation. The maximum sampling rate per channel is 10 S/s. Therfore only 2 channels were used for the experiment, in order to keep to all over sampling rate at 3 S/s.

The reflow experiment is shown in figure 10c and was carried out in a Mistral 260

reflow oven with three heating zones. Since the datalogger cannot pass through the oven, the T-type extension wires have a length of 2 m in order to reach from one end of the oven to the other. The accuracy of these manufactured T-type thermocouples was compared to a standard K-type thermocouple (wire) in a temperature controlled furnace (Binder FP53). The temperature difference between both thermocouples did not exceed ±0.4 K for all investigated temperatures (20 °C, 90 °C, 140 °C, and 180 °C).

A pilot reflow experiment was carried out using the specimen shown in figures 9 and 10a. The results of the temperature measurement on SnAg3.0Cu0.5 solder are given in figure 11. The diagram contains the profiles of the reference temperature measured by a K-type thermocouple (wire), the solder joint temperature measured through one of the five solder joints acting as the tip of a T-type thermocouple, and the internal TC-08 cold junction temperature. The approximate measurement positions are given in figure 10b. The resulting profiles for reference and solder joint are very similar (see figure 11a). The profile clearly shows the different zones of the reflow oven and one can see that the solder joint temperature recording rises earlier during transition into a new zone. For the preheat zones both temperatures recordings (Reference, Solder) approach each other by the middle of each zone.

The melting and the solidification can be detected by the crossing of the red (reference temperature) and the black line (solder temperature) enlarged in figures 11b and 11c, which is similar to the obeservations depicted in figure 8b. The melting of the solder material is indicated by the horizontal course of the solder joint temperature in figure 11b. It becomes obvious that the supposed melting temperature at approx. 237 °C is significantly above the solder's melting range between 217 °C … 220 °C. The melting onset known from the Differential Scanning Calorimetrie (DSC) measurements reported in [7], which were carried out with very slow heating rates (5 K/min), is very different from the current results. It is not completely clear what causes the higher melting temperature recording. Therefore further experiments need to be carried out to understand this phenomenon.

The solidification of the solder joint is depicted in the enlarged view of the reflow profile in figure 11c. The crossing of the red (reference temperature) and the black line (solder temperature) indicates the beginning of the solidification. Moreover a characterisitic rise in temperature can be observed, which can be related to the release of latent heat during the solidification. The solidification temperature of approx. 145 °C shows significant undercooling which is well known for these solder alloys [8].

481

These two characterisitic shapes in the course of the solder temperature compared with the reference temperature indicate supposedly the onset of melting or solidification. Therefore the application of T-type thermocouples provides valuable additional information compared to the standard profile acquisition by K-Type thermocouples.

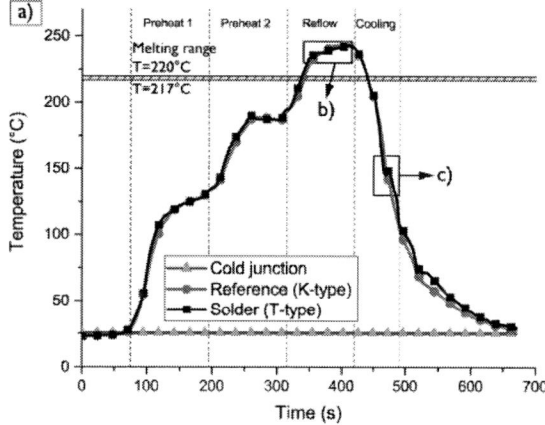

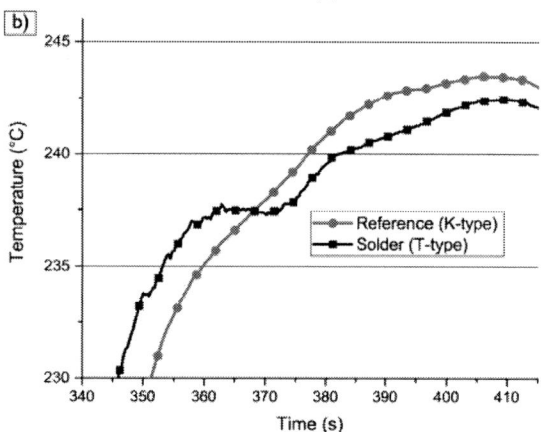

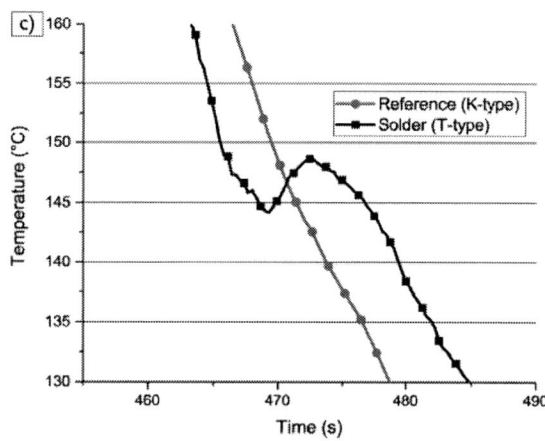

Figure 11: Temperature measurements during reflow of BGA solder balls forming the tip of a T-type thermocouple: a) complete reflow profile with positions of the different reflow zones, detail enlargements of b) the melting, and c) the solidification of the solder joint

Conclusions

T-type thermocouples were evaluated concerning the temperature profile measurements for solder reflow processes. A pilot experiment was carried out on a standard laboratory batch reflow oven, which was equipped with a four channel temperature recorder designed for K-type thermocouples. K-type and T-type thermocouples were connected to the temperature recorder in different ways. The results show, that T-type thermocouples are feasible to record solder temperatures during the reflow process. In order to record temperatures on the component level, a BGA type component was manufactured by soldering a PCB with Cu-Ni metallization on top of a PCB with Cu metallization. The solder joint itself acts as the tip of the resulting T-type thermocouple. The temperature profile therefore reliably shows the temperature of the solder joint with its characteristic melting and solidification behavior.

Acknowledgements

The authors like to thank Dr. Gerald Hielscher, Dr. Krysztof Nieweglowski, Dr. Christian Wenzel and Dr. Martin Straub for their contributions in the specimen manufacturing and Dr. Andreas Ruh for his support in experiment preparation and evaluation.

References

[1] H. Bell, "Reflowlöten - Grundlagen, Verfahren, Temperaturprofile und Lötfehler", Eugen G. Leuze Verlag, first edition, Bad Saulgau, 2004.

[2] A. Rahn, "Erfassung von Lötprofilen - Methodik, Fehlerquellen, Messtoleranzen", Eugen G. Leuze Verlag, first edition, Bad Saulgau, 2008.

[3] M. X. Maida, "IC Temperature Sensor Provides Thermocouple Cold-Junction Compensation", National Semiconductor Linear Applications Handbook: Application Note 225, Santa Clara, CA: National Semiconductor 1994.

[4] J. Marcin, "Thermocouple Signal Conditioning Using the AD594/AD595", Application Note 274, Norwood, MA: Analog Devices 2016

[5] M. Mueller, S. Wiese, K.-J. Wolter, "Influence of cooling rate and composition on the solidification of SnAgCu solders", IEEE-Proceedings of the 1st Electronics System-integration Technology Conference (ESTC), 5-7 Sept., Dresden (D) 2006.

[6] D. Čáp, "In situ Temperaturmessung einzelner Lotkontakte während des Lötens", Studienarbeit in der Studienrichtung Geräte-, Mikro- und Medizintechnik, Technische Universität Dresden, Dresden 2018

[7] S. Schindler, S. Wiese, "Solidification processes of SnCu, SnAg and SnAgCu solder alloys and interface reactions to charactize solar cell interconnections processes", IEEE-Proceedings

of the 18th European Microelectronics & Packaging Conference, 12-15 Sept., Brighton (GB) 2011.

[8] S. Schindler, M. Mueller, S. Wiese, " Investigation of the undercooling of SnCu solder", IEEE-Proceedings of the 5th Electronics System-integration Technology Conference (ESTC), 16-18 Sept., Helsinki (FIN) 2014.

High Frequency Substrate Technologies for the Realisation of Software Programmable Metasurfaces on PCB Hardware Platforms with Integrated Controller Nodes

D. Manessis[1], M. Seckel[1], L. Fu[2], O. Tsilipakos[3], A. Pitilakis[3], A. Tasolamprou[3], K. Kossifos[4], G. Varnava[4], C. Liaskos[3], M. Kafesaki[3], C. M. Soukoulis[3], S. Tretyakov[2], J. Georgiou[4], A. Ostmann[1], R. Aschenbrenner[1], M. Schneider-Ramelow[1], and K-D. Lang[5]

[1]Fraunhofer Institute for Reliability and Microintegration (IZM), Gustav-Meyer-Alle 25, 13355 Berlin, Germany
[2]Aalto University, Dept. of Electronics & Nanoengineering, FI-00076 Aalto, Finland
[3] Foundation for Research & Technology – Hellas /FORTH, 71110 Heraklion, Crete
[4] University of Cyprus, Dept. of Electrical & Computer Engineering, 1678 Nicosia, Cyprus
[5]Technical University of Berlin, Gustav-Meyer-Alle 25, 13355 Berlin, Germany

Corresponding Author: +49-30-46403788, Dionysios.Manessis@izm.fraunhofer.de

Abstract

The proposed work is performed in the framework of the FET-EU project "VISORSURF", which has undertaken research activities on the emerging concepts of metamaterials that can be software programmable and adapt their properties. In the realm of electromagnetism (EM), the field of metasurfaces (MSF) has reached significant breakthroughs in correlating the micro- or nano-structure of artificial planar materials to their end properties. MSFs exhibit physical properties not found in nature, such as negative or smaller-than-unity refraction index, allowing for EM cloaking of objects, reflection cancellation from a given surface and EM energy concentration in as-tight-as-possible spaces.

The VISORSURF main objective is the development of a hardware platform, the Hypersurface, whose electromagnetic behavior can be defined programmatically. The key enablers for this are the metasurfaces whose electromagnetic properties depend on their internal structure. The Hypersurface hardware platform will be a 4-layer build-up of high frequency PCB substrate materials and will merge the metasurfaces with custom electronic controller nodes at the bottom of the PCB hardware platform. These electronic controllers build a nanonetwork which receives external programmatic commands and alters the metasurface structure, yielding a desired electromagnetic behavior for the Hypersurface platform.

This paper will elaborate on how large scale PCB technologies are deployed for the economical manufacturing of the 4-layer Hypersurface PCB hardware platform with a size of 9"x12", having copper metasurface patches on the top of the board and the electronic controllers as 2mmx2mm WLCSP chips at 400μm pitch assembled at the bottom of the platform. The PCB platform designs have stemmed from EM modeling iterations of the whole stack of high frequency laminates taking into account also the electronic features of the controller nodes. The manufacturing processes for the realization of the selected PCB architectures will be discussed in detail.

Key words: Metasurfaces, metamaterials, software defined materials, High frequency substrates.

1. Introduction

Metasurfaces (MSs), the two dimensional versions of metamaterials, are ultrathin periodic structures with designed, subwavelength building blocks enabling exotic functionalities [1-2]. The building blocks can consist of metallic, dielectric, semiconducting, and 2D material inclusions. Tunable metasurfaces can be realized by modifying the properties of the meta-atoms via an external stimulus, leading to adjustable and, in some cases, reconfigurable functions. The available tuning schemes can be classified as "global" or "local" depending on the ability to tune the unit cells collectively or independently. Global control of the

unit cells can, for example, enable tunable perfect absorption, whereas local control can provide more advanced functionalities such as wavefront manipulation, steering or focusing [3,4, 10-11]. An efficient control mechanism naturally suited to a *local* tuning scheme is that of *voltage-controlled* lumped electronic elements incorporated inside the meta-atom to provide control over the MS properties [5,6, 10-11]. In the framework of the VISORSURF project this is accomplished by specifically-designed integrated circuits (chips), embedded in the meta-atoms. The control of the circuits can be software-driven and the behavior of the metasurface can be defined programmatically; this is the concept of the HyperSurface (HFS) [7]. In addition, the controllers can be nano-networked [7-9] enabling the broad vision of smart devices in the emerging Internet of Things paradigm. In this work, we present the concept of the Hypersurfaces with a focus on their electromagnetic aspects [10-11].

2. The Hypersurface Concept of The Visorsurf project

The functional and physical architecture of the Hypersurface tile is presented in Fig. 1(a) and it consists of a metasurface layer, an intra-tile control layer and a tile gateway controller. This paper focuses on the metasurface layer which realizes the electromagnetic functionalities. In the overall concept of the HFSs, there is a switch fabric design for operation in the microwave (GHz) regime.

The unit cell of the switch fabric is presented in Fig. 1(b). It consists of four copper patches (size is in millimeter range) residing on a low-loss dielectric substrate backed by a copper plate. A chip is attached in each unit cell, lying behind the copper plate, shown in Fig. 1(c), and its physical dimensions are also subwavelength. The chips are interconnected by means of communication tracks. The four patches are connected to the RF ports of the chip by means of through vias. When microwave radiation impinges on the chip-loaded unit cell, it induces local currents in the patches and the metallic substrate; in turn, the induced currents act as secondary electromagnetic sources modifying the scattered field which leads to the desired operation, when the aggregate effect of all unit cells of the metasurface is accounted for. Consequently, for the modification of the metasurface response, one needs to adjust dynamically the complex-valued surface impedance in each unit cell. This is achieved with the controller chip [Fig. 1(c)], which is judiciously placed behind the backplate in order to minimize interference with the impinging electromagnetic wave. By controlling the resistive and reactive contributions in each chip we can demonstrate an angle-tunable perfect absorber that can operate inside the 4.5-5.5 GHz range, and more advanced functionalities of wavefront manipulation such as anomalous reflection [10-11]. This paper describes in detail the manufacturability of such large metasurface panels based on standard industrial PCB processes conducted in the Substrate line of Fraunhofer IZM.

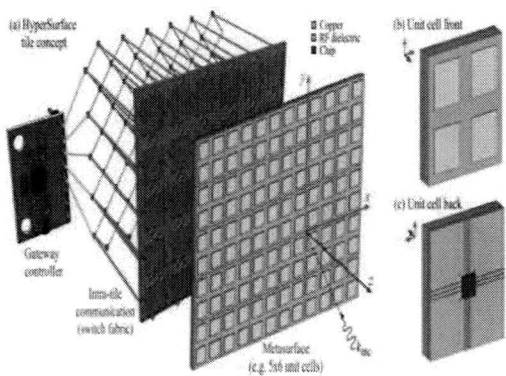

Figure 1:(a) The HyperSurface tile. The switch state configuration setup provides the desired function. A controller intra-network communicates the relevant commands and the inter-tile and external communication are handled by standard gateway hardware, such as an FPGA. Unit cell of the metasurface: (b) front side with periodic square 2x2 patches pattern and (c) back side with controller chip (dark grey box) and communication lines.

3. Manufacturing feasibility study on 3-layer HSF substrate

This Section presents the first manufacturing results of a 3-layer HSF substrate using a stack of a HF Rogers laminate material and a high-Tg FR4 material. The first 3-layer HSF substrate was constituted of a thick HF laminate based on Rogers RT5880 laminate and an underlying conventional high-Tg PCB prepreg. Figure 2 shows a perspective of a single unit cell from its top side which is the patch layer and its bottom side which is the chip layer. The copper patch is 4.2mm scquare; the space between them is 0.8mm. The HSF substrate consisted of 8 single unit cells with a final size of 96mmx96mm, and is shown in the bottom of Figure 2.

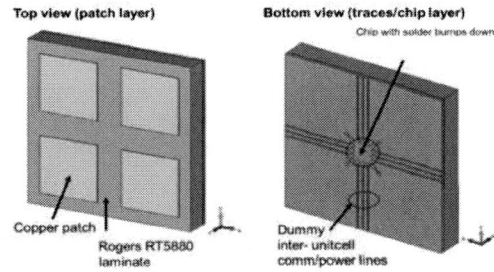

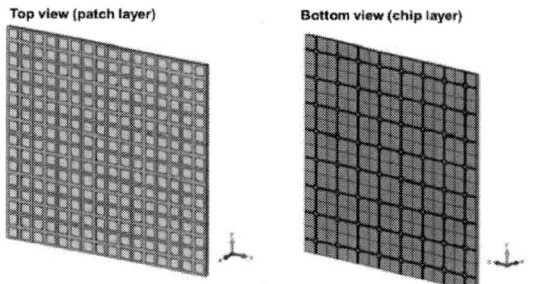

Figure 2: Perspective of a single unit cell and the whole HSF consisted of 8 single unit cells (96mmx96mm).

The geometrical design of the 3-layer board is shown in Figure 3 where the copper layer (1 st layer), the ground backplane layer (2nd layer) and the chip/copper trace layer (3rd layer) can be distinguished. The copper layers were 35μm thick and the high frequency RT5880 laminate was 1.575 mm whereas the underlying Hitachi 679FGS prepreg was 0.254mm. The whole thickness of the HSF substrate was targeted to be 1.934mm. There were 4 through vias from the patch layer to the chip layer which were mechanically drilled and had a diameter of 300μm.

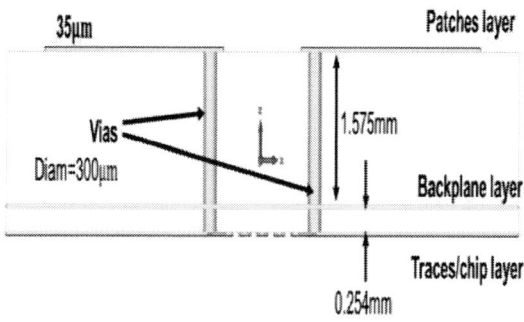

Figure 3: Geometrical cross section of the 3-layer HSF stack-up. Total estimated thickness is 1.934mm.

After the drilling of the vias, a typical permanganate desmear process was followed to clean the via from rest of the RT5880 material. However, such desmear process is not appropriate for RT5880 material and the copper has not really stuck to the via walls. No copper electroplating has taken place in the through vias.

Although repeated desmear processes were run with various parameters, the vias could not be cleaned efficiently and could not be electroplated at all. Therefore, the recommended process for Teflon based laminates was followed which is a plasma process. After 20 min plasma desmear, the through vias were cleaned efficiently and copper electroplating had been successful. Figure 4 shows cross sections of the 3-layer HSF after plasma desmear and electroplating. The copper wall

thickness was increased to 20μm as shown in Figure 4. Figure 5 shows the corresponding X-Ray from the 3-layer HSF substrate from the chip layer with the vias filled with copper (dark contrast).

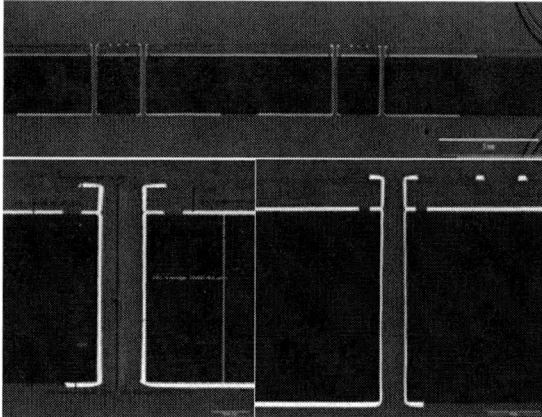

Figure 4: Cross-sections of the 3-layer HSF board with the bottom RT5880 laminate layer and vias filled with copper. Vias were cleaned with plasma desmear.

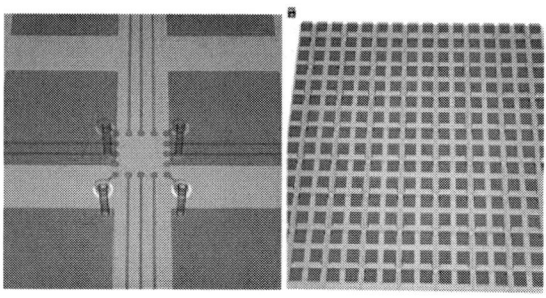

Figure 5: X-Ray of the 3-layer HSF stack-up from the chip layer side view. The Chip and the 4 vias can be seen filled with copper (dark contrast) due to used plasma desmear processes in the RT5880 laminate. On the right side, a X-Ray perspective of the whole HSF substrate (96mmx96mm) is provided from the chip side.

After structuring the first layer (patch layer) and the third layer (chip layer), 5μm Ni/80nm Au was applied on both layers whereas solder mask was also coated on the chip layer. Figure 6 shows pictures of the copper patches and the chip bottom side.

Figure 6: 3-layer HSF substrate 96mmx96mm in size.

The results provide ample evidence that the processing of RT5880 Teflon based laminate is quite challenging with conventional PCB processes and other alternative processes like plasma desmear should be employed, raising the cost of the process significantly. Furthermore, due to usage of different prepreg and laminate materials with respect to their thermal expansion coefficient, the 3-layer HSF board has shown a warpage of about 2mm, making the assembly of the controller chips on the third layer quite challenging. For these reasons, it was decided to deploy the Megtron 7N family of materials which are available both as laminates and prepregs. That would ensure the usage of uniform materials and the elimination of warpage at large production panels of 9"x12" (22cmx30cm) or 12"x12" (30cmx30cm) during the course of the project.

4. Manufacturing of 4-layer HSF substrate

After the experience gained in the feasibility studies with the 3-layer stack and having evaluated the pros and cons of employing the Rogers RT5880 Teflon based material, the Consortium decided to switch to Megtron 7N materials which could become readily available both as prepregs and laminates and could be potentially processed with conventional PCB processes. The Megtron 7N materials were available as R-5785(N) laminates with a thickness of d_1:750μm and d_2: 100μm and as prepreg R-5680(N) with a glass fiber of type 2116 and a thickness of 100μm. It was decided to build a symmetric and a slightly asymmetric stack-up to evaluate the final warpage effect on a 9"x12" board. The actual PCB process sequence is not differentiated whether a symmetric or asymmetric stack up is chosen.

○ *Symmetric and asymmetric laminate designs for 4-layer substrate*

Figure 7 shows the symmetric and asymmetric stack up next to each other for better comparison. The only slight difference between the two designs are the missing laminate (d_2) and a prepreg (t_1) on top of the asymmetric design which makes it eventually slightly thinner than the symmetric one. Figure 7 shows the stack-ups used and the main vias for interconnectivity among layers; namely the L2-L3 and L1-L4 through vias as well the L4-L3 blind vias.

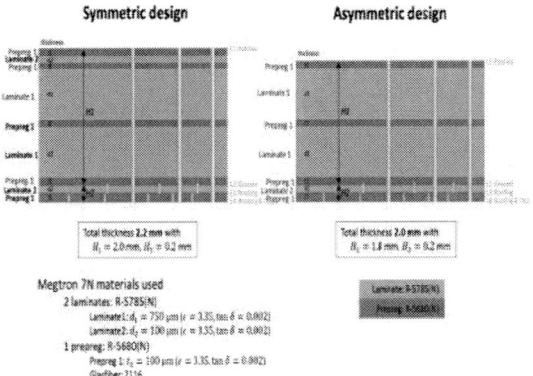

Figure 7: Symmetric and asymmetric stack-ups in comparison. Megtron 7N materials are being used.

○ *Process developments and flow for 4-layer substrate*

The process flow was decided based on the design requirements and especially the fine line L/S for the chip layer where a thin copper should exist at the end of the electroplating process so as to achieve the fine L/S under the chip area. Figure 8 provides the process flow and some dimensional via details.

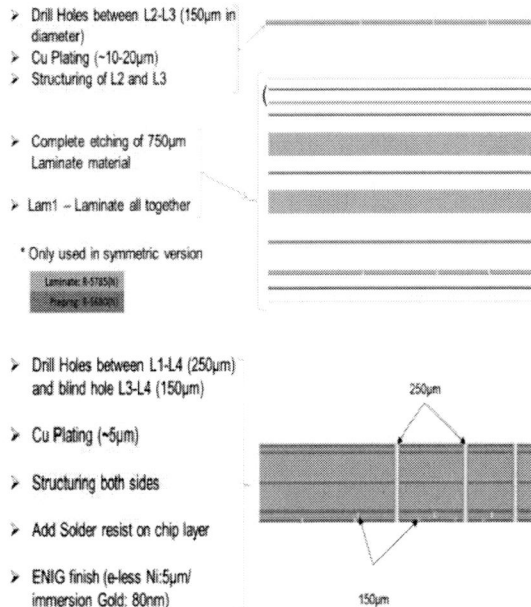

Figure 8: Process flow for 4-layer HSF substrate.

○ *Manufacturing process results and challenges*

The manufacturing of the 4-layer HSF substrate provided the first processing experience with Megtron 7N materials and yielded very good results and revealed the processing challenges that need further developments before the launch of the next HSF substrates. Figure 9 shows a cross section of the symmetric 4-layer substrate after final processing. The thicknesses of the symmetric and asymmetric are 2.4 mm and 2.2 mm, respectively. Both values are 200µm higher than the estimated ones, mainly attributed to deviation in prepreg and lamination thicknesses in the stack compared to nominal ones and from slightly higher copper thicknesses after electroplating on the copper layers.

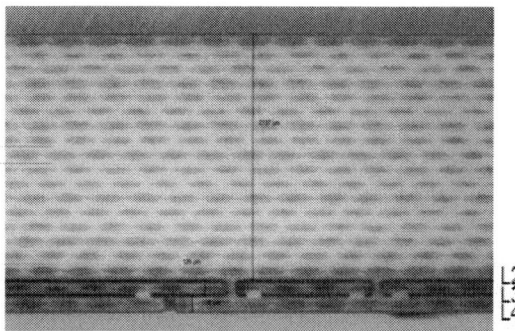

Figure 9: Cross section of the 4-layer stack.

The major challenges for the 4-layer HSF substrate are the fine line structuring of the bottom chip layer, the opening of the L4-L3 blind vias and the electroplating of all vias. As shown in Figure 10, very fine L/S of 45µm/55µm was successful with a copper thickness of 15µm. The application of solder resist and the deposition of 5µm Ni/80nm Au metallisation is also shown.

Figure 10 shows on the top the chip position where 45µm/55µm Line/Spacing was achieved and at the bottom the L3-L4 blind vias and the structuring around it. A 100µm line was structured. The results provide ample evidence that such fine copper structuring especially under the chip area is feasible.

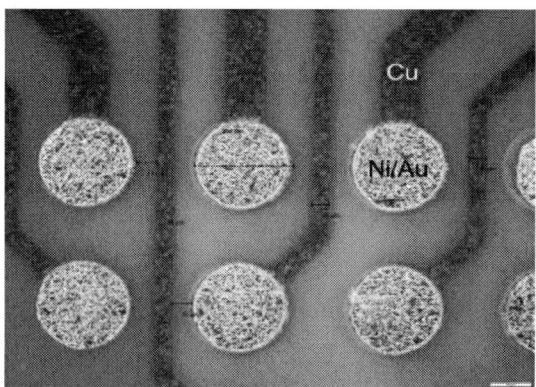

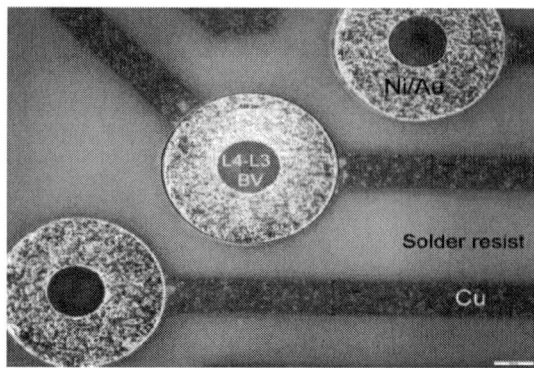

Figure 10: Bottom layer of HSF board at chip position. Fine line structures of 45µm/55µm L/S achieved. Structured bottom layer in proximity to L3-L4 blind vias.

Figure 11 shows X-Ray pictures of the 4-layer substrate where the copper filling of the L1-L4 vias, the L2-L3 vias and L4-L3 vias is witnessed with dark contrast. Via electroplating is successful.

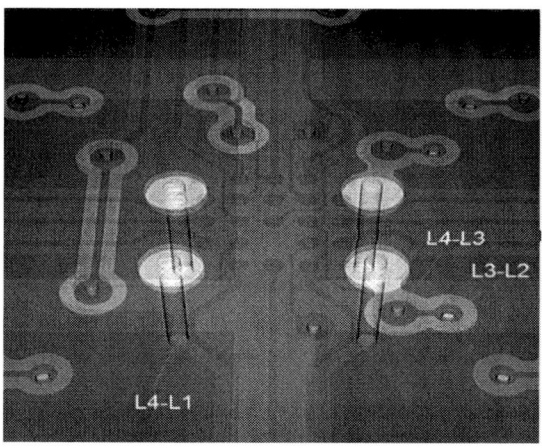

Figure 11: X-Ray of the 4-layer substrate. The dark contrast in all vias comes from the copper coating in the vias.

The through vias L4-L1 were successfully electroplated, as shown in Figure 12. It can be seen that inside the via 23µm were deposited and on the L1 and L4 layers the copper ring thickness was about 40µm and away from the ring the copper remained at about 11µm which is the ideal thickness for L1, L4 fine line structuring.

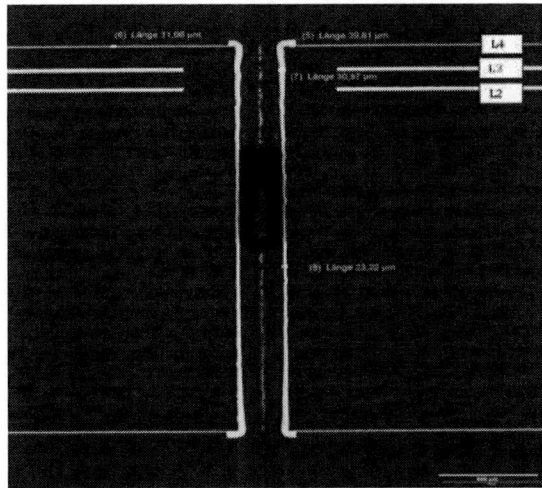

Figure 12: L4-L3-L2-L1 copper layers in a drilled through via of 250μm. Reinforced copper filling to achieve about 23μm in the via.

The blind vias L4-L3 were tried with the pico UV laser to reopen the blind vias which were initially opened by mechanical drilling. Some of the L4-L3 vias could not be efficiently opened by mechanical drilling. Few microns, around 10μm, were remaining to touch the L3 copper pad. That would be an ad-hoc solution for the boards, and mechanical drilling of all vias would be in any case preferable as a quick and economical industrial process. The laser was used to re-open and overshoot the blind vias and subsequently desmear and copper electroplating was performed. The results were very successful using the laser for the blind vias and the drilled board. In turn, the board was successfully electrically tested for continuity. Figure 13 shows the blind via, 150μm in diameter, opened with the UV laser to be also copper filled.

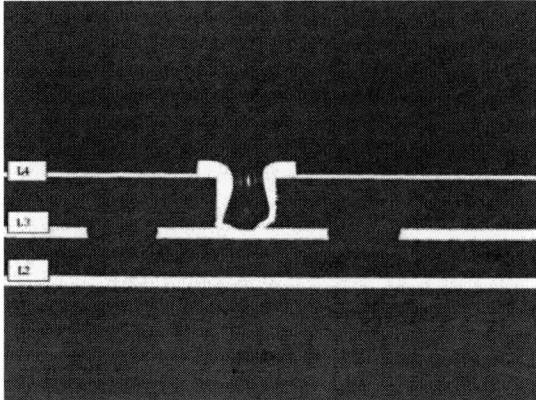

Figure 13: Blind via (L4-L3), 150μm in diameter, opened by UV laser and then copper coated.

The laser technology looks promising to solve the problem of drilled L4-L3 blind vias and is also compatible with filled Megtron 7N materials.

Figure 14 shows the 4-layer HSF substrate finished with solder resist on the bottom side. The warpage of the board, due to usage of uniform Megtron 7N materials, was less than 100μm, much smaller than the warpage of the 3-layer RT5880 board and is considered minimal from EM considerations.

The resultant HSF prototype has a net area of 6.2"x9.4" and containts 384 unit cells. It is electrically functionable and proves that the 4-layer HSF is manufacturable.

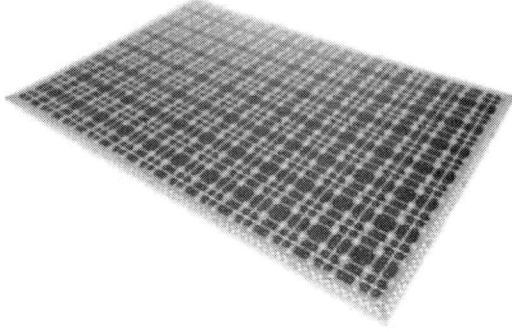

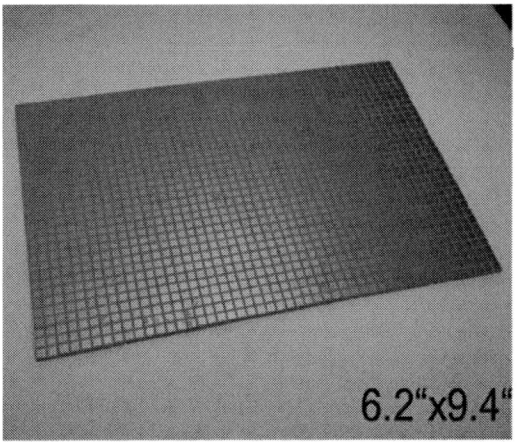

Figure 14: Bottom side (chip layer) and HSF unit cell area with Ni/Au metallization (top layer) of the 4-layer HSF substrate (net area of 6.2"x9.4", 384 unit cells)

Conclusions

This paper has presented the concept of Hypersurfaces, i.e., the approach of project VISORSURF to software-driven metasurfaces whose complex surface impedance can be locally modified with a set of programmable commands. The control is enabled by a network of voltage-driven electronic chips; they provide variable resistance and reactance values that modify the complex surface impedance of the metasurface in a local or a global manner. This paper also provides the first manufactured Hypersurfaces on panels employing industrial PCB processes. Hypersurfaces were made on a 3-layer and a 4-layer PCB consustruction which will be finally the HSF design for the project. The 3-layer substrate was made of a RT5880 Teflon laminate outstanding for its HF properties, but it was difficult to process with conventional chemical desmear processes and therefore the electroplating did not work. The usage of plasma desmear has solved the problem. The 3-layer HSF substrate has shown high warpage due to usage of non-uniform laminates. Due to warpage and processability problems, it was decided to use Megtron 7N materials available as laminates and prepregs. A first version of the VISORSURF 4-layer substrate is already manufactured. Very fine structuring of 45μm/55μm L/S was successfully demonstrated at the chip level. Through vias were also successfully drilled and copper coated. Drilling of blind vias was not consistent and the results were not reliable. As a solution to solve the problem, pico UV laser has been employed successfully for the blind vias which were also copper coated. The warpage with the usage of uniform Megtron 7N materials remained under 100μm. Panels of 9"x12" were manufactured with a net area of 6.2"x9.4", consisted of 384 unit cells. The 4-layer HSF design is slightly modified and will be manufactured again in the next 2 months in the final size of 300mmx300mm (12"x12").

Acknowledgements

This work was supported by European Union's Horizon 2020 Future Emerging Technologies call (FETOPEN-RIA) under grant agreement no. 736876 (project VISORSURF).

References

[1] C. M. Soukoulis and M. Wegener, "Past achievements and future challenges in the development of three dimensional photonic metamaterials," *Nat. Photonics*, 5, pp. 523–530, 2011.

[2] S. B. Glybovski, S. A. Tretyakov, P. A. Belov, Y. S. Kivshar and C. R. Simovski, "Metasurfaces: from microwave to visible," *Phys. Rep.*, 634, 1-72, 2016.

[3] O. Tsilipakos, A. C. Tasolamprou, Th. Koschny, M. Kafesaki, E. N. Economou and C. M. Soukoulis, "Pairing toroidal and magnetic dipole resonances in elliptic dielectric rod metasurfaces for reconfigurable wavefront manipulation in reflection," *Adv. Opt. Mater.* 6, 1800633, 2018.

[4] F. Liu, O. Tsilipakos, A. Pitilakis, A.C. Tasolamprou, M. S. Mirmoosa, N. V. Kantartzis, D. H. Kwon, M. Kafesaki, C. M. Soukoulis, and S. A. Tretyakov, "Intelligent Metasurfaces with Continuously Tunable Local Surface Impedance for Multiple Reconfigurable Functions," *Phys. Rev. Appl.*, accepted, 2019. *arXiv:1811.10082*, 2018.

[5] T. J. Cui, M. Q. Qi, X. Wan, J. Zhao and Q. Cheng, "Coding metamaterials, digital metamaterials and programming metamaterials," *Light Sci. Appl.*, vol. 3, pp. 1-9, 2014.

[6] H. Yang, X. Cao, F. Yang, J. Gao, S. Xu, M. Li, X. Chen, Y. Zhao, Y. Zheng and S. Li, "A programmable metasurface with dynamic polarization, scattering and focusing control," *Sci. Rep.*, vol. 6, 35692, 2016.

[7] C. Liaskos, A. Tsioliaridou, A. Pitsillides et al., "Design and development of software defined metamaterials for nanonetworks," *IEEE Circuits Syst. Mag.*, vol. 15, no. 4, pp. 12–25, 2015.

[8] S. Abadal, C. Liaskos, A. Tsioliaridou, S. Ioannidis, A. Pitsillides, J. Solé-Pareta, E. Alarcón and A. Cabellos-Aparicio, Computing and Communications for the Software-Defined Metamaterial Paradigm: A Context Analysis, *IEEE Access*, pp. 6225–6235, 2017.

[9] H. Taghvaee, S. Abadal, J. Georgiou, A. Cabellos-Aparicio, E. Alarcón, Fault Tolerance in Programmable Metasurfaces: The Beam Steering Case, *arXiv:1902.04509*, 2019.

[10] O.Tsilipakos et al., "Software-Defined Metasurfaces: The VISORSURF Project Approach", to be presented in 13th International Congress on Artificial Materials for Novel Wave Phenomena – Metamaterials 2019, Rome, Italy, Sept. 16th – Sept. 21st, 2019.

[11] A. Tasolamprou et. al., "The Software-Defined Metasurfaces Concept and Electromagnetic Aspects", to be presented in META 2019, Lisbon - Portugal, July 23 – 26, 2019.

Highly Thermal Conductive and Light-weight Graphene-based Heatsink

Nan Wang[1], Ya Liu[2], Lilei Ye[1], Johan Li[2]†

[1]SHT Smart High Tech AB, Kemivägen 6, SE-412 58 Göteborg, Sweden
Nan.wang@sht-tek.com and lilei.ye@sht-tek.com
&
[2]lectronics Materials and Systems Laboratory (EMSL), Department of Microtechnology and Nanoscience (MC2), Chalmers University of Technology, Kemivägen 9, SE-412 96 Göteborg, Sweden
johan.liu@chalmers.se

Abstract

With the developing trend of miniaturization and integration of modern electronic devices, commercial heatsinks materials, like copper and aluminum, are facing more and more challenges, such as inefficient cooling performance, large size and heavy weight. Here, we solve the problem by developing a novel highly thermal conductive and light-weight graphene heatsink. Composed by vertically-aligned and continuous graphene structures, heat transport was highly efficient from the base to fin structures inside the heatsink. The maximum through-plane thermal conductivity of graphene heatsink can be up to 1000 ~ 1500 W/mK, which is over 7 times higher than aluminum, and even outperforms copper about 4 times. Graphene heatsink demonstrated outstanding cooling performance which was superior to copper heatsink with the same dimension and same power input. Noticeably, the graphene heatsink also has important advantages of light-weight and high emissivity. The measured density (1.1 g/cm³) is only one-eighth of copper and less than half of aluminum and emissivity is about ten times higher than pure copper and aluminum. The resulting graphene heatsink thus opens new opportunities for addressing large heat dissipation issues in weight driven electronics and other high power systems.

Key words: graphene, heatsink, thermal conductivity, cooling

Introduction

The life-time performance and reliability of power devices such as power transistors, lasers and light emitting diodes (LED) are strongly related to their working temperature[1]. Efficient cooling is therefore highly essential. A heatsink is designed to maximum the heat exchange between power components and surrounding cooling mediums, and thereby preventing overheating, premature failure, and improving the reliability and performance of the components. Commercial heatsinks are mainly made by copper and aluminum alloys due to their relatively high thermal conductivity and good ductility. However, their applications are facing more and more challenges in power electronics due to the tremendously increased power density induced by miniaturization and high integration. In addition, commercial heatsink materials are suffering some other issues related to material properties, such as large size, rigid structure and heavy weight[2]–[5].

Here, a highly thermal conductive and light-weight heatsink based on graphene composites was developed to address the above issues. Composed by vertically-aligned and continuous graphene structures, heat transport was highly efficient from the base to fin structures inside the heatsink. The maximum through-plane thermal conductivity of

graphene heatsink can be up to 1000 ~ 1500 W/mK, which is over 7 times higher than aluminum, and even outperforms copper about 4 times. Graphene heatsink demonstrated outstanding cooling performance which was superior to copper heatsink with the same dimension and same power input. Noticeably, the graphene heatsink also has important advantages of light-weight and high emissivity. The measured density (1.1 g/cm3) is only one-eighth of copper and less than half of aluminum and emissivity is about ten times higher than pure copper and aluminum. The resulting graphene heatsink thus opens new opportunities for addressing large heat dissipation issues in weight driven electronics and other high power systems.

Experimentals

Graphene heatsink was fabricated from vertically aligned graphene composites. It was customerized to certain shapes according to different applications. As shown in Figure 1, the graphene heatsink used in this study has a size of 27*27*13mm. It includes 25 cooling pins acting as main heat dissipation areas. The cooling pin has a size of 3*3*10 mm. A copper heatsink with the same size was also fabricated as the reference.

Bulk thermal conductivity of graphene heatsink was measured by Laser Flash equipment (LFA 447). A home-made testing platform was built to show the

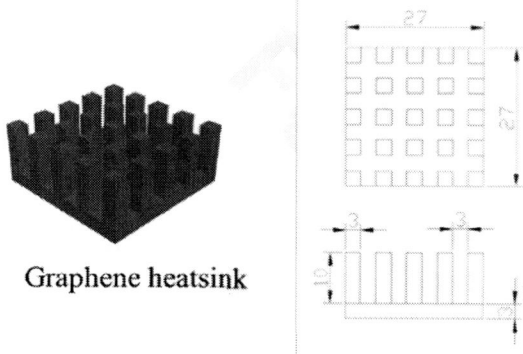

Figure 1: Schematic graphene heatsink design.

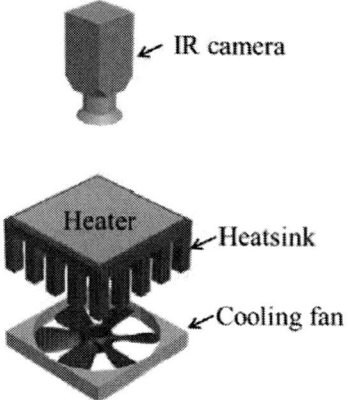

Figure2: Schematic graphene heatsink design.

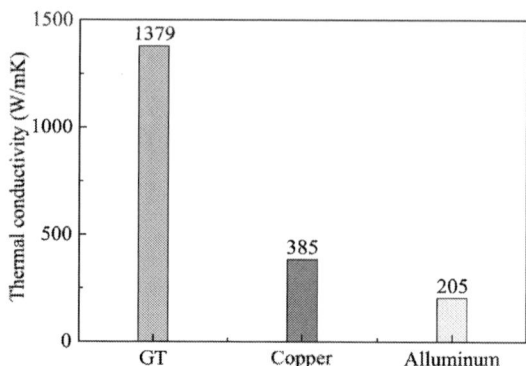

Figure 3: Bulk thermal conductivity comparison between graphene heatsink and common heatsink materials including copper and aluminum.

cooling performance of graphene heatsink. Copper heatsink was used as comparison. As shown in Figure 2, a heater was fixed on the top surface of the heatsink. To ensure good heat conduction from heater to heatsink, thermal grease with a thickness of 150μm was applied between them. A cooling fan (KF0210C1MR) was used to take away heat from heatsink. A thermal imaging camera (Therm CAM PM595 NTSC) was used to record the temperature change on the heater. The surface of the heater was coated by a thin layer of graphite to improve the emmisivity. The working power of the heater was varied from 1 to 6 W. The maximum working temperature of the heater in full power is 150°C. The input power of the cooling fan is 0.84 W. The air flow rate is 282 LMF (linear feet per minute). The environmental temperature is 20°C.

Results and Discusssions

The cooling performance of heatsinks is mainly dominated by the heat convection process from heatsink to surroundings[6]. In the same size and environmental conditions, bulk thermal conductivity of heatsink materials determins the efficiency of such a process. Figure 3 shows the comparison of

bulk thermal conductivity between the graphene heatsink and heatsinks made by common materials, such as copper and aluminum. Composed by vertically aligned graphene structure, graphene heatsink shows extraordinary bulk through-plane thermal conductivity, which is over 4 times and 6 times higher than copper and aluminum, respectively. The superior thermal conductivity of graphene heatsink can greatly benefit fast heat dissipation from power components to surroundings.

Besides heat convection, radiation is the other important factor to affect the heat dissipation of heatsink to surroundings, especially in the case of natural convection[7], [8]. Previous studies have shown that surface emissivity of heatsink materials plays a key role on determining the heat radiation performance[9]. For common heatsink materials without surface treatment, emissivity values are lower than 0.1. Differently, the graphene heatsink has a very high emissivity about 0.98, which can be considered as an ideal radiator.

Noticeably, the graphene heatsink also has important advantage of light-weight. The measured density (1.1 g/cm^3) is only one-eighth of copper and half of aluminum.

The cooling performance of graphene heatsink was tested on a home-made testing platform. The average temperature rising of the flexible heater was recorded by IR camera upon different applied powers both in natural convection and forced convection with a fan. As shown in Figure 4 and Table 1, the power increase led to linear increase of the working temperature of the heater. Upon the highest applied power (6W), the average temperature rises of the heater were 75.2 °C and 33.8 °C in natural convection and forced convection, respectively. According to the product datasheet, the maximum temperature rising of the heater is limited within 130°C. Therefore, the use of graphene heatsink successfully reduced the working temperature of heater and kept it from overheating both in natural and forced convection.

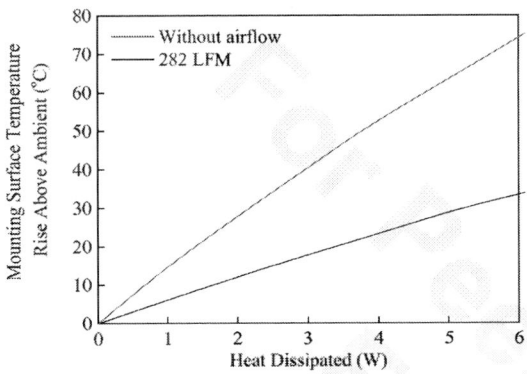

Figure 4. Heater Surface temperature rises above ambient at different powers both in natural convection and forced convection.

Table 1. Temperature rise list of the heater above ambient at different powers both in natural convection and forced convection.

Power (W)	Heater Temperature Rise Above Ambient ($\Delta T = T_{hs} - T_a$) (°C)	
	Natural Conv.	282 LMF
0	0	0
1	15.7	6.4
2	27.8	12.0
3	40.4	17.9
4	53.1	23.2
5	63.7	29.2
6	75.2	33.8

The cooling performance of the graphene heatsink was also compared to that of copper heatsink. The tests were carried out in forced convection with the maximum air flow rate of 282 LMF. The working power of the heater is 6 W. Figure 5 shows the temperature changes of the two heaters from the start of heating to the equilibrium state. It was found that graphene heatsink showed a higher temperature than copper heatsink at the beginning of the heating. The reason for that is the relatively large weight of the copper heatsink which took more energy for the same temperature rise. With the same volumetric size, the copper heatsink consumes 15.3 J of energy for every degree increase in temperature, which is about 4 times higher than that of graphene heatsink (3.9 J). Therefore, graphene heatsink showed a faster temperature rise at the beginning of the heating. The fast increase of temperature in the graphene heasink can lead to a large temperature gradient, which accelates the following heat convection process in the equilibrium state. The surface temperatures of graphene heasink and copper heatsink reached to the same at about 6 minutes after the start of heating. After that, graphene heatsink started to show a lower surface temperature than that of copper heatsink. It is attributed to the ultrahigh thermal conductivity of

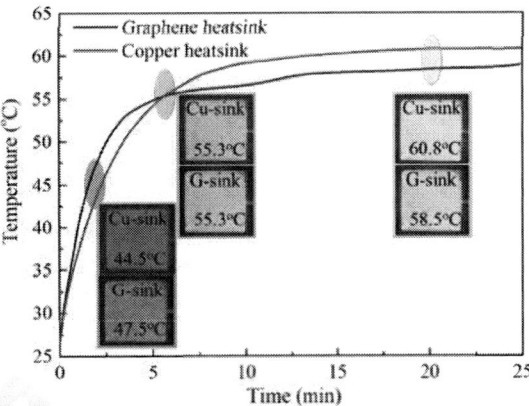

Figure 5. Surface temperature changes of graphene heatsink and copper heatsink as a function of time. The inserts show the IR images at different states of heating.

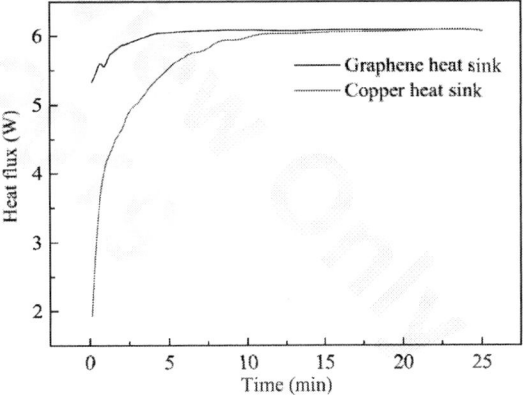

Figure 6. The change of heat flux that dissipated to surroundings as a function of time.

vertically aligned graphene structures. At the equilibrium state, the surface temperature of the graphene heatsink kept 2.3 °C less than that of copper heatsink, showing a superior cooling performance. Figure 6 shows the change of heat flux that passed through heatsinks and dissipated to surroundings. It was found that graphene heatsink can reach to the maximum heat flux much faster than copper heatsink due to its ultra high thermal conductivity and extremely low density. It means a much faster heat dissipation speed than copper.

Conclusions

In summary, a highly thermal conductive and light-weight heatsink based on graphene composites was developed to address the overheating issues in power electronics. The uique vertically-aligned graphene structure offers the graphene heasink many advantages, including fast heat dissipation, high surface emissivity, and low density. The resulting graphene heatsink thus opens new opportunities for addressing large heat dissipation issues in weight driven electronics and other high power systems.

Acknowledgements

We thank for the financial support from the Swedish National Board for Innovation (Vinnova) Graphene SIO-Agenda Program, Formas program on graphene enhanced composite. We also acknowledge the financial support from the Swedish Foundation for Strategic Research (SSF) (No SE13-0061), National Swedish Science Foundation as well as from the Production Area of Advance at Chalmers University of Technology, Sweden.

References

[1] L. Yang, J. Hu, and M. W. Shin, "Degradation of high power LEDs at dynamic working conditions," *Solid-State Electron.*, vol. 53, no. 6, pp. 567–570, Jun. 2009.

[2] J. Park, H. Ham, and C. Park, "Heat transfer property of thin-film encapsulation for OLEDs," *Org. Electron.*, vol. 12, no. 2, pp. 227–233, Feb. 2011.

[3] C. Oshman, Q. Li, L.-A. Liew, R. Yang, V. M. Bright, and Y. C. Lee, "Flat flexible polymer heat pipes," *J. Micromechanics Microengineering*, vol. 23, no. 1, p. 015001, Nov. 2012.

[4] X. Yang, Y. Y. Yan, and D. Mullen, "Recent developments of lightweight, high performance heat pipes," *Appl. Therm. Eng.*, vol. 33–34, pp. 1–14, Feb. 2012.

[5] T. Icoz and M. Arik, "Light Weight High Performance Thermal Management With Advanced Heat Sinks and Extended Surfaces," *IEEE Trans. Compon. Packag. Technol.*, vol. 33, no. 1, pp. 161–166, Mar. 2010.

[6] Y. Peles, A. Koşar, C. Mishra, C.-J. Kuo, and B. Schneider, "Forced convective heat transfer across a pin fin micro heat sink," *Int. J. Heat Mass Transf.*, vol. 48, no. 17, pp. 3615–3627, Aug. 2005.

[7] C. E. Brench, "Heatsink radiation as a function of geometry," in *Proceedings of IEEE Symposium on Electromagnetic Compatibility*, 1994, pp. 105–109.

[8] V. Rammohan Rao and S. P. Venkateshan, "Experimental study of free convection and radiation in horizontal fin arrays," *Int. J. Heat Mass Transf.*, vol. 39, no. 4, pp. 779–789, Mar. 1996.

[9] S.-H. Yu, D. Jang, and K.-S. Lee, "Effect of radiation in a radial heat sink under natural convection," *Int. J. Heat Mass Transf.*, vol. 55, no. 1, pp. 505–509, Jan. 2012.

Electronic Packaging for MEMS Infrared Sensor With Filtered Optical Window

A. Maierna (1)*, M. Del Sarto (1), N. Manca (1), G. Bruno (2), M. E. Castagna (2)
A. La Malfa (2), A. Gritti (3)

STMicroelectronics

(1) Via Tolomeo, 1 - 20010 Cornaredo (MI) - Italy
(2) Stradale Primosole, 50 - 95121 Catania (CT) – Italy
(3) Via C. Olivetti, 2 - 20041 - Agrate Brianza (MB) - Italy

* +39 0293517666, amedeo.maierna@st.com

Abstract

Sensor developments involving semiconductors techniques are even more becoming a typical vehicle for high-level of integration and in many cases are constituting a solid base for a dedicated category of Micro Electro Mechanical Systems (MEMS).
In the present paper the package structure along with the System in Package (SIP) assembly details for an optical MEMS sensor are shown. An infrared sensitive structure is involved, based on semiconductor technology. The contribution of the package presence around the sensor chip and the physical interactions with sensor constitutes one of the key aspects for the overall system performance and these physical aspects are focused in the present paper.
The package structure is a cavity Land Grid Array (LGA). Structural and physical characteristics of the involved materials points to match the optical signal handling as well as the electrical signal processing as output of the integrated Application-Specific Integrated Circuit (ASIC) controller.
A dedicated organic substrate design, a molded-cavity structure and a Si-based IR filtering window constitutes the main features of the sensor system, in terms of concept and design boundary. Measured sensor performance and opto-electrical characterizations are finally reported, including IR opto-geometrical considerations as FFOV (Full Field Of View) results for two different package versions.

Key Words: Infrared Sensor, Package, Optical Window, IR filter.

Introduction

Thermal detectors are today used for a wide range of functions including motion detection, temperature measurement, people counting, and fire & gas detection within numerous markets such as building, security, appliances, industry and consumer.
Five applications will drive the thermal detector market revenue growth: spot thermometry in mobiles devices, motion detection, smart building, heating,

ventilation, and air conditioning (HVAC) and other medium array applications and people counting.
Every object emits thermal radiation, depending on its temperature. There is a fixed relation between the temperature of an object and the radiated energy, as expressed by the Stefan-Boltzmann law [1, 2]. The peak of the radiation shifts to smaller wavelengths with rising temperature: the emission peak at 300K (room temperature) is at 10 um, while the sun (6000K) has a peak at 500nm, which is visible light.
A thermal detector, when incident IR radiation is absorbed, converts the wave energy into electrical

signal by means of different temperature dependent mechanisms such as thermoelectric voltage, Seebeck Effect [3], resistance or pyroelectric voltage [4].

Nowadays, modern semiconductor technology, especially adopting MEMS manufacturing techniques, makes possible to produce very efficient uncooled IR detectors, thanks to the possibility to achieve thermal insulation: the sensor is extremely sensitive, shows a very fast response time due to its smallness, and it is additionally inexpensive, because of the employment of semiconductor mass production means [5, 6]. The MEMS sensor performances have to be coupled to package optical elements in order to fabricate an efficient sensor-system.

Physical components, as the package-body housing and the optical window that allows IR radiation to reach the sensor, has also function to protect neighboring circuitry and interconnecting wires. In some cases a filtered window is used to modify the responsivity spectrum of the device and hide the sensor from visible radiation. This window material is typically silicon based with interferential filters.

The optical interface is physically located on the top of the package body, on the opposite surface of the leads interconnections to PCB.

In the present paper is described a system in package (SIP) composed by an IR sensor, an ASIC included in a package with optical features. The attention is focused on the package-related aspects in terms of materials, optical performance and total system sensitivity. The package is a cavity Land Grid Array (LGA) with an optical IR-filtered window has been drawn, fabricated and characterized. The sensor field of view ranges from 80° to 110° depending on window geometry. The package impact on sensor sensitivity has been also investigated.

MEMS-IR sensor devices

The innovative package has been drawn for a MEMS-IR sensor based on a micromachined thermopile, moreover the package structure can be generally applied to different IR sensors, also for different active areas just recalculating the geometrical package dimensions, maintaining the concept and the materials. Thermopiles consist of N thermocouples connected in series: this allows increasing the sensor output voltage, that becomes N times the one of the single thermocouple element. Thermocouples consist of two different conductor materials joined at one end. The two junctions are referred to as hot and cold junction. When the junctions are at different temperatures, a

voltage difference ΔV originates between the two conductors thanks to the Seebeck effect [3], as shown in the following expression:

$$\Delta V = N\alpha\Delta T \qquad (1)$$

where ΔT is the temperature difference between the hot and the cold junction, and α the Seebeck coefficient which depends on the conductor materials.

In micromachined thermopiles, the thermocouples legs are embedded in a dielectric membrane: the hot junction is positioned in the suspended part of the membrane, while the cold junction is in correspondence of the silicon substrate in order to optimize the temperature difference between old and hot junction, and then maximize the output voltage, that is usually of the order of hundreds of μV, a few mV at most: hence, an appropriate amplification is required so that the subsequent circuitry can correctly handle the signal.

The proposed miniaturized micromachined thermopile sensor consists of p/n polysilicon thermocouple elements placed in series. A central metal plate in aluminum, covered with a dielectric material, acts as the absorbing membrane for the radiation, and the active area is 600μmX600μm. A schematic layout of the sensor is reported in Fig. 1. The actual die presents a test area to monitoring the sensors parameter during the characterization, but will be eliminated in an advanced version reducing the die dimensions, and the optical window positioning.

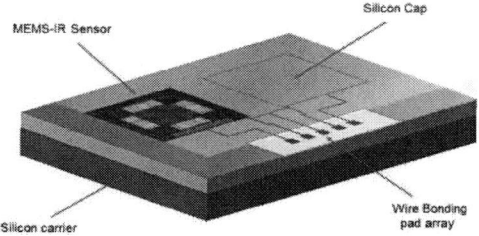

Figure 1: IR sensor body with thermopile IR-sensing area and an area used to integrate testing-reference devices.

The MEMS-IR sensor is usually electrically connected to an application-specific integrated circuit (ASIC) to drive the sensor and amplify the signals [7], for this reason a System in Package assembly has been evaluated. To guarantee the incident IR radiation to reach the sensor active area, avoiding in this way the radiation noise due to flash lamps in visible light, an

IR-selective filter, for the selected application a long pass with λ>5.5 μm, has been integrated at package level.

Total losses introduced by infrared long-pass filter is limited to about 20% in the wavelength region interested by this presence sensor system requirement; this level of energy loss originated by the filtering elements was considered sufficiently limited for the main application purposes, as example presence sensor or IR detection for temperature monitoring inside a populated PCB assembled system. For other future kind of possible application this interferometric filter shall be changed with a different one presenting other transmission spectra.

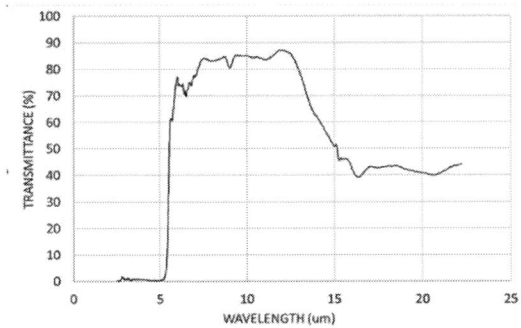

Figure 2: Transmittance spectra for the integrated IR long-pass filter on the top of the package

The filter is silicon adopted for this application is based with interferential layers integrated usually on both sides. Different filters can be mounted, making the device suitable for different application like for example NDIR spectrometer [8, 9].

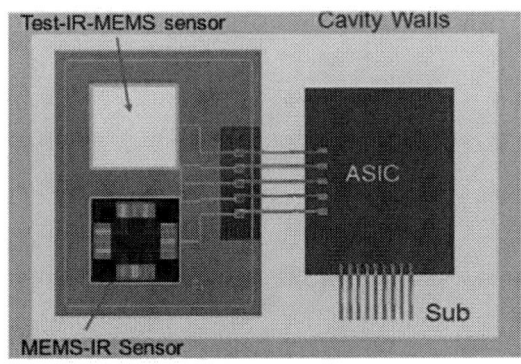

Figure 3: Assembly structure of a MEMS-IR sensor and ASIC layout

The package for this IR sensor has been designed and implemented taking present a typical side-by-side layout: the sensor and the ASIC are encapsulated in side by side configuration (Fig. 3).

On the top package surface an optical window is integrated in order to select the radiation's wavelength components. This solution for temperature detection can prevent environmental radiation to reach the detector with the result to reduce the total system noise[8]. The polymer constituting the package top and cavity walls can be considered totally opaque to visible-IR radiation, and can be classified as Liquid (Chrystal Plastic LCP) material. Different filters can be then mounted for different application, as example for NDIR spectrometer [10]. Structural elements as two dice and wire bonding are connecting the sensing and signal processing dice to an interconnecting substrate, as shown in Fig. 3.

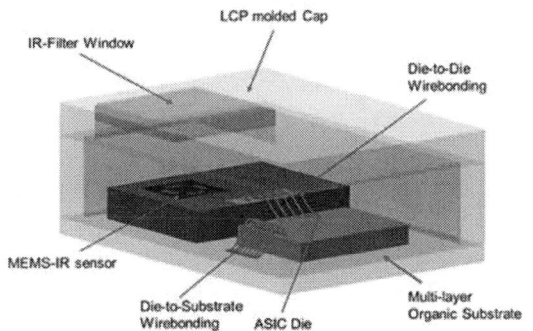

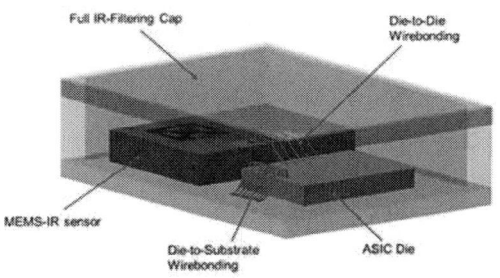

Figure 4: Package with "Small IR window" or "Full IR filtering cap", rendering images

Experimental Setup and Measurements

A MEMS-IR sensor electro-optical characterization has been performed using a calibrated black body radiator ranging from -20C to 160C as target object. The employed black body radiator is the SR-800R 4D/A model by CI Systems [11], which features a 4 in x

4 in area and emissivity equal to 0.99. The characterization has been performed by positioning the device at 5.0 cm from the black body surface in order to entirely cover the sensor view area.

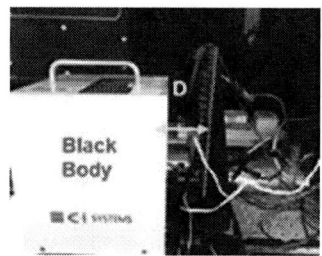

Figure 5: Experimental setup.

The acquisition has been done with and without the filter and a signal to noise of 1.6 and 2.36 has been respectively observed. The reduction of the signal to noise for the sample with the filter is due to the filter attenuation and is fully compatible with spectrum reported in Fig. 2.

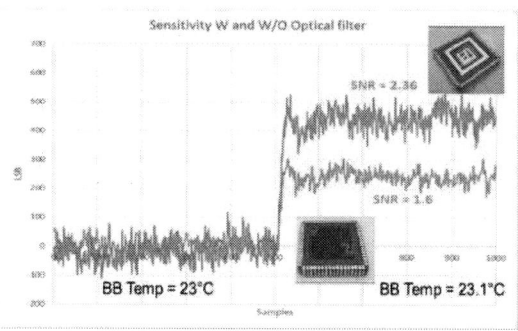

Figure 6: Sensitivity characterization of the sensor in ceramic package with and without IR filter.

System output is constituted by a digital signal and the number of bit obtained as lsb (lower significant bit) difference under IR radiation exposure is a measure of the output change of the system. By fixing the geometrical conditions and ensuring the black body is entirely covering the window area view, a total sensitivity of about 2000lsb/°C has been estimated for the sensor under test, with a noise evaluable in 150lsb. The IR long pass filter can be choice mainly in order to match the desired detection selectivity and nature and dimensions of the detectable bodies to be identified in terms of presence in front of the sensor window.

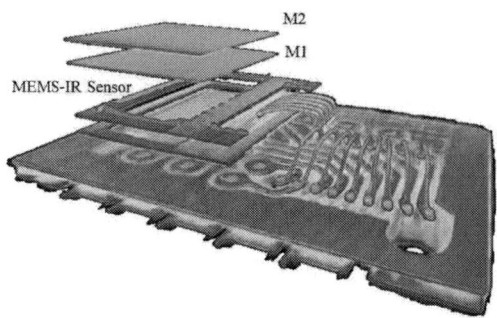

Figure 7: image obtained by 3D-Xray tomography of the assembly with filter metallization M1 and M2 of IR silicon filter

In above Fig. 7 are shown two metallization, layer M1 and M2 placed on the MEMS-IR sensor, with function of IR-filtering for the incident radiation on the package surface. In the 3-D picture are also shown wire bonding structures and substrate metal traces for the sensor and ASIC interconnections.

Field of View (FOV) Calculation

For an optical system, in general we can define a field of view (FOV), in order to evaluate the portion of a geometrically defined space that can be detected by the system. The field of view (FOV) of any optical device can be defined as either the half field-of-view (HFOV), where FOV = $\pm\theta$, or the full field-of-view (FFOV), where FOV = θ. In this paper, the HFOV definition of FOV = $\pm\theta$ is used. In our geometrical evaluation we assumed n=3.44 for Si; n=1 for air and vacuum. The following picture shows the FOV calculation scheme on the section structure considered

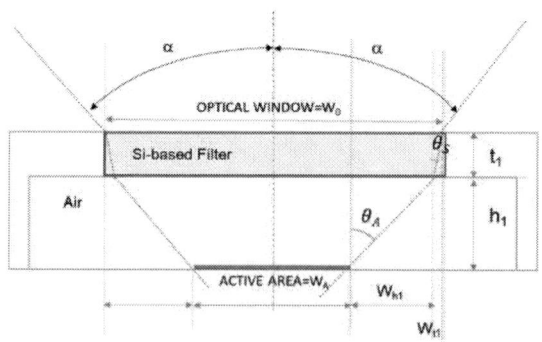

Figure 8: FOV calculation schematic section
When making calculations for the field of view, consider the refraction (or bending) of light as occur when the light passes through the window.

Using basic relations from trigonometry, we can find:

$$W_O = W_A + 2\,(W_{tl} + W_{h1}) \qquad \text{(eq. 1)}$$

Where W_O is the width of the package's optical window, W_A the width of the sensor active area, $W_{tl} + W_{h1}$ the widths defined by the optical path in air and silicon calculated by eq. 2 following set:

$$W_{tl} = t_1 \cdot tg\theta_S; \qquad \text{(eq. 2a)}$$
$$W_{h1} = h_1 \cdot tg\theta_A; \qquad \text{(eq. 2b)}$$

where, t_1, and h_1 are geometrical vertical parameters of the package and the device itself, θ_A and θ_S are respectively the angle for IR propagation in air and in silicon. The Snell's law, reported in below eq. 3, links the two angles:

$$n_1.\sin(\theta_1) = n_2.\sin(\theta_2) \qquad \text{(eq. 3)}$$

With n_1 and n_2 are indicated the refractive indexes of each material, θ_1 and θ_2 are the propagation angles (counterclockwise considered) from the surface normal in each material, assuming value for Si ($n=3.44$) and for Air/Vacuum ($n=1$). With the above assumptions in terms of geometry the expected value was ranging FFOV=80°-82°. A tentative concept-design of a cavity package was then started and two lots of prototypes were manufactured on an assembly pilot line lab. With the purpose to have different FFOV two different windows design was proposed. Molding resin materials was tested at IR transmission measurements, in order to verify "T%=0" condition for the package wall material in the wavelength range 1.0um -13.0um. The assembly structure for this system, referring to the following drawings, is a SIP with ASIC die placed side by side with MEMS-IR sensor, die-to-die wire bonding (WB).

Figure 9: Package with IR window (on the left) or full IR filter (on the right) pictures Surface Mounting Technology (SMT) on a DIL 24 test board.

The two systems in-package above described have been characterized by using the previously described black body as radiating object at a distance of 22cm from the top package surface.

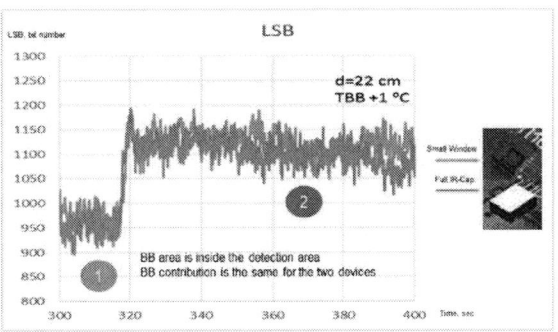

Figure 10: Graph of sensitivity of MEMS-IR sensor in package with "Small window" on cap compared to "Full IR filtering" cap

After this experimental activity no measurable sensitivity difference between small-window and full-IR cap was observed at 22cm, the same for response time. The distance has been chosen in order to approximate the beam direction to an IR planar incident wave on the device top surface. To perform the FOV characterization the device has been mounted in a rotational stage from -90° to +90° degree respect the normal condition where the sensor active area is positioned in front the black body.

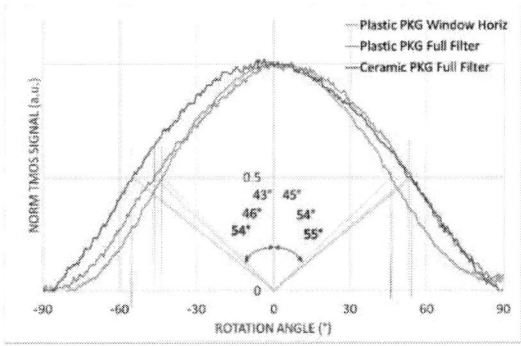

Figure 11: FOV characterization results for IR sensor in package with IR small window, full IR filter and large ceramic package assembly

Electronic Packaging for MEMS Infrared Sensor With Filtered Optical Window

In the ceramic large package, the IR sensor showed a FFOV of 109°±2°, smaller than expected for a Lambert's distribution (theoretically 120° then) probably due to the MEMS Si-embedding structure. For the small window a FFOV of 88° has been measured. The full IR-cap filter of the plastic package presented a FFOV of 100°, acquired by the same package-rotating technique. In the last case an asymmetric effect has been observed, due to the package plastic walls adjacent to the sensor's active area.

Package Stress Simulation

An important factor to obtain a large output voltage from a thermopile is to have a high thermal isolation in to maximize the temperature difference between hot and cold junctions for a specific absorbed power. By using a MEMS package, the gas inside the cavity can be selected as well its pressure in a range from 100μBar to 100mBar: the gas thermal conductivity impacts on the temperature propagation and on the difference between hot and cold thermopile junctions and then on the output voltage variation and then increasing the sensor efficiency.

The MEMS package is obtained by wafer-to-wafer bonding technique. MEMS sensors system mainly consists of a silicon microstructure manufactured through surface micro-machining process and encapsulated in a Si package (the die) usually made of two or more wafers stacked and bonded together with glass-frit compound.

The presence of a Si protection cap of thickness about 150um on the sensor introduces by itself an IR-wavelength dependent natural filtering on the radiation reaching the sensor surface. Of course the optical performance, due to IR transmission spectra of the Si-cap is reduced in dependence of silicon properties in the 1-13um wavelength infrared region [12].

The sensor evolution involves to integrate the MEMS cap on the sensor wafer. The whole sensor system constituted by the IR sensor with the silicon cap, the ASIC and the package has been than simulated. Being the sensor die stacked on the package substrate, the sensing microstructure is coupled with package structures. It follows that the signal performance of the device is affected by its package. In addition to the stress related to the operative life, criticalities can arise during manufacturing processes, especially during the cooling following the soldering of the package on PCB. Since the package is made of materials having different Coefficient of Thermal Expansion (CTE), thermal gradients induce warpage causing stress transfer to the sensing microstructure, thus affecting sensing performance.

3D finite element model is established in SolidWorks Simulation to simulate the warpage of the silicon holding the sensing microstructure. Post-soldering cooling simulation considers the package bonded on a reference PCB. Table 3 summarizes thermal loads and boundary conditions. Figure 12 shows the finite element model.

Table 2 lists the material properties used for the simulation.

Although know that simulation results strongly depend on the material model and proprieties used [13], considering common practices in package simulation literature, the comparative aim of the analysis, the available material data and the static nature of the simulation performed, elastic isotropic materials have been assumed [14, 15, 16,17].

In order to reduce the computational time, a simplified model has been considered. However, due to the asymmetry of the ASIC placement inside the package and the cap window, the entire model needs to be simulated. For the top and bottom substrate layers, equivalent mechanical proprieties have been calculated as follows [14]:

$$E_{eff} = \frac{\sum E_i V_i}{\sum V_i}$$

$$\alpha_{eff} = \frac{\sum \alpha_i E_i V_i}{\sum E_i V_i} \qquad \text{(eq. 3)}$$

where E_{eff} is the effective Young's Modulus, α_{eff} the effective CTE, E_i, α_i, V_i the Young's Modulus, the CTE and volume or area percentage of the constituents materials. Figure 12 shows the FEM model and Figure 13 shows the warpage simulation on sensor, ASIC and the Substrate resulting from simulation. The warpage w of the substrate holding the sensing microstructure is defined as the difference between the maximum and the minimum z displacement along the frame itself.

Table 2. Material proprieties

Material	E [GPa]	α [ppm/°K]
Silicon	131	2.8
Solder mask	6.2 @30°C 0.23 @260°C	α1 = 60 α2 = 130 Tg = 114°C
Core substrate	21.0	α1 = 19 α2 = 5 Tg = 230°C
Copper	117.0	17.0
Cap mold material	8.1@30°C	α1 = 24 α2 = 42 Tg = 281°C
Die Attach Film	1.7@30°C 0.04@260°C	α1 = 80 α2 = 170 Tg = 128°C
Epoxy Cap-Filter	7.5-8.5@30°C	α1 = 40-60 (<Tg) α2 = 150-170 (>=Tg)
FR4	25.0	16.0

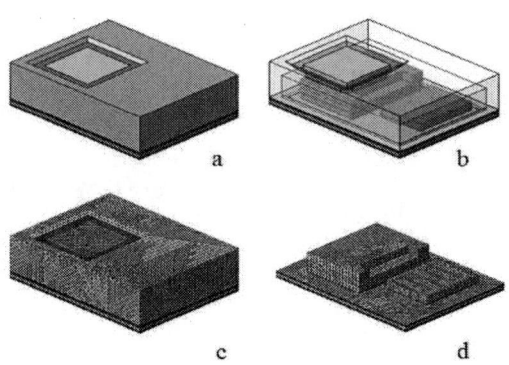

Figure 12: Finite Element Model for thermo-mechanical simulation. a,b) CAD model, c,d) finite element models with and without package cap. In the pictures aren't shown the PCB board for post-soldering simulation.

Table 3. Thermo-mechanical FEA boundary conditions and loads

Post-soldering Condition
217°C→25°C

- T_{ref} = 217°C (zero strain)
- T_{unif} = 25°C

Figure 13: package substrate, ASIC and MEMS (without top wafer) warpage (w).

Conclusions

Presented package for IR sensor was designed, manufactured and characterized with promising results. The angle of FFOV was measured, ranging from 80° to 110°, depending from optical window dimensions. A silicon-based IR filter has been mounted on top-package and characterized in order to eliminate flash lamps effects and environmental noise. Stress simulation did not show critical situations on material interfaces. Package reliability in terms environmental stress are preliminary matching JEDEC L3 requirements.

Acknowledgments

Special thanks to ST Microelectronics Analysis Labs direct by Daniela Morin, thanks to Alexandra Colombo and Luca Privileggi for system in-package physical analysis and 3D-tomography. Thanks also to Angelo Recchia and Michele Vaiana for system's electrical characterizations. For first prototypes manufacturing the author team want to thanks Michael Chen, Chris YC Huang from ASE Chung-Li, Sharon Liu and Christophe Zinck from ASE Europe.

Bibliography

[1] J. R. Mahan, "Radiation Heat Transfer: A Statistical Approach, Volume 1", pp. 7-8, John Wiley & Sons Inc., 2002.

[2] Y. Houdas, E.F.J. Ring, "Human Body Temperature: Its Measurement and Regulation", pp. 24-27, Springer, 1982.

[3] A. Rogalski, Infrared Detectors, 2nd Edition, CRC Press, Boca Raton, Florida - 2010.

[4] A. Rogalski, History of infrared detector, Opto-Electro Review, 20, n.3, Varsaw 2012.

[5] Eran Socher, Oflr Bochobza-Degani, Yael Nemirovsky, "Tmos-infrared uncooled sensor and focal plane array", US20060244067A1, 2003.

[6] L. Gitelman, S. Stolyarova, S. Bar-Lev, Z. Gutman, Y. Ochana, and Yael Nemirovsky, "CMOS-SOI-MEMS Transistor for Uncooled IR Imaging", IEEE Transactions on Electron Devices, Vol 56, No. 9, September 2009.

[7] E. Moisello, M. Vaiana, M. Castagna, G. Bruno, E. Bonizzoni, P. Malcovati, "A Chopper Interface Circuit for Thermopile-Based Thermal Sensors", IEEE International Symposium on Circuits and Systems (ISCAS), 2019.

[8] Jane Hodgkinson et al., "Non-Dispersive Infrared (NDIR) measurement of carbon dioxide at 4.2um in a compact and optically efficient Sensor", Sensors and Actuators B: Chemical, Volume 186, Pages 580–588, Sep. 2013.

[9] A. Graf et al. "System Design and Analysis Concept of a Highly Adaptable NDIR Sensor for Gas Analysis", 14th International Conference on Solid-State Sensors, Actuators and Microsystems, Lyon, France, June 10-14, 2007 (2007).

[10] M. Ebermann1 et al. "A Fast MEMS Infrared Micro-spectrometer for the Measurement of Hydrocarbon Gases", Transducers 2015, Anchorage, Alaska, USA, June 21-25 (2015).

[11] Online doc http://www.laseroptronic.it/doc/SR-800R.pdf.

[12] M. Geddo, B. Pivac, A. Sassella, A. Stella, A. Borghesi, A. Maierna, "Infrared Determination of Interstitial Oxygen behavior during epitaxial silicon growth on Czochralski Substrates", J. Appl. Physics, 72, (9), American Institute of Physics, Nov 1992.

[13] S. Fischer, J. Wilde, E. Deier, E. Zukowski, "Influence of materials data on the performance modelling in the design of MEMS packages" In: 9th International Symposium on Advanced Packaging Materials: Processes, Properties and Interfaces. 2004 Proceedings. IEEE, 2004. p. 57-62.

[14] X. Zhang, T. Y. Tee, J. Luan, "Comprehensive warpage analysis of stacked die MEMS package in accelerometer application", In: Electronic Packaging Technology, 2005 6th International Conference on. IEEE, 2005. p. 1-6.

[15] S. Wei, J. Tang, J.Song, "FEM study on the effects of flip chip packaging" 2009 International Conference on Electronic Packaging Technology & High Density Packaging", In: 2009 International Conference on Electronic Packaging Technology & High Density Packaging. IEEE, 2009. p. 403-406.

[16] M. Li, J. Song, Q.A. Huang, F.X Chen, J.Y. Tang, "The thermomechanical coupling effect in multi-layered microelectronic packaging structures", In: 2006 8th International Conference on Solid-State and Integrated Circuit Technology Proceedings. IEEE, 2006. p. 2135-2137.

[17] M. Lishchynska, C. O'Mahony, O. Slattery, O. Wittler, H Walter, "Evaluation of packaging effect on MEMS performance: simulation and experimental study", IEEE Transactions on Advanced Packaging, 2007, 30.4: 629-635.

Ag and Si particles sintering technology for SiC power device

M. Ueshima[1], T. Mototsuji[2], Y.Isono[3] and M. Haga[1]

Tomoaki Mototsuji(Hirata Lab.), Yusuke Isono(Koizumi Lab.)

[1]Daicel Corpration, 805 Umaba, Ibogawa-cho, Tatsuno-shi, Hyogo 671-1681, Japan

[2]Osaka University, Graduate School of Engineering

Phone +81-791-72-5411, and E-mail Address mn_ueshima@jp.daicel

Abstract

Ag sintering technology is one of the best candidates for SiC die attach because of good mechanical and thermal properties at high temperature. However cracks propagates into SiC dies or DBC substrates at thermal fatigue test at more than 200 °C because Ag is too hard for die attach materials and 10ppm/ °C of CTE mismatch between SiC dies and Ag layers is too big. Therefore a development of low CTE sinterng materials is dispensable for SiC power device operated at high temperature. In this study, a method to coat Si particles with Ag layer at room and its sintering properties are researched. Si particles with φ30μm and Ag flakes with φ5μm are prepared and Si-20,40and 60 vol%Ag particles are produced by utilizing high-speed 3D motion of a patented 3D Ball Mill produced by Nagao System Inc.. It is revealed that their Si particles after the ball milling are almost covered by Ag layer except for Si-20vol.%Ag particles and furthermore the coated particles are crushed to 5μm by SEM observation. Their Si-Ag particles are mixed with a solvent to formulate into pastes. These Si-Ag pastes are printed onto Ag electroplated Cu plats, and then the bonding experiments are carried out at temperatures of 250°C for 60 min without pressure at air atmosphere. It is revealed that some Ag layers are jointed on the surface of Si particles and others are delaminated by FE-SEM observation. Si particles are connected via Ag layers on their surface. The strength between Ag layer and Si particles must be improved in order to adopt Ag-Si sintering paste to die-attach technology. .

Key words: Si, sintering ,Ball mill die bonding

Introduction

ELV (End-of Life Vehicles Directive) and RoHS (Directive on the Restriction of the Use of Certain Hazardous Substances in Electrical Equipment) have been requiring replacement of Pb-rich high temperature solders to suitable alternatives. Pb-rich solders have been widely used for die-attach of Si dies on substrates or higher temperature soldering in multiple step soldering. Some of Pb rich alloys [6, 7] had been already replaced by Sn-3Ag-0.5Cu alloy even though the components with Sn-Ag-Cu should not be heated beyond 200 °C at second reflow. However, many components are actually soldered by the conventional reflow oven and cracks sometimes penetrate into the solder joints of the components, because volume of solders increase by more than 5 % when they melt.

Several types of Pb-free high temperature solders have also been proposed [8, 9]. For example, there are Au alloys [10, 11] and Zn alloys [12]. However, most of these alloys cannot be applied in the temperature range above 250 °C. Moreover, these Pb-free solders have typical disadvantages such as high cost for the Au alloys, poor electrical

and thermal conductivities for the Bi alloys, and brittleness for most of the alloys, except for Zn–Sn..

Furthermore, a low carbon society requires new devices that efficiently utilize electric energy. Today, high power devices applied widely such as in electric trains, hybrid and electric vehicles, elevators, and renewable energy systems, as well as in many consumer electronics. Howerver, conventional Si-based power devices have reached near theoretical efficiency and operation temperature. In addition, the miniaturization requirement for these devices leads to shift from the water cooling system to air cooling system. Therefore, new devices with wide bandgap semiconductors have recently gained considerable attention as a breakthrough technology for new high power moduless. Moreover, there have been many reports on wide bandgap semiconductors such as SiC, GaN and GaO [1, 2, 3, 4, 5]. Among these, SiC has not only excellent physical properties such as resistance to high temperature and to high voltage operations, but also excellent electrical properties such as low power loss and high power density. In addition, the excellent heat-resistance above 250 °C at SiC semiconductors can make the water cooling system eliminated on their power modules [3]. In such cases, the packaging materials, particularly the die-bonding materials, for these new high power devices must be required not only high solidus line above 260 °C, but also high thermal fatigue resistance and thermal conductivities.

Ag sinter joining has recently been gaining much attention as candidates for Pb-free high temperature solder for die attach due to its high thermal conductivity, high electrical conductivity, and high temperature stability. This metal-bonding technology is expected to have various applications to high-power semiconductor devices and LED. There have been several reports on Ag sinter joining technologies by using Ag nanoparticles, hybrid Ag particles, and Ag oxide particles [13, 14, 15, 16, 17, 18, 19, 20]. Certainly, Ag sintering technology is one of the best candidates for SiC die attach. However cracks propagates into SiC dies or DBC substrates at thermal fatigue test at more than 200 °C because Ag is too hard for die attach materials and CTE mismatch 10ppm/ °C between SiC dies and Ag layers is too big. Therefore a development of low CTE sinterng materials is dispensable for SiC power device operated at high temperature.

In this study, a method to coat Si particles with Ag layer at room and its sintering properties are researched. Si particles with φ30μm and Ag flakes with φ5μm are prepared and Si-20,40and 60 vol%Ag particles are produced by utilizing high-speed 3D motion of a patented 3D Ball Mill producted by Nagao System Inc.. It is revealed that their Si particles after the ball milling are almost covered by Ag layer except for Si-20vol.%Ag particles and furthermore the coated particles are crushed to 5μm by SEM observation. Their Si-Ag particles are mixed with a solvent to formulate into pastes. These Si-Ag pastes are printed onto Ag electroplated Cu plats, and then the bonding experiments are carried out at temperatures of 250°C for 60 min without pressure at air atmosphere. It is revealed that some Ag layers are jointed on the surface of Si particles and others are delaminated by FE-SEM observation. Si particles are connected via Ag layers on their surface. The strength between Ag layer and Si particles must be improved in order to adopt Ag-Si sintering paste to die-attach technology.

Experimantal Procedure

AgC-239 (Fukuda Metal Foil and Powder Co., Ltd.) flakes with average diameters of 5.0 μm compsed of Ag and and Si particles with φ30μm were used as the starting materials. The AgC-239 flakes had an average thickness of 260 nm and a high specific surface area of 1.2m2/g. The appearance of Ag flackes and Si particles are shown in Fig. 1 and 2 respectively. Si-20, 40 and 60 vol%Ag particles are produced by utilizing high-speed 3D motion of a patented 3D Ball Mill producted by Nagao Systems tem Inc.

Figure 1:SEM image of Ag flake

Figure 2:SEM image of Si particles

Three types of Si-Ag pastes were prepared by mixing the three Si-Ag paricles with DPNP soluvent. Their weight ratio of the soluvent to the Si-20, 40 and 60 vol.%Ag particles pastes were respectively 21, 25 and 30wt.%, at which proportion the pastes had a viscosity suitable for use as joining pastes.

Furthermore four types of pastes composed of Si-60 vol.%Ag paritcles and Ag flakes are prepared. Their weight ratio of Si-60 vol.%Ag paritcles to Ag flakes were 0.2:0.8, 0.4:0.6, 0.6:0.4 and 0.8:0.2 respectivly.

Figure 3: Picture of test piece where Si-Ag paste was printed and Si die was mounted on the paste

These Si-Ag pastes are printed onto Ag electro-plated Cu plats by metal mask and Si dies with Ag metallization with 3 mm × 3 mm × 0.45 mm are mounted on their pastes. The picture of test piece where Si-Ag paste was printed and Si die was mounted on the paste is shown figure 3. The bonding experiments are carried out at temperatures of 250°C for 60 miniutes with pressure 0 and 2.0 MPa at air atmosphere. This joint sample was used for measurement of the joining strength and microstructural observation. To examine the joining strength, the joint samples were sheared using a die shear tester (DAGE, XD-7500). The shear strength was measured at a shear head speed of 50 μm/s. Field emission scanning electron microscopy was performed in order to observe the joint interfaces with Si and Ag particles after die bonding.

Results and Disucussion

The appearance of Si-20, 40 and 60 vol%Ag particles produced by utilizing high-speed 3D motion of a patented 3D Ball Mill are shown in Fig. 4, 5 and 6 respectively. The surfaces of Si particles are almost covered with Ag flakes and some Ag flakes are torn with less than 1μm width by sharp edges of Si particles during 3D ball mill operation. Many Ag flake aggregates included Si particles are formed after 3D ball mill. The grainsize of their aggregates is the distribution of the range between 20 μm and 50 μm, mainly. Furthermore, the volume of their small aggregates isn't enough to fill the gaps among their big aggregates. Ag coated Si particles can be produced by utilizing high-speed 3D motion of a patented 3D Ball Mill but the small Ag flakes are almost consumed by formation of Ag-Si aggregates before sintering.

As show figure 4, 5 and 6, the ratio of Si particles surface coverd with Ag flakes increases, as Ag content of Si-Ag particles increasing. Some Si particles are not almost covered on the surfaces with Ag flakes at Si-20 vol%Ag particles.

Figure 4: SEM image of Si-60 vol%Ag particles produced by 3D Ball Mill

Their surfaces without Ag flakes can not be sintered with Ag and Si surfaces at low temperature reflow process. Therefore Ag-Si particles with more than 40 vol.% Ag are required for Ag-Si particles sintering process at least.

The joining strengths with their pastes of Si-20, 40, 60 vol%Ag particles were 2.9, 13 and 23N respectively. The strength of Si-20 vol%Ag particles paste is too low, because it is difficult that the big Ag aggregates exposing Si particles surface sinter each other.

Four kinds of pastes composed of Si-60vol.% Ag particles and Ag flakes are prepared in order to fill the gap between the big Ag aggregates. The joining strength with their paste composed of Si-60vol.%Ag particles and Ag flakes, where their weight ratio of Si-Ag paritcles and Ag flakes were 0.2:0.8, 0.4:0.6, 0.6:0.4 and 0.8:0.2 , were 43, 84, 146 and 149 N respectively. Their joining strengths can be quite improved by addition of free Ag particles because free Ag paricles connected each Si particle covered with Ag flakes.

Figure 5: SEM image of Si-40 vol%Ag particles produced by 3D Ball Mill

Figure 6: SEM image of Si-20 vol%Ag particles produced by 3D Ball Mill

The typical interface microstructures for the joint sintered with 2MPa puressure at the paste composed of Si-60vol.%Ag particles and Ag flakes ,where the weight ratio of Si-Ag paritcles to Ag flakes were 0.6:0.4 is shown in Fig. 7. As shown in figure 7, Ag particles sintered with Ag plating on Si die and Cu plate and Si particles with black cSSLonstrast homogeiously disperse at the joining layer. Many Si particles don't joint with Ag particles, nevertheless Si-60vol.%Ag particles are covered with Ag flakes before sintering. However Si particles don't interrupt to sinter Ag flakes each others. Therefore the Si-Ag particles sintering can be improved, if Si particles could keep to be covered with Ag flakes during the reflow. The interface between Si particle and Ag flake is show in figure 8. As shown in figure 8, the interface between Si and Ag is formed only heating at 250°C for 60 miniutes. However the surface of Si particle isn't almost covered with Ag.

In this study, it is revealed that some Ag layers are jointed on the surface of Si particles and others are delaminated by FE-SEM observation. Si particles are connected via Ag layers on their surface with each other. The strength between Ag layer and Si particles must be improved in order to adopt Ag-Si sintering paste to die-attach technology.

As the next step of the current Si-Ag particles die-attachment study, Ag layer delamination on the Si particles during sintering process would be improved and as a results, die shear strengths would be reach to that of Ag sintering pastes.

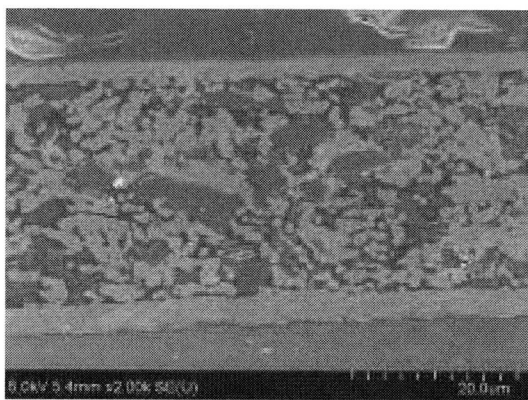

Figure 7: SEM image of the cross section at the joint at the paste composed of Si-60vol.%Ag particles and Ag flakes (0.6:0.4)

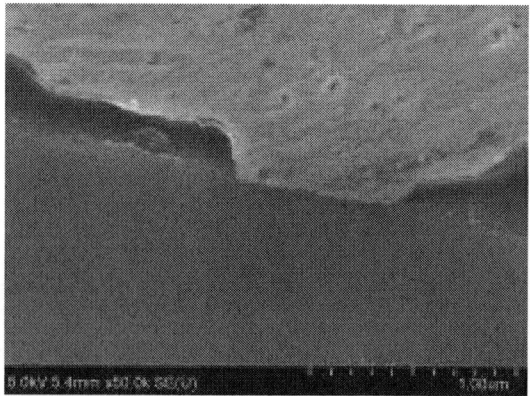

Figure 8: SEM image of the cross section at the interface between Si particle and Ag flake at the

paste composed of Si-60vol.%Ag particles and Ag flakes (0.6:0.4)

Conclusion

1. Ag coated Si particles can be produced by utilizing high-speed 3D motion of a patented 3D Ball Mill and big Ag-Si aggregates with the distribution of the range between 20 µm and 50 µm are formed at the sam time.
2. The small Ag flakes are almost consumed by formation of Ag-Si aggregates by 3D ball mill.
3. Si particles are almost covered on the surfaces with Ag flakes at Si-40 and 60 vol%Ag particles but not Si-20vol.%Ag paricles.
4. The joint strength can be improved at the paste composed of Si-Ag particles and Ag flakes, where the weight ratio of Si-Ag paritcles to Ag flakes were 0.6:0.4 and 0.4:0.6.
5. Many Si particles don't joint with Ag particles, nevertheless Si-60vol.%Ag particles are covered with Ag flakes before sintering.
6. The interface between Si and Ag is formed only heating at 250°C for 60 miniutes at the atmosphere.

References

Articles from conference proceedings

[1] A. Katz, C.H. Lee, K.L. Tai, Mater. Chem. Phys. 37, 303–328, 1994.
[2] P.G. Neudeck, R.S. Okojie, L.Y. Chen, Proc. IEEE 90(6), 1065–1076, 2002.
[3] H.S. Chin, K.Y. Cheong, A.B. Ismail, Metall. Mater. Trans. 41B, 824–832, 2010.
[4] W. Wondrak, R. Held, E. Niemann, U. Schmid, IEEE Trans. Ind. Electron. 48(2), 307–308, 2001.
[5] Y. Sugawara, Mater. Sci. Forum 457–460, 963–968, 2004
[6] K.S. Kim, C.H. Yu, N.H. Kim, N.K. Kim, H.J. Chang, E.G. Chang, Microelectron. Reliab. 43, 757–763, 2003
[7] W.D. Zhuang, P.C. Chang, F.Y. Chou, R.K. Shiue, Microelectron. Reliab. 41, 2011–2021, 2001
[8] K. Suganuma, Curr. Opin. Solid State Mater. Sci. 5, 55–64, 2001
[9] K. Suganuma, S.J. Kim, K.S. Kim, JOM 61(1), 64–71, 2009
[10] R.W. Johnson, C. Wang, Y. Liu, J.D. Scofield, IEEE Trans. Electron. Packag. Manuf. 30(3), 182–193, 2007
[11] P. Hagler, R.W. Johnson, L.Y. Chen, IEEE Trans. Compon. Packag. Manuf. Technol. 1(4), 630–639, 2011
[12] S. Kim, K.S. Kim, S.S. Kim, K. Suganuma, G. Izuta, J. Electron. Mater. 38(12), 2668–2675, 2009
[13] J.G. Bai, G.Q. Lu, IEEE Trans. Device Mater. Reliab. 6(3), 436–441, 2006

[14] E. Ide, S. Angata, A. Hirose, K.F. Kobayashi, Acta Mater. 53, 2385–2393, 2005

[15] M. Knoerr, A. Schletz, in Proceedings of 6th IEEE CIPS Conference, 2010, pp, 1–6

[16] H. Alarifi, A. Hu, M. Yavuz, Y.N. Zhou, J. Electron. Mater. 40(6), 1394–1402, 2011

[17] J.G. Bai, Z.Z. Zhang, J.N. Calata, G.Q. Lu, IEEE Trans. Compon. Packag. Technol. 29(3), 589–593, 2006

[18] T. Morita, E. Ide, Y. Yasuda, A. Hirose, K. Kobayashi, Jpn. J. Appl. Phys. 47(8), 6615–6622, 2008

[19] T. Morita, Y. Yasuda, E. Ide, Y. Akada, A. Hirose, Mater. Trans. 49(12), 2875–2880, 2008

[20] K. Suganuma, S. Sakamoto, N. Kagami, D. Wakuda, K.S. Kim, M. Nogi, Microelectron. Reliab. 52(2), 375–380, 2011

Behaviour of Silver-Sintered Joints by Cyclic Mechanical Loading and Influence of Temperature

Zeynep Gökdeniz[1], Golta Khatibi[1], Johann Nicolics[2], Andreas Steiger-Thirsfeld[3]

[1] Christian Doppler Laboratory for Lifetime and Reliability of Interfaces in Complex Multi-Material Electronics, Inst. for Chemical Technologies and Analytics, TU Wien, Getreidemarkt 9, 1060 Vienna, Austria

[2] Dep. of Electronic Materials, Inst. of Sensor and Actuator Systems, Gusshausstraße 27-29, TU Wien, Austria

[3] USTEM, TU Wien, Wiedner Hauptstr. 8-10, A-1040 Wien

Phone: +43-1-58801-164391, E-mail: zeynep.goekdeniz@tuwien.ac.at

Abstract

Devices in power electronics operate at high temperatures and demand controlled power loss density, while the trend leads to low temperature bonding methods in the packaging procedure. As an alternative to using Pb-alloys and Cu-Sn transient liquid phase soldering, Ag-sintering is a promising bonding technology. Silver with its high thermal and electrical conductivity is an advantageous joining material for different applications (e.g. die attach in power modules). One of the main advantages of this joining technology is the low processing temperature in combination with the high melting point of silver. It is also compatible with multilayer processing. The features of the bonding silver layer affect the elastic and thermo-mechanical behavior of the whole devices. A thorough investigation of the properties of Ag-sintered joints in relationship with their microstructure at various temperatures is essential for an appropriate lifetime prediction. This work investigates the low cyclic fatigue of Cu-Ag-Cu la joints at room and elevated temperature prior and subsequent to thermal treatment. Furthermore, the ratcheting behaviour of the joints as a part of the plastic-creep phenomenon has been evaluated. SEM investigations were performed to determine the porosity of the silver bonding layer and to analyze the fracture surface of the failed samples.

Key words: low temperature silver sintering, temperature effect, shear strength, low cycle fatigue, lifetime

INTRODUCTION

While high temperature Pb-solders are still exempted from RoHS regulations for high power semiconductors with specific applications, industry is seeking for reliable alternative solutions. A leadfree alternative is silver sintering, which has the advantage of being a low temperature bonding method. Additional, regarding the device operation temperature, Ag sintered joints are supposed to be insensitive with respect to high temperature (melting point of Ag 960 °C) than conventional solder. Pure silver having high thermal (430 W/(m·K)) and electrical conductivity ($60 \cdot 10^6$ S/m) is an outstanding material for different applications in the field of power electronics.

Formulations of silver paste compositions from various suppliers have a broad range containing Ag-particles from flake-shaped to drop-form and from micron to nanosize. They are used to form silver joints by pressureless sintering and sintering under pressure [1] and under atmosphere such as air or nitrogen. The thermo-mechanical behavior of the sintered joints depends on parameter such as their geometry (e.g. thickness of the paste and applied area) and sintering conditions, which affect the porosity and hence the performance of the bonding layer. Based on the microstructure, thermal and mechanical properties of Ag-sintered components are topics of ongoing studies for lifetime prognosis. In this study a commercial silver paste from Heraeus (specified particle size 1 µm – 20 µm) developed for pressure sintering on copper substrate [4] was used.

EXPERIMENTAL

Sample preparation

To perform tests on silver sintered Cu-Ag-Cu lapjoints, samples were prepared as followed: copper stripes of 1 mm thickness, 50 mm length and 3 mm width were grinded and polished to remove the surface oxidation, put in isopropyl for rinsing by ultrasonic bath and dried by air blowing. First a silver layer of approximately 500 nm was sputtered on both mating copper surfaces. Then the Ag-paste was applied by stencil printing (mask thickness

100 μm) on both strips, forming a pad area of ca. 10 mm² on each side. The Ag-paste was dried at 130 °C under nitrogen flow. Sintering was performed under pressure of 70 MPa for 30 minutes at 230 °C in air (Figure 1).

After sintering some samples were heat treated at 250 °C up to 250 h in air and in protective atmosphere in order to study the effect of aging on the mechanical properties and evolution of pores and grain size.

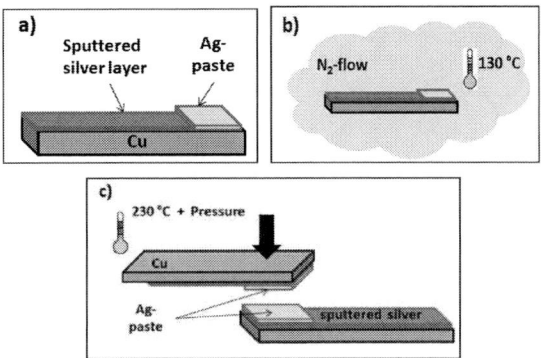

Figure 1: a) Preparation and application of Ag-paste by stencil printing b) Drying under nitrogen-flow c) Sintering of the lapjoints.

Microstructure

Mainly the sintering conditions including temperature, pressure, and time lead to a difference in the microstructure hence in mechanical behavior of the joint. Typical overview of an Ag-sintered lapjoint in Figure 2.a shows that sintering for 30 minutes under 70 MPa pressure leads to a relative dense Ag-layer. The thickness of the silver layer in the lapjoint was reduced from 200 to 70 μm after the sintering process. However, due to the geometrically induced lower bonding pressure at both sides of the joints, the porosity of a marginal area was around 40 % (Figure 2.b) [5]. Nanoindentation tests were performed to determine the Young's modulus of the samples, leading to a value of about 70 GPa for the Ag-joint. The E-Modulus of the porous area was around 45 GPa.

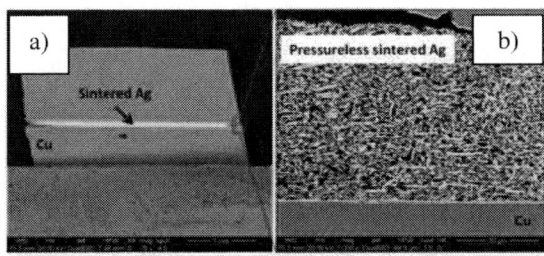

Figure 2: SEM image of the cross-section of a sintered Cu-Ag-Cu lapjoint and the microstructure of the margin area which was sintered with a lower applied pressure.

In order to reveal the microstructure of the joints and to analyze the size and distribution of the pores samples were prepared by using focused ion beam (FIB) technique as shown in Figure 3. The image analyses result in a porosity of around 6 % for the samples prior to heat treatment (for the 0-hour). Not only the total porosity but also the shape and connectivity of the pores affect the properties of a sintered part [9]. The porosity of the sintered layer seemed not to be isolated but formed an interconnected network. After heat treatment at 250°C for 250 hours a considerable change of the microstructure including rearrangement and coarsening of the pores as well as grain growth is observed (Figure 3.b).

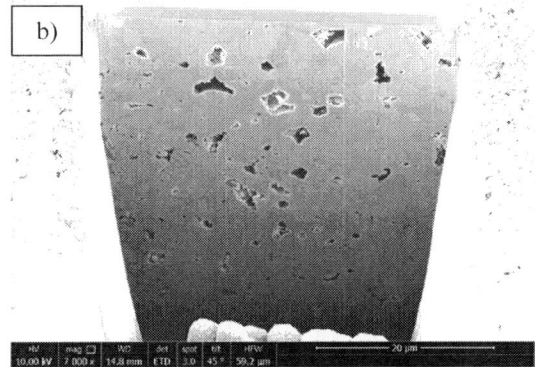

Figure 3: FIB milled section of the Ag-layer which was sintered under 70 MPa pressure a) in non-aged state and b) after aging for 250 hours at 250 °C in air.

Aging in air leads to copper-oxide formation at the Cu-Ag interface as shown in Figure 4. In our study a series of samples were protected from oxidation during the heat treatment by encapsulation in hermetically sealed glass tubes.

Tensile and fatigue test set-up

A universal testing machine from Messphysik Austria equipped with a load cell capacity of 1 kN was used for determination of mechanical properties.

509

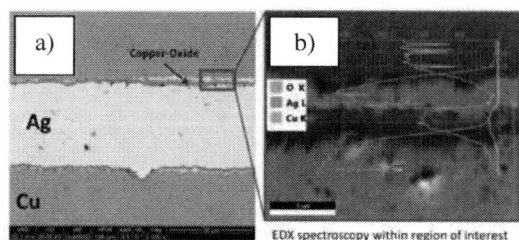

Figure 4: SEM image from the cross-section of a sample after heat treatment in air with the EDX analysis showing copper-oxide.

A thermal chamber was part of the set-up to implement the experiment also at elevated temperatures. A Laser Speckle Extensometer (LSE) system was used for contactless strain measurement of the joint area (Figure 5).

Figure 5: Tensile machine with LSE for measurements of shear strength and low cycle fatigue experiments of Cu-Ag-Cu lapjoints.

RESULTS

The effect of surface roughness of copper substrate for pressureless sintered silver on the shear strength has been investigated by Meiyu Wang et al. [7]. The pre-metallization with silver of the surface leads to a stronger interface connection compared to electroplating [6]. In a previous study, we found that using Ag-sputtered strips results in an increase of the joint strength comparing with those sintered on blank copper [8]. Therefore, in this study only sputtered samples were prepared for the fatigue investigation. All samples were half an hour and sintering pressure of 70 MPa were chosen for the samples to be investigated for mechanical cycling.
For the pre-characterization shear tests were performed on several samples resulting in a maximum shear strength of 45 MPa and a mean value of 35 MPa. A rather high scatter of shear strength in all samples was observed with 44 % of non-aged samples and 26 % of aged samples showing shear strength values below 40 MPa. These results indicate an improvement of the shear strength of Ag-sintered joints subsequent to aging. As described in the next section, low cycle fatigue tests were conducted with a maximum stress of 40 MPa.

Low cycle mechanical loading and ratcheting

Low cycle fatigue tests were performed with a cross head speed of 2 mm/min which corresponds to a testing frequency of about 0.1 Hz. The tests were conducted in tensile-tensile loading mode with a minimum stress of 8 - 10 MPa and maximum stress levels of 30, 35 and 40 MPa.

Figure 6 shows the $\sigma - \varepsilon$ relationship of a Cu-Ag-Cu lapjoint corresponding to the strain data captured with the LSE directly from the silver joint area (Figure 5) in comparison with the response of the whole system including the testing machine and the joint. For an example at a stress level of 30 MPa, the strain of the joint is about 4.5-5 % for the first five cycles with LSE, while the strain value of the whole system is around 14 %.

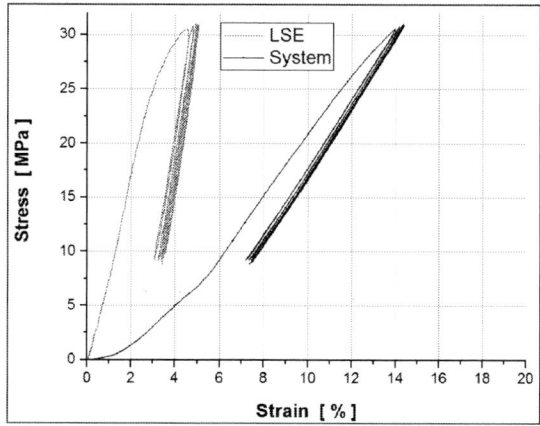

Figure 6: $\sigma - \varepsilon$ relationship of the Ag-interface captured with LSE (red plot) and of the total setup (black plot).

Ratcheting is a creep-phenomenon which occurs when the maximum of the exerted stress exceeds the elastic limit of the material. By the

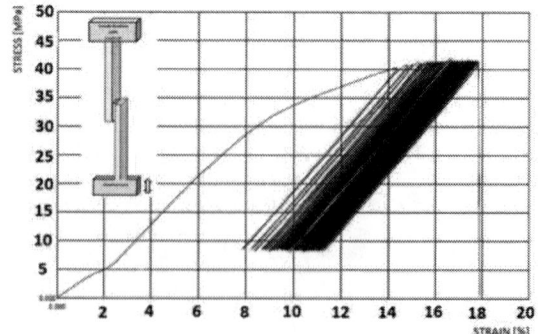

Figure 7: An example for the $\sigma - \varepsilon$ –curve of the Ag-sintered lapjoint by cycling loading.

application of cyclic stress in the tensile-tensile mode, a shift of the stress-strain hysteresis is observed due to plastic deformation. The rate of ratcheting is affected by the stress level (mean and maximum), temperature, loading history and other factors [2]. An example of this behavior for the Cu-Ag-Cu lapjoints is shown in Figure 7.

Figure 8 shows an increase of the ratcheting strain (ε_r) of the non-aged samples for stress levels of 30 and 40 MPa. A comparison of the ε_r - N_f, curves up to 300 cycles shows that a higher stress amplitude results in an increase of ratcheting strain from 14.8 % to 21.5 % (Figure 8.a). The strain rate at the first 50 cycles in both cases is higher, which gradually decreases with increasing the cycles to failure (N_f). Figure 8.b shows that the curve corresponding to a σ_{max} of 30 MPa tends to saturate after a couple of thousand cycles.

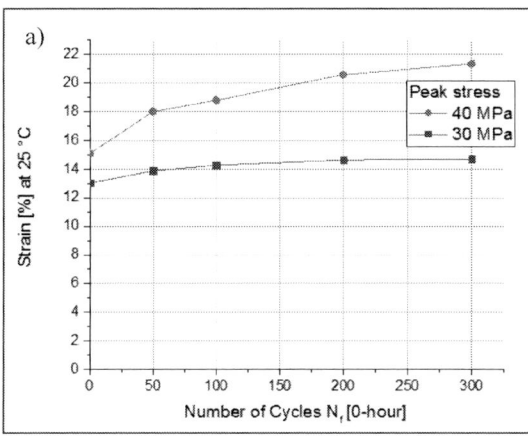

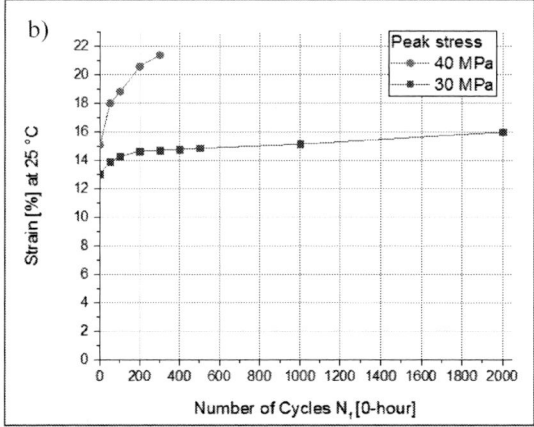

Figure 8: Average ratcheting strain at two stress levels for non-aged samples at room temperature up to N_f = 300 (a) and N_f = 2000 (b).

Influence of temperature on lifetime N_f

A comparison of the S-N curves of the non-aged lapjoints tested at 25 °C and 130 °C in the low

cycle regime is presented in Figure 9. The curves are obtained in the range of maximum stress 40 to 30 MPa up to ~ 1e4 cycles to failure. Though a high scatter of data is observed in both cases, a clear shift of S-N curve with increasing the testing temperature is observed resulting in the reduction of lifetime around one order of magnitude.

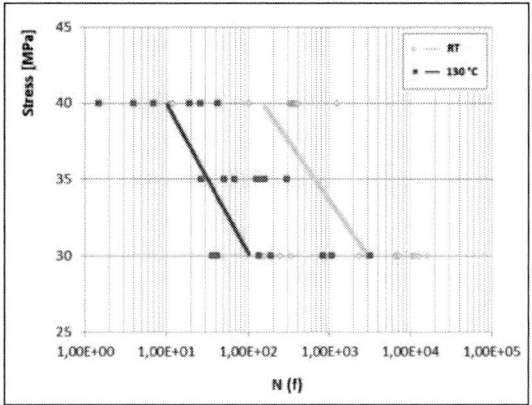

Figure 9: S-N curve for sintered Cu-Ag-Cu lapjoints

The fracture probability curves of non-aged joints at 25 °C and 130 °C in comparison with those of in air aged samples tested at 25 °C is displayed in Figure 10. All samples were tested at a σ_{max} of 30 MPa (mean stress of 20 MPa). The plots show that the aged samples have a 50 % lifetime expectation of 5.000 cycles, whereas it is 3.000 cycles for non-aged ones (Figure 10.a). Aging in a protective atmosphere results in a clear improvement of lifetime as shown in Figure 10.b for samples investigated at σ_{max} of 40 MPa (mean stress of 25 MPa). The 50 % lifetime expectation is 150 cycles versus 4.000 cycles for the non-aged and aged samples respectively.

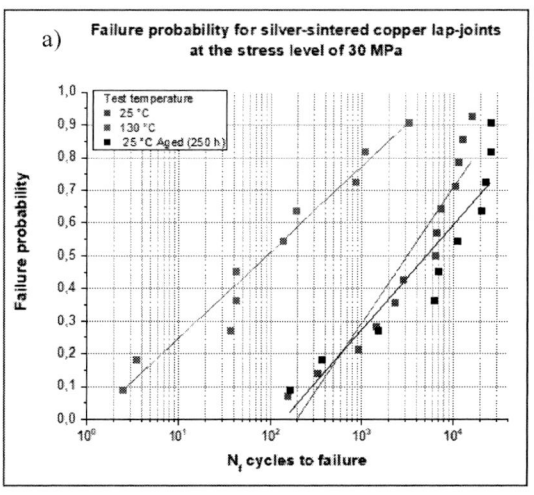

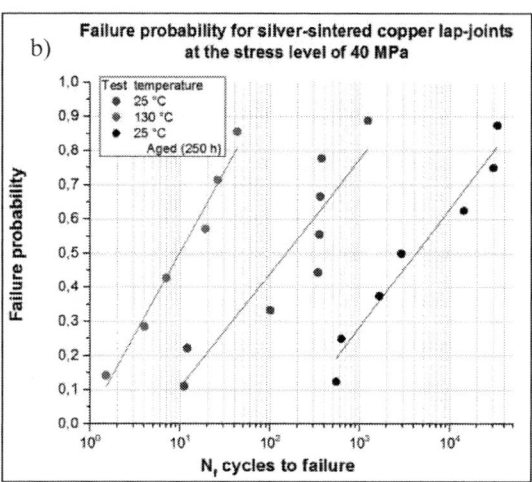

Figure 10: Failure probability plots of the Cu-Ag-Cu joints tested at elevated (red plots) and room temperature for two stress levels as shown in a) and b) and for both, 0-hour and aged sample series (blue or violet and black plots).

Fracture surface analysis

SEM investigations were performed to analyze the fracture surface of the aged and non-aged samples after shear and fatigue testing at room and elevated temperatures. Basically, two types of crack propagation were observed: One near the copper interface but mainly between sputtered and sintered silver layers (Figure 11.a), the other mainly through the Ag-sintered layer (Figure 11.b). Joints without heat treatment (0-hour samples) tested at room temperature, tend to break near Cu/Ag bonding interface. Heat treated samples which were tested at higher temperatures as well as aged samples tend to break rather in the bulk of silver sintered layer. This type of fracture was observed in the case of samples either subjected to static or cycling loading.

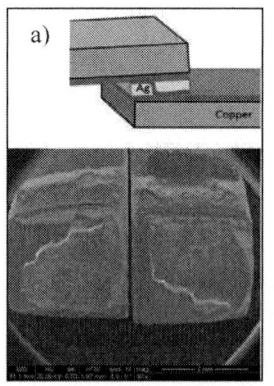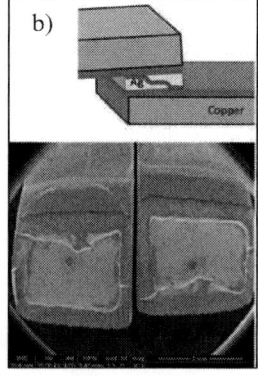

Figure 11: Exemplary SEM images and schematic crack path of fracture propagation near interface (a) and through sintered silver bulk (b).

CONCLUSION

Primarily the sintering conditions such as temperature, pressure and duration influence the microstructure of the silver sintered joints. Optimization of the manufacturing parameter includes surface treatment of the interconnecting components such as surface roughness and metallization. Mechanical properties e.g. static shear strength or fatigue behavior of the joint is affected by the porosity of the sintered bonding layer.

In this study an increase of porosity from 6 % to 40 % within the Ag-sintered joints resulted in a decrease of Young's Modulus from 70 to 45 GPa. The ratcheting behavior of the lapjoints under constant stress cyclic loading was studied. The ratcheting strain which describes the creep phenomenon of the material was found to be strongly dependent on the maximum cyclic stress at room temperature.

For non-aged samples fatigue testing at elevated temperature of 130 °C lead to a significant reduction of lifetime compared to testing at 25 °C. Long term thermal exposure in air at 250°C/250h resulted in formation of copper-oxide at the Cu/Ag interface which is believed to have a weakening effect on the joint. Furthermore a considerable grain growth and alteration of the pore shape and distribution was observed during the aging up to 250 hours. Thereby the volume percentage of the pores remained almost unchanged. Long term aging did not result in a decrease of fatigue response of the Ag-joints at room temperature. Independent of aging at air or protective atmosphere, an improvement of the lifetime was observed. In the latter case even a considerably higher lifetime was obtained. In addition, aging resulted in a change of failure mode from near interface fracture to failure in the bulk of the sintered layer due to the higher capacity of plastic deformation in the recrystallized material.

In a study on the effect of aging on the tensile behavior of sintered silver a clear trend between the mechanical response in the plastic regime and the ageing time or density was not found. This unpredictable behavior of the aged porous silver was rather related the statistically inhomogeneous distribution of the pores leading to a variation of the local density of the samples [3]. In the present work, the improvement of the fatigue strength of the aged samples at low cycle regime and high stress amplitudes seems to be related to several concurrent factors. These may include a possible completion of the sintering process during the aging, increased plasticity as result of recrystallization, increase in the mean pore spacing and improvement of the interfacial bonding strength as a result of diffusion. For a better understating of the low cycle fatigue behavior of silver sintered joints further studies are ongoing.

Acknowledgements

The financial support by the Austrian Federal Ministry for Digital and Economic Affairs and the National Foundation for Research, Technology and Development is gratefully acknowledged.

We would like to thank Prof. J. Fleig and A. Opitz from Div. Electrochemistry CTA–TU Wien for providing the sputter facility.

References

[1] Kim S. Siow, "Are Sintered Silver Joints Ready for Use as Interconnect Material in Microelectronic Packaging?" Journal of Electronic Materials, Volume 43, Issue 4, 2014, pp. 947-961.

[2] Tao Wanga, Gang Chena, Yanping Wangb et al., "Uniaxial ratcheting and fatigue behaviors of low-temperature sintered nano-scale silver paste at room and high temperatures", Materials Science and Engineering A, Volume 527, Issue 24-25, September 2010, pp. 6714-6722.

[3] P. Gadaud et al. "Ageing Sintered Silver: Relationship between Tensile Behavior, Mechanical Properties and the Nanoporous Structure Evolution" Materials Science & Engineering A 669 (2016) 379–386

[4] Ly May Chew et al., "Micro-silver sinter paste developed for pressure sintering on bare Cu surfaces under air or inert atmosphere", 2018 IEEE 68th Electronic Components and Technology Conference, 29 May - 1 June 2018, San Diego - California, USA.

[5] Wolfgang Schmitt et al., "Silver sinter paste for SIC bonding with improved mechanical properties", EMPC 2017, September 10th to 13th 2017, Warsaw University of Technology, Poland.

[6] Chen, C. et al. "High-temperature reliability of sintered microporous Ag on electroplated Ag, Au, and sputtered Ag metallization substrates", Journal of Materials Science: Materials in Electronics, Volume 29, Issue 3, 1 February 2018, pp. 1785-1797.

[7] MeiyuWang et al., "How to determine surface roughness of copper substrate for robust pressureless sintered silver in air", Materials Letters, Volume 228, 1 October 2018, pp. 327-330.

[8] Z. Goekdeniz, et. al. "Temperature Dependent Mechanical Properties of Sintered Silver-Copper Joints" ISSE 2019 1-6. 10.1109/ISSE.2018.8443659

[9] Herbert Danninger et al., "Automatic measurement of the effective load bearing cross section Ac in sintered steel", Praktische Metallographie / Practical Metallography, Volume 39, Issue 8, August 2002, pp. 414-425.

The Challenge of a Self-Soldering PCB

Arne Neiser[1], Dirk Seehase[2] and Andreas Reinhardt[1]

[1] Research and Development, SEHO Systems GmbH,
Kreuzwertheim, Germany
[2] Institute of Electronic Appliances and Circuits, University of Rostock,
Rostock, Germany

corresponding author: arne.neiser@seho.de

Abstract

In the most commonly used soldering process, reflow soldering, it is necessary to heat the system to stably transfer the required energy to the product during the soldering process. 70 percent of the heat energy of a machine is used for self-heating, the rest for soldering. In a current project, the possibility of realizing the heating necessary for soldering by embedding a Joule heating element in the printed circuit board is investigated. To realize this, the board must be mechanically contacted. These contacts must meet the following requirements: The necessary energy must be transferred and the handling of the board must be robust and simple. There are several contact methods that have been developed and tested. The experiments mainly relate to contact resistance, reliability and flexibility of the contact method.

Key words: embedded component, heating layer, self-soldering PCB, reflow soldering

Introduction

The soldering of electronic components has hardly changed in the last 20 years with respect to the use of convection. The energy transfer is done by a gaseous medium, which is heated by means of a mostly electric heater. This heated air is applied to the product by forced convection. From an energetic point of view, this process has an efficiency of approx. 30% when the performance of the heating element is considered in relation to the heating used to solder the product.

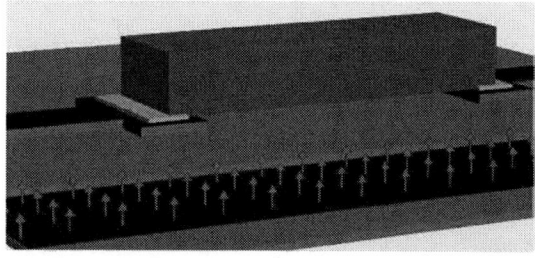

Fig. 1: Schematic representation of an embedded heating layer with red arrows to show the heat flow

Taking this percentage into account, a higher efficiency was sought. The focus was on the heat conduction. The result is to embed a heating layer in the circuit board and use this layer to provide the required energy for soldering. [1]

The methods presented here for contacting with this heating layer depend on the current status of the project. This condition is as follows: There is a usable heating material that can be embedded in the current printed circuit board technology. This material has a resistance of approximately 20 ohms / square. This layer is for soldering a euro card (100 mm x 160 mm) at max. 60 V and 10 A. First printed circuit board demonstrators were developed and promising tests were carried out.

The tests are supported by simulations made by research partners and thus certain criteria are worked out. In addition, two prototype machines were built, which are currently being completed. For these systems, various contact methods have been developed, which are presented in this article. It focuses on 2 different contact methods. The first, in which a contact for a continuous soldering oven is to be realized, the second in a batch oven with contacts on the small side of the board.

The target temperature profile of the board must correspond to a soldering profile and reach and maintain the required temperature ranges. For this purpose, it seems to be necessary that the temperature is monitored by a sensor / sensor. For this, a thermocouple or thermal imager can be used which survey the surface of the board during the profile. [2]

State of the art

One factor for evaluating the different contact methods is contact resistance. To determine the contact resistance, the touch areas must be defined and all dependent factors determined. In the following, the surface texture (roughness), the temperature of the contact surfaces, the force and the material properties (conductivity, hardness, etc.) are considered. [3]

There are simple formulas in the literature to approximate contact resistance. The contact resistance (R_E) results from the specification. electrical resistance (ρ) divided by twice the radius is assumed here a circular area as a contact surface

$$R_E = \frac{\rho}{2a}. \text{ [3]} \qquad (1)$$

This consideration is considered to be very ideal and is not due to a change in resistance due to temperature or other roughness contact surface. A better approximation results can be archived from the following formula [3]

$$\Lambda \approx \frac{3,7}{E^* \rho l} F_N , \qquad (2)$$

$$E^* = \frac{E}{2(1 - v^2)}, \qquad (3)$$

$$R_E = \frac{1}{\Lambda}. \qquad (4)$$

Here, the symbols stand for: the elastic modulus (E), the root mean square roughness height distribution (l), the force to which the contact surfaces are connected, and the Poisson's number (v). With eq.(4) we can compare the calculated value with the measured one in the evaluation chapter.

Test Setups

For the tests, a test board was used, which has lateral, front and rear contact elements, see Fig. 2. Embedded in the board are 12 heating surfaces of equal size, which can be controlled individually or in a complete row via a contact pair on the front and rear side [source]. In addition, a copper side contact is applied to test a contac tmethode near the transport. The test board is also used to investigate the solderability with the embedded heating layer. For this reason, there are food prints with DaisyChain to recognize.

Fig. 2: above: test board in a continuous soldering oven with contact points on the front and back (not used) as well as on the printed circuit board side (contacted here) bottom: 3D illustration of the contact elements

The first experimental setup is designed for a continuous soldering oven, a prototype system can be seen in Fig. 4. The contact measurement is designed in such a way that each transport contacts on each side. They consist of bronze (CuSn6) and contact the test board on the side contacts (see Fig. 2). During the passage of the test board through the soldering oven, the contact resistance between the potential of the connected contacts (blue cable) and the contact surfaces of the test plate was determined.

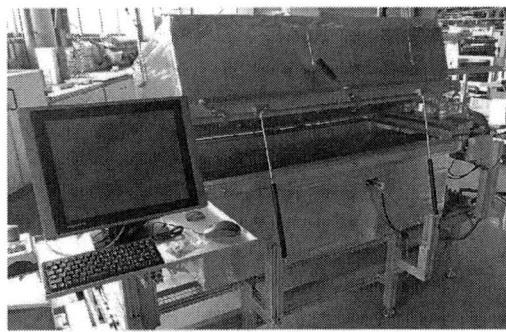

Fig. 3: Continuous soldering oven (front view)

The second contact methode here presented is a for use in a batch oven. The contacts are made of the same material as in the continuous version. There are 15 contacts for the 12 heating surfaces, which can be heated separately or in series by the interconnection on the board. The Tesboard is located here in a chamber at the end of which the contact is located (see Fig. 4 and Fig. 5). [4]

Fig. 4: 3D model of contact measurement for use in a batch soldering oven

The test chamber for the second contact methode of a batch soldering oven is shown in Fig. 5. It can be equipped with the board from the front. The chamber is also planed with a heater and a fan and capable to soldered under nitrogen. This paper deals only with the contact, but not with the chamber itself.

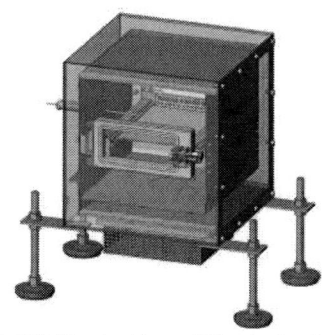

Fig. 5: 3D illustration of the test soldering chamer of a oven in the batch variant

The measurement setup for the second contact type for the batch oven is shown in Fig. 6. It uses a power source, oscilloscope and thermocouple multimeter. The resistance is measured with the four-wire method. For each separate area on the test board, the associated contact resistances were measured.

Fig. 6: Test setup for the contact method of a soldering batch oven

Results

The results of 25 series of measurements of the contact resistance are shown in Fig. 7 for the continuous soldering oven. The maximum and

minimum values of the resulting resistance of the contacts are shown. The mean is 0.04 Ω and has a deviation of up to 0.006 Ω.

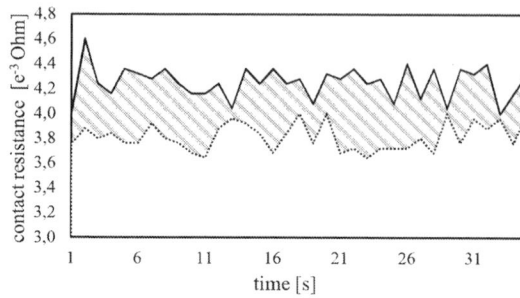

Fig. 7 Contact resistance of continuous measurements of a moving test PCB over time

Fig. 8 shows the contact resistances of the contact pairing per heating zone (12 zones) for 10 measurements. Here, values between 0.016 Ω and 0.029 Ω are measured. The deviation of the individual segments varies between 5% and 20%.

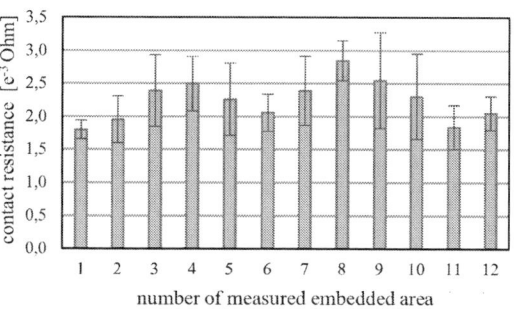

Fig. 8: Contact resistance for the batch contact method per heating segment

When calculating the contact resistance (with eq.(4)) for the comparison with the measured results, the values in TABLE I were used. The result is a contact resistance of 1.25 mΩ. [5]

TABLE I: Value for calculation of the contact resistance

size	unit	value
E	GPa	100
l	m	2e-6
v	-	1/3
F_N	N	3
ρ	Ωm	1,111e-7

Evaluation

The results show that the contact resistance is in the lower mΩ range as expected. By the use of bronze which has a factor of 10 lower conductivity than copper and the contact resistance is higher by a factor of 10. This only plays a subordinate role for energy transfer, as this factor causes only a loss of 0.1 Volt (1 Watt) at a maximum of 10A. The calculated result lies in the same size range and can thus be used in the first approximation in the future. There is no need to worry about the roughness of the contact surface because the influence of the contact resistance is negligible.

When measuring the board in the continuous seoldering oven, some challenges had to be overcome: The contact force of the contacts was greater than the friction to the transport chain, which led to the stopping of the board. The contact material thickness of the contacts was too thick, which led to damage of the board and contact with light oblique installation difficult. By reducing the thickness of the contact material, these challenges could be solved.

When measuring the second method for the batch oven, the thickness of the contacts was too high, which meant that the circuit board had contact with only a few contacts. Again, it can be solved by reducing the material thickness.

Both methods can be used to contact the test board. The biggest advantage of the batch version is the increased number of contacts and thus the possible higher number of heating segment. At the moment no investigations have been carried out here in relation to the wear of the contacts.

Conlcusion and Future Plans

It was shown how a PCB with embedded heating layers can be contacted in a continuous pass of a soldering oven and a batch soldering one. In the case of using these heating layers with resistance at 10 Ω and more, the contact resistance can be neglected. But if the contacts will be used for other applications the results can be important.

In the future, further contact methods will be investigated, which may open up new possibilities in terms of playing the contact on the PCB and using the room inside the soldering machine. One is shown in Fig. 9. This is also intended for use in a batch oven. Furthermore, it will be investigated how the temperature measurement on the PCB can be realized. As soon as the batch process is completed, the first tests under nitrogen are carried out, as well as control algorithms which require a stable solder profile.

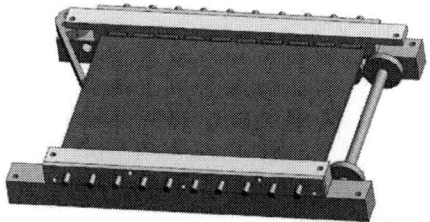

Fig. 9: 3D representation of another contact method in a batch oven

Acknowledgements

The authors acknowledge the financial support by the Federal Ministry of Economy Affairs and Energy of Germany in the framework of the research project ERFEB (03ET1533A).

Also, the authors want to acknowledge the following companies: B & B Sachsenelektronik GmbH, Siemens Corporate Technology, FutureCarbon GmbH, Heraeus Materials Technology GmbH C.KG, Neue Materialien Bayreuth, Lackwerke Peters, University of Bayreuth and University of Rostock.

References

[1] D. Seehase, A. Novikov, and M. Nowottnick, "Resistance Development on Embedded Heating Layers during Climatic Test," in *41st International Microelectronics and Packaging Conference (IMAPS)*, 2017.

[2] A. Neiser, D. Seehase, P. Koschorrek, and A. Reinhardt, "Control a Joule-Heating Embedded Layer within a Printed Circuit Board," in *2018 7th Electronic System-Integration Technology Conference (ESTC)*, 2018, pp. 1–4.

[3] E. Vinaricky, A. Keil, W. A. Merl, K. H. Schröder, and J. Weiser, *Elektrische Kontakte, Werkstoffe und Anwendungen*. 2002.

[4] A. Neiser, D. Seehase, and A. Reinhardt, "Self-Heating Printed Circuit Board - A Challenge to Contact and Control for a Reflow Soldering Process," in *2018 41st International Spring Seminar on Electronics Technology (ISSE)*, 2018, pp. 1–4.

[5] D. Kupferinstitut, "German copper institute: CuSn6," 2005. [Online]. Available: https://www.kupferinstitut.de/fileadmin/use r_upload/kupferinstitut.de/de/Documents/S hop/Verlag/Downloads/Werkstoffe/Datenbl aetter/Bronze/CuSn6.pdf.

Investigation of ScAlN for piezoelectric and ferroelectric applications

R. Petrich[1], H. Bartsch[1], K. Tonisch[1], K. Jaekel[1], S. Barth[2], H. Bartzsch[2], D. Glöß[2], A. Delan[3],
S. Krischok[1], S. Strehle[1], M. Hoffmann[1] and J. Müller[1]

[1] Technische Universität Ilmenau, Institut für Mikro- und Nanotechnologien MacroNano®, Ilmenau

[2] Fraunhofer-Institut für Organische Elektronik, Elektronenstrahl- und Plasmatechnik FEP, Dresden

[3] Technische Universität Dresden, Institut für Festkörperelektronik, Dresden

+49 3677 69 3440, rebecca.petrich@tu-ilmenau.de

Abstract

$Sc_xAl_{1-x}N$ is a promising material to expand the application range of nitride materials, since scandium increases the piezoelectric constants while retaining the crystalline wurtzite structure. In this work, stationary reactive pulse magnetron sputtering is used for the deposition of functional layers with a scandium content $x = 0...0.44$. Layer morphology, piezoelectric properties and breakdown voltage are studied. XRD measurements reveal that high scandium content yields to a weak wurzite formation. The lattice constant dependent on the parameter x is calculated on the base of XRD data and the curve characteristic agrees with the density functional theory. The highest piezoelectric coefficient d_{33} was observed at a scandium content of 36.6 %, it amounts to 27.5 pC/N. Parallel capacitor structures are generated by means of chlorine-based ICP etching and lift-off structuring of the top electrodes. Technological details of the structuring process are presented. The etch rate depends strongly on the scandium content. The permittivity was determined on the base of these test structures. It increases significantly with increasing scandium content. High capacitance values up to 7.4 nF were measured. The adequate breakdown voltage of 51 V for scandium concentrations of $x = 0.22$ or higher suggest a use of such layers for integrated thin film capacitors in addition to well-tried piezoelectric applications.

Key words: Scandium aluminum nitride, piezoelectricity, ferroelectricity, magnetron sputtering

Introduction

Autonomous sensor applications require new concepts for ultra-low energy electronics fed by energy harvester components to guarantee an independent and continuous operation. $Sc_xAl_{1-x}N$ based piezoelectric harvester structures promise high performance due to high piezoelectric constants and power generation figure of merit [1, 2] and exhibit outstanding thermal, mechanical and piezoelectric properties. The material can expand the application range of nitride materials, since scandium increases the piezoelectric constants while retaining the crystalline wurtzite structure and it shows ferroelectric properties for high Sc contents [3].

This paper investigates the properties of pulsed magnetron sputtered layers $Sc_xAl_{1-x}N$, $x = 0...0.44$ and presents crystallographic, electric and piezoelectric properties.

Layer deposition

The AlN and $Sc_xAl_{1-x}N$ films were deposited by stationary reactive pulse magnetron sputtering from a Double Ring Magnetron DRM 400 developed by Fraunhofer FEP [4]. This type of magnetron combines two concentric discharges, which enables the uniform coating of substrates with a diameter of 200 mm. Pure Al-targets (purity 5N) or an Al outer target (5N) and Sc inner target (3N5) were used to deposit AlN or $Sc_xAl_{1-x}N$, respectively. The film composition of $Sc_xAl_{1-x}N$ was varied by changing the power ratio of the two targets. The deposition rate was 200 nm/min (AlN) or $100 - 200$ nm/min ($Sc_xAl_{1-x}N$, depending on applied target powers).

The reactive sputtering in an argon-nitrogen mixture (gas purities 5N) was carried out in unipolar/ bipolar hybrid pulse mode using a pulse unit UBS-C2 developed by Fraunhofer FEP and standard DC power supplies [5]. A closed loop reactive gas control was applied to stabilize the process in the transition mode between metallic and full reactive sputtering. The base pressure of the cluster tool equipment was $2.0 \cdot 10^{-7}$ mbar.

Piezoelectric coefficients

Piezoelectric properties were determined using a PM300 piezometer (Piezotest Ltd., London, UK) on a simple ultrasound transducer layout. It is a three-layer structure, consisting of a circular AlN or $Sc_xAl_{1-x}N$ layer with a diameter of 13 mm, sandwiched between two aluminum electrodes with 10 mm diameter serving as contact pads.

The scandium content was determined on Si substrates at the center position of deposition by energy dispersive spectrometry of X-rays (Apollo

XV, EDAX) using a silicon drift detector with an energy resolution of 125 eV. For the analysis, an acceleration voltage of 10 keV was used, which allows a good excitation of Aluminum and Scandium K_α-lines. The relative measurement uncertainty is in the range of ± 5 %. The dependency of the piezoelectric coefficient on Sc concentrations x is shown in Table 1. The piezoelectric coefficient d_{33} increases with higher Sc concentration up to $Al_{0.57}Sc_{0.43}N$. Above 50 % Sc, the films exhibit nearly no piezoelectric properties [6].

Table 1: Piezoelectric coefficient d_{33} [pC/N] of $Sc_xAl_{1-x}N$ ($x = 0...0.57$)

x	0	0.24	0.29	0.33	0.35	0.37
d_{33}	6.5	21	22	23.5	23.8	27.5

x	0.40	0.41	0.42	0.51	0.54	0.57
d_{33}	26	27.2	26.9	1.5	0.4	0

Crystal structure of the layers

X-ray diffraction has been done in Bragg-Brentano configuration using a Bruker AXS D8. The XRD diffractograms of the $Sc_xAl_{1-x}N$ layers are depicted in Figure 1. They reveal a pure c-axis orientation of wurtzite $Sc_xAl_{1-x}N$ as can be expected for the columnar growth being typical for reactively sputtered AlN and $Sc_xAl_{1-x}N$.

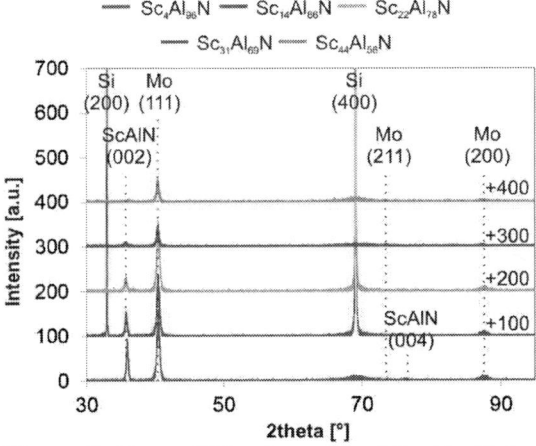

Figure 1: X-ray diffractograms of $Sc_xAl_{1-x}N$ ($x = 0.04...0.44$)

The (002) peak was used for calculating the lattice constant c. The values are depicted in Figure 2. For very high Sc amounts ($x = 0.44$) only a weak wurtzite (002) peak is visible, since the shift from hexagonal AlN towards cubic ScN leads to an increasing share of the layer being amorphous. The lattice constant increases with increasing content of Sc and decreases again for higher Sc contents. This behavior has also been described theoretically by the density functional theory (DFT) [7], calculated values are also displayed in Figure 2 for comparison. Theoretical and measured values agree qualitatively.

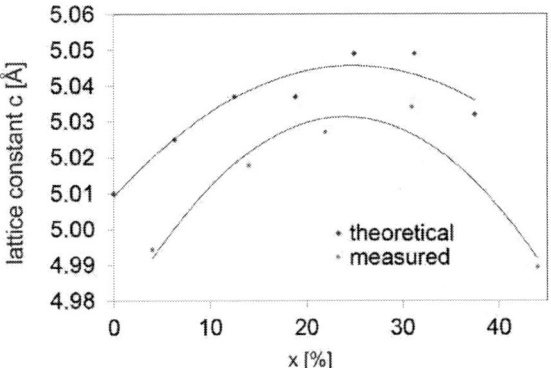

Figure 2: Comparison of lattice constants as measured and theoretical DFT values

Layer composition

Energy dispersive X-ray spectroscopy (EDX) is carried out at 12 kV acceleration voltage in order to determine the chemical composition (Thermo - SDD-Detektor, NORAN 7). The error for the method is estimated to be 2 %. K_α-lines intensity was used for the element ratio calculation and the molybdenum and silicon portions representing the contact layer and substrate were not considered in the percentage analysis of the spectrum. The obtained values are summarized in Table 2.

Table 2: Sample composition measured by EDX of $Sc_xAl_{1-x}N$ ($x = 0.04...0.44$)

Sample	Sc [at%]	Al [at%]	N [at%]	O [at%]
AlN	0	40.0	57.3	2.7
$Sc_4Al_{96}N$	1.7	36.8	57.3	4.2
$Sc_{14}Al_{86}N$	5.2	32.7	54.7	7.4
$Sc_{22}Al_{78}N$	8.1	29.4	54.4	8.1
$Sc_{31}Al_{69}N$	11.9	26.2	52.7	9.2
$Sc_{44}Al_{56}N$	17.0	21.8	50.0	11.2

Oxygen was present in all samples. Its content increases with increasing scandium content. The origin lies most probably not in the sputter process, since Auger electron spectroscopy with depth profiling (not shown here) showed an oxidized surface with a depth of $10 - 15$ nm in total. It is assumed to be a consequence of exposure to air.

The thickness of all layers was determined evaluating scanning electron microscopy (SEM) images, (SEM Hitachi 4800) captured at 5 kV and a working distance of 8.2 mm. Figure 3 depicts the cross section of the layer sequence and illustrates the measurement of the dielectric thickness d. The transition region between the molybdenum layer and $Sc_xAl_{1-x}N$ layer is visible as a dark line. Single points at different layer positions were selected for measurement. The average relative error of the thickness measurement E_d was evaluated on the base of a sample size of 17 measurement points and amounts to 7%.

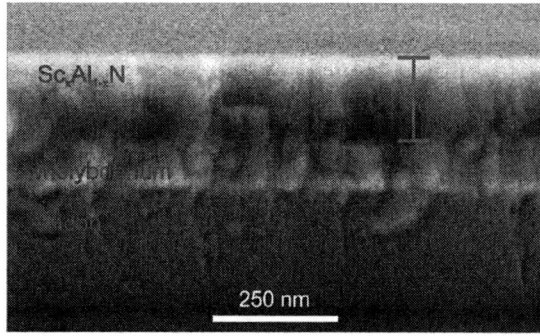

Figure 3: SEM image of a cross section through the Sc14Al86N layer

Roughness studies were carried out using laser scanning microscopy (LSM, OLS 4100, Olympus, Germany) considering a scanning area of 128 x 128 μm^2. The arithmetic mean height R_a was determined without filtering using the device software and the obtained values are summarized in Table 3.

Table 3: Layer thickness d [nm] and surface roughness R_a [nm] of $Sc_xAl_{1-x}N$ (x = 0.04...0.44)

x	0	0.04	0.14	0.22	0.31	0.44
d	171	189	165	151	139	153
R_a	12	10	11	17	23	29

Preparation of test structures

To determine the permittivity, plate capacitor structures were realized. The main process steps and a schematic view of the structures are depicted in Figure 4. 200 nm molybdenum were sputtered on a 300 μm (100) p-type Si wafer. Subsequently, $Sc_xAl_{1-x}N$ was deposited with a nominal layer thickness of 200 nm. The structuring of the $Sc_xAl_{1-x}N$ was done by chlorine-based ICP etching at a gas flow of 15 sccm Cl_2, 5 sccm BCl_3 and 5 sccm Ar. A mask consisting of photoresist (AZ®1518) is used.

The etch rate of the layers was determined in a preliminary test. Masked samples of $Sc_xAl_{1-x}N$, were etched for 50 seconds each at 25 W CCP power at 20°C. The etch depth was subsequently determined by profilometry (Dektak 150 stylus, Veeco Instruments Inc., USA). The results are shown in Table 4.

The test revealed that the etch rate decreases strongly with increasing scandium content. At a scandium portion of x = 0.44, no etching was detectable with the measurement device. Processing of $Sc_xAl_{1-x}N$ layers with scandium content up to x = 0.04 is reasonable applying a CCP power of 25 W. For higher scandium concentrations it is necessary to increase CCP power to 50W. In order to guarantee the chemical selectivity when approximating to the Mo electrode, the remaining layers with a thickness of

approx.70nm are finally etched with reduced CCP power of 25W. The process temperature was reduced from 20°C to 10°C to counteract the erosion of the resist mask. Table 5 summarizes the etch parameters for structuring of $Sc_xAl_{1-x}N$ layers with x = 0 to x = 0.44.

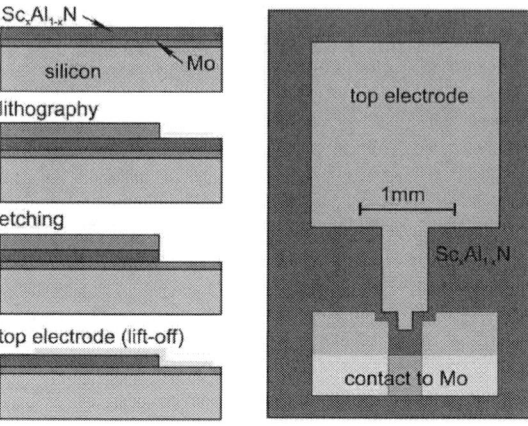

Figure 4: Schematic view of test structures and main process steps for their preparation

Table 4: Etch rates r_{etch} [nm/min] of $Sc_xAl_{1-x}N$ (x = 0.04...0.44)

x	0	0.04	0.14	0.22	0.31	0.44
r_{etch}	104.1	90.7	29	17.3	4.5	n.a.

Table 5: Etch parameters of $Sc_xAl_{1-x}N$ (x = 0.04...0.44)

x	0	0.04	0.14	0.22	0.31	0.44
t_{25W}[min]	5	5	3	5	9	12
t_{50W}[min]	0	0	2.5	5	7	13

After etching of the $Sc_xAl_{1-x}N$ layers, an oxygen plasma (5 sccm O_2, 20W CCP) was applied to improve the release of the photoresist (AZ®1518). Subsequently, the photoresist was dissolved with acetone.

Finally, the top contact is made with 10 nm Ti and 50 nm Au using the photoresist AZ® 5214E and a lift-off process.

Capacitance measurements

A digital LCR meter (VOLTCRAFT LCR-9063) was used for capacitance measurements at a frequency of 250 Hz. The relative error E_C of the measurement was determined considering eight values and it is lower than 5%. The capacitance values are depicted in Figure , the error bars indicate the relative error E_C. The permittivity was calculated using equation (1):

$$\varepsilon_r = \frac{C \cdot d}{A \cdot \varepsilon_0} \qquad (1)$$

Here, ε_r describes the relative permittivity of the dielectric, C is the measured capacitance, d the thickness of the dielectric, A the capacitor area and ε_0 the vacuum permittivity. The capacitor area A amounts to 2 x 2 mm with a deviation of 0.25 % due to process tolerances of the lithography and lift-off step. These tolerances are considered in the calculation of the relative permittivity. The obtained values are presented in Figure 5.

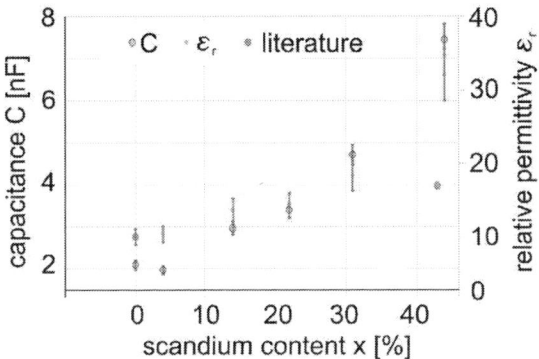

Figure 5: Capacitance and permittivity dependent on the Sc content; ε_r (AlN) = 10 [1]; ε_r (Sc$_{0.43}$Al$_{0.57}$N) = 17 [2]

The permittivity for x = 0 agrees good with the literature value. The literature value for x = 0.43 is much lower than the values measured in this study (compare Figure 5).

The uncertainty of layer thickness measurement and capacitance measurement are taken into account for the estimation of the error of the relative permittivity ε_r. It is carried out on the base of the errors E_C and E_d using Eq. 2 and 3, respectively:

$$\varepsilon_{rmin} = \frac{(C - E_C) \cdot (d - E_D)}{A \cdot \varepsilon_0} \quad (2)$$

$$\varepsilon_{rmax} = \frac{(C + E_C) \cdot (d + E_D)}{A \cdot \varepsilon_0} \quad (3)$$

C represents the capacitance and d the layer thickness and the respective mean values obtained from the measurements were used. To these values, the respective errors $E_c = 5\%$ and $E_d = 7\%$ were added for maximum calculation and subtracted for minimum calculation. The obtained values present the errors bars in Figure .

To determine the dielectric strength, the Sc$_x$Al$_{1-x}$N layers were contacted with 100 nm-thick molybdenum electrodes and characterized using a Keithley 4200 parameter analyzer (Keithley Instruments, USA). The measurement results are shown in Table 6.

Table 6: Breakdown voltage U_d [V] of Sc$_x$Al$_{1-x}$N (x = 0.04...0.44)

x	0.04	0.14	0.22	0.31	0.44
U_d	112	65	51	57	51

The breakdown voltage U_D was found to be 112 V for x = 0.04 and decreases with increasing Sc concentration. At x = 0.14 it drops to 65 V and at x = 0.22 to 51 V. From here it remains relatively constant despite increasing concentration of scandium.

Conclusions

Sc$_x$Al$_{1-x}$N layers were produced by stationary reactive pulse magnetron sputtering, varying x from 0 to 0.44. Morphology, layer thickness and roughness were investigated. The real thickness of the layers deviated from the nominal value. The roughness of the layers increases with increasing scandium content. XRD studies show, that (002) c-axis orientation decreases with increasing scandium content.

Test structures were processed by means of chlorine-based ICP etching. The etch rate of the layers decreases strongly with increasing scandium content. Therefore, higher power and longer etch time is required to remove high-doped layers. Capacitor elements are obtained by subsequent deposition of Ti/Au electrodes. The relative permittivity of the layers was calculated analytically on the base of measured geometric data and capacitance.

The results show that the relative permittivity depends strongly on the scandium content. While the relative permittivity of undoped AlN agrees good with values found in the literature, those of high-doped Sc$_{44}$Al$_{56}$N layers is significantly higher than values found in the literature. It is assumed that the reason for the deviation is the increasing surface roughness, which causes inhomogeneity of thickness and field distribution, affecting the effective relative permittivity, which is calculated on the base of ideal capacitor structures. The capacitor structures of high-doped Sc$_{44}$Al$_{56}$N show high values up to 7.4 nF and a breakdown voltage of 51 V.

Acknowledgements

This research was funded by the German Federal Ministry for Economic Affairs (BMWi) within the project "BiSWind, Integrated Sensor Technolgies for Wind Turbines" (Ref. 0325891B, 0325891C und 0325891E) and supported by the Free State of Thuringia and the European Social Fund (2017 FGR 0060).

Special thanks go to the technologists of the Institute of Micro- and Nanotechnologies MacroNano® Christian Koppka, David Venier, Birgitt Hartmann, Joachim Döll, Manuela Breiter, Henry Romanus and Andrea Knauer.

References

[1] N. Dutoit, B. Wardle, and S.-G. KIM, "design considerations for mems-scale piezoelectric mechanical vibration energy harvesters," *Integrated Ferroelectrics*, vol. 71, no. 1, pp. 121–160, 2006.

[2] M. Akiyama, K. Umeda, A. Honda, and T. Nagase, "Influence of scandium concentration on power generation figure of merit of scandium aluminum nitride thin films," *Appl. Phys. Lett.*, vol. 102, no. 2, p. 21915, 2013.

[3] S. Fichtner, N. Wolff, F. Lofink, L. Kienle, B. Wagner, "AlScN: A III-V semiconductor based ferroelectric", *Journal of Applied Physics,* vol. 125, p. 114103, 2019

[4] P. Frach, C. Gottfried, H. Bartzsch, and K. Goedicke, "The double ring magnetron process module —a tool for stationary deposition of metals, insulators and reactive sputtered compounds," *Surface and Coatings Technology*, vol. 90, no. 1-2, pp. 75–81, 1997.

[5] S. Barth *et al.,* "Adjustment of plasma properties in magnetron sputtering by pulsed powering in unipolar/bipolar hybrid pulse mode," *Surface and Coatings Technology*, vol. 290, pp. 73–76, 2016.

[6] S. Barth *et al.,* "Magnetron sputtering of piezoelectric AlN and AlScN thin films and their use in energy harvesting applications," *Microsyst Technol*, vol. 22, no. 7, pp. 1613–1617, 2016.

[7] S. Zhang, W. Y. Fu, D. Holec, C. J. Humphreys, and M. A. Moram, "Elastic constants and critical thicknesses of ScGaN and ScAlN," *Journal of Applied Physics*, vol. 114, no. 24, p. 243516, 2013.

Investigation of Pressureless Sintered Interconnections on Plasma Based Additive Copper Metallization for 3-Dimentional Ceramic Substrates in High Temperature Applications

Christian Schwarzer[1], Alexander Hensel[2], Frederik Roth[1], Corinna Merz[2], Joerg Franke[2]
Michael Kaloudis[1]

[1]Aschaffenburg University of Applied Sciences, Aschaffenburg, Germany

[2]Institute for Factory Automation and Production Systems (FAPS),

Friedrich-Alexander-University Erlangen-Nuremberg, Nuremberg, Germany

christian.schwarzer@th-ab.de

Abstract

Conventional Molded interconnect device technology with spatial circuit boards offer freedom of design as well as potential for rationalization and miniaturization by electronics integration. By integrating the third dimension into the circuit design, material and cost savings can be realized. Due to the limited application temperature of thermoplastics and thermosets, the potentials of spatial assemblies cannot be fully exploited. In the field of power electronics and LED technology, ceramic substrates are increasingly gaining acceptance for thermally demanding conditions. This publication investigates the interaction of a copper interface produced in an additive metallization process of a ceramic substrate by using the Plasmacoat® method and interconnection joints formed in a pressureless silver sintering process. Therefore, copper was deposited on different ceramic substrates and different particle temperatures as well as subsequent thermal treatment were evaluated. Silver sintering on selective coated interface was investigated for later use on a three-dimensional metallized ceramic substrate. Cross sections and Shear tests reveal that forming pressureless sintered interconnections on an additively formed copper surface is more demanding compared to common direct copper bonded (DCB) substrates.

Key words: additive manufacturing, plasma coating, silver sintering, high temperature application

Introduction

Spatial circuit boards in conventional Molded interconnect device (MID) technology offer freedom of design as well as potential for rationalization and miniaturization by electronics integration. Furthermore, material and cost savings can be realized by integrating the third dimension into the circuit design [1]. Due to the limited application temperature of thermoplastics, the potentials of spatial assemblies cannot be fully exploited. By using more thermally stable thermosets, the thermal range of applications for substrate can still be slightly shifted towards higher temperatures. For automotive and demanding industrial applications the requirements on the maximum operating temperature for electronic modules and their degree of spatial integration are increasing [2]. In the field of power electronics and LED technology, ceramic substrates are increasingly gaining acceptance for thermally demanding conditions. Current state of the art processes for the production of suitable metallization on ceramic substrates like Direct Copper Bonding (DCB) or Active Metal Bracing (AMB) are proven options for the realization of circuit layouts on flat ceramic substrates, however there are strong limitations in the implementation of spatial circuit layouts.

In the following a plasma-based coating method for the application of copper on different substrates is evaluated for the adaption in demanding electronic applications. This additive method has shown the potential of a copper metallization of semiconductors for enabling a copper bonding process on the applied functional layers [3]. The technology allows an easy to use, flexible and efficient alternative without the risk of environmental impacts or necessary follow-up processes like back-etching. The generated surface can then be used as an interface layer on which electronic devices like bare dies are mounted by interconnection materials [4, 5].

To meet the increasing requirements regarding the operating temperature or mechanical connection, a silver sinter process is used to generate the interconnection. Common solder materials are not suitable for high temperature applications due to their low melting temperature [6 - 9]. In recent years, silver sinter material has attracted attention as high reliable interconnection material especially for power electronics devices with demanding requirements on efficiency and lifetime [10, 11]. Silver sinter material is adequate for high temperature application due to the high melting temperature (961 °C) and possess excellent properties such as high thermal and electrical conductivity [12 - 15].

In this investigation the results of the additive manufacturing process for metallizing Al_2O_3 substrates by using a plasma coating technology and the results of a silver sinter process on this additive manufactured surface metallization will be discussed and compared to common DCB-substrates.

Methodology

This investigation focusses on the interaction of the copper surface produced in an additive metallization process by using the Plasmacoat® method and an interconnection layer of pressureless sintered silver material.

Plasma-Based Coating Techniques

A plasma is described as the 4th aggregate state and represented by partly ionized gas. The atomic and molecular excitement in this case, is achieved by an electric discharge caused by a high voltage impulse in the carrier gas between the nozzle, used as the anode, and the cathode. Once the plasma is ignited, it is stabilized by high a current at low voltage level. Because of the usage of argon as a carrier gas, an overall operating voltage of less than 40 V can be used and therefore a possible voltage flashover, which would cause the destruction of the semiconductor, is avoided. To operate a high-power plasma at low voltages, the carrier gas has to be chosen respectively. When applying argon as a plasma gas, the required ionization energy is in comparison to molecular gases like nitrogen relatively low. This is because of the missing side effects like dissociation or energy migration into molecular vibrational or rotational effects [16].

In Figure 1 the function of an atmospheric plasma gun is displayed.

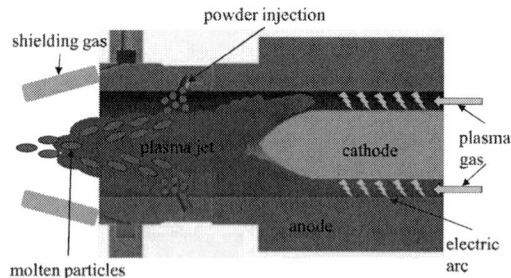

Figure 1: Principle of an atmospheric plasma gun [5]

A fine spherical copper powder with an average diameter of 12 μm is injected in the ignited plasma flame. The temperature of the plasma flame can be approximated at about 12.000 K [17]. Thus, the particles are complete or superficially molten and accelerated by the gas jet. A shielding gas at the nozzle helps to focus the particle beam. The solidification occurs when the particles strike the substrate and are cooled down immediately because of the high temperature difference.

The process and the workpiece handling are shown in Figure 2 where the substrate handling is done by an industrial robot while the gun is mounted fixedly in the chamber. This prevents a disturbance of the process due to unexpected movements in the process medium supply lines.

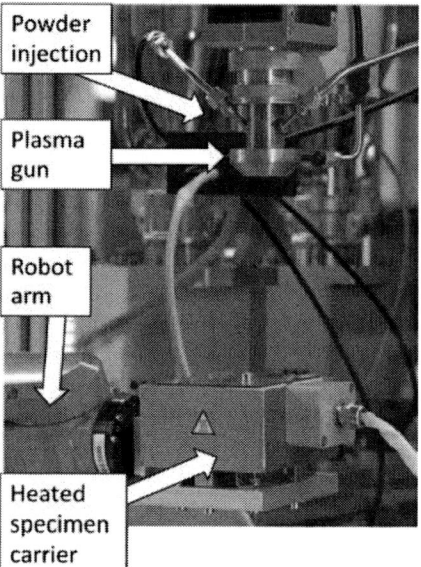

Figure 2: Construction of the process chamber

Silver Sintering

Sintering generally defines the formation of a dense body from a powder which is characterized by the elimination of pores and grain growth and occurs under the driving force of lowering the free energy (Figure 3). The sintering material used for the Low-temperature joining technique is a sub-micron or micron silver powder.

The silver powder consists of flake-shaped silver particles whose structural dimension is approximately 1 μm. The silver flakes have a coating of organic acids which are decomposed at temperatures above 200 ° C. This coating was formed in the manufacturing process from the added grinding media by chemical bonding. Such an organic protective shell is indispensable for the Ag powders, since otherwise the Ag particles could already sinter together under storage conditions (room temperature, normal pressure). In addition, non-chemically bound residues of grinding media may remain in the particle

524

pores. The decomposition of the coating initiates sintering, because only then an oxide-free surface of the particles is provided. Such surfaces have a strong tendency for surface self-diffusion at process temperatures of 230 °C to 250 °C. As a result, contact necks corresponding to (Figure 3) are formed at the points of contact of the particles, which produce the macroscopic connection. At a temperature of about 250 ° C - 260 ° C and a pressure of about 35 MPa sinter the silver particles within a few seconds between two surfaces and thereby form a solid compound with a high melting point of 961 °C.

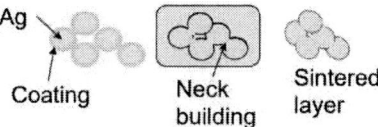

Figure 3: Scheme for silver sintering process

Nevertheless, silver sintering process can also be conducted without additional external pressure. This "pressureless" sintering process requires a longer process time due to the absence of the main process driver external sintering pressure. Pressureless silver sintering, which can be done in a conventional convection oven, results in a die attach layer with a higher porosity, due to the lack of pressure driven densification of the silver particles. However, this interconnection layer benefits from the excellent material properties of silver and the pressureless process enables the use of sintering interconnection technology for pressure sensitive devices such as LEDs, sensors, thin semiconductors or three-dimensional ceramic substrates.

Experimental

The aim of this study is to investigate the feasibility of forming a sinter interconnection on additively manufactured copper surfaces on Al_2O_3 substrates. Therefor first additive metallized ceramic substrates were built up and the parameters of the additive process were investigated regarding the bonding towards the substrate. In a second step silver sinter material was applied on those additive formed surface metallization and semiconductor chips were attached by using a pressureless sintering process. The processes were mainly investigated by using destructive testing such as shear tests and metallography but also X-ray was used to evaluate the quality of the sintered joint.

Substrate metallization

Initially the metallization of the substrates was tested on planar Al_2O_3 substrates from CeramTec GmbH with thicknesses of 63 µm and 100 µm.
The additive metallization process was realized by using a plasma torch, control unit and powder conveyor with a heated specimen tray on a six-axis robot arm.
The copper powder used for the plasma-induced coating process was a spherical copper powder (D_{50} 12µm) with silica shell to prevent agglomeration on the conveyor. The plasma gas was argon/argon with hydrogen and nitrogen sheath gas, which allows an optimized focus of the particle beam and a more uniform coating layer.
The Al_2O_3-substrates were preheated to 245 °C – 260 °C to lower the thermally induced stress by the hot plasma flame, which was driving over the ceramic in 30 mm distance during the metallization process. For each passage of the plasma flame a copper deposition of 25 µm to 30 µm was realized, so within six runs the intended surface metallization of 150 µm thickness was achieved. To prevent a certain amount of overspray and allow a fine structuring the substrates were masked with a lasercut Polymid layer (see Figure 4a), which is removed after the spray coating process. In Figure 4b a spray coated copper layout on a ceramic substrate can be seen. During the experimental testing three temperatures for the deposited copper particles (1200 °C, 1400 °C, 1600°C) were evaluated regarding the influence on the adhesion of the metallization on the ceramic substrates.

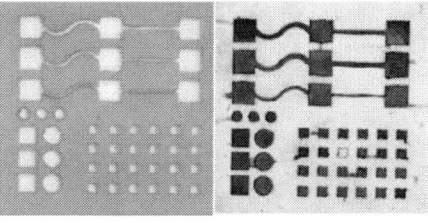

a)   b)

Figure 4: a) Substrate w. lasercut Polyimid-mask, b) Spray coated copper layout on planar Al_2O_3 substrate

After the additive metallization a thermal fellow-up treatment was investigated as well. Therefor some of the metallized specimen were put in a tube furnace system with maximum temperature of 1060 °C under vacuum. The purpose of the thermal process was mainly the reduction of the copper oxides which form during the copper deposition process. Therefor three heating stages were used. The bonding of the achieved copper metallization on the Al_2O_3 substrate was evaluated by measuring the maximum force for tearing of the copper layer during a shear test (see Figure 5). Prior to the die attach process the surface of the additive manufactured copper layer was examined by using scanning electron microscope (SEM) and energy dispersive X-ray spectroscopy (EDX) in addition to this the roughness of the surface of the specimen was measured using White light interferometry (WLI) and compared to common direct copper bonded substrates (DCB).

Figure 4: Shear chisel and test pads for shear strength measurements of the additive manufactured copper metallization on Al₂O₃ substrate

Silver sintering

During the second part of this investigation silver sinter material was applied by dispensing silver paste on the spray coated copper surfaces of the Al₂O₃ substrates as well as on DCB substrates as reference. Commercially available sinter pastes from Heraeus Deutschland GmbH & Co. KG was used for attaching Semikron 4*4 mm² diodes with silver backside metallization on the additive metallized copper surface.

The Si-chip was placed in the wet sinter paste. The average thickness of the sintered layer was 60 μm. The sintering process was conducted in a convection oven at 230 °C for 60 min under nitrogen (heating rate 5 K/min). To prevent the oxidization of the copper substrates the oxygen rate was set lower than 50 ppm. Figure 5 shows the test specimen, which was later used for testing.

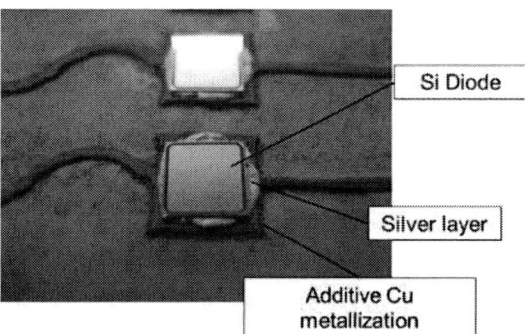

Figure 5: Test specimen with Si-diode silver sintered on additive copper surface

After the sintering process the test specimen were evaluated by using X-ray as well as destructive test methods like Die Shear Test and metallography. Cross sections were prepared for further examination of the sintered interconnection layer by using SEM.

Results

The investigation of the bonding strength of the selective metallized copper layer (Figure 6) reveal a very significant rise of shear strength after the thermal treatment at 1060 °C. For test specimen without an additional heat treatment almost no shear

strengths were measurable. Furthermore, the measured bonding strength unfolds that the different material thicknesses of the ceramic substrates (63 μm and 100 μm) have no significant effect and can be neglected. Nevertheless, the results show that the particle temperature has an influence on the bonding strength. The highest shear values could be measured for particle temperatures of 1200 °C, while the measured values for 1400°C and 1600°C were slightly lower, for both material thicknesses.

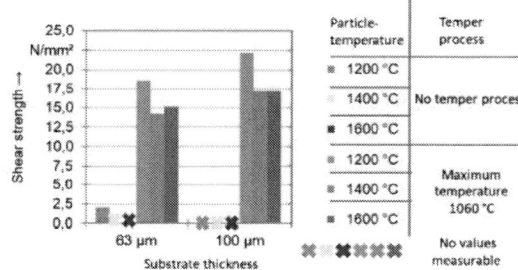

Figure 6: Results of the shear strength measurement of the additive manufactured copper metallization on Al₂O₃

The cross sections of the additive metallization (Figure 7a) show a lamellar structure of the deposited copper particles which corresponds mainly to transformation of kinetic energy to plastic deformation of the heated copper particles after the plasma driven impact on the Al2O3 substrate. After the heat treatment near the eutectic temperature the cross sections show a change of the copper structure (see Figure 7b).

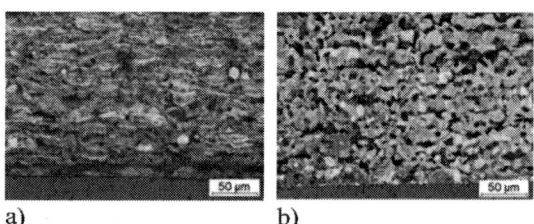

a)   b)

Figure 7: Cross section of the copper metallization initial a) and after thermal treatment at 1060 °C b)

The elevated temperatures lead to visible grain growth and further diffusion-based connection between the individual copper layers but larger pores. The restructuring of the additive manufactured copper layer during the heat treatment can also be seen in SEM images.

The results of the EDX analysis show for all test specimen that several residues besides copper can be detected. For the untreated surface mainly copper-oxides as well as some silica, carbon and aluminum residues. The element analysis on the surface after the oven process reveals less copper-oxides.

After the thermal treatment the color of the copper surface changed from brownish to pinkish-reddish (Figure 8a and 8b). Nevertheless, the porous layer still contains copper-oxides, mainly inside the pores. In contrast to this the detected amount of silicon rises significantly. Surprisingly the analysis also found residues of chrome on every and manganese on more than 50 percent of the tested samples. The detected silicon may origin from the copper particles themselves, due to the silica coating of the copper powder. The measured carbon may remain from the attached polyimide-mask. The other residues which were only detected after thermal treatment may result from sedimentation during the oven process.

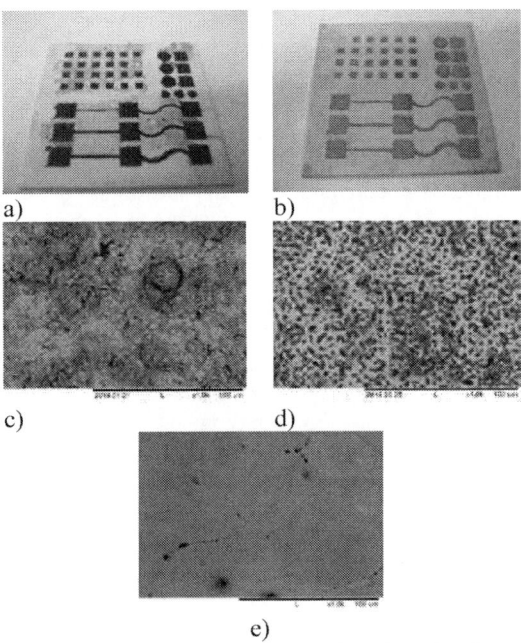

Figure 8: Copper surface initial a); b) after heat treatment; c) SEM image of copper surface initial (1000x), d) copper surface after heat treatment (1000x) and e) DCB-surface (1000x)

By comparing SEM images of the additively manufactured copper surface with the DCB substrate of the reference specimen (Figure 8e), the higher surface roughness of the plasma-coated specimen becomes visible. This finding is enforced by the results of the roughness measurements by using white light interferometry (see Table 1).

The measured maximum height difference for the DCB substrate was 72 µm with an average roughness of 2.1 µm, while the maximum height of the additively formed copper layer was between 75.7 µm and 160 µm. These results show also and higher average roughness value S_a between 5.3 µm and 15.3 µm for the untreated surface and between 5.49 µm and 18.7 µm for the surface after thermal storage at 1060 °C. S_q corresponds to the standard deviation of the height.

Table 1: Measured surface roughness

	DCB substrate	Copper layer (untreated)		Copper layer (therm. treated)	
	[µm]	[µm]		[µm]	
		Min.	Max.	Min.	Max.
S_z	72.12	56.9	168.70	75.76	160.90
S_a	2.10	5.30	15.37	5.49	18.70
S_q	2.66	5.30	19.25	7.27	23.82

In Figure 9 the result of the second phase, the pressure-less sintering of silicon diode on the additive applied copper layer, are shown in an SEM image of a cross sections. The SEM image of the cross section shows the fine structure and porosities of the sintered silver layer and the larger pores of the copper metallization as well.

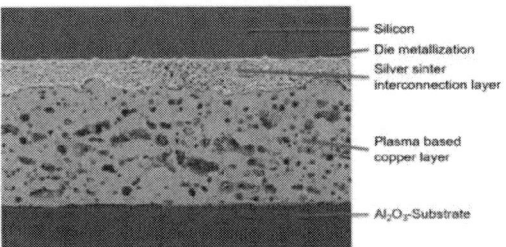

Figure 9: Cross section of a silver sintered Si-diode on the additive manufactured copper metallization (200x)

High resolution images of the sections show the silicon chip, which is attached to the reference DCB substrate Figure 10a and on the additive copper surface Figure 10b by a sintered silver layer which is 50 µm to 60 µm thick.

By comparing the visible appearance of the interconnection layer, it can be seen that the dispensed and sintered silver is compensating the higher surface roughness of the plasma-coated substrates. Nevertheless, some unfilled areas and holes were detected mainly near steep angles or protruding edges of the copper metallization (see Area A in Figure 10). It is presumed that the holes occur during the placement of the Si-chip in the wet paste, mainly due to the overlapping of some deformed copper particles and the rheological behavior of the silver sinter paste, which may prevent wetting of the surface in these areas.

The results of X-ray analysis (Table 2) reveal a significant difference between common DCB metallization and the additively applied copper surface regarding the measured void rate. By calculating the 2D void ratio inside the sintered interconnection layer it appears that the plasma-based copper layer shows a mean void rate between 19.5 % voids for particles, which were applied at 1200°C, and respectively 24.2 % for particles at 1400 °C and 24.9 % at 1600 °C.

In contrast to this the void rate for the commercially available DCB substrate was only 5.4%. The X-ray images show more and larger voids as well as drying channels inside the sinter layer (see Figure 11b).

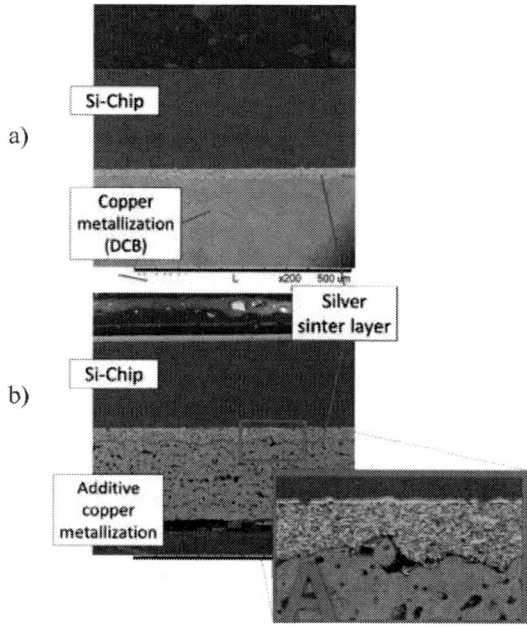

a)

b)

Figure 10: a) Cross section of a silver sintered Si-diode on the DCB-copper, b) additive manufactured copper

Table 2: Measured void ratio

Substrate	Particle tempera ture [°C]	Tested devices	Mean void ratio [%]	Standard deviation
DCB	-	8	5.4	0.56
Plasma- coated copper	1200	18	19.5	1.69
	1400	51	24.2	0.89
	1600	9	24.9	1.25

These drying channels as well as the fact that test specimen were sintered in one batch allows an assumption that the higher surface roughness of the plasma-based copper surface obstructs the evaporation of the solvent during the sintering process. Parts of the solvents as well as some of the coating remain inside the silver layer. The voids in the silver layer reduce the connecting area of the attached surfaces and solvent as well as organic residues may prevent the interparticle diffusion of the silver particles. The high voiding rate may not only influence the mechanical stability inside the sintered joint but also the thermal conductivity and reliability.

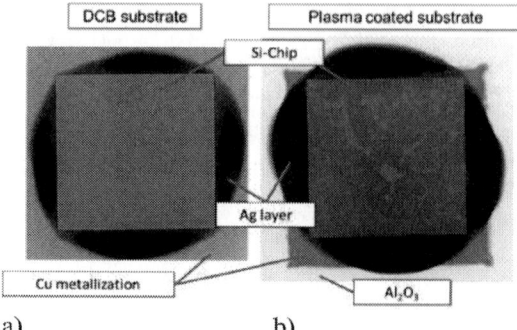

a) b)

Figure 11: a) X-ray images with voids inside sinter layer for Si-chips on DCB substrate and b) additive manufactured copper

During the evaluation of the sintering process it became necessary to sinter under nitrogen atmosphere to prevent oxidization of the copper substrates. The results of the die shear test (Figure 12) show a median of 7.9 N/mm² for the reference test samples with conventional DCB substrates.

For the plasma-coated specimen the measured shear values were significantly lower, with a median strength of 3 N/mm² for particles applied at 1200 °C and 1400 °C as well as 2.8 N/mm² for a particle temperature of 1600 °C. The breakage patterns of the sheared interconnections show good adhesion of the silver sintered layer towards the semiconductor chip but less signs of bonding at the substrate metallization.

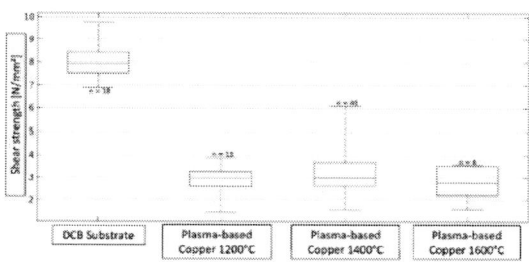

Figure 12: Boxplot of measured die shear strength

By evaluating the backside of the sheared off semiconductors (see Figure 13), the dried and well sintered areas appear on the edges, while the center of the semiconductor remains in a darker gray, which is an indication of remaining solvent and organics.

It is assumed that besides the trapped thinner and the coating residues, due to the higher porosity and roughness of the copper layer, the surface impurities like silicon from the shell of the copper particles as well as the external contaminations during the oven process lead to the low adhesion of the sinter material. Unlike regular solder materials sinter pastes do not contain surface activating fluxes, which usually allow an optimized joining.

Figure 13: Breakage pattern after die shear test and backside of a Si-chip with remaining silver layer

These findings reveal that forming pressureless sintered interconnections on the additive formed copper surface is more demanding than on regular DCB substrates. Further experiments will be conducted with more advanced substrates, which will be using the selective plasma coating process with particle temperatures of 1200°C and thermal treatment due to the good adhesion towards the Al_2O_3 substrate. The results showed that additional non-copper depositions on the surface need to be prevented or removed due to their negative influence on the sintered joints. For the next samples a sinter paste with an optimized and more homogenous drying properties will be evaluated. Also, an infrared assisted sinter process will be tested regarding lower voiding and better sintering results.

Finally, the metallization and the joining processes will be transferred onto a three-dimensional technology demonstrator for high temperature applications (Figure 14a). Currently the metallization process is tested on the pyramidal Al2O3 substrate (see Figure 14b).

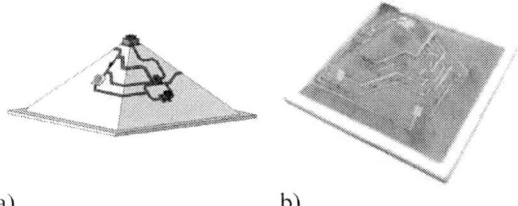

a)   b)

Figure 14: a) Early model of a 3D-demonstrator and b) pyramidal Al₂O₃ substrate with three-dimensional masking

Conclusion

This contribution focused on the combination and compatibility of a selective metallization technology with a highly reliable joining technology. Therefor planar ceramic substrates were metallized by using a spray coating process in a second step silver sintering was used to mount semiconductor chips on the additive copper surface of the Al_2O_3 ceramic.

The though fabricated copper metallization as well as the results of the silver sintering process were analyzed and compared with commercially available direct copper bonded substrates.

This publication showed that a particle temperature of 1200 °C is sufficient in combination with a thermal treatment at 1060 °C, which has a major impact on the bonding strength of the copper on the ceramic and is also required for a proper reduction of the copper oxides.

The bonding strengths for silver sintering on a copper interface deposited from a plasma coating process, measured in this investigation, are currently not on par with values measured on common DCB substrates, despite the larger surface of the additive metallization. The remaining silica coating of the copper particles, interfering depositions on the surface as well as poor sintering results due to solvent and coating residues were identified as root causes.

This publication demonstrates that the combination of two individually successful technologies like plasma-coating and silver sintering requires fine-tuning and alignment of the processes as well as the materials.

Further research will be conducted to optimize the results and transfer the processes onto a three-dimensional technology demonstrator.

Acknowledgment

The results presented in this publication have been achieved within the research project 'AgOn3D'. The IGF-project 20132 N was supported by the German Federal Ministry for Economic Affairs and Energy (via AiF) and the research association 3D-MID (Molded Interconnected Device). The authors would like to thank all members of the committee accompanying the project for the contribution and the valuable support of the project.

References

[1]   J. Franke, *"Räumliche elektronische Baugruppen (3D-MID): Werkstoffe, Herstellung, Montage und Anwendungen für spritzgegossene Schaltungsträger."* München: Hanser, 2013

[2]   I. Kriebitzsch, *"3-D-MID Technologie in der Automobilelektronik,"* Bamberg, 2002.

[3]   H. Herman, *"Plasma-Sprayed Coatings.,"* in Scientific American, vol. 259, no. 3, 1988, pp. 112–117. JSTOR, JSTOR.

[4]   A. Hensel, M. Müller, J. Franke, and K. Kohlmann-von Platen: *"System Concept of a Robust and Reproducible Plasma-Based Coating Process for the Manufacturing of Power Electronic Applications"*; 7. WGP Jahreskongress (RWTH Aachen, 05.10.2017 - 06.10.2017)

[5]   Hensel A, Schwarzer C, Scheetz M, Kaloudis M, Franke J; *"Investigations of Silver Sintered Interconnections 3-*

Dimensional Ceramics with Plasma Based Additive Copper Metallizations", 20th IEEE Electronics Packaging Technology Conference (EPTC) (Singapore, 04.12.2018 - 07.12.2018)

[6] J. Lutz, "*Halbleiter-Leistungsbauelemente: Physik, Eigenschaften, Zuverlässigkeit*", 2nd ed. Berlin, Heidelberg: Springer, 2012.

[7] S. Klaka, "*Eine Niedertemperatur-Verbindungstechnik zum Aufbau von Leistungshalbleitermodulen*", 1st ed. Göttingen: Cuvillier, 1997.

[8] T. Krebs et al., "*Breakthrough in Power Electronics Reliability – New Die Attach and Wire Bonding Materials*," in 63. Electronic Components & Technology Conference (ECTC): May 28-31, 2013, 2013.

[9] S. Duch, T. Krebs, Y. Loewer, W. Schmitt and M. Thomas, "*Novel interconnect materials for high reliability power converters with operation temperatures above 150°C*," in 2012 IEEE 62nd Electronic Components and Technology Conference, San Diego, CA, USA, 2012, pp. 416–422.

[10] Roth and W. Schmitt, "*Improving the bond strength of sinter joints by modifying the DBC without noble finishes and modified silver sinter pastes*," in Integrated Power Systems (CIPS), 2014 8th International Conference on: Date 25-27 Feb. 2014, 2014.

[11] C. Weber, M. Hutter, C. Ehrhardt, and K.-D. Lang, "*Failure Analysis of Ag Sintered Joints After Power Cycling under harsh temperature conditions from + 30°C up to + 180°C*," in European Microelectronics Packaging Conference 2015.

[12] W. Schmitt, L. M. Chew and R. Miller, "*Pressureless sintering of large dies by infrared radiation*," CIPS 2018; 10th International Conference on Integrated Power Electronics Systems, Stuttgart, Germany, 2018, pp. 1-5.

[13] L. M. Chew, W. Schmitt and M. Dubis, "*High bonding strength of silver sintered joints on non-precious metal surfaces by pressure sintering under air atmosphere using micro-silver sinter paste*," 2018 IEEE 20th Electronics Packaging Technology, Conference (EPTC), Singapore, Singapore, 2018, pp. 125-131.

[14] L. M. Chew; W. Schmitt; C. Schwarzer; J. Nachreiner, "*Mirco-Silver Sinter Paste*

Developed for Pressure Sintering on Bare Cu Surfaces Under Air or Inert Atmosphere"; – ECTC2018, SanDiego, CA, USA

[15] M. Beierlein and M. Kaloudis, "*Evaluation of Pressureless Silver Sintered High Power Semiconductor Devices by Measurement of thermal Impedance*," Dresden, Oct. 8 2013.

[16] M. Boulos, P. Fauchais, J. Heberlein, "*Thermal Spray Fundamentals: From Powder to Part.*", New York, Springer Verlag, 2014 pp. 17-110, 383 – 467C.

[17] P. Fauchais, G. Montavon, and G. Bertrand, "*From Powders to Thermally Sprayed Coatings*," Journal of Thermal Spray Technology, vol. 19, no. 1, pp. 56–80

Alternative Heating Methods for Printed Circuit Boards and a Practical Comparison of Direct Resistance and Inductive Heating Processes

Dirk Seehase, Arne Neiser, Jacob Maxa, Fred Lange, Andrej Novikov
and Mathias Nowottnick

Institute of Electronic Appliances and Circuits, University of Rostock, Albert-Einstein-Str. 2, Rostock, Germany
dirk.seehase@uni-rostock.de

Abstract

Heating processes on printed circuit boards (PCBs) are required for various manufacturing steps and also for different tasks throughout the operation of electronic assemblies. This paper aims to give a brief summary of the state of the art for alternative heating methods which can be applied for PCB heating, besides external convective processes. Also, three distinctly different methods of providing various amounts of heating energy directly on (or within) the PCB, which were experimentally studied, are presented here. For one, the usability of exothermic heat coming from a reactive material, which can be applied to the PCB, is described. Furthermore the inductive heating of suitable susceptor structures integrated in a PCB material is shown. And finally, the technological applications of integrated resistive layers, which can produce Joule heating, were evaluated.

Key words: PCB, soldering, exothermic, induction, Joule, heating

1. Introduction

Heat and temperature are omnipresent in electronics and at the same time both curse and blessing. During production as well as in the operating state, PCBs and its electronics are subject to heating processes. These cause changes in their physical and functional parameters and are widely undesirable (e.g. thermal deformation, electrical drift, temperature induced corrosion ...). In order to minimize or eliminate these negative influences, a wide variety of cooling concepts has been established which can be partially integrated into an electronic assembly. [WIT08], [KÖT17] At the same time, however, there are a number of applications which require controlled heating of a PCB or assembly. Such applications are lamination processes, soldering processes, drying processes and many more.

With all these applications, there is a need to carry out heating processes, both in the production process and during operation with an integrated, ready-to-use-system. Furthermore, there is an increasing demand for more energy efficient heating processes, especially with regard to today's reflow soldering applications. It was therefore the goal of previous research studies to produce and characterize heating systems which can be integrated into a PCB. In this paper, alternative heating methods, which have been researched and reviewed, as well as comparing studies for practical heating solutions will be presented.

2. State of the Art

In this part, a brief summary of the state of the art for various heating methods and their implementation in PCB technology will be given. Here, priority is given to those methods of interest which allow a certain degree of integration into the PCB system and thus enable an energetically more efficient heat transfer.

2.1. Electro-Thermal Heating

There are many electrothermal processes that are widely used, both in everyday life and in the industrial environment. Using the conversion of electrical energy into heat for the heating of PCBs is therefore an obvious process. Inductive heating and resistance heating have been identified as the most suitable methods for general heating on PCBs.

2.1.1 Induction Heating

Inductive heating is primarily used in the metalworking industry. It is utilized, for example, for inductive hardening [NEU16] or inductive melting [BAA18]. Also, in the electronics industry, the use of inductive heating methods is not uncommon and is often utilized for the brazing or welding of metal components.

Inductive selective soldering systems are presented for example in [EUT14] and [WOL15]. Here the component connection and/or the solder itself serves as a susceptor, which is heated by the electromagnetic alternating field of a surrounding inductor loop. However, these systems are unsuitable for general soldering applications.

By [IRI13] the use of special field-coupling structures for heat energy conversion is presented. These structures (susceptors) are intended to be embedded in the circuit board material and can then be used for repair soldering of components mounted above it.

In the description of [TRU01], the inductors themselves are inserted into the circuit board material. Each inductor then generates an electromagnetic field which is intended to heat an overlying joint. Also, the use of high-permeable magnetic field concentrators are proposed, which allow a partial shielding, so that unwanted heating in other places of the assembly can be prevented.

The patents [MCG92] and [SUP96] describe how ferri- or ferromagnetic materials can be added to the solder. These will have a high permeability, small geometric dimensions and are equally distributed in the solder in order to release the inductive heat there. Since these susceptive particles are not melted when the solder materials do, the additives would remain as part of the solidified solder joint after the inductive soldering process. Thus theoretically, reusability for inductive re-melting would be possible. An application for this can be found in [HAB10]. Here FeCo particles ($d \sim 38\ \mu m$) in various concentrations (2 wt.-% - 10 wt.-%) are mixed into an SnAg3Cu0.5 solder paste.

2.1.2 Direct Resistance Heating

On the circuit board, the direct resistance heating of electrically loaded components and thus the indirect heating of the surroundings are almost omnipresent and sometimes require elaborate cooling concepts. The publication [NEU14] shows how these power losses can be put to use. Here, the direct heating of SMD chip resistors is used to melt a low melting point bonding material on the opposite side of the substrate.

The simplest way of implementing a heating layer in circuit boards is by the use of copper heating coils. In [MER99], microfluidic applications are realized by the use of copper meander structures, which simultaneously serve as a heating element and as lateral channel boundary for a non-conductive fluid flowing through. Further heater variants for use in microfluidics are presented in [WEG01] by means of heating wire and in [GAS08] by means of thin-film heating elements for the operation of membranes on micro pumps.

Various patents also describe PCB heaters based on copper, Constantan (CuNi44) or another metallic layer as an embedded element in the substrate material [WOL02], [HEN91] or embedded in a special interposer [AOK19]. Also their use for soldering applications is discussed.

2.2. Exothermic Heating

Chemical energy is a form of internal energy. The energy conversion goes hand in hand with the expiration of chemical reactions. A distinction is made between exothermic reactions in the course of which chemical energy in the form of other types of energy is released and endothermic reactions, for the course of which energy must be continuously supplied. In many cases, this other form of energy is thermal energy.

This is used, for example, in thermochemical heat stores, such as the sorption storage. These storages are capable of reversibly converting and storing heat energy into chemical energy. In [BRE14] a method is presented in which a water-storing coating produces a cooling effect on electronic assemblies by means of sorption heat. Conversely, when externally cooled, adsorption of water vapour from the environment takes place under exothermic release of energy (evaporation enthalpy). However, for a significant heating of printed circuit boards or assemblies, the achievable energy densities are too low.

Another example for a release of exothermic energy are reactive multilayer-foils as described in [HEM14], [LEV04] and [MAN97]. Such foils consist of thousands of nano-scale layers (25 nm - 90 nm), which are constructed of two alternating materials (e.g., Ni and Al). In such layer structures, heat is released as a result of thermally induced atomic diffusion. In [LEV04] special joining applications on electronic assemblies are described by the use of solder preforms based on reactive multilayer-systems. However, such reactive solder foils are unsuitable for large-scale use on a full assembly.

3. Alternative Heating Methods

This article presents alternative options for heating on PCBs and assemblies. Here, the summarized results of various research activities in which novel heating methods were investigated and applied, should be presented.

3.1. Reactive Paste Material

In the research project ThermoFlux [NOW15] reactive paste materials have been developed which serve as an additional source of energy during the reflow soldering process. The reactive components were adjusted so that the exothermic release of energy occurs 10 K to 20 K before reaching the melting range of a standard SnAg3Cu0.5 solder paste (T_S: ~ 217 °C). This way, an ignition can be controlled via the temperature profile in a reflow oven, for example. The binding of the reactive components in a pasty mixture also allows the application by means of printing or dispensing processes.

3.1.1. Basic Principle

The basic principle of the desired solution is based on the fact that the amount of energy required for the soldering process is obtained in part from a chemical reaction. The main energy supplier is a sacrificial metal in the reactive mixture, the oxidation of which gives off exothermic heat. Furthermore a specific metal compound is added to the system as an reaction initiator. This can decompose at relatively low temperatures with a release of energy and thus support or initiate the oxidation reaction of the sacrificial metal. Since the availability of oxygen required for the oxidation of the sacrificial metal is limited in time, an oxygen supplier is incorporated as a third component into the system. Above a defined temperature this oxygen source becomes unstable and decomposes with a release of oxygen, thus providing the necessary amount of oxygen for the reaction (even in inert atmosphere).

3.1.2 Materials

For the initiation of the chemical reaction, it was decided to use carbonyls. Considering the results of [NOW11], the metal compounds, iron pentacarbonyl [$Fe(CO)_5$], cobalt tetracarbonyl [C_4HCoO_4] and nickel tetracarbonyl [$Ni(CO)_4$] were primarily considered. After initial practical measurements, the use of cobalt carbonyls was favoured.

The sacrificial metal powder used herein must have the property of oxidizing at elevated temperature, while being kicked-off by the initiator compound. By means of TGA, (thermal gravimetric analysis) different metals were therefore examined for their reaction kinetics. Regarding their oxidation behaviour and reaction starting temperature, the elements Fe, Zn and Mn were shortlisted. However, the use of Zn involves increased laboratory expense because of its tendency to react with atmospheric moisture and atmospheric oxygen. The investigated metals Fe and Mn are both suitable. Although, the investigations have shown that the reaction in Fe runs more smoothly. Therefore, a better heat transfer was expected. While comparing different particle sizes of the sacrificial metal powders, it was observed that smaller grain sizes produced a larger energy release. In addition, the starting temperature for the reaction can be reduced with smaller particle sizes. Therefore, for most experiments a Fe powder with a particle size of 4 µm - 6 µm was used.

Theoretically, various organic and inorganic peroxides can be used as an oxygen supplier within the reactive system. Many kinds of peroxides have been excluded due to their hazardous potential or for a decomposition temperature (T_D) which was too low/too high. Shortlisted for practical testing were zinc peroxide (ZnO_2, $T_D = 225\ °C$), calcium peroxide (CaO_2, $T_D = 275\ °C$) and potassium permanganate ($KMnO_4$, $T_D = 240\ °C$). The $KMnO_4$ has been proven to be the most suitable because of higher storage stability.

A sketched composition of the most commonly used reactive paste is shown in Figure 1. Finally, the most balanced mixing ratio of the different reaction components was determined as 20 wt.-% of the reaction initiator, 10 wt.-% of the sacrificial metal powder and 24 wt.-% of the oxygen supplier, which were all mixed into a flux matrix.

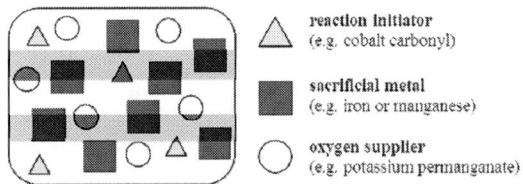

Figure 1: Composition of a reactive paste

3.1.3 Applications

The developed reactive paste system was applied for the thermal support of reflow soldering processes. Initially, the approach was to mix the reactive components into the solder paste material used. However, here a proper solder joint could not be formed with any mixing ratio. Either not enough additional energy could be generated or the amount of additives prevented the formation of a coherent solder ball. As a good alternative, the energy release in a reaction area which is separate from the solder material, has proven itself. For this purpose, a good thermal conductivity between the reaction depots and the respective solder paste deposits was required. On a PCB, copper pads and conductor tracks are ideal for the thermal connection, since copper has a very good thermal conductivity of approx. 380 W/m·K. Depending on the layout, these heat-conducting structures can then also be used for electrical signal transmission during the later operation of an assembly.

Figure 2 shows different sketches for several investigated variants for the use of reactive material in separate reaction areas.

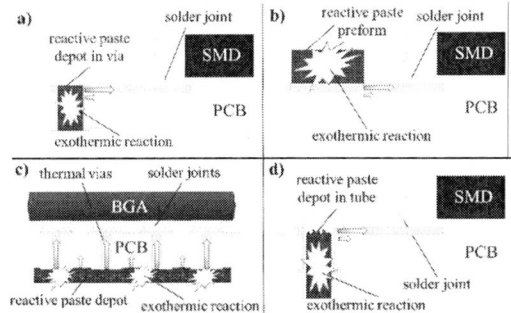

Figure 2: Variants for the use of reactive materials for soldering support; a) in VIA; b) as preform; c) depot on opposite side; d) in metallic tube

One possibility is the use of VIAs which are filled by paste-in-hole printing. (Figure 2 a)) By varying the diameter of the VIA and its distance from the solder paste deposit, the intensity of the heat energy can be controlled. If the energy from one VIA is not sufficient, several VIAs can be connected to a paste depot (if space is available). For special applications it is also possible to insert a copper-plated cavity around the solder pad and then fill it with reactive paste. By this, particularly large

533

reaction depots can be generated and a simultaneous ignition is guaranteed.

In addition to the use of paste, the use of reactive preforms represents another application option. These can be generated beforehand in the desired geometry and then placed on the board like a component before the soldering process. (Figure 2 b)) Since the preforms have no fluidic binder (like a paste), the use of a primer is useful for safe placement on the thermal pads.

Another variant is the thermal connection between screen or stencil printed reactive paste deposits and the solder material. Here, the reactive paste is applied like solder paste onto thermal Cu pads. It would be possible to print the reactive paste on the same side as the solder paste, but this would require the use of special stencils for the second printing process. Alternatively, the reactive paste can be printed to the opposite side of the board and the heat path can be realized by thermal vias. (Figure 2 c))

As an alternative to the filled VIA, it is also possible to fit the plated-through holes with tubes that are filled with reactive material. (Figure 2 d)) These inserted tubes can be removed from the VIAs after the soldering process and have the advantage that contaminations of the assembly by reaction process residues are minimized.

Figure 3 shows a temperature measurement during a forced convection soldering process in which a BGA was soldered with a reactive paste deposit on the opposite side of the PCB. The solder material used was a SnAg3Cu0.5 paste. The heating zones were adjusted so that the external heating of the assembly was approx. 10 K lower than the melting temperature of the solder. The achieved melting of the solder deposits could therefore only be done by the additional energy input provided by the exothermic reaction. It should be noted that the exothermic temperature peak was well above 300 °C (maximum value of the measuring system).

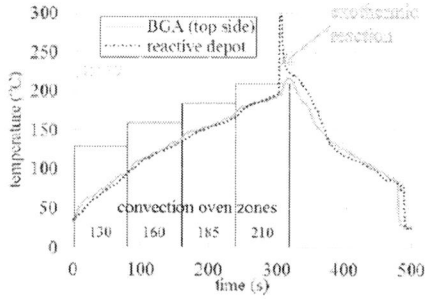

Figure 3: Example of measured reaction peak and component temperature during convective reflow below melting temperature of the solder

3.2. Induction Heating with Susceptors

Here are the possibilities that can be used for a heating of PCBs by using the energy transfer of electromagnetic alternating fields to field-coupling material structures (susceptors). The induction offers the advantage of a contactless energy transfer into the PCB, whereby external/internal leads can be avoided and separate susceptor structures are possible.

3.2.1. Basic Principle

The basic concept for the use of induction for the heating of PCBs is shown in Figure 4. In this work, two types of inductor geometries were investigated. On the one hand a planar coil, in which the transverse field component (in axis of the coil), was used for heating susceptors (horizontal to the coil). And on the other hand, a cylindrical coil was used for longitudinal field heating of corresponding susceptors.

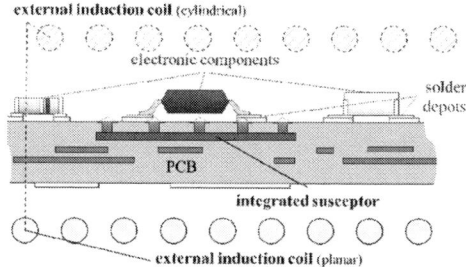

Figure 4: Basic principle for endogenous heating of PCBs with induction heat

The energy conversion is based on various loss mechanisms (eddy currents, hysteresis, excess) and takes place in susceptor structures whose material, shape and size must be tuned to the external magnetic field parameters (e.g., field orientation, frequency). In this case, the susceptor can be positioned both on the outside and in the interior of the base material. The transport of the heat energy to the desired areas (e.g., solder joints) takes place primarily by heat conduction in the PCB material. The susceptor and circuit board must be placed as close as possible to the inductor coil or inside the coil, without coming into physical contact with it. The more distance between the coil and the susceptor the weaker the magnetic field strength (exponential drop) gets.

A detailed description of the physical basics for inductive heating in metals was omitted in this summary. Further descriptions can be found in [SEE18] or in textbooks about induction.

3.2.2 Materials

For suitable susceptor materials, various metallic materials were investigated. The selection of materials was based on their availability, but above all on the estimation of their penetration depth with regard to the respective magnetic field frequency. Condition for a significant heating is that the dimensions of the electrical conductor amount to many times of its penetration depth. An important material parameter for this is the permeability. Figure 5 shows the penetration depth for different metals as a function of the magnetic field frequency. The transverse field inductor used here was operated at a frequency of 20 kHz and the longitudinal field inductor at a frequency of 100 kHz.

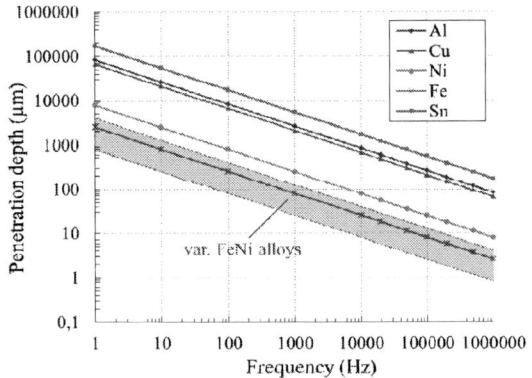

Figure 5: Penetration depth for different metals in dependence of the magnetic field frequency

For the susceptors in the transverse magnetic field, copper was primarily used, being the standard material in the electronics and printed circuit board technology. This material can be used either as an extra foil or as part of the base material of an FR-4 substrate. Alternatively, aluminium can also be used here. Thin ferromagnetic susceptors are unsuitable for heating in the transverse field because they direct the magnetic flux due to their high permeability.

However, materials such as Fe, Ni and various FeNi alloys could be heated in the longitudinal magnetic field. Here, the susceptor structures couple in the field primarily with their thickness.

3.2.3 Applications

The suitability of the process for the melting of solder material on printed circuit boards and the associated production of solder joints was demonstrated by means of small test samples. The demonstrators used for this purpose are approx. 50 mm x 20 mm. Larger sized structures could not be used in these inductor setups. Figure 6 shows a test sample that was used. In order to perform an electrical function test following the inductive soldering process, an inverter (SO-14) and a voltage regulator (SO-8) were used as components.

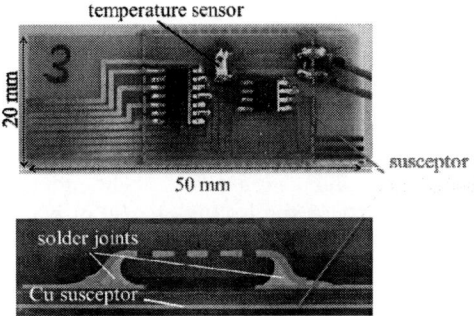

Figure 6: Test sample for induction heating (top); cross section of an test sample with IC and susceptor (bottom)

The susceptors were placed on the backside of the base material to allow for an easy sample preparation. For the transverse field soldering demonstrator, 100 μm thick copper susceptors with dimensions of 20 mm x 30 mm were used. The solder demonstrator for the longitudinal field coil inductor had a 60 μm thick nickel layer measuring 20 mm x 25 mm. The solder material used was a SnAg3Cu0.5 solder paste.

In order to perform the inductive soldering tests reproducible and as close to typical soldering processes as possible, a LabVIEW software was used for control of specified temperature profiles. It should be noted that it was not the goal to find an optimal soldering profile. Instead, the question was answered as to whether a successful soldering process can be carried out by a sufficiently precise control according to specification with the used susceptor / inductor combinations. An example for the set and the measured solder profiles, within the transverse and the longitudinal field heating, is shown in Figure 7. With these a successful soldering process could be achieved at the tested components.

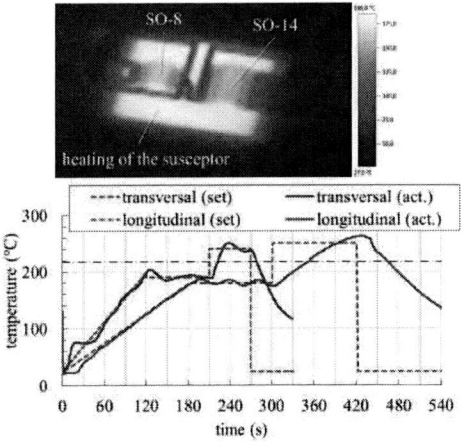

Figure 7: IR image of a heated sample (top); temperature profiles for soldering (bottom)

With both specified profiles, the required time of 60 s to 100 s above liquidus temperature of the solder, could be achieved during the soldering process. By using thicker (Ni) susceptors, the control behaviour, in terms of tracing the set profile, can be improved.

On the components used here, no damage or errors due to the electromagnetic field could be detected. Nevertheless, there is the possibility of unwanted coupling of the field in certain components, which can lead to thermal or electrical errors. Larger components, such as inductors or electrolytic capacitors, have shown rapid self-heating in experiments in the electromagnetic field. In ICs, the tendency for self-heating in the induction field is determined by its respective lead frame material. Ferromagnetic structures, such as e.g. Alloy 42 or Alloy K are commonly used here and may get too hot in the field or generate false electrical signals in larger structures (depending on the frequency).

3.3. Resistive Heating Layers

In addition, resistive layers in printed circuit boards whose electrical resistance is high enough to achieve Joule heating with predefined current and voltage values (extra-low voltage) were investigated. Unlike inductive heating, physical electrical contacting of the resistive layers to an external DC voltage source is required here.

3.3.1. Basic Principle

Resistance or conductive heating describes the generation of Joule heat on an electrically conductive, current-carrying body. Here, charge carriers which are accelerated by an electrical conductor give their kinetic energy to the lattice atoms. This energy is then converted into heat. The basic principle for the heating of PCBs with resistive layers is shown in Figure 8.

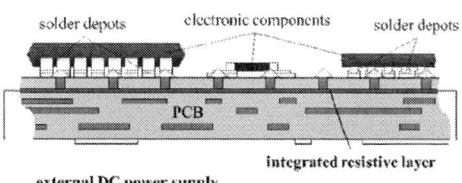

Figure 8: Basic principle for endogenous heating of PCBs with direct resistance heating

The required magnitude of the thermal losses of a heating layer, integrated in a printed circuit board, was estimated to be 1 W/cm² to 2 W/cm². With these power densities it was possible to achieve temperatures of up to 250 °C on the board (depending on the present heat capacities) in numerical studies as well as in experiments. For examinations, PCB dimensions were assumed to be no greater than double Eurocard (160 mm x 233 mm). For the electrical supply maximum values of 60 V and 10 A were specified. With these, a reasonable surface resistance range of 5 Ω/sq to 80 Ω/sq was determined for a full-faced heating layer.

3.3.2 Materials

A detailed material description has already been presented in [SEE17]. Therefore, only the most important features should be summarized here. For the characterization of integrated heating layers and their processability in a technological production process for PCBs, the following materials were selected to be investigated as heating structures:

- a polyimide foil (d ~ 25 μm) with a conductive carbon/polyimide layer (d ~ 25 μm)
- various carbon/polymer compounds (lacquer, paste)
- a copper foil (d ~ 18 μm) with a nickel-phosphorus coating (d ~ 400 nm)

The polyimide layer is filled with nanoscale carbon black particles and has a relatively high sheet resistance of (~ 100 Ω/sq). Polyimide has the advantage of a high thermal resistance. By lamination processes, the foil could be applied to or in base materials. Limitations have been found in the adhesion between foil and prepreg and in the application of an additional metallization layer (electroplated copper) on the carbon layers as electrodes.

As a further class of materials, various conductive carbon composites were investigated, whose electrical conductivity results from a mixture of nanoscale carbon fillers and carbon particles in the micrometre range. Different application methods were investigated for producing a layer of the different mixtures on the substrate material. Here, screen printing was identified as a suitable method. However, restrictions have resulted from a very limited pot life of the compounds and an increased contamination of the screens after several consecutive printing steps. The printing processes have thereby an insufficient reproducibility and require a thorough cleaning of the printing equipment. Differences have also been found in the homogeneity of the layer thickness between the individual material systems. For relatively large heating layers, as investigated here, local differences in layer thickness have a direct effect on the homogeneity of the temperature distribution during resistive heating.

Another investigated resistance foil consists of a nickel-phosphorus alloy, which comes deposited electrolytic on a copper foil. First, the application of the foil to the substrate material was conducted by means of lamination processes. Subsequently, the copper can be selectively etched. As a result, the desired NiP heating structure is uncovered and can be used. By using a special $CuSO_4$ etching solution, it was also possible to structure the NiP layer. Figure 9 shows an SEM image of the thin NiP layer (and a Cu electrode) on the base material of a PCB.

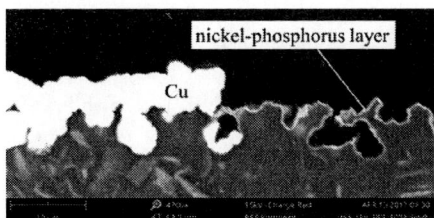

Figure 9: Cross section of NiP Layer on PCB

3.3.3 Applications

Here, soldering tests were also performed to demonstrate the usability of resistive heating layers. In the example presented here, a SnBi58 solder paste ($T_S = 138$ °C) was used. As heating material, a NiP layer, which was applied to the backside of the boards, was used. The use NiP heating layers was preferred over the carbon materials, since they exhibited for a more homogenous temperature distribution and a more stable temperature coefficient.

Several resistive layer segments were created on the test boards, which can be heated separately from each other. By this, a better process control was achieved. Segmented heating layers can also perform selective heating tasks. Figure 10 (top) shows two heating segments during endogenous heating. Controlling of the soldering profiles was also done here with LabVIEW. A measured temperature profile on a soldered segment is also shown in Figure 10 (bottom).

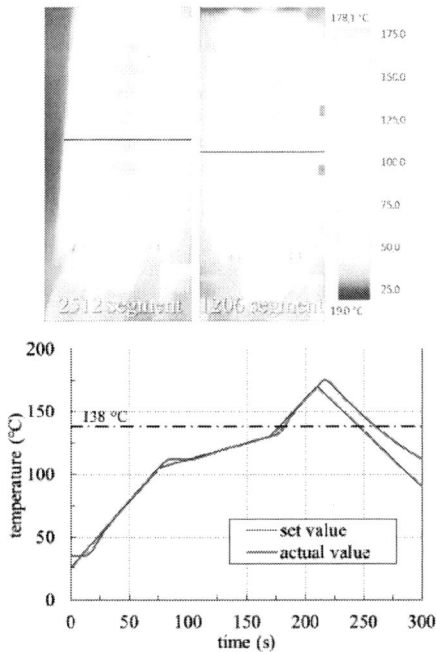

Figure 10: Selectively heated solder-segments (top); temperature profile for soldering (bottom)

The soldering tests have shown that it is possible to carry out soldering processes with an integrated resistive heating layer, where the heat energy is merely provide by its Joule heating. The required energy consumption (3.8 kJ per test segment) is much lower than with other conventional soldering methods. Comparative measurements of shear strengths on the soldered test components revealed no significant difference between endogenously soldered specimens and convectively soldered ones.

4. Conclusion

In addition to a brief summary of the state of the art for alternative heating methods on printed circuit boards, this paper presented three process examples with which a specific thermal energy release in printed circuit boards can be accomplished.

The various applications of exothermically reacting pastes, preforms or filled tubes can be used with skilful application for the general or selective soldering of a number of components on electronic assemblies. However, this technology can only be a one-off as well as a supporting component for the soldering process. Igniting the reaction also requires a powerful additional external energy supply. In addition, there remains the need to thoroughly clean any residues of the reaction process.

With the use of integrated susceptors within an electromagnetic field, it is possible to perform a heating of PCBs that is independent of furnaces. Which materials and structure sizes couple in depends strongly on the orientation of the used field as well as its frequency. Here, with the fields used, most oft the standard PCB structures (tracks, pads, pins,...) were avoided by the field-coupling. However, larger structures or 'antenna areas' could still be undesirably affected by the electromagnetic field. Therefore the general heating of typical electronic assemblies seems impractical. Although for selective process this method could be adapted.

The best results in terms of general heating of PCBs were obtained with the direct resistance heating of full-faced resistive layers on or within the substrate. The energy control of the process by an external DC voltage is easy enough to implement. Segmented heating structures make it possible to choose whether a general or a selective heating should take place. As with susceptors, the heating layer remains useable throughout the life cycle of the assembly and could be used for various maintenance or repair tasks.

Further studies which are currently conducted by the use of resistive layers are the construction of reliability test boards with integrated heating. These would allow for more flexibility throughout combined testing methods which require heat (e.g.: thermal cycling + vibration or shear force measurement at higher temperature).

Acknowledgements

Parts of these research are conducted within the research project "ERFEB - energy- and resource-efficient manufactured electronic assemblies", which is financially supported by the german ministry for economic affairs and energy (BMWi) and the project management Jülich (PtJ).

References

Articles from conference proceedings

[HEM14] G. Hemken, C. Walz, Einsatz von reaktiven Multischichten zum Fügen von Elektronikkomponenten, *Proceedings EBL 2014*, Fellbach, 2014.

[LEV04] J. P. Levin, T. R. Rude, J. Subramanian, et.al, Room Temperature Lead-Free Soldering of Microelectronic Components using a Local Heat Source, *ASM Conference Proceedings: Joining of Advanced and Specialty Materials*, pp. 75 - 79, 2004.

[SEE17] D. Seehase, F. Lange, A. Novikov, et al., Material study for full-faced heating layers, integrated in printed circuit boards, *Proceeding of ISSE 2017*, Sofia, 2017.

Journal articles

[HAB10] A. H. Habib, M. G. Ondeck, et al., Novel Solder-Magnetic Particle Composites and Their Reflow Using AC Magnetic Fields, *IEEE TRANSACTIONS ON MAGNETICS*, Vol. 46, No. 6, pp. 2187-2190, 2010.

[MAN97] A. Mann, A. Gavens, M. Reiss, et al., Modeling and characterizing the propagation velocity of exothermic reactions in multilayer foils, *Journal of Applied Physics*, Vol. 82, No. 3, pp. 1178-1188, 1997.

[MER99] T. Merkel, M. Graeber und L. Pagel, A new technology for fluidic microsystems based on PCB technology, *Sensors and Actuators A: Physical*, Vol. 77, No. 2, pp. 98-105, 1999.

[NEU14] J. Neubert, A. Rost und H. Lipson, Self-Soldering Connectors for Modular Robots, *IEEE Transactions on Robotics*, Vol. 30, No. 6, 2014.

[NEU16] J. Neumeyer, C. Groth, J. Wibbeler et al.; FE-Simulation des induktiven Härtens am Beispiel einer Kalanderwalze, *Journal of Heat Treatment Materials*, Vol. 71, pp. 43-50, 2016.

[SEE18] D. Seehase, C. Kohlen, A. Neiser, et al., Selective Soldering on Printed Circuit Boards with Endogenous Induction Heat at Appropriate Susceptors, *Periodica Polytechnica Electrical Engineering and Computer Science Journal*, Vol.62 No. 4, pp. 172 – 180, 2018.

Books, PhD Thesis

[BRE14] F. Bremerkamp, Integration eines Sorptionsspeichers in das Wärmemanagement von elektronischen Baugruppen, Rostock , 2014.

[GAS08] S. Gaßmann, Mikrosystem zur Fließ-Injektions-Analyse in Leiterplattentechnologie, Rostock, 2008.

[KÖT17] T. Kötter, Hochleistungsdichte Phasenwechsel-Komposit-Materialien für das thermische Management elektrotechnischer Systeme, Erlangen, 2017.

[NOW11] M. Nowottnick, J. Wilde und U. Pape, Flussmittel mit nanochemisch aktiven Metallverbindungen zur Stabilisierung von Weichloten, *AVT in der Elektronik*, Vol. 12, Templin, 2011.

[NOW15] M. Nowottnick, J. Wilde und U. Pape, ThermoFlux - Energieeffiziente Lötprozesse durch autonom schmelzende Lotpasten, *AVT in der Elektronik*, Vol. 21, Templin, 2015.

[WEG01] A. Wego, Entwicklung einer thermopneumatischen Mikromembranpumpe auf Basis der Leiterplattentechnologie, Rostock, 2001.

[WIT08] W. Wits; Integrated Cooling Concepts for Printed Circuit Boards, Twente, 2008.

Patents

[AOK19] R. S. Aoki, J. W. Thibado, J. L. Smalley, et al., Rework Grid Array Interposer with direct Power, *US 10,211,120 B2*, 2019.

[HEN91] H. E. Henschen, M. J. McKee und J. M. Pawlikowski, Self Regulating Temperature Heater as an Integral Part of a Printed Circuit Board, *US 5,010,233, 23*, 1991.

[IRI13] S. Iriguchi, K. Hatanaka, S. Watanabe, et al., Method of repair of electronic device and repair system, *US 8,456,854 B2*, 2013.

[MCG92] T. H. McGaffigan, Method, System and Composition for Soldering by Induction Heating, *US 5.093.545*, 1992

[SUP96] A. B. Suppelsa, Auto-Regulating Solder Composition, *US 5.573.859*, 1996.

[TRU01] H. A. Trucco, Inductive Self-Soldering Printed Circuit Board, *US 6229124 B1*, 2001.

[WOL02] P. D. Wolfarth, Self-Heating Circuit Board. *US 6,396,706 B1*, 2002.

Others

[BAA18] E. Baake, Induktives Schmelzen von Eisen- und Nichteisenwerkstoffen, *www.prozesswaerme.net*, pp. 77-80, 2018.

[EUT14] EUTECT GmbH, Prospekt: Induktionslöten Schnell u. Konrolliert, *www.eutect.de*, Dusslingen, 2014.

[WOL15] Wolf Produktionssysteme GmbH, Prospekt: Speziallöttechnik, *www.wolf-produktionssysteme.de*, Freudenstadt, 2015.

Mechanical behavior of SAC305 lead-free alloy

Julien Vieilledent [a], Wilson Maia [a], Pascal Retailleau [b], Marie Gabrielle Ameil-Ferbos [b],
Frédéric Dulondel [c], Philippe Milesi [c], Catherine Munier [d], Catherine Jephos [e]

[a] Thales, Group Industry, Vélizy-Villacoublay 78140, France
[b] MBDA France, 1, Avenue Réaumur 92358 Le Plessis Robinson, France
[c] Safran Electronics & Defense, 21, avenue du gros Chêne 95610 Eragny sur Oise, France
[d] Airbus Group, 1 Boulevard Jean Moulin, 78990 Élancourt, France
[e] DGA, Maîtrise de l'information, 35170 Bruz, France

Phone: +33 (0)1 40 83 24 02, Mobile: +33 (0)7 86 67 93 82 and E-mail: julien.vieilledent@thalesgroup.com

ABSTRACT

Components solder joint reliability is a continuous challenge for Aerospace Defense and High Performance (ADHP) equipment as they operate in harsh environments, combining different types of loadings. Numerous studies address either thermal, thermomechanical or mechanical stress and fewer focuses on combined impact. This paper presents results from "COSAC" PEA (Plan d'Etude Amont), which was aimed at assessing the mechanical strength of lead-free SAC305 considering the impact of temperature and pre-conditioning. The results were compared to tin-lead assemblies, used as reference. Mechanical loadings were applied at room temperature, high temperature (125°C), low temperature (-20°C, -40°C and -55°C) and after thermal cycling on BGA and QFN, with continuous monitoring. A significantly high quantity of around 2000 components was tested. This paper focuses on vibration endurance testing as they represent the greatest amount of data. Tin-lead and lead free SAC305 results are compared and discussed concluding on solder joint reliability testing methodology impact, considering the mission profile.

Key words: Lead-free, SAC305, harsh environment, mechanical stress, fatigue, thermal cycling, temperature, interaction.

Introduction

Aerospace, Defense and High Performance (ADHP) equipment are still out of scope, for reliability or safety reasons, from RoHS 2011 directive which restrict the use of lead in electronic equipment. Electronics boards are consequently still mainly assembled using tin-lead alloys. ADHP equipment requires high reliability associated to harsh environments and long mission profile. They are mainly exposed to temperature, vibration and shocks loadings [1]. On the other hand, Electronic industry innovations, most of the time only dedicated to lead-free assembly, are driven by the consumer market, which has less stringent constraints regarding mission profile and reliability over time. As a consequence, solder joint reliability is a continuous challenge. Regarding tin-lead alloy, ADHP equipment designers can rely on a significant return of experience. Concerning lead-free alloys, return of experience is limited especially considering the effect of combined loadings such as mechanical and thermomechanical stresses. Numerous studies address either one or the other, and fewer focus on combined effects with comparison of tin-lead and lead-free alloys.

Thermal cycling generates plastic deformations in solder joints and failures by low-cycle fatigue. Mechanical vibrations representative of real mission profile conditions can generate low-cycle and high cycle fatigue failures. Solder joint reliability for mission profile combining thermal cycling ageing and mechanical stresses consequently relies on the proper understanding the relationship between stress level and number of cycles to failure, by experimentally obtain a fatigue curve (number of cycles to failure as function of stress, S-N curve).

1. Experimental setup description

In order to plot fatigue curve, the test methodology used was developed to generate failures in solder joints with continuous monitoring of the load intensity and time to failure. The mechanical loading conditions applied were sinusoidal vibrations at the first resonance frequency of the electronic board. This methodology was already applied in previous studies [2, 3]. Figure 1 shows the deflection profile of the board in that condition. The maximum board displacement at the center is calculated from the measured acceleration.

$$Z_{max} = \frac{a_{max}}{4\pi^2 f_0^2}$$

Z_{max}: Maximum displacement [mm]
a_{max}: Acceleration at the center of the board [m.s^{-2}]
f_0: 1st resonance frequency [s^{-1}]

Random vibrations are nevertheless a more common type of loading in operation and also commonly used [4]. The choice of applying sinusoidal vibration was motivated to only address the 1st natural resonant frequency of the board. It enables to assume that "Z_{max}" is proportional to the solder joint stress level. Per opposition, random vibrations are susceptible to address several deflection modes generating a more complex loading of the solder joint which cannot be considered anymore as proportional to "Z_{max}:" as previously defined.

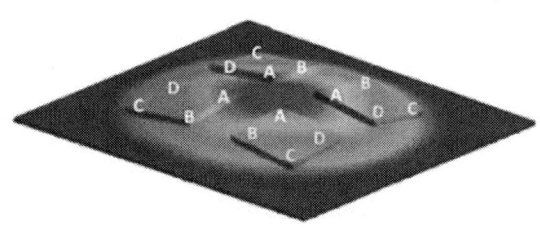

Figure 1: Board deflection at 1st resonant frequency, modeled with finite element analysis

Another key point of the methodology is its ability to generate the targeted failure mode which is the crack of the solder joint. To do so and being representative of typical ADHP application boards, the test vehicle 1st resonant frequency was designed to be between 300 and 400 Hz. Targeting the 1st natural resonant frequency participates to optimize the probability to generate solder joint cracks as the deflection mode generates maximum board displacement compared to subsequent modes.

Test vehicles were designed to host a single component reference. In order to optimize the number of testing trials, the components were placed on board symmetry axis, to be subjected to the same constraint, and experimental set-up enabled to test simultaneously 4 boards (16 components).

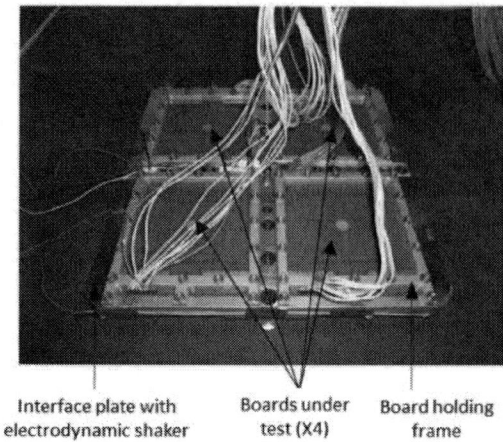

| Interface plate with electrodynamic shaker | Boards under test (X4) | Board holding frame |

Figure 2: Experimental set-up

Four identical references of test vehicles tested simultaneously present slight differences of resonant frequency. In order to be able to address the 4 resonant frequencies a sweep of 2 Hz per second was made around the mean value. Experimentally the four resonance frequencies were within a frequency window of ±5% around the mean frequency, achieved when necessary by adding little mass in the center of the boards.

Figure 3 is an example of acceleration measurement during a sweep window of 120 Hz (60 s) around the 1st natural frequency of a BGA1156 test vehicle at ambient temperature with an input acceleration of 19 G (186 m.s^{-2}). Greatest acceleration and board displacement occur when vibration frequency is f$_0$.

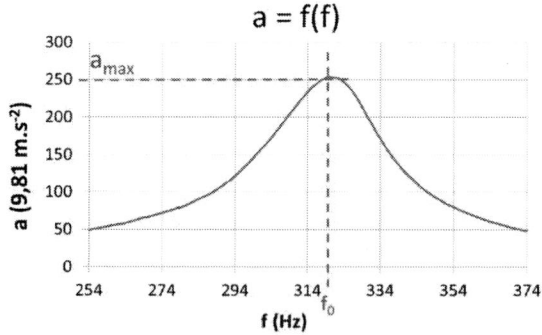

Figure 3: Acceleration measurement on a BGA1156 test vehicle during a sweep window of 120 Hz, at ambient temperature under constant 19 G sinusoidal vibration

The continuous monitoring of the daisy-chained components was performed with an event detector according to IPC standard test method [5].

Three different daisy chained packages were tested: BGA1156, QFN68 and BGA288. Components were assembled using either tin-lead

($Sn_{62}Pb_{36}Ag_2$) or SAC305 ($Sn_{96,5}Ag_3Cu_{0,5}$) solder pastes. BGA packages were either fitted with tin-lead or SAC305 balls. Table 1 summarizes packages main properties.

Package type	BGA	QFN	BGA
Termination count	1156	68	288
Pitch [mm]	1,0	0,5	0,8
Package dimension [mm]	35x35	10x10	19x19
Termination finish	SAC305 Or SnPb	Sn	SAC305 Or SnPb

Table 1: Components description

QFN and BGA were chosen for the study as they are widely used in equipment and because they have different behavior regarding mechanical and thermo-mechanical damage accumulation. BGA solder joint loading distribution is different in thermal cycling and vibration. Mechanical deformations are concentrated under silicon die during thermal cycling and on the corners for vibrations. For QFN, for both tests, deformations are higher in the corners.

2. Results

Some dispersion of results for reliability tests of solder joints is expected due to the solder joint bulk structural heterogeneity, surface imperfection, variation of dimensions and experimental setup. This scattering generally increases with lower level of stress. Taking the expected dispersion of mechanical tests into account, a population of 8 or 16 (SAC305 at ambient temperature) components was tested for each loading condition (alloy / package / temperature / pre-conditioning). Up to five different stress levels were tested per condition in order to be able to plot the different fatigue curves. Time-to-failure was characterized by a 2-parameters Weibull law, with "η" being the scale factor (characterizing order of magnitude of time-to-failure) and "β" the shape factor (characterizing the scattering of the distribution). The fatigue curve X-axis is the statistical time-to-failure ($Tre_{50\%}$) or number of cycles to failure ($Nf_{50\%}$) coming from Weibull distribution. The Y-axis "Zmax" is the maximum board deflection measured with an accelerometer at the center of the board at its 1st natural resonance frequency. Results are fitted with exponential laws.

Figure 4and Figure 5 respectively show test results for BGA1156 and QNF68 both assembled either with tin-lead or SAC305. Each dot is a component failure. It enables analyzing the time-to-failure distributions.

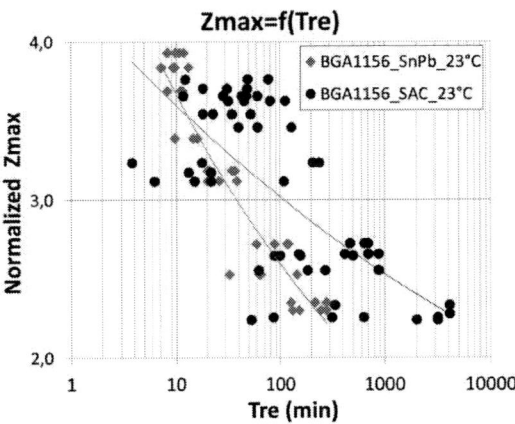

Figure 4: BGA1156 SAC305 and SnPb endurance testing at room temperature

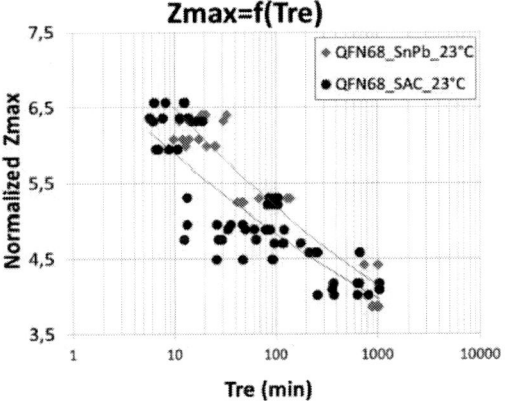

Figure 5: QFN68 SAC305 and SnPb endurance testing at room temperature

For BGA1156 and QFN68, results show that SAC305 can exhibit a higher dispersion regarding time to failure. No significant difference is noticed on the average values.

3 different packages were tested at room temperature both with SAC305 and tin-lead. Figure 6 shows that the impact of the component type is much more significant than the alloy type for a same component package. The alloy impact is also different depending on the package type: for QFN68 average mechanical strength was better for tin-lead assemblies whereas for BGA1156 SAC305 showed better results and for BGA288 results tend to be equivalent.

541

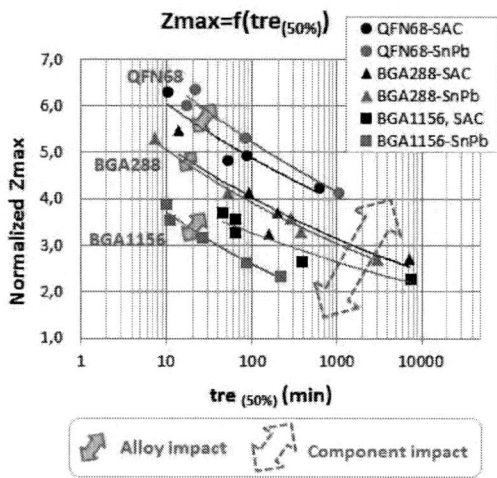

Figure 6: QFN68, BGA288 and BGA1156, fatigue curve with tin Lead and SAC305 assemblies

Table 2 indicates pre-conditioning parameters.

Max. temp.	Min. temp.	Dwell time max. temp.	Dwell time min. temp.	Temp. variation rate
125°C	-55°C	20 min	20 min	< 20 °C/min

Table 2: Thermal cycling preconditioning parameters

Table 3 indicates the number of cycles per preconditioning condition.

CT1bis	CT1	CT2	CT3
250	1015	1450	SnPb: 4306 SAC: 3579

Table 3: Definition of the different preconditioning conditions

Figure 7 and Figure 8 show SAC305 and tin-lead QFN68 fatigue curves with and without pre-conditioning.

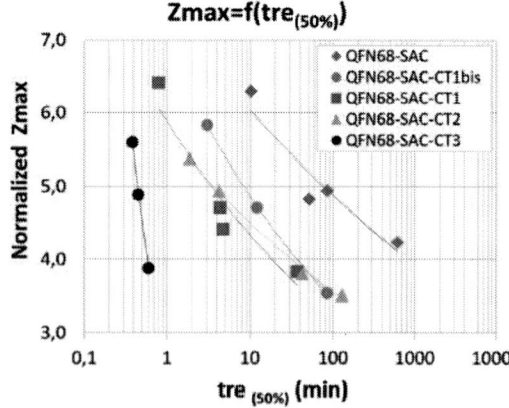

Figure 7: SAC305 QFN68 fatigue curve with and without thermal cycling pre-conditioning

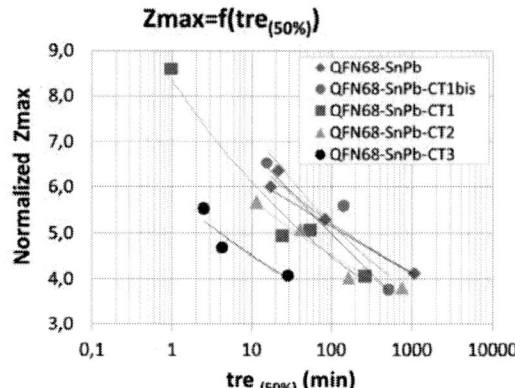

Figure 8: Tin lead QFN68 fatigue curve with and without thermal cycling pre-conditioning

Results indicate the impact of the wear-out of the terminations generated by thermal cycling. Vibration tests time-to-failure decrease with number of thermal cycles of pre-conditioning. For SAC305, it is significant after 250 cycles. The decrease for tin-lead exists as well but is more moderated. It is important to note that only the QFN68 was tested at 250 cycles and that SAC305 QFN68 thermal cycling results exhibited a different behavior compared to tin-lead QFN68, as illustrated by Figure 9.

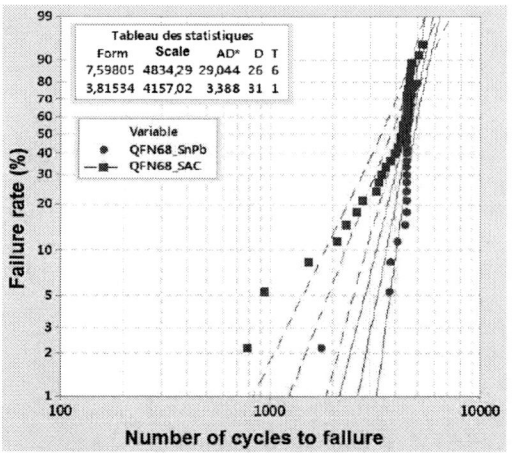

Figure 9: Thermal cycling Weibull curve for QFN68

The impact of temperature was also assessed. Figure 10 and Figure 11 show the results for BGA1156 which is the only package for which tests were performed at both high and low temperature for endurance vibrations.

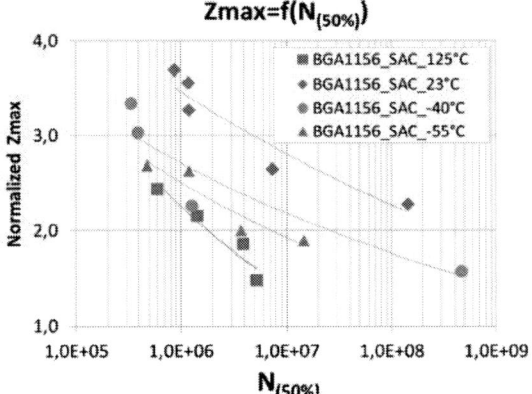

Figure 10: SAC305 BGA1156 fatigue curve at ambient temperature, 125°C, -40°C and -55°C

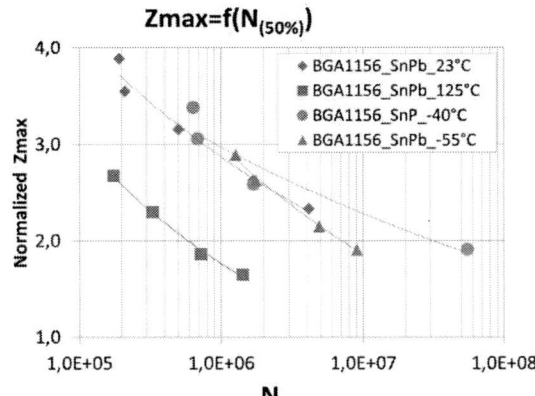

Figure 11: SnPb BGA1156 fatigue curve at ambient temperature, 125°C, -40°C and -55°C

Results indicate for both tin-lead and SAC305 a decrease in time-to-failure of equivalent order of magnitude at high temperature, compared to ambient temperature. At low temperatures there is a decrease of the performance for SAC305 whereas for tin-lead no obvious decrease can be observed.

Considering two tests conditions, Z_{max1} and Z_{max2}, the ratio between their time-to-failure is called an acceleration factor (AF).

$$AF = \frac{tre_{(50\%)1}}{tre_{(50\%)2}} = \left[\frac{Z_{max2}}{Z_{max1}}\right]^{\frac{-1}{B}}$$

Table 4 summarizes the value of the acceleration factor (1/B exponent).

Package	Temperature	SAC	SnPb
		-1/B	-1/B
	-55°C	8,62	4,85
	-40°C (1)	3,65	4,18
BGA1156	-40°C (2)	16,13	11,49
	125°C	4,83	4,22
	23°C (1)	6,71	6,37

	23° (2)	20,00	–
BGA288	23°C	8,00	9,09
	-55°C	–	–
QFN	-40°C (1)	–	–
	125°C	8,00	8,26
	23°C	11,76	10,20

Table 4: Value of exponent deduced from 50% failure rate fatigue curve fitted with exponential laws, for SAC305 and tin lead at different temperature.

The exponent values, representative of the slopes of the fatigue curves between SAC305 and tin-lead are equivalent. Acceleration factor for tin-lead and SAC305 can be considered the same.

For each test condition either 8 or 16 components (SAC305 at ambient temperature) were tested. In order to validate that the electrical failures were due to the solder joint cracks, one cross-section was performed for each test condition on the last failing component. The analysis revealed that cracks occurred almost exclusively in the solder joints. At component level, cracks were mainly localized in the corner solder joints both for BGA and QFN, whichever the alloy was. Corner A, B and D exhibited solder joints with cracks whereas corner C did not (cf **Figure 1**, for corner and localization on board). At solder joint level, cracks were localized either on component side or PCB side in equivalent proportions for SAC305 and tin-lead assemblies. Figure 12 indicates which the proportions were for QFN68 and BGA1156.

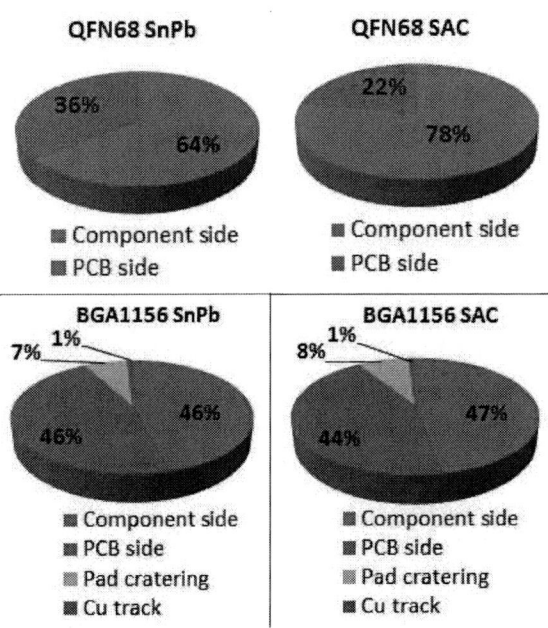

Figure 12: Cracks localizations on cross sections

Figure 13 shows examples of representative failure modes observed on the cross sections.

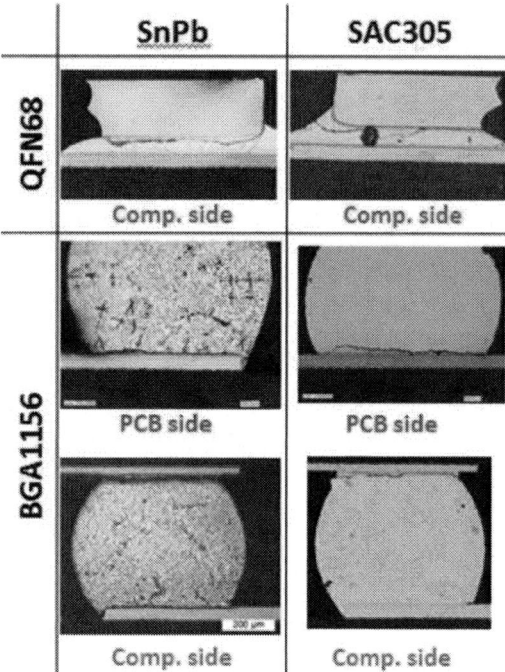

Figure 13: Cross sections performed after mechanical tests showing typical cracks in the solder joints.

Discussion

Three fatigue domains are commonly used by material sciences for metal and metal alloys, as illustrated in Figure 14. Firstly, low-cycle fatigue, where stress-levels are high (higher than yield strength) leading to plastic strain, the number of cycles to failure in this domain is lower than 10 or 100 thousands. Secondly, limited endurance, or high-cycle fatigue domain, where the number of cycles to failure are below 10 million. In this domain, apparent stress levels in the elastic domain generate failures by local concentration of stresses. Finally, the endurance domain where the number of cycles to failure is too high to be measured.

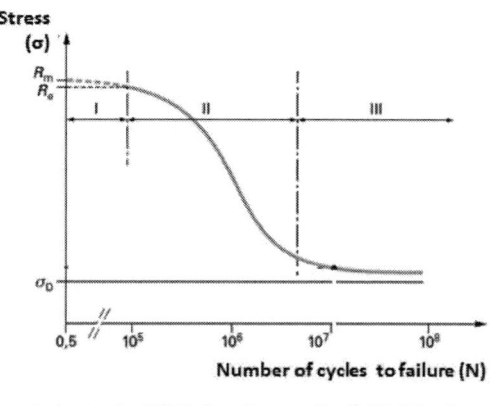

I	Low cycles fatigue domain	R_e	Yield strength
II	High cycles fatigue domain	R_m	Tensile strength
III	Endurance	σ_D	Fatigue limit

Figure 14: Fatigue curve [6]

Tested stress levels are representative of high cycles fatigue domain. For some conditions we observed the beginning of a transition with the endurance domain.

Conclusions

The presented study enabled to characterize the mechanical behavior of SAC305 solder joints considering also combined impact of temperature and thermal cycling which are typical harsh environment applications loadings. A comparison with tin-lead as reference used in harsh environment was made. Equipment are qualified (test methods) to verify reliability considering their mission profiles. The stakes of this test campaign was beyond mechanical characterization, evaluating if the use of SAC305 would necessitate reconsidering test methods ensuring solder joint reliability regarding environments combining mechanical, thermal and thermo-mechanical loadings in their mission profiles.

The presented results do not include all tests performed during this study. It does not include vibration step stress and shocks step stress tests. The hereafter conclusions are the study conclusions which take into account all test results and not only restricted to endurance testing.

The main differences observed between the both alloys are the following ones:
- SAC305 can exhibit larger dispersion regarding time to failure. This was not systematically observed.
- For BGA156, the impact of low temperature (-40°C, -55°C) regarding ambient temperature, on average time to failure, was greater on SAC305 (No endurance test were performed on QFN68 at low temperature).
- The impact of thermal cycling pre-conditioning on time to failure for a same constraint level tends to be greater for SAC305. It is important to note that the only tested condition below 1000 cycles was 250 cycles only applied to QFN68.
- Tin-lead has an overall lower mechanical strength than SAC305 at high temperature (+125°C).

On the other hand, similarities have been observed:
- At ambient temperature, average behavior.
- Acceleration factor
- Package relative strength: BGA1156 < BBGA288 < QFN68 (influence of size)
- Failure mode (crack location on component and within the solder joint)

Another noticeable observation is that the effect of package type between QFN and BGA is more significant than the effect of the solder alloy.

This study enabled to assess impacts of different loadings on time to failure. Designing application for mechanical high cycles fatigue domain requires taking into account the previous observations considering the specificity of the considered mission profile, and develop tests ensuring sufficient reliability.

This study did not enable to measure and conclude on potential impact on fatigue limit (σ_D). If considering that design enables to guarantee keeping the stress level below the fatigue limit, previous considerations regarding impact of temperature and thermal cycling should not apply. For a given constraint, the stress intensity at solder joint level mainly depends on components types and placements, and boards deflections.

References

[1] Steinberg D., "Vibration Analysis for Electronic Equipment", 3rd edition, John Wiley & Sons, Inc. 2000.

[2] M.Brizoux , W.C.Maia Filho, A.Grivon, Thales Global Services, Corporate Engineering. "BGA behavior under harsh mechanical environment using Backward SnPb+". In Brasage 2010, Brest, France. May 19-20st 2010.

[3] Da Yu, and al. "High-cycle fatigue life prediction for Pb-free BGA under random vibration loading" Microelectronics Reliability 51 (2011) 649–656

[4] "NASA-DoD Lead-Free Electronics Project" In IPC/JEDEC International Conference on Reliability, Rework, and Repair of Lead-Free Electronics, Mars 2008.

[5] IPC Test Standards, IPC-9701. Performance test methods and qualification requirements for surface mount solder attachments. IPC-association connecting electronics industries; 2002. p. 13.

[6] P.Rabbe, and al. "Essais de fatigue partie I" (m4170), Techniques de l'ingénieur, 10/03/2000.

Low-temperature fine-pitch flip-chip bonding by using snap cure adhesives and Au stud bumps

Ali Roshanghias[1*] and Augusto Daniel Rodrigues [1, 2]

[1] Silicon Austria Labs GmbH, Europastr. 12, A-9524 Villach, Austria
[2] Federal Institute of Santa Catarina, Florianopolis, Brazil
* Email: Ali.Roshanghias@silicon-austria.com

Abstract

Fine-pitch flip-chip bonding with a low-temperature budget can be considered not only as the key enabler of hybrid and flexible electronics, but also as a requirement for 3D packaging platforms with temperature-sensitive components. In this study, low-temperature flip chip bonding by using snap cure non-conductive paste (NCP) and anisotropic conductive paste (ACP) was investigated. By exploiting the compliancy of soft gold-stud-bumps and compressive stress induced by snap cure adhesives, flip-chip bonding with both adhesives were successfully conducted in a thermode station at 170 °C within a few seconds. It was also found out, that for fine-pitch flip-chip bonding with Au-stud bumps, snap cure NCA is the better choice, since it can handle finer pitches, provide sufficient underfilling, has lower cost and simplified process.

Keywords: Flip-chip bonding; Fine-pitch; Anisotropic conductive paste; NCP; Au stud-bump

Introduction

Interconnection technologies are one of the major gears for innovation in electronics areas, and for a quite long time, wire bonding and soldering techniques have been the main tools to create the interfaces between chips, substrates and boards. However, with the upcoming trends of system integrations such as internet of things (IoT), lab on chip, hybrid electronics and flexible electronics, alternative interconnect schemes are extensively demanded. Here conventional semiconductor components and circuit boards are required to be connected directly to materials with much low temperature and pressure resistance such as PET, PDMS and paper [1- 3].

Flip chip bonding in comparison to chip-on-board bonding offers more variety of materials and therefore is currently the mainstream in hybrid electronics. Different types of bump structures such as solder bumps (Sn, In, Sn-Ag), Cu pillars, SLID, plated Au bumps and Au stud bumps are commonly used for flip chip technology. Except the stud bumps, all other bumping processes yield a flat or curvy surface. During these processes, a metal layer is deposited over the bond pad using processes such as plating and sputtering [2-7].

Stud bumping is a very similar process to wire bonding where a wire is melted into a sphere and attached to a bond pad. After the attachment, the wire is removed at the end to leave a stud on the bond pad. Stud bumped flip chips are bonded to a substrate using either adhesives or thermosonic process. Since Au stud bumping can be done directly on Al pads of the chips without under-bump-metallurgy (UBM) processes, it is a flexible and inexpensive process in comparison to other bumps. High-speed stud bumpers can nowadays bump a whole wafer in minutes.

Flip chip bonding with Au-stud bumps and adhesives uses the shrinkage of paste to maintain good physical contact between the Au stud bumps and the corresponding copper pads of the Ni/Au surface finish.

The most critical steps in developing a low-temperature flip-chip bonding technology with Au-stud bump and adhesive are to define:

1) the right morphology for the Stud bumps (as-bumped or coined)
2) the Pad metallization on the counterpart
3) the right adhesive in terms of polymerization time and shrinkage
4) the optimal bonding profile (temperature, pressure and time) [7-16]

Concerning adhesive, advanced snap cure adhesives has been introduced and commercialized recently. Not only non-conductive paste (NCP) can be snap cured within a few seconds, but also isotropic conductive pastes (ICP) and anisotropic conductive adhesive (ACP) can be produced with similar snap cure epoxies. Snap cure adhesives are engineered with appropriate viscosity to minimize excessive flow and prevent undesirable bleeding. Snap cure compositions are frequently employed in high volume production processes. In another study,

546

Cu pillars were flip-chip bonded with a snap cure ACP [17].

Materials and methods

The chip used in this experiment was a 4.75 mm x 4.75 mm silicon die with two nested daisy-chains and kelvin sensing structures. The Chip had 72 peripheral Al pads; each was 100 µm* 100 µm in size and bumped with an Au-stud bump. The pitch size in the chip was 250 µm. As seen in 3D surface profile and the corresponding 2D-line scan in Fig. 1, the stud bumps have an average height of 30 µm and have a spiky morphology.

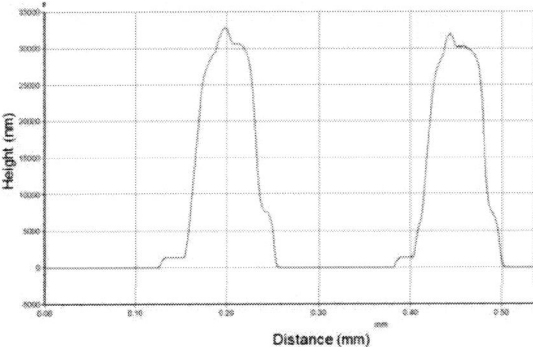

Fig.1 Surface profilometry of Au-stud bumps

The respective printed circuit board with the complementary daisy chain structures had cu pads with electroless nickel immersion gold (ENIG) metallization. A non-conductive and an anisotropic snap cure adhesive with identical epoxy matrix were used in this study for bonding. The ACP had additionally conductive Ni particles.

The bonding process was carried out by means of a micro-assembly station (Fineplacer ®, Finetech GmBH). The snap cure adhesives were dispensed in a controlled volume on top of the board by a digital dispenser. The die was picked by the mechanical arm of the machine, and placed precisely above the adhesive deposited area. When the die was in full contact with the substrate via the adhesive, the low bonding forces ranging from 8 to 30 N (1.85

– 7 MPa) were exerted for 10 to 15 seconds. Simultaneously, the stack was being heated from bottom and top up to 170 °C and cooled down in the air.

The microstructure and elemental composition of the samples were characterized by a scanning electron microscope (SEM, FEI) equipped with energy-dispersive X-ray spectroscopy (EDS). A white-light interferometer (WLI, Polytec) was used for topography characterization of the stud-bumps.

Results and Discussion

The average contact resistance per bump versus the bonding force is plotted in Fig. 2. As inferred from this graph, ACP and NCP bonded samples yield to the similar contact resistance at higher bonding forces. In other words, under 20 N and 30 N bonding forces, ACP and NCP bonded samples revealed to have almost equal contact resistance.

In essence, the electrical contact in the NCP and ACP adhesives comes from physical contact and not the metallurgical bonding mechanism such as in thermo-compression bonding. Concerning NCP, the electrical contact is established by physical contact of stud bump and chip pad, in contrast to the ACA bonding process, where the electrical contact is established by conductive particles which are clamped between the planarized Au studs and the pads.

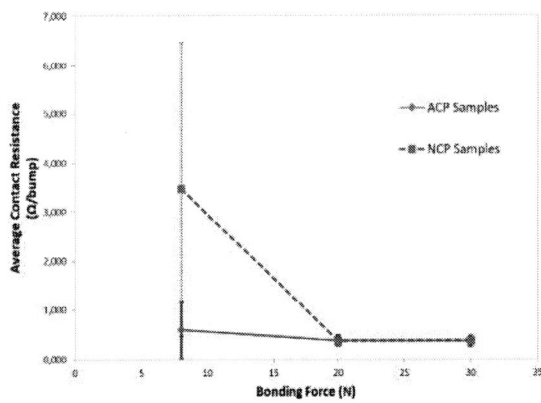

Fig. 2 Average contact resistance per bump versus bonding force at 170 °C

Similar contact resistance at higher bond force implied that conductive particles contributed barely to the contact resistance. However, at lower bond force, the difference in the contact resistance of ACP and NCP could be attributed to the conductive particles. Therefore, a certain threshold force seems to be necessary for NCP to provide good physical contact between the Au studs and the bonding pads due to the co-planarity of the bumps.

The corresponding cross-sectional SEM pictures of the ACP and NCP samples are given in Figs. 3 and 4 respectively.

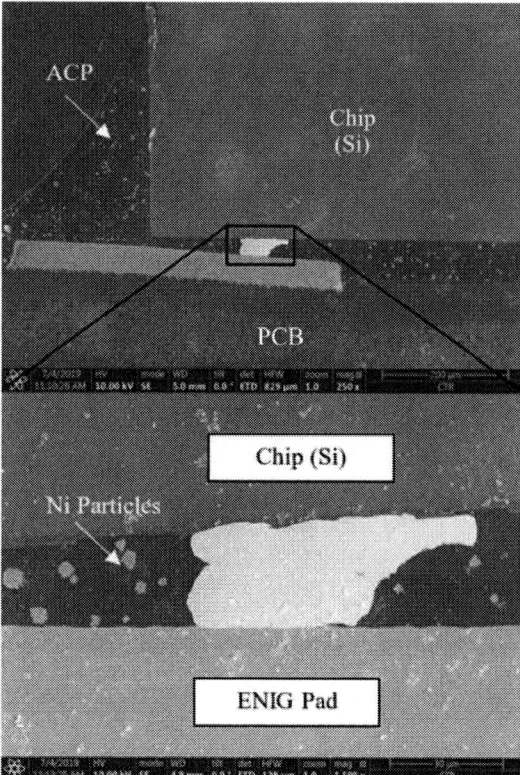

Fig. 3. Cross-sectional SEM image of ACP bonded sample under 30 N bonding force

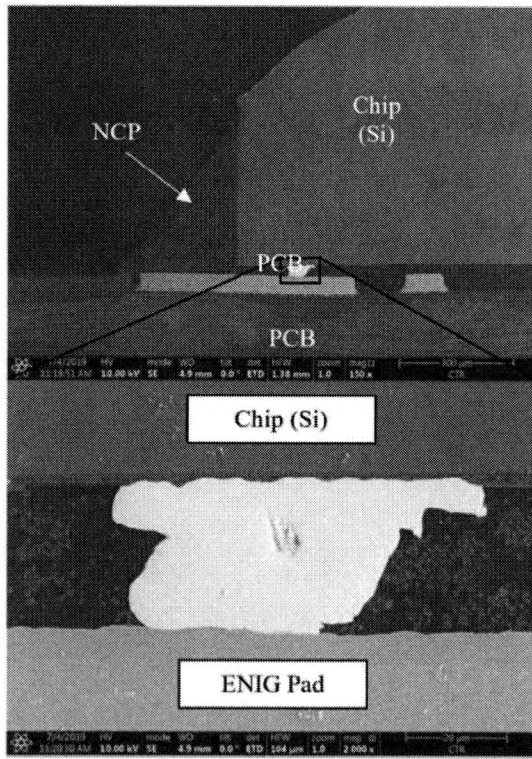

Fig. 4. Cross-sectional SEM image of NCP bonded sample under 30 N bonding force

Bumps topography and compliance

The Au studs provide a compliant interconnect between chip and substrate, which is due to the softness of pure gold. Gold has a Vickers hardness of 118-226, whereas copper has a Vickers hardness of 343–369 and Nickel has a hardness of 638. As a result, Au stud bumps can deform and accommodate to the bond-line much easier than Cu pillars. Additionally, Au stud bumps have a spiky shape which locally increases the contact pressure.

As seen in Figs. 3 and 4, Au stud bump is uniformly bonded to the ENIG pad metallization of the PCB. By comparing Fig.3 and 4, no detectable differences between the bonded Au- bumps are observed. That could explain the similar contact resistance of ACP and NCP samples.

The elemental mapping of the ACP bonded sample (Fig. 5) also indicates that due to the sharp tip of the Au-stud bump, short processing time and the softness of Au in comparison to Ni, no Ni particles were trapped in between bump and the Pad. Hence, one can conclude that for fine pitch flip-chip bonding with as-bumped Au stud bumps and snap cure adhesive, NCP bonding can be more beneficial since it can handle the finer pitch bonding potentially. However, for the coined Au-stud bumps with larger surface area, trapped conductive particles in ACP might essentially result in better electrical conductivity, higher yield and reliability of the bond.

It is noteworthy to mention that for fine pitch flip-chip bonding with Au-stud bump and NCP, the co-planarity of the two parts is a determining factor, especially when low bonding force and temperature are applied such as in this study. When higher bonding force is permitted and the substrate is rigid, then tilting can be partially compensated by squeezing stud bumps.

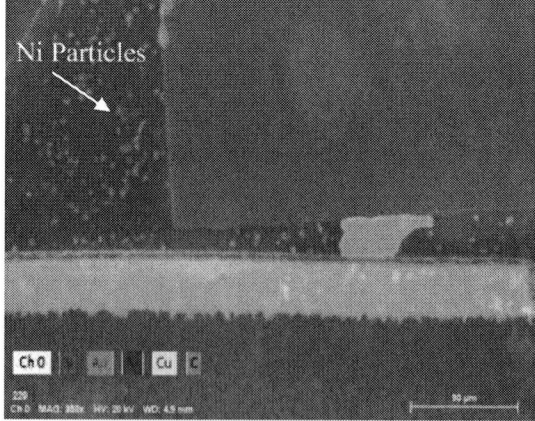

Fig. 5 EDS elemental mapping of ACP bonded Au stud bump

Conclusion

In this study, low-temperature fine pitch flip chip bonding by using snap cure adhesives and Au stud-bumps was conducted. Contact resistances of the bumps were analyzed for both ACP and NCP bonded daisy chain Si chips. The results revealed that for Au-stud bumps with spiky shapes (as-bumped) NCP and ACP bonded samples yield to similar values. It was concluded that NCP bonding with Au-stud bumps can be the optimal technology which can handle finer pitch and promises simplified process, and lower cost. The concurrent polymerization of the snap cure adhesive during stud bump deformation is the main promoter of this low-temperature bonding technology which can induce a permanent compressive force on the bonded area and ensure the reliability of the bond. The proposed technology with snap cure adhesive can be promising for high throughput flip chip bonding process, especially for hybrid and flexible electronics.

Acknowledgments

This work was carried out within the framework of the project "Flex-Si-sense", financed by the Austrian Research Promotion Agency (FFG).

References

[1] Lu, D. and Wong, C.P. eds., 2009. Materials for advanced packaging (Vol. 181). New York: Springer.

[2] Blackwell, G.R., 2017. The electronic packaging handbook. CRC Press.

[3] C. Chuang, Q. Liao, H. Li, S. Liao, G. Huang, Increasing the bonding strength of chips on flex substrates using thermosonic flip-chip bonding process with nonconductive paste, Microelectronic Engineering 87, pp. 624-630, 2010.

[4] Zhong, Z., 1999. Assembly and reliability of flip chip on boards using ACAs or eutectic solder with underfill. Microelectronics International, 16(3), pp.6-14.

[5] K. Harr, S. Kim, Y. Kim, Y. Kim, Development of chip-on-flex bonding using Sn-based bumps and non-conductive adhesive, Microelectronics Reliability 55, pp. 12411247, 2015.

[6] S. Kim, Y. Kim , Low temperature chip on film bonding technology for 20 um pitch applications, J Mater Sci: Mater Electron27, pp. 36583667, 2016.

[7] Lee, T.K., Lua, E., Low, K.C., Ng, A. and Ng, H.W., 2005, December. Bonding development for non-conductive paste (NCP). In 2005 7th Electronic Packaging Technology Conference (Vol. 2, pp. 6-pp). IEEE.

[8] X. Cai , A. Zhai , C. Cheng, The preparation and the reliability study of anisotropic conductive paste (ACP) for RFID tag inlays packaging, 14th International Conference on Electronic Packaging Technology, pp. 368-371, 2013.

[9] K. Saarinen-Pulli, S. Lahokallio, L. Frisk, Effects of different anisotropically conductive adhesives on the reliability of UHF RFID tags, International Journal of Adhesion and Adhesives 64, pp 5259, 2016.

[10] Min, T.A., Lim, S.P.S., Yeo, A. and Lee, C., 2005, December. Influence of bump geometry, adhesives and pad finishings on the joint resistance of Au bump and A/NCA flip chip interconnection. In 2005 7th Electronic Packaging Technology Conference (Vol. 2, pp. 5-pp). IEEE.

[11] J. Choi, T. Oh, Contact Resistance Comparison of Flip-Chip Joints Produced with Anisotropic Conductive Adhesive and Nonconductive Adhesive for Smart Textile Applications, Materials Transactions, Vol. 56, No. 10, pp. 1711-1718, 2015.

[12] J. Kim, J. Kim, K. Paik, Effects of the Types of Anisotropic Conductive Films on the Bending Reliability of Chip-in-Plastic Packages, IEEE transactions on components, packaging and manufacturing technology, vol. 9, no. 3, pp. 405-411, 2019.

[13] Leppänen, L., 2016. Bendability of Flip-Chip Attachment on Screen Printed Interconnections (Master's thesis).

[14] Chuang, C.L. and Jiang, Y.R., 2018, August. Investigation on the Reliability of Chips onto the Flex Substrate With a Non-Conductive Paste for Flex Substrates Pretreated by Argon Plasma. In 2018 19th International Conference on Electronic Packaging Technology (ICEPT) (pp. 338-342). IEEE.

[15] Min, T.A., Lim, S.P.S., Yeo, A. and Lee, C., 2005, December. Influence of bump geometry, adhesives and pad finishings on the joint resistance of Au bump and A/NCA flip chip interconnection. In 2005 7th Electronic Packaging Technology Conference (Vol. 2, pp. 5-pp). IEEE.

[16] Yoneda, Y., Kuramochi, T., Sohara, T. and Liao, J.M., 1999. A novel flip chip bonding technology using au stud bump and lead-free solder. In Pan-Pacific Conference.

[17] Ma, Y., Roshanghias, A. and Binder, A., 2018. A comparative study on direct Cu–Cu bonding methodologies for copper pillar bumped flip-chips. Journal of Materials Science: Materials in Electronics, 29(11), pp.9347-9353.

Can Electrolytic Capacitors Meet the Demands of High Reliability Applications?

Greg Caswell

DfR Solutions 9000 Virginia Manor Road, Suite 290, Beltsville, MD USA 20705
301-640-5825 – gcaswell@dfrsolutions.com

Abstract

Electrolytic capacitors have long been identified as a weak link for long term high reliability applications. However, capacitor manufacturers have made significant improvements to the materials and manufacturing processes to enhance their reliability. This paper will discuss those changes, provide insight into the various failure mechanisms for electrolytic capacitors and describe appropriate accelerated tests to validate performance.

We will take a deeper dive into the methodologies utilized to improve capacitor performance, e.g. foil purity and electrolyte volume. We will also discuss, from a reliability perspective, the impact of changing to a higher temperature electrolyte (from ethylene glycol to DMF, DMA and GBL) and also changes in the bung material (from butyl to EPDM).

There are several environmental factors involved in the aging of electrolytic capacitors. Electrolyte loss due to drying out and leakage current due to oxide degradation are thermally related as is the self-heating associated with ripple current. The impact of the applied voltage level is also a driver as it can cause leakage current increases as well. All of these issues result in a capacitance decrease, an increase in ESR, and a change to the dissipation factor. Many other failure mechanisms associated with manufacturing will also be discussed.

Key Words: Life Expectancy, Electrolytic Capacitors, Reliability

Introduction

Aluminum electrolytic capacitor manufacturers state the endurance lifetime of their products in their datasheets, supported by testing that applies rated voltage and ripple current at the maximum rated temperature to the capacitor. In service, electrolytic capacitors are rarely run at their rated values, so the endurance lifetime is not directly applicable. The industry uses a rule of thumb approach of lifetime doubling for every 10°C decrease below rated temperature to estimate lifetime at lower temperatures. However, there is no explicit approach to estimate lifetime at lower applied ripple currents, which is also known to increase lifetime. So, how can we examine the different suppliers to ascertain their differences in reliability? Evaporation prediction has been based on a widely held standard aging relationship that ties back to the 10°C previously noted.

$$L_x = L_o \times 2^{(T_o - T_x)/10}$$

However, there are differences in the way that the manufacturers make this determination. This paper will discuss these differences as the ripple current, voltage, and can size play a large part.

This paper will then identify which parameters have a distinct impact on capacitor lifetime and how a Reliability Physics approach offers the best methodology for ascertaining the most reliable capacitor for your application. The process of determining the capacitor's ESR, then calculating its core temperature, leading to the calculation of its vapor pressure and finally calculating the loss of electrolyte will be presented as will an approach to shorten test time when validating electrolytic capacitors.

Examples of calculations for life expectancy will be shown to demonstrate the effects of applied voltage, rated temperature, ripple current, and the endurance factor coupled with the application usage profile. Finally, a best in class qualification methodology will be presented.

Changes in Electrolytic Capacitors to Fix Past Reliability Issues

Capacitor manufacturers have done significant work to improve the reliability of electrolytic capacitors. They have improved their quality control with respect to the foil purity, electrolyte volume etc. in manufacturing. They have also migrated to higher temperature electrolytes (from ethylene glycol to dimethylformamide (DMF), gamma-butyrolactone (GBL) and Dimethylacetamide (DMA) and made changes to the bung material from butyl to ethylene propylene diene monomer rubber (EPDM). These materials have enhanced the ability of the rubber compound to minimize the effects of ultraviolet rays and to operate in moderate heat.

Even with these changes the manufacturers are still making empirically based life calculations They are also dealing with the balance between high capacitance (thin dielectric) and high voltage (thick dielectric).

Electrolytic Capacitor Failure Mechanisms

There are several factors that can drive the aging of electrolytic capacitors. For example, temperature can affect the loss and dry out of electrolyte and also increase leakage current due to oxide degradation. Issues like thermal stress, wrong polarity installation, overvoltage and rapid charge/discharge can drive these failures. Ripple current, through self-heating, can also increase the operating temperature and exacerbate the loss of electrolyte, while increased voltage levels can also impact the oxide degradation. These effects result in decreased capacitance, increased ESR and a change in dissipation factor.

Issues during manufacture can also result in failures. These include impurities of metal particles, contamination by chloride, deficiencies in the oxide layer, burred foil, poor sealing and poor connections. These issues can lead to short circuits or increases in leakage current.

Capacitor Life Prediction Approaches

Different capacitor manufacturers have created their own models for determining the life expectancy of their capacitors. For example, the following equations are utilized by Nichicon for small devices (with and without ripple current). The 3rd formula is used for large can electrolytics:

$$L = L_r \times 2^{\frac{T_r-T}{10}} \times \frac{1}{B_n} \quad \text{Miniature w/o ripple}$$
$$B_n = 2^{\alpha \times \left(\frac{I_r}{I}\right)^2 \times 2^{-\left(\frac{T_r-T}{30}\right)}}$$

$$L = L_r \times 2^{\frac{T_r-T}{10}} \times 2^{\alpha\, 1-\left(\frac{I_r}{I}\right)^2 \times 2^{-\left(\frac{T_r-T}{30}\right)}}$$
Miniature w/ ripple

$$L = L_r \times 2^{\frac{T_r-T}{10}} \times 2^{1-\frac{\Delta t_r \times \left(\frac{I_r}{I}\right)^2}{K}}$$
Large can

Where, L_r is rated lifetime, T_r is rated temperature, T is ambient temperature, I_r is rated ripple current, I is actual ripple current, Dt_r is the temperature rise due to rated ripple current, Dt is the temperature rise due to actual ripple current and a and K are coefficients. To obtain the coefficients, you have to contact the manufacturer.

Nippon Chemi-Con utilizes a different set of equations for their capacitors and again has coefficients that are part of their internal data.

$$L = L_r \times 2^{\frac{T_r-T}{10}} \times 2^{-\frac{\Delta t}{5}}$$
Miniature w/o ripple

$$L = L_r \times 2^{\frac{T_r-T}{10}} \times 2^{\frac{\Delta t_r-\Delta t}{5}}$$
Miniature w/ ripple

$$L = L_r \times 2^{\frac{(T_r+5)-(T-25)}{10}} \times 2^{\frac{25-\Delta t}{4}} \times Kv$$
Large can

Where, L_r is rated lifetime, T_r is rated temperature, T is ambient temperature, I_r is rated ripple current, I is actual ripple current, Dt_r is the temperature rise due to rated ripple current, Dt is the temperature rise due to actual ripple current and a and K are coefficients.

Reliability Physics Approach

A reliability physics approach to determining the life expectancy of electrolytic capacitors is derived from electrolyte vapor pressure, the volume of electrolyte, the critical volume level, the evaporation or leak rate and the embrittlement of the bung material. Unfortunately, most capacitor manufacturers consider this information proprietary. There are 4 steps to this process, 1) determine the Equivalent Series Resistance (ESR) of the capacitor, 2) calculate the core temperature of the device, 3) calculate the vapor pressure and 4) calculate the loss of electrolyte.

The following example illustrates the calculation for determining ESR.

$$ESR_T/ESR_{25C} = A \times (B - (C \times TANH((T-T_0)/D)))$$

Where A=6, B=1.8, C=1.7, $T_0 = 240$, and D=25. Figure 1 shows the change in the ESR ratio as a function of the change in temperature.

551

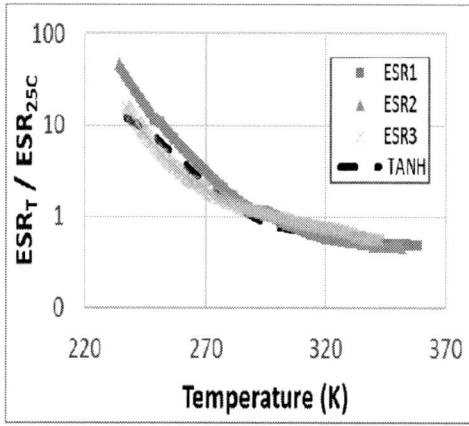

Figure 1 – Ratio of ESR_t to ESR_{25C} With Respect to Temperature

Month	Max Temp	Min Temp
Dec, Jan, Feb	20	5
Mar, Nov	25	10
Apr, Oct, May, Sep	30 35	15 20
Jun, Jul, Aug	40	25

Figure 2 – Temperature Variability in Phoenix, Arizona

Next, the core temperature is determined by the following calculation:

$$\Delta T = \frac{ESR_T \times I^2}{H \times (2\pi r h + 2\pi r^2)}$$

Where H is the heat transfer per surface area

Once determined you add this value to the ambient temperature to determine the core temperature of the capacitor. The next step is to determine the vapor pressure which utilizes the Antoine Equation.

$$[\log P = A - \frac{B}{C + T}]$$

For example, for ethylene glycol/99%H_2O A=9.19, B=3103 and C=309.7.

Finally, you calculate the loss of electrolyte.

$$V_{t0+Dt} = V_{t0} - (k \times P \times Dt)$$

Where k is leak rate based on vapor pressure (empirically determined, ml/mmHg/hr) and Δt is the time step.

Then it is necessary to select whether your calculations will utilize a constant temperature or temperature variation. Figure 2 shows the variability in temperature for Phoenix, Arizona.

The electronics industry tends to be reliant on supplier ratings. However, some things to consider are that failure definition can vary from supplier to supplier; lifetime calculations can be with or without ripple current; and the probability of failure after lifetime is never defined (e.g. test to 0 failures). There are also different ways to extend lifetime. Manufacturers can increase the volume of electrolyte, use higher boiling point electrolytes, provide better bung seals, as well as the ability of the capacitor to operate with lower electrolyte volumes.

To illustrate this overall concept let's do an example of a calculation to determine the life expectancy of an electrolytic capacitor using the simplest formula used in the industry.

$$L = L_r \times 2^{\left(\frac{T_r - T_E}{10}\right)}$$

Where, L_r = rated lifetime of the capacitor at rated ripple current, T_r = rated temperature of the capacitor and T_E = actual temperature of the electrolyte.

For our purposes we set the field environment as follows:
1) ~10C rise due to ripple current
2) 65C for 12 hours each day for thermal environment (1/2 year)
3) Voltage applied is 80% of rated

The charts in Figure 3 below show the difference in life expectancy with respect to the manufacturers rated life based on their endurance testing.

		Nichicon
Rated voltage		63
Op voltage		50.4
rated temp		105
op temp		65
rated life		8000
delta T ripple		10
		worst case
calculated life hours		80000
hours/year	4337.6	
Years of life estimated		18.44338

		Nichicon
Rated voltage		63
Op voltage		50.4
rated temp		105
op temp		65
rated life		6000
delta T ripple		10
		worst case
calculated life hours		60000
hours/year	4337.6	
Years of life estimated		13.83253

Figure 3 – Life Expectancy Calculations for Nichicon Capacitors

Figure 4 shows the same calculations for Su'scon electrolytic capacitors where their endurance testing is a lower value. The impact on their life expectancy is obvious.

		Su'scon
Rated voltage		63
Op voltage		50.4
rated temp		105
op temp		65
rated life		4000
delta T ripple		10
		worst case
calculated life hours		40000
hours/year	4337.6	
Years of life estimated		9.221689

		Su'scon
Rated voltage		63
Op voltage		50.4
rated temp		105
op temp		65
rated life		3000
delta T ripple		10
		worst case
calculated life hours		30000
hours/year	4337.6	
Years of life estimated		6.916267

Figure 4 - Life Expectancy Calculations for Su'scon Capacitors

Finally, you can further control your calculations by determining the ripple current in the capacitor which is derived from:

$$I_{cap}=P_{pv}V_{pv}*\sqrt{2}*n$$

Where I_{cap} is the RMS capacitor ripple current in A, P_{pv} is the PV Power in W, and V_{pv} the PV voltage in V and n is the number of capacitors in parallel. Note that this equation is valid because there is no significant power conversion induced high frequency ripple current in the electrolytic capacitors.

There are two ways to calculate thermal conductance (and therefore the temperature rise due to ripple current). The first is per EIA-ECA-797 assumes that applying the rated ripple current will result in a 5C temperature rise. Let's look at 2 examples:

- UCC: Power = $I^2R = (3.8)^2$ x 0.014 = 0.202 W
- Nichicon: Power = $I^2R = (3.4)^2$ x 0.02 = 0.231 W

- Thermal conductance can therefore be determined

- UCC = 5 / 0.202 = 24.7 °C/W
- Nichicon = 5 / 0.231 = 21.6 °C/W

The second way, also per EIA-ECA-797 is to assume a convection coefficient of .0001 W°C/cm². Thermal conductance is the product of the convection coefficient and the surface area. So, for our example:

- Length = 40mm
- Width = 18mm
- Area = $2pr^2 + 2prh = 509 + 2260$ $= 2769$ mm^2 = 27.7 cm^2

- Thermal conductance is therefore 36.1 °C/W

Temperature rise will be different depending on the approach chosen. Also, the analytical equations being used to calculate temperature rise assume heat loss primarily through convection. However, with some designs, heat loss could be primarily through conduction. Actual heat rise will likely need to be through more sophisticated modeling or measurement.

Conclusions

This paper has presented a methodology for users of electrolytic capacitors to determine the life expectancy of the next general components for use in their applications.

References

1) Menzel, Stephan, "Aluminum Electrolytic Capacitors – Failure Modes," APEC 2018 Capacitor Workshop, San Antonio, Texas
2) Lohrber, Pierre, " High Voltage in Aluminum Capacitors," APEC 2018 Capacitor Workshop, San Antonio, Texas
3) Gulbrandsen, S. Cartmill, K., Arnold, J., Kirsch, N., Caswell, G., "A New Method for Testing Electrolytic Capacitors to Compare Life Expectancy," 2014 IMAPS Symposium

Tin Whisker Growth Mechanism on Tin Plating of MLCCs Mounted with Sn-3.5Ag-8In-0.5Bi Solder in 30°C60%RH

Akira SAITO[*1*2], Hiroshi NISHIKAWA[*3]

[*1]Murata Manufacturing Co., Ltd., Japan [*2] Graduate School of Engineering, Osaka University, Japan

[*3] Joining and Welding Research Institute, Osaka University, Japan

+81-(0)77-586-8290, and akirasaito@murata.com

Abstract

Low-temperature soldering using Sn-3.5Ag-8In-0.5Bi solder, Sn-8Zn-3Bi solder, 42Sn-58Bi eutectic solder cause the tin whisker growth on the tin plating in the testing for tin whisker. Especially, the tin whisker on the tin plating of MLCC (multi-layer ceramic capacitor) mounted by Sn-3.5Ag-8In-0.5Bi solder grew up fast, and its length became 160 μm at 8000 h in 30°C60%RH. But the mechanism of the whisker growth was unknown. We suggest a new mechanism of whisker growth related to the shape of ternmination and reflow temperature.

Key words: Tin whisker, Indium, Low-temperature soldering, Termination, MLCC

Introduction

On July 1, 2006 the Restriction of Hazardous Substances Directive (RoHS) came into effect [1], restricting the inclusion of lead in most consumer electronics sold in the EU, and having a broad effect on consumer electronics sold worldwide. Lead-free solders in commercial use may contain tin, copper, silver, bismuth, indium, and zinc.

Tin-silver-copper (SAC) solders are used by almost Japanese manufacturers for reflow. Sn-3.0Ag-0.5Cu (SAC305) is the most popular Lead-free solder and RoHS, REACH and JEIDA compliant [2].

However, SAC305 has two problems. One is the melting point of SAC305 that is about 30°C higher than 60/40 Sn-Pb eutectic solder. Therefore, Sn-Ag-In-Bi solder, Sn-Zn-Bi solder, and Sn-Bi eutectic solder are used for the assembly of temperature-sensitive components [3]. Another is whisker growth on the pure tin plating of components mounted with Lead-free solders, such as SAC305.

For preventing tin whisker, three testing conditions which were dump heat testing, thermal cycling testing and ambient storage testing, have been established as international standerds, such as

IEC and JEDEC [4-5]. By the enforcement of the standerds, there became little short circuit fault of electronics by tin whisker. However, we found that low-temperature soldering with low-temperature solder caused tin whisker growth [6].

In this paper, we show the results of tin whisker testing of low-temperature soldering and a new mechanism of tin whisker growth on the tin plating of MLCC mounted by 220°C reflow process with Sn-3.5Ag-8(6)In-0.5Bi solder.

Experiment

Table 1 shows sample prepareration for tin whisker testing. $BaTiO_3$ and $CaZrO_3$ were prepared as base materials. All terminations had same structure, Cu/Ni/Sn. MLCCs (1.6 mm × 0.8 mm) were suppied from two makers. MLCC mounted with Sn-3.5Ag-8(6)In-0.5Bi solder on 1.6 mm glass-epoxy board using 220°C and 245°C reflow process. MLCCs mounted with SAC305 using 220°C and 245°C reflow process were prepared as comparison samples. These samples were held in 30°C60%RH, and their whisker length observed using optical microscopre and scanning electron microscope (SEM) at 500, 1000, 2000, 4000, and 8000 h. A whisker was cross-sectioned by focused ion beam (FIB).

Table 1: Sample prepareration for tin whisker testing

No.	Bace material	Termination	MLCC	Solder	Reflow
1-1	$BaTiO_3$	Cu/Ni/Sn	A	Sn-3.5Ag-8In-0.5Bi	N_2 / 220°C
1-2	$CaZrO_3$	Cu/Ni/Sn	A	Sn-3.5Ag-6In-0.5Bi	Air / 220°C
2-1	$BaTiO_3$	Cu/Ni/Sn	A	Sn-3.5Ag-8In-0.5Bi	Air / 220°C
2-2	$BaTiO_3$	Cu/Ni/Sn	B	Sn-3.5Ag-8In-0.5Bi	Air / 220°C
2-3	$BaTiO_3$	Cu/Ni/Sn	B	Sn-3.5Ag-8In-0.5Bi	N_2 / 245°C
3-1	$BaTiO_3$	Cu/Ni/Sn	A	Sn-3.0Ag-0.5Cu (SAC305)	N_2 / 245°C
3-2	$BaTiO_3$	Cu/Ni/Sn	A	Sn-3.0Ag-0.5Cu (SAC305)	N_2 / 220°C

Results

As our previous study [6], figure 1 shows maximum tin whisker length on MLCCs mounted with three low-temperature solders after three kinds of tin whisker testing conditions. Tin whiskers on MLCCs mounted with Sn-3.5Ag-8(6)In-0.5Bi solder, Sn-8Zn-3Bi solder, 42Sn-58Bi eutectic solder grew up than 40 μm after three tin whisker testing of IEC 60068-2-82. The whiskers on MLCCs mounted with Sn-3.5Ag-8(6)In-0.5Bi solder were longer than other solders. Especially, we focused on the tin whisker growth in 30°C60%RH.

No whisker was observed on MLCC soldering with SAC305 in 220°C and 245°C reflow process after three kinds of tin whisker testing.

Figure 2 shows the time dependency of maximum tin whikser length on MLCC mounted with Sn-3.5Ag-8(6)In-0.5Bi solder in 30°C60%RH. These whiskers linearly grew up with time. Especially, the tin whisker of samle 2-2 (▲) and sample 1-2 (●) grew up fast. The tin whisker length was 160 μm at 8000 h about 0.9 year. If the tin whisker grows with same rate, the tin whisker length will become about 360 μm at two years and abnout 530 μm at three years. It is very risky for short circuit. These growth rate can be classified into two groups. Fast rate group includes sample 2-2 (▲) and sample 1-2 (●), and slow rate group includes sample 2-1 (△).

Figure 3 shows the optical images of tin whisker growth point of MLCCs. The tin whiskers of sample 2-2 and sample 1-2 in fast rate group grew up at the edge of termination. On the other hand, the tin whisker of sample 2-1 in slow rate group grew up at the point except the edge of terminartion. These results indicate that the acceleration factor for tin whisker growth exists at the edge of termination.

Figure 4 shows SIM image of the cross-section of the tin whisker at termination of sample 2-2. Tin whisker grew on grain boundary of tin plating at the flexure point of tin plating. SIM means scanning ion microscope. Ni-Sn intemetalic compound (IMC) was observed at Ni/Sn boundary. The thickness of Ni-Sn IMC was uniform, no Ni-Sn IMC that caused tin whisker growth was observed.

Figure 5 shows SEM/EDS images of cross-section at the growth point of tin whisker. No oxidation of Cu/Ni/Sn termination, no diffusion of copper to tin layer and no corrosion was observed.

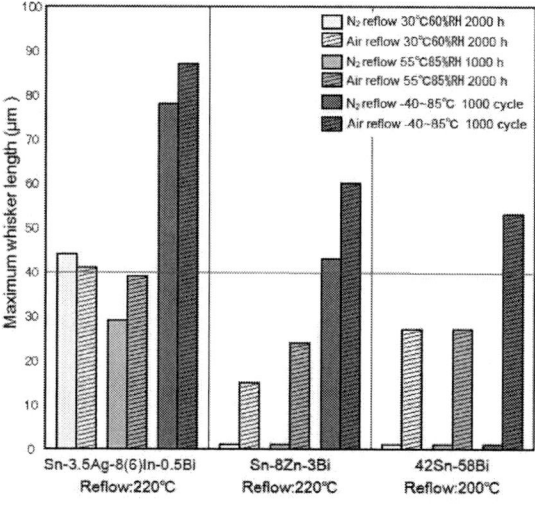

Figure 1: Maximum tin whisker length on MLCCs mounted three low-temperature solders after three whisker testing

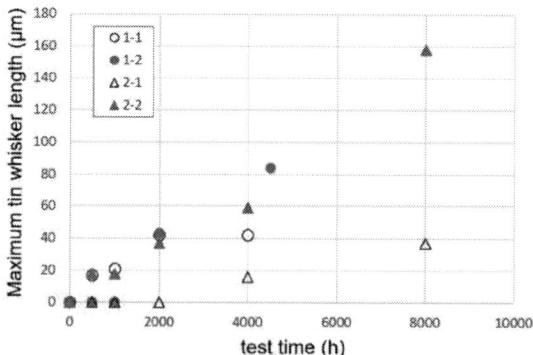

Figure2: Time dependency of maximum tin whisker length in 30°C60%RH

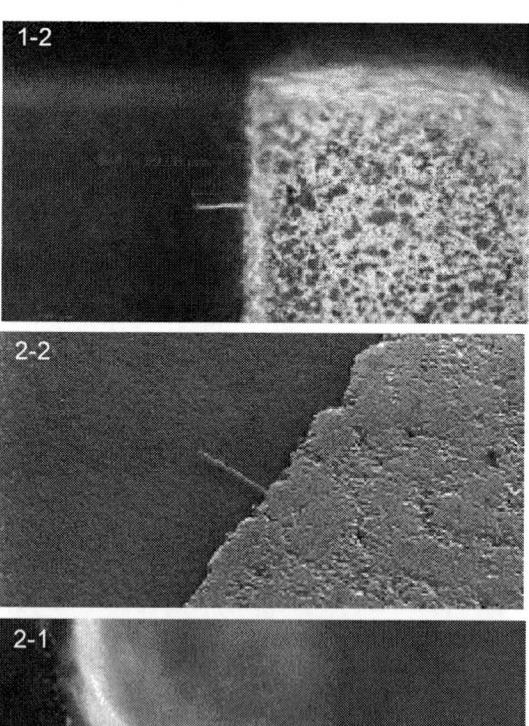

Figure 3: Growth points of tin whiskers

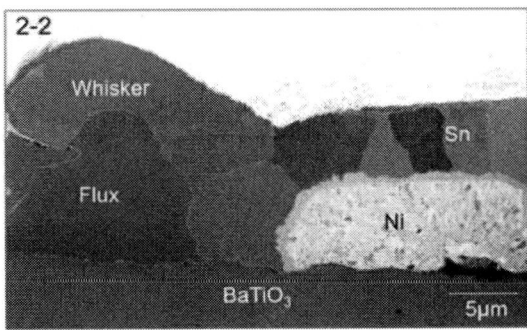

Figure 4: SIM image of the cross-section at the growth point of tin whisker

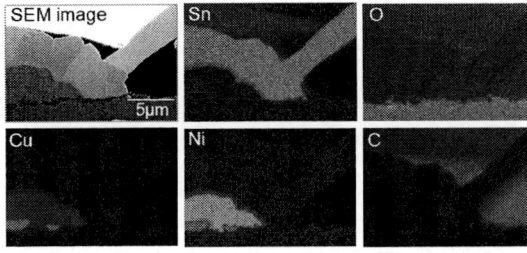

Figure 5: SEM and EDS images of the cross-section at the growth point of tin whisker

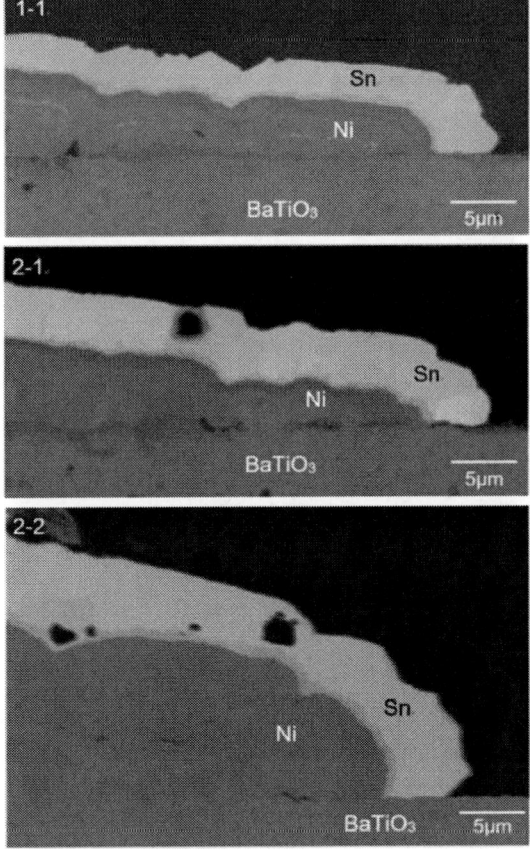

Figure 6: Optical images of the cross-section of terminations

These results indicates that no factor of the whisker growth was found.

Figure 6 shows the shape of terminations of sample 1-1, sample 2-1 and sample 2-2 before reflow. Tin plating layers of sample 1-1 and sample 2-2 in fast rate group of tin whisker growth were sharply curved at the edge of termination. On the other hand, tin plating layer of sample 2-1 in slow rate group of tin whisker growth was slightly curved at the edge of termination.

Discussion

Tin whisker growth mechanism was studied by many researchers. Figure 7 shows the well-known mechanism of tin whisker growth and whisker testing conditions. Compressive stress causes tin whisker growth. Oxidation and corrosion of termination, mismatch of CTE (coeffient of thermal expansion) and formation of IMC causes compressive stress. However, tin whiskers in this report can not be explained by existing mechanism.

We examined two possibility about tin whisker growth mechanism. One is diffusion of tin atom through flux redsidual. Another is convert from tensile stress to compressive stress of tin plating layer at the edge of termination.

The mechanism about flux residuals includes dissolution of tin oxide to flux resuiduals, difussion of tin atom in flux residuals and separation of tin from flux residuals to whisker. Due to check this mechanism, MLCC removed flux residuals after reflow was held in 30°C60%RH. The result shown

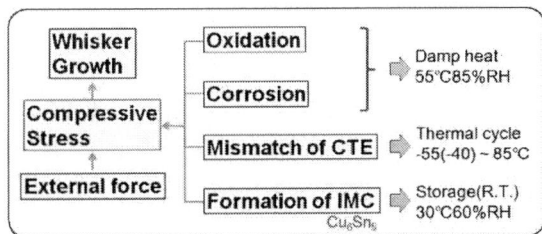

Figure 7: Tin whisker growth mechanism and whisker testing conditions

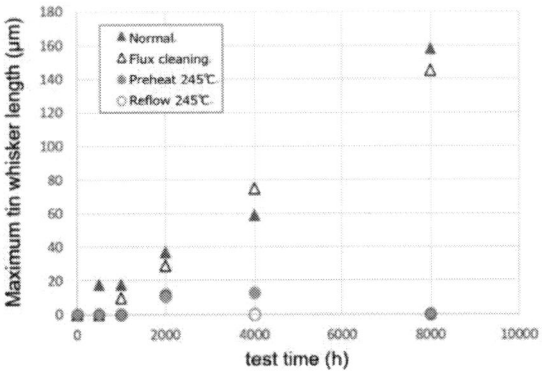

Figure 8: Influence of flux cleaning and reflow temperature to tin whisker growth rate

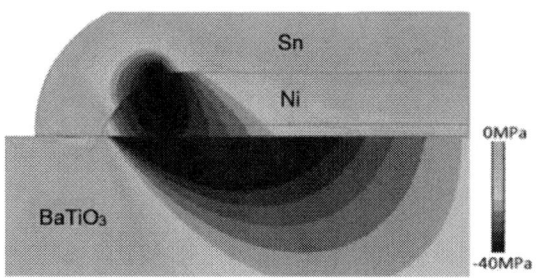

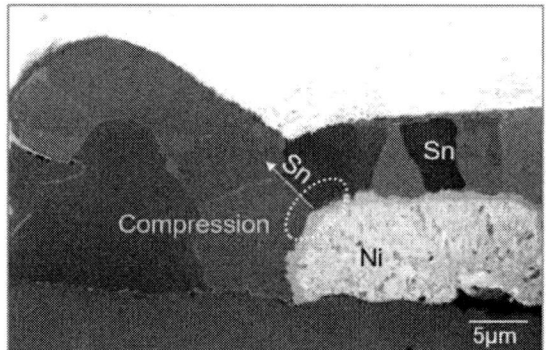

Figure 9: Stress simulation and SIM image of cross-section at the edge of termination

figure 8 indicated that flux residuals has no relation to tin whisker growth.

To investigate the convert of the stress, 2D stress simulation and SIM image of cross-section at the edge of termination is shown in figure 9. In this figure, compressive steress (the negative side of tensile stress) was displayed as a black area, when cooling to 25°C from 217°C. The cooling causes tensile stress of tin plating, because tin has the largest CTE in materials of MLCC. The tensile stress pulled tin plating to the right side in the figure. As a result, Tin plating layer is pushed to the corner of Ni plating. Tin atom migrated from compressive area (in dashed line area) to tin whisker.

Against a purpose of low-temperature assemblies, no tin whisker was observed on MLCC mounted with Sn-3.5Ag-8In-0.5Bi solder in 245°C reflow process in figure 8. No tin whisker was observed on 245°C preheated MLCC mounted with Sn-3.5Ag-8In-0.5Bi solder in 220°C reflow process. These results indicate that the melting of tin plating prevents the growth of tin whisker.

Conclusion

Tin whisker grew up on the tin plating of MLCC mounted at 220°C reflow using Sn-3.5Ag-8In-0.5Bi solder. Main results are as follows;
Tin whiskers linearly grew up with time in 30°C60%RH.
1. Tin whiskers at edge of termination grew up faster than others.
2. Whisker growth becomes fast so that the curve of tin plating is sharply.
3. Compressive stress of tin plating occurs at the edge of termination in cooling after reflow.

4. Tin whisker grew up on the surface of tin plating at the edge of termination in 30°C60%RH.
5. No tin whisker grew up on the tin plating of MLCC heated 245°C.

International standard for whisker testing condition has not yet supported this phenomenon.

Acknowledgements

This work has been supported in part by whisker study working group in JEITA. We also thank the member of whisker study working group.

References

[1] EUR-Lex – 32002L0095.
[2] Ma, H., Suhling, J. C., "A Review of Mechanical Properties of Lead-Free Solders for Electronic Packaging", Journal of Materials Science, Vol. 44, pp.1141-1158, 2009.
[3] Morgana Ribas et al.,"Comprehnesive Report on Low Temperature solder alloys for portable electronics", Proceedings of SMTA International, Rosemont, IL, Sep. 27-Oct. 1, 2015.
[4] IEC 60068-2-82.
[5] JEDEC JESD201C.
[6] A. Saito, et al., "Influence of Low Temperature Soldering on Tin Whisker Formation", Mate2018, pp.251-256, 2018.

Characterization of Impact of Vertical Stress on FinFETs

T. Furuhashi[1,*], M. Haneda[1], T. Sasaki[1], Y. Kagawa[1], Y. Ooka[1], T. Hirano[1], M. Saito[1], K, Ohno[1], H. Iwamoto[1],

Y. Liu[2,3], G. Hiblot[2], K. Vanstreels[2], M. Gonzalez[2], D. Velenis[2], G. Beyer[2], G. Van der Plas[2], I. De Wolf[2,3,*], and E. Beyne[2]

[1]Sony Semiconductor Solutions Corporation, 4-14-1, Asahi-cho, Atsugi-shi, Kanagawa, 243-0014, Japan

[2]imec, Kapeldreef 75, Leuven, Belgium

[3]Dept. Materials Engineering, KU Leuven, Belgium

+32-478-91-08-06, and Takahisa.Furuhashi@sony.com,

+32-16-281463, and Ingrid.DeWolf@imec.be

Abstract

We investigated the vertical stress impact on FinFETs using in-situ electrical measurements in a nano-indenter setup. We found that the impact of vertical stress on I_d for N and P-type FinFETs increases for longer gate lengths. According to mechanical simulations, if vertical stress is applied to the sample surface, the FinFET feels not only vertical compressive stress but also non-negligible horizontal compressive stress. Furthermore, we confirmed that the in-plane compressive stresses have different values in directions perpendicular and parallel to the Fin using S-Device simulation. The experimental results of I_d variations can be explained by a change of electron (hole) effective mass and electron scattering considering the vertical and horizontal piezo resistivity coefficients.

Key words: Vertical mechanical stress, Nano-indenter, in-situ probing, FinFET

Introduction

The necessity of multi-functionality in Large Scale Integration (LSI) systems has been rising due to the increase in demand for IoT (Internet of Things). One of the solutions is 3D packaging, such as 3D stacking of heterogeneous chips and FO-WLP (Fan Out-Wafer Level Package) [1]. The more the number of stacked chips is increasing, the more one needs to consider CPI (Chip Package Interaction) in details when designing the circuit.

It is expected that in such 3D Systems in Package (3D SiP), the impact of stress on devices also becomes more significant. So far, the in-plane stress induced by, e.g. wafer warpage or the differences of CTE (Coefficient of Thermal Expansion) have been extensively studied [2-4]. However, even though 3D SiP can possibly introduce vertical (out-of-plane) stress, there are less reports of vertical stress impact on devices and it is necessary to investigate this impact [5-7]. Therefore, we are investigating the stress sensitivity, especially vertical stress sensitivity of different types of MOSFETs using in-situ measurements in a nano-indenter setup [8].

In this work, we discuss the change of FinFET electrical characteristics induced by vertical stress. It is confirmed that the variations of I_d for N and P-type FinFETs depend on the gate length, and that their sensitivity to stress is increasing for longer gate length. These results are attributed to the change of electron (hole) effective mass and carrier scattering rates considering the vertical and horizontal piezo resistivity coefficients.

Experimental

For this study, bulk FinFETs were used with gate lengths (L_g) varying from 26nm to 1μm for N-type and from 30 to 100nm for P-type, respectively. The vertical force was applied to the sample surface by the flat-end tip of a nano-indenter. A picture of the in-situ nanoindentation probing set-up is shown in Fig. 1. It combines the nano-indenter (Bruker TI950 Triboindenter), to apply the vertical stress, with remotely controlled movable probes (miBots[TM], Imina Technologies SA, red cubes) to measure in-situ the FinFETs response to this stress. As shown in Fig. 2, the size of the tip needs to be larger than the size of the devices to ensure a uniform vertical stress distribution in the indentation region [8]. This set-up enables us to measure reliably the electrical characteristics of devices while applying vertical stress simultaneously. The force applied to the devices varies from 0 to 200mN. Before and after applying stress, the electrical characteristics were also measured to verify that the device was not damaged by the applied forces. The values of the effective stresses in the vertical and horizontal directions at 200mN were estimated using mechanical simulations (Finite Element Modeling: FEM) and the I_d of each gate length with or without stress was calculated using S-Device simulation (TCAD).

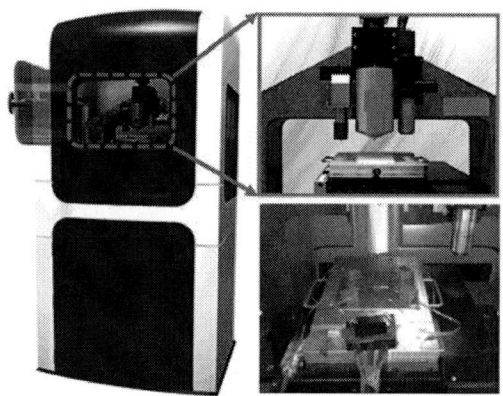

Figure 1: Images of nano indentation set-up with miBots for electrical probing.

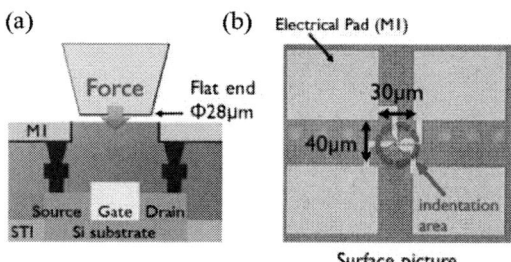

Figure 2: Schematic of the studied samples. (a) Cross section of FinFET and nano indentation-tip, (b) Sample surface (top view). The red circle indicates the indentation area.

Results and discussions

Figure 3 shows the I_d-V_g curve of an N-type FinFET with L_g = 100nm with and without 200mN applied on the surface. The I_d-V_g curves show normal behavior if the force is applied, which means the device was not damaged by the indentation. However, there are differences at higher gate voltage. The value of I_d at 200mN is larger compared to the unstressed curve (0mN).

The I_d changes with the applied force of an N and P-type FinFET whose gate length is 100nm are shown in Fig. 4. From this result, it can be seen that I_d clearly changes when the force increases. The N-type I_d increases with increasing force. On the contrary, the P-type I_d decreases with increasing force. Furthermore, we confirmed that I_d was restored back to the original value without stress

after indentation. (triangle symbols in Fig. 4). This demonstrates that the FinFET is affected by the force, but not damaged by indentation during the experiment.

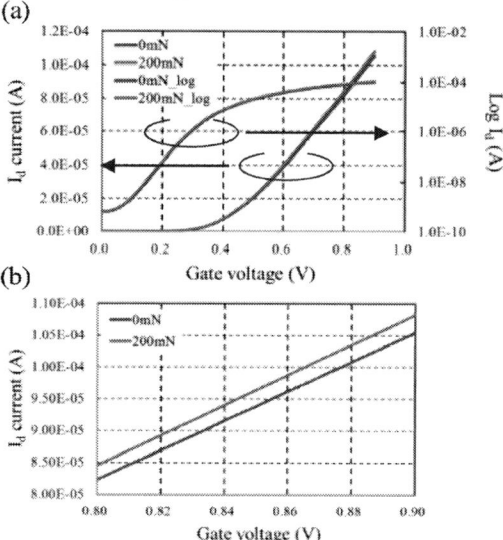

Figure 3: I_d-V_g curve with and without vertical force. (a) V_g: 0 to 0.9V, (b) V_g: 0.8 to 0.9V.

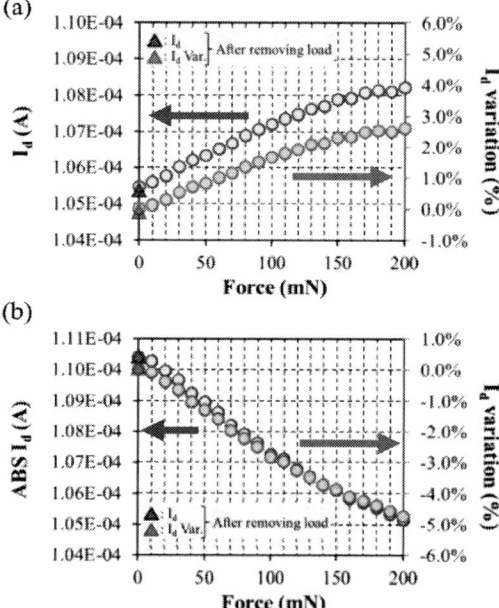

Figure 4: I_d changes under the applied vertical force. (a) N-type of L_g = 100nm, (b) P-type of L_g = 100nm.

We next applied this method for FinFETs of different gate lengths and obtained the gate length dependence of the stress-induced I_d variation. Figure 5 shows the I_d variations of N and P-type at different applied forces versus L_g. These results indicate that the sensitivity of I_d of N and P-type to the force is increasing for longer L_g. Moreover, we

560

found that at the same L_g, I_d increases with increasing force applied by the nano-indenter, and that P-type FinFETs are more sensitive to force than the N-type ones.

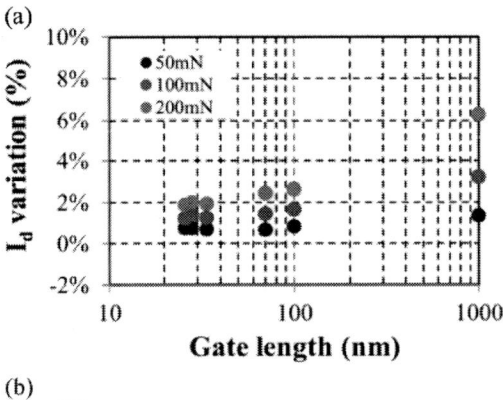

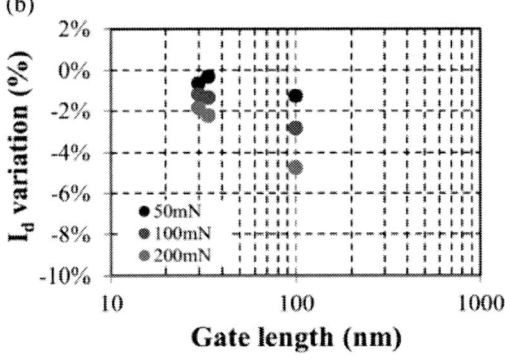

Figure 5: Gate length dependence of I_d variation. (a) N-type, (b) P-type.

Following the experiments, we validate quantitatively the experimental results using mechanical and S-Device simulations in order to consider the cause of I_d change under the force.

Figure 6 shows schematically the simulation result of the stress induced by a vertical force (from the nano-indenter). It shows that this force not only introduces vertical stress, but also in plane compressive stress. The device region under the nano-indenter tip expands laterally due to Poisson ratio when the vertical stress is applied. By reaction, the surrounding material resists the in-plane push of the device region, hence inducing in-plane compressive stress on the device. From this result, it is confirmed that when a vertical nano-indenter push is applied to the sample surface, the FinFET is submitted not only to a vertical compressive stress but also to in-plane compressive stresses. Based on the mechanical simulation and the piezo resistivity coefficient (shown in Table 1) [9], we investigated the effect of these additional horizontal stress components on FinFETs when 200mN is applied on the sample surface, using Synopsis S-Device simulation.

Figure 7 shows the S-Device simulation results at $L_g = 70$nm. In both N and P type FinFETs,

considering that the in-plane stresses in x and y directions have the same value (blue dots), the resulting I_d variation is not in agreement with the experimental values. Next, we assume that the in-plane stresses are not the same, i.e., we change the values in x and y directions and reduce the contribution of stress in the z direction (red dots). It is then observed that the I_d variations of N and P-type are more consistent with the experimental values. Furthermore, the larger the stress in y direction becomes (green dots), the smaller the difference of I_d variations between the simulation and the experimental values are. We also evaluated the I_d variations at other gate lengths using the stress field x/y/z = 50/130/100MPa. The simulation results were in very good agreement with the experiment values in both N and P-type. This result suggests that the imposed stress perpendicular to the Fin is larger than the one parallel to the Fin due to the RMG gate. That is because the RMG gate (y-direction) is much stiffer than the Fin (x-direction). The remaining mismatch of P-type devices between S-Device and Experiment can be attributed to unknown or uncalibrated device parameters such as boron active concentration in SiGe of the S-Device simulations.

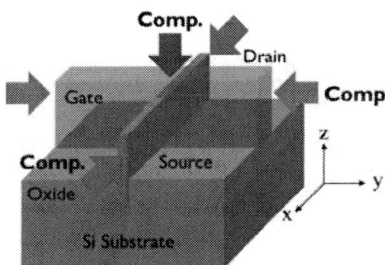

Figure 6: Mechanical simulation result of effective stress for FinFET.

Table 1: Piezoresistive coefficient of sample device based on Matsuda's value [9].

%/100MPa	N-type	P-type
X direction	-2.6	+5.4
Y direction	-1.2	-5.9
Z direction	+3.9	+0.1

We can explain compressive stress induced I_d variations of N and P-type by the theory of electron (hole) mobility change. As shown in Fig. 8, the electron mobility increases under vertical compressive stress due to density transfer towards the Δ2 valley, while the hole mobility decreases due to the increase of the effective mass of the heavy hole valley, hence reducing the current.

Finally, it was also observed in this work that the stress sensitivity of the N and P-type FinFETs increases with gate length. This is explained by different mechanisms, such as the contribution of the series resistance in short devices which are less sensitive to stress due to their high doping. In addition, the carrier velocity tends to saturate in shorter devices, and the saturation velocity itself is less sensitive to stress than the mobility. Finally, at very short gate lengths, the carrier transport mechanism changes from diffusion to ballistic, whose sensitivity to stress is generally lower [10].

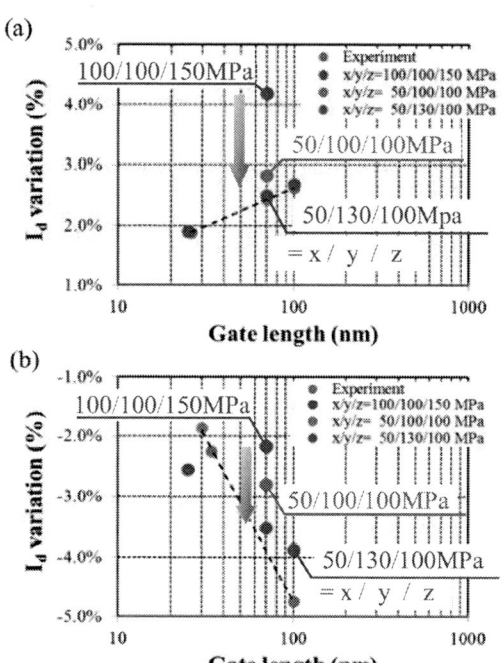

Figure 7: S-Device simulation results. (a) N-type, (b) P-type.

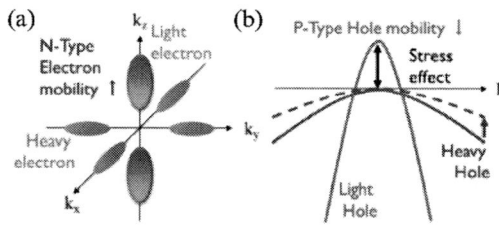

Figure 8: Subband models of FinFETs. (a) N-type, (b) P-type.

Conclusions

We investigated the vertical stress impact on FinFET transistors using in-situ measurements with electrical probes (miBots) inside a nano-indenter setup. We found that the variations of I_d for N and P-type transistors depend on the gate length and the sensitivity of I_d of N and P-type to stress is increasing for longer gate length. Using mechanical and S-Device simulations, we confirmed that in-

plane compressive stresses are also imposed on FinFET when the vertical stress is applied, and the in-plane compressive stresses have different values in directions perpendicular and parallel to the Fin. The results of I_d variations can be explained by a change of electron (hole) mobility and electron scattering.

Acknowledgments

The authors would like to thank Dr. N. Horiguchi and Dr. Y. Kikuchi from Logic Technologies for providing the samples.

References

[1] E. Beyne, "The rise of the 3rd dimension for system integration", 2006 IITC, Burlingame, California, June 5-7, pp.1–5, 2006.

[2] J. C. Suhling, et al., "Silicon Piezoresistive Stress Sensors and Their Application in Electronic Packaging", IEEE sensors Journal, Vol. 1, pp. 14-30, June 2001.

[3] J. H. Lee, et al., "Reliability Aging and Modeling of Chip-Package Interaction on Logic Technologies Featuring High-k Metal Gate Planar and FinFET Transistors", 2015 IEEE IIRW, Stanford, California, October 11-15, pp. 63-67, 2015.

[4] J. P. Prayer, et al., "Study of the piezoresistive properties of NMOS and PMOS Ω-Gate SOI Nanowire transistors: scalability effects and high stress level", 2014 IEEE IEDM, San Francisco, California, December 15-17, pp. 20.5.1-20.5.4, 2014.

[5] V. Cherman, et al., "3D stacking induced mechanical stress effects", 2014 IEEE 64th ECTC, Lake Buena Vista, Florida, May 27-30, pp. 309-315, 2014.

[6] V. Cherman, et al., "Effect of packaging on mechanical stress in 3D-ICs", 2015 IEEE 65th ECTC, San Diego, California, May 26-29, pp. 354-361, 2015.

[7] Y. Liu, et al., "Study of Out-of-plane Mechanical Stress Impact on Si BJT and Diffusion Resistor Using In-situ Nanoindentation Probing", 2019 30th ESREF, Toulouse (France), September 23-26, to be published, 2019.

[8] Y. Liu et al., "In-situ investigation of the impact of externally applied vertical stress on III-V bipolar transistor". 2018 IEDM, San Francisco, California, December 1-5, pp. 408-411, 2018.

[9] Matsuda et. al., "Nonlinear piezoresistance effects in silicon", Journal of Applied Physics, Vol. 73, pp.1838-1847, 1993.

[10] K. Uchida. et. al., "Physical mechanisms of electron mobility enhancement in uniaxial stressed MOSFETs and impact of uniaxial stress engineering in ballistic regime", 2005 IEDM Technical Digest., Washington, DC, pp. 129-132, 2005.

Characterization of Tin Pest Phenomenon in a Low Ag Content SAC Solder Alloy

Balázs Illés[1], Agata Skwarek[2], Tamás Hurtony[1], Olivér Krammer[1], Gábor Harsányi[1],

and Krzysztof Witek[2]

[1] Budapest University of Technology and Economics, Department of Electronics Technology, Budapest, Egry József u. 18, 1111 Hungary

[2] Łukasiewicz Research Network - Institute of Electron Technology, Kraków Division, Kraków, ul. Zabłocie 39, 30–701 Poland

+36-1-4632755, billes@ett.bme.hu

Abstract

In the electronics technology, the metallic β-Sn (white tin) is the basic material of solder alloys and surface finishes. The "tin pest" phenomenon is the spontaneous allotropic transition of β-Sn to the semiconductor α-Sn (gray tin) under the temperature of 13.2°C. In this work, the tin pest susceptibility of the widely used Sn99Ag0.3Cu0.7 solder alloy was investigated and compared to Sn99Cu1 alloy (as a well know reference). Bulk solder bars were prepared by metal casting and the samples were inoculated by InSb, CdTe and α-Sn powders to enhance the allotropic transition process. The inoculator materials were pressed onto the surfaces of the samples by a mechanic lamination. The samples were stored at -10 °C temperature for 8 weeks. The allotropic transition was monitored by optical inspection and by electrical resistance measurements. The microstructural changes of the samples – caused by the transition of crystal structure of Sn – were investigated by metallurgical cross-sections. The results showed, that in all cases the Sn99Ag0.3Cu0.7 solder alloy is much less susceptible to tin pest development than the Sn991Cu, which might be explained by the suppression effect of the Ag alloying. Furthermore, it was found that the process of transition highly depends on the applied inoculator material.

Key words: soldering; tin pest; SAC alloy; inoculator; allotrope transition;

1. Introduction

In the electronics technology, the metallic β-Sn (white tin) is the basic material of solder alloys and surface finishes [1]. β-Sn has BCT crystal structure, and it is stable temperature above 13.2°C (with the melting point at 231.9 °C). Sn has further three more allotropes the α-, γ- and σ-Sn. From the aspect of microelectronics applications, the γ- and σ-Sn are less known, since they can from only at harsh conditions, temperature above 161°C and pressures above several GPa in high pressure [2]. α-Sn (gray Sn) has diamond cubic crystal structure and it is stable temperature below 13.2°C. α-Sn is a nonmetallic material, since its atoms are in covalent structure, in which electrons are not freely movable. It is a semiconductor and powdery material.

So called "tin pest" phenomenon marks the allotropic transition of β to α-Sn below 13.2°C [3]. The first sign of the transition is the appearance of discolored spots / warts on the surface of the objects, (Fig. 1). Further result of the transition is the decomposition of the objects into powder, since the transition causes ~27% volume increase [4]. The allotropic transition can have three main phases: "the nucleation", when new phases starts to develop via self-organization or new thermodynamic phases form; "the growth", when the transition proceeds and "the saturation", when the transition stops, and some α-Sn remains in metastable state [5]. The speed of the transition highly depends on the temperature and the presence of materials with the same crystal structure with similar lattice parameters to the α-Sn, like CdTe or InSb [6].

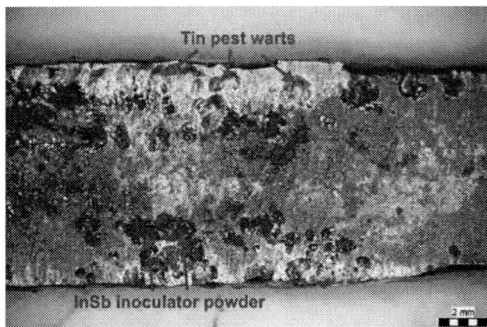

Figure 1: Tin pest warts on the surface of Sn99Ag0.3Cu0.7 alloy inoculated with InSb after 10 weeks of storage at -10°C.

Tin pest can occur not only in pure Sn objects, but in high Sn content alloys as well [3, 6]. However, alloying of electropositive metals (like Pb, Bi, Sb), which are soluble in the solid Sn, usually suppress the transition [7]. After the application of RoHS directives since 2006, mostly the lead-free SnAgCu (SAC) solder alloys are applied in microelectronics [8]. In these alloys, the Sn content is over than 95 wt%, so the risk of the tin pest phenomenon might be increased considerably in microelectronic appliances [3, 6].

In addition, a further trend in the soldering technology is to reduce more and more expensive Ag from SAC solder alloys [9]. Among the so called "low Ag content SAC alloys", the Sn99Ag0.3Cu0.7 (SAC0307) alloy is the most widespread in the electronics industry. The solder joints made from SAC0307 have the similar quality and reliability properties (electrical conductivity, mechanical stability, long-term durability, corrosion susceptibility, etc.) as the solder joints made from the widely applied SAC305 or SAC405 alloys [10]. The price of SAC0307 is ~20-30% lower than the prices of SAC305, however, the even 99 wt% Sn content could increase further the possibility of the allotropic transition of tin.

Consequently, the study of tin pest susceptibility of low Ag content SAC solder alloys is important for electronic devices working in sub-zero temperature in aerospace, automobile and aeronautical applications.

2. Materials and Methods

In this work, the tin pest susceptibility of the SAC0307 solder alloy was investigated and compared to Sn99Cu1 alloy. The tin pest susceptibility of the Sn99Cu1 alloy is well researched [5, 6]. Bulk solder bars were casted with the size of 45x6x3 mm (Fig. 2). For shortening the nucleation phase different inoculator powders: InSb, CdTe and α-Sn itself were used.

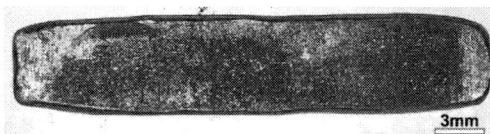

Figure 2: The inoculated sample.

Inoculating powders were pressed onto the surfaces of the samples by a mechanic laminator, applied 30kN force. Fifteen - fifteen samples were prepared from the six different sample types. The samples were stored at -10 °C temperature for 8 weeks.

Basically, the tin pest development was monitored by optical inspections. After the first tin pest sign appeared, the progress of the transition was characterized by electrical resistance measurements

as well, since the transition (from metal to semiconductor) increases the electrical resistance of the samples. The measurements were carried out by 4-wire method with an AGILENT 4338B mΩ meter. At mΩ range the measurement accuracy of the instrument is under 3% and the repeatability error was under 2%. The initial average resistance value of the samples was between 0.23 – 0.25mΩ.

Metallurgical cross-sections were also prepared and analyzed by an Olympus BX51 metal microscope in order study the microstructural changes of the samples caused by the changing of the crystal structure of Sn.

3. Results and Discussion

Fig. 3 shows the average electrical resistance changes of Sn99Cu1 samples inoculated with InSb, CdTe and α-Sn in logarithmic scale. During the first 3 weeks of the study, no detectable resistance change was observed. However, minor tin pest warts (see Fig. 1) already appeared on the surface of the samples after 2-3 weeks storage at -10 °C. At 4 weeks, the samples inoculated with InSb and α-Sn showed resistance increases, between 30–150% for InSb and 5–20% for α-Sn inoculation. In the case of CdTe inoculation, 5-20% resistance increase was detected only after 4 weeks of storage. Hence, the nucleation phase of the transition has been found to be 3 and 5 weeks, at InSb, α-Sn and CdTe inoculation, respectively.

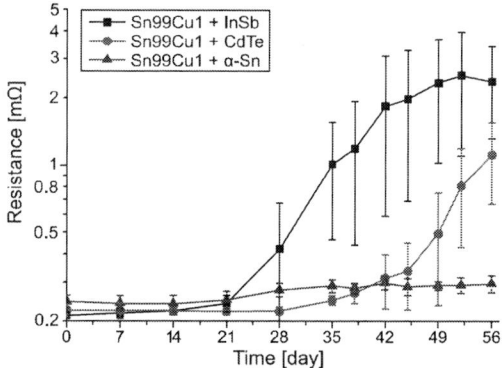

Figure 3: Electrical resistance changes of the Sn99Cu1 samples during 8 weeks of storage at -10°C.

During the rest of the study, the samples with different inoculation showed very different behavior. The samples with InSb inoculation showed a considerable resistance increase between 3–6 weeks (in the growth phase). After 6 weeks, the changes of the resistance increase reached the saturation phase. After 8 weeks, the samples started to decompose (Fig. 4). The decomposition of the samples usually occurred when the electrical resistance had increased of 25-30 times as compared to the initial value. The highest detected resistance increase of the samples with InSb inoculation was ~33 times (8.3mΩ) after 8 weeks of storage at -10 °C.

In the case of Sn99Cu1 samples with CdTe inoculation, considerable increase of the electrical resistance was noticeable between the 5 – 8 weeks of the study. In this case no saturation phase was observed (Fig. 3). Di Maio and Hunt found similar results in the case of different SnCu alloys inoculated with CdTe [11]. Here, the highest resistance increase was ~32 times (8.13mΩ) after 8 weeks of storage at -10 °C. After 8 weeks the decomposition of the samples begun.

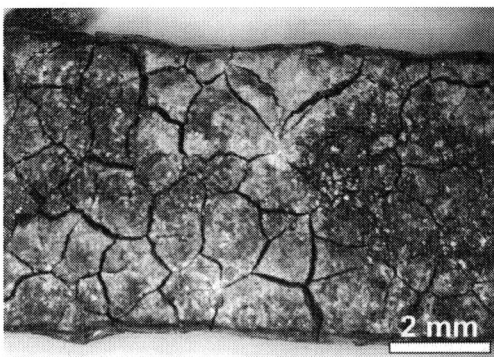

Figure 4: Decomposed Sn99Cu1 sample with InSb inoculation after 8 weeks of storage at -10°C.

Generally, the high deviation of the electrical resistances of the samples with InSb and CdTe inoculation is the result of the not totally even inoculation, and the autocatalytic nature of the transition. The differences in the appearance of α-Sn on the samples result in differences in the transition rates.

In the case of α-Sn inoculation, after the moderate (5–20%) resistance increase at 4 weeks, no further changes were detected. The highest resistance increase was ~0.4 times (0.351mΩ) after 8 weeks of storage at -10 °C. Here, the decomposition of the samples was not observed at all, the transition could not develop in the whole sample volume and it was limited only on the surface of the samples.

Fig. 5 shows the average electrical resistance changes of SAC0307 samples inoculated with InSb, CdTe and α-Sn in logarithmic scale. The samples inoculated with InSb and CdTe were totally resistant to the transition, no any visible tin pest signs or electrical resistance changes during the study were observed. In the case of SAC0307 samples inoculated with α-Sn, similar effect was observed as at the Sn99Cu1 samples inoculated with α-Sn. The first tin pest warts appeared after 2 weeks of storage at -10°C (Fig. 6) and after 3 weeks 15-25% resistance increase was detected, without any considerable changes later (Fig. 5).

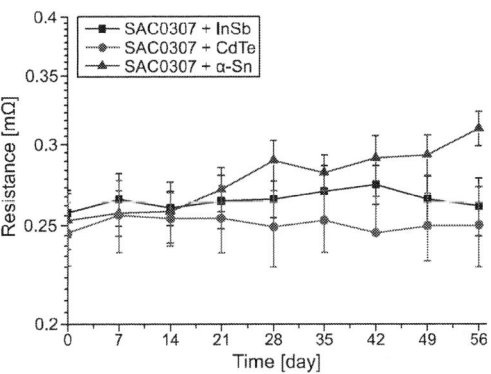

Figure 5: Electrical resistance changes of the SAC0307 samples during 8 weeks of storage at -10°C.

Figure 6: Tin pest warts on the SAC0307 samples inoculated α-Sn after 3 weeks of storage at -10°C.

According to the obtained results, two main findings can be concluded: i) the α-Sn inoculation resulted in the allotropic transition at both alloy, however the resistance changes were moderate; ii) the InSb and CdTe inoculation resulted in considerable and fast changes but only in the case of the Sn99Cu1 alloy. In order to explain the followings, metallurgical cross-sections were prepared from the samples.

Fig. 7 shows the cross-sections of Sn99Cu1 samples inoculated with α-Sn (Fig. 7a) and InSb (Fig. 7b) after storage at -10°C for 8 weeks. In the case of α-Sn inoculation, the transition does not develop into the whole sample volume, it was limited to the surface of the samples, causing the peeling of the outer layer of material after a while. In the case of InSb or CdTe inoculation, the inoculators can diffuse into the sample body, so the allotropic transition starts at several places inside the sample body and not only on the surface (Fig. 7b), finally leading to the decomposition of the sample. The brittle α-Sn is not visible in Fig. 7b, due to falling out from the sample during the polishing.

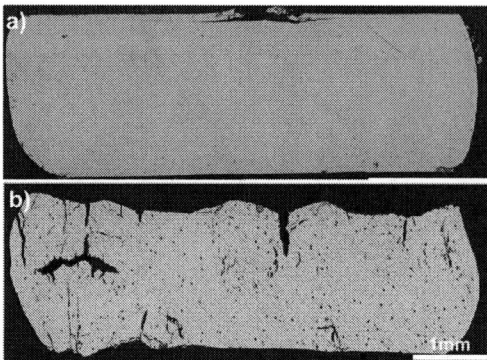

Figure 7: Cross-sections of Sn99Cu1 samples after 8 weeks of storage at -10°C: a) α-Sn inoculation; b) InSb inoculation.

Different tin pest susceptibility of the similar solder alloys (Sn99Cu1 and Sn99Ag0.3Cu0.7) might be explained by the Ag content of the SAC0307 alloy. Addition of Ag to the solder alloy, enhances the creation of Ag_3Sn intermetallic compounds (IMCs) in the joint after its solidification. Although in the low Ag content solder alloys, the amount of Ag_3Sn IMCs is relatively lower compared to the classical SAC305 or SAC405, but their dispersion around the grain boundaries of the Sn grains is usually even [12]. The presence of the Ag_3Sn at the Sn grain boundaries can decrease the grain-boundary diffusion which results in larger grains and lower number of the grain boundaries. This could decrease the allotropic transition rate. Similar effect of the Sn grain structure on Sn whisker growth was already found [13]. However, further research is necessary to prove this theory.

4. Conclusions

In this paper, the tin pest susceptibility of the Sn99Ag0.3Cu0.7 (SAC0307) solder alloy was investigated and compared to Sn99Cu1 alloy with different inoculator materials application. The main findings are the following: the SAC0307 alloy is much less prone for the Sn transition than the Sn99Cu1. This might be explained with the presence of Ag in the solder alloy, since Ag_3Sn IMCs can decrease the grain-boundary diffusion. Therefore, from reliability aspects, silver cannot be totally eliminated from solder alloys. Furthermore, process of the allotropic transition highly depends on the inoculator material. The non-Sn inoculators can diffuse into the body of the samples and start the transition inside the samples as well, which can lead to the fast decomposition of the samples. In the case of α-Sn itself inoculation, the process is limited to the surface of the samples which results in much slower degradation of the samples.

5. Acknowledgement

This research was partially supported by the National Research, Development and Innovation Office of Hungary – NKFIH, project ID: FK 127970 and by the Higher Education Excellence Program of the Ministry of Human Capacities in the frame of Nanotechnology and Materials Science research area of Budapest University of Technology and Economics (BME FIKP-NAT).

References

[1] A. Pietriková, M. Kravčík, "Boundary Value of Rheological Properties of Solder Paste", Proceedings of the 34th Int. Spring Seminar on Electronics Technology (ISSE), Tatra Lomnic, Slovakia, pp. 94-97, (2011).

[2] A. M. Molodets, S. S. Nabatov, "Thermodynamic Potentials, Diagram of State, and Phase Transitions of Tin on Shock Compression", High Temperature, Vol. 38, No. 5, pp. 715–721, (2000).

[3] W. Plumbridge, "Recent Observations on Tin Pest Formation in Solder Alloys", Journal of Electronics Materials, Vol 37, No. 2 pp. 218–223 (2008).

[4] S. Gialanella, F. Deflorian, F. Girardi, I. Lonardelli, S. Rossi, "Kinetics and microstructural aspects of the allotropic transition in tin", Journal of Alloys Compounds, Vol. 474 pp. 134–139, 2009.

[5] A. Skwarek, B. Illés, B. Horváth, A. Géczy, P Zachariasz, D. Bušek, "Identification and caracterization of ß→α-Sn transition in SnCu1 bulk alloy inoculated with InSb", Journal of Materials Science: Materials in Electronics, Vol. 28 pp. 16329–16335, (2017)

[6] A. Skwarek, P. Zachariasz, J. Kulawik, K. Witek, "Inoculator dependent induced growth of α-Sn", Materials Chemistry and Physics, Vol. 166, pp. 16–19, (2015).

[7] D. Giuranno, S. Delsante, G. Borzone, R. Novakovic, "Effects of Sb addition on the properties of Sn-Ag-Cu/(Cu, Ni) solder systems", Journal of Alloys Compounds, Vol. 689, pp. 918-930, (2016).

[8] A. Pietriková, M. Kravčík, Investigation of Rheology Behavior of Solder Paste, Proceedings of the 35th Int. Spring Seminar on Electronics Technology (ISSE), Bad Aussee, Austria, pp. 138-143, (2012).

[9] B. Medgyes, P. Tamási, F. Hajdu, R. Murányi, M. Lakatos-Varsányi, L. Gál, "Corrosion investigations on lead-free micro-alloyed solder alloys used in electronics", Proceedings of the 38th Int. Spring Seminar on Electronics Technology (ISSE), Egerszalók, Hungary, pp. 296-299, (2015).

[10] X Zhong, L Chen, B Medgyes, Z Zhang, S Gao, L Jakab, "Electrochemical migration of Sn and

Sn solder alloys: a review", RSC Advances, Vol. 7, No. 45, pp. 28186-28206, (2017).

[11] D. Di Maio, C.P. Hunt, "Monitoring the Growth of the α Phase in Tin Alloys by Electrical Resistance Measurements", Journal of Electronic Materials, Vol. 38, No. 9, pp. 1874-1880, (2009).

[12] O. Krammer, T. Garami, B, Horváth, T. Hurtony, B. Medgyes, L. Jakab, "Investigating the thermomechanical properties and intermetallic layer formation of Bi micro-alloyed low-Ag content solders", Journal of Alloys and Compounds, Vol. 634, pp. 156-162, (2015).

[13] B. Horváth, B. Illés, T. Shinohara, G. Harsányi, "Whisker Growth on Annealed and Recrystallized Tin Platings", Thin Solid Films, Vol. 520, pp. 5733-5740, (2012).

Paving the way for the replacement of solder interconnections in power electronics by silver-sinter using pulsed infrared thermography

Dan R. Wargulski[1], Battist Rabay[2], Adrian Stelzer[2], Jacek Rudzki[3], Armin Hindel[4],
Markus Bast[4], Dirk Busse[5], Ellen Auerswald[6], Mohamad Abo Ras[1]

[1]Berliner Nanotest und Design GmbH, Berlin, Germany
[2]Nano-Join GmbH, Berlin, Germany
[3]Danfoss Silicon Power GmbH, Flensburg, Germany
[4]Fachhochschule Kiel, Kiel, Germany
[5]Budatec GmbH, Berlin, Germany;
[6]Fraunhofer Institut für Elektronische Nanosysteme ENAS, Chemnitz, Germany

Corresponding author: wargulski@nanotest.eu

Abstract

In this work we show the capabilities of non-destructive testing by pulsed infrared thermography (PIRT) on large-area silver-sinter layers. This method can achieve a similar reliability in detecting delamination as SAM and X-ray analysis. The great advantage of PIRT is its potential to become a tool for 100% in-line inspections in the production line of power modules, whereas SAM and X-ray are preferred as offline tools for reasons of costs, time or the need of a coupling medium. PIRT could help to replace vulnerable solder layers in power modules by much better performing silver-sinter layers to enhance the reliability of those modules. By PIRT analysis on sample series with varied sinter process parameters and different sinter pastes, which will be compared to conventional failure analysis such as SAM and x-ray, this work can show the quality of this method for the developing process of new sinter applications and for the following quality assessment in production lines.

Key words: pulsed IR thermography, non-destructive testing, power modules, silver-sintering

Introduction

Since the emergence of the fields of renewable energies and e-mobility, power modules are playing an increasing role for industry. For the controlling and switching of large electrical loads these electronics must handle huge thermal loads. The state-of-the-art method to conduct away the heat from the semiconductors to the heat sinks are large-area solder interconnections. These solder layers can be prone to form flaws such as tilts and voids and, furthermore, just show medium thermal conductivities [1]. The replacement of the solder layers by silver-sinter layers would be very beneficial since they show higher thermal conductivities compared to solder [2] and can be applied with smaller bond line thicknesses, which further decreases the thermal resistance. Therefore, low pressure sinter processes and pastes are necessary, because to reach pressures used for conventional sinter processes enormous forces would be needed. In this study we can show that pulsed infrared thermography (PIRT) is a powerful tool for the evaluation and testing during the development of such pastes and sinter processes and, furthermore, for a 100% in-line inspection in later production lines.

Methods

Failure analysis by pulsed infrared thermography

Pulsed infrared thermography (PIRT) is a method for non-destructive testing. Most common applications are the inspection of macroscopic components such as parts made of composites for aerospace [3] or the inspection of concrete structures [4]. The application of PIRT for failure inspections of thermal interfaces in electronics is a relatively new field and more commonly used in laboratories than in industry [5].

The principle is that the surface of the device under test (DUT) gets thermally excited by a short pulse. An infrared (IR) camera records a sequence of images during the measurement. In this work the excitation by flash lamps has been used. After the pulsed excitation, as part of the surface cooling process, the applied heat starts to diffuse through the sample. In case of a defect such as a void or delamination in the DUT the heat path is disturbed. This disturbance can be detected by the IR camera as hot spots or areas since the heat is trapped above those defects. The setup is schematically shown in Figure 1.

The samples investigated in this work are direct copper bonded substrates (DCBs), which are bonded to a copper base plate using a silver sinter layer.

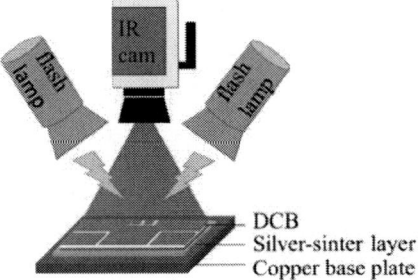

Figure 1: Schematic of the pulsed infrared thermography failure analysis

The IR camera is an Infratec IRimager 8300 combined with two 1.2 kJ flash lamps. Currently. it is still necessary to coat the DCBs by black camera varnish to ensure a suitable flashlight absorption at the one hand and a high emissivity for higher signal strength at the other hand, but for the in-line capabilities the development of a method with black foils to avoid the blackening just started and will be a topic in future works. The resulting images are difference images. This means an initial image before the flash excitation is subtracted from the resulting recordings after the flash [6].

Failure analysis by scanning acoustic microscopy and X-ray analysis

Scanning acoustic microscopy (SAM) and X-ray analysis were used as reference methods for comparison and evaluation of the quality of the results obtained by PIRT in this work. Scanning acoustic microscopy is a well-established method for failure analysis of electronic packages. It is a reliable method for the detection of voids and delamination in thermal interface materials. The principle is based on reflections of acoustic sound waves on defects. The disadvantage and a cause for the incapability to be a tool for in-line inspections in most of the production lines is the need for a coupling medium such as water.

X-Ray analysis such as computer tomography (CT) is also a non-destructive method. It requires an expensive setup and is in general more suitable to detect voids but is almost unusable to detect delamination, which is more commonly found in sinter layers than voids. The principle here is based on the transmissibility of x-rays.

More details about these two conventional techniques can be found in literature [7].

Silver-sinter layers in power modules

Power modules consist of semiconductors which are soldered or sintered on a DCB. The structure of a typical power module is shown schematically in Figure 2. The DCB is responsible for an optimal heat dissipation from the semiconductors and simultaneously shows a similar coefficient of thermal expansion as silicon which ensures a good reliability during thermal cycling. The state-of-the-art technique to thermally connect the DCB to the copper base plate, which is connected to the heat sink, is soldering. The use of sinter technology instead of soldering could provide a lot of benefits. As shown in Table 1 silver-sinter layers are able to dissipate heat much better since the thermal conductivity of silver-sinter is by a multiple higher and the bond line thicknesses is thinner. This reduces the thermal resistance enormously.

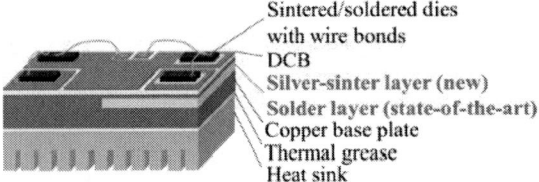

Figure 2: Schematic of a power module

Additionally, sinter layers are less prone to show flaws such as voids and tilts, which is expected to result in an increased reliability of power modules. Today's applications of sinter technology are limited to areas of up to 200 mm². The current goal is to achieve areas of up to 5000 mm².. The issue of larger sinter areas is the need of enormous forces to reach sinter process preassures, which most of the current sinter materials still need.

Table 1: List of advantages of the sinter technology over the state-of-the-art solder technology in power modules

Properties DCB on Base Plate	State-of-the-Art (Soft Solder)	Silver-Sinter Technology (goals)
Thermal Conductivity	50 W/(m·K)	>250 W/(m·K)
Bond Line Thickness	200-400 µm	>30 µm
Voids	frequent and large	no voids
Tilts	often	no tilts
Reliabilty, -40°C to +125°C temperature changes	<500 cycles	>1000 cycles
Area of todays inter layers	<200 mm²	>200 mm² to 5.000 mm²

To achieve the goal of sintering large areas new sinter materials and processes need to be developed. In best case the new sinter paste should be capable to be sintered without pressure or at least at low preassures.

Silver-sinter sample designs

For the tests of new sinter pastes, sinter processes and the PIRT inspection, samples have been designed and built in three stages as shown in Figure 3.

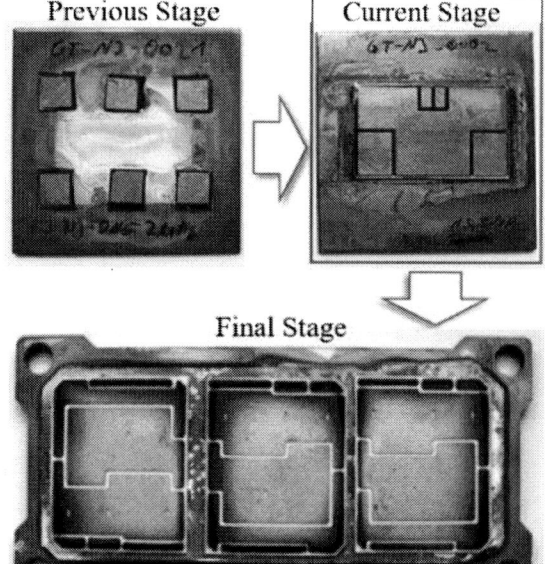

Figure 3: Photographs of three stages of sample designs for testing the sinter process and the PIRT failure inspections

The first stage consists of a 6 x 6 cm² copper base plate with six 1 x 1 cm² DCBs sintered on it. Four of six DCBs have been sintered on just half their area to simulate a delamination at a defined area.

The second and current sample stage is the sample design this work will report about in general. It consists of a 6 x 6 cm² copper base plate and a 2.5 x 4.5 cm² DCB sintered on it. Series of samples with this design have been built with two different silver-sinter pastes, a commercial paste and a modified paste with increased silver content and flake shaped silver particles, and varied sinter pressures of 0, 2, 5, 10 and 13.3 MPa.

The final stage and just an outlook in this work has the appearance of a power module with three large DCBs on a conventional power module base plate.

Results

Previous sample stage

First PIRT analyses have been performed on the first sample stage with 1 x 1 cm² DCBs which are completely, and half sintered on a copper base plate, shown in Figure 4. The DCBs were blackened for a suitable absorption and emissivity. These samples were designed to test and calibrate the system because it is visible by eyes where the heat dissipation should be good (sintered area in grey) and where the heat dissipation is disturbed (non-sintered area). The PIRT images have been overlaid over the sample image and as expected the DCBs completely connected to the copper plate by silver-sinter show a suitable and homogeneous heat dissipation indicated by the blue color (low temperatures) whereas the half sintered DCBs show a disturbed heat dissipation indicated in yellow and red (increased temperature) at approximately half their area due to heat accumulations above the non-sintered area.

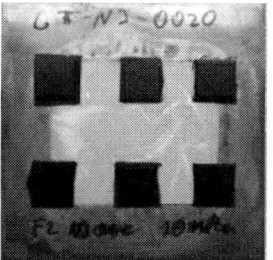 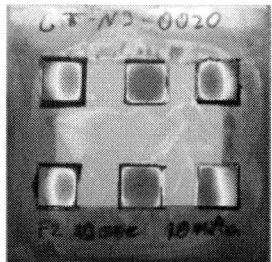

Figure 4: First sample stage with small DCBs, left: photograph of a sample with black coating and right: photograph of the same sample with overlaid PIRT images.

Current sample stage

After the confirmation that PIRT is working very well on sintered DCBs, a series of samples with the second sample design was built to test two different silver-sinter pastes at different sinter process pressures to figure out how theses two pastes are performing at certain pressures. The PIRT measurement of one sample takes approximately one second.

Since the sizes and positions of delamination or whether there are any defects at all are unknown for this type of sample compared to the previous sample stage, it is more difficult to find the optimal temperature scaling settings in the thermograms, which are indicating the true sample state. The issue of the temperature scaling in PIRT thermograms is demonstrated in Figure 5. The given temperatures in this figure are temperature differences, since an initial image before the flash was subtracted from the recorded PIRT image sequence.

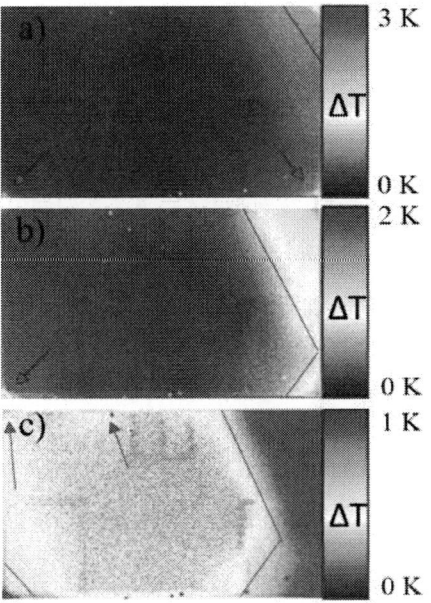

Figure 5: PIRT analyses on one sample shown with three different temperature scaling settings, a temperature scale from 0 to 3 K (a), 0 to 2 K (b) and 0 to 1 K (c). The arrows indicate small flaws and the red lines the apparant boundaries of delamination

Depending on the scaling settings defects can appear larger, smaller or can even be invisible. The reason for this effect is the heat spreading in the material above the defects which causes non-discrete contrast boundaries. Therefore, verification and calibration by conventional methods such as SAM and X-ray are often necessary. But once the optimal settings have been found other methods than PIRT are not required if the boundary conditions are kept constant.

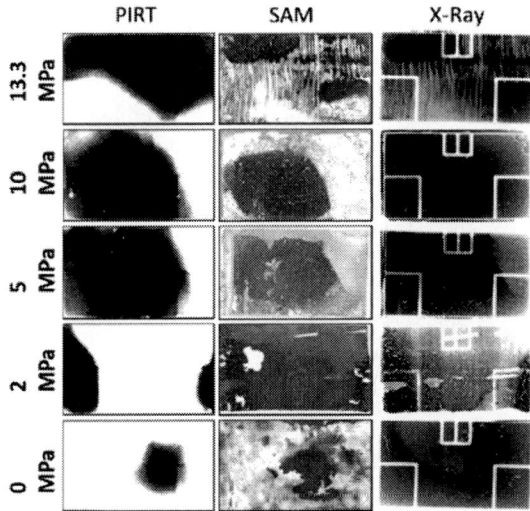

Figure 6: Result images, one exemplary sample per pressure, of the PIRT, SAM and X-ray analysis

Figure 6 shows results of the PIRT, SAM and X-ray analysis of one exemplary sample of each sinter pressure. The defect signal in the PIRT images is shown as white areas (high temperatures), black areas (low temperatures) indicate a suitable thermal interconnection by a better heat dissipation. Quite similar, but less distinct it is in the SAM and X-ray images In the SAM and X-ray images white areas correspond to voids or delamination. Four samples were sintered and analysed per pressure, these results can be seen in the attachment. All of them show a similar agreement between results obtained by PIRT, SAM and X-ray. The scaling of the PIRT images was adjusted to match the defect patterns in the SAM images. A black and white color palette has been chosen to enhance the comparability to the other measurements and for the image analysis which was used to obtain how many percent of the sintered area is delaminated. After calibration, the results of PIRT and SAM were matching very well. But there are differences which need to be mentioned. Whereas the PIRT results solely give information about the thermal behavior of the sintered layer and its interfaces, the SAM and the X-ray images are showing also other detected effects, which can be seen as a disturbing issue in this case. SAM and X-ray are also detecting structural information of the sinter material which do not necessarily influence the heat transport. However, the thermal behavior is the desired information, because it is the crucial property for the functionality of the sinter layer between DCB and heat sink. The 13.3 MPa sample in Figure 6 is a good example for such an issue. In SAM and X-ray vertical scratches are visible which could be a result of the sinter paste application, but obviously they do not show an influence on the thermal behavior and even cover the delamination-caused signal in the SAM image. Such an effect could trigger a false-positive failure alarm in in-line inspections. Another example is the 2 MPa sample, where a grainy structure is visible in the SAM and X-ray images. The PIRT image clearly shows that this grainy region has a relatively good thermal interconnection, whereas the rest seems to be delaminated. It is often hard to interpret the SAM and X-ray images compared to PIRT. This simplicity and clarity of PIRT results could enable a more reliable automated defect detection in future 100% in-line inspections.

One reason for the low quality of the SAM images is that the applied SAM equipment reached its limits by sensing through a mm-thick copper base plate. Whereas with X-ray it is difficult to detect delamination. This is a general issue by using this method, which is more suitable for void detection.

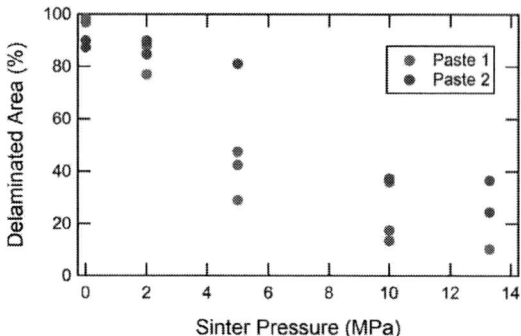

Figure 7: Results of the PIRT image analysis showing the measured percentage of delaminated area of a sample over the sinter pressure

A series of 20 samples (4 per pressure) have been measured by PIRT and subsequently analyzed by a graphic program to quantify the results. The resulting values are the percentage of the DCB area, which was measured as delaminated (Figure 7). It is not surprising that the quality of the sintered thermal interconnection decreases with decreasing sinter pressure. The pressureless sintered samples showdelamination of 87.4 - 98.4%, where paste 2, a paste with increased silver content and flake shaped particles, shows a slightly better performance than the standard paste. From 10 MPa to 13.3 MPa the delamination reached values in a range of 10.4 – 37.4%. Those values are still high, but it should be mentioned that these were the first attempts to sinter such large areas without any optimizations and the samples can be rather seen as a proof-of-concept for the failure analysis system. The main reason for the delamination here is the strong warpage and deformation of the 2 mm thick copper base plate due to the heating to up to 250°C during the sinter process and the following cooling process back to room temperature. Another reason is the relatively low sinter process pressure, which can require up to 30 MPa in conventional die to substrate sinter processes. It is expected that after process optimizations 5-10 MPa can already be enough for completely non-delaminated samples.

Conclusions

In this study it is shown that PIRT can act as an efficient and fast tool for the inspection of sintered interconnections. This was demonstrated on 20 samples with DCBs sintered at different pressures onto copper base plates, similar to the assemblies in power modules. The results obtained by PIRT show a good agreement to results obtained by SAM and X-ray techniques for the same samples. Delamination is the most common type of defect in such samples and the investigations have shown that it can be reduced and avoided by applying pressure during the sinter process. The delamination was most probably caused by warpage and deformation of the copper base plate, which could be avoided by

fixation during the sinter process or specifically preformed base plates. The comparison of the two pastes has shown a slightly better result for paste 2 (flakes, increased silver content) under pressureless sinter conditions. For more detailed results a larger sample amount would be necessary for more statistical relevance.

Once the correct setting was found and a calibration completed, the images of the pulsed infrared thermography have shown a good match to the SAM images, and are easier and clearer to interpret, which will make automated analyses in production lines more reliable. A clear advantage of PIRT is the direct measurement of the thermal behavior of the sinter interconnection without the detour via sonic-based or x-ray-based methods. The speed, non-destructivness and accuracy of PIRT can help to accelerate development processes of new sinter pastes and sinter processes for power electronics and can further be a tool for 100% in-line inspections in power module production lines.

Acknowledgements

The presented work has been performed within the project Grotherm funded under grant number 16ES0665 by the Federal Ministry of Education and Research (Germany) which is gratefully acknowledged by the authors.

References

[1] J. A. King, "Material Handbook for Hybrid Microelectronics," Artec House, Norwood, MA, 1988

[2] M. Abo Ras et al., "LaTIMA - an innovative test stand for thermal and electrical characterization of highly conductive metals, die attachs, and substrate materials," in 21st International Workshop on Thermal Investigations of ICs and Systems (THERMINIC), 2015, pp. 1–6.

[3] F. Ciampa, P. Mahmoodi, F. Pinto and M. Meo, "Recent Advances in Active Infrared Thermography for Non-Destructive Testing of Aerospace Components", Sensors (2018), 18:609

[4] F. Weritz, R. Arndt, M. Röllig, C. Maierhofer, H. Wiggenhauser, "Investigation of concrete structures with pulse phase thermography", Mat. Struct. (2005) 38: 843

[5] M. Schaulin, M. Oppermann, T. Zerna, "Thermographic inspection method for quality assessment of power semiconductors in the manufacture of power electronics modules", 7th Electronic System-Integration Technology Conference (ESTC), 2018

[6] D. May et al., „Transient thermal response as failure analytical tool – a comparison of different techniques'', Thermal, Mechanical and Multi-Physics Simulation and Experiments in Microelectronics and Microsystems (EuroSimE), 14th International Conference on, 2013, S. 1/8-8/8.

[7] R. J. Ross, C. D. Hartfield, T. M. Moore, Microelectronics Failure Analysis, ASM International, pp. 362, 2011

Supplementary Information

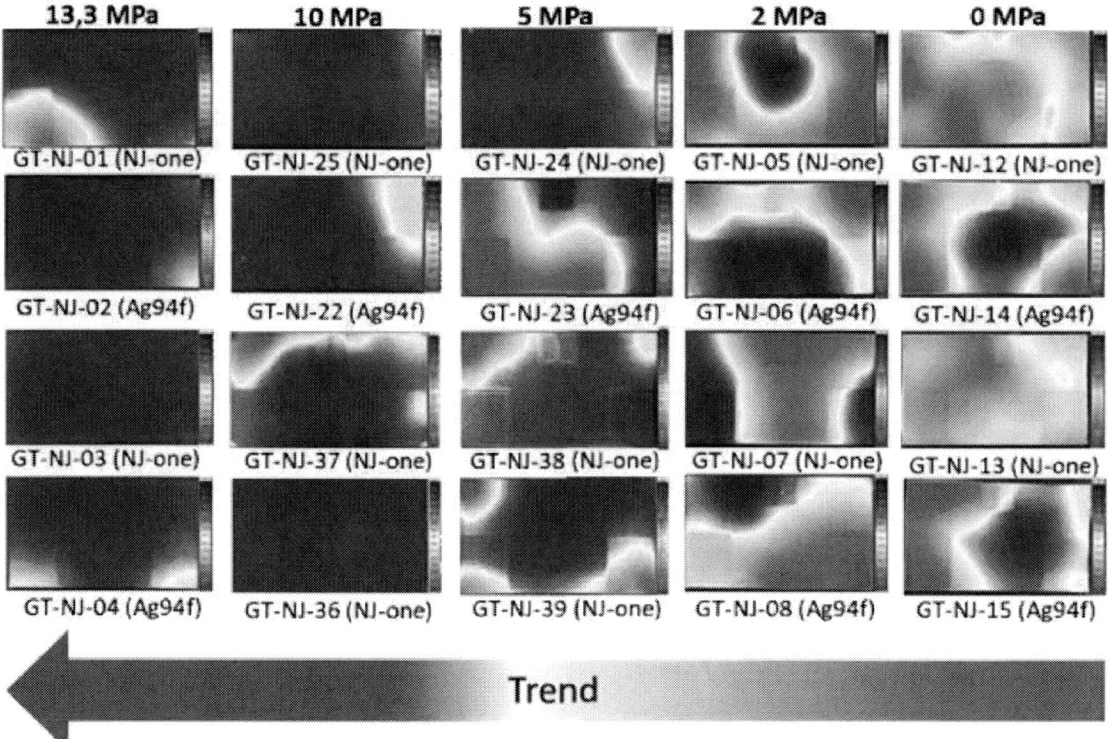

Figure 8: PIRT measurements of 20 samples without calibration by SAM

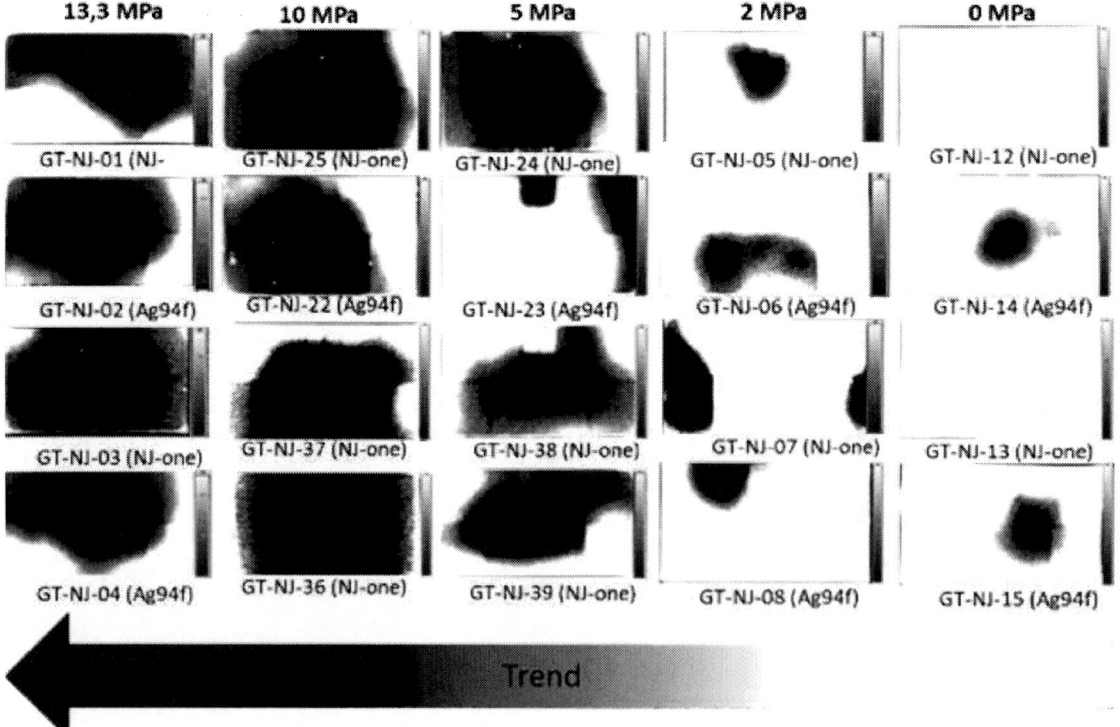

Figure 9: PIRT measurements of 20 samples with calibration by SAM

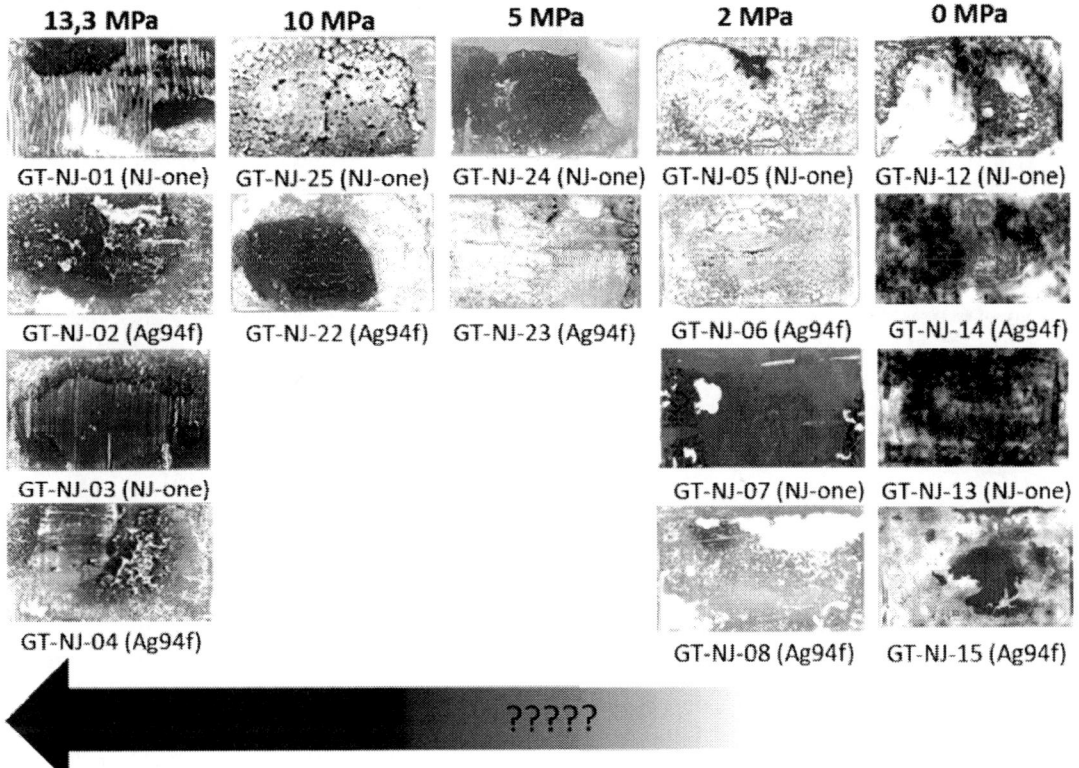

Figure 10: SAM measurements of 16 samples. The four missing samples have not been measured until the time of writing

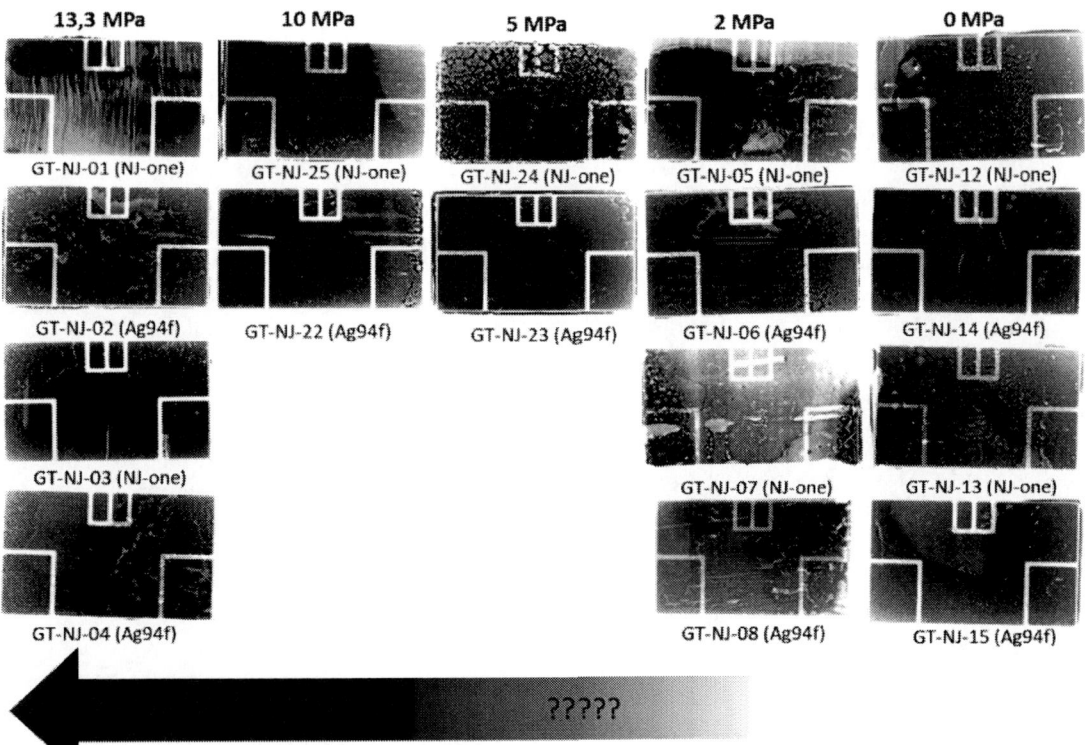

Figure 11: X-ray measurements of 16 samples. The four missing samples have not been measured until the time of writing

Optical method for validation of changes in the cleaning process and cleaning process optimization

VLADIMÍR SÍTKO

PBT Works s.r.o., 756 61 Rožnov pod Radhoštěm, Lesní 2331, email: info@pbt-works.com,

www. pbt-works.com

IVAN SZENDIUCH

Brno Faculty of electrotechnical and communication technologies, Department Microelectronics, 61600 Brno, Technicka 10,

szend@feec.vutbr.cz

INTRODUCTION

Currently, electronic assemblies are more complex and compact. Requirements on the reliability of electronic assemblies in a harsh environment are rapidly growing. Assemblies become dense, with small distances between poles. Processing of low and very high-frequency signals requires high values of surface insulation. Experiences confirm that using No-clean soldering technology has only limited efficiency in protection against leakage current, ion migration, and corrosion.

Cleaning high reliable electronic designed for the harsh environment becomes to be an important option. Due to new solder paste composition, bottom terminated components (with a very low gap under components), special surface protection, and strict requirements on final cleanliness; cleaning is not anymore a simple process. Improper setup can damage the assembly. SIR values of incompletely cleaned parts are not high enough. Too long cleaning can damage components, surface plating layers, destroy solder resist or damage marking.

Properly chosen soldering material and cleaning material combination can avoid many troubles. Easiness of cleaning can not only increase capacity and reducing direct cost. Often, difficult to clean solder residues prevent to clean assembly safely at all.

The assembly must be cleaned completely, i.e., also under components. This cannot be easily checked and often is not done, even if residue under the component represent the big risk. Also, not completely cleaned residues are more dangerous as uncleaned ones.

One of the main concern of all manufacturers of reliable electronic is the absence of modern testing methods. Those which are available was developed and standardized already in the last century, some of them are 50 years old. Such methods cannot reflect the situation in modern SMT assemblies, also new methods for obtaining evidence for protection against electrochemical corrosion of electronic assemblies are nedessary.

CURRENT SITUATION IN STANDARDS

Already several years ago, after introducing No-clean and SMT processes, it became commonly known, that measuring of ionic contamination is probably not the best option to qualify new electronic assemblies.

Finally, in November 2018, an addendum to IPC – J-STD 001, add 01 documents were released which *bans using ionic contamination as evidence for electronic assembly cleanliness qualification.*

This situation made us develop an easy and reproducible method for testing the efficiency of the cleaning process. The Glass Test Board and small automatic optical inspection instrument (AOI) to evaluate results of cleaning under components. We have enhanced its precision and speed of measurement so that it enables to make even bigger comparison studies.

OPTICAL MEASUREMENT OF FLUX RESIDUES UNDER COMPONENTS USING PRESISE GLASS TEST BOARD.

This methodology has several applications:

- Can be used for soldering residue and cleaning to chemistry matching, as the first step of the cleaning process introduction, (see also IPC CH 65B) In this paper, we concentrate on the flux residue solubility testing and show how easy and precise the method can be. We were ourselves surprised with some data, resulted from processing a large number of samples. Probably such data were not available ever before.It is suitable for validation of cleaning process changes acc. to J-STD 001G, Am 1,

- Cleaning machine capability studies, process capability studies and, optimization of the cleaning process can be done easily and effectively by this method. We present shortly such application

- It proves to be much more sensitive than measuring ionic contamination. We show a comparison of these measurements and explain why this is so.

- Can be used for monitoring the cleaning process instead of PICT (process Ionic Contamination Testing) using Glass Test Boards.

GLASS TEST BOARD

The glass test board (Fig.1) is a precise model of PCB with 400 pcs of 0805 chips models, made of ceramic and sealed with patented technology on the glass substrate. This design assures very precise geometry, especially the gap thickness under the chip. Test boards are recyclable, and it can be used many times w/o losing its properties. Glass test board dimension are in Fig.1.

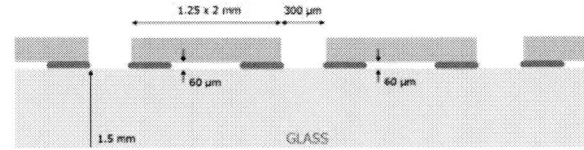

Fig. 1: Glass Test Board dimension

Gaps under chips can be underfilled with flux, centrifuged from solder paste. The test board is then reflowed to simulate soldering, cleaned. AOI than measures rest residue amount on the Test Board. Many of paste manufacturers already agreed to sell flux from solder paste separate to do such tests.

GLASS TEST BOARD PREPARATION

During the underfill process by 75 – 100°C it can be recognized, how sensitive some fluxes are to slump at increased temperature. Some samples flow nicely; others needs some mechanical agitation with a small spatula.

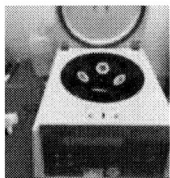
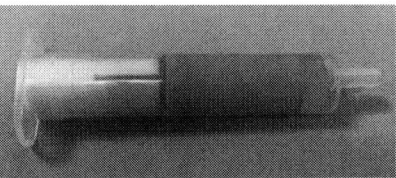

Fig.2: Paste centrifuging to get jelly flux

Finally, the upper surface of chips is wiped with a squeegee to ensure as much as possible defined thin layer of flux on top (Fig.2, Fig.3).

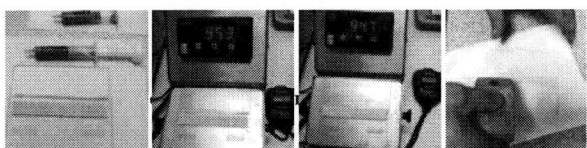

Fig.3: Jelly Flux underfill

After standard reflow process of glass coupon test boards are cleaned. Cleaning parameters are dependent on the type of test we have to do. For some test (flux to cleaner matching) require a machine with very uniform cleaning space. Otherwise, test boards have to be mounted only on the same specific place of the machine (Fig.4).

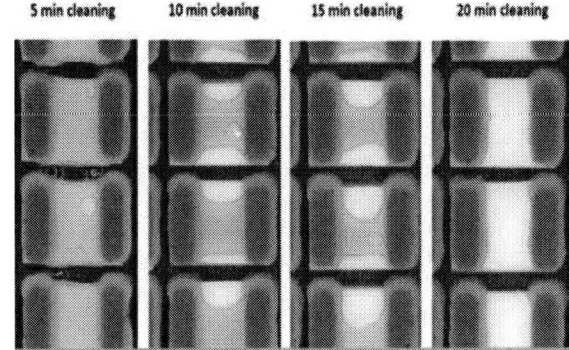

Fig.4: Test Board details in different stage of cleaning process

MEASURING AND DATA EVALUATION

For Glass Test Boards measuring we use dedicated AOI tester shown in the Figure 5.

Fig.5: AOI tester for Glass Test Board measuring

Tester makes pictures in a modified color scanner with 6400x9600 resolution, enough to follow details even under smallest chips. Special software was developed for this purpose processes pictures. Measuring of Glass Test Board gives 400 values of flux residues separately under each ship. The precision of the entire method is about +- 3%. Test time for the one Glass Test Board is 180 sec.

Result of the test is an automatically generated test protocol with all statistical data from processing the 400 measured data under each chip (Fig.6).

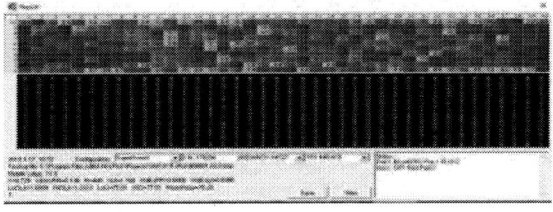

Fig.6: Test protocol from AOI tester for Glass Test Board

SOLDER RESIDUES AND CLEANER MATCHING STUDY

We made this type of study, not only to get results on solubility (cleaning) resistance of different flux residues but also to demonstrate easiness and speed of such studies.

We studied differences in cleaning resistance between differently prepared solder residues. We made 3 Test Boards from each material sample.

1: sample which was reflowed once

2: sample which was reflowed twice (as double side boards in production do)

3: sample, which was reflowed once and left 30 days at room temperature and about 60% of relative humidity – "aged" to simulate usual process delays)

We tested 25 different solder pastes in this study.

TEST BOARD CLEANING

For cleaning, we used the cleaning machine with Direct Linear Spray in Air, which has the best possible spray uniformity across the whole chamber (Fig.7). The position of substrates in the machine was, therefore, not important. This is essential because not every type of cleaning machine can perform this way.

Fig.7: Spray in air machine used for this study

We used one type of cleaner chemistry, microemulsion, neutral with very low saponification. Concentration 20%

Concentration was measured by titration always at the beginning and the end of each cleaning session. Rinsing with DI water in two steps, (each equipped with DI water reclaim) and drying was necessary for each cycle, for precise evaluation of coupons.

We split the estimated cleaning time to short sequences (cycles) and measured cleaning result between sequences. The values of residue removing under chips during the wash step of cleaning gave a plot of the

"washing curve."- This data makes a record of flux residue removing during the entire cleaning until the contamination under all chips of coupon reaches 0% Process times were defined by the movement cycle of spray and blow arm. (nozzle system drives from left to right and back along the frame with Glass Test Boards). Cleaning parameters are shown in Tab.1 and test boards in the cleaning machine are shown in the Fig.8.

Tab.1: Cleaning parameters: (set values on the machine)

Cycle	Wash	Temp	Pres	Rinse 1	Rinse 2	T	Dry	T
	min	°C	Bar	min	min	°C	min	°C
1-4	4,4	60	2,9	2,2	2,2	40	8,8	110
5-8	8,8	60	2,9	2,2	2,2	40	8,8	110

Fig.8: Test Boards in cleaning machine. Mention diferent discoloration of diferent samples

MEASURING AND DATA EVALUATION

In this study, we processed 103 samples on Test boards. We put samples with 1 reflow and 1 reflow + aging always on identical coupons. Coupons are fully recyclable.

We have made 635 test protocols, which means, we processed 254000 readings under individual chips to get all washing curves and to be able to compare it. Even this seemed to be a high number, the time needed for this study was 12 men/days only, including paste centrifuging and test boards preparation.

TEST RESULTS PROCESSING

For each material (in all three conditions), we generated " washing curves.", a record of test board cleaning (Fig.9).

Then, we defined a simple method for comparison. Taking the total cleaning as the comparison is not precise enough because, due to the size of the study, we had to step the cycle time in 4,4 and 8,8 min. Such stepping cannot express actual differences between flux residues.

578

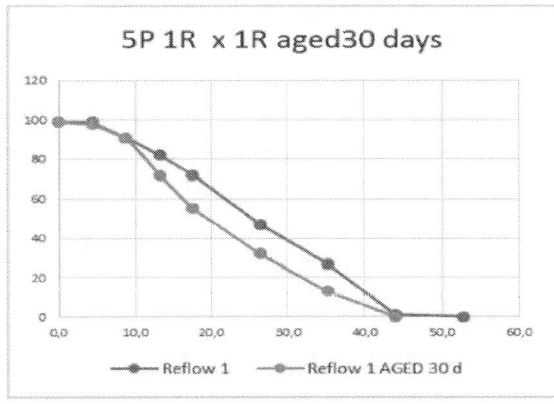

Fig.9: sample of washing curve

In practical cleaning, we cannot always rely on the speed of removing contamination under components, because some gaps are very small (single microns size) and some flux residues need extra effort (time) at the finale of cleaning (near to 0%), while others do not show such effect. The solubility during the entire cleaning is, therefore, important. Therefore, we have chosen for comparing solubilities the area under the washing curve.

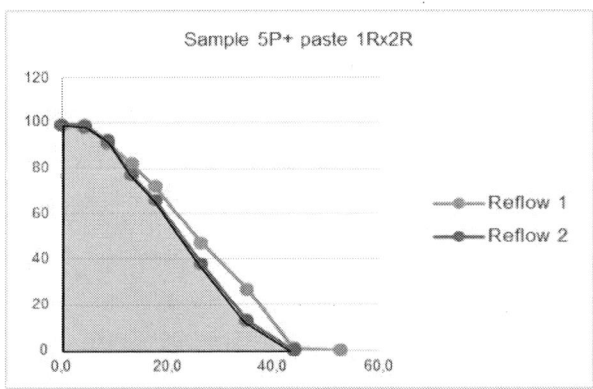

Fig.10: Definition of cleaning resistance

Cleaning resistance (equal to the area under the washing curve) depends on many factors (Fig.10). Cleaning parameters and power of cleaning machine are the biggest. Cleaning resistance can be according to our experience recalculated from machine to machine using testing and comparing two machines with the same parameters and same flux with constant data.

RESULTS OF EXPERIMENT

Differencies in cleaning resistance between different types of flux residues fluxes are high (see Fig.11).

We can generally distinguish pastes as Easy cleanable (green), Cleanable (yellow) and Difficult to clean (red). Users, which have to clean, should avoid cleaning difficult to clean pastes, especially, when assemblies

have very thin gaps under component (too high thickness of solder mask, small chips, QFN packages, etc.), sensitive surface plating, labels, marking.

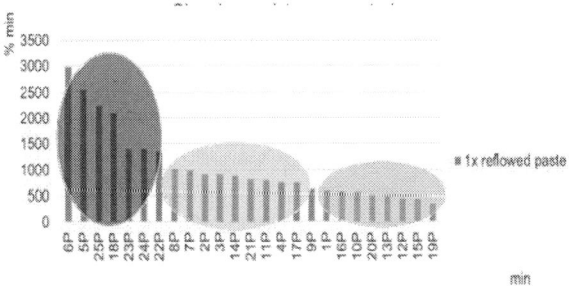

Fig.11: Diferent clearing resistence demonstration

Comparing the effect of 1 and 2 reflow cycles, we have got results, which we not completely expected (Fig.12). A common rule says that all flux residues solubility is getting worse with higher heat exposure. It is, however, not fully true!

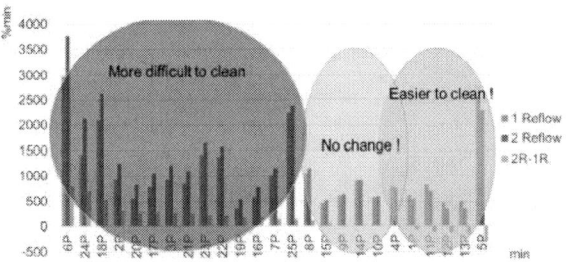

Fig.12: Cleaning resistence sorted after 1 reflow (left column) and 2 reflows (right column)

As seen in the comparison of 25 different tested materials, almost 50% does not change it´s cleaning resistance with the second reflow or is even easier cleanable. This is the first surprising result of our study.

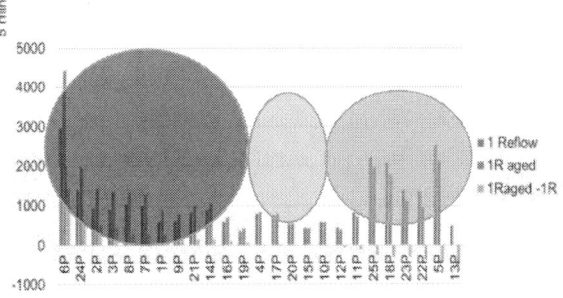

Fig.13: Cleaning resistence after 1 reflow and and after 30 days ageing

Cleaning resistance is growing also after aging (see Fig.13). This sentence belongs to main empirical postulates of cleaning knowledge. Our results show that this is not necessarily always true!

We found, among 25 different materials, that about 60% of them does not show any change in cleaning resistance after 30 days of aging and, some of them was even more easy to clean. This is a second surprise.

CENTRIFUGED SOLDER PASTE AND JELLY

Centrifuging solder paste is not a difficult process, but not all people have the necessary equipment, skill, and time to do it.

There was a question if flux jelly taken directly from the manufacturer of paste and centrifuged solder paste do give the same results. We had, unfortunately, no chance to get flux from manufacturers to all samples, we were able to make only 10 pairs of materials.

Results in Fig.14, 15 show good consistency except for three materials from ten (33%). Even with 2 reflow exposure, the situation does not change.

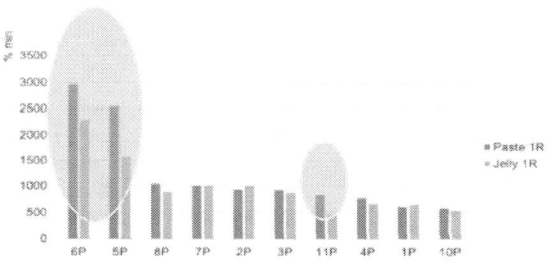

Fig.14: Paste and Jelly after 1 reflow

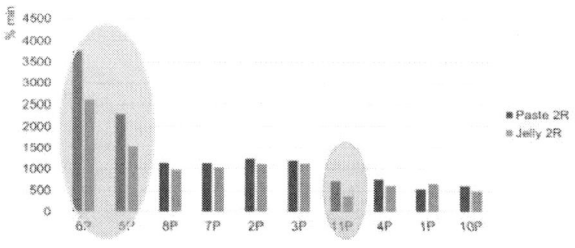

Fig.15: Paste and Jelly after 2 reflow

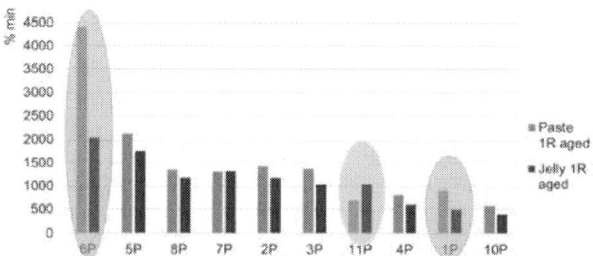

Fig.16: Paste and Jelly after 30 days ageing

With aging, the centrifuged paste and jelly differences are changing, but still, the belonging to easy, cleanable, or hard to clean category does not change (Fig.16).

We consider using jelly as good for shop floor testing approach, both for paste to cleaner selection (matching) and, especially for cleaning process changes validation and process stability monitoring.

PARTIAL CONCLUSION

Glass Test Board AOI test technology is capable of making cleaning resistance comparison even on big groups of samples easy and with reasonable time investment.

Unfortunately, process specialists, which are responsible for cleaning, must still behave like alchemists in middle age. There is a lack of data on solubility and sensitivity to heat exposure and aging after soldering.

Information about solubility was not that important in the past. Today, solder pastes got different properties due to demands on printing. Cleaning chemistry manufacturers have to react to that development. Such progress changes the cleaning process significantly.

To build an efficient and not expensive cleaning process, cleaning people need new kind of data, which were in the past and even until now, not available.

Those, which are available for the solubility, are not commonly published - due to the competition awareness.

However, many other data from pastes are public and are a part of paste classification. Why not then the solubility data under different conditions?

CLEANING MACHINE CAPABILITY STUDY

This topic is a never-ending conflict story caused by a misunderstanding of the meaning of machine capability. It may come from mostly evaluated mechanical lines, like pick&placer, where it is relatively easy to measure the precision of placement, by defined speed and to get many values in a short time. By cleaning, it is not practical, to think this way. For instance: 90 ltr bath in the machine can be heated to process temperature in 20 min if needed, but cooling down can take even 24 hrs (at compact machines no cooling system is installed). Moreover, the temperature is only one of many parameters. Any other parameter measured separately will not describe cleaning stability.

Here using precise Glass Test Board, is valuable. With glass test board, at first, the machine has to be checked on its uniformity of cleaning conditions — some cases we can see already in a comparison of uniformity of cleaning within one substrate (Fig.17, 18 and 19).

We can make a similar test through the entire process chamber. In this case, it is better to run the complete "cleaning curve" with several substrates, put on places with the expected fastest and slowest cleaning speed.

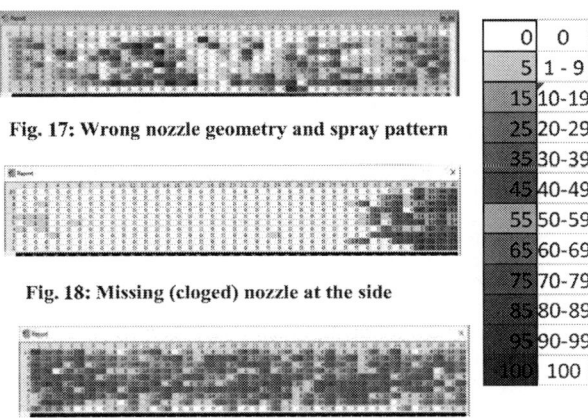

Fig. 17: Wrong nozzle geometry and spray pattern

Fig. 18: Missing (cloged) nozzle at the side

0	0
5	1 - 9
15	10-19
25	20-29
35	30-39
45	40-49
55	50-59
65	60-69
75	70-79
85	80-89
95	90-99
100	100

Fig. 19 Proper nozzle distribution – uniform cleaning

Results are very much dependent on the optimization grade of spray mechanics of the cleaning machine.

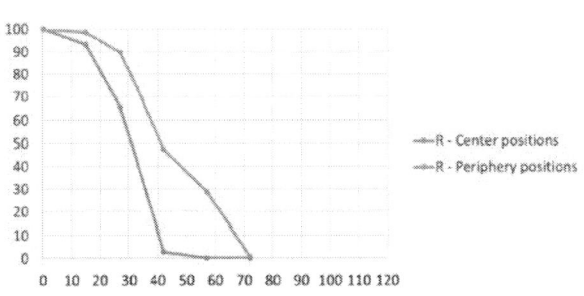

Fig. 20: Rotational direct spray

As an example, we measured cleaning field uniformity of cleaning machine with rotational direct spray nozzles (Fig.20). We have used 8 pcs of Glass Test Board placed 4 pc near the center of rotation of nozzle bar and 4 pcs at the corners of spray working area.

Fig. 21: Cleaning speed – center/periphery position

Results are on the Fig.21, showing a big non- uniformity between center and periphery. Still, the machine can clean, but results from the center are not comparable with results from corners. This problem, especially when cleaning assemblies with hard to clean flux residues and very small gap under components, leads to incomplete cleaning or failures in component compatibility.

The next step on cleaning machine capability study can be repeated cleaning of Glass Test Board with an evaluation of each result. From this data, calculating of C_{pk} is simple. Cleaning parameters should be set so that the result on Glass test board is near to 50% of residues; in this range, the method is most sensitive.

COMPARISON OF IONIC CONTAMINATION AND GLASS TEST BOARD OPTICAL CHECK

We made a study with different solder residues where, besides measuring cleaning curves, we measure after each cleaning sequence also the ionic contamination of the Glass Test Board. The detailed process was: cleaning step, ionic contamination evaluation, optical testing on AOI. This approach does not distinguish between cleaning on cleaning machine and some, additional cleaning on the contamination measuring instrument (which is a kind of cleaning " machine" too). This additional cleaning effect is between 3 to 1% of the result, and we can ignore it. Results are in Fig.22.

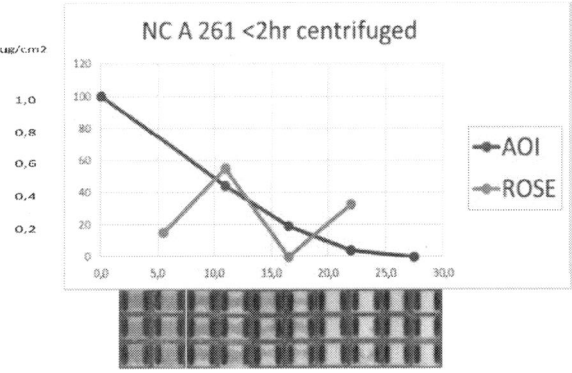

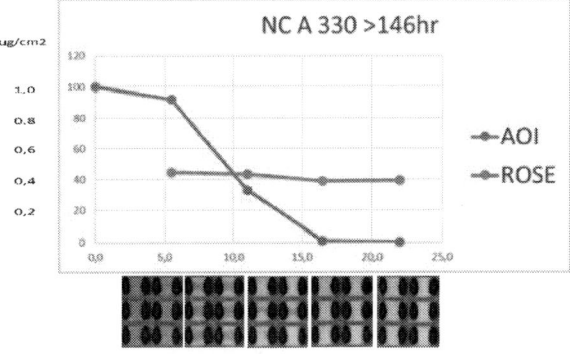

Fig. 22: Ccomparation of ionic contamination and AOI

Graphs together with photos of details (see Fig.23) from appropriate test boards show that just after the open surface of the |Test Board is clean, ROSE method claims that " PCB is clean. Due to low impingement of ROSE test equipment with cleaning mixture IPA + DI water (weak solvent for the majority of today's solder paste formulations) measuring instrument for Ionic contamination cannot see residues in low gaps under components.

So far, the gap under component gets open, even not yet all residues are cleaned, the ROSE value can grow again. The explanation is simple: some flow under component has been established again, which can transport ions to instrument sensor.

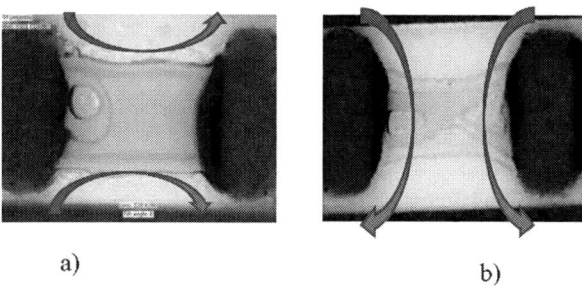

a)   b)

Fig.23: Gap under component a) clogged with flux b) open

The conclusion from this study shows that modern electronic assemblies, full of SMT and leadless components cannot be checked on cleanliness by ionic contamination measuring methods.

CLEANING PROCESS MONITORING

The cleaning process is dependent on many parameters, which not everyone can be strictly separated and measured. Beside of clear, measurable parameters, like time, temperature, concentration, there are plenty of other (loading of cleaner, organic contamination of rinsing steps, distance from nozzles, nozzle shape, angle, flow speed under immersion and other details) which cannot be easy monitored and comparably measured.

Monitoring of displayed parameters on the machine is certainly useful, but cannot reflect all possible changes, which can happen during the process. Therefore we tried to use Glass Test Board to monitor all changes in the cleaning process. Such monitoring can be, of course, used as validation of process changes according to J-STD 001 Add. 01.

In this experiment, we run a long term simulation test of a complete cleaning machine with all maintenance activities automated (automatic refilling of cleaner concentrate and DI water in the washing step, refilling some part of DI water in first rinsing loop and refilling of DI water in final rinsing step).

As the machine was equipped with adaptive cleaner refilling, after some defined volume of cleaner was used, the system measured the concentration and put the exact amount of concentrate + DI water to refill and keep the concentration stable. We always run the same dummy PCB assemblies and simulated the real flux residue load with one reflowed board with the equivalent of total mass solder paste, on all dummy boards. We monitored a complete function of the filtering system on the machine and, besides that, also a quality of cleaning process. (especially washing). We set the cleaning time and flux on the Glass Test |Board so, that final result of cleaning was near to 50%. As a part of the study, we have stopped the function of automatic adaptive refilling for some time (until the level of cleaner in the washing bath did not reach the minimum.

We monitored concentration. All other parameters of cleaning were kept constant. In each cycle, we cleaned two Glass Test Boards to get more data.

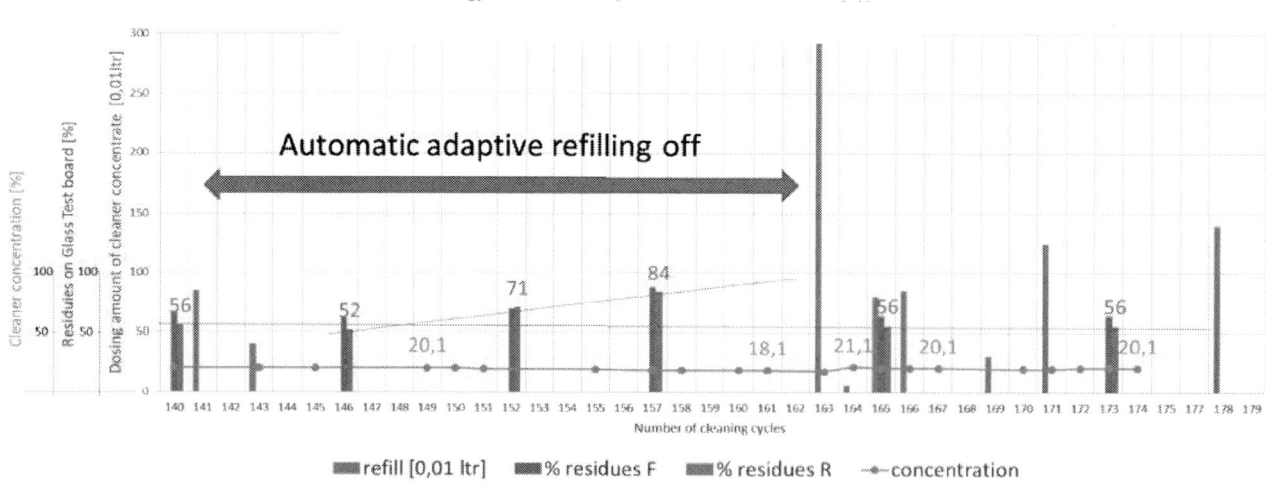

Fig.24: Influence of automatic refilling

Result of an automatic refilling stop was very small concentration dropping. From initial 20,1% to 18,1% after 19 cleaning cycles of machine. Two percent difference is a value comparable with the standard precision and repeatability of any concentration measuring method (either titration or sensor measuring.) So, not a big change.

When observing Glass Test Boards cleaning results, the cleaning result response on automatic refilling is much more dramatical. During 11 cycles, the average cleaning result, which was 56%, got changed to 84%, and this value grew probably further. Testing of cleaning with Glass Test Board clearly shows here that automatic adaptive cleaner refilling is a very useful option on the cleaning process.

Also, this experiment shows that Glass Test Board method is a good tool for process changes monitoring.

Response on the cleaning of Glass Test Board can detect any irregularity of the cleaning process. Of course, this only can monitor cleaning as a complex result. Monitoring all possible physical parameters is also important to explain the reason for changes.

CONCLUSION

We have described some possibilities of Glass Test Board inspection method for cleaning process introduction, optimization, and monitoring.

Experiences with practicing such measurements show good repeatability, easiness of interpretation, and test preparation.

The fact that Glass Test Boards are re-usable is an important factor. Some measurement can require high precision of reading and using the same Glass board for repeated measurements (especially in Cleaning process

monitoring) excludes any doubts on influence by dimensional allowances of that Glass Test Board.

ACKNOWLEDGEMENT

The arcticle was supported by project FEKT-S-17-3934 „Utilization of novel findings in micro and nanotechnologies for complex electronic circuits and sensor applications".

REFERENCES

[1] Bixenman, M., Sitko, V., & Hub, P., Cleaning Flux Residue under Leadless Components using Objective Evidence to Determine Cleaning Performance. IPC APEX Technical Conference, San Diego, CA, 2017

[2] Sitko, V., & Bixenman, M., Visual Method for Determining Time to Clean Flux Residues under Leadless Components. IPC/SMTA High-Performance Cleaning and Conformal Coating Conference. Chicago, I, 2018

[3] Sitko, V., Cleanliness process development. SMTA Electronics in Harsh Environments Conference. Amsterdam, 2019

[4] Technical report Project NOMEN, a joint project of PBT Works s.r.o. and Technical University Brno, sponsored by EU., 2011–14

[5] Patent PV 2012-940, Method of creating interlayer on glass testing substrate intended for sticking chips, and applying device for carrying out the method, 13446038, 2013

An approach for failure prediction in H³TRB-tests

Elisabeth Kolbinger[a*], Felix Wuest[a], Marius van Dijk[a], Stefan Trampert[a],

Klaus-Dieter Lang[b]

[a] Fraunhofer Institute for Reliability and Microintegration, Gustav-Meyer-Allee 25, 13355 Berlin,
Germany
[b] Technical University Berlin, Gustav-Meyer-Allee 25, 13355 Berlin, Germany
*elisabeth.kolbinger@izm.fraunhofer.de / Tel: +49 30 / 46403272

Abstract

*Applications for power modules under harsh environmental conditions have gained importance in recent years, e.g. in offshore wind turbines. One of the critical conditions that leads to failures in these applications is humidity. Therefore, it is important to subject the power electronics to appropriate load tests in advance, in order to generate these failures. An established test method for the qualification of power electronics is the **H**igh **H**umidity **H**igh **T**emperature **R**everse **B**ias (H³TRB)-test. This paper deals with the prediction of failures in H³TRB-tests. The goal in this paper is to use the currents noise to improve the method of forecasting time to failure in H³TRB-testing. This is based on the assumption that there may be a correlation between the time to failure and the point in time at which the electrical parameters begin fluctuating.*

Key words: humidity, diffusion processes, power modules, reliability, H³TRB-Test

1. Introduction

1.1. General Motivation

In order to predict the lifetime of semiconductor power modules in a manageable way, it is necessary to perform tests that are able to simulate accelerate the environmental conditions in the field. The humidty robustness is one of the most important factors for outdoor applications. In addition to humidity, high voltages must also be taken into account in a test. The combination of the applied voltages and geometries determines the dominant field strengths. These have an influence on the degradation reaction, since corrosion processes start in the areas with the highest field strength.

In addition to external conditions, the choice of materials plays an important role. Especially when using polymers as e.g. encapsulants, moisture plays a critical role. The moisture can diffuse into and through the polymer, causing potentially hydrolysis at the bonding interface, which can lead to delamination effects. Due to the possible delamination, corrosion processes can occur at the interface. Besides the possible mechanical stress due to the encapsulation process, the corrosion processes represent another critical risk for the reduction of the reliability of power modules [1][2].

In order for a moisture-induced defect to occur, the diffusion processes of moisture through the test substrates must be observed. The diffusion coefficients of the individual components can differ significantly, based on the molecular structure of it. Braun et al. have shown that moisture absorption at 60 °C / 60 % r.H. in silicone is very fast in comparison to epoxies. But the maximum moisture absorption is significantly lower compared to these epoxides and is only in the range of less than 0.1 wt-% [1]. Based on these investigations, the moisture absorption of the silicone is considered in this paper in addition to the electrical measurements.

The diffusion of moisture can be described in one dimension with the 2nd Fick's law. In fickian diffusion, it is assumed that the moisture flux is directly proportional to the concentration gradient in a material [2]. Whereby t is the time, c the concentration, x space coordinate in the sample and D the diffusion coefficient.

$$\frac{\partial c}{\partial t} = D \frac{\partial^2 c}{\partial x^2} \qquad (1)$$

1.2. H³TRB-Test

The H³TRB-test offers the possibility to combine high humidity and high stress in one test. Recently there has been much discussion about the relevance of this test. Especially the choice of the test voltage is an important basis for discussion [3][4]. The advantage of the H³TRB-test compared to the THB (temperature humidity bias) test previously used is that higher voltages are applied. Both test methods promote charge or ion movement due to increased humidity in conjunction with relatively higher temperatures and voltages, enabling the detection of instabilities due to process variations or inadequate design margins. However, by using the higher voltages in the H³TRB-test, corrosion mechanisms are accelerated by electrochemical mechanisms that could play a more dominant role in field operation than the classical effects of charge or ion movement. In the presence of high electric fields and sufficient humidity of the reactant ions on the chip surface, electrochemical corrosion may occur [4]. Furthermore, the use of a higher test voltage leads to a more realistic lifetime estimation. The test voltages occur in a range that will also be used in later operation.

One of the most discussed degradation phenomena after the H³TRB-test takes place in the guard rings of the PI insulation layer [4][5]. This results in corrosion of aluminium. By using different metals in the power modules, there is always a risk of contact corrosion, with the most base metal most likely to corrode [6]. Due to the different field strengths available, there are areas that are more prone to corrosion than others. Nevertheless the literature shows that in most cases aluminium corrosion is present.

2. Experiments

2.1. Determination of the diffusion coefficient

The diffusion coefficient of the silicone gel is determined by the moisture uptake. The silicone, potted in an aluminium dish with a defined height, was first dried for 48 hours at 100 °C, followed by storage at 85 °C and 85 % r.H. The weight was intermittently measured until saturation occurred.

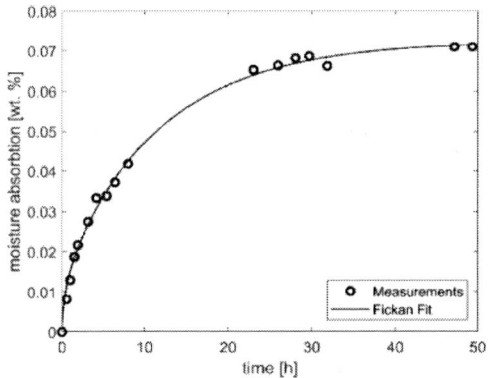

Figure 1 – Measurement and fit of humidity uptake of the silicone gel at 85°C / 85 % r.H.

The measured data, Figure 1, was fitted occording to the analytical solution for Fickian diffusion [7]:

$$\frac{M_t}{M_\infty} = 1 - \frac{8}{\pi^2} \sum_{n=0}^{\infty} \left\{ \frac{exp\left[-(2n+1)^2\pi^2\left(\frac{D\,t}{s^2}\right)\right]}{(2n+1)^2} \right\} \quad (2)$$

Where M_t and M_∞ are the weight at time t and at full saturation respectively, D the diffusion coefficient and s being twice the diffusion path length. The diffusion coefficient at 85 °C / 85 % r.H. is determined to be 0.0085 mm²/s.

2.2. H³TRB-Test

For the investigations under bias and humidity a CS-70/300/S climate chamber of CTS Clima Temperatur Systeme GmbH was used. The test is carried out under climatic conditions of 85 °C and 85 % r.H. at a rated blocking voltage of 80 % $((1360 \pm 66)V)$. During the test the leakage current of all samples is measured every 1.5 minutes with an accuracy of $\pm 35 \mu A$. The current was limited to a maximum leakage current of 3.5 mA.

Figure 2 shows the functional diagram of the electrical measurement setup. Current and output voltage are sensed on high voltage side of the powersupply and sent through two isolated amplifiers to provide a galvanic isolation between high voltage (secondary side) and low voltage (primary side). A passive attenuator is used to reduce the output voltage to a suitable lower voltage before fed to the isolated amplifier.

The V/I conversion for the current measurement setup is done via a passive resistor, and then amplified by the isolated amplifier. The output voltage of an unregulated HV DC/DC converter is controlled by an error amplifier which corrects the derivation between V_{in} (reference voltage) and V_{sense} (equivalent output voltage) in order to correct the gain errors.

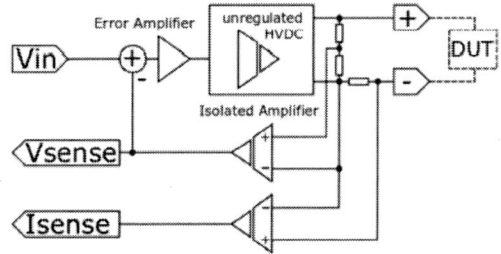

Figure 2 – Functional diagram of the electrical measurement setup.

The specially designed modules consist of four parallel diodes with a rated blocking voltage of 1700 V and a nominal current of 150 A. Eleven modules were tested within two separate runs using the same setup and equipment.

3. Results and Discussion

3.1. Simulation

Numerical simulations of the diffusion process are performed in order to estimate the time necessary for complete saturation. Due to symmetry, only one half of the the power modul was simulated. At $t = 0$ the module is complete dry (concentration 0). Since the diffusion coefficient of the silicone gel is much higher than that of the housing plastic, only the diffusion in the silicone gel was simulated. The cover of the power module was neglected, which is a valid assumption, since large gaps are present between the cover and side frame. This will cause similar conditions both outside the power module as inside above the silicone gel.

The transient diffusion simulation was performed with the commercial finite element software COMSOL Multiphysics®. The resulting concentration distribution after 15 min. at 85 °C and 85 % r.H. is displayed in Figure 3. Here, the normalized concentration is shown ranging from completely dry at "0" untill complete saturation at "1".

Figure 3 – Distribution of normilazed concentration after 15 min. exposure at 85 °C / 85 % r.H.

Due to the high diffusion coefficient of the silicone gel, combined with the low thickness of it, the time for complete saturation is relatively short. The concentration at the die attach of an IGBT is shown in Figure 4. After only one hour an almost complete saturation is reached.

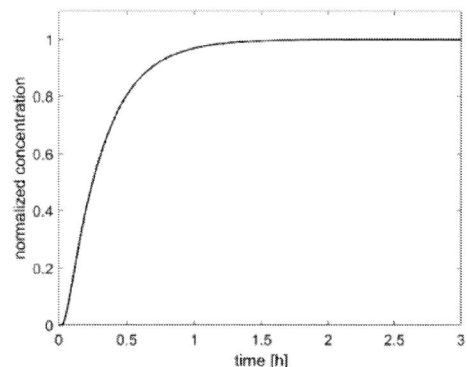

Figure 4 – Normalized concentration at the position of the 1st level die attach of an IGBT.

The simulation results indicate cleary that effects caused by moisture can occur almost directly after exposing the power module to harsh conditions such as 85 °C / 85 % r.H.

3.1. H³TRB-Test Results

During the tests it was noticeable that in the early test phase the current stays stable with only slight changes in amplitude. After a while the current signal tends to rise and fall quite frequently, as can be seen in Figure 5. This behavior is expected to be caused by corrosion effects. The noise of the leakage

current is used to find a precursor for an early failure of the power modules during H³TRB-testing.

For easier identification of the noise of the leakage current signal a floating average filter with a Gaussian window and a width of 50 values and a σ of 5 has been used (Figure 5). For this filter a rectangular filtering window was used with zero phase shift filtering according to [8].

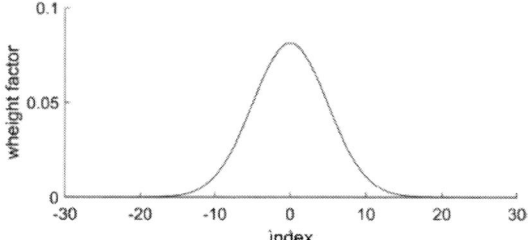

Figure 5 – filter window for floating average filtering of leakage current signal.

This floating average represents the signal whereas the difference between the signal and its floating average represents the noise. As can be seen in figure 6 the noise signal rises well before the current increases and reaches the failure criterion of 3.5 mA. For more reliable evaluation a floating root mean square filter was used on the noise signal (equally weighted from 50 values) with s being the original signal, n the number of values observed, i the index of the signal currently processed and s_{RMS} the resulting floating RMS signal. Again, zero phase shift filtering was used.

$$s_{RMS} = \sqrt{\frac{1}{n} \sum_{i}^{i+n} s_i^2} \qquad (3)$$

Table 1 – Overview of the investigated samples with the determined times between exceeding the effective value and time-to-failure or end of test.

Sample	Time between first effective value exceeded and failure time / end of test [h]	Testresult pass / fail
1	152.09	fail
2	272.38	fail
3	177.69	fail
4	103.84	fail
5	202.10	fail
6	167.19	fail
7	809.13	pass
8	572.79	pass
9	781.45	pass
10	-	pass
11	707.56	pass

A threshold level of 0.038 V was selected to identify a point at which a failure might be predicted. From all modules that failed during the H³TRB-test

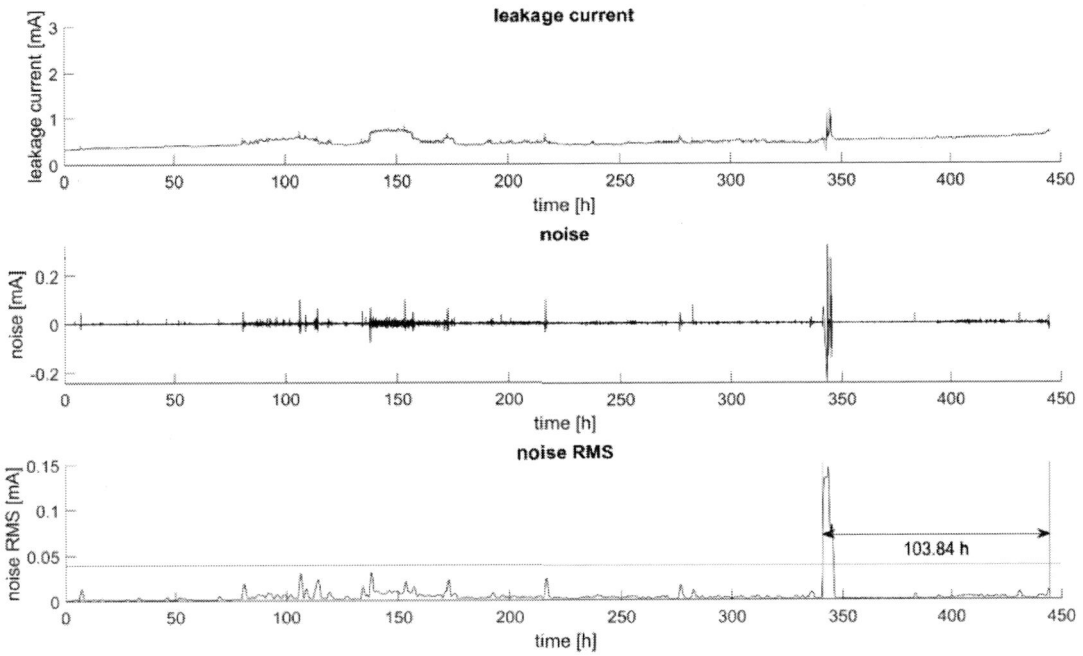

Figure 6 – leakage current, noise and floating root mean square of noise over time with threshold (exemplary graph from sample 4).

the forecast data was taken (Table 1). The mean value for all failed modules is 179.2 hours with a mean absolute deviation of 38.7 hours. This method has been used on 6 of 6 failed modules in the test. Additionally five modules passed the test without reaching the failure criterion of 3.5 mA. Nonetheless, four of these modules should not have passed according to the newly developed forecast algorithm, since the noise RMS rises above threshold level quite early.

Figure 7 – corrosion on the guard rings in the corner area of a diode after finishing the H³TRB-test.

After the H³TRB-test, the modules were optically examined using microscope imaging. Figure 7 shows an example image for the guard rings of the corner region on a diode. Black discolorations are visible, which are more or less pronounced depending on the field strengths. The discolorations can, as already known in the literature [4][5], explained by corrosion of aluminium.

Also on some modules black discolorations can be found on the edge and corners of the diode, as can be seen in Figure 8. The reason of this effect is not been determined. Probably it is not due to corrosion. It being residues of a burnt dendrite is possible, but not confirmed.

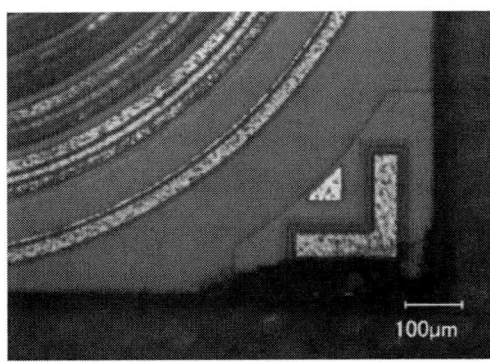

Figure 8 – Corrosion at edge and corner of a diode after finishing the H³TRB-test.

All modules, even the ones passing the H³TRB-test show traces of corrosion. The levels of visual corrosion have been summarized in Table 2. The overall corrosion impression is based upon the visual appearance of the diodes surface and its surroundings. The corroded guard rings determine which of the ten guard rings show corrosion traces. Sample 1 for example shows corrosion effects on the guard rings 1 through 9 (counting outwards). The discoloring effects on the edges can be seen only on modules that have failed.

Table 2 – Visible corrosion effects on all tested modules.

	Sample	Overall corrosion impression	Corroded guard rings [of 10]	Edge area discoloration
fail	1	Strong	1-9	None
	2	Medium	5-9, 4-9	Partially
	3	Low	6-8, 3-5	Partially
	4	Strong	1-9, 1-8	Partially
	5	Medium	4-8, 5-9	Partially
	6	Medium	4-8, 4-9	None
pass	7	Medium	1-7, 5-7, 1-5, 6	None
	8	Medium	3-8	None
	9	Medium	1-8, 1-7	None
	10	Medium	1-8, 1-7	None
	11	Medium	1-8, 1-7	None

4. Outlook and Conclusions

The results of the moisture simulation show that a complete saturation of the silicone is already present after approx. one hour at 85 °C / 85 % r.H. Due to the fast absorption of moisture, moisture swelling can occur, which can lead to delamination at the silicone substrate interface. These defects can be the source of corrosion mechanisms.

The time divergence between time-to-failure and saturation of the silicone can have several causes. On the one hand the reaction kinetics of the corrosion reaction must be taken into account and on the other hand due to the PI layer above the aluminium there is an additional diffusion path, which must also be taken into account. The test results show that a failure only occurs after more than 300 hours, which clearly shows that not only the moisture absorption of the silicone gel is a factor to be included in the considerations.

The electrical signal processing of the H³TRB-test lead to the same conclusion: the noise of the leakage current, which was initially intended to be used as a precursor for failure of the modules, can now clearly not be used in such a way anymore.

Guard ring corrosion can be seen on all modules whereas the edge effects can only be found on failed modules. Therefore it is assumed that the edge effects caused the modules to reach failure criterion of 3 mA. The algorithm developed in this paper might be sensitive to the aluminum corrosion. Further investigations are to be done to correlate this corrosion to the measurements in upcoming tests. A life evaluation will be run and interrupting the test when the noise RMS reaches the threshold at increasing levels. It is expected that different levels of corrosion can be observed.

Acknowledgements

The research leading to these results was funded by the Federal Ministry of Education and Research as part of the KorSikA project.

References

[1] T. Braun, J. Bauer, K.-F. Becker, O. Höck, H. Walter, M. van Dijk, R. Aschenbrenner, K.-D. Lang „Influence of Humidity on reliability of Plastic Packages", Proceedings of the **2015** European Microelectronics Packaging Conference (EMPC), 14-16 September 2015, Friedrichshafen, Germany.

[2] M. Akram, K. M. B. Jansen, S. Bhowmik, Leo J. Ernst, „Moisture absorption analysis of high performance polyimide adhesive", SAMPE Fall Technical Conference **2011**, 17 – 20 October 2011, At Fort Worth, Texas.

[3] J. Jormanainen, E. Mengotti, T. B. Soeiro, E. Bianda, D. Baumann, T. Friedli, A. Heinemann, A. Vulli, J. Ingman, "High Humidity, High Temperature and High Voltage Reverse Bias - A Relevant Test for Industry Application", PCIM Europe **2018**, 5 – 7 June 2018, Nuremberg, Germany.

[4] C. Papadopoulos, C. Corvasce, A. Kopta, D. Schneider, G. Pâques, M. Rahimo, „The influence of humidity on the high voltage blocking reliability of power IGBT modules and means of protection", Microelectronics Reliability, Vol. 88-90, pp. 470-475, **2018**.

[5] S. Kremp, O. Schilling, „Humidity robustness for high voltage power modules: Limiting mechanisms and improvement of lifetime", Microelectronics Reliability, Vol. 88-90, pp. 447–452, **2018**.

[6] C. Zorn, M. Piton, N. Kaminski, „Impact of Humidity on Railway Converters", PCIM Europe 2017, May 16-18, Nuremberg, Germany.

[7] C. Shen and G. Springer, "Moisture Absorption and Desorption of Composite Materials", J. Composite Materials, vol. 10, **1976**.

[8] J Kormylo, V Jain, „Two-pass recursive digital filter with zero phase shift", IEEE transactions on Acoustics, Speech, and Signal Processing 22(5), 384-387, **1974**.

Risk Assessment Study of Package Warpage by using Bayesian Networks

Kazuaki Ano

Dialog Semiconductor
8F, W Bldg., 1-8-15 Konan, Minato-ku, Tokyo, 108-0075 Japan

Phone: +81-3-5769-5100, kazuaki.ano@diasemi.com

Abstract

Heterogeneous IC packaging is being increased its complexity of the structure year by year. In such complex packaging, package warpage is a major technical challenge for Design For Manufacturing (DFM). The warpage brings troubles in manufacturing process for both semiconductor supplier and customer. Until today, various warpage studies have been done and key parameters of the warpage have been identified and verified. In such studies, Finite Element Analysis (FEA) modeling technique is used. However, it's not easy to predict how it gets better or worse without using the technique when packaging materials or process parameters are changed. And the change risks can be predicted only by expertise modeling engineers. Therefore, it must be beneficial, if there is an effective risk assessment method of package warpage against the changes. This is a motivation of this study and it aims to predict warpage risk level and its effectiveness by using Bayesian Networks (BN) as a risk assessment tool. In this study, an idea is to distinguish the warpage parameters into Concave (CC) and Convex (CV), then calculate the warpage rate. Meta-analysis was done to determine contribution rate of the warpage parameters, and then a BN model was examined its effectiveness for Ball Grid Array (BGA) packages. As a conclusion, a warpage risk model by using BN suggested proper warpage risk level and resulted a good co-relation to FEA model.

Key words: Risk Assessment, Bayesian Networks, IC Package Warpage,

Introduction

Package warpage is not a new issue but chronic since package is increased pin count and body size, decreased its thickness. Further, recent complex heterogeneous package and/or Fan-Out area array package with wafer or panel level packaging have also faced the issue. Therefore, it's not only single package level but panel or wafer level because of the manufacturing in large size dimension. Warpage brings some troubles during handling or testing not only in IC supplier's manufacturing but customer's Surface Mount Technology (SMT) process. Prime cause of the warpage is due to the use of various packaging materials which have different physical properties such as thermal expansion, elastic modulus, glass transition temperature and so on.

In a package design phase, risk assessment is an important event and package warpage is assessed by Failure Mode and Effective Analysis (FMEA). In the analysis, if there is a potential warpage risk, it's conducted further analysis by using FEA. Most of the case, FEA results are provide by an expert modeling engineer. Because the analysis needs special skills such as creating a model, meshing, proper parameter definitions, and so on, even though it uses matured modeling software. Therefore, it's a limited analysis method and risk analysis is also limited by the engineer. Without the FEA results, it is difficult to know how much of the warpage will improve or get

worse. If there is a model which can suggest certain warpage risk level without special skills or modeling techniques, it will be beneficial for quality engineers who are involved into a package development project. This is a motivation of this study how to assess warpage risk with an uncomplex method and a simple model. This paper studies a warpage risk of BGA package by establishing a BN model and discuss its effectiveness compared to FEA model results.

BN model as a Risk Assessment tool

As introduced briefly, package warpage is primary induced by packaging material properties. In this causal relation, warpage is a result and material properties can be defined as causal factors. These factors lead to improve or worsen package warpage. This relation can be fit well into a BN model. BN is based on Bayes theorem and it's a type of probabilistic graphical model to represent causal relationship and presumes quantitative probability of the relationship. It's also a powerful tool for reasoning under uncertainty. And introduction of various BN models in manufacturing area have studied broadly. For examples, [1] and [2] adopted BN model as a manufacturing design decision support tool. Manufacturing process conditions, fault diagnostics, process robustness and process maintenance were examined by [3]. Fault root cause analysis and preventive maintenance (PM) area have also studied its effectiveness of BN [4]. Reliability assessment with BN model was also studied by [5],[6].

In the Bayes theorem, if 3 of 4th event probabilities have known already, rest of probability can be calculated. This feature is also applicable to BN. Figure 1 shows a directed graph of simple BN models. In this figure, A, B, and C are called as "node". Arrows indicates a direction of influence. Figure 1(a) shows a probable model of event B after event A. In this model probability, P(B|A), is given by an equation (1). Similarly, probability of P(C) in Figure 1(b) is given by equation (2). In this equation, probability of P(C|A,B) means; event C is determined by conditions of event A and B. If causal event increases like A, B, C... n, the condition also increases in power of 2 (2^n). Conditional Probability Table (CPT) of model (b) is described in Table 1. A probable result can be calculated by some causal event factors with its conditions. And these events and results chain can be established a complex chain system by influencing probable numbers which come out from previous node. However, BN model has some constraints in its model. A chained system must be acyclic. It cannot be a looped system. And causal factors should be independent. Those should not have any relations among the factors.[7]

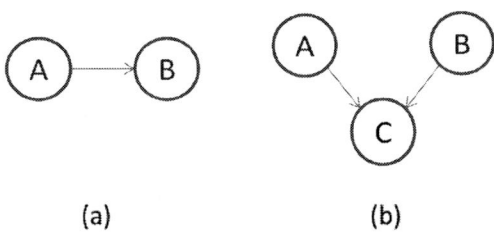

(a)   (b)

Figure 1: BN example models

$$P(B|A) = \frac{P(A|B)P(B)}{P(A)} \quad \cdots (1)$$

$$
\begin{aligned}
P(C) &= P(C|A,B)P(A)P(B) \\
&+ P(C|A,\overline{B})P(A)P(\overline{B}) \\
&+ P(C|\overline{A},B)P(\overline{A})P(B) \\
&+ P(C|\overline{A},\overline{B})P(\overline{A})P(\overline{B})
\end{aligned} \quad \cdots (2)
$$

Table 1: Conditional Probability Table

A	B	P(C) 0	P(C) 1
0	0	$P(\overline{C}|\overline{A},\overline{B})P(\overline{A})P(\overline{B})$	$P(C|\overline{A},\overline{B})P(\overline{A})P(\overline{B})$
0	1	$P(\overline{C}|\overline{A},B)P(\overline{A})P(B)$	$P(C|\overline{A},B)P(\overline{A})P(B)$
1	0	$P(\overline{C}|A,\overline{B})P(A)P(\overline{B})$	$P(C|A,\overline{B})P(A)P(\overline{B})$
1	1	$P(\overline{C}|A,B)P(A)P(B)$	$P(C|A,B)P(A)P(B)$

IC Package Warpage factor analysis

In order to establish a package warpage model by BN, meta-analysis was conducted by reviewing prior articles. The analysis results are summarized in Table 2. The articles were extracted from IEEE Digital Library with keywords of "Package" and "Warpage". About 200 articles could be matched with its published last 15 years from 2003 to 2018. Then, narrowed down by removing special packaging case such as TSV or multiple stacked die and so on. Since this study intended to confirm effectiveness of a BN model, simple package warpage studies were chosen.

With the analysis, it was confirmed that the warpage behavior was different with the package structure. For example, Package-on-Package case is well known about opposite behavior on top and bottom packages.[15] Top package is a full molded Ball Grid Array (BGA) structure and bottom package is a partially molded BGA mostly. This molded area difference makes opposite warpage behavior. So finally, 7 articles which are mostly for full molded BGA package are referenced for warpage factor analysis. [10], [14-18], [21]

The warpage factors were normalized with the specimen's mechanical dimension which used in the study. For example, if the warpage value is 20μm for 12x12mm package, normalized warpage rate is given by 20/(12/2x√2). Highest warpage value is seen at the center of package hence warpage rate is calculated with a half of diagonal dimension of a package.

Warpage direction is defined by each article, but it's not standardized. In this study, it's defined that concave (CC) as positive (smile) direction and convex (CV) as negative (cry) direction. (Figure 2)

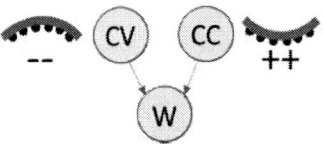

Figure 2: Warpage direction

Normalized factors in this study are shown in Table 3. From the analysis, convex factors are majority of the warpage and contributed about 77% of the total. It's seen that material properties of Epoxy Mold Compound (EMC) is a major factor of the warpage behavior and Mold process as well. This is because difference of Mold process condition, especially Mold temperature, is the prime contributor of the warpage because mechanical properties are determined by the process parameters such as temperature and time. [16], [23-26]

Die thickness and size are also major contributors of the warpage. Relative X-Y-Z size ratio against package dimensions behaves either CC or CV directions. [13] predicted the warpage behavior change by die size ratio. The ratio around 0.2 to 0.5 changes the warpage direction. In this study, the ratio doesn't include because of the model simplification.

Table 2: Package warpage factors by meta-analysis

Ref.	PKG Type	Dimension	Studied by	Warpage Direction	Factors	W rate / param	unit	W rate (um/mm)
[8]	FO-PLP	Panel: 370x470 Chip: 10x10 PKG: 12.25x12.5	FEA	Pos: Smile Neg: Cry	PI thickness	9.08	um	
					Cu thickness	-1.389	um	
					Cu ratio	-1.060	%	
					PI elastic modulus	13.060	GPa	
					PI C.T.E.	42.700	ppm	
					Die-Mold Gap	0.007	um	
					Mold Temperatureerature	-0.019	°C	
[9]	FC BGA	PKG: 42.5x42.5 Chip: 21x21x0.775	FEA	Pos: Smile Neg: Cry	Substratestrate thickness	0.374	um	
					Chip thickness	-0.113	um	
					Substratestrate:Chip ratio	-7.813	%	
[10]	PoP (Top)	PKG: 12x12 Stacked die	FEA	Pos: Cry Neg: Smile	Cure shrinkage ratio	-580	%	-6.835366
					EMC C.T.E. <Tg	-8.5	°C	-8.013877
					EMC C.T.E. >Tg	3.1286	°C	9.217608
					EMC Elastic modulus < Tg	-1.987	GPa	-4.683050
					EMC Elastic modulus > Tg	25	GPa	2.946278
					Substrate C.T.E. <Tg	8	ppm	8.662058
[11]	FOP	Panel: 320x320x0.9 (carrier) EMC: 299.8x299.8x0.6 Chip: 6x6 PKG: 8x8x0.86	Shadow moire FEA	Pos:Smile Neg:Cry	Die thickness	35.967	um	
					EMC thickness	-15.25	um	
					Carrier thickness	-17.65	um	
					IMC Temperature	35	um	
					Inplane C.T.E.	927.97	um	
[12]	PoP Modeling (PCR) Principal Component Regression	PKG: 12x12x0.6mm 14x14mm Chip: 8.47x8.24x(0.06-0.1)mm	FEA	Pos:Smile Neg:Cry	Die Elastic modulus	0.22	GPa	
					Mold Elastic modulus < Tg	-1.08	GPa	
					Mold Elastic modulus > Tg	-20	GPa	
					Mold C.T.E. >Tg	-1		
					Substratestrate Elastic modulus < Tg	-1.08	GPa	
					Die width	16	mm	
					Die thickness	0.5	um	
					Pkg Size	44	mm	
					EMC Thickness	0.146	um	
					EMC Elastic modulus < Tg	0.475	GPa	
					EMC C.T.E. <Tg	2.633	ppm	
					Die area	-11.476	ratio	
					Die thickness	-0.1857	um	
					Substrate Elastic modulus < Tg	0.28	GPa	
					Substrate C.T.E. <Tg	-1.1	ppm	
[13]	FOWLP	PKG: 15×15×0.2 mm 2 dies 8x9, 2x3mm Wafer: 12"(R=150mm)	FEA	Pos:Smile Neg:Cry	EMC Elastic modulus < Tg	16	GPa	
					EMC C.T.E. <Tg	125	ppm	
					EMC Tg	-17.5	deg.c	
					Mold Temperature	25	deg.C	
					PMC Temperature	0.28	deg.C	
[14]	BGA strip	BLK: 42x52 PKG: 12x12x1.0mm Chip: 1.25x1 25x0.33mm	Actual measurement	Pos:Smile Neg:Cry	EMC Shrinkage	565.7	0.001	17.012078
					EMC C.T.E. >Tg {W 175c-W250c)	1.13	ppm	--
[15]	PoP (Top,Bottom)	Top BGA PKG: 12x12x0.7mm Chip: 8x8x0.05mm 6x6x0.05mm Bot: PoP PKG: 12x12x0.45mm Chip: 5x5x0.05mm	FEA	Pos:Cry Neg:Smile	Substrate C.T.E. <Tg	5.8	ppm	7.086700
					EMC C.T.E. <Tg	-13.7	ppm	-15.590500
					PKG thickness ratio (EMC/Chip)	-24.6	1x	-7.479100
					Substratestrate C.T.E. <Tg	15.8	ppm	19.337000
					EMC C.T.E. <Tg	-11.4	ppm	-13.569400
					PKG thickness ratio (MC/Chip)	-32		-15.084900
					Die Substratestrate thickness ratio (Substrate / Die)	29.6	1x	3.582700
					Die Substratestrate size ratio (Substrate/Die)	29	1x	4.025600
					Die Substrate thickness ratio (Substrate / Die)	-4.2	1x	-1.806700
					Die Substrate size ratio (Substrate/Die)	-18.7	1x	-3.832800
[16]	EMC/Cu	Coupon board 40x10x1.7	FEA	Pos:Smile Neg:Cry	Mold Temperature	30	deg.C	-232.834200
					Mold pressure	--	kgf/cm2	39.897100
[17]	BGA Strip	Strip 240x74(0.265+0.182)mm	FEA	Pos:Cry Neg:Smile	Mold Temperature	25.7	deg.C	299.905405
					Close cure time	3.5	sec	-2.702703
[18]	EMC/ Substrate	PKG: 13x9x0.74mm	Shadow moire	Pos:Smile Neg:Cry	Die thickness	-0.827	um	-17.7530
					Substrate core thickness	-0.327	um	-6.7736
					D/A Tg	--	--	
[19]	FC BGA	PKG: 35x35x2.0mm Chip: 0.775umt Substrate: 1.2mmt	FEA Shadow moire	Pos:Smile Neg:Cry	Under-Fill Elastic modulus	8.16	GPa	
					Under-Fill C.T.E.	--	--	
[20]	PoP (Bottom)	PKG: 8x10.5mm Chip:5x9.7mm PKG:10x10.5mm Chip:9.3x9.5mm PKG(Bot): 15x15m Chip: 4.1x4.1mm, 8.3x8.3mm 12x12mm Chip thickness: 0.1, 0.05mm	FAE Therm Moire	Pos:Smile Neg:Cry	EMC Elastic modulus < Tg	4.17	GPa	
					EMC C.T.E. <Tg	1.4	ppm	
					Chip thickness	1.19	um	
					Chip Size	20	mm	
[21]	PoP (Top)	PKG: 12x12x1.0mm Chip: 1.25x1.25x0.33mm	FEA Laser or S/M	Pos:Smile Neg:Cry	EMC C.T.E. <Tg	21	ppm	27.537095
					EMC Elastic modulus < Tg	--	--	
[22]	FOWLP	PKG: 12x15mm Chip: 8x9mm 4x9mm (12x9mm)	FEA	Pos:Smile Neg:Cry	Chip thickness	-0.075	um	
					EMC thickness	-0.2	um	

Table 3: Normalized warpage factors

	Norm. rate	Contribution %
CC++ factors	**100.5698**	**23.150%**
EMC a1 (C.T.E, =<Tg)	17.0472	3.924%
EMC E1 (Elastic modulus, =< Tg)	33.0503	7.608%
EMC Shrinkage ratio	0.3934	0.091%
EMC/Die thick ratio (MC/Chip)	7.4791	1.722%
Mold Close cure time	2.7027	0.622%
Mold pressure	39.8971	9.184%
CV-- factors	**-333.8559**	**76.850%**
Die Substrate size ratio (Sub/Die)	-4.0256	0.927%
Die Substrate thickness rario (Sub	-3.5827	0.825%
Die thickness	-17.7530	4.087%
EMC a2 (C.T.E. >Tg)	-3.6298	0.836%
EMC E2 (Elastic modulus, >Tg)	-2.9463	0.678%
EMC Tg	-20.9007	4.811%
Mold temp	-266.3698	61.315%
Substrate a1 (C.T.E., =<Tg)	-7.8744	1.813%
Substrate thickness	-6.7736	1.559%
Warpage (absolute)	**434.4257**	**100.000%**

Warpage Model by BN

BN's outcome is a value of probability with a range from 0.0 to 1.0. While warpage is a bidirectional mechanical change and it's defined as CV(+) and CC(-). Therefore, neutral value of BN outcome is set at 0.5. If the outcome is more than 0.5, it warps toward CV direction. Oppositely, it warps toward CC direction in case of the value is less than 0.5. Figure 3 illustrates the warpage direction by BN outcome value.

Warpage behavior is a result of combination of CV and CC results. Therefore, the result is defined

Figure 3: Warpage direction by BN outcome value

as warpage (W) and causal factors can be defined as CV and CC respectively as shown in Figure 2. Each CV, CC factors have second level factors. In the case of CV, secondary factors are Die, EMC-cv, Mold Temp, Substrate (Sub) and similarly CC's secondaries are EMC and Die thick ratio (EMC/Die), EMC-cc, Mold. The model prefers to be a hierarchy structure to minimize the combinations in a CPT. As it's described previously, the table combination increases by 2^n, if the factors are defined in a sole layer. As the initial value of each factors, it's set at 0.5 which is the neutral value. When a factor value is changed, the warpage level is influenced by its contribution rate which is shown in Table 3.

Established BN model of this study is shown in Figure 4.

The nodes which only have output arrows such as Die thickness (thk), EMC E2, Mold Temp, Cure time, and so on in this model are called as "parent node". These nodes define 2 states value as True and False respectively. In this model the value is set as a change value by increasing or decreasing into its neutral value of 0.5.

The value range should be within 0.0 to 1.0, of course. Other secondary nodes have each CPT. For instance, node named "Die" has a CPT with 3 parent nodes; Die/Sub Size, Die/Sub thk, and Die thk. Therefore, the table contains 2^3 conditions by True/False cases. Similarly, Warpage node also has a CPT which contains 2^2 conditions with CV and CC. The table used in this study is shown in Table 4.

Table 4: CPT of P(W|CV, CC)

		P(W)	
CV	CC	0	1
0	0	0.2315	0.7685
0	1	0	1
1	0	1	0
1	1	0.7685	0.2315

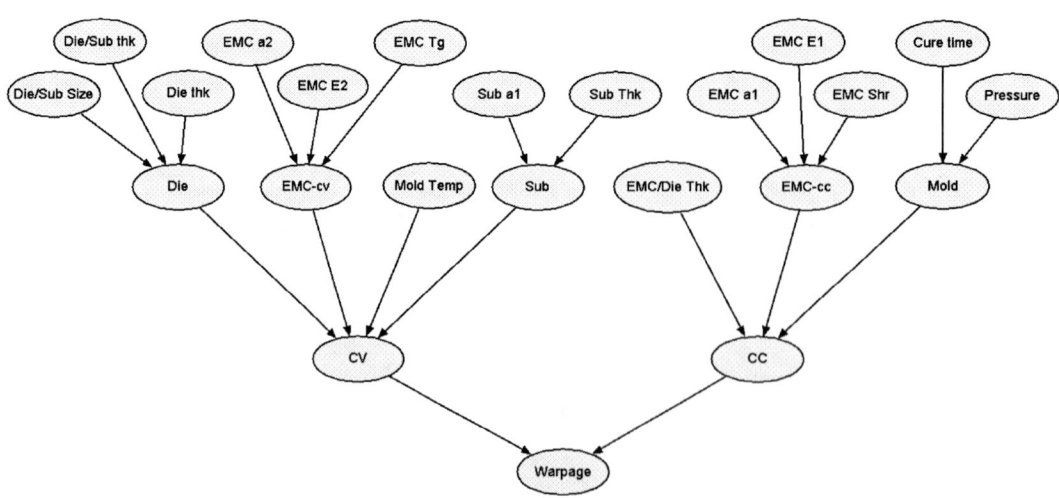

Figure 4: BN conceptual model of Package Warpage

At the initial state, all BN nodes' value is set at 0.5. Therefore, the warpage outcome value becomes also 0.5. By changing the factor values, the outcome value changes accordingly. And the change value for all factors should be within 0.0 to 1.0. For example, when EMC-a2 increases the value from 0.5 to 0.6, following EMC-cv, CV value changes then Warpage outcome value changes accordingly. In this case, the Warpage value changes to less than 0.5 because EMC a2 belongs to CV factor.

How CC and CV value influences W value, can be seen a chart in Figure 5. CC and CV values influence W value with its warpage contribution rate as it's shown in Table 3.

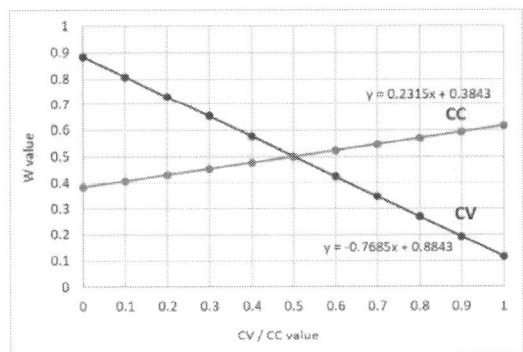

Figure 5: CV / CC curve in this BN model

Case Study

A BN model for this package warpage study was created as shown in Figure 4. Each nodes' results can be calculated by using spread sheet software but for future portability and extensibility, it's suggested use a BN software. In this conceptual BN model study, Hugin Lite v8.5 software was used to verify effectiveness and validity of the established model.

In order to verify the effectiveness, 8 articles were selected. [10], [14-18], [21], [27] Those articles were for full molded BGA package cases. As it described previously, the warpage behavior is different with the package structure. In this study, a BN model was created for full molded BGA.

Warpage factors were extracted from each article and examined into the model. The factors can be checked in Table 2. How to examine with the model is, for example, when the EMC's modulus (EMC E1) is going to change from 8GPa to 14GPa, the change rate is 1.75x. The change rate reflects into the factor value in the model. Therefore, the given value becomes 0.875 which is 1.75x of its neutral value 0.5. Then, 0.875 is set in True side value, 0.125, which is given by 1 - 0.875, is set in False side value of the factor value. In this example, warpage (W) value, 0.5285, is given as a result from the model. This W number is 0.0285 increase from the neutral value of 0.5. Therefore, this E1 change makes 0.0285x increase of concave warpage risk from its original state. This W value is able to convert to an actual predicted warpage

value. For example, predicted warpage of case [10] is given by following equation.

$(0.5285 - 0.5) \times 434.4257 \times 12/2 \times \sqrt{2} = 210.1$ (µm)
Where 434.4257 is the total warpage value of this model and $12/2 \times \sqrt{2}$ is half length of diagonal of 12x12mm Package.

As a typical example of the case study, Case [27] is introduced how it was examined. The study was how EMC influenced the warpage. The study introduced 11 types of EMCs and those material properties were given as elastic modulus below Tg (E1), CTE below Tg (a1), CTE above Tg (a2), glass transition temperature (Tg), EMC shrinkage rate (Shr). Table 5 shows the summary of the given material properties. In the table, EMC type 1A is used as a reference material. Other EMC types are compared with this 1A material. With this, the warpage result has either CC or CV direction from an origin warpage of 1A. Table 6 shows converted ratio based on the properties of 1A. These values use to determine a set value of warpage factors in the BN model. The determined values for each factor are shown in Table 7. Those values are simply multiple of 0.5 for the converted ratio. Then, these values are set into the BN model so that warpage change value, which is relative to 1A, can be obtained by each EMC type.

Table 5: Material properties of the studied EMCs

	E1	a1	a2	Tg	Shr
1A (ref)	12.3	12	29	209	0.034
1B	13.1	11	37	196	0.022
1C	11.4	11	31	187	0.033
1D	12.8	11	35	177	0.051
1E	12.6	12	35	168	0.081
MCA	12.6	9	35	189	0.054
2A	13.5	11	45	180	0.011
2B	12.4	10	38	171	0.020
2C	11.3	11	42	161	0.030
2D	13.5	11	41	154	0.026
2E	13.3	12	34	147	0.049

Table 6: Converted value ratio against 1A

	E1	a1	a2	Tg	Shr
1A (ref)	1	1	1	1	1
1B	1.065	0.917	1.276	0.938	0.647
1C	0.927	0.917	1.069	0.895	0.971
1D	1.041	0.917	1.207	0.847	1.500
1E	1.024	1.000	1.207	0.804	2.382
MCA	1.024	0.750	1.207	0.904	1.588
2A	1.098	0.917	1.552	0.861	0.324
2B	1.008	0.833	1.310	0.818	0.588
2C	0.919	0.917	1.448	0.770	0.882
2D	1.098	0.917	1.414	0.737	0.765
2E	1.081	1.000	1.172	0.703	1.441

Table 7: Set values for the BN model and W results for each EMCs

	E1	a1	a2	Tg	Shr	W
1A (ref)	-	-	-	-	-	-
1B	0.5325	0.4583	0.6379	0.4689	0.3235	0.5008
1C	0.4634	0.4583	0.5345	0.4474	0.4853	0.4980
1D	0.5203	0.4583	0.6034	0.4234	0.7500	0.5033
1E	0.5122	0.5000	0.6034	0.4019	1.1912	0.5101
MCA	0.5122	0.3750	0.6034	0.4522	0.7941	0.4983
2A	0.5488	0.4583	0.7759	0.4306	0.1618	0.5028
2B	0.5041	0.4167	0.6552	0.4091	0.2941	0.5000
2C	0.4593	0.4583	0.7241	0.3852	0.4412	0.4993
2D	0.5488	0.4583	0.7069	0.3684	0.3824	0.5066
2E	0.5407	0.5000	0.5862	0.3517	0.7206	0.5096

W values in Table 7 show the warpage level and direction relative to 1A. The values resulted less than 0.5 implies the warpage direction toward CV. Oppositely, it implies CC direction when the value is greater than 0.5. Those W values are used to calculate its predicted warpage change value and it's added or subtracted into the warpage value of 1A. The predicted warpage value can be seen in a summary table, Table 8.

With this same manner, other 7 cases are also examined, and predicted warpage value was given accordingly. Table 8 shows a summary of predicted warpage value of BN and FEA value which was extracted from the articles.

The BN model does not intend to be a replaceable model with FEA, so the outcome is not going to chase the gap of FEA outcome value. However, it should not be far from the FEA results. Figure 6 shows a plot chart of BN vs FEA outcome values. How the BN is far from FEA can be seen a distance from BN=FEA line on the chart. It's observed that the BN model is prone to increase difference with FEA results in greater than 100µm warpage region. In the SMT process, greater than 100µm warpage cannot be acceptable, hence, for risk assessment purpose, values within a range of +/-100µm is considered as predictable.

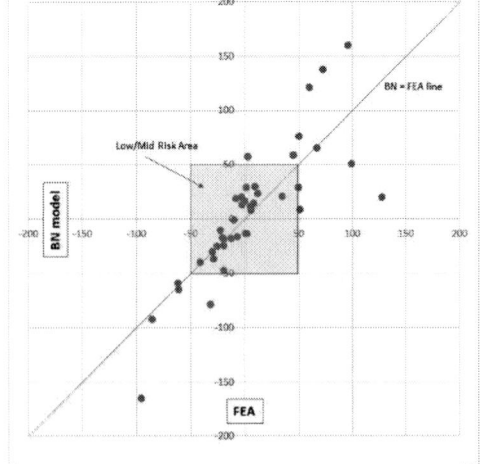

Figure 6: BN model vs FEA

This BN model intends to assess the warpage level when some material or process parameters are changed. Therefore, the BN outcome is used as an assessment value. The risk assessment is often classified into 3 level of Low, Middle, and High. In this model, the risk is distinguished into 3 levels with the predicted value. Table 8 shows a risk classification by coloring in green, yellow, and red respectively. Green indicates, the warpage result was in a range within 0 to +/-25µm. Yellow is within greater than +/-25µm and less equal +/-50µm. Red is greater than +/-50µm. In case of BGA package, the warpage needs to be less than 100µm in terms of its safe SMT process, hence, warpage within 25µm can be state as safe region. More than 50µm change will become a potential risk region of SMT issue. That is a reason of the color setting and risk classification. The risk assessment results are summarized in Table 9. Some of assessment results showed gap between BN and FEA in the classification. This gap is discussed in following paragraph.

If it's considered as a conservative judgement, 3 cases (1 and 2) on 0~25 column of FEA result may be acceptable. And 2 cases on >50 column of FEA can be barely captured the risk as Middle level risk in the BN model. With this consideration, as a result, in the total 40 assessment cases, 36 out of 40 (90%) were fit well with this BN model though the gap needs to be clarified.

Table 8: Warpage value and Risk level summary of BN model and FEA (µm)

	BN	FEA		BN	FEA
Case[10]	1.5	29.0	Case [17]	-22.5	-10.0
	2.9	58.0	Case [18]	-20.6	-17.5
	49.4	29.0		-41.2	-39.5
	98.8	51.0		-61.8	-59.5
	-6.6	-16.0		-19.2	-24.0
	-19.2	-47.0	Case [21]	8.1	14.7
	-31.7	-78.0		11.8	23.7
	50.9	9.0		59.7	121.6
	127.5	20.0		72.2	138.0
	-12.5	-18.0		95.8	160.0
	-25.8	-25.0	Case [27]	-2.4	13.3
	-61.2	-65.0		5.9	8.0
Case [14]	34.7	21.0		-9.8	-0.5
	-2.9	20.0		-29.9	-29.9
	9.6	30.0		5.0	11.6
Case [15]	66.7	65.6		-8.3	18.7
	-85.3	-92.6		0.0	16.4
	-95.9	-165.0		2.1	-13.5
	50.2	76.0		-19.5	-18.3
	44.7	59.0		-28.4	-36.1
			Case [16]	1058.6	940.0

Note: Case [16] is strip warpage value.

Table 9: Risk Summary Table

N=40 (µm)		FEA		
		0 ~ 25	>25 ~ 50	>50
BN	0 ~ 25	16	3	1
	>25 ~ 50	1	5	2
	>50	2	0	10

Low	Mid	High

Discussion

Case studies showed assessment results of the package warpage with the BN model and 90% of the cases were fit well into the model. However, there were some gaps between the BN model and FEA results. Especially in Table 9, 4 cases which are from 2 cases on a row of FEA 0~25, 1 and 2 cases on FEA >50 row, those are critical when it's evaluated the risk level. Because the risk may be overlooked due to the result as Green from the BN model. In the table, colored in yellow can be marginal case so it may be neglected. In those gaps, most of them were seen in Case [10]. The case was a study of EMC's mechanical properties such as elastic modulus and CTE. The FEA results were given with a full molded BGA package but it was stacked die case. In the stacked die package, EMC and die thickness ratio in a package is changed and warpage behavior changes by its ratio. While EMC thickness is thicker than Die, the warpage is prone to change CC direction. Opposite case the warpage changes to CV direction. Therefore, the ratio should have been input as one of CC factor separately. Because of this behavior difference, the BN value in Case [10] could be higher than FAE.

In other case review, Case [21] showed almost 2x difference in BN and FEA. The study was for EMC properties and similar with Case [10]. CTE was originally calculated and called as "effective CTE". A reason of this introduction was, CTE varied by process conditions. This CTE value made the warpage value difference compared to other study results. This was a reason of the gap and it could be seen that the value gap became bigger when the outcome value became high. Similar result could be seen in a result in Case[15]. BN and FEA result showed 2x difference, as shown the value, -95.9, -165.0 respectively. This result was because of EMC and Die thickness ratio and the original FEA curve changed at around 4.0. When die thickness becomes 4x thicker than remained EMC thickness in a package, the warpage value and direction changed. Other studies and BN model results showed a reasonable relation.

In reviews of the BN model results, a gap from the warpage result comes from modeling type of FEA. The BN model shows good fit with FEA, when FEA model uses linear modeling and material properties. As shown in Figure 5 and the equation (2), BN results were given by the proportional equation. This is a limitation of this BN model. On the other hands, as [23] described in their warpage study, non-linear model which used viscoelastic properties improved predicted modeling results for the warpage. Further, [13] and [15] also predicted non-linear warpage behavior with functions of die size and die thickness. If the BN model is modified with those non-linear properties and equations, the model may be able to close gaps with the FEA results.

Conclusion

Risk assessment by using BN model was studied by comparing FEA results. The BN model could demonstrate about 90% of fit in 3 level risk analysis. There were some gaps in the assessment results, but it was a limitation of linear modeling feature of this model. In terms of risk assessment, the BN model could confirm its effectiveness and conceptual methodology, though the model was examined in limited number of BGA package cases. Non-linear properties and extensive examinations with various package types will be able to improve the model quality. With this study, BN model was verified its feasibility and effectiveness to be used as a risk assessment tool.

Reference

[1] W. P. Cheah, K. Kim, H. Yang, S. Choi, and H. Lee, "A Manufacturing-Environmental Model Using Bayesian Belief Networks for Assembly Design Decision Support," 20th International Conference on Industrial, Engineering and Other Applications of Applied Intelligent Systems, Kyoto, Japan, June 26-29, pp. 374–383, 2007.

[2] Y. Zhu and A. Deshmukh, "Application of Bayesian decision networks to life cycle engineering in Green design and manufacturing," Eng. Appl. Artif. Intell., vol. 16, no. 2 SPEC., pp. 91–103, 2003.

[3] L. Yang and J. Lee, "Bayesian Belief Network-based approach for diagnostics and prognostics of semiconductor manufacturing systems," Robot. Comput. Integr. Manuf., vol. 28, no. 1, pp. 66–74, 2012.

[4] S. Dey and J. A. Stori, "A Bayesian network approach to root cause diagnosis of process variations," Int. J. Mach. Tools Manuf., vol. 45, no. 1, pp. 75–91, 2005.

[5] H. Langseth and L. Portinale, "Bayesian networks in reliability," Reliab. Eng. Syst. Saf., vol. 92, no. 1, pp. 92–108, 2007.

[6] N. Li and Z. Lu, "Reliability Assessment Based on Bayesian Networks for Full Authority Digital Engine Control Systems", 11th International Conference on Reliability, Maintainability and Safety (ICRMS), 2016.

[7] Richard Neapolitan, "Learning Bayesian Networks", Pearson Prentice Hall, NJ, Chapter 1, pp. 1-62, 2004

[8] P. B. Lin, C. Ko, W. Ho, C. Kuo, K. Chen, Y. Chen, T. Tseng, and U. T. Corp, "A Comprehensive Study on Stress and Warpage by Design , Simulation and Fabrication of RDL-First Panel Level Fan-Out Technology for Advanced Package", Electronic Components and Technology Conference, 2017.

[9] K. M. Chen, K. K. Ho, P. C. Kuo, and C. Y. Wu, "Bill of Material and Geometry on FCBGA

Packaging Warpage Impacts", IMPACT, pp. 345–348, 2013.

[10] M. J. Yim, R. Strode, R. Adimula, C. Yoo, and B. V Numonyx, "Effects of Material Properties on PoP Top Package Warpage Behaviors", Electronic Components and Technology Conference, pp. 1071–1076, 2010.

[11] F. Hou, T. Lin, L. Cao, F. Liu, J. Li, X. Fan, and G. Q. Zhang, "Experimental Verification and Optimization Analysis of Warpage for Panel-Level Fan-Out Package", TRANSACTIONS ON COMPONENTS, PACKAGING AND MANUFACTURING TECHNOLOGY, vol. 7, no. 10, pp. 1721–1728, 2017.

[12] P. Warpage, P. Lall, K. Patel, and V. Narayan, "Model for Inverse Determination of Process and Material Parameters for Control", Transactions on components, packaging and manufacturing technology, vol. 5, no. 9, pp. 1358–1375, 2015.

[13] F. X. Che, D. Ho, M. Z. Ding, and X. Zhang, "Modeling and Design Solutions to Overcome Warpage Challenge for Fan-Out Wafer Level Packaging (FO-WLP) Technology", Electronics Packaging Technology Conference, 2015.

[14] T. Ahsan, H. Tang, and H. Corporation, "Mold Compound Properties for Low Warpage Array Package," in 10th Electronics Packaging Technology Conference, 2008, pp. 1421–1426.

[15] C. Ren and F. Qin, "Parametric Study of Warpage in Package-on-Package Manufacturing", ICEPT-HDP, pp. 339–343, 2009.

[16] L. Hong, S. Hwang, H. Lee, and D. Hung, "Simulation of Warpage Considering Both Thermal and Cure Induced Shrinkage during Molding in IC Packaging", Polytronic 2004, pp. 78–84, 2004.

[17] M. Chae and E. Ouyang, "Strip Warpage Analysis of a Flip Chip Package Considering the Mold Compound Processing Parameters", Electronic Components & Technology Conference, pp. 441–448, 2013.

[18] C. Y. Huang, T. D. Li, and M. Y. Tsai, "Warpage Measurement and Design of wBGA Package under Thermal Loading", IMPACT, pp. 415–418.

[19] M. Y. Tsai and H. Y. Chang, "Warpage Measurement and Simulation of Flip-Chip PBGA Package under Thermal Loading", Int. Conf. on Thermal, Mechanical and Multiphysics Simulation and Experiments in Micro-Electronics and Micro-Systems, Euro Sim E, pp. 145–148, 2008.

[20] W. Sun, W. H. Zhu, C. K. Wang, A. Y. S. Sun, and H. B. Tan, "Warpage Simulation and DOE Analysis with Application in Package-on-Package Development", 2008.

[21] H. Tang, J. Nguyen, J. Zhang, and I. Chien, "Warpage Study of a Package on Package Configuration", Proceedings of HDP, 2007.

[22] H. Li, A. Chen, S. Peng, G. Pan, and S. Chen, "Warpage Tuning Study for Multi-chip Last Fan Out Wafer Level Package", Electronic Components & Technology Conference, pp. 1384–1391, 2017.

[23] M. Amagai and Y. Suzuki, "A Study of Package Warpage for Package on Package (PoP)", Electronic Components & Technology Conference, pp. 226–233, 2010.

[24] M. Chae and E. Ouyang, "Strip Warpage Analysis of a Flip Chip Package Considering the Mold Compound Processing Parameters", Electronic Components & Technology Conference, pp. 441–448, 2013.

[25] D. Y. Gui, L. J. Ernst, D. G. Yang, L. Goumans, and P. Semiconductors, "Study of the Effects of Molding Pressure on the Warpage of HVQFN Packages", 6th International Conference on Electronic Packaging Technology, 2005.

[26] J. W. Y. Kong, J. Kim, S. Member, M. M. F. Yuen, and S. Member, "Warpage in Plastic Packages: Effects of Process Conditions, Geometry and Materials," vol. 26, no. 3, pp. 245–252, 2003.

[27] K. I. Y. Chien, J. Zhang, L. Rector, and M. Todd, "Low Warpage Molding Compound Development for Array Packages," pp. 1001–1006, 2006.

Flex Cracking of Multilayer Ceramic Capacitors:
Experiments on Fracture Propagation

Erik Wiss, Vlad Serea, Steffen Wiese

Saarland University, 66123 Germany

+4968130271825, s9erwiss@stud.uni-saarland.de

Abstract

This paper presents a comparison between a three point and a four point flexural board test. Multilayered ceramic capacitors were soldered onto the board using SAC305 solder while force and capacitance measurements were acquired during the bending. Additionally pictures of the board were taken to determine the curvature radius of the board. The capacitors are then metallographic prepared for microscopy of the cross sections, which offer a precise picture of the origin and propagation of the crack.

Key words: MLCC, reliability

1. Introduction

Multilayer Ceramic Capacitors are the backbone of most electronic circuits and are widely used in numerous applications due to their large capacity range and high volumetric efficiency. MLCC reliability has become of more concern in recent years, mainly because of increasing demand of capacitors. Depending on the application field they have to sometimes withstand large mechanical stresses or temperature variations which led to standardized testing procedures.

These capacitors usually have a brittle ceramic body which makes them prone to cracking especially while undergoing a bending load. Mechanical stresses induced by bending or large temperature variations combined with remaining stresses from the manufacturing process and board integration often lead to component failure [1].

When a MLCC fails due to cracking, there are more possible outcomes that lead to the functional break down of the electric circuit. One of the more harmless consequences would be the loss of capacity. A more dangerous outcome is the formation of hot spots on the board, which leads to damage in the vicinity of the capacitor [2].

Figure 1: Top view of a MLCC 1206 specimen, that was soldered on PCB stripe having the same width as the capacitor. The layout of the copper structures consisted on two well defined soldering pads for the MLCC and on two test pads that allow a capacitance measurement during the bending test

2. Experiment

Reliability tests are usually divided into tests on the component itself and tests with components mounted on PCBs. IPC/JEDEC and ASTM qualification tests [3,4] proposed standard four point test configurations for soldered MLCCs. However, standards introduced for components in the area of automotive electronics proposed a three point bending method where the pin that displaces the board has a radius of 340 mm. This provides a different stress distribution when compared to a sharp center pin.

A three point bending apparatus was built in order to compare the results with a four point technique in order to better understand the influence of the testing method on the failure of capacitors. Whereas three point bending delivers the maximum stress right underneath the capacitor, four point bending offers constant stress along a longer portion of the board.

Both bending fixtures are driven by a stepper motor (Nanotec AR59-H-50) while force and capacitance measurements are taken during the bending with a digital multimeter (Rigol DM3061) and a force sensor (KD24S). In addition, a Canon camera is mounted parallel to the plane of bending in order to be able to measure the curvature radius of the board at the time of failure. Figure 2 presents the equipment used.

The ram used to displace the board was an aluminum cube of 20 mm side length with the lower faced curved at a radius of 140 mm [5] while the fixed supports were placed 90 mm away from each other. This radius was chosen in order to achieve higher displacement of the boards. The 320 mm ram proposed by [6] would only allow a displacement of 2 mm before starting to push the board with its edges turning it into a four point test. Such a small displacement is not enough to generate any MLCC

597

failure. For the four point fixture, the bending rolls were placed at a 45 mm distance.

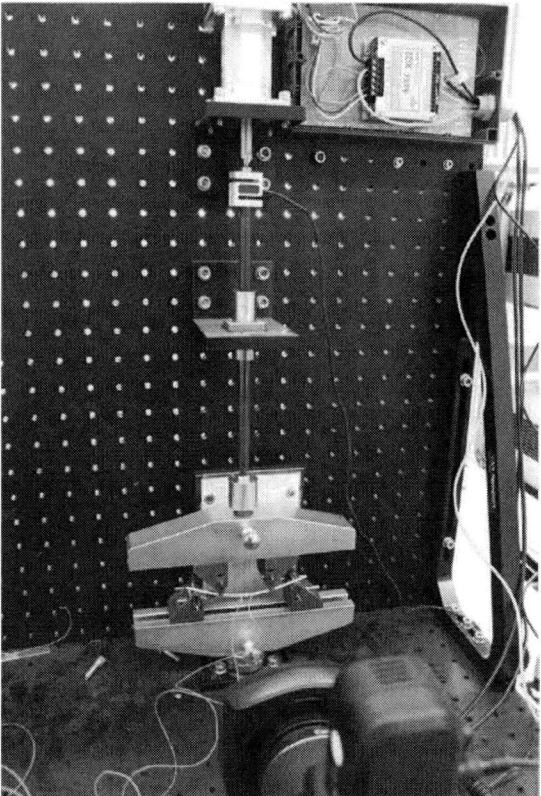

Figure 2: Four point bending apparatus

The width of the PCB is 1.6 mm and corresponds to the width of the capacitor. The thickness is also 1.6 mm and the copper layer is 40 μm thick. The specimen chosen in this experiment is a MLCC with 1206 size and 0.1 μF capacitance. The cross section can be seen in Figure 3.

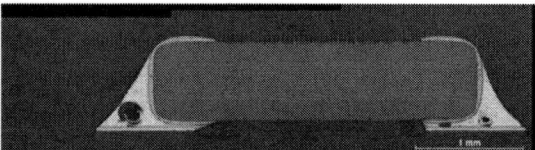

Figure 3: Cross section of the MLCC before bending, showing the alternating electrode layers throughout the ceramic body as well as the metallization and the solder shape

Even though it has not been subjected to any tests, the capacitor shows small horizontal cracks at the ceramic-termination interface which can be the effect of thermal shock when the capacitor cools down after soldering.

The ceramic body is made out of BaTiO3 dielectric with X7R temperature characteristics. The solder used was SAC305 (Sn96.5Ag3.0Cu0.5) with a melting point at 217 °C. A total of 20 capacitors were tested in each fixture.

Control of experiment and data acquisition was done by a LabVIEW script which also triggered

the camera at specific time points. A capacitor was considered failed when a 10% decrease in capacitance was measured. The experiment is then stopped and the curvature radius of the board at that time is measured, as seen in Figures 4 and 5.

3. Results

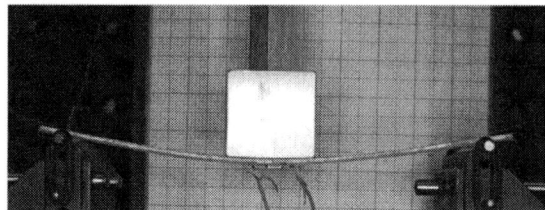

Figure 4: Measured radius of 220 mm at a displacement of 4742 μm using the three point ram

The radius was used to calculate the strain of the board at the moment of failure and the acquired data is subjected to a statistical analysis. A two parameter Weibull distribution was calculated with the maximum likelihood method using the following distribution and density functions:

$$F(x) = 1 - e^{\left(\frac{x}{\eta}\right)^{\beta}}$$

$$f(x) = \frac{\beta}{\eta} * \left(\frac{x}{\eta}\right)^{\beta-1} * e^{-\left(\frac{x}{\eta}\right)^{\beta}}$$

where η is the scale parameter and β the shape parameter. The likelihood function [7] is described by

$$L(\eta, \beta) = \prod_{i=1}^{n} [f(x_i)^{\delta_i} * [1 - F(x_i)]^{\delta_i - 1}$$

Some of the capacitors did not show a voltage drop even after the PCB broke, that data being censored from the Weibull distribution.

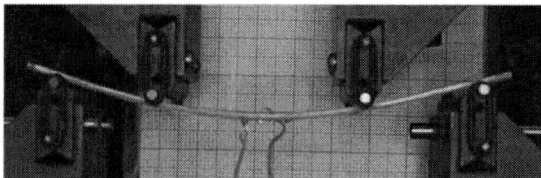

Figure 5: Measured radius of 148 mm at a displacement of 4200 μm using the three point ram

The results can be seen in the Weibull distribution plots for the two cases. Using the three point bending fixture, the capacitors break at much lower strain compared to the four point mechanism.

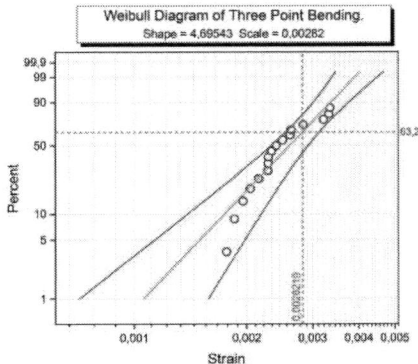

Figure 6: Weibull probability plot of the three point bending experiment

Microscopy analysis of the cross section shows the crack propagation throughout the ceramic. The cracks originate at the interface between the metallization and the ceramic body where the stress transmitted through the board is at its highest. Most of the cracks are K shaped and have a large angle. An example of such a crack is presented in Figure 7.

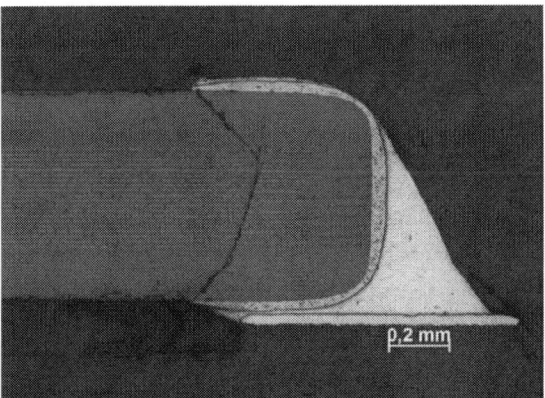

Figure 7: Crack in one of the MLCC specimens used with the three point bending fixture

While using the four point fixture we observed that more capacitors do not show a drop in capacitance (5 compared to 2 in the three point fixture). Microscopy analysis confirms the absence of any cracks (Figure 8).

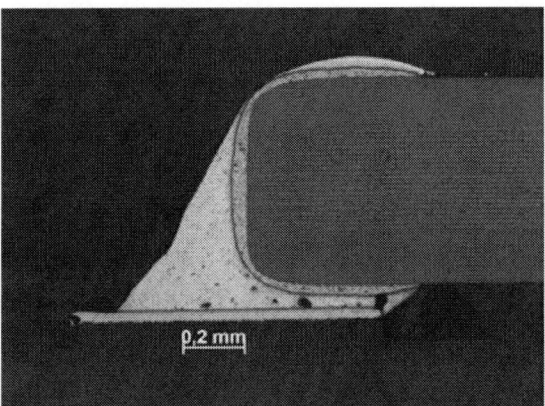

Figure 8: One of the MLCC specimens that did not show a capacitance drop during bending

The shape parameter is also lower, indicating that the distribution is more spread out, as seen in Figure 9.

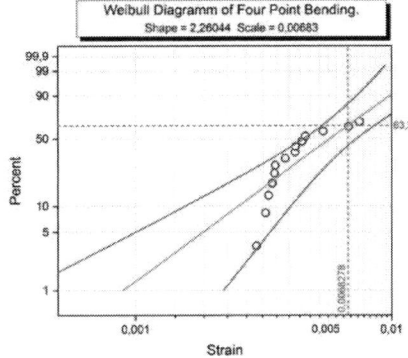

Figure 9: Weibull probability plot of the four point bending experiment

Figures 10 and 11 present typical cracks observed after bending with the two mechanisms. We notice that the crack path goes throughout the whole height of the capacitor between the ends of the terminal metallization. We also notice more crack lines originating next to the main crack which are usually shorter and do not reach the end of the ceramic body.

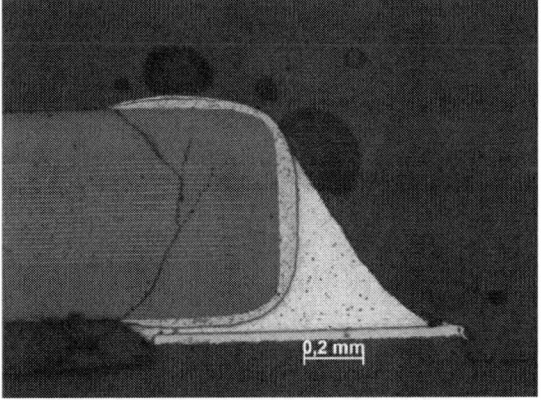

Figure 10: Crack in one of the MLCC specimens used with the three point bending fixture

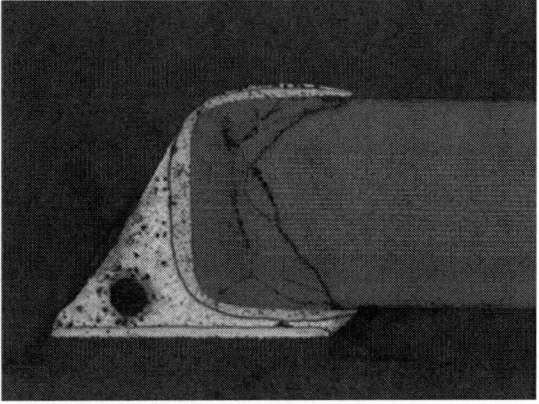

Figure 11: Crack in one of the MLCC specimens used with the four point bending fixture

4. Conclusions

The bending experiments done in this study present the impact of the ram on capacitor bending strength. Comparison with the three point mechanism indicates that using a four point ram generates less stress at the same displacement, which leads to capacitor failure at a higher strain.

Acknowledgements

We would like to thank Dr. Andreas Ruh for the metallographic preparation of the samples.

References

[1] Teverovsky, A., "Guidelines for Selection, Screening and Qualification of Low-Voltage Commercial Multilayer Ceramic Capacitors for Space Programs", Rev. A, Technical Report GSFC.DPW.7499.2012, NASA Goddard Space Flight Center; Greenbelt, MD 2012.

[2] Steward, M., "AVX MLCC FlexiTermTM: Guarding Against Capacitor Crack Failures", AVX technical paper, http://www.avx.com/docs/techinfo/mlccflex.pdf , visited on February 2018.

[3] IPC/JEDEC-9702, "Monotonic Bend Characterization of Board-Level Interconnects", IPC, Bannockburn, Illinois (2004).

[4] ASTM C1161 - 02a "Standard Test Method for Flexural Strength of Advanced Ceramics at Ambient Temperature", American Society for Testing and Materials, West Conshohocken, Pennsylvania (1990).

[5] EIA Proposed Draft PN-2271, Rev. G, 1993, Electronic Industries Association.

[6] Industries Association.AEC-Q200-005 REV A, „Board Flex Test", June 1, 2010.

[7] W. Q. Meeker and L. A. Escobar, "Statistical Methods for Reliability Data" John Wiley & Sons, 2014.

Investigating the Overdischarge Failure on Copper Dendritic Phenomenon of Lithium ion Batteries in Portable Electronics

Chengcheng Chen*, Guanghui He, Jinbao Cai, Zhenbo Zhao and Daojun Luo

China Electronic Product Reliability and Environmental Testing Research Institute (CEPREI), Guangzhou 510610, China

13920693795, and chenchengcheng@ceprei.com

Abstract

Lithium-ion batteries (LIBs) have been widely applied in portable electronics as a power source. Nevertheless, LIBs also become the potential safety hazard for their high activity and flammable organic component, which threatens the safety of consumers. Failure analysis of LIBs is useful to discovery their defective technology and explore the failure mechanism, guiding the improvement of LIBs reliability. Herein, we report on the investigation of overdischarge failure phenomenon of LIBs and its internal short-circuit mechanism. A nondestructive X-ray computed tomography strategy in conjunction with microzone analysis has been employed to study the failure mechanisms. Firstly, nondestructive test shows that the failure initiating point is located at the bottom center of cell core with the most serious wrinkles. Secondly, microzone analysis and simulated tests display that dissolution of copper foil on anode occurs when cell voltage below 2.5 V. Then many copper dendrites generate during the recharge process. Additionally, growth of copper dendrite is the main reason leading to the internal microshort-circuit, the decomposition of SEI layers and closure of separator. As condition of LIBs continuously deteriorates, thermal runaway of LIBs occurs when the heat accumulates to a critical point. Moreover, the relationship between copper dissolution and cell voltage have been further investigated. Result offers information that designing cut-off voltage protection is an effective way to avoiding the overdischarge failure of LIBs.

Key words: Overdischarge, Failure analysis, Lithium-ion battery, Nondestructive analysis, Copper dendritic

1. Introduction

Rechargeable lithium-ion batteries (LIBs) are widely applied in portable electronic devices owing to their high energy density, long cycling life and environmental friendly, which is the main recyclable power source for mircorelectronics products. With the development of portable electronic devices, LIBs with the highly integrated, high rate capability and high power density are urgently required. This may bring about the potential safety hazard as well, which threatens the safety of consumers.[1-2] Therefore, enhancing the investigation of reliability evaluation and failure analysis of LIBs is an effective way to discovery the defects and increase quality of electronic products. Although the importance of the failure analysis of LIBs is obvious, researches of failure analysis is still in infancy.[3-4] Facing the various failure modes of LIBs in portable electronics, it is significant to deeply study the mechanism failure of LIBs and provide a practical strategy to avoid their security issues.

Failure modes of LIBs could be divided into explicit behaviors and implicit behaviors. Explicit behaviors are the phenomenons that could be obseverd through surface shape change or damage, including explosion, buring, swelling, weeping and breakage. Implicit behaviors are the phenomenons that could not be directly discovered without the further disassembly, nondestructive detection or microscopic analysis, including interal micro-short, generation of Li dendrite, separator aging, electrolyte drying and cell core deformation.[5-6] Therefore, failure analysis of LIBs is comprehensive project owing to their variable failure modes. It generally includes background research, visual examination, nondestructive test, electrochemical evaluation, disassemble analysis, micro-zone analysis, environmental tests and failure reinspection. This strategy could estimate the failure modes and causes, forecast and prevent the unprepared failure behaviors, control and manage the failure extension.[7-8] Although researches on failure analysis of LIBs have been studied and obtained some achievements, the investigations are always focus on part materials of LIBs instead of the whole device or battery, which ignores the applied background of LIBs and the working state of electric device.[9-11] Thus, researches that based on the application of LIBs to judge their failure modes are urgently needed, which may be a high-effective and facile way to understand the failure mechanism and improve their reliability.

Herein, we report on the analysis of burning failure of LIBs in a used portable electronic device, which is attributed to the implicit behaviors leading to the explicit failure during its long-time service. We have carefully explored the investigation of overdischarge failure phenomenon of LIBs and its internal short-circuit mechanism, and tried to understand their relationship. Nondestructive X-ray computed tomography strategy combined with microzone analysis are employed to locate the failure site and explore the failure mechanmise. Additionally, overdischarge simulation experiment is applied to verify the Cu dissolution and Cu dendrite deposition phenomenon, implying the relation between copper dissolution and cell voltage.

2. Experiment and methodology

The failure analysis of LIBs was composed of electrical test, structure analysis, nondestructive testing and microzone analysis. Then, the verification test was employed to review the analysis results.

The interial structure of invalid battery was measured by nondestructive 3D X-ray computed tomography (CT) with a Y. CT Precision S instrument at 100 kV and 7 mA for the X-ray studies. The morphology and elemental compositions of the electrode materials in a disassembled battery were determined by scanning electron microscopy coupled with energy dispersive X-ray diffraction spectroscopy (SEM and EDS, S-4300 & Genesis-60, operating at 5 kV), and the Cu^{2+} content was determined by an Agilent 5100 inductively coupled plasma - optical emission spectrometry (ICP-OES). Electrochemical discharge process was operated by a CHI660E electrochemical work station.

3. Resluts and Discussions

The background information of invalid battery has been carefully inquired. Firstly, the invalid battery is placed in a portable electronic device, which have been used for about one year, and then disused for about half a year. Secondly, the failure accident of battery occurs on the recharging process, and the device had never been charged for about half a year. Thrid, the failure phenonmen of battery is burning, which is an explicit behavior due to the long-time implicit interal breakdown. The failure analysis of battery should exmain two aspects, battery protection module and unit cell.

The battery protection module has been disassembled, and carried out the functional tests and electrical test. The functional testing results reveal that all the parameters of battery protection module in the failed LIB comply with its specifications shown in Table 1. Besides, the electrical test (Figure 1) shows that the battery protection modules of noraml battery and invalid battery perform the same V-I profile. These tests verify the functional integrity of battery protection modules in invalid battery, which could reasonably eliminate the failure of battery protection modules leading to the burning of LIB.

Table1. The functional tests result of battery protection module.

No.	Function	Tested parameter	Specifications
1	Overcharge protection	4.28 V	4.28±0.025 V
2	Over discharge Protection	2.997 V	3.0±0.075V
3	Over discharge current protection	2.0 A	1.3~4.0A
4	Short-circuit protection	OK	OK
5	Current comsumption (Operation)	5.2 µA	Max 6.0 µA
6	Current comsumption (Power down)	0.16 µA	Max 6.0 µA
7	Internal Impedance	83 mΩ	Max 100 mΩ

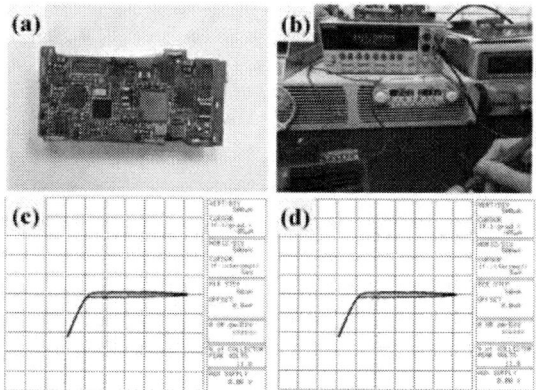

Figure 1: (a) image of battery protection module (BPM). (b) The comparsion of electric tests between the BMP in (c) noraml battery and (d) invalid battery.

Then, the invalid LIB has been carefully analyzed through the nondestructive testing, disassembled experiment, micro-analysis and simulation testing. To clearly observe the electrode piece distorted location and extent without any damages, computed tomography (CT) has been employed and typical three views of the battery displayed in Figure 2. Top view of battery (Figure 2a) shows the serious distortion of core and peeling of active materials, which located on the line A and B sections. Front view images (Figure 2b) clearly exhibit the drastic stratification and wrinkles in the center of battery, where is near lead layer. Side view images of A and B sections (Figure 2(c, d)) displays that the serious delamination of electrode pieces concentrates on below of battery. It implies that failure locus located on the bottom area. The nondestructive testing gives a valid information that the bottom of battery has occurred severe wrinkle, delamination and peeling of active materials, which could be reasonably attributed to the extreme deterioration of electrode in this area before the failure. Thus, the internal structure and materials at bottom of battery should be paied more attentions.

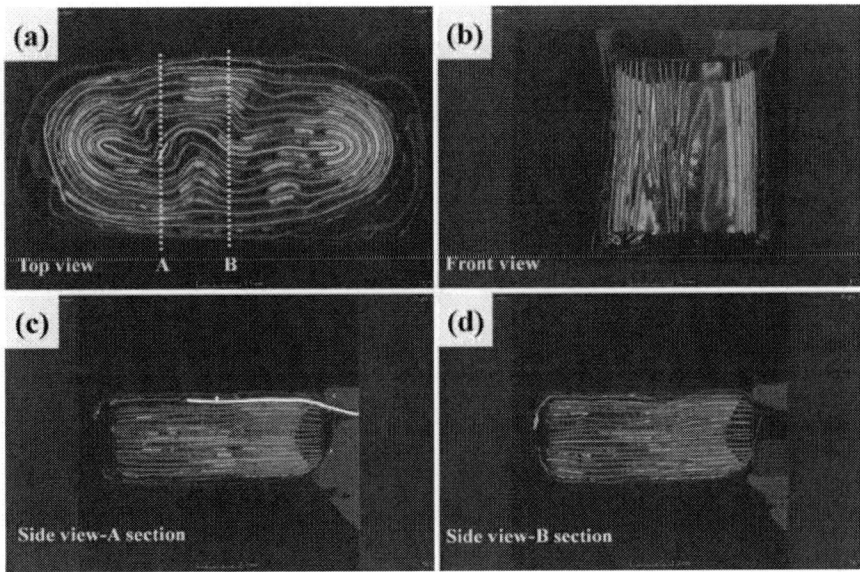

Figure 2: CT images of the invalid battery with three section views: (a) top view, (b) front view, (c) side view-A section, (d) side view-B section.

The inner change of anode, cathode and current collectors should be analyzed through battery disassembly. Figure 3 shows that the failed battery occurs serious swelling compared with the normal battery, and the aluminum film is burnt through. More importantly, the bottom of battery has a big hole owing to the rapid release of thermal stress. Besides, there are some silver metal precipitate on the edge of electrode, which could be ascribed to the melt of Al current collector during the thermal runaway process. Disassembled battery displays that both the Al current collector and part of graphic anode have been burnt out, and only $LiCoO_2$ cathode and Cu current collector with severe destruction remain. Therefore, $LiCoO_2$ cathode and Cu current collector become the only clue to find the reason of failure.

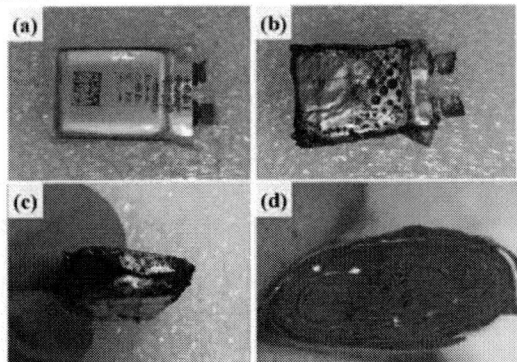

Figure 3: Images of (a) normal battery and (b) invalid battery. (c) The most serious failure point of battery. (d) image of invalid cell core.

To further analysis the change of cathode, anode materials and Cu current collector, scanning electron microscope (SEM) is employed. Compared the morphology and particle size change between normal battery and invalid battery (Figure 4), it reveals that the anode material of normal battery is relatively flat with compacted structure, and the binder holds up all the particles together. In contrast, the anode of invalid battery displays slack and undulating surface without linking between blocks and dispersive active particles. Similar phenomenon has been observed in the cathode, and there are some side reaction products adhering to the surface of active materials. It implies the serious pulverization and structural construction of anode and cathode during repeated Li^+ insertion/extraction in invalid battery.

Then, the selected Cu current collector located in the failure triggering area has been carefully analyzed shown in Figure 5. Notably, a lot of dendritic crystals are observed on the surface of Cu current collector. The dendritic crystals exhibit the dense needlelike morphology with the length of 2 μm. EDS result (Figure 5c) indicates the main elelment of dendritic crystals is also Cu. Moreover, there are numerous pits on the surface of Cu current collector. This could be ascribed to the dissolution of Cu. The anode Cu metal loss electrons convering to the Cu^{2+} during the over discharge process, which are free in the electrolyte ($Cu \rightarrow Cu^{2+} + 2e^-$). When the battery is recharged, the free Cu^{2+} would obtain the electrons on the surface of anode, and deposit into the dendritic Cu crystals ($Cu^{2+} + 2e^- \rightarrow Cu$). The dendritic crystals will continuously grow up and keep the needlelike morphology, which may pierce the separator and cause the internal short. This may be root of the safety hazard.

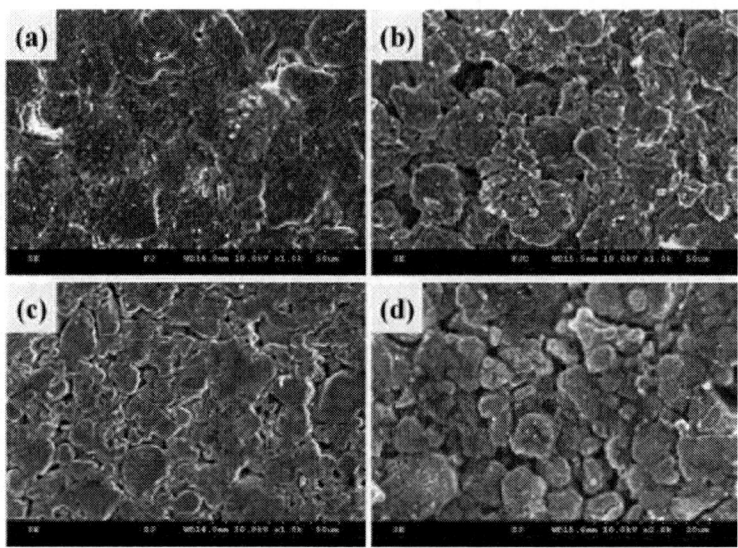

Figure 4: SEM images of anode materials in (a) normal battery and (b) invalid battery. SEM images of cathode materials in (c) normal battery and (d) invalid battery.

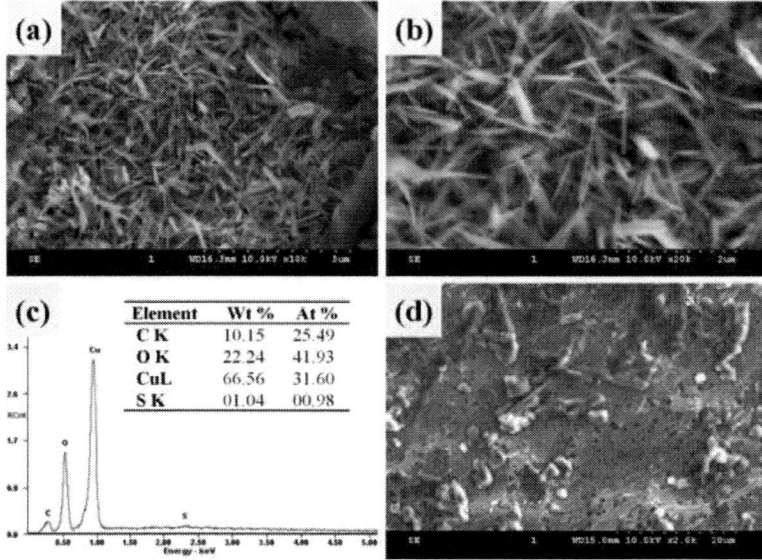

Element	Wt %	At %
C K	10.15	25.49
O K	22.24	41.93
CuL	66.56	31.60
S K	01.04	00.98

Figure 5: (a-b) SEM images of copper dendritic on Cu current collector and (c) their EDS image. (d) SEM image of the surface of anode current collector

Nevertheless, it is significant to discovery the reason of overdischarge and potential of battery leading to the Cu dissolution.

According to the capacity data of battery, the remanent capacity of battery is about 60 mAh when the potential of battery below 2.8 V. Meanwhile, based on the specification data of battery protection module, its current consumption is about 20 μA. Background informations tell us the device disused for about half a year, it means that the battery has never been charged for about 4380 hours. The energy consumption of battery protection module is approximately 87.6 mAh, which is more then the remanent capacity of battery. That is to say the

potential of battery should reduce to 0 V before the device reused. Therefore, the simulation test of overdischarge has been operated using the same type battery (Figure 6). It reveals that the concentration of Cu^{2+} gradually increase. When the potential of battery is 1.0 V, the concentration of Cu^{2+} is 300 mg/L, However, when the potential blew 1.0 V, the dissolution rate of Cu^{2+} rapidly enhance. When the potential decrease to 0 V, the concentration of Cu^{2+} reaches as high as 1100 mg/L. This results indicate the main reason of overdischarge is the energy consumption of battery protection module. When the potential of battery reaches 0 V, free Cu^{2+} with high concentration has generated in electrolyte. They may deposit on anode and become the dendritic crystals after recharging process, which may cuase the internal short of battery leading to the burning of battery.

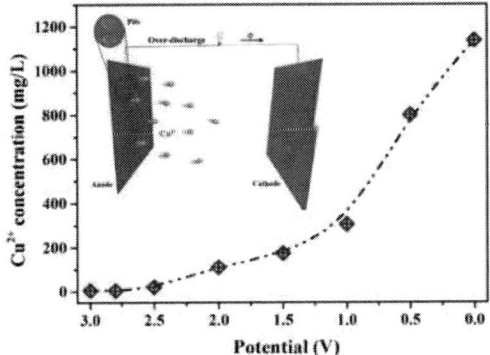

Figure 6: Cu^{2+} concentration curve related with the battery potential, insertion image: schematic diagram of Cu dissolution in discharging process.

Conclusion

In conclusion, analysis shows that the burning failure of battery in a used portable electronic device is owing to the implicit behaviors leading to the explicit failure during its long-time service. Firstly, nondestructive test shows that the failure initiating point is located at the bottom center of cell core with the most serious wrinkles. Secondly, microzone analysis and simulated tests display that dissolution of copper foil on anode occurs when cell voltage below 2.5 V. Then many copper dendrites generate during the recharge process. Additionally, growth of copper dendrite is the main reason leading to the internal microshort-circuit, the decomposition of SEI layers and closure of separator. As condition of LIBs continuously deteriorates, thermal runaway of LIBs occurs when the heat accumulates to a critical point. Therefore, it is significant to design overdischarge cut protection mechanise, and use the thicker separator to enhance the safeft of battery.

Acknowledgements

This work was financially supported by the National Natural Science Foundation of China (21801048), Material Testing and Evaluation Platform-Electronic Materials Industry Center (by MIIT).

References

[1] S. K. Babu, A. I. Mohamed, J. F. Whitacre, S. Litster, "Multiple imaging mode X-ray computed tomography for distinguishing active and inactive phases in lithium-ion battery cathodes", Journal of Power Sources, Vol. 283, pp. 314-319, February, 2015.

[2] R. Cartera, B. Huhmanc, C. T. Lovea, I. V. Zenyukd, "X-ray computed tomography comparison of individual and parallel assembled commercial lithium iron phosphate batteries at end of life after high rate cycling", Journal of Power Sources, Vol. 381, pp. 44-55, 2018.

[3] L. Zielke, T. Hutzenlaub, D. R. Wheeler, I. Manke, T. Arlt, N. Paust, R. Zengerle, S. Thiele, "A Combination of X-Ray Tomography and Carbon Binder Modeling: Reconstructing the Three Phases of LiCoO$_2$ Li-Ion Battery Cathodes" Adv. Energy Mater. Vol. 4, pp. 1301617, 2014.

[4] C. Chen, Y. Wei, Z. Zhao, Y. Zou, D. Luo, "Investigation of the swelling failure of lithium-ion battery packs at low temperatures using 2D/3D X-ray computed tomography", Electrochimica Acta, Vol. 305, pp. 65-71, 2019.

[5] S. Saxena, Y. Xing, M. Pecht, "A Unique Failure Mechanism in the Nexus 6P Lithium-Ion Battery", Energies, Vol 11, pp. 841, 2018.

[6] X. Lai, Y. Zheng, L. Zhou, W. Gao, "Electrical behavior of overdischarge-induced internal short circuit in lithium-ion cells", Electrochimica Acta, Vol 278, pp. 245-254, 2018.

[7] H. Maleki, J. N. Howard, "Effects of overdischarge on performance and thermal stability of a Li-ion cell", Journal of Power Sources, Vol 160, pp. 1395-1402, 2006.

[5] M. Ouyang, Z. Chu, L. Lu, J. Li, X. Han, X. Feng, G. Liu, "Low temperature aging mechanism identification and lithium deposition in a large format lithium iron phosphate battery for different charge profiles", Journal of Power Sources, Vol. 286, pp. 309-320, 2015.

[8] C. Chen, H. Guo, Z. Zhao, S. Li, Z. Jiang, D. Luo, Y. Wang, "A robust strategy for engineering Li4Ti5O12 hollow micro-cube as superior rate anode for lithium ion batteries", Electrochimica Acta, Vol. 293, pp. 141-148, 2019.

[9] L. Kong, C. Li, J. Jiang, M. Pecht, "Li-Ion Battery Fire Hazards and Safety Strategies", Energies, Vol 11, pp. 2191, 2018.

[10] C. Fear, D. Juarez-Robles, J. A. Jeevarajan, P. Mukherjee, "Elucidating Copper Dissolution Phenomenon in Li-Ion Cells under Overdischarge Extremes", Journal of The Electrochemical Society, Vol. 165 (9), pp. A1639-A1647, 2018.

[11] R. Guo, L. Lu, M. Ouyang, X. Feng, "Mechanism of the entire overdischarge process and overdischarge-induced internal short circuit in lithium-ion batteries", Scientific RepoRts, Vol 6, pp. 30248, 2016.

Time-dependent Behaviour of Bonded Silicon Dies under Subcritical, Constant Shear Loading

N. Pflügler[1*], R. Pufall[1], M. Goroll[1], S. Breitenreiner[1], and B. Wunderle[2]

[1]Infineon Technologies AG, Am Campeon 1-15, 85579 Neubiberg, Germany
[2]Technische Universität Chemnitz, Reichenhainer Str. 70, 09126 Chemnitz, Germany

*Phone: +49 89 23426205, E-Mail: nadine.pfluegler@infineon.com, ORCID: 0000-0002-7415-7532

Abstract

We explore the capabilities of creep testing for the reliability and robustness assessment of die attach layers within this paper. To this end we conducted finite element simulations and experiments. We optimised our setup in terms of thermal stability and measurement accuracy with the introduction of a temperature monitoring system, and with the installation of an active cooling system of our load cell. Sample alignment to the chisel was optimised with the introduction of a new rotational Degree of Freedom and a rotation system. Finite element simulation results and experiments revealed that it is possible to detect defects within the die attach layer using the die creep test, but the investigation of thin die attach layers is challenging, as displacements are within the submicron range. The two biggest influences on the measurement results are the bond line thickness and the temperature.

Key words: electronics packaging, reliability, adhesion, creep testing, shear testing, die attach

Introduction

When conducting reliability assessments of semiconductor packages it is important to consider the validity and sensitivity of testing. This is particularly true for the study of failure mechanisms and failure modes on dedicated test vehicles. Reliability of electronic components depends on the application profile and on the desired lifetime. The ambient temperature inside the engine compartment, for example, is around 125 °C once the operating temperature of the car is reached. If a semiconductor product is operated within this environment the junction temperature can easily reach 150 °C and beyond. Electro-thermal simulations show that the die attach material is also subjected to temperatures close to junction temperature in case of thin silicon dies. In order to assess the reliability or even robustness of connected materials against delamination, dedicated experiments and finite element simulations are utilised within the development phase of the semiconductor package. Experiments used to study adhesion are typically conducted with short-time loading of the sample under test (e.g. die shear testing, button shear testing). Little experience exists in regards to the behaviour of the interfaces under long-lasting, constant subcritical loading, which is also closer to the actual application conditions.

We are studying the time-dependent behaviour of semiconductor dies attached to copper alloy substrates using arbitrary die attach materials with the method proposed in [1]. In earlier investigations we detected "spontaneous" delamination under constant loading for a polyimide-moulding compound interface, once we subjected it to a subcritical loading condition for several hours [1]. The investigation of time-dependent behaviour under subcritical loading is therefore important for a better understanding of this failure mode and the driving failure mechanism.

In this paper we are reporting on our results of silicon dies bonded to a lead frame substrate by an adhesive. Especially, the impact of defects inside the die attach layer (e.g. voids, delamination, pores) on the mechanical behaviour under constant loading is of interest. The presence of two interfaces (silicon/adhesive and adhesive/lead frame) in combination with the typically minimised bond line thickness of adhesive die attach layers renders testing and the interpretation of measurement results into a challenging task.

Finite Element Simulation Results – Potentials and Limits of Die Creep Tests

A finite element simulation model aids to explore potentials and limits of the application of creep testing, especially, in terms of detecting and distinguishing defects inside the die attach layer. Defects are voids or delaminations, for example, which can be introduced due to improper choice of process parameters or dispensing patterns, or due to contaminated adherend surfaces. Simulations were conducted using Abaqus [2]. Figure 1 depicts the simplified geometrical model and the boundary conditions. The substrate (lead frame) is omitted, and the silicon die is replaced by a 50 μm thick, stiff tungsten carbide plate, because we wanted to focus on the behaviour of the die attach material. Interfaces are assumed to show perfect adhesion, i.e. damage criteria are not implemented. The bond between the visco-elastic die attach material and substrate is represented by a (pinned) boundary condition, prohibiting displacements in all three translational directions (x, y and z). Both the plate and the die attach layer are 2 mm long and 2 mm wide. We applied a constant force of 40 N (reasonable force value for tests under shear loading in our experimental setup) to one side of the plate for one hour, similar to the creep experiment.

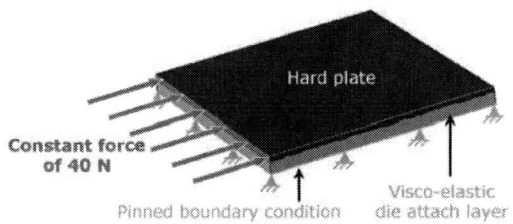

Figure 1: Simplified simulation model of the creep test.

The simulation study is qualitative, therefore, we used an arbitrary material model for the die attach layer, and residual stresses were omitted. The die attach layer's visco-elastic behaviour is modelled using a Prony series of 30 terms with a glass transition point T_g around 35 °C. Simulations were

conducted with a defect-free die attach layer as reference, and varying defective states, e.g. interfacial delamination, voiding, and insufficient wetting. Furthermore, we conducted our simulations at two temperatures, 25 °C (below T_g) and 50 °C (above T_g), to study the impact of temperature. Creep curves (displacement-time curves) were extracted from the simulation for die attach material thicknesses of 10 µm (Figure 2) and 100 µm (Figure 3).

A comparison of maximum values of the displacement axis in graphs a) and b) in Figure 2 and Figure 3 respectively reveals that the displacement does increase with increasing temperature, as expected. Polymer chains can move and detangle more easily with increasing temperature, due to the rise in free volume. Moreover, we see a strong influence of the die attach layer thickness on the possibility to differentiate between defective states. For the thick die attach layer at room temperature (Figure 2a) the creep curves can be clearly separated. This is not the case at the higher temperature (Figure 2b). This behaviour is inverse for the thin layer (Figure 3). At room temperature most defective states result in a creep curve that overlaps with that of the reference, e.g. the defect-free state. An increase in temperature on the other hand, leads to a fan-out of the curves.

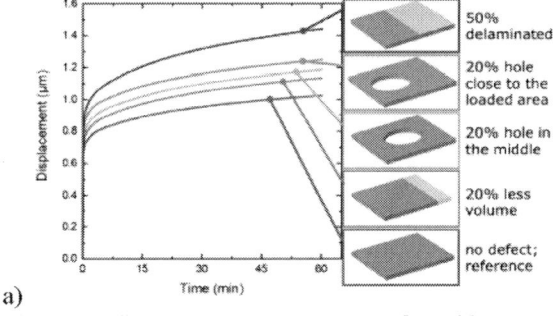

a)

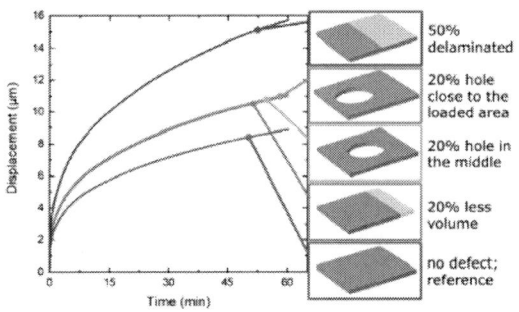

b)

Figure 2: Simulated creep curves of a 100 µm thick die attach layer under 40 N shear loading a) at 25 °C, and b) at 50 °C. Both graphs depict the behaviour of die attach layers with different states of defect. Pictures on the right of each graph give information on the type of defect.

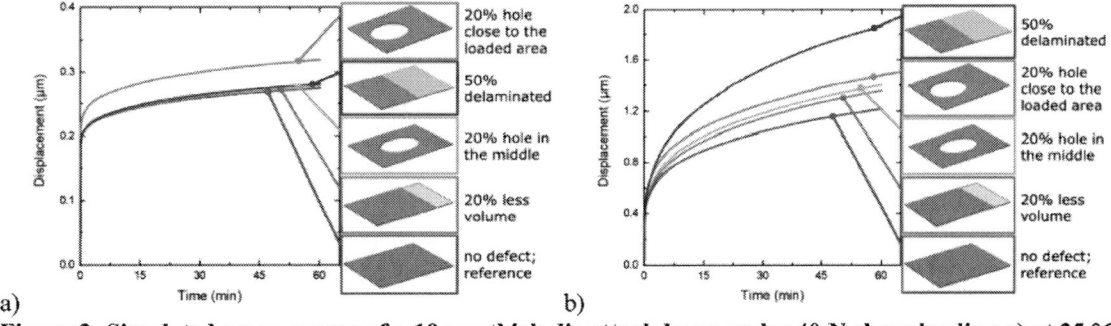

Figure 3: Simulated creep curves of a 10 µm thick die attach layer under 40 N shear loading a) at 25 °C, and b) at 50 °C. Both graphs depict the behaviour of die attach layers with different states of defect. Pictures on the right of each graph give information on the type of defect.

Both can be explained with the strength of the coupling between plate and substrate (Figure 4). The die attach material acts as the coupling agent. A smaller volume of coupling material leads to a stronger coupling effect, and therefore to a decrease in the total deformation. Furthermore, coupling can be weakened in case of defects, which are within the deforming volume part of the die attach layer. Thus, defects located further away from the loaded edge of the die might not show a significantly different behaviour compared to the reference sample for thin die attach layers (Figure 3a). Above T_g the Young's modulus is reduced, the displacement is therefore higher, and aforementioned defects can be distinguished more easily (Figure 3b).

In terms of total displacement values the simulation results for the thin die attach layers (Figure 3) point to another limit of the method concerning metrology. The measurement equipment needs to be able to detect displacements in the submicrometre range. This is especially important with the drive for thinner die attach layers, with the intention to reduce the thermal resistance of the whole package. Our simulation results show, on the contrary, the thinner the die attach layer, the lower the possibility to trigger creep behaviour inside the adhesive layer.

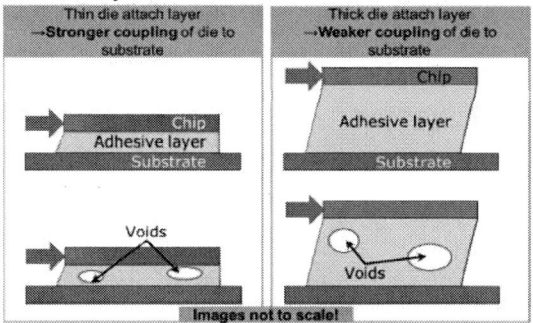

Figure 4: Differences in the loading condition of the die attach layer based on its thickness (thin: left, thick: right). Schematics depict a side view of a bonded die under shear loading with and without defects. Images are not to scale.

Experimental Setup

To accurately measure in the micrometre scale, a tester with very high machine stiffness is needed. The tester used for the experiments is a specifically designed combined shear and creep tester by Kammrath & Weiss GmbH (Figure 5). In addition, the test vehicle behaviour is monitored from the side using a high-resolution camera system in combination with the digital image correlation software VEDDAC [3]. An extensive study on the main influences during testing and information about the calibration procedure can be found in [4].

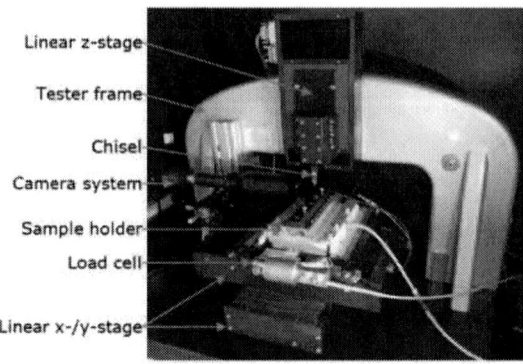

Figure 5: Photograph of the combined shear and creep tester highlighting the important components.

We further optimised our experimental setup in order to increase the measurement accuracy [5]. This comprises the installation of a temperature monitoring system, the installation of an active cooling system close to our load cell, and the introduction of a new rotational degree of freedom to compensate for small sample misalignments. These measures are described in detail within the following sections.

Thermal Stability of the Experimental Setup

In a first step we verified the thermal stability of our experimental setup described above. Creep tests are conducted over several hours and usually at temperatures above room temperature (up to 200 °C in our case). By implementing an Arduino-based temperature monitoring system, we are able to track temperatures at different locations of the equipment using calibrated type K thermocouples. We set our heating stage to 180 °C and monitored the temperature at the heat stage, at the linear stage, and at the load cell for 20 hours. Heat stage and load cell are thermally decoupled using ceramic plates. Nonetheless, this decoupling is not perfect, which leads to an increase in temperature at the load cell of roughly 25 °C within eight hours. After that a thermal equilibrium is reached and its temperature remains constant. We noticed an influence on the measurement results when conducting shear tests within this time frame. The temperature compensation of the load cell was insufficient for measurement durations of creep tests.

We were able to solve this problem with the implementation of an active liquid cooling system. Thermal energy, which is conducted from the heat stage, through the ceramic plates and the linear stage to the load cell, is now absorbed by three cooling plates. The plates are custom-made to fit at each side of the linear stage and directly underneath the load cell (Figure 6). Thermal conductivity between cooling plates and linear stage is improved with thermal grease. The temperature of the cooling water is controlled by a 200 W chiller.

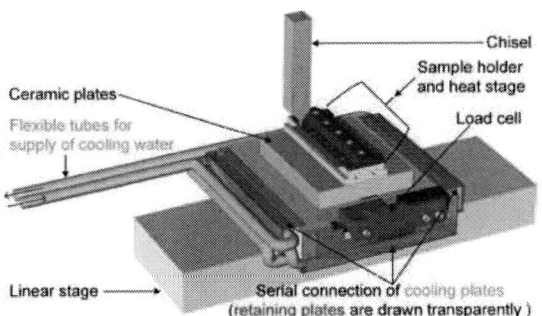

Figure 6: Three-dimensional drawing of the experimental setup including the newly installed components of the water cooling system.

Introduction of a New Rotational Degree of Freedom

We reported previously on the impact of small angular sample misalignments between the chisel and a sample of rectangular shape [4], i.e. a semiconductor die or a cuboidal moulding compound button. Misalignments lead to an asymmetric loading condition of the interface or of the die attach material, and hence, to an earlier damage initiation. It was therefore necessary to introduce a rotational degree of freedom to enable a defined sample rotation.

The linear x- and y-stages are installed on top of a rotary stage, which was set tight using four locking levers. We added a possibility for a defined rotation with the installation of two screws with fine threads (Figure 7). They are used to press one of the linear stages to the side. The chisel is the axis of rotation. An angular scale with nonius allows us to control the alignment of the sample to the chisel in steps of 0.02°. The alignment can be checked using a second camera system that is directed at the sample from the top.

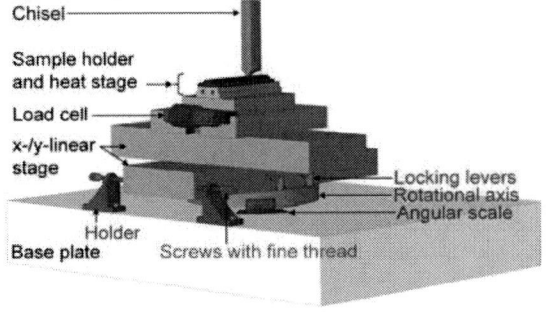

Figure 7: Three-dimensional drawing of the experimental setup including the newly installed components of the rotational axis.

Sample Preparation and Characterisation

We bonded unstructured silicon dies (2x2 mm² in size, 150 µm thick) to a 200 µm thick copper alloy lead frame material. The adhesive used in this study is a non-conductive half epoxy half acrylic resin with a glass transition temperature T_g of approximately 40 °C. It contains organic spacer

particles to produce a uniform bond line thickness of 20 µm, and silica filler particles for improved thermal conductivity and adaption of the Coefficient of Thermal Expansion to silicon and typical copper alloy substrates. The total volume percentage of filler particles is in the range of 40% according to its data sheet. The curing took part at 175 °C for 30 minutes in nitrogen atmosphere to avoid strong oxidation of the substrate surface.

Moreover, we want to investigate the impact of defects inside the die attach layer. Thus, we reduced the volume of the dispensed adhesive to create insufficient wetting. Figure 8 depicts infrared images (top views) of two samples at room temperature. Silicon is transparent in this wavelength range, and therefore, infrared light can be used here to characterise the die attach layer. The grey scale value gives information on the emissivity of the sample. On the left a sample with little defects inside the die attach layer is visible; the emissivity is nearly constant over the whole area of the silicon. The sample on the right was processed with insufficient wetting of the adhesive to the silicon and to the substrate. The black area stands for a lower emissivity in this non-wetted area.

Voids inside the adhesive layer

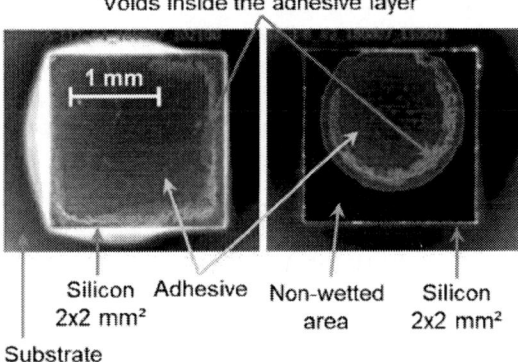

Figure 8: Infrared images (top view) of two samples. Left: Reference sample with little defects in the die attach layer. Right: Defective sample with insufficient wetting.

Experimental Results – Advanced die shear test

Based on the simulation results reported above, and our die attach layer thickness of 20 µm, we chose to run the creep tests at 45 °C, i.e. above the glass transition point. We conducted an initial advanced die shear test to determine the strength of the bonded sample. We did this for one die, where the adhesive wetted the complete area of the silicon back-side ("reference"), and for one sample with insufficient wetting ("defective"). Experimental parameters are listed in Table 1, and the resulting load-displacement curves are depicted in Figure 9.

For the reference sample, the maximum shear force F_{max} is 106 N, and the shear force at crack opening F_{crack} was determined to be 14 N using the image sequence. The crack is initiated early, and the failure mode is a cohesive failure of the die attach material. The defective sample failed at 47 N. Here, it is not possible to determine the crack initiation point, because the die attach layer is not visible in the image sequence, due to the insufficient wetting of silicon and substrate. The reason for the 56% decrease in maximum shear force is the smaller interconnection area, as material and processing conditions are identical. The diameter of the connected area of the defective shear sample is roughly 1.6 mm, which converts to a round interconnect area of 1.2 mm². This is a drop of 70% in bonded area in relation to the reference sample (4 mm²), which correlates to the decrease in maximum shear force.

The difference in the stiffnesses of the samples, i.e. the slopes of the force-displacement curves, can be explained with the interconnection shape. 30% of the silicon die's area, precisely the area at the edges of the silicon, does not have a connection to the substrate for the defective sample. The silicon is "floating" with the only connection being a small volume of cured adhesive in the middle of the die. When applying load from one edge the first reaction of the sample is the opening of the existing gap, resulting in a linear slope. On the contrary, the reference sample's die attach layer starts directly underneath the loaded edge of the silicon, which results in a higher stiffness.

Table 1: Experimental parameters of the advanced die shear test

Shear velocity	1 µm/s
Shear height	50 µm
Frame rate	1 fps
Temperature	45 °C

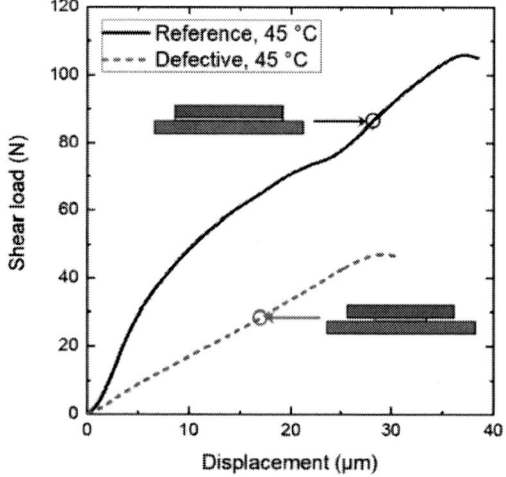

Figure 9: Advanced die shear test results of two samples at 45 °C. The schematics depict side views of the samples (not to scale).

Experimental Results – Advanced die creep test

The creep load needs to be subcritical to prevent damage to the sample, especially, during the load ramping phase. We chose a creep load of 7 N for both samples, which is lower than the shear load at crack initiation F_{crack} during the advanced die shear tests described above ($F_{creep} \cong 65\% \cdot F_{crack}$). Experimental parameters are listed in Table 2, and the creep curves are depicted in Figure 10.

Both curves do not show a steady increase in displacement, but rather a "bumpy" response to the constant force. The declining curve of the reference sample even suggests that the sample is moving away from the chisel. We assume both phenomena are connected partly to the control circuit of the creep tester. The force stayed constant during the measurement with small deviations in the range of less than ±0.1 N (Figure 11). Nevertheless, we see a correlation between the "bumps" in the force signal to the ones in the displacement curves, i.e. it is a measurement artefact due to our control system.

This also implies that we are not detecting a time-dependent reaction of the reference sample under the chosen conditions. The most probable reason is that the displacement in the range of 1 µm is at the lower limit of the measurement capabilities of our creep tester. Another possibility is that the die attach material, that we chose for the experimental part of the study, does not show excessive time-dependent behaviour at all or within the boundaries of the given layer thickness and the volumetric ratio of elastic filler and spacer particles to visco-elastic polymer matrix. The reaction can also be superimposed with the damage initiation and propagation at one of the interfaces or within the adhesive layer, i.e. creep is not the dominant effect under the chosen experimental conditions. We will investigate these topics further using different analysis and measurement methods. Furthermore, we are questioning and revising the utilisation of bulk material data used within finite element simulations of thin adhesive layers.

Nonetheless, the difference in behaviour between the reference sample and the defective sample is instantaneously visible. We will use these results for a more extensive study on the possibility to employ the creep test method for the detection of defects. This could aid with the assessment of reliability and mechanical robustness of die bonded systems.

Table 2: Experimental parameters of the advanced die creep test

Creep load	7 N
Shear height	50 µm
Frame rate	1/30 fps
Temperature	45 °C

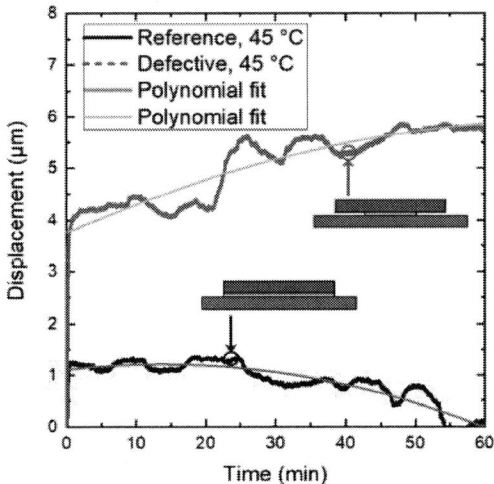

Figure 10: Advanced die creep test results of two samples at 45 °C. The schematics depict side views of the samples (not to scale).

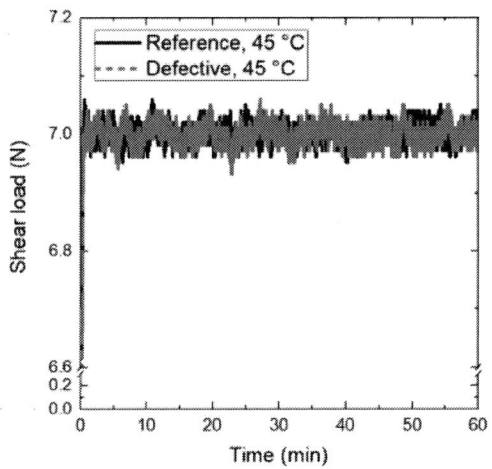

Figure 11: Force-time signal of the creep measurements.

Conclusion

In this paper we showed finite element simulations and results from experimental investigations of creep tests on silicon dies bonded to lead frame substrate using an adhesive. We also reported on the improvements of our test setup that were necessary to refine the measurement accuracy. We installed a temperature monitoring system to characterise the thermal stability of our experimental setup, to be able to keep a constant temperature at the sample for several hours. The investigation revealed that the activation of our heating stage influenced the load cell during extended measurement times. Therefore, we installed an active liquid cooling system to keep a constant temperature at the affected load cell. Moreover, we introduced a new rotational Degree of Freedom and a second camera system at the tester to be able to align our samples in reference to the chisel within angular steps of 0.02°.

The finite element simulation results of the die creep test show a clear influence of temperature and bond line thickness on the measurement results. Thin die attach layers (thickness around 10 µm) are better to be investigated at elevated temperatures (above the glass transition point). This has two advantages. First, it increases the total displacement due to the reduced Young's modulus of the adhesive. Second, it eases the detectability of defects. This is because the actively loaded volume of the die attach layer increases at elevated temperatures. Thick die attach layers (thickness around 100 µm) are, on the contrary, best investigated at room temperature. The actively loaded volume is already high at room temperature, making it easy to detect defects. Elevated temperatures increase the effect here as well, but the differentiation between defects (void close to or further away from the loaded edge) decreases.

Furthermore, simulation results show that a low total displacement in the submicron range is to be expected. This is challenging in terms of metrology, which we could also confirm during our creep experiments on die bonded samples. We investigated samples with a bond line thickness of 20 µm. The adhesive we used is a non-conductive adhesive with a filler content of 40%vol. At first we conducted advanced die shear tests to learn about the strength of the interconnect. We then chose the creep load to be in a subcritical range to not damage the sample by exceeding the ultimate strength of the interfaces or the adhesive. We investigated reference samples with full wetting of the adhesive on the silicon's and the substrate's surface. In contrast, we also prepared samples with a defined defect (insufficient adhesive volume), and compared it to the reference samples. The creep tests show that it is possible to clearly distinguish between samples using the displacement-time curves. Nevertheless, more though needs to be put into the control circuit, which introduced measurement artefacts. Furthermore, the reference sample did not show significant creep behaviour. On the one hand, this can be due to the limitations of our test setup, as we are already at the lower limit of the displacement measurement capabilities with 1 µm displacement of the sample. On the other hand, the adhesive itself might be the reason. It might not show excessive time-dependent behaviour at all or within the boundaries of the given layer thickness and the volumetric ratio of elastic filler and spacer particles to visco-elastic polymer matrix. Possible damage at one of the interface or within the adhesive layer can also be the reason for the behaviour observed. We will investigate these topics further with different analysis and measurement methods.

Acknowledgements

We thank Dr. Daniel May, Technische Universität Chemnitz, for assistance with infrared measurements.

References

[1] N. Pflügler, R. Pufall, M. Goroll, B. Wunderle, "Non-Destructive Die Attach Test Method for Future Robust Packages - a Proposal", Proceedings of the 2016 6th Electronic System-Integration Technology Conference (ESTC), Grenoble, September 13-16, pp. 1-4, 2016.

[2] 3DS SIMULIA, "Abaqus Unified FEA", Retrieved May 21, 2019, URL: https://www.3ds.com/products-services/simulia/products/abaqus/abaquscae/

[3] Chemnitzer Werkstoffmechanik GmbH, "VEDDAC", Retrieved May 21, 2019, URL: http://www.cwm-chemnitz.de/produkte/software.

[4] N. Pflügler et al., "Advanced Risk Analysis of Interface Delamination in Semiconductor Packages: A Novel Experimental Approach to Calibrating Cohesive Zone Elements for Finite Element Modelling," Proceedings of the 19th International Conference on Thermal, Mechanical and Multi-Physics Simulation and Experiments in Microelectronics and Microsystems (EuroSimE), Toulouse, April 15-18, pp. 1/11–11/11, 2018.

[5] S. Breitenreiner, "Optimization of an experimental setup for a more precise determination of adhesion parameters", Bachelor Thesis, University of Applied Sciences Munich, Munich, 2019.

Acceleration factors for reliability analysis of electronic assemblies

Enrico Galbiati

GESTLABS S.r.l.
Via lecco 61, Vimercate (MB), Italy
enrico.galbiati@gestlabs.it

Abstract

There are several models for the calculation of the acceleration factor for the electronic assemblies subjected to stress tests. A correct reliability assessment requires an accurate analysis of these models and the evaluation of their real effectiveness and adequacy. As regard to the solder joints subjected to the thermomechanical stresses given by temperature variation, applicable models are the original Coffin-Manson model, the Coffin-Manson model modified by Norris and Landzberg, and the model described in the standard IPC-SM-785. The last one is the most complete model to evaluate the acceleration factors given by slow temperature swings. However, the most effective method for the estimate of the solder joint aging is based on the analysis of the microstructure of the alloy. Regarding the acceleration factors of the electronic components, the international standard IEC 61709 can be used. This standard provides the equations to determine the intrinsic failure rates of the electronic components. Those equations allow to calculate the acceleration factors for the different types of stress (caused by temperature, voltage, current, etc.). This standard is more linked to the actual characteristics of the components than MIL-STD-217F, which was widespread in the past. IEC 61709 is also a useful tool for the design of electronic assemblies. A new method for the calculation of the acceleration factors is the analysis of the variation of the electrical parameters of the assembly when the stress is applied. This new method allows to evaluate directly the effects of the stress on the functionality of the assembly.

Key words: Acceleration factor, reliability, Coffin-Manson, microstructure, failure rate, electronic assembly

Introduction

The reliability assessment is a very critical topic, because it often requires the "acceleration" of the product life. This acceleration is obtained by specific tests, called "accelerated tests", where the use of appropriate stresses (e.g., caused by temperature, voltage, etc.) accelerates the product life. More precisely, the objective of the stress tests is to accelerate specific failure mechanisms. This acceleration can be quantified through the "acceleration factor". To be effective and avoid unwanted effects, the accelerated tests must not cause alterations of the chosen failure mechanisms, nor trigger failure mechanisms that do not occur in the product life.

The acceleration factor depends on both the failure mechanism and the component subjected to the stress. Therefore, for a given failure mechanism, each component has its own acceleration factor, and, for a given component, each failure mechanism has its own acceleration factor.

It is very difficult to know the real acceleration given by the stress tests. Therefore, a lot of effort has been made also in the past to develop models for the calculation of these acceleration factors. However, caution should be used in the choice and application of those models, because they sometimes are improperly used or extended to areas outside their applicability.

Models for solder joints reliability

One of the most important topics in reliability analysis of electronic assemblies is the aging of solder joints during the product life. This problem stimulated the development of different models to evaluate the effect of the accelerated thermal cycling (ATC), where the temperature is varied slowly.

The effect of ATC on the solder joints is a thermomechanical fatigue, which accelerates the aging of the alloy, i.e. its thermodynamic evolution. For lead-free alloys, this process results in the precipitate coarsening. This phenomenon causes the embrittlement of the solder joints, leading to microfractures and, finally, cracks. The most general law describing the damage of a metal subjected to fatigue is given by the Morrow equation:

$$\overline{N}_f = C\,(\Delta W)^{1/c} \qquad (1)$$

where
$\overline{N}_f$ = mean number of cycles to fail (cyclic life)
C = material constant
ΔW = visco-plastic strain energy
c = fatigue ductility exponent

Starting from (1), the original Coffin-Manson plastic strain fatigue life was developed:

$$\overline{N}_f = C\left(\Delta \gamma_p\right)^{1/c} \qquad (2)$$

where

$\Delta\gamma_p$ = cyclic plastic strain range

Several models were developed from the original Coffin-Manson model. The most widespread is the modified Coffin-Manson model, developed by Norris and Landzberg. Its equation is

$$\overline{N}_f = C_0 \frac{f^\alpha}{(\Delta T)^\beta} e^{\frac{E_a}{k\,T_{max}}} \qquad (3)$$

where

C_0 = constant
f = cycling frequency
α = cycling frequency exponent
β = temperature swing exponent
E_a = activation energy
ΔT = cycling temperature swing
T_{max} = maximum cycling temperature
k = Boltzmann constant

The general equation for the acceleration factor, AF, determined by a stress test is

$$AF = \frac{\overline{N}_{f,u}}{\overline{N}_{f,s}} \qquad (4)$$

where

$\overline{N}_{f,u}$ = mean number of cycles to fail in use (operating) conditions
$\overline{N}_{f,s}$ = mean number of cycles to fail in stress conditions

Therefore, the acceleration factor for the modified Coffin-Manson model is

$$AF = \frac{\overline{N}_{f,u}}{\overline{N}_{f,s}} = \left(\frac{f_u}{f_s}\right)^\alpha \left(\frac{\Delta T_s}{\Delta T_u}\right)^\beta \\ \times e^{\frac{E_a}{k}\left(\frac{1}{T_{max,u}} - \frac{1}{T_{max,s}}\right)} \qquad (5)$$

where

$T_{max,u}$ = maximum temperature in use conditions
$T_{max,s}$ = maximum temperature in stress conditions
f_u = cycling frequency in use conditions
f_s = cycling frequency in stress conditions

In this model, α, β and E_a have default, fixed values. Normally, they are set at the following values: α = 1/3, β = 1,9 and E_a = 0,122 eV. The parameters of this model do not take into consideration characteristics that are really influent on the reliability of the solder joint: alloy type, coefficient of thermal expansion (CTE) of the component case and the printed circuit board (PCB), geometrical characteristics of solder joint and pad, etc. Therefore, despite being widely used for ATC and often also for thermal shock, this model is not adequate to provide realistic acceleration factors and does not allow a correct reliability analysis. In fact, this model was developed more than 20 years ago for a tin-lead alloy, and only for specific components, solder joints and PCB material. Therefore, it has not a general validity and should no longer be used.

A more adequate model to evaluate the aging of solder joints was developed by Werner Engelmaier (USA) and published in a specific standard, the IPC-SM-785. In this standard, issued in 1992, Engelmaier's model is described in detail, with focus on the effect of ATC on tin-lead alloy. This model is complete and realistic, because it considers all factors that can influence the reliability of solder joints: materials, CTE of component case and PCB, distance between the farthest solder joints of the component, geometric characteristics of solder joint and pad, temperatures, etc. The equation describing this model is complex, because of the quantity of parameters influencing the effect of ATC.

The number of cycles, $N_f(x\%)$, corresponding to x% failure probability for leadless SMT solder joints is given by (6), while the value of $N_f(x\%)$ for leaded solder joints is given by (7).

$$N_f(x\%) = \frac{1}{2}\left(\frac{2\varepsilon_f}{F}\frac{h}{L_D\,\Delta\alpha\,\Delta T_e}\right)^{-1/c} \\ \times \left[\frac{\ln(1-0,01\,x)}{\ln 0,5}\right]^{1/\beta} \qquad (6)$$

$$N_f(x\%) = \frac{1}{2}\left[\frac{2\varepsilon_f}{F}\frac{200\,\text{psi}\,A\,h}{K_D\,(L_D\,\Delta\alpha\,\Delta T_e)^2}\right]^{-1/c} \\ \times \left[\frac{\ln(1-0,01\,x)}{\ln 0,5}\right]^{1/\beta} \qquad (7)$$

where

ε_f = fatigue ductility coefficient
F = empirical factor accounting for effects such as cyclic warpage, cyclic transients, non-ideal solder joint geometry, brittle intermetallic compounds
K_D = stiffness of component lead
A = solder joint area
h = solder joint height
L_D = half the maximum distance between component pads
$\Delta\alpha$ = absolute difference in CTE between the component and PCB
ΔT_e = equivalent cycling temperature swing, accounting for component power dissipation effects, external temperature variation and CTE values of component and PCB
c = fatigue ductility exponent, depending on the alloy characteristics, the mean cycling solder joint temperature and the average time at the extreme cycling temperatures
β = slope of Weibull statistical failure distribution, typically greater than 1

As also clearly stated in IPC-SM-785, this model applies only when the creep fatigue caused by grain structure coarsening is the dominant failure mechanism. For example, the model does not apply when:

- the temperature is below 0°C
- the temperature ramp is greater than 30°C/minute
- the cycling frequency is high (e.g., greater than 0,5 Hz)
- the intermetallic compound layer is too thick and brittle
- the global CTE mismatch is very small compared with the local CTE mismatch, as in case of ceramic-on-ceramic, or silicon-on-silicon.

IPC-SM-785 model gives us a very important information: each solder joint of the electronic assembly has his own acceleration factor, which may differ from the other solder joints. In fact, as shown in (4), the acceleration factor depends on $N_{f,n}$ and $N_{f,s}$, which in turn depend on the geometrical and thermomechanical characteristics of the solder joint, pad and component. Therefore, as far as the solder joint reliability is concerned, it is absolutely incorrect to think that the effects of a thermomechanical stress can be described using only a single acceleration factor applicable to the whole electronic assembly, because the stress test really gives different acceleration factors to the different solder joints. This fundamental result gives a further evidence of the inadequacy of the modified Coffin-Manson model, which provides a single acceleration factor applicable to any solder joint of any electronic board, irrespectively of the peculiarity of each single solder joint, component and PCB. Instead, IPC-SM-785 model has a correct approach.

IPC-SM-785 model was originally developed for tin-lead alloys, but, with the necessary changes in the parameter values, it can be applied also to lead-free alloys. This model applies to slow thermal variations, typical of the ATC, not to fast thermal changes, typical of the thermal shocks. In fact, in this last case, the dominant failure mechanism is the warpage of the PCB, causing shear and especially tensile stress in the solder joints. This failure mechanism is outside of the scope of IPC-SM-785 model.

This model is complex and contains some parameters (e.g., ε_f, F, K_D, c and β) that are very difficult to know or measure. Moreover, some of them also depend on the manufacturing process, and particularly on the thermal profile of the reflow. For these reasons, the values of these parameters, which can be found in literature for some alloy, can be taken as a reference only. Those values can be used for a rough analysis applicable to the comparison of solder joints or the detection of possible critical situations. In fact, the uncertainty affecting those parameters makes impossible a realistic estimate of the time to fail and the acceleration factor.

From a practical point of view, the best way to evaluate the reliability of the solder joints is the analysis of the microstructure. Basing on this analysis, it is possible to have an objective method to evaluate the evolution of the microstructure and thus the acceleration of the solder joint life. For lead-free alloys, the size of the intermetallic precipitates (Ag_3Sn, Cu_6Sn_5, etc.) can be used to "measure" the aging of the current lead-free alloys (SAC305, SAC387, etc.). In fact, the precipitate coarsening process is an indicator of the aging of the alloy. The reason is that the surfaces of the precipitates give a big contribution to the Gibbs free energy of the alloy. Therefore, the Gibbs free energy decreases when the total extension of the surfaces of the precipitates decreases. According to the thermodynamics, the evolution of the alloy is towards the reduction of the Gibbs free energy, and therefore towards the reduction of these surfaces, which occurs during the precipitate coarsening. This process, which is accelerated by the thermal or thermo-mechanical stresses, can be described by the following model, based on the diffusion of solute atoms (Ag, Cu, etc.) in Sn:

$$d(t)^m - d_0^m = K\, C_{sol}\, D_{sol}\, t \qquad (8)$$

where
$d(t)$ = precipitate diameter after the time t
$d_0 = d(0)$ = initial precipitate diameter
C_{sol} = solubility of solute atoms in Sn
D_{sol} = diffusivity of solute atoms in Sn
K = constant
m = 2 (for plated-shaped precipitates) or 3 (for spherical or three-dimensional precipitates)

Even if C_{sol}, D_{sol}, and K are not known, the acceleration factor can be determined comparing the growth of the precipitates in the stress test conditions and in the use conditions after the same time t. For example, applying this approach to the three-dimensional growth of spherical precipitates, we obtain:

$$AF = \frac{d(t)_s^3 - d_0^3}{d(t)_u^3 - d_0^3} \qquad (9)$$

where subscript "s" refers to stress conditions and subscript "u" refers to use (i.e., operating) conditions.

Of course, the alloy inside the solder joint is not perfectly uniform, so the shape and size are not the same for all the precipitates. This problem can be handled using the equivalent diameters to take into considerations non-spherical precipitates, then averaging these diameters to have a value that represents the mean precipitate size inside the volume of alloy under analysis. Obviously, special attention should be given to the measure of such small particles. These problems cause some uncertainty in the evaluation of the precipitate size. However, this approach provides a value of the acceleration factor that is approximated but realistic, because directly related to the aging of the alloy.

Since this method is based on experimental data coming from stress tests, it requires more effort than the use of models simply based on a formula. Moreover, this approach requires specific and expensive equipment to perform cross-sections of solder joints and, especially, to analyse them with the X-ray spectroscopy. In fact, this technique requires the use of a scanning electron microscope (SEM). However, the obvious advantage of using this approach is that this method is more linked to the reality than a mathematical equation using parameters (e.g., activation energies) that are not determined experimentally from the specific material/component under test.

In any case, if the measure of the size is too difficult or has an excessive uncertainty, the simple visual inspection and qualitative evaluation of the microstructure allow a better and more realistic reliability evaluation than the application of formulas coming from too generic models.

The reliability analysis and the acceleration factor evaluation based on the microstructure analysis, and especially on the grain and precipitates coarsening, can be applied not only to the thermal cycling, but also to all types of stress (e.g., high temperature) that cause the increase of alloy aging.

Obviously, generally it is not possible to analyse all solder joints in the assembly. The solder joints to be analysed should be selected basing on the amount of stress that they can receive. For example, the solder joints of ceramic components are subjected to more stress than solder joints of plastic components, because of the big difference between the CTE of the ceramic case and the PCB material. Another factor increasing the stress is the length of the component. In fact, greater lengths and the consequent greater distances among the pads of the same component give more increase in the shear strain and stress to the solder joints.

Models for electronic components and assemblies

The acceleration factors for electronic components and assemblies are generally based on the Arrhenius model referred to the temperature stress and, less frequently, to models for other types of stress (humidity, voltage, current, etc.). Regarding the temperature stress, the well-known Arrhenius model is clearly a good model, adequate to describe the effect of the temperature in the acceleration of chemical reactions and chemical-physical processes. Regarding the stresses other than temperature, there are many models that often disagree with each other. Therefore, a considerable uncertainty affects these models.

Independently from the model used, the parameters present in the formulas, including Arrhenius equation, are generic or, in the best cases, referred to the component family (e.g. electrolytic capacitor, Zener diode, etc.). Moreover, even assuming that the model is correct, to obtain realistic estimates of the acceleration factor, the parameters should take into account all the peculiarities of the component, i.e. all the factors than can influence the effect of the stress (materials, shape and size, manufacturing process, etc.); otherwise, the acceleration factor just remains a number disconnected from the real behaviour of the component. Moreover, the acceleration factor referred to a given stress often depends on other stresses, because the different stresses generally are not independent from each other because of possible interactions among them. Therefore, for example, the activation energy of the Arrhenius model, which is related to the temperature stress, may depend on the voltage stress or humidity stress; sometimes, the activation energy also depends on the temperature. Really, it is very difficult to find theoretical models that consider all these aspects.

One of the most complete and up-to-date standards containing models to calculate the acceleration factors for electronic components is the international standard IEC 61709 "Electric components – Reliability – Reference conditions for failure rates and stress models for conversion". IEC 61709 treats the intrinsic failure rate (IFR) of the electronic components, in the hypothesis of constant failure rate. In this standard, the IFR, λ, of the electronic components under operating conditions is calculated multiplying the reference IFR (i.e. the IFR declared by the manufacturer in specified reference conditions), λ_{ref}, by the factors, π_x, related to the specific stresses (subscript "x" indicates the type of stress: "T" for temperature, "U" for voltage, "I" for current, etc.) present in the actual application. Therefore, each π_x, called "stress factor" in the standard, can be considered an acceleration factor between the operative conditions and the reference conditions, regarding to a specific stress (temperature, voltage, current, etc.).

IEC 61709 has structure and models similar to those of Military Handbook MIL-STD-217F "Reliability prediction of electronic equipment", an old standard that was widely used in the past years. Of course, IEC 61709 is more up-to-date than MIL-STD-217F, but the most important difference is that the values of λ_{ref} in IEC 61709 come from the component manufacturers, while in MIL-STD-217F all values come only from its own internal tables. Therefore, the IFRs of MIL-STD-217F do not have any connection with the actual specific characteristics of the component (construction feature, quality, etc.).

The general formula for acceleration factor, AF, is the following:

$$AF = \frac{MTTF_u}{MTTF_s} \qquad (10)$$

where $MTTF_u$ and $MTTF_s$ are the mean time to failure (MTTF) in use conditions and stress conditions, respectively.

In the assumption of constant failure rate, the MTTF is also the mean time between failure (MTBF), and (9) becomes

$$AF = \frac{MTTF_u}{MTTF_s} = \frac{MTBF_u}{MTBF_s} = \frac{1/\lambda_u}{1/\lambda_s} \quad (11)$$

where $MTBF_u$ and $MTBF_s$ are the MTBF in use conditions and stress conditions, respectively, and λ_s and λ_u are the IFRs in the stress conditions and in the use conditions, respectively.

Therefore, the acceleration factor, AF, given by a stress test can be calculated using the values of π_x, as follows:

$$AF = \frac{MTBF_u}{MTBF_s} = \frac{1/\lambda_u}{1/\lambda_s}$$
$$= \frac{\lambda_s}{\lambda_u} = \frac{\lambda_{ref} \prod_{i=1}^{n} \pi_{s,i}}{\lambda_{ref} \prod_{i=1}^{n} \pi_{u,i}} = \frac{\prod_{i=1}^{n} \pi_{s,i}}{\prod_{i=1}^{n} \pi_{u,i}} \quad (12)$$

where all $\pi_{s,i}$ and $\pi_{u,i}$ are all the applicable π_x in the stress conditions and in the use conditions, respectively.

As an example, we can analyse the equation for acceleration factor of the temperature, π_T, which is

$$\pi_T = \frac{A\,e^{Z\,E_{a,1}} + (1-A)e^{Z\,E_{a,2}}}{A\,e^{Z_{ref}\,E_{a,1}} + (1-A)e^{Z_{ref}\,E_{a,2}}} \quad (13)$$

$$Z = \frac{1}{k}\left(\frac{1}{T_0} - \frac{1}{T_u}\right) \quad (14)$$

$$Z_{ref} = \frac{1}{k}\left(\frac{1}{T_0} - \frac{1}{T_{ref}}\right) \quad (15)$$

where
A = constant
$E_{a,1}$, $E_{a,2}$ = activation energies in eV
k = Boltzmann constant
T_0 = 313 K
T_u = temperature in use (i.e. operating) conditions
T_{ref} = temperature in reference conditions

T_u (often indicated with "T_{op}") and T_{ref} have different meanings depending on the component type: they are the junction temperatures for integrated circuits and semiconductor components, the average case temperatures for capacitors and resistors, and the winding temperatures for inductors.

The use of two activation energies, both referred to the component family, makes this model more complete than models based on a single activation energy. However, also this model has important critical points to be considered:
a) The failure mechanisms accelerated by the temperature may be more than two, making the use of two activation energies not enough to take into account the effects of all failure mechanisms.
b) The activation energies of the different failure mechanisms may depend on the construction features of the component, thus they may depend on the specific manufacturer or model of the component; consequently, the use of fixed values of activation energy for a whole component family may result in a wrong estimate of the acceleration factor.

The critical issues described in a) and b) also apply to the acceleration factors for the other types of stress (voltage, current, etc.), and consequently to the factors π_U, π_I, etc.

Moreover, this standard does not provide models for all types of stress. For example, no model is given for humidity and mechanical stress, which indeed are very common in the real life.

Only the manufactures of the electronic components would have the possibility to perform deep studies of the failure mechanisms of their own components and carry out extended tests to identify the correct models of the failure mechanisms and calculate the true values of the related activation energies. However, these types of studies are rare, and normally they are not shared with the user of the components.

Based on the considerations described above, even if IEC 61709 is one of the most accurate and up-to-date publications on this matter, this standard is more adequate to identify weaknesses or criticalities in an electronic assembly, rather than to obtain realistic estimates of acceleration factors. IEC 61709 is a very useful tool for the design of electronic assemblies, but the IFR values calculated by its models should not be used to estimate the actual value of the defectiveness in field.

The calculation of the acceleration factor referred to the whole electronic boards is even more critical. The acceleration factor given by a stress test to an assembly should be calculated basing on the acceleration factors and the IFRs of the individual components, as shown below.

$$AF_{ass} = \frac{\lambda_{s,ass}}{\lambda_{u,ass}} = \frac{\sum_{i=1}^{N} \lambda_{s,i}}{\sum_{i=1}^{N} \lambda_{u,i}}$$
$$= \frac{\sum_{i=1}^{N} AF_i \lambda_{u,i}}{\sum_{i=1}^{N} \lambda_{u,i}} \quad (16)$$

where
AF_{ass} = acceleration factor for the whole assembly
$\lambda_{s,ass}$, $\lambda_{u,ass}$ = IFR of the assembly in stress conditions and in use conditions, respectively
$\lambda_{s,i}$, $\lambda_{u,i}$ = IFR of the i-th component in stress conditions and in use conditions, respectively

AF_i = acceleration factor for the i-th component
N = total number of components in the assembly

According to (16), the acceleration factor for the assembly is not simply the average of the acceleration factors of the individual components, but it is their weighted average, where the weights are the IFRs in the use conditions. Of course, for the same reason, in case of thermal stress, where the Arrhenius equation is used, the acceleration factor for the whole assembly must not be calculated averaging the activation energies of the individual components. The same considerations apply to the parameters used in the other models for the acceleration factors related to other types of stress (voltage, current, humidity, etc.).

There is also the practice to calculate AF_{ass} considering the assembly as a single unit, using predetermined parameters (for example, using a fixed value of activation energy for Arrhenius model). This practice is sometimes used because it is obviously very simple to apply, but it is incorrect. In fact, this practice does not consider the peculiarity of each single component of the assembly, treating the assembly as made of identical components. This simplistic approach results in wrong reliability estimates.

New method for electronic assemblies

A new method, proposed by the present article for the definition of acceleration factors to be assigned to an electronic assembly, is based on the analysis of the degradation of selected electrical parameters of the assembly, after the assembly has been subjected to a certain stress. These parameters may be values of voltage, current, electrical power, or signals with specific frequency and duty cycle, etc. As parameters to be analysed, it is convenient to take the functional parameters or electrical outputs of the assembly, because they are strictly related to the functionality of the assembly or the product containing the assembly. However, in some cases it is not possible to use the electrical outputs, because they are too difficult to measure, or because their variation caused by the stress is too irregular. In these situations, it is suggested to choose other parameters, selecting, where possible, those that are the most linked to the functions of the assembly.

Of course, to correctly evaluate the degradation compared to the initial situation taken as a reference, the parameters shall be measured in the same environmental conditions (temperature, humidity, etc.) of the initial situation. This is necessary because the electrical parameters may be sensitive to the environmental conditions during the measurement, showing a drift.

The stresses (temperature, humidity, thermal cycling, overvoltage, etc.) applied to the assembly generally cause the degradation of the electrical parameters. However, not all parameters are varied in the same way, because each individual parameter is the result of electrical interaction of different components, and the different components may have their own specific degradation, even if they are subjected to the same stress. Therefore, it is possible that some parameter will have big changes, while some others will have small changes or no change at all. For this reason, it is not correct to assign a single acceleration factor to the whole assembly; each parameter will have his own acceleration factor.

As an example, indicating with Y an electrical parameter of the assembly, we can assume that the applied stress leads to a decrease of the value of Y over time. To calculate the acceleration factor, AF_Y, for this parameter, we can compare the time in the use conditions after which the parameter reaches a chosen value, Y_1, with the time in the stress conditions after which the parameter reaches the same value, Y_1, starting in both cases from the same initial value, Y_0. The value of the acceleration factor is given by

$$AF_Y = \frac{t_{u,0,1}}{t_{s,0,1}} \qquad (17)$$

where
$t_{u,01}$, $t_{s,0,1}$ = time to reach Y_1 starting from Y_0 in use conditions and in stress conditions, respectively

Of course, all measurements of the parameter shall be performed in the same environmental conditions.

In the hypothesis that the acceleration factor does not vary with time, as assumed in all models concerning the acceleration factor for the electronic components (Arrhenius, Coffin-Manson, Peck, Eyring, etc.), AF_Y is independent from Y_0 and Y_1.

Of course, since the effects of the applied stress are different for the different parameters, each single electrical parameter has his own acceleration factor, which may be different from the others. For this reason, it is not possible to assign a single acceleration factor to the whole assembly.

The evaluation of the effects of the applied stress is easier and more effective if any or all these parameters have specification limits that are related to the functionality of the assembly or product.

The advantage of this method is the possibility to refer the accelerated aging, caused by the applied stress, directly to the electrical characteristics or the functionality of the assembly, and consequently to the performance of the assembly or the product. Moreover, the acceleration factor determined in this way is not a result of a theoretical model only, because the application of (17) requires experimental data coming from specific tests to determine $t_{u,01}$ and $t_{s,0,1}$.

Summary and conclusion

There are several methods to calculate the acceleration factor for solder joints of electronic assemblies. The most known method is the Coffin-Manson model modified by Norris and Landzberg.

This model is very simple to use, but it is not linked to the characteristics of the solder joint, component and the PCB. Therefore, it is not adequate to provide realistic estimates and should not be used.

A more complete model is the one developed by Engelmaier and described in IPC-SM-785. It contains many parameters, which take into account all the necessary characteristics of the solder joint, component and PCB. However, from a practical point of view, it is generally not feasible to determine the values of all the parameters. In fact, for some of those, the knowledge of the actual values would require an extensive analysis supported by specific tests. Therefore, for those parameters, reference values are normally used instead of the actual ones. According to above considerations, this model should be considered very useful to make comparisons among solder joints and to detect possible weaknesses, but not adequate to estimate the real values of the time to fail or the acceleration factor.

An alternative approach to determine the acceleration factor is based on the evaluation the solder joint aging, which is carried out analysing the microstructure of the alloy. This approach requires complex equipment and skill to perform the measurement and to evaluate the results, but it has the advantage to be based on experimental data and to be addressed directly to a root cause of the solder joint failures: the aging of the microstructure of the alloy. Moreover, this approach, unlike the modified Coffin-Manson model and IPC-SM-785 model, can be applied to the aging caused by any type of stress, like the pure thermal stress given by the exposure to a high constant temperature, not only in case of thermal cycling.

Regarding the reliability analysis of electronic components, one of the most accurate and complete standards is IEC 61709. This standard is very useful to verify and improve the design of electronic assemblies. It allows to identify critical situations related to the design, which cause too much electric or thermal stress to the component, or due to a poor component quality, resulting in a too high reference IFR. However, since this standard is based on theoretical models using predetermined parameters, the calculated IFRs should not be used to estimate the actual MTBF or the defectiveness of the assembly.

Other approaches using simplified models with default values for the activation energy or other parameters should not be employed because they lead to wrong reliability estimates.

A new alternative method to evaluate the effects of the stress on the electronic assemblies is the analysis of the behaviour of some selected electrical parameters, in order to assign to each of them a specific acceleration factor. This objective is obtained determining the acceleration factors from experimental data. This approach gives two advantages: it is tailored on the specific assembly, and

it is based on the real effects of the stress on the functionality of the assembly.

To have a more effective and complete reliability analysis, some of the above described method can be used together, since they are complementary to each other. In particular, a complete reliability analysis of an electronic assembly should include:

- the analysis of the microstructure, which gives an effective evaluation of the reliability and acceleration factors of selected solder joints;
- the application of IEC 61709, which provides useful information to identify weaknesses and criticalities in the design of electronic assembly or in the choice of the electronic components;
- the analysis of selected electric parameters, which allows to perform the reliability analysis of the whole electronic assembly in relation to its functionality and the final use of the product containing the assembly.

References

[1] IPC-SM-785 "Guidelines for accelerated reliability testing of surface solder attachments", 1992.

[2] E. Galbiati, "Reliability evaluation of solder joints in electronics assemblies", 21th European Microelectronics and Packaging Conference (EMPC), Warsaw, Poland, September 10-13, 2017.

[3] P. Kumar, Z. Huang, I. Dutta, G. Subbarayan, and R. Mahajan, "Influence of microstructure on creep and high strain rate fracture of Sn-Ag-based solder joints", K.N. Subramanian, Wiley, 2012.

[4] E. Galbiati, and A. Vailati, "Evaluation and prediction of lead-free solder joints reliability and robustness", International Conference of Electrical and Electronic Technologies for Automotive, Milan, Italy, July 9-11, 2018.

[5] IEC 61709 "Electronic components - Reliability - Reference conditions for failure rates and stress models for conversion", 2017.

High frequency effects on the PBGA stress developments at random vibration

Yeong K. Kim, Seok M. Lee, Dosoon Hwang and Sun Kim

[1] Inha University, Dept. of Mechatronics, 100 Inha-ro, Incheon, South Korea
[2] Korea Aerospace Research Institute, Daejun, South Korea
Tel.: + 82328607334, fax: + 823287565261
E-mail: ykkim@inha.ac.kr

Abstract

Large size commercially available plastic ball grid array chip packaging was tested and analyzed under random vibration to access its application feasibility on satellite electronics. Two extreme levels of the random vibrations were applied sequentially in the vibration tests to investigate the sustainability of the PBGA chips mounted on the polyimide PCB with aluminum frame. The test results did not show any solder failure under the test conditions, indicating the robust structural integrity and providing the evidences justifying the PBGA packaging to the aerospace applications. Numerical analyses were also performed for the solder stress developments, demonstrating that the first natural mode was not necessarily the dominant source for the maximum solder stress, and the usual assumption that the center of the PCB center could be bad choice for the chip mounting might be wrong. The calculations represented the locations away from the PCB center could be worse due to the higher frequency natural mode contributions to the stress developments.

Key words: PBGA reliability, Random vibration, COTS, Aerospace applications, Solder failure.

Introduction

The electronics packaging techniques employed for the satellite have been traditionally based on legacy, which usually lead to relatively heavy, bulky and expensive due to customized manufacturing. To overcome the pitfall, the applications of Commercially-Off-The-Shelves (COTS) such as PBGA chips have been suggested as a solution to the issues [1-3]. This study is to investigate the PBGA packaging application to the satellite by using big chips and PCB installation structure. A polyimide PCB was designed to a realistic size of actual electronics devices, and the large size PBGA chip was mounted to induce the solder joint stress. The entire specimen was tested under the two levels of powewr spectrum density, and the solder joint failures were monitored by in-situ resistance measurements to examine the sustainability of the chips under the vibrations. Generally, it is widely accepted that the maximum stress is generated by the first natural mode. To verify the predication, numerical calculations were also performed for the detail solder stress development mechanism under the random vibrations by changing the chip sizes and their locations.

Sample and tests

Fig. 1 shows the test sample with PBGA chips mounted on the printed circuit board. Two different sizes of the PBGA of 16 mm x 16 mm (type-1) and 10 mm x 10 mm (type-2) with daisy chains were employed with Sn/Pb (63/37) eutectic solder full array, the number of which were 361 and 100, and the ball size 0.45 mm and 0.4 mm, respectively. No underfills were applied. The power spectrum density of the random tests is shown in Fig. 2 (a).

Two steps of the mean square acceleration levels per unit bandwidth of 0.35 g²/Hz (acceptance level) and 0.7 g²/Hz (qualification level) were employed, and the signal's root mean square values were 22.48 grms and 31.78 grms, respectively, in the frequency range of 20 ~ 2000Hz.

The sample was mounted on a pheumatic shaker following the power spectrum density levels.

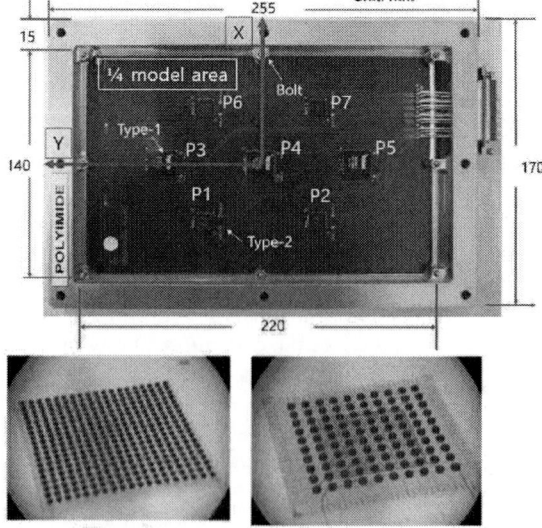

Type-1   Type-2

Figure 1: The sample configuration and the type-1 and type2 sodlers by X-ray.

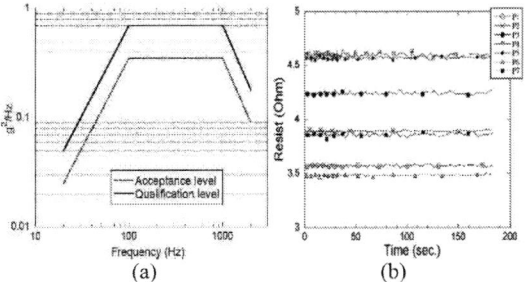

(a)   (b)

Figure 2: (a) Power spectrum densities of the two levels. (b) The solder resistance measurements during the qualification level.

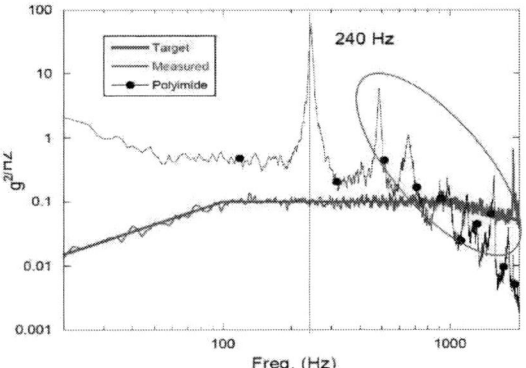

Figure 3: The PCB vibration behavior measurements under the random vibration.

Two minutes of the acceptance level and three minutes of qualification level were applied for the tests. Fig 2 (b) reprensts the resistance measurements data at the qualification level. There was no failure found, which demonstrated the robust structural integrity of the PBGA chip packaging. To examine the PCB vibration behavior, an accelerometer was attached at near the chip P4, and the power spectrum density was applied again. Fig. 3 showes the measurement data, showing the multiple peaks were captured according to the natural modes.

Fig. 4 is for the 1/4 finite element modeling of the sample structure. To cover the entire matural frequencies, the PCB boundaries were changed to four different boundary conditions (BC) of Xsymmetry-Ysymmetry (Xs-Ys), Xsymmetry-Yasymmetry (Xa-Ya), Xa-Ys and Xa-Ya. With the Xs-Ys BC, the first calculated mode is the first natural mode of the entire PCB. When the BC is Xs-Ya, the first calculated mode is the second natural mode of the entire PCB, etc.

Fig. 5 (a) represents the calculation results of the sample under the Xs-Ys boundary condition applied. The figure represents only the solder model of the chips at the 1/4 scale. In this calculation, the maximum stress location was found at the corner of the center chip (P4) due to the effect of the first natural mode. At that maximum node, the rest of three boundary condition results were calculated and the RMS stress values of each BC were drawn in Fig. 5 (b). As seen, the stress of each BC was increased by

accumulating the stress development at the frequencies from 20 to 2000 Hz considered in this study. The total maximum stress value was 102.5 MPa when all the values of each BC were added at the 2000 Hz.

The calculation results clearly indicated that the major stress was developed when the BC was Xs-Ys, where the first natural mode was affiliated. This results are in accordance with usual assumption that the maximum stress is occured at the edge of the center mounted chip due to the first natural mode.

For further investigation, the finite modeling was modified by replacing the P6 chip with the type-1 as represented in Fig. 6 (a), rendering all the chips were same size. And the same calcualtions by changing the BC's were performed. Fig. 6 (b) represents the stress results when the boundary condition was Xs-Ys. Again, the maximum stress was found at the corner of the center chip, and the total stress was obtained as 102.5 Mpa.

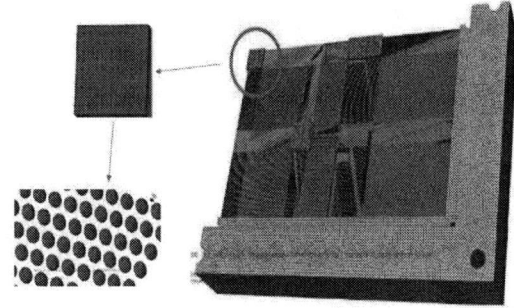

Figure 4: 1/4 model for the test sample.

Next, the maximum stress location was searched and found when the BC was Xs-Ya at the corner of the P6 chip as shown in Fig. 7 (a). Unlike the Xs-Ys case, the first natural frequency calaulated in this BC was the second natural mode. Same process to add the stresses under the four different boundary conditions were carried out, and the results are presented in Fig. 7 (b).

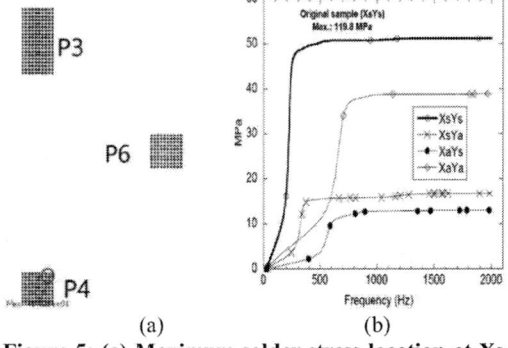

(a)   (b)

Figure 5: (a) Maximum solder stress location at Xs-Ys of the sample using 1/4 model. (b)The stress developments by changing the boundary conditions.

As seen, the total maximum stress in this location was found to be 109.2 MPa, which was

621

actually higher than the maximum stress at the center chip under Xs-Ys BC in Fig. 6. The graph also represented that the major stress development profile was when the BC was Xs-Ya, implying that the second natural mode contributes mainly to the stress development.

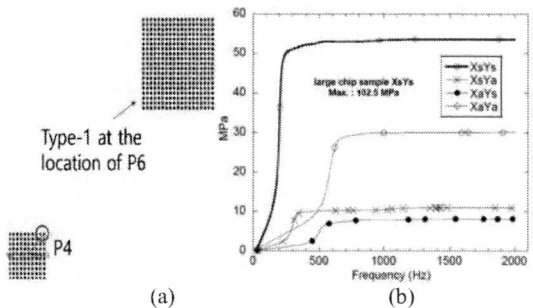

(a)   (b)

Figure 6: (a) Maximum solder stress location at Xs-Ys of the modified sample using 1/4 model. (b) The stress developments by changing the boundary conditions.

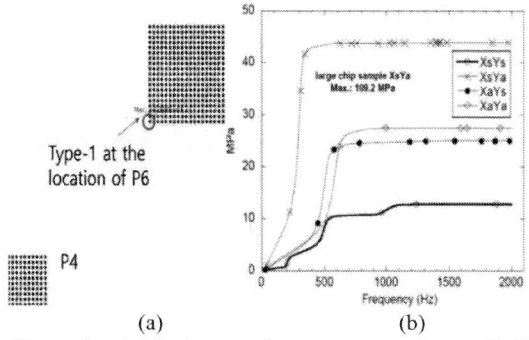

(a)   (b)

Figure 7: (a) Maximum solder stress location at Xs-Ya of the modified sample using 1/4 model. (b) The stress developments by changing the boundary conditions.
\

The results indicated that, the effects of the natural modes higher than the first one were so significant that higher solder stress might occur at the chip located at nearby the PCB center.

Conclusions

The strong structural integrity of the PBGA chip packaging was verified by testing the PCB with sizable chips under the harsh random vibration tests. The results showed a possibility of the COTS part applications to the satellite electronics devices.

The 1/4 finite element model was effectively used for the numerical calculations. The results represented that the maximum stress of the sample was found at the center chip solder due to the major stress development contribution of the first natural mode. However, when the model was modified in the different chip size, the maximum stress was found at the chip located away from the PCB center.

The analyses of the calculation results represented that the second natural mode was the dominant source of the stress development at that location. The results demonstrated that, unlike usual assumption, the first natural mode is not necessarily the dominant mode for the maximum stress development. Depending on the chip size and location, the mode higher than the first one was significant in the random vibration, which induced the maximum solder stress at the chip located aside from the PCB center.

Acknowledgements

This research was supported by Korea Aerospace Research Institute (NRF-2017M1A3A4A04037651)

References

[1] Qi H, Osterman M, Pecht M. Plastic ball grid array solder joint reliability for avionics applications. IEEE Transaction of Components and Packaging Technology. Vol. 30, pp. 242–7. 2007.

[2] Liu F, Lu Y, Wang Z, Zhang Z, "Numerical simulation and fatigue life estimation of BGA packages under random vibration loading," Microelectronics Reliability, Vol. 55, pp. 2777–85, 2015.

[3] Kim YK, Hwang DS, "PBGA packaging reliability assessments under random vibrations for space applications," Microelectronics Reliability, Vol. 55, pp. 172–9, 2015.

Formatting Advice

For various word processor files the layout appears to be printer dependent. As a consequence a word processor file that is well formatted for your printer can be badly formatted for our PDF-printer. Using a PDF-Writer as your default printer, when creating your document, can circumvent this problem. Printing to a PDF-Writer or savibg as PDF will create a PDF-file of your document. PDF-files can be opened by PDF-reader. The layout then appears to be printer independent.

For authors having a PDF-Writer installed:

- Set PDF-Writer as your default printer before formatting your document.
- Format the document as previously described
- Please check the following settings of the PDFWriter:
  Page set-up:
  Page size: A4
  Format: Portrait
  Graphics: Resolution: 600 dpi
                     Scaling: 100%
     PDFWriter font embedding:
     — Embed all fonts
     PDFWriter compression:
     — General settings:
     — Compress text and vector graphics
     — Recalculate graphics
     Colour / greyscale images:
     — Compression with JPEG low
     B/W images:
     — Compression with CCITT group 4

Remark:

Your document should be formatted for A4 paper portrait layout before creating the PDF file of the document.

Chip Package Interaction with Cu Pillar Interconnects – Impact of Die Warpage

Bjoern Boehme[1], Maricel Sy[2], Simone Capecchi[1], Dirk Breuer[1], Ta-Chien Cheng[3], Christian Goetze[1], Frank Kuechenmeister[1]

[1]GLOBALFOUNDRIES Dresden Module One LLC & Co. KG, Wilschdorfer Landstrasse 101, D-01109 Dresden, Germany;
Phone: +49-351-277-4004, bjoern.boehme@globalfoundries.com
[2]ASE Group Kaohsiung, 26. Chin 3rd Rd., Nantze Export Processing Zone, Kaohsiung, Taiwan 811
[3] GLOBALFOUNDRIES Taiwan Ltd. 15F-1, No. 289, Kuang Fu Road, Section 2, Hsinchu 30071, Taiwan

Abstract

This paper summarizes the work which was done to understand the die warpage at different temperatures and the influence on the assembly yield in a FCCSP package in combination with Cu pillar first level interconnects. A Cu BEoL stack with two ultra-thick metal (UTM) layers was selected as one of the most critical learning vehicles to understand and quantify the effect of a high effective Cu thickness in combination with a reduced die thickness to match the package form factor requirements. With this vehicle, the "BEoL to bulk silicon thickness"-ratio can reach up to 10% or more, depending on die thickness.

Based on a bi-material strip working hypothesis (bulk silicon + BEoL stack) an experiment design was devised with the aim of maximizing the assembly yield and minimizing the die warpage. Die thickness, Cu pillar solder height, BEoL metal density and assembly parameters have been selected as experiment design factors. The FC assembly was run in multiple experiments in a standard reflow process. The assembly yield was determined by inline x-ray inspection for non-wet failures and electrical test on packaged samples.

The Shadow Moiré technique was used to collect substrate and die warpage hysteresis data in the temperature range in between RT and 260°C reflecting the behavior during the solder reflow process. The die warpage hysteresis was identified as a key factor for high assembly yields: the lower the hysteresis the lower the assembly yield loss.

Additional cross sections were prepared to confirm the results of the X-Ray inspection and electrical tests. The critical fails occurred at the die corners showing non-wetting of the solder and matched the working hypothesis.

The collected die warpage hysteresis data matched well with a prediction of the warpage hysteresis during the solder reflow using the rule of mixture for the BEoL CTE. With the collected results a forecast of the assembly yield loss in dependency of the different factors is possible.

After solving the assembly challenges the packages were stressed in standard package reliability tests and passed the requirements. Additionally, assembly robustness tests were run (Multiple reflow, quick temperature cycle) and passed with margin.

Keywords: CPI; BEoL; warpage, Shadow Moire

I. INTRODUCTION

Advanced silicon nodes support multiple backend of line (BEoL) stacks to address customer needs for functionality, cost and performance. Especially for mmWave including 5G applications BEoL stacks with ultra-thick metal (UTM) are needed. Ultra thick metal structures enable better RF performance for mmWave applications in terms of low loss requirements (e.g. interference, resistance) and through the opportunity of passive integration on the chip level (e.g. on chip inductors or antennas).

GLOBALFOUNDRIES is dedicated to understand the chip package interaction. Several systematic studies and qualification have been completed during recent years [BOE16, BOE17]. A strong cooperation with ASE as OSAT was established and led to a mutual understanding of CPI related aspects.

In the figure 1 an overview of identified key factors impacting the CPI for FCCSP packages is shown. The three highlighted factors "BEoL stack", "Assembly Process parameters" and "Interconnect material" are the focus of this paper.

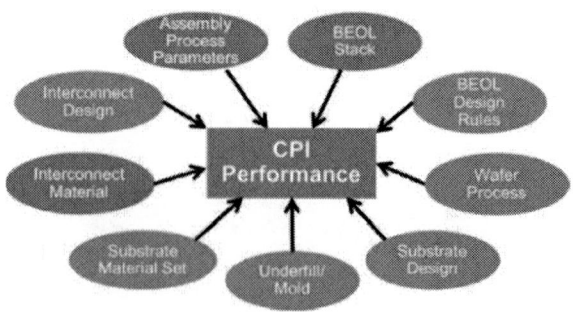

Fig. 1. Overview of important factors for the Chip Package Interaction of advanced silicon node products with FCCSP package with highlighted focus of this paper in blue

II. TEST VEHICLE DESCRIPTION

A. BEoL Stack, Die, Interconnect, Package

The TV (test vehicle) for this study was a 7x7mm² die of an advanced silicon node. In order to investigate the die warpage at elevated temperatures (e.g. reflow for die attach) a very thick Cu-BEoL stack was selected for this TV. The BEoL stack contained a total of 10 Cu metal layers which includes two UTM modules. Due to the two UTM modules, the BEoL stack is significant thicker than stacks without UTM module. Due to the die thinning for packaging and the UTM Cu (trace, vias and fill patterns), an interaction during high thermal loading was expected on die level.

For the FCCSP package configuration, Cu pillar first level interconnects were selected (oblong, 45μm width, 75μm length). The final height was 30μm Cu height, 3μm Ni height, and 25μm SnAg Cap height. The die pattern of Cu pillar was distributed uniformly. It had a 3-row dense pattern at the die periphery and a 350μm in the die center. The minimum pitch was 100μm in the periphery. The package size was 14x14mm². A coreless ETS substrate with less than 200μm thickness was used as a common substrate technology for advanced nodes packages especially in cost sensitive product portfolios.

Fig. 2. Schematic of FCCSP Package

A critical configuration was used to understand impact and marginality for assembly processes and assembly yield. A standard high volume manufacturing (HVM) assembly flow at the OSAT ASE was executed. Multiple Reflow testing and quick temperature cycling followed by CSAM analysis was used to prove manufacturing robustness.

In figure 3 the standard assembly flow is shown.

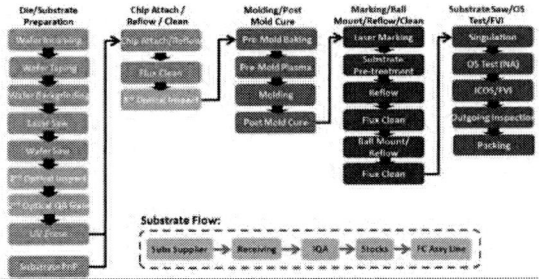

Fig. 3. General process flow of the assembly

B. Electrical CPI Test Structures

The TV contained multiple electrical test structures to assess the chip package interaction on critical locations of the die in the package. For this purpose perimeter structures, under bump structures and stitch structures cover especially the die corner and periphery region to have an electrical check on package level after assembly.

These structures are checked initially on wafer level, to assemble only known good dies. By routing the structures in the 3 layer ETS substrate to BGA balls these can be measured on package level (on leakage and resistance). Detailed electrical analysis of CPI critical location are possible before and after packages level reliability (Temperature Cycle, High Temperature Storage or unbiased HAST, etc.). At the T0 package level tests, assembly related fails could be detected. This is important to check for potential non-wets or substrate defects. In case of severe yield loss at T0, focused root cause analysis with short turnaround times is possible. Furthermore these structures are used to monitor the health of the packages during the reliability stress tests to qualify the technology.

III. WARPAGE AND MEASUREMENTS

A. Bi-material strip working hypothesis

Multiple types of materials are used in dies and package and interact with each other. These can be grouped in polymers, metals, ceramics/glasses and others. These have typically different mechanical properties. For thermal-mechanical load conditions (e.g. assembly reflow, temperature cycle test) the Young's modulus "E" and the coefficient of thermal expansion (CTE) are two of the key material property parameter (simplified concept).

This concept can be found in FCCSP packages everywhere in local or global scale. One example is the ETS substrate as a combination of Cu traces/pads in combination with the polymer prepreg. A second example, which is the key focus of this paper, is the DIE itself as combination of bulk silicon and the BEoL stack (Fig 4). It will be explained in the following section in more details.

B. Bi-material strip working hypothesis for die

In figure 4, the general behavior of the thinned (backgrinded) and singulated (Laser groove, Dicing) die, as a bi-material strip of BEoL stack and bulk silicon, is shown. This is a simplified concept which ignores the polyimide layer for now.

The CTE of bulk silicon is about 2.6ppm/K and the Young's modulus is about 130GPa. For the BEoL stack as a combination of Cu structures (traces, vias and fill patterns) and inorganic dielectric material (TEOS, ULK, ...) the CTE, based on a simple rule of mixture, is expected to be higher than 2.6ppm/K. This means, during a significant temperature increase, the lateral expansion of the BEoL stack layers will be higher than for the bulk silicon. For a reflow profile from 20°C to a peak of 260°C, there is a temperature delta of 240K.

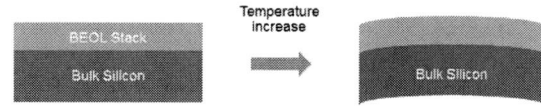

Fig. 4. Simplified model of bi-material strip warpage during temperature increase

In [BER04] this bi-material strip warpage is proposed to be modeled as "thermal warpage" with a formula containing the delta CTE, the temperature change, the strip length L and the strip thickness d. Additional the thicknesses d_{BEOL} and $d_{Bulk\ Si}$ and the Young's modulus E_{BEOL} and $E_{Bulk\ Si}$ is considered. The combination of the formulas results in a nonlinear equation which can be used to predict the die warpage in dependence of temperature change, total die thickness (BEoL stack+Bulk Si) and die size.

For the TV in this study with a 7mmx7mm die size the diagonal length is 9.9mm (which is the bi-material strip reference). For a typical product package configuration a final die thickness of 150-200µm is common. Since the total die thickness can easily varied by the backgrinding process in the assembly process, this was the focus parameter for the data collection in this study. Multiple runs of wafer preparation on different die thicknesses were done.

C.   Bi-material strip hypothesis for ETS Substrates

For completeness of the discussion in this paper, the warpage behavior of the substrate, especially ETS, need to be introduced. Also for this combination of Cu and polymer it is expected, that a warpage during temperature increase can occur. (e.g. assembly reflow, temperature cycle). The risk is higher compare to normal substrates, because for the trace embedding there is always a so called recess of trace up to 5µm in reference to the substrate surface. In combination with a 20µm SnAg cap of a Cu pillar, this can already be a challenge for large dies.

For the FC assembly reflow process, it increases the risk of local non-wets of the first level interconnects (here Cu pillar) since the substrate is not flat anymore. Most critical mode of the substrate deformation is, if the warpage is in the opposite direction to the die warpage. In figure 8 a schematic is show for the reflow peak. A mitigation scenario is to increase solder cap volume or use SnAg balls as first level interconnects (if possible).

Chen et al published key learning in [CHE15]. Details on managing the warpage of the substrate strips are given. The key learning from design perspective is that the balancing the top /bottom copper volume plays the significant role. This is not only possible by pattern densities but also by thickness optimization.

Additionally, the prepreg selection with high modulus prepregs is essential. To understand the actual performance of the design for the test vehicle

Shadow Moire measurements were completed and compared to assembly technology baseline.

For the measured ETS substrate strip the absolute warpage (Z-direction, out of plane) was determined. It was analyzed for the full package area 14mmx14mm and for the actual under die area of 7mmx7mm. This is important, because due to routing complexity variations in the local metal content can cause significant differences. Additional, in the area of the die, it is more critical for potential non-wet fails during reflow FC attach.

D.   Shadow Moire measurements

In [CHE15] the Shadow Moire methodology is introduced. It is in general a methodology *"...which employs a collimated light beam pointed at an angle to a 100 line/mm master grating in front of ..."* the object to be measured. During heating up the object, the fringe patterns will change and the algorithm calculates the deformation in z-direction. This methodology can be applied to many different objects and the key challenges are typically the sample preparation and the control of the temperature increase. The smaller the object or area to be analyzed, the more complicate the measurement. To check the consistency, always multiple objects should be measured. For larger objects the uniform temperature control plays an important role.

For this study here it was done on ETS substrate strip level in the assembly boat to assess the actual warpage behavior during the assembly reflow process. The strip contained 84 single substrates in a 14x6 matrix.

On die level it was thinned dies on a baseplate on the die backside. Typically 5 dies were measured to proof repeatability. The accuracy for z-direction warpage accuracy is about 2.5µm.

E.   Inline non-wet check for strip assembly

As the assembly was executed on selected splits the first inline parameter to assess the potential non-wet fail due to very high die warpage hysteresis was the x-ray. Based on the hypothesis that the warpage hysteresis in the reflow avoids good contact between Cu pillar SnAg cap and ETS landing trace, the die corner area and edge were the focus. 100% engineering inline inspection was required for the ETS substrate strip. After completion the assembly flow was continued until the complete package.

Subsequently, the electrical tests on package level were done to double confirm passing and failing units and statistics on percentage of fails. Finally cross sections were made to analyze the shape of interconnects. For some split legs with stretched interconnects have been observed and indicated a marginality for this specific configuration.

IV. BEoL MANUFACTURING ROBUSTNESS

A. Multiple Reflow Prove Point

The split legs with a sufficient assembly margin without critical non-wet on the die corner were examined by a standard qualification approach of multiple reflow experiments. The dies were assembled on substrate with the initial assembly reflow. Without applying the mold or any underfill, the substrate strips with dies were stressed by multiple reflow cycles (up to 20x) to over stress the BEoL of the die. The thermo-mechanical stress is induced by the mismatch of the coefficients of thermal expansion (CTE) between the silicon die and polymer substrate. In Figure 3 the general process flow is shown. In regular intervals the BEoL was checked by scanning acoustic microscopy (CSAM). For ULK BEoL stacks in combination with Cu pillar first level interconnects, there is a risk of delamination in the BEoL (white bumps) for critical configurations. The highest risk occurs in the die corner since this is the region of the highest stress.

In Figure 5 a typical reflow profile is shown, which was used for assembly and MR stress. The process was executed on HVM tools in order to assure the repeatability.

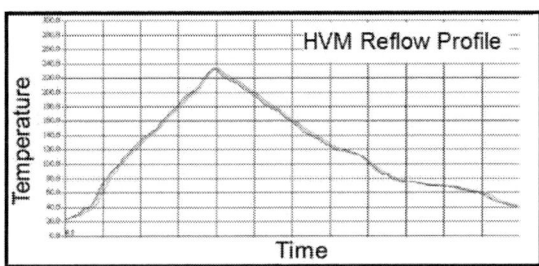

Fig. 5. Reflow profile for assembly and multiple reflow stress of FCCSP sample strip with assembled dies

V. PARAMETER FOR EXPERIMENTS ON PACKAGE

A. BEoL stack variations and design

To prove the working hypothesis of the dependency of the warpage hysteresis from the effective Young's modulus E_{BEoL} and CTE_{BEoL}, 3 different reticles sets were designed for the 2 UTM modules. It allowed to cover a metal density range of "low", "medium" and "high". Based on the hypothesis that the higher Cu ratio increases the effective CTE_{BEoL}, the "high" metal density split lot was expected to show a significant higher die warpage hysteresis during the reflow process. CTE of bulk Cu is 17ppm/K and therefore much higher than Silicon and TEOS. Accordingly the lower Cu ratio in the UTM layer should decrease the warpage hysteresis.

The wafer material was processed through standard Fab processes and inserted in the standard assembly process flow.

B. BOM and Assembly Process Parameter

The key parameter which influences the assembly process is the die thickness. Wafers were grinded to different thicknesses in the range from 100μm to 780μm (full wafer thickness). Based on the working hypothesis and the literature, a nonlinear dependency was expected. The thinnest dies were expected to show maximal die warpage hysteresis during assembly reflow process and a high risk for non-wet yield loss was expected.

Additional, variations for the reflow profile, the flux process and the strip warpage control tooling have been done to further understand the significance and options to increase the margin for given die thickness and die size. Furthermore a thermal pre-treatment procedure of the ETS substrate strips was investigated on its effect on non-wet risk.

VI. RESULTS

A. Die Warpage vs. die thickness Results

The Shadow Moire measurements quantified the typical warpage hysteresis in the standard reflow profile. For the data representation the warpage for the diagonals of the 7mmx7mm was extracted from the Shadow Moire data.

In Figure 6 the die warpage during the temperatures in the assembly reflow process is shown for 4 different die thicknesses. For all die thicknesses the warpage is increased during the heating process and reached its maximum at the peak temperature. During the cooling the warpage is decreased and reaches its minimum at final temperature of 32°C.

The lower the die thickness is, the higher the maximum warpage during the reflow process is. Thin thicknesses corresponds to a high "BEoL to bulk silicon thickness"-ratio. For the experiments, the ratio ranged from 16% (very aggressive, very thin die) to 1% (very conservative, very thick die). The used design for these experiments had a "medium" metal density value (see next section)

For the most aggressive ratio (BEoL/Bulk=16 %), the warpage hysteresis ranges up to 45μm (at 260°C peak). Considering a Cu pillar interconnect with a POR height of 58μm and a SnAg cap of 25μm, it is clear, that a high risk of non-wet fails in the substrate assembly occurs. Even tuning assembly parameter like flux process or reflow heat up/cool down ramp will barely mitigate the risk.

On the other hand, for the medium and low BEoL/Bulk-ratios (up to 11%) the warpage hysteresis is acceptable because the typical values do not exceed 20μm. The relationship between die thickness and warpage hysteresis is nonlinear, which means, that minor changes in the die thickness can have a significant impact (for thinner dies).

Additionally the warpage hysteresis between heating and cooling was compared and matched well (arrows in Figure 6).

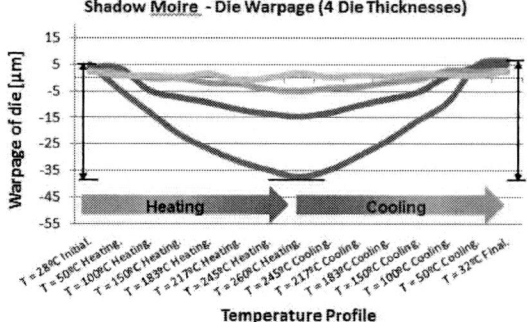

Fig. 6. Bare die warpage for 4 different die thicknesses during reflow process measured by Shadow Moire (thinner die – more warpage)

B. Die Warpage vs. Metal density Results

Three different metal density rations in the two UTM modules were modified (low, medium, high). Based on the bi-material working hypothesis, the effective CTE and E are different and will cause different warpage hysteresis behavior during reflow. The reference was "medium" density for a 150µm die thickness. Two die thicknesses 200µm and 150µm were investigated for consistency check. A result summary is shown in the in Figure 7.

For the 150µm thick die, it was confirmed that higher Cu ratios can increase the warpage hysteresis significantly for about 70%. On the other hand, lower Cu rations decrease the warpage hysteresis for 29%.

For 200µm thick dies, which have lower absolute warpage hysteresis and are less critical in die warpage, the "low" metal density leg has only 34-40% of die warpage hysteresis of the POR leg.

In order to exclude errors caused by thermal inertia, the heat up and cool down process was checked. A good consistency was seen (Delta: 1% … 9%) was seen across all 6 legs (3 metal densities, 2 die thicknesses).

DOE Leg (Condition)	Metal Density	Die Thickness	Warpage Hysteresis (Relative Values)	
			Heat up (RT-->260°C HT)	Cool down (260°C HT --> RT)
2xUTM, High, 150µm	High (109%)	150µm	169%	170%
2xUTM, Med, 150µm	Med (100%)	150µm	100%	100%
2xUTM, Low, 150µm	Low (85%)	150µm	73%	71%
2xUTM, High, 200µm	High (109%)	200µm	104%	109%
2xUTM, Med, 200µm	Med (100%)	200µm	79%	70%
2xUTM, Low, 200µm	Low (85%)	200µm	40%	34%

Fig. 7. Impact of UTM metal density on the die warpage hysteresis in Shadow moire analysis

C. ETS Substrate Strip Warpage Results

To understand the ETS substrate warpage deformation during the assembly reflow (Flip Chip attach process), the complete substrate strip (14x6 single substrates) was measured individually with the Shadow Moire methodology. The warpage at different temperatures during heating and cooling is shown in figure 8. The region of interest was the single substrate die areas (7mmx7mm).

An example of the overall deformation mode "convex" is visible in Figure 9 (one single substrate).

It does not change its mode during heating and cooling (always convex, Cu pillar landing traces are up).

Based on Figure 8 the (profile) the warpage decrease slightly area below the die was evaluated (Fig). At solder peak temperature, when the highest die warpage is seen the warpage is about 10µm.

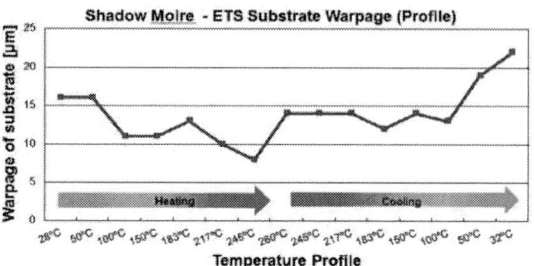

Fig. 8. Substrate warpage profile (diagonal) during reflow process measured by Shadow Moire

Substrate Deformation Mode vs. Temperature

T = 28°C Initial.	T = 50°C Heating.	T = 100°C Heating.	T = 150°C Heating.	T = 183°C Heating.
T = 217°C Heating.	T = 245°C Heating.	T = 260°C Heating.	T = 245°C Cooling.	T = 217°C Cooling.
T = 183°C Cooling.	T = 150°C Cooling.	T = 100°C Cooling.	T = 50°C Cooling.	T = 32°C Final.

Fig. 9. Substrate deformation mode during reflow process profile measured by Shadow Moire

For the complete 14mmx14mm single substrate area much higher values up to 35µm occur. This is the minimum to maximum delta on the measurement area.

D. Assembly risk and general failure mode margin

During die attach reflow the die and the substrate strip are undergoing individual warpage deformation. The individual values have been presented in the previous section. Especially for thin dies (with high BEoL/Bulk ration) and dies with a high Cu metal density in the BEoL (UTM layers) a significant die warpage (Concave, Cu pillar down) can occur. This overlays with the individual ETS substrate warpage. The critical worst case scenario is that the warpage occurs in the opposite direction (substrate convex, die concave) and is shown in figure 10. The risk of non-wets at the die corner is increased.

	Room Temperature	Reflow Temperature
Ideal case		
Critical case		

Fig. 10. Worst case scenario where substrate and die have a warpage deformation in the opposite direction

In Figure 11 the failure mode of non-wet is shown in comparison to a good wetting result. The criticality is a combination of die size, die thickness, BEoL thickness and finally also Cu pillar SnAg cap high.

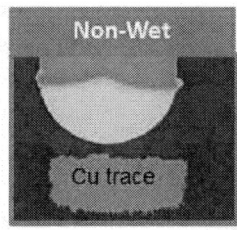 

Fig. 11. Non-wet as failure mode post assembly reflow which can be detected by x-ray and verified by cross section

The inline detection of non-wets is done with x-ray inspection for the die corner and edges. For the CPI work presented here, the electrical test structures were designed in a way to identify such case electrically on package level and matched well with the x-ray findings.

VII. PACKAGE RELIABILITY VERIFICATION

After the comprehensive study of die and substrate warpage multiple package level studies were executed.

The thinnest die option (Figure 6) could not be assembled successfully without non-wet. So it is proposed to avoid very thin dies in combination with very thick (2xUTM) BEoL stacks for this die size. If very thin dies are needed, a much larger SnAg volume for the Cu pillar or even SnAg balls as first level interconnects are proposed (for die size 7mmx7mm). For smaller dies, the die warpage is less and the non-wet risk is reduced. In order to optimize an assembly setup, the die warpage hysteresis needs to be collected and analyzed. If it is less than 20μm it is expected to be passing. If it is higher a risk for low margin due to non-wet fails can occur.

In figure 12, a cross section of a typical package without non-wet fails is shown. At the die corner region the x-ray result is confirmed no no-wets.

Multiple Reflow tests (without underfill) were used to prove the BEoL robustness for several legs (e.g. die thickness variation, metal density variation). Additional packages for component level reliability test were assembled and checked for any non-wet units by x-ray and electrically. All data passed.

Fig. 12. Confirmation of package assembly without non-wetting first level interconnects in the die corner

Package level reliability tests were executed and the passing results are summarized in Table 1. The package level tests include Temperature Cycling (TC - 55°C/125°C), High Temperature Storage (HTS 150°C) and uHAST.

TABLE I.   OVERVIEW OF PASSED RELIABILITY TESTS FOR QUALIFICATION OF NEW TECHNOLOGIES

Test item	Test Criteria	Summary
Manufacturing Robustness	Multiple reflow (MR), 5x	Passed
	Quick Temperature cycle (QTC), -40°C / 60°C, 20x	Passed
Preconditioning	MSL3	Passed
Temperature Cycling (TC)	-55°C / 125°C 1000x / 0 fails	Passed
High Temperature Storage (HTS)	150°C 1000h / 0 fails	Passed
unbiased Highly Accelerated Stress Test (uHAST)	130°C/85%rh 96h / 0 fails	Passed

VIII. SUMMARY AND CONCLUSION

A systematic study of the influence of the die warpage during Flip Chip attach reflow process was carried. A standard FCCSP package configuration was used. The die attach must be controlled well especially for products, which require UTM stacks in combination with low die thickness. Shadow Moire measurements should be applied to generate the die and substrate warpage data. These data can be used to assess the risk of non-wet failures in assembly. For a Cu pillar with 25μm SnAg cap, the non-wet risk is controllable for thick BEoL stacks with high Cu metal density in BEoL.

REFERENCES

[CHE15] C. Chen, M. Lin, G. Liao, Y. Ding and W. Cheng, "Balanced embedded trace substrate design for warpage control," 2015 IEEE 65th Electronic Components and Technology Conference (ECTC), San Diego, CA, 2015, pp. 193-199

[BER04] Frank Bernhard, Technische Temperaturmessung, ISBN 978-3-642-62344-8, Springer Verlag Berlin Heidelberg 2004, DOI 10.1007/978-3-642-18895-4

[BOE16] Bjoern Boehme, Christian Goetze, Sebastian Dej, Po-Hsiang Wang, Frank Kuechenmeister, Dirk Breuer, Jens Paul and Michael Thiele, "Advanced chip package interaction qualification for critical stacks in combination with Cu pillar interconnect technology," 2016 6th Electronic System-Integration Technology Conference (ESTC), Grenoble, 2016, pp. 1-6. ; doi: 10.1109/ESTC.2016.7764456

[BOE17] Bjoern Boehme, Dirk Breuer, Christian Goetze, Al Rhea Estoque, Falk Tischer, Frank Kuechenmeister, Jens Paul, Michael Thiele, "Chip package interaction with Cu Pillar interconnects - systematic study of key factors impacting the qualification", Microelectronics Packaging (NordPac) 2017 IMAPS Nordic Conference on, pp. 68-73, 2017, doi: 10.1109/NORDPAC.2017.7993167

Integrated Condition Monitoring by Measuring the Delay of Gate Turn-off

Felix Wuest[a], Stefan Trampert[a], Frederic Sehr[a], Klaus-Dieter Lang[b]

[a] Fraunhofer Institute for Reliability and Microintegration, *Gustav-Meyer-Allee 25, 13355 Berlin, Germany*
[b] Technical University Berlin, *Gustav-Meyer-Allee 25, 13355 Berlin, Germany*

* felix.wuest@izm.fraunhofer.de / Tel: +49 30 / 46403-686

Abstract

Since renewable energies are becoming increasingly important in today's energy supply, improving the reliability of components in the energy conversion chain, namely power inverters, is becoming a major field of interest in current research. Reliability issues of these power inverters have been discussed thoroughly in previous research. An in-depth exploration of condition monitoring on their insulated gate bipolar transistors (IGBTs) is, however, still outstanding. To predict age-related failures it is necessary to implement condition monitoring into the inverters. Therefore an integrated condition monitoring system for IGBT power electronic devices is developed. The chip's junction temperatures are derived from measuring the miller plateau duration during turn-off. A circuit is integrated into the gate driver board for easy acquisition.

Key words: condition monitoring, miller plateau, turn-off, IGBT, reliability

1 Introduction

There are numerous electrical parameters that show a dependence on the chip temperature and thus indicate the degradation of the die attach of IGBTs. One of these is the gate turn-off duration, also known as the Miller Plateau. In previous research [1] the possibility of condition monitoring of IGBT devices using the gate turn-off duration has been investigated utilizing a sampling oscilloscope for data acquisition which leads to high power consumption and acquisition of large amounts of data. The goal of this research is to implement a more convenient and accurate way by integrating the measuring of the delay into a gate driver board.

2 Condition Monitoring

There are three concepts of condition monitoring (Figure 1). The first concept is parameter monitoring, where a parameter inherent to the system that is sensitive to the aging process is constantly monitored (Figure 1a). The second concept is known as the canary device (Figure 1b). This refers to a component unnecessary to the devices functionality that is loaded just as the regular components but has a weaker structure. Therefore it will fail prematurely. Different structural layouts can be used to implement different warning stages so that a prediction becomes quite accurate. The third concept, the so-called life cycle unit (Figure 1c), constantly monitors the load upon a device and predicts the remaining life-time via an underlying life-time model [2].

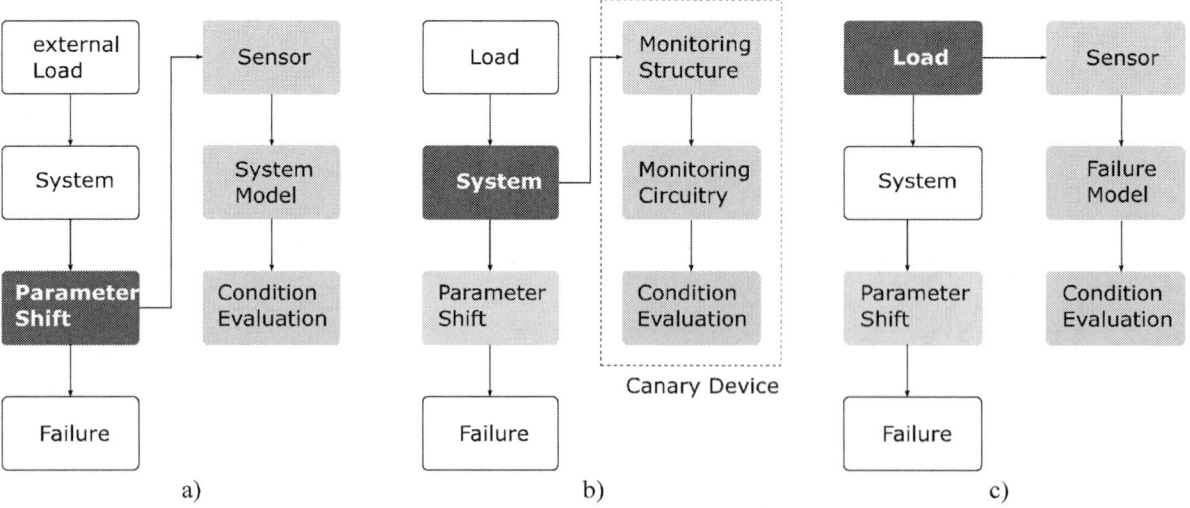

Figure 1: Concepts of condition monitoring: a) parameter monitoring b) canary device c) life cycle unit [2]

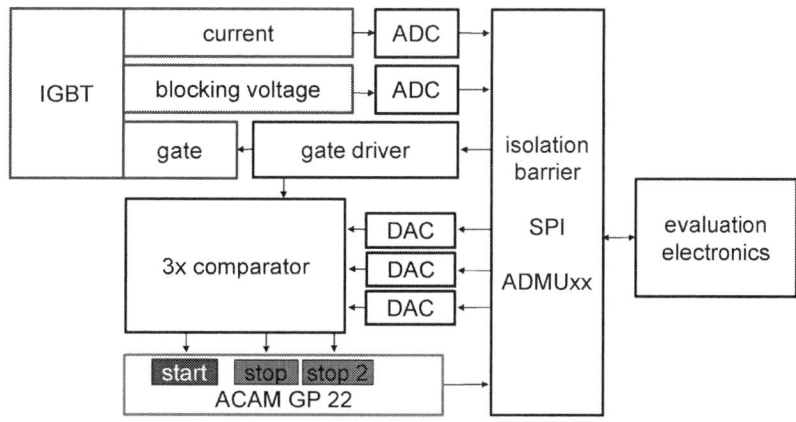

Figure 2: Concept for measuring the gate turn-off duration.

As the IGBT's die temperature is dependent on both load and state of damage, a combination of the first and the third concept is feasible for usage in this research. Temperature swings, as being one of the most critical aspects regarding lifetime [3][4], can be counted as well as a continuously increasing die temperature under simmilar working conditions might be observed and related to a progressed damage.

3 Measurement Circuit

Figure 2 depicts the concept of the measurement circuit. The gate driver takes care of supplying sufficient energy to the IGBT's gate. Three high speed comparators monitor the gate voltage. These define the start and two stop events, of which only one is evaluated at the moment. The time to digital converter (TDC), an Acam GP 22, measures the time between start and stop with a sampling accuracy of up to 22 ps. The resulting time is communicated via SPI over an isolation barrier to

Figure 3: Image of prototype board of integrated delay measuring on gate driver.

the evaluation unit. In addition, current and collector-emitter voltage are measured by analog-digital-converters (ADC) and communicated to the evaluation unit. Figure 3 depicts the prototype board. To ensure that the measurements taken by the digital acquisition unit (DAQ) are within reasonable limits, the temperatures on the heatsink and the module will also be measured using thermocouples.

4 Test Setup

A simplified setup circuit is depicted in Figure 4. It consists of a current source with a voltage limit of 4V and an inductor (L) that forces a high voltage during turn-off. Additionally a snubber circuit of a parallel resistor (R) and capacitor (C) is implemented to achieve the desired turn-off voltage and limit the voltage transient. The passives were selected to have a low temperature coefficient. These values are also shown in Table 1. With the low values given, a rather small impact from thermal influences is expected during the test.

Table 1 shows the values chosen for the passives.

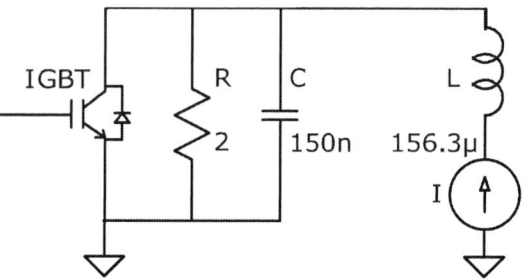

Figure 4: A simplified circuit used for testing the measurement method in application.

The passives were selected to have a low temperature coefficient. These values are also shown in Table 1. With the low values given, a rather small impact from thermal influences is expected during the test.

Table 1: Values for the passives in the test.

Part	Value	Thermal coefficient
L	156.3 μH	cooled at 22 °C
R	2 Ω	100 ppm
	4 Ω	
C	150 nF	-200 ppm

The selected IGBT is an IXGN 200N60A2 in a SOT-227B package with a current rating of 200 A and a maximum blocking voltage of 600 V. Its field of application is motion control, DC choppers, uninterruptible power supplies and power supplies.

The IGBT is turned on for a pulse duration of 30 ms and turned off afterwards. This allows a test with relatively low self-heating.

This setup allows testing the measurement method with close to real life application blocking voltages and high currents.

Figure 5 shows an exemplary development of the gate emitter voltage during turn-off with threshold voltages ($V_{start} = 12.8\,V$ and $V_{stop} = 0.73\,V$) and time measurement.

The parasitic inductivity of the gate circuit is quite high due to the fact that the board is not directly mounted onto the IGBT, but connected via wires of approximately 6 cm length. This leads to a short oscillation peak between 10 ns and 150 ns. The Miller plateau ends approximately at 300 ns where the gate voltage decreases again. The second threshold voltage has to be lower than the Miller plateau level to trigger the end reliably. The starting threshold voltage however may be set in a greater range due to the high dv/dt during the first nanoseconds.

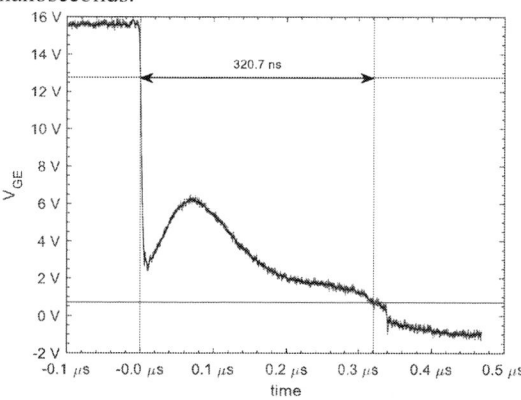

Figure 5: Exemplary gate emitter voltage during turn-off and corresponding time measurement at threshold voltages.

To compare and benchmark the TDC measurement system, the electrical parameters are also monitored with a 12-bit oscilloscope. The evaluation is done equally with the same threshold voltages.

The corresponding collector-emitter voltage is shown in Figure 6. For a current of 50 A the turn-off voltage should result in 200 V with a snubber resistor of 4 Ω. Due to the parasitic inductance of the snubber circuit the switch-off voltage overshoots above 300 V.

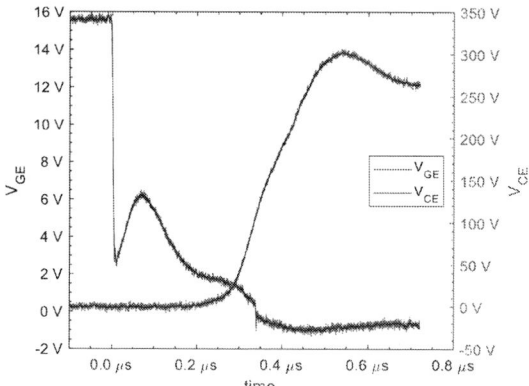

Figure 6: Exemplary collector-emitter-voltage with corresponding gate-emitter-voltage during turn-off.

The switching losses, i.e. the non-linear behaviour of the semiconductor, result in a current-dependent attenuation of the snubber. This results in a non-linear curve between the maximum breaking voltage and the breaking current as shown in Figure 7. However, a linear function is still a good approximation.

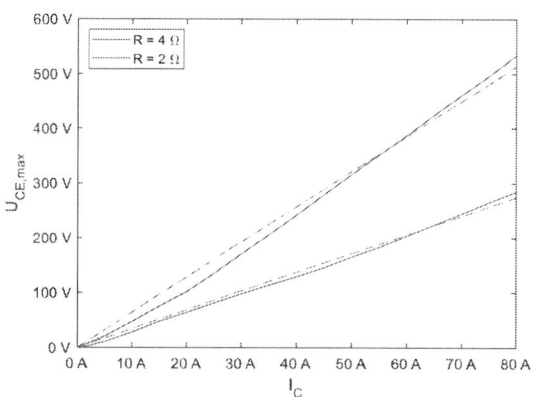

Figure 7: Maximum collector-emitter-voltage peak during turn-off with linear fit (dashed line) at room temperature.

5 Results

Figure 8 depicts the results for the TDC measurement for two different snubber resistors for various currents and its corresponding oscilloscope measurements at room temperature. Clearly, there is no linear correlation between current and turn-off duration for low currents. At approximately 20 A, which is 10 % of the nominal IGBT current, the curves start to linearize. Therefore, the proposed measurement method should only be used for currents above this limit. TDC and oscilloscope measurement show similar behavior, apart from having an offset.

Figure 8 also shows that the standard deviation for the TDC measurement is rather low for each of the 17 measurement points. At each point 100 measurements were taken to acquire enough statistical information. The oscilloscope data, on the other hand, shows quite a large standard deviation which reaches its maximum at 10 A. It is about 8.26 ns for the same number of measurements. The TDC data also shows its biggest standard deviation at 10 A, but with only 0.268 ns it is more than one order of magnitude smaller than corresponding oscilloscope data.

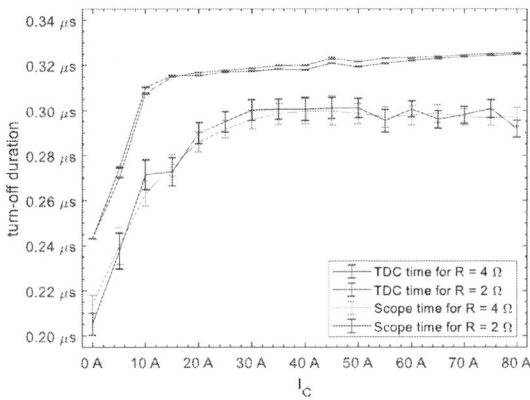

Figure 8: Time measurement by scope and TDC circuit in comparison for two different snubber resistors at room temperature

Both curves with different snubber resistors nearly overlap, however the standard deviation of the TDC measurement is so low that it is feasible to identify the difference between both. At 50 A both differ about 2.2 ns which is about 1 % of the absolute value.

The high deviation of the oscilloscope data becomes even more visible when a steady temperature slope is monitored continuously by both methods and compared against each other.

In Figure 9 each measurement is plotted at its corresponding leadframe temperature. The TDC measurement (blue dots) shows great uniformity within itself, producing quite a good curve. A linear fit gives little error. The scope data (orange dots) however is scattered a lot. The linear fit is parallel to the TDC fit but gives greater error. Compared to the opperation of different snubber resistors, the temperature error would be less than 2.7 °C, which would be acceptable. However, with increasing temperature the error increases, as Table 2 and Figure 10 show. For the interesting temperatures at approximately 150 °C the error will be unacceptably high.

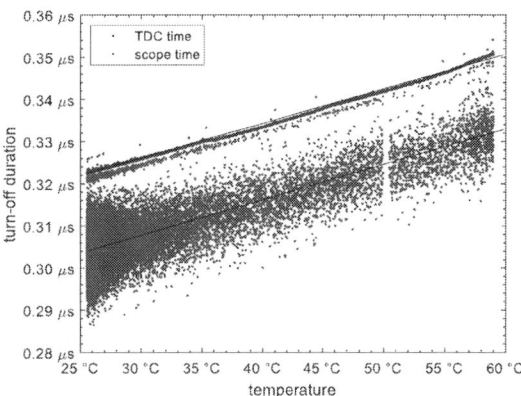

Figure 9: Comparison of TDC and scope measurement on a temperature slope for 50 A with 4 Ω snubber resistor.

Table 2: Temperature coefficients for different currents

Current	R	Gradient	Offset
25 A	2 Ω	0.9429 ns/K	301.62 ns
50 A	2 Ω	0.9193 ns/K	301.11 ns
70 A	2 Ω	0.8516 ns/K	310.08 ns
50 A	4 Ω	0.8022 ns/K	302.02 ns

Nevertheless, the comparison of the turn-off duration versus temperature under different DC currents with a snubber resistor of 2 Ω in Figure 10 shows obvious differences. The linear fit is depicted with a strong line wheras the 95 % error estimates are painted with the corresponding lighter color. Table 2 sums up the gradients of the different curves.

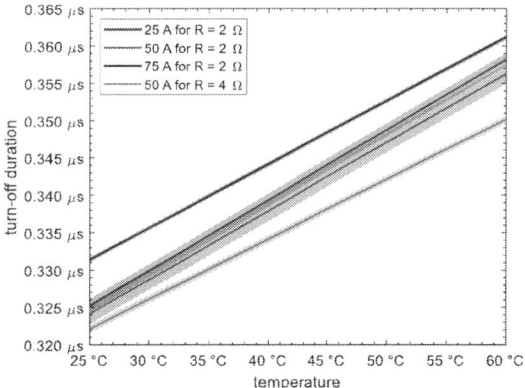

Figure 10: Comparison of influences of different currents onto the temperature dependency for two snubber resistor values.

This shows that there is no simple way to derive the temperature directly from the measurement of the turn-off duration. Some assumptions and simplifications have to be made at this point.

6 Conclusion and Outlook

For most applications, the DC link voltage is quite stable with only small changes in amplitude during operation. Therefore, the proposed method may be used, only being calibrated for currents and temperatures even though the parameter of interest shows a dependence from voltage. In addition, the voltage dependence is small compared to those on current and temperature. Using a double pulse method for calibration, a characteristic curve map can be obtained and embedded into the microcontroller. However, this needs to be done for each semiconductor type that is used with a few parts, so that sufficient statistics can be used.

With this method, detailed information of the chip temperature and therefore its load can be recorded. The temporal temperature resolution is very important as previous research has shown [1]. With the proposed method integrated into a gate driver board the chip temperature can be monitored continuously during operation. Life-time models can be implemented using this information.

With continuous work and more measurements taken for various currents and voltages a numerical approach for the junction temperature being a function of turn-off duration, current and voltage may be possible.

7 Acknowledgements

As part of the AMWind project, the Federal Ministry of Economics and Technology funded the research leading to these results.

8 References

[1] F. Wuest, S. Trampert, S. Janzen, S. Straube und M. Schneider-Ramelow, „Comparison of temperature sensitive electrical parameter based methods for junction temperature determination during accelerated aging of power electronics,“ in Microelectronics Reliability, 2018

[2] K. Jerchel, M. Krüger, A. Middendorf, N. F. Nissen und K.-D. Lang, „Reliability Improvements in Electronic Systems by Combining Condition Monitoring Approaches,“ in *International Conference on Electronics Packaging (ICEP) 2014*, Toyama, Japan, 2014

[3] M. Ciappa, Selected failure mechanisms of modern power modules, Microelectron. Reliab. 42 (2002) 653–667.

[4] U. Choi, et al., Study and handling methods of power IGBT module failures in power electronic converter systems, IEEE Trans. Power Elec. 30 (5) (May 2015) 2517–2533.

Numerical study on the influence of material models for tin-based solder alloys on reliability statements

R. Metasch, R. Schwerz, K. Meier*, M. Roellig

Fraunhofer Institute for Ceramic Technologies and Systems, Branch Materials Diagnostics, Dresden, Germany

*Technische Universität Dresden, Institute of Electronic Packaging Technology, Germany

Phone: +49 351 88815 581, and E-mail: rene.metasch@ikts.fraunhofer.de

Abstract

This paper presents a comparison of three different material models, which are currently used to describe creep behaviour of solder alloys in thermo-mechanical use cases. The first model is the commonly utilized Garofalo approach. This approach is based on a hyperbolic-sine equation to describe the secondary (stationary) creep rates dependent on the mechanical stress in combination with the Arrhenius equation to consider temperature dependencies. The other two material models cover the primary and the secondary creep stage. Here, unified viscoplastic constitutive models initially proposed by Anand and Chaboché et al. are introduced. In the presented FEM study result comparison has been done for the accumulated solder joint strains of a SMT component exposed to thermal cycling. Subsequently, the influence on creep strain progression and service life time estimation has been investigated.

Keywords: lead-free solder, thermo-mechanical properties, Unified, Anand, Garofalo, SinH, stationary, secondary, primary, creep, life-time, electronics

Motivation

The service life of electronic circuits is mainly determined by the degradation behaviour of the interconnections between the printed circuit board and the electrical components. This degradation process is influenced by the applied solder alloy and its individual time and temperature dependent deformation behaviour. But it is also affected by the occurring temperature change, the thermal mismatch and the complex stiffness situation of the adjacent geometries and materials. On the other hand, the overall trend is towards smaller solder joints and solder alloys with increased strength. Hence, temperature driven effective strains are decreased. This leads to the fact that stresses evolving within the solder volume are increasingly dominated by primary creep mechanisms rather than by secondary creep. Though this is well known, the creep data input of solder material models in numerical simulations, which cover primary creep behaviour and enable quantitatively accurate analysis, are not available as of today. Additionally, efficient experimental characterisation approaches to determine the necessary material property input are not at hand even though they are a prerequisite to accomplish successful material, packaging and system design comparisons with numerical analysis.

Material modelling of tin-based solder

For years solder material has proposed significant challenges to material characterization, because the actual mechanical response is rather complex. Soon it became clear, that the time dependent behaviour cannot be neglected. State of the art implementations in today's virtual design include the creep behaviour, which is most often based on stationary creep and rate dependent deformation properties. The primary and tertiary creep regimes are generally not considered.

The authors proposed an advanced modelling of the solder response by superposition of the hardening effects, the viscous effects and relaxation. The result is a unified plastic representation of the material response, which also models the hardening and primary creep deformation regimes. Further, the Anand model has been fitted to the measured stress-strain curves. This approach is also able to model the primary and the secondary creep behaviour with some restrictions.

Different to other studies, the used experimental procedure allows the extraction of parameters for all of the aforementioned material models, based on the same set of characterization measurements. This significantly improves the evaluation of the addressed material models by preventing the drawbacks of having different measuring setups and their individual influences during the data acquisition process before the parameter extraction even begins.

The solder characterisation process has been performed in the temperature range from -40 °C to 150 °C while strain rates were varied from 1e-3 1/s to 1e-6 1/s and multiple relaxation steps were applied. The detailed characterisation procedure has been presented in a previous publication [1], [2] and is shown exemplary for 25 °C in Fig. 1. The hysteresis for 25 °C and 150 °C is given in Fig. 2.

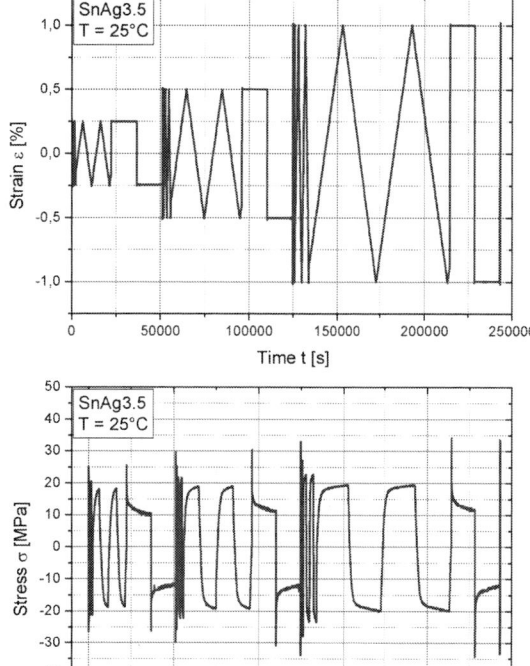

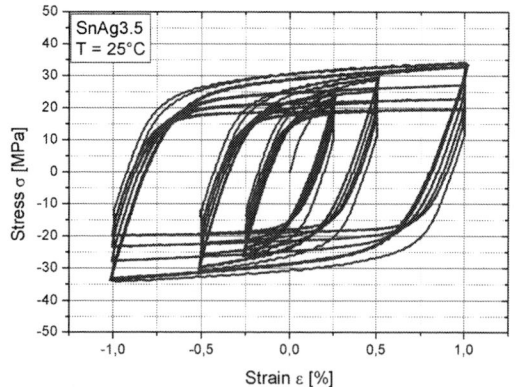

Fig. 1: Example of measured strain and stress progress by variation of strain amplitudes and strain rates under tensile-compression for 25 °C [1]

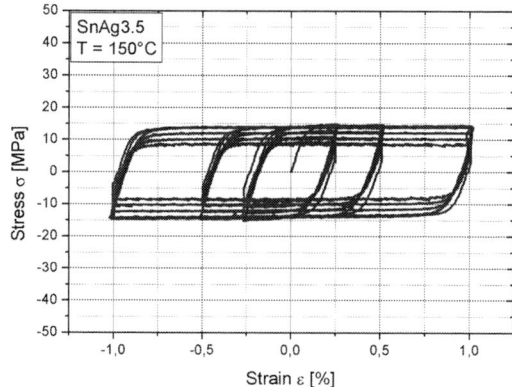

Fig. 2: Example of measured strain-stress hysteresis for 25 °C and 150 °C [1]

This approach enables a ranking of the material models and a general answer to the question whether primary creep modelling should be considered for a thermo-mechanical load case.

Tin-based solder modelling with Garofalo (sinh)

The Garofalo material model describes stationary creep rate behaviour as a function of stress and temperature. This model cannot represent the non-linear stress evolution at a given strain rate in the primary creep region, since material hardening effects are not considered within the mathematical function.

Material model parameters have been determined based on the end stresses for the measured strain rates at temperatures over 25 °C, because only here the stresses reach a stationary state. The result fits this measured stresses in this limited temperature range. At -40 °C the model does not fit the measured end stresses, since stationary creep is not reached within the strain amplitudes in the experiment. Instead, stresses are extrapolated based on the stationary data for the -40 °C (see Fig. 3).

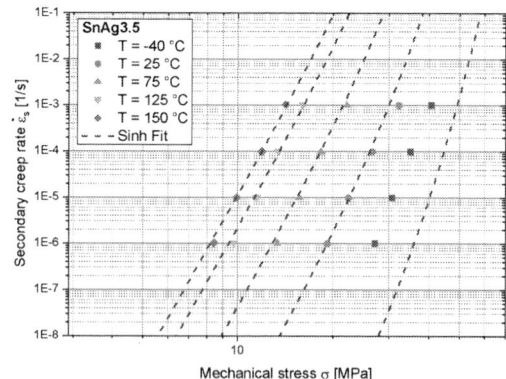

Fig. 3: Sinh-fit to model stationary creep behaviour [1]

Tin-based solder modelling with Anand model

The Anand model is a constitutive material model, which is able to describe the inelastic deformation dependent on strain rate and temperature, see Fig. 4. To determine the Anand model parameters the tensile-compression progresses have to separate into single curves beginning at zero stress for every strain-rate and temperature constellation (5 temperatures with 4 strain rates each). This results in 20 stress-strain master curves which are analytical approximated in two steps. After that a numerical model of the tensile-compression solder specimen was used for further parameter optimization.

In this first step only, the stationary stresses are fitted in comparison to the Sinh model and their restriction, that the stresses at -40 °C are extrapolations based on the value of the other temperatures (see Fig. 4). The results are very similar to the Sinh result (compare to Fig. 3).

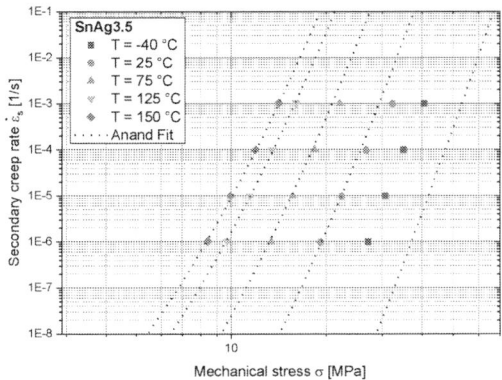

Fig. 4: First Anand-fit to model only the stationary creep behaviour [1]

The next fitting step is to determine the Anand model parameters, which calculate the progress to the stationary stress state, which represent the beginning of the curves or the primary creep parts.

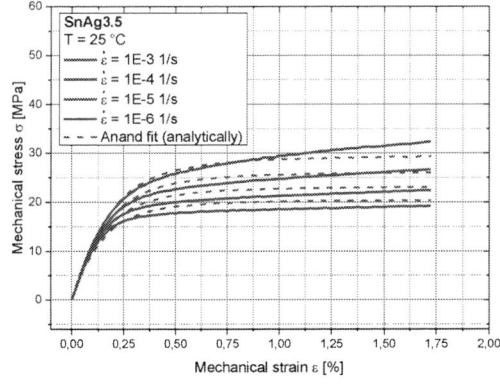

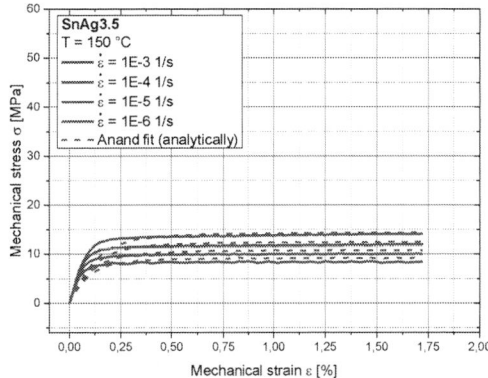

Fig. 5: Strain-rate dependent stress-strain master curves with analytical optimized Anand model fit at 25 °C and 150 °C

The last step is a validation and optimization step to improve the Anand model parameters. For that, a numerical model of the real specimen geometry was used. Displacement measurement data were transferred to FEM-loading model and the parameters of the Anand material model were optimized under continuously comparative evaluation of deformation behaviours.

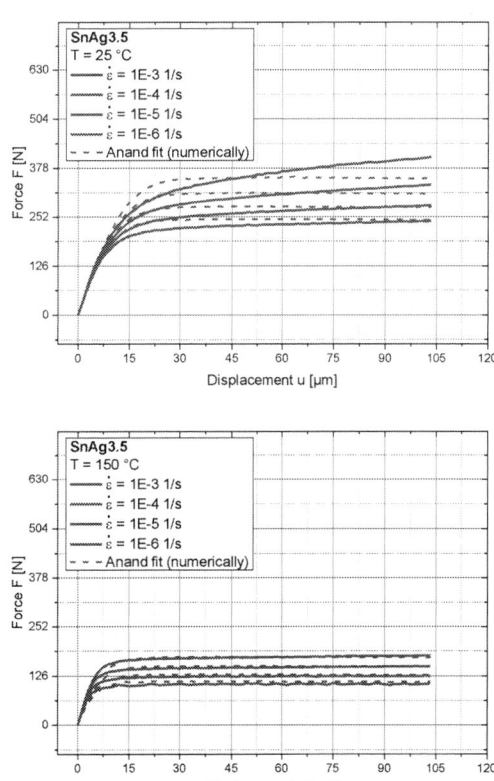

Fig. 6: Strain-rate dependent force-displacement master curves with numerical optimized Anand model fit at 25°C and 150°C

Tin-based solder modelling with Unified model

The last material model utilized is extracted differently than the two aforementioned. For the unified approach the measurements for multiple temperatures are fitted independent from each other. Therefore, the parameters don't have to be built around a compromise between data at low temperatures and data at high temperatures. On the other hand, the effort to extract all necessary coefficients is higher. The complete material description consists of a parameter set for each part of the model at every temperature.

The base of the modelling approach is a formulation for the kinematic hardening portion, which is mostly responsible for the primary creep. This is super-positioned with visco-hardening terms, which enable the rate dependant plasticity. Finally, this is supplemented with kinematic static recovery behaviour to achieve the correct relaxation. The following figure gives a representation of the distinct model parts based on the measurement at 75 °C (Fig. 7). The extraction of the material parameters has been published previously by Kabakchiev et al. [2].

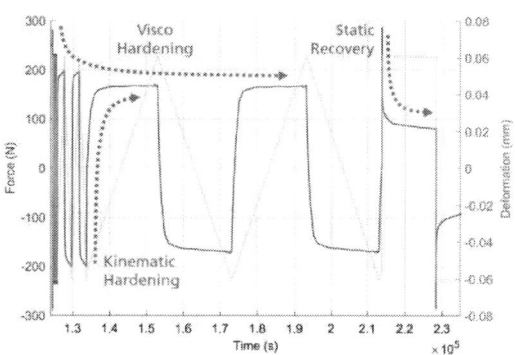

Fig. 7: Representation of the Unified model portions to represent the complete plastic behaviour including primary creep

Comparison of the mechanical behaviour

To compare the Anand and unified plasticity modelling (further called Unified) with the current state of the art stationary creep modelling (further called SinH) a FEM based parameter extraction with the virtual representation of the tensile-pressure specimen has been conducted. The model utilizes quarter symmetry and is given in Fig. 8. For the invers extraction the reactionary forces measured during the experiment were repeatedly compared to the simulated forces while the material parameters undergo variation. Once the difference becomes very small the correct material model parameters have been identified.

The results for the temperatures 25 °C and 150 °C are given in the following figures (Fig. 9 and Fig. 10).

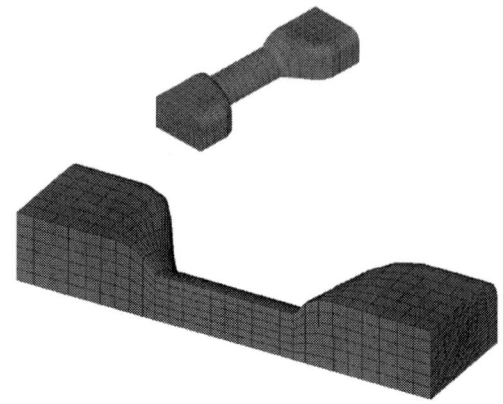

Fig. 8: FEM-Quartermodel of the tensile-pressure specimen for the invers material properties extraction.

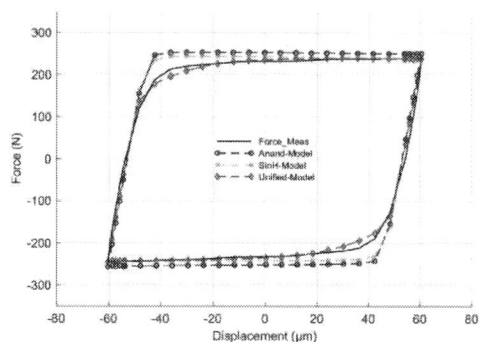

Fig. 9: Comparison of constitutive models and measurement for one mechanical cycle (25 °C, 1e-6 1/s)

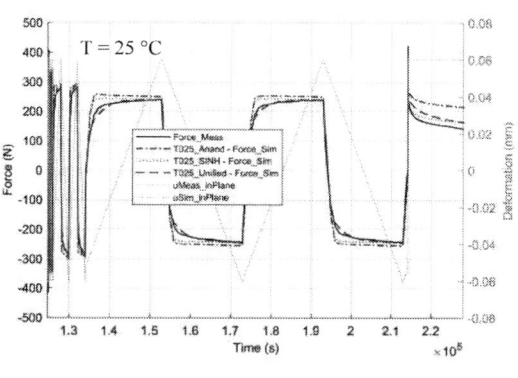

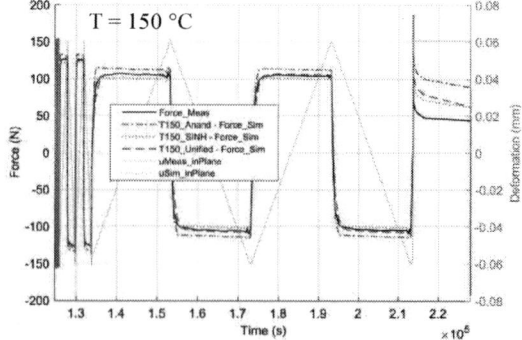

Fig. 10: Comparison between the measurement data and constitutive modelling results

Within the stress-strain behaviour it becomes obvious that the steady state SinH-model reaches its limitation for creep resistant solder alloys. The primary part of creep strongly dominates the mechanical behaviour. Either solders at low temperature or applications of advanced high temperature solder material result in such behaviour.

As a result of the comparison one has to conclude that only the unified model is able to fully describe the primary and the secondary creep state throughout the whole temperature range. The unified model almost identically follows the measured behaviour. The potential for higher congruence for this model is also increased at low temperatures, where primary creep dominates. In consequences a better fit for stiffer and more creep resistant solder is given by the application of the unified plasticity modelling approach. It is therefore recommended to utilize it for future advanced solders as Innolot, Indium, Bismuth, Nickel alloyed solder or similar solder compositions, which are created for higher creep resistivity.

Numerical influences on the life time prediction

In order to establish a product specific comparison of effects for the 3 different constitutive models, additional numerical simulations have been done and evaluated (SinH, Anand, and Unified). The utilized FEM model is shown in the following figure and consists of a single resistor on PCB (Fig. 11). The mechanically relevant features pad, metallization, ceramic body and solder contact have been included to appropriately calculate the mechanical response under thermal loading. A thermal cycling profile using 60 minutes cycling time in temperature range -40 °C to 125 °C has been applied to the structure.

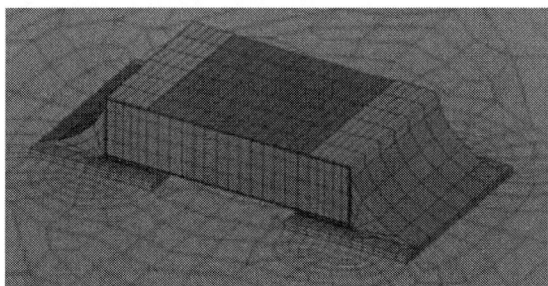

Fig. 11: Geometric details and mesh of the SMD resistor assembled on the PCB

For the evaluation the loading stresses and the solder joint's plastic strain response has been extracted. The results show the Mises-stress and accumulated plastic/creep strain during the structural loading scenario over three cycles (see Fig. 12, Fig. 13). Naturally the accumulated strain is increasingly larger for the unified model, because the primary creep strain fractions are considered and add strain while simultaneously reducing the stress.

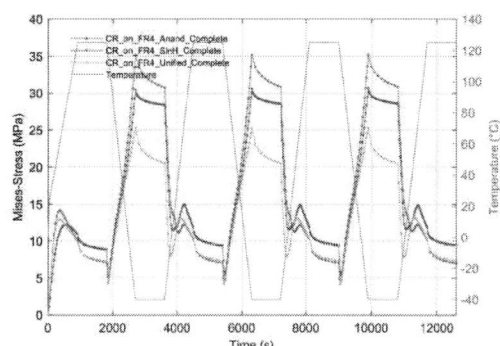

Fig. 12: v. Mises-Stress of SnAg3.5 over simulation time for the complete solder volume

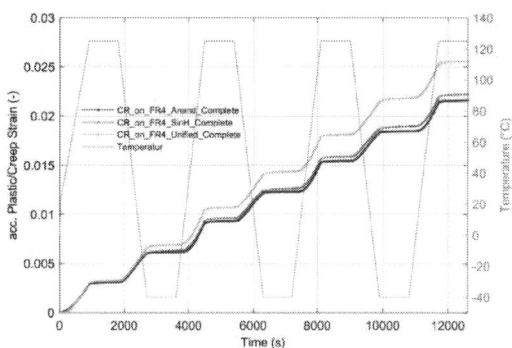

Fig. 13: Accumulated plastic and creep strain of SnAg3.5 over simulation time for the complete solder volume

As it can be seen in the equivalent strain distribution at maximum temperature (Fig. 14), the standoff region is more severely loaded compared to the meniscus. This is very typical behaviour for a two-pol component.

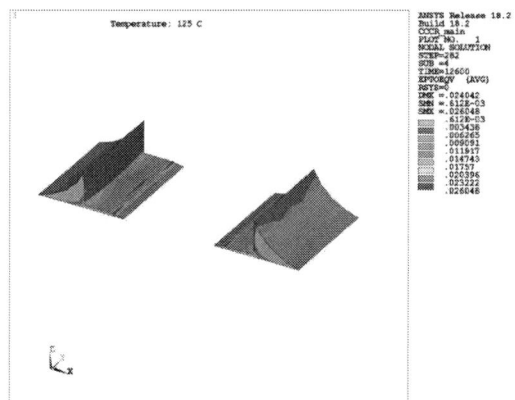

Fig. 14: Equivalent strain at the high temperature point

Lifetime assessment is usually based on the cyclic increment of the plastic response. This characteristic value can then be correlated to real measurements through a lifetime-model (e.g. Coffin-Manson). To evaluate the difference for the constitutive models this inelastic strain per cycle has been extracted.

Since the strain distribution has also been analysed and has shown significant deviation throughout the joints, the extraction has been expanded for three distinct volumes of the solder joint:

1) the complete solder contact,
2) the standoff region and
3) the meniscus region.

The results in relative strain increments is given in the Fig. 15. Here the hyperbolic sine calculations have been taken as reference and show very similar accumulation as the Anand model. Unsurprisingly the Unified model comes out with significant higher inelastic strain per cycle. Obviously, the stationary creep allows for higher stress in the solder, whereas the Unified approach calculates higher strains and lower stress due to primary and tertiary creep considerations (see Fig. 15). Consequently, the same trend can be observed in the cycle-based accumulations, which is usually utilized to extrapolate and estimate lifetimes. Here the results indicate, that the SinH and Anand models would result in overestimated lifetimes, which might lead to incorrect assumptions for the design process.

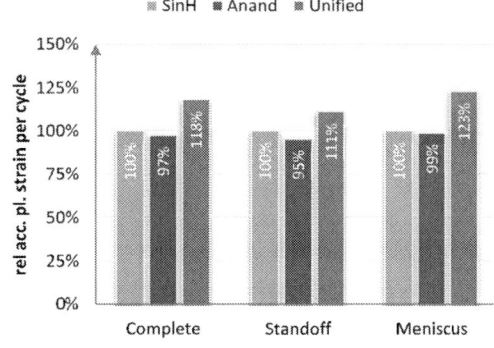

Fig. 15: Relative accumulated plastic & creep strain during on cycle shown for different solder volume parts

Summary

The finite element simulation on chip resistor demonstrates, only the unified model is able to describe the primary deformation behaviour under cyclic thermo-mechanical loading. The comparison of the resulting strain portions per cycle on a typical chip resistor shows that the accumulated strain increases by approx. 18 % for the unified approach.

The inelastic strain calculation of Anand is similar to the Sinh. Even if, the Anand is able the model a primary portion of creep strain, it flops under non-stationary and reverse loading conditions. Starting from 2nd temperature cycle the models represents the secondary creep behaviour only.

The intention for the distinction between the stand-off and the meniscus region is to compare the influence on a more as well as less stressed region.

Here one can see that the modelling with the Unified model differs more strongly on less stressed regions were primary creep dominates the deformation.

In general, the choice of a Unified plasticity model is to favour, since the accuracy of calculated strain-stress condition is increasing compared to historical models. The primary creep relevance increases strongly by the use of creep resistance solder alloys and by increasing temperature gradients in thermos-shock loadings.

References

[1] Metasch, R.; Roellig, M.; Kabakchiev, A.; Metais, B.; Ratchev, R.; Meier, K.; Wolter, K.-J., "Experimental investigation of the visco-plastic mechanical properties of a Sn-based solder alloy for material modelling in Finite Element calculations of automotive electronics," *Thermal, mechanical and multi-physics simulation and experiments in microelectronics and microsystems (eurosime), 2014 15th international conference on* , vol., no., pp.1,8, 7-9 April 2014

[2] Kabakchiev, A.; Metais, B.; Ratchev, R.; Guyenot, M.; Buhl, P.; Hossfeld, M.; Schuler, X.; Metasch, R.; Roellig, M., "Description of the thermo-mechanical properties of a Sn-based solder alloy by a unified viscoplastic material model for Finite Element calculations," *Thermal, mechanical and multi-physics simulation and experiments in microelectronics and microsystems (eurosime), 2014 15th international conference on* , vol., no., pp.1,6, 7-9 April 2014

[3] Xu Chen, Gang Chen and M. Sakane, "Prediction of stress-strain relationship with an improved Anand constitutive Model For lead-free solder Sn-3.5Ag," in *IEEE Transactions on Components and Packaging Technologies*, vol. 28, no. 1, pp. 111-116, March 2005

[4] Muller, M.; Wiese, S.; Wolter, K.-J., "Influence of Cooling Rate and Composition on the Solidification of SnAgCu Solders," *Electronics Systemintegration Technology Conference, 2006. 1st* , vol.2, no., pp.1303,1311, 5-7 Sept. 2006

[5] Wiese, S.; Schubert, A.; Walter, H.; Dukek, R.; Feustel, F.; Meusel, E.; Michel, B., "Constitutive behaviour of lead-free solders vs. lead-containing solders-experiments on bulk specimens and flip-chip joints," *Electronic Components and Technology Conference, 2001. Proceedings., 51st* , vol., no., pp.890,902, 2001

[6] Metasch, R.; Boareto, J.C.; Roellig, M.; Wiese, S.; Wolter, K.-J., "Primary and tertiary creep properties of eutectic SnAg3.8Cu0.7 in bulk specimens," *Thermal, Mechanical and Multi-Physics simulation and Experiments in Microelectronics and Microsystems, 2009.*

EuroSimE 2009. 10th International Conference on , vol., no., pp.1,8, 26-29 April 2009

[7] H. J. Frost, M. F.Ashby, "Deformation-Mechanism Maps "Plasticity and Creep of Metalls and Ceramics", Pergamon Press 1982

Effect of Test Fixture on the Drop Reliability of Solid State Drive

Tae-Min Kang[1], In-Jun Jung[2], Jae-Young Jang[2], Dong-Kil Shin[2]*

[1]SKHynix, R. of Korea (South)
[2]School of Mechanical Engineering, Yeongnam University, R. of Korea (South)
*E-mail: dkshin@yu.ac.kr

Abstract

Dropbehavior of M.2 SSD module was investigated. Speciment mounting fixtures to evaluate drop reliability were designed considering the SSD's real application. Socket and screw were crucial parts to be properly designed to mount a specimen on a drop tester. Acceleration, deflection, and strain were measured by a real time data acquisition system. It was observed that the dynamic behavior of M.2 module was strongly dependent on the types of fixture. And the fixture should be carefully designed to obtain appropriate reliability data.

Key words: solid state drive, drop, reliability, acceleration, deflection, strain

Introduction

Solid state drive (SSD) has various configurations depending on their applications. Next generation form factor, commonly known as M.2 form factor, is widely used for ultrabooks or tablet computers [1]. The M.2 form factor minimizes usage of the card space, while minimizing the foot print. Slender shape of M.2 weakens the reliability under the drop test because large deflection is induced in the transversal direction. Electronic packages on the M.2 module need to endure the impact load. In general, solder ball and copper pad are the weakest locations under the drop test. Solder ball absorbs impact energy. Copper pad should be of the shape and type that can survive during the test [2, 3].

In order to evaluate the drop reliability, the effect of inertia force and the stress concentration at the crack are very important [4]. In addition, the deflection behavior of M.2 is strongly dependent on the vibration characteristic of the module printed circuit board (PCB). Analytical solutions for the shock and vibration responses of a board ere summarized in references [5, 6]. Board level drop reliability of electronic package has been investigated by many researchers [7-10].

In this study, the effects of mounting fixture on the drop characteristics of M.2 module were investigated.

Specimen Fixtures for Drop Test

An M.2 module has slender rectangular shape, with an edge connector on one side and a semicircular mounting hole on the other side. The module is installed on a main board as shown in Fig. 1. To install it on the main board, firstly, insert the edge connector into a socket, and then join the other end with mounting post using a screw.

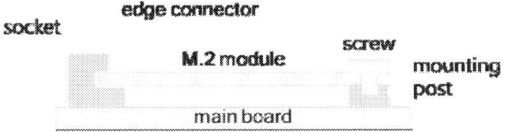

Figure 1: Schematic of solid state drive (M.2 module) on a main board

When one considers shock energy during the drop test, the energy reaches the electronic package on the M.2 module through outer case of a system– main board – M.2 module – package. In this path, socket and screw are crucial joints to design an accelerated test equipment. As it can be seen in the figure, the socket holds the edge connector by spring force of terminals in the socket. And the screw holds only semicircular parts of the edge. In addition, M.2 can not be tested by standard test method (JESD22-B111) in which it has 4 screw holes to fix the board on a drop table.

One method is to design a special test board which has the same design concept as the JESD22-B111 carring M.2 module instead of conventional package. In this case, many design variables should be carefully considered, such as size of a board, thickness, number of M.2 module on a board, location of M.2 modulle on a board, direction of laying, and so on.

Another method is to install a M.2 module on a drop table directly. In this case, it needs a special fixture to hold the module. Therefore, we investigated the effect of fixtures when the M.2 module is directly mounted on a drop table. We designed two types of fixture. Fig. 2 (a) shows socket / screw fixture, which uses the same socket and screw as a real product. The socket needs to be positioned on the fixture. Fig. 2 (b) shows clamp / clamp fixture. Both sides of the module are clamped.

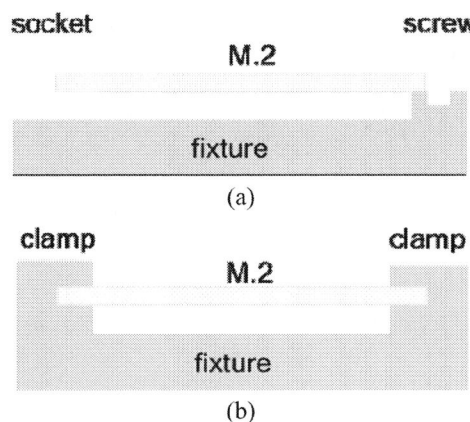

(a)

(b)

Figure 2: Types of mounting fixture

The characteristics of a M.2 module were evaluated by a drop tester. Applied shock level was 1500 G peak acceleration and 0.5 msec duration time, which is one of energy levels in JESD22-B104C standard [11].

Acceleration, strain, and bending deflection were measured by accelerometer, strain gage, and gap sensor, respectively. Fig. 3 shows schematic of the location of sensors. A small accelerometer was attached at the center of a package as shown in Fig. 3 (a). A strain gage was attached at the center of module where was flat region as shown in Fig. 3 (b). Deflection was measured by a gap sensor which measured a gap far from the metal layers in the PCB as shown in Fig. 3(b). Note that the strain and deflection were measured not simultaneously but individually. Measured signals were stored in a personal computer through a data acquisition system (NI Co.) with 50 kHz of sampling rate.

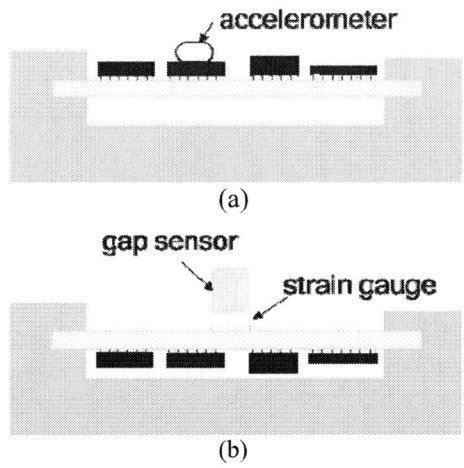

(a)

(b)

Figure 3: Sensoring positions; (a) acceleration, and (b) deflection and strain

Measured Acceleration, Deflection and Strain

Fig. 4 shows typical results of the acceleration at a flash memory package on M.2 module. A socket/screw fixture was used. In the graph, a black colored half sine curve shows an acceleration at drop table. The maximum acceleration at the package is about a half of the peak of table acceleration. Full range of time history shows a combination of short period (table acceleration) and long period (about 4.0 msec).

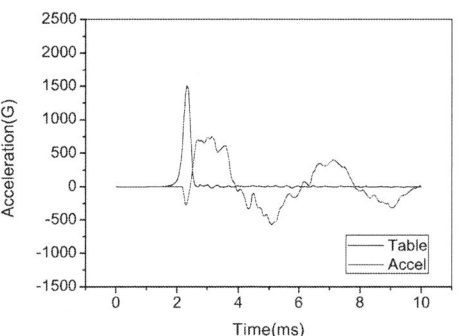

Figure 4: Measured data of accelerations at drop table and a package under socket / screw fixture

Fig. 5 shows typical results of the acceleration at the same package as that of Fig. 4 when a clamp / clamp fixture was used as shown in Fig. 2(b). The peak acceleration at the package is about 2050 G which is much higher than that of table acceleration. Overcall fluctuation period is about 2.0 msec.

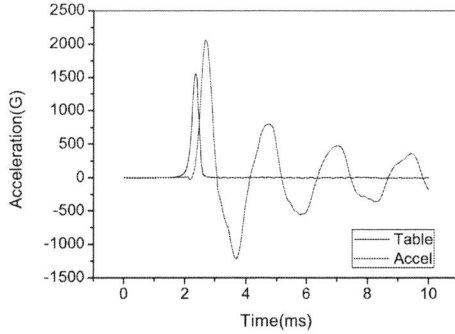

Figure 5: Measured data of accelerations at drop table and a package under clamp / clamp fixture.

Fig.6 shows a typical result of the deflection at the center of M.2 module as shown in Fig. 3 (b). Minus sign means the module move away from the initial position (downward deflection), and plus sign means vise versa. It shows a large deflection. Time history shows about 2.0 msec of period.

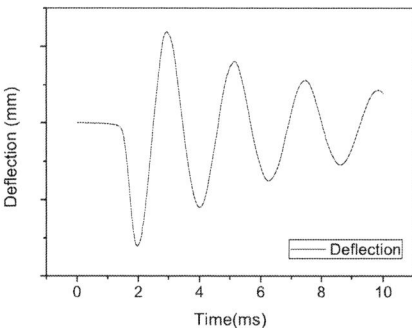

Figure 6: Measured data of deflection at the center of module with clamp / clamp fixture

Fig. 7 shows a typical result of strain at the center of M.2 module as shown in Fig. 3 (b). Minus means compressive strain and plus means tensile strain. As expected in Fig. 6, it shows about 2.0 msec of period.

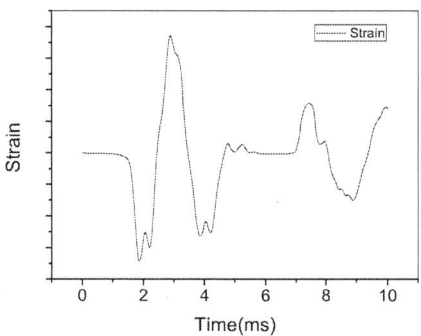

Figure 7: Measured data of strain at the center of module with clamp / clamp fixture.

Reliability Test on M.2 module

Reliability of M.2 module was evaluated by a drop tester. Designed fixtures were used. It was appeared that the socket / screw fixture showed large deviation of drop life compared with the clamp / clamp fixture.

Finite Element Analysis

Mechanical behaviors during the short time of the drop were analyzed by numerical simulation, finite element analysis. Commercial software, Simulia (Dassault Co.), was used. Implicit dynamic analysis scheme was applied. Linear elements (C3D8R and C3D6) were used. Table acceleration was applied at the two short edges of the module

Fig. 8 shows contour plot of deflection (z direction) and deformed shape at maximum deflection time.

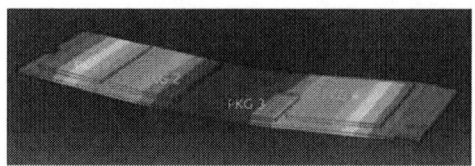

Figure 8: Deflection contour of M.2 module at maximum deflection time

Numerical results of acceleration at four packages on the M.2 module were extracted and plotted in Fig. 9. The accelerations at 4 packages are very different depending on the positions on the module. The peak acceleration of PKG3 is the first, followed by PKG2 and PKG4. In the initial period in which acceleration occurs, it appears a small and short acceleration with duration time of 0.5 msec.

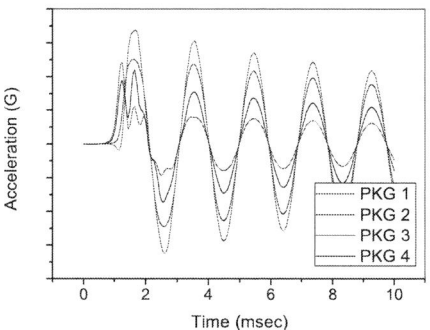

Figure 9: Analyzed accelerations at 4 packages on M.2 module.

Analytic Analysis of the Natural Frequency of M.2 bare PCB

The shape of M.2 module is a rectangular beam. Therefore, the vibration behavior is analyzed analytically. Eigenvalue problem to obtain natural frequency and normal modes of a uniform beam is [6]

$$\frac{d^4 Y(x)}{dx^4} - \beta^4 Y(x) = 0 \qquad (1)$$

where,

$$\beta^4 = \frac{\omega^2 m}{EI} \qquad (2)$$

and ω, m, E, I are circular natural frequency, mass per unit length, elastic modulus, moment of inertia, reapectively. Applying boundary conditions and obtaining the solution of frequency equation, natural frequency of nth mode is

$$\omega_n = \kappa_n^2 \sqrt{\frac{EI}{mL^4}} \qquad (3)$$

When a bare PCB without any package is considered, the first two natural frequencies and periods depending on boundary conditions are listed in Table 1. Note that, the natural frequenciy of M.2 module becoms slightly lower than that of bare PCB, because the addition of package weight increases total mass of the rectangular beam.

Table 1. Natural frequency of bare PCB

B.C. (left / right)	1st mode		2nd mode	
	(Hz)	(msec)	(Hz)	(msec)
Cantilever	68.9	14.5	432.0	2.31
Clamp/Clamp	438.7	2.27	1209.2	0.83

Conclusion

Drop behavior of M.2 SSD module was investigated. Installoation fixtures were designed considering real product. Acceleration on a package, deflection at the center of module, and stain at the center of module were measured. It was observed that the behavior of M.2 module was strongly dependent on the installation condition. Socket / screw fixture showed large deviation of drop life compared with clamp / clamp fixture.

Acknowledgements

(1) This work was supported by SKHynix Co.
(2) This research was supported by Basic Science Research Program through the National Research Foundation of Korea (NRF) funded by the Ministry of Education (NRF-2018R1D1A1A09083672).

References

[1] J. Handy et al.,"SNIA Webcast: All about M.2 SSDs" SNIA, 2015.
[2] D.A. Shnawah, M.F.M. Sabri, and I.A. Badruddin, "A review on thermal cycling and drop impact reliability of SAC solder joint in portable electronic products", Microelectronics reliability, 2012.
[3] T.M. Kang et al., "A study on the correlation between experiment and simulation board level drop test for SSD", 18th International Conference on Thermal, Mechanical and Multi-Physics Simulation and Experiments in Microelectronics and Microsystems, 2017.
[4] L.B. Freund, "Dynamic fracture mechanics", Cambridge University Press, Cambridge, 1990.
[5] D.S. Steinberg, "Vibration analysis for electronic equipment", John Wiley & Sons, Inc, New York, 1988.

[6] L. Meirovitch, "Analytical methods in vibrations", Macmillian Publishing Co., 1967.
[7] D. K. Shin et al., "Development of multi stack package with high drop reliability by experimental and numerical methods", Proc 56th ECTC Conf., San Diego, 2006.
[8] E. Suhir, "Could shock test adeqeately mimic drop test conditions," Proc 52th ECTC Conf., 2002.
[9] A. Syed, et al., "A methodology for drop performance prediction and application for design optimization of chip scale packages," Proc 55th ECTC Conf., 2005.
[10] F. Yang, S.A. Meguid, "Efficient multi-level modeling technique for determining effective board drop reliability of PCB assembly", Microelectronics Reliability, 2013.
[11] JEDEC Standard JESD22-B104C, Mechanical shock
[12] JEDEC Standard JESD22-B111, Board level drop test method of components for hanheld electronic products

High-Temperature Inter-Cavity Silicon Interposer Substrate for Multi-Chip Module Assembly

Erick M. Spory, T.S. Kalkur, Ph.D.

Global Circuit Innovations, Inc., University of Colorado at Colorado Springs
Colorado Springs, CO USA

Erick.Spory@GCI-Global.com +1 (719) 649-0947, TKalkur@UCCS.edu

Abstract

Many of the Integrated Circuit (IC) solutions generated by GCI involve multi-chip module (MCM) types of layouts within the package cavity to emulate the original monolithic device (single die in package cavity), ultimately providing a form, fit, and functional drop-in replacement. These MCM architectures provide the greatest degree of flexibility when ensuring that the obsolete component's electrical requirements are met[1]. This paper will address the design, development, and manufacture of inter-cavity silicon interposer substrates, by GCI and the University of Colorado at Colorado Springs Electrical and Computer Engineering department (UCCS-ECE), that provide signal routing of the passive and active devices within the cavity to meet the functional requirements of older obsolete devices.

In this manner, a wide array of newer generation components can be configured within the package cavity on these silicon interposer substrates to architect obsolescence solutions while also meeting the high-temperature requirements of the Au-Sn (gold/tin) lid solder seal processes (~340°C). Silicon substrates can also replace ceramic interposer versions[2], while meeting or exceeding performance requirements, decreasing lead times, and manufactured with more competitive costs.

Key Words: Silicon Interposer Substrate, High-Temperature Electronics, Die Extraction, IC Obsolescence, ENEPIG Plating, Multi-Chip Module

INTRODUCTION & BENEFITS OF SILICON INTERPOSERS

Global Circuit Innovations, Inc. (GCI) has recently been awarded a variety of government contracts to provide solutions to otherwise obsolete IC's supporting DoD requirements. These contracts have been in the form of Small Business Innovative Research (SBIR) and Rapid Innovation Funding (RIF) projects, although recently GCI has also been able to work directly with the Air Force Sustainment Center (AFSC) Strategic Alternate Sourcing Program Office (SASPO) to reverse engineer IC solutions for the Defense Logistics Agency (DLA). In all cases, the final product is designed, manufactured with franchised components, and qualified to support the requirements of MIL-STD 883.

IC obsolescence has become a significant problem preventing continued manufacturing, or even sustainment, of long lifetime military electronic platforms. Specifically, many individual ICs incorporated into legacy Department of Defense (DoD) electronic systems are military qualified versions of

their commercial counterparts. The demand for these higher-end components is typically insufficient to maintain a -55°C to +125°C screened ceramic package line by the original component manufacture (OCM) for any length of time. Further, commercial demand for even the -40°C to +85°C plastic packaged components can change quickly due to industry migration to faster, more complex, or higher memory density versions of these older devices.

The DoD has routinely tasked GCI with innovative solutions for these obsolete IC devices in order to avoid expensive and time-consuming system redesigns. One example of the benefit of inter-cavity interposer substrates manufactured by UCCS is seen below when GCI was requested to develop an IC solution to an original high-voltage 16K (2Kx8) Electrically Erasable Programmable Read Only Memory (EEPROM) requiring a 21V input signal (see Figure 1 below). No currently available EEPROM replacement devices are designed to tolerate these voltages.

Figure 1. 2018 GCI 16K EEPROM Equivalent of 1982 SEEQ EEPROM Using FR-4 Cavity Interposer/Substrate.

GCI then was required to design a replacement device using multiple die and passive components within the cavity to emulate the form, fit, and function solution within the targeted 24-Pin Ceramic Package. The original 2018 GIC component solution using an organic FR-4 interposer substrate solution can be seen below in Figure 2.

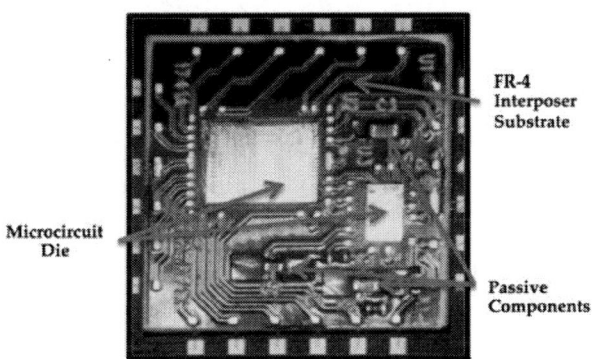

Figure 2. Higher Magnification of Cavity for Figure 1.

The FR-4 interposer substrate allows the ability to mount and connect multiple active devices (microcircuits) with passive components (resistors, capacitors, and inductors) to complete the required circuit. However, due to solder lid seal thermal requirements of 280°C to 340°C, FR-4 interposers are incompatible because of their temperature exposure limitation of 125°C and outgassing at higher temperatures. Even more expensive, higher-temperature substrates such as Polyimide that can withstand assembly temperatures of 260°C to 320°C, can outgas hydrocarbons and moisture within the cavity during the lid seal operation due to their organic nature.

Silicon interposers, however, do not outgas and can tolerate temperature exposure to 400°C. They are compatible with die stacking and can be easily die attached to ceramic substrates[3]. Another benefit of the silicon interposers is that although they are typically manufactured at 0.020" thick, they can be thinned to as

little as 0.004" (100 um) to accommodate lower profile stacking within the cavity when height restrictions are present. GCI's partnership with UCCS ECE has effectively allowed for silicon interposer lead times that rival FR-4 and Polyimide board shops, while providing inter-cavity component connectivity that can route amps of current at high-speeds through the traces, and with identical thermal coefficients of expansion that are equivalent to the silicon die mounted to the interposer.

Equivalent device architectures to those seen in Figures 1 and 2 can be seen below in Figures 3 and 4 using the inter-cavity silicon interposers. Silicon interposer substrates are generally high yielding during manufacture due to relaxed line, space, and pitch layouts, which are generally much larger than the minimum line and space capability of the process.

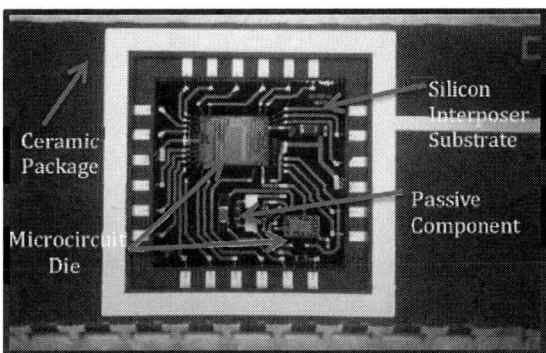

Figure 3. GCI's 16K EEPROM Equivalent of the SEEQ EEPROM Using a Silicon Cavity Interposer/Substrate.

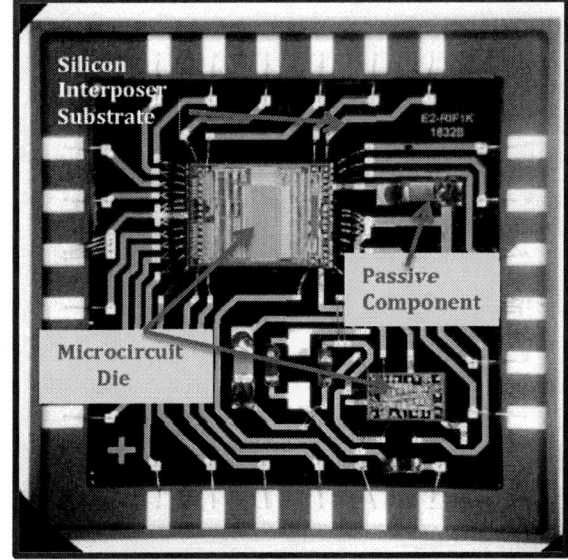

Figure 4. Higher Magnification of Cavity for Figure 3.

Global Circuit Innovations has patented die extraction and re-packaging (or, *DER™)* processes (see Figure 5 below) readily provide bare die for MCM manufacturing using silicon interposers.

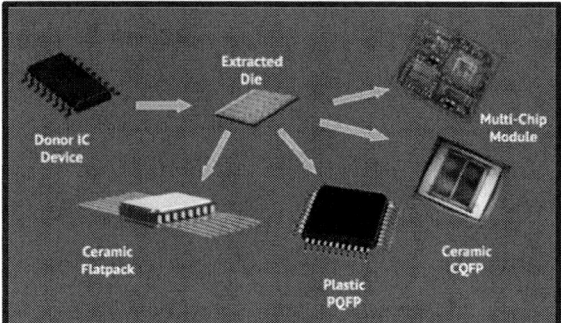

Figure 5. GCI's *DER™* Process Illustration.

Another GCI process routinely used on the silicon interposers is electroless nickel, electroless palladium, and immersion gold (ENEPIG) die pad plating (DEER™). This plating process readily allows for gold ball or aluminum wire bonding, as well as solderability of the silicon substrate for assembly and mounting of passive components (see Figure 6 below).

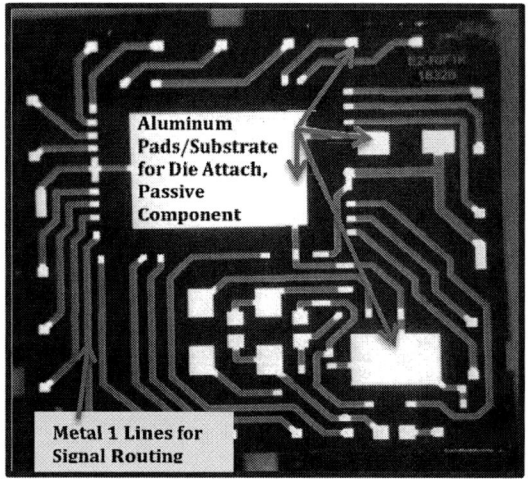

Figure 6. Diced Silicon Interposer Routing Substrate Following ENEPIG Plating of Aluminum Die Pads.

The aluminum die pad plating process was developed to provide a reliable extraction and re-packaging solution, following original gold ball removal, that eliminates undesired compound wire bonding (new bond on original bond) during reassembly, particularly for military applications. This process also provides a new die pad surface medium for either aluminum or gold wire bonding that ultimately prevents subsequent gold/aluminum interface degradation (Kirkendall or

Horsting Voiding) at temperature exposure greater than 150°C.

A close-up optical photo of the aluminum die pads following ENEPIG processing can be seen below in Figure 7. Typical total plate-up is on the order of 4 – 5 um. A scanning electron microscope (SEM) image of plated die pads on a silicon microcircuit can be seen in Figure 8. Subsequent gold ball bonding to ENEPIG plated pads can be seen in Figure 9. The ENEPIG process for re-conditioning aluminum die pads can be summarized by the illustration within Figure 10 on the following page. None of these steps require masking.

Figure 7. Optical Photo of ENEPIG Plated Die Pad.

The thicknesses of these plated layers can be determined by a number of analytical techniques such as cross-sectional analysis and X-ray Fluorescence Spectroscopy (XFS). All of the plating sequences and benefits for conventional silicon microcircuit die pad plating apply relative to the silicon interposer substrates discussed in this paper.

Figure 8. SEM Photo of Die with ENEPIG Plated Pads.

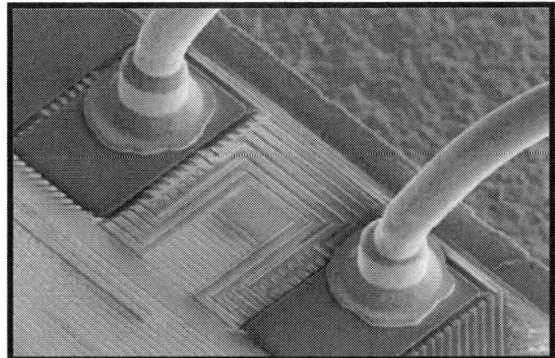

Figure 9. SEM Photo of Bonded Die with ENEPIG Plated Pads.

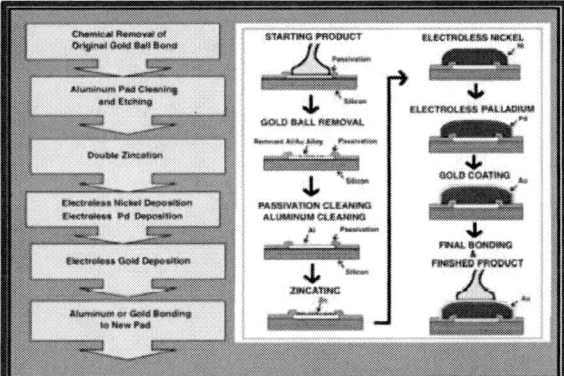

Figure 10. GCI's *DEER™* Process Flow Including Gold Ball Removal and ENEPIG Pad Re-conditioning. Target: 4 μm Ni, 0.25 μm Pd, & 0.04 μm Au

One of the most significant benefits of aluminum pad plating, aside from superior high-temperature connectivity reliability, is the increase in manufacturing window for optimized bonding parameters relative to bonding force and power (see Figure 11 below).

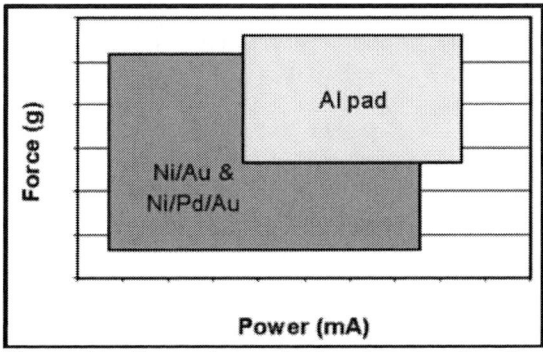

Figure 11. Comparison of Manufacturing Bonding Window for Current/Power and Force.

Relative to general harsh environment applications, the new *DEER™* (ENEPIG Plating) process has already been proven to be capable of meeting the stringent requirements involved with high-temperature integrated circuit (IC) reliability for the oil and gas drilling industry, including 30,000 g-force shock and vibrational testing, high-temperature reliability at 250°C beyond 6,000 hours, and tens of thousands of hours of field functionality. Figure 12 represents the benefit of DEER™ processing for high-temperature reliability relative to conventional plastic or ceramic assembly flows.

Ultimately, the use of the silicon substrate interposers discussed in this paper enables identical bonding parameters for the interposers as the active IC devices used. Figures 13 and 14 show oblique optical photos of the package cavity for finished product. Bonds from the package posts to the silicon interposer substrate, as well as the active ICs to the silicon substrate, can be clearly seen.

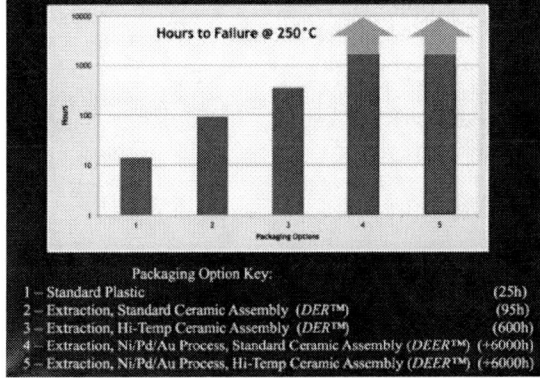

Figure 12. DER™ and DEER™ Processing Reliability Package Assembly Flow Comparison Study @ 250°C.

Figure 13. Oblique Optical Photo of Silicon Interposer Substrate Mounted in Cavity Following Bonding.

Figure 14. Same as Above, Except Main Microcircuit Bonding Region Highlighted.

SILICON INTERPOSER SUBSTRATE MANUFACTURE AND SPECIFICATIONS

The original interposer substrate design layout was proven functional in FR-4 prior to conversion in silicon. The appropriate AutoCAD file was submitted to the mask shop to produce a 5" mask plate for the 4" wafer process. Each individual interposer substrate was designed at 0.362" on a side with a 0.012" street (300 um to accommodate the 250 um die saw thickness for individual interposer substrate dicing), for a total pitch of 0.374" as can be seen in Figure 15 below.

The final mask set was configured in a matrix of 69 interposer substrates as seen in Figure 16. Only two masks were generated: Metal 1 and Passivation Cut. The Metal 1 mask was configured as 'digitized data chrome' (blocking the exposure where the metal is in place), and the Passivation Cut mask was configured as 'digitized data clear' (thereby allowing exposure only over the aluminum pad portions requiring subsequent oxide removal). The reason for these designations is that the resist used was a 'positive' type. A **positive resist** is a type of **photoresist** in which the portion of the **photoresist** that is exposed to light becomes soluble to the **photoresist** developer and is removed. Thus, the Metal 1 mask was printed with opaque chrome where the Metal 1 was to be maintained for routing, and the Passivation Cut areas were made to be transparent such that the Passivation could be etched (no resist) in the defined regions.

The following characteristics of the processing can be listed below:

Wafer Size and Type: 4", N-Type
Metal Deposition Type and Thickness (Resistivity is a function of Thickness): Pure Aluminum, ~14,000 angstroms (1.4 um)

Overlying Oxide Type and Thickness: Spin on Glass (SOG), ~5000 angstroms (0.5 um)
Original Wafer Thickness: 0.020"
Contact Mask: Metal 1 – Chrome, Passivation Cut – Clear
Layer Etch: Argon Mill (Tech Ion Mill, Beam Voltage: 900V, Beam Current: 20 mA).

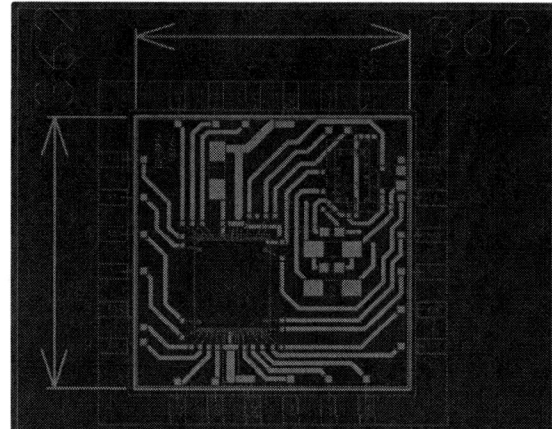

Figure 15. Individual Interposer Substrate Design/Layout Within Ceramic Cavity to Show Dimensions in mils.

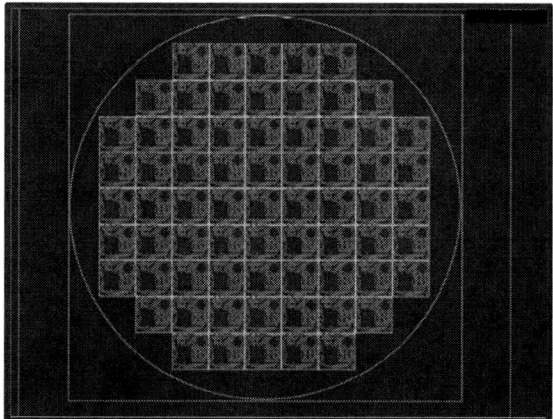

Figure 16. Layout of Silicon Interposer Substrates on 4" Wafer to Show Leverage (69 Interposers).

To date, there has not been a reliability performance variation seen between packaging flows 4 and 5 of Figure 12 using *DEER*™ extracted devices, thereby indicating that conventional ceramic packaging of *DEER*™ product is just as reliable as using much more complex and expensive packaging techniques requiring bake-outs and desiccants for moisture control. Thus, *DEER*™ bare die can be made available for conventional processing with multi-chip <u>module (MCM) assemblies</u> as well as monolithic packaging.

The entire MCM referenced within this paper is expected to pass all standard Department of Defense (DoD) MIL-STD 883 tests, including Groups A, B, C, and D, 1000 hour +125°C dynamic life test, and 1000+ hour 250°C unbiased baking.

CONCLUSION

The final product can be seen below in Figure 17, which includes the 24-pin ceramic package, the silicon interposer, the two mounted and bonded microcircuits,

Figure 17. Overall Oblique Optical View of Ceramic Package, Silicon Interposer Substrate, Mounted Passive Components, and Bonded Microcircuits.

and the 5 passive components (2 capacitors and 3 resistors), all of which were tested functionally to pass relative to the end customer system requirements.

The silicon interposer ultimately allows for a higher temperature gold/tin solder lid seal process of 340°C for 2 minutes, while eliminating the presence of organics within the sealed cavity. Future silicon interposers may have two or more metal layers routed on the interposer for greater routing complexity, depending on the number of bonded pins and microcircuit devices within the cavity.

REFERENCES

1. Smithsonian Chips/Integrated Circuit Engineering Corporation, Chapter 12. Retrieved October 8, 2018, from http://smithsonianchips.si.edu/ice/cd/PKG_BK/CHAPT_12.PDF

2. AdTech Ceramics: AdTech Introduces the Addition of an ENEPIG Plating Line. 13-Jul-2018. Retrieved October 5, 2018 from http://www.adtechceramics.com/post/adtech-introduces-the-addition-of-an-enepig-plating-line.

3. Chip Stack MCMs. Retrieved October 5, 2018, from https://en.wikipedia.org/wiki/Multi-chip_module.

Laser Frit Sealing Approach for Ultrathin Glass

M. Bedjaoui*, J. Brun, and J. Amiran

Univ. Grenoble Alpes, CEA-LETI, Minatec Campus
*00334338782926, messaoud.bedjaoui@cea.fr

Abstract

This work aim to adapt the laser sealing technique to the ultrathin glass materials for the encapsulation of thin-film batteries using alkali free glass. An adapted sealing tool is necessary to prevent the thermal stress within the ultrathin glass to achieve crack free sealing.

Key words: Ultra thin glass, laser sealing, frit glass, thin film batteries, encapsulation

Introduction

Packaging is one of the major challenges in the fabrication of moisture and oxygen sensitive device (e.g lithium microbatteries, solid and organic state lighting, liquid crystals displays...). These kinds of miniature electronic devices require hermetic and ultrathin seal with long-term stability as a packaging system. For the particular case of lithium-based thin film batteries and due to the very great sensitivity of lithium with water and oxygen, the required barrier performances is generally smaller than 10^{-4} g/m^2/day and 10^{-4} cc/m^2/day for the water permeability and the oxygen permeability, respectively [1]. These packaging requirements are critical to enable shelf life of thin film batteries for sevreal years. Likewise, the energy requirements (autonomoy) of wide range of miniaturized microlectronic applications such as medical implants, hearing aids and smart lens involve the recourse to thin-solid lithium microbatteries with important volumetric energy density. Thus, the packaging approach is an important factor that requires attention for the realization of high-performance microbatteries [2].

Meanwhile, the advent of ultrathin glass materials (thickness<100µm) can provide acceptable encapsulation using new joining techniques. Overall, glass frit bonding is the most common approach employed in wafer level encapsulation and packaging. It's a thermo-compressive process based on the use of an intermediate layer positioned between two thick substrates (>200 µm) and subjected to pressure and heating. However, the hermetic joining techniques generally require the heating of the whole device at high temperature (hundreds of degrees). To cope with this issue, the laser sealing is a localized joining technique typically adapted for temperature-sensitive devices.

In this paper, a novel packaging approach based on the laser glass frit sealing of ultrathin glass materaisl has been adapted to thin film battery devices. More particularly, the main objective of this work consists to evaluate the impact of different sealing process parameters on sealing quality (such as frit composition, thermal treatment, laser power, sealing speed...).

Conventional packaging of thin film batteries

To date, the most common approach of packaging relies on the deposition of housing structures made of adhesive layer and barrier film [3, 4] using lamination or sealing approaches. As such, the barrier performances of the standard encapsulation model presented in Figure 1 are widely related to the adhesive thickness and the edge sealing. Indeniably, the high barrier level required for lithium thin film batteries is achieved through barrier films with large sealing margin values (>1mm). Therefore, the conventional packaging for thin flim batteries consumes a significant part of the lateral area. The edge sealing area may account for a significant portion of battery area. Thus, the inactive content can drastically outweighs the active volume. This fact is more prounced for small size batteries (few mm^2) in comparison to large size batteries (few cm^2).

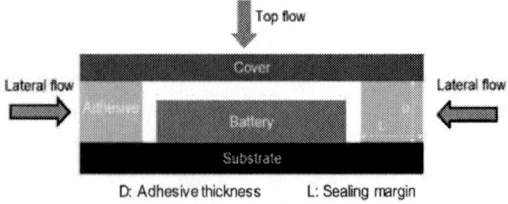

D: Adhesive thickness   L: Sealing margin

Figure 1: Schematic of standard packaging of thin film batteries using laminated-barrier films

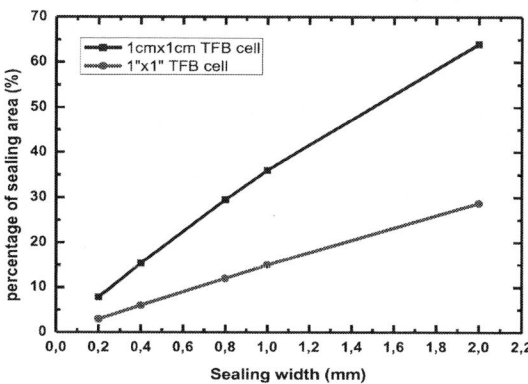

Figure 2: Percentage of sealing area vs sealing margin width corresponding to 1"x1" TFB-cell and 1cmx1cm TFB-cell

As such, we plot in Figure 2 the percentage estimation of sealing area relative to two square battery cells (1"x1" and 1cmx1cm) as a function of the edge sealing width values. If the sealing width is 2 mm (typical sealing margin value for the lamination technique), the percentage of sealing area exceeds 28% for 1"x1" cell and 60% for 1cmx1cm of the total battery area. Accordingly, alternative packaging solutions allowing optimized sealing width combined with high barrier level are highly recommended. In this study, we propose particular housing of the thin film battery using ultrathin glass materials sealed by laser technique through intermediate frit glass layer. The aim of this approach is to reduce the packaging volume part, and therby, to achieve the anticpated high performance of microbatteries.

Materials and methods of the study

The concept of laser-assited frit glass sealing applied to ultra thin glass is evaluated through the encapsulation of thin film batteries. In practice, the proposal approach involves four main steps: (i) deposition of lithium-based layers on ultrathin glass substrate, (ii) screen printing of the glass frit paste onto the periphery of ultrathin glass cover, (iii) thermal treatment of the glass frit paste, (iv) laser sealing of the ultrathin glass UTG cover on the ultrathin glass UTG substrate.

Size	1"x2"
Thickness	50 µm
Transformation temperature Tg	715°C
Coefficient of thermal expansion CTE (0 to 300°C)	3-5 ppm/°C
Transmission (λ=550nm-1000nm)	92%

Table 1: Main characteristics of AF32 UTG material employed in this study

Firstly, our approach exploits the advent and properties of ultrathin glass UTG materials (thickness <100µm) in the design of the packaging scheme. Among a serie of properties, UTG has excellent barrier performances regarding the water vapor and the oxygen. Additionally to the flexibility and the relative low CTE, the optical transmission is attractive for laser sealing techniques. Therefore, the packaging module assumes symmetrical AF32 alkaline-free borosilicate glass proposed by Schott for either the microbattery substrate or the encapsulation cover. This material is suitable for the fabrication of thin film microbatteries as precdently reported [4]. Table 1 reports the general properties of the UTG materials employed on this study. Other UTG references (G-Leaf[TM] from Nippon Electric Glass, Willow® from Corning) may fulfill the requirements of laser-based glass frit sealing.

Secondly, to achieve a succeful sealing process, it is important to select the proper glass glass frit. For instance, the matching of coefficient of thermal expansion (CTE) of glass frit and UTG materials is a critical criterion to overcome thermomechanical stress in the bonding interface. On the other hand, the laser absorption efficiency of the glass frit is for a great importance to enable a realiable joint formation. In this paper, we evaluate the sealing quality of two differents lead-free glass frit materials: the Vita® paste from Corning and the 5115HT1 paste from AGC-Asahi Glass Company, Ltd). Table 2 summrizes the elementary properties of the used frit glass pastes according to the suppliers datas. The viscosity values of the two paste candidates are compatible with the most common printing methods. Here, the screen-printing technique with a 300 mesh/in stainless steel screen allows the deposition of the glass frit onto the surface to be joined of the UTG cover. The wet screen-printed pastes consist of two close loop forms along the edge of the UTG cover (external dimension 23mmx23mm). The screen-printing technique enables regular and high repatability of cords at different widths: 400µm and 800µm.

	5115HT1	Vita®
Viscosity (Pa.s)	50-200	57
Transformation temperature Tg	358°C	303°C
Coefficient of thermal expansion CTE (0 to 300°C)	7.5 ppm/°C	7 ppm/°C
Transmission (λ=550nm-1000nm)	<15%	<15%
Composition	Bi_2O_3-SiO_2	Borosilicate

Table 2: Main characteristics of frit glass materials employed in this study

Process step	Temperature (°C)		Duration (min)	
	5115HT1	Vita®	5115HT1	Vita®
Drying	120	120	30	30
Organic Burn-Out	290	380	30	60

Table 3: Firing profiles of frit glass materials employed in this study

After deposition, the glass frit pastes was subjected to thermal conditioning in order to obtain compact and free-voids glass. Indeed, the glass paste initially consists of glass powders (frit) incorporated in an organic matrix. The thermal treatment schedule includes two elementary stages: solvent drying step and organic binder burn out step. Table 3 compiles the recommandations for the firing profiles of the glass frit pastes employed in this work. Consequently, the thickness of the 5115HT1 and the Vita® frit cords is reduced after thermal treatment from 26 μm to 19 μm and from 25 μm to 12 μm, respectively. This fact should be related to the evacuation of the different intruders including solvents and organics. Thus, it is important to consider the frit past shrink limitation on the design of the packaging system of TFB stack.

Commonly, the temperature schedule inside a furnace is of a paramount importance to achieve excellent adhesion of the paste to the cover as well as on the sealing quality. Morever, the temperature-cooling rate plays a critical role not only on the sealing quality but also on the warp control of the UTG cover. According to the 5115HT1 and Vita® manufacturers, the recommended cool down rate to room temeprature is fixed at 10°C/mn. Based on this, the UTG samples with frit glass cords presents a warp aspect after annealing treatment as shown in Figure 3(a). Evidently, the handling of warped UTG samples complicates the flattening as well as the alignment of UTG cover and UTG substrate during laser bonding process. The control of the cooling down step at 2°C/mn offers the advatnatge to overcome the UTG warp as shown by Figure 3(b). Figure 3 photos correspond to the Vita® paste but the warp phenomen characterizes the UTG covers indepently of the frit glass paste.

Thirdly, the laser sealing process can took place to defintly encapsulate the active layers. In this way, the assembly of the UTG cover and the UTG substrate is done in an in-house made sealing cell whose principle is described elsewhere [5]. Briefly, the laser-sealing cell consists of a coper block characterized by a receiving zone, which enables an accurately positioning for UTG materials. The copper block equipped with vacuum facilities is able to support an upper cap window in order to enclose UTG materials. This design ensures that the UTG materials stays in close contact during the laser scan.

Figure 3: Pictures of rectangular 1"x2" UTG cover with 23mmx23mm close loops frit glass paste after thermal annealing schedule with two room temperature-cooling rates: 10°C/mn (a) and 2°C/mn (b).

Figure 4 shows the laser system used in all our experiments. The laser bonding setup comprises an ytterbium fiber-delivery diode with continuous wave (CW) output at wavelength of 800-820nm, controlled heating plate and XY workpiece (200mmx200mm with speed capabilities up to 300mm/s). All the part of the laser system are monitored by LabVIEW software application except the heating plate, which works independtly. The laser power capability of the laser diode was up to 100 W.

Prior to the laser irradiation, the UTG samples were positioned and stacked inside the sealing cell before performing void pumping under inhert glove box atmosphere. A description of the former sealing cell, the requirements and the development of the dedicated cell, the features are described in recent paper [5]. The sealing cell is transferred from the glove box to the laser system to complete the sealing of UTG materials. The laser beam is directly focused on the frit glass line with a circular shape spot (diameter around 800μm). Based on precedent results [5], we privilege separate laser scan segments instead commonly closed loops of the laser movement along the glass frit line.

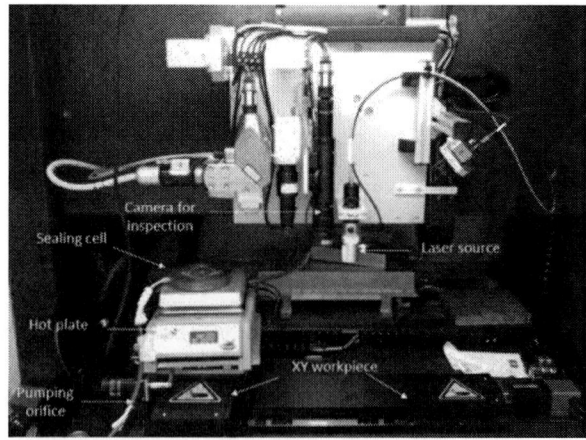

Figure 4: Photograph of experimental laser station with dedicated-sealing cell used in the glass frit bonding trials of UTG materials.

Laser sealing experiments

The main objective of this work aim to adapt the laser sealing technique to the ultrathin glass UTG (thickness of 50µm) materials. This process can be exploited for the packaging of thin lithium-based and sensitive-temperature devices. Depending on the nature of batteries family, the maximum temperature is 260°C for lithium-free batteries and 180°C for lithium metal batteries. In comparison to conventional thermocompression bonding process, the thermal heating induced by laser irradtion is exclusively limited to the frit glass cord and sparing the battery layers. The process is based on the principle of laser transmission welding which requires one material (UTG cover or UTG substrate) to be highly transmittive but the other material (frit glass paste) must be strongly absorbed at the wavelength of the laser beam. As a result, the frit glass of the UTG cover locally heats up, melts and is bonded to the UTG substrate without affecting the active layers.

In an earlier paper [5], we have shown the importance of the force applied during the bonding process of UTG materials. The convenience of the proposed sealing cell guarantees uniforme force distributed along the frit line during the laser irradiation. The heat excess is drawn away by the copper block at the bottom of the sealing cell. In addition to the indicated points in this work, the key parametrs that we have identified as critical to avoid thermal stress during laser sealing process of UTG materials are mainly laser power output and laser scan velocity. Indeed, the heat flow generated by the laser irradiation depends on one hand on the laser power output and on the other hand on the laser scan speed. To evaluate the influence degree of these parametrs on the sealing quality, we use microscope inspection with bright and dark field illumination. Furthermore, we employ Scanning Electron Microscopy (SEM) after adequate preparation of the sample to control the joint quality of the UTG bonding.

In the rest of this document, we will exclusively focus on the use of AGC frit glass paste. Despite the difference of the firing profiles corresponding to 5115HT1 and Vita® pastes, the obtained results after laser sealing are comparable. Moreover, the transmittance and reflectance of the pastes measured with a spectrometer to extract the absorbance of the glass frit materials are quite similar. Thus, the 5115HT1 and Vita® materials are suitable for laser-based packaging. The only drawback concerns the smaller thickness of Vita® paste; contrary to 5115HT1 paste where the thickness is compatible with geometrical topography of thin film battery application (thickness of active layers in the 10-15 µm range). This explains why the 5115HT1 is prefered for the laser sealing experiments.

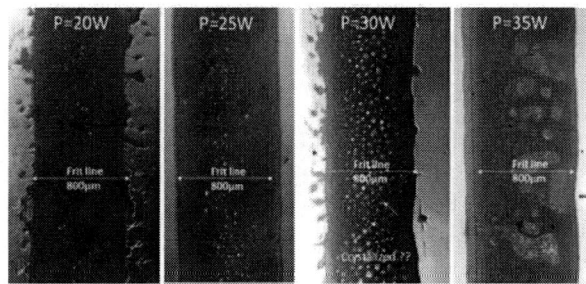

Figure 5: Microscope inspections of the joining area of UTG seals using different laser power values: fixed scan velocity 20mm/s; 5115HT1 glass frit line, UTG thickness 50µm.

To adjust the laser power to the required melting point of the glass frit paste we vary the power of the diode laser in the 10W to 40W range, while keeping the laser sealing speed at 20 mm/s. Figure 5 shows differents visual inspections using optical microscope of the frit lines corresponding to some laser power conditions. At low power values (≤20W), the glass frit partially melts and the mating of the UTG substrate and the UTG cover is unreliable. Indeed, the UTG materials are readily separable. In opposition, the use of high power values (≥30W) generates cristalized points on the frit glass line, which is generally highly undesirable for hermetic seals. Thus, intermediate power values around 25 W allows reliable and reproducible sealing as the frit cord in Figure 5 (P=25 W) seems free-defects, such as possible cracking or delimantion.

In the same way, the moving speed of the focused laser source can affect the temperature field and the width of the heat-affected area for a fixed power value (P=25W). Figure 6 shows that intermediate laser scan speed (20 mm/s) was necessary in order to avert the void formation and cracks on the frit line and the UTG materials. At low sealing speed (<15 mm/s), the residence time of laser spot is highly localized at fixed point on the frit line for a relative long time in comparison to high seeling speed (>20mm/s). Consequently, the reduction of thermal stresses is more difficult than for high speed processing. This fact generates voids inclusions on the glass frit line as shown by Figure 6 (10mm/s and 15 mm/s conditions). In parallel, the increase of the sealing speed (Figure 6 at 30 mm/s condition) results in cracks on the frit line. Obviously, the residual stress is an important parameter, which depends on the laser-moving rate for a fixed laser power value. These results may originate from the residual stress defined by the cooling rate to room temperature of the whole parts of the assemably structure (UTG cover, frit glass and UTG substrate). The best operating sealing speed was 20 mm/s avoiding the formation of micro-cracks across the frit glass cord and the UTG materials as shown by Figure 6.

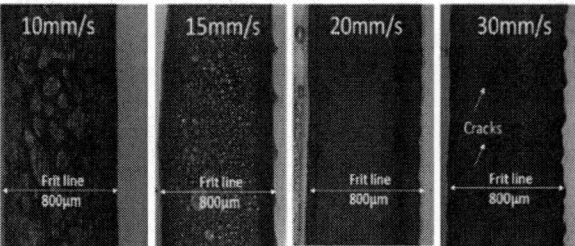

Figure 6: Microscope inspections of the joining area of UTG seals using different laser power values: fixed laser power 25 W; 5115HT1 glass frit line; UTG thickness 50μm.

As seen from the above, the key factor guiding succeful bonding is thermal history resulted from laser heating. Conceptualy, the duration time of the laser sealing process is composed from three phases: heating, wetting and cooling [6, 7]. Indeed, the perfect form without voids or cracks of the joining parts (UTG cover, UG substrate and frit glass cord) depends on a smooth simultaneous heating and cooling of the whole system. The sealing results of UTG materials are in total accordance with conventional laser sealing of thick glass materials (>200 μm). Indeed, S. Logunov et al. [8], highlight the impact of laser speed and power on the peak temperature at the frit interface. It was shown that the increase of the laser power or the decrease of the laser speed amplifies the temperature along the frit cord. Additionnaly, in the laser sealing process the glass thickness starts to play a role when the heat diffusion distance becomes comparable to the thickness of the glass. It is clearly the case for ultrathin glass (50 μm thick). Hence, the UTG materials require an adapted thermal management of the laser process in order to avoid the formation of significant temperature gradients between the different parts of the assembly. In summary, the use of appropriate sealing cell combined with adapted process for the frit line preparation and the laser sealing are the major factors to enable a reliable joint formation between UTG materials.

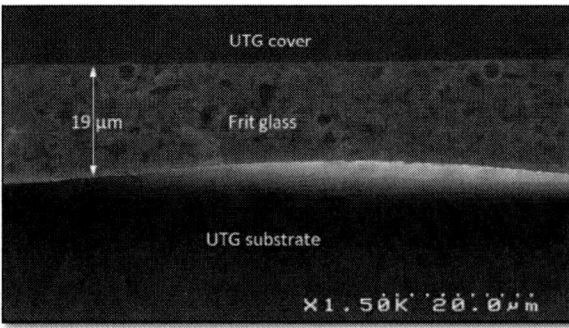

Figure 7: SEM micrograph of a cross-section of the joining area of UTG seals: laser power 25 W; scan velocity 20mm/s, 5115HT1 glass frit line (width 800μm, thickness 19μm); UTG thickness 50μm.

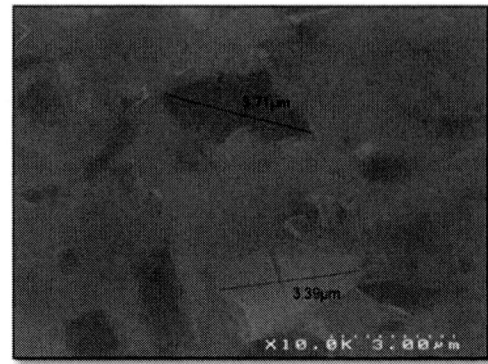

Figure 8: Magnified SEM micrograph corresponding to the Figure 7 observation.

Specifically, a sealing speed of 20 mm/s and a power of 25 W parametrs allow typical seals of UTG materials through 5115HT1 frit line as shown by Figure 7. Closer inspection of this sample showed that the morphology of the glass frit cord is similar to that obtained by thermos-compressive process [9]. This inspection shows an excellent contact of the frit cord (19 μm thick) with UTG substrate and UTG cover. As shown in Figure 8, the glass frit cord is composed of a main matrix containing filled particles (mean size diameter<5μm). According to the manufacturer guidelines, the main matrix is a bismuth oxide matrix while the composition of the filled particles is on silica and aluminum oxide.

An important deal for the microbattery technology concerns the design of the packaging system regarding the terminal connections. In order to test the effectiveness of the UTG laser sealing, dummy TFB cells fabricated on UTG substrate were subjected to the encapsulation process using UTG cover with 5115HT1 frit line (800μm width and 19 μm thick). As shown in Figure 9(a), the laser irradiation using typical conditions (25W, 20mm/s) generates several cracks on the UTG materials. The platinum layers influence the thermal equilibrium elaborated with UTG materials and frit glass cord. Thus, the presence of metal layers on the UTG substrate modify the interface of the frit line. Indeed, the metal layers can affect the laser sealing parametrs during laser irradiation to achieve successful bonding. In the laser frit glass-sealing concept, the temperature distribution is the key factor affecting the bonding quality as preveiously discussed. To overcome this issue, the adjustement of the laser parameters when the laser spot pass through the metal zone allows free-crack sealing as exhibited in Figure 9(b). In this case, the laser power was reduced to 20 W acrross the metallic zone and kept at 25 W for the other parts. Preliminary hermeticity tests of the UTG laser sealing process in the presence of platinum therminals and lithium layers are very promising. Thus, the lithium encapsulated using this process is not damaged by the storage at harsh conditions (1 week at 60°C, 85%RH) indicting a succeful UTG bonding.

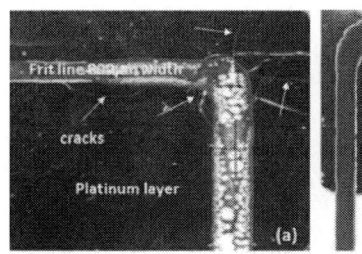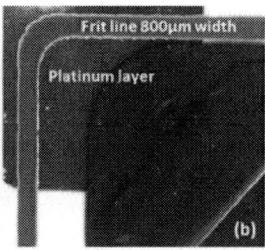

Figure 9: Microscope inspection of UTG to UTG bonding laser in the presence of TFB terminal layers (a) unsecusseful (b) successful.

Summary

This paper concerned on the laser sealing of ultrathin glass materials and the effect of the key parametrs on the bonding quality using adapted sealing cell. The investigations shows that the laser glass frit sealing process is suitable for the ultrathin glass materials and the glass frit materials Vita® from Corning and 5115HT1 from AGC. The process window is mainly related to the glass frit preparation (thermal schedule) and the laser source parametrs (power output, scan speed). The most highlighted outcome is the temperature monitoring (temperature distribution, duration, cooling rate) during the laser process to overcome the frit glass and the UTG cracks.

The given laser-assisted glass frit bonding process using UTG materials is a promising technique for the packaging of temperature sensitive devices, which requires a high-density integration, such as lithium thin film batteries.

Acknowledgements

The authors thank EnSO project for financial support. EnSO has been accepted for funding within the Electronic Components and Systems for European Leadership Joint Undertaking in collaboration with the European Union's H2020 Framework Program (H2020/2014-2020) and National Authorities, under grant agreement n° 692482.

References

[1] R. Salot, S. Martin, S. Oukassi, M. Bedjaoui, J. Ubrig, "Microbattery Technology Overview and Associated Multilayer Encapsulation Process", Applied Surface Science 256S, pp. S54-S57, 2009.

[2] N. A. Kyeremateng, R. Hahn, "Attainable Energy Density of Microbatteries", ACS Energy Letters, Vol.3, pp. 1172-1175, 2018.

[3] F. Gasco and P. Feraboli, "Manufacturability of Composite Laminates with Integrated Thin Film Li-Ion Batteries", Journal of composite materials, vol.48 (8), No 1-2, pp. 899-910, 2014.

[4] M. Bedjaoui and S. Poulet, "Direct Bonding and Debonding Approach of Ultrathin Glass Substrates for High Temperature Devices", Proceedings of the 2017 Electronic Components and Technology Conference (ECTC), Orlando, Florida, May 31-June 2, pp. 725-732, 2017.

[5] M. Bedjaoui, J. Amiran, J. Brun, "Ultrathin Glass to Ultrathin Glass Bonding Using Laser Sealing Approach", Proceedings of the 2019 Electronic Components and Technology Conference (ECTC), Las Vegas, Nevada, May 28-31, pp.995-1001, 2019.

[6] Y. Xiao, W. Wang, J. Zhang, "Accurate Predetermination of the Process Parametrs for Glass/Glass Laser Bonding Based on the Temperature Distribution Analysis", Journal of Electronic Packaging, Vol. 138, pp 0210061-0210067, 2016.

[7] H. Kind, E. Gehlen, M. Aden, A. Olowinsky and A. Gillner, "Laser glass frit sealing of vacuum insulation glasses", Physics Procedia, vol.56, pp. 673-680, 2014.

[8] S. Logunov, S. Marjanovic, J. Balakrishnan, "Laser Assisted Frit Sealing for High Thermal Expansion Glasses", Journal of Laser Micro/Nanoengineering, Vol.7, n°3, pp. 326-333, 2012.

[9] R. Cruz, J. Ranita, J. Maçaira, F. Ribeiro, A. da Silva, J. M. Oliveira, M. Fernandes, H. Ribeiro, J. Mendes and A. Mendes "Glass-Glass Laser-assisted glass frit bonding", IEEE Transactions on Components, Packaging and Manufacturing Technology, vol.2, No 12, pp. 1949-1956, 2012

Picosecond Laser Structuring Technology for LTCC – the Improvement of Fine Line Structuring

N.Gutzeit[1], A. Schulz[1], M. Fischer[1], T. Thelemann[2], J. Müller[1]

[1] Institute of Micro- and Nanotechnologies MacroNano®
Electronics Technology Group
Technische Universität Ilmenau
Gustav-Kirchhoff-Str. 1
Ilmenau, Germany, 98693

phone: +49 3677 693453, fax: +49 3677 693350, e-mail: nam.gutzeit@tu-ilmenau.de

[2] Micro-Hybrid Electronic GmbH
Gustav-Kirchhoff-Straße 5
Ilmenau, Germany, 98693

Abstract

The paper presents actual results and applications of the developed picosecond laser structuring technology for low temperature cofired ceramics (LTCC). Standard screen-printed thick films on LTCC are ablated in the green and fired state for buried structures and structures at the top layer using a pulsed picosecond UV laser. Resolutions of <20 µm lines and spaces with an accuracy and a reproducibility of below 1 µm are demonstrated. Micro vias with diameters down to 30 µm are presented. Furthermore laser drilling of micro nozzles in sintered LTCC membranes are shown. Related to high resolutions for buried structures a process for the minimized positional tolerance of structures in different layers is described.

Key words: picosecond laser structuring, LTCC, thick film

Introduction

The resolution of screen-printed structures in LTCC is limited due to the screen printing process itself [1-3]. Accompanying with smaller lines and spaces the reproducibility and the accuracy of these structures decreases. Laser structuring of screen-printed thick films was done for some years [4-8]. At the Electronics Technology Group at the Ilmenau University of Technology a picosecond laser structuring technology for LTCC was developed which meets the demands regarding to higher resolution, higher precision and a higher reproducibility. It is based on standard screen-printed thick films on LTCC which are ablated using a pulsed picosecond UV laser to remove not required parts of the thick films [3,9-13]. For the ablation a picosecond laser with a pulse length of <10 ps and a wavelength of 355 nm was used. These parameters offer the possibility to ablate thick film materials avoiding a heat affected zone in the regions nearby. This so-called cold ablation [4,8] enables precice laser structuring also in the unsintered or green state.

The developed picosecond laser structuring technology of screen-printed thick films on LTCC offers three big advantages. Firstly, the resolution of structures in LTCC can be increased. Lines and spaces with a width of 20 µm and lower are possible.

The accuracy and reproducibility of the structures is increased to 1 µm and below. Secondly, the laser structuring process can be directly integrated in the standard LTCC manufacturing process. The screen-printed thick films can be laser structured after sintering or in the green state, which enables high resolution structures for buried structures as well as for structures at the top layer. Also, laser drilling of miniaturized vias can be easily implemented. Thirdly the positional tolerances of structures in different layers, induced by the applied stacking method (optical, mechanical), can be circumvented by using the laser structuring technology and specific fiducials.

Experimental

The picosecond laser ablation machine at the Institute of Micro- and Nanotechnologies of the TU Ilmenau (3D Micromac, Chemnitz, microstruct C, laser source: Lumentum PicoBlade) uses a pulsed UV laser beam (wavelength: 355 nm). The maximum mark speed is 1500 mm/s. The maximum average laser power is 14 W. The pulse length is smaller than 10 ps at a pulse repetition rate of 200 kHz and the focus spot diameter is approximately 7 µm (100 mm laser optics).

For all investigations DuPont DP951 in different thicknesses as substrate material, DuPont

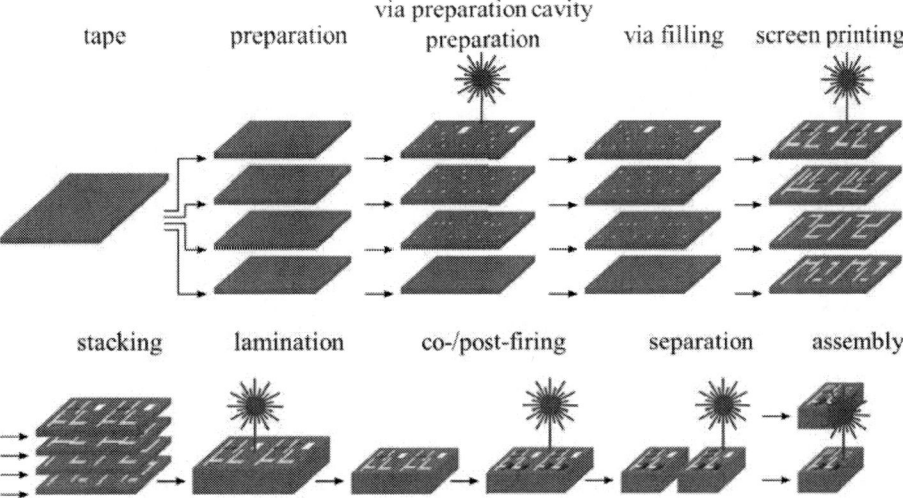

Figure 1: LTCC standard process with illustrated applicability of the picosecond laser structuring

5740A as conductor paste and DuPont 5738R as viafill paste are used. All substrates were fired with the recommended sintering profile. Figure 1 shows the standard LTCC manufacturing process, with illustrated applicability of the picosecond laser structuring technology. The first step is the production of microvias if necessary. Then standard screen-printed thick films are applied on the LTCC with oversized lateral dimensions in regions intended for laser structuring. For buried patterns the dried thick films must be laser structured in the green state.

First the gold thick film must be ablated. Due to inhomogenities within the dried thick films the ablation rate varies. This leads to areas where the gold thick film is totally ablated and areas with residual gold at the same time. To prevent shorts all residual gold within the ablation area must be removed.

The ablation rate of the green DP951 tape material is higher than the ablation rate of the gold thick film paste. Due to this fact the needed repetitions of the laser ablation process to reach spaces with no shorts, results in a relatively high ablation depth into the tape material. The ablation of the dried, green thick films needs less energy compared with the fired ones. Due to this fact and to prevent a deep ablation of the sheet material the laser power is reduced to 0.12 W.

After the structuring of the green thick film paste it is important to clean the surface from the residual particles with compressed air. This step dereases the risk of shorts. Thereafter, the different layers are laminated to each other.

To reach smaller positional tolerances of structures in different layers additional laser structuring processes can be implemented into the LTCC manufacturing process. For this several process steps are added. The first laser structuring is done on the top side of a LTCC sheet. This sheet is laminated with the top side onto the layers below,

mostly supporting layers. After removing the mylar tape from the back side of the laserstructured sheet a further standard screen printing process is done on the laminate. By using the same specific fiducials like the first laser structuring processalso on the back side, smaller positional tolerances of structures in the different layers can be achieved. This step sequence can be repeated several times to reach small positional tolerances for three or more layers. This can be easily done by the alignment on vias in buried layers and a sequential lamination. This process is called advanced sequential structuring process.

After the sintering process thick films at the top layer can be laser structured. Because the ablation rate of the fired LTCC substrate is lower compared to green tape mateiral, a higher laser power of up to 0.8 W can be used. This leads to a lower number of repetitions of the ablation process and a smaller ablation depth into the LTCC substrate.

Postfired laser structuring offers a wide range of further processing opportunities. Examples are the separation of samples, the production of cavities and micro nozzles or the shaping of resistors.

Results

Micro Vias

The minimum diameter of punched vias in LTCC is 50 µm. The picosecond laser structuring offers the possibility to decrease the diameter of the vias down to 30 µm. This enhances the application of multi layered high resolution patterns. Figure 2 shows a microscopic image of the top and bottom side of a laser drilled micro via in DuPont 951 C2. The diameter of the laser inlet and outlet side is 42 µm and 32 µm respectively in the green state. These vias can be filled with DuPont 5738R using standard filling processes.

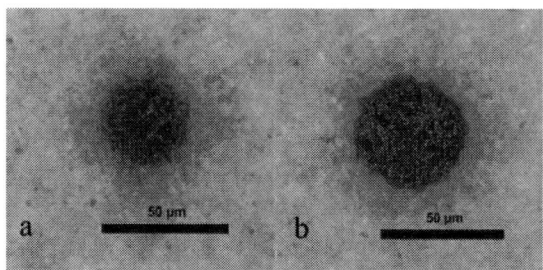

Figure 2: Microscopic image of a micro via: a) laser outlet side, b) laser inlet side

Green State Laser Structuring

Picosecond laser structuring in the green state can be used to reach a high resolution in buried layers. Applications for this are substrates with a high density of electrical contacts [11], microwave filters with buried miniaturized inductors [12] and capacitors or displacement sensors with buried flat inductors.

Figure 3 illustrates the advantages of the picosecond laser structuring for substrates with a high density of electrical contacts. Small line widths provide the fan-out capability of an array of small pads with a small pitch. In combination with micro vias the inner pads can be routed through buried layers. The achieved accuracy and reproducibility is 1 μm and 1.6 μm respectively (see Table 1).

Microwave filters based on lumped elements need inductors and capacitors with very precise geometrical dimensions in order to achieve high self resonance frequencies. Figure 4 shows an example of a finger capacitor with a line width of 27.8μm which is close to the desired value of 28.7 μm. which corresponds to 24.2 μm after the sintereing process.

The smallest continuous line for green state structuring with the used picosecond laser ablation mashine is around 24 μm ± 1.0 μm in the green state. Spalling parts of conductor material have a significant impact on the lower limit of the line width. These parts are peeled off during the laser ablation process, due to the pressure of the evaporating material in the ablation area.

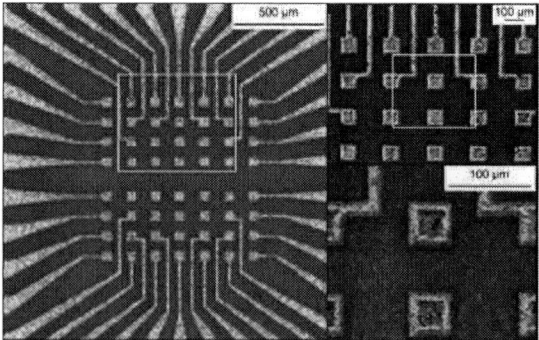

Figure 3: LSM image of a LTCC substrate for a silicon chip with high electrical contact density

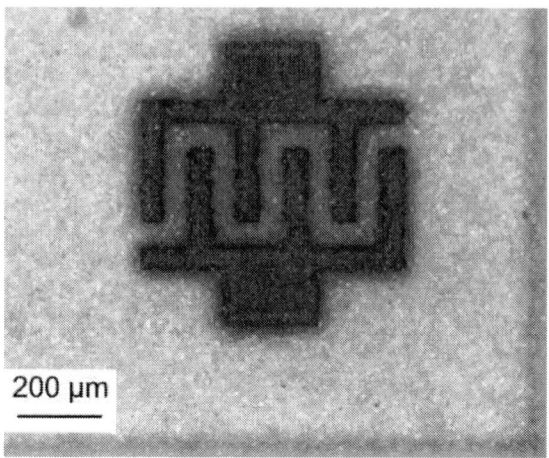

Figure 4: Green structured finger capacitor

Because of the same fact the smallest space for green state laser structuring without shorts is 25 μm ± 1.0 μm in the green state.

The ablation depth into the tape material at the ablation areas of spaces covered with gold thick films (dried thickness of about 8 μm) is 16 μm or higher (cf. Table 1).

Figure 5 shows an X-RAY image of a buried flat inductor in LTCC with 26 windings. It was manufactured using the lowest possible line and space width in green state of 24 μm and 25 μm respectively. After the sintering the lines and spaces are 21 μm and 22 μm respectively. This inductor can be used for displacement sensors. The measured inductance of the samples is 1.75 μH ± 0.11 μH. The octagonal geometry leads to a shorter processing time of the laser structuring process compared to a fully circular spiral design.

Stacking Tolerance

In addition to small lines, spaces and micro vias, multi layered electrical elements with an increased resolution require very small layer-to-layer positional tolerances. These can be decreased by the desribed advanced sequential structuring process. Through this process the same specific fiducials for different laser structuring steps at different layers are used. During the sequential lamination process a lateral shift of the position and an increased line width can be observed. This must be taken into account

Table 1: Measured values of green and post fired laser structured lines and pads

μm	buried (green state laser structuring)	top (post fired laser structuring)
line width	24 ± 1.0 (before firing) 20 ± 1.0 (after firing)	24 ± 0.4
pad size	50 ± 1	59 ± 0.5
ablation depth	24 (16 μm into green tape)	9.5

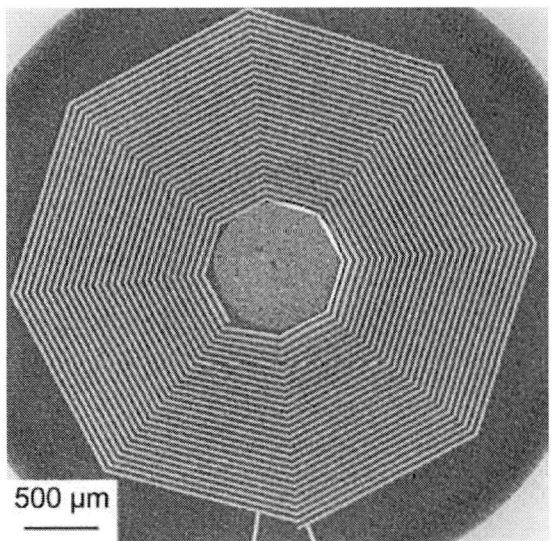

Figure 5: XRAY image of a bruried inductor

during the calculation of the laser design (cf. Figure 6). The reached positional tolerances are 5 μm and below.

Post Fired Laser Structuring

The postfired picosecond laser structuring of thick films provides a high accuracy and reproducibility of 0.4 μm and 0.6 μm repectively (see Table 1).

The smallest continuous line for post fired laser structured screen-printed thick films is around 14 μm ± 0.5 μm, the smallest space without shorts is 13.5 μm ± 0.5 μm (cf. Figure 7). The minimum line width is determined by the probabiliy of inhomogeneities within the screen-printed thick film. Film thickness, small voids or microscopic variations of the adhesion between the thick film and the LTCC substrate material are likely to cause to open circuits.

Another example for a very precise post fired picosecond laser structuring process is the laser drilling of micro holes into LTCC membranes (thickness 135 μm). These holes are used as micro

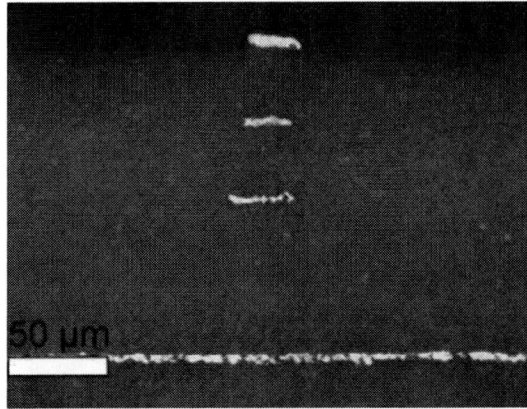

Figure 6: Microscopic image of a cross section of an inductor with minimized positional tolerance

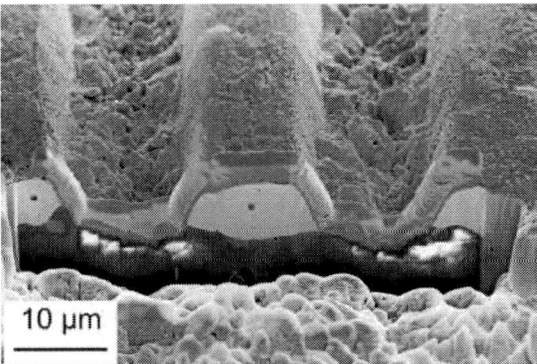

Figure 7: SEM Image of a FIB preparated post fired laser structured gold line

nozzles within a micro plasma source [13]. The holes have a diameter on the laser inlet side of 18 μm and on the laser outlet side of <5 μm. Figure 8 shows a microscopic image a micro nozzle array.

Conclusion

The developed picosecond laser structuring technology enables high resolution structuring of screen-printed thick films on LTCC. Buried structures with minimum lines and spaces of 24 μm and 25 μm respectively in the green state can be manufactured. After the sintering the green structured lines and spaces are 21 μm and 22 μm respectively. Post fired laser structuring offers a minimum line width of 13.5 μm and a minimum space of 14 μm. Laser drilled vias in LTCC with a diameter of 32 μm at the laser outlet side can be filled reliably with standard via filling processes. Laser drilling of micro nozzles with a diameter of <5 μm at the laser outlet side can be drilled in thin LTCC membranes easily by using an picosecond UV laser machine.

Acknowledgements

The authors would like to thank Mrs. Ina Koch for the fabrication of the LTCC substrates. The investigations were funded by the Thüringer Aufbaubank within the project KerMuSens (funding code: 2016 FE 9050)

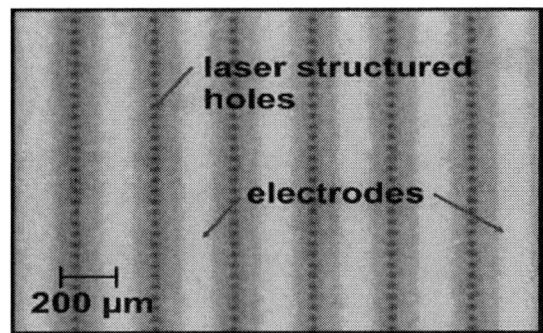

Figure 8: Microscopic image of the micro nozzle array for the use in the micro plasma source

References

[1] J. Müller, R. Perrone, H. Thust, K.-h. Drue, C. Kutscher, R. Stephan, J. Trabert, M. Hein, D. Schwanke, J. Pohlner, G. Reppe, R. Kulke, P. Uhlig, A. F. Jacob, T. Baras, A. Molke, "Technology Benchmarking of High Resolution Structures on LTCC for Microwave Circuits", 2006 1st Electronic Systemintegration Technology Conference, Dresden, Germany, pp. 111–117, 2006.

[2] D. Stöpel, K.-H. Drüe, S. Humbla, M. Mach, T. Mache, A. Rebs, G. Reppe, G. Vogt, M. Hein, J. Müller, "Fine-line structuring of microwave components on LTCC substrates", 2010 3rd Electronic System-Integration Technology Conference (ESTC), Berlin, Germany, pp. 1–6, 2010.

[3] A. Schulz, N. Gutzeit, D. Stöpel, T. Welker, M. Hein, J. Müller, "High resolution patterning of LTCC based microwave structures for Q/V-band satellite applications", 2016 German Microwave Conference (GeMiC), Bochum, Germany, March 14 - 16, pp. 19–22, 2016.

[4] J.Bliedtner, H.Müller, A.Barz, "Lasermaterialbearbeitung: Grundlagen - Verfahren - Anwendungen - Beispiele", Hanser, München, 2013

[5] D. Nowak, A. Dziedzic, M. Hrovat, J. Cilenšek, "Miniaturization of Thick–Film Resistors by Laser–Shaping", Johann Nicolics (Ed.), 33rd International Spring Seminar on Electronics Technology (ISSE), 2010: 12 - 16 May 2010, Warsaw, Poland , conference proceedings, IEEE, Piscataway, NJ, 2010.

[6] M. F. Shafique, K. Saeed, D. P. Steenson, I. D. Robertson, "Laser of Microwave Circuits in LTCC Technology", IEEE Transactions on Microwave Theory and Techniques, Vol. 57, No. 12, pp. 3254–3261, 2009.

[7] J. Zhu, W. K. C. Yung, "Studies on laser ablation of low temperature co-fired ceramics (LTCC)", The International Journal of Advanced Manufacturing Technology, Vol. 42, 7-8, pp. 696–702, 2009.

[8] R. Hull, John Dowden, R. M. Osgood, J. Parisi, H. Warlimont (Eds.), "The Theory of Laser Materials Processing: Heat and Mass Transfer in Modern Technology", Springer, Dordrecht, Heidelberg, 2009.

[9] N. Gutzeit, T. Welker, K.-H. Drüe, J. Müller, "High resolution LTCC laser processing in the green and fired state for future technologies", 12th International Conference and Exhibition on Ceramic Interconnect and Ceramic Microsystems Technologies, CICMT, Denver, CO, USA, Apr 19-21, 2016.

[10] J. Müller, N. Gutzeit, T. Welker, „Laser-Processing an grünen und gesinterten Folien", DKG-Herbstsymposium 2016 Keramische Mehrlagentechnik – Herstellverfahren und Anwendungen, 30.11.-01.12.2016, Heinrich-Lades-Halle, Erlangen, 2016.

[11] N. Gutzeit, A. Schulz, T. Welker, C. Wagner, E. Schäfer, J. Müller, "High-precision picosecond laser structuring on LTCC for silicon chip assembly with high electrical contact density", European Microelectronics and Packaging Conference, EMPC, Warschau, Polen, 10.-13. September 2017.

[12] N. Gutzeit, A. Schulz, J. Müller, T. Thelemann, "Miniaturized laser structured components in LTCC for 5G applications", 14th International Conference and Exhibition on Ceramic Interconnect and Ceramic Microsystems Technologies, CICMT 2018, April 18-20, 2018, University of Aveiro, Aveiro, Portugal

[13] M. Fischer, M. Stubenrauch, A. Naber, N. Gutzeit, M. Klett, S. Singh, A. Schober, H. Witte, J. Müller, "LTCC-Based Micro Plasma Source for the Selective Treatment of Cell Cultures", European Microelectronics and Packaging Conference, EMPC, Warschau, Polen, 10.-13. September 2017..

Author Family Name	Author First Name	Topic	Paper
Aasmundtvei	Knut	InT-09	High-energy X-ray Tomography for 3D Void Characterization in Au–Sn Solid-Liquid Interdiffusion (SLID) Bonds
Acharya	Sarthak	Post-19	Realization of Embedded Passives using an additive Covalent bonded metallization approach
Ali Roshanghias	Ali	Post-32	Low-temperature fine-pitch flip-chip bonding by using snap cure adhesives and Au stud bumps
Altenbockum	Silvia	MT-06	High throughput two-stage bonding technique for advanced wafer level packaging
Ano	Kazuaki	R&Q-08	Risk Assessment Study of Package Warpage by using Bayesian Networks
Bakr	Mona	MT-07	Effect of overmolding process on the integrity of electronics circuits
Bartsch	Heike	EST-01	LTCC as substrate - enabling semiconductor and packaging integration
Bedjaoui	Messaoud	Sub-02	Laser Frit Sealing Approach for Ultrathin Glass
Bertocci	Francesco	InT-02	Corrosion of Tin-Indium Solder during the Manufacturing Process of Biomedical Ultrasound Transducers
Bhogaraju	Sri Krishna	InT-10	Hybrid Cu particle paste with surface-modified particles for high temperature electronics packaging
Bickel	Jan	MT-03	Increasing the productivity of the novel atmospheric pressure sputtering technology for 3D chip interconnection
Billaud	Mathilde	MT-08	Process Flow and Cost Modelling for Fan-Out Panel Level Packaging
Boehme	Bjoern	R&Q-14	Chip Package Interaction with Cu Pillar Interconnects – Impact of Die Warpage
Cadalen	Eric	MT-10	Dispensing Improvements with Drop on Demand Technology
Caswell	Greg	R&Q-01	Can Electrolytic Capacitors Meet the Demands of High Reliability Applications?
Chen	Chengcheng	R&Q-10	Investigating the Overdischarge Failure on Copper Dendritic Phenomenon of Lithium ion Batteries in Portable Electronics
Cherman	Vladimir	Post-08	High spatial resolution measurements of thermo-mechanical stress effects in flip-chip packages
Del Carro	Luca	InT-15	Sintering of oxide-free copper pastes for the attachment of SiC power devices
Delrue	Jean-Pol	AP-03	Glass Wafer Level Packaging Enabled by Laser Induced Deep Etching of Closed Cavities
Derakhshandeh	Jaber	AP-02	Novel embedded microbump approach for die-to-die and wafer-to-wafer interconnects with variable microbump diameters and down to 5 μm interconnect pitch scaling
Do	Duyen Nu Bich	Mod-02	New Encapsulation Concepts for Medical Ultrasound Probes – A Heat Transfer Simulation Study
Dohle	Rainer	MT-04	New Assembly Technology for VCSEL arrays comprising ultra-thin diodes
Eom	Yong-Sung	Post-01	Advanced Interconnection technology with Laser Assisted Bonding Process for PET Substrate
Ernst	Daniel	InT-13	Heterogeneous Integration of Surface Mounted Devices for Bendable Electronics
Fischer	Michael	MEMS-02	Active cooling using fluid channels in a silicon-ceramic composite substrate
Fumita	Yusuke	MT-02	A novel DAF mount method for SDBG process
Furuhashi	Takahashi	R&Q-03	Characterization of Impact of Vertical Stress on FinFETs
Gadhiya	Ghanshyam	Mod-06	Assessment of FOWLP process dependent wafer warpage using parametric FE study
Galbiati	Enrico	R&Q-12	Acceleration factors for reliability analysis of electronics assemblies
Géczy	Attila	GrE-04	Challenges of SMT assembling on biodegradable PCB substrates
Gökdeniz,	Zeynep	Post-26	Behaviour of Silver-Sintered Joints by Cyclic Mechanical Loading and Influence of Temperature
Goldberg	Adrian	MEMS-03	Monolithic Ceramic IR-Emitter in Multilayer Technology
Griffart	Aurélien	MT-13	Opto-electronics flip-chip bonding Automation and *in-situ* quality monitoring
Gutzeit	Nam	Sub-03	Picosecond Laser Structuring Technology for LTCC – the Improvement of Fine Line Structuring
Hirman	Martin	Post-12	Comparison of QFN Chips Glued by ACA and NCA Adhesives on the Flexible Substrate
Hlina	Jiri	Post-02	Advanced Application Capabilities of Thick Printed Copper Technology
Hsieh	Sheng-Chi	AP-04	mmWave Antenna Design in Advanced Fan-Out Technology for 5G Application
Hunstig	Matthias	MT-11	Thermosonic wedge-wedge bonding using dosed tool heating
Hurtony	Tamás	InT-04	Investigation of the Microstructure of Mn-doped Tin-Silver-Copper Solder Alloys Solidified with Different Cooling Rates
Ihle	Martin	InT-11	Functional Printing of MMIC-Interconnects on LTCC Packages for sub-THz applications
Illés	Balázs	R&Q-04	Characterization of Tin Pest Phenomenon in a Low Ag Content SAC Solder Alloy
Inoue	Masahiro	InT-06	Design of Interfacial Chemistry for Inducing Low Temperature Sintering of Silver Micro-fillers within Epoxy-based Binders
Jeong	Haksan	Mod-03	Thermomechanical Properties of Fan-Out Wafer Level Package Fabricated with Various Epoxy Mold Compounds
Kadija	Igor	NT-01	Flexible Electronics Printing by Electroplating
Kahle	Ruben	MT-05	Evaluation of adaptive processes for the embedding of bare dies in IC substrates
Kim	Yeong K	R&Q-13	High frequency effects on the PBGA stress developments at random vibration
Klengel	Sandy	InT-03	Influence of copper wire material to corrosion resistant packages and systems for high temperature applications
Kolbinger	Elisabeth	R&Q-07	An approach for failure prediction in H^3TRB-tests
Koscielski	Marek	GrE-03	Recovery of valuable BGA components from used electronic mobile devices and their application in new electronics products
Lee	Jae Hak	Post-03	Study on the Embedded Flexible Hybrid Stack Package using Polymer Elastic Bump Interconnection Method
Li	Zhaorong	Mod-05	Evaluation Method of Electronic Performance Margins for SiP Based on Field-Circuit and Multiple Physical Cooperative Simulation
Lorenz	Georg	Post-07	Investigation of mechanical and microstructural properties of a new, corrosion resistant gold-palladium coated copper bond wire
Lundén	Heidi	AP-05	Hermetic and Radiation Tolerant Glass Package for VCSELs Using Novel Micro Bonding Method
Maierna	Amedeo	Post-24	Electronic Packaging for MEMS Infrared Sensor With Filtered Optical Window
Mailliart	Olivier	InT-14	Assembly of very fine pitches Infrared focal plane array with indium micro balls
Manessis	Dionysios	Post-22	High Frequency Substrate Technologies for the Realisation of Software Programmable Metasurfaces on PCB Hardware Platforms with Integrated Controller Nodes
Manessis	Dionysios	EST-03	Embedding Technologies for the Manufacturing of Advanced Miniaturised Modules toward the Realisation of Compact and Environmentally Friendly Electronics Devices
Matuskova	Bozena	Post-20	Ultra-thin polymer spray coating for advanced adhesive bonding applications
Metasch	René	R&Q-16	Numerical study on the influence of material models for tin-based solder alloys on reliability statements
Meuser	Carmen	EST-02	Printed Functional Applications: Batteries, Communication Elements, Antennas and Conductive Paths on Technical Textiles
Mitra	Kalyan	Post-17	Manufacturing of bio-compatible & degradable devices using inkjet technology for transient electronics
Moise	Madalin Vasile	P&T-02	Implementation of a prototype embedded system for in-car multipoint temperature measuring
Natta	Lara	MEMS-01	Aluminium Nitride based bio-MEMS for vascular graft monitoring
Neiser	Arne	Post-27	The Challenge of a Self-Soldering PCB
Nguyen	Hoang-Vu	Post-10	Reworkable Anisotropic Conductive Adhesive for Assembly of Medical Devices
Nishikawa	Hiroshi	MT-12	Effect of bonding temperature on shear strength of joints using micro-sized Ag particles for high temperature packaging technology
Passlack	Ulrike	MT-09	Hybrid Systems-in-Foil (HySiF) – Low Stress CFP Process for Biomedical Application
Petrich	Rebecca	Post-28	Investigation of ScAlN for piezoelectric and ferroelectric applications
Pflügler	Nadine	R&Q-11	Time-dependent Behaviour of Bonded Silicon Dies under Subcritical, Constant Shear Loading
Piallat	Fabien	Post-11	ALD Coatings to Mitigate Against Tin Whiskers and Upgrade the Environmental Durability of Electronic Circuit Boards
Pradhan	Mamta	EST-04	Ultra-thin Capacitors in Silicon for 3D-Integration and Flexible Electronics
Reinhardt	Kathrin	P&T-03	Printpower – Paste systems for multifunctional copper power modules
Richter	Jessica	Post-13	Tilting Behavior and Phase Formation of Sn-Cu Composite Solder for Large Area Baseplate Solder Joints
Rovitto	Marco	Post-05	Transfer Molding Simulation to Predict Filling Flaws and Optimize Package Design
Rusanen	Outi	Apl-01	Injection Molded Structural Electronics Brings Surfaces to Life
Saito	Akira	R&Q-02	Tin Whisker Growth Mechanism on Tin Plating of MLCCs Mounted with Sn-3.5Ag-8In-0.5Bi Solder in 30°C60%RH
Schein	Friedrich-Leonhard	AP-07	High density fan-out panel level packaging of multiple dies embedded in IC substrates
Schellscheidt	Benjamin	GrE-01	Life-Cycle Assessment for Power Electronics Module Manufacturing
Schischke	Karsten	GrE-02	Embedding as a key Board-Level Technology for Modularization and Circular Design of Smart Mobile Products: Environmental Assessment
Schneider	Andreas	InT-08	Stencil Printing and Flip-Chip Bonding for Assembly of Pixelated X-ray Detectors using PCB-type Interposer and Flexible Printed Circuit Boards
Schober	Jakob	InT-07	Copper Pillar Bumps on Acoustic Wave Components
Schulz	Alexander	Apl-04	Laser structured passive Components and RF Filters in LTCC Technology with Operating Frequencies up to 40 GHz focusing on 5G Mobile Applications
Schwarzer	Christian	Post-29	Investigation of Pressureless Sintered Interconnections on Plasma Based Additive Copper Metallization for 3-Dimentional Ceramic Substrates in High Temperature Applications
Seehase	Dirk	Post-30	Alternative Heating Methods for Printed Circuit Boards and a Practical Comparison of Direct Resistance and Inductive Heating Processes
Serea	Vlad	Post-16	A Finite Element Modelling Approach of Test Setups for Multilayer Ceramic Capacitors
Shaw	Mark	AP-06	High Capacity Silicon Photonics Packaging
Shehzad	Adil	Post-06	New approach to apply 1,2,3-benzotriazole as a capping layer on UBMs for 3D TCB stacking and investigation of oxidation protection and solder wetting
Shi	Xiangyang	Mod-01	Electrical Modeling and Analysis of Through-Silicon-Via Crosstalk Based on Scalable Physical Lumped Circuit Model for 3D Packaging
Shin	Dong-Kil	R&Q-17	Effect of Test Fixture on the Drop Reliability of Solid State Drive
Sítko	Vladimír	R&Q-06	Optical method for validation of changes in the cleaning process and cleaning process optimization
Somma	Cristina	Post-04	Digital Core Supply in Automotive FC-BGA Package: a Smart Modularity Solution for Design and Validation
Spory	Erick Merle	Sub-01	High-Temperature Inter-Cavity Silicon Interposer Substrate for Multi-Chip Module Assembly
Steenmann	Anna	InT-05	TLPB Improves Solder Connections by On Chip Creation of Intermetallic Phase Precursors
Stenzel	David	InT-12	Characterization of Alternative Sinter Materials for Power Electronics
Stojek	Krzysztof	P&T-04	Metalization impact on heat transfer through sintered nanosilver based thermal joints
Stoyanov	Stoyan	Mod-04	Packaging Challenges and Reliability Performance of Compound Semiconductor Focal Plane Arrays
Su	James	InT-01	Flip Chip Assembly on Coreless Substrate Challenge with Die Bond Solution
Trulli	Susan	Apl-03	Multi-material Printed Microwave Element for Phased Array Applications
Tsai	Mike	AP-01	Advanced Antenna Integration of 3D System in Package Solutions for IoT and 5G Application
Ueshima	Minoru	Post-25	Ag and Si particles sintering technology for SiC power device
Vasile	Daniel Ciprian	Apl-02	Innovative Authentication Method for IoT Devices
Vieilledent	Julien	Post-31	Mechanical behaviour of SAC305 lead-free alloy
Vijay	Karthik	Post-18	Substrate to Baseplate Attach: A Novel Solder Solution with an Embedded Metal Matrix for Enhanced Reliability
Wang	Nan	Post-23	Highly Thermal Conductive and Light-weight Graphene-based Heatsink
Wargulski	Dan	R&Q-05	Paving the way for the replacement of solder interconnections in power electronics by silver-sinter using pulsed infrared thermography
Wiese	Steffen	Post-21	Temperature evaluation of solder joints for adjusting reflow profiles
Wiss	Erik	R&Q-09	Flex Cracking of Multilayer Ceramic Capacitors: Experiments on Fracture Propagation
Wright	Daniel	MT-01	Development of mechanically compliant flip chip interconnect using single metal coated polymer spheres
Wuest	Felix	R&Q-15	Integrated Condition Monitoring by Measuring the Delay of Gate Turn-off
Ye	Suiying	P&T-01	Anisotropic Composite Core Material for Inductor-based Fully Integrated Voltage Regulator
Yuile	Adam	Post-15	Simulation of temperature profiles in reflow ovens for soldering area array components
Zehri	Abdelhafid	NT-02	Graphene-coated copper nanoparticles for thermal conductivity enhancement in water-based nano-fluid

2025 IEEE International Conference on Electron Devices and Solid-State Circuits (EDSS 2025)

Yinchuan, China
13-15 June 2025

IEEE Catalog Number: CFP25754-POD
ISBN: 979-8-3315-2209-4